TITANIUM '92

SCIENCE AND TECHNOLOGY

VOLUME II

TITANIUM '92

SCIENCE AND TECHNOLOGY

VOLUME II

Proceedings of a Symposium sponsored
by the Titanium Committee of the Minerals, Metals & Materials
Structural Metals Division

Held at the Seventh World Titanium Conference
June 29 - July 2, 1992
in San Diego, California

Edited by

F.H. FROES and I.L. CAPLAN

A Publication of

TMS

Minerals • Metals • Materials

A Publication of The Minerals, Metals & Materials Society
420 Commonwealth Drive
Warrendale, Pennsylvania 15086
(412) 776-9000

Printed in the United States of America
Library of Congress Catalog Number 93-78482
ISBN Number 0-87339-222-1

TABLE OF CONTENTS

VOLUME II

INTERMETALLICS

FORGING, FORMING, JOINING, AND CASTING

PHYSICAL AND MECHANICAL PROPERTIES

INTERMETALLICS

MICROSTRUCTURES AND MECHANICAL BEHAVIOR OF GAMMA TITANIUM ALUMINIDES

Masaharu YAMAGUCHI

Department of Metal Science and Technology, Kyoto University
Sakyo-ku, Kyoto 606, Japan

Abstract

This is a review of the current research and development activities on the γ/α_2 two-phase TiAl compounds which have recently received a great deal of research interest because of their refractory nature with a very low density. Particular emphasis is laid on their microstructure and mechanical behavior. Recent work on the environmental embrittlement in TiAl is also briefly reviewed.

Titanium '92
Science and Technology
Edited by F.H. Froes and I. Caplan
The Minerals, Metals & Materials Society, 1993

Introduction

In the last few years, there has been an enormous increase in the research and development activity on titanium aluminides, in particular, on the TiAl compounds since they have immense potential as new high-temperature light-weight structural materials. In Japan, a national project aiming at the development of high temperature intermetallic compounds was launched in 1989. TiAl was chosen as a target compound for the project (see review [1]).

The TiAl phase possesses a wide composition range. However, it extends primarily on the Al-rich side and the TiAl compounds, with nearly stoichiometric or Ti-rich compositions, exhibit a two-phase microstructure composed of the TiAl phase (γ) and a small volume fraction of the Ti_3Al phase (α_2). The so-called TiAl compounds to which a recent enormous increase in the research and development activity has been devoted, are such γ/α_2 two-phase compounds rather than the Al-rich TiAl compounds with a γ single phase structure. In the last few years, marked improvements in the mechanical properties of the two-phase compounds have been achieved. However, there are still remaining rood-blocks to the commencement of their structural applications. In particular, their mechanical properties are often inappropriately balanced. For example, elongation up to 3~4% has been achieved at room temperature through microalloying of ternary elements and controlling their microstructure by thermomechanical processing. However, improvements in room temperature elongation often lead to reductions in other mechanical properties such as toughness and high-temperature strength.

In this paper, the characteristics of the microstructure and mechanical behavior of the γ/α_2 two-phase compounds will be discussed as will their current status and applications. The readers are also recommended to refer to recent reviews on the microstructure/property relationships in the two-phase compounds and their applications [1-7].

Microstructures

As-Solidified Lamellar Structure

When the two phase compounds are prepared by usual ingot-metallurgy methods, the γ and α_2 phases constitute a lamellar structure. Such a lamellar structure in the two-phase compounds with nearly stoichiometric compositions is formed during the reaction $\alpha(\alpha\text{-Ti}) \rightarrow \alpha + \gamma$ [8] and the ordering of the α phase to the α_2 phase occurs after the formation of lamellar structure [9,10]. For Ti-rich compositions such as Ti-39~43 at%Al, however, the α phase becomes ordered and then γ-lamellae are formed via the reaction, $\alpha_2 \rightarrow \alpha_2 + \gamma$ [11].

Whichever reaction occurs, γ lamellae are formed in such a way that close-packed planes and directions in the γ phase are parallel to the corresponding planes and directions in the α or α_2 phase [12]. However, close-packed $[\bar{1}10]$ and $[\bar{1}01]$ directions on the close-packed (111) plane in the γ phase are not equivalent because of the tetragonality of the $L1_0$ structure of the γ phase. Thus, the γ phase can be formed in six orientation variants corresponding to the six possible orientations of

$[\bar{1}10]$ on (111) in the γ phase with respect to $\langle 11\bar{2}0\rangle$ on (0001) in the α or α_2 phase. This suggests that a domain structure corresponding to these six orientation variants results in the γ phase in the two-phase compounds with the lamellar structure [13-18].

The growth of γ lamellae in the direction perpendicular to the basal plane of the α or α_2 phase gives rise to the occurrence of domain boundaries parallel to the lamellar boundaries. In other words, the lamellar structure of the γ/α_2 two-phase compounds is composed of γ lamellae separated by such domain boundaries and α_2 lamellae. Each γ lamella also has a domain structure. Recently, using an etching technique to remove the γ phase and to leave the α_2 phase, large internal stresses have been suggested to exist in the α_2 lamellae in the two-phase compounds with the lamellar structure [19]. Such internal stresses are supposed to be due to the lattice mismatch between the γ and α_2 phases.

Microstructure Control

Microstructure types The microstructure of two-phase compounds can be adjusted to be either lamellar or equiaxed, or a mixture of both microstructures through appropriate hot-working and heat treatments [3-6]. Primary hot-working, usually isothermal forging is typically conducted near the eutectoid temperature. Hot-worked ingots are then subjected to heat-treatments for further microstructure control. The resulting microstructures can be classified into four groups. The characteristics of microstructures belonging to the four groups and the corresponding heat treatments are summarized as follows (see reviews [4,6]).

i) Near-gamma microstructure
- generally non-uniform and consisting of coarse γ grains and banded regions composed of fine γ grains and α_2 particles.
- obtained by heat-treatments at temperatures just above the eutectoid temperature.

ii) Duplex microstructure
- consisting of fine γ grains, lamellar grains and some α_2 plates.
- obtained by heat treatments at temperatures where the volume fractions of α and γ are roughly equal.

iii) Nearly lamellar microstructure
- consisting mostly of coarse lamellar grains and a small amount of fine γ grains.
- obtained by heat treatments at temperatures near the α-transus.

iv) Fully lamellar microstructure
- consisting of large lamellar grains such as those observed in as-cast ingots of the two-phase compounds.
- obtained by heat treatments at temperatures above the α-transus.

Although isothermal forging is commonly employed as a primary hot-working method, rolling and extrusion are another possible primary hot-working routes. Rolling at temperatures in the $\alpha + \gamma$ field has been reported to lead to a weaker texture than isothermal forging in the $\alpha + \gamma$ field [20].

Alloying effects Additions of alloying elements such as Cr and Mo result in the formation of a third phase [21-23]. In the case of the additions of Cr, the electron diffraction patterns for the third phase have been reported to be consistent with a B2-type structure (nominally Ti_2AlCr) [21]. Several γ-base ternary alloys containing 45 to 54 at%Al and 1 to 4 at%Cr have been investigated and the third phase has been found to tend to form with a decreasing Al or an increasing Cr concentration [21]. In Ti-48.4Al-0.6Mo (at%), thermomechanical processing has been reported to result in a microstructure composed of very fine equiaxed grains of TiAl with Ti_3Al and βTi present in the form of thin plates or fine particles in TiAl grains and on their grain boundaries [23]. Chromium or molybdenum modified ternary TiAl compounds having such a fine microstructure have been reported to deform superplastically [22,23].

Recrystallization As mentioned above, isothermal hot-working is very effective in obtaining a fine microstructure, however, the microstructure of isothermally forged two-phase TiAl ingots is often inhomogeneous [4,24]. The deformation and recrystallization of the lamellar structure occur simultaneously in the ingots during their isothermal forging and their recrystallization behavior plays an important role in the homogeneity of the resulting microstructures.

When the lamellar boundaries lie at about 45° from the rolling direction, two-phase TiAl crystals with the lamellar structure can be rolled to about 50% reduction in thickness even at room temperature [25,26]. Using such lamellar TiAl specimens (PST crystals, see the footnote in the section of Mechanical Properties), the recrystallization mode and behavior of the lamellar structure have been studied as a function of reduction in thickness [26]. Figure 1 shows softening isotherms of specimens, rolled to 15, 30 and 40% at room temperature and all annealed at 900℃. It can be seen that the decrease in microhardness

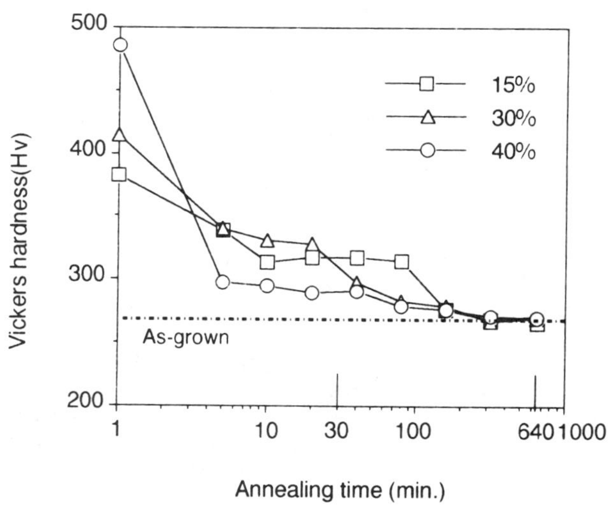

Fig. 1 Change in microhardness of cold-rolled PST crystals of TiAl as a function of annealing time at 900℃.

occurs in two stages with increasing annealing time. The first and second stages of softening have been found to correspond to the rearrangement and annihilation of dislocations and the annealing-out of deformation twins, respectively. When the reduction in thickness is smaller than 20%, TiAl lamellae are deformed by shear deformation parallel to the lamellar boundaries. However, since TiAl lamellae are composed of domains of the six different types [13-18], some domains deform by slip and the others by twinning. Thus, the deformation structure in a TiAl lamella is expected to be like Fig. 2(a). As the amount of reduction in thickness is increased, shear deformation via both slip and twinning across the TiAl lamellae becomes more and more predominant. Such a deformation structure could be described schematically as in Fig. 2(b).

The rearrangement and annihilation of dislocations occur much faster than the annealing-out of deformation twins. Therefore, in specimens cold-rolled to less than 20%, i.e. with such a deformation structure as that shown in Fig. 2(a), the growth of domains being mostly free of dislocations occurs at the expense of domains still containing deformation twins without disturbing the lamellar structure.

Figures 3(a) and (b) show microstructures observed on the rolling plane of a specimen before rolling and a specimen rolled to 15% and annealed for 640 min at 900℃. Figures 4(a) and (b) show (111) pole figures corresponding to Figs. 3(a) and (b), respectively. The (111) pole normal to the lamellar planes, which is indicated by an arrow, and other {111} poles are correspondingly located in the two pole figures. This indicates that recrystallization of the specimen rolled to 15% occurs without disturbing the lamellar structure. By contrast, on annealing of specimens with a deformation structure like Fig. 2(b), any significant

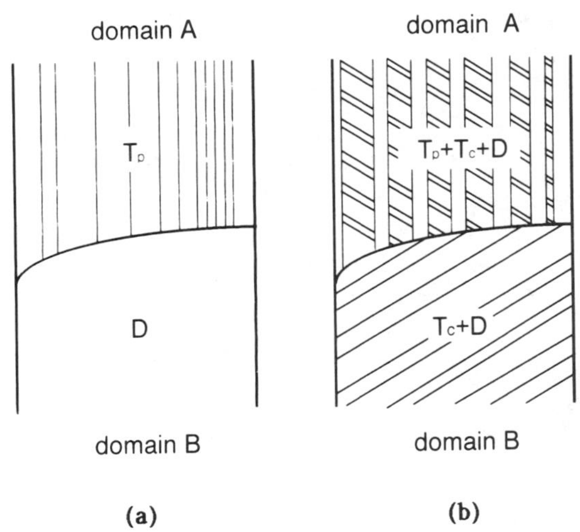

Fig. 2 Deformation structures of PST crystals of TiAl (schematic). (a) cold-rolled to less than 20%, (b) cold-rolled to more than 40%. D, T_P and T_C mean dislocations, deformation twins parallel to the lamellar boundaries and those lying across the TiAl lamellae.

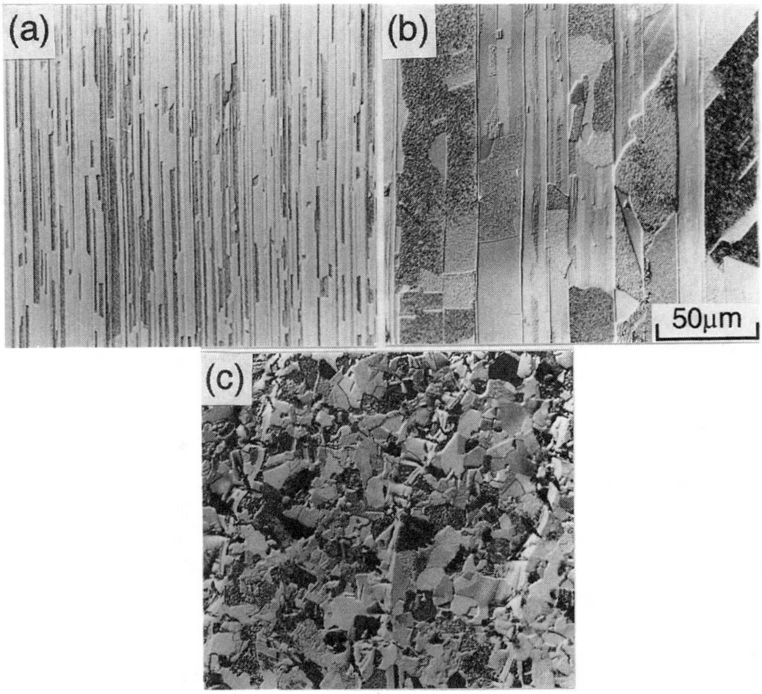

Fig. 3 Microstructures of PST crystals of TiAl. (a) cold-rolled to 15%, (b) cold-rolled to 15% and annealed for 640 min at 900℃, (c) rolled to 50% through successive roll-and-anneal process. The final annealing was made for 40 min at 900℃.

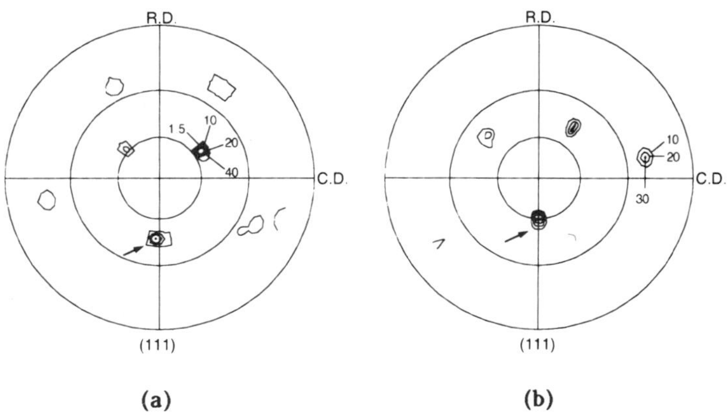

Fig. 4 (111) pole figures for PST crystals of TiAl. (a) as-grown, (b) cold-rolled to 15% and annealed at 900℃ for 640 min.

difference in the stored energy may not arise between neighboring domains since each of the domains contains dislocations and twins more or less evenly. Therefore, recrystallization occurs in a general manner of nucleation of randomly oriented new grains and their growth (Fig. 3(c)). The occurrence of such two different modes of recrystallization may lead to an inhomogeneous microstructure. Thus, deformation should be applied uniformly and at a degree above a certain threshold to obtain a homogeneous microstructure consisting of equiaxed grains by isothermal forging of ingots with the lamellar structure.

Mechanical Properties

Ductility of the TiAl Phase in the Two-Phase TiAl Compounds

The strength and ductility of the TiAl phase coexisting with the Ti_3Al phase in the lamellar TiAl compounds have been investigated using "single crystals†" having a structure composed of a single set of lamellae with a specific orientation [2,28-30]. The results of the work can be summarized as follows.

1) Yield stress of PST TiAl is highly dependent on lamellar orientation with respect to the loading axis ; it is high when the lamellae are parallel or perpendicular to the loading axis, but very low for intermediate orientations.

2) When the lamellae are parallel or perpendicular to the loading axis, shear deformation proceeds across the lamellar boundaries (hard-type), whereas shear deformation occurs parallel to the lamellar boundaries (easy-type) for intermediate orientations.

3) For intermediate orientations, tensile strain can be as high as 20% at room temperature.

Since the easy-type of deformation is thought to correspond to the deformation of the TiAl phase itself, it can be inferred that the TiAl phase in equilibrium with the Ti_3Al phase is soft and easy to deform.

Microstructure-Mechanical Property Relationships

The room-temperature tensile properties of the two-phase TiAl compounds are directly related to their microstructure. The fine duplex microstructures result in a tensile elongation ranging between 2% and 4% depending on alloy composition and microstructure characteristics such as grain size [4-6]. Recently, in a binary TiAl compound of composition Ti-50at%Al, thermomechanical processing has been reported to result in a fine duplex microstructure (grain size : ~30μm) showing 4.6% room-temperature elongation [31]. Fully lamellar microstructure generally exhibits poor ductility and low strength at room temperature. A decrease in the size of lamellar grains has been shown to increase both strength

---

†We call these "single crystals" polysynthetically twinned (PST) crystals since the TiAl phase in such "single crystals" are composed of numerous thin twin-related lamellae and this is analogous to the phenomenon known as polysynthetic twinning which is often observed in mineral crystals [27].

and ductility [4]. However, the room-temperature tensile elongation of fully lamellar TiAl has been currently reported to range from 0 to 2% [4-6,32], usually less than 1%. Considering that PST TiAl can show a tensile elongation as large as 20% at room temperature, the room-temperature ductility of the two-phase TiAl compounds would be expected to achieve higher values.

Variations in tensile properties with temperature are collectively shown for several different microstructures in [4,6]. Both ductility and yield strength for the thermomechanically processed duplex structures are generally better than those for the coarse lamellar structures [4-6]. However, creep and fracture properties of fully lamellar microstructure are much better than those of other microstructures [33-36]. In particular, serrated or interlocked boundaries of lamellar grains have been found to be fracture tolerant and resistant to high-temperature creep [36]. Treatments to improve the ductility of two-phase lamellar TiAl are required. If refining of lamellar grains, together with refining the microstructural constituents of the lamellar structure [2], can be achieved in the fully lamellar two-phase compounds, their mechanical properties would be expected to be more appropriately balanced [1,2,4].

Alloy development efforts have been devoted to improve mechanical properties of the Ti-rich two-phase compounds which are more ductile than the Al-rich single-phase compounds [37,38]. An early alloying study [39] on TiAl has shown that an addition of 1 at%V to Ti-48at%Al results in a ductility of 2% and a good combination of mechanical properties such as ductility, strength and creep resistance. More recently, additions of Mn [40,41], Cr [21,42,43] and Mo [23] to Ti-rich two-phase compounds have been shown to increase their room-temperature ductility. However, none of these alloying elements shows any positive effects on the ductility of Al-rich single-phase compounds. Therefore, ductility improvements achieved in the two-phase TiAl compounds by additions of ternary elements such as V, Mn, Cr and Mo must be strongly associated with the change in the microstructural constituents of the α_2/γ two-phase structure. We recently found that the domain size in the lamellar structure of Mn modified compounds was much finer than that in the lamellar structure of the binary compounds.

Environmental Embrittlement

A number of intermetallic compounds including FeAl, Fe_3Al, Co_3Ti and Ni_3Al show distinctly lower ductility at room temperature when tested in air than in vacuum (see review [44]). Recently, we have found that TiAl also displays less ductility in air than in vacuum [45]. Figure 5 shows the effect of test environment on the ductility of PST TiAl tested at room temperature. The lamellar boundaries in these PST crystals are inclined at 31° with respect to the tensile axis. Therefore, the easy-type deformation occurs. At the lowest strain rate of $2.0\times10^{-4}s^{-1}$, the PST TiAl shows about 55% elongation in vacuum, but about 20% and 10% in air and in hydrogen, respectively. The environmental loss in ductility decreases with increasing strain rate. This is consistent with hydrogen embrittlement. When tested in air or in hydrogen at lower strain rates, the PST TiAl fails by cleavage-like mode along (111) in the TiAl phase which is

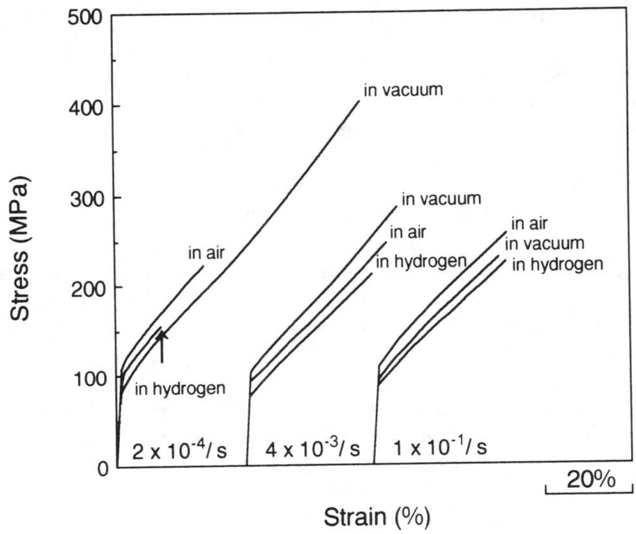

Fig. 5 Effects of test environment and strain rate on the room-temperature ductility of PST crystals of TiAl.

parallel to the lamellar boundaries. However, in vacuum or at higher strain rates in air or in hydrogen, fracture occurs across the lamellar boundaries. The environmental embrittlement was observed also in PST specimens which were so oriented that the hard-type deformation occurred. The Cr or Mn modified PST crystals of TiAl have been found to show a small environmental loss in ductility as compared with the binary PST crystals [46]. Further studies on the behavior and mechanism of environmental embrittlement in PST TiAl are now in progress.

Applications

TiAl has widespread potential utility both in air- and land-base applicability because of its refractory nature with a very low density. Investment casting has been applied to Ti-48Al-M(M = V, Mn, Mo), Ti-48Al-2Nb-M(M = Cr, Mn, V) [47], Ti-48Al-2V-TiB_2 [48] and some new castable TiAl compounds [1], and various components of aero engines such as turbine blade, vane and compressor case have been manufactured [1,2,4,7,47-49]. Components of automobile engines such as valve and turbocharger wheel were also manufactured from TiAl compounds [1,2,32]. The turbocharger wheel in [32] was reported to show better performance than the Inconel 713C wheel tested for comparison. Forging of larger components such as turbine rotor has been also investigated [50,51].

Concluding Remarks

In recent years, there has been an enormous increase in the research and development activity on the two-phase TiAl compounds. However, more work is needed for the commencement of the structural applications of the compounds. Refining of lamellar grains, together with refining of the microstructural constituents of the lamellar structure in the fully lamellar compounds would be of promise to struck a balance between the room-temperature ductility, room-temperature toughness and elevated temperature strength [1,2,4,6]. The author believes PST crystals of TiAl are extremely useful to study the fundamental mechanical properties including the fracture mechanisms of the lamellar structure which is so common and persistent after thermal treatments of TiAl compounds.

With regards to alloying elements, we should clarify the reasons why additions of alloying elements such as V, Mn, Cr and Mo improve the ductility of two-phase TiAl, but are not effective in improving the ductility of single-phase TiAl. TiAl is susceptible to test environment at ambient temperatures. Therefore, the effects of alloying additions to TiAl should be tested at least in air and in vacuum. Some alloying elements might be effective mainly in reducing the environmental embrittlement.

Acknowledgements

This work was supported by Grant-in-Aid from The Ministry of Education, Science and Culture for Scientific Research on the Priority Area "Intermetallic Compounds" and in part by the research grant from R & D Institute of Metals and Composites for Future Industries.

References

1. M. Yamaguchi, Materials Science and Technology, 8, April (1992) 1.
2. M. Yamaguchi and Y. Umakoshi, Prog. Mater. Sci., 34(1) (1990) 1.
3. Y.W. Kim and F.H. Froes, High Temperature Aluminides and Intermetallics, ed. S.H.Whang, C.T.Liu, D.P.Pope and J.O.Stiegler (Warrendale, PA : TMS, 1990) 465.
4. Y.W. Kim and D.M. Dimiduk, JOM, 43(8) (1991) 40.
5. S.C. Huang and D.S. Shih, Microstructure/Property Relationships in Titanium Aluminides and Alloys, ed. Y.W.Kim and R.R.Boyer (Warrendale, PA : TMS, 1991) 105.
6. D.M. Dimiduk, D.B. Miracle, Y.W. Kim and M.G. Mendiratta, ISIJ International, 31(10) (1991) 1223.
7. F.H. Froes, C. Suryanarayana and D. Eliezer, ISIJ International, 31 (10) (1991) 1235.
8. C. McCullough, J.J. Valencia, C.G. Levi and R. Mehrabian, Acta Metall., 37 (1989) 1321.
9. M. Yamaguchi and H. Inui, Ordered Intermetallics-Physical Metallurgy and Mechanical Behavior (Proc. NATO Advanced Research Workshop), ed. C.T. Liu, R.W. Cahn and G. Sauthoff, to be published by Kluwer Academic, Boston.
10. P.A. MoQuary, P.A. Dimiduk and S.L. Semiatin, Scripta Metall. Mater., 25 (1991) 1689.
11. Y.W. Kim, High Temperature Ordered Intermetallic Alloys IV, MRS Symposia Proceedings Vol.213, ed. L.A.Johnson, D.P.Pope and J.O.Stiegler

(Pittsburgh, PA, MRS, 1991) 777.
12. M.J. Blackburn, Technology and Application of Titanium, ed. R.T.Yaffee and N.E.Promisel (London, Pergamon) 633.
13. C.R. Feng, D.J. Michel and C.R. Crowe, Scripta Metall., 22 (1988) 1481 ; 23 (1989) 241, 1135.
14. D.S. Schwartz and S.M.L. Sastry, Scripta metall., 23 (1989) 1621.
15. Y.S. Yang and S.K. Wu, Scripta Metall. Mater., 24 (1990) 1801 ; 25 (1991) 255.
16. H. Inui, A. Nakamura, M.H. Oh and M. Yamaguchi, Ultramicroscopy, 39 (1991) 268.
17. H. Inui, M.H. Oh, A. Nakamura and M. Yamaguchi, Philos. Mag. A, in press.
18. D.S. Schwartz and W.O. Soboyejo, the same proceedings as [5], 65.
19. N. Aritomi, K. Ogawa, K. Honma, K. Sato and T. Tsujimoto, Mater. Sci. Eng. A, 149 (1991) 41.
20. P.A. Beaven, T.H. Pfullmann, J. Rogalla and R. Wagner, the same proceedings as [11], 151.
21. S.C. Huang and E.L. Hall, Met. Trans. A, 22 (1991) 2619.
22. K. Hashimoto, N. Masahashi, Y. Mizuhara, H. Fujii and M. Matsuo, the same proceedings as [5], 253.
23. T. Maeda, M. Okada, and Y. Shida, the same proceedings as [11], 555 ; Proc. 1989 Sapporo Meeting of JIM (Sendai, Japan Institute of Metals, 1989) 238.
24. J. Seeger, C. Hartig, A. Bartels and H. Mecking, the same proceedings as [11], 157.
25. S.R. Nishitani, M.H. Oh, A. Nakamura, T. Fujiwara and M. Yamaguchi, J. Mater. Res., 5 (1990) 484.
26. M.H. Oh, H. Inui, A. Nakamura and M. Yamaguchi, Acta metall. mater., 40 (1992) 167.
27. C.S. Barrett and T.B. Massalski, Structure of Metals, 3rd revised edition (Oxford, Pergamon Press, 1980) 406.
28. T. Fujiwara, A. Nakamura, M. Hosomi, S.R. Nishitani, Y. Shirai and M. Yamaguchi, Phil. Mag. A, 61 (1990) 591.
29. H. Inui, M.H. Oh, A. Nakamura and M. Yamaguchi, Acta metall. mater., in press.
30. Y. Umakoshi, T. Nakano and T. Yamane, Scripta metall., 25 (1991) 1525.
31. M. Hosomi, T. Maeda, M. Okada, H. Inui and M. Yamaguchi, Proc. 1992 Chiba Meeting of JIM (Sendai, Japan Institute of Metals, 1992) 317.
32. Y. Nishiyama, T. Miyashita, S. Isobe and T. Noda, the same proceedings as [3], 557.
33. D.S. Shih, S-C. Huang, G.K. Scarr, H. Jang and J.C. Chesnutt, the same proceedings as [5], 135.
34. S.L. Kampe, P. Sadler, D.E. Larsen and L. Christodoulou, the same proceedings as [5], 313.
35. K.S. Chan and Y.W. Kim, the same proceedings as [5], 179.
36. S. Mitao, S. Tsuyama and K. Minakawa, the same proceedings as [5], 213, 297.
37. E.L. Hall and S.C. Huang, J. Mater. Res., 4 (1989) 595.
38. S.C. Huang and E.L. Hall, Acta metall. mater., 39 (1991) 1053.
39. M.J. Blackburn, US Patent No.4294615, (1981).

40. T. Tsujimoto and K. Hashimoto, High Temperature Ordered Intermetallic Alloys III, MRS Symposium Proceedings Vol.133, ed. C.T.Liu, A.I.Taub, N.S.Stoloff and C.C.Koch (Pittsburgh, PA : MRS, 1989) 391.
41. T. Hanamura, R. Uemori and M. Tanino, J. Mater. Res., 3 (1988) 656.
42. T. Kawabata, T. Tamura and O. Izumi, the same proceedings as [40], 329.
43. W. Wunderlich, T. Kremser and G. Frommeyer, Z. Metallkde., 81 (1990) 802.
44. C.T. Liu and C.G. McKamey, the same proceedings as [3], 133.
45. M. Misaki, M.H. Oh, H. Inui and M. Yamaguchi, Proc. 1992 Chiba Meeting of JIM (Sendai, Japan Institute of Metals, 1992) 322.
46. M. Kobayashi, H. Inui, M.H. Oh and M. Yamaguchi, Proc. 1992 Chiba Meeting of JIM (Sendai, Japan Institute of Metals, 1992) 318.
47. B. London and T.J. Kelly, the same proceedings as [5], 285.
48. G. Colvin, M. Cianci, B. Kleyn and L. VanderLeest, the same proceedings as [5], 361.
49. H. Lipsitt, High Temperature Ordered Intermetallic Alloys I, MRS Symposia Proceedings Vol.39, ed. C.C.Koch, C.T.Liu and N.S.Stoloff (Pittsburgh, PA : MRS, 1985) 351.
50. Y. Nonaka, T. Mochizuki, M. Shinohara, Proc. 1992 Chiba Meeting of JIM (Sendai, Japan Institute of Metals, 1992) 328.
51. M. Shinohara, T. Matsumoto and Y. Nonaka, Proc. 1992 Chiba Meeting of JIM (Sendai, Japan Institute of Metals, 1992) 328.

Environmental Effects on Fatigue Properties of Alpha-2 Titanium Aluminides

Prasad S. Godavarti, Mark D. Lipschutz *, John G. Snow, and James A. Hall

Allied-Signal Aerospace Company, Garrett Engine Division,, 111 S. 34th Street, P.O. Box 5217, Phoenix, AZ 85010

Abstract

The use of titanium alloys, in particular the titanium aluminide alloys, is inhibited by degradation of properties following exposure to the service environment. Preexposures were performed in the 538°C to 760°C temperature range and post-exposure degradation of 25°C LCF and HCF life was quantified for a beta-processed and beta solutioned Ti-25Al-10Nb-3V-1Mo (atomic %) alpha-2 titanium aluminide alloy. Samples were tested following preexposure in hot air, hot salt, and molten sulfate salt environments. Air and molten sulfate salt preexposures at temperatures > 649°C resulted in a 100X drop in LCF life but caused little change in HCF strengths. Hot salt effects were severe, both under LCF and HCF conditions. LCF debit following exposures in hot air and molten salt could be attributed to the poor oxidation resistance of this alloy and a tendency to form brittle strain-intolerant alpha-2 stabilized outer case. Hot salt sensitivity can be attributed to pitting-type attack on the sample surface which was similar in nature to that observed in conventional alpha+beta titanium alloys.

Introduction

The fatigue capabilities of titanium aluminides in various environments must be considered when designing these alloys into gas turbine engines. Earlier published work has documented debits in tensile and creep life [1], as well as debits in fatigue crack growth rates [2], that were attributable to the presence of aggressive air environments at temperatures as low as 427°C. A USAF-funded program [3] was conducted by the authors that aimed at developing a better understanding on the mechanisms of environmental degradation in tensile, creep and fatigue properties of titanium aluminides. Alpha-2 and gamma titanium aluminides were evaluated. The effectiveness of two coatings: sputter-deposited Ti_5Si_3 and ion-implanted Al-Nb, in reducing the environmentally-induced strength debits was also studied and the results are reported elsewhere [3].

For the fatigue testing phase of this contract, a series of tests were conducted to investigate the effect of environment on the mechanisms of fatigue initiation, fatigue crack growth, and fatigue failure. Fatigue initiation in low cycle fatigue (LCF) and high cycle fatigue (HCF) regimes was studied using smooth specimens. In this paper, the effects of pre-exposure to typical gas turbine engine environments on residual 25°C fatigue life are discussed.

---

* Formerly with Garrett Engine Division, Allied-Signal Aerospace Company

Titanium '92
Science and Technology
Edited by F.H. Froes and I. Caplan
The Minerals, Metals & Materials Society, 1993

Experimental Methods

Alloy Chemistry and Microstructures

A Ti-25Al-10Nb-3V-1Mo (atomic %) alloy ingot was double-melted by Timet, Henderson, NV. The alloy ingot was forged above the beta-transus and then processed via a series of extrusion steps and heat treatments to a beta-processsed /beta solutioned and double-aged microstructure (Figure 1).

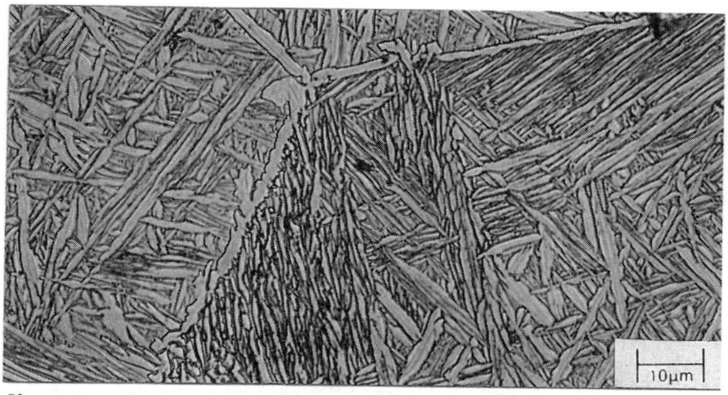

Figure 1. Microstructure of Fully Heat Treated Ti-25-10-3-1 Alloy Consists of a Transformed Beta Widmanstatten Structure with a Large Beta Grain Size (=685 microns). A Significant Amount of Feathery Alpha-2 was also Present at the Grain Boundaries.

Preexposure Conditions

12.7 millimeter diameter button head fatigue specimens were exposed in several environments. Exposures in laboratory air were performed in a muffle furnace. Pre-exposures in molten salt environments were performed by locating the samples in a flowing (SO_2+Ar+O_2) environment with a resultant oxygen level of 0.001 atmospheres. These samples were coated with a Na_2SO_4-$MgSO_4$ salt solution just prior to the pre-exposure. Hot salt sensitization comprised on application of ASTM D1144 sea-salt solution on the test bars followed by a 649°C/ 1 hour exposure in an air furnace. Vacuum pre-exposures were conducted by double-bagging the samples in cp-titanium foils followed by annealing in a vacuum furnace at $5x10^{-6}$ torr .

Fatigue Testing

LCF tests were performed on MTS 810 load frames in strain control modes per ASTM E-606. Strain was monitored by an MTS 632.516.04 extensometer attached to the center section of the bar. The button heads were gripped on both ends using specially designed spilt collet fixturing. Specimen alignment better than 3% at maximum load was achieved using this arrangement. All LCF tests reported in this paper were conducted at 0.33 Hz. frequency, at room temperature (25°C) at a R(strain)-ratio of 0.05. Tests were terminated either at failure or at $>10^5$ cycles (designated as run-out).

Load controlled HCF tests were conducted on MTS 810 load frames using a frequency of 10 Hz., and an R (stress)-ratio=-1.0 at 25°C. All tests were tested to failure or to run-out (10^6 cycles).

Post-Test Examination

Selected samples were examined using SEM and optical microscopy (OM) techniques. OM samples were generally electroless nickel plated to improve edge retention during mounting and cross-sectional metallographic examination. Cross-sections were made such that the optical section intersected the initiation site.

Results

Air Preexposures and 25°C LCF Life Debit

A short-time, 1.3 Hours, exposure to laboratory air at 760°C resulted in a significant debit in LCF life(Figure 2). Tests on samples preexposed under identical conditions, but in vacuum, did not result in loss of LCF life when compared to the as-received, unexposed alloy. These results confirm that the debit in LCF life is attributable to the exposure to the environment.

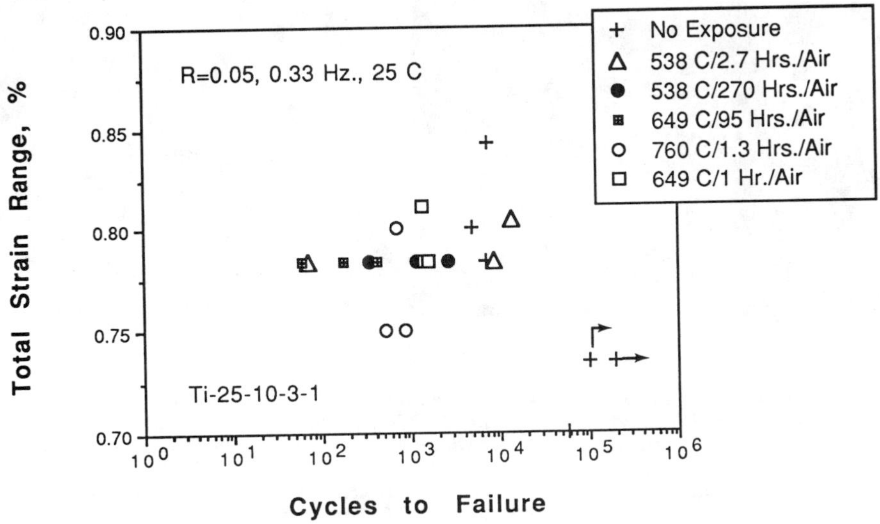

Figure 2. Effect of Air Preexposure on 25°C LCF Life of Ti-25-10-3-1. Long Time Preexposures at 649°C and 538°C, as well as Short Time Preexposures at 760°C Result in a 10^2 Cycles Life Debit at 25°C. Run Out Life = $1X10^5$ Cycles

On exposure to high temperature air, Ti-25-10-3-1 formed a brittle, oxygen rich/ alpha-2 stabilized outer core,as shown in Figure 3a . Fatigue crack initiation appeared to be form the presence of an outer mantle (3-5 microns thick) at an aligned alpha-2 colony. The presence of cracks only in this outer mantle belies the brittle nature of the mantle. Cross-sectional microscopy close to the initiation site, confirmed the presence of an alpha-case wherein cracks perpendicular to the stress axial, were localized(Figure 3b). These cracks are microstructure-independent.

Molten Salt Preexposures and 25°C LCF Life Debit

The effects of molten salt preexposure on room temperature LCF life were dramatic. These results are shown in Figure 4. Preexposure in SO_2/SO_3 gaseous environments at 760°C for 100 hours with and without Sulfate-bearing salts on the surface resulted in a dramatic reduction in strength. Samples tested at strain range = 0.784% following preexposure, failed upon loading. The breaking stress measured was significantly lower than the ultimate tensile strength of the unexposed alloy at 25°C (570 MPa compared to 725 MPa).

The strength reduction attributable to the presence of molten salt during preexposure was greater than the corresponding debit in the absence of salt. Samples that were preexposed in the SO_2/SO_3 gas environment but without any applied sulfate salt on the surface, a life of 840 cycles was recorded at strain range =0.665%, compared to a life of 0-26 cycles following identical exposures, but with sulfate salts on the surface. The molten salt thus tends to further increase the level of attack and resulting LCF debit life.

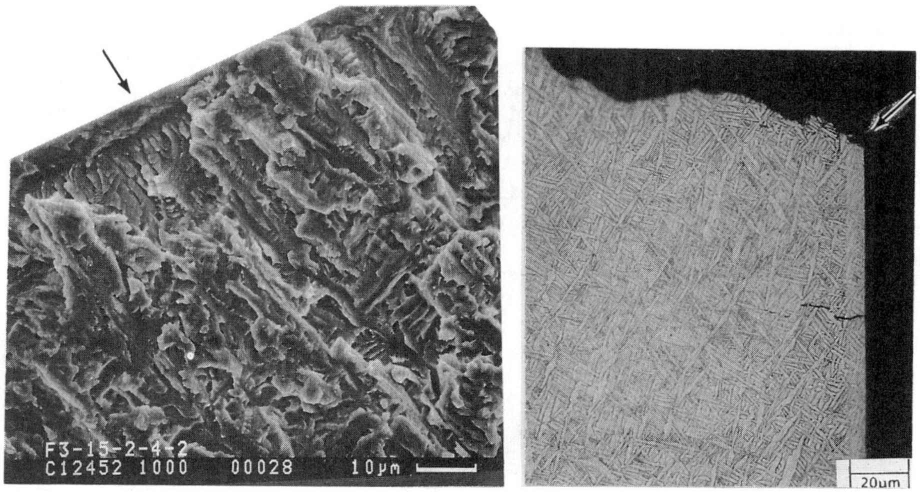

(a) (b)

Figure 3. Failure Analysis of Sample Preexposed at 649°C for 95 Hours in Laboratory Air, and Tested at 25°C Under LCF Conditions (a) SEM of Fracture Surface.(b) Cross Sectional Optical Microscopy in the Vicinity of the Initiation Site Showing Surface Cracks Normal to the Stress Axis and a 10 Microns Thick Alpha-Case. Kroll's Etch. Initiation Site Indicated by Arrow.

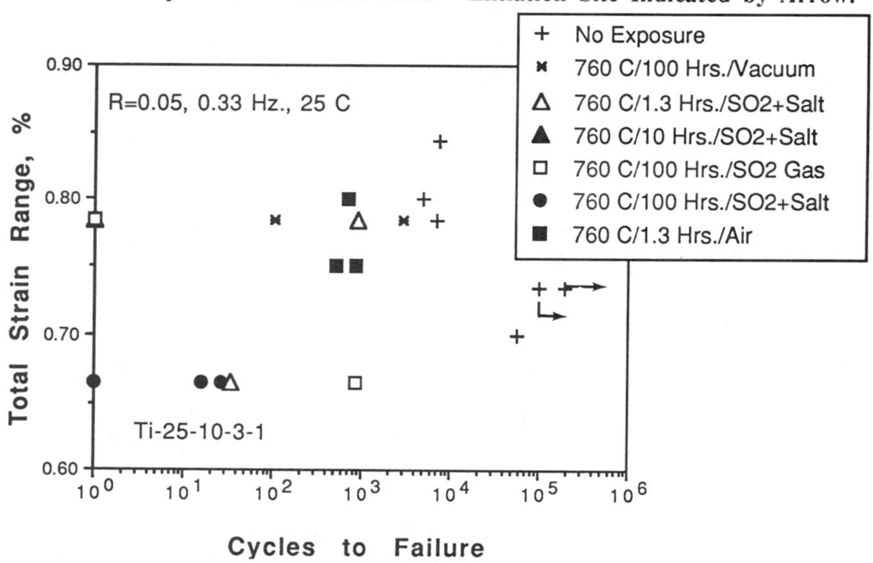

Figure 4. Effect of Molten Salt Preexposure on 25°C LCF Life of Ti-25-10-3-1. Preexposures Result in Significant Debit in LCF Life Compared to Unexposed Condition. Run Out Life = 1x10^5 Cycles.

Figure 5 shows features similar to that seen in air exposed samples: a brittle outer mantle,as well as the presence of lateral cracks in the alpha-case. This sample was exposed at 760°C for 1 hour. Increasing preexpsoure time resulted in deeper alpha cases and a further debit in LCF life.

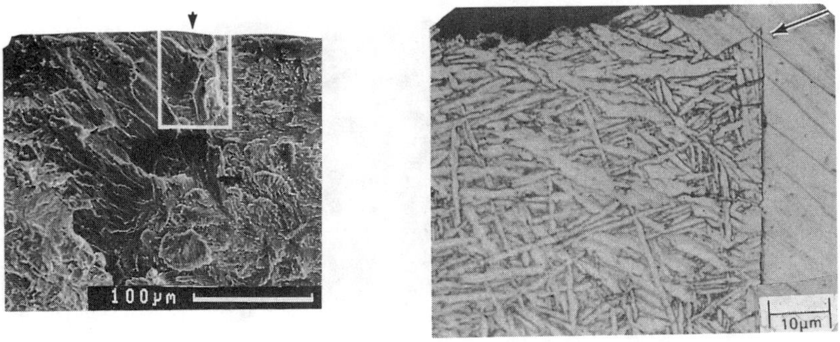

(a) (b)

Figure 5. Failure Analyses of Sample Which was Preexposed at 760 C for 1 Hour, with Salt on the Surface, and Tested at 25°C Under LCF Conditions (a) SEM Fractography, (b) Cross Sectional OM on The Vicinity of the Origin . Initiation Site Indicated by Arrow.

Hot Salt Preexposure and 25°C LCF Life Debit

Hot salt preexposure reduced room temperature LCF life by two orders of magnitude (Figure 6). Microstructural examination revealed that the high susceptibility to hot salt attack occurred by pitting- type attack along alpha-2/beta interfaces and at alpha-2 bearing grain boundaries (Figure 7).

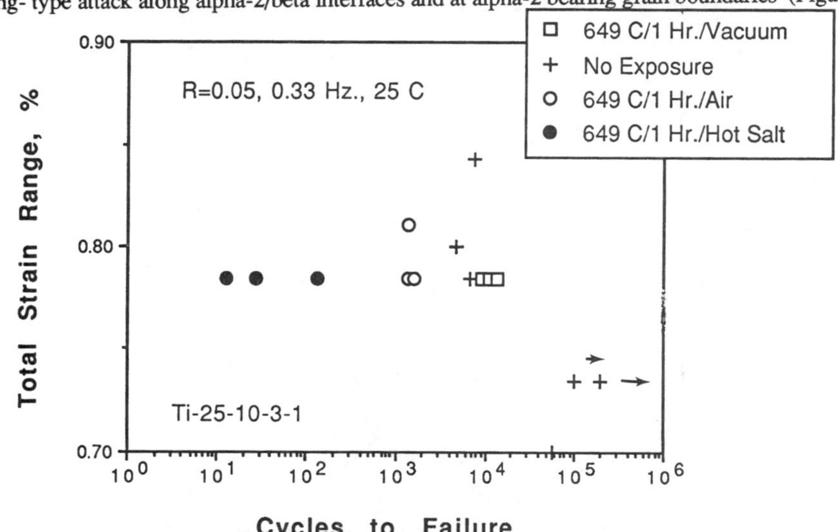

Figure 6. Effect of ASTM Sea-Salt Preexposure on 25°C LCF Life of Ti-25-10-3-1. Hot Salt Effects Were Severe and Exceed Debits Attributable to Exposure to Hot Air. Run Out Life = 1x10^5 Cycles.

Air Preexposures and 25°C HCF Life

Unlike under LCF conditions, air preexposures resulted in little change in 10^6 runout stresses (Figure 8). The scatter in the data was within 10^2 cycles, which is typical to engineering alloys in the HCF regime.

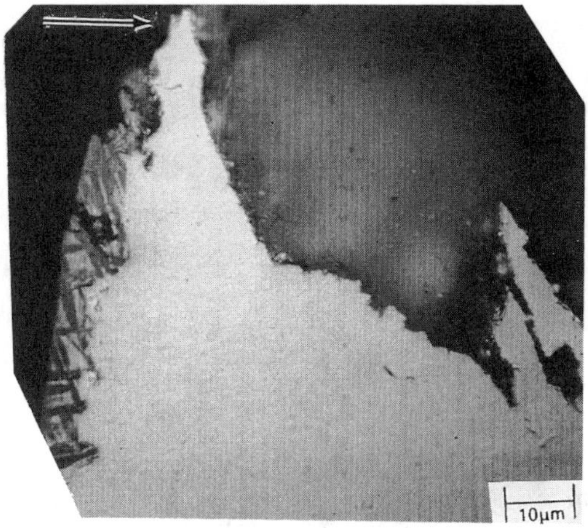

Figure 7. Cross Sectional OM of Sample Preexposed Under Hot Salt Conditions and Tested at 25°C Under LCF Conditions. As-Polished Structure. Initiation Site Indicated by Arrow. Pitting of Sample is Detected.

All initiation sites were at the sample surface. All samples were characterized by a single initiation site. As shown in Figure 9, the initiation sites were generally associated with locations where a grain boundary containing "feathery" alpha-2 intersected the sample surface.

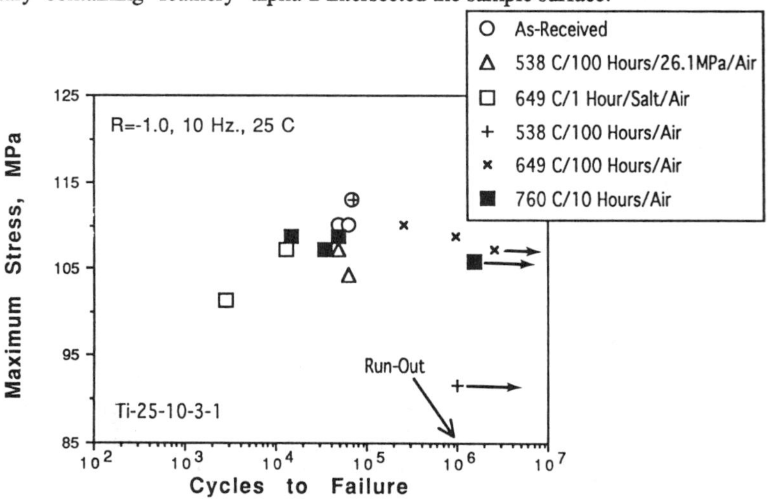

Figure 8. Effect of Preexposure in Various Environments on 25°C HCF Behavior of Ti-25-10-3-1. Preexposures in air at 538°C, 649°C, and 760°C did not Cause any Significant Loss in 25°C HCF Lives Under Conditions that Resulted in 10^4-10^5 Cycles Life. Run Out Stress = $1x10^6$ Cycles Life.

(a)

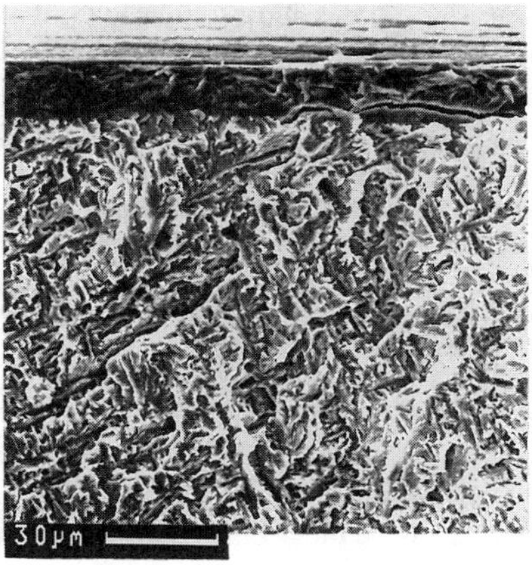

(b)

Figure 9. SEM Fractographs of a Sample Preexposed at 760°C for 10 Hours in Laboratory Air and Tested at 25°C Under HCF Conditions. Initiation Site Indicated by Arrow in (a). Cracks in Brittle Alpha-case are Shown in (b).

Hot Salt Preexposures and 25°C HCF Life Debit

Hot salt preexposure significantly reduced HCF life compared to baseline unexposed samples (Figure 8). Failure analyses of the sample revealed severe attack along alpha-2/beta interface and grain boundaries akin to that seen in the LCF test sample shown in Figure 7.

Discussion

The total fatigue life, N_f, of a smooth sample is composed of two constituents, N_i, the number of cycles to initiation of a fatigue crack, and N_p, the number of cycles required to propagate the crack. The primary effect of preexposures, in air or other aggressive environments, is to reduce N_i. On exposures to high temperature air, Ti-25-10-3-1 formed a brittle alpha-case. Increasing exposure times and/or temperatures resulted in deeper alpha cases as well as a higher oxygen/nitrogen content within the alpha-case. These higher interstitial levels decrease the strain tolerance of the alpha-case. The deeper alpha-case permits the initiation of deeper cracks during the fatigue tests, depending on the imposed strain levels. Hot salt attack is severe and produces a notch-like crack at microstructural features that are surface connected, resulting in the formation of a defect zone that is notched and can reduce life.

LCF Conditions

Under LCF conditions, the mechanism of air-induced debit at 25°C is predicated upon the formation of brittle, deep alpha-cases. The high strain levels imposed during the test induce cracks in this alpha-case, probably in the early stages of the LCF test. These cracks act as very effective notches for the subsequent rapid crack propagation stage. The reduced LCF lifes, in Ti-25-10-3-1, are thus a reflection of the reduced level of N_i. Reduced LCF lives following preexposures in molten salt environments are also a result of formation of alpha-cases at the surface of the specimen. The mechanisms for LCF life debit, following molten salt preexposures, are thus similar to that discussed for the air preexposures.

Hot salt attack in Ti-25-10-3-1 was severe. The attack causes significant damage to the sample surface (formation of pits) so as to embrittle the surface. The embrittled layer, and the severe notch sensitivity of this alloy, results in the dramatic LCF life debit.

HCF Conditions

The formation of a brittle alpha-case did not play a significant role in defining post air-exposure HCF life at 25°C. This is most likely a result of the low total strain imposed on the surface during HCF testing. Thus the strain-tolerance of the alpha-case was not exceed in the test. N_i is not reduced, thus N_t is not severely affected.

Preexposures in hot salt environment resulted in significant reduction in HCF life. The surface damage increased the notch sensitivity of the sample locally, resulting in a reduced life when compared to a smooth, damage-free sample.

Conclusions

The poor oxidation resistance and hot salt resistance of Ti-25-10-3-1 play a major role in defining the extent of fatigue life debits when subsequently tested at 25°C. The following conclusions are drawn from this study:

1. Two decades of LCF life debits were measured following short time preexposures at 760°C, and 270 hours preexposures at 538°C in air. This is attributable to the formation of brittle alpha-cases on the sample surface during preexposures, and the poor crack growth behavior in this alloy. The alpha-cases crack due to their low strain tolerance, thereby reducing N_i.

2. The lower total strains imposed on the sample during a HCF test results in no crack formation in these alpha-cases in the early stages of the fatigue tests. This resulted in minimal effect of air preexposures on 25°C HCF life.

3. Pre-exposures in molten sulfate environments resulted in dramatic reductions in fatigue life. Mechanisms of fatigue life debit were similar to those due to hot air pre-exposures.

4. A short time preexposure to hot salt, without an externally applied creep stress, was sufficient to cause severe embrittlement of the alloy measured as a dramatic reduction in LCF and HCF life at 25°C.

Acknowledgements

This work was supported by WL/MLLN, Dayton, OH as USAF Program No. F33615-87-C-5295. Paul Smith is the Program Monitor at WL. The support of the Mechanical Properties Test Laboratory staff at GED (Mr. C. Desadier and Mr. D. Chamblee) in the fatigue testing is gratefully acknowledged. The contributions of the Metallography Laboratory at GED are appreciated. The molten salt preexposures were conducted at Materials Engineering Associates, Lanham MD.

References

1. C.H. Ward and S.J. Balsone, "The Effect of Salient Microstructural Constituents on Tensile and Creep Deformation of Ti-25Al-10Nb-3V-1Mo", in "Microstructure/Property Relationships in Titanium Aluminides and Alloys", (Eds. Y-W. Kim and R.R. Boyer), TMS, Warrendale, PA, pg.373, 1991.

2. P.B. Aswath and S. Suresh, "Fatigue Crack Growth Behavior in a Titanium Aluminide Intermetallic", Materials Science and Engineering, Vol. 114, pg. L5, 1989.

3. P.S. Godavarti and M.D. Lipschutz, "Environmental Effects on Titanium Aluminides", USAF Contract F33615-87-C-5295, Final Report, In Press.

The Isothermal Forging of the Ti_3Al Based Alloy, Superalpha 2.

J.C.F.Millett, J.W.Brooks*, I.P.Jones.

Interdisciplinary Research Centre for Materials in High Performance Applications, University of Birmingham, Elms Road, Edgbaston, Birmingham, B15 2TT, United Kingdom.

*INCO Engineered Products LTD, 28-30 Derby Road, Melbourne, Derbyshire, United Kingdom.

Abstract

The Ti_3Al based alloy Superalpha 2 has been deformed in compression at constant strain rate, over a range of temperatures and strain rates, 900°C - 1050°C and $0.0003s^{-1}$ - $0.035s^{-1}$ respectively. Derivation of the values of activation energy of deformation and strain rate sensitivity have shown that dynamic recrystallisation and grain boundary sliding have approximately equal roles, and metallographic examination of the forged specimens shows that primary α_2 break up during deformation is also a significant factor. The aforementioned material values have also been used in a constitutive model so that experimental and theoretical flow curves can be compared.

Introduction

The development of intermetallic materials has reached the stage where commercial quantities are available for process development. The work presented in this paper is part of a wider study of the hot deformation behaviour of intermetallics which is aimed at defining production routes for isothermally forged parts using process modelling. Ti_3Al based alloys are seen as potentially useful high temperature materials, whilst isothermal forging has been chosen since the effects of temperature and strain rate, and their effects on microstructure and properties, can be rigorously examined.

Titanium '92
Science and Technology
Edited by F.H. Froes and I. Caplan
The Minerals, Metals & Materials Society, 1993

Experimental Procedure

The alloy, composition Ti-25Al-10Nb-3V-1Mo, used in this investigation was supplied by TIMET as forged 150 mm diameter billet in the alpha-beta condition. Cylindrical compression specimens (15 mm dia.x26 mm) were machined from a longitudinal orientation, and coated with a boro-silicate glaze to prevent oxidation and alpha case formation during deformation. Compression tests took place on a Zwick 1484 tensile machine, to a reduction in height of 50%, with a decreasing crosshead speed to provide a genuinely constant strain rate. Test temperatures were in the range 900°C to 1050°C and strain rates $3.0x10^{-4}s^{-1}$ to $3.5x10^{-2}s^{-1}$. After deformation, each specimen was quenched into water, and then sectioned, ground and polished using standard metallographic techniques. Etching was done by immersion in an aqueous solution of 2% HF and 10% HNO_3 by volume. Micrographs were taken using a JEOL 5200 Scanning Electron Microscope using secondary electron imaging. As forged grain sizes were revealed by exposure of polished surfaces to air at 500°C for 15 minutes. A comparative series of tests, using heat treatments alone was carried out at 1025°C to assess the effect of the heat treatment necessarily included in the hot deformation.

Results

The microstructure from the as-received billet is shown in figure 1. The microstructure consists of primary α_2 particles (marked a) in a transformed beta matrix. The original beta grain size was observed to be of the order of 2-3 mm. In this respect, superalpha 2 seems to be similar to a forged conventional titanium alloy.

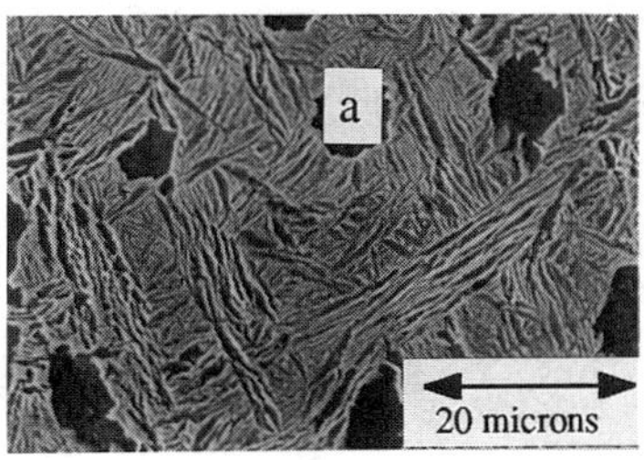

Figure 1. Microstructure of As Received Billet.

True stress-strain curves in figure 2 were produced by assuming conservation of volume throughout each compression test. Arrhenius plots of strain rate and temperature (figure 3) give an activation energy of deformation of 436.5 kJ mol^{-1}. A log-log plot of stress versus temperature compensated strain rate, Z (figure 4) gives a strain rate sensitivity, m, of 0.268.

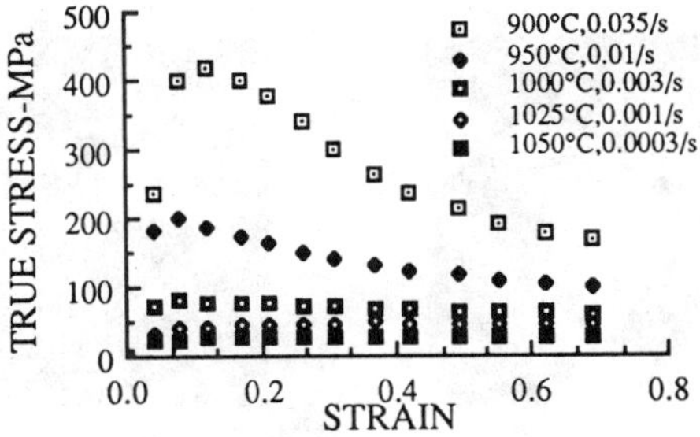

Figure 2. True Stress-Strain Curves for Superalpha 2 at Elevated Temperatures.

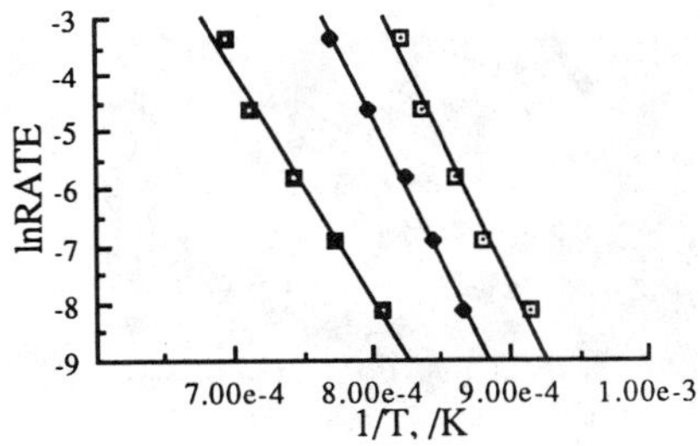

150 MPa, Q=426 kJ mol^{-1}.

100 MPa, Q=418 kJ mol^{-1}.

50 MPa, Q=338 kJ mol^{-1}.

Figure 3. Strain Rate vs Temperature For Constant Stress Levels.

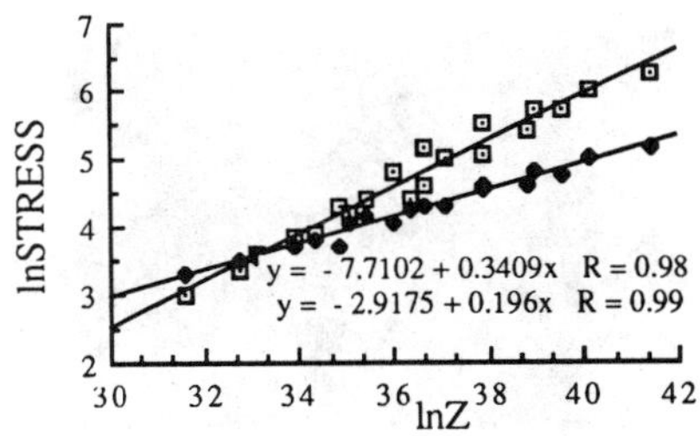

Figure 4. Stress vs Temperature Compensated Strain Rate.

The values of Q and m displayed in the above figures were obtained for the steady state condition where stress becomes constant in the flow curve. The actual values quoted above take into account an "initial" stress associated with the "starting" microstructure.

Forged microstructures for T=950°C and 1025°C and $\dot{\varepsilon}$=3.5x10^{-2}s^{-1} and 3.0x10^{-4}s^{-1} are shown in figure 5 and 6. These show that as temperature increases the amounts of both primary and secondary α_2 particles decrease. With decreasing strain rate, the amount of secondary α_2 decreases. Microstructures of specimens held at 1025°C (figure 7) show that whilst the amount of primary α_2 remains about the same, there is virtually no secondary α_2 at all.

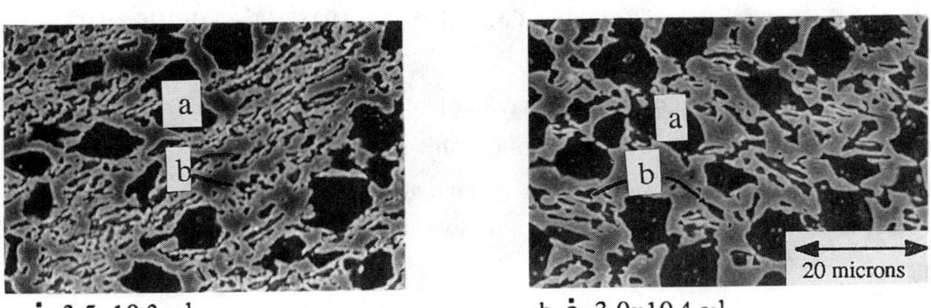

a. $\dot{\varepsilon}$=3.5x10^{-2} s^{-1}.

b. $\dot{\varepsilon}$=3.0x10^{-4} s^{-1}.

Figure 5. Microstructures of Forged Superalpha-2 at 950°C.

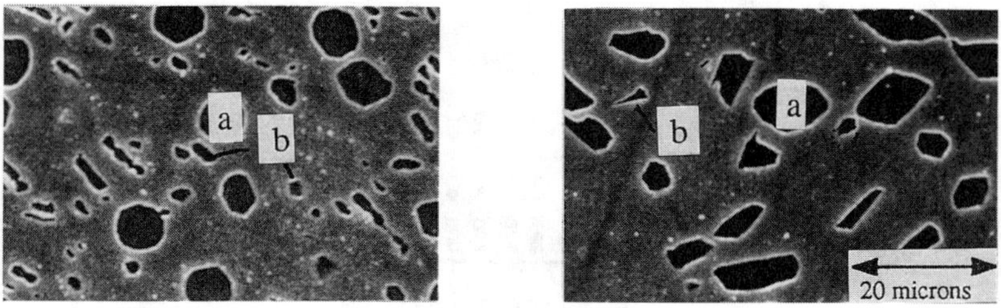

a. $\dot{\varepsilon}=3.5\times10^{-2}\ s^{-1}$.   b. $\dot{\varepsilon}=3.0\times10^{-4}\ s^{-1}$.

Figure 6. Microstructures of Forged Superalpha 2 at 1025°C.

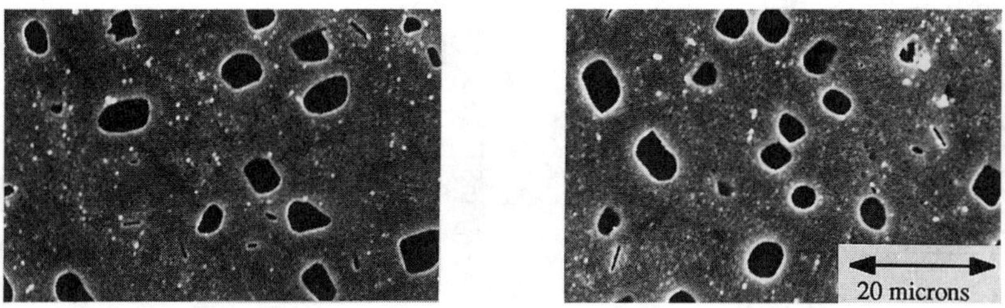

a. T=1025°C, ($\dot{\varepsilon}=3.5\times10^{-2}\ s^{-1}$.)   b. T=1025°C, ($\dot{\varepsilon}=3.0\times10^{-4}\ s^{-1}$.)

Figure 7. Microstructures of Specimens Held at 1025°C for the Equivalent Time of Strain.

Figure 8. shows the forged grain size from the conditions 1025°C, $3\times10^{-3}s^{-1}$. This shows a grain size of the order of a few millimetres, and is typical for the entire temperature/strain rate range. This indicates that the forged grain size is independent of forging conditions.

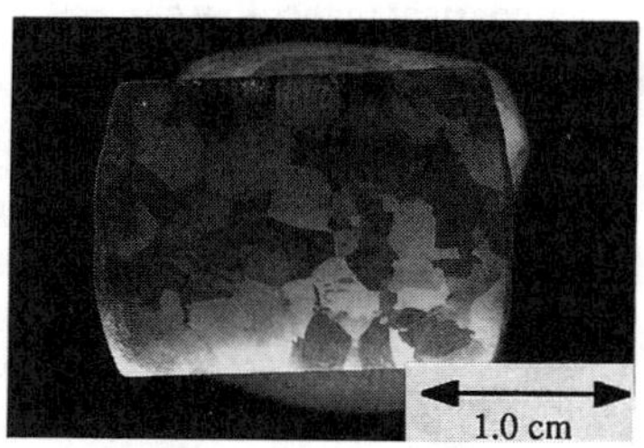

Figure 8. Typical as Forged Grain Size. (T=1025°C, $\dot{\varepsilon}=3.0\times10^{-3}\ s^{-1}$.)

The algorithm used to describe flow behaviour has been developed by Bate (1), and is used by IEPL to model successfully a wide range of metallic materials.

The precise nature of the flow curve for a given material is dependent on the temperature T and strain rate, $\dot{\varepsilon}$. It is convenient to combine the two into a single variable, the Zener-Holloman Parameter, Z:-

$$Z = \dot{\varepsilon} \exp\left[\frac{Q}{RT}\right] \qquad (1)$$

where Q is the activation energy of deformation and R is the universal gas constant.

The flow curve also depends upon the microstructural state. Usually, this is taken to be grain size, but factors such as dislocation density, second phase distribution etc are involved. Therefore, a single parameter λ, is used to cover all of these and hence direct comparison between λ and any one microstructural feature is precluded. It is assumed that flow stress has a linear relationship with λ. Therefore, the equilibrium stress σ, at any point during the test may then depend on Z by a simple power law:-

$$\sigma = K\lambda Z^{m} \qquad (2)$$

where K is a constant and m is the strain rate sensitivity.

Since equilibrium stress is expected to depend on microstructure, it may reasonably be assumed that equilibrium microstructure, λ_{ss} is dependent on Z only, again by a simple power law:-

$$\lambda_{ss} = \lambda_0 Z^{q} \qquad (3)$$

where λ_o is a constant and q is the sensitivity of microstructure to Z.

Figure 2 shows that stress (and presumably microstructure) tend towards a steady state value with increasing strain, most likely in an exponential way.

Hence:-

$$\frac{d\lambda}{d\varepsilon} = \frac{\lambda_{ss} - \lambda}{\alpha} \qquad (4)$$

and

$$\frac{d\sigma}{d\varepsilon} = \frac{\sigma_{ss} - \sigma}{\beta} \qquad (5)$$

where λ_{ss} and σ_{ss} are the steady state values of microstructure and stress respectively, and α and β are exponential damping factors.

The experimental and predicted flow curves presented in figure 9 show that there is a reasonable correlation between the two, indicating that the algorithm derived above can outline the deformation of superalpha 2.

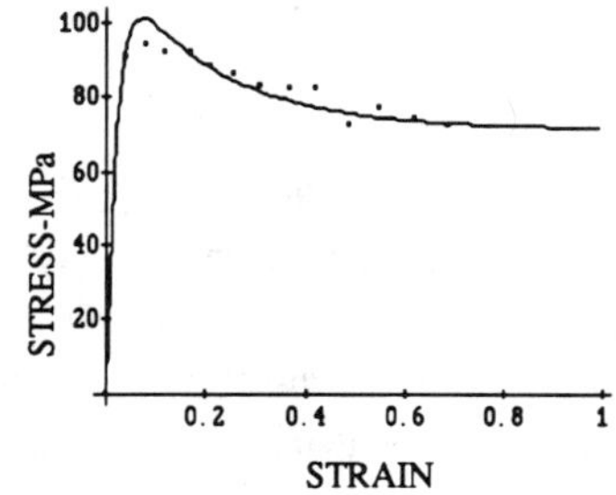

a. 1000°C, 0.01/s

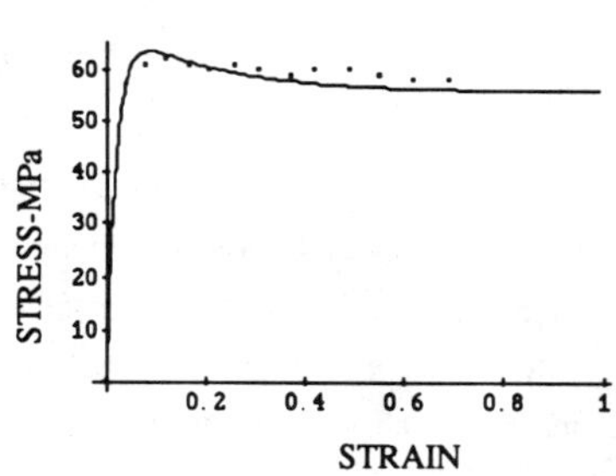

b. 1025°C, 0.003/s

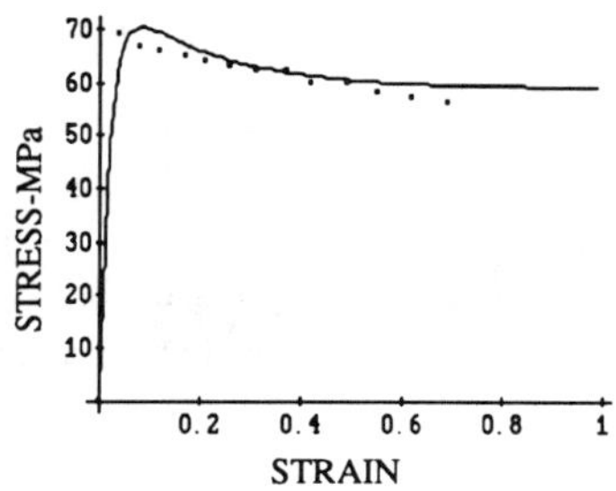

c. 1050°C, 0.01/s

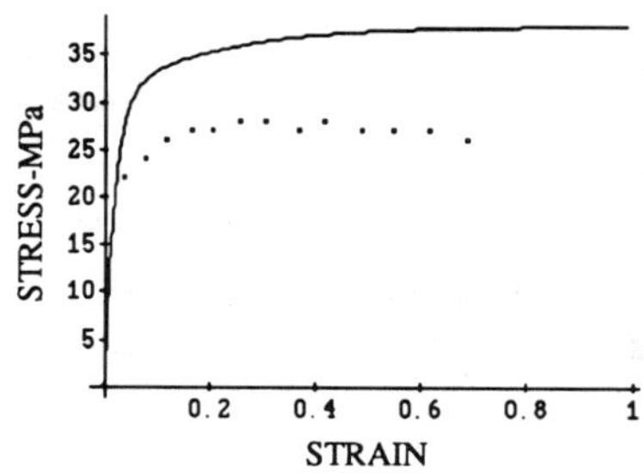

d. 1050°C, 0.0003/s

Figure 9. Comparison of Experimental (dotted) and Predicted (solid) Stress-Strain Curves.

Discussion

The flow curves in figure 2 show that the degree of flow softening decreases with increasing temperature and decreasing strain rate. This is consistent with a wide variety of materials, including conventional titanium alloys (2) and nickel base superalloys (3), in which dynamic recrystallisation plays an important role in hot deformation. How this material deforms can be understood better by examining various material constants such as activation energy of deformation, Q and strain rate sensitivity, m. From the results in figure 3, an activation energy of deformation of 436.5 kJ mol^{-1} was calculated. This is not the activation of any one process, and must be analysed by comparing with other values obtained by different methods. From background work (4) carried out for this investigation, the boundaries of the α_2-B2 phase field, where the majority of the forging trials took place, were placed at 937°C and 1080°C. From this an activation energy of α_2 dissolution in the B2 phase of 606 kJ mol^{-1}was calculated. A recent study of the superplasticity of superalpha 2 shows an activation energy of deformation of 308 kJ mol^{-1} (5), which can be assumed to be associated with a grain boundary diffusion process. This value is aproximately half that for the α_2 dissolution in the B2 phase which is assumed to be the bulk diffusion activation energy. Since the activation energy of bulk diffusion is usually considered to be twice that for grain boundary diffusion (7), the present activation energy

(436.5 kJ mol^{-1}) of deformation was initially thought to be the result of roughly equal proportions of bulk diffusion process (dynamic recrystallisation) and grain boundary sliding (superplastic flow). However, closer examination of the strain rate/temperature plots in figure 3 suggest that the situation may be more complex. Values of activation energy were calculated by assuming that a curve of $\ln\dot{\varepsilon}$ against 1/T would give a straight line, using standard linear regression techniques, the gradient being the quantity Q/R. This method therefore gives an average Q/R covering the entire temperature/strain rate matrix, and does not take into account that Q/R may vary according to the forging conditions. From figure 3, it can then be seen that there are, in fact two gradients or Q/R values for each plot, and so instead of dynamic recrystallisation and grain boundary sliding taking place concurrently, there is a transition point between the two.

Strain rate sensitivity, m, was calculated to be 0.268. This can be compared to values greater than 0.3, necessary for complete superplastic deformation, and around 0.1 for both conventional deformation and creep based processes (8). The intermediate value of m calculated in this investigation is an indication that no single deformation process is operating, but rather a combination of processes. Combined with the low strain rates used in this study, it suggests that there is a transition between a superplastic-dynamic recrystallisation process and a creep based process.

Examination of the forged microstructures (figures 5 and 6) shows that as temperature increases, the proportion of α_2 in the microstructure decreases, as would be expected. A more interesting feature is the behaviour of the secondary α_2 with decreasing strain rate. An overall decrease indicates that at lower strain rates, secondary α_2 has more time to dissolve and reprecipitate on the larger primary α_2 particles. If a comparison is made with samples that have received a heat treatment alone (figure 7) it can be seen that there is very little, if any, secondary α_2. This shows that α_2 breakup is possible during deformation, and hence must have a significant effect upon flow properties. Comparing this to a conventional titanium alloy (2), no break up of the analogous alpha phase can be seen. Therefore, instead of the softer "beta" phase deforming around the harder "alpha" particles, in superalpha 2, at least some of the deformation is localised in the α_2 particles. This appears to be consistent with the recent report (6) that when quenched from the α_2-B2 phase field, the B2 phase was the harder phase.

From figure 8, it is clear that the α_2 phase does not prevent B2 grain growth in this alloy. Thus the short time required to remove the samples from the testing rig was sufficient to allow grain growth to occur, and as a consequence the forged grain sizes are of the order of a few millimetres, indicating that the B2 grain size cannot be controlled easily by hot working. It is possible that prior working lower in the α_2-B2 field could provide some control due to the higher α_2 volume fraction, but this is not likely.

The modelling of the flow curves using the algorithm described in the previous section would seem to describe adequately the deformation of superalpha 2 although at higher temperatures and lower strain rates predicted stresses are rather higher than those gained experimentally. This can be understood by the fact that creep becomes more important at higher temperatures and lower strain rates, but was not taken into account in the model.

Conclusions

Hot deformation of superalpha 2 takes place via a combination of dynamic recrystallisation, grain boundary sliding and primary α_2 breakup. At higher temperatures and lower strain rates, creep plays an increasingly important role. Rapid grain growth immediately after forging makes grain size control difficult in superalpha 2.

Acknowledgements

The authors would like to thank the Science and Engineering Research Council and INCO Engineered Products Limited for financial support during this project, Rolls Royce for supplying the material and Mr. Simon Day for doing most of the compression tests.

References

1. P.Bate "Modelling of Recovery and Recrystallisation During Hot Working". (Report No.9 Project No. 1528 "Isothermal Forging Development". IEPL. 1987.)

2. P.S.Bate, P.L.Blackwell and J.W.Brooks "Thermo-Mechanical Processing of Titanium IMI 834" Sixth World Conference on Titanium, France (1988) 287-292

3. J-P.A.Immarigeon and P.H.Floyd "Microstructural Instabilities During Superplastic Forging Of a Nickel-Base Superalloy Compact" Met Trans A 12 (1981) 1177-1186

4. J.C.F.Millett "The Isothermal Forging of the Ti_3Al based Alloy, Superalpha 2" (M.Phil Thesis, University of Birmingham 1992)

5. H.S.Yang, P.Jin, E.Dalder and A.K.Mukherjee "Superplasticity of a Ti_3Al-Base Alloy Stabilized With Nb, V and Mo." Scripta Met 25 (1991) 1223-1228

6. A.K.Gogia, D.Banerjee and T.K.Nandy "Structure, Tensile Deformation and Fracture of a Ti_3Al-Nb Alloy" To be published.

7. R.E.Smallman "Modern Physical Metallurgy" Fourth Edition.

8. G.J.Davis, J.W.Edington, C.P.Cutler, K.Padmanabhan "Superplasticity: A Review" J. Mat. Sci 5 (1970) 1091-1102.

SPRAY FORMING OF TI-ALUMINIDE-BASED ALLOYS

J. W. Sears**, G. Itoh*, and M. H. Loretto*

*IRC in Materials for High Performance Applications
The University of Birmingham, Edgbaston, England B15 2TT,

**ManLabs, A Division of Alcan Aluminium Corporation
21 Erie Street, Cambridge, Massachusetts, 02139

ABSTRACT

The aim of this programme is to spray form Ti Aluminides free from porosity and interstitial contamination. Initial work was carried out by atomising IMI-318 (Ti-6-4) and Ti-aluminide alloys to assess the operation of the cold-wall induction bottom-pour (CWIBP) as a clean melting system. Ti-aluminide-based alloys have been sprayed onto flat and tubular substrates. The IRC plasma melting system at The University of Birmingham, Edgbaston, England, was used to provide ingot feed stock for the spray deposition and atomisation. The plasma system, used for casting ingots, was able to be converted for atomisation or spray deposition. Spray forming was facilitated by the use of a scanning or centrifugal atomiser. The operation of this process will be discussed and the effects of various operating parameters on the spray deposit product will be presented. The chemistry, microstructure, homogeneity and macrostructure of the as sprayed material has been examined.

INTRODUCTION

The Ti aluminides form a technologically important group of intermetallic compounds. TiAl (γ) is one of these which exhibits great promise for commercial exploitation, being strong at elevated temperatures while also exhibiting superior oxidation resistance. However, the compound is brittle at ambient temperatures and this limits its application. Possible sources of this brittleness include compound purity, planarity of slip and the nucleation of voids at the points of intersection of deformation twins. Some attempts have been made to improve the room temperature ductility by alloying additions, the introduction of a dispersion of hard second phases, and the control of grain size. Consequently, there has been considerable interest in the application of powder metallurgy (PM) and rapid solidification processing (RSP) to the preparation of these materials. Research has been focused on IM route by isothermal forging and the powder metallurgy (PM) approach through powder consolidation via HIP, plasma spraying and mechanical alloying. Although, there has been considerable success in the IM and PM techniques, they have proved to be very expensive[1]. The spray forming of a deposit directly from the melt offers a more economical process to produce such difficult to form materials. As will be shown, deposits of Ti-aluminides can be produced directly from a stream of molten metal without the handling problems associated with the P M route.

Titanium '92
Science and Technology
Edited by F.H. Froes and I. Caplan
The Minerals, Metals & Materials Society, 1993

It has been pointed out that there are as many as five different solid phases possible for an alloy of composition Ti-52Al which has been undercooled ≈200 K below the equilibrium liquidus, these being β (bcc), α (hcp), and the intermetallic compounds γ ($L1_0$), B2, and α_2 ($D0_{19}$)[2]. It is known that when processing with the cold-wall induction bottom-pour (CWIBP), described below, that C and O_2 can be introduced into the powders. Carbon is highly insoluble in TiAl, and the compound Ti_2AlC may precipitate during consolidation and/or heat-treatment of the powder[3]. Part of the investigation, involving the powders and spray deposits, will be the examination of Ti-alloy particles before and after heat treatment to determine the extent of the C contamination. Oxygen affects the beta transus temperature of these alloys and adds to the material inherent brittleness.

COLD-WALL INDUCTION BOTTOM-POUR (CWIBP)

The CWIBP system includes a melt chamber capable of being pressurised to 170 kPa. This allows for the flexibility to ramp an overpressure in the melt chamber to account for the change in metalistatic head during the metal pour. A controlled overpressure allows for a constant metal flow rate which is critical in producing uniform spray deposits. The cold-wall crucible is a segmented, water-cooled, induction heated device. An induction-heated graphite nozzle at the base of this crucible constricts the molten metal into a stream for atomisation and spray forming.

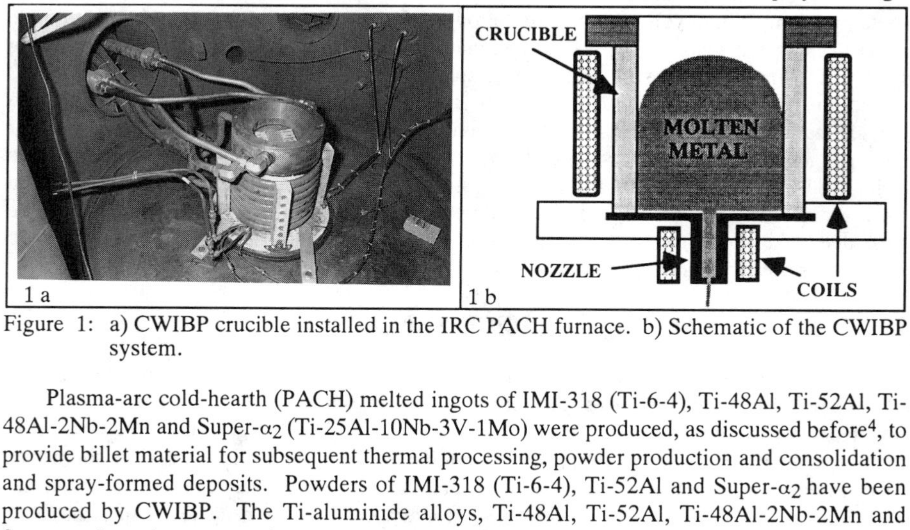

Figure 1: a) CWIBP crucible installed in the IRC PACH furnace. b) Schematic of the CWIBP system.

Plasma-arc cold-hearth (PACH) melted ingots of IMI-318 (Ti-6-4), Ti-48Al, Ti-52Al, Ti-48Al-2Nb-2Mn and Super-α_2 (Ti-25Al-10Nb-3V-1Mo) were produced, as discussed before[4], to provide billet material for subsequent thermal processing, powder production and consolidation and spray-formed deposits. Powders of IMI-318 (Ti-6-4), Ti-52Al and Super-α_2 have been produced by CWIBP. The Ti-aluminide alloys, Ti-48Al, Ti-52Al, Ti-48Al-2Nb-2Mn and Super-α_2 have been spray formed onto sheet and tubes. The spray system utilises the same CWIBP crucible that was used in powder production to provide a controlled, molten stream of metal to a scanning atomiser for the production of flat or drum preforms or to a centrifugal atomiser (CSD) for the production of rings. A photograph and schematic of the CWIPB device is shown in Figure 1(a) and (b). The operation of this process has been discussed before[5]. The effects of various operating parameters on the spray deposit product will be presented. The chemistry, homogeneity and microstructure of the as sprayed material has been examined and compared with material made through both the PM and IM routes. The plasma-melted, as-cast and heat-treated microstructures of Ti aluminides have been described elsewhere[6,7].

ATOMISATION

Alloys of IMI-318 (Ti-6-4), γ-Ti (Ti-52Al) and Super α_2-Ti were atomised. CWIBP served as the melt source to provide molten metal to the atomisation die for these alloys. Metal temperature was measured using an IRCON, two-colour pyrometer. Superheat of ≈ 30°C was obtained when melting IMI-318 (Ti-6-4) and ≈ 20°C for most of the Ti-aluminides. When the charge is completely melted, the nozzle is heated and molten metal begins to flow when the metal directly above the nozzle melts. Low powder yields, due to nozzle blow back and stream instability, were experienced during these early runs with lots of 3 to 6 kgs being produced from 25 kg charges (6 to 8 kgs retained in skull). Metal to gas flow rates of about 1:1 were planned. The details of the atomisation system has been described in detail before[3].

Shown in Table 1 is the data from the first atomisation runs. Mean particle sizes for these runs were higher than expected. Based on the data collected from these runs, it was found that the over pressure compensation was incorrect, resulting in higher than expected metal flow and therefore, low gas to metal flow rates resulting in larger powder particle size. Nozzle blockage by metal blow back was a main contributor to low yields. It is anticipated that a change to an lower nozzle angle (<15°) will correct the blow back problem.

Table 1. Operating conditions from the initial atomisation runs.

RUN #	ALLOY	CHARGE WEIGHT	SKULL WEIGHT	GAS/MET RATIO*	YIELD %**	MEAN SIZE@	NOZZLE DIAMETER	JET ANGLE
1	IMI-318	25 kgs	8.1 kgs	0.72	5.3	123 μm	5.0 mm	7.5°
2	IMI-318	24 kgs	7.9 kgs	0.41	2.0	133 μm	4.0 mm	10°
4	Ti-52Al	14 kgs	4.5 kgs	0.86	41.6	158 μm	5.0 mm	15°
5	Super-α2	27 kgs	7.2 kgs	0.62	9.3	144 μm	5.0 mm	15°

*Gas / Met Ratio = kgs min^{-1} Ar Gas ÷ kgs min^{-1} Metal, **Yield % = Total weight powder < 250 μm produced ÷ Weight of metal poured through atomiser, @Mean size is at the 50% cumulative weight for all powders < 250 μm.

IMI-318 (Ti-6-4) Powder

The powder particles shown in Figure 2(a) are representative of the material produced by this process. Initial chemistry results on IMI-318, Ti-52Al and Super-α_2 powder are shown in Table 2. Analysis was performed on several size fractions of powder with the larger size fractions consistently having less C and O_2 contamination. Results on IMI-318 powders indicated C levels from 0.06 to 0.08 wt.%,. This indicates that C pickup is from 0.04 to 0.06 wt.%, with 65 μm powder about an 0.06 wt.% increase and 150 μm powder an 0.04 wt.% increase. In the main, the increase in C contamination in the particles is most likely due to graphite nozzle erosion. Upon examination after these runs, it was found that some C-based insulating material broke-down as a result of the nozzle blockage. A C-based film was discovered coating the chamber walls, and would have coated the powders also. This would explain why the 65 μm particles have a higher concentration of C than the 150 μm powder, owing to the fact that smaller particles have a higher surface to volume ratio. For a similar reason, smaller particle have higher O_2 levels due to surface oxides.

Table 2: Ti Alloy Powder O_2 and C Levels from the Initial Atomisation Runs as a function of particle size.

RUN #	ALLOY	45-90 μm Oxygen ppm	125-180 μm Oxygen ppm	45-90 μm Carbon wt%	125-180 μm Carbon wt%
1	IMI-318	2900	2600	0.08	0.06
2	IMI-318	4100	2500	0.08	0.06
4	Ti-52Al	1050	900	0.02	0.01
5	Super-α2	1850	1000	0.09	0.04

A TEM micrograph of IMI-318, ≈50 μm powder, Figure 2(b), shows a martensitic structure, common in rapidly solidified conventional Ti alloys. It was noted that the microstructures coarsen progressively from the surface to the centre of the particle. This observation agrees with the radiation cooling model presented by Sastry, et. al.[8], further discussion of this topic is outside the context of this paper and will be presented later. The microstructure was devoid of second phase particles (e. g., TiC present in Ti alloys contaminated by C) and similar to other reports on Ti with low levels of C[9]. This is to be expected, since the C concentration in the powder was less than the solubility limit in Ti.

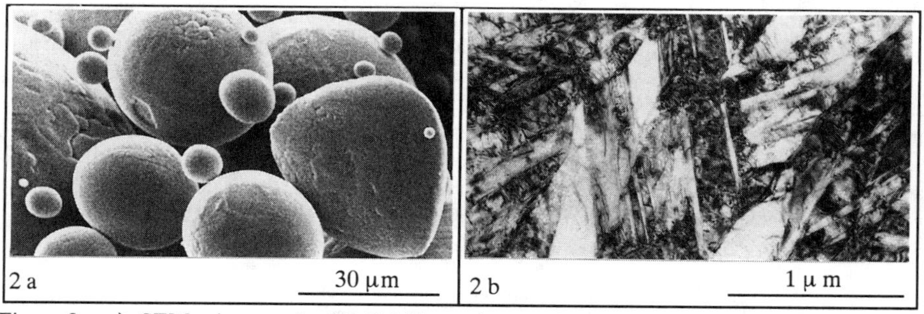

Figure 2: a) SEM micrograph of IMI-318 powders produced by CWIBP gas atomisation. b) TEM micrograph showing martensitic microstructure of IMI-318 gas atomised powder [10].

<u>Ti-52Al Powder</u>

The Ti-52Al powder, which was the best atomisation run (41.6 % yield), contained low C (0.01 to 0.02 wt.%) and O_2 (900 to 1050 ppm). These results indicate very low interstitial pickup during atomisation (See Table 2), considering that the billet material was PACH double-melted with blend levels at 600 ppm O_2 and 0.01wt.% C.

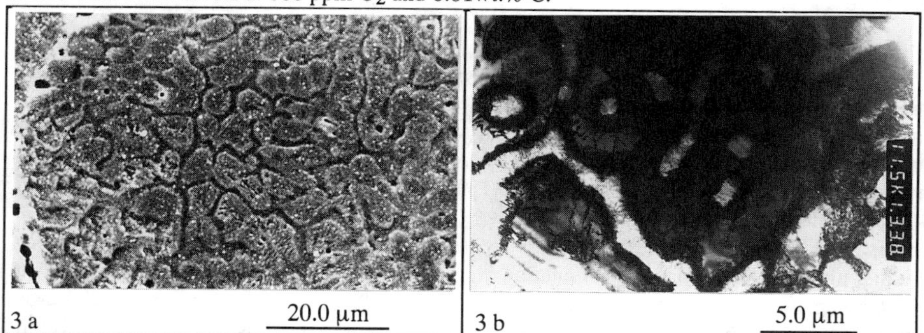

Figure 3: a) SEM micrograph of Ti-52Al as-atomised powder. b) TEM micrograph of Ti-52Al as-atomised powder[10].

Typical microstructures of gas atomised powders of Ti-52Al are shown in figure 3(a) and (b). As can be seen, the microstructure is dendritic, with the dendrites being Ti_3Al and the interdendritic regions being TiAl. These two phases have been identified by microdiffraction and X-ray studies[11]. When hot isostatically pressed (HIP'd) at 950°C for 4 hours under a pressure of 100 MPa, the dendritic microstructure transforms to equiaxed grains of TiAl. It is important to note that the microstructure consists of only one phases, namely TiAl, and that no carbides or nitrides, such as Ti_2Al(C or N) or TiC, are present.

<u>Super-alpha-2 powders</u>

The super-α_2 powders had slightly higher levels of O_2 and C than did the Ti-52Al powder, as shown in Table 2. It has been reported that O_2 content has a significant effect on the beta-transus of this alloy[12], therefore, maintaining low O_2 interstitial content is very important when processing this alloy. The powder produced in this run contained higher O_2 levels than would be desired. These powders were most likely contaminated due to a burnt o-ring that failed due to the nozzle blockage. O_2 levels similar to the Ti-52Al or better should be attainable.

A TEM micrograph of the super-α_2 powder, Figure 4(a), shows APB's indicating an ordered structure. In Figure 4(b), a SEM micrograph of a 60 μm powder cross-section shows the distribution of primary α_2 precipitates distributed through a β matrix. These observations are similar to other work in progress[13].

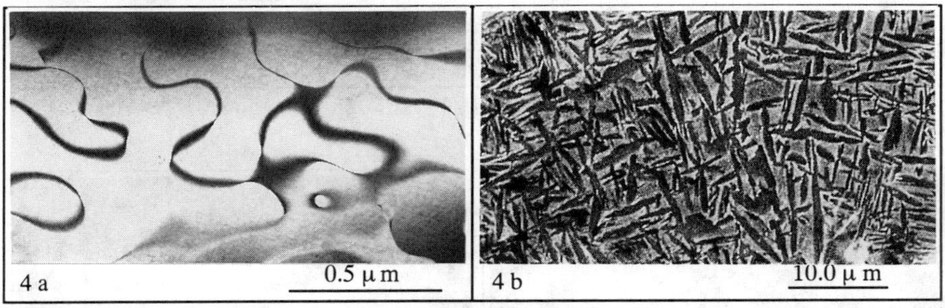

Figure 4: a) TEM micrograph of super-α_2 as-atomised powder particle showing APB's, b) SEM of a cross-section of a 60 μm as-atomised super-α_2 powder particle.

SPRAYFORMING

The sprayforming part of the system provided by Sprayforming Developments Limited (SDL), Swansea, U. K., is a novel three in one system. The three configurations of the system, shown in figure 5, allow the formation of Ti spray deposits in the form of thick sheet, tubular-sprayed on outside of a drum and tubular-sprayed on the inside of a mould. The first two methods use a pneumatic scanner coupled with a gas atomiser to spray metal. Spray formed sheet is produced by moving a flat substrate under the spray in a direction perpendicular to the scanning motion. Tubular shapes are formed by rotating a drum under the spray with the rotational direction perpendicular to the scanning motion. Spraying inside of a mould is accomplished by centrifugal spray deposition (CSD). CSD consists of a rotating disk which sprays molten metal on the inside of a mould. The mould is programmed to move up and down as the spray deposit builds up on the inside surface.

Figure 5: Photographs of the spray forming systems, sheet and tubes are formed using a scanning atomiser and tubular shapes are formed from Centrifugal Spray Deposition (CSD).

Drum Preform

The first spray-forming operations at the IRC were performed using Ti-aluminide ingot material (150 mm diameter), previously double-plasma melted, scanning onto a rotating drum preform. Objectives of these trials were to establish operating parameters. The data from Ti-aluminide spray-forming experiments (1 to 3) are shown in Table 3. A more detailed description of the spray forming operation has been presented elsewhere[3]. An example of a deposit is shown in figure 6(a), this preform was produced during run #1, scanning Ti-48Al onto the drum substrate, shown in Figure 5.

Table 3: Data from the first three spray-forming experiments using the 200 mm diameter drum substrate.

RUN #	ALLOY	CHARGE WEIGHT	SKULL WEIGHT	GAS/MET RATIO*	AT/SC GAS**	DRUM SPEED	NOZZLE DIA.	SUB. PRE-HT
1	Ti-48Al	12 kgs	3.5 kgs	0.19	2.1	300 rpm	5.0 mm	NONE
2	Ti-48-2-2	16.5 kgs	4.2 kgs	0.20	5.6	300 rpm	5.0 mm	200°C
3	Ti-52Al	16.5 kgs	11 kgs	0.30	6.2	100 rpm	4.0 mm	250°C

* Gas / Met Ratio = kgs min^{-1} Ar Gas ÷ kgs min^{-1} Metal, ** AT/SC Gas = Atomiser Gas Flow (kgs min^{-1}) ÷ Scanner Gas Flow (kgs min^{-1})

Figure 6: a) Drum deposit of Ti-48Al, b) CSD deposit of Ti-52Al. Note the variation in size and surface texture.

The optical micrographs shown in figure 7, are examples of the microstructures obtained in Runs # 2. The deposition can be divided into three regions. The first is a splat layer where the first particles run together on the surface of the substrate (Figures 7(a)). Cooling is faster in this region, resulting in high porosity and evidence of many surviving particle boundaries. Region II begins as this layer builds-up, temperatures increase at the surface, allowing arriving particles to splatter and flow together before solidifying (Figures 7(b)). In region II the optimum spray conditions are obtained, which are characterised by a uniform microstructure and low porosity. The transition to region III (Figures 7(c)) begins when the metal flow rate becomes unsteady and breaks-up near the end of the run. Larger particles, at a wider size distribution, are sprayed during this time as the spray conditions deteriorate.

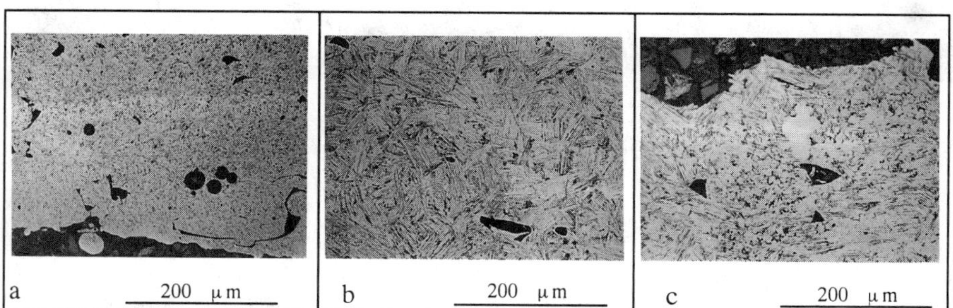

Figure 7: a) Splat-Substrate interface from Run #2 Ti-48-2-2 showing Region I, b) Run #2 - Region II, c) Run #2 - Region III.

Interstitial chemistries from the first two spray forming experiments (Table 4) were similar to those obtained from ingot material, as discussed above. O_2 and C concentrations, in the spray-formed deposits, are only slightly higher than in the double-melted plasma ingot material.

Table 4: Interstitial chemistry Data from Spray-Forming Runs 1 and 2.

RUN #	ALLOY	DEPOSIT THICKNESS	OVERSPRAY OXYGEN	OVERSPRAY CARBON	DEPOSIT OXYGEN	DEPOSIT CARBON
1	Ti-48Al	4 mm	1500 ppm	600 ppm	900 ppm	200 ppm
2	Ti-48-2-2	6 mm	3400 ppm	1000 ppm	850 ppm	200 ppm

Centrifugal Spray Deposition (CSD)

CSD is of particular interest in the spray forming of Ti alloys, since it can be operated under vacuum to minimise entrapped gas which would be detrimental for these alloys in rotating parts. The use of a centrifugal disk to atomise and distribute particles precludes the need for atomising and scanning gases. As of the present, the IRC system is not configured to operated under vacuum, but could be modified in the future. The experiments that have been run using CSD, indicate that higher deposition rates than with gas scanning are possible. Also, initial results showed that the porosities are much lower in CSD deposits than in the gas scanned material. The substrate-deposit interface produced by CSD is superior to the gas scanned devices, as shown in Figure 6(b). A super-α_2 alloy was sprayed by this method with the resultant microstructures shown in Figure 8. There is minimal porosity in this deposit, and the grain structure is relatively coarse (compared to PM) but smaller than cast material and more uniform. The grains size is uniform from substrate surface to free surface (Compare Figure 8(a) to 8(b)). Note that the microstructure is devoid of porosity.

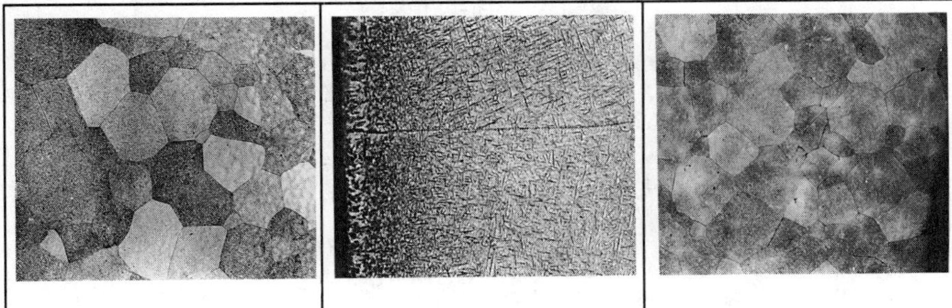

8 a 1.0 mm | 8 b 100 µm | 8 c 1.0 mm

Figure 8 CSD spray formed Super-α_2. a) Micrograph of the deposit showing the substrate-surface interface, b) Higher magnification from the same area as (a) showing fine precipitates of primary α_2, c) The deposit free surface.

Plate Preform

The flat preform offers a source of thick billet-like material. This configuration is also the easiest to operate. The initial experiment using the plate preform consisted of spraying Ti-48Al alloy. The scanning conditions were such that a deposit of about 2 cm thick was built up on the 20 x 25 cm plate, with little oversprayed deposit. Metal flow rate for this run approached a singular value (6 kgs/min) due to improved overpressure controls. The plate was transversed under the spray three times at a speed of ≈ 40 cm/sec. The microstructures shown in Figure 9 give a cross-sectional view from near the centre of the deposit. The deposit has three dense areas separated by bands of porosity. The porosity is present at the interface of the spray passes. One can infer from this that a singular pass or faster transverse speed may improve the deposit density. Mechanical test were being performed at the time of this writing.

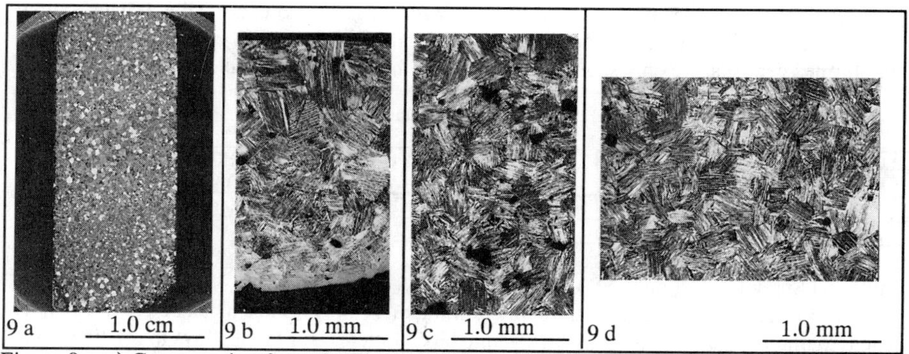

Figure 9 a) Cross-section from the Ti-48Al sheet deposit, b) The substrate deposit interface, c) Porosity at the pass interface, d) Bulk microstructure showing the lamella γ and α_2.

CONCLUSIONS

It has been show that it is possible to spray form Ti-aluminides into various shapes without compromising the alloy's interstitial content. High density preforms have been able to be produced. Porosity will be a challenge to prevent in traditional gas-scan spray forming but with CSD, operating under a vacuum, a totally dense material may be achievable after thermo-mechanical processing (i. e., HIP'ing). Grain size refinement has been shown to be much better than IM. Spray forming of Ti-aluminides presents an economical alternative to PM processing.

ACKNOWLEDGEMENTS

The author would like to thank the SERC for providing funds through the foundation grant with the IRC at The University of Birmingham, Edgbaston, England. Many thanks to the people at IMI Titanium for their contributions of time and materials. The assistance of the IRC staff, especially T. Perry, J. Boots, M. Glywn and A. Fellows made this project possible. Special thanks to T. Johnson and K. O'Rielly for their contributions.

---

1 Advanced Materials & Processes, ASM Publication, February 1992, p. 9.

2 J. L. Murray, "Binary Titanium Alloy Phase Diagram Compilation," Final Technical Report to the Office of Naval Research, Project No. N00014-85-F-0069, 1985.

3 M. J. Kaufman, D. G. Konitzer, R. D. Shull and H. L. Fraser, Scripta Metallurgica, Edited by Pergamon Press plc., Vol. 20, p. 103-108, (1986).

4 J. W. Sears, "Clean Melting Techniques for the Production of Titanium Alloy Powders and Spray Deposits", Proceedings: Advances in Powder Metallurgy - 1991, by APMI, Chicago, IL, Vol. 6, p. 335-346.

5 J. W. Sears, "Titanium Spray Formed Structures", Proceedings: 1992 PM World Congress, by APMI, San Francisco, CA, June 21- 26.

6 R. V. Ramanujan, I. P. Jones, J. M. Young and J. W. Sears. "The Influence of Heat Treatment on the Microstructure of Plasma-Melted g-based Titanium Aluminides," This conference.

7 C. Huang, T. A. Dean and Z. Chen, "The Influence of Thermomechanical Processing on the Microstructure and Properties of Plasma Melted Ti Aluminides, "This conference.

8 S. M. L. Sastry, T. C. Peng and B. London, "Rapidly Solidified Ti-Alloy Powders Produced by Plasma-Arc-Melting / Centrifugal-Atomisation (PAMCA)," Advances in Powder Metallurgy -1989, Editors: T. G. Gasbarre and W. F. Jandeska, Jr., MPIF, Princeton, N.J., 1989, Vol. 3, p. 387-400.

9 J. W. Sears, et. al., "The Production of Rapidly Solidified Ti Alloy Powers", Processing of Structural Metals by Rapid Solidification, Editors: F. H. Froes and S. J. Savage, ASM International, Metals Park, OH, 1987, p. 223-229.

10 T. T. Cheng, Work in progress at IRC, University of Birmingham, Edgbaston, England, 1992.

11 G. Itoh, Work in progress at IRC, University of Birmingham, Edgbaston, England, 1992.

12 A. Szaruga, M. Sajib, R. Omlor and H. A. Lipsitt, Scripta Metallurgica, Edited by Pergamon Press plc., Vol. 26, p. 787-790, (1992).

13 Z. Chen, Private conversation, University of Birmingham, Edgbaston, England, 1992.

MICROSTRUCTURAL AND TEXTURAL CHANGES DURING

ROLLING OF ALPHA-2 TITANIUM ALUMINIDE FOILS

Ravi Verma and Amit K. Ghosh

Department of Materials Science and Engineering
The University of Michigan
Ann Arbor, MI 48109

Abstract

Foils of high strength-to-weight ratio materials, such as titanium aluminides, may be useful for high temperature structural applications in composites and honeycomb configurations. Processing of foils by rolling can lead to different microstructures and textures depending upon the processing method used. This research was undertaken to study the effect of different rolling and heat treatment schedules on the development of these metallurgical features, and to examine the resulting strength or hardness. A hot pack-rolling procedure has been developed, and a cold rolling/annealing process has been optimized for maximum allowable rolling strain between annealing steps. Microstructural development through various stages of foil processing has been investigated. Hot pack-rolling produces foils with highly refined microstructures (average grain size 1 μm). Cold rolling, on the other hand, produces very smooth, uniformly thick foils with little oxygen contamination. Rolled foils typically show a duplex rolling texture: {1120}<1010> "prism", and (0001)<1010> "basal". While the strength of the "prism" texture is highly dependent on temperature, manner of rolling, and initial sheet microstructure, the "basal" texture is nearly independent of these factors. Microhardness measurements on rolled foils show anisotropy effects due to crystallographic texture.

Introduction

Cost-effective sheet or foil processing methods are needed for utilization of Ti_3Al (α_2) alloys in advanced aerospace structural applications. The α_2 phase has an ordered hexagonal closed packed structure, DO_{19}. The structure is characterized by limited slip systems, super dislocations and near absence of "c+a" or c/2 dislocation activity (1), which result in low ductility of Ti_3Al alloys. Generally, attempts to roll sheets can cause strain localization and cracking in these alloys. The design of alloys with some room-temperature ductility (2) has centered around stabilizing the ductile high temperature Ti phase, β, into the α_2 structure. Niobium has found favor as a β stabilizer, since it increases the non-basal slip activity in the α_2 phase (3). Niobium also enhances oxidation resistance of Ti_3Al (4) which is important in hot processing of these alloys. The alloy Ti-24Al-11Nb (at.%) is a typical α_2 alloy composition developed for possible high temperature applications (5).

Hot pack-rolling (6-7) as well as cold rolling/annealing (8) processes have been developed for α_2 alloys, and production of thin foils with good surface finish is a commercial reality. However, little effort thus far has been spent in understanding evolution of the microstructure through the processing steps and using this knowledge to manipulate foil microstructure through changes in heat treatment and thermomechanical processing. Further, since

Titanium '92
Science and Technology
Edited by F.H. Froes and I. Caplan
The Minerals, Metals & Materials Society, 1993

crystallographic texture in h.c.p. alloys has an important bearing on the alloy properties, and since the same has not been examined in detail in the α_2 alloys, it was also an object of this study to determine texture in rolled α_2 foils.

We have studied such a hot pack-rolling process and a cold rolling/annealing procedure for producing foils of Ti-24Al-11Nb $\leq$ 0.25 mm in thickness. We have used these two rolling procedures to produce a variety of α_2 foils. Microstructural evolution through various stages of foil processing has been studied. The texture in rolled foils was examined by the pole figure method. Influence of microstructural features on mechanical response of the foils has also been examined.

Processing

The starting α_2 stock was a cast and hot forged plate of Ti-24Al-11Nb produced by TIMET. The β-transus temperature for this alloy is reported to be 1125°C (9). This high value of β transus is consistent wth an oxygen content of 840 ppm in our alloy. A schematic of the processing steps involved in reducing the initial 10 mm thick plate to .25 mm thick foil is shown in Figure 1.

Forging. Sections cut from the initial α_2 plate were upset forged in high purity argon to a 4 mm thick pancake form. Two different forging temperatures were chosen: 1130°C (> β transus) to obtain a "β-forged" microstructure, and 1060°C (< β transus) to obtain an "(α_2 + β)-forged" microstructure. Thickness reduction in forging was 50-60%.

Pack for Hot Rolling. An evacuated hot-rolling pack, developed specifically for this study, was used to hot roll the forged α_2 pancakes. The pack was designed with the following objectives in mind: a) near isothermal rolling of the workpiece, b) an oxygen and nitrogen free rolling environment, c) a good surface finish of the rolled product, d) easy separation of the workpiece from the pack, and e) a 75-80% reduction of the workpiece without failure of the outer envelope of the pack. The maximum thickness reduction attainable with the pack rolling procedure is limited only by the integrity of the vacuum envelope of the pack. The key to achieving a high level of rolling reduction, without failure of the outer envelope, is a differential thickness reduction between the workpiece and the envelope as shown in Table I.

Table I. Thickness Reductions in a Typical α_2 Hot Rolling Pack

Layer	Thickness (mm)		Reduction %
	Before	After	
α_2 Workpiece	3.91	1.02	74
Protective Layers *	1.58	0.36	77
Outer Envelope †	1.06	0.86	19
Total	6.55	2.24	66

* 304 SS with Nb foil, grafoil, and Y_2O_3 coating
† 304 SS

Sheet Forming by Pack Rolling. Forged α_2 pancakes, contained in the evacuated pack, were preheated to 985°C and rolled in a two-high mill with 10-12% reduction per pass and intermediate reheating between passes. The rolling temperature was estimated (7) as 950°C. After 75-80% thickness reduction of the workpiece, the packs were cut open to recover the workpiece. Both "β-forged" and the "α_2 + β-forged" materials were processed to obtain 1 mm thick sheets, which served as the starting stock for foil rolling as described below.

Hot Pack-rolling to Foil Gauge. To hot roll the 1 mm thick sheets to foil gauge, sections were cut from both the "β-forged" and the "α_2 + β-forged" sheet stocks, re-packed, and hot rolled using the procedure described in the previous section. Typically, a thickness reduction of 75-80% was needed to produce a 0.20 - 0.25 mm thick foil.

Cold Rolling to Foil Gauge. To cold roll the 1 mm thick sheets to foil gauge, a cold rolling/annealing process was developed with the objective to: 1) introduce maximum possible amount of cold work prior to annealing, and 2) obtain maximum softening prior to introducing further reduction.

Initial cold rolling experiments were used to investigate the microstructure and the process variables that limit the rollability of foils. These experiments were conducted on an $\alpha_2 + \beta$ processed stock. Coupons, 10 x 20 mm^2 in size, cut from this sheet were vacuum sealed in quartz capsules, annealed at 900-1100°C for 2 hr and then water quenched. The annealed coupons were cold rolled in small reduction increments (1-2% per pass) till they developed cracks. Figure 2 shows the possible net rolling reduction prior to cracking as a function of the annealing temperature. The plot shows a maximum cold-rolling reduction of 38% at 1000°C. It was concluded that a heat treatment of 1000°C for 2 hr followed by water quenching provides an optimum microstructure for cold rolling of the α_2 sheet.

The intermediate annealing treatment was designed to relieve accumulated cold working strain and restore sheet ductility, before introducing further reduction. A temperature of 900°C was considered suitable for this purpose because of the risk of altering the optimum microstructure at temperatures greater than 900°C, and to keep the temperature high enough in order to restore sheet ductility in a reasonable time. Coupons cut from an $\alpha_2 + \beta$ processed stock were cold rolled with 8% reduction, annealed at 900°C in vacuum sealed quartz capsules for 10-50 minutes and air cooled. Knoop hardness measurements were made on the annealed coupons, and the data is plotted as a function of the annealing time in Figure 3. The plot shows that the hardness of a cold rolled α_2 sheet can be fully recovered by annealing at 900°C for 30 minutes or more. This treatment should be adequate to restore ductility of cold rolled sheets as well.

Based on these initial experiments, the following procedure evolved for cold rolling of α_2 foils:
Step 1: Microstructure optimization - Anneal at 1000°C/2 hr/WQ
Step 2: Cold roll with 25% reduction
Step 3: Strain relieving - Anneal at 900°C/.5 hr/WQ
Step 4: Cold roll with 25% reduction
Step 5: Repeat steps 3 and 4, as needed.

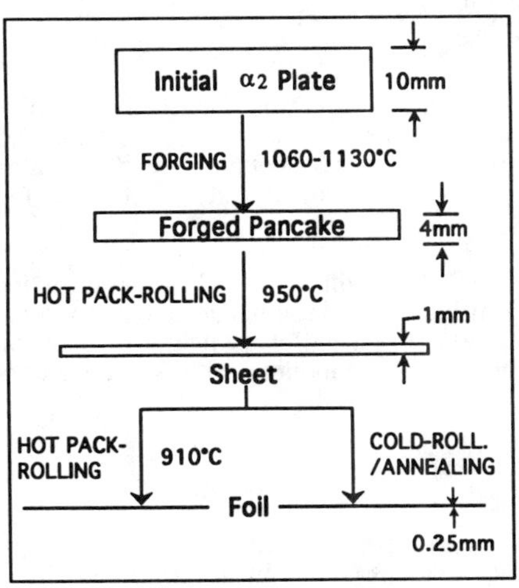

Figure 1 - Schematic of foil processing.

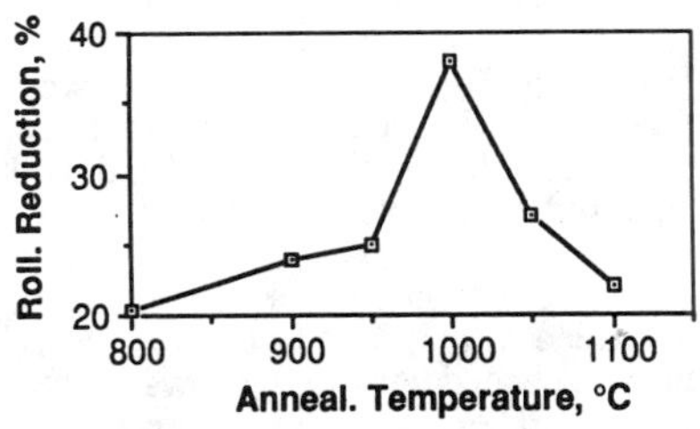

Figure 2 - Net cold rolling reduction, prior to cracking of annealed sheets, annealed for 2 hr and followed by water quenching.

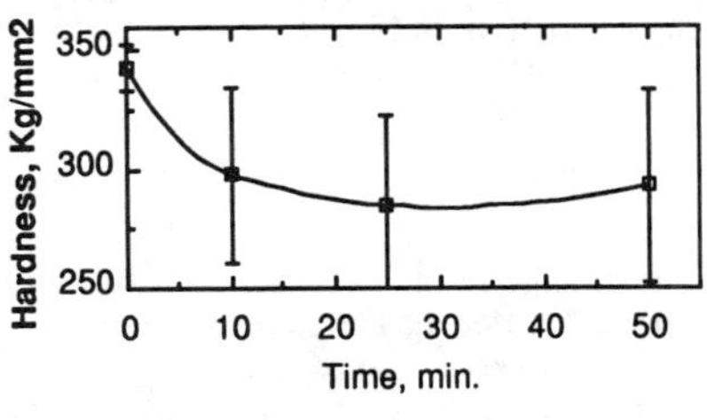

Figure 3 - Drop in Knoop hardness with time of annealing in cold rolled sheet, annealed at 900°C.

Microstructure Evolution

The Ti-24Al-11Nb alloy contains a mixture of two phases: the ordered α_2 phase, based on Ti_3Al, and a high temperature phase β, stabilized by niobium (2). Microstructure manipulation in Ti-24Al-11Nb, as in other titanium alloys, centers around the $\beta \rightarrow \alpha$ transformation. The transformation can occur martensitically or by nucleation and growth. The microstructure is further complicated by ordering of the β phase to β_o, a B2 structure (10), and by phase separation in the α_2 phase (11).

"β-forged" Material

Forged Pancake. Figure 4a shows the microstructure of a pancake forged at 1130°C, with 60% reduction, followed by furnace cooling. This shows a colony microstructure with very large prior-β grain size of 150-300 μm.

Hot Rolled Sheet. The forged pancake was hot pack-rolled with 77% reduction to form a sheet. Microstructure of the hot-rolled sheet, in Figure 4b, shows severe bending of laths of the colony structure. The laths tend to align parallel to the rolling plane and are thinned out in the sheet thickness direction. This results in a fine microstructure in the LS and TS planes. Furthermore, the β_o laths show a tendency to break down into fine equiaxed β_o grains.

Unidirectionally Hot Rolled Foil. A section of the hot rolled sheet was re-packed and further hot rolled with 80% reduction, to a 0.22 mm thick foil. The foil microstructure, in Figure 4c, shows highly refined β_o and α_2 grains in the prior lamellar regions and heavy flattening of the isolated primary α_2 grains. This gives a near homogeneous microstructure comprised of very fine (≤ 1 μm) β_o particles dispersed in an α_2 matrix.

"α_2 + β-forged" Material

Forged Pancake. Figure 5a shows the microstructure of a plate section forged at 1060°C with 50% reduction, followed by furnace cooling. The microstructure largely consists of equiaxed primary α_2 grains, 5-10 μm in size, with small regions of β_o phase located at grain-boundary triple points.

Hot Rolled Sheet. The forged plate was hot pack-rolled with 77% reduction. Microstructure of the hot-rolled sheet, in Figure 5b, shows pancake-shaped α_2 grains flattened along the short transverse direction. The surrounding β_o phase region shows development of a fine (1-2 μm) equiaxed grain structure.

Hot Unidirectionally Rolled Foil. A section of the hot-rolled sheet was re-packed and further hot rolled with 75% reduction to give a .25 mm thick foil. The foil microstructure in Figure 5c, in as-quenched condition, shows further grain refinement and loss of continuity in the β matrix region. There was not a significant difference between as-rolled and as-quenched foil microstructures.

Cold Unidirectionally Rolled Foil. Figure 6 shows a 0.28 mm thick cold rolled (net reduction 65%) foil, produced from a "β-forged" sheet stock, using the cold rolling/annealing procedure described earlier. The foil has very smooth surfaces and a uniform thickness. The foil microstructure in Figure 7 shows highly flattened primary α_2 grains along with agglomerates of small equiaxed β_o grains.

Grain Size

The change in the average grain size with these processing steps is plotted in Figure 8. In "β-forged" material, the lamellar transformed-β structure breaks down by hot rolling to give a highly refined equiaxed microstructure at the foil level. In "α_2 + β-forged" material, the primary α_2 grains hardly show any refinement due to rolling, and the large primary α_2 volume fraction keeps the average grain size relatively large throughout foil processing.

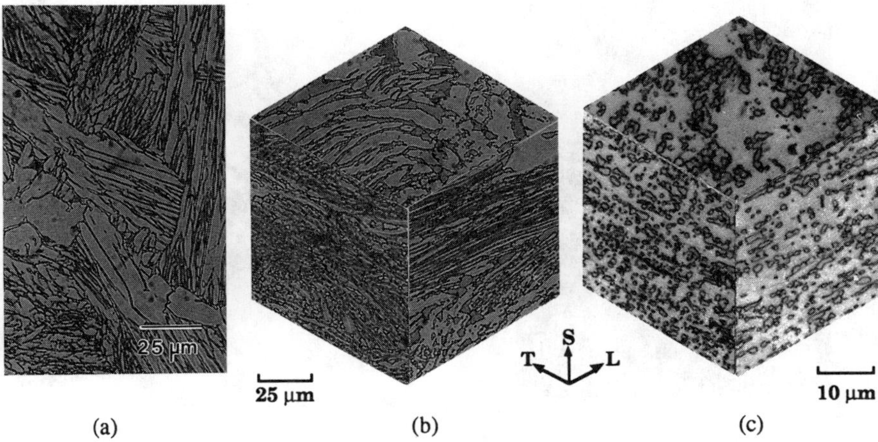

Figure 4 - Microstructures of (a) "β-forged" pancake (T = 1130°C; reduction = 60%; furnace cooled), (b) hot-rolled sheet from the "β-forged" stock, (c) unidirectionally hot-rolled foil, 0-22 mm thick, from the "β-forged" stock.

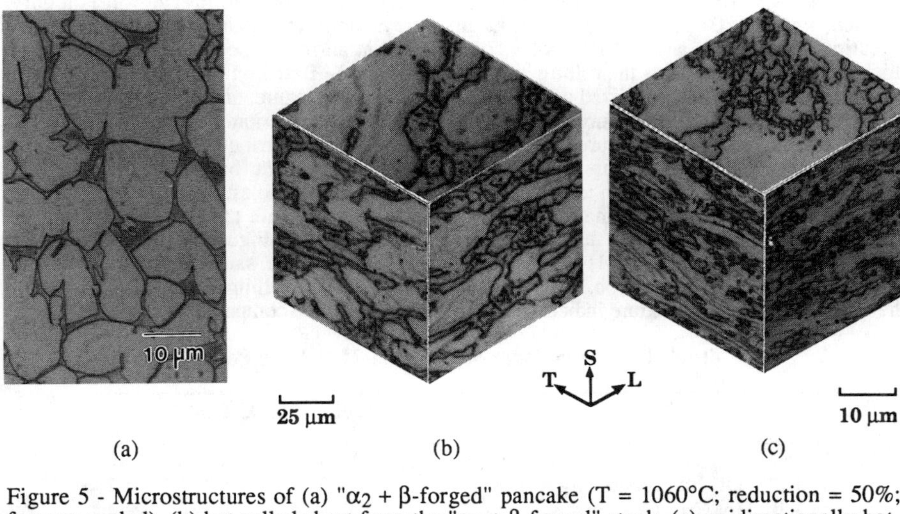

Figure 5 - Microstructures of (a) "α_2 + β-forged" pancake (T = 1060°C; reduction = 50%; furnace cooled), (b) hot-rolled sheet from the "α_2 + β-forged" stock, (c) unidirectionally hot-rolled foil, 0-25 mm thick, from the "α_2 + β-forged" stock.

Figure 6 - Cross-section of 0.28 mm thick unidirectionally cold-rolled foil from "β-forged" sheet stock.

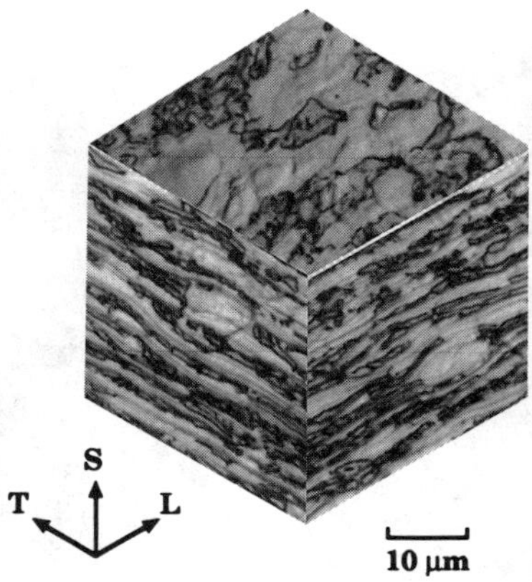

Figure 7 - Microstructure of 0.28 mm thick unidirectionally cold-rolled foil from "β-forged" sheet stock.

Texture

Crystallographic texture in rolled foils was recorded by the X-ray pole figure method. Pole figures were recorded for four important crystallographic directions of the hexagonal α_2 lattice. Figure 9 shows pole figures recorded from a "β-forged" hot-rolled foil. The foil is highly textured, with $\langle 11\bar{2}0 \rangle$ poles concentrated, 12 times the random strength, along the foil thickness direction (Figure 9a), and $\langle 10\bar{1}0 \rangle$ poles concentrated at an inclination of 30° from the foil thickness direction, towards the rolling direction (Figure 9b). Based on these two pole figures, the rolling texture can be idealized as $\{11\bar{2}0\}\langle 10\bar{1}0 \rangle$ "prism" texture, and the same is depicted by a hexagonal unit-cell in Figure 10a. The $\langle 11\bar{2}1 \rangle$ pole distribution in Figure 9c is largely consistent with the "prism" texture. However, the $\langle 0001 \rangle$ pole distribution in Figure 9d is not consistent with the "prism" texture and thus represents an independent "basal" texture in the foil. In Figure 9c, out of the four high intensity locations of the $\langle 11\bar{2}1 \rangle$ poles, the two lower strength points, placed on the longitudinal direction, are not consistent with the "prism" texture. If, however, they are attributed to the "basal" texture represented by Figure 9d, the latter can be completely described as $(0001)\langle 11\bar{2}0 \rangle$ rolling texture, and the same is depicted by the hexagonal unit-cell in Figure 10b. This duplex "prism" + "basal" rolling texture is typical of all the rolled foils studied. Texture indices for various rolled foils are compared in Table II.

Table II. Texture Index of Rolled Ti-24Al-11Nb Foils

Foil	Texture Component	
	"Prism"	"Basal"
β / Hot Unidirectionally Rolled	12.0	3.0
α_2 + β / Hot Unidirectionally Rolled	9.6	3.2
α_2 + β / Hot Cross Rolled	2.6	3.4
β / Cold Unidirectionally Rolled	5.6	3.5

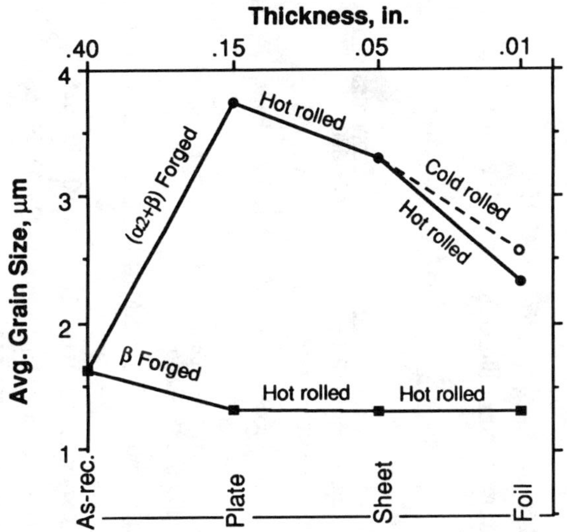

Figure 8 - Change in average grain size through foil processing.

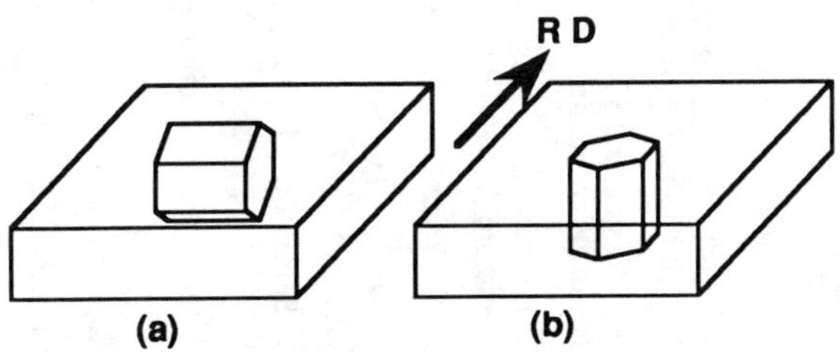

Figure 10 - Ideal rolling textures (a) {1120}<1010> "prism" type, (b) (0001)<1120> "basal" type.

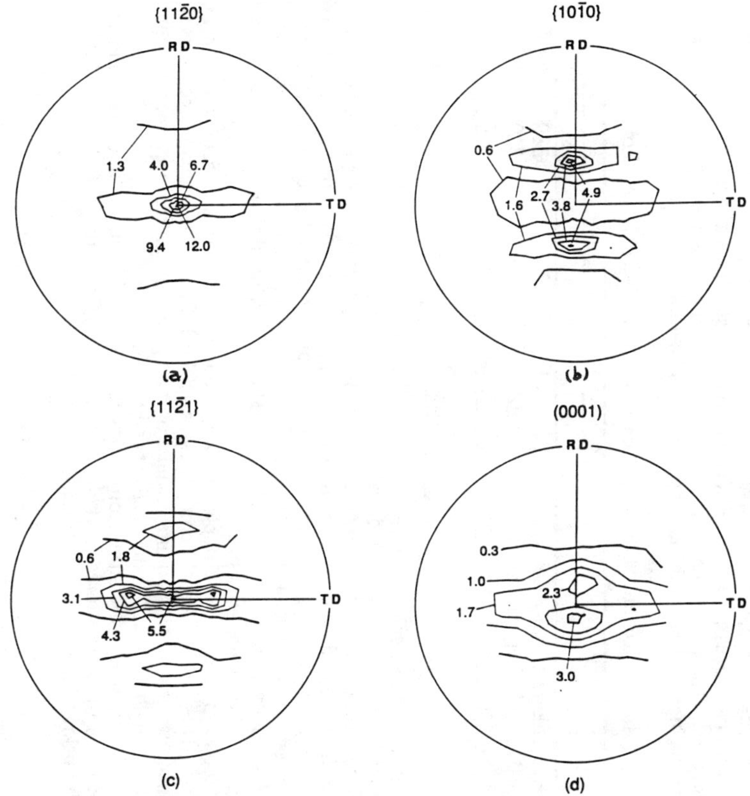

Figure 9 - Pole figures from "β-treated" unidirectionally hot-rolled foil for crystal directions (a) <1120>, (b) <1010>, (c) <1121>, and (d) (0001).

Microhardness

Table III gives Knoop hardness values measured on the rolled foils. The foils show significant anisotropy, with the through-thickness direction being the strongest and the rolling direction being the weakest. The cross rolled foil (with alternate rolling passes at right angles to each other) shows the highest through-thickness strength of all the foils.

Table III. Knoop Hardness (kg/mm^2) of Rolled Ti-24Al-11Nb Foils

Foil	Direction		
	L	T	S
β / Hot Unidirectionally Rolled	279	343	389
$\alpha_2 + \beta$ / Hot Unidirectionally Rolled	280	293	320
$\alpha_2 + \beta$ / Hot Cross Rolled	301	295	440

Conclusions

1. Hot pack-rolling and cold rolling/annealing procedures for processing foils of Ti-24Al-11Nb alloy have been developed and studied.
2. Foil microstructure depends on the pre-rolling microstructure of the α_2 stock used. Hot rolled foils processed from "β-forged" sheet stock, have highly refined and equiaxed microstructures (average grain size 1 µm). This highly refined structure results from breaking down of the lamella of the "β-forged" structure during hot rolling while hot or cold rolled foils processed from "$\alpha_2 + \beta$-forged" sheet stock exhibit relatively coarse foil microstructures (average grain size 2-3 µm).
3. Rolled foils typically show a duplex rolling texture comprising a "prism" component, {1120}<1010>, and a "basal" component, (0001)<1120>. The strength of "prism" component varies with temperature, manner of rolling, and the initial sheet microstructure while the strength of "basal" texture is practically constant.
4. Rolled foils show significant anisotropy, with through thickness direction being the strongest and transverse direction being the weakest. Cross rolled foil shows highest through-thickness strength of all the foils.

Acknowledgments

The authors wish to thank Texas Instruments for the financial support, and GM Allison Gas Turbine for providing Ti-24Al-11Nb material.

References

1. S.M.L. Sastry and H.A. Lipsitt, Metall. Trans. A, 8A (1977), 1543-1552.
2. H.A. Lipsitt, High Temperature Ordered Intermetallic Alloys, eds. C.C. Koch, C.T. Liu, and N.S. Stoloff (Pittsburgh, PA: MRS, 1985), 351-64.
3. S.M.L. Sastry and H.A. Lipsitt, Ti' 80, Science and Technology, eds. H. Kimura and O. Izumi (Warrendale, PA: TMS-AIME, 1980), 1231-1243.
4. N.S. Chowdhury, H.C. Graham, and J.W. Hinze, Proc. Symp. on Properties of High Temperature Alloys, (Princeton, NJ: Electrochemical Society, 1977), 668-680.
5. M.J. Blackburn and M.P. Smith, (Report AFWAL/TR-81-4046, Air Force, 1981).
6. C. Bassi, J.A. Peters, and J. Wittenauer, "Processing Titanium Aluminide Foils," Journal of Metals, 9 (1989), 18-20.
7. S.L. Semiatin, M. Ohls, and W.R. Kerr, "Temperature Transients during Hot Pack Rolling," Scripta Metallurgica, 25 (1991), 1851-1856.
8. S.C. Jha et al., "Titanium Aluminide Foils," Advanced Materials & Processes, 4 (1991), 87-90.
9. A.K. Gogia, D. Banerjee, and T.K. Nandy, Metall. Trans. A, 21A (1990), 609-625.
10. D. Banerjee, T.K. Nandy, and A.K. Gogia, Scripta Metallurgica, 21 (1987), 597-600.
11. D. Banerjee et al., Acta Metallurgica, 36 (1988), 871-882.

The Role of Silicon in a Ti_3Al Based Alloy

D.J.Arrell[+], H.M.Flower[+] and S.Kerry[*]

[+] Dept. of Materials, Imperial College, London SW7 2BP, England

[*] Defence Research Agency, Aerospace Division, RAE Pyestock, Hampshire, GU14 0LS, England.

In the drive for lighter weight gas turbine engines operating at higher temperatures the titanium aluminides, Ti_3Al and TiAl, are being considered as potential replacement materials for nickel superalloys and conventional titanium alloys. However, the stoichiometric alloys are brittle at ambient temperature and therefore alloying with these intermetallic materials is being studied in order to improve this lack of ductility.

Alloys based on the Ti-Al-Nb (α_2) based system are currently being developed at the DRA. An important alloying addition to these alloys is that of silicon. Results are presented which demonstrate the importance of silicon to the development of good mechanical properties, particularly ambient temperature ductility and elevated temperature creep rupture life. An optimum silicon content has been found, above which the ductility is reduced markedly.

These results will be discussed with reference to the influence of silicon on microstructural development and phase chemistry in the Ti-Al-Nb system, where silicon modifies the partitioning between the β_o and α_2 phases, and alters the stability of the β phase with respect to martensitic transformation on quenching. The α_2 phase in the alloy with optimum silicon content can be accurately represented by $(Ti+Nb)_3(Al+Si)$.

Titanium '92
Science and Technology
Edited by F.H. Froes and I. Caplan
The Minerals, Metals & Materials Society, 1993

Experimental

A series of quaternary alloys based on the ternary composition Ti-20Al-11Nb (atomic%), in which silicon has been substituted for titanium at levels up to 1.5at%, has been examined. Each alloy was prepared commercially, and then forged in air (in the $\alpha_2 + \beta$ phase field) by DRA into approximately 200g blocks.

Samples of each alloy were then vacuum heat treated in the single phase β field at 1150°C for 1hr and argon quenched. An ageing treatment at 625°C for 2hrs was then applied, again followed by an argon quench. The cooling rate, an estimated 50K.sec^{-1}, was equivalent to the rate achieved by a conventional air cooling method. D.S.C was used on samples of each alloy to ascertain the temperatures of the phase transformations which occur in the system. Tensile specimens of each alloy in this heat treated condition were produced, and room temperature ductility and tensile strength measured. Small sections of each alloy were subsequently aged at 1050°C for 168hrs in an argon atmosphere, and water quenched in order to examine the equilibrium phase compositions.

Results and Discussion

Samples of each alloy in the 625°C aged condition were examined after polishing and etching in Kroll's reagent, and compared with the microstructure of the base Ti-20Al-11Nb alloy. As can be seen in Fig 1a, the structure consists of a network of α_2 platelets embedded in a small fraction of retained β_o matrix. The prior β_o grain boundaries are marked by a continuous film of α_2. This microstructure is apparently unaffected by additions of silicon of up to 0.9at% (figs 1b and c). There is, however, a marked change in samples with a composition containing greater than approximately 1.0at% Si (fig 1d). At this level, the microstructure is noticeably coarsened, though the basic morphology of platelets in a matrix of a second phase remains. However, the prior grain boundaries, differ in that they do not appear clearly delineated, having platelets apparently crossing them in large numbers, and do not have the α_2 film decorating them.

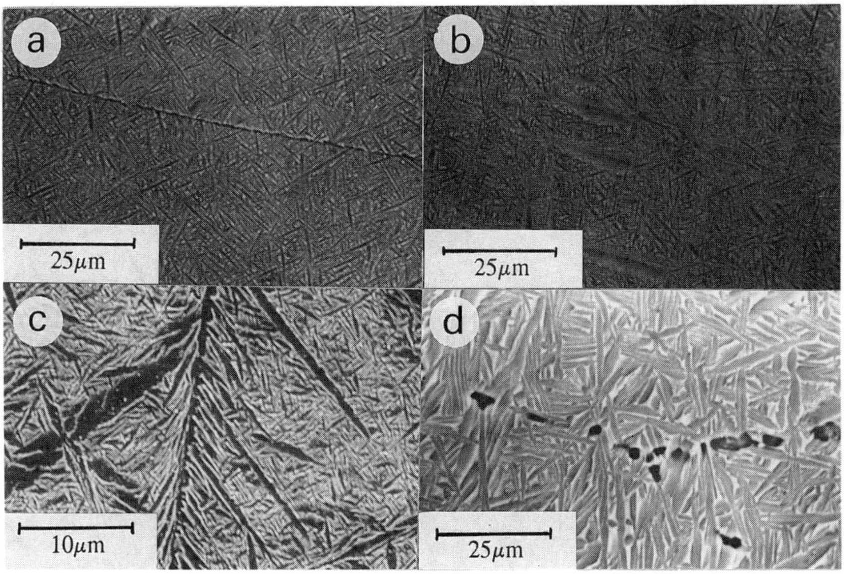

Fig 1. SEI images of structures etched in Kroll's reagent, (a) α_2 + retained β_o structure in Ti-20Al-11Nb, aged at 625°C/2hrs. (b) Microstructure unchanged in Ti20Al-11Nb-0.5Si. c) Detail of α_2 coating on grain boundary in Ti-20Al-11Nb-0.9Si. (d) BEI image of silicides on prior β boundaries in α_2 + O microstructure in Ti-20Al-11Nb-1.5Si alloy.

TEM and X-ray confirm identification of $\alpha_2 + \beta_o$ in low Si alloys. However, in high Si alloys, the β_o is replaced by an orthorhombic phase (O phase), which is similar to that first reported by Banerjee et al[1] with lattice parameters of a=585pm, b=975pm and c=466pm. As can be seen in Table I, the b lattice parameter in the orthorhombic phase is noticeably distorted from that of the α_2 phase (if α_2 is indexed as an orthorhombic structure, then the b parameter would be 1003pm). This distortion is greater than that which has been reported in most other orthorhombic structures of similar composition[2-4], which is probably explained by the effect of the silicon atom diameter being smaller than that of aluminium .

Table I. Lattice parameters of alloys (in pm)

Si Content	α_2		β_o	O		
	a	c	a	a	b	c
0	579	466	326	-	-	-
0.5	579	465	326	-	-	-
0.9	579	465	326	-	-	-
1.2	579	465	-	585	975	466
1.5	579	465	-	585	975	466

Thin foil X-ray EDS analysis (on a JEOL 2000FX) of the two different structures (Table II), shows that the chemical composition of the β_o and O phases are very similar. Examination of the phase transformations using D.S.C techniques indicates that the O phase is stable at temperatures of up to approximately 1000°C. Figure 2, shows the transformation of O -> β_o on heating using D.S.C. in Ti-20Al-11Nb-1.2Si, in the 625°C aged condition. A trace produced by similarly aged Ti-20Al-11Nb-0.5Si, is included for comparison.

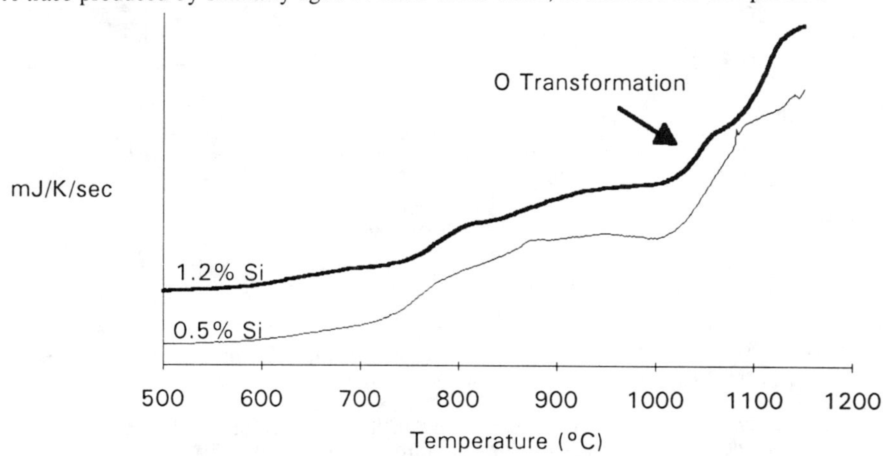

Fig 2. D.S.C. Curve for alloys in aged condition heated at 30°C/min.

When the silicon level is raised further still (~1.2at%Si), a phase based on a composition of $(Ti+Nb)_5(Si+Al)_3$ is produced. Most of this forms by precipitation during the heat treatment at 1150°C, though in the 1.5% Si alloy, there is evidence that some of the silicide is in the form of a eutectic structure, indicating its formation in the initial solidification process. In the 1150°C/1hr + 625°C/2hrs heat treated condition, the ductility of the base alloy is improved markedly when the silicon level is raised to the 1.0 at% level.

Table II. Phase Composition of Alloys

Alloy	Structure	Composition (atomic%) Ti	Al	Nb	Si
Ti-20Al-11Nb-0.9Si	α_2	67.6	21.1	10.3	1.0
	β_o	66.0	11.1	22.1	0.0
Ti-20Al-11Nb-1.2Si	α_2	66.6	21.5	10.2	1.7
	O	67.6	11.7	21.8	0.4

This can be attributed to the strengthening of the prior β_o grain boundaries due to the removal of the film of brittle α_2 structure, which in lower silicon level alloys is a marked feature. The role of O phase in its own right has not yet been determined, but may further contribute to the ductility. As can be seen in Figure 3, this is achieved with only a small reduction in tensile strength relative to that achieved at the 0.5at% silicon level. It is likely that the coarser microstructure found in this alloy is a contributing factor to this drop in strength. It is also possible that the presence of O is also influences the observed strength. However, as the silicon level is increased further, the ductility drops off quite sharply. This is due to the presence of the silicide particles, which form preferentially, as grain boundary precipitates, causing a weakening of the grain boundary, and hence a drop in ductility.

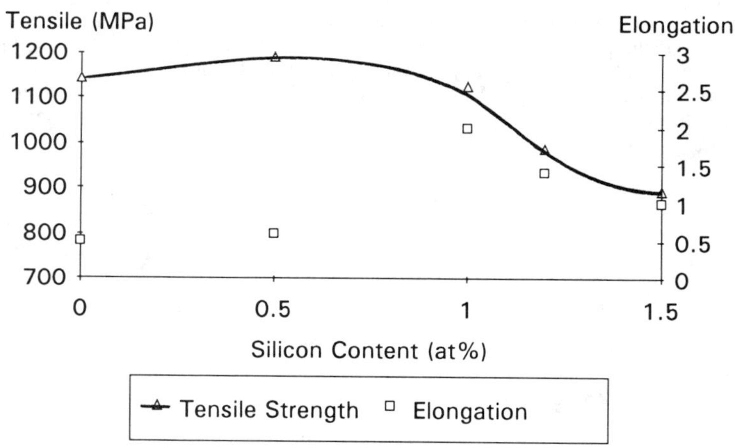

Fig 3. Mechanical Properties of Ti-20Al-11Nb-XSi in 1150°C/1hr + 625°C/2hrs heat treated condition.

Subsequent heat treatment of these alloys in the $\alpha_2 + \beta_o$ phase field at 1050°C for 168hrs, water quenched, produces a microstructure which consists of a large number of α_2 needles in a β_o matrix (figure 4a). The prior β_o grain boundaries are coated by a series of overlapping α_2 plates, surrounded by a precipitate free zone. The thickness of this boundary layer increases with increasing silicon content. At silicon levels of 1.2at% and above there are a small number of silicide particles which nucleate in the matrix, and grow to a size of approximately 15μm diameter. These form with a surrounding "α_2 free" zone (fig 4b). There is also a modification to the grain boundary regions, which also contain a series of small (~1μm) silicide precipitates (figure 4b). If these particles are close to an α_2 / β_o grain boundary, then on the intersection of the boundary they grow into the β_o phase to produce much larger (~10μm) particles (Figure 4c).

In these, and other related alloys studied in the present work, but not discussed here, if it is assumed that niobium substitutes on titanium lattice sites[4-6], and silicon on aluminium ones, then the α_2 phase in the two phase field is consistently highly stoichiometric (see Table III).

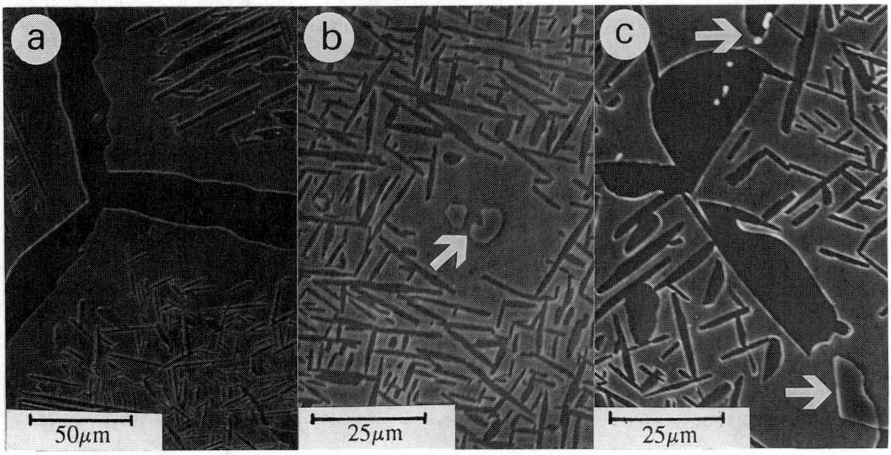

Fig 4. SEI images of $\alpha_2+\beta_o$ aged (1050°C/168hrs) alloys etched in Kroll's reagent, (a) α_2 film on prior β_o grain boundary in Ti-20Al-11Nb. (b) Precipitate free zone surrounding silicide particle (indicated). (c) Silicide particles on grain boundaries (indicated).

Table III. Elemental distribution in α_2 + β_o phase field at 1050°C.

Alloy	Phase	Atomic % Ti	Al	Nb	Si	Element Ratio+
Ti-20Al-11Nb-0.5Si	α_2	67.9	24.2	7.4	0.5	3.0:1
	β_o	68.1	18.5	12.9	0.5	-
Ti-20Al-11Nb-0.9Si	α_2	68.2	23.6	7.1	1.1	3.0:1
	β_o	68.4	17.7	12.1	0.8	-
Ti-20Al-11Nb-1.2Si	α_2	67.8	23.4	7.7	1.5	3.0:1
	β_o	67.9	18.4	12.9	0.8	-
	ζ^*	55.3	5.9	7.7	31.1	5.1:3

+ Ti+Nb : Al + Si
* $(Ti+Nb)_5(Si+Al)_3$

This adherence to stoichiometric behaviour leads to a modification in the behaviour of silicon from that which occurs in the ternary Ti-Al-Si system. In the binary Ti-Si system, at low silicon alloy levels, silicon acts as a eutectoid forming element [7]. The addition of aluminium transforms this into a ternary peritectoid [8] at aluminium contents of greater that ~2wt%. Evidence provided by the present work suggests that the effect remains the same when α_2 is substituted for αTi. Thus when silicon content in a ternary Ti_3Al - Si alloy is increased in the α_2 + β_o heat treated condition, it would be expected from considerations of isothermal data that it would increase the volume fraction of α_2 in the alloy substantially, as the composition of the α_2 is effectively on a line from Ti_3Al to Ti_3Si due to the stoichiometric alignment evident in the current work. Thus, the overall effect of the silicon is to increase the fraction of α_2 in the alloy on an *isothermal* basis. As can be seen in the vertical phase diagram section in Figure 5, this does not affect its overall peritectoid forming effect. Other work carried out by the authors, but not discussed here, shows that on long term ageing, all of these alloys will have a microstructure containing O phase. It is suggested that the effect of the silicon in causing the β_o to orthorhombic transformation at this temperature is due to a shifting of the O curve on a CCT diagram to the left of that for the α_2 phase, thus promoting the relatively coarse O precipitates which nucleate during air cooling through the $\alpha_2+\beta_o$ phase field in the higher silicon content alloys.

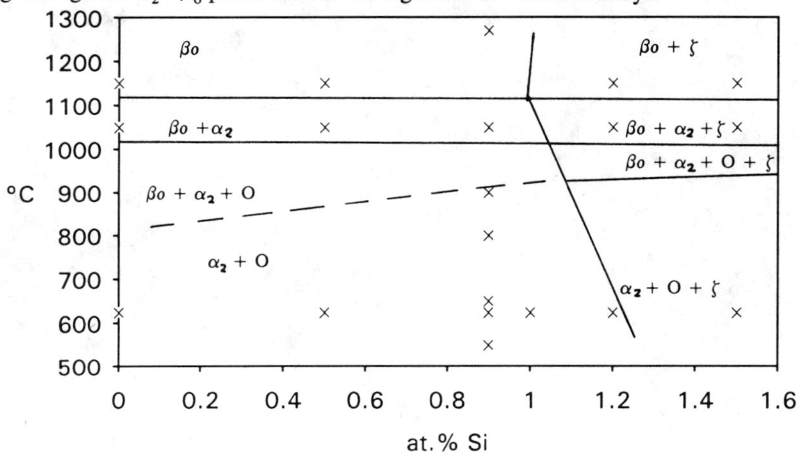

Fig 5. Pseudo-vertical section with constant Al and Nb held at 20at%Al and 11at%Nb.

Conclusions

1) In the solution treated plus aged condition, the addition of >0.9at% silicon to the alloy causes an orthorhombic structure to be formed, which significantly enhances room temperature ductility with minimal reduction in tensile strength.
2) At silicon levels of 1.2at% and above, the precipitation of titanium silicide particles produces a degradation in both tensile strength and ductility.
3) Silicon appears to substitute for aluminium in the α_2 structure in the $\alpha_2 + \beta_o$ phase field at 1050°C. This results in an increase in the volume fraction of α_2 at that temperature.
4) The addition of niobium at levels of up to 11at% does not affect the peritectoid forming effect of the silicon.

References

1. D. Banerjee et al., "A New Ordered Phase in a Ti_3Al-Nb Alloy", Acta Met., 36 (4) (1987), 871-882.
2. K. Muraleedharan et al., "Orthorhombic Distortions of the α_2 Phase in Ti_3Al-Nb Alloys: Artifacts and Facts", Scripta Met., 24 (12) (1990), 27-32.
3. H.T. Kestner-Weykamp et al, "Microstructures and Phase Relationships in the Ti_3Al - Nb System", Scripta Met., 23 (10) (1989), 1692-1702.
4. B. Mozer et al, "Neutron Powder Diffraction Study of the Orthorhombic Ti_2AlNb Phase", Scripta Met., 24 (12) (1990), 2363-2368.
5. D.G. Konitzer, I.P. Jones and H.L. Fraser, "Site Occupancy of Solid Solutions of Nb in the Intermetallic Compounds of TiAl and Ti_3Al", Scripta Met., 20 (1986), 265-268.
6. T.K. Nandy, D. Banerjee and A.K. Gogia, " Site Substitution Behaviour in Ti_3Al (DO_{19}) and Ti_2AlNb (B2) Intermetallics" (Paper presented at the Sixth World Conference on Titanium, Cannes, 1988), 943-948.
7. J.L. Murray, Phase Diagrams of Binary Titanium Alloys, 289-294.
8. F.A. Crossley and D.H. Turner, "Titanium Rich Corner of the Ti - Al - Si System", Met. Trans., 1959, no 2:60-63.

Acknowledgements

Thanks to Defence Research Agency, RAE Pyestock for funding for this project.

THE INFLUENCE OF DEVIATION FROM STOICHIOMETRY ON MICROSTRUCTURE AND TENSILE PROPERTIES OF A Ti_3Al BASE ALLOY

Isabelle Ehrhart, Alain Vassel

ONERA, BP 72, 92322 Chatillon Cedex, France

Abstract

Alloys of nominal compositions Ti-xAl-10V (at.%), with x between 19 and 27, were vacuum arc-melted in the shape of 200g buttons, isothermally forged and heat treated. Microstructural analysis on thin foils revealed that only α_2 and ordered β phases are present in the alloys. Tensile tests have shown that the yield strength displays a minimum and that the elongation to rupture steadily increases with the decrease in aluminium concentration. The improvement in ductility in sub-stoichiometric alloys is attributed to the activation of an additional slip system in the α_2 phase.

Introduction

Low ductility and fracture toughness properties are the main obstacle to the use of intermetallics in general, and to that of Ti_3Al based titanium aluminides in particular. Deviation from stoichiometry is one of the parameters that may enhance the plasticity of intermetallic compounds. Indeed, the introduction of lattice defects may change the deformation mechanisms and the fracture behaviour. For instance, unit dislocations are responsible for plastic deformation in sub-stoichiometric Fe_3Al whereas paired dislocations are found in Fe-31Al (at.%). It follows that sub-stoichiometric Fe_3Al is more ductile than aluminium rich compositions [1]. Similarly, the highest elongation to rupture in the TiAl compound has been observed at an aluminium concentration around 48at.% [2,3]. In this case, the ductility improvement is attributed to the refinement of the microstructure due to the precipitation of a second phase and to variations in the deformation mechanisms.

A slight increase in the ductility when lowering the aluminium content in Ti_3Al+Nb alloys has been reported in the literature but no interpretation is given [4]. The present study was aimed at getting a better understanding of the influence of deviation from stoichiometry on microstructure, deformation mechanisms and room temperature tensile properties of Ti_3Al based alloys.

Titanium '92
Science and Technology
Edited by F.H. Froes and I. Caplan
The Minerals, Metals & Materials Society, 1993

Experimental

This study was performed on the Ti_3Al+V model system. Previous investigations on forged buttons revealed that about 10at.% of vanadium has to be added to obtain both a significant refinement of the microstructure and some ductility at room temperature [5]. Alloys of nominal compositions Ti-xAl-10V (at.%), with x between 19 and 27, were vacuum arc-melted in the shape of 80x50x15mm buttons. These buttons were melted four times to ensure a good chemical homogeneity. The measured compositions of the alloys together with their beta transus temperature (T_β) are listed in Table I. The oxygen level was kept below 850ppm (wt.%) in all alloys.

Table I List of alloys studied

Chemical composition (at.%)	T_β (±5°C)
Ti-19.4Al-10.3V	1030
Ti-20.8Al-10.3V	1040
Ti-23.3Al-10.3V	1090
Ti-25.1Al-10.3V	1100
Ti-27.1Al-10.3V	1110

The buttons were isothermally forged at T_β-50°C to 80% reduction in one step ($\varepsilon=\ln(H_o/H)=1{,}6$). The forging operation was carried out under high vacuum in a 30 tons press at a constant strain rate of $10^{-3}s^{-1}$. Post-forging heat treatments were performed in an argon atmosphere. Flat specimens 2mm in thickness with a 16mm long x 4mm wide gauge section were used to determine tensile properties at room temperature.

Identification of phases and determination of slip systems were performed on thin foils in a JEOL 200CX electron microscope.

Results

Initially, an annealing treatment (850°C, 4 hours) was applied to the forged buttons. Optical micrographs (Fig. 1) show differences in structure morphology with the aluminium content. In sub-stoichiometric alloys, the microstructure is composed of nodular α_2 phase surrounded by a continuous β phase layer whereas a fine decomposition of the β phase is observed in alloys containing 25%Al or more. Examination on thin foils revealed that only α_2 and β phases are present in the alloys ; this result shows that the two-phase α_2+β field exists in a wide interval of aluminium content. Besides, superlattice spots appears on diffraction patterns of the β phase indicating an ordering reaction. There is evidence in the literature of ordering of β in the Ti-Al-Mo [6] and Ti-Al-Nb [7] systems around the Ti_2AlX composition. It is to be noted that ordering of the β phase has been observed in the Ti_3Al+V system at vanadium concentrations >8at.% [5].

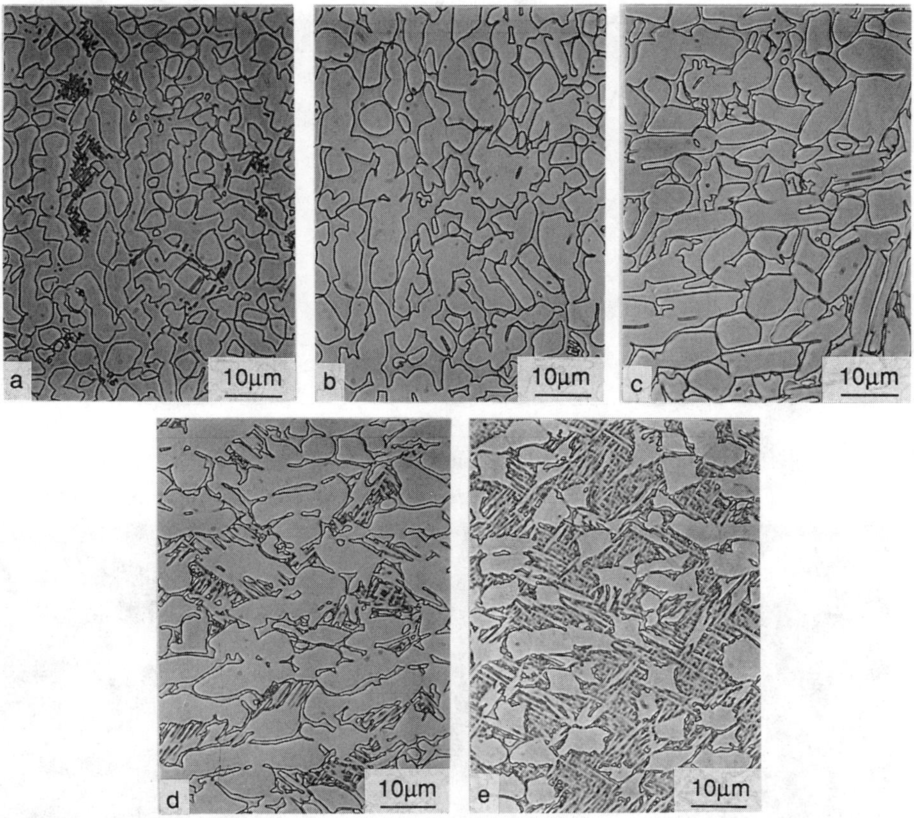

Figure 1 Alloys microstructure in the annealed condition : a) Ti-19.4Al-10.3V, b) Ti-20.8Al-10.3V, c) Ti-23.3Al-10.3V, d) Ti-25.1Al-10.3V, e) Ti-27.1Al-10.3V.

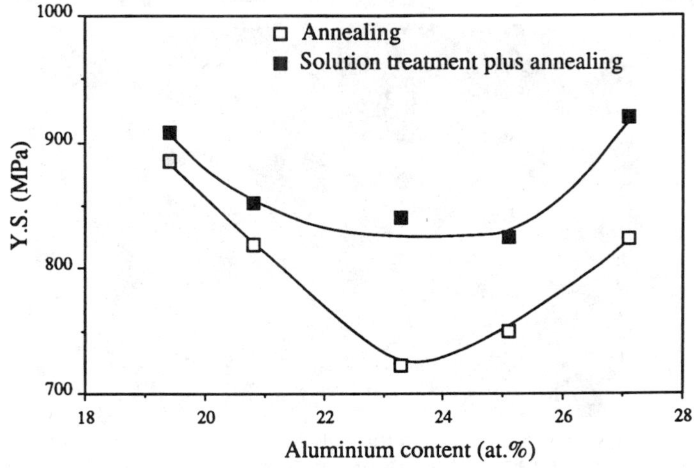

Figure 2 Yield strength versus aluminium alloy content

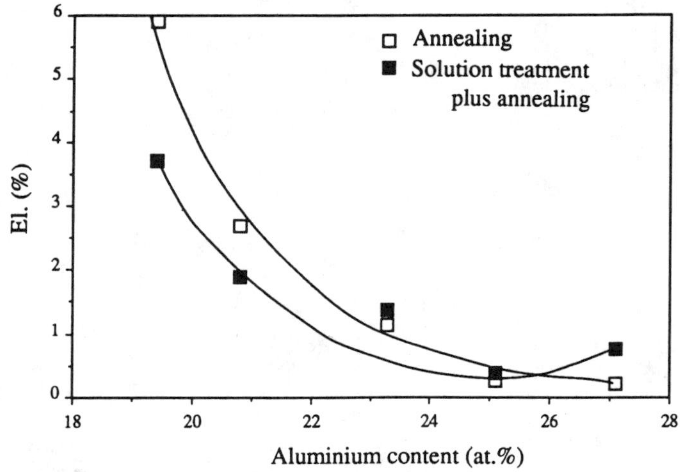

Figure 3 Elongation to rupture versus aluminium alloy content

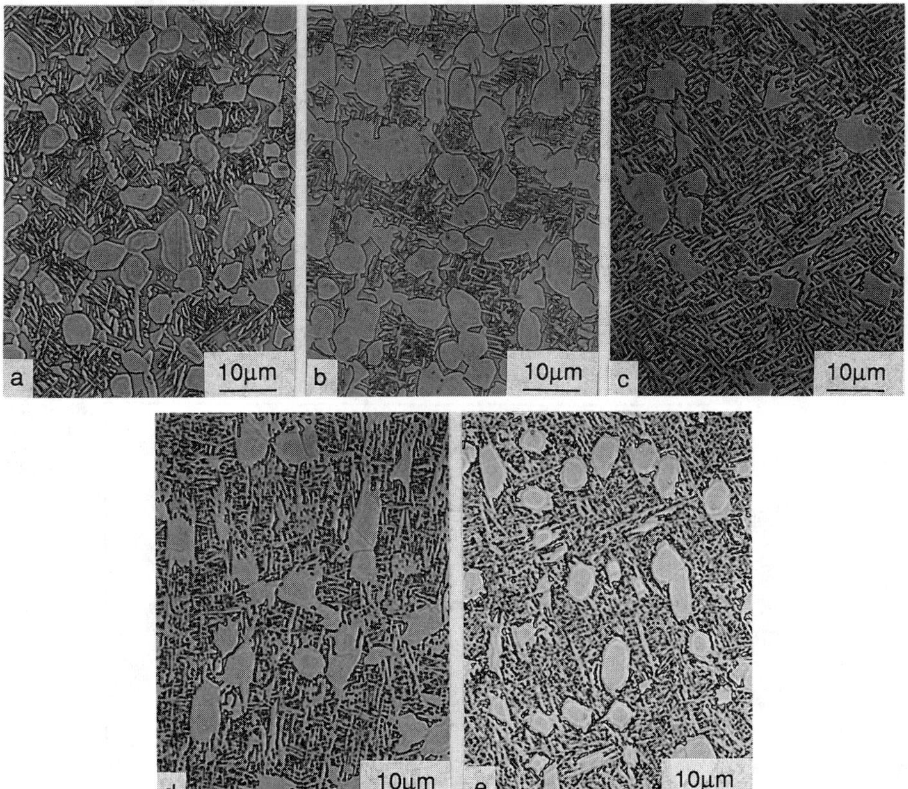

Figure 4 Alloys microstructure in the solution treated plus annealed condition : a) Ti-19.4Al-10.3V, b) Ti-20.8Al-10.3V, c) Ti-23.3Al-10.3V, d) Ti-25.1Al-10.3V, e) Ti-27.1Al-10.3V.

The yield strength and the elongation to rupture at room temperature versus aluminium alloy content are presented in Figures 2 and 3 respectively. The yield strength in the annealed condition varies significantly from one alloy to another and exhibits a minimum around 23%Al. As far as the ductility is concerned, it is observed that elongation to rupture steadily increases with decrease in aluminium content and reaches 6% in the alloy with 19%Al. The variation in tensile properties will be discussed in the next section.

Considering the micrographs in Figure 1, it cannot be rejected that the change in the mechanical behaviour of the α_2 phase with the aluminium concentration may be hidden by microstructural differences between alloys. For this reason, a solution treatment at T_β-30°C followed by the preceding annealing treatment (850°C, 4 hours) was applied to all alloys in order to obtain similar microstructures. In that condition, the microstructure of the different compositions consists of a low volume fraction of equiaxed α_2 phase in a transformed β matrix (Fig. 4).

Results of tensile tests on the solution treated plus annealed condition are plotted in Figures 2 and 3. The variation in yield strength versus aluminium content follows the same trend as in the annealed condition but with a smaller difference between extreme values. Again, a minimum is observed near 23%Al. The elongation to rupture is improved when decreasing the aluminium concentration ; a value of 4% has been measured in the alloy with 19%Al. It is worth noting that in both heat treated conditions a simultaneous increase in strength and ductility is observed in sub-stoichiometric compositions.

Discussion

The large variation in yield strength with the aluminium concentration in the annealed condition arises partly from changes in microstructure (Fig. 1). Indeed, the coarsening of the structure between 19%Al and 23%Al can account for the decrease in yield strength and the fine decomposition of the β phase beyond 25%Al may contribute to the further increase in strength. This explanation cannot be put forward in the case of the solution treated plus annealed condition as the microstructures are very similar. So, it is assumed that deviation from stoichiometry has an influence on the mechanical behaviour of the α_2 phase.

Microprobe analysis was performed on the α_2 phase of the different alloys and it was found that the minimum in yield strength which appears in Figure 2 corresponds to a stoichiometric composition of that phase (Table II). This result is in agreement with a previous study on binary Ti-Al alloys within the composition range 17-60at.%Al [8]. Measuring the Vickers hardness as a function of the aluminium content, the authors observed two minima at the stoichiometric compositions Ti_3Al and TiAl. It is to be mentioned that a similar effect has been reported in other intermetallic compounds such as Ni_3Al and Ni_3Ga [9]. Further work is required to give a detailed explanation for the aluminium dependence of the yield strength.

The ductility is improved when decreasing the aluminium alloy content for both heat treatments (Fig. 3). In particular, it is interesting to point out that the strength level is equivalent for both structures at the lowest aluminium concentration (Fig. 2) whereas the elongation to rupture is higher in the annealed condition. We have shown elsewhere that the morphology of the β phase influences the ductility of Ti_3Al based

Table II Microprobe analysis on the α_2 phase in the solution treated plus annealed condition (at.%).

Alloy	Ti	Al	V
Ti-19.4Al-10.3V	72.1 ± 0.1	23.2 ± 0.4	4.7 ± 0.4
Ti-20.8Al-10.3V	70.6 ± 0.1	24.0 ± 0.3	5.4 ± 0.4
Ti-23.3Al-10.3V	69.7 ± 0.1	25.2 ± 0.1	5.1 ± 0.1
Ti-25.1Al-10.3V	68.2 ± 0.1	26.2 ± 0.1	5.6 ± 0.1
Ti-27.1Al-10.3V	67.0 ± 0.1	27.2 ± 0.1	5.8 ± 0.1

alloys and that the presence of this phase is the most beneficial when it forms a continuous layer surrounding the α_2 phase [10]. Precisely, this type of microstructure has been developed during annealing on the Ti-19.4Al-10.3V alloy (Fig. 1).

As in the case of strength, differences in microstructure cannot account alone for the variation in ductility with the aluminium alloy content ; this is the reason why deformation mechanisms of the α_2 phase have been investigated. Thin foils of Ti-19.4 Al-10.3V and Ti-27.1Al-10.3V alloys have been examined after plastic deformation. In both alloys, the dominant slip system consists of "a" type dislocations with a $1/3\langle 11\bar{2}0\rangle$ Burgers vector moving on $\{10\bar{1}0\}$ prismatic planes (Fig. 5). No evidence of glide on (0001) basal planes was found here although this slip system has been identified in Ti_3Al [8,11]. Besides, "c+a" dislocations with a $1/6\langle 11\bar{2}6\rangle$ Burgers vector have been observed on $\{11\bar{2}1\}$ pyramidal planes in both alloys (Fig. 6). Finally, some "c" dislocations are visible at the α_2/β interface but, according to Yang [12], this type of dislocation does not contribute to deformation. The main difference between the two alloys is the much higher density of "c+a" dislocations in the sub-stoichiometric com-

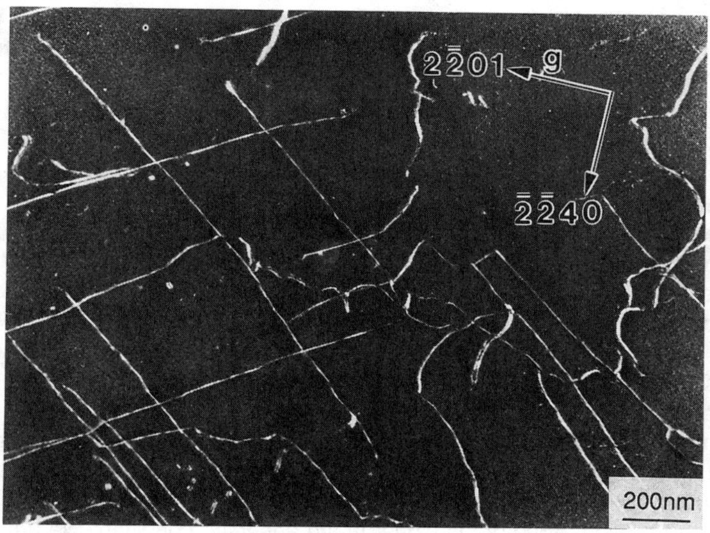

Figure 5 "a" type dislocations in α_2 phase of Ti-27.1Al-10.3V alloy ($g=2\bar{2}01$).

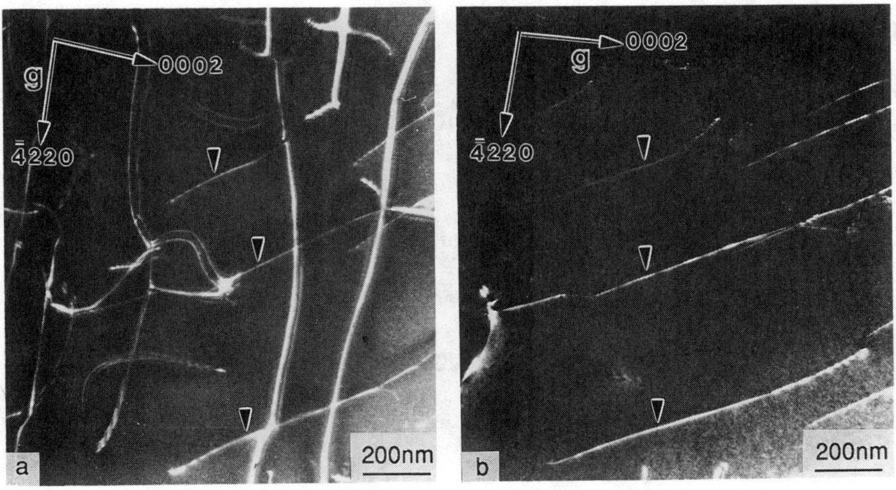

Figure 6 "c+a" type dislocations (arrows) in α_2 phase of Ti-19.4Al-10.3V alloy : a) g=$\bar{4}$ 220, b) g=0002.

position. It is assumed that deviation from perfect order when diminishing the aluminium content may reduce lattice frictional stresses thereby enhancing dislocations mobility and improving plasticity. Also, it is to be noted that the volume fraction of the ductile β phase slightly increases when lowering the aluminium concentration and this parameter may contribute to the better ductility.

Conclusions

The influence of deviation from stoichiometry on microstructure and room temperature tensile properties of Ti_3Al+V alloys has been investigated.

In the range of aluminium concentration considered, the only phases that have been identified are α_2 and ordered β.

In both annealed and solution treated plus annealed conditions, the yield strength versus aluminium content exhibits a minimum which corresponds to a stoichiometric composition of the α_2 phase.

The improvement in ductility in sub-stoichiometric alloys is mainly attributed to an increased activity of "c+a" dislocations on $\{11\bar{2}\,1\}$ pyramidal planes. Another parameter that can account for the better plasticity is the slight increase in the volume fraction of the ductile β phase with the decrease in aluminium content.

Acknowledgements

The authors would like to acknowledge DRET for financial support.

References

[1] S.K. Ehlers, M.G. Mendiratta, J. of Metals, Dec. 1982, 80.

[2] Y.W. Kim, J. of Metals, July 1989, 24.

[3] S.C. Huang, E.L. Hall, M.F.X. Gigliotti, Sixth World Conference on Titanium, vol.2 (Les Editions de Physique, Paris, 1989), 1109.

[4] M.J. Blackburn, M.P. Smith, U.S. Patent 4 292 077, Sep. 1981.

[5] I. Ehrhart, Doctorat Thesis, University of Orsay, March 1991.

[6] T. Hamajima, G. Lutgering, S. Weissmann, Met. Trans., 4 (1973), 847.

[7] D. Banerjee, T.K. Nandy, A.K. Gogia, Scripta Met., 21 (1987), 597.

[8] J.C. Williams, M.J. Blackburn, Ordered Alloys (Claitor's Press, Baton Rouge, 1970), 425.

[9] O. Noguchi, Y. Oya, T. Suzuki, Met. Trans. A, 12A (1981), 1647.

[10] A. Vassel, I. Ehrhart, M. Thomas, 1990 Int. Conf. on Titanium Products and Applications, vol.1 (T.D.A., Dayton, 1990), 131.

[11] S.M.L.Sastry, H.A. Lipsitt, Titanium'80 Science and Technology, vol.2 (The Metallurgical Society of AIME, 1980) 1231.

[12] W.J.S. Yang, Met. Trans. A, 13A (1982), 324.

TRANSFORMATION KINETICS IN $\gamma+\alpha_2$ TiAl-BASED ALLOYS

A. Denquin, S. Naka and T. Khan

Office National d'Etudes et de Recherches Aérospatiales,
29, Av. de la Division Leclerc, 92322 Châtillon CEDEX France

Abstract

Transformation kinetics during heat treatments in five two-phase $\gamma+\alpha_2$ alloys were examined in order to establish a better understanding of the relationship between microstructure and mechanical properties of TiAl-based alloys. Various transformation modes were observed and three types of competitions were identified. (1) The formation of the lamellar structure can take place both in the $\alpha+\gamma$ and $\alpha_2+\gamma$ phase fields, depending on the cooling rate; in this case, the propagation of Shockley partials promoting the γ lamella formation competes with the $\alpha \Rightarrow \alpha_2$ ordering reaction. (2) Quenching from the single phase α field can hinder the formation of lamellae to promote either the ordering $\alpha \Rightarrow \alpha_2$ reaction or a "massive" $\alpha \Rightarrow \gamma$ transformation, depending on the alloy composition. (3) The lamellar structure of lower Al-content alloys tends to show a discontinuous coarsening while higher Al-content alloys give rise to an easy formation of monolithic γ grains. Finally, the present study shows that it is possible to classify the two-phase TiAl-based alloys into different categories based on the above competitions.

Introduction

Titanium aluminide alloys based on TiAl are now receiving an increasing attention because of their low density and attractive properties at high temperatures, such as high mechanical strengh and good oxidation resistance. However, these alloys suffer from a low ductility up to about 600°C, which limits their application. Two phase $\gamma+\alpha_2$ TiAl-based alloys containing about 48 atomic percent of aluminium exhibit a higher ductility than the single phase γ alloys [1]. A fine lamellar structure is generally believed to promote a certain amount of ductility. This two phase lamellar structure can be modified by thermomechanical and heat treatments. In view of the great industrial potential of these intermetallic materials, the relationship between microstructure and mechanical properties has to be carefully evaluated. It is therefore essential to have a precise control of the microstructure. The purpose of the present study is to examine the kinetics of various transformation processes which take place during heat treatments on various TiAl-based alloys.

Titanium '92
Science and Technology
Edited by F.H. Froes and I. Caplan
The Minerals, Metals & Materials Society, 1993

Experimental procedures

The present study was carried out on five two-phase alloys. Their atomic concentrations are as follows : $Ti_{54}Al_{46}$, $Ti_{52}Al_{48}$, $Ti_{51}Al_{47}Cr_2$, $Ti_{50}Al_{46}Cr_2Nb_2$ and $Ti_{48}Al_{48}Cr_2Nb_2$. These alloys were prepared using the vacuum arc melting technique. An argon furnace was used for heat treatments below 1250°C but a vacuum furnace allowing oil quenching was employed for heat treatments at higher temperatures. Microstructures were observed both by optical and transmission electron microscopy (TEM) on a JEOL 200CX. Since the ternary and quaternary phase diagrammes are unknown and because of the small additions of chromium and niobium, the binary phase diagramme proposed by McCullough et al. [2] was used as a reference for understanding phase transformations in all the compositions. Solution heat treatments in the α phase field of such alloys have not been extensively studied, both because of an inevitably large grain size resulting through such heat treatments and due to the difficulty of performing them at temperatures as high as 1400°C on an industrial scale [3]. Therefore, most of the previous microstructural examinations were made only after heat treatments in the $\alpha_2+\gamma$ and $\alpha+\gamma$ phase fields without any solution heat treatment. In the present work, heat treatments were performed in three different phase fields, i.e. solution heat treatments in the α phase field as well as heat treatments in the $\alpha+\gamma$ and $\alpha_2+\gamma$ phase fields either after solution heat treatment or directly from the as-cast state.

Results

1. The lamellar microstructure

In the as-cast conditions, all the alloys examined showed a lamellar-grained structure (Fig.1a). The lamellar grains are formed of alternate lamellae of α_2 and γ phases. Higher Al-content alloys, i.e. $Ti_{52}Al_{48}$, $Ti_{51}Al_{47}Cr_2$ and $Ti_{48}Al_{48}Cr_2Nb_2$, also showed monolithic γ regions corresponding to the interdendritic segregation of aluminium during solidification. The α_2

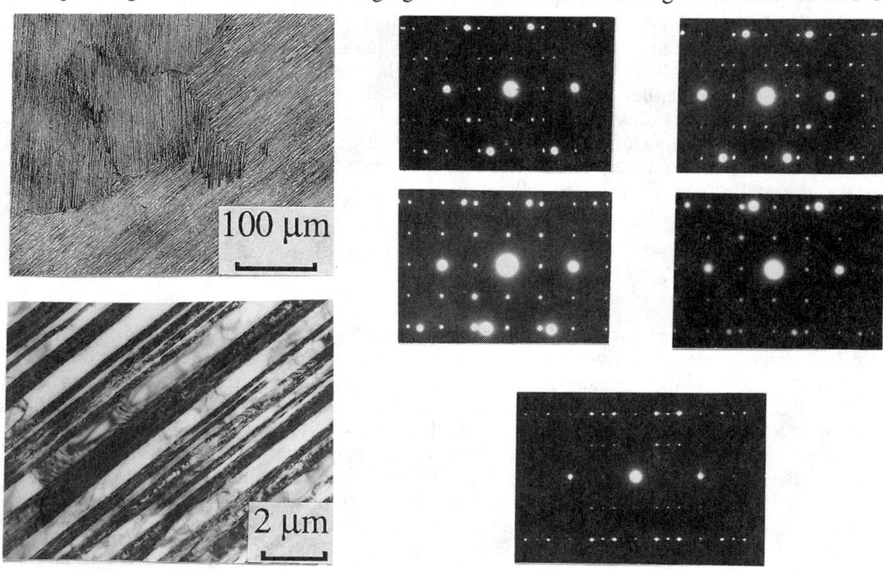

(a) (b)

Fig.1 (a) Optical and TEM observations of the lamellar structure.
(b) Orientation relationship revealed by selected area diffraction.
The zone axis of the α_2 and γ phases are $<11\bar{2}0>$ and $<1\bar{1}0>$, respectively.

and γ phases of the lamellar structure have the orientation relationship: $<11\bar{2}0>\alpha_2$ // $<1\bar{1}0>\gamma$ and $\{0001\}\alpha_2$ // $\{111\}\gamma$. The habit plane is the {0002} plane of α_2 [4]. A TEM study of the lamellae revealed the presence of small domains in each γ lamella. The α_2 phase has a DO_{19} structure in which the $<11\bar{2}0>$ directions are equivalent. Because of the atomic positions of Ti and Al in the $L1_0$ structure of γ, $<1\bar{1}0]$ and $<10\bar{1}]$ or $<0\bar{1}1]$ are not equivalent. A $<1\bar{1}0]$ zone axis diffraction pattern is different from $<10\bar{1}]$ or $<0\bar{1}1]$ ones. It is then possible to distinguish four orientation relationships between one γ domain and the α_2 phase of the lamellar structure by selected area diffraction. As shown in fig.1b, the selected area diffraction pattern generally obtained from several lamellae corresponds to a composite of those four patterns [5,6]. It can be noticed that two γ domains are twin related, order-fault related or pseudo-twin related [6,7], and two domains with the same orientation can be antiphase domains. Domains separated by an antiphase boundary or an order-fault boundary have the same stacking sequences of the {111} planes, i.e. ...ABCABC... and those separated by a twin boundary or a pseudo-twin boundary have the opposite stacking sequences ...ABCABC... vs. ...ACBACB...[8]. The results of extensive analyses made in the present study indicate that the domains included in a given γ lamella are of the same stacking sequence, while two adjacent γ lamellae are often twin or pseudo-twin related, although the explanations of such observations are still subject to some controversies [5,6,7,8,9,10].

2. Competition between the $\alpha \Rightarrow \alpha_2$ ordering and the formation of γ lamellae

The 46 at.% Al alloy directly quenched from the α phase field showed large single-phase grains accompanied by the formation of some lamellae at grain boundaries (Fig.2a). TEM observations of the same sample (Fig.2b) indicate that the matrix is ordered (α_2) containing numerous antiphase boundaries (APB's) and a few isolated γ lamellae. These lamellae always cut through antiphase domains. These results lead to the conclusion that the nucleation of the γ lamellae starts subsequently to the ordering of the α phase. As proposed by Blackburn [4], this nucleation probably proceeds from the movement of Shockley partials, locally giving rise to an ordered FCC type stacking sequence in the ordered hexagonal structure. The stress induced by this reaction favours the growth by the ledge mechanism.

(a) (b)

Fig. 2 a and b $Ti_{54}Al_{46}$ quenched from the α phase field.

On the contrary, if a sample of the same alloy was cooled down from the α phase field to 1230°C (α+γ phase field) and held for 72 hours before quenching, an entirely lamellar microstructure was obtained (Fig.3a) in which both the α_2 lamellae containing small APB's and γ lamellae are well developped (Fig.3b). These observations indicate that the nucleation and growth of γ lamellae can precede the ordering of the α phase, suggesting that the explanations based on the γ lamella nucleation and growth (propagation of Shockley partials and ledge mechanism) are also valid in the disordered hexagonal structure.

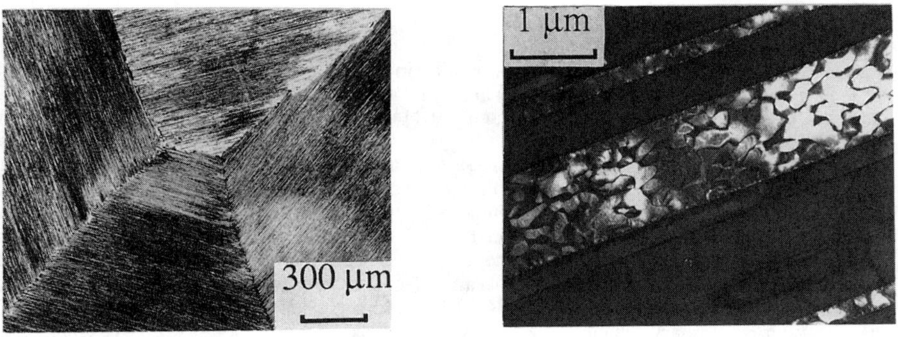

(a) (b)

Fig. 3 a and b : $Ti_{54}Al_{46}$ cooled down from the α phase field and hold 72 h at 1230°C before quenching.

Finally, the existence of large order-fault domains within the γ lamellae formed in this phase field rules out the explanation given by Yang and Wu [8] in which the origin of the order-fault domains are attributed to the pre-existing antiphase domains of α_2.

3. Competition between ordering ($\alpha \Rightarrow \alpha_2$) and massive formation of the γ phase ($\alpha \Rightarrow \gamma$)

The 48 at.% Al alloy quenched from the α phase field showed a grain size similar to that of the 46 at. % Al alloy, but the optical microstructure of the former alloy (Fig.4a) was quite different from that of the latter alloy (Fig.2a). Apart from some lamellar regions observed around grain boundaries, the grains in the 48 at.% Al alloy are extensively occupied by "martensitic-like" or "massively transformed" regions. TEM observations indicate that these regions were entirely constituted by the γ phase, often faulted and exhibiting numerous subgrains (Fig.4b). Twins, pseudo-twins, order faults and antiphase boundaries can be observed in some cases (Fig.5a and 5b). Thus, the second type of competition takes place between the $\alpha \Rightarrow \alpha_2$ ordering of the 46 at.% Al alloy and the "massive" formation of the γ phase in the 48 at.% Al alloy.

(a) (b)

Fig. 4 a and b $Ti_{52}Al_{48}$ quenched from the α phase field.

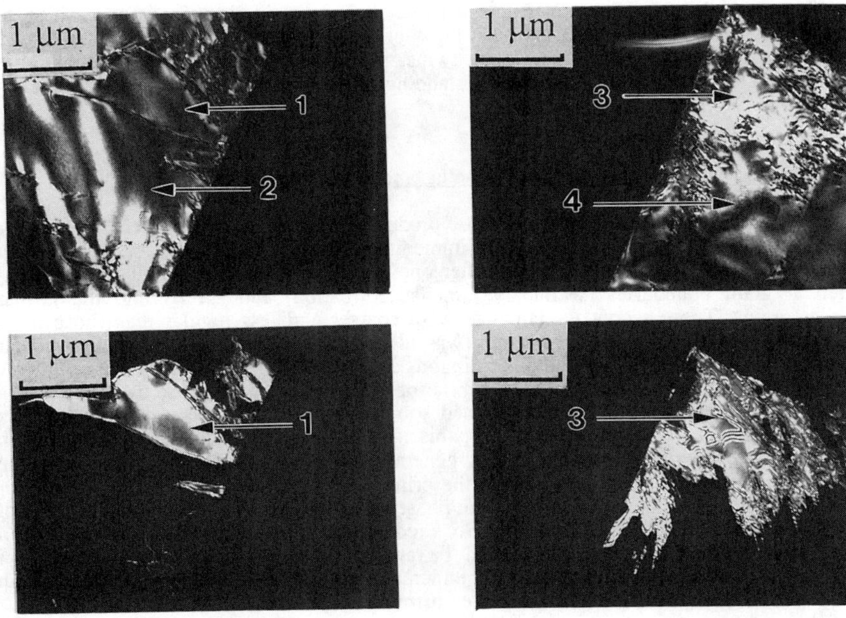

Fig. 5a Dark field observation of various domains respectively twin related (1 and 3, 2 and 4), pseudo-twin related (1 and 4, 2 and 3) and order-fault related (1 and 2, 3 and 4)

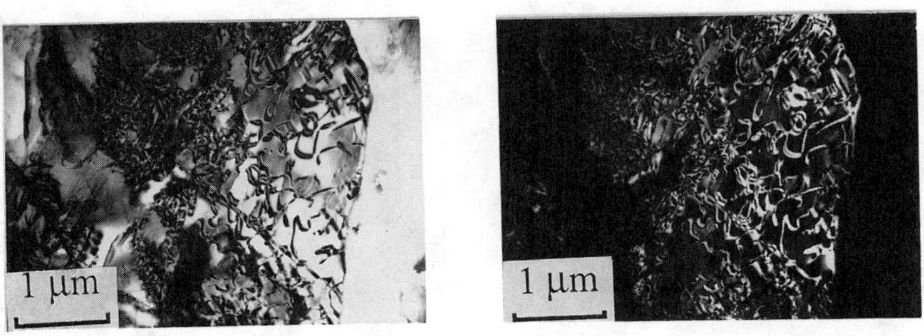

Fig.5b Observation of antiphase boundaries in the alloy massively transformed. (dark field image with g = 001)

High rate transformations from one phase to another can be either of a martensitic type or a massive type. Both these modes are composition-invariant. The martensitic transformation occurs by a shearing of specific atomic planes of the parent phase without any diffusion, often producing a metastable phase. A specific orientation relationship is generally observed between the two phases and a surface tilt can be seen on the sample surface. On the contrary, in a massive transformation, a high temperature phase is suddenly transformed in another phase of the phase diagramme. The phase interface moves probably by thermally activated jumps [11]. This transformation usually proceeds with approximately the same rate in all directions and can easily cross the grain boundaries. Moreover, the massive reaction can take place inside the two-phase field [11].

Work is now in hand both to determine whether there is a simple orientation relationship between the parent α phase and the resulting γ phase and to characterize the nature of the above mentioned crystal defects such as twins, pseudo-twins, order faults and antiphase boundaries. This will lead to a better understanding of the observed massive formation of the γ phase.

4. Competition between coalescence of lamellae and discontinuous coarsening

When the sample quenched from the α phase producing a single α_2 phase microstructure was maintained at 1000°C, all the grains were immediately transformed into a very fine lamellar structure (primary lamellae formation). After one hour, however, new lamellar domains appeared at grain boundaries (secondary lamellae formation) and spread into the primary lamellar grains. These secondary lamellae were coarser and less regular than those of the primary reaction (Fig.6). This coarsening was always accompanied by the grain boundary migration. Such a reaction is called discontinuous coarsening and proceeds by a dissolution of the precipitates of the adjacent grain. A similar observation has been made by Yamabe et al. [12], but this phenomenon can be explained somewhat differently. As generally accepted, grain boundaries have a greater solubility; this may promote the dissolution of the second phase. The free energy of the reaction is generally composed of chemical and interfacial contributions. The high formation rate of the primary lamellar structure may lead to a non-equilibrium chemical state of the constituent phases (i.e. their compositions are different from those predicted by the phase diagramme). So, a redistribution of solutes should operate during the discontinuous coarsening. This could be an as important factor as the decrease in interfacial energy in order to compensate the increase in grain boundary energy. Studies using atom probe are now in progress in order to confirm this explanation.

Fig.6 $Ti_{54}Al_{46}$ quenched from the α phase field and held at 1000°C for 24 h.

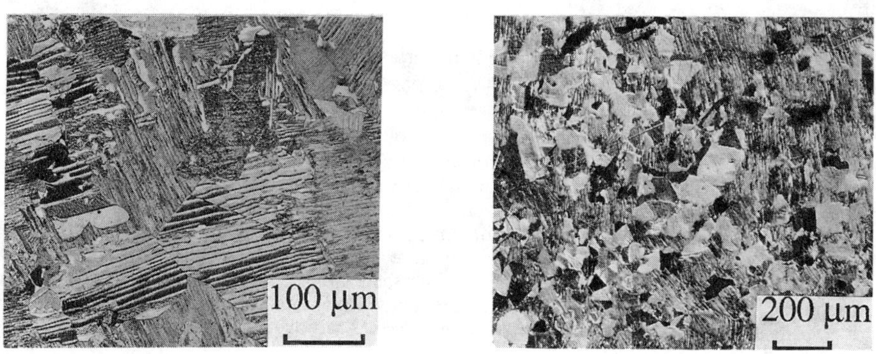

Fig.7 (a) $Ti_{54}Al_{46}$ held at 1150°C for 72 h, (b) $Ti_{52}Al_{48}$ held at 1150°C for 1 h.

The same alloy, held at 1150°C (just above the "eutectoid" temperature) without prior solution treatment, frequently showed discontinuous coarsening. Modifications at grains boundaries can be noticed in optical microscopy after 24 hours (Fig.7a). On the contrary, higher Al-content alloys tend to produce monolithic γ grains without any significant discontinuous reaction (Fig.7b). Thus, these two phenomena, i.e. formation of monolithic γ grains and discontinuous coarsening, constitute the third competition.

Discussion

Three types of competitions have been identified during the microstructural analysis of the two binary TiAl-based alloys :

- a competition between the $\alpha \Rightarrow \alpha_2$ ordering reaction and the formation of the γ lamellae,
- a competition between the $\alpha \Rightarrow \alpha_2$ ordering reaction and the "massive" formation of the γ phase,
- a competition between the coalescence of the lamellae and the discontinuous coarsening.

These three competitions, and in particular the last two, allow to classify the two-phase TiAl-based alloys into various categories. The major distinction can be made between the alloys showing the $\alpha \Rightarrow \alpha_2$ ordering reaction such as the 46 at.% Al binary alloy and those showing the "massive" formation of the γ phase such as the 48at.% Al binary alloy. All the ternary and quaternary alloys examined in this study undergo the "massive" transformation upon quenching from the α phase field (Fig.8a). It is however worth noting that the $Ti_{51}Al_{47}Cr_2$ and $Ti_{50}Al_{46}Cr_2Nb_2$ alloys exhibit more lamellar regions than the $Ti_{52}Al_{48}$ one. On the contrary, if a heat treatment is performed at 1150°C, two types of behaviour can be distinguished (Fig.8b). The $Ti_{50}Al_{46}Cr_2Nb_2$ alloy involves a discontinuous reaction while the higher Al content alloys (i.e. $Ti_{51}Al_{47}Cr_2$ and $Ti_{48}Al_{48}Cr_2Nb_2$) produce monolithic γ grains. With these two heat treatments, it is thus possible to evaluate the nature of the microstructural modifications without knowing the phase diagrammes corresponding to these alloys. This is an important conclusion, because the determination of a phase diagramme is generally a very time-consuming task.

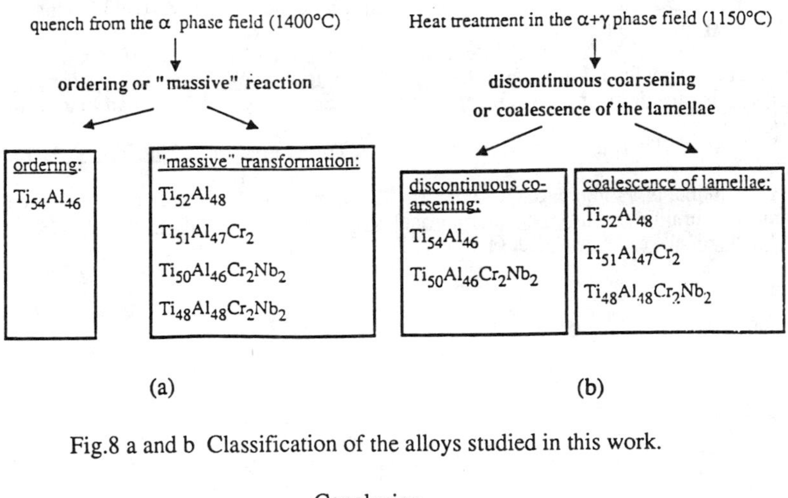

Fig.8 a and b Classification of the alloys studied in this work.

Conclusion

Microstructural modifications are more complex than those previously described by other investigators and the kinetics of transformations are greatly dependant on the alloy composition. Investigations made on the two binary alloys allowed us to identify various transformation modes, obtained essentially after two heat treatments (Quenching from the α single phase field and heat-treatment near the "eutectoid" temperature); these transformation modes can be understood in terms of competitions. By checking these competitons, ternary and quaternary alloys can then be classified into various categories.

Acknowledgements

This work was conducted with a financial support of SNECMA and TURBOMECA. The authors are grateful to A. Bachelier-Locq for her technical assistance and P. Thévenin for the alloy preparation.

References

1.T. Kawabata, T. Tamura and O. Izumi, "Parameters for Ductility Improvement in TiAl", in Proc. Mat. Res. Soc. Symp. on "High Temperature Ordered Intermetallic Alloys", Boston, MRS, 1989, Vol.133, 329.

2. C. McCullough, J.J. Valencia, C.G. Levi and R. Mehrabian, "Phase Equilibria and Solidification in Ti-Al Alloys", Acta Metall., 37 (1989) 1321.

3. S.C. Huang and D.S. Shih, "Microstructure - Property Correlation in TiAl-Base Alloys", paper presented at the TMS Symp., Detroit, 1990.

4. M.J. Blackburn, "Some Aspects of Phase Transformations in Titanium Alloys", in "The Science, Technology and Applications of Titanium", Ed. R.I. Jaffee and N.E.Promisel, Pergamon Press, Oxford, 1970, 633.

5. C.R. Feng, D.J. Michel and C.R. Crowe, "Twin Relationships in TiAl", Scripta Metall., 22 (1988) 1481.

6. Y.S. Yang and S.K. Wu, "Orientation Relationships between γ Plates and α_2 Matrix in a Ti-39at.%Al Alloy", Scripta Metall. et Mater., 24 (1990) 1801.

7. M. Yamaguchi and Y. Umakoshi, "The Deformation Behavior of Intermetallic Superlattice Compounds", Progress in Materials Science, 34 (N°1) (1990) 59.

8. Y.S. Yang and S.K. Wu, "Orientation Faults of γ Lamellae in a Ti-40at.% Al Alloy", Scripta Metall. et Mater., 25 (1991) 255.

9. C.R. Feng, D.J. Michel and C.R. Crowe, "Antiphase Domains and {011} Twins in TiAl", Phil. Mag. Let., 61 (N°3) (1990) 95.

10. M. Kikuchi and Y. Yamabe, "Mecanism of Formation of Primary γ/α_2 Lamellar Microstructure in Titanium Aluminides", in Proc. International Symp. on "Intermetallic Compounds. Structure and Mecanical Properties", Ed. O. Izumi, The Japan Inst. of Metals, 1991, 815.

11. M. Hillert, "Thermodynamics of the Massive Transformation", Met. Trans. A, 15A (1984) 411.

12. Y. Yamabe, N. Honjo and M. Kikuchi, "Discontinuous Coarsening of γ/α_2 Lamellae in Titanium Aluminides", in Proc. International Symp. on "Intermetallic Compounds. Structure and Mechanical Properties", Ed. O. Izumi, The Japan Inst. of Metals, 1991, 821.

A NOVEL INERT GAS ATOMIZATION FACILITY FOR THE PRODUCTION OF INTERMETALLIC TITANIUM-BASED ALLOY POWDERS

R. Gerling, F.P. Schimansky and R. Wagner

GKSS Research Center, Institute for Materials Research,
Max-Planck-Str.,
D-2054 Geesthacht, Germany

Abstract

For the production of pure, ceramic-free Titanium-based intermetallic alloy powders with a high fraction of undersize powder (<45μm), a new type of atomization facility has been developed and put into operation. This laboratory scale plant combines a cold crucible plasma torch melting technique with gas atomization. A special melt stream transfer system has been developed which consists of an induction heated water-cooled copper funnel for forming a thin melt stream and guiding it into the gas nozzle. Technical details of this plant and results of first runs concerning the particle size distribution, the impurity content and the microstructure of intermetallic Ti-based alloy powders are presented in this paper.

Introduction

In the field of research and development of new high temperature alloys, intermetallic Titanium-based alloys such as Titanium-Aluminides (γ-TiAl, α_2-Ti_3Al) and Titanium-Silicides (based on Ti_5Si_3) play an important role. These alloys are characterized by high strengths at low densities. However, other properties such as room temperature ductility and corrosion resistance need to be improved prior to industrial application of these materials.

Titanium and Titanium-based alloy melts and powders are known to be very reactive, and the mechanical properties of components consisting of such alloys/alloy powders depend strongly on the respective impurity content. Research and development of these alloys must therefore be based on a production technique which on the one hand keeps the impurity content as low as possible and on the other hand allows the microstructure of these alloys to be controlled with respect to small grain sizes, phase compositions and chemical homogeneity.

The Institute for Materials Research of the GKSS-Research Center, Germany, is engaged in the development of high temperature intermetallic Ti-based alloys. For the powder metallurgical route of these alloys an inert gas atomization facility has been developed together with Leybold-Durferrit. This pilot-plant meets all the requirements for the production of pure, ceramic free

Titanium '92
Science and Technology
Edited by F.H. Froes and I. Caplan
The Minerals, Metals & Materials Society, 1993

Ti-based intermetallic alloy powders with a high fraction of undersize powder.
The specifications of this gas atomization facility and properties of powders from first test runs are described in this paper.

The Plasmamelting Inert Gas Atomization Facility PIGA

For this pilot production plant for intermetallic Ti-based alloy powder a new technical concept has been developed and realized. The concept consists of a plasma-skullmelting-inert-gas-atomization technique (PIGA) with an induction heated water cooled copper funnel for forming and guiding the melt stream into the gas nozzle.
The most important part of PIGA is the melting chamber (schematically shown in Fig. 1) which houses the plasma torch, the water cooled copper crucible, the transfer system for the melt stream and the gas nozzle.
As heat source a plasma torch was chosen, which can be operated with either Helium or Argon gas. The use of a plasma torch has three advantages:

1. During melting, the inert gas pressure inside the melting chamber is about normal pressure. This reduces the evaporation of alloy constituents during melting.
2. With the hot plasma gas mechanical energy is transferred into the melt. This supports a thorough mixing of the alloy constituents.
3. Using a plasma torch, the injection of impurities into the melt - e.g. electrode material - is negligibly small.

Since Ti-based alloy melts are very reactive, a skull melting technique using a water cooled copper crucible (volume ~ 1.4 l) has to be chosen to keep the impurity content of the melt at low values.
As atomization technique, gas atomization has been chosen, since this technique can yield a high fraction of undersize powders and high cooling rates which are required for the production of supersaturated or metastable fine-grained alloy powders. The atomization nozzle can be operated with either Argon or Helium.
In order to obtain high quality powder, a thin, homogeneous melt stream has to be fed into the middle of the gas nozzle. The transfer of the melt from the crucible into the gas nozzle is one of the most critical technical problems in designing a gas atomization facility. In particular the following problems have to be solved:

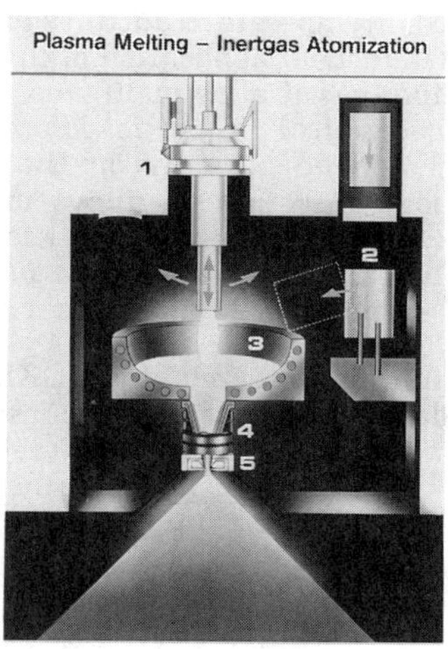

Fig. 1:
Schematic drawing of the melting chamber showing the plasma torch (1), the feeder (2), the watercooled copper crucible (3), the water-cooled induction heated melt stream transfer system (4) and the gas nozzle (5).

1. A thin, homogeneous and continuous melt stream has to be formed.
2. The overheat of the melt has to be maintained.
3. The melt stream has to be guided into the middle of the nozzle.
4. Contamination of the melt stream with impurities has to be avoided.

For PIGA a new transfer system has been developed. It consists of an induction heated watercooled copper funnel, which forms a thin melt stream and guides it into the middle of the gas nozzle (Fig. 1). The induction coil used has three functions: The electromagnetic forces are thought to keep the melt away from the funnel walls such that contamination of the melt with copper can be avoided. The electromagnetic forces influence the mass flow of the melt stream yielding a thinner melt stream and a high fraction of undersize powder. Finally, the induction heating allows the overheat of the melt, generated by the plasma torch, to be maintained.

The product of a gas atomization run, the metal powder, is collected in a powder can which is mounted below a cyclone separator. The powder can can be closed gas-tight so that the pure metal powder does not come into contact with air or moisture when the can is dismounted and carried to glove boxes for further handling in an Argon-atmosphere.

Further technical details of PIGA concerning the powder production process and safety precautions are discussed in detail in previous papers [1,2].

Cooling Rates

In order to achieve small grain sizes, a homogeneous microstructure and to be able to produce metastable or highly supersaturated alloy powders, short solidification times and high cooling rates for the subsequent cooling of the powder particles are required.

In order to check which parameters dominate the cooling rates and the solidification times and to get an idea of what can be realized with PIGA, we have calculated cooling rates (dT/dt) and solidification times using equations and thermodynamic data for TiAl reported in the literature [3,4].

Fig. 2 shows a summary of our calculations of the cooling rate as a function of powder particle size for $Ti_{51}Al_{49}$ (at%) alloy powder at 1750 K. As variables are shown the surrounding medium (Vacuum, Argon, Helium) and the difference velocity V_R between the powder particle and the surrounding atmosphere. In principle we are not able to atomize under vacuum, however these data allow comparison with other techniques e.g. centrifugal atomization.

V_R becomes important only for the cooling under Ar and He. V_R was measured by Bauckhage et al. [5] and their results show that V_R has its maximum in the vicinity of the gas nozzle in the zone of desintegration of the melt stream. As a consequence of the slowing down of the gas and the acceleration of the particles by the gas, V_R decreases with increasing distance from the nozzle and soon becomes about zero. However, as pointed out by Bauckhage et al. V_R does not only depend on the distance from the nozzle but also on the respective particle size.

We have assumed V_R to be identical for all particle sizes, and in Fig. 2 we show the results of our calculations for V_R = 0 m/s for He and Ar gas, and V_R = 100 m/s for He only. The assumption V_R = 0 m/s is conservative and indicates the lower limit for the cooling rates. It also allows a direct comparison with vacuum atomization. In contrast, V_R = 100 m/s is seen to be an upper limit, and realistic cooling rates should lie between the boundaries given by V_R = 0 and 100 m/s.

Fig. 2:
The cooling rate as a function of powder particle size for a $Ti_{51}Al_{49}$ alloy at 1750 K under vacuum, in Argon and Helium atmosphere and for different relative velocities between powder particle and atomization gas.

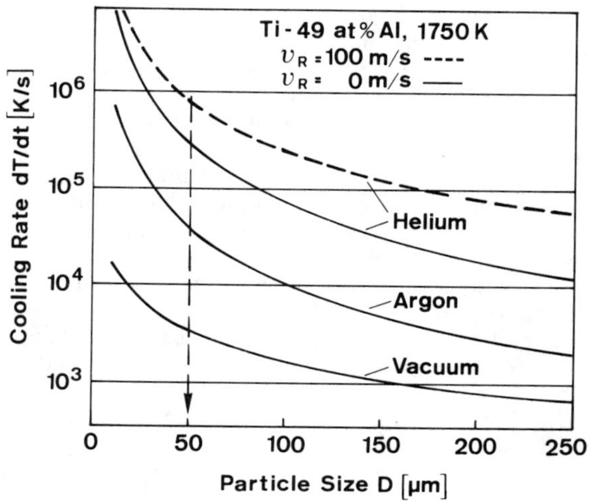

As can be seen from Fig. 2, dT/dt increases strongly with decreasing particle size regardless of the respective medium. On the other hand dT/dt increases with respect to the surrounding medium from vacuum to Ar to He. The respective cooling rates for a 50 µm particle are: $3 \cdot 10^3$, $3 \cdot 10^4$ and $3 \cdot 10^5$ K/s at $V_R = 0$ m/s. If we allow for $V_R = 100$ m/s, dT/dt reaches a value of $8 \cdot 10^5$ K/s on He atomization.

It may be concluded, that the highest cooling rates can be achieved using He as atomization gas. For a 50 µm $Ti_{51}Al_{49}$ particle dT/dt should lie at 1750 K in the range $3 \cdot 10^5 \leq dT/dt \leq 8 \cdot 10^5$ K/s.

Our calculations of the solidification times reveal similar results. The solidification time is the shorter the smaller the powder particle is; it is shorter for He - than for Ar - atomization and it decreases with increasing V_R.

Powder Particle Size Distributions

As shown in the previous section the high cooling rates required for optimized microstructures of the powder particles are given for small powder particles. One of the aims of optimizing the powder production process should therefore be to yield a high fraction of undersize powder. In Fig. 3 the results of sieve analyses of two test runs with PIGA (January 1992) are shown. The test runs were performed using Helium and Argon for atomization. The other conditions, e.g. alloy composition ($Ti_{48}Al_{50}Nb_2$), inert gas pressure (1.5 MPa) and the nozzle geometry were kept unchanged. The Helium (#2) atomization yielded the higher fraction of undersize powder (15% < 45 µm). The reason may be seen in the higher gas velocity with Helium compared to Ar (#1) atomization. The particle size distribution aimed for is shown by the curve indicated by (*). This size distribution ~40% < 45 µm is thought to be attainable after further optimization runs, which may include increased inert gas pressures and/or optimized nozzle geometries.

Scanning electron microscopy images indicated the powder particles to be spherical to a high degree, irrespective of whether He or Ar was used for atomization.

Fig. 3:
The particle size distribution for Helium (#2) and Argon (#1) atomized $Ti_{48}Al_{50}Nb_2$ alloy powder as given by test runs in January 92. The size distribution indicated by (*) appears to be possible after further process optimization.

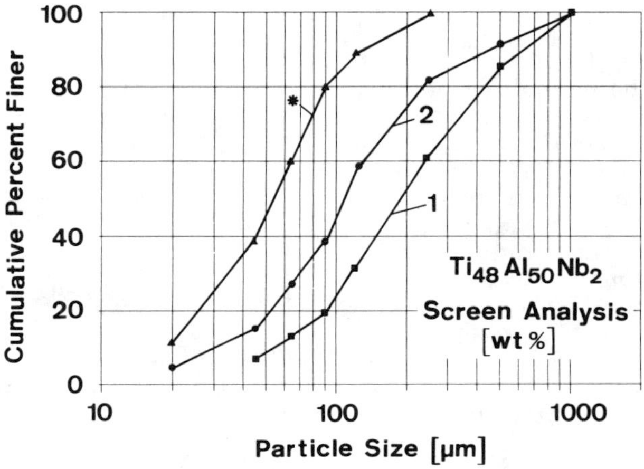

Microstructural Effects of Different Cooling Rates

As shown in the previous sections, there is a strong influence of the type of atomization gas and the powder particle size on both the solidification time and the cooling rate in the solid state. In this section we want to show the way in which the microstructure - as far as it can be characterized by the ratio of the amounts of the two phases and their distribution - is affected by the cooling conditions achieved with PIGA.
Using light microscopy and x-ray diffraction, several powder size fractions of a $Ti_{48}Al_{50}Nb_2$ powder atomized with Helium and Argon were investigated.
As reported by McCullough et al. [6] upon solidification of alloys having about these compositions, in the first stage α-dendrites are formed which then become surrounded by interdendritic γ-phase. With decreasing temperature α-dendrites transform to α_2 and undergo solid state decompositions to $\alpha_2 + \gamma$. Upon high cooling rates, the formation of the γ-TiAl phase can be suppressed to a certain extent. The ratio of the volume fractions of the α_2-Ti_3Al and the γ-TiAl phase may therefore be seen as an indication of the cooling conditions the powder particles have been exposed to. Since we only need an indirect measure of the volume fraction of the two phases, we have calculated the normalized ratio of the peak areas of the most intense diffraction lines of the α_2- and the γ-phase: $I\alpha_2(201)/[I\alpha_2(201)+I\gamma(111)]$. These normalized intensity ratios as a function of the powder size are shown in Fig. 4. For both gases the α_2-fraction increases with decreasing particle size, and for each particle size the α_2-fraction is larger for He than for Ar atomization. These results indicate that:

1. With decreasing particle size the solidification time decreases and the cooling rate in the solid state increases.
2. For identical particle sizes He atomization results in higher cooling rates/shorter solidification times than Ar atomization.

In conclusion, these results are in good qualitative agreement with the predictions of the cooling rate-calculations (Fig. 2).

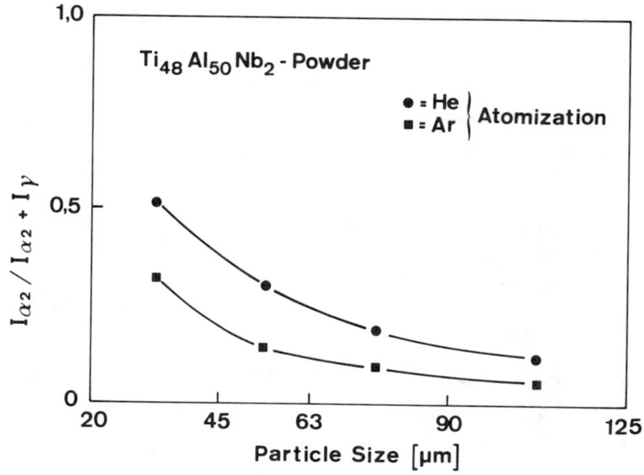

Fig. 4:
The normalized peak intensity ratios of the α_2-Ti_3Al and the γ-TiAl phase as a function of powder particle size for Helium and Argon atomized $Ti_{48}Al_{50}Nb_2$.

In Fig. 5 two light microscopy images of $Ti_{48}Al_{50}Nb_2$ powder particles atomized with He and Ar are shown. Both particles have their origin in the size fraction 45-63 μm. The microstructure is characterized by dark regions separated by bright channels. The dark regions represent the primary α-dendrites which have transformed upon cooling to $\alpha_2+\gamma$. This phase mixture however cannot be resolved by light microscopy. The bright channels represent the interdendritic γ-phase. Due to the short solidification times and high cooling rates, both particles consist of a very fine and homogeneous phase distribution. The phase distribution of the He-atomized particle appears to be finer compared to the Ar-atomized one. This again indicates that higher cooling rates can be achieved using He as atomization gas.

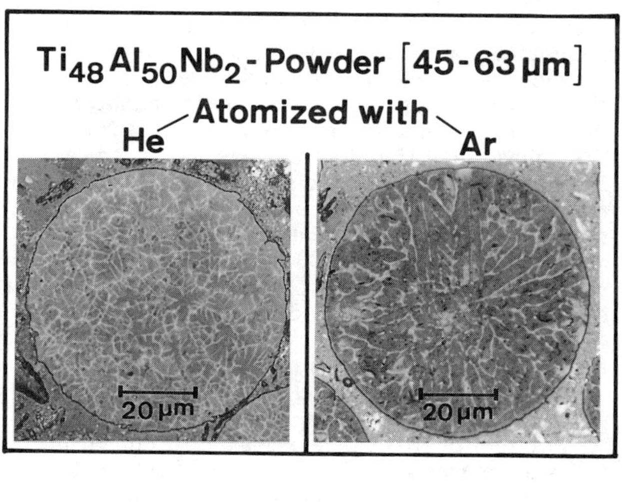

Fig. 5:
Light microscopy images of $Ti_{48}Al_{50}Nb_2$ powder particles from the 45-63 μm size fraction atomized with Helium and Argon.

Contamination During Melting and Atomization

The objective of designing a new type of gas atomization facility was the production of pure, ceramic-free Titanium-based intermetallic alloy powder. One task therefore was to keep the additional contamination during melting and atomization as low as possible. In order to get a first idea of the additional contamination which may occur during the process we have focussed on two elements: Oxygen and Copper. Oxygen is well known to be a crucial element with respect to the mechanical properties of the powder, and copper is one of the most probable metal-contaminations since both the crucible and the melt stream transfer system consist of copper.

The oxygen and copper contents were analysed in three stages of the whole production process of TiAlNb-powder:

1. Content of these impurities in the elements Ti, Al and Nb prior to melting.
2. Impurity content after melting and re-melting several times for homogenization.
3. Cu- and O_2-content of the final TiAlNb-powder.

As shown in table 1, the oxygen and the copper content of the mixture of elements is found to be 345 and 17 µg/g respectively. Chemical analyses of the melted $Ti_{48}Al_{50}Nb_2$ alloy revealed the oxygen and the copper concentration to be 350 and 23 µg/g respectively. Thus, within the experimental error neither the oxygen nor the copper content of this alloy increases upon melting the elements and homogenizing the alloy.

The copper content of the alloy powder (90-125 µm) was analysed to be 30 µg/g. This indicates that during forming and guiding the melt stream by the transfer system a small amount of copper may have been taken up by the melt. However since such copper concentrations are near to the limits of detection for these alloys, it can not be excluded, that all three concentrations as shown in table 1 lie within the experimental errors. It may, therefore, be concluded that the copper impurity content picked up during the whole process of melting and atomizing is only very small or even zero.

The procedure of the powder adapted for subsequent O_2-analysis was as follows: The plant was thoroughly cleaned and flushed serval times with Helium. Afterwards the first atomization run of $Ti_{48}Al_{50}Nb_2$ was performed using He as atomization gas. After this run, the powder can was replaced by an empty can, without opening the plant, and the second run using Ar as atomization gas was performed.

	O_2 [µg/g]	Cu [µg/g]
Mixture of elements prior to melting	345	17
TiAlNb-alloy	350	23
Alloy powder 90-125 µm	420[a]/350[b]	30[a]
Alloy powder 20-45 µm	1100[a]/600[b]	

a: first run with He gas, b: second run with Ar gas.

Table 1: The oxygen and copper impurity content after melting and gas-atomizing intermetallic $Ti_{48}Al_{50}Nb_2$.

Subsequently two powder size fractions of each run were analysed with respect to the oxygen impurity content [tab. 1]. The oxygen content of the two size fractions 90-125 μm and 20-45 μm were found to be 420 and 1100 μg/g for the first He-run and 350 and 600 μg/g for the second Ar-run. For both runs the oxygen content increases with decreasing powder particle size i.e. with increasing surface area indicating that the oxygen is rather attached to the surface than distributed inside the powder particles. For the He-atomized powder the O_2-content is found to be considerably higher than for the second Ar atomization. Since the oxygen content of both gases is about the same, the lower O_2-content of the Ar-powder is thought to be a consequence of the first run acting as a 'cleaning run' during which a considerable amount of oxygen is picked up i.e. from the walls of the plant.
In conclusion, an O_2-content of 600 μg/g found in the 20-45 μm fraction of intermetallic $Ti_{48}Al_{50}Nb_2$ powders produced during one of the first test runs of PIGA appears to be an excellent value. Moreover further improvements appear to be possible

- by heating the whole plant up to 150°C during vacuum pumping prior to flushing with inert gas.
- by increasing the numbers of flushing of the plant.
- by using atomization gases with further reduced oxygen content.
- by 'pre-atomization runs' used as cleaning runs.

References

1. R. Gerling, F.P. Schimansky and R. Wagner, "Spezifikationen der Pulververdüsungsanlage zur Herstellung von intermetallischen Titan-Basislegierungspulvern" (paper presented at the 2nd Symposium Materialforschung 1991, Dresden, Germany, 26-29. August 1991), 1, 659-671.
2. R. Gerling, F.P. Schimansky and R. Wagner, "Specification of the Novel Plasma-Melting Inert Gas Atomization Facility PIGA 1/300 for the Production of Intermetallic Titanium-Based-Alloy Powder" (paper presented at the 1992 Powder Metallurgy World Congress, San Francisco, California, 21.-26. June 1992), in press.
3. R.M. German, Powder Metalurgy Sience, (Princeton, J.J.: Metal Powder Industries Federation, 1984), 88.
4. E. Klar and J.W. Fesko, Metals Handbook Ninth Edition, Volume 7 Powder Metallurgy, (Metals Park, OHIO: American Society for Metals, 1984), Atomization 33.
5. K. Bauckhage et al., "Spray forming of liquid steel; local size and velocity distributions of particles in the spray cone and their reference to varying process parameters" (paper presented at PM'90 World Conference on Powder Metallurgy, London 2.-6. July), (1) 207-215.
6. C. McCullough et al., "Phase Equilibria and Solidification in Ti-Al Alloys", Acta Metall. 37 (5) (1989), 1321-1336.

Microstructural Development in Ingot and Forged XD™

Ti-48Al-2Mn-2Nb (+2% TiB_2) Near - Gamma Titanium Aluminide.

M Jezioro, S Moorhouse, M Cope and P Postans

Rolls-Royce plc, PO Box 31, Derby DE2 8BJ, UK.

Abstract

XD and monolithic near gamma titanium aluminides have been identified as candidate materials for high pressure compressor stators and low pressure turbine blade applications in future gas turbine engine developments. It is envisaged that weight savings of over 100 kgs are possible by replacement of Ni base superalloys.

As part of a wider programme aimed at evaluating process - structure - property relationships in both alpha-2 and near gamma titanium aluminides, RR have carried out an evaluation of Ti-48Al-2Mn-2Nb + 2% TiB_2 in ingot and forged form. Microstructural characterisation has established the presence of a number of solidification related features which may be detrimental to mechanical properties. Some of the segregation effects observed are specific to the XD (TiB_2 dispersion) process, others are related to the alloy chemistry). Phase diagram modelling is utilised to explain partitioning within this alloy system.

Finally the effect of forming and heat treatment on microstructural development is discussed. By utilising a thermomechanical process route it is possible to generate a variety of microstructures from lamellar to equiaxed. The effect of these structures on tensile properties is then discussed.

Introduction

This paper describes one aspect of the Rolls-Royce investigation of titanium aluminides for gas turbine applications. The overall programme covers alloy chemistry development, manufacturing assessments and various engineering aspects such as design studies and structure/property optimisation. Components manufactured from near gamma titanium aluminides have been identified as a potential replacement for low pressure turbine blades currently produced in the nickel based casting alloy, IN713LC. The specific strength and modulus characteristics of the near gamma material offer significant advantages which translate to a possible 100Kg weight saving compared to the current design.

Titanium '92
Science and Technology
Edited by F.H. Froes and I. Caplan
The Minerals, Metals & Materials Society, 1993

Although investment casting has been shown to be a viable processing method for gamma aluminides, thermomechanical processing (TMP) appears to offer greater freedom to manipulate the microstructure and hence mechanical properties. Both casting and TMP and electro-chemical machining of monolithic and XD™ alloys are being evaluated in the early stages of the development programme prior to definition of the optimum treatment. However, the present work concentrates on structural and mechanical property evaluation of cast plus HIP ingot and forged XD™ Ti-48Al-2Mn-2Nb. The microstructural and property information is supported by phase diagram and forging modelling programmes, both of which are described in the current paper.

Experimental Procedure and Materials

As part of the preliminary evaluation of XD gamma aluminide, an ingot of Ti-48at%Al-2at%Mn-2at%Nb + 2vol% TiB_2 has been purchased from Martin Marietta, the composition of which is given below.

Table I

Composition of XD gamma ingot.

	Al at%	Nb at%	Mn at%	Si at%	O ppm	N ppm	C ppm	H ppm	B ppm
Aim	48	2	2						
Mid Centre	47	1.9	1.9	300	800	105	250	40	6600

Although Rolls-Royce has its own alloy development programme, this alloy was selected as a model system on which information on process structure property relationships can be determined.

The work reported in this paper was carried out using a 60lb ingot, VAR melted by Alta Group, Fombell PA and HIP'ed by IPS, New London, OH. One third of this ingot was supplied to Ladish Co. Inc. to forge a pancake (= 14"0 x 1"thick), the remaining 40lbs being retained for microstructural evaluation, small scale compression testing etc.

Optical, SEM, EPMA and image analysis techniques were used to evaluate the material. Phase diagram modelling (Ref 1) was used to enhance understanding of phases present during both melting and thermomechanical processing.

Small scale compression testing was carried out at Swansea University using 6mm x 8mm cylinders taken from the HIP'ed ingot. These tests were used to generate an understanding of the effect of deformation on microstructure. In addition, flow stress data and a preliminary constitutive equation was generated which will be used for modelling of extrusion and blade forging.

XD Processing

The TiB_2 dispersion was added to the ingot using the XD™ process which has been developed by Martin Marietta (Refs 2-5). XD gamma titanium aluminides are melted in a conventional manner using CEVAM with the XD process aiming to introduce a uniform dispersion of TiB_2. However, evaluation of HIP'ed ingot product has highlighted a number of segregation effects.

Results and Discussion

a) Segregation Effects in CEVAM Melted XD-Near Gamma Ti-48at%Al-2at%Mn-2at%Nb.

TiB_2 Cellular Structure

Figure 1 shows a cellular distribution of TiB_2 particles surrounding islands of lamellar gamma + alpha-2. This feature was approximately 1mm x 2mm in size. It is probable that this feature is a remnant of undissolved TiB_2 master alloy. Previous reports of TiB_2 segregation (Ref 6) have been a result of microsegregation during solidification. Such segregation effects clearly illustrate the problems associated with producing a fine distribution of TiB_2 using the XD™ process.

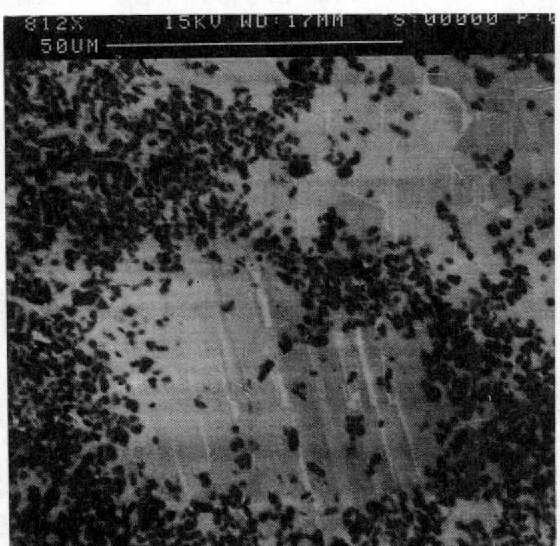

Figure 1 - Cellular distribution of titanium diboride particles.

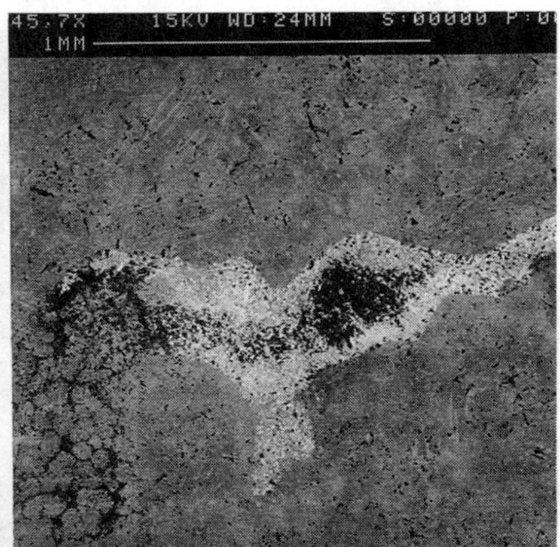

Figure 2 - Agglomeration of particles surrounded by lighter needles.

TiB_2 agglomerations + $(Nb, Ti)_x B_y$ needles.

An agglomeration of TiB_2 particles surrounded by lighter needles of $(Ti,Nb)_xB_y$ is shown in Figure 2. This TiB_2 agglomeration is likely to be a remnant of the original Al + TiB_2 sponge that has not dissolved on melting. This is similar to the TiB_2 agglomeration in Figure 1. However Nb has reacted with TiB_2 to form mixed Nb + Ti boride needles as shown in Figure 3. It is possible that the mixed boride needles have encapsulated the original Al + TiB_2 sponge not allowing the TiB_2 to dissipate during melting. Clearly, such a feature is both XD and alloy chemistry related.

Figure 3 - Higher magnification view of $(Nb,Ti)_x B_Y$ etching feature in Figure 2.

Mn Segregation

Figure 4 shows areas of single phase gamma grains surrounded by dark fully lamellar regions. The gamma stabilised zone was found to be rich in Al and Mn. At the centre of the gamma stabilised region, darker particles were seen (Figure 5). These were found to be Ti-Al-Mn intermetallics of composition, 50.9 at % Al, 43.95 at % Ti, 1.31 at % Nb and 3.68 at % Mn.

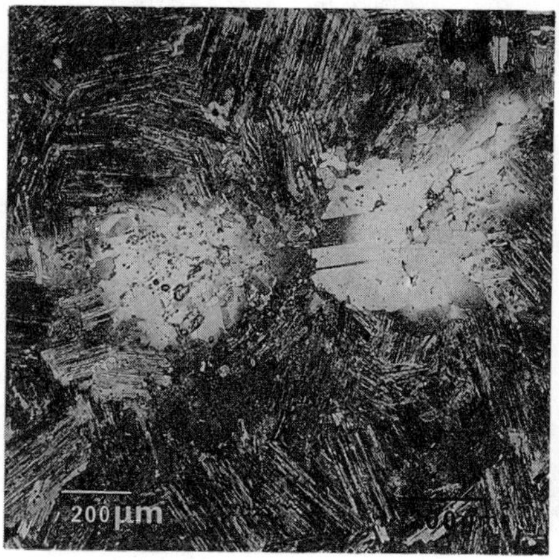

Figure 4 - Single phase gamma grains surrounded by fully lamallar regions

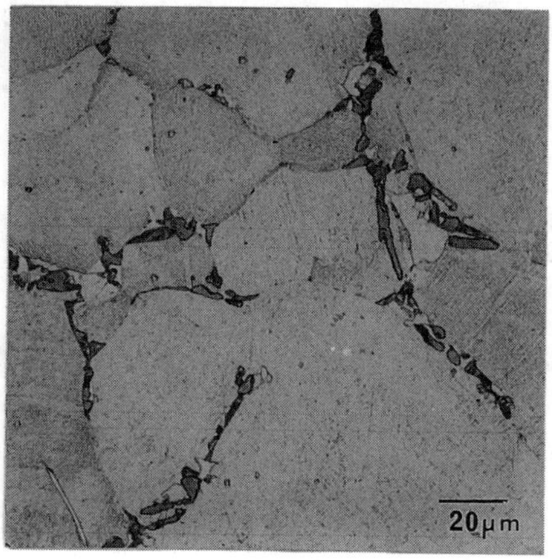

Figure 5 - Ti-Al-Mn intermetallics within the gamma stabilised region.

These effects were observed predominantly in the hot top region of the ingot. As a result of the local solidification conditions, the low melting point Ti-Mn-Al intermetallic forms within to the last liquid to solidify. This mechanism is supported by phase diagram modelling predictions (Ref 7) which shows Mn strongly partitioning in the last liquid to solidify (Figure 6). Use of the binary Ti-Al phase diagram is meaningless for understanding the Ti-Al-Mn-Nb system. Thus a phase diagram model has been used to predict the phases present at both the forging and heat treatment temperatures. A phase diagram prediction for the ternary Ti-Al-Mn system is shown in Figure 7 (Ref 7).

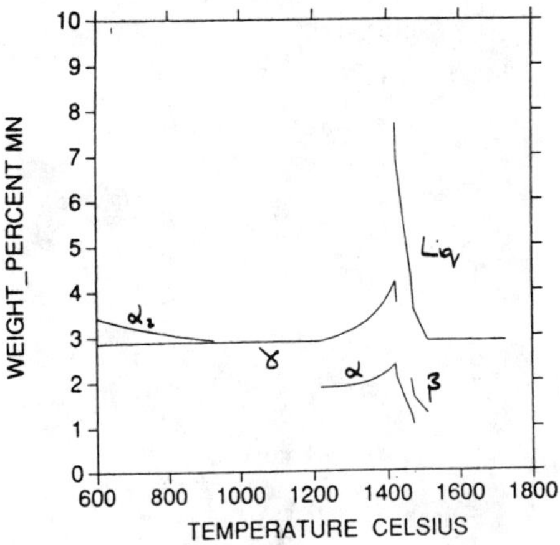

Figure 6 - Solidification sequence for Ti-48at%Al-2at%Mn.

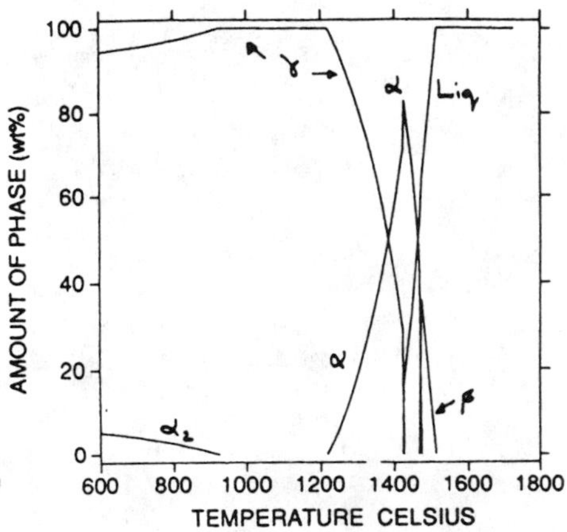

Figure 7 - Phase Prediction for the Ti-Al-Mn system.

b) Thermo Mechanical Processing

As described in Section 1 extrusion + ECM of HIP'ed ingot is a possible option to produce LPT blades. However, in order to produce the required microstructures and hence properties it is necessary to understand the deformation and heat treatment behaviour of XD gamma aluminides. As a result Rolls-Royce have carried out a number of small scale compression tests (Ref 8) at a strain rate of $5 \times 10^{-3}\ s^{-1}$ and at forging temperatures of 1000, 1050, 1100 and 1150°C in order to assess microstructural development.

The microstructures generated by small scale compression testing and associated control specimens are shown in Figure 8. Compression at 1000°C (Figure 8a) shows both deformation of the prior alpha grains and the laths of alpha-2 and gamma within them, indicating that the material has been worked below the gamma + alpha transus. Little or no equiaxed gamma phase was observed in either the small scale compression specimen or the control heat treated sample.

As the compression temperature is increased to 1050°C it is apparent from Figure 8b that a necklace type structure of deformed prior alpha, containing alpha-2 and gamma lamellae, surrounded by broken up alpha-2 and gamma is beginning to occur. This trend is continued by forging at 1100°C (Figure 8c), until forging at 1150°C (Figure 8d) develops a fully necklace structure. This consists of 200um diameter deformed prior alpha surrounded by recrystallised equiaxed gamma, 10um in diameter.

Thus isothermal forging can significantly refine the as cast ingot structure of XD gamma titanium aluminides to produce a 'necklace' type structure of recrystallised 20µm gamma and broken up alpha-2 + gamma lamellae.

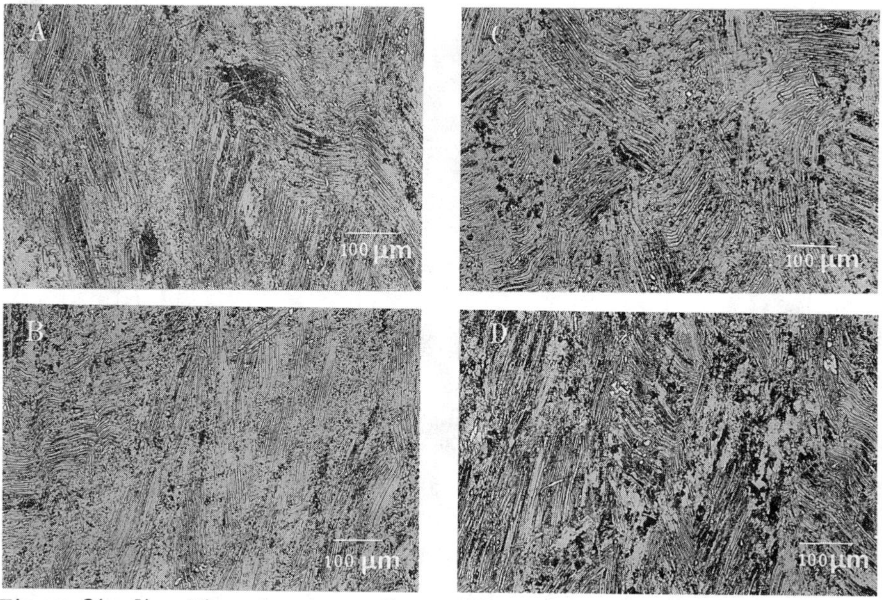

Figure 8(a-d) - Microstructure after small scale compression testing at 5 x 10^{-3} s^{-1} and temperatures of a) 1000°C b) 1050°C c) 1100°C and d) 1150°C.

c) Heat Treatment

The response of such structures to heat treatment are shown in Figures 9 and 10 where heat treatments in the nominally single phase alpha and duplex gamma + alpha region have been carried out. This shows that either a fine equiaxed structure (~10µm grain size) or a fine ~80µm grain sized lamellar structure can be produced.

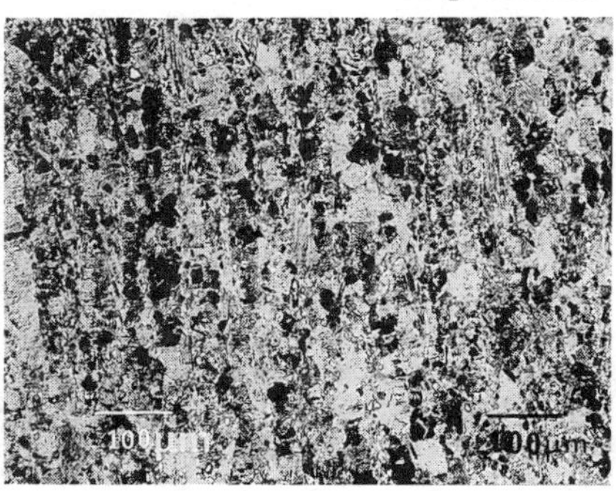

Figure 9 - Micrograph of material forged and heat treated in the gamma + alpha region.

Figure 10 - Micrograph of material forged and heat treated in the single phase alpha region, as predicted by the Ti-Al-Mn-Nb phase diagram model (Ref 7).

d) Mechanical Testing

Tensile tests were performed on both the cast plus HIP ingot and on the heat treated, isothermally forged pancake at 20°C and 750°C (2 per condition). In addition the forging was also tested at 850°C. Test specimens were manufactured by electrodischarge machining cylindrical blanks from the bulk material and then grinding the gauge length and threads. Mechanical test data generated within the current programme is listed in Table II.

TABLE II

MATERIAL CONDITION	TEST TEMP (°C)	0.2% PS (MPa)	UTS (MPa)	El (%)
CAST PLUS HIP INGOT	20	377 384	430 425	0.5 0.4
	700	323 333	462 468	2.5 2.0
HEAT TREATED FORGING	20	379 382	465 484	0.9 1.0
	700	322 325	517 509	10 20
	850	245 209	311 270	26.6* 103.2

* Specimen failed outside gauge length hence lower elongation

It can be seen that the R.T. 0.2% proof stress values were unaffected by the thermomechanical process route. However, higher elongations were observed after TMP. The effect of the higher plasticity was to allow greater strain hardening and hence UTS values. A more important advantage is that the ductile to brittle transition temperature is lowered in the forged plus heat treated condition. Ductility of the forged material at 700°C is within a similar range to that measured for IN713LC (Ref 9).

Whilst the proof stresses decreased with increasing temperature, as expected, the UTS values increased up to a test temperature of 700°C. This is because the measured 20°C values were in fact fracture stresses rather than true UTS values. Observations such as this are common for intermetallics tested below the brittle/ductile transition temperature (Ref 10).

Compared to the ductility claim for the matrix alloy (Ref 11), the elongations measured during the present study were significantly lower. This is partly due to the presence of hard, brittle TiB_2 particles in the XD™ processed material which act as stress raisers in the low ductility matrix. Additionally, the heat treatment was developed to give an optimum property balance rather than maximum ductility which would be achieved in a fine duplex microstructure, compared to the fully lamellar structure tested here (Ref 12).

Conclusions

This work has shown that there are a number of processing obstacles to be overcome before the XD process can be considered as a viable production route for the manufacture of LP turbine blades. Some of these problems are alloy chemistry related, some unique to the XD process. Further development is underway to minimise the segregation effects of the alloying additions. The use of phase diagram modelling has proved a powerful tool in the prediction of not only segregation effects during melting but also structural development during thermo-mechanical processing.

Clearly the aero engine utilisation of near gamma titanium aluminides can only be achieved once a fundamental understanding of process-structure-property relationships has been generated.

Acknowledgements

The authors would like to acknowledge Professor R.Evans at Swansea University for carrying out small scale compression testing, and Dr N.Saunders and Mr C.Small for generating phase diagram modelling predictions. Also our thanks to Ash Thakker at RR Inc. and to Pam Sadler of Martin Marietta Laboratories for helpful discussions.

References

1. C Small, Rolls-Royce Report MTL081335, December 1991.

2. United States Patent - 5,015,534 14th May 1991 - Rapidly solidified intermetallic second phase composites.

3. L.Christodoulou, P.A.Parrish, and C.R.Crowe, "XD™ Titanium Aluminide Composites", Materials Research Society Symposium Proceedings, Vol. 20, 1988.

4. S.L.Kampe, J.A.Clarke, and L.Christodoulou, "Structure/Property Comparisons in Particulate and Short Fibre Gamma Based Titanium Aluminide Composites", Materials Research Society Symposium Proceedings, Vol. 194, 1990.

5. J.D.Bryant, S.L.Kampe, P.Sadler, and L.Christodoulou, Effect of Phase Morphology on the Mechanical Behaviour of Two Titanium Aluminide Composites", Metallurgical Transactions A, Vol 22A, (1991), 2009-2019.

6. J.D.Bryant and S.L.Semiatin, "Segregation in Multicomponent Ingot Metallurgy Gamma Titanium Aluminides", Scripta METALLURGICA et MATERIALIA, Vol. 25 (1991), 449-453.

7. N.Saunders and C.Small, Rolls-Royce Report MTL081306, August 1991.

8. R.Evans, private communication with authors, Swansea University, September 1991.

9. IN713 LC Mechanical Property Database (Unpublished), Rolls-Royce.

10. M.T.Cope, Rolls- Royce Report MEG034236, May 1990.

11. S.C.Huang, European Patent Application, 0,362,470, 26th May 1989.

12. Y.W.Kim and D.M.Dimiduk, Journal of Metals, August 1991, 40-48.

THE EFFECT OF CHROMIUM ON THE KINETICS OF ALPHA

DECOMPOSITION IN GAMMA TiAl ALLOYS

P. A. McQuay, D. M. Dimiduk, H. A. Lipsitt†, and S. L. Semiatin

Wright Laboratory Materials Directorate
WL/MLLM
Wright-Patterson Air Force Base, Ohio 45433

†Wright State University,
Dayton, Ohio 45435

Abstract

Chromium (Cr) additions (1-3 at.%) have been found to improve the room temperature strength and ductility of two phase $\alpha_2+\gamma$ alloys. However, little is understood about the mechanism of ductility improvement offered by Cr additions, and the effect of Cr on the microstructure and kinetics of microstructural evolution in gamma alloys. In the present work, the effects of Cr additions on the kinetics of microstructural evolution in gamma alloys within the range Ti - (46-49)Al -(1-3)Cr have been investigated. Continuous cooling and isothermal transformation studies were conducted on several gamma alloys to establish the broad aspects of the kinetics of decomposition of the disordered alpha during heat treatment above the eutectoid temperature. Optical and electron microscopy were used to characterize the resulting microstructures. A schematic continuous cooling diagram is presented and discussed for Ti-48Al-2Cr and Ti-48Al-3Cr alloys.

Introduction

Two phase gamma ($\alpha_2+\gamma$) titanium aluminides have been under development for several years as potential aerospace propulsion or high temperature structural materials [1]. Due to the overall good oxidation and burn resistance of gamma titanium aluminides, compared with more conventional titanium alloys or the alpha-2 class of titanium aluminides, gamma alloys offer real promise as a light weight compressor material for gas turbine engines. However, concerns with fabricability, damage tolerance, and strength have precluded their use. In particular, room temperature ductility and fracture toughness are of concern. Additions of transition metals (1-3 atomic %), such as Cr, Mn, and V, have been shown to improve the room temperature ductility of near-gamma alloys [2-4], and there are also data which suggest that Cr improves the room temperature fracture toughness of these alloys [5]. These advantages have led to the development of several alloys containing Cr, the most notable, Ti-48Al-2Cr-2Nb.

Yet, within this progress on developing a useful balance of properties in gamma alloys, the chemical effects on the intrinsic alloy response versus their influence on processing and the kinetics of microstructure development cannot be separated. This is especially important in light of the strong influence of microstructure on properties which has been demonstrated in near-gamma alloys [6-7]. Within this view, the present manuscript summarizes the status of known effects of Cr on phase transformations and microstructural evolution. Work is also underway to establish the phase stability of Cr containing gamma alloys, and this work will be presented elsewhere.

Titanium '92
Science and Technology
Edited by F.H. Froes and I. Caplan
The Minerals, Metals & Materials Society, 1993

Experimental Procedure

The alloys used in the study were produced as 200 gram "cigars" by vacuum arc melting in Ar. The nominal and analyzed compositions of the alloys are shown below in Table I.

Table I. Aim and Analyzed Wet Chemistry of Alloys

Aim Chemistry (at %)	Analyzed Chemistry (at %)			
	Ti	Al	Cr	O(ppm)
Ti-48 Al-2Cr	50.1	48.0	1.9	800
Ti-48Al-3Cr	49.2	47.9	2.9	1040

A careful determination of the alpha transus and a homogenization study in each alloy was conducted by means of an annealing and quench study. The alpha transi temperatures (T_α) were found to be approximately 1370°C and 1360°C for the 2 and 3% Cr alloys respectively. This was also confirmed by DTA and a subsequent isothermal heat treatment and quench study of alloys in the homogenized (fully lamellar) condition.

The second part of the present investigation comprised a determination of continuous cooling and isothermal transformation kinetics. Specimens were glass coated and heat treated in air at T_α+ (10-20°C)/15 min to ensure the specimens were single phase alpha at temperature, and then subjected to various continuous cooling treatments or sub-transus isothermal annealing treatments[8]. This procedure was chosen to minimize the time the specimens were annealed above the transus, and to maximize the control of the cooling rates. Following the thermal treatments, the specimens were sectioned and characterized as-polished via Back-Scattered Electron Imaging (BEI) imaging on a JEOL 840 SEM, and optical polarized light microscopy.

Results and Discussion

A. Continuous Cooling Studies

Other studies have shown a significant effect of cooling rate on the decomposition of gamma alloys heat treated in the single phase alpha field. At relatively slow cooling rates, such as a furnace cool (dT/dt < 100°C/min), the alpha phase will decompose into the classical $\alpha_2 + \gamma$ lamellar structure, shown in Figure 1a, via the following reaction (for alloys where C_{Al}>47) [8,11]:

$$\alpha^{ss} \rightarrow (\alpha+\gamma) \rightarrow (\alpha_2+\gamma)_L \qquad \{1\}$$

Also, it has been demonstrated that the lamellar decomposition reaction can be superceeded at higher cooling rates by several different reactions which yield very different morphologies of α_2 and γ, such as the massive gamma microstructure (Figure 1b) [8,9-10]. Between these extremes of slow furnace cooling and water quenching, a range of unique microconstituents can be developed in a fashion akin the heat treatment of carbon steels. The purpose of this study was to identify those different morphologies and microconstituents, and to begin to develop a CCT curve in order to characterize the effect of 2 and 3% Cr on the decomposition of alpha.

The 2 and 3% Cr alloys were subjected to a number of cooling rates starting from an annealing temperature above the alpha transus. Table II lists the conditions and methods employed in the cooling rate experiments, and summarizes the microstructures which were developed.

Table II. Effect of Continuous Cooling During Alpha Decomposition

Cooling Rate	Microconstituents*	
	Ti-48Al-2Cr	Ti-48Al-3Cr
WQ: ~ 10^5°C/min	100% γ_M	100% γ_M
AC: ~ 10^3°C/min	85%$(\gamma+\alpha_2)_A$ + 15%$(\gamma+\alpha_2)_L$	80%$(\gamma+\alpha_2)_A$ + 20%$(\gamma+\alpha_2)_L$
100°C/min	60%$(\gamma+\alpha_2)_L$+40%$(\gamma+\alpha_2)_W$	40%$(\gamma+\alpha_2)_W$+30%$(\gamma+\alpha_2)_A$+ 30%γ_L
10°C/min	75%$(\gamma+\alpha_2)_L$+ 25%γ_W	NA
1°C/min	$(\gamma+\alpha_2)_L$ – Some DCL	$(\gamma+\alpha_2)_L$ – Some DCL
0.1°C/min	$(\gamma+\alpha_2)_L$ – Heavily DCL	$(\gamma+\alpha_2)_L$ – Heavily DCL

*A - Accicular; L - Lamellar; W - "Widmanstatten-like"; M - "Massive"; DCL- Discontinuously Coarsened Lamellar
Note: All microconstituent percentages and cooling rates are approximate.

As is noted in Table II, there appears to be at least four distinct microstructural morphologies obtainable in the range of cooling rates employed in this study. As mentioned above, the lamellar structure is commonly formed at relatively slow cooling rates such as a furnace cool (FC) at cooling rates of between 10 and 100°C/min, in the range of 1400°C to 1100°C (Figure 1a). At slower cooling rates the plates coarsen considerably, as can be seen in Figures 1c of the specimens cooled at 0.1°C/min. The coarsening of the lamellar plates has been attributed by some to be a discontinuous coarsening process, and is most pronounced in the grain boundary region [12]. The driving force for the coarsening can be attributed to the higher equilibrium volume fraction, and hence driving force for formation, of gamma as the temperature is dropped near the eutectoid temperature. In addition to a discontinuous coarsening process, many of the plates may have originally nucleated and grown as coarse plates at temperatures high in the $\alpha + \gamma$ field, a supposition which seems reasonable in light of the isothermal heat treatments discussed in the next section.

a 200 µm b 200 µm c 300 µm

Figures 1a-c. OPL micrographs of Ti-48Al-3Cr. (a) Furnace cooled at 100°C/min. (b) Water quenched. (c) Furnace cooled at 0.1°C/min.

At high furnace cooling rates, such as an 100°C/min, several other morphologies of gamma and alpha-2 may be seen in addition to the classic lamellar microstructure, and differences between the 2 and 3% Cr alloys become noticeable. At least three morphologies can be seen in the 3% Cr alloy with a cooling rate of 100°C/min: the classic lamellar microstructure (Figure 1a); lamellar plates with many orientations within a single prior alpha grain, termed Widmanstatten-like plates (Figure 2b); the last is the accicular, or needle-like microconstituent (Figure 2b).

a 300 μm b 100 μm

Figures 2a and b. OPL micrographs of Ti-48Al-3Cr, furnace cooled at 100°C/min.

The Widmanstatten-like microconstituent may be another morphological variation of the classic lamellar structure which reveals large regions or colonies of lamellar structure having more than one habit plane within a single alpha grain. In the classic lamellar structure, gamma forms on the basal plane of the hexagonal alpha crystal with the following relationship: $(0001)\alpha_2 \| \{111\}\gamma$, $<11\bar{2}0>\alpha_2 \| <1\bar{1}0>\gamma$. This relationship has been confirmed many times by numerous studies [6-7, 9,10]. With this relationship in mind, when gamma forms only on the basal plane of alpha, there should only be one habit plane for the plates formed on each prior alpha grain (ie. only one direction of plates within a prior alpha grain). This unique morphology has also been observed by Wang et al [10] in a similar study, in which they refer to the multi-directional plates as Widmanstatten, or having a Widmanstatten morphology. The Widmanstatten-like plates are usually found extending from a grain boundary or other disruption in the lamellar structure, but often do not extend all the way across a prior alpha grain. Interestingly, these Widmanstatten-like plates appear to form on distinct crystallographic planes, similar to the "Type III" lamellar structure described by Kim [11].

The accicular microconstituent is the alpha decomposition product found in this study which does not apparently have an orientation relationship with the parent alpha phase, as can be seen from Figure 2b. This microconstituent dominates the microstructure of the air cooled samples in the 2 and 3% Cr alloys, as described below. In BSI, very little contrast can be seen in this microconstituent at the highest contrast levels, indicating that the structure is single or nearly single phase gamma. Upon aging at 1100°C, thin alpha-2 needles will form between some of the gamma needles, but the overall accicular microconstituent appears stable.

In comparison to the 3% Cr alloy, the microstructure from the 2% Cr alloy (not shown) is dominated by the lamellar and Widmanstatten-like microconstituents, with little of the accicular microconstituent. This indicates that the higher level of Cr may inhibit the kinetics of formation of the lamellar structure. This effect has also been seen in alloys containing Ta or W, elements which have a relatively low solubility in gamma [8].

At cooling rates above 1000°C/min, such as in an air cool (AC) between 1400 and 1100°C, the accicular microconstituent dominates, with a small volume fraction of lamellar plates emanating from the grain boundaries, as shown in Figures 3a-b. The microstructure for the 2 and 3% Cr alloys given an air cool are quite similar, with the 3% Cr alloy exhibiting a larger amount of the Widmanstatten-like plates, with the accicular microconstituent being the dominant morphology in both alloys. It is not clear whether the accicular morphology is a degenerate lamellar structure, or a distinct morphology which results from a different transformation mechanism from the lamellar structure. In BSI, very little contrast is evident, making the structure very difficult to see, indicating it may be a variant of the massive gamma structure.

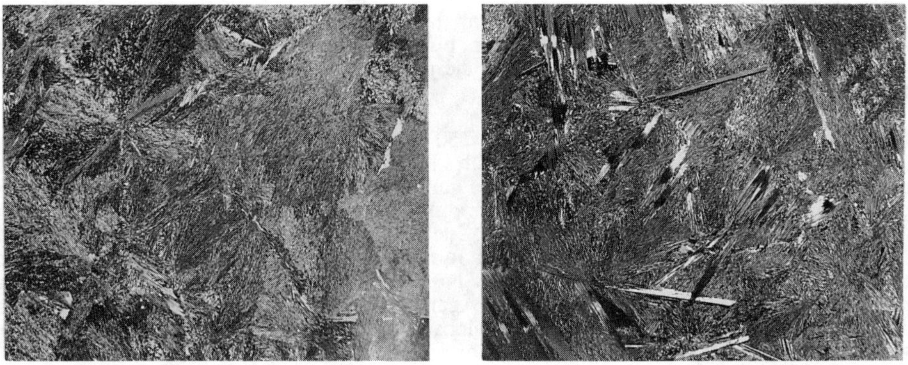

Figures 3a-b. OPL micrographs of Ti-48Al-2Cr and Ti-48Al-3Cr respectively, air cooled.

The structure which is formed at the highest cooling rate, a water quench, in both alloys is the massive gamma structure mentioned earlier. It should be noted that although this structure is refered to as "massive" [9,10], it still remains to be quantitatively proven if the transformation is in fact massive. In OPL microscopy, it appears as a mottled structure (see Figure 1b) which has very few plates or needles. Very little structure is discernable at normal contrast levels in BEI, and EDS analysis showed no discernable partitioning of Ti, Al or Cr. SAD has also demonstrated that the structure is $L1_0$, with a slightly lower c/a ratio than is normally found in gamma [10]. If the prior alpha phase cools below T_0 (the critical transformation temperature where the free energy of formation of alpha and gamma are equal) it is expected that any supersaturated alpha which has not transformed to the lamellar structure may transform in a massive fashion via the reaction path (for alloys where C_{Al}>47):

$$\alpha^{ss} \rightarrow \gamma^{ss}_{Massive} \qquad \{2\}$$

during quenching. In binary alloys where $C_{Al} \leq 47$, it has been demonstrated that the alpha or alpha-2 phase may be quenched in, or a combination of alpha-2 and massive gamma may result from quenching [9]. The massive gamma structure, because it is 100% gamma, is obviously a non-equilibrium structure in nominally two phase $\alpha_2+\gamma$ alloys, and will decompose upon further annealing. The quantitative aspects of this transformation remain to be investigated.

Referring again to the accicular microconstituent, it may also be possible that this structure is a secondary product of the massive gamma structure, but in light of the apparent non-crystallographic nature of the microconstituent, this seems unlikely. Even the massive gamma microconstituent has been demonstrated to be coherent or semi-coherent with itself and the parent phase. Instead, it is suggested that the accicular structure may have a different transformation mechanism than the lamellar or Widmanstatten-like microconstituents.

B. Development of Microstructure - Isothermal Annealing Study

In addition to the continuous cooling study, several isothermal annealing experiments were performed to verify the continuous cooling data. The experiments and results are summarized below in Table III. The first direct isothermal annealing treatment from the alpha field was conducted at a slight undercooling below the transus temperature, where nucleation rates would be minimized, but diffusion and growth rates very high. The microstructures from the 2 and 3% Cr sample are shown in Figures 4a-b. The 2% Cr sample exhibited very coarse gamma plates (up to 40 µm), with large but irregular spacing. The microconstituent between the gamma plates appears to be massive gamma, indicating that nearly half of the alpha transformed into coarse gamma plates during the initial furnace cool and isothermal hold, while the remainder of the alpha phase underwent the massive transformation to gamma during the subsequent water quench,. This unusual microstructure should then be composed entirely of two morphologies of gamma, albeit unstable forms which decompose upon low temperature heat treatment. The 3% Cr sample, however, contains only the massive form of gamma, indicating again that the higher level of Cr significantly retards the process of forming lamellar plates. Interestingly, the accicular microconstituent is also seen in this specimen, but is restricted only to the end of gamma plates which have not grown continuously across the prior alpha grain boundary, or at some grain boundaries.

Results from the direct isothermal anneals at 1150°C are shown in Figures 5a-b. Specimens from both alloys exhibit a fine lamellar structure with a high volume fraction of the Widmanstatten-like plates, and a small amount of equiaxed gamma grains, both at the grain boundaries and within the prior alpha grains. These microstructures are difficult to interpret in light of the previous heat treatments and proposed transformation paths. One possible explanation is that the Widmanstatten-like plates formed during the furnace quench to the lower aging temperature (expected slower cooling rate compared to an air cool since ΔT is smaller). The remaining super-saturated alpha phase might then massively transform to gamma, and subsequently form the lamellar microconstituent during the 1150°C age, as alpha is precipitated out into the massively formed gamma. However, other possible transformation paths are being considered.

Schematic Continuous Cooling Transformation (CCT) curves based on the continuous cooling and isothermal annealing studies are presented in Figure 6a and b for the 2 and 3% Cr alloys respectively. Several assumptions have been made in constructing the CCT diagrams, the largest assumption being that the Widmanstatten-like microconstituent is a derivative of the classic lamellar structure, formed at higher cooling rates, and as such, is not shown explicitly. Further, an assumption has been made that the start of the massive gamma transformation occurs at $\sim T_0$. Alternative constructions are being considered. The start (γ_L^s)and finish (γ_L^f) curves for the lamellar transformation were in part determined from continuous cooling data from DTA.

Table III. Effect of Continuous Cooling and Isothermal Anneals

Heat Treatment†	Microconstituents*	
	Ti-48Al-2Cr	Ti-48Al-3Cr
T_α+30°C/15min$\xrightarrow{FC}$ T_α-20°C/1hr→WQ	40%γ_{CP} + 60%γ_M	10%γ_{CP} + 90%γ_M
T_α+30°C/15min$\xrightarrow{FQ}$ 1150°C/1hr→WQ	50%$(\gamma+\alpha_2)_L$ + 40%$(\gamma+\alpha_2)\gamma_W$ + 10%γ_G	40%$(\gamma+\alpha_2)_L$ + 50%$(\gamma+\alpha_2)\gamma_W$ + 10%γ_G

* A - Acicular; L - Lamellar; W - "Widmanstatten-like"; G - Equiaxed Gamma Grains; CP - Coarse Gamma Plates; M - "Massive" Gamma

† FQ: Direct quench to lower temperature furnace; FC: 100°C/min to the lower temperature

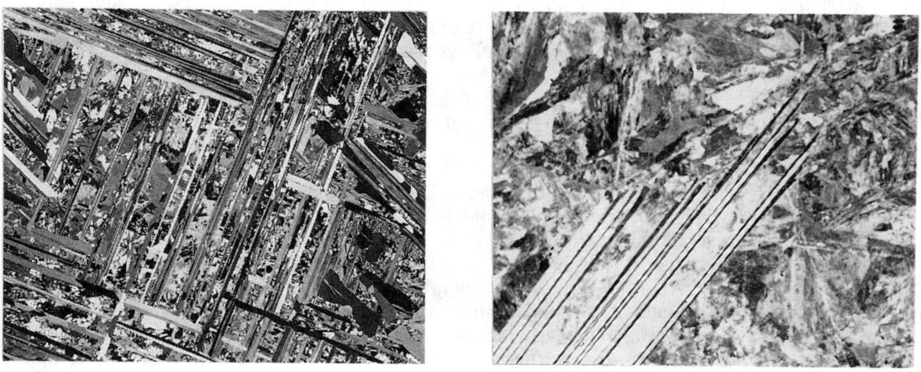

a 300 µm b 300 µm

Figures 4a and b. OPL micrographs of Ti-48Al-2Cr and Ti-48Al-3Cr respectively, T_α+30°C/15min→ T_α-20°C/1hr→WQ.

Figures 5a and b. OPL micrographs of Ti-48Al-2Cr and Ti-48Al-3Cr respectively, T_{α}+30°C/15min→ 1150°C/1hr→WQ

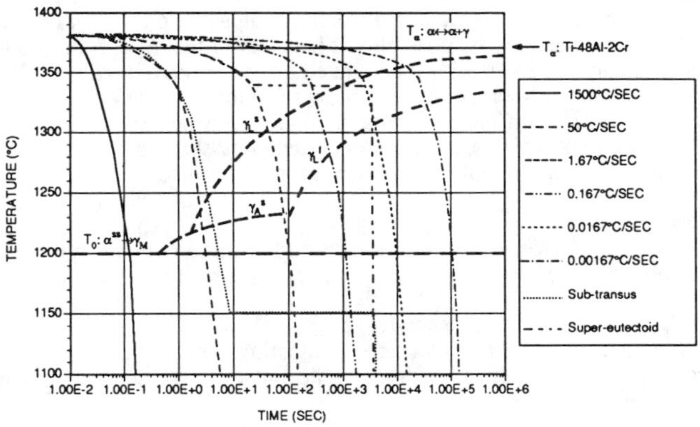

a

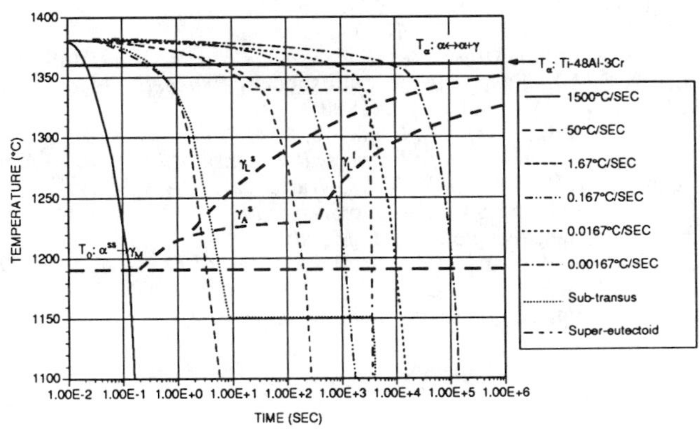

b

Figures 6a-b. Continuous Cooling Transformation (CCT) curves for the 2 and 3% Cr

Summary and Conclusions

The present investigation has demonstrated that the addition of Cr to near gamma alloys has a strong effect on both the phase equilibria and the kinetics of microstructural evolution. It has been demonstrated that Cr acts as an alpha stabilizer, decreasing the alpha transus temperature by approximately 10°C per atomic percent Cr, up to 3% Cr. Evidence also indicates that Cr restricts the $\alpha_2+\gamma$ phase field by shifting the gamma transus toward the Al-lean region, thereby decreasing the equilibrium solubility of Al in gamma in the presence of alpha or alpha-2.

Characterization of the microstructural evolution indicates the Cr-rich B2 phase in cast Cr containing gamma alloys is metastable between 1100 and 1400°C in near-gamma alloys with 1-3 at% Cr. The B2 phase is most likely a ordered derivative of the BCC Ti-Cr phase which should solidify first from the melt. It has been demonstrated that the B2 phase will go into solution following hot working and annealing, or a heat treatment in the single phase alpha region. Below 1100°C, in the alloys containing more than 2 atomic percent Cr, two Cr rich phases are present which are not alpha-2. The compositions of these phases are roughly Ti_2AlCr and TiAlCr respectively. AEM analysis of these phases is continuing, and the results will be presented elsewhere.

At least four distinct microconstituents containing unique morphologies of alpha-2 and gamma have been observed in this study in addition to the typically observed fully lamellar microstructure. These are: coarsened fully lamellar; accicular; Widmanstatten-like plates; and massive gamma. Further distinctions within these may be possible. It is believed that a discontinuous coarsening mechanism is responsible for the interlocking undulated grain boundaries seen in alloys cooled at slow cooling rates. The accicular microstructure appears to be the only varient which may not have a crystallographic relationship with the parent alpha phase. Further, the addition of Cr to near gamma alloys retards the rate of alpha decomposition, effectively shifting the CCT domains for the lamellar, Widmanstatten-like, and massive gamma microstructures to the longer times or slower cooling rates.

Careful Analytical Electron Microscopy studies are needed to further elucidate the decomposition transformations of alpha in two-phase gamma alloys, and the mechanisms by which they occur.

Acknowledgements
The authors would like to thank Mr. Mark Dodd for conducting heat treatments, Ms. René Roesner and Mr.'s Mike Scott and Eric Fletcher for metallographic and TEM specimen preparation.

References

1. H.A. Lipsitt, *High Temperature Ordered Intermetallic Alloys*, MRS, Pittsburgh, 351, (1985).
2. S-C. Huang and E.L. Hall, *Metall. Trans.A,* **22A**, 2619 (1991).
3. S-C. Huang and E.L. Hall, *Acta Met.,* **39**, 1053 (1991).
4. T. Tsujimoto and K. Hashimoto, *Mater. Res. Soc. Symp. Proc.*, **133**, 391 (1989).
5. W.O. Sobeyejo and S.M.L. Sastry, to be published in *Proceedings of Seventh World Conference on Titanium* , TMS, Warrendale, PA (1992).
6. Y-W. Kim and D.M. Dimiduk, *Journ. of Metals,* **43**, 40 (1991).
7. S-C. Huang and D.S. Shih, *Microstructure/Property Relationships in Titanium Aluminides and Alloys*, eds. Y-W. Kim, and R.R. Boyer, 105, TMS, Warrendale, PA, (1991).
8. P.A. McQuay, D.M. Dimiduk and S.L. Semiatin,*Scripta Metall.*, **25**, 1689 (1991).
9. S.A. Jones and M.J. Kaufman, accepted to *Acta Metall.*.
10. P.W. Wang, G.B. Viswanathan and V.K. Vasudevan, *Met all.Trans. A*, **23A**, 690 (1992).
11. Y-W. Kim, *Acta Metall.*, **40**, 1121 (1992).
12. D.S. Shong and Y-W. Kim, Scripta Metall., **23**, 257 (1989).

AN INVESTIGATION OF THE EFFECTS OF HEAT TREATMENT ON MICROSTRUCTURE AND FRACTURE BEHAVIOR OF EXTRUDED Ti-48Al-1.5Cr

W. O. Soboyejo

Edison Welding Institute
1100 Kinnear Road
Columbus, Ohio 43212

S.M.L. Sastry

Department of Mechanical Engineering
Washington University
St. Louis, Missouri 63130

Abstract

The results of a recent study of the effects of heat treatment on the microstructural evolution and fracture behavior of an extruded ingot metallurgy gamma-based Ti-48Al-1.5Cr alloy (compositions quoted in atomic % unless stated otherwise) are presented. These include room- and elevated-temperature tensile property data, room-temperature fracture toughness values and fatigue crack growth-rate data. The paper assesses the influence of heat treatment in the $\alpha + \gamma$ and/or $\alpha_2 + \gamma$ phase fields on microstructural evolution and mechanical properties. The addition of 1.5 at.% Cr to Ti-48Al is shown to slow down the microstructural transformation kinetics and improve damage tolerance. A heat treatment schedule that results in attractive combinations of room- and elevated-temperature properties is presented.

Introduction

Recent efforts to develop hypersonic vehicles have stimulated renewed interest in gamma-based titanium aluminides for intermediate-temperature (700-100°C) structural applications (1). The strong interest in this class of alloys has been due largely to their attractive strength retention and creep resistance at elevated temperatures (1,2). Unfortunately, however, the use of gamma-based alloys has been limited by the room-temperature fracture properties, in spite of recent improvements in ductility (3-9) and damage tolerance (10-12). The engineering of a good balance of room- and elevated-temperature has also been difficult due to the often conflicting microstructural/composition requirements for strength, ductility and toughness (13,14).

This paper explores the potential for engineering "balanced" properties in an extruded Ti-48Al-1.5Cr alloy. Balanced properties are engineered by systematic heat treatment in the $\alpha + \gamma$ and/or $\alpha_2 + \gamma$ phase fields after extruding near the α-transus with an extrusion ratio of 14:1. The alloy is shown to have attractive combinations of room- and elevated-temperature strength, fracture toughness and better fatigue crack growth resistance than mill annealed Ti-6Al-4V after careful control of microstructure via extrusion and heat treatment.

Materials and Heat Treatment Schedules

The ingots employed in the study were procured from Duriron, Dayton, OH, sealed in Ti-6Al-4V cans, and extruded at 1343°C at the Wright Patterson Air Force Base, Dayton, OH, with a reduction ratio of 14:1. The resulting extrusions contained 35.2 wt. % Al, 0.011 wt. % C, 0.044 wt. % 0, 42 ppm N and 19 ppm 0.

Titanium '92
Science and Technology
Edited by F.H. Froes and I. Caplan
The Minerals, Metals & Materials Society, 1993

Two sets of heat treatments were carried out. The first set of heat treatments involved two stage annealing (TSA) in the $\alpha + \gamma$ and $\alpha_2 + \gamma$ phase fields. The first stage of the TSA was done at different temperatures in the $\alpha + \gamma$ phase field to introduce different volume fractions of lamellar and equiaxed grains. This was followed by a stabilization anneal at 982° C which was designed to improve chemical homogeneity and promote further transformation/diffusion until equilibrium conditions were reached at the maximum potential service temperature of 982° C. Stabilization annealing was carried out at 982° C for 4 and 100 h. The second set of heat treatments involved annealing solely in the $\alpha_2 + \gamma$ phase field. This included annealing at 982° C for 4 or 100 h and an empirical three-stage heat treatment (13) which henceforth will be referred to as HTC.

The tensile tests were conducted on smooth cylindrical button-heated specimens with a gage diameter of ~3.18 mm. These were loaded monotonically to failure at a ramp rate of 5×10^{-4} s^{-1}. The fracture toughness and fatigue-crack growth rate tests were conducted on single edge notched (SEN) specimens with a square cross-section (6.35×6.35 mm^2) and a length of ~50 mm. The SEN specimens were pre-cracked under far-field compression loading (13) prior to fatigue or fracture toughness testing under three-point bending loading. The fracture toughness specimens were loaded monotonically to failure at a ramp rate that corresponded to a stress intensity factor increase rate of 0.92 MPa√m s^{-1} at the crack-tip. The fatigue crack growth rate specimens were subjected to constant amplitude sinusoidal loading at a cyclic frequency of 10 Hz. Initial loads that corresponded to a stress intensity factor range, ΔK, of 7 MPa√m were employed, and the load range was maintained constant, i.e., $\Delta P = P_{max} - P_{min}$, during the tests which were stopped prior to specimen fracture for microscopic examination of the interaction of the crack path with the underlying microstructure. SEN specimens were then fractured under monotonic loading, and the fracture surface morphologies were examined by scanning electron microscopy (SEM).

Results and Discussion

(a) Microstructural Evolution

Ternary alloying with 1.5% Cr and heat treatment have very significant effects on microstructure, as shown in Figures 1 and 2. The as-extruded alloy has a coarse (150-200 µm) nearly lamellar microstructure with small recrystallized equiaxed γ grains at some colony/grain boundaries (Figure 1a). Three-stage annealing (HTC) results in the recrystallization of small (3-5 µm) equiaxed γ grains (Figure 1b), while single stage annealing at 982° C results in progressive transformation to a microstructure with ~80% lamellar volume fraction after 100 h (Figures 1c and d).

Two-stage annealing of Ti-48Al-1.5Cr results in relatively slow transformation from a nearly lamellar as-extruded microstructure to nearly lamellar or duplex microstructure, with a high incidence of microstructural twins (Figures 2a-d). In fact, further annealing studies revealed that long annealing durations (up to 100 h each at 1150 and 982°C) are needed to reach microstructural equilibrium conditions in Ti-48Al-1.5Cr (Figure 3). Alloying with Cr therefore promotes microstructural stability apparently as a result of slow transformation kinetics. A coarse (100-200 µm) $\alpha_2 + \gamma$ packet size is also promoted by the addition of Cr. It is also interesting to note that a high incidence of twinning is observed in the equilibrium microstructures produced after prolonged annealing of Ti-48Al-1.5Cr (Figure 3). Further details of the possible phase transformations in ternary Ti-Al-Cr gamma alloys are provided in Ref. (15).

(b) Fracture Properties

Ternary alloying of Ti-48Al with 1.5 at% Cr has a very significant effect on fracture properties, as shown in Table I. At room temperature, Cr promotes increased strength and toughness in all the microstructural conditions that were examined. The highest fracture toughness of

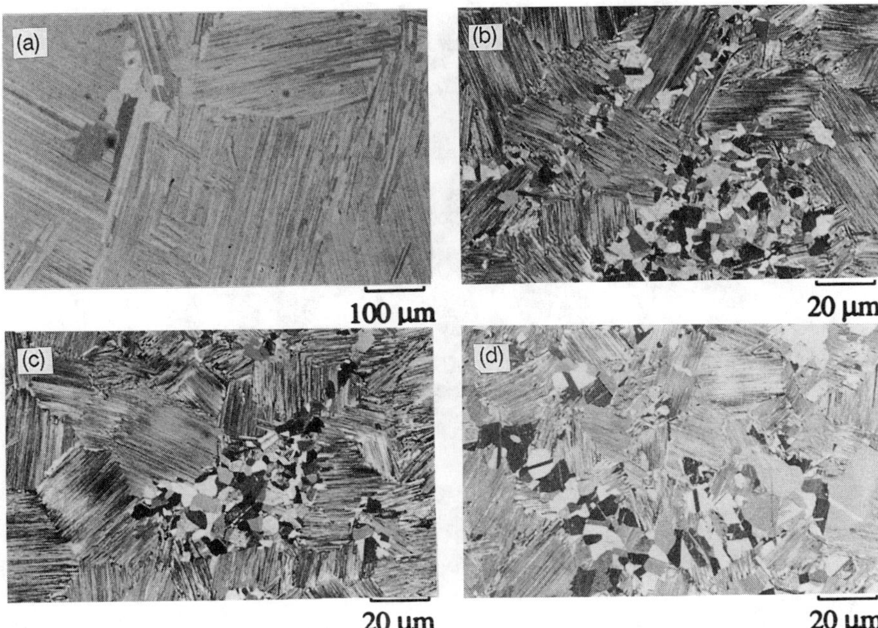

Figure 1. Effect of annealing in the α_2 + γ phase field on the microstructure of Ti-48Al-1.5Cr at (a) as-extruded (included for comparison), (b) 982°C/4 h/AC + 704°C/8 h/AC + 815°C/24 h/AC, (c) 982°C/4 h/AC, (d) 982°C/100 h/AC (AC = Air Cool; FC = Furnace Cool).

Figure 2 - Effect of annealing in the α_2 + γ phase field on the microstructure of Ti-48Al-1.5Cr at (a) 1150°C/0.5 h/AC + 982°C/4 h/AC, (b) 1150°C/0.5 h/AC + 982°C/100 h/AC, (c) 1250°C/0.5 h/AC + 982°C/4 h/AC, (d) 1250°C/0.5 h/AC + 982°C/100 h/AC (AC = Air Cool).

Figure 3 - Slow microstructural transformation in Ti-48Al-1.5Cr at (a) 1150°C/2 h/AC, (b) 1150°C/8 h/AC, (c) 1150°C/24 h/AC, (d) 1150°C/100 h/AC + 982°C/100 h/AC (AC = Air Cool).

32 MPa √m is greater than that of most conventional structural aluminum alloys. The Ti-48Al-1.5Cr alloy also retains its strength up to 815°C. The "best" balance of fracture properties is again obtained after three-stage heat treatment in the $\alpha_2 + \gamma$ field (HTC). Prolonged annealing in the $\alpha_2 + \gamma$ field at 982°C results in lower strength and fracture toughness which again is associated with lower lamellar volume proportions after prolonged annealing (Figures 1 and 2). The drop in strength observed after annealing may also be attributed partly to the annealing out of stored work due to extrusion (13).

(c) Fatigue Crack Growth

The fatigue crack growth rate data for the Ti-48Al-1.5Cr alloy are compared with previously obtained data for mill annealed (MA) Ti-6Al-4V (16) in Figure 4. The alloy has better fatigue crack growth resistance than MA Ti-6Al-4V in the as-extruded condition, and after heat treatment (HTC). Very little difference was observed between the fatigue crack growth rates in as-extruded and heat treated (HTC) material. This suggests that the improvements in fatigue crack growth resistance are due largely to intrinsic effects of Cr. However, more work is needed to distinguish between solid solution and microstructural effects on the fatigue crack growth behavior of this alloy. Nevertheless, the results are very significant since they indicate the potential to engineer damage tolerant gamma alloys with better fatigue crack growth resistance than MA Ti-6Al-4V which is currently used as a structural material in aerospace structures.

(d) Engineering Balanced Properties

It is clear from the above discussion that the "best" balance of properties was obtained using the empirical heat treatment (HTC). This heat treatment has been found to promote the best balance of room- and elevated-temperature properties in a wide range of gamma alloys (14) although the precise reasons for its effects are not fully understood at present. More recently, Ti-48Al-1.5Cr has been shown to have better primary and secondary resistance than the quarternary Ti-48Al-2Cr-2Nb alloy subjected to HTC and similar extrusion conditions (14). The alloy, therefore, has significant potential for future elevated-temperature structural applications although its oxidation resistance may require some improvement.

TABLE I. - EFFECTS OF HEAT TREATMENT AND ANNEALING ON THE FRACTURE PROPERTIES OF Ti-48Al-1.5 Cr

25°C

Heat Treatment	Fracture Toughness MPa √m (ksi √in.)	Plastic Elongation to Failure (%)	0.2% Offset Yield Stress MPA (ksi)	Ultimate Tensile Stress MPA (ksi)	Young's Modulus GPa (Msi)
1300°C/2 h/AC + 982°C/4 h/AC	--	--	--	--	--
1250°C/0.5 h/AC + 982°C/4 h/AC	28.6 (26.0)	0.3	598 (86.7)	625 (90.6)	140 (20.4)
1250°C/0.5 h/AC + 982°C/100 h/AC	24.5 (22.2)	0.3	559 (81.1)	578 (83.4)	140 (20.4)
1250°C/2 h/AC + 982°C/4 h/AC	--	--	--	--	--
1200°C/0.5 h/AC + 982°C/4 h/AC	27.5 (25.0)	1.3	590 (85.6)	670 (97.2)	173 (25.2)
1200°C/0.5 h/AC + 982°C/100 h/AC	23.6 (28.5)	0.4	554 (80.4)	577 (83.7)	157 (22.8)
1200°C/2 h/AC + 982°C/4 h/AC	--	--	--	--	--
1150°C/2 h $\xrightarrow{FC}$ 982°C/100 h/AC	--	--	--	--	--
1150°C/0.5 h/AC + 982°C/4 h/AC	29.1 (26.5)	0.7	585 (84.9)	644 (93.4)	201 (29.2)
1150°C/0.5 h/AC + 982°C/100 h/AC	30.0 (27.2)	0.3	530 (76.8)	556 (80.6)	175 (25.4)
982°C/4 h/AC	31.8 (28.9)	1.0	625 (90.6)	710 (103.0)	214 (31.1)
982°C/100 h/AC	21.2 (19.3)	1.8	523 (75.8)	584 (84.8)	180 (26.1)
982°C/4 h/AC + 704°C/8 h/FC + 815°C/24 h/AC	30.7 (27.9)	1.4	607 (88.0)	704 (102.1)	160 (23.2)
As-Extruded[(a)]	32.2 (29.3)	1.0	630 (91.4)	809 (117.3)	147 (21.3)

TABLE I. - (Continued)

815°C

Heat Treatment	Plastic Elongation to Failure (%)	0.2% Offset Yield Stress MPa (ksi)	Ultimate Tensile Stress MPa (ksi)
1300°C/2 h/AC + 982°C/4 h/AC	--	--	--
1250°C/0.5 h/AC + 982°C/4 h/AC	13.8	549 (79.6)	670 (97.1)
1250°C/0.5 h/AC + 982°C/100 h/AC	1.4	496 (72.0)	425 (61.7)
1250°C/2 h/AC + 982°C/4 h/AC	--	--	--
1200°C/0.5 h/AC + 982°C/4 h/AC	28.3	476 (69.0)	614 (89.1)
1200°C/0.5 h/AC + 982°C/100 h/AC	0.2	521 (75.6)	552 (80.1)
1200°C/2 h/AC + 982°C/4 h/AC	--	--	--
1150°C/2 h $\xrightarrow{FC}$ 982°C/100 h/AC	--	--	--
1150°C/0.5 h/AC + 982°C/4 h/AC	17.1	554 (80.4)	663 (96.2)
1150°C/0.5 h/AC + 982°C/100 h/AC	30.4	461 (66.9)	591 (85.7)
982°C/4 h/AC	18.4	549 (79.6)	648 (94.0)
982°C/100 h/AC	29.6	491 (71.2)	601 (87.1)
982°C/4 h/AC + 704°C/8 h/FC +	--	--	--
815°C/24 h/AC	25.6	555 (80.5)	659 (95.6)
As-Extruded[(a)]	--	--	--

AC = air cool
FC = furnace cool
(a) = included for comparison

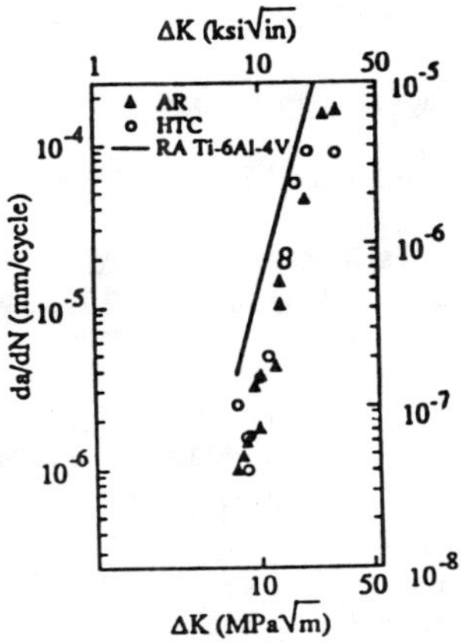

Figure 4 - Effects of heat treatment on the fatigue crack growth rates of Ti-48Al-1.5Cr.

Conclusions

1. A high incidence of microstructural twinning is observed in the microstructures of Ti-48Al-1.5Cr produced by two-stage annealing in the $\alpha + \gamma$ and $\alpha_2 + \gamma$ fields or prolonged annealing in the $\alpha + \gamma$ field after extrusion at 1343°C. Two-stage annealing also generally results in lower strengths (presumably as a result of the annealing out of stored work due to extrusion), ductility and fracture toughness.

2. Ternary alloying of Ti-48Al with 1.5% Cr results in the formation of coarse (100-200 μm grain size) microstructures with very slow transformation kinetics. The nearly lamellar microstructure produced by extrusion at 1343°C followed by three-stage annealing (HTC) has the "best" balance of microstructural stability, strength, toughness and fatigue crack growth resistance.

Acknowledgements

The authors would like to acknowledge useful discussions with Dr. Y-W. Kim, Dr. K. K. Sankaran, Lt. Paul McQuay, and 2Lt. David Lee. Appreciation is also extended to Messers. J. D. Keyes, J. J. Evans, II, D. A. Deuser, and Ms. C. Sabinash for their technical assistance. The research was conducted under the McDonnell Douglas and Edison Welding Institute Independent Research and Development programs.

REFERENCES

1. H. A. Lipsitt, "Titanium Aluminides-Future Turbine Materials," Advanced High-Temperature Alloys: Processing and Properties, S. M. Allen, R. M. Pelloux, and R. Widner (eds.), ASM, Metals Park, OH, 1985, pp. 157-164.

2. Y-W. Kim, J. Metals, 1989, Vol. 41, pp. 24-30.

3. M. J. Blackburn and M. P. Smith, United States Patent No. 4294615, October 13, 1981.

4. S. C. Huang and D. S. Shih, in Microstructure/Property Relationships in Titanium Aluminides and Titanium Alloys, eds. Y-W. Kim and R. R. Boyer, TMS, Warrendale, PA, 1990, pp. 105-122.

5. S. C. Huang and M. F. X. Gigliotti, United States Patent No. 4842891, June 1989.

6. S. C. Huang and M. F. X. Gigliotti, Untied States Patent No. 4923534, May 1990.

7. T. Kawabata, M. Tadano and O. Izumi, Scripta Metall. 1988, Vol. 22, pp. 1725-1730.

8. Hashimoto, H. Dohi and T. Tsiyimoto, Proc. Symp. Japanese Institute of Metals, "Plastic Deformation of Ordered Alloys and Intermetallic Compounds, July 1986 (in Japanese), p. 17.

9. T. Kawabata, T. Tamura, and O. Izumi, in High-Temperature Intermetallic Ordered Alloys III, Vol. 133, eds. C. T. Liu, et al., Materials Research Society, Pittsburgh, PA, 1989, pp. 373-383.

10. D. Larsen, M. L. Adams, S. L. Kampe, L. Christodolou and J. D. Bryant, Scripta Met. Mater., 1990, Vol. 24, pp. 851-862.

11. S. L. Semeatin, D. C. Vollmer, S. El-Soudani and C. Su, Scripta Metall. Mater., 1991, Vol. 24, pp. 1409-1413.

12. W. O. Soboyejo, P. B. Aswath and J. E. Deffeyes, Mater. Sci. Engng., 1991, Vol. A138, pp. 95-101.

13. W. O. Soboyejo, D. S. Schwartz and S. M. L. Sastry, "An Investigation of the Fracture Behavior of Gamma-Based Titanium Aluminides: Effects of Annealing in the $\alpha + \gamma$ and $\alpha_2 + \gamma$ Phase Fields," Metall. Trans. A, Vol. 23A, pp. 2039-2059, 1992.

14. W. O. Soboyejo, R. J. Lederich and S.M.L. Sastry, "An Investigation of the Effects of Alloying and Heat Treatment on the Mechanical Behavior of Extruded Gamma Alloys," Metallurgical Transactions, 1992 (submitted).

15. P. A. McQuay, D. M. Duniduk, H. A. Lipsitt and S. L. Semiatin, "The Effect of Cr on the Phase Equilibria and Alpha Decomposition in Gamma TiAl Alloys," Proceedings of the 7th World Conference on Titanium (to appear).

16. W. O. Soboyejo, unpublished research, McDonnell Douglas Research Labs, 1991.

EFFECT OF REFINING GRAINS ON MECHANICAL PROPERTIES IN Ti-RICH TiAl

T.Maeda, M.Okada and Y.Shida

Advanced Technology Research Laboratories,
Sumitomo Metal Industries, Ltd., Amagasaki 660, Japan

Abstract

The effects of refining grains in Ti-rich TiAl(γ) on its basic mechanical properties such as tensile properties, creep strength and fracture toughness, been studied by comparing two microstructures, duplex of coarse γ grains with lamellar(γ + α_2) and fine equi-axed grains structures. The refinement of grains strongly affects the mechanical properties, improving strength below the DBT temperature and ductility at room to elevated temperatures, while deteriorating high temperature strength including creep strength and fracture toughness. It is suggested that crack initiation, crack propagation and hot deformation behaviours relating with the microstructures determine the mechanical properties. It was emphasized that the selection of microstructure is necessary depending on the application, and microstructures for well balanced mechanical properties can hardly be obtained by controlling grain size only.

Introduction

Most intermetallic compounds with ordered structure maintain strong atom bonding to elevated temperatures, which is attractive as heat resistant materials. TiAl, designated as γ, has a large potential to be used as a high temperature structural material, especially for aerospace industries, because of its light weight and mechanical stability at elevated temperatures up to about 1073K. However, it suffers from poor ductility and low fracture toughness, which are essential characteristics in most intermetallics[1]. Ti-rich compositions of TiAl exhibits better room temperature ductility than the stoichiometric composition and small amounts of ternary elements are often added to the composition[1][2]. In addition to alloy design, microstructural control may be an important technique to improve the above properties. It is known that coarse grained lamellar structures obtained in as-cast ingot, or in material heat treated in the high temperature single α phase region, exhibits poor ductility[2]. Refining the microstructure can often be an effective way to improve ductility in conventional disordered alloys and intermetallic compounds as with Ni_3Al doped with boron[3].

The purpose of the present study is to investigate the effect of refining grains on basic mechanical properties in Ti-rich TiAl, such as tensile properties, fracture toughness and creep strength, by comparing two microstructures, coarse and fine grains structures.

Titanium '92
Science and Technology
Edited by F.H. Froes and I. Caplan
The Minerals, Metals & Materials Society, 1993

Experimental

The alloy composition examined was Ti-33.3Al-1.7Mo-0.06O(in wt%) equivalent to $Ti_{52}Al_{47.3}Mo_{0.7}$, in which Mo was doped into Ti-rich TiAl in order to improve room temperature ductility(4) and oxidation resistance(5). A large ingot (250kg) was obtained by double melting using vacuum arc remelting(VAR). The ingot was subsequently HIPed at 1373K in Ar atmosphere to eliminate cavities formed during solidification and to ensure homogeneity in the microstructure. Refinement of microstructure was achieved by isothermal forging. Small blocks machined from the ingot were isothermally forged to 15 or 30mm thickness plates, which were both forged at 1223K with an initial strain rate of $5x10^{-4}s^{-1}$, by an 80% reduction in thickness. The forged plates were then annealed at 1273K for 3.6ks followed by air cooling (designated as IFA). Microstructures of as-HIPed and IFA materials are shown in Figure 1. As-HIPed material exhibits the mixed structures of large coarse γ grains and lamellar structure consisting of thin γ plates and thin α_2 layers, in which lamellar structure in as-cast condition partly transformed to single γ phase. The structure of IFA material shows fine equi-axed γ grains up to 30 μm with a small amount of $Ti_3Al(\alpha_2)$.

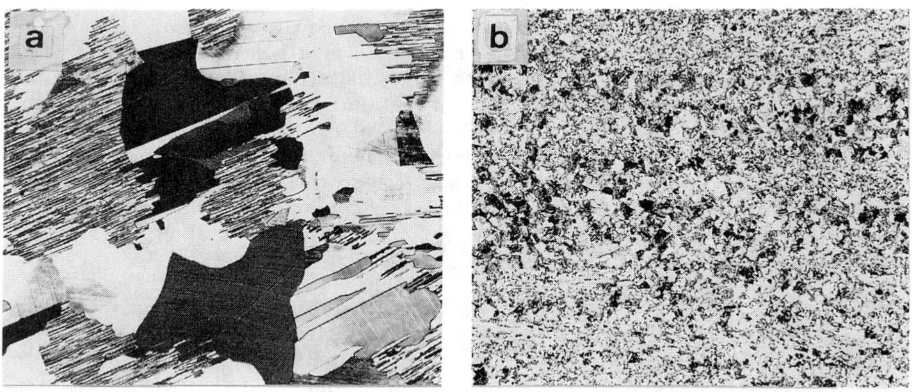

Figure 1 Microstructures of $Ti_{52}Al_{47.3}Mo_{0.7}$ examined. (a)As-HIPed (b)IFA.

Tensile properties from room temperature to 1273K were examined in air using cylindrical specimens of 4 or 6.25mm diameter. Creep rupture tests were carried out at 1073K and 1173K using cylindrical test pieces with 6mm in diameter. Fracture toughness was measured at room temperature with standard size CT test pieces with 25.4mm in thickness in which fatigue crack was introduced. Microstructures were characterized by optical microscope and TEM and fracture surfaces were examined by SEM.

Results

Tensile Properties

Strength Variation of tensile properties with temperature are shown in Figure 2. It shows that IFA material exhibits higher 0.2% proof stress by 100MPa than As-HIPed material, showing refining grain size substantially increases the yield strength. This result supports the previous reports that TiAl obeys the Hall-Petch relationship(6)(7). 0.2% proof stress in both materials continuously decreases with the temperature and no anomalous increase of strength with

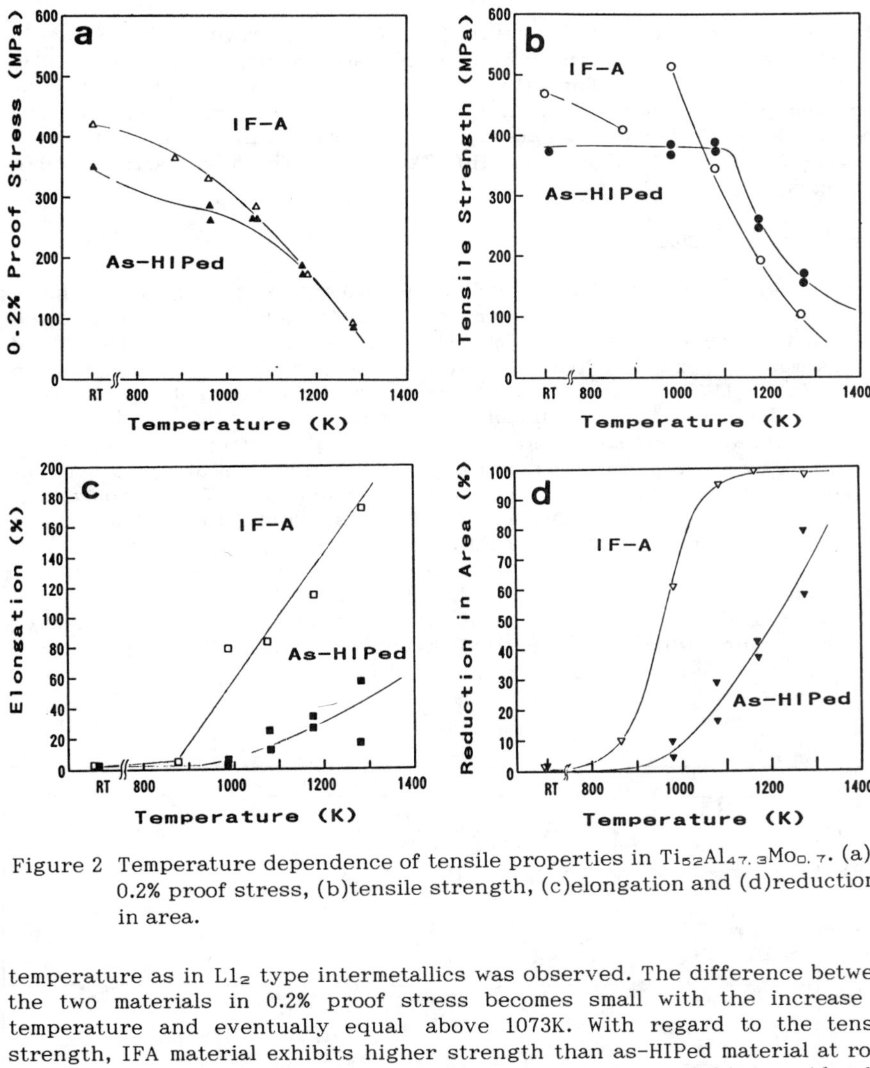

Figure 2 Temperature dependence of tensile properties in $Ti_{52}Al_{47.3}Mo_{0.7}$. (a) 0.2% proof stress, (b)tensile strength, (c)elongation and (d)reduction in area.

temperature as in $L1_2$ type intermetallics was observed. The difference between the two materials in 0.2% proof stress becomes small with the increase of temperature and eventually equal above 1073K. With regard to the tensile strength, IFA material exhibits higher strength than as-HIPed material at room temperature just as 0.2% proof stress. However, they exhibit considerably different behaviour at elevated temperature. As-HIPed material maintains the room temperature strength up to 1073K, then it abruptly decreases above the temperature. In IFA material, on the contrary, it decreases even at 873K, then it shows a discontinuous increase at 973K, and decreases again rapidly above this temperature. This discontinuity is due to the substantial plastic deformation above 973K, whereas it fractures before exhibiting sufficient ductility below 973K, implying that ductile-brittle transition(DBT) temperature exists between and 873 and 973K in IFA material.

Ductility As-HIPed material exhibited less than 0.2% tensile elongation at room temperature, whereas distinct elongation more than 0.5% was observed in IFA material, showing that refining grains improves room temperature ductility. As-HIPed material shows substantial elongation above 1073K, whereas IFA materials does so above 973K(Figure 2c), which are reverse behaviour in strength stated above. Furthermore, the increase gradient in IFA material is considerably higher than as-HIPed material to reach 180% at 1273K. The appearance of superplasticity in fine grains structure is consistent with the previous result[8]. Reduction in area follows same manner in elongation, in which it attains close to 100% above 1073K in IFA material. This is attractive for hot working of TiAl.

Creep Rupture Strength

Creep rupture curves at 1073 and 1173K in the two materials are shown in Figure 3. IFA material with fine grains structure shows substantially lower creep strength than as-HIPed material. 360ks(100hr) creep rupture time in as-HIPed and IFA materials obtained from Figure 3 were 200 and 135MPa at 1073K and 185 and 93MPa at 1173K respectively. It should also be noted that the slopes of curves in IFA material are significantly larger than those in as-HIPed material. These results are as expected from the high temperature strength stated above, although, strictly speaking, effects of microstructure are not always consistent between high temperature strength and creep strength[9]. A TEM micrograph of creep specimen of IFA material ruptured at 1073K is shown Figure 4. A number of twins were observed in a grain and they transmit to neighbour grains and a lot of dislocations were also found in twin plates, suggesting that both twinning and slip dominate creep deformation in TiAl. It is also interesting to note that fine grains without deformation structure at grain boundaries in the bottom right of the photograph, which are expected to have newly formed by dynamic recrystallization during creep deformation.

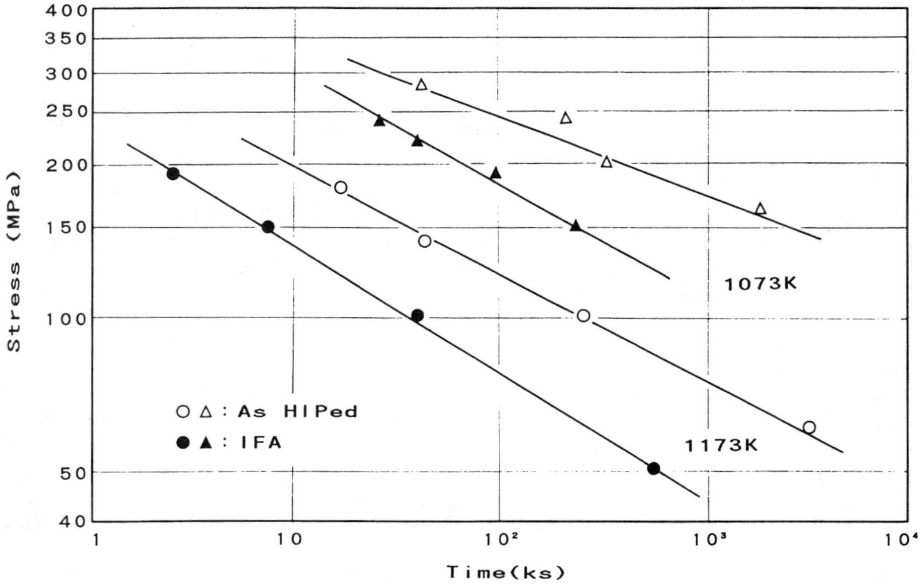

Figure 3 Creep rupture curves in $Ti_{52}Al_{47.3}Mo_{0.7}$ at 1073 and 1173K.

Figure 4 TEM micrograph of creep specimen in isothermally forged and annealed $Ti_{52}Al_{47.3}Mo_{0.7}$ ruptured at 1073K under 150MPa for 222ks.

Fracture Toughness

Fracture toughness of As-HIPed and IFA materials were 21MPa·$m^{-1/2}$ and 11 MPa·$m^{-1/2}$, respectively. The load-distance curves showed the stable crack growth to fracture. The former did not satisfy the condition for the validity of K_{IC}, $P_{max} \leqq 1.1P_Q$, on the other hand, the validity was confirmed on IFA material. K_Q of the former is higher than those in ceramics, however the latter is close to those in ceramics which have high fracture toughness. These values of $K_{IC}(K_Q)$ are consistent with previously reported values of 10-30MPa·$m^{-1/2}$[2]. This result shows the refining grains in TiAl substantially decrease the fracture toughness. Fractographs of the two materials are shown in Figure 5. As-HIPed material exhibits rough cleavage fracture surfaces containing steps and flat regions, which corresponds to crack propagation passing across lamellar plates and passing along lamellar interfaces or grain boundaries respectively. On the contrary, IFA material exhibited very flat fracture surface macroscopically indicating that cracks propagate straight. The large magnification of the fractograph(Figure 5c) reveals the layers which reflect the fine and coarse grain bands of microstructure(Figure1b) attributed to unidirectional isothermal forging, and no grain boundary (only cleavage fracture) was observed.

Discussion

The above results show that the refinement of grains strongly influences the mechanical properties in TiAl based alloys as follows. The refinement of grains improves proof stress and tensile strength below DBT temperature, ductility at room and elevated temperatures, while deteriorates tensile strength at elevated

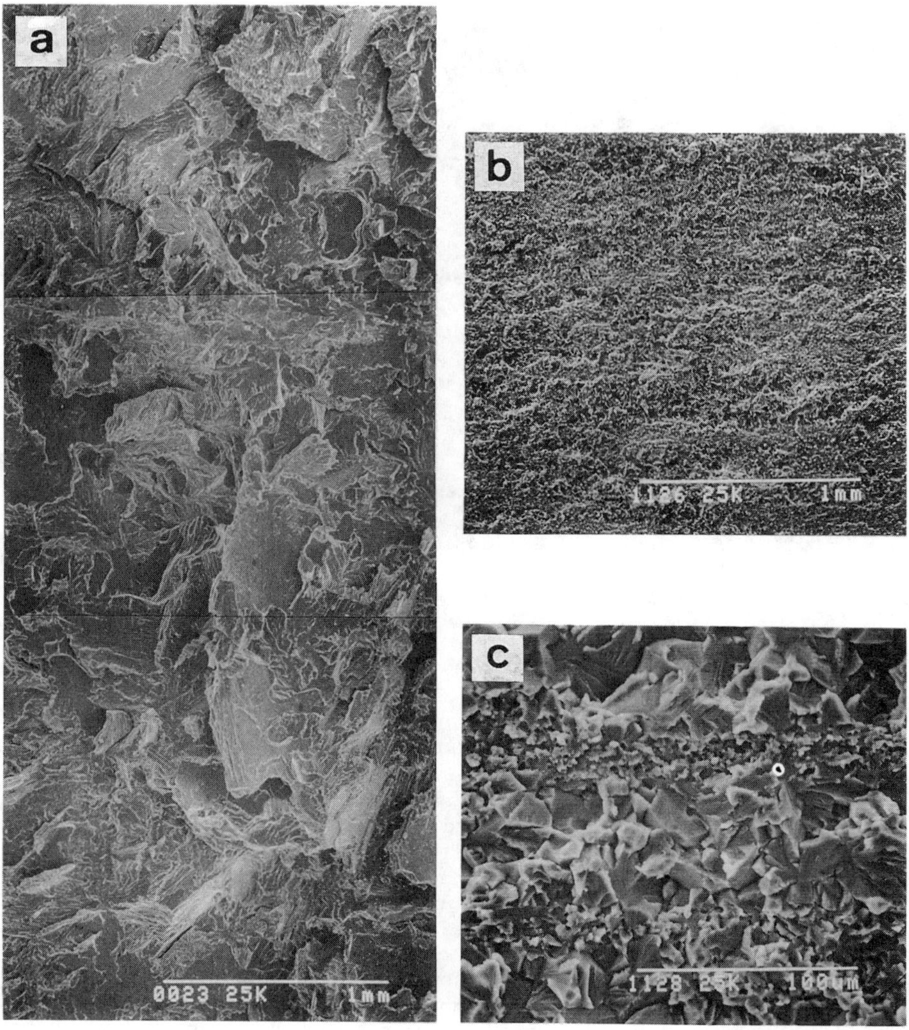

Figure 5 Fractographs of CT specimens in $Ti_{52}Al_{47.3}Mo_{0.7}$ examined. (a)As-HIPed (b) and (c)IFA.

temperatures, and fracture toughness at room temperature. It is quite important to note that the fracture toughness of the fine grain microstructure is half of that in coarse grains structure and the value is close to those in high fracture toughness ceramics. Furthermore, the decrease in high temperature strength, including creep strength, by grain refinement is a serious problem for a high temperature material. These facts limit the advantages of this material for engineering use when compared to ceramics or conventional high temperature alloys such as superalloys.

The improvement of room temperature ductility by refining grains may be due to the reduction of stress concentration at the grain boundaries, which results in crack initiation at the grain boundaries in coarse grain structures. In the deformation at elevated temperatures, dynamic recrystallization easily occurs in TiAl based alloys resulting in stress relief and furthermore grain boundary sliding, which is enhanced in fine grains structure leading to the reduction in high temperature strength associated with substantial ductility. For crack propagation in fracture toughness tests, once cracks are formed, it was found that cracks do not preferentially propagate at grain boundaries in fine grain structures. This implies that the minimum stress for crack propagation in taransglanular cleavage mode is substantially lower than that for crack initiation at the grain boundaries leading to crack blanching. This means that the stress for cleavage fracture is considerably low so that grain boundary fracture does not occur and eventually cracks propagate straight, regardless of grain boundaries, leading to very low fracture toughness in fine grain structures. In coarse grain structures, stress concentration easily occurs at grain boundaries or the lamellar boundaries, especially α_2 layers interfere with crack propagation passing across lamellar plates, resulting in higher fracture toughness.

The results suggest that it is not easy to achieve high ductility without sacrificing fracture toughness or high temperature strength only by refining grain size. These imply that the optimum microstructure which exhibits well balanced properties may need modification of morphology in grains such as lamellar or precipitates coupled with the control of grain size. It was reported that lamellar structure shows high strength at elevated temperatures (2) (9) and that, even in fully lamellar structure, finer grain size exhibits better room temperature ductility, although ductility is small probably because of coarse grain size due to heat treatments at high temperatures(10). The refinement of lamellar grains requires thermomechanical treatments(11). Although considerably coarse duplex structure of equi-axed γ and lamellar in the present study showed very little room temperature ductility, refined duplex microstructure shows high room temperature tensile elongation more than 3%(10). It must be noted that these these designed microstructures based on fine grains exhibit relatively high ductility but the data on high temperature strength are very limited. Ti-rich TiAl with two phase microstructures involve phase transformations at high temperatures as understood from the phase diagram, so that it can give a wide variety in modification of two phase structures. Further development of optimum microstructures in TiAl based alloys is one of the most important subjects for improving mechanical properties for applications as engineering materials.

Conclusion

The refinement of grains strongly influences the mechanical properties of $Ti_{52}Al_{47.3}Mo_{0.7}$ as follows. The refining grains;

- -improves : proof stress and tensile strength both below the DBT temperature, ductility at room and elevated temperatures
- -deteriorates : tensile strength at elevated temperatures above the DBT temperature, Fracture toughness at room temperature

Improvement of ductility by refining grain size sacrifices high temperature strength and fracture toughness which are important properties in TiAl based alloys. The optimum microstructure which exhibits well balanced properties may need the modification of morphology in grains such as lamellar or precipitates in addition to grain size .

Acknowledgment

The authors gratefully acknowledge Mr.Y.Iwata for his technical assistance.

References

1.Y.Yamaguchi and Y.Umakoshi, "The Deformation Behaviour of Intermetallic Superlattice Compounds," Progress in Materials Science , 34, No.1 (1990), 1-148.

2.Y-W.Kim, "Progress in the Understanding of Gamma Titanium Aluminides," JOM, August,(1991), 40-47.

3.M-S.Kim, S.Hanada, S.Watanabe and O.Izumi, "Effect of Grain Size on Deformation and Fracture in Recrystallized Ni_3Al Polycrystals Doped With Boron," Trans. Japan Inst.Met., 29(1988), 274-283.

4.T.Maeda, M.Okada and Y.Shida, "Ductility and Strength in Mo Modified TiAl," Intermetallic Compounds (Proc. of JIMIS-6) ed. by L.A.Jhonson, D. P.Pope and J.O.Stiegler,Mater. Res. Soc. (1991), 555-560.

5.T.Maeda, H.Anada, M.Okada and Y.Shida, "Total Improvement of γ TiAl by the Addition of Molybdenum," Proc. of Intermetallic Compounds (JIMIS-6)" ed. by O.Izumi, Japan Inset. Met., (1991), 463-467.

6.V.K.Vasdevan, S.A.Court, P.Kurath and H.L.Fraser, "Effect of Grain Size on the Yield Stress of the Intermetallic Compound TiAl," Scripta Metall., 23(1989), 467-469.

7.W-Y.Chu and A.W.Thompson, "Effect of Grain Size on Yield Stress in TiAl," Scripta Metall., 25(1991), 641-644.

8.T.Maeda, M.Okada and Y.Shida, "Superplasticity in Ti-rich TiAl" Proc. of Superplasticity in Advanced Materials, ed. by S.Hori, M.Tokizane and N.Furushiro, (1991), 311-316.

9.T.Maeda, M.Okada and Y.Shida, "Effects of Microstructure on the Tensile and Creep Properties in TiAl," Proc. of "Mechanical Behaviour of Materials- Ⅳ no.3, Pergamon Press, (1991), 199-204.

10.Y-W.Kim, "Microstructural Evolution and Mechanical Properties in Gamma Titanium Aluminides," Microstructure/property relationships in Titanium Aluminides and Alloys, ed. by Y-W.Kim and R.R.Boyer, TMS,(1991), 91-103.

11.T.Maeda and M.Hosomi, " Refinement of Lamellar grains through Hot Working Conditions in α Region in an $\alpha_2+\gamma$ Type Titanium Aluminide," (Paper presented at the 108th Japan Inst. Met. Meeting, Tokyo, 3 April,1991), 189. Submitted to J. Japan Inst. Met..

The Effects of Al, Cr, Nb, and Ta on the Tensile Properties of Cast Gamma Titanium Aluminide

C. M. Austin, T. J. Kelly and S. C. Huang*

GE Aircraft Engines, Cincinnati, OH 45215
*GE Corporate Research, Schenectady NY 12301

Abstract

Al, Cr, Nb and Ta were varied in 25 near-gamma titanium aluminide alloys prepared as investment castings. The effects of these elements on room temperature tensile behavior were determined through graphical and statistical analysis. Microstructures were quantified and related to composition and properties. The relationship between Al and strength/ductility is similar to that found by others in wrought material. Cr has little effect at levels of 0 and 2%; at higher levels, it increases strength but sharply lowers ductility. Nb and Ta both increase yield strength at the rate of 17MPa per percent added, but reduce log(elongation) at the rate of 0.06 and 0.10, respectively. Oxygen, not intentionally varied, increases strength without lowering ductility at levels between 500 and 900 wppm.

Performed under NAWC Contract N00140-90-C-01742, A. S. Culbertson, Monitor.

Background

Gamma alloys appear to be capable of substituting for Ni and Ti alloys in gas turbine engines [1]. At the temperature limit of Alloy 718, gamma has comparable strength and environmental resistance. Gamma also has 50% greater specific stiffness than conventional Ti and Ni alloys. Current-generation gamma alloys, notably 48Al-2Cr-2Nb [2,3], meet the requirements of many components; however, to fully exploit the unique properties of gamma, we are attempting to identify a second-generation alloy optimized for cast applications.

The components for which gamma is best suited are presently cast of either Ti or Ni alloys. Gamma has been shown to be amenable to conventional Ti foundry practice [4], while wrought processing is presently limited to simple shapes via extrusion or isothermal forging [5]. Ductility of 2% can be produced reliably in investment castings. Tensile strength is about 20% lower than for some forms of wrought gamma, but creep capability is greater.

The first task of the program has been completed, which involved the evaluation of 64 investment cast alloys with variations in Al, Cr, Nb, Ta, W, Mo, Ni, Co, Fe, B, Si and C [6]. The evaluation included castability, heat treatment response, tensile properties, creep resistance and environmental stability. In this paper, discussion is limited to the effects of Al, Cr, Nb and Ta on room temperature tensile behavior based on a subset of 25 alloys. While these elements have been studied extensively in wrought material, precise measurements of their effects in cast material are necessary to choose optimum levels for a cast alloy meeting aerospace requirements.

Table I

Composition, Heat Treatment, Structures and Tensile Data

Alloy	Composition (a/o, analysed)							Specimen 1		Specimen 2		Structure			
	Al	Cr	Nb	Ta	O	Other	HT	YS	El	YS	El	Type	Col	Col	Gam
							°C	ksi	%	ksi	%		%	µm	µm
1B-01	47.7	0.0	4.1	0.0	0.12		1330	58	0.9			D	60	80	60
1B-02	45.8	2.0	3.8	0.0	0.12	0.16Ni, 0.05B	1300	56	0.9	54	0.9	D	70		
							1330	50	0.3	56	1.6	Cn	100	500	
1B-14	46.6	2.1	0.0	3.8	0.23		1300	71	0.3			D	50		
							1330	70	0.4	67	0.3	Cn	100	1000	
1C-3	45.6	2.1	3.9	0.0	0.13	0.1Si	1275	57	0.3			D	90	600	30
							1300	64	0.6			Cn	98	1000	50
							1330	73	0.8			Cn	98	500	50
1C-4	45.7	4.1	3.8	0.0	0.13		1330		0.1			Cn	100		
							1360		0.1			Cn	95	800	
1C-6	45.7	3.2	4.1	0.0	0.13	0.14Ni, 0.04B	1385	60	0.2			Cn	95	500	
1D-1	45.8	2.0	0.0	4.0	0.15		1300	59	1.0	58	0.9	D	90	400	30
							1315	66	0.9	62	0.5	Cn	98	500	40
1D-2	45.4	2.0	2.1	2.0	0.14		1300	54	0.8	53	0.8	D	90	800	100
							1315	56	0.6			D	90	500	20
							1330	58	0.7			Cn	99	800	40
1D-3	46.4	2.0	1.9	1.9	0.16		1300	51	1.0	54	0.7	D	95	500	100
							1315	55	1.2			D	80	400	40
							1330	65	0.7			Cn	98	400	40
1D-4	45.5	2.0	0.0	3.8	0.21	0.23B	as-HIP	78	1.9			Cn	98	1000	20
							1300	67	0.6	73	0.2	Cn	98	1000	20
							1315	71	0.7			Cn	98	1000	30
							1330	69	0.3			Cn	100	800	
1D-6	46.1	2.0	0.0	3.2	0.15	0.18Si	1300		0.4	65	0.7	D	50	200	50
							1315	56	0.3			D	80	300	50
							1330	66	0.4			D	70	200	50
B1	47.2	0.0	0.0	0.0	0.18	0.08B	1300	51	2.5	51	2.3	D	70	100	40
Bin1	45.0	0.0	0.0	0.0	0.12		1300	50	0.5	51	0.2	Cn	100	500	
Bin2	45.6	0.0	0.0	0.0	0.13		1300	62	1.0	65	0.8	D	90	500	
Co2	46.6	0.0	0.0	0.0	0.14	0.65Co	1300	50	1.9	49	1.5	D	80	300	75
Cr1	47.7	2.0	0.0	0.0	0.12		1300	45	1.9	46	2.4	D	20	30	60
Cr2	47.6	3.9	0.0	0.0	0.17		1300	65	0.8	68	1.1	D	80	200	
HT1	46.9	1.9	0.0	2.0	0.17		1300	56	2.1	58	1.7	D	90	250	40
							1300	55	1.8			D	90	250	40
Nb1	45.2	0.0	3.8	0.0	0.14		1300	63	0.4	61	0.4	Cn	100	300	
Nb2	44.5	0.0	8.1	0.0	0.13		1300	80	0.4	81	0.3	Cn	100	300	
Nb3	45.6	0.0	11.3	0.0	0.17		1300	81	0.4	81	0.4	D	90	200	
							1350		0.1		0.1	D	30	200	
Nb5	43.8	0.0	7.9	0.0	0.13		1300	82	0.3		0.2	Cn	100	300	
Ni2	47.2	0.0	0.0	0.0	0.13	0.63Ni	1300	43	2.0	44	2.1	D	80	200	80
Ta1	45.2	0.0	0.0	3.9	0.24		1300		0.2	71	0.2	D	90	300	
Ta2	48.1	0.0	0.0	8.1	0.11		as-HIP	68	0.2	68	0.2	D	50	60	
							1300	63	0.4	65	0.3	D	50	100	

Notes: Heat treatments were 20 hours at indicated temperature. All material was HIPed at 1260°C/20h/25ksi prior to heat treatment.

Approach

A quantitative approach to alloy development was emphasized. This includeed the use of designed experiments, statistical analysis, and mathematical models of compositional and structural effects. Some alloys were chosen and evaluated in a more ad hoc manner, since designed experiments can be compromised by unanticipated behavior.

The alloys considered here included binaries and more complex alloys that offer information on the effects of Al, Cr, Nb and Ta, Table I. They are a subset of alloys from a larger program that evaluated many other elements; some of the alloys contained small quantities of these elements that we believe have a negligible effect on tensile properties. Aluminum levels varied from about 44 to 48% (all compositions are given in atomic percent). Most alloys had Al levels such that heat treatment could produce the duplex structure, i.e., a mixture of gamma grains and colonies of alpha-two/gamma lamellae. Cr was either 0, 2, 3 or 4%. Nb ranged from 0 to 12%, though most alloys contained between 0 and 4%. Ta ranged from 0 to 8%, mostly from 0 to 4%. Oxygen content ranged between about 0.12 and 0.24% (500 and 900 ppm by weight).

Procedures

All alloys were produced by Howmet Corporation, Whitehall MI, by standard arc melting and investment casting methods. Starting ingots were made by cold hearth vacuum induction melting by Duriron, Dayton OH. High purity Ta was used in some heats to prevent correlation between Ta and oxygen. Chemical analysis was performed by Sherry Laboratory. Castings were HIPed at 1260°C/25ksi/4h. Heat treatments were performed in a vacuum furnace backfilled with flowing Ar at various temperatures (often 1300°C) for 20h and furnace cooled at about 1°C/s.

The structure resulting from each composition/heat treatment combination was quantified by visual inspection; the number of such combinations precluded more rigorous methods. Many samples exhibited significant variation in structure as a result of local solidification rate and segregation. Individual data points may be in considerable error.

Tensile testing was performed on specimens with 4mm gage diameters at 0.005/min strain rates. Plastic elongation was measured from extensometry recordings.

The statistical models provided here were developed after extensive trials of different combinations of variables, examination of outliers, and tests of validity. Residual analysis and "partial" regression plots were used to examine more closely the individual effects of variables, as explained in the text associated with Figures 3, 4 and 5.

Results and Discussion

This section addresses structures, strength and ductility. There are numerous potential interactions between composition, structure and properties, leading to several valid views of the data. In other words, composition, heat treatment and structural data "over-specify" the problem, such that including all parameters in an analysis can be redundant. We have selected views that seek composition-property relationships as a first priority. Potential sources of remaining variation is examined in plots of residuals versus certain microstructural measurements.

Structures

The variations in structure were largely those expected from variations in Al levels and heat treatment temperatures. Structures were designated as shown below. All plots in this paper are coded to identify these structures (see Figure 2).

G: All gamma grains
D: Duplex (gamma grains plus colonies)
Cn: Essentially all colonies (small)
Cs: Large colonies (from above-transus heat treatment)

A small amount of B2 phase was identified in alloys having 2Cr; much larger amounts were found in 3Cr and 4Cr alloys. Nb and Ta had little effect on the type of structure, but produced sharply different segregation patterns as shown in Figure 1. The initial segregation and slow diffusion of Ta appears to lead to a smaller colony size, even after full heat treatment.

Ductility

The variation in plastic elongation for different types of structures is shown in Figure 2. The highest levels were recorded for the duplex structure. Full colony structures rarely provided more than 1%.

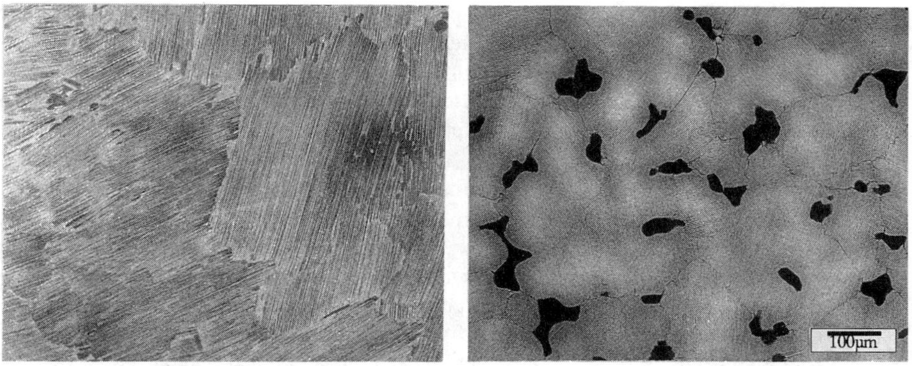

Fig. 1: Segregation of Nb (left) and Ta (right) in 45Al-4Nb and 45Al-4Ta alloys after HIP and heat treatment at 1300°C for 20 hours.

Regression analysis of tensile data from fine colony and duplex structures produced the following expression and error statistics:

$$\log(\text{elong}) = -6.5 + 0.142[\text{Al}] - 0.31[\text{Cr}>2\%] - 0.06[\text{Nb}] - 0.10[\text{Ta}]$$

t-values: 4.4   5.1   5.8   7.0

$r^2 = 62\%$ F = 26   D.F. = 63

This model and related plots use log(elongation), a more linear quantity that produces a normal residual distribution. The variable "Cr>2%" is the amount of Cr exceeding 2%. Variations in Al, Cr, Nb and Ta "explain" 62% of the variation in log(elongation) observed in the data; 38% remains unexplained, a consequence of data error, non-linear behavior and factors not included in the model. The t values are high and imply significance of each element (oxygen, if included, yields a t value indicating insignificance). The F value is high and indicates validity for the model as a whole.

The differences between the measured values and the values predicted by expression (1) are known as the residuals. They collectively represent the unexplained 38% of the variation. These are plotted against predicted values in Figure 3. Inspection of these values allows identification

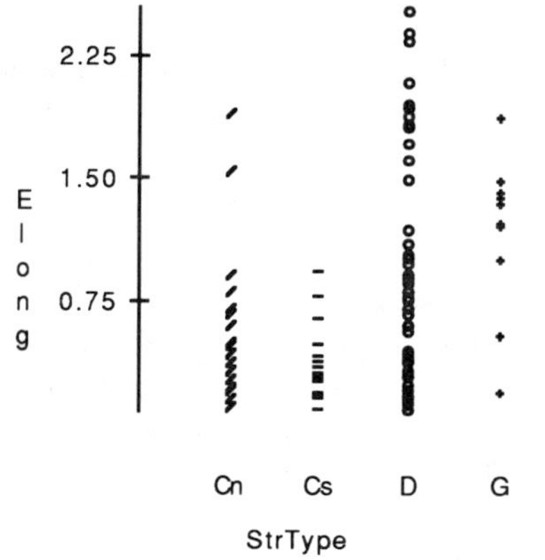

Fig. 2: Variation in measured elongation for different types of structures (see text). The symbols coded in this plot are used throughout this paper.

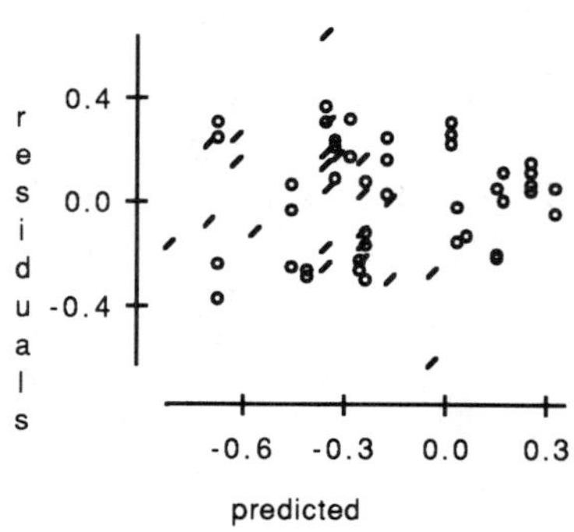

Fig. 3: The errors, or residuals, in the ductility model plotted against predicted values; duplex and colony structures are equally distributed about zero.

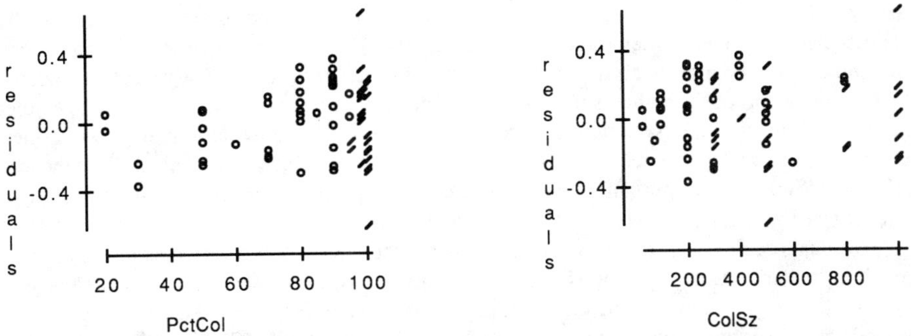

Fig. 4: Residual plots useful for detecting potential additional effects of structural characteristics on ductility (sizes in μm).

of additional effects, the major linear effects having been accounted for. For example, in Figure 3 the distribution of duplex and fine colony residuals is uniform about zero, suggesting that structure type cannot explain ductility variations beyond that explained by composition.

Figure 4 shows residuals plotted against colony percentage and colony size. In other words, we attempt here to explain the remaining 38% variation in terms of microstructural measurements. (All the real error in the data is concentrated in such plots; considerable scatter is inevitable.) While the trends in these plots are not distinct, there is a suggestion that if one chooses the duplex structure,

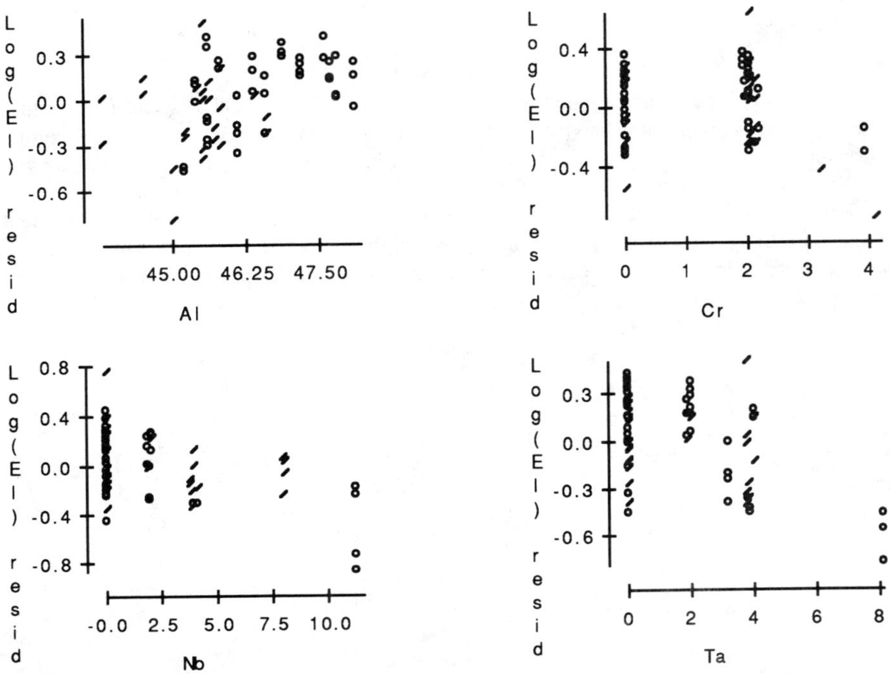

Fig. 5: Partial regression plots showing the underlying effect of each element in the ductility model (e.g., Al is plotted against the residuals of a model including Cr, Nb and Ta).

then 80-90% colony is most desirable. Examination of same-alloy trends is not conclusive, but also suggests that high colony structures are best (but not 100%). Overall, however, these plots suggest a structural independence that is difficult to accept. This may reflect a high degree of dependence of structure on composition that is indirectly accounted for in the model, though it must be noted that many alloys are represented here in several heat treatment/structural conditions. A linear model using colony fraction, colony size and gamma grain size can explain only 23% of the total variation. However, an important set of structural characteristics were not measured in this study—lameller spacing and alpha-two/gamma proportions.

Residuals can also be used to view more closely the effects of the variables included in the model. So-called "partial regression" plots can be formed by omitting the variable of interest (e.g., Al) from the original model, and then plotting the residuals of this reduced model against the omitted variable (e.g., residuals versus Al). This has been done for Al, Cr, Nb and Ta in Figure 5. The underlying Al effect after accounting for Cr, Nb and Ta reflects the expected behavior [7], perhaps showing a peak at 47% Al (a non-linearity). The plot for Cr shows that there is little difference between 0 and 2% Cr, but the four points from high-Cr alloys all had low ductility; this is the plot that justifies the use of "Cr>2%" as a variable in the model. The Nb and Ta plots show the trends suggested by the linear model but it is not clear that the effects are linear, particularly for the case of 2% Ta, which does not reduce ductility according to this view of the data.

Strength

The trend of 0.2% yield strength with structure type is shown in Figure 6. A model was developed for yield strength based on data from duplex or fine colony structures:

$$YS = 155.4 - 2.56[Al] + 7.0[Cr>2\%] + 2.5[Nb] + 2.5[Ta] + 95[O]$$

t-values:	3.2	3.5	7.9	7.4	4.1

$r^2 = 72\%$ $F = 28$ $D.F. = 55$

In this case, 72% of the variation is explained by Al, Cr, Nb, Ta and oxygen. The residuals are plotted against predicted values in Figure 7. Again, no clear distinction can be made according to structure type. The two high duplex points are from Bin2, a 46Al binary alloy; it is not clear why this alloy behaves differently than the others, being stronger than the model predicts.

Potential additional effects from structure are shown in the residual plots in Figure 8. Variation in strength is very wide among alloys with high colony fraction. The strength of duplex specimens varies inversely with colony size (ignoring Bin2 data); specimens with fine colony structures appear insensitive to colony size.

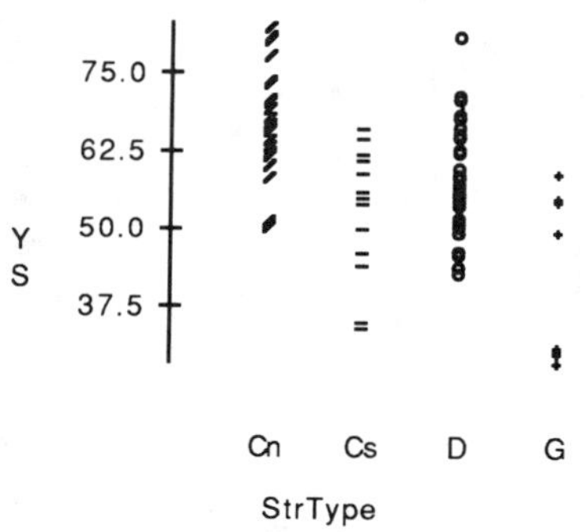

Fig. 6: Variation in yield strength for different types of structures.

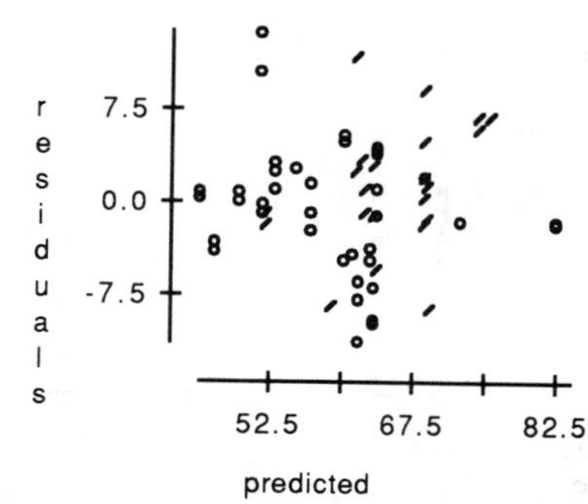

Fig. 7: The residuals of the strength model plotted against predicted values; no additional effect of structure is apparent.

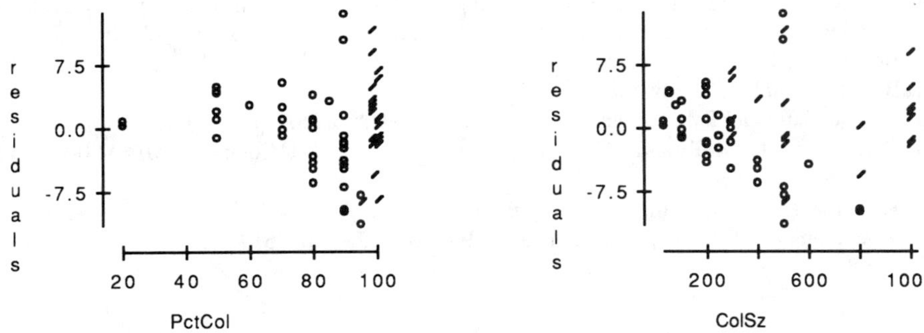

Fig. 8: The potential effects of structure on strength after accounting for Al, Cr, Nb, Ta and oxygen content as shown in residual plots (sizes in μm).

The underlying effects of Al, Cr, Nb, Ta and oxygen are shown in the partial regression plots in Figure 9. A non-linear effect of Al is again apparent. Cr at 2% may reduce strength; data above 2% is sparse, but suggests strengthening. Ta and perhaps Nb show a potential dip in strength at 2%; this may correspond to elongation behavior (see Figure 5). Oxygen is the only element found in this study to strengthen without reducing ductility.

Conclusions

Al: As has been observed previously [7], Al improves ductility until structures become mostly gamma. Peak ductility is obtained at about 47%. Yield strength is reduced by Al at the rate of 17 MPa (2.5 ksi) per atomic percent added.

Cr: Tensile behavior of cast material is essentially unaffected by 2% Cr, which is at or slightly greater than the solubility limit. Levels above 2% lead to copious B2 phase formation and poor ductility. The ductilizing effect of 2% Cr observed for wrought gamma [8] may be related to recrystallization or other deformation-related effects that are unimportant in cast material.

Nb and Ta: Both reduce ductilty; Ta is 60% more deleterious than Nb. Both strengthen at the same rate (17 MPa per percent added), meaning that Nb may be a better choice for improving tensile properties. Non-linear behavior may be present at low levels, however, most notably for the case of 2Ta. The overall study has not yet resolved whether Ta improves creep strength more than 60% faster than Nb, which would allow greater creep strength at constant ductility. Greater density-adjusted creep strength becomes possible if Ta is 100% more effective than Nb. These are important issues in our alloy development program and are the subject of current effort.

Acknowledgments

The authors are grateful for the support of the Naval Air Warfare Center-Trenton, AS Culbertson, Program Monitor. The authors wish to acknowledge the contributions of WA Baeslack, JC Chesnutt, DA Eaken, HL Lipsitt, B London, PL Martin, PA McQuay, DS Shih, JS Schaeffer and D Wheeler.

References

1. H.A. Lipsitt, Mater. Res. Soc. Symp. Proc. 1985, vol. 39, p.p. 351-364.
2. S.C. Huang, D.W. McKee, D.S. Shih and J.C. Chesnutt, in Intermetallic Compounds, edited by O. Izumi, The Japan Institute of Metals, 1991, p.p. 363-370.

3. D.S. Shih, S.C. Huang, G.K. Scarr, H. Jang and J.C. Chesnutt, in Microstructure/Property Relationships in Titanium Alloys and Titanium Aluminides, edited by Y.W. Kim and R.R. Boyer, TMS, 1990, p.p. 135-148.
4. B. London and T.J. Kelly, ibid, p.p. 285-295.
5. S.L. Semiatin and P.A. McQuay, Metall. Trans., 1992, vol 23A, p.p. 149-161.
6. C.M. Austin, T.J. Kelly and S.C. Huang, Phase 1 Report, NAWC Contract N00140-90-C-1742, 1992.
7. S.C. Huang and E.L. Hall, Metall. Trans., 1991, vol. 22A,p.p. 427-439.
8. S.C. Huang and E.L. Hall, Metall. Trans., 1991, vol. 22A,p.p. 2619-2627.

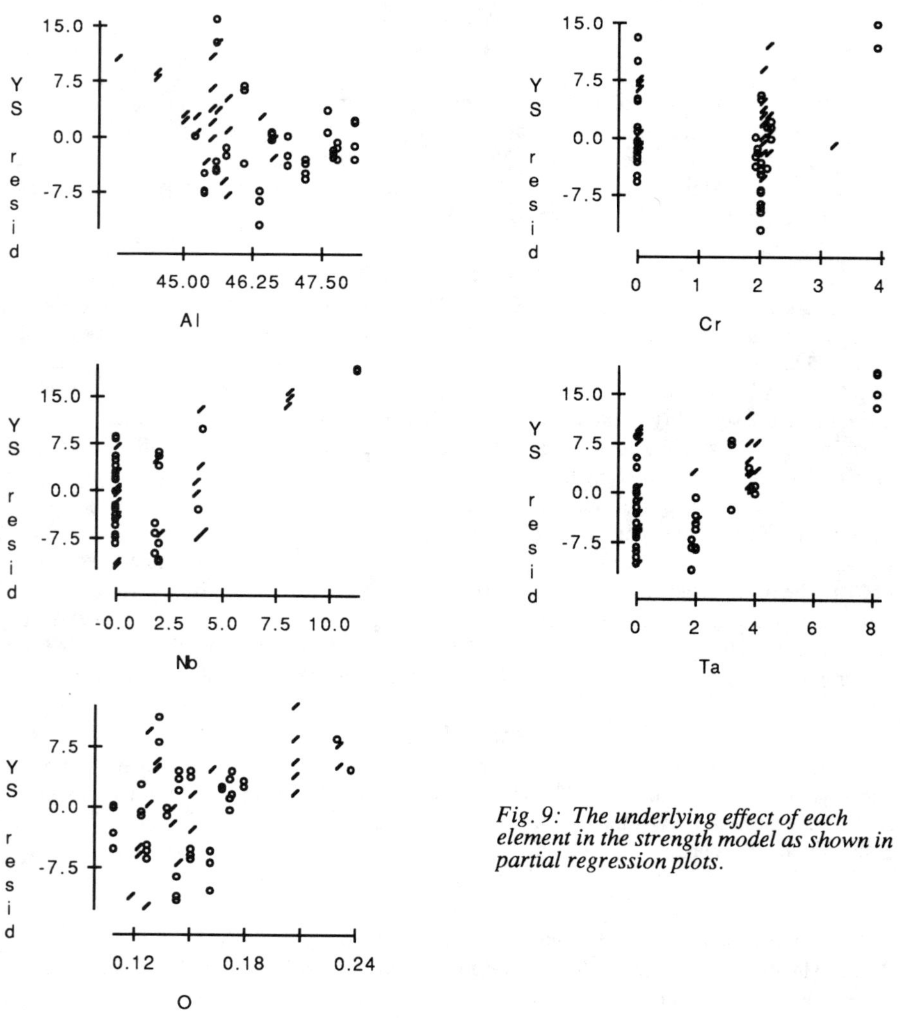

Fig. 9: The underlying effect of each element in the strength model as shown in partial regression plots.

PROCESSING AND PROPERTIES OF DUCTILE PHASE REINFORCED GAMMA TITANIUM ALUMINIDE ALLOYS

Mahyar S. Dadkhah and Jeffrey A. Graves

Rockwell International Science Center
1049 Camino Dos Rios
Thousand Oaks, CA 91360

Abstract

Gamma Ti aluminide (γ, TiAl) alloys offer significant potential for a variety of aerospace applications, however, their use is limited in part by low ductility and fracture toughness. A powder metallurgy (P/M) composite approach, involving the blending of a γ alloy with a ductile/tough Ti-Nb alloy powder has shown potential for increasing the extrinsic fracture toughness of γ. In this study, moiré interferometry was used to measure fracture toughness behavior for monolithic and phase blended γ alloys. This whole-field measurement technique was used to record both vertical and horizontal displacements associated with crack growth and to evaluate the crack opening displacement (COD) for compact tension specimens. The results showed that the fracture toughness of phase blended γ sheet was noticeably increased, while ductility remained similar to that of as-HIP'ed monolithic γ. The increase in fracture toughness in the phase blended γ is due to formation of crack face bridging, crack path deviation, blunting and multiple crack formation.

Titanium '92
Science and Technology
Edited by F.H. Froes and I. Caplan
The Minerals, Metals & Materials Society, 1993

I. Introduction

Phase blending has been proposed as a method of improving the fracture toughness of Ti aluminides [1]. Phase blending involves incorporating a ductile/tough phase into the matrix of a brittle material to serve as a crack arrestor or to restrict crack propagation by forming ductile ligaments to bridge the crack front [2]. Traction produced by the bridges reduce crack tip stress intensities. Ductile phase reinforced composites often experience large-scale bridging effects [3], especially for short cracks or in components with small dimensions. The overall load displacement characteristic $P(\delta)$ of a reinforced component is governed by traction law $p(u)$ for the ductile phase, as well as the effective crack tip fracture resistance of the composite, Γ_t and the elastic modulus E' [4]. In addition, important secondary toughening mechanisms have been demonstrated, which arise from crack particle interaction processes [4]. These mechanisms include crack trapping and renucleation leading to increases in Γ_t and crack branching and deflection. The toughening contributions from these mechanisms, depend upon the combined properties of the ductile phase, interface reaction layer and the matrix. Numerous studies of the fracture behavior of brittle matrices containing ductile particles show that, if strongly bonded to the matrix, the ductile particles act to toughen the matrix. The degree of toughening increases with volume fraction and ductile particle size. Difficulty in applying phase blending to γ-TiAl alloys has stemmed from excessive interdiffusion and reaction of the constituents, during both consolidation and mechanical working at elevated temperature. Typically, the high temperatures required to consolidate, extrude, or roll phase blended γ-TiAl alloys results in a variable composition within the microstructure (i.e., between the matrix and the toughening constituent) and reaction product formation at the γ-TiAl / ductile phase interface. In addition, deformation of the powder mixture produces a nonuniform distribution and morphology of the ductile phase particles. These factors limit the effects of phase blending. If, however, these limitations can be reduced through proper selection of the constituent chemistries and thermomechanical processing route, the resulting γ alloy, having an improved balance of engineering properties, would prove very attractive for light weight, high temperature structural applications. It is the primary purpose of this study to explore microstructural development and crack propagation behavior in a γ system, using a moiré interferometry, a whole-field strain mapping technique, in concert with a detailed microstructural analysis.

A collaborative effort at Rockwell International Science Center (RISC) and Nuclear Metals, Inc. (NMI) has been carried out to explore powder metallurgy techniques for the fabrication of "phase blended" γ-TiAl sheet. The NMI PREP™ powders exhibit excellent compositional uniformity with low interstitial content, and may be useful in producing a sound composite with good fracture toughness. In this study, the microstructures and properties of monolithic γ-TiAl were compared to HIP-consolidated sheet comprising the γ alloy blended with a β-Ti alloy.

II. Experimental

Materials and Processing

The "γ" alloy selected for this study was a two-phase, $\gamma + \alpha_2$, alloy of composition Ti-48Al-2.5Nb-0.3Ta (at.%). The toughening powder constituent was an alloy of composition Ti-30at.%Nb, having primarily a β-Ti microstructure with a minor volume fraction of α at low temperature. These powders were produced by NMI using a plasma rotating electrode (PREP™) process. The powders were sieved prior to blending, with -250 μm (-60 mesh) γ powder and 180-500 μm (+80/-35 mesh) Ti-Nb powder being used for the study. Both monolithic γ and a mixture of 80 volume % γ + 20 volume % Ti-Nb powder were HIP consolidated at 1000°C/300 MPa for characterization. This low temperature/high pressure condition was selected to minimize interdiffusion and extensive reaction at the γ/Ti-Nb interface.

Tensile Testing

Machining of the specimens was performed at MET-CUT Research Associates in Cincinnati, Ohio. Flat tensile specimens, having a nominal gauge section of 2 mm x 9.4 mm x 25.4 mm long, were tested at room temperature. Tensile test pieces were machined using low stress grinding techniques. Two tensile specimens were tested for each material. Tests were conducted at room temperature on a 20 kip closed loop servo-hydraulic test machine per ASTM E8. Strain control was set so that a strain rate of 0.005 / sec was maintained through 0.2% yield and the head rate was 0.05 in./min. after that. Extensometers were used on all specimens.

Crack Growth Experiments (Resistance Curves and Fracture Toughness)

Cracks were grown stably under measured loading conditions in compact tension specimens of dimension 30 x 32 x 2 mm. A loading fixture that could attach to the stage of an optical microscope was used, to allow in situ observation of the crack tip and wake zones. A four-beam moiré interferometry system was used to measure two orthogonal in-plane displacement fields simultaneously [5]. This technique involves replicating a diffraction grid to the surface of the compact tension specimen before growing the crack, and then, during loading to extend the crack, illuminating the surface with an Argon-ion laser (wavelength 514 nm). In this set up, part of the incident beam impinged directly onto the specimen surface, while the other part, after reflection from three plane mirrors, was incident in a symmetrical direction to produce a virtual reference grating. The combination of this reference grating and the diffraction grating bonded to the specimen produced the moiré pattern, which was recorded photographically. The moiré pattern consists of a set of fringes, which represent contours of constant displacement in the x_1 and x_2 directions, the increment between adjacent fringes being equal to the period of the reference grating (0.417 μm). In the present experiments, this technique allowed strain resolution of 10^{-3} over gauge lengths as small as 40 μm. The testing environment for toughness testing was ambient air. Digitized recordings of the moiré patterns were converted to strains automatically [6]. The evaluation of the J-Integral is essentially a numerical integration along a loop encompassing the crack where the three strain components must be evaluated at identical points along the chosen path [6]. J-Integral requires the strain components, the stress components and the strain energy density. The three stress components are calculated using J_2-deformation theory of plasticity for multi-axial states with a power hardening stress-strain relation. Stress intensity factors were evaluated using $k=(JxE)^{0.5}$. Fracture toughness values were obtained by marking the maximum in K_R vs. crack extension curves.

III. Results

The HIP-consolidated γ alloy formed a fine grained (~20 μm) equiaxed microstructure, while the γ + Ti-Nb phase blended microstructure comprised spherical γ-TiAl regions (of approximately the same 20 μm size) encompassed by a nearly continuous β-Ti phase. This is most easily seen in the back-scattered electron image of Figure 1a, in which the Ti-Nb region is lighter in contrast with the darker γ particles. In addition, an approximately 10 μm thick interdiffusion/reaction zone formed between the phase blended constituents, as shown in Figure 1b. The monolithic γ and phase blended γ + Ti-Nb sheets had measured densities of 3.76 and 3.97 g/cc, respectively.

The uniaxial stress-strain relations for monolithic and phase blended alloys with the two coefficients for the power hardening relations are shown in Figure 2a and 2b, respectively. Room temperature elongation's were below 1% for both the monolithic and phase blended sheet. All fracture surfaces showed occasional flat cleavage facets and ductile tearing. There was no evidence of intergranular fracture. Although the phase blended fracture surface was more rough, grain boundary facets and prior powder boundary facets were not observed.

Crack opening profiles for monolithic and composite γ-TiAl alloys, measured by moiré interferometry, are shown in Figures 3a and 3b, respectively. Comparison of the profiles show that in the composite material the crack is bridged. In this case bridging is supplied by ductile ligaments that remain connected to both fracture surfaces behind the crack front. The bridging traction could be determined from measurements of such crack opening profiles [7]. Figure 4 shows crack mouth opening displacement (CMOD) vs. crack length. The phase blended alloy shows larger CMOD, because it has lower elastic modulus and higher ductility than the monolithic alloy.

K_R-Δa resistance curves are shown in Figure 5. For the monolithic alloy crack extension has practically no effect on the toughness, when the crack is fully grown out of the semi-chevron zone. The average toughness for monolithic alloy is ≈11 MPa√m. In contrast, some resistance curve behavior is observed for the phase blended composite. K_R increases from an apparent initiation value of ≈ 15 MPa√m to ≈19 MPa√m at $\Delta a \geq 10$ mm.

Figures 6a and 6b show micrographs taken along the crack path in the compact tension specimen. Ductile ligaments along the fracture surface remain intact as shown in Fig. 6a. These ligaments simply pull out when the bridging zone is saturated. With reference to work of Ashby [8], by allowing for the constraint of the surrounding matrix, the average flow stress of the ductile particles increases significantly. The crack growth sequence in the phase blended alloy is: crack grows in γ phase toward the reaction zone, crack blunts when it grows into ductile zone, crack renucleates behind the ductile ligament, (at this point crack is bridged), and series of microcracks develop parallel to the second crack and grow toward Ti-30Nb phase (Fig. 6b). Blunting and microcrack development allow the particles to stretch out as the crack faces separate, forming bridging ligaments that hinder the crack advance. Due to these ligaments, the value of K_{Ic} will increase as the crack length increases up to the point at which the Ti-30Nb particle pulls out by separation along the reaction zone.

IV. Discussion

Bridging is the most puissant way of improving the toughness of brittle materials. The physical mechanism of toughening is straightforward: if ductile particles span the advancing crack, they must stretch as the crack opens until they fracture or decohere: the work-of-stretching contributes to the overall toughness of the solid. If the particle is so weakly bonded to the matrix that it easily pulls free as the crack approaches, then it is not stretched and there is almost no contribution to the toughness. But if it is strongly bonded, it is constrained. Its force-displacement curve is then very different from that of the unconstrained material as measured in an ordinary tensile test. This is an important difference because the energy absorbed in stretching the particle, crucial in calculating the contribution to the toughness, depends strongly on the degree of constraint. Based upon the microstructural observations, the reaction zone boundaries in the phase blended sheet are the weakest link between the two constituents. Therefore, optimization of the HIP cycle for minimization of interdiffusion between the γ and Ti-30Nb constituents is crucial. In addition, altering the composition of the matrix and/or the toughening constituent may modify the kinetics of interdiffusion/reaction and thereby modify the interface strength characteristics.

The sheet ductility's were all below 1% demonstrating the need for subsequent thermomechanical processing, as has been found in other studies on compaction of Ti aluminide powders. Thus as-HIP'ed ductility of the matrix appears to be a controlling factor in the overall behavior of this material. The yield stress of the phase blended sheet was slightly lower than the monolithic sheet, but the ductility was somewhat improved. The nearly continuous ductile Ti-30Nb phase would account for this. The toughness of phase blended sheet was greatly improved demonstrating the effectiveness of the Ti-30Nb phase. The quantitative contribution of different mechanisms of fracture toughness enhancement is now being investigated.

V. Conclusions

It has been demonstrated that PREP™ powders of γ and a ductile binary Ti-30Nb alloy can be physically blended and compacted into thin sections using hot-isostatic-pressing.

Compared to monolithic γ sheet, the phase blended sheet showed a slight increase in ductility and a noticeable increase in fracture toughness. Techniques to assure homogeneous distribution of the ductile phase and improve the as-HIP'ed ductility must be developed to realize the full potential of this material.

The increase in toughness in the phase blended composite is due to formation of crack face bridging, crack path deviation, blunting and multiple cracking.

VII. References

1. M.A. Ohls, W.T. Nachtrab and P.R. Roberts, "Processing and Properties of Gamma Titanium Aluminide Sheet Produced from PREP Powder," P/M N Areo. and Def. Tech. Symp., Metal Powder Indus. Fed., March 4-6, 1991.

2. C.K. Elliott, G.R. Odette, G.E. Lucas and J.W. Sheckard,"Toughening Mechanisms in Intermetallic γTiAl Alloys Containing Ductile Phases," MRS Proc. 120, p. 95 (1988).

3. F. Zok & C. Hom "Large Scale Bridging in Brittle Matrix Composites," Acta Met. Mat.

4. G.R. Odette, H.E. Deve, C.K. Elliott, A. Harigowa and G.E. Lucas, "The Influence of the Reaction Layer Structure and Properties on Ductile Phase Toughening in Titanium Aluminide-Niobium Composites," in Interfaces in Ceramic Metal Composites, R.J. Arsenault, R.Y. Lin, G.P. Martins and S.G. Fishman (eds), TMS-AIME, Warrendale, PA, 443 (1990).

5. Dadkhah, M.S., Wang, F.X. and Kobayashi, A.S., "Simultaneous On-Line Measurement of Orthogonal Displacement Fields by Moire Interferometry," Experimental Techniques 12 (1988) 28-30.

6. Dadkhah, M.S., Kobayashi, A.S. Wang, F.X. and Graesser, D.L., "J-integral Measurements Using Moire Interferometry," Proc. of the VI Int'l Congress on Experimental Mechanics (1988) 227-234.

7. B.N. Cox and C.S. Lo, "Load Ratio, Notch, and Scale Effects for Bridged Cracks in Fiberous Composites," Acta Metall. Mater. Vol. 40, No. 1, pp. 69-80, 1992.

8. M.F. Ashby, F.J. Blunt and M. Bannister, "Flow Characteristics of Highly Constrained Metals Wires," Acta Metall. 37, 1847 (1989).

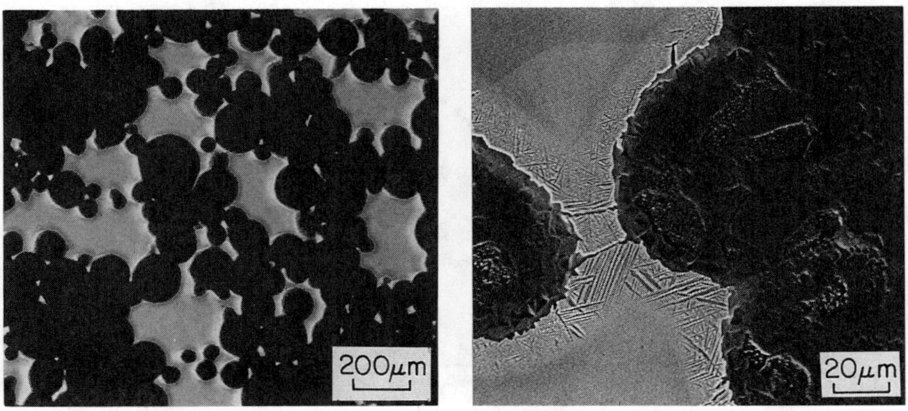

(a) (b)

Figure 1. Micrographs of the as-HIP'ed phase blended γ + Ti-30Nb alloy showing the morphology and distribution of the toughening constituent (1a) and the interface microstructure development (1b).

(a)

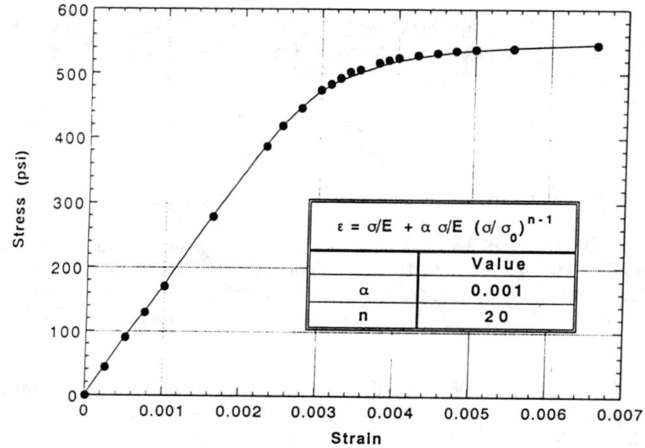

(b)

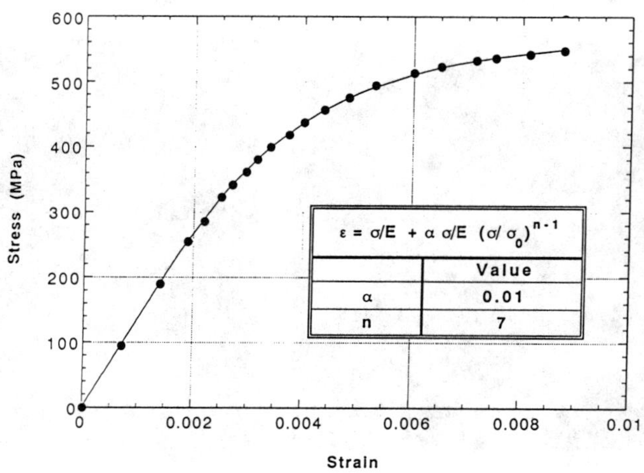

Figure 2. The uniaxial stress-strain relations for a) monolithic and b) phase blended alloys.

(a)

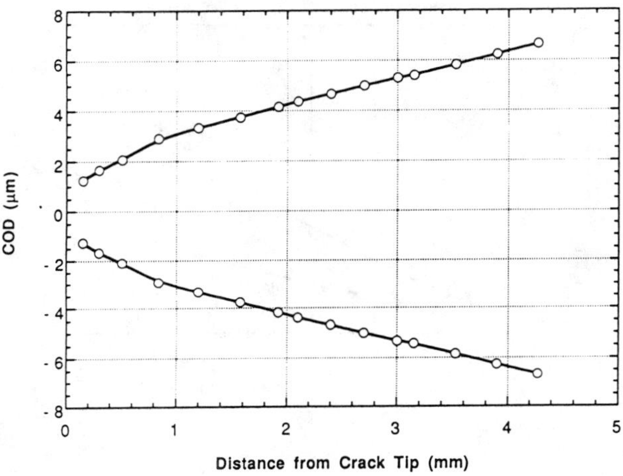

(b)

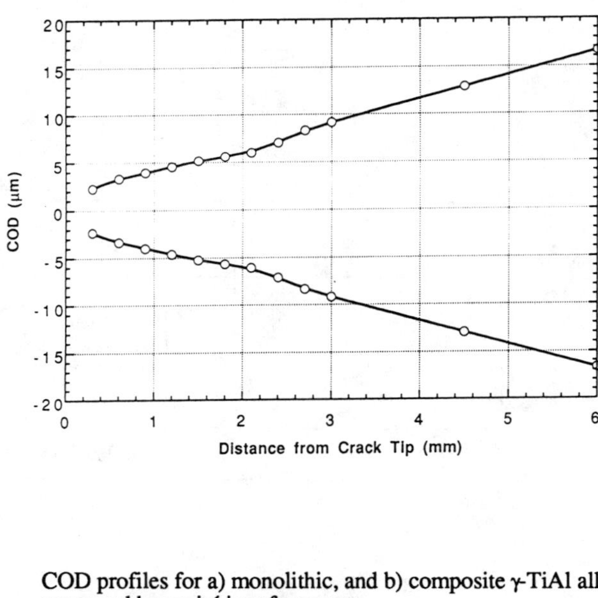

Figure 3. COD profiles for a) monolithic, and b) composite γ-TiAl alloys, measured by moiré interferometry.

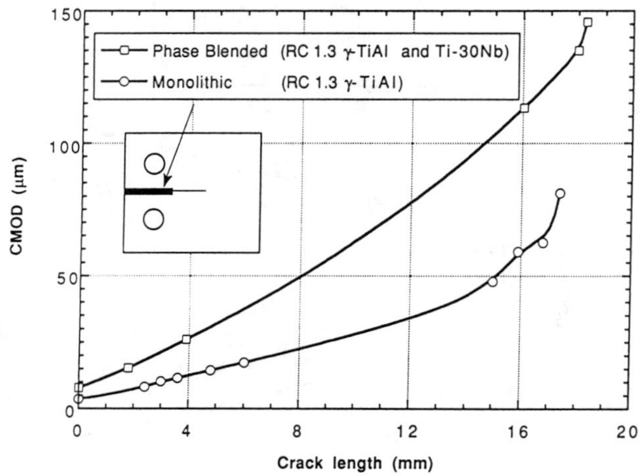

Figure 4. Crack mouth opening displacement (CMOD) vs. crack length.

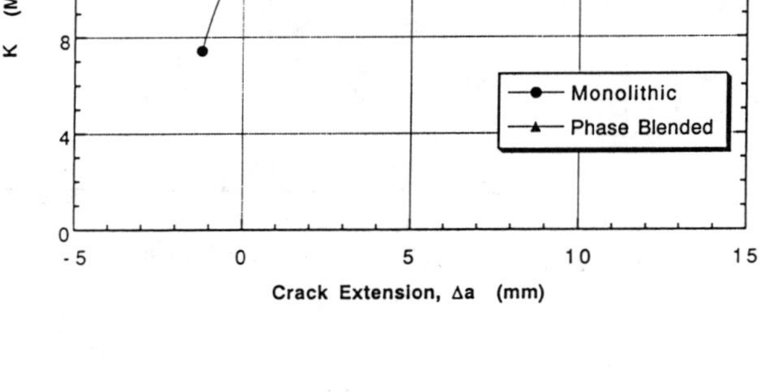

Figure 5. K_R-Δa curves for Monolithic and Phase blended alloys.

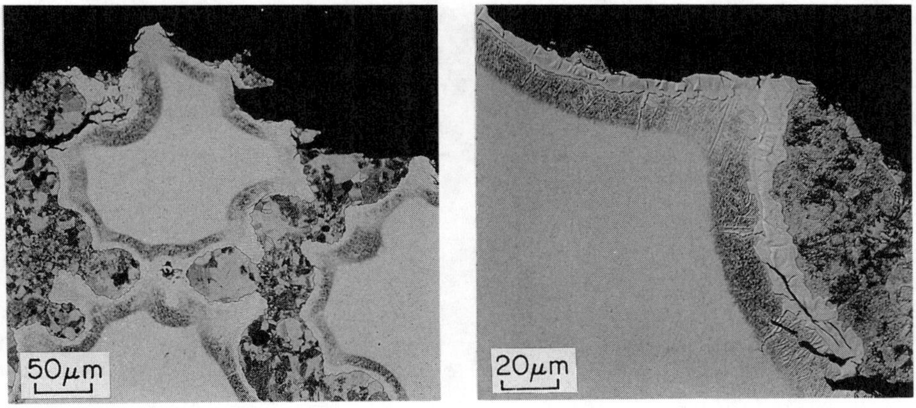

(a) (b)

Figure 6. Micrographs of phase blended alloy taken along the crack path.

Effects of Processing and Heat Treatment on Microstructure and Mechanical Properties of Gamma Titanium Aluminide Powder Compacts

C. F. Yolton* and D. Eylon**

* Crucible Materials Corporation, Crucible Research, P.O. Box 88, Pittsburgh, PA 15230-0088 USA

** University of Dayton, Graduate Materials Engineering, KL407, Dayton, OH 45469-0240 USA

Abstract

Titanium aluminide alloys based on the gamma (TiAl) intermetallic compound are desirable for advanced aerospace applications. The properties which make these alloys attractive are low density, good elevated temperature tensile and creep strength, and good oxidation resistance. The main drawback to using TiAl alloys is their low ductility at temperatures up to approximately 760C (1400F) and the difficulty in forging and machining, the result of their low ductility. Powder metallurgy (PM) processing potentially minimizes the problems associated with conventional processing while maintaining the attractive properties. This paper will present the results of a study to determine the effect of hot isostatic pressing (HIP) conditions, and heat treatment on the microstructure and mechanical properties of consolidated Ti-48Al-2Cr-2Nb gamma titanium aluminide alloy powder made by gas atomization. Mechanical testing included room and elevated temperature tensile and stress rupture tests.

Introduction

Gamma titanium aluminide alloys have received considerable attention in recent years, particularly for high temperature applications in advanced aircraft. These alloys are based on the intermetallic compound TiAl (gamma). Gamma titanium aluminide alloys are of interest for high temperature aircraft applications because they have a number of attractive properties. These properties include low density, high strength-to-weight ratio at elevated temperature, good elevated temperature creep-rupture strength and good oxidation resistance. Although alloys based on gamma titanium aluminides have attractive elevated temperature properties, room temperature ductility and fabrication are areas which require additional work. One means of minimizing the difficulties associated fabrication of titanium aluminides into finished products is through powder metallurgy (P/M). The production of complicated fully dense

Titanium '92
Science and Technology
Edited by F.H. Froes and I. Caplan
The Minerals, Metals & Materials Society, 1993

shapes by P/M which require only minimal machining to achieve finished dimensions has been demonstrated.(1) For conventional titanium alloys the mechanical properties of P/M product made from prealloyed powder are generally comparable to cast and wrought product.(2,3) The work described in this paper was conducted to determine the mechanical properties of P/M Ti-48Al-2Cr-2Nb which is one of the more recently developed gamma alloys. The alloy was produced by gas atomization and consolidated by hot isostatic pressing (HIP).

Experimental Procedure

Powder Production

The Ti-48Al-2Cr-2Nb (at%) prealloyed powder used in this work was produced by gas atomization. The gas atomization unit is shown schematically in Figure 1. In this unit the charge to be atomized is melted in a cold wall induction crucible. The charge is typically in the form of titanium ingot, master alloy and elemental additions. The molten alloy is atomized by tilt pouring into a non-ceramic tundish. At the bottom of the tundish is a refractory metal nozzle which produces a well formed stream of the molten alloy. The stream is atomized by free fall through high pressure jets of argon gas. The atomized powder is carried by the gas via a transfer tube from the atomization tower to a cyclone where it is collected in a cannister. Figure 2 shows the base of the atomization tower, the cyclone and the collection cannister.

Powder cleanliness is very important for critical aerospace applications. The atomizer has several features which promote powder cleanliness. As shown in Figure 3, all internal surfaces of the atomizer which may come in contact with the powder are polished stainless steel. The atomizer tower, transfer tube and cyclone collector contain multiple flanges to allow disassembly and thorough cleaning of all internal surfaces every time the alloy is changed. This is done to prevent cross contamination of alloys. The atomizer tower and all sections of the system below the melting furnace are housed in positive pressure room which has a dedicated air handling system to prevent airborne contamination from the ambient shop atmosphere. A powder processing room is located adjacent to the atomizer room where operations such as screening, blending and loading of cans for shipment or consolidation is done.

Melting and atomization are conducted from a control room located beside the atomization facility for operator safety. The controls are centralized on one console as shown in Figure 4. Melting is viewed with one video camera and atomization is viewed with a second camera. Both views can be displayed simultaneously on one monitor. Approximately 30 sensors are used to collect information during melting and atomization which is automatically logged on a personal computer.

The as-atomized powder is typically screened to -35 mesh (500 micron) and processed through a magnetic separator to remove any magnetic particles which may have been picked up during atomization.

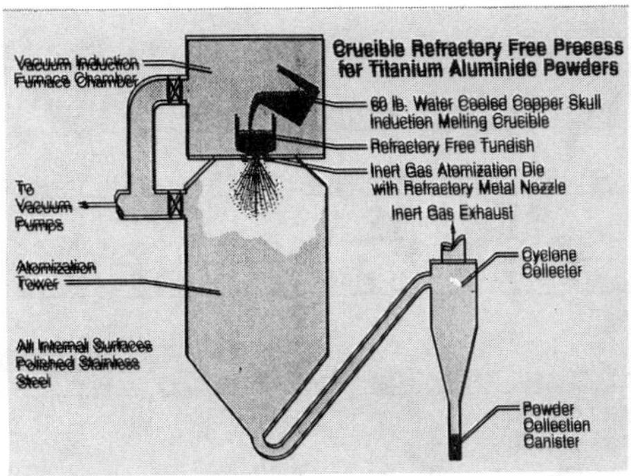

Figure 1. Schematic diagram of titanium gas atomization unit.

Figure 2. Photograph of base of atomization tower and cyclone.

Figure 3. Photograph of titanium gas atomization unit disassembled for cleaning.

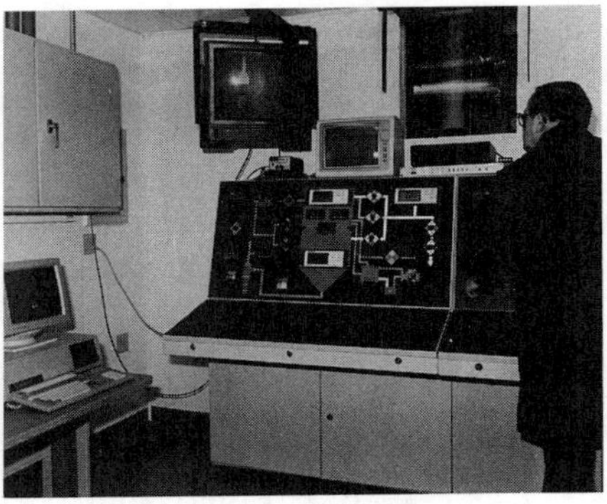

Figure 4. Photograph of atomizer control room.

The size distribution of the powder is given in Table I. Mean particle size is approximately 160 microns. Chemical analysis results on the powder are given in Table II. The analyzed composition is close to the nominal composition and the oxygen content is well below 0.1 wt% which is generally considered necessary for optimum mechanical properties.

Powder Consolidation

The powder for mechanical property testing was consolidated in commercially pure titanium containers. After thorough cleaning, the containers were loaded with powder, evacuated, first at room temperature then heated to 65C (150F) and sealed by pressure welding the stems. Compacts were consolidated by hot isostatic pressing using two conditions: 1177C (2150F) and 103 MPa (15 ksi) for 4 hours, and 1000C (1832F) and 200 MPa (29 ksi) for 4 hours.

Table I. Size Distribution of Ti-48Al-2Cr-2Nb Powder

	Percent Undersize						
U.S. Standard Mesh	35	45	60	80	100	200	325
Micron	500	354	250	177	149	74	44
	100	89	72	55	45	16	5

Table II. Chemical Composition of Ti-48Al-2Cr-2Nb Powder

Weight Percent (Atomic Percent)

Al	Nb	Cr	Fe	C	O	N	H	Ti
32.5	5.0	2.8	.050	.007	.082	.006	.0010	Bal
(47.0)	(2.1)	(2.1)	(.035)	(.023)	(.200)	(.017)	(.0387)	Bal

Mechanical Testing

Specimen blanks for mechanical testing were cut from the hot isostatically pressed compacts and heat treated as individual blanks. The heat treatments used are listed in Table III. Tensile tests and stress rupture tests were conducted using round specimens with a 6.3mm (0.25in.) diameter x 25mm (1.0in.) long gage length in accordance with ASTM-E8. All tests were conducted in air. Elevated temperature tensile specimens were held at temperature for 30 minutes prior to testing.

Table III. Heat Treatments Used For Ti-48Al-2Cr-2Nb Compacts

Heat Treatment
a.) 1177C (2150F) 4 hr AC
b.) 1290C (2354F) 1 hr AC
c.) 1290C (2354F) 4 hr AC
d.) 1290C (2354F) 4 hr AC + 871C (1600F) 8 hr AC

Results and Discussion

Figure 5 shows an SEM micrograph of the as-atomized powder. The powder is predominantly spherical and has a minimum of satellite particles. Tap density of the powder is 2.63 grams/cc or about 66% of solid density. The good packing characteristics of the gas atomized powder make it well suited for making near net shapes via HIP. The as-atomized microstructure of the powder is shown in Figure 6. Fine structure analysis by other workers (4) has shown that as-atomized gamma titanium aluminide powder may consist of varying amounts of alpha two phase and gamma phase depending on the particle size which determines cooling rate.

Figures 7a and 7b show the microstructure of the Ti-48Al-2Cr-2Nb alloy after HIP consolidation at 1000C (1832F) and 1177C (2150F) respectively. Both HIP condiditions resulted in a bimodal grain size distribution which has been observed in other gamma alloys. As indicated in the photomicrographs, a much finer grain size is produced by HIPing the powder at the lower temperature. The heat treated microstructures of the material HIPed at 1000C are shown in Figure 8. All heat treatments were above the HIP temperature which produced some grain growth. Heat treating at 1290C results in a small amount of an intergranular phase which is presumed to be alpha two. Room temperature tensile results for the Ti-48Al-2Cr-2Nb alloy

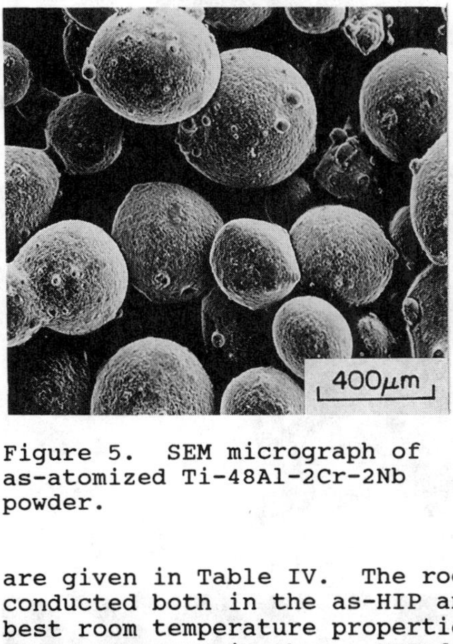

Figure 5. SEM micrograph of as-atomized Ti-48Al-2Cr-2Nb powder.

Figure 6. Microstructure of as-atomized Ti-48Al-2Cr-2Nb powder.

are given in Table IV. The room temperature tests were conducted both in the as-HIP and heat treated conditions. The best room temperature properties were obtained with material HIPed at 1000C in the as-HIPed condition. The material HIPed at the higher temperature or heat treated after the 1000C HIP exhibited lower strength associated in some cases with lower ductility. Elevated temperature tensile results are given in Table V. At 649C the strength is approximately equilivant to the room temperature strength with slightly higher ductility. Yield strength remains at about 340MPa at 760C but elongation is significantly higher at 12%. At 871C yield strength decreases about 10ksi from the room temperature value and ductility is over 50%. Increasing the test temperature to 1000C results in a decrease in strength to about 45% of the room temperature value and ductility remains very high.

Stress rupture test results for the Ti-48Al-2Cr-2Nb alloy are given in Figure 9. The material HIPed at 1177C has improved stress rupture strength over the material HIPed at 1000C. The improved stress rupture properties would be expected with the larger grained material.

Figure 7. Microstructures of consolidated Ti-48Al-2Cr-2Nb powder: (a) as-HIP 1000C (1832F)/200MPa (29ksi)/4 hr (b) as-HIP 1177C (2150F)/103MPa (15ksi)/4 hr

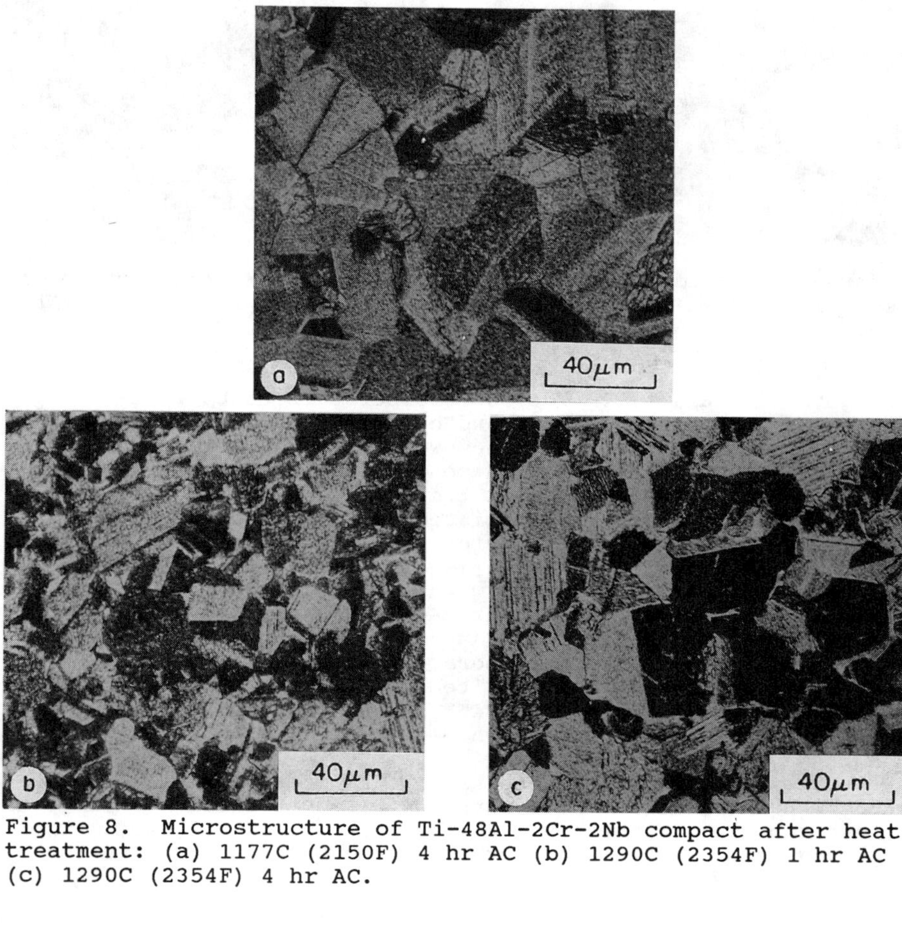

Figure 8. Microstructure of Ti-48Al-2Cr-2Nb compact after heat treatment: (a) 1177C (2150F) 4 hr AC (b) 1290C (2354F) 1 hr AC (c) 1290C (2354F) 4 hr AC.

Table IV. Room Temperature Tensile Properties of HIP Consolidated P/M Ti-48Al-2Cr-2Nb

Heat Treatment(*)	Test Temp C (F)	Tensile Strength MPa (ksi)	Yield Strength 0.2% Offset MPa (ksi)	Elong. in 4D %	Rdn of Area %
As-HIP 1000C	RT	551 (80)	482 (70)	2.0	2.0
HIP 1000C + (a)	RT	482 (70)	413 (60)	2.0	3.6
HIP 1000C + (b)	RT	475 (69)	400 (58)	2.1	2.0
HIP 1000C + (c)	RT	468 (68)	386 (56)	1.8	1.8
HIP 1000C + (d)	RT	441 (64)	344 (50)	1.0	2.4
As-HIP 1177C	RT	496 (72)	420 (61)	1.5	3.6

* Refer to Table III

Table V. Elevated Temperature Tensile Properties of HIP Consolidated P/M Ti-48Al-2Cr-2Nb

Heat Treatment(*)	Test Temp C	(F)	Tensile Strength MPa (ksi)	Yield Strength 0.2% Offset MPa (ksi)	Elong. in 4D %	Rdn of Area %
HIP 1000C+(d)	649	(1200)	427 (62)	344 (50)	2.0	3.6
HIP 1000C+(d)	760	(1400)	524 (76)	338 (49)	12.0	13.4
HIP 1000C+(d)	871	(1600)	351 (51)	282 (38)	66.0	52.5
As-HIP 1000C	1000	(1832)	193 (28)	138 (20)	32.3	29.0
HIP 1000C+(b)	1000	(1832)	186 (27)	158 (23)	61.3	52.6
HIP 1000C+(c)	1000	(1832)	179 (26)	152 (22)	81.6	69.0

* Refer to Table III

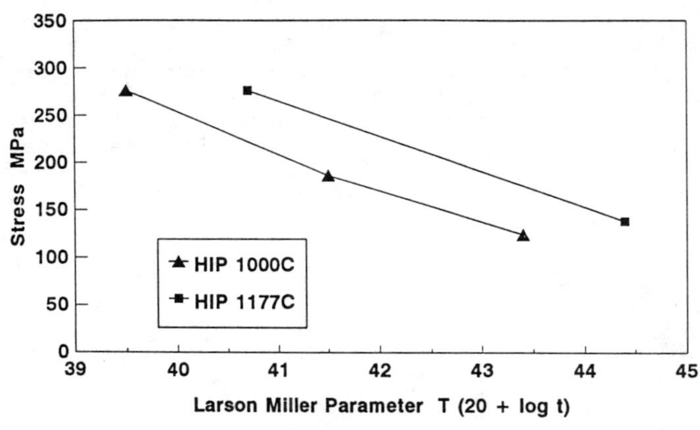

Figure 9. Stress rupture properties of HIP consolidated Ti-48Al-2Cr-2Nb.

Summary

Gas atomized Ti-48Al-2Cr-2Nb powder was consolidated to full density by HIP at 1000C and 1177C. A very fine grain size (5 micron) was obtained by HIP at 1000C. The fine grain size is projected to yield a material with very good hot workability.

Tensile and stress rupture properties of the P/M Ti-48Al-2Cr-2Nb compacts were determined. The material HIPed at 1000C gave the best combination of room temperature properties, highest strength and highest ductility. The larger grain material HIPed at 1177C gave the best elevated temperature stress rupture properties.

References

1. C. F. Yolton, T Lizzi, V. K. Chandhok and J. H. Moll, "Powder Metallurgy of Titanium Aluminide Components", Titanium Rapid Solidification Technology, F. H. Froes and D. Eylon Eds., TMS-AIME Publication, Warrendale, PA, 1986.

2. A. A. Sheinker, "Prealloyed Titanium P/M Parts in the F-18 Aircraft", Presented at the 1884 AIME Spring Meeting, Los Angeles, CA, February, 1984.

3. R. H. Witt and I. G. Weaver, "Titanium PM Components for Airframes", Titanium Net Shape Technologies, F. H. Froes and D. Eylon Eds., TMS-AIME Publication, Warrendale, Pa, 1984.

4. J. A. Graves, J. H. Perepezko, C. H. Ward and F. H. Froes, "Microstructural Development in Rapidly Solidified TiAl", Scripta Metallurgica, Vol 21, 567, 1987.

α/γ PERITECTIC OSCILLATIONS IN DIRECTIONALLY SOLIDIFIED Ti-52at%Al

David B. Snow, Charles T. Burilla, and Ben F. Oliver*

United Technologies Research Center
411 Silver Lane, East Hartford, CT 06108 USA

*Department of Materials Science and Engineering
The University of Tennessee, Knoxville, TN 37996-2200 USA

Abstract

A Ti-52at%Al alloy exhibited bands of segregation perpendicular to the growth direction after directional solidification under planar interface conditions. The phases within these bands consisted of either an α_2/γ lath structure (≈51±1at%Al), or γ-TiAl (≈54±1at%Al) containing small regions of α_2, indicative of cellular instability during solidification. The banded microstructure was caused by an oscillation of the solidifying alloy composition, so that the specimen alternated between solidification as α and solidification as γ. This in turn implied that solidification occurred via an $\alpha + L \rightarrow \gamma$ peritectic transformation, and that the 52at%Al alloy composition was hypoperitectic.

Introduction

Current research efforts to examine the fundamental mechanical properties of γ TiAl have engendered a related interest in the production of TiAl single crystals. One of the processes which has been successfully utilized to grow single crystals is containerless directional solidification [1-3]. However, particular combinations of alloy composition and processing conditions can produce polycrystalline TiAl specimens with unusual microstructures, characterized by distinct bands oriented perpendicular to the growth direction [3]. Similar microstructures have been observed after peritectic solidification under planar interface conditions [4]. Consequently, it appeared that quantitative analysis of the banded microstructure observed in directionally solidified Ti-52at%Al might augment the current efforts to determine the high temperature phase relationships in near equiatomic Ti-Al alloys [5-8].

Experimental Procedure

The Ti-52at%Al specimen was consolidated by arc melting high purity Ti and Al (99.999%) under purified argon. The concentrations of the specific impurities in the Ti were Al 10, Cu 8.6, Fe 5, Si 8, C 40, N 3, and O 90 (wt ppm). The arc melted alloy was drop cast to form a 2.5cm diameter rod. The rod was directionally solidified under planar interface conditions by containerless, levitated zone melting [1-3], using a protective atmosphere of palladium purified hydrogen. The turbulence produced by induction melting, together with the rotation of the lower rod section at 100rpm, combined to create a highly mixed molten zone. The interface velocity (solidification rate) was maintained at 3.53μm/s (0.5in/h).

Titanium '92
Science and Technology
Edited by F.H. Froes and I. Caplan
The Minerals, Metals & Materials Society, 1993

Longitudinal sections of the specimen were etched for light microscopy with a solution of 5%HF/5%HNO_3 in glycerin. Thin foils for examination by TEM were electropolished in 50ml HClO (70%)/300ml butanol/500ml methanol with a double jet apparatus at 70VDC and -50°C. Backscattered electron images and Al concentration profiles were obtained with a Cameca "Camebax" electron microprobe operated at 15kV, using wavelength dispersive X-ray (WDX) spectroscopy with pure element standards. Thin foils were examined with a Philips EM400T TEM/STEM equipped with a 30mm^2, ultra-thin window, energy dispersive X-ray spectrometer.

Results

Chemical Analyses

The chemical composition of the as-solidified Ti-52at%Al rod was determined by various analytical techniques, as presented in Table I. Samples for these analyses were cut so as to incorporate enough segregation layers to determine the overall composition of the specimen.

Table I Post-Solidification Chemical Analysis

Element	Concentration (weight)	(atomic)	Analytical Procedure
Aluminum	37.96 ± 0.33%	51.9%	ICP/AES*; mean of 3 samples
Hydrogen	45 ± 47 ppm	1650 ppm	gas fusion analysis
Carbon	247 ± 65 ppm	760 ppm	gas fusion analysis
Nitrogen	275 ± 57 ppm	725 ppm	gas fusion analysis
Oxygen	367 ± 8 ppm	850 ppm	neutron activation analysis (18.8g)

*Inductively Coupled Plasma/Atomic Emission Spectroscopy

Microstructure

Light Micrographs. The as-solidified microstructure of Ti-52at%Al viewed at low magnification is shown in Figure 1A. The area in the right third of this photomicrograph was directionally solidified once; the left two thirds was solidified a second time in the opposite direction. The microstructure exhibited prominent bands of varying thickness, which extended across the entire specimen diameter. The shape of the boundaries between these bands delineated the location of prior liquid/solid interfaces. The macroscopic grain structure was columnar, aligned with the direction of solidification.

Examination of this banded microstructure at higher magnification revealed somewhat greater complexity (Figure 1B). Some of the wider bands contain fine scale banding. In addition, small rod-shaped regions of α_2 were present throughout the γ bands, indicative of cellular instability during solidification (Figures 2-3). Interspaced between the γ bands are regions which had solidified as α, subsequently transformed to lamellar α/γ, and finally to lamellar α_2/γ .

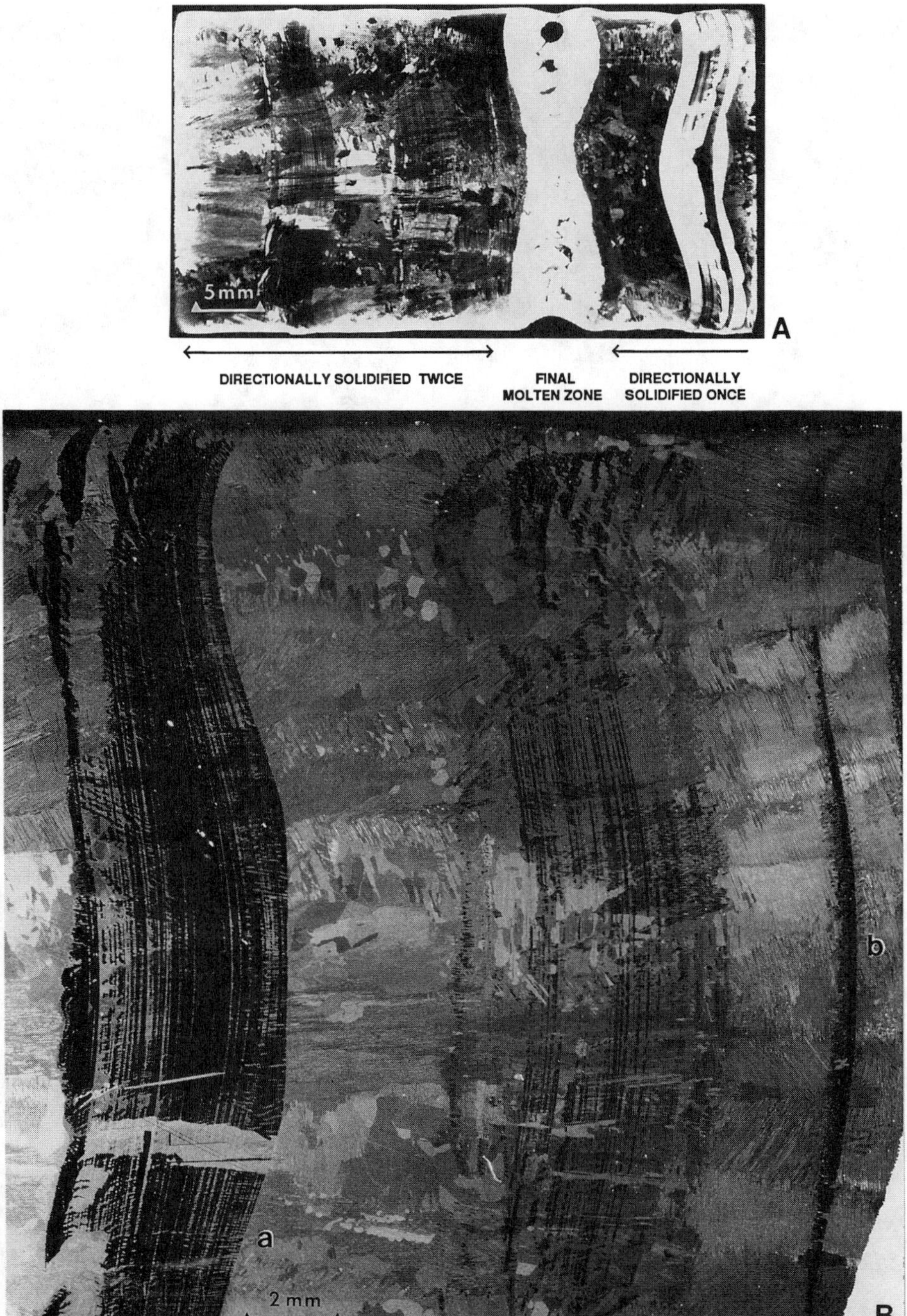

Figure 1 - Banded microstructure, Ti-52at%Al, directionally solidified under purified hydrogen. Etched surface; photographed with polarized light. (A) Low magnification overview. (B) Left side of overview at higher magnification.

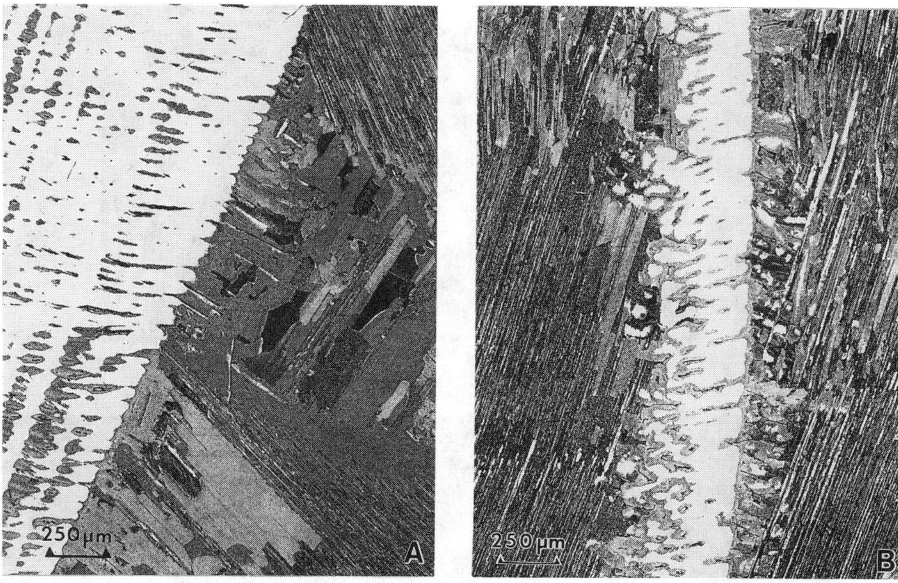

Figure 2 - (A) Interface between γ band containing α_2 cellular instability (left) and α_2/γ band (right); area "a", Fig. 1B. (B) Narrow γ band (center), with α_2/γ regions on either side; area "b", Fig. 1B. Etched surfaces, photographed with polarized light.

<u>Backscattered Electron Images</u>. Phase identification over large areas, both between and within bands, was most easily achieved by backscattered electron imaging of unetched surfaces, which allowed unambiguous distinction between α_2 (light contrast) and γ (darker contrast) (Figure 3). These images confirmed that the rod-shaped areaswere α_2, indicative of cellular instability in the γ bands during solidification; and that the microstructural banding observed by light microscopy was the manifestation of alternating changes in Al content as the solidification front progressed.

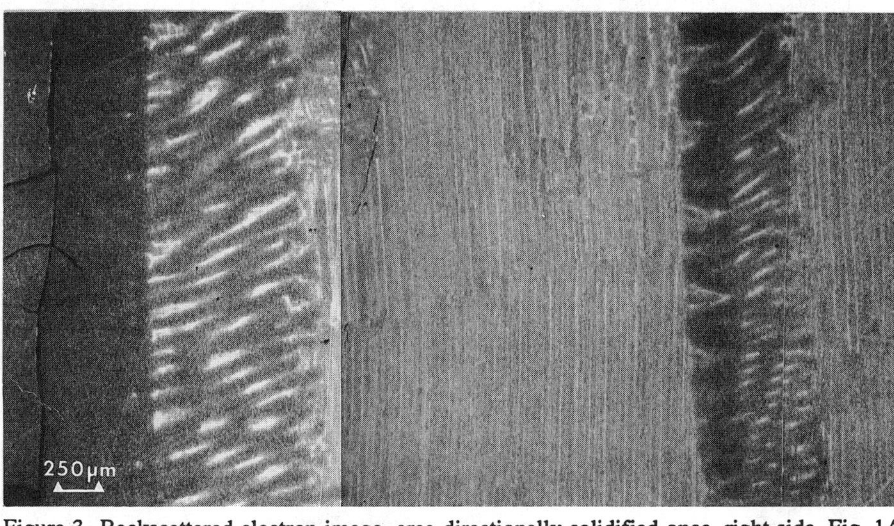

Figure 3- Backscattered electron image, area directionally solidified once, right side, Fig. 1A. Final molten zone is at far left.

Aluminum Concentration Profiles

The Al concentration was measured by electron microprobe along a linear trace on an unetched surface near the centerline of the specimen (Figure 4). Both the region directionally solidified once (Figure 4A) and that solidified twice (Figure 4B) display macroscopic Al segregation on a scale of 0.5-3mm, corresponding to the banding shown in Figure 1. However, on a smaller scale there is extensive variation in both Al concentration, and in the width of the bands.

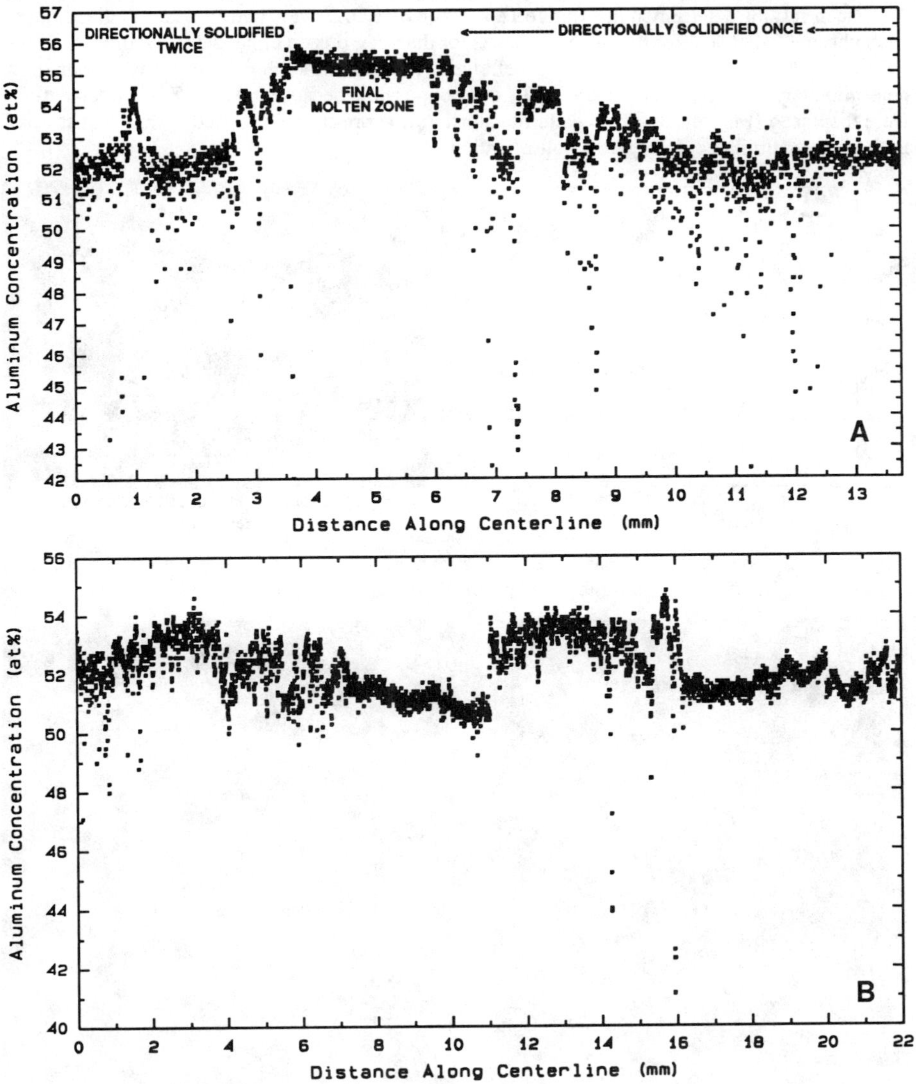

Figure 4 - Electron microprobe data, Al concentration vs. distance along centerline, trace parallel to growth direction, 5µm spacing between sample points. (A) Data on right is from region directionally solidified once. (B) All data from within region directionally solidified twice, Fig. 1A.

Cellular α_2

Some of the fine scale instances of lower Al concentration shown in Figure 4 can be attributed to the presence of cellular α_2. This was confirmed by microprobe analysis within a single γ band (Figure 5), and by transmission electron microscopy (Figure 6). The steep concentration gradients associated with the α_2 (Figure 5) may be artifacts derived from the intersection of the electron beam activation volume with regions of λ lying below the free surface. By contrast, the Al concentrations determined in a thin foil specimen by EDXS were constant across the width of each phase: 42±3at%Al for the α_2, 54±3at%Al for the γ regions on either side (Figure 6A; 10nm spot size; 120kV). A cross-check of the cellular α_2 Al concentration from its lattice parameters, determined by CBED pattern analysis [9], was not possible because the HOLZ line intersections were displaced (Figure 6B). This pattern distortion was presumably caused by residual stresses in the α_2, accumulated during post-solidification cooling.

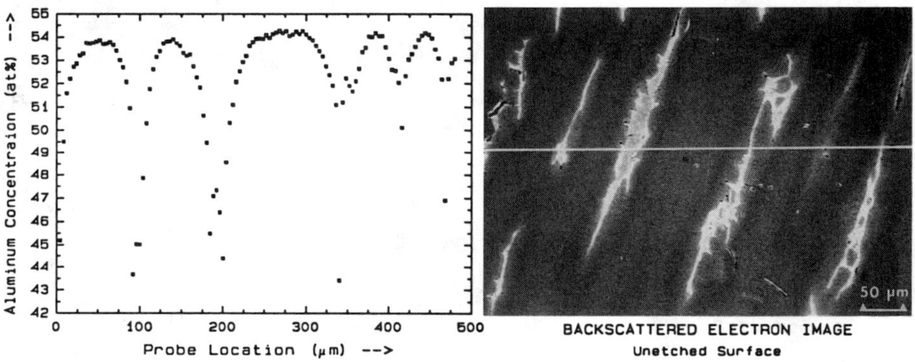

Figure 5 - Al concentration along linear trace shown at right; area within γ band which contains cellular α_2; region directionally solidified once.

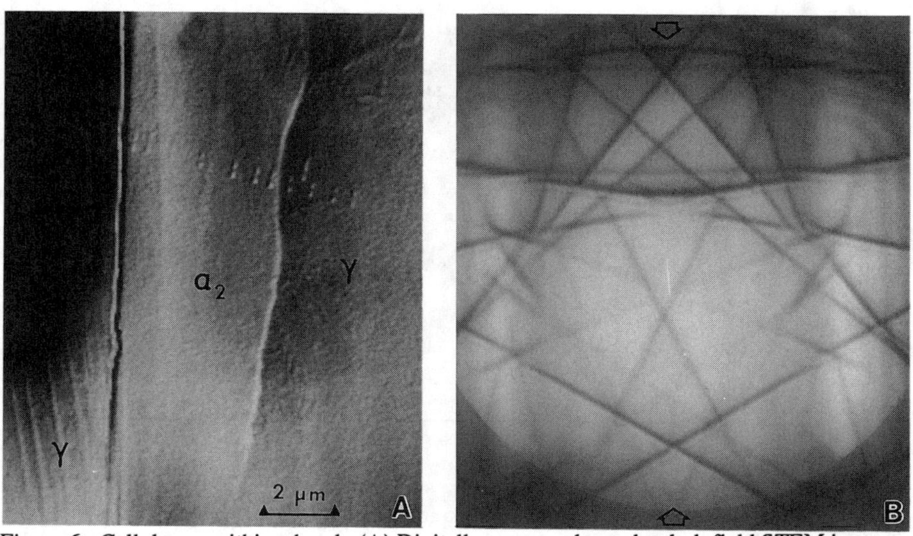

Figure 6 - Cellular α_2 within γ band. (A) Digitally processed annular dark field STEM image. (B) $[3\bar{3}01]$ CBED from the α_2 shown at left. The HOLZ line intersections are asymmetric about a vertical mirror plane, indicating the presence of residual stress.

Discussion

The presence of macroscopic segregation bands within this directionally solidified alloy was interpreted as evidence that the solidifying composition had oscillated from end to end of a $L + \alpha \rightarrow \gamma$ peritectic transformation. This occurred because the mass transfer requirements of the peritectic transformation could not be continuously fulfilled under (nearly) planar interface conditions. The alternate solidification between the α and γ phases implied that the overall alloy composition (52at%Al) was hypoperitectic. Similar banded microstructures have been observed to form during directional, planar interface solidification of a Sn-Cd hypoperitectic alloy [4]. The electron microprobe determination of the average %Al with the segregation bands suggest that the boundaries of the $L + \alpha \rightarrow \gamma$ peritectic are 51±1at%Al and 54±1at%Al, respectively. The existence of this peritectic is consistent with the results of several recent investigations [5-8].

In the case of the cellular instability which was manifested as small regions of α_2 within the γ bands (Fig. 6A), it is difficult to understand why the α_2 composition should be as low as 43±2at%Al. The relatively high concentration of hydrogen, a strong β stabilizer, retained in the specimen (Table I) suggests that the use of hydrogen as a protective atmosphere may have expanded the β phase field, and thus promoted the formation of β during the oscillation of the peritectic mass transfer instabilities. Cellular instability during solidification of a γ band could then be manifested by the formation of hydrogen stabilized, intercellular β of relatively low Al content, which would subsequently transform to α, and then to α_2. By comparison, when helium was used instead of hydrogen as the protective atmosphere during the directional solidification of Ti-52at%Al, banding of the microstructure was largely suppressed [3].

Conclusions

1. Analysis of the banded microstructure observed in directionally solidified Ti-52at%Al is consistent with its formation by a $L + \alpha \rightarrow \gamma$ peritectic transformation.

2. The bands in the microstructure were formed as the alloy oscillated between solidification as α of ≈51±1at%Al composition and solidification as γ of ≈54±1at%Al composition, since the mass transfer requirements for the peritectic transformation were not continuously met under planar interface conditions.

3. The oscillation between the α and γ phases during $L + \alpha \rightarrow \gamma$ solidification implies that Ti-52at%Al is a hypoperitectic composition.

4. Based on the current understanding of the Ti-Al phase diagram, it is not evident why the occurrence of cellular instability during solidification of the γ bands should produce α_2 with an Al content as low as ≈43at%. One possible explanation is that the hydrogen protective atmosphere promoted the formation of intercellular β phase, which subsequently transformed to α and then to α_2.

Acknowledgment

The authors would like to acknowledge the assistance of the following UTRC staff members: Mr. Donald W. Tucker for preparation of the thin foil TEM specimens; Dr. David A. Condit and Mr. Paul A. Vozzella for the bulk chemical analyses.

References

1. B. F. Oliver, B. Y. Huang, and W. C. Oliver, "Containerless Growth of Large Single Crystals of TiAl," Scripta Metall., 22 (1988), 1405-1408.

2. R. D. Reviers, B. F. Oliver, and D. D. Burns, "Computer Controlled Containerless Processing of High Temperature Intermetallic Compounds," Materials and Manufacturing Processes, 4 (1) (1989), 103-131.

3. B. F. Oliver and B. Kad, "A Study of α/γ Peritectic Oscillations Using Containerless Liquid Directional Solidification of γ-TiAl Compositions," J. Less-Common Metals, 168 (1991), 81-90.

4. W. J. Boettinger, "The Structure of Directionally Solidified Two-Phase Sn-Cd Peritectic Alloys," Metall. Trans., 5 (1974), 2023-2031.

5. C. McCullough, J. J. Valencia, C. G. Levi, and R. Mehrabian, "Phase Equilibria and Solidification in Ti-Al Alloys," Acta Metall., 37 (1989), 1321-1336.

6. S. C. Huang and P. A. Siemers, "Characterization of the High-Temperature Phase Fields near Stoichiometric γ TiAl," Metall. Trans. A, 20A (1989), 1899-1906.

7. J. C. Mishurda and J. H. Perepezko, "Phase Equilibria in Ti-Al Alloys," in Microstructure/Property Relationships in Titanium Aluminides and Alloys, ed. Y-W. Kim and R. R. Boyer, (Warrendale, PA: The Minerals, Metals & Materials Society, 1991), 3-30.

8. E. L. Hall and S. C. Huang, "Microstructures of Rapidly-Solidified Binary TiAl Alloys," Acta Metall., 25 (1991), 1805-1809.

9. J. W. Steeds, "Convergent Beam Electron Diffraction," in Introduction to Analytical Electron Microscopy, ed. J. J. Hren and D. C. Joy, (New York: Plenum Press, 1979), 387-422.

Wrought Processing of a Gamma Titanium Aluminide Alloy Ti-48Al-2Cr-2Nb

J.C. Chesnutt and D.S. Shih*

GE Aircraft Engines
One Neumann Way, MD M89
Cincinnati, Ohio, USA 45212-6301
*now at McDonnell Douglas Research Laboratories
Box 516
St. Louis, Missouri 63166-0516

Abstract

Considerable research on and development of the alloy, Ti-48Al-2Cr-2Nb (at %), has occurred in the past several years, and microstructural conditions with an attractive balance of elevated temperature strength, low temperature ductility and toughness have been achieved. A number of microstructural conditions with this desirable balance of properties were produced by either isothermal forging or extrusion of as-cast ingots followed by heat treatment. This paper will describe the processing sequences for both forging and extrusion and will compare the degree of microstructural refinement and control available from each process. Input material for isothermal forging trials was produced both by vacuum induction skull melting (VIM) and vacuum arc re-melting (VAR). Both laboratory scale (approximately 140 mm diameter) and production scale (approximately 485 mm diameter) forgings have been produced. Extrusion process development has been concentrated on VIM material with starting billet diameter of 70 mm. Extrusion ratios ranging from 6:1 to 20:1 have been utilized to develop a range of microstructures. Microstructures and mechanical properties of extruded and forged material have been characterized

Introduction

The objective of the process development activities was to produce wrought material in sufficient size and quantity to allow heat treatment development and mechanical property characterization that would be representative of production material. In this paper we will discuss the processing of extrusions from vacuum induction melted (VIM) ingot and isothermal forging of both VIM and vacuum arc re-melt (VAR) ingot. Microstructural development in both extruded and isothermally forged material will be discussed, and the mechanical properties of extruded and forged material will be compared.

Titanium '92
Science and Technology
Edited by F.H. Froes and I. Caplan
The Minerals, Metals & Materials Society, 1993

Material Production

The material characterized in this study was produced as VIM or VAR ingots which were subsequently extruded or forged and heat treated.

Ingot Material

Ingot material was obtained in several forms including 70 mm diameter, 16 kg, VIM ingot; 152 mm, 16 kg, double VAR ingot, and 203 mm, 55 kg, triple VAR ingot. The smaller VAR ingots were initially ordered to conserve resources, but it was later found that the larger, triple-melted VAR ingots insured improved chemical homogeneity. The extrusion and small-scale forging studies were conducted on the VIM ingots, the large-scale forgings studies were conducted on the 203 mm diameter VAR ingot. VIM ingots were hot isostatically pressed at 1200°C and 105 MPa for 3 hours and homogenized at 1175°C for 24 hours prior to machining the extrusion or forging multiples. The larger VAR ingots were hot isostatically pressed at 1175°C and 105 MPa for 4 hours. In addition, the forging multiples for the large scale forgings were homogenized for 10 hours at 1300°C prior to forging

Extrusion

The extrusion studies were conducted using the 8.9 MN press at Wright-Patterson AFB. Extrusions were produced using a variety of extrusion ratios from 6:1 to 20:1 at a ram speed of ~13 mm/s on billets heated to 1315°C. With the exception of one extrusion trial of a bare billet, all billets were encapsulated in a titanium alloy can (either CP or Ti-6Al-4V). A "standard" can configuration was used in which the gamma billet was surrounded by a can of 5 mm wall thickness. The actual gamma alloy billet was approximately 63 mm diameter by 130 mm long, and the overall extrusion billet diameter was 73 mm. Extrusion pressures were generally a function of extrusion ratio; the maximum extrusion pressure was ~1000 MPa. Material processed at three different extrusion ratios (6:1, 9:1, and 20:1) was subjected to fairly extensive heat-treatment and mechanical property evaluation. Results of the heat treatment and mechanical property evaluation of material extruded at the highest ratio will be summarized in the following sections.

Forging

Isothermal, small-scale and large-scale forging trials were conducted. The small-scale forging trials resulted in forging dimensions of 140 mm diameter by 12.7 mm thick. The large-scale forging trials resulted in forging dimensions of ~485 mm diameter by 30 mm inch thick. Prior to these forging trials, a determination of flow-stress as a function of temperature and strain-rate for cast and hot isostatically pressed Ti-48Al-2Cr-2Nb was made using compression specimens 5 mm diameter x 10 mm high. The material used for these tests had been heat treated to produce a roughly equal volume fraction of primary gamma and transformed alpha (lath gamma and alpha-2) as will be discussed in the microstructure section. The flow-stress as a function of temperature and strain-rate is shown in Figure 1.

The flow stress for Ti-48Al-2Cr-2Nb has a strong temperature and strain-rate dependence. To achieve adequate forging reductions, forging temperatures need to be relatively high and forging strain-rates need to be relatively low. This fact is illustrated by the three bars on the plot that occur at the lowest strain-rate and relatively high temperatures. These three bars represent the forging conditions for the three forgings described in the following paragraphs.

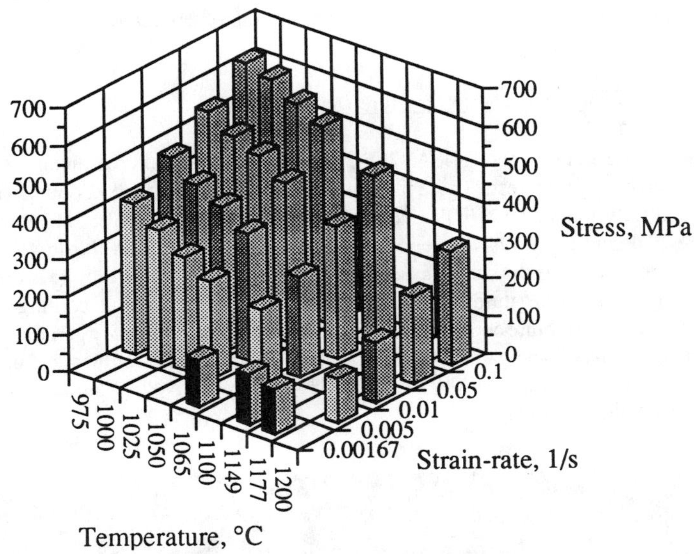

Figure 1. Flow stress as a function of temperature and strain-rate for Ti-48Al-2Cr-2Nb.

Large scale forgings on production, isothermal presses are likely to be limited to this portion of the temperature/strain-rate envelope by press capacity and die material.

Small-Scale Forgings - The small scale forgings were made on the Wyman-Gordon, 1.8 MN isothermal press. Specimens 70 mm diameter by 51 mm high were forged to ~140 mm diameter by 12.7 mm high at 1175°C and a strain-rate of $1.67 \times 10^{-3}\ s^{-1}$. Due to the limited capacity of the press, a number of the specimens failed to reach the desired minimum thickness of 12.7 mm. The forgings were subsequently subjected to several heat treatment schedules selected to produce a range of microstructures. In general, the mechanical properties for these forgings when compared to an equivalent microstructure in the extrusions, showed only a small decrease in strength and little change in ductility. In a set of tests designed to evaluate directionality in the plane of the forging, it was found that the properties in the radial and tangential directions were essentially equivalent.

Large-Scale Forgings - The large scale forgings were made on the Wyman-Gordon, 71.2 MN press. Two forging trials were conducted on this press, one at 1149°C and a second at 1066°C. The forging trial at 1149°C involved two forging steps at a strain-rate of $1.67 \times 10^{-3}\ s^{-1}$. In the first step, a 184 mm diameter by 229 mm high multiple was forged from 229 mm to 76 mm and in the second from 76 mm to 30 mm. The forging was removed from the press and conditioned after the first step. The final diameter of the forging was 483 mm. Material from this forging was subsequently heat treated to a duplex microstructure and tested; the microstructure and tensile results are discussed in a following section. The forging trial at 1066°C was conducted in an attempt to get work into the material at a lower temperature and thereby obtain a finer recrystallized grain size upon subsequent heat treatment. The forging involved a single step from 184 mm diameter by 114 mm high to 304 diameter by 43 mm high at a strain-rate of $1.67 \times 10^{-3}\ s^{-1}$. Attempts to recrystallize the microstructure of this forging at sub-transus temperatures met with little success in that the recrystallized microstructure was very non-uniform and comprised areas of fine, recrystallized grains and large regions of highly worked, lamellar microstructure. Heat treatment above the alpha transus resulted in a fully transformed structure with large grain size.

Microstructure Development

Early in the development cycle for Ti-48Al-2Cr-2Nb at GEAE, three microstructural conditions were identified as being of interest. These previously have been referred to as duplex (DP), sub-transus (ST), and fully-transformed (FT). Details of the heat treatments utilized to obtain two of these structures, DP and FT are have been reported by Shih, et al. [1]. A third microstructure studied has been referred to as the sub-transus microstructure. This microstructure is obtained by holding the alloy at a temperature slightly below the alpha transus and cooling from that temperature. That microstructure comprises a small volume fraction of primary gamma and a large fraction of lamellar (alpha-2 and gamma) structure. The heat treatments for the three microstructures were conducted for 2 hours at 1300, 1400, and 1350°C, respectively. Examples of the three microstructures are shown in Fig. 2.

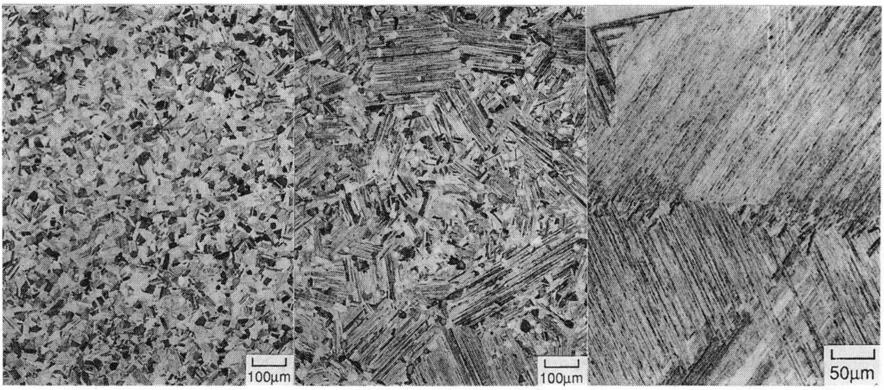

(a) (b) (c)

Figure 2. Representative microstructures of extruded Ti-48Al-2Cr-2Nb. (a) Duplex, (b) Sub-Transus, and (c) Fully-Transformed.

Mechanical Property Characterization

Extensive mechanical property characterization of the numerous microstructural conditions of both extrusions and forgings has been conducted. In the following section, the tensile properties for a high extrusion ratio (20:1) extrusion will be compared with data from one of the large scale forgings.

Extrusion

The material used for this comparison was extruded at an extrusion ratio of 20:1 as discussed previously. Tensile and creep properties for the alloy in both a duplex and fully-transformed microstructural condition are shown in Figures. 3 and 4.

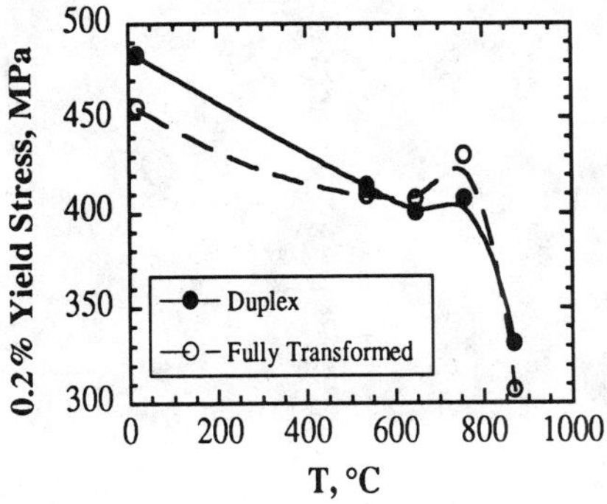

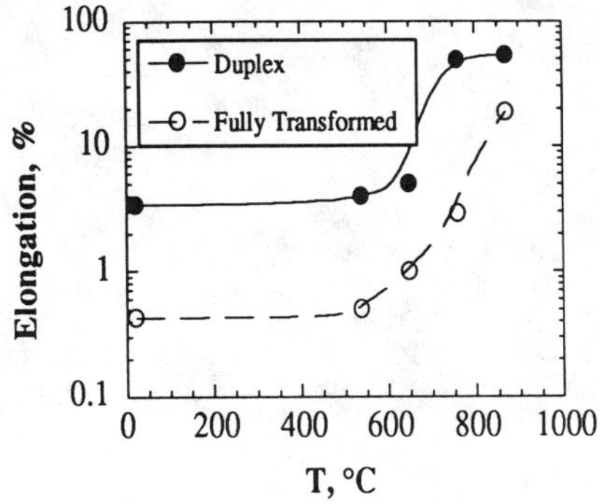

Figure 3. Tensile properties as a function of temperature for extruded Ti-48Al-2Cr-2Nb

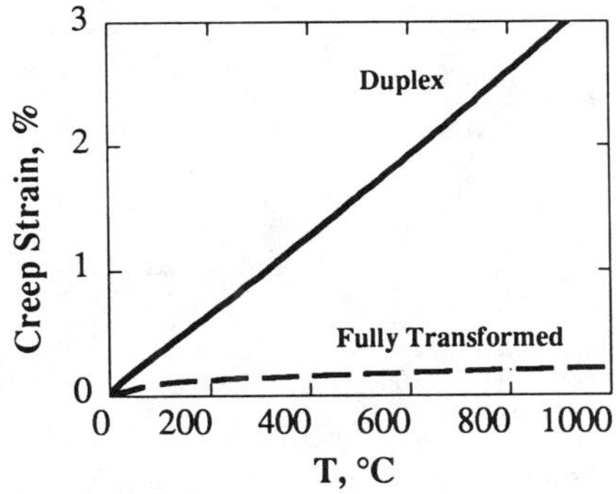

Figure 4. Creep of extruded Ti-48Al-2Cr-2Nb at 760°C and 103 MPa.

It can be seen from Figures 3 and 4 that the duplex microstructure provides a good combination of strength and ductility, but does so at the expense of creep strength. The sub-transus microstructure provides a compromise between the duplex and fully transformed microstructures, in that the creep strength is intermediate to that of the duplex and fully transformed microstructures and the tensile ductility is generally greater than 2 percent.

Large-Scale Forging

The large scale forging was heat treated (10 hours at 1300°C in argon, and furnace cooled) to a microstructural condition equivalent to the duplex microstructure in the extrusion. The microstructures for both are shown in Figure 5. The mechanical properties for the two processing conditions are shown in Figure 6.

(a) (b)

Figure 5. Duplex microstructure of Ti-48Al-2Cr-2Nb. (a) Extruded, (b) Forged

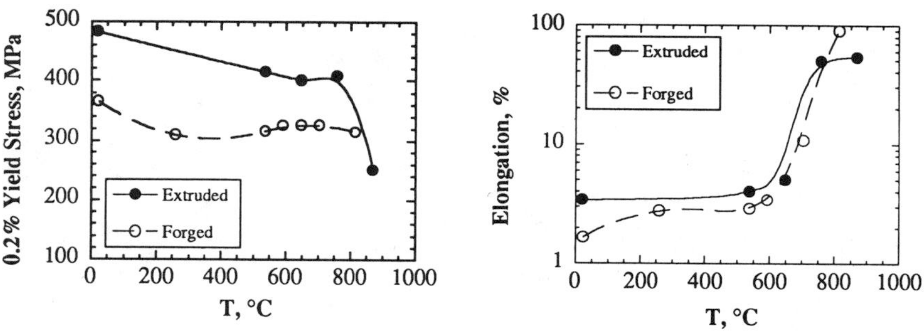

Figure 6. Comparison of the tensile properties of extruded and forged Ti-48Al-2Cr-2Nb with the duplex structure

Discussion and Conclusions

A range of microstructural conditions can be produced in both extruded and forged Ti-48Al-2Cr-2Nb by heat treatment subsequent to the working operation. Of the microstructures produced in this material, the duplex structure with approximately equal volume fractions of primary gamma and transformed microstructure, Fig. 5 (a), provides the best balance of properties. Microstructures that are fully lamellar, or which contain a small volume fraction of primary gamma, appear to have improved creep resistance, but at the expense of tensile ductility. The difference in the mechanical properties of the extruded and the forged material when heat treated to a duplex microstructure does not appear to be a function of grain size, but may be related to one or more of several factors. These factors could include: 1) the degree of deformation imparted during the processing step, 2) specimen size, and/or 3) microstructural and crystallographic texture. The 20:1 extrusion ratio yielded a true strain of 3 while the forging was conducted in two steps each of which yielded a true strain of ~ 1 (or a total of 2 with an intermediate heating to forging temperature). While this difference might have been expected to provide a significantly different grain size, it appears not to have and, therefore, cannot be directly correlated with the difference in strength and ductility.

The difference in tensile ductility (plastic elongation) may be related to specimen size in that the material from the extrusion was tested using a specimen with a 4.1 mm gage diameter while the material from the forging was tested using a specimen with a 6.4 mm gage diameter. It has been reported in tensile testing of materials of low ductility, that apparent ductility often decreased as the size of the test bars is increased. Finally, although the texture (either microstructural or crystallographic) of the two product forms was not characterized, the possibility of either form of texturing providing an increase in the tensile properties of the extrusion should be considered. In conclusion, it should be noted that the heat treatment for the forging was not optimized, but was based on an extensive heat treatment study on the extruded material the goal of which was to provide the best balance of properties in a duplex microstructure. The need for optimization of microstructural development in forged near gamma alloys exists and is being pursued in both in-house research and contract funded activities at GEAE and other engine and airframe contractors.

Acknowledgments

The use of the extrusion press at Wright-Patterson AFB and the technical assistance of Mr. J. Brown of UES in the extrusion work is greatly appreciated. The assistance of Dr. T. Howson of Wyman-Gordon during both the small-scale and large-scale forging studies contributed significantly to the success of those studies. The authors are grateful to Mr. R. Grant and Ms. K. Schumacher for their experimental assistance. We also wish to thank Dr. L. Johnson and Dr. H. Lipsitt for valuable discussions.

This work was supported in part by GE Aircraft Engines Independent Research and Development program and in part by Army Contract DAAJ02-89-C-0030.

References

1. D. S. Shih, et al., "The Microstructural Dependence of Mechanical Properties of Ti-48Al-2Cr-2Nb", Microstructure/Property Relationships in Titanium Aluminides and Alloys, eds. Y-W. Kim and R. R. Boyer, (Warrendale, PA: The Minerals, Metals & Material Society, 1991) 135-148.

The Influence of Forging on the Microstructure of Plasma-melted γ-based Titanium Aluminides

X. D. Zhang, R. V. Ramanujan, T. A. Dean, R. Dewes,
M. H. Jacobs and M. L. H. Wise

IRC in Materials for High Performance Applications, The University of Birmingham,
Edgbaston, B15 2TT, UK

Abstract

The influence of forging conditions on the workability, stress-strain relation and microstructures of plasma-melted γ-TiAl based intermetallics have been studied and sound forgings have been produced using traditional hot forging processes. The good workability of alloys of nominal composition Ti-48Al-2Mn-2Nb and Ti-48Al (all compositions in atomic percent) was resulted when a high forging temperature (1350 °C) and a high die temperature (750 °C) were used. The stress-strain relationship largely depended on billet temperature. The lamellar structure of Ti-48Al-2Mn-2Nb alloy was very stable and existed at all forging temperatures.

Introduction

Titanium aluminides based on γ-TiAl are being considered as potential aerospace and automotive materials for high temperature applications. Recent interest in these alloys has been concentrated on the two-phase γ alloys which are composed of the γ-TiAl phase and a small volume fraction of the α_2-Ti_3Al phase since better mechanical properties (such as improved ductility, room temperature fracture toughness and good creep resistance) can be obtained over the single phase γ-TiAl alloys [1-3]. Thermomechanical processing, including HIPping, forging and subsequent heat treatment is critically important to optimise microstructure and mechanical properties of these alloys [4]. In forging, the quality of the products (e.g., soundness) and microstructure are very sensitive to processing parameters such as temperature and strain rate. Since the mechanical properties of γ-based alloys are strongly dependent on their microstructure, it is important that this is optimised by the appropriate thermomechanical treatment [5].

The aim of this paper is to present primary experimental results on the hot workability of two-phase γ-TiAl alloys using a traditional forging process at higher billet temperatures and strain rates than those currently used in isothermal forging for conventional titanium alloys.

Experimental Procedure

Alloys of nominal composition Ti-48Al and Ti-48Al-2Mn-2Nb and actual composition Ti-49Al and Ti-50Al-2Mn-2Nb, weighing about 30 kg, were cast in a Retech plasma arc melting furnace as cylindrical ingots with a diameter of 10 cm. The oxygen contents for

Titanium '92
Science and Technology
Edited by F.H. Froes and I. Caplan
The Minerals, Metals & Materials Society, 1993

both alloys ranged from 650 wt ppm to 700 wt ppm. Cylindrical billets for forging, of size Ø19 x 25 mm, Ø35 x 35 mm and Ø35 x 50 mm, were cut from the cast ingots (using wire electrical discharge machining) with their axes orientated parallel to that of the original ingot .

Forging was carried out on a hydraulic press with a load capacity of 150 ton. Three sets of flat forging tools made from alloys with different temperature capabilities were used: a 5% Cr hot-work tool steel (BS BH13) which could be heated to 200 °C; a Ni 718 alloy which could be heated to 600 °C and an IN100 alloy which could be heated to 1000 °C. Forging load was measured by a strain-gauged load cell situated in the tool stack and tool displacement measured by an induction transducer placed between the press slide and bed. Outputs from these instruments were recorded simultaneously against time on an oscilloscope and photographed. Before heating the billets were coated with a thick layer of glaze and wrapped in a layer of ceramic wool and the working surfaces of the dies were coated with a thick layer of colloidal graphite. The cold billets were then placed in an electric resistance muffle furnace and heated in air to the required temperature for 30 minutes while the dies were raised to their working temperature in a separate furnace. When the dies were at their working temperature they were removed from the furnace and positioned on the press; the billet was then removed from the furnace, placed between the dies and forged to a chosen reduction, the whole sequence of events being undertaken as quickly as possible.

The area of a forging at any stage of deformation was calculated by dividing its volume by its instantaneous height, obtained from the ram displacement. Stress was calculated by dividing the instantaneous load by the instantaneous calculated area. Ram velocity was calculated by measuring the shape of the recorded displacement-time relation and strain rate computed by dividing the ram velocity by the instantaneous height of the workpiece.

Results and Discussion

Workability

The forging tests were undertaken using different conditions. The billet samples after forging are shown in Fig.1 and Fig.2 with ram speed 1~ 2 mm/sec. Both the temperature during forging and the cooling rates after deformation appear to be critical parameters in controlling the quality of forged materials. The temperature during forging has been shown to be influenced by several factors including: the surface area of contact with the die, the contact time, the billet volume, the die temperature, the amount of heat generated by deformation and the friction between die and billet [6].

Three initial die temperatures were used and it was found that increasing die temperature improved the quality of the forged billets by reducing the formation of surface cracks, as shown in Fig. 1. Chilling of the billets and low forging temperatures produced poor quality forgings, so billet temperatures ranging from 1150 °C to 1350 °C were used. Fig. 2 shows billets forged at different temperatures from which it is obvious that forging at 1350 °C can give a sound sample only when the die temperature is 750°C.

Since surface cracks also developed during air cooling, the forgings were furnace cooled from 900 °C off the press. Nonuniform deformation due to the formation of a dead zone between the workpiece and die surface was apparent in all forged billets (Fig. 3). This was caused by constraints at the billet-die interfaces due to friction and heat loss.

Stress-Strain Relations

Figs. 4 and 5 show the effect of forging temperatures on stress-strain and strain-strain rate behaviour derived during forging for Ti-48Al-2Mn-2Nb and Ti-48Al alloys, respectively, using die temperatures of between 150 °C and 200 °C. Due to the characteristics of the hydraulic press the initial strain rate for each forging temperature varied slightly and also changed throughout the deformation. It can be seen that flow stress increased rapidly to a maximum then remained almost constant with increasing strain, but the forging carried out at 1250 °C exhibited an increase in flow stress at strains above about 0.7. This behaviour can be explained in terms of the balance of heat generated and lost within the workpieces. At the higher forging temperatures, the heat resulting from a combination of plastic deformation, friction between the billet and die and dynamic recrystallization probably balanced the heat loss during forging once a certain strain value had been exceeded. In the case of the 1200 °C forging, the combination of lower initial temperature and chilling of the interfacial region may have caused additional frictional losses. Similar behaviour was found with the Ti-48Al alloys.

At the higher preheat temperatures, the initial forging temperature was always greater than 1100 °C (measured by optical pyrometer) and only dropped to about 950 °C at the end of deformation, a range of temperature which would include that typically used in isothermal forging. Since there would have been little heat loss in the initial forging stage (i.e. strain less than about 0.2) the maximum stress values, shown in Table 1 for different forging conditions, can be compared with the data obtained from isothermal forging by Nobuki et al.[4,5]. For example, their values of maximum stress obtained when isothermally forging at 1210 °C, at a strain rate of 0.1 sec^{-1}, were 231 MPa for Ti-47Al-2V and 250 MPa for Ti-48.2Al, respectively. Our values were similar; i.e. 205 MPa and 201 MPa for Ti-48Al-2Mn-2Nb and Ti-48Al, respectively, forged at 1250 °C at strain rates between 0.05 sec^{-1} and 0.5 sec^{-1}, despite the fact that our forging was carried out with decreasing billet temperature and varying strain rate. From their extensive work on flow stress and deformability maps, Nobuki et al. predicted that these materials could suffer brittle fracture when forged conventionally (1000°C to 1200°C, strain rate ~50 sec^{-1}). In the present study this problem was overcome by increasing the temperatures of both billet (1200 °C to 1350 °C) and die (400 °C to 750 °C) and by using a moderate strain rate of 0.05 sec^{-1} to 0.5 sec^{-1} on a hydraulic press. The significance of this achievement is that traditional hot forging has potential for TiAl intermetallics.

Microstructures of As-cast and As-forged Samples

The as-cast Ti-48Al and Ti-48Al-2Mn-2Nb alloys had predominantly lamellar microstructures with a few equiaxed γ grains formed around the lamellar grain boundaries near the centres of the ingots. In the Ti-48Al-2Mn-2Nb alloy, the microstructure was more heterogeneous than that of the binary alloy and a duplex microstructure (lamellae and equiaxed γ grains) was apparent at the centre of the ingot (Fig. 6). SEM EDX analysis from the edge to the centre of the ingot did not reveal any significant variation in composition, indicating that this nonuniform microstructure should have been caused by the solidification rate differences.

The microstructure of as-forged Ti-48Al-2Mn-2Nb (billet temperature 1350 °C, die temperature 150 °C to 250 °C and 60% reduction) consisted of networks of fine recrystallised grains and lamellae (Fig. 7). The lamellae with original orientations parallel or almost parallel to the compression axis were rotated through 90°. Similar microstructures were found in all other billets, irrespective of forging temperature; this was due to the high stability of the lamellar structure over the range of temperatures used in the present study. Although some recrystallized grains were observed, large-scale dynamic recrystallisation typically reported in isothermal forging appeared to have been restricted by high strain rates used for the hot-die forging.

Conclusions

1. Sound forgings have been achieved using a conventional hot forging process at medium strain rate (0.05 sec^{-1} to 0.5 sec^{-1}); increasing both die and billet temperatures improved the quality of the forged billets.

2. Flow stress increased with decrease in forging temperature; maximum flow stress values obtained from measurements made during conventional hot forging were comparable with those obtained from isothermal forging.

3. The microstructure of forged Ti-48Al-2Mn-2Nb consisted of networks of fine recrystallised grains and lamellae.

Acknowledgements

The financial support of SERC through the grant to the IRC is gratefully acknowledged.

References

1. M. J. Blackburn and M. P. Smith, , Rep. AFWAL-TR-79-4056, 1979 (U.S. Air Force Wright Aeronautical Laboratories).

2. S. Mitao, S. Tsuyama and K. Minakawa, Mater. Sci. Eng., **A143**, 1991, 51-62

3. Y-W. Kim and D. M. Dimiduk, JOM, August, 1991, 40-47

4. M. Nobuki, K. Hashimoto, J. Takahashi and T. Tsujimoto, Mater. Trans., JIM, **31**, No. 9, 1990, 814-819

5. M. Nobuki and T. Tsujimoto, ISIJ International, **31**, No. 8, 1991, 931-937

6. Metals Handbook, 9th Edition, vol. 14, " Forming and Forging ", 1988, 36

(a) (b)

Figure 1 Ti-48Al samples forged at 1250°C, a) die temperature 400°C and 70% reduction, b) die temperature 750°C and 80% reduction (viewed along forging direction).

(a) (b)

Figure 2 Ti-48Al samples forged at 1350°C, a) die temperature 400°C and 75% reduction, b) die temperature 750°C and 80% reduction (viewed along forging direction).

Figure 3 Macrostructure of forged Ti-48Al sample (billlet temperature 1250°C, die temperature 400°C and 70% reduction.

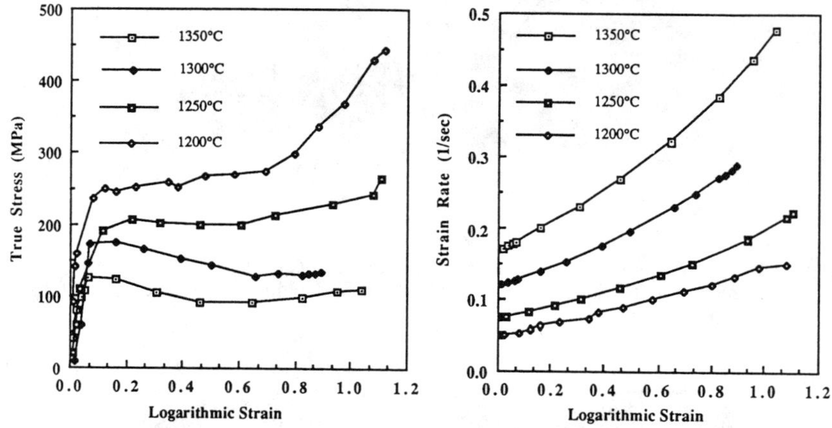

(a) (b)

Figure 4 Flow stress - strain and strain - strain rate curves of Ti-48Al-2Mn-2Nb alloys forged at die temperature 150°C.

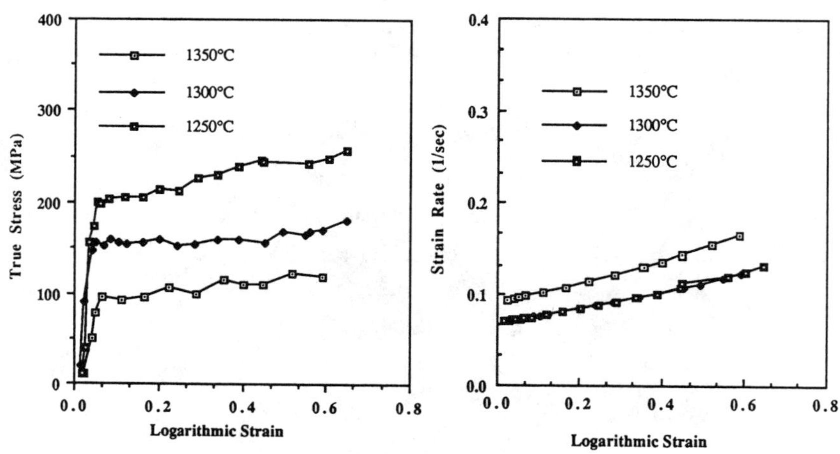

(a) (b)

Figure 5 Flow stress - strain and strain rate curves of Ti-48Al alloys forged at die temperature 150°C.

Table 1. Maximum stress (at initial period, strain < 0.2) on forging tests (die temperature 150°C to 200°C) and total forged strain of γ - TiAl alloys

	Ti-48Al		Ti-48Al-2Mn-2Nb	
Temp.	Stress (MPa)	Strain	Stress (MPa)	Strain
1200°C	—	—	258	1.1
1250°C	201	0.6	205	1.1
1300°C	159	0.6	174	0.9
1350°C	95	0.6	124	1.0

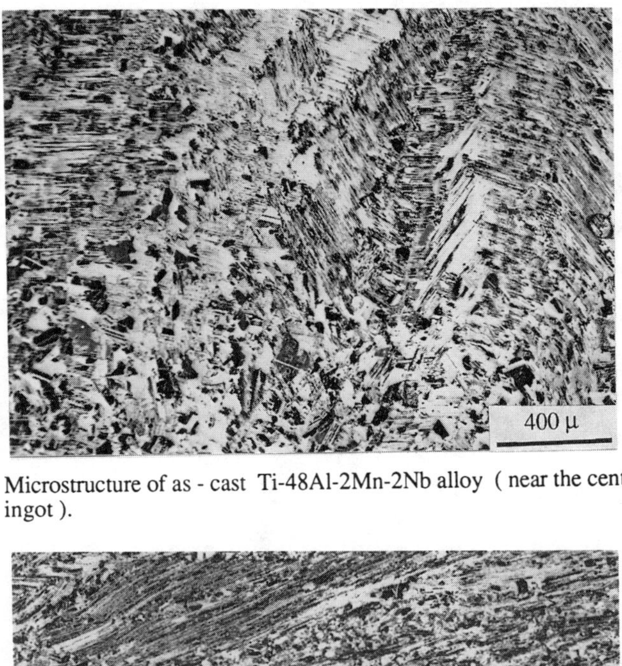

Figure 6 Microstructure of as - cast Ti-48Al-2Mn-2Nb alloy (near the centre of the ingot).

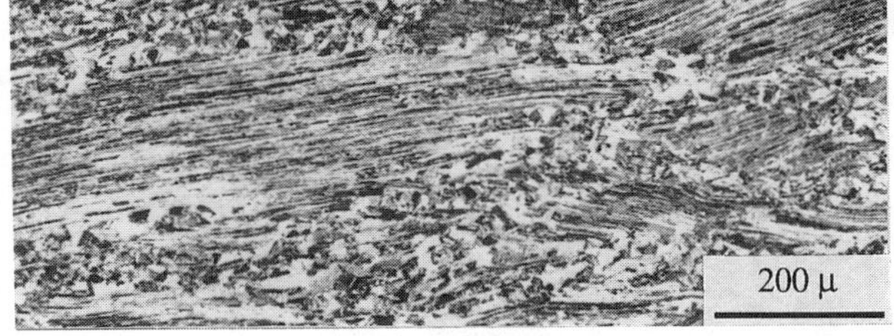

Figure 7 Microstructure of as - forged Ti-48Al-2Mn-2Nb at die temperature 150°C, billet temperature 1350°C and 60% reduction.

GAS TUNGSTEN-ARC WELDING OF A Ti-48Al-2Cr-2Nb

GAMMA TITANIUM ALUMINIDE

L. C. Mallory,* W. A. Baeslack III,* D. Phillips,** T.J. Kelly**

*Department of Welding Engineering
The Ohio State University
Columbus, OH 43210 USA

** GE Aircraft Engines,
Cincinnati, OH 45212 USA

Abstract

Ti-48Al-2Cr-2Nb (at%) is a second-generation gamma titanium aluminide which has recently been the subject of appreciable investigation both in the wrought and in the cast product forms. The effective application of Ti-48Al-2Cr-2Nb and future generation gamma titanium aluminides in structural gas turbine engine applications will require joining both to themselves and to other titanium-based materials. Gas tungsten arc (GTA) welding with preheat is one suitable process for joining gamma alloys. The purpose of the present study was to investigate evolution of weld fusion zone and heat-affected zone (HAZ) microstructures in preheated GTA welds of Ti-48Al-2Cr-2Nb and to develop initial relationships between these structures and mechanical behavior. This work was accomplished through the detailed characterization of both simulated and actual GTA welds. The microstructures of both the fusion zone and near-HAZ were primarily comprised of $\gamma + \alpha_2$ lamellae. Fusion zone microstructures of GTA welds differed from those of the near-HAZ due to solidification behavior and microsegregation. Weld HAZ simulation studies performed on a Gleeble® 1500 system showed a significant influence of weld heating and cooling rates and peak temperature on the microstructure and hardness, especially for high peak temperatures within the alpha phase field.

Introduction

Alloys based on the gamma titanium aluminide TiAl are potentially valuable for structural gas turbine engine applications because of their low density and good elevated-temperature tensile properties. The effective use of gamma alloys in structural gas turbine engine applications will require their joining, both to themselves and to other titanium based materials. In addition, production of gamma-alloy castings becomes more cost effective when weld repair may be performed as necessary. A previous investigation of welding gamma alloys has shown a principal limitation to gamma titanium aluminide weldability to be solid-state cracking associated with the alloy's low room temperature ductility [1]. Patterson and Damkroger showed that cracking in electron beam welds of Ti-48at%Al could be controlled by adjusting welding parameters to reduce the cooling rates in fusion zone and HAZ to 100 °C/s. Similarly, for a composite material Ti-48at%Al-6.5vol%TiB_2, cracking was avoided by using a preheat of 335 °C to 470 °C to reduce the cooling rate to 300 °C/s. Decreasing the weld cooling rate decreases the magnitude of thermal contraction stresses on cooling. The weld cooling rate may also control weldment mechanical properties and cracking tendency by affecting the microstructures of the fusion zone and HAZ.

Since GTA welding with preheat may thus be an appropriate process for joining gamma alloys, the microstructures of gamma alloy welds must be characterized and the effects of thermal cycles on gamma alloys should be understood. In this study, crack-free GTA welds in Ti-48at%Al-2at%Cr-2at%Nb were produced using preheat and were characterized using standard techniques.

Titanium '92
Science and Technology
Edited by F.H. Froes and I. Caplan
The Minerals, Metals & Materials Society, 1993

Gleeble® 1500 thermal simulation was used to study the effects of peak temperature and cooling rate on HAZ microstructure.

Experimental Procedures

The material used in this study was nominally Ti-48at%Al-2at%Cr-2at%Nb, provided in the form of 2.5 cm (1 in) thick forged pancakes. Forging was performed at 1165 °C, followed by a 4 hour heat treatment at 1300 °C, furnace cooling to 870 °C for 8 hours, and furnace cooling to room temperature. The actual chemical composition of the material studied was 47.4at%Ti-48.7at%Al-2.0at%Cr-1.9at%Nb, slightly richer in aluminum than the nominal 48%, with 540 ppm/wt oxygen. The alpha transus temperature increases with aluminum content for gamma alloys with 2%Cr, 2%Nb, and aluminum contents between 45.9%Al and 48.1%Al [2]. Extrapolation from this data predicted an alpha transus of 1358 °C for Ti-48.7Al-2Cr-1.9Nb.

Welding was performed using the manual gas tungsten-arc (GTA) process on coupons 1.7 mm (0.067 in) thick electro-discharge machined from the pancake, using filler wire of matching composition. Welding current was 60 A, welding voltage was 12 V, and welding speed was 1.3 - 1.7 mm/s (3 - 4 in/min). Preheating was effected by means of quartz lamps directed at an argon-filled chamber, and the chamber temperature was measured by a thermocouple attached to the weld specimen. Using these preheating lamps, the specimen was heated to 800 °C before welding. After the arc was extinguished, the heating lamps were left on for 5 min, allowing slow cooling to 800 °C. The lamps were then shut off and the welded specimen cooled slowly to room temperature. Inspection for surface defects was carried out by fluorescent particle examination.

Heat-affected zone thermal simulation and other controlled thermal cycling were performed on specimens 2.0 mm (0.080 in) thick in a Gleeble® 1500 thermo-mechanical system. Thermal cycles experienced in the HAZ of partial-penetration GTA welds in gamma alloys were calculated using an analytical solution to the conduction heat transfer equation, and thermal properties measured for a Ti-48Al-1V-0.1C gamma alloy [3]. For a peak temperature of 1400 °C and welding conditions similar to those used in the preheated welds of this study, the cooling rate from 1400 °C to 1200 °C was calculated to be 100 °C/s. Figure 1 shows the calculated time-temperature history for a peak temperature of 1400 °C. Thermal cycles with a range of peak temperatures were calculated, in order to investigate the microstructural variation across the near HAZ. These thermal cycles were performed on specimens of the wrought Ti-48Al-2Cr-2Nb under vacuum, in the Gleeble® 1500. As well, comparison was made with specimens heated to 1400 °C, held 30 s, and cooled from 1400 °C at slower rates of 10 °C/s and 1 °C/s.

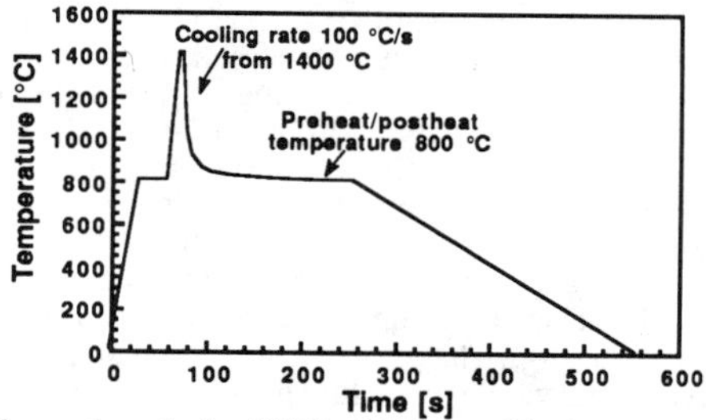

Figure 1: Calculated thermal cycle for HAZ of GTA weld of gamma alloy used for thermal simulation.

Metallographic sections were cold-mounted in epoxy, ground to 2400 grit using silicon carbide paper, and polished first with 1 μm diamond or 0.3 μm alumina and then with neutralized colloidal silica. Etching was performed using Kroll's reagent (3 ml HF + 6 ml HNO_3 +100 ml H_2O). Metallographic characterization was done using light microscopy on etched specimens, optical polarized light microscopy (OPL) on unetched specimens, scanning-electron microscopy using backscattered electron imaging (BSE), energy-dispersive X-ray spectroscopy (EDS), transmission-electron microscopy (TEM), and DPH microhardness measurement.

Results and Discussion

Base Metal Microstructure

The base metal structure of this forged Ti-48Al-2Cr-2Nb alloy was primarily equiaxed gamma, with α_2 present at grain boundaries (Figure 2). The α_2 was not visible under OPL, but may be seen in backscattered electron images as a light-imaged grain boundary phase. Figure 2b also reveals the presence of some fine lamellae. The base metal contained bands of large twinned gamma grains (100 - 200 μm in diameter) and bands of finer grains. This banded structure was similar to the near-gamma (NG) structure described by Kim for forged Ti-47Al-1Cr-1V-2.5Nb after heat treating low in the $\alpha + \gamma$ field (1210 °C) for 2 h [4]. He observed regions of coarse (100 μm) gamma grains, and bands of fine (3 - 15 μm) gamma grains with α_2 particles (1 - 4 μm diameter) pinning the gamma grain boundaries.

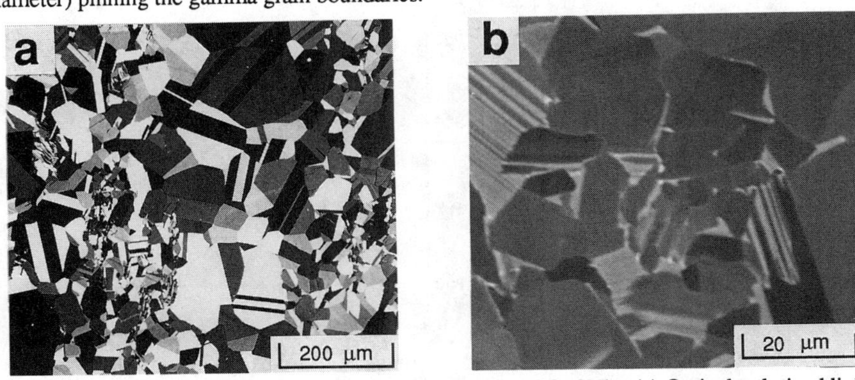

Figure 2: Base metal microstructure of wrought Ti-48Al-2Cr-2Nb: (a) Optical polarized light micrograph; (b) Backscattered electron image (BSE), showing presence of α_2 at grain boundaries.

GTA Weld Microstructure

In the preheated GTA welds of this study, no cracking was detected in fluorescent-dye penetrant examination or in metallographic sections. Interestingly, weld beads produced under identical welding conditions but without preheat all contained cracks. Macroscopic examination of preheated GTA welds showed the near-gamma structure of the base metal and the fine transformed structures of near HAZ and fusion zone (Figure 3a). In a light macrograph of a section etched with Kroll's reagent, the fusion boundary (indicated with an arrow) was readily distinguished by the change in etching contrast. The transformed HAZ (between the two arrows) was approximately 0.3 mm wide at the bottom of the weld and as narrow as 0.12 mm wide at the side.

In the near-HAZ of preheated GTA welds, the equiaxed gamma grains had gradually transformed to a fine $\gamma + \alpha_2$ structure with mixed lamellar and acicular morphologies and some small equiaxed gamma grains (Figure 4). At the outer edge of the transformed HAZ may be distinguished some partial transformation as evidence of a far-HAZ. The fusion zone was shown by OPL to have a similar transformed structure to the HAZ, but the morphology appeared more fully lamellar, with larger grains. Near the fusion boundary, in Figure 4a, some light etched regions free of lath structure had a shape characteristic of interdendritic regions. Occasional evidence of small gamma grains was seen in the fusion zone. McQuay *et al.* described three morphologies of $\gamma + \alpha_2$ lamellae, classic lamellar (one orientation per grain), Widmanstatten-like plates (several orientations per prior-α grain), and an acicular morphology (no apparent orientation relationship with prior alpha grains) [5]. A higher cooling rate or a higher Cr content reduced the fraction of classic lamellar morphology. The HAZ and fusion zone structures observed in the present work were consistent with observations from Ti-48Al-2Cr and Ti-48Al-

3Cr held at 1400 °C for 15 min and furnace cooled at 100 °C/s, about the same rate as that of HAZ.

Further examination of the fusion zone by scanning-electron microscopy with backscattered-electron imaging revealed a dendritic solidification structure, with evidence of interdendritic gamma (Figure 5). In this backscattered electron image, contrast was indicative of a change in composition. Thus, Figure 5 shows the solidification segregation which occurred from the white dendrite core to the black interdendritic regions. Under close inspection (see arrow in Figure 5b), a discontinuity in contrast was seen in many dendrites. This would be consistent with a peritectic reaction. Superimposed on the solidification structure, the $\gamma + \alpha_2$ lamellae may be seen.

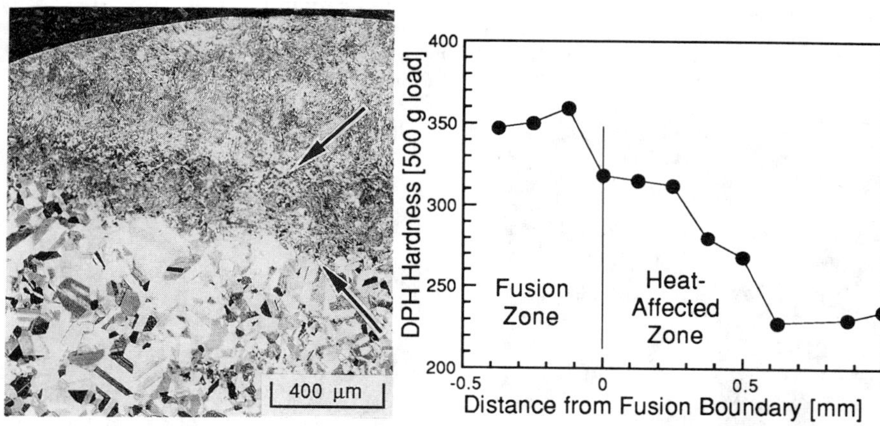

Figure 3: Partial-penetration GTA weld bead in wrought Ti-48Al-2Cr-2Nb: (a) Macro-photograph of section etched with Kroll's reagent; (b) Microhardness traverse.

Analysis of the compositional segregation by EDS showed that the first-to-solidify dendrite cores, shown on the BSE micrograph as thin white regions, were enriched in Cr and depleted in Al, relative to the overall composition, and the interdendritic regions which solidified last were enriched in Al relative to the overall composition. Consistent with the observations of Shih *et al.*, negligible segregation of Nb was observed [6].

Recent revisions to the understanding of the binary Ti-Al equilibrium diagram give three possible primary solidification phases for near equiatomic alloys: β, α, and γ. Two peritectic reactions occur: $\beta + L \rightarrow \alpha$ and $\alpha + L \rightarrow \gamma$ [7]. The primary solidification phase has been identified as BCC beta for all binary alloys with less than 49%Al [5, 7]. When the primary solidification phase was β, a microstructure of fully lamellar $\gamma + \alpha_2$, possibly with interdendritic gamma, was seen. No microstructural evidence of the $\beta + L \rightarrow \alpha$ peritectic reaction was observed by McCullough *et al.*, due to the subsequent solid-state transformations. For binary alloys with between 49% and 55% Al, α was the primary solidification phase. The commonly observed room temperature microstructure was $\gamma + \alpha_2$ laths surrounded by cellular gamma, believed to form in the solid state, with interdendritic gamma solidified from enriched liquid. The sequence of solidification reactions has not yet been identified for Ti-48Al-2Cr-2Nb, although Shih *et al.* observed about 1% β present in fully lamellar Ti-48Al-2Cr-2Nb and suggested that β might thus be the primary solidification phase [6]. BSE and EDS observations of the current work indicated the presence of interdendritic gamma, and showed patterns of compositional variation consistent with some off-peritectic solidification sequence. The precise nature of the fusion zone solidification sequence and its influence on subsequent on-cooling phase transformations is a subject of continuing investigation.

Transmission-electron microscopy of fusion zone confirmed the presence of $\gamma + \alpha_2$ structures in classic lamellar morphology and possibly in acicular morphology as well as some fine gamma grains (Figure 6a). Transmission electron micrographs of the near (transformed) HAZ also showed $\gamma + \alpha_2$ lamellae and fine gamma grains (Figure 6b).

Figure 4: Microstructure of GTA weld in wrought Ti-48Al-2Cr-2Nb: (a) Light micrograph of section etched with Kroll's reagent; (b) Optical polarized light (OPL) micrograph of unetched section.

Microhardness of fusion zone and transformed HAZ was increased relative to the base metal microhardness of approximately 230 DPH, as shown in the microhardness traverse down the weld centerline of Figure 3b. The fine lamellar and acicular structures of near HAZ (315 DPH) were harder than the near-gamma structure of unaffected base metal, while the fusion zone hardness exceeded 340 DPH.

Characterization of the near-HAZ by TEM is difficult because one cannot easily correlate microstructural observations on a particular thin foil sample such as Figure 6b, with a known thermal cycle. The gradient of peak temperature is high in the HAZ and the resulting microstructure is strongly influenced by peak temperature. For this reason, in order to learn more about the microstructures produced in wrought Ti-48Al-2Cr-2Nb by HAZ thermal cycles, thermal simulation was performed using a Gleeble® 1500.

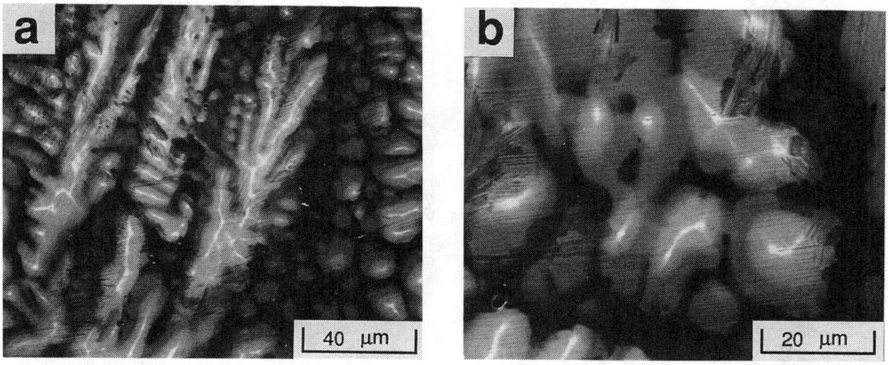

Figure 5: Backscattered electron images (BSE) of GTA weld fusion zone showing solidification segregation.

Figure 6: Transmission electron microscope images of GTA weld in wrought Ti-48Al-2Cr-2Nb: (a) Fusion zone; (b) Near heat affected zone.

Simulated HAZ

Using a Gleeble® 1500 thermo-mechanical system, HAZ thermal cycles with peak temperatures between 1300 °C and 1434 °C, as shown in Figure 1, were conducted on specimens of wrought Ti-48Al-2Cr-2Nb. With peak temperatures of 1375 °C and below, and a 4 s hold at peak temperature, the structure of the cycled specimen resembled the base metal microstructure, equiaxed gamma with grain boundary α_2. After thermal cycles with peak temperatures between 1387 °C and 1434 °C the transformed microstructure consisted mainly of $\gamma + \alpha_2$ in acicular and lamellar morphologies (Figure 7). Although 1375 °C is believed to exceed the alpha transus, the time above the transus, about 5 s, was insufficient for the transformation to occur. The kinetics for the transformation of γ to α above the alpha transus have been measured in a similar alloy, Ti-46.6Al-2.7Nb-0.3Ta, to require 300 minutes for complete transformation at 20 °C above the alpha transus, and 4 minutes at 95 °C above the transus [8]. Cr has been shown to retard the kinetics of the $\alpha \rightarrow \gamma$ transformation on cooling because Cr stabilizes alpha; its effect on the kinetics of the $\gamma \rightarrow \alpha$ transformation on heating have not been reported.

Microhardness of the untransformed specimens averaged 271 DPH. The fully transformed specimens were significantly harder than the untransformed specimens, with an average microhardness of 324 DPH. Figure 8 shows the hardness of simulated HAZ specimens as a function of peak temperature. Comparison with the weld hardness traverse of Figure 3b shows that this was consistent with the relatively high hardness of the transformed HAZ (about 315 DPH) and lower hardness of untransformed HAZ.

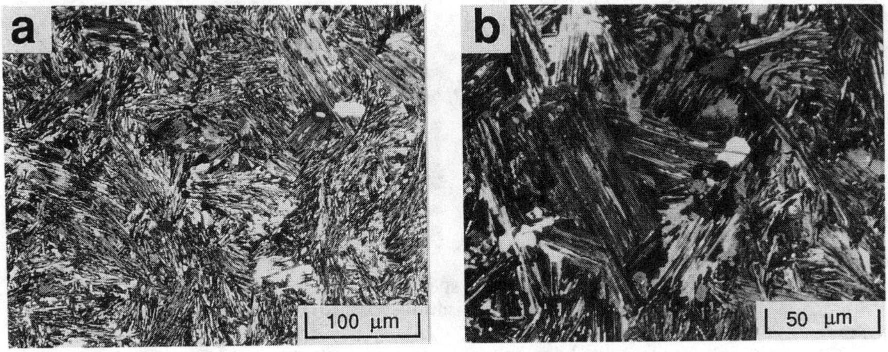

Figure 7: Microstructure of simulated HAZ specimen with peak temperature of 1387 °C.

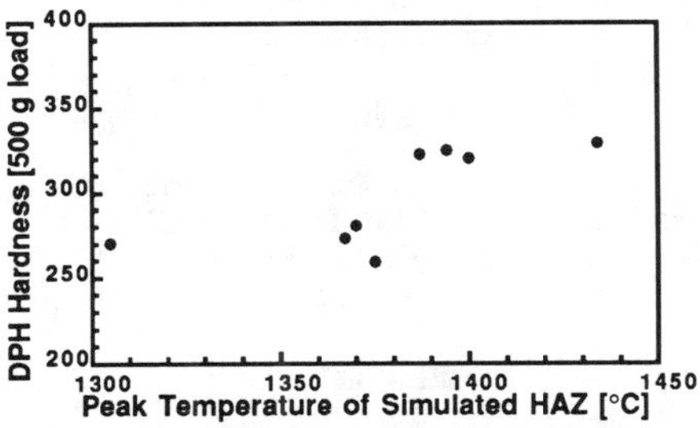

Figure 8: Hardness versus peak temperature for HAZ-simulation thermally cycled specimens of Ti-48Al-2Cr-2Nb.

Specimens which were held at 1400 °C for 30 s before cooling at slower constant cooling rates of 10 °C/s and 1 °C/s exhibited coarse lamellar structures consistent with slow furnace cooling results of McQuay *et al.* (Figure 9) [5]. Microhardness of the specimen cooled at 10 °C/s was 269 DPH; of the 1 °C/s specimen it was 246 DPH. In comparison, the hardness of transformed HAZ in preheated GTA welds was 315 DPH and base metal hardness was about 230 DPH. This indicated the possibility of reducing HAZ hardness and cracking susceptibility by reducing weld cooling rates.

Conclusions

GTA welding with preheat has been used to produce weld beads without cracking in wrought Ti-48Al-2Cr-2Nb. The base metal structure of this Ti-48Al-2Cr-2Nb was primarily equiaxed gamma with α_2 at grain boundaries.

The near HAZ of the GTA welds was comprised of mixed-morphology (lamellar and acicular) $\gamma + \alpha_2$ and some small gamma grains. The fusion zone of the GTA welds contained lamellar $\gamma + \alpha_2$ with interdendritic gamma and some small equiaxed gamma grains. The fusion zone solidification structure of cored dendrites with interdendritic gamma was visible in BSE.

Thermal simulation of HAZ indicated that transformation to a primarily $\gamma + \alpha_2$ structure occurred with a peak temperature exceeding some temperature between 1375 °C and 1387 °C, for thermal cycles typical of the HAZ of GTA welds. Slower cooling rates and longer hold times above the alpha transus produced softer structures of coarser $\gamma + \alpha_2$ lamellae.

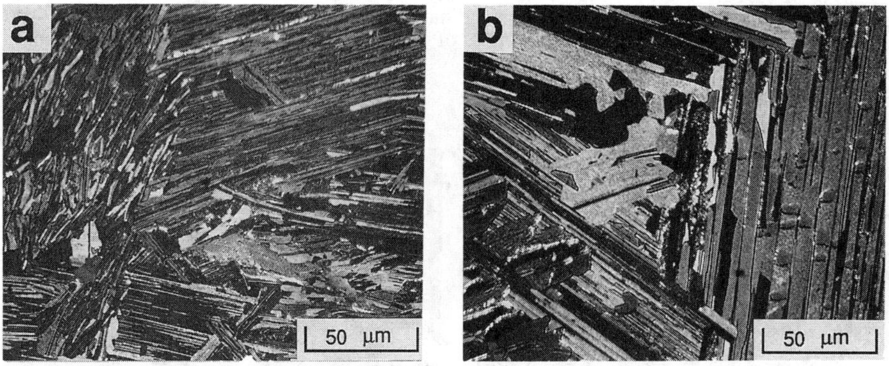

Figure 9: Microstructures of Ti-48Al-2Cr-2Nb specimens held at 1400 °C for 30 s and cooled with fixed cooling rates: (a) Cooling rate 10 °C/s; (b) Cooling rate 1 °C/s.

As GTA welding with preheat has been shown to be a suitable process for joining gamma alloys, more thorough microstructural characterization of gamma alloy welds should be performed. Future work should include efforts to characterize the far HAZ of gamma alloy welds, investigations of transformation kinetics for the reactions on heating and on cooling in the HAZ, and the identification of solidification reactions in the fusion zone.

Acknowledgements

Financial support of this study was provided by General Electric Aircraft Engines. The authors gratefully acknowledge the assistance of Mr. Jim Barber of GE with GTA welding, Mr. Eric Fletcher of Universal Energy Systems, WPAFB, Ohio with optical polarized light microscopy and Mr. Kuang-Hua Hou of OSU with transmission electron microscopy.

References

1. R.A. Patterson, B.K. Damkroger, "Weldability of Gamma Titanium Aluminide," *Weldability of Materials*, ed. R.A. Patterson, K.W. Mahin (Materials Park, OH: ASM International, 1990), 259-267.
2. H.A. Lipsitt, private communication, 1992.
3. M.J. Blackburn, M.P. Smith,"R&D on Composition and Processing of Titanium Aluminide Alloys for Turbine Engines," AFWAL-TR-82-4086, 1982.
4. Y.-W. Kim, "Microstructural Evolution and Mechanical Properties of a Forged Gamma Titanium Aluminide Alloy, " *Acta Metall Mater*, 40 (6) (1992), 1121-1134.
5. P.A. McQuay *et al.*, "The Effect of Chromium on the Phase Equilibria and Alpha Decomposition in Gamma TiAl Alloys," to be published in *Proceedings of Seventh World Conference on Titanium*, (Warrendale, PA: The Metallurgical Society,1992).
6. D.S. Shih *et al.*, "The Microstructural Dependence of Mechanical Properties of Ti-48Al-2Cr-2Nb," *Microstructure/Property Relationships in Titanium Alloys and Titanium Aluminide*, ed. Y.W. Kim and R.R. Boyer (Warrendale, PA: The Metallurgical Society,1990), 135-148.
7. C.McCullough *et al.*, "Phase Equilibria and Solidification in Ti-Al Alloys," *Acta Metall*, 37 (5) (1989), 1321-1336.
8. R.M. Nekkanti, P.A. McQuay, S.L. Semiatin, "Measurement of Homogenization Kinetics for a Near-Gamma Titanium Aluminide," *Scripta Metall Mater*, 26 (7) (1992), 1089-1094.

CREEP BEHAVIOR OF Ti-27 mol% Al INTERMETALLICS

Makoto OHTSUKA * and Hiroshi OIKAWA

Department of Materials Science, Faculty of Engineering,
Tohoku University, Sendai 980, JAPAN
* Now with Institute for Advanced Materials Processing,
Tohoku University, Sendai 980, JAPAN

Abstract

Constant-stress compressive creep tests were carried out on a single-phase Ti-27 mol% Al intermetallic compound, whose average grain diameter was 75 μm, at the temperature between 1050 and 1250 K under the stress ranging from 100 to 500 MPa. Stress and temperature-dependence of the minimum creep rate was analyzed using Dorn-type equation and they were divided into two groups. In addition, the structural change during creep was investigated using an optical microscope. At the stage of minimum creep rate, in the high stress region, small new grains are formed along original grain boundaries, whereas in the low stress region, the original smooth grain boundaries become irregular but new grains are rarely formed. The restoration process by the formation of dynamic recrystallization is thought to be one of the causes of the acceleration creep. The results observed can be understood in view of dynamic recrystallization that takes place depending on the magnitude of stress.

Introduction

Intermetallic compounds of Ti-Al system are frequently cited as one of the most promising structural materials for use at high temperature, mainly because of their low density and high yield strength at elevated temperatures [1]. The mechanical behavior of Ti_3Al intermetallic compounds at elevated temperature, however, has been investigated only to a limited extent. For examples, tensile (short-time) properties of a single-phase Ti_3Al binary intermetallic compound have been studied at elevated temperatures [2]. When a material is considered as a heat resisting material, creep behavior is the fundamental characteristic. However, experimental information about the creep behavior of Ti_3Al-based intermetallic compounds is available to a very limited extent. Only one study on creep behavior of a single-phase Ti_3Al binary intermetallic compound has been reported [3].

Titanium '92
Science and Technology
Edited by F.H. Froes and I. Caplan
The Minerals, Metals & Materials Society, 1993

In the present paper, some experimental results of the creep behavior obtained on a single-phase Ti_3Al binary intermetallic compound are reported, with a special reference to effects of stress and temperature.

Experimental Procedure

Material

The material used in the present study was a single-phase Ti-Al binary intermetallic material whose chemical composition was Ti-27.5 mol% (17.6 mass%) Al. Main impurities (in mass%) were Fe : 0.055, O : 0.092, N : 0.063 and H : 0.0018. The ingot was prepared by consumable electrode vacuum arc remelting, and then hot-extruded up to about 80 % reduction in cross-section at 1473 K. The extruded bars were annealed in vacuum at 1400 K for 100 ks. The resulting microstructure is shown in Fig. 1. The grain size was fairly uniform and the average grain diameter determined by the linear intercept method was about 75 μm. Compressive creep specimens having a rectangular parallelepiped shape were prepared with dimensions of 2 mm × 2 mm × 3 mm.

Creep Test

Constant-stress compressive creep tests were carried out in an argon atmosphere at the temperature between 1050 and 1250 K under the stress ranging from 100 to 500 MPa. The reduction in specimen height was detected by a linear variable differential transformer (LVDT) and recorded continuously. The temperature of a specimen was monitored with K-type thermocouples and kept constant within ± 1 K of the set value. The applied true stress was kept constant within ± 0.8 % error from the nominal value.

Fig. 1 Optical micrograph of the equiaxed structure of Ti-27 Al. The average grain diameter is 75 μm.

Results and Discussion

Creep Curves

Typical examples of the compressive creep curves observed in a single-phase Ti-27 Al intermetallic compound are shown in Fig. 2. Some examples of the variation in creep rate $\dot{\varepsilon}$ with strain ε are shown in Fig. 3.

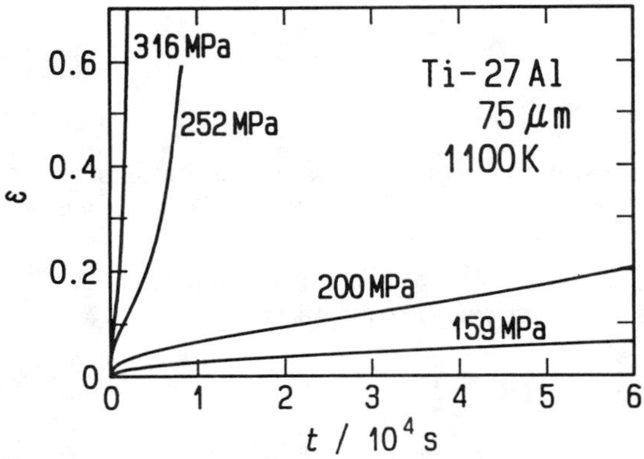

Fig. 2 Typical compressive creep curves observed in Ti-27 Al at 1100 K under various stresses.

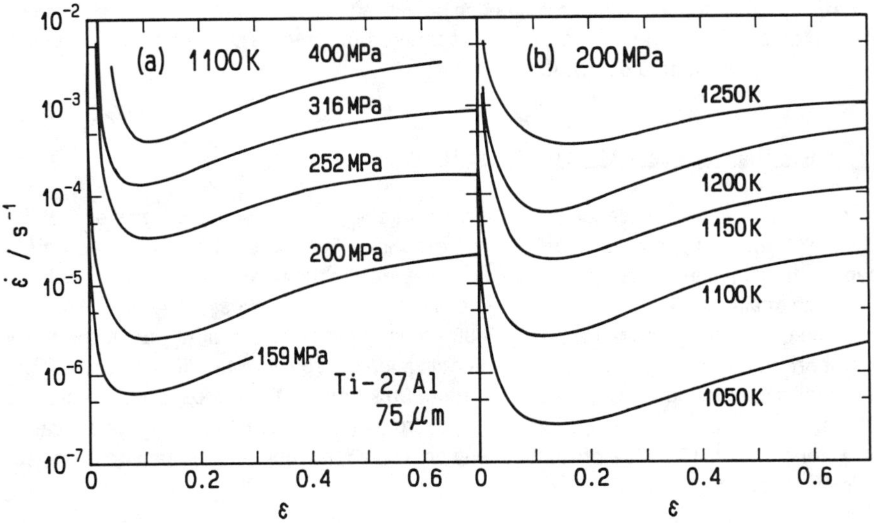

Fig. 3 Creep rates $\dot{\varepsilon}$ as a function of strain ε in Ti-27 Al (a) at 1100 K and (b) under 200 MPa.

Under the present testing conditions, the primary stage is always of the normal type. A minimum in the creep rate appears at about 0.1 (true) strain. The strain indicating the minimum creep rate ε_m is little affected by the testing conditions. The creep rate increases gradually soon after it reaches a minimum value : the steady-state stage does not appear. In all stresses and temperatures examined, these general features are observed, although the shape of the increasing creep rate stage is affected by the testing conditions. The acceleration creep becomes more pronounced as the testing stress increases. For example, the creep rate becomes twice the minimum creep rate with increasing the strain of 0.18 under 159 MPa and 0.13 under 400 MPa. The absolute value of creep rate increases with testing stress σ (Fig. 3(a)) and temperature T (Fig. 3(b)), as in cases of metals and alloys.

Stress Dependence of Minimum Creep Rate

The minimum creep rate, instead of the steady-state creep rate, will be used hereafter as one of the fundamental characteristics of high-temperature creep of Ti_3Al intermetallic compound.

The relationship between the minimum creep rate $\dot{\varepsilon}_m$ and stress σ is shown in Fig. 4. The stress dependence of $\dot{\varepsilon}_m$ is analyzed using the Dorn-type equation, i.e.,

$$\dot{\varepsilon}_m = A' \cdot \sigma^n \cdot \exp(-Q_m / RT).$$

Here, A' is constant, n the stress exponent, Q_m the apparent activation energy of creep, R the gas constant and T the testing temperature.

As shown in Fig. 4, the $\dot{\varepsilon}_m$ vs. σ relation can be divided into two groups. In the high stress and/or high strain-rate region, $n^H \sim 7$ independent of temperature, whereas in the low stress and/or low strain-rate region, $n^L \sim 9$.

Temperature Dependence of Minimum Creep Rate

The temperature dependence of minimum creep rate $\dot{\varepsilon}_m$ shown in Fig. 5 indicates that the $\dot{\varepsilon}_m$ vs. $1/T$ relation can be divided into two groups. The apparent activation energy of creep for the minimum creep rates Q_m is evaluated from the relations. In the high stress and/or high strain-rate region, $Q_m^H \sim 340$ kJ/mol independent of stress, whereas in the low stress and/or low strain-rate region, $Q_m^L \sim 480$ kJ/mol. These values of Q_m are significantly larger than that reported for diffusion. The activation energy for interdiffusion of the Ti_3Al intermetallic compounds is reported to be about 90 kJ/mol [4]. In most pure metals and simple alloys, the apparent activation energy of creep is found to be close to the activation energy for diffusion [5]. The present case is an exceptional one and the reason is unclear at this moment.

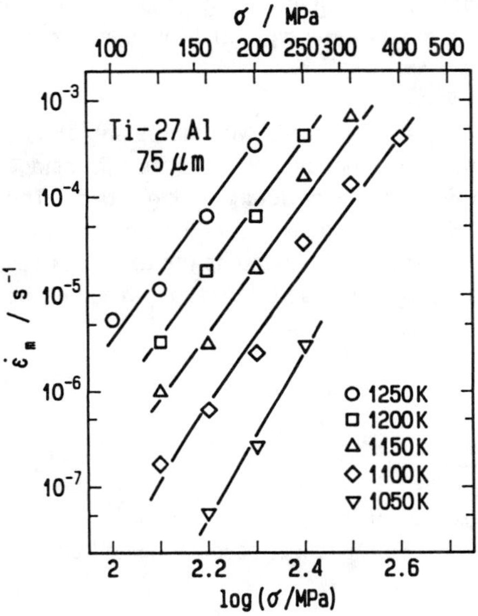

Fig. 4 Stress dependence of the minimum creep rate $\dot{\varepsilon}_m$ in Ti-27 Al at various temperatures.

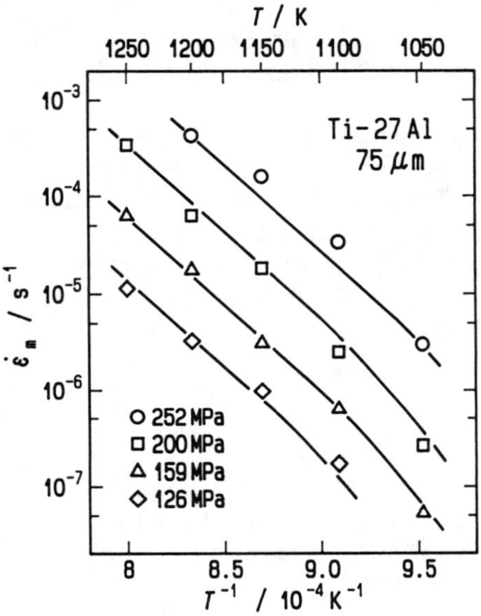

Fig. 5 Temperature dependence of the minimum creep rate $\dot{\varepsilon}_m$ in Ti-27 Al under various stresses.

Structural Change during Creep

The deformation structure was observed with an optical microscope on the section of specimens after creep deformation. Examples are shown in Fig. 6 for the specimens which were deformed up to the strain indicating the minimum creep rate ε_m at 1100 K under various stresses. In the high stress region, small new grains are always observed around original grain boundaries, whereas in the low stress region, the original smooth grain boundaries become irregular but new grains are rarely observed. The formation of new small grains becomes more pronounced as the testing stress increases.

The structural change during creep in the high stress region is shown in Fig. 7. Original grain boundaries are irregular in an early stage of deformation. Some new small grains appear around original grain boundaries at ε_m. The area occupied by recrystallized grains widens continuously with increasing strain. These results observed in the high stress region can be understood in view of dynamic recrystallization as the rate-controlling restoration process.

The structural change during creep in the low stress region is shown in Fig. 8. Original grain boundaries exhibit a very irregular feature in an early stage of deformation. The wave length of original grain boundaries corresponds roughly to the diameter of recrystallized grains in a later stage of deformation, although new small grains are rarely observed at ε_m. Some new small grains appear along original grain boundaries at the stage of a gradual increase in creep rate. The area occupied by recrystallized grains widens gradually with increasing strain.

The strain where new small grains are first observed decreases with increasing stress, though the strain indicating the minimum creep rates does not depend on stress. The stage of an increase in creep rate is related to the refinement of structure during creep. The restoration process by dynamic recrystallization is thought to be one of the causes of the acceleration creep. The acceleration creep becomes pronounced as the testing stress increases. This result can be understood in view of dynamic recrystallization which may be significantly accelerated with increasing stress.

Summary

Constant-stress compressive creep tests were carried out in argon atmosphere on a single-phase Ti-27 Al at temperatures between 1050 and 1250 K under stresses ranging from 100 to 500 MPa.

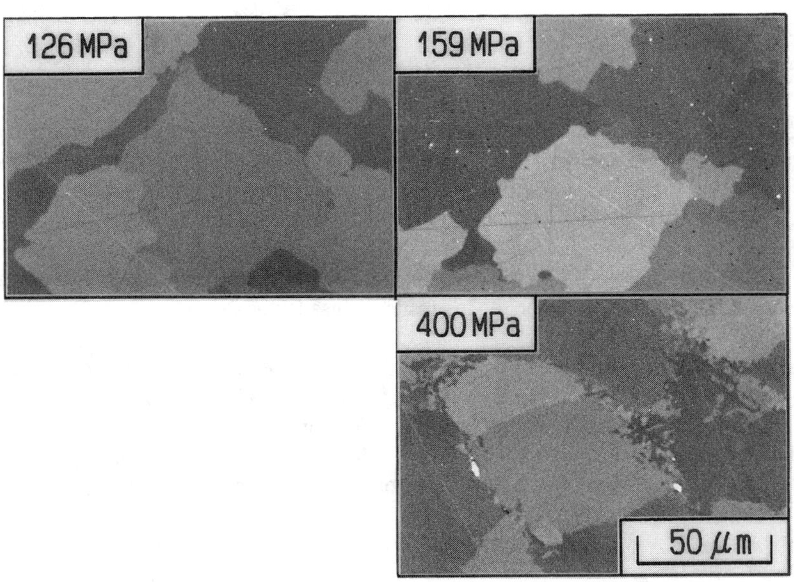

Fig. 6 Optical micrographs deformed up to the strain indicating the minimum creep rate ($\varepsilon_m \cong 0.1$) under various stresses.

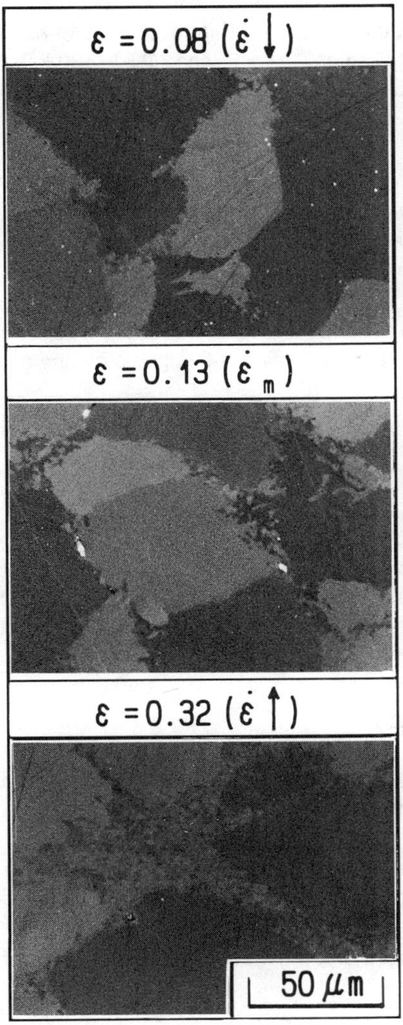

Fig. 7 Optical micrographs showing the structural change during creep at 1100 K under 400 MPa.

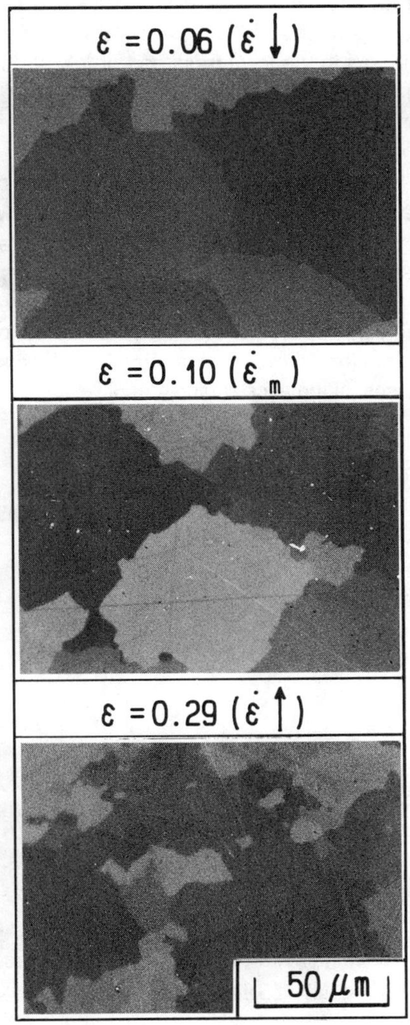

Fig. 8 Optical micrographs showing the structural change during creep at 1100 K under 159 MPa.

The results show that a minimum in the creep rate appears at about 0.1 strain which is independent of testing conditions. The stage of a gradual increase in creep rate were observed in all stresses and temperatures examined.

Stress and temperature-dependence of the minimum creep rate $\dot{\varepsilon}_m$ can be divided into two groups. In the high stress and/or high strain-rate region, $n^H \sim 7$ and $Q_m{}^H \sim 340$ kJ/mol, whereas in the low stress and/or low strain-rate region, $n^L \sim 9$ and $Q_m{}^L \sim 480$ kJ/mol. These Q_m values are significantly larger than that for interdiffusion in this intermetallic compound.

At the strain indicating the minimum creep rate, small new grains are formed along original grain boundaries in the high stress region, whereas in the low stress region new grains are rarely formed, though they appear in a later stage. The area occupied by recrystallized grains widens continuously with increasing strain.

The stage of acceleration creep is related to the refinement of structure during creep. The restoration process by the formation of dynamic recrystallization is thought to be one of the causes of the acceleration creep.

Acknowledgments

The authors would like to express their appreciation to Kobe Steel, Ltd. for donating the materials used in this investigation.

References

1. H. A. Lipsitt, High-Temperature Ordered Intermetallic Alloys, ed. C. C. Koch, C. T. Liu and N. S. Stoloff, MRS Symp. Proc., vol. 39 (Pittsburgh, PA: MRS, 1985), 351-364.

2. H. A. Lipsitt, D. Shechman and R. E. Schafrik, "The Deformation and Fracture of Ti_3Al at Elevated Temperatures", Metall. Trans., 11A (1988), 1369-1375.

3. M. G. Mendiratta and H. A. Lipsitt, "Steady-state Creep Behaviour of Ti_3Al-base Intermetallics", J. Mater. Sic., 15 (1980), 2985-2990.

4. K. Ouchi, Y. Iijima and K. Hirano, Titanium '80, ed. H. Kimura and O. Izumi, (Warrendale, PA: TMS-AIME, 1980), 599-568.

5. O. D. Sherby and P. M. Burke, "Mechanical behavior of crystalline solids at elevated temperature", Prog. Mater. Sci., 13 (1968), 325-390.

THERMAL STRAINS IN TITANIUM ALUMINIDE AND

NICKEL ALUMINIDE COMPOSITES*

A. Saigal
Department of Mechanical Engineering
Tufts University
Medford, MA 02155

and

D. S. Kupperman
Materials and Components Technology Division
Argonne National Laboratory
Argonne, IL 60439

Abstract

Neutron diffraction was used to measure residual thermal strains developed during postfabrication cooling in titanium aluminide and nickel aluminide intermetallic matrix composites. Silicon carbide/titanium aluminide, tungsten and sapphire/NiAl, and sapphire and SiC-coated sapphire/$NiAl_{25}Fe_{10}$ composites were investigated. The thermal expansion coefficient of the matrix is usually greater than that of the fibers. As such, during cooldown, compressive residual strains are generated in the fibers and tensile residual strains are generated in the matrix, parallel to the fibers. Liquid–nitrogen dipping and thermal cycling tend to reduce the fabrication–induced residual strains, measured at room temperature, in silicon carbide–fiber–reinforced titanium aluminide matrix composites. However, matrix cracking can occur as a result of these processes. The axial residual strains in the matrix were lower in the nickel aluminide matrix than in the titanium aluminide matrix. As the matrix undergoes plastic deformation, residual thermal strains are related to the yield stress of the matrix.

* Work has benefitted from the use of the Intense Pulsed Neutron Source at Argonne National Laboratory and was supported by the U. S. Department of Energy under Contract W–31–109–Eng–38.

Titanium '92
Science and Technology
Edited by F.H. Froes and I. Caplan
The Minerals, Metals & Materials Society, 1993

Introduction

Silicon carbide–fiber–reinforced titanium matrix composites, along with tungsten and sapphire–fiber–reinforced nickel aluminide matrix composites, are currently being evaluated for structural applications in jet engines and propulsion systems because of their low weight, high strength, and high stiffness potential at high temperatures. These composites are usually fabricated at temperatures around 1000°C. As there is a significant mismatch between the coefficients of thermal expansion of the fibers and the matrix, the thermally induced residual strains and stresses can be significant. In many engineering composites, frictional forces at the interfaces often provide the necessary link between the reinforcements and the matrix because chemical bonding is either weak or nonexistent (1,2). In addition, subsequent mechanical behavior of high–temperature composites is strongly influenced by the process–induced residual strains and stresses that are locked in the constituents during cooling after fabrication. Therefore, it is important to be able to measure and estimate the residual strains and stresses that exist in composite materials.

A number of programs have been conducted to measure residual strains and stresses in composite materials. For crystalline structures, including crystalline constituents in composite systems, one of the most widely used method involves X-ray diffraction. X–rays have been used to determine microstresses in systems with relatively light elements, such as aluminum/silicon carbide composites (3). However, X-rays typically measure stress fields in the near-surface region. Neutron diffraction is another powerful tool for measuring bulk elastic residual strains, from which residual stresses can be calculated in much the same way as done with X–ray diffraction (4,5). Neutron diffraction offers significant advantages over X–ray diffraction. Neutrons can penetrate deep into the interior of most engineering materials, typically by a factor of 1000 deeper than can X–rays. This provides for an excellent volume sample, a true bulk measurement of residual strains, and independence from surface–related effects.

Allen et al. used a neutron diffraction method to measure residual stresses and load–induced bulk stresses in a metal matrix composite (6). Krawitz et al. measured residual stress as a function of temperature in a high volume fraction tungsten carbide–nickel cemented carbide composite (7). Saigal and Kupperman have used neutron diffraction to measure residual thermal strains in NiAl matrix composite systems (8). In addition, Majumdar et al. have used the Intense Pulsed Neutron Source (IPNS) and the General Purpose Powder Diffractometer (GPPD) at Argonne National Laboratory to measure residual strains in a number of engineering composite materials (9).

In the present study, the neutron diffraction technique was used to measure and compare residual thermal strains, developed during cooling, in silicon carbide–fiber–reinforced titanium aluminide, tungsten and sapphire–fiber–reinforced nickel aluminide (NiAl), and sapphire and SiC coated sapphire-fiber-reinforced nickel aluminide ($NiAl_{25}Fe_{10}$) high–temperature composites. The effect of thermal processing, such as liquid–nitrogen dipping and thermal–cycling, on residual strains in SiC/titanium aluminide composites was also studied.

Neutron Diffraction Measurements

Thermal neutrons with wavelengths on the order of the lattice spacings are used in the experiments. Bragg's law of diffraction can be applied to neutrons as follows:

$$2\ d_{hkl}\ \sin\theta = \lambda_{hkl} \qquad (1)$$

where d_{hkl} is the lattice spacing, 2θ is the angle between the incident and the scattered neutron beams when a Bragg peak is detected, λ is the de Broglie wavelength of the neutron, and *hkl* are the Miller indices of the diffracting planes.

In neutron diffraction, the lattice spacings in various crystallographic directions of stress–free powders and/or fibers (which are used to fabricate the composite) are determined first. For any {*hkl* } diffraction peak, the lattice strain is given by:

$$\varepsilon_{hkl} = \frac{d_{hkl} - d_o}{d_o} \qquad (2)$$

where d_{hkl} and d_o represent the average interplanar spacings in the stressed and unstressed lattice, respectively.

The data were collected using the IPNS and the GPPD at Argonne. During a single measurement at IPNS, many diffraction peaks, i.e., crystallographic directions, of each phase are recorded simultaneously in various spatial directions (9). Residual strains in the fibers and the matrix, parallel and perpendicular to the fibers, are measured simultaneously by aligning the fibers at an angle of 45° to the neutron beam and analyzing neutron diffraction data with detectors $\pm 90^{o}$ from the neutron beam direction, as shown in Fig. 1.

Three different composite systems were investigated. They consist of:

1. 35 volume percent silicon carbide (SCS–6) fibers in a titanium aluminide matrix. Columnar grains of the 140 μm diameter fibers were oriented normal to the fiber axis.

2. 35 volume percent tungsten and 30 volume percent sapphire fibers in nickel aluminide (NiAl) matrix. The tungsten fibers and sapphire fibers were 140 μm and 200 μm in diameter, respectively.

3. 17 volume percent sapphire and 10 volume percent SiC coated sapphire fibers in nickel aluminide ($NiAl_{25}Fe_{10}$) matrix. The fibers were 150 μm in diameter.

Results and Discussions

Lattice parameters for various crystallographic directions were measured parallel and perpendicular to the fiber axis in the different composite systems. For the silicon carbide–fiber–reinforced titanium aluminide samples, the diffraction peak that provided the most useful data for silicon carbide fibers was that for the {220} plane. For the titanium aluminide matrix, the 1.75 Å line provides the average residual strain in the matrix. In order to reduce residual strains and stresses in the composites as compared to those in the as-fabricated condition, the fabrication procedures and the thermal history of the composites can be altered. Two commonly used procedures are liquid–nitrogen dipping (LND) and thermal cycling.

Figure 2 compares the residual strains, measured at room temperature, in the as–fabricated, liquid–nitrogen–dipped, and thermally cycled silicon carbide–fiber–reinforced titanium aluminide samples. In the as-fabricated condition, the residual strain in the fibers, parallel to the fibers, is compressive (–0.0019) and that in the matrix is tensile (+0.0042). The tensile strains in the matrix and compressive strains in the fibers are lower in the LND and thermally cycled specimens than in the as–fabricated sample. Thermal cycling appears to reduce the residual strains more than LND. The transverse strains in the fiber and matrix are both compressive (–0.0005 and - 0.0007, respectively). It was found that the matrix and fiber strains perpendicular to the fibers did not change significantly with processing.

Based on figure 2, there is no doubt that the axial strains (parallel to fibers) in the fibers and the matrix are reduced by the processing methods discussed. The limited ductility of the matrix is not sufficient to produce the observed reductions in residual strains. It is possible that matrix cracking occurs because of thermal processing; this is further suggested by elastic moduli measured by the velocity of ultrasound propagating perpendicularly to the fibers. Decreases in Young's modulus for liquid–nitrogen–dipped (134 GPa) and thermally cycled (131 GPa) samples were measured as compared to the as–fabricated (137 GPa) sample.

Table 1 shows the measured residual strains in the matrix in various composites. The residual strains (parallel to the fibers) are tensile in the matrix and are higher in the titanium aluminide matrix than in the nickel aluminide matrix. The strains in the nickel aluminide (NiAl) matrix due to 35 vol.% tungsten fibers and 30 vol.% sapphire fibers are similar. Based on 3-D elastoplastic finite element analyses, the thermally induced effective stress in the matrix exceeds the yield stress and the matrix undergoes plastic deformation (8,10). This suggests that the residual matrix strains are primarily controlled by the yield stress of the matrix. For the composite systems investigated, it was found that as the yield stress of the matrix increases, the axial residual strain in the matrix increases.

Conclusions

Neutron diffraction was used to measure residual thermal strains in titanium aluminide and nickel aluminide intermetallic matrix high–temperature composites. It has been shown that significant residual strains develop in these composites during postfabrication cooling. As the thermal expansion coefficient of the matrix is usually greater than that of the fibers, the fibers are under compressive residual strains and stresses. Liquid–nitrogen–dipping and thermal cycling tend to reduce the fabrication–induced residual strains in the silicon carbide–fiber–reinforced titanium aluminide matrix composite. However, matrix cracking can occur as a result of these processes. Strains in the matrix (parallel to fiber axis) are tensile and are higher in the titanium aluminide matrix than in the nickel aluminide matrix. As the matrix undergoes plastic deformation, the residual matrix strains are primarily controlled by the yield stress of the matrix. For the composite systems investigated, it was found that as the yield stress of the matrix increases, the axial residual strain in the matrix increases.

Acknowledgments

This work was supported by the U.S. Department of Energy through Argonne National Laboratory discretionary funds, and by General Electric Co. The work benefited by the use of the Intense Pulsed Neutron Source at Argonne National Laboratory. The authors wish to thank P. K. Wright for supplying the samples and R. L. Hitterman for assistance in data acquisition.

References

1. D. C. Phillips, "Interfacial Bonding and Toughness of Carbon Whisker Reinforced Glass and Glass–Ceramics," Journal of Materials Science, Vol. 9, No. 11, 1974, pp. 1847–54.

2. K. M. Prewo and J. B. Brennan, "High–Strength Silicon Carbide Fiber–Reinforced Glass–Matrix Composites," Journal of Materials Science, Vol. 15, No. 2, 1980, pp. 463–68.

3. P. Predecki, A. Abuhasan, and C. S. Barrett, "Residual Stress Determination in Al_2O_3/SiC (Whisker) Composites," Advances in X–Ray Analysis, Vol. 31, 1988, p. 231.

4. M. T. Hutchings and C. G. Windsor, "Industrial Application of Neutron Scattering," Ch. 25 in Neutron Scattering, Edited by K. Skold and D. L. Price, Academic Press, New York, 1986.

5. I. C. Noyan and J. B. Cohen, Residual Stress, Springer–Verlag, New York, 1987.

6. A. J. Allen, M. Bourke, M. T. Hutchings, A. D. Krawitz, and C. G. Windsor, "Neutron Diffraction Measurements of Internal Stress in Bulk Materials:–Metal Matrix Composites," Residual Stresses in Science and Technology, E. Macherauch and V. Hauk, Eds., DGM Informations–gelsellschaft Verlag, Oberursel, Vol. 1, 1987, pp. 151–57.

7. A. D. Krawitz, D. G. Reichel, and R. L. Hitterman, "Residual Stress and Stress Distribution in a WC–Ni Composite," Materials Science and Engineering, Vol. A119, 1989, pp. 127–34.

8. A. Saigal and D. S. Kupperman, "Residual Thermal Strains and Stresses in Nickel Aluminide Matrix Composites," Scripta Metallurgica, Vol. 25, No. 11, 1991, pp. 2547-52.

9. S. Majumdar, J. P. Singh, D. S. Kupperman, and A. D. Krawitz, "Application of Neutron Diffraction to Measure Residual Strains in Various Engineering Composite Materials," Journal of Engineering Materials and Technology, Vol. 113, January 1991, pp. 51–59.

10. R. P. Nimmer, R. J. Bankert, E. S. Rusell, and G. A. Smith, "Micromechanical Modeling of Fiber–Matrix Interface Effects in SiC/Ti Metal Matrix Composites," Proceedings of 13th Annual Conference on Composite Materials and Structures, Cocoa Beach, Florida, Jan. 18–20, 1989.

Table 1 Residual strains in the matrix (parallel to the fibers)

	Residual strain in the matrix (±0.0003)	Yield stress of the matrix (MPa)
Titanium Aluminide (35 vol.% SiC fibers)	+0.0042	620
Nickel Aluminide ($NiAl_{25}Fe_{10}$) (17 vol.% sapphire fibers)	+0.0022	414
Nickel Aluminide ($NiAl_{25}Fe_{10}$) (10 vol.% SiC-coated sapphire fibers)	+0.0017	414
Nickel Aluminide (NiAl) (30 vol.% sapphire fibers)	+0.0014	240
Nickel Aluminide (NiAl) (35 vol.% tungsten fibers)	+0.0013	240

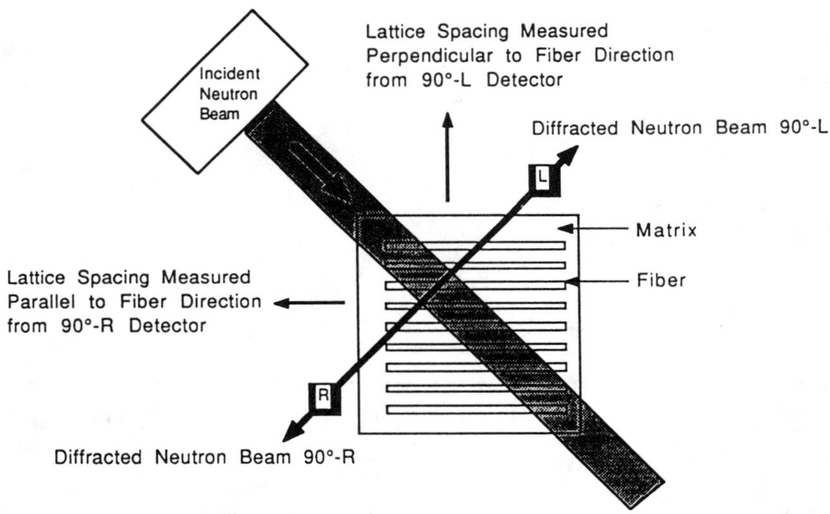

Figure 1. **Schematic representation of experimental set-up.**

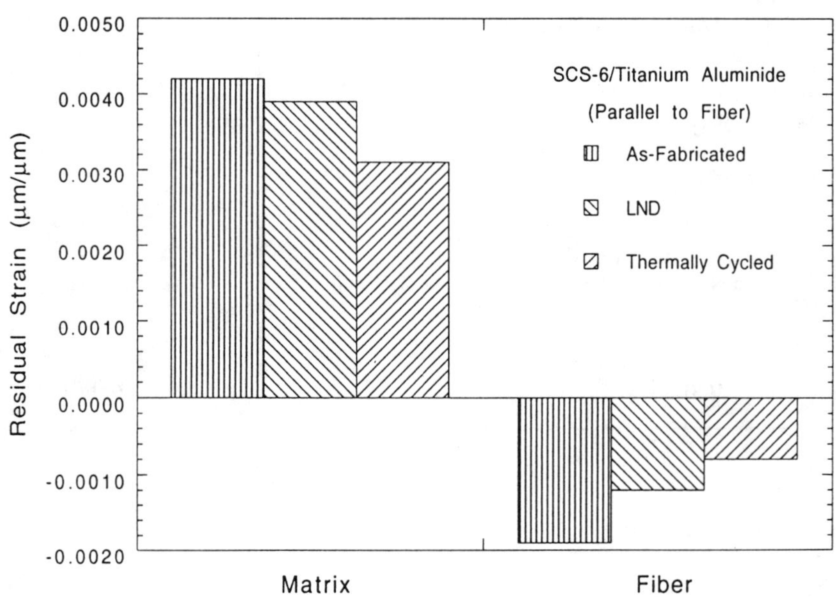

Figure 2. Residual strains in SiC (SCS–6) fibers and titanium aluminide matrix, parallel to fiber axis, in as–fabricated, liquid–nitrogen–dipped, and thermally cycled specimens.

Fracture and Fatigue of TiAl Based Aluminides

A.W. James and P.Bowen

IRC in Materials for High Performance Applications /School of Metallurgy and Materials
The University of Birmingham,
Edgbaston, Birmingham, B15 2TT, UK.

Abstract

Effects of microstructural orientation on ambient temperature fracture toughness; and effects of environment and orientation on fatigue crack growth rates at ambient and elevated temperatures have been studied in a gamma based titanium aluminide. A cast ingot of the binary intermetallic compound Ti-48Al with a fully α_2/γ two phase lamellar microstructure has been shown to exhibit both highly anisotropic fracture toughness and resistance to fatigue crack propagation. A marked increase in fatigue crack growth rates and a large reduction in fracture toughness is observed when microstructural orientation and testpiece loading favour crack propagation in an interlamellar direction.

Introduction

An attractive combination of elevated temperature oxidation and creep resistance, coupled with high specific strength and modulus has led to intense interest in gamma-based titanium aluminides as potential materials for high temperature applications. However, at present low ductility and poor fracture toughness are major factors limiting their use, and much work remains to be done to establish their fitness for purpose especially at envisaged operating temperatures. In particular, there is little published data concerned with the fatigue crack growth of gamma-based titanium aluminides(1,2). In this present study the effects of microstructure, environment and test temperature on the fatigue crack growth rates and fracture toughness of binary Ti-48Al (atomic %) have been investigated.

Titanium '92
Science and Technology
Edited by F.H. Froes and I. Caplan
The Minerals, Metals & Materials Society, 1993

Experimental

Single edge notch bend (SENB) testpiece blanks were cut using electro-discharge machining from a cast ingot of 100 mm diameter and 675 mm length, in three orientations, at a distance of approximately 100 mm from the top of the ingot (Figure 1).

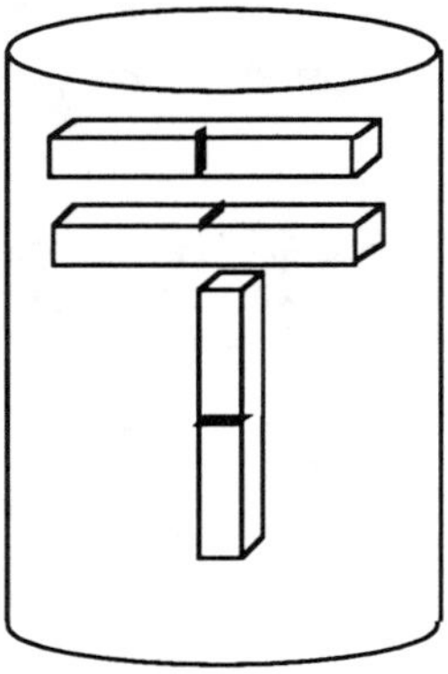

Figure 1 - Schematic representation of the ingot showing testpiece orientation and position of the machined notches.

Testpieces in the transverse direction measured 8 x 8 x 70 mm and those in the longitudinal direction 10 x 10 x 60 mm. The ingot was produced by cold hearth overflow melting within the IRC at Birmingham using a large-scale plasma furnace equipped with twin computer controlled, servo-hydraulic 150 kW helium transferred arc torches, and manufactured by Retech Ltd., California. Note that the ingot was cooled relatively slowly to ambient temperature. The material was of nominal composition Ti-48Al (atomic %), and compositional details of the ingot are shown in Table 1.

Table I - Compositional analysis of the ingot

Position	Al ($^{a}/_{o}$)	O(ppm)	N_2(ppm)	H_2(ppm)	C($^{w}/_{o}$)
Top	48.8	750	20	10	0.010
Middle	49.53	650	20	10	0.015

All specimens were notched using a diamond wafering blade of 150μm thickness, to a depth of approximately 0.25 the width of the specimen (i.e. a/W = 0.25 where a = depth and W = specimen width) . Holes to facilitate the attachment of current carrying leads were drilled in the ends of the specimens using carbide tooling.

Fatigue crack propagation tests were performed in three-point bending at 700° and 800°C under vacuum of better than $5x10^{-4}$ mbar; at 700°C in air and argon; and at ambient temperature both in air and in vacuum. All tests were carried out at a frequency of 10 Hz and at a low mean stress R = 0.1, (where R is the ratio of the minimum applied stress intensity, K_{min}, to the maximum applied stress intensity, K_{max}, over the fatigue cycle i.e.

R=Kmin/Kmax) on an ESH servo-hydraulic testing machine equipped with a custom built vacuum chamber. The initiation and growth of fatigue cracks were monitored by a direct current potential difference (d.c.p.d.) technique. Tests were performed at constant cyclic load range i.e. increasing cyclic stress intensity range, ΔK (where $\Delta K = K_{max} - K_{min}$) with crack growth.

Some testpieces were fatigued in three point bending at room temperature to produce pre-cracks for measuring fracture toughness at room temperature according to BS 5447:1977(3). Other fracture toughness tests were also performed on specimens manufactured from the broken halves of the fatigue specimens, but these were pre-cracked in cyclic compression. In these specimens surfaces were polished to 1µm in order to measure pre-cracks using a surface replication technique. The specimens were notched to a depth, a/W = 0.25 using a diamond wafering blade of 150µm thickness and pre cracked using uniaxial cyclic compression with an initial stress intensity range, ΔK, of between 18 and 22 MPa√m.

The fracture and fatigue surfaces were examined using an ISI scanning electron microscope operating at 10° tilt and 20 kV.

Sections for optical examination were taken from three orientations and prepared by grinding to 1200 grit and polishing with colloidal silica. The specimens were then swab etched with Krolls II reagent (2% HF, 1% HNO_3, 97% H_2O) prior to optical examination.

Results

Microstructure

The cast microstructure comprised large elongated lamellar grains consisting of alternating plates of TiAl (γ) and Ti_3Al (α_2). The lamellar plates were stacked in an orientation normal to the length of the grains and thus the majority of grains had lamellae arranged in a similar orientation. The plates of TiAl in the lamellar regions are approximately 1-2 µm in thickness. Testpiece notch orientations A, B and C in relation to the lamellar plates are shown schematically in Figure 2.

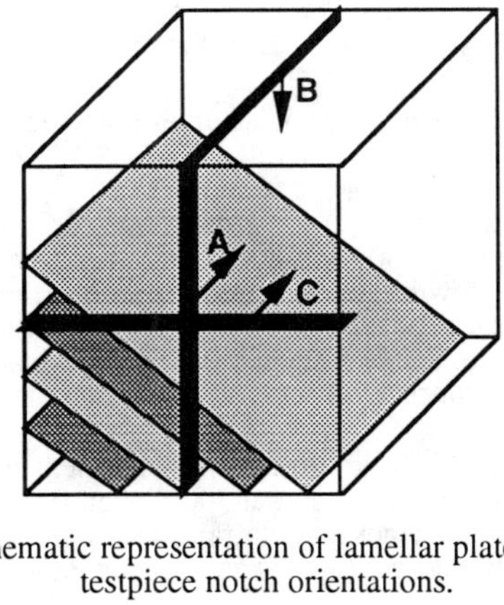

Figure 2 - Schematic representation of lamellar plates in relation to testpiece notch orientations.

Fracture toughness

The effect of testpiece orientation on fracture toughness at ambient temperature is shown in Table II. Marked differences are observed for testpieces of 'A', 'B' and 'C' orientations. It should be noted that transverse testpieces notched in the 'B' orientation often failed by interlamellar failure in a direction up to 40° away from the mode I direction.

Table II - Ambient temperature fracture toughness values for the various testpiece orientations.
(The results marked * conform to BS5447:1977)

Specimen orientation	Notch orientation	Fracture toughness (MPa√m)
Transverse	A	20.4/24.2/23.8*
Transverse	B	13.1*
Longitudinal	C	12.2

Fatigue crack growth

At ambient temperature, marked effects of testpiece orientation on fatigue crack growth rates are once again observed, see Figure 3, for 'A' and 'B' testpiece orientations.

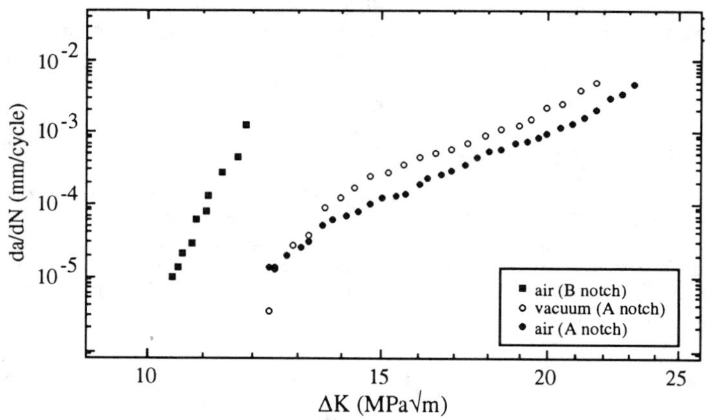

Figure 3 - Fatigue crack growth resistance curves obtained at ambient temperature.

In the optimised 'A' orientation stable crack growth rates are maintained to high values of applied stress-intensity range. At 700° and 800°C in vacuum, however, there appears to be little effect on fatigue crack growth rates of transverse orientations, 'A' and 'B', see Figure 4. In sharp contrast, higher crack growth rates are observed for testpieces of orientation 'C', as shown in the Figure, (To date, stable fatigue crack growth has not been obtained for this orientation at ambient temperature).

For the optimised 'A' orientation, little effect of environment is observed on fatigue crack growth at ambient temperature, Figure 3, but at 700°C significant effects of environment are observed, Figure 5. In particular, crack growth rates at lower applied stress intensity ranges increase in air relative to both inert gas (argon) and vacuum environments.

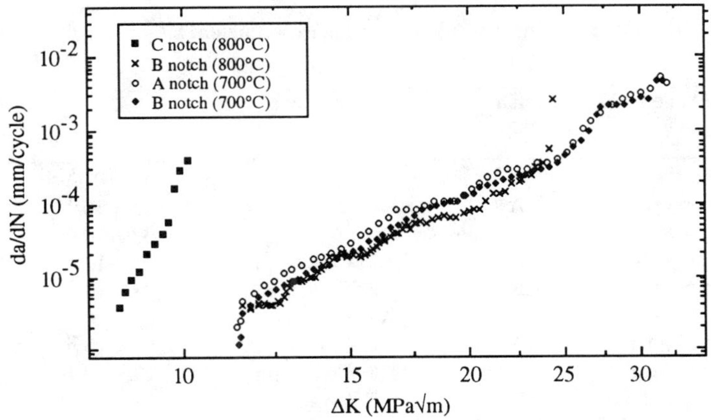

Figure 4 - Fatigue crack growth resistance curves obtained in vacuum at elevated temperatures.

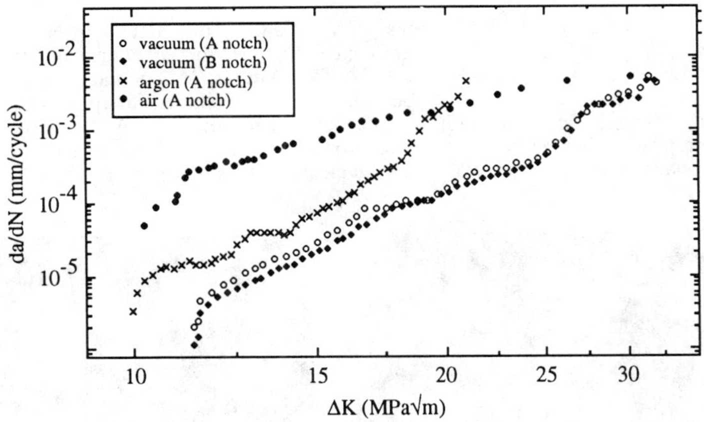

Figure 5 - Fatigue crack growth resistance curves obtained at 700°C

If the crack growth rates can be represented by a Paris-Erdogan law, of the type:

$$da/dN = A\Delta K^m \qquad (1)$$

where da/dN is the crack increment per cycle, and A, m are materials constants, then the values of A and m are given in Table III for all tests reported here. Note that in the determination of these constants, no data are included for crack growth within 200μm of the machined notch (of root radius approximately 100μm). The steep dependence of da/dN on ΔK is clear in the high values of m obtained for testpieces of 'B' orientation (ambient temperature) and 'C' (800°C).

Table III - Materials constants obtained from experimental fatigue crack growth rates.

Specimen orientation	notch orientation	Test conditions	A	m (mm/cycle)
Transverse	A	R.T/vacuum	1.4×10^{-15}	9.5
Transverse	A	R.T./air	4.33×10^{-14}	8.0
Transverse	B	R.T./air	9.32×10^{-44}	37.4
Transverse	A	700°C/vacuum	3.27×10^{-13}	6.8
Transverse	B	700°C/vacuum	1.28×10^{-13}	7.0
Transvarse	A	700°C/argon	8.09×10^{-15}	8.6
Transverse	A	700°C/air	2.91×10^{-8}	3.7
Transverse	A	800°C/vacuum	5.27×10^{-12}	5.7
Longitudinal	C	800°C/vacuum	1.51×10^{-31}	27.8

Fractography

Fatigue surfaces are shown for testpieces of 'A', 'B' and 'C' orientations in Figures 6, 7 and 8 respectively. They indicate clearly the influence of underlying lamellar plate morphology on micromechanisms of fatigue crack growth. For the optimised 'A' orientation, fracture surfaces produced by fatigue crack growth at 700°C in vacuum are shown for applied stress intensity ranges of 11 and 26 MPa√m in Figures 6(a) and (b) respectively. At the higher ΔK range, inter-lamellar decohesion becomes more pronounced, Figure 6(b). In sharp contrast, the lamellar plates themselves are often observed clearly for testpieces of 'B' and 'C' orientations.

(a) ΔK ~ 11 MPa√m 30 μm (b) ΔK ~ 26 MPa√m 30 μm

Figure 6 - Scanning electron micrographs showing the fatigue surfaces (a and b) of a transverse testpiece notched in the 'A' orientation, tested at 700°C.

For testpieces of 'B' orientation, tested at ambient temperature, fracture surfaces exhibit large smooth facets, see Figure 7(a). Note that crack growth was deflected parallel to and along such plates, away from the direction of maximum mode I opening, at angles of up to 40°. At elevated temperatures, such deflection was not observed, Figure 7(b) and fatigue cracks grew in the mode I direction.

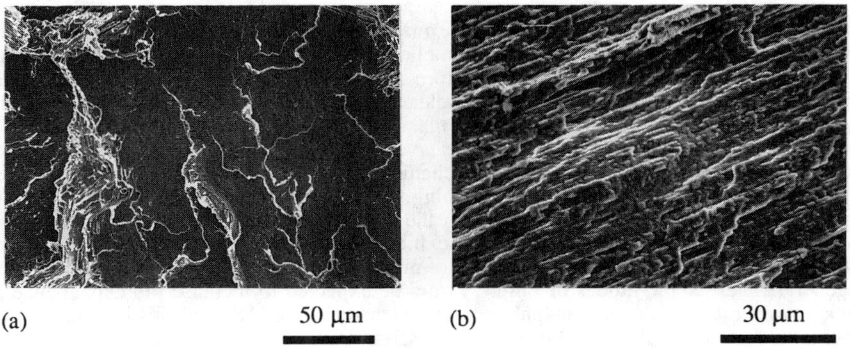

Figure 7 - Scanning electron micrographs showing the fatigue surfaces of transverse testpieces notched in the 'B' orientation, (a) tested at ambient temperature and (b) tested at 800°C

For orientation 'C', the influence of the plates themselves are observed on fracture surfaces even for tests performed at 800°C in vacuum, see Figure 8. Note that for this orientation the fatigue surface exhibits large facets and 'steps' between different lamellar layers.

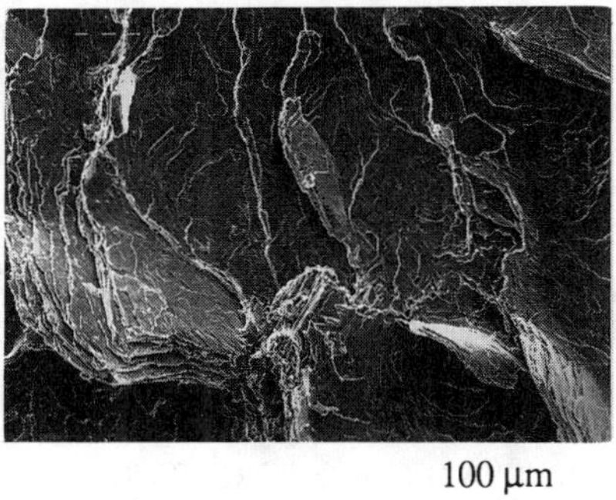

Figure 8 - Scanning electron micrograph showing the fatigue surface of a longitudinal testpiece notched in the 'C' orientation, tested at 800°C.

Discussion

For the orientation optimised for fatigue crack growth resistance, and ambient temperature fracture toughness, condition 'A', little effect of environment is observed at ambient temperature. Significant effects are observed at an elevated temperature of 700°C, where crack growth rates in air can be up to one order of magnitude greater than those observed in argon and in vacuum environments. Such effects would be expected to increase at lower frequencies of cyclic loading and hence for use at elevated temperature attention must be given to such environmental interactions. Further work is required to examine the mechanisms of environmental attack.

The α_2/γ lamellar microstructure has been shown previously to exhibit improved crack growth resistance and fracture toughness over microstructures containing significant amounts of allotriomorphic or twin related gamma(4,5,6). However, this present study has shown that not only is a fully lamellar microstructure important, but also that the orientation of the fully lamellar microstructure may be critical. Low values of fracture toughness, comparable to those obtained from a fully gamma grained microstructure(4) were measured when the crack propagated in a direction parallel to the lamellar plates. The mechanism by which the crack propagates in a direction parallel to the lamellar plates has not yet been determined precisely, but it would appear at ambient temperatures that cleavage of the plates occurs, see Figure

7(a), even if the plates are inclined to the maximum mode I opening direction. In the longitudinal orientation, condition 'C', the lamellae plates are aligned so as to allow the crack to advance without impediment. The fracture and fatigue surfaces of the longitudinally oriented specimens exhibit large regions of cleavage fracture separated by 'steps' where the growing crack has changed from one particular pair of lamellae to a parallel set, Figure 8. Thus in such a suitably oriented testpiece, cleavage can persist even to test temperatures of 800°C, Figure 8, but it appears that this mechanism cannot operate if the plates are inclined steeply to the axis of maximum mode I opening, as occurs for the 'B' orientation, see Figure 7(b). Such observations have implications for the definition of a ductile to brittle transition in two phase fully lamellar aluminides which exhibit a strong microstructural orientation and may also be taken to illustrate the influence of mixed mode failure on transition temperature in general. These static modes of failure would be expected to increase the dependence of crack growth rate, da/aN, on the applied stress intensity range, ΔK, and this is observed see Figures 3, 4, 5 and Table III. Further work is required to establish whether cleavage is occurring through the γ plates or through the thinner constrained plates of α_2. For the purpose of the development of titanium aluminides for structural use at elevated temperatures, such considerations are of an academic nature only because such anisotropic crack growth resistance exhibited in these as-cast structures is clearly unacceptable. If fully lamellar microstructures are to be used in practice, it is essential that these well aligned packets of α_2/γ plates are distributed randomly throughout the material.

Conclusions

Strong microstructural orientation effects have been observed on both fracture toughness and fatigue crack growth resistance. Rapid fatigue crack propagation and low values of fracture toughness result if crack growth occurs in a direction parallel to the lamellar plates, even if these plates are aligned away from the direction of maximum mode I opening.

Crack growth rates are independent of test temperature between 700 and 800°C in vacuum, but increased crack growth rates (by approximately one order of magnitude) may be obtained from tests in air compared with those in vacuum at a test temperature of 700°C such effects of environment are not observed at ambient temperature.

Acknowledgements

Support for one of the authors (AWJ) from a SERC CASE award with AEA Technology, Harwell is gratefully acknowledged, together with useful discussions with Dr M. Strangwood and Dr C.A. Hippsley (AEA Technology, Harwell). Thanks are also due to Dr M. Kearns, IMI Titanium, plc for the provision of both feedstock materials and compositional analysis.

References

1. W.O. Soboyejo, P.B. Aswath and J.E. Deffeyes, "An Investigation of Room- and Elevated Temperature Fatigue Crack Growth in Ti-48Al" to be published in Mat. Sci. and Eng. A.

2. S.M.L. Sastry and H.A. Lipsitt, "Fatigue Deformation of TiAl Base Alloys"Met. Trans. A, 8A, (1977), 299-308.

3. BS5447 "Plane strain fracture toughness testing of metallic materials", 1977.

4. A.W. James and P. Bowen, "Elevated Temperature Crack Growth Resistance of TiAl Under Monotonic and Cyclic Loading", paper presented at the International Conference on High Temperature Aluminides and Intermetallics, 16-19 September 1991, San Deigo, USA.

5. Y.W. Kim, "Recent Advances in Gamma Titanium Aluminide Alloys" High Temperature Ordered Intermetallics IV, Vol. 213, ed J.O. Stiegler et al., (MRS, Pittsburgh, PA 1991).

6. N.J. Rogers and P. Bowen, "Effects of Microstructure on Toughness in a Gamma Titanium Aluminide", Conf. Proc. Euromat '91, July 22-24, University of Cambridge, The Institute of Metals, U.K.

DEFORMATION OF THE ORTHORHOMBIC PHASE IN Ti-Al-Nb ALLOYS

AT HIGH TEMPERATURES

D. Banerjee* and R.G. Rowe**

* Defence Metallurgical Research Laboratory,
Hyderabad-500258, India
** General Electric Corporate Research and Development,
Schnectady, NY-12301, USA

Abstract

The deformation structure of the O phase in a Ti-26Al-21Nb alloy has been characterised by transmission electron microscopy after compressive strains at 650C. Both 'a' slip and 'c' component slip have been observed. Both types of 'a' dislocations, <100> and <110>, are operative. While <110> dislocations slip predominantly on {110} planes, the slip planes for <100> dislocations could not be unequivocally identified although cross slip on to (001) planes was recorded. 'c' component dislocations, <114> and <102>, are present as screw dipoles on {220} and {041} planes. Extensive cross slip of these is observed. The dislocation structures are compared to those generated by deformation at room temperature.

Introduction

A new class of Ti-Al-Nb alloys has been recently developed (1-3) in which the major constituent phase is the O phase (4,5) which is based on the Ti_2AlNb composition with a Cmcm space group. These alloys possess superior combinations of strength and toughness as compared to alpha2 based compositions, and therefore an analysis of deformation behaviour of the O phase, especially in comparison to alpha2, is of considerable interest. We have earlier carried out a comparative study of dislocation arrangements generated by room temperature compressive strains in the alpha2 and O phases (6,7), and now report a similar study on the O phase at 650C.

Materials and Methods

A Ti-26Al-21Nb alloy (atom%) composition with 600 wppm oxygen was quenched from 1200C and heat treated at 900C for 300 hours to produce equiaxed grains of a single phase O structure. Samples were deformed at 650C to 2% strain in compression and sliced perpendicular to the stress axis for examination by transmission electron microscopy.

Titanium '92
Science and Technology
Edited by F.H. Froes and I. Caplan
The Minerals, Metals & Materials Society, 1993

Results

Fig.1 shows a comparison of slip vectors in the alpha2 and O phases. The unit cell shown for the O phase has been chosen to be the same as that of alpha2 to illustrate equivalent vectors and changes in the nature of the antiphase vectors. Note that, in the O phase unit cell, the 1/2[100] and 1/4[110] vectors differ slightly in magnitude as do the 1/2[102] and1/4[114] vectors, and are also associated with different types of antiphase boundaries (6).

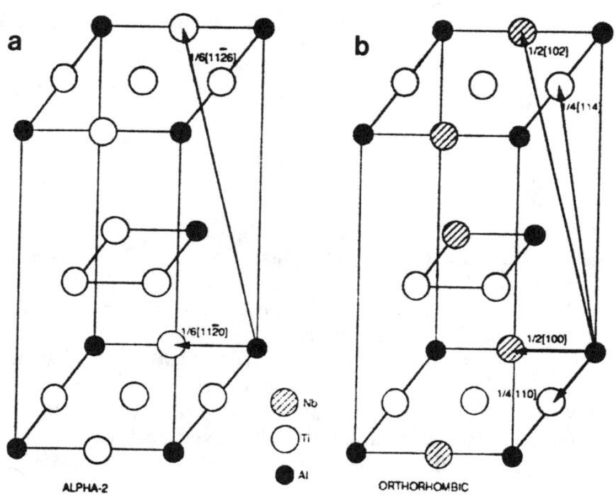

FIG.1 (a) The unit cell and Burgers vectors of dislocations in the alpha2 structure (b) equivalent Burgers vectors of dislocations in the O phase. The unit cell has been chosen to be same as that of alpha2 for ease of comparison.

The broad features of the arrangements of 'a' dislocations are essentially unchanged on deformation at 650C from that at room temperature (6,7). Weak beam microscopy was used to examine the dissociation of 'a' dislocations. Fig.2 shows an example of [110] dislocations imaged with the (1$\bar{1}$0) plane inclined and edge on, respectively. The dissociation of the [100] superdislocation can clearly be seen in the former case, but not in the latter, indicating that these dislocations are dissociated on the (1$\bar{1}$0) planes and therefore glide on them. Thus, no differences are seen in the case of <110> dislocations between room temperature and 650C. Similar analyses for [100] dislocations, however, reveal a different picture. Figs. 3a and 3b show examples of two different [100] dislocations imaged with the prismatic (010) and (001) planes nearly edge on, respectively. The behaviour of both these dislocations is quite distinct from that observed for the <110> dislocations. In the case of the first example, the dislocation appears to be dissociated on the (001) plane since the superpartials can barely be seen when the (001) plane is nearly edge on. However, with the (010) plane edge on, the dislocation is seen to consist of two segments, one showing clear dissociation into superpartials while the other shows a comparatively thick image in which the superpartials cannot be distinguished. In the second example, the images suggest that many local cross slip events have occured along the dislocation line between the basal and (010) planes as shown by the arrows. In general, it was difficult to detect [100] dislocations, their density being far lower than <110> dislocations.

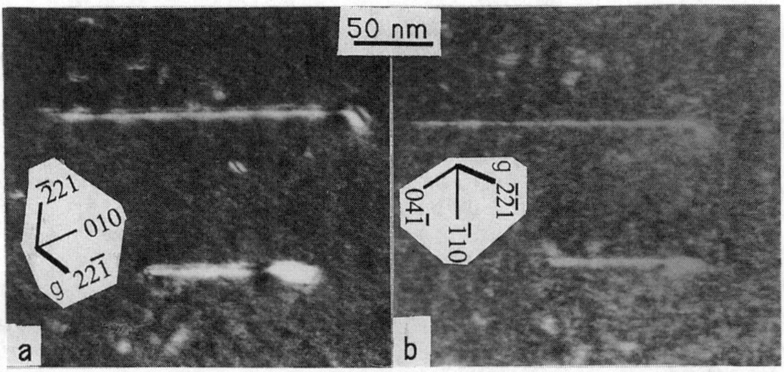

FIG.2: (a) [110] dislocations dissociated into 1/4[110] superpartials, weak beam with a $22\bar{1}$ reflection near the [102] zone axis. (b) The same dislocations near the [114] zone axis imaged with the same reflection with the $(1\bar{1}0)$ plane edge on.

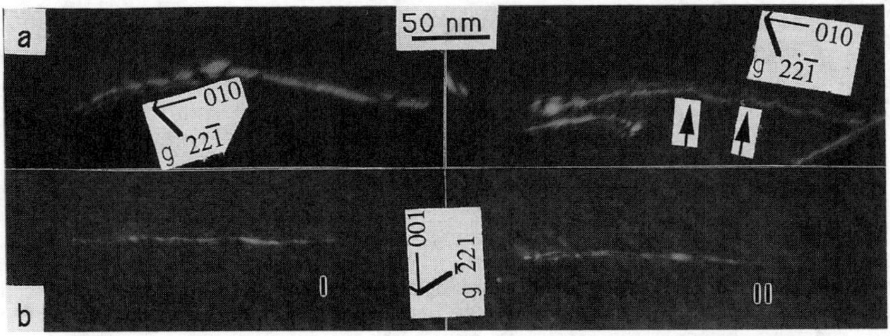

FIG.3 Examples of [100] dislocations. (a) dislocations I and II imaged with a $22\bar{1}$ reflection near a [102] zone axis with the the (010) plane edge on. (b) The same dislocations near a [110} zone axis with a $\bar{2}21$ reflection, and the (001) plane nearly edge on. Arrows mark locations where cross slip may have occured.

Considerable 'c' component dislocation activity is observed at 650C, as at room temperature (6,7). However the arrangement of these dislocations differed considerably in detail. The dislocations are localised in slip bands as shown in Fig.4. Within each slip band, the dislocations are arranged in pairs (Fig.4b) which careful contrast experiments show to be dipoles (Fig.4c).

Burgers vector analysis shows that both <102> and <114> are operative. Typical contrast experiments are given in Fig.5 . An analysis of the line directions shows these dipoles to be essentially in screw orientations. Tilting experiments were conducted which unambiguously show the <102> dipoles to lie on {041} planes (Fig.5,), and <114> dipoles to lie on {221} planes. An analysis of the contrast behaviour of individual dislocations constituting the dipoles with varying excitation parameters (8) shows that the images are consistent with a magnitude of g.b equal to 4 for (002) and (221) reflections, so that these dislocations are <102> and 1/2<114> superdislocations. Weak beam microscopy shows the dislocations to be dissociated (Fig.5e) with the separation of the superpartials in the projection plane of the image to be as low as 1.5nm which just borders the expected resolution in weak beam microscopy. The very small superpartial separation precludes the determination of the dissociation plane with any great degree of accuracy by

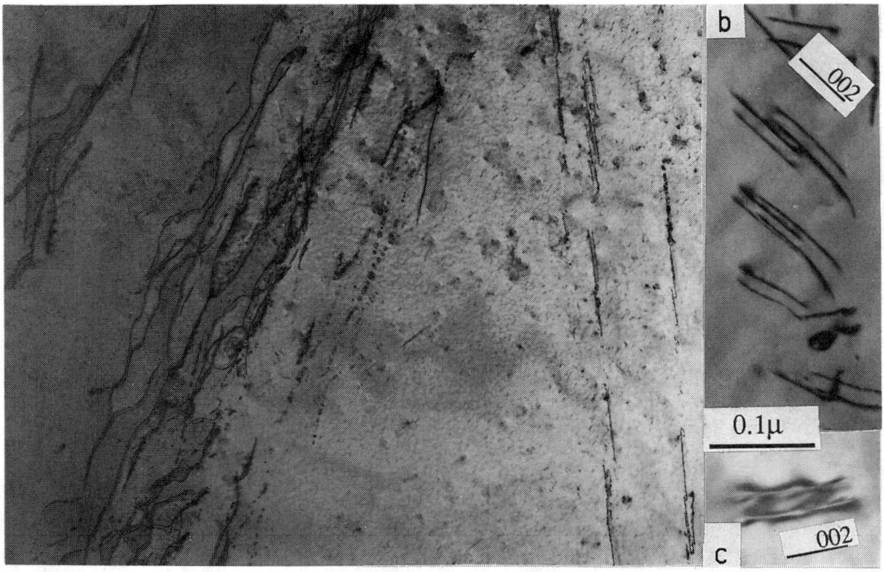

FIG.4: (a) 'c' component dislocations in a slip band. (b) The dislocations within the slip bands are paired. (c) The opposite sense of contrast of the paired dislocations show them to be dipoles. Each dislocation in the pair is associated with a double image.

tilting experiments of the type described in Figs.3 and 4. While the dipoles are essentially in screw orientations the individual dislocations constituting the dipoles are seen to cross slip, as seen in locations marked 'A' in Fig.5 . Dislocation loops are seen to trail the dipoles (Fig.6) probably formed by pinching off the long dipole segments by cross slip.

Discussion

The discussion of the results is confined to comparing the differences observed in the deformation structures generated at 650C and at room temperature (6,7). An analysis of the causes underlying these differences will be left to subsequent publications.

The macroscopic nature of dislocation arrangements remain similar to that observed on deformation at room temperature (6,7). 'a' dislocations are distributed randomly while 'c' component dislocations are confined to widely spaced slip bands. Of the two types of 'a' dislocations, [100] and <110>, the <110> dislocations are dissociated into 1/4<110> superpartials on {110} planes, again as observed on deformation at room temperature. These planes are equivalent to the prismatic planes of the closely related alpha2 structure. A considerably lower density of the [100] dislocations is, however, seen on deformation at 650C, and the few dislocations that are observed show very localised cross slip along their length with some segments dissociated into 1/2[100] superpartials on the (001) planes (equivalent to the basal planes of the alpha2 structure). Since the difference in the magnitude of the Burgers vectors of the [100] and <110> dislocations is very small (6), the difference in their behaviour must be attributed to the nature of the antiphase boundary structure bound by the respective superpartials. At this time we do not understand why this difference is manifested at 650C and not at room temperature.

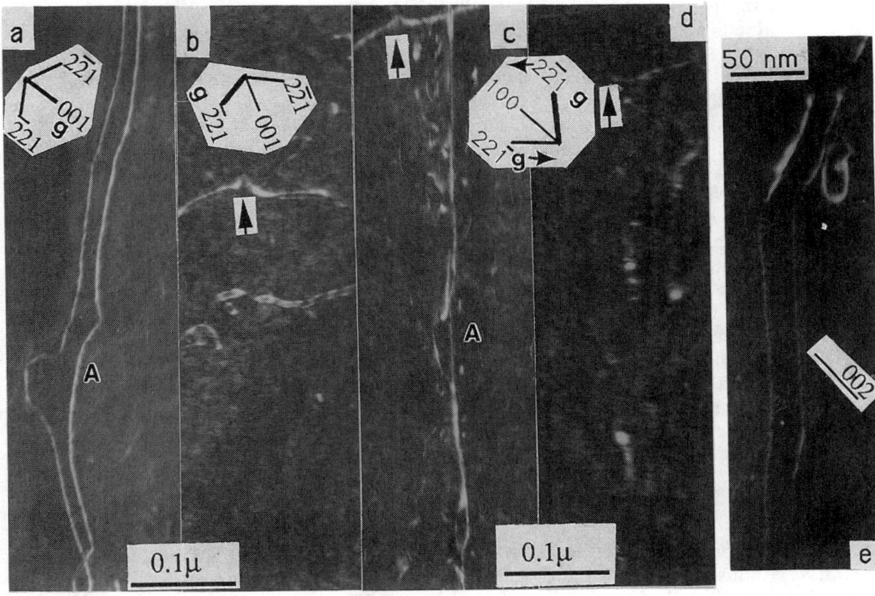

FIG.5 Burgers vector and slip plane analysis of 'c' component dislocations.

(a) weak beam image with a 002 reflection near a [110] zone axis.

(b) invisibility condition in the [110} zone axis with the $\bar{2}21$ reflection.

(c) Weak beam image with the $2\bar{2}1$ reflection in the [012] zone axis. The dipoles now overlap with the $(22\bar{1})$ plane edge on.

(d) invisibility condition with the $22\bar{1}$ reflection in the [012] zone axis. The invisibilty conditions confirm that the Burgers vector is [102].

(e) Dissociation of a [102] dislocation into superpartials.

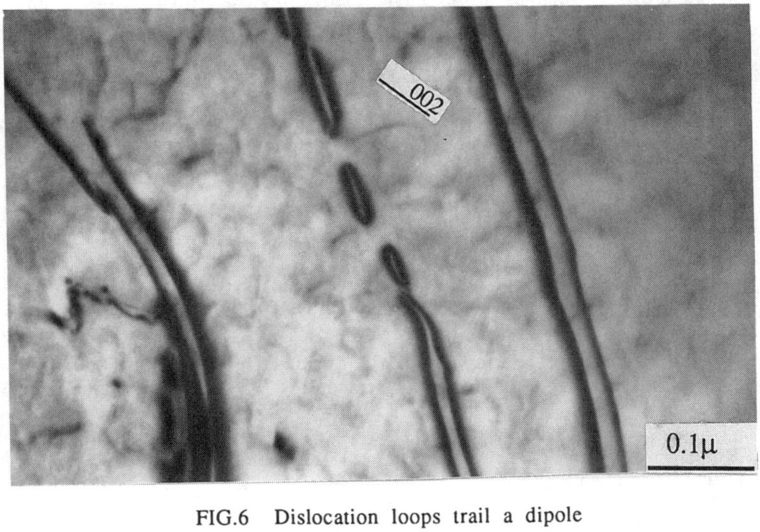

FIG.6 Dislocation loops trail a dipole

'c' component dislocations are confined to slip bands, but their arrangement within the slip bands is quite distinct from that observed at room temperature. In the latter case, <114> and <102> dislocations were observed to exist in curved, mixed orientations on {221} and {041} planes, respectively (these are equivalent to the {2201} planes of the alpha2 structure). These dislocations were dissociated on these planes into superpartials, and the trailing superpartial was observed to be strongly pinned by unspecified obstacles. At 650C, however, these dislocations are observed as screw dipoles on the same planes. The individual dislocations constituting the dipoles are superdislocations which are dissociated into superpartials with much smaller separations than those seen at room temperature. No evidence of superpartial pinning was observed at this temperature; on the contrary, cross slip of the superdislocations as a whole appears to occur. These features suggest that (a) the obstacles to trailing superpartials which exist at room temperature are not operative at 650C (b) edge <114> and <102> dislocations are extremely mobile at 650C since they are not observed at all (c) the glide of these dislocations generates screw superdislocation dipoles by a mechanism as yet unknown and (d) individual superdislocations constituting the dipole can cross slip, in the process generating superdislocation loops. It is emphasised that 'c' component dislocations are rarely observed in the alpha2 phase, both at room temperature and high temperatures, and that too only in grains for which the ratio of the resolved shear stress for 'c' component slip to that for 'a' slip is very low (9).

Conclusions

1. The dislocation structure in a Ti-26Al-21Nb alloy has been examined after compressive strains of 2% at 650C. [110] dislocations are observed to be randomly distributed, and dissociated to 1/4<110> superpartials on {110} planes. Very few [100] dislocations are seen, and these show evidence of profuse localised cross slip along their line lengths, with segments dissociated into 1/2[100] superpartials on (001) planes.

2. 'c' component slip occurs in the form of <114> and <102> dislocations which are localised in slip bands on {221} and {041} planes, respectively. These are present in the form of screw dipoles. Individual superdislocations constituting the dipoles are seen to be dissociated into superpartials with separations of about 1.5 nm. Cross slip of the superdislocations is observed with trailing superdislocation loops.

Acknowledgements

One of us, DB, acknowledges funding of the Defence Research and Development Organisation, India, for the TEM characterisation, and is grateful to the Director, DMRL for permission to publish this work.

References

1. R.G. Rowe, High Temperature Aluminides and Intermetallics, ed. S.H. Wang et al., (Warrendale PA: The Metallurgical Society, 1990), 375-401.

2. R.G. Rowe, Microstructure-Property Relationships in Titanium Aluminides and Alloys, ed. Y-W. Kim and R. Boyer, (Warrendale PA: The Metallurgical Society, 1991), 387-398.

3. R.G. Rowe et al., High Temperature Ordered Intermetallic Alloys IV, ed. L.A. Johnson, D.P. Pope and J.O. Stiegler, (Pittspurgh PA: MRS, 1991), 703-708.

4. D. Banerjee et al., Acta Metall., 36 (1988), 871-882.

5. B. Mozer, L. Bendersky and W.J. Boettinger, Scripta Metall., 24 (1990), 2363-2368.

6. D.A. Koss et al., High Temperature Aluminides and Intermetallics, ed. S.H. Wang et al., (Warrendale, PA: The Metallurgical Society, 1990), 175-196.

7. D. Banerjee, R.G. Rowe ands E.L. Hall, High Temperature Ordered Intermetallic Alloys IV, ed. L.A. Johnson, D.P. Pope and J.O. Stiegler, (Pittspurgh PA: MRS, 1991), 285-290.

8. D. Banerjee and D. Mukherjee, unpublished research, DMRL, India (1992).

9. B.A. Marquardt et al., "Research and Development for Improved Toughness Aluminides", (Report WRDC-TR-89-4133, Wright Research and Development Center, 1989).

MICROMECHANISMS OF FRACTURE IN TITANIUM ALUMINIDES

Anthony W. Thompson

Dept. of Materials Science and Engineering
Carnegie Mellon University
Pittsburgh, PA 15213

Abstract

Low ductility and brittle fracture modes near room temperature are familiar in titanium aluminide alloys based on both Ti_3Al and TiAl. The fracture mode is often characterized as "cleavage," despite the fact that not all the fracture surface is cleavage-like, nor does it appear that the "cleavage" is either catastrophic in nature or normal-stress controlled. Instead, ductile regions may be present among brittle areas, and these correlate well with relative ductility and toughness data. Micromechanisms of cleavage fracture can be quantitively compared to experimental results.

Introduction

Extensive development work has taken place in recent years because of the technological potential [1,2] of alloys based on titanium aluminides, either the DO_{19} Ti_3Al phase (α_2) or the $L1_0$ TiAl phase (γ). Considerable study of the physical metallurgy of these alloys has established that alloying additions which stabilize the bcc β phase in α_2 alloys, or which give rise to a certain amount of α_2 phase in γ alloys, are a major benefit to ductility and toughness at low temperatures [2-5]. It also appears that mechanical properties exhibit a marked sensitivity to microstructure, as is true for many structural alloys.

There is as yet little fundamental understanding of the fracture processes in these alloys, particularly in terms of their dependence on microstructure. Fracture processes operating on the scale of the microstructure, i.e. of the order of micrometer size scales, have come to be called "micromechanisms." As in other structural materials, extant micromechanism concepts can be applied to these alloys, and should be relevant not only for fundamental fracture issues [6], but also for technological uses such as NASP (National Aerospace Plane) applications [7]. However, many questions remain about the mechanism(s) and details of these effects, and the need for additional work to answer such questions is highlighted in this summary.

Mechanical Properties

Mechanical properties of titanium aluminide alloys have been summarized in several recent overviews [2-5,8,9]. There is a pronounced strain rate effect, much like what has been reported in conventional Ti alloys containing hydrides [10]. The dependence of mechanical properties on

Titanium '92
Science and Technology
Edited by F.H. Froes and I. Caplan
The Minerals, Metals & Materials Society, 1993

temperature for a particular microstructure of Ti-24-11 [11-13] is shown in Fig. 1. The ductility shown is the uniform true strain ε_p, rather than the fracture ductility ε_f. Notched specimens were tested to investigate fracture processes [14,15]. When results are converted to the local, true fracture stress below the notch [14,16-18], the results were as shown in Fig. 2, with a temperature-independent stress σ_F [13,14], consistent with the original postulate of Orowan [19]. Both toughness and crack growth resistance or R-curve behavior have also been measured for this material, as reported in more detail elsewhere [12,13,20,21]. Fig. 3 shows that the local fracture stress σ_F and the plane strain fracture toughness K_{Ic} both have a similar dependence on temperature.

Micromechanisms. There is considerable current interest, now that the basic mechanical properties are being established, to turn to the issues of mechanisms of behavior, for example, deformation, on which there has been extensive work [4,5]. In fracture, work on mechanisms is beginning. The fractographic appearance of both α_2 and γ fracture surfaces is reminiscent of cleavage, and that term is widely used in the literature on these alloys. Yet some care in use of this term may be appropriate [13,22,23]. It remains to be shown that the fractures obey all the criteria for classical cleavage [24,25], particularly in γ alloys, where fracture facets are commonly observed to be curved or conchoidal [22,23]. Curved fracture facets are also evident in α_2 alloys when the α_2 phase is sufficiently large, e.g. $\geq 10\ \mu$m. Fracture surfaces of many microstructures in titanium aluminide alloys exhibit regions of ductile tearing between brittle facets. This tearing evidently arises [3-5,12] from retained or transformed β phase in α_2 alloys and from α_2 regions in γ alloys. It remains to be explored whether a "toughest link" model [26] can be applied to such aluminide fractures.

For the sake of discussion, however, we may consider that fracture in aluminide alloys is in fact a cleavage process. On that basis, comparisons are possible to extant theories [6,13]. In α_2 alloys, for example, the appropriate mechanism to compare would be the Smith model for homogeneous cleavage [27], that is, cleavage which is not nucleated by a second-phase particle or other microstructural inhomogeneity. For that model, the initiation condition for a crack coinciding with the slip plane at the head of the pile-up, that is, for slip-initiated cracking in the slip plane, as appears to occur [13] in Ti-24-11, has been written as [27]

$$\tau_F = \tau_f + \{2\mu\gamma_{eff} / \pi L\,(1 - \nu)\}^{1/2}, \qquad (1a)$$

where τ_F stands for a critical shear stress for fracture, τ_f for the friction shear stress, μ for the shear modulus which can be calculated from Young's modulus E as $\mu = E/2(1+\nu)$, γ_{eff} for the effective energy required for fracture (expected to have the approximate magnitude of the true surface energy γ_s), L for the pile-up length, and ν for Poisson's ratio. In terms of tensile stresses, Eq. (1a) can be rewritten as

$$\sigma_F = \sigma_f + g\{2\mu\gamma_{eff} / \pi L\,(1 - \nu)\}^{1/2}, \qquad (1b)$$

where g is a geometric factor resolving the flow stress into the appropriate shear component (a Taylor factor of roughly two should reasonably approximate g), and σ_F and σ_f are tensile analogs of the values in Eq. (1a). The model is sketched in Fig. 4, where L is identified with grain size d. For the present circumstances, the dimension d should be interpreted as the critical dimension of the microstructural constituent which cracks, i.e. the width of an α_2 lath in an α_2-base alloy, or a γ plate in a γ alloy.

Equation (1) is suitable for initiation-controlled cleavage [6,27], a process in which microstruc-

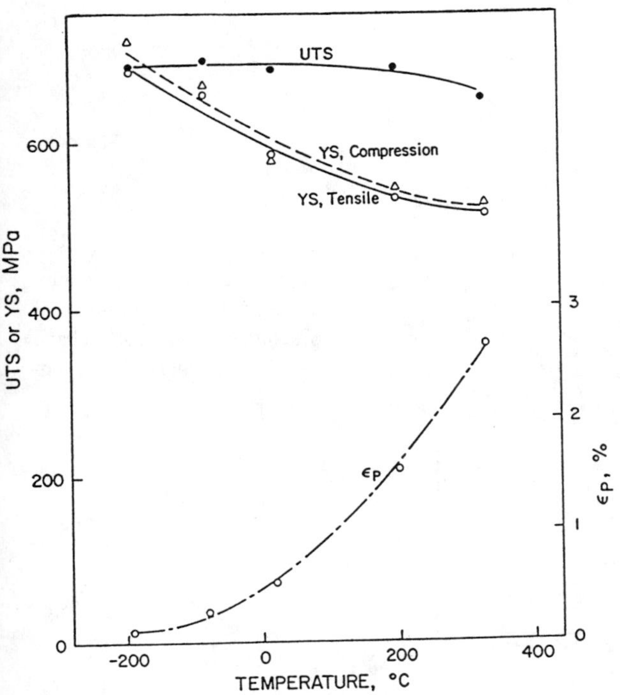

Figure 1. The dependence on temperature of compressive and tensile yield strength (YS), failure stress (either UTS, ultimate tensile strength, or fracture stress), and ductility (uniform true strain ε_p). Each point averages 2 to 5 tests. From ref. 13.

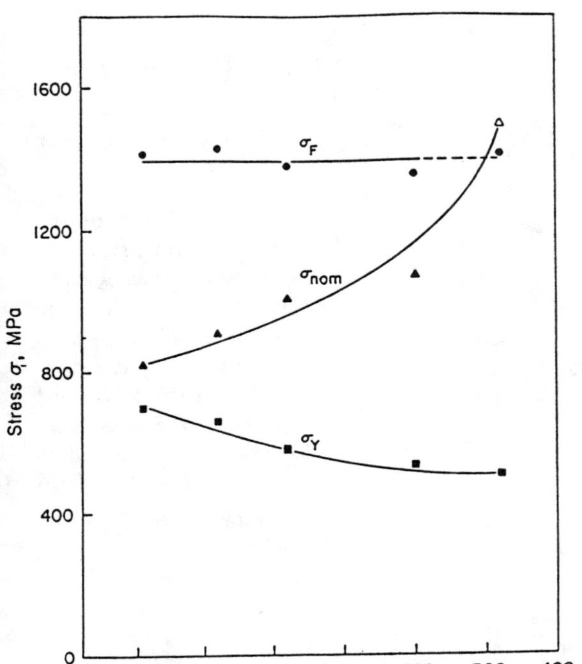

Figure 2. Determination of temperature-independent local fracture stress σ_F in bend-bar tests, with YS and nominal stress σ_{nom} values included. From ref. 13.

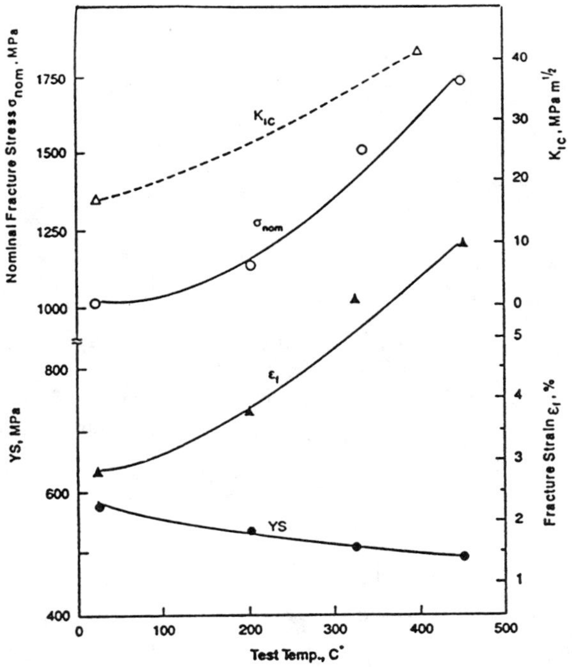

Figure 3. Dependence of YS, fracture ductility ε_f, fracture toughness K_{Ic} and nominal fracture stress in notched bars, $\sigma_{nom,}$ on temperature. From ref. 12.

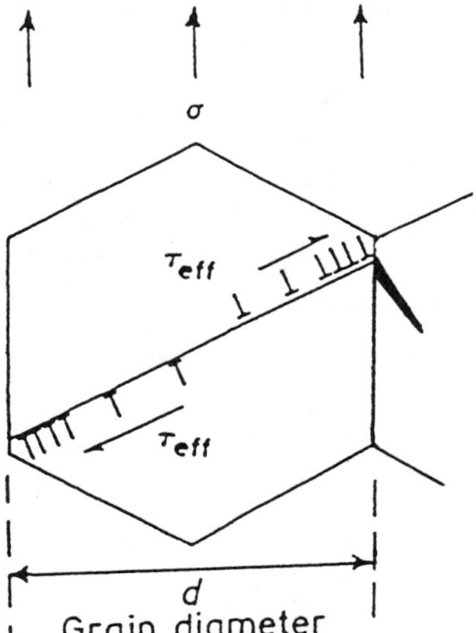

Figure 4. A sketch of the microstructural situation envisioned for development of Equation (1), text, with the effective shear stress τ_{eff} given by the difference between the applied shear stress τ_{app} and the internal or friction stress τ_i. The pile-up is imagined to nucleate a Zener (or Stroh) crack [6] in the adjacent grain. Adapted from ref. 6.

tural nuclei such as the carbides which dominate cleavage in mild steel [25,28-31] are absent or ineffective. In this situation, the value of γ_{eff} controls fracture. Since in all real materials $\gamma_{eff} \geq \gamma_s$, where any "surplus" irreversible energy above γ_s is designated γ_{irr}, one can write that $\gamma_{eff} = \gamma_s + \gamma_{irr}$. However, since a major contributor to γ_{eff} is plastic work γ_p, the term γ_{irr} is usually [32] reserved for minor effects such as heating from relaxation of stretched atomic bonds as the crack tip passes, deviations of the fracture surface from perfect flatness, and so forth. The total amount of plastic work, however, comprises both "precursor" work, i.e. plastic work which is done in enlarging the crack-tip plastic zone prior to crack extension, and "growth" work associated with crack propagation. Experimental techniques are chosen [6,14,17,31] to measure only the growth part of γ_p. Thus the γ_{eff} term is interpreted as $\gamma_{eff} = \gamma_s + \gamma_{p\,(growth)} + \gamma_{irr}$.

The measured value [13] of γ_{eff} for Ti-24-11 was 27 J/m^2, at least an order of magnitude greater than the expected value of the true surface energy of the lattice, about 0.9 J/m^2 for Ti_3Al [13]. Such an "effective" energy value is broadly consistent with results on iron [25,30,31], where measured values of γ_{eff} are about 14 J/m^2 and the known value of γ_s is about 2 J/m^2. It should be reiterated that this γ_{eff} corresponds to crack propagation [6,14] and thus is not directly related to the total fracture energy $\gamma = 2\gamma_s + \gamma_p$, where $\gamma_p = \gamma_{p\,(growth)} + \gamma_{p\,(precursor)}$. In Ti-24-11, γ has a value [13] of about 3000 J/m^2 because of the large amount of $\gamma_{p\,(precursor)}$ work associated with plastic zone development, typical of metallic alloys with tougnesses of 10 MPa$\sqrt{m}$ or more [6,13,33].

Another approach to this fracture interpretation [13] is to estimate the fracture toughness K_{Ic} from σ_F and the other parameters shown above. The estimated value calculated was $K_{Ic} = 18.5$ MPa$\sqrt{m}$, which is very close to the experimental average value [13] of 17.2 MPa$\sqrt{m}$. However, because g, γ_{eff}, and L in Eq. (1) are uncertain, this agreement may be fortuitous. Nevertheless, it indicates that a Smith-like model may apply to cleavage fracture in Ti-24-11. More work is needed to investigate the details of the fracture process, relevant elastic-plastic fracture mechanics [33,34], operative micromechanism(s) [6,25,33], and crack propagation behavior in Ti-24-11 to place this conclusion on a firm basis.

The alternative to the foregoing estimates from a cleavage model would be a "critical distance" approach, as first postulated by Ritchie, Knott and Rice [31]. This approach uses the measured ratio of σ_F to the yield strength, YS, to determine the distance l^* ahead of the crack tip at which that stress ratio must be maintained to propagate fracture [14,,31,33]. For the Ti-24-11 on which the above measurements were made [13], $\sigma_F / YS = 2.37$, and the calculation [13] then yields a value for l^* of about 16 μm. This estimated l^* value, equalling about three α_2 lath widths, is broadly consistent with the microstructural dimensions found in other determinations [31,33] of l^*. It would be of interest to examine microcracking in α_2 laths to permit application of a statistical model [34] to Ti-24-11, and such work is now in progress.

Microstructure effects. The calculations just described are for a single microstructure of α_2 material. Microstructural variation could introduce a variety of effects. For example, the lath or plate size of the α_2 phase is readily varied in α_2 alloys, and it may be essential to refine understanding of the factors in Eq. (1) to interpret fracture within laths. It is evident that variation in lath dimensions could significantly affect fracture, through the dimension L in Eq. (1), but heat treatment to vary lath width could also alter composition within α_2 laths, so that both the slip resistance τ_f in Eq. (1), and also γ_{eff} could be affected. These aspects of local plasticity appear to require a fracture model with a strain component [17,35]. On the other hand, should it prove the case that normal-stress cleavage of laths is dominant, as appears to be the case in many hydride-forming metals [10,36], then the "particle cracking" version of Smith's model [28] would require more detailed examination. This could prove difficult, because elastic property effects from crystallographic texture, or variation in the parameters which control surface energy, may be essential to the process. Additional metallurgical study, of phase transformations and microstructural dependence of mechanical properties of aluminides [3-5,8,37], would also pay dividends. Regardless of how this "cleavage" fracture is nucleated, it appears that considerable work remains to be done to elucidate the processes of brittle fracture in titanium aluminide alloys.

Acknowledgements

I am grateful for discussions with J.J. Lewandowski on aluminide fracture, and with H.A. Lipsitt and R. Strychor on aluminide metallurgy, as well as the experimental work performed by W.-Y. Chu during his research visit to Carnegie Mellon University. Preparation of this manuscript has been supported by the David Jackson Fund for Scholarly Endeavors and by the Alcoa Center for Engineered Materials at Carnegie Mellon, and has benefited from the warm hospitality of the Materials Department of the University of California, Santa Barbara, during my sabbatical visit.

References

1. H.A. Lipsitt, in *Ordered Intermetallic Alloys* (MRS Symp. Vol. 39), ed. C.C. Koch, C.T. Liu and N.S. Stoloff (Pittsburgh: Mater. Res. Soc., 1985), pp. 351-364.
2. R. L. Fleischer, D.M. Dimiduk and H.A. Lipsitt, *Ann. Rev. Mater. Sci.*, 19 (1989), 231-263.
3. C.H. Ward, J.C. Williams, A.W. Thompson, D.G. Rosenthal, and F.H. Froes, in *Proc. 6th World Conf. on Titanium*, Vol. 2, ed. P. Lacombe, R. Tricot and G. Beranger (Paris: Les Editions de Physique, 1989), pp. 1103-1108.
4. D.A. Koss, D. Banerjee, D.A. Lukasak and A.K. Gogia, in *High Temperature Aluminides and Intermetallics*, ed. S.H. Whang, C.T. Liu, D.P Pope and J.O. Stiegler (Warrendale, PA: TMS-AIME, 1990), pp. 175-196.
5. J. M. Larsen, K. A. Williams, S. J. Balsone and M. A. Stucke, in *High Temperature Aluminides and Intermetallics*, ed. S.H. Whang, C.T. Liu, D.P Pope and J.O. Stiegler (Warrendale, PA: TMS-AIME, 1990), pp. 521-556.
6. A.W. Thompson and J.F. Knott, *Metall. Trans. A*, in press.
7. H.G. Nelson, *SAMPE Quarterly*, 20, No. 1 (1988), 20-23.
8. Y.-W. Kim and F.H. Froes, in *High Temperature Aluminides and Intermetallics*, ed. S.H. Whang, C.T. Liu, D.P Pope and J.O. Stiegler (Warrendale, PA: TMS-AIME, 1990), pp. 465-492.
9. W.-Y. Chu and A.W. Thompson, *Scripta Metall. Mater.*, 25 (1991), 641-644.
10. N.E. Paton and J.C. Williams, in *Hydrogen in Metals*, ed. I.M. Bernstein and A.W. Thompson (Metals Park, OH: Amer. Soc. Metals, 1974), pp. 409-431.
11. W.-Y. Chu, A.W. Thompson, and J.C. Williams, in *Hydrogen Effects on Material Behavior*, ed. N.R. Moody and A.W. Thompson (Warrendale, PA: TMS-AIME, 1990), pp. 543-553.
12. A.W. Thompson, in *Environmental Effects on Advanced Materials*, ed. R.H. Jones and R.E. Ricker (Warrendale, PA: TMS-AIME, 1991), pp. 21-33.
13. W.-Y. Chu and A.W. Thompson, *Metall. Trans. A*, 23A (1992), 1299-1312.
14. J.F. Knott, *Fundamentals of Fracture Mechanics*, Revised Ed. (London: Butterworths, 1979).
15. A.W. Thompson, *Mater. Sci. Technol.*, 1 (1985), 711-718. 16.
16. J.R. Griffiths and D.R.J. Owen, *J. Mech. Phys. Solids*, 19 (1971), 419-431.
17. J.J. Lewandowski and A.W. Thompson, *Acta Metall.*, 35 (1987), 1453-1462.
18. I.-G. Park and A.W. Thompson, *Metall. Trans. A*, 22A (1991), 1615-1626.
19. E. Orowan, *Trans. Inst. Engrs. Shipbuilders Scotland*, 89 (1945), 162-215.
20. A.W. Thompson and W.-Y. Chu, in *Summary Proceedings of the 3rd Workshop on Hydrogen-*

Materials Interaction, NASP Joint Program Office Workshop Pub. 1007, ed. H.G. Nelson (Moffett Field, CA: NASA-Ames, 1990), pp. 125-130.

21. W.-Y. Chu and A.W. Thompson, in *High Performance Composites for the 1990's*, ed. S.K. Das, C.P. Ballard and F. Marikar (Warrendale, PA: TMS-AIME, 1991), pp. 143-157.
22. A.W. Thompson and W.-Y. Chu, in *Microstructure/Property Relationships in Titanium Alloys and Titanium Aluminides*, ed. Y.-W. Kim and R.R. Boyer (Warrendale, PA: TMS-AIME, 1991), pp. 165-177.
23. W.-Y. Chu and A.W. Thompson, *Scripta Metall. Mater.*, 31 (1991), 2133-38.
24. C.D. Beachem and R.M.N. Pelloux, in *Fracture Toughness Testing and Its Applications*, STP 381 (Philadelphia: A.S.T.M., 1965), pp. 210-244.
25. J.F. Knott, in *Atomistics of Fracture*, Proc. NATO Advanced Research Institute, ed. R.M. Latanision and J.R. Pickens (New York: Plenum, 1983), pp. 209-234.
26. A.W. Thompson, in *Fracture 1977*, Proc. 4th Int. Conf. on Fracture, Vol. 2, ed. D.M.R. Taplin (Waterloo, Ont.: Univ. Waterloo Press, 1977), pp. 237-242.
27. E. Smith, *Acta Metall.*, 14 (1966), 985-989 and 991-996.
28. E. Smith and J.T. Barnby, *Met. Sci. J.*, 1 (1967), 56-64.
29. C.J. McMahon and M. Cohen, *Acta Metall.*, 13 (1965) 591-604.
30. J.F. Knott, *J. Iron Steel Inst.*, 205 (1967), 288-291.
31. R.O. Ritchie, J.F. Knott and J.R. Rice, *J. Mech. Phys. Solids*, 21 (1973), 395-410.
32. J.P. Hirth, *Phil. Trans. Roy. Soc. A*, 295A (1980), 139-149.
33. R.O. Ritchie and A.W. Thompson, *Metall. Trans. A*, 16A (1985), 232-248.
34. T. Lin, A.G. Evans and R.O. Ritchie, *Acta Metall.*, 34 (1986), 2205-2216.
35. R.O. Ritchie, W.L. Server and R.A. Wullaert, *Metall. Trans. A*, 10A (1979), 1557-1570.
36. H.K. Birnbaum, in *Hydrogen Effects on Material Behavior*, ed. N.R. Moody and A.W. Thompson (Warrendale, PA: TMS-AIME, 1990), pp. 639-658.
37. R. Strychor, J.C. Williams and W.A. Soffa, *Metall. Trans. A*, 19A (1988), 225-234.

INFLUENCE OF STRAIN RATE ON THE STRUCTURE/PROPERTY

BEHAVIOR OF THE ALPHA-2 ALLOY Ti-24.5Al-10.5Nb-1.5Mo*

George T. Gray III, Sun Ig Hong, and B.J. Marquardt+

Materials Science and Technology Division
Mail Stop G730
Los Alamos National Laboratory
Los Alamos, NM 87545

+ GE Aircraft Engines, 1 Neumann Way
Evendale, Ohio 45215-6301

Abstract

In this paper, a study of the effect of strain rate and temperature on the substructure evolution and mechanical response of a Ti_3Al (α_2) alloy, Ti-24.5Al-10.5Nb-1.5Mo, is presented. The compressive true stress-true strain response and substructure evolution were both found to depend on both the applied strain rate and test temperature. The strain-rate sensitivity of the flow stress at 2% strain was 0.008. The rate of strain hardening was seen to increase slightly with increasing strain rate. The calculated ambient temperature strain-hardening rate for the quasi-static (0.001 s^{-1}) data was 3300 MPa / unit strain and 4200 MPa / unit strain at 6000 s^{-1}. Calculation of shear modulus normalized hardening rates yielded μ / 165 for the quasi-static rate data and μ / 130 for the high-rate data at room temperature. Increasing temperature at high strain rates was observed to decrease the flow stress but have no effect on the rate of strain hardening. Dislocation g•b analysis revealed that following room temperature deformation at low strain rate the dislocations were **a**-dislocations lying on basal planes, 2nd order pyramidal (**a/2** + **c**) slip on $\{1\bar{2}11\}$, and 1st order pyramidal **a**-slip on $\{10\bar{1}1\}$. Increasing the rate of deformation at 25°C increased **a**-slip on prism planes and decreased basal slip. Higher strain-rates and elevated temperatures increased the amount of basal slip. The defect generation and the rate sensitivity of Ti-24.5Al-10.5Nb-1.5Mo are discussed as a function of strain rate and temperature.
*Work performed under the auspices of the U.S. Department of Energy

Introduction

Although the influence of strain rate on the microstructure / property relationships of pure titanium and a variety of titanium alloys has been extensively studied[4], the effect of strain rate on the stress-strain and deformation response of titanium aluminides has received limited attention[1-6]. Dynamic constitutive behavior is however crucial to understanding the high-speed deformation and impact performance of titanium aluminides under high-rate loading histories such as might be encountered during foreign object damage, high-speed forging, or machining. To date, only two studies have investigated the mechanical response of α_2-based alloys to high-rate deformation[3,5]. Results of strain-rate change tests on Ti-25Al-9Nb and Ti-24Al-11Nb revealed that the strain hardening and strain-rate hardening responses of the two alloys were very similar[3]. The strain-rate hardening exponents were measured to be 0.008 and 0.010 at 27°C for the Ti-25Al-9Nb and Ti-24Al-11Nb, respectively. The rate sensitivity

Titanium '92
Science and Technology
Edited by F.H. Froes and I. Caplan
The Minerals, Metals & Materials Society, 1993

of the Ti-24Al-11Nb alloy was found to be reasonably constant with increasing temperature; an "m" value of 0.007 was measured at 650°C[3].

Tests on the β-phase alloy Ti-11Al-23Nb found a rate sensitivity of 0.020 at room temperature, strain hardening in the β-phase to be significantly lower than in α_2, and a strong dependence of the yield strength on temperature[3]. Based on these α_2 and β-phase results, Gittis and Koss[3] concluded that the macroscopic deformation behavior of Ti-24Al-11Nb is controlled by the α_2 phase when the alloy is in a basketweave microstructural condition. High-rate compression tests conducted on Ti-24Al-11Nb studied with an equiaxed α_2 plus discontinuous β-phase microstructure exhibited somewhat different results[5]. While the rate sensitivity in the equiaxed Ti-24Al-11Nb alloy was also relatively low (m <0.01), the quasi-static work-hardening of the equiaxed alloy was approximately a factor of two higher than when the alloy was heat-treated to a basketweave structure[5]. Under dynamic-loading conditions with strain rate = 10^2 s^{-1}, the equiaxed Ti-24Al-11Nb alloy was also seen to show a pronounced decrease in strain hardening compared to quasi-static loading and the occurrence of localized plastic flow[5]. This paper reports the results of a study of the effect of strain rate and temperature on the substructure evolution and mechanical response of Ti-24.5 Al-10.5 Nb-1.5 Mo.

Experimental Procedures

The material used for this investigation was Ti-24.5Al-10.5Nb-1.5Mo produced by TIMET (composition in at.% of 23.8 Al, 10.6 Nb, 1.43 Mo, 0.19 O, 0.05 Fe and bal. Ti). The Ti-24.5Al-10.5Nb-1.5Mo alloy studied was produced by triple vacuum-arc-remelting a 410 kg, 35.5 cm-dia. ingot. The processing was started with upsetting and drawing operations above the β-transus temperature. Processing was initiated at a diameter of approximately 23 cm and continued until the diameter of the bar had been reduced to 10 cm. The alloy was heat-treated at 1085°C for 1 hour and then in a salt bath at 815°C for 30 minutes followed by aging at 704°C for 8 hours. Cylindrical compression samples (7.5 mm in dia. by 4.6 mm long) were electro-discharge machined (EDM) from the forged bar section. The in-plane microstructure of the heat-treated Ti-24.5Al-10.5Nb-1.5Mo alloy, Figure 1a and1b, consist of equiaxed primary α_2 and fine secondary α_2 plates and ordered B2(β_0)[2] forming a basketweave structure. Quasi-static compression tests were conducted at strain rates of 0.001 and 0.1s^{-1}. Dynamic tests were conducted as a function of strain rate, 1000-6000 s^{-1}, and temperature, between 25 and 700°C, utilizing a Split-Hopkinson Pressure Bar. The inherent oscillations in the dynamic stress-strain curves and the lack of stress equilibrium in the specimens at low strains make the determination of yield inaccurate at high strain rates. Specimens for optical metallography and TEM were sectioned from the heat-treated and deformed samples. TEM foils were prepared using conventional jet polishing and observed using a Phillips CM-30.

Results and Discussion

Mechanical Response

The compressive true stress-true strain response of Ti-24.5Al-10.5Nb-1.5Mo was found to depend on both the applied strain rate, which ranged from 0.001 to 6000 s^{-1}, and the test temperature, which was varied between 25 and 700°C (Figures 2 and 3). The flow stress of Ti-24.5Al-10.5Nb-1.5Mo increased with increasing strain rate. The strain-rate sensitivity, "m" ($= \delta\ln\sigma / \delta\ln\dot{\varepsilon}$), of Ti-24.5Al-10.5Nb-1.5Mo at 2% strain is 0.008. This rate sensitivity is less than one third that measured in Ti-48Al-1V[6] but higher than Ni_3Al[7] which displays no strain-rate sensitivity of the yield stress at 25°C.

The quasi-static mechanical behavior of Ti-24.5Al-10.5Nb-1.5Mo (to true strains of 0.20 in compression) exhibits stress-strain curves with falling work-hardening rates approaching

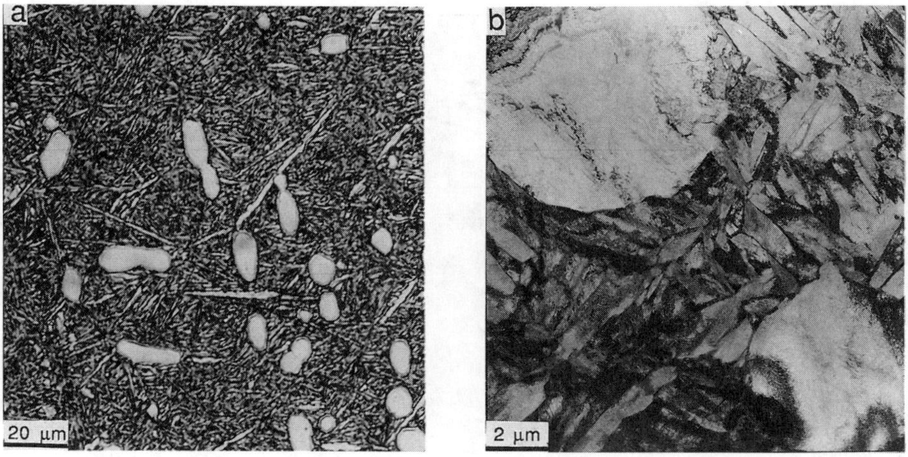

Figure 1 - a) optical micrograph, and b) TEM brightfield of the microstructure of Ti-24.5 Al-10.5 Nb-1.5 Mo showing equiaxed primary α_2 in a matrix of fine secondary α_2 plates and β_0 phase (basketweave).

saturation at approximately 1700 MPa (μ/100) similar to Ni_3Al[7] which saturates at a stress level of ~μ/105 but contrary to the almost linear stress-strain sustained hardening response of Ti-48Al-1V[6] to similar stress levels (assuming a polycrystalline shear modulus). Based on this assumption, Ti-24.5Al-10.5Nb-1.5Mo , Ti-48Al-1V, and Ni_3Al all harden to stress levels roughly a factor of three higher than pure metals such as copper and tantalum which saturate at stress levels of ~μ/333 and μ/450 at low strain rates[6]. It is likely that a portion of the ~3x factor in the saturation stress levels is due to anisotropy, such as a higher Taylor factor in the low symmetry intermetallic systems. The majority of the sustained hardening, however, is believed to result from a lack of dynamic recovery. The sustained plasticity exhibited by this alloy in compression is similar to the crack-free deformation in Ti-48Al-1V tested in compression to 20% strain[6]. These results suggest that the limited tensile ductility exhibited in Ti-aluminides may be due to the intervention of cleavage or intergranular fracture at stresses lower than that required to activate sufficient slip to accommodate polycrystalline plasticity rather than solely an inherent lack of sufficient slip systems[8].

In disordered metals the decrease in work hardening at high stresses and strains is generally linked to dynamic recovery. This is usually associated with the loss of dislocations at cell walls, tangles, etc. It is worth noting that in the case of intermetallics, however, the absolute stress levels are unusually high, i.e., ~ μ/100 compared with ~μ/300 in disordered metals. This suggests that: 1) dynamic recovery mechanisms in intermetallics are physically different than those in disordered metals, and 2) it is possible that other mechanisms are limiting (saturating) the flow stress such as mutual annihilation by mechanical recombination of dislocations[8]. These results show that while hardening in disordered metals saturates at modest stress levels, some ordered systems continue Stage-II hardening to very high stress levels. The details of how dynamic recovery processes would occur in ordered intermetallics will most probably depend on the local dislocation configuration. Independent of the controlling mechanisms these results clearly suggest that radically different dynamic recovery processes appear to be operative between α_2 and γ-based titanium aluminides.

The rate of strain hardening in Ti-24.5Al-10.5Nb-1.5Mo is seen to increase slightly with increasing strain rate. The calculated average strain-hardening rate, θ , for the quasi-static (0.001 s^{-1}) data at 25^0 C is 3300 MPa / unit strain. The hardening rate at 6000 s^{-1} at 25^0 C is 4200 MPa / unit strain. Normalizing the work hardening rates with the Taylor Factor for a random polycrystal, [θ / $(3.07)^2$], yields work hardening rates of μ / 165 for the quasi-static

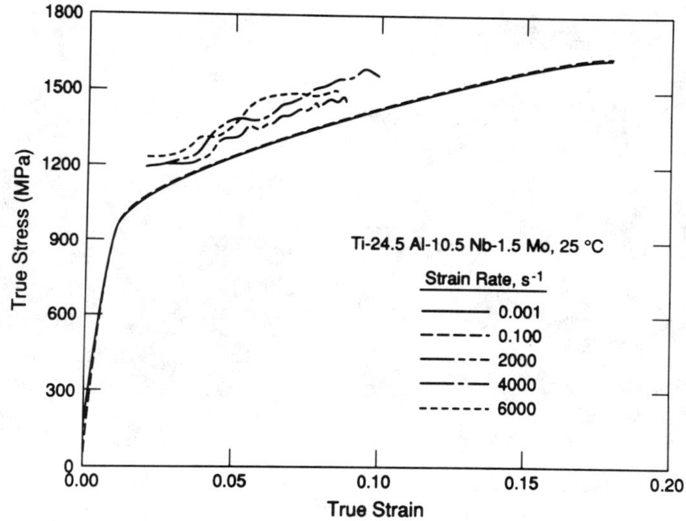

Figure 2 - Compressive Stress-Strain response of Ti-24.5Al-10.5Nb-1.5Mo as a function of the applied strain rate at 25⁰ C.

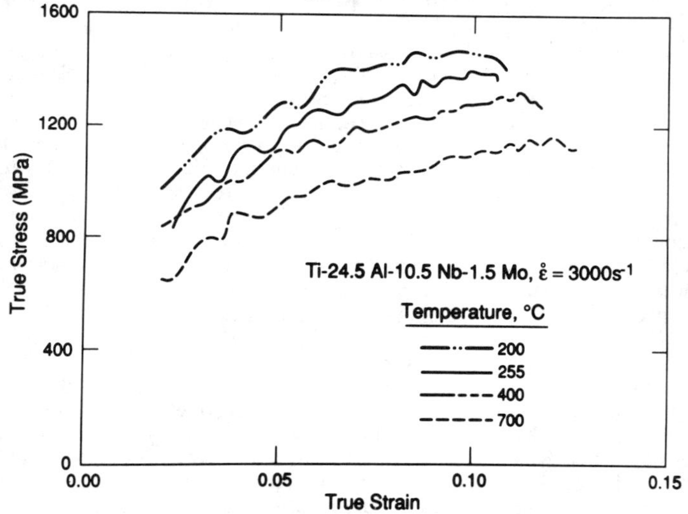

Figure 3 - Compressive stress-strain response of Ti-24.5Al-10.5Nb-1.5Mo as a function of test temperature at a strain rate of ~3000 s^{-1}.

rate data and μ / 130 for the high-rate data assuming a shear modulus of 57 GPa for Ti_3Al at room temperature[9]. In comparison to a strain hardening rate of $\mu/175$ for Ti-48Al-1V at 4500 s^{-1} at 25⁰C[6], the shear modulus normalized strain hardening rate of Ti-24.5Al-10.5Nb-1.5Mo is seen to exhibit sustained work-hardening at high-strain-rate similar to Ti-48Al-1V rather than that exhibited by conventional Ti-alloys such as Ti-6Al-4V as discussed below. The observation of stable work hardening in Ti-24.5Al-10.5Nb-1.5Mo, in contrast to previous high-rate data on Ti-24Al-11Nb[5], is thought to be a direct result of the finer-scaled crack-free basketweave microstructure of the current study.

The rate sensitivity and strain hardening rates of Ti-24.5Al-10.5Nb-1.5Mo exhibit somewhat different trends when compared to those seen in Ti-48Al-1V[6] or Ti-6Al-4V[4]. While the rate sensitivities of the yield and flow stresses of Ti-6Al-4V in both tension and compression are also high, the rate sensitivity "m" is 0.015[4]. This value is roughly twice that of Ti-24.5Al-10.5Nb-1.5Mo while the yield stress levels are all in the range of 1000 MPa for Ti-6Al-4V at low strain rate compared to approximately 950 MPa for Ti-24.5Al-10.5Nb-1.5Mo in this study. The strain hardening of Ti-6Al-4V as a function of strain rate, using a shear modulus of 44.5 GPa at 25^0C[9] is μ / 146 at a strain rate of 0.0001 s^{-1} and μ / 436 at a strain rate of 3000 s^{-1}. The normalized quasi-static strain-hardening rate for Ti-6Al-4V is similar to the high and low-rate normalized hardening rates for Ti-24.5Al-10.5Nb-1.5Mo. Both of the normalized rates, i.e. ~ μ/120 to μ/200, are consistent with Stage II-type hardening behavior such as that seen in disordered FCC polycrystals. The drastic decrease in the strain hardening in Ti-6Al-4V at high strain rate reflects a significant influence of adiabatic heating[4]. In contrast, Ti-24.5Al-10.5Nb-1.5Mo displays no temperature-induced decrement in the strain hardening but rather an increased hardening rate at high strain rates. Increasing temperature at high strain rates is observed to decrease the flow stress of Ti-24.5Al-10.5Nb-1.5Mo (Figure 3). Increasing temperature shows only a small effect on the strain hardening rate; the calculated hardening rate at 700^0 C at 4000 s^{-1} is 4178 units / unit strain. This observation is similar to the lack of a temperature-sensitivity on strain hardening seen in Ti-24Al-11Nb[3] up to 650^0C. The overall decrease in flow stress with increasing temperature is also consistent with several previous studies[2,3,10] on the temperature dependence of the yield stress for Ti_3Al.

Substructure Evolution

The fine scale of the as-heat-treated basketweave microstructure in the current study is consistent with the previous observation that Mo or V additions appear to promote finer-scale microstructures through their influence on retarding the $\beta - \alpha_2$ transformation[3]. The substructure evolution of Ti-24.5Al-10.5Nb-1.5Mo was observed to depend on the applied strain rate and temperature of deformation. The substructure of Ti-24.5Al-10.5Nb-1.5Mo deformed to ε=0.10 at 0.001 s^{-1} at 25^0C consisted of long straight screw dislocations.(Figure 4) The observation of straight screws suggests a greater degree of edge mobility and reduced cross-slip of screw dislocations. Dislocation g•b analysis revealed that following room temperature deformation at low strain rate the majority of the dislocations are **a**-dislocations lying on basal planes and 1st order pyramidal slip on $\{10\bar{1}1\}$, and 2nd order pyramidal (**a/2** + **c**) slip on $\{1\bar{2}11\}$. Increasing the rate of deformation at room temperature to 6000 s^{-1} resulted in increased **a**-slip on prism planes, and 1st order pyramidal planes, and a lessor amount of basal slip. In Figure 4a, dislocations are seen to lie on $(1\bar{1}01)$ and (0001) traces at low strain rates whereas they lie on $(\bar{1}010)$ and $(1\bar{1}0\bar{1})$ plane traces following high-strain-rate deformation (Figure 4b).

In Ti-24.5Al-10.5Nb-1.5Mo deformed simultaneously at high-strain rates and elevated temperatures the substructure was seen to be generally similar to that observed following high-rate deformation at room temperature except for an increased amount of basal slip and a somewhat higher incidence of 2nd order pyramidal slip. An interesting finding of this study is that extensive slip occurred on $<1\bar{2}10>$ $\{10\bar{1}1\}$. Court et al.[11] observed similar slip in Ti_3Al-4 at.%Nb, although less extensively than in the current study, and did not observe this slip system in unalloyed α_2. Figures 5a and 5b show **a/2** + **c** dislocations observed following high-rate deformation at both 25 and 700^0C. Figures 6a and 6b show the substructure of Ti-24.5Al-10.5Nb-1.5Mo following deformation at 700^0 C to ε=0.10 at a strain rate of 3000 s^{-1}. In Figure 6a, the trace of $(10\bar{1}1)$ planes is clearly seen; labeled "A". The slip traces labeled "B", are parallel to the traces of $(1\bar{2}11)$ or (0001), however the sharpness of the slip traces suggest that they are (0001) planes since the basal planes are parallel to the electron beam direction. The groups of dislocations labeled "C" are shown to lie along the traces of $(11\bar{2}1)$ or $(11\bar{2}\bar{1})$ planes. At elevated temperatures at high rate, particularly above 400^0 C, evidence of an increased amount of shearing of the β_0 laths was also seen. This observation is consistent with previous observations[3] that above 650^0C the

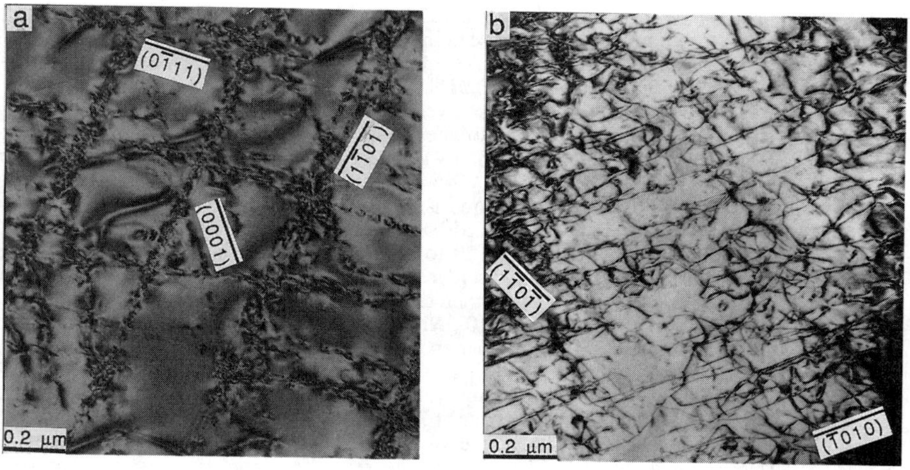

Figure 4 - Substructure of Ti-24.5Al-10.5Nb-1.5Mo following deformation at 25^0 C to $\varepsilon = 0.10$, a) 0.001 s^{-1} showing basal slip and 1st order pyramidal slip, [1$\bar{2}$10] zone axis, and b) 6000 s^{-1} showing prism slip and 1st order pyramidal slip, [1$\bar{2}$16] zone axis.

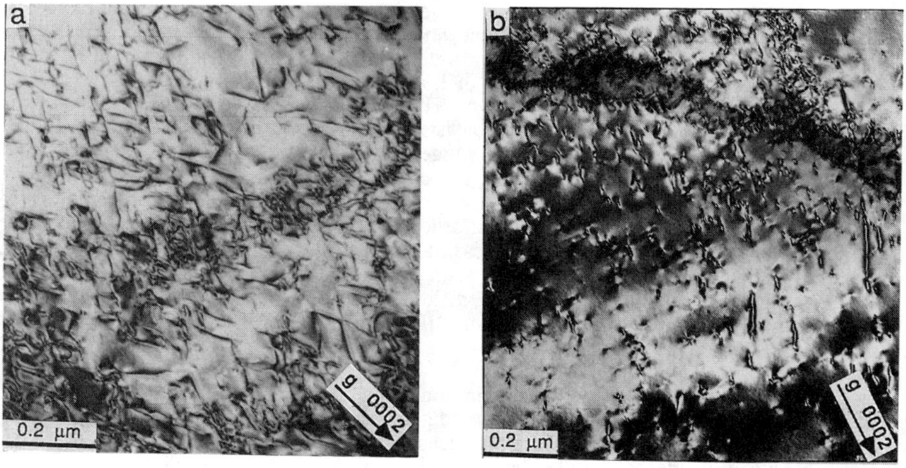

Figure 5 - Brightfield micrograph showing the presence of **a/2 + c** dislocations: a) 25^{0}C, 6000 s^{-1} and b) 700^{0}C, 3000 s^{-1}.

yield strength of β_0 decreases rapidly. Finally, contrary to previous observations of microtwinning at elevated temperatures[1,2], no deformation twins were observed in Ti-24.5Al-10.5Nb-1.5Mo, regardless of the temperature or applied rate of deformation. This observation is in total contrast to the pronounced propensity for deformation twinning in Ti-48Al-1V[6] which increases with increasing temperature and/or strain rate. This observation is important in terms of identification and modeling of the controlling mechanisms of work-hardening between the two Ti-aluminides. The similarity in normalized quasi-static work-hardening rates between Ti-24.5Al-10.5Nb-1.5Mo (μ/165) and Ti-48Al-1V (μ/175) shows that while twinning occurs readily in Ti-48Al-1V and not in Ti-24.5Al-10.5Nb-1.5Mo it is not readily obvious what role twinning plays in the overall defect storage and thereby work hardening.

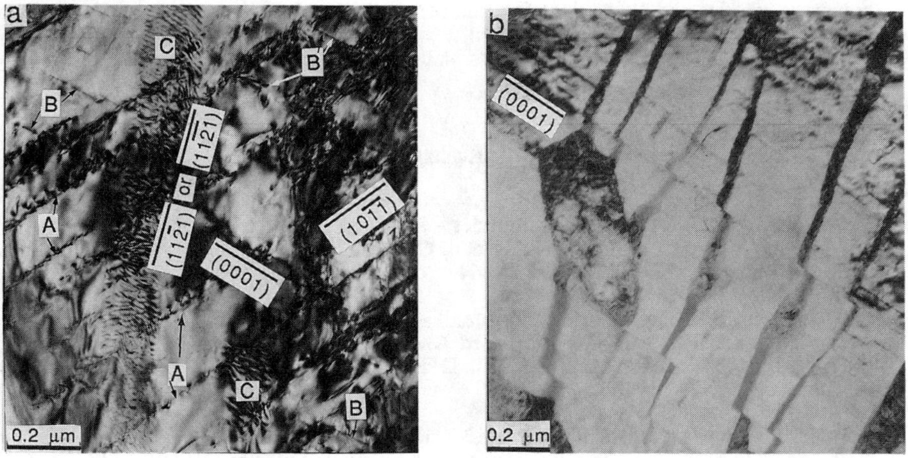

Figure 6 -Brightfield micrograph of dislocation debris in Ti-24.5Al-10.5Nb-1.5Mo following deformation to ε=0.10 at 3000 s^{-1} at 700^0C showing: a) pyramidal and basal slip, [1$\bar{2}$10] zone axis and b) increased incidence of shearing of β_0 lath at high temperature.

Summary and Conclusions

Based on a study of the influence of strain rate and temperature on the mechanical response and substructure evolution of Ti-24.5Al-10.5Nb-1.5Mo, the following conclusions can be drawn:

1. The compressive true stress-true strain response of Ti-24.5Al-10.5Nb-1.5Mo was found to depend on both the applied strain rate and the test temperature. The strain-rate sensitivity, m, of Ti-24.5Al-10.5Nb-1.5Mo at 2% strain was calculated to be 0.008.

2. The rate of strain hardening in Ti-24.5Al-10.5Nb-1.5Mo is seen to increase slightly with increasing strain rate. Stage-II work-hardening behavior in this ordered alloy is shown to continue to very high stress levels, ~μ/100, contrary to disordered metals. Increasing temperature is seen to have no effect on the rate of strain hardening; the calculated hardening rate at 700^0 C at 4000 s^{-1} is 4178 units / unit strain.

3. The substructure evolution of Ti-24.5Al-10.5Nb-1.5Mo was observed to depend on the strain rate and temperature of deformation. Following room temperature deformation at low strain rate the majority of the dislocations are **a**-dislocations lying on basal planes and 1st-order pyramidal planes, and 2nd order pyramidal (**a/2** + **c**) slip on {1$\bar{2}$11}. Increasing the rate of deformation at room temperature to 6000 s^{-1} is seen to result in increased **a**-slip on prism planes and a lessor amount of basal slip. At high-strain-rates and elevated temperatures the substructure was seen to be similar to that following high-rate deformation at room temperature except for an increased amount of basal slip and a somewhat higher incidence of 2nd order pyramidal slip. At elevated temperatures at high rate evidence of an increased amount of shearing of the β_0 laths was also seen.

Acknowledgments

The authors acknowledge the assistance of M.F. Lopez for conducting the quasi-static compression tests and W. Wright for conducting the Hopkinson-Bar tests. The authors would like to acknowledge stimulating discussions with J.D. Embury, A.D. Rollett, and M.F. Ashby. This work was performed under the auspices of the U.S. Department of Energy.

References

1. H.A. Lipsitt, "Titanium Aluminides - An Overview", High-Temperature Ordered Intermetallic Alloys, vol. 39, ed. C.C. Koch, C.T. Liu, and N.S. Stoloff, (Pittsburgh, PA: Materials Research Society, 1985), 352-364.

2. D.A. Koss, D. Banerjee, D.A. Lukasak, and A.K. Gogia, "A Review of the Deformation and Fracture of Ti_3Al-Based Alloys", in High Temperature Aluminides and Intermetallics, ed. S.H. Whang, C.T. Liu, D. Pope, and J. Stiegler, (Warrendale, PA: The Metallurgical Society, 1990), 175-196.

3. S.J. Gittis and D.A. Koss, "On the Macroscopic Flow Behavior of the Phases Present in Ti-24Al-11Nb", in High-Temperature Ordered Intermetallic Alloys III, vol. 133, ed. C.T. Liu, A.I. Taub, N.S. Stoloff, and C.C. Koch, (Pittsburgh, PA: Materials Research Society, 1989), 323-328.

4. P.S. Follansbee and G.T. Gray III, "An Analysis of the Low Temperature, Low and High Strain-Rate Deformation of Ti-6Al-4V," Metall. Trans., 20A (1989), 863-874.

5. L.S. Harbison, R.J. Bourcier, and D.A. Koss, "Flow Behavior of Ti-24Al-11Nb at High Strain Rates", in Microstructure / Property Relationships in Titanium Alloys and Titanium Aluminides, ed. Y-W. Kim and R.R. Boyer, (Warrendale, PA: The Metallurgical Society, 1991), 437-445.

6. G.T. Gray III, "Influence of Strain Rate on the Structure/Property Behavior of Ti-48Al-1V", in Microstructure / Property Relationships in Titanium Alloys and Titanium Aluminides, ed. Y-W. Kim and R.R. Boyer, (Warrendale, PA: The Metallurgical Society, 1991), 263-274.

7. H.W. Sizek and G.T. Gray III, "Mechanical and Microstructural Response of Ni_3Al at High Strain Rates and Elevated Temperatures," Shock-Wave and High-Strain-Rate Phenomena in Materials, ed. M.A. Meyers, L.E. Murr, and K.P. Staudhammer, (New York, NY: Marcel Dekker, Inc., 1992), 117-125.

8. J.D. Embury and G.T. Gray III, "Plasticity, Damage Accumulation and Fracture at Large Strains," 13th Riso Int. Sym. on Matls. Sci.: Modeling of Plastic Deformation and its Engineering Applications, 1992, in press.

9. R.E. Schafrik, "Dynamic Elastic Moduli of the Titanium Aluminides," Metall. Trans., 8A (1977), 1003-1006.

10. D.A. Lukasak and D.A. Koss, "The Flow and Fracture of a Ti_3Al-Nb Alloy", Metall. Trans., 21A (1990) 135-143.

11. S.A. Court, J.P.A. Loflander, M.H. Loretto, and H.L. Frazer, "The Influence of Temperature and Alloying Additions on the Mechanisms of Plastic Deformation of Ti_3Al", Philos. Mag. A, 61(1990) 109-139.

Microstructure, Tensile Ductility, and Fracture Toughness

of Ti-25Al-10Nb-3V-1Mo

C.H. Ward[a], I. Roman[b], J.C. Williams[c], and A.W. Thompson[d]

[a] Wright Laboratory, Wright-Patterson AFB, OH 45433 USA
[b] Hebrew University of Jerusalem, 91904 Jerusalem, Israel
[c] GE Aircraft Engines, Cincinnati, OH 45215 USA
[d] Carnegie Mellon University, Pittsburgh, PA 15213 USA

Abstract

Both the tensile ductility and fracture toughness of titanium aluminide alloys based on Ti_3Al are typically lower than those observed for conventional titanium alloys at room temperature. Six different microstructures were produced to evaluate their effects on these two properties. The microstructures were manipulated through thermomechanical processing to achieve various primary α_2 volume fractions and secondary α_2 morphologies. Room temperature tensile and fracture toughness tests were performed. These were monitored for acoustic emission during mechanical testing to aide in the understanding of the deformation and failure mechanisms. Additionally, flat, pre-polished tensile bars were tested and examined to determine the deformation behavior of the microstructures. This information, together with examination of sectioned round tensile bars, helped identify the microstructural components most critical to failure initiation. The correlation of crack nucleation and propagation to observed properties and microstructure is discussed.

Introduction

The room temperature ductility and fracture toughness of α_2 titanium aluminides have been shown to be dependent on microstructure.[1-8] However, there does not appear to be a significant correlation between these two properties when comparing microstructures. Measurement of acoustic emission (AE) during tensile testing has been used in a recent study to examine the flow and fracture behavior of Ti-25Al-10Nb-3V-1Mo (atomic percent) at room temperature.[9] In that study, it was determined that the events that lead to tensile failure in most microstructures occurred very near the fracture point. Since the nucleation of critical flaws occurs so close to failure, the effect of fracture toughness in extending tensile elongation by delaying crack propagation is negligible. Thus, it must be considered that crack nucleation is the controlling event in the room temperature tensile fracture for most microstructures in α_2 alloys. This study follows the previous report of acoustic emission response during tensile testing and attempts to specifically identify the controlling fracture initiation sites in six different microstructures.

Titanium '92
Science and Technology
Edited by F.H. Froes and I. Caplan
The Minerals, Metals & Materials Society, 1993

Experimental

Six microstructures were prepared by thermomechanical processing (TMP) using hot-die forging and heat treatment. Room temperature tensile tests were performed for each microstructure. Both round and flat geometries were used in tensile testing. The details of the alloy chemistry, TMP, round tensile bar geometries and testing conditions can be found elsewhere.[9] The flat tensile bars had nominal dimensions of 100 mm in length, a 25.4 mm reduced section length, and a 2.75 mm x 9.35 mm reduced cross-section. The flat tensile specimens were polished, with the final polish being accomplished with a colloidal silicate suspended in a basic solution (Beuhler MasterMet). A 12 μm (center to center) square grid (2 μm wide lines) was then applied using the same photo-resist and mask technique used in the production of semiconductors. The grid was etched into the flat tensile bars by the reactive ion technique. Application of the 12 μm grid on the flat tensile bars was generally successful, with a large portion of the gage length containing the pattern. The flat bars were tested in tension at a crosshead speed of 0.02 mm/sec and with a nominal extensometer gage length of 18 mm. Longitudinal cross-sections of the round tensile bars and the surfaces of the flat pre-polished tensile bars were examined using scanning electron microscopy (SEM). Room temperature fracture toughness tests were performed on five of the microstructures according to ASTM E399-89. Compact tension specimens were used having a nominal W=20 mm and B=5 mm. Acoustic emission (AE) response during mechanical testing was monitored for both the round tensile and fracture toughness tests. Specifics regarding the AE monitoring methods have been described in another paper.[9]

Results and Discussion

Tensile Testing

The six microstructures examined in this study are found in Figure 1. A summary of the tensile results are shown in Table I. Significantly lower elongation to failure in the flat tensile specimens is quite noticeable. Failure of at least one of the specimens was caused by a large, apparently machining induced, defect at the surface. The other specimens generally exhibited failure initiation near a specimen corner. This indicates that the sharp corners may have acted as a stress concentrator. Note that the microstructures can be divided into two levels of yield and fracture stress, this is the direct result of of secondary α_2 lath size and morphology.

SEM examination of the flat tensile bars after room temperature tensile testing produced fruitful results. Distortions in the etched grid indicated the scale of deformation on the surface. These distortions can be described in two ways: (1) dislocation slip bands interacting with the grid pattern; (2) the macroscopic effect of slip on causing larger distortions to the pattern.

Table I. Room temperature tensile test results

Microstructure	Type	E (GPa)	YS (MPa)	UTS (MPa)	% Elong.	% ROA	ε_f	σ_f (MPa)
3	Round	94.4	563	806	7.0	9.9	0.104	895
	Flat	100.2	672	NA	NA	6.8		
4	Round	115.1	726	958	6.2	9.4	0.098	1058
	Flat	113.7	745	896	1.6	10.7		
7	Round	104.8	581	821	6.5	8.7	0.091	900
	Flat	104.5	632	725	0.8	9.2		
8	Round	105.9	753	967	5.9	6.7	0.070	1037
	Flat	106.1	711	904	2.1	4.0		
11	Round	90.2	556	745	3.5	6.7	0.069	799
	Flat	88.6	618	656	0.3	1.1		
14	Round	104.0	809	996	3.6	7.6	0.079	1078
	Flat	101.6	727	812	0.6	4.5		

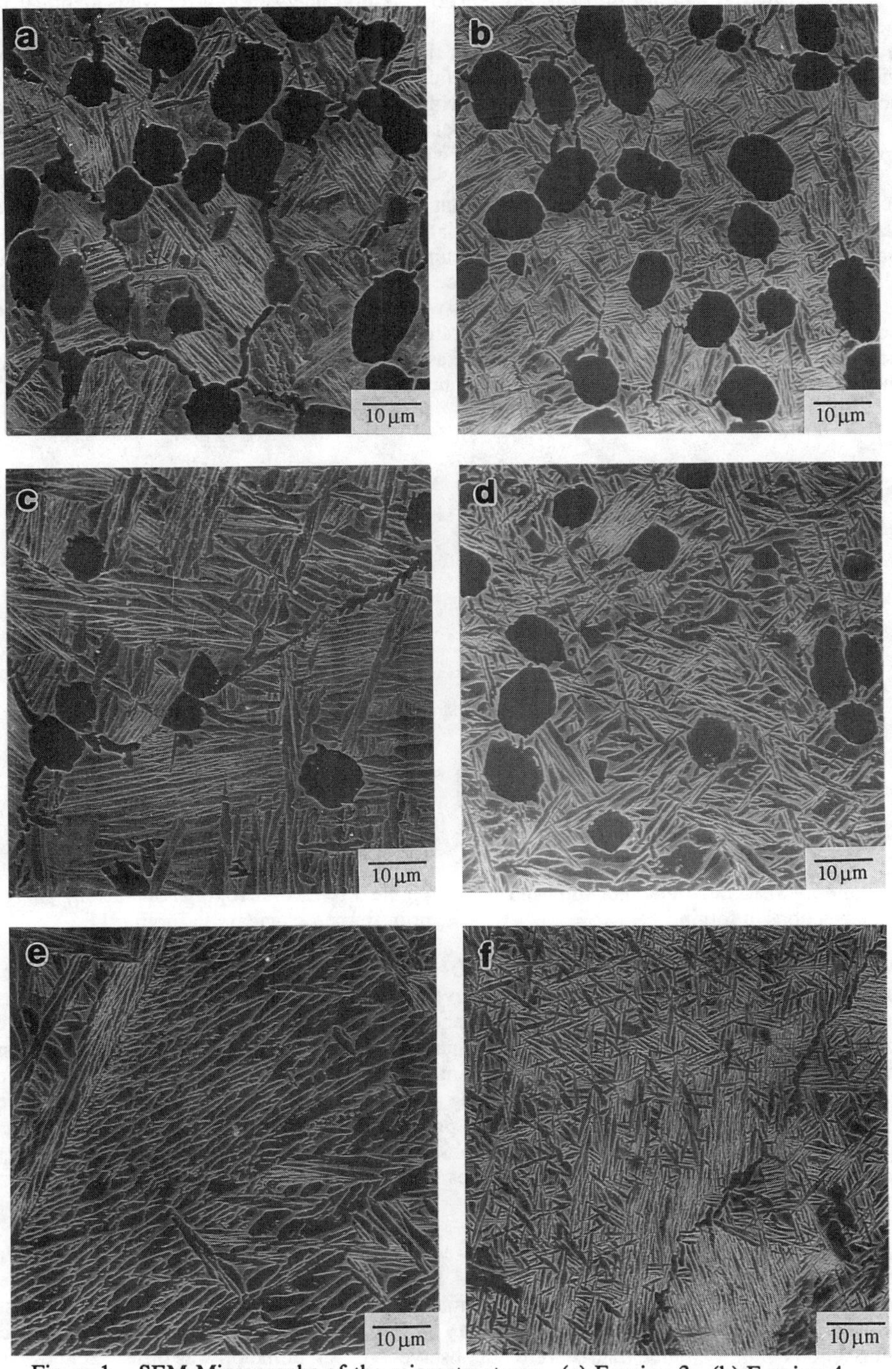

Figure 1 - SEM Micrographs of the microstructures. (a) Forging 3. (b) Forging 4. (c) Forging 7. (d) Forging 8. (e) Forging 11. (f) Forging 14.

The results of AE monitoring of round tensile bars presented previously were identical in behavior between the duplicate tests performed for each microstructure.[9] The output of the transducer was generally quiet from the start of the tensile test, excepting the instances where a yield peak was detected, until very near final failure of the test specimen. Increased acoustic activity was typically observed at strains only within approximately 0.5% of catastrophic failure. This increase in activity is believed to be associated with the nucleation of microcracks in the specimen.

Examination of both the surfaces of the failed flat tensile bars and the cross-sections of the round tensile bars has shown differences in failure initiation sites amongst the different microstructures. Starting with the coarse secondary α_2 lath structure in a β processed material, microstructure 11, slip band intersections on the specimen's surface were profuse with some of these intersections resulting in crack nucleation, Figure 2. Cracks were also observed to occur at the prior β grain boundaries within grain boundary α_2 due to strain incompatibility, Figure 3. This behavior has been previously observed and described by Lukasak and Koss in Ti-24Al-11Nb.[2] Additionally, there was evidence of limited shear crack nucleation.

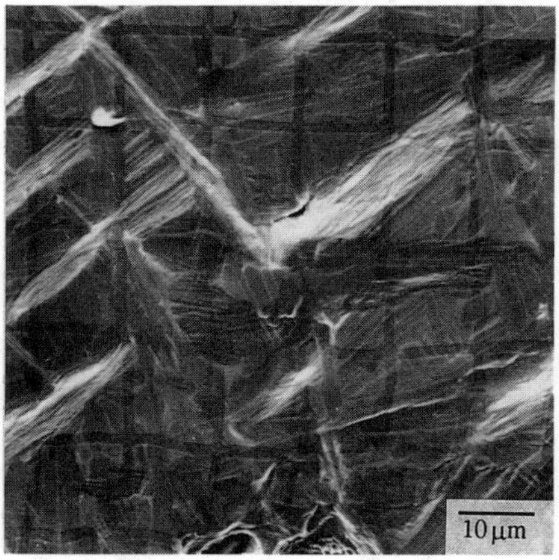

Figure 2 - SEM micrograph of slip band intersections in forging 11.

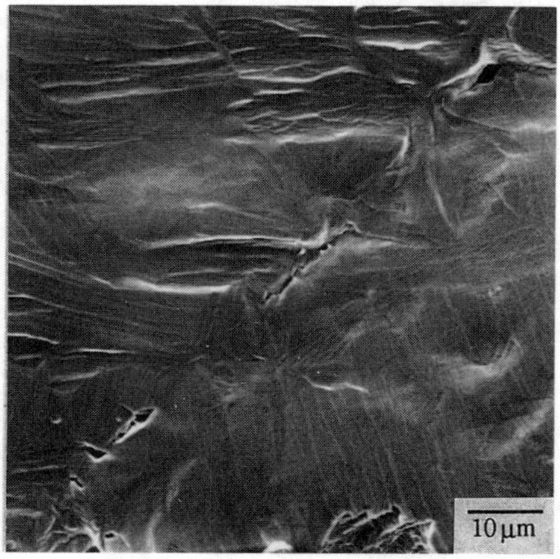

Figure 3 - SEM micrograph of cracking along prior β grain boundary in forging 11.

With the small addition of primary α_2, 0.10 volume fraction, in a coarse microstructure, microstructure 7, some of the cracking observed on the surface of the flat tensile bars was associated with the primary α_2 grains. Slip bands in the matrix were often times seen to have intersected with a failed primary α_2 particle, Figure 4. A large fraction of the cracks observed on the surface were still associated with slip band intersections in the matrix. Increasing the volume fraction of primary α_2 to 0.25 while maintaining a coarse secondary α_2 size, as in microstructure 3, still resulted in long slip bands being a prominent feature on the specimen's surface. However, a larger proportion of cracks were observed to be nucleating at or within primary α_2 grains. The slip band length in microstructures 3 and 7 appeared to be shorter as compared to microstructure 11.

Slip bands were still observed in the β processed microstructure 14, but were on a much finer scale as compared to microstructures 3, 7 and 11. In the fine Widmanstätten microstructure containing 0.25 volume fraction primary α_2, microstructure 4, the initiation sites were almost exclusively associated with the primary α_2 grains. In fact, cracking due to slip band intersections within the fine matrix were not observed. Crack initiation was observed to occur within the individual primary α_2 grains, as in Figure 5, or at the primary α_2/matrix interfaces, as in Figure 6. The cracks within the primary α_2 grains appeared to be nucleated by two mechanisms. First, as seen in previous studies, some cracks were observed to lie along the operating slip planes and were not oriented perpendicular to the stress axis, Figure 5.[5] The second case found cracks nucleating perpendicular to the stress axis with no apparent slip within the α_2 particles, Figure 7. Thus, the first case appears to be described by shear band decohesion and the second cleavage.

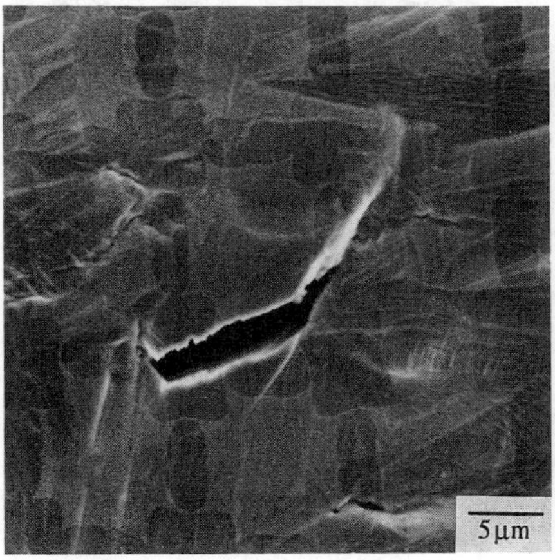

Figure 4 - Matrix slip band intersection with primary α_2 particle in microstructure 7.

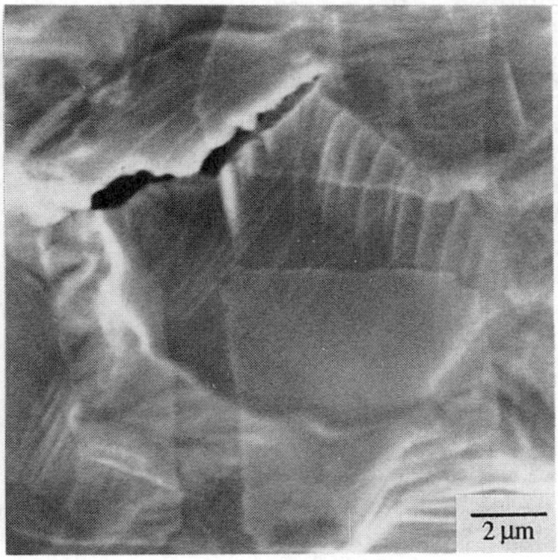

Figure 5 - Failure along active slip planes in a primary α_2 particle in microstructure 4.

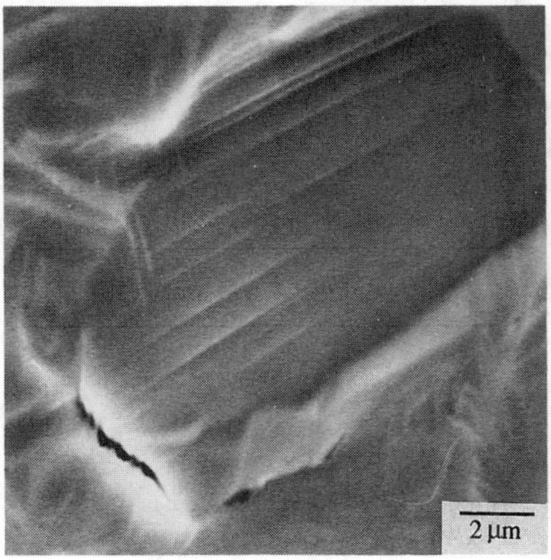

Figure 6 - Crack nucleation at the interface of a primary α_2 particle in microstructure 8.

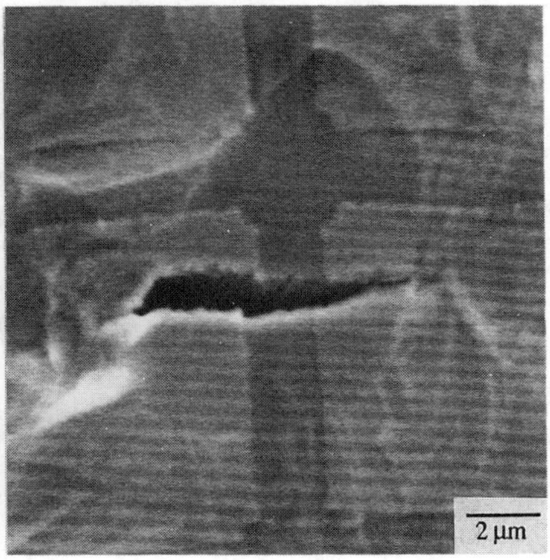

Figure 7 - Cleavage crack nucleated in a primary α_2 particle in microstructure 4.

Longitudinal sectioning of the failed tensile specimens reveals similar crack nucleation sites to those observed on the surface of the pre-polished tensile bars. This result helps validate and generalize the surface observations noted previously. Figure 8 shows crack nucleation at primary α_2/matrix interfaces as well as within primary α_2 particles in microstructure 7. Figure 9 shows crack nucleation associated with grain boundary α_2 and the aligned α_2 laths at prior β grain boundaries in microstructure 11. An enlargement of the crack tip region reveals the mechanism by which the crack must be propagating, Figure 10. The α_2 laths failed along a plane at an apparent orientation near that of the basal plane. Furthermore, the B2 phase was observed to be pulled up around the α_2 laths as if it had behaved as a plastic bridge or ligament during crack extension.

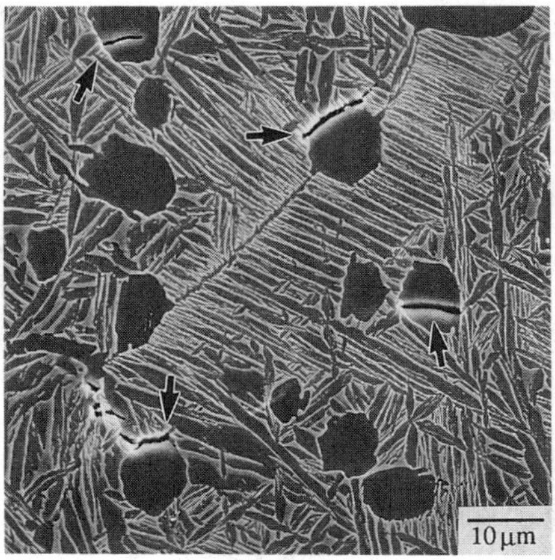

Figure 8 - Sectioned tensile bar showing cracks at interfaces and within primary α_2 particles in microstructure 7.

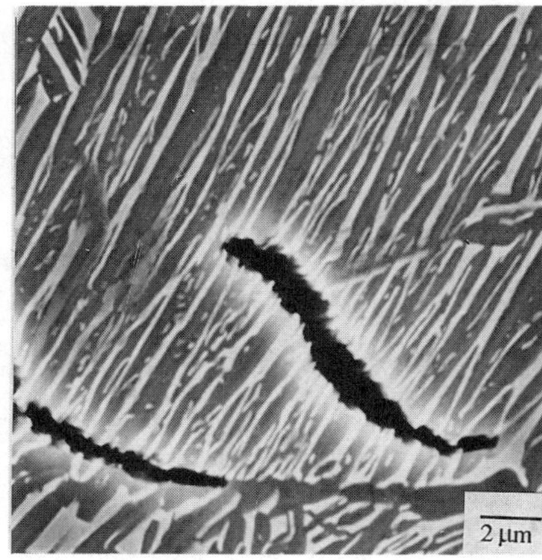

Figure 9 - Cracking associated with prior β grain boundaries in microstructure 11.

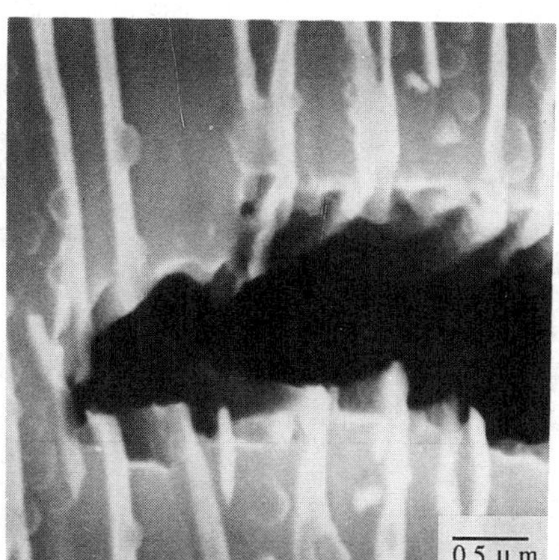

Figure 10 - Enlarged crack tip region of crack in Figure 9.

The behavior of these six microstructures can be grouped into four categories based on stress and strain to failure. Microstructure 11 fails at low stress and strain, whereas microstructure 14 fails at high stress and low strain. Microstructures 3 and 7 fail at low stress but high strain and microstructures 4 and 8 fail at high stress and high strain. Stroh demonstrated that crack nucleation, and hence failure, depends on slip length in brittle fracture.[10] This corresponds to a critical stress at the end of a slip band, which is directly related to the applied tensile stress. As seen above, microstructures 3 and 7 exhibit similar slip band lengths, while those in microstructure 11 are longer, particularly near the prior β grain boundaries. Microstructures 4, 8 and 14, on the other hand, show a much finer slip band length.

Failure in microstructure 11 is dominated by crack nucleation at the prior β grain boundaries. This microstructure has significant amounts of aligned secondary α_2 plates at the prior β grain boundaries which lead to strain incompatibility across the boundaries.[2] The very large effective slip length of these regions due to the alignment of slip planes means rapid strain accumulation at the boundaries and a correspondingly high stress at the end of the resulting slip bands. The high stress at the end of the slip band eventually leads to Stroh crack nucleation at a relatively low applied stress. Rapid propagation of this crack along the prior β grain boundary results in a correspondingly low strain to failure.

Smaller microstructural units of deformation, and thus smaller effective slip lengths, in microstructures 3 and 7 take greater applied stresses than microstructure 11 to nucleate a Stroh crack. These cracks are nucleated at the intersection of a slip band with either another slip band or with a primary α_2 particle. A greater stress to nucleate a crack will obviously result in a larger strain to failure for materials of similar yield stress. Microstructures 4 and 8 having a fine, strong basketweave morphology of secondary α_2 avoid the the formation of slip bands within the matrix and incompatibility across prior β grain boundaries. Thus, an absence of long slip bands intersecting primary α_2 grains means that the applied stress must be even larger to reach a critical cleavage stress. β processed microstructure 14 has a low incidence of aligned secondary α_2 plates at the prior β grain boundary and has a fine, basketweave matrix morphology. It is evident from the tensile data that it started out with a higher yield strength and failed at a similar fracture stress as microstructures 4 and 8, as would be expected for microstructures of similar scale.

As is evident in the micrographs in Figure 1, primary α_2 grains can lie relatively close to one another and have a good probability of being oriented similarly to the tensile axis within a single prior β grain. When one primary α_2 grain fails, the local stress field surrounding this particle is very likely to cause the almost immediate failure of a nearby particle. Thus, it is not surprising that a critical crack size for failure can be reached rapidly by this process. Based on Chan's model of tensile fracture in a brittle material, the low fracture toughness of this material at room temperature is not able contain a moderately sized crack from propagating once one is nucleated.[11] Therefore, tensile fracture occurs very shortly after crack nucleation.

Fracture Toughness Testing

Room temperature fracture toughness tests were performed on all but microstructure 14. The results of testing showed that all five microstructures had a fracture toughness of approximately 20 MPa√m, ± 1 MPa√m. The first thing to note is that these are amongst the highest fracture toughness values reported for Ti-25-10-3-1, which are typically closer to 15 MPa√m.[12,13] No significant trends were observed in the fracture toughness results. Most fracture surfaces were observed to exhibit very flat fracture surfaces, although that of microstructure 11 had a visually more tortuous appearance. The coarse aligned microstructure of microstructure 11 would have been expected to have a higher fracture toughness value than the other microstructures based on results obtained by Marquardt.[6] However, while the coarse secondary α_2 laths are aligned in regions, this microstructure cannot be termed 'colony', as was the case in Marquardt's study. Additionally, the prior β grain size in this microstructure is found to be smaller than for all other microstructures, 0.6 mm *vs.* 1 mm. This is due to the one hour β anneal after forging; evidently the time allowed for solution treatment was insufficient to permit substantial β grain growth after recrystallization. No significant differences in AE response amongst the six microstructures were found during fracture toughness testing.

Summary and Conclusions

The behavior of these microstructures at room temperature was studied by monitoring the acoustic emission response during mechanical testing. Failure was generally found to be preceded by very little acoustic emission activity, indicating failure primarily controlled by the crack nucleation event. The length of slip bands formed in the secondary α_2+B2 matrix during plastic deformation was found to control the fracture stress of these microstructures. SEM examination showed the differences in crack nucleation sites within the six microstructures. Failure initiation ranged from crack nucleation at slip band intersections and shear cracking in coarse β processed microstructure, to cleavage, slip band decohesion and interfacial failure associated with primary α_2 grains in $\alpha_2+\beta$ processed microstructures. A β processed microstructure with a coarse microstructure exhibited lower ductility due to the presence of aligned secondary α_2 laths at prior β grain boundaries, leading to strain incompatibility across the boundaries at a low applied stress. Coarse $\alpha_2+\beta$ processed microstructures were found to have nucleation sites associated with both slip band intersections in the matrix and slip band intersections with primary α_2 particles. Conversely, fine $\alpha_2+\beta$ processed microstructures appear to nucleate failure solely at primary α_2 particles. A β processed microstructure with a fine microstructure failed at a lower strain but a similar fracture stress as in the fine $\alpha_2+\beta$ processed microstructures. Thus, failure does appear to be controlled by a critical cleavage stress which is determined by the microstructural unit controlling slip band formation.

Although the fracture toughness values reported for the material in this study are relatively high for this alloy, ~20MPa$\sqrt{m}$, no significant effect of microstructure on toughness was observed. It should also be noted that the strength levels reported in this study are lower than those previously reported for this alloy, and likewise the ductility is higher than seen before. Thus, the relatively subtle effect that microstructure has been shown to have on the fracture toughness of these alloys may be masked by an increased capacity for plastic work in the present material.[5,7,8]

Acknowledgements

The authors greatly appreciate the experimental assistance received from J. Eblin, J. Ruschau and J. Eschweiler. One author (CHW) would like to thank Dr. M. Peters for providing the facilities of the German Aerospace Research Establishment (DLR) where part of this work was completed.

References

1. C.H. Ward, J.C. Williams, A.W. Thompson, D.G. Rosenthal, and F.H. Froes, Sixth World Conference on Titanium, ed. P. Lacombe *et al.*, (Les Ulis Cedex, France: les éditions de phisique, 1989), 1103.

2. D.A. Lukasak and D.A. Koss, Met. Trans., 21A (1990), 135.

3. J. Kumpfert, K.-J. Grundhoff, H. Schurmann, Y.T. Lee, C.H. Ward, and M. Peters, Proc. Conf. 'Euromat '91', (London: The Institute for Metals, in press).

4. C.H. Ward and S.J. Balsone, Microstructure/Property Relationships in Titanium Aluminides and Alloys, ed. Y-W. Kim and R.R. Boyer, (Warrendale, PA: TMS, 1991), 373.

5. K.S. Chan, Met. Trans., 21A (1990), 2687.

6. B.J. Marquardt, "Fracture Toughness of Alpha-2 Titanium Aluminides" (Paper presented at AeroMat '90, Long Beach, CA, 23 May 1990).

7. K.S. Chan, Met. Trans., 23A (1992), 183.

8. W.O. Soboyejo, Met. Trans., 23A (1992), 1737.

9. I. Roman and C.H. Ward, Scripta Met., 24 (1992).

10. A.N. Stroh, Adv. Phys., 6 (1957), 418.

11. Kwai S. Chan, Scripta Met., 24 (1990), 1725.

12. M.J. Blackburn and M.P. Smith: "R&D on Composition and Processing of Titanium Aluminide Alloys for Turbine Engines," AFWAL Technical Report 82-4086, Wright-Patterson AFB, 1982.

13. D.P. DeLuca and B.A. Cowles, "Fatigue and Fracture of Titanium Aluminides," WRDC Technical Report 90-4075, Wright-Patterson AFB, 1990.

MICROMECHANISMS OF FRACTURE IN ALPHA TWO BASED TITANIUM ALUMINIDES

R.A.Chave, S.Kerry†, C.J.Beevers and P.Bowen

School of Metallurgy and Materials / IRC in Materials for High Performance Applications, University of Birmingham, Edgbaston, Birmingham, B15 2TT, UK.
† Defence Research Agency, Aerospace Division, Royal Aerospace Establishment, Pyestock, Hampshire, GU14 0LS, UK.

Abstract

Micromechanisms of fracture have been studied in experimental alloys based on Ti_3Al (α_2) based titanium aluminides, with and without the addition of molybdenum (Mo). At ambient temperature, failure occurs by transgranular cleavage. The critical failure event appears to be the fracture of individual α_2 laths in the base alloy, but of a group or packet of laths in the Mo containing alloy. Thus, although the lath size is refined by the addition of Mo, the fracture toughness is not increased. At the test temperature of 500°C, the base alloy fails by plastic collapse, and failure occurs by a process of micro-void coalescence. In contrast, the Mo containing alloy fails in a macroscopically brittle manner, and by a process of intergranular micro-void coalescence.

Introduction

In recent years titanium aluminides have shown good potential for use within the aerospace industry as low density materials possessing high temperature strength [1]. The drive behind the development of these materials comes from the considerable weight advantages to be gained if titanium alloy or nickel-based superalloy components within a gas turbine engine can be replaced by low density materials with comparable mechanical properties. The reduction in engine weight combined with the increase in operating temperatures that titanium aluminides offer would assist in the development of engines with greatly increased speed and efficiency.

The ordered intermetallic compounds Ti_3Al (α_2) and TiAl (γ) have received considerable attention over the past two decades due to their good strength and modulus retention at high temperatures. However, both these compounds suffer from poor low temperature ductility and poor fracture toughness. The addition of niobium (Nb) to Ti_3Al has a significant effect on the ductility [2] due both to the stabilisation of the high temperature β phase and to a reduction in the planarity of slip. Commercial alloys based on Ti_3Al with Nb additions, such as Ti–24Al–11Nb and Ti–25Al–10Nb–3V–1Mo, have been identified as possessing a good balance of elevated temperature rupture strength and room temperature ductility.

Oxidation resistance currently limits the potential use of α_2 based alloys to the compressor section of gas turbines where titanium alloys are currently used. Alloys based on the γ phase show greater promise at higher temperatures as replacement materials for superalloys, but suffer severely at present from a lack of room temperature ductility or

Titanium '92
Science and Technology
Edited by F.H. Froes and I. Caplan
The Minerals, Metals & Materials Society, 1993

toughness. Alloys based on α_2 are at a more advanced stage of development than γ alloys, and fabricated components have successfully undergone engine trials [3]. The use of α_2 alloys is also being considered as a matrix material for advanced composites. This paper forms part of a fundamental study into the mechanisms of fracture of a range of experimental alloys based on Ti_3Al.

Experimental

The materials used in the work presented here form part of an alloy development programme being carried out at RAE Farnborough [4]. Extensive work on alloys with a range of aluminium and niobium contents identified a base alloy composition of Ti–23Al–11Nb–0·9Si (atomic %) as showing potential for reasonable room temperature elongation (up to 3%) and excellent creep rupture strength (in excess of 1000 hours at 625°C and 250MPa stress). The further substitution of molybdenum (Mo) in this base alloy, resulting in the composition Ti–23Al–9Nb–2Mo–0·9Si, has shown increased high temperature strength and creep resistance [5]. For the purpose of this present study these two alloys underwent thermo-mechanical processing in the form of a β-extrusion followed by controlled heat treatments in the β phase field. Three different alloy conditions have been studied: the base alloy was solution treated at 1150°C for 1 hour followed by two different ageing treatments (625°C for 2 hours, or 800°C for 1 hour), and the Mo alloy was solution treated at 1150°C for 1 hour followed by an age at 800°C for 1 hour.

Single edge notched (SEN) specimens of dimensions 50x10x10mm were cut from lengths of 50mm diameter extruded bar by electrical discharge machining. Each specimen was notched to a depth of 2·5mm using a 150μm thick diamond blade. In addition a small number of notched bend specimens of a specific notch profile [6] were also machined from the base alloy in the 625°C aged condition. Notches were ground to the required notch root radius of 200μm and notch included angle of 45°. SEN bend specimens were pre-cracked to an a/W of 0·45–0·55 using an Amsler Vibrophore electro-magnetic resonance machine as described elsewhere [7]. Fracture toughness testing was carried out at room temperature in accordance with BS 5447:1977[8], using a four point bend geometry on an ESH servo-hydraulic testing machine. At elevated temperatures the constraints of the fixtures necessitated the use of three point bending, and toughness values were derived solely from a cross-head displacement versus load trace and so do not qualify as K_{IC} values.

Fractographic examination was carried out on an ISI 100A scanning electron microscope (SEM), operating at 0° tilt, 20kV. Additionally, some metallographic sections cut transverse to fracture surfaces were also examined.

Results

Optical and SEM micrographs of Ti–23Al–11Nb–0·9Si and Ti–23Al–9Nb–2Mo–0·9Si in the β heat treated conditions are shown in Figure 1. These show the 'basket weave' microstructures which result from air cooling from the single phase β field. Qualitative energy dispersive X-ray (EDX) analysis on the 625°C aged condition has shown the thin regions surrounding each lath (the light regions of Figure 1*b)* to be richer in niobium but to have a lower aluminium content compared to the centre of the laths. Examination of foils in a transmission electron microscope (TEM) has shown this surrounding phase to be retained β. The microstructures of the two ageing conditions of the base alloy appear to be basically the same, see Figure 1*c)*, but Mo has the effect of refining the 'basket-weave' structure significantly, Figure 1*d)*.

The average values of fracture toughness exhibited by the alloys under investigation are given in Table I. The room temperature values of the base alloy are relatively low, with marginally the best toughness exhibited by the alloy aged at 625°C. The addition of Mo decreases the toughness of the alloy. The SEM micrographs in Figure 2 show the fracture surfaces of specimens tested at room temperature. Fracture occurred by transgranular cleavage in all conditions. Moreover, at high magnification, for the base alloy, see Figure 2*a)*, river lines can be seen clearly within an individual α_2 lath. Thus for the base condition, the failure of

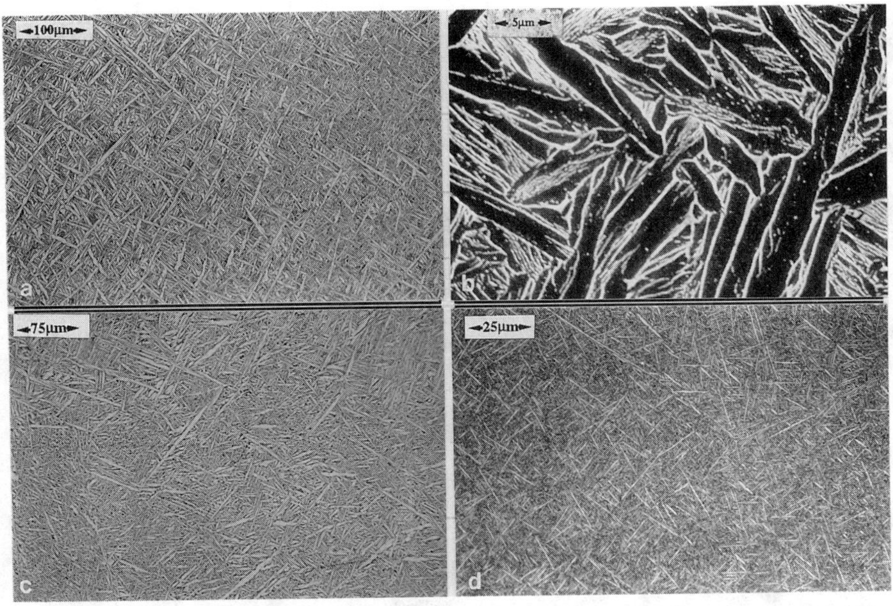

Figure 1 - Microstructures of *a), b)* base alloy in the 625°C aged condition, *c)* base alloy in the 800°C aged condition, *d)* Mo containing alloy in the 800°C aged condition.

an individual lath appears to be a discrete failure event. This is in contrast to the fracture of the Mo containing alloy where cleavage appears to radiate out through a group of individual laths, Figure 2*b)*. Ductile tear ridges are present around the cleavage facets on the surfaces of both base alloy conditions, but appear less prevalent on the fracture surface of the Mo containing alloy. The appearance of the fracture surface of the Mo containing alloy is also smoother with a larger apparent cleavage facet size.

Table I - Fracture toughness results.

Alloy Condition	Room temperature K_{IC} (MPa$\sqrt{m}$)	500°C - Nominal toughness (MPa$\sqrt{m}$)
Base Alloy, β+625°C age	15.7	39.3
Base Alloy, β+800°C age	14.5	35.9
Mo Alloy, β+800°C age	12.6	44.9

At 500°C the base alloy conditions fail by a process of plastic collapse. The fracture surfaces in Figure 3*a),b)* show extensive micro-void coalescence and large featureless areas of localised shear. Fracture in the Mo alloy at this temperature is intergranular, as shown in Figure 3*c)*. The intergranular failure mechanism is one of ductile micro-void coalescence rather than cleavage as is evident from Figure 3*d)*. The load/displacement curves for the base alloy and the Mo alloy in the same condition are shown schematically in Figure 4. Failure in the Mo alloy was characterised by a series of 'pop-ins' (determined from a sudden reduction in load), whereas in the base alloy the failure occurred by plastic collapse. The values of toughness at 500°C quoted in Table I are nominal since the use of a peak load/cross-head displacement trace rather than a clip gauge at elevated temperatures does not provide a valid K_{IC} under the conditions given in BS 5447. Also, due to the substantial plasticity exhibited at

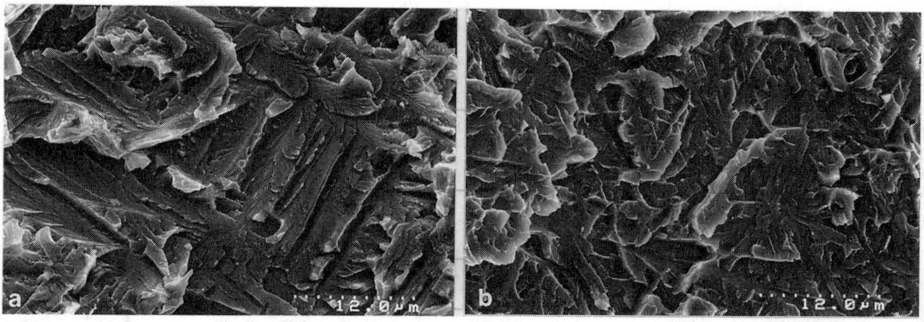

Figure 2 - SEM fractographs of room temperature fracture in *a)* base alloy, aged at 800°C, *b)* Mo containing alloy.

elevated temperatures, it would be expected that the use of elastic-plastic fracture mechanics parameters to characterise failure would result in an increased level of toughness for the base alloy conditions.

Notched bend tests on the base alloy in the 625°C aged condition gave an average cleavage fracture stress, σ_F*, value of 1280MPa at room temperature which is a little over twice the yield stress. Methods of evaluating σ_F* have been described in detail elsewhere, see for example [9].

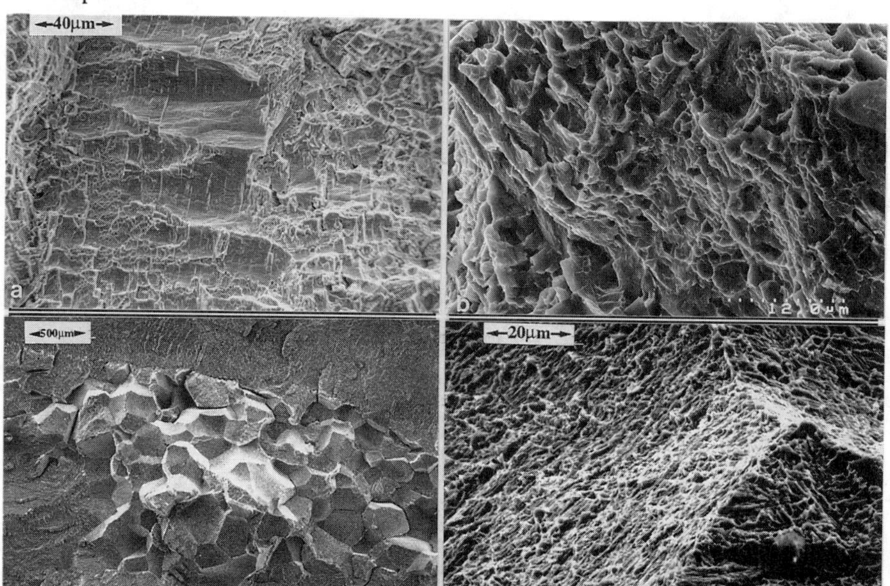

Figure 3 - SEM fractographs of fracture at 500°C in *a)* base alloy, aged at 625°C, *b)* base alloy, aged at 800°C, *c),d)* Mo containing alloy.

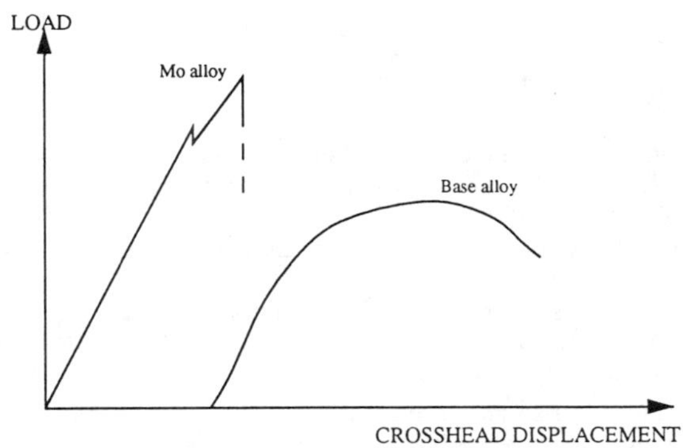

Figure 4 - Schematic of the load versus crosshead displacement curves at 500°C.

Discussion

Failure of these alloys at room temperature occurs by a process of transgranular cleavage. From the observation of secondary microcracks behind the main crack front in the sectioned fracture surfaces (Figure 5), fracture at room temperature in the base alloy would appear to proceed by cleavage across α_2 laths followed by tearing of the more ductile intervening β phase. Ligaments of this β phase can be seen left bridging some microcracks, providing "crack tip shielding". Similar observations have also been made in other studies on alpha-2 based alloys [10-12].

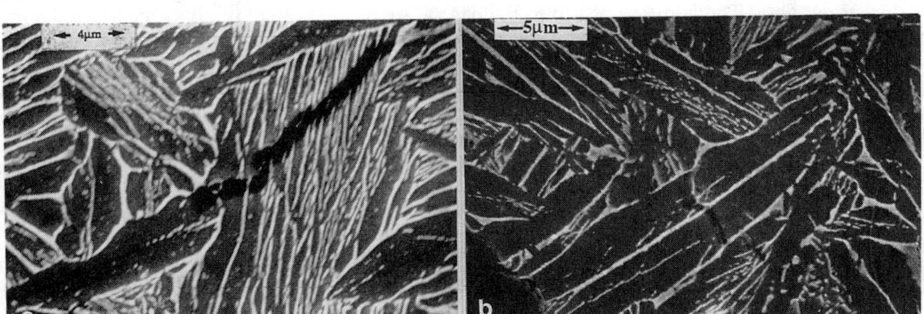

Figure 5 - SEM micrographs of the sectioned room temperature fracture surface of *a)* base alloy aged at 625°C, *b)* base alloy aged at 800°C.

It is accepted generally that in many body centred cubic metals transgranular cleavage occurs under a critical value of tensile stress. Moreover, some recent work on "super α_2" in an α_2+β heat-treated condition has shown that the cleavage fracture stress, σ_F*, is independent of temperature [13], indicative of a critical tensile stress controlled fracture criterion. Such failures are associated with a growth controlled mechanism where a microcrack propagates catastrophically at a critical value of tensile stress (σ_F*). If a growth controlled process is assumed in this case, then critical microcrack size can be linked to the cleavage fracture stress through a modified Griffith relationship:

$$\sigma_F^* = \sqrt{\frac{4E\gamma_p}{\pi(1-\nu^2)d}}$$

where E is Young's modulus, γ_p is the effective surface energy, ν is Poisson's ratio, and d is the critical microcrack size. Values can be assigned to most of the variables in this equation for the base alloy 625°C aged condition (which has been the subject of the most extensive study so far [14]). Measurements of cleavage fracture stress are in the range 720–1280MPa, with values being calculated from uniaxial tensile results and notched bar tests in pure bending respectively. The modulus is taken as 130GPa and Poisson's ratio as 0·26[15]. Estimation of the effective surface energy is, however, more problematic. Values for bcc metals range from 2–200Jm^{-2} [16]. It is likely that a brittle hexagonal intermetallic will exhibit a value towards the bottom of this range, and so an estimate of between 2–20Jm^{-2} has been made for γ_p, and would not seem too unreasonable. With these values applied, microcrack sizes which would result in the fracture stresses observed are of the order of 0·2–7μm. Although this range is relatively large, it should be noted that these values are estimated outer limits. Even so, the microstructural feature of the β-heat treated base alloy which corresponds most closely in order of magnitude to this critical microcrack size is the α_2 lath width.

This analysis, although only of a preliminary nature, when combined with the observations of cleavage across α_2 laths made on sectioned fracture surfaces, Figure 5, and the river markings on individual α_2 laths, Figure 2*a)*, would suggest that the measured fracture stress is consistent with the propagation of a crack across a single α_2 lath being the controlling factor in the fracture of this alloy at room temperature. If this is so, then since a smaller microcrack size leads to an increased fracture stress, it should thus be possible to raise this fracture stress and improve the toughness of the alloy by reducing the lath width size. It has indeed been reported elsewhere [17] that a decrease in the size of the α_2 phase can result in substantial increases in room temperature tensile strength, fracture strength, ductility, and fracture toughness. The addition of Mo to the base alloy has the effect of significantly reducing the lath width size and keeping yield stress values the same, yet as can be seen from Table 1, this does not appear to improve the fracture toughness of this alloy. Most important, fractographic observations suggest that after the addition of Mo, the critical failure event is now the fracture of a group or packet of α_2 laths, see Figure 2*b)*. This new unit size appears to be somewhat larger and hence remains consistent with the use of a modified Griffith equation. The precise role of the Mo is yet to be determined but it may plausibly change the nature of the surrounding β phase. Indeed bridging ligaments such as are observed in the base alloy have not yet been observed in the Mo alloy. Analytical TEM will be used to confirm this suggestion.

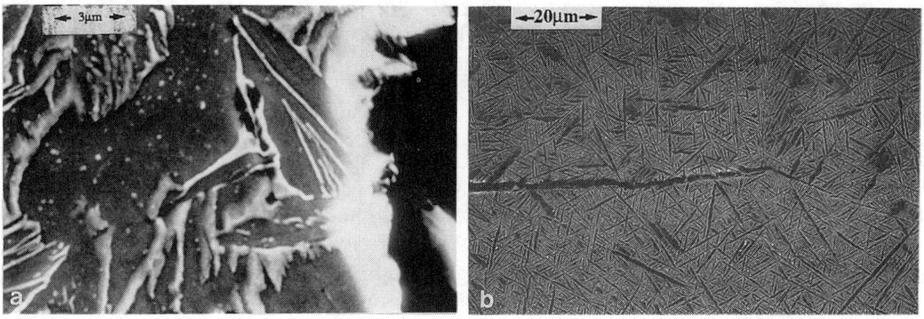

Figure 6 - SEM micrographs of the sectioned fracture surface at 500°C of *a)* base alloy aged at 625°C, *b)* Mo containing alloy.

At a test temperature of 500°C the base alloy is above its brittle to ductile transition and failure occurs primarily by a process of micro-void formation, growth and coalescence. Void formation would appear to occur at the boundary between the α_2 and β phases as revealed in the sections through fracture surfaces shown in Figure 6*a)*. The Mo alloy appears to be macroscopically brittle at 500°C, see Figure 4. In section, the fracture surface suggests that failure occurs along prior β grain boundaries, see Figure 6*b)*, and this is confirmed fractographically, Figure 3*c)*. However, the failure micromechanism is one of intergranular micro-void coalescence, see Figure 3*d)*. These voids are of the order of 1μm and are relatively shallow. The precise reason why the addition of Mo promotes intergranular failure at 500°C is at present unclear. Further work is required to understand the fracture behaviour of this alloy over this range of test temperatures.

Conclusions

1. Fracture at room temperature occurs by transgranular cleavage. The room temperature fracture toughness is relatively low in all alloy conditions studied to date. Ageing temperature appears to have little effect on the value of fracture toughness or the mechanism of failure in the base alloy. It is postulated that the mechanism of failure in the base alloy can be characterised by microcrack initiation within individual α_2 laths and their subsequent propagation through more ductile β regions to give catastrophic failure. The addition of Mo refines the lath structure but does not promote increased toughness. Fractographic observations suggest that the critical failure event is now the simultaneous failure of a group or packet of α_2 laths.

2. At the test temperature of 500°C the base alloy is above its ductile-brittle transition and failure occurs by a process of micro-void initiation, growth, and coalescence. Void initiation appears to occur at the boundary between the α_2 and β phases. The Mo containing alloy fails in an intergranular manner at this temperature with void formation and growth occurring along prior β grain boundaries.

Acknowledgements

This work has been carried out with the support of the Defence Research Agency. One of the authors (RAC) is grateful for a SERC CASE award with Defence Research Agency, Aerospace Division, RAE Pyestock.

References

1. H. A. Lipsitt:, High Temperature Ordered Intermetallic Alloys, C. C. Koch, C. T. Liu, and N. S. Stoloff, eds., (MRS, Pittsburgh, PA, 1985), 351-64.

2. S. M. L. Sastry and H. A. Lipsitt, Metall. Trans. A, 8A, (1977), 1543-52.

3. J. C. Chestnutt and J. C. Williams, Metals and Materials, 1990, 509.

4. S.Kerry: Royal Aerospace Establishment Technical Report 89049, 1989.

5. S. Kerry: Defence Research Agency, Aerospace Division, Private Communication, 1991.

6. J. R. Griffiths and D. R. J. Owen, J. Mech. Phys. Sol., 19, (1971), 395-415.

7. R. A. Chave, et al, (Paper presented at the Seventh World Conference on Titanium, 1992).

8. British Standard Institution BS5447, 1977.

9. P. Bowen, S. G. Druce, and J. F. Knott, Acta Metall., 34, (1986), 1121-31.

10. K. S. Chan, Metall Trans A, 21A, (1990), 2687-99.

11. D. A. Lukasak and D. A. Koss Metall Trans A, 21A, (1990), 135-43.

12. A. K. Gogia, D. Banerjee, and T. K. Nandy, Metall Trans A, 21A, (1990), 609-25.

13. R. J. T. Penton and P. Bowen: (Proc. ICM-6, Kyoto, Japan, August 1991).

14. R. A. Chave, C. J. Beevers, and P. Bowen, (Proceedings of the 2nd Eoropean Conference on Advanced Materials and Processes, 1991), 333-9.

15. R. E. Schafrik, Metall Trans A, 8A, (1977), 1003-6.

16. P. Bowen and J. F. Knott, Met. Sci., 18, (1984), 225-235.

17. W. Y. Chu and A. W. Thompson: Metall Trans A, 22A, (1991), 71-81.

ON THE INFLUENCE OF THE PHASE MORPHOLOGY ON THE MECHANICAL PROPERTIES OF THE TWO-PHASE INTERMETALLIC COMPOUND SUPER ALPHA 2

G. Proske[1], G. Lütjering[1], J. Albrecht[1], M.A. Däubler[2], D. Helm[2]

1) Technical University Hamburg-Harburg, 2100 Hamburg 90, FRG
2) MTU Munich, 8000 Munich, FRG

Abstract

This investigation focussed on the improvement of the fracture toughness of the titanium alloy "Super Alpha 2" without major loss in room temperature ductility, yield stress, and creep resistance. For this purpose four different bi-modal microstructures and for comparison one lamellar structure were investigated. The microstructure, the fracture surfaces, and the crack front profiles were examined using light microscopy, transmission electron microscopy, and scanning electron microscopy. The best combination of the different mechanical properties was found for a bi-modal structure with a primary α_2 volume fraction of about 25 - 30% within a lamellar matrix containing a combination of large and very small secondary α_2 plates.

Titanium '92
Science and Technology
Edited by F.H. Froes and I. Caplan
The Minerals, Metals & Materials Society, 1993

Introduction

Titanium alloys based on the intermetallic compounds Ti_3Al and TiAl are potential aerospace engine materials because of their superior thermal performance, lower density, greater strength, and lower cost compared to current nickel-base alloys (1). The drawback of these materials is their poor room temperature ductility and the low toughness. One of the most promising developments is the commercial alloy "Super Alpha 2" (Ti-14Al-20Nb-3.2V-2Mo in wt.-%), which is based on the intermetallic compound Ti_3Al. The addition of Nb and other β stabilizing elements leads to a two phase microstructure consisting of the ordered hexagonal α_2 phase (Ti_3Al) and the ordered bcc β_2 phase. The geometrical arrangement of the two phases can be varied by thermomechanical treatments leading to equiaxed, lamellar or bi-modal (equiaxed primary α_2 in a lamellar matrix consisting of α_2 and β_2 lamellae) microstructures. Earlier investigations (2,3) have shown that the bi-modal microstructure shows the best combination of strength, room temperature ductility, and creep resistance. The investigations described in this paper make an attempt to modify the bi-modal microstructure towards an improved fracture toughness without major loss in strength, room temperature ductility, and creep resistance. For this purpose four different bi-modal structures were generated to examine the influence of the cooling rate after the recrystallization treatment and an additional annealing treatment on the above described mechanical properties. For comparison the same mechanical tests were performed using a lamellar structure.

Experimental Procedures

The material used for these studies was delivered by RMI, Niles, Ohio, USA. The material in the as-received condition had a fine lamellar microstructure. To generate the different microstructures the material was homogenized in the β phase field and cross rolled in the $\alpha_2 + \beta$ region.
The bi-modal microstructures were all recrystallized in the two phase field at the same temperature and subsequently cooled at one of the following cooling rates: Fast cooling (Bi-M./Fast), intermediate cooling (Bi-M./Medium), and slow cooling (Bi-M./Slow). To obtain the fourth bi-modal structure the Bi-M./Medium microstructure received an additional annealing treatment at a temperature below the recrystallization temperature followed by fast cooling. This microstructure is designated as Bi-M./Annealed.
The lamellar structure was obtained by β annealing followed by controlled cooling with a cooling rate of 0.017 Ks^{-1}. All microstructures received an identical final heat treatment in the $\alpha_2 + \beta_2$ phase field to generate the same total volume fraction of α_2 and β_2 for all microstructures.

The microstructure was evaluated by light microscopy (LM) and transmission electron microscopy (TEM). TEM specimens were prepared using the twin-jet electropolishing technique in a solution of perchlorid acid. Scanning electron microscopy (SEM) was used for the examination of the fracture surfaces of the fracture toughness specimens. Additionally, the crack front perpendicular to the direction of crack propagation was investigated using light microscopy.

Tensile tests were performed at room temperature and 600° C with a crosshead speed of 1mm/min (initial strain rate: $8 \times 10^{-4}s^{-1}$) using round specimens. The creep tests were conducted under constant tensile load with an initial applied stress of 210 MPa at a test temperature of 600° C. The creep resistance was defined as the plastic deformation after a test period of 100 hours. Fracture toughness tests following the ASTM standard E 399-81 were performed for the determination of the fracture toughness of the different microstructures. These tests were carried out at room temperature using three point bending specimens.

Results and Discussions

Microstructural Development

Rolling in the $\alpha_2 + \beta$ region produced a uniform distribution of the equiaxed primary α_2 grains. The volume fraction of primary α_2 and the size of the α_2 and β_2 plates within the lamellar matrix are influenced by the $\alpha_2 + \beta$ recrystallization temperature and by the cooling rate.

Light micrographs and transmission electron micrographs of the different bi-modal conditions are presented in Figure 1 - Figure 4. The bi-modal structure with the highest cooling rate (Bi-M./Fast, Figure 1a) has a volume fraction of 10% equiaxed primary α_2. The volume fraction increases slightly with decreasing the cooling rate. The highest volume fraction (25-30%) of primary α_2 was determined for Bi-M./Slow (Figure 3a). The latter and the Bi-M./Annealed condition (Figure 4a) have roughly the same amount of primary α_2. The size of the primary α_2 grains is about 3 μm for Bi-M./Fast and Bi-M./Medium (Figures 1a and 2a) and 4 to 5 μm for Bi-M./Slow and Bi-M./Annealed (Figures 3a and 4a). The light micrographs of Bi-M./Slow (Figure 3a) and Bi-M./Annealed (Figure 4a) show that these two microstructures consist of equiaxed primary α_2 and of large secondary α_2 plates. The volume fraction of these large secondary α_2 plates is higher for the Bi-M./Slow condition compared to Bi-M./Annealed. The width of these lamellae is about 1 μm. The TEM micrograph of the Bi-M./Fast microstructure (Figure 1b), the structure with the highest cooling rate, shows a very fine lamellar matrix consisting of α_2 and β_2 lamellae. Due to the fineness of the lamellar matrix it is not possible to distinguish between α_2 and β_2 plates. Decreasing the cooling rate leads to a coarsening of the α_2 and β_2 plates (Figures 2b and 3b). The TEM micrograph of Bi-M./Annealed (Figure 4b) shows that the lamellar matrix between the equiaxed primary α_2 grains consists of a mixture of large secondary α_2 plates and very fine α_2 and β_2 plates.

Figures 5a and 5b show the LM and TEM micrographs of the β annealed structure. The recrystallization in the β phase field leads to a considerable grain coarsening, resulting in a β grain size of 1-1.5 mm (Figure 5a). During cooling, α_2 plates begin to grow heterogeneously from the prior β grain boundaries when the temperature falls below the β-transus. The average width of these plates is about 3 to 4 μm and the length up to several hundred microns (Figure 5a). The width of the β_2 plates is with 0.5 to 1.0 μm comparatively small (Figure 5b).

Tensile Properties

The results of the tensile tests performed at room temperature are presented in Table I. The bi-modal microstructure with the finest lamellar matrix (Bi-M./Fast) has the highest yield stress (1010 MPa). A decrease of the cooling rate leads to a reduction in the yield stress (Bi-M./Medium: 935 MPa), Bi-M./Slow: 765 MPa). The Bi-M./Annealed condition has a yield stress (970 MPa) ranging between Bi-M./Fast and Bi-M./Medium. A pronounced influence of the cooling rate and the additional annealing treatment on the room temperature ductility was not detected. The highest value was measured for the Bi-M./Medium condition with 6.4%, while the Bi-M./Slow structure has the lowest room temperature ductility (2.8%).

At 600° C the bi-modal structures show the same sequence in the yield stress compared to the tests at room temperature (Table II). The Bi-M./Fast and the Bi-M./Annealed condition have the highest yield stress with about 790 MPa. The lowest yield stress was again measured for the Bi-M./Slow condition (550 MPa). Again no distinct influence of the cooling rate and the additional annealing treatment on the ductility was evident. The highest tensile elongation was measured for the Bi-M./Slow condition (11.2%) and the lowest value for Bi-M./Annealed (4.4%).

The lamellar microstructure has a much lower yield stress at room temperature (585 MPa) and at 600° C (370 MPa) compared to the bi-modal structures. The room temperature duc-

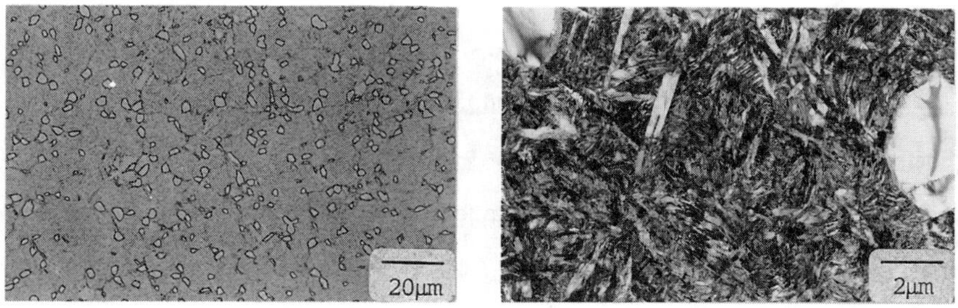

a) LM b) TEM

Figure 1: Microstructure of the Bi-M./Fast condition.

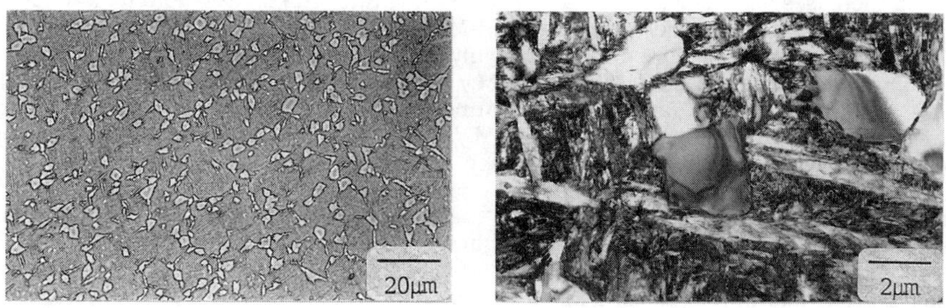

a) LM b) TEM

Figure 2: Microstructure of the Bi-M./Medium condition.

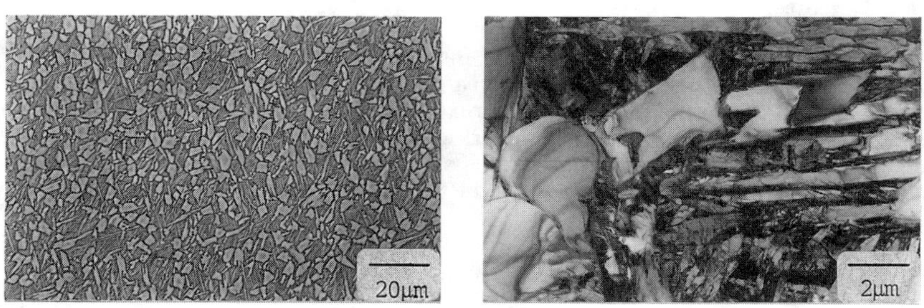

a) LM b) TEM

Figure 3: Microstructure of the Bi-M./Slow condition.

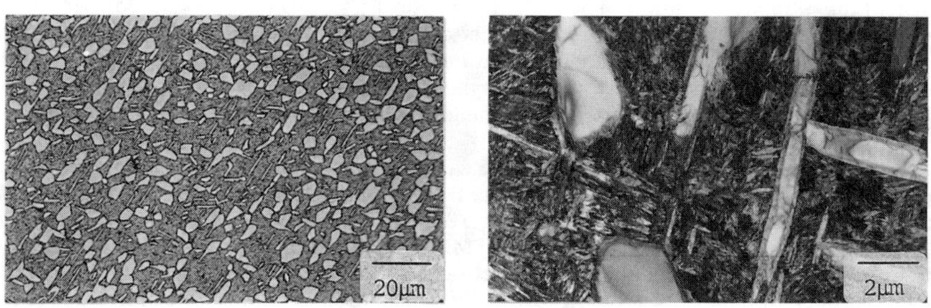

a) LM b) TEM

Figure 4: Microstructure of the Bi-M./Annealed condition.

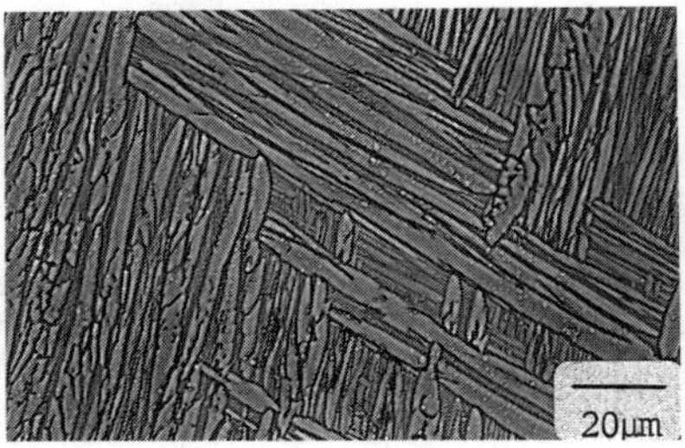

a) LM

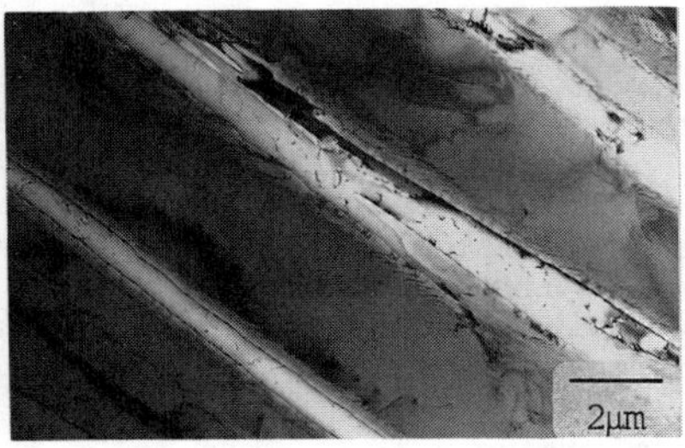

b) TEM

Figure 5: Microstructure of the lamellar condition.

Table I: Results of the tensile tests performed at room temperature.

Condition	$\sigma_{0.2}$ (MPa)	UTS (MPa)	σ_F (MPa)	Tens.Elong. (%)	R.A. (%)
Bi-M./Fast	1010	1235	1350	5.1	8.6
Bi-M./Medium	935	1180	1315	6.4	10.9
Bi-M./Slow	765	930	1015	2.8	8.6
Bi-M./Annealed	970	1225	1415	4.4	13.5
Lamellar	585	615	625	0.3	1.3

Table II: Results of the tensile tests performed at 600° C.

Condition	$\sigma_{0.2}$ (MPa)	UTS (MPa)	σ_F (MPa)	Tens.Elong. (%)	R.A. (%)
Bi-M./Fast	790	1045	1275	7.8	19.3
Bi-M./Medium	690	965	1200	7.0	21.3
Bi-M./Slow	550	825	1045	11.2	20.9
Bi-M./Annealed	785	1050	1200	4.4	13.1
Lamellar	370	560	610	6.7	16.9

tility is very low (0.3%), but a sharp increase of the ductility (6.7%) was measured at 600° C.

These investigations have shown that the strength values are influenced by the size of the α_2 plates in the lamellar matrix. The same result was also found by Ward, Williams et al. (4). A refinement of the α_2 lamella size was found to increase the yield stress significantly (Tables I and II). This implies that the interlamellar boundary strengthening, respectively the reduction of the slip length (5) is the dominant factor for the bi-modal structures. The yield stress of Bi-M./Annealed is mainly determined by the small secondary α_2 plates, a distinct influence of the large secondary α_2 plates is not evident.

In general, it can be concluded that a fine lamellar matrix in combination with a small prior β grain size improves the resistance to crack nucleation, thereby resulting in good ductility and indirectly high true fracture stress values for the bi-modal structures.
The yield stress and the ductility of the lamellar structure is also determined by the size of the α_2 plates. Additionally, the α_2 plates within one packet of α_2 lamellae have a similar crystallographic orientation facilitating slip transfer over wide distances. The resulting large slip length is considered to be the reason for the low yield stress and the low ductility of the lamellar structure.

Creep Behavior

The results of the creep tests are listed in Table III. They show that the creep resistance decreases with decreasing cooling rate. The bi-modal structure with the highest cooling rate (Bi-M./Fast) has the best creep resistance (0.15% plastic strain) while Bi-M./Slow (lowest cooling rate) has the lowest creep resistance (0.32%). Bi-M./Medium and Bi-M./Annealed have almost an identical creep resistance (Bi-M./Medium: 0.19%, Bi-M./Annealed: 0.20%). The creep resistance of the lamellar structure is comparatively low (0.32%).

Former investigations (2,3,6) have suggested that the plastic deformation during the creep tests at 600° C and an applied stress of 210 MPa is primarily determined by the motion of dislocations within the softer β_2 phase. These studies have also shown that the creep resistance depends on the volume fraction and the size of the β_2 phase. The creep resistance of the four bi-modal and the lamellar structure depends mainly on the size of the β_2 phase, because the volume fraction of the β_2 phase is the same for all these microstructures due to the identical final heat treatment. The TEM examinations of the microstructures have revealed that the α_2 and β_2 plates coarsened when the cooling rate decreased. The growth of the β_2 plates leads to an increase in the slip length which is the cause of the decreasing creep resistance. The low volume fraction of large secondary α_2 plates in the Bi-M./Annealed condition (Figure 4b) has only little influence on the creep resistance. A higher volume fraction of secondary α_2 plates (Bi-M./Slow, Figure 3b) leads to a distinct decrease of the creep resistance (Table III) due to the large size of the β_2 plates.

Table III: Results of the creep tests, 600°C, 210 MPa.

Condition	$\varepsilon_{pl.100h}$ (%)
Bi-M./Fast	0.15
Bi-M./Medium	0.19
Bi-M./Slow	0.32
Bi-M./Annealed	0.20
Lamellar	0.32

Table IV: Results of the fracture toughness tests.

Condition	K_5 (MPa $m^{1/2}$)
Bi-M./Fast	11.8
Bi-M./Medium	14.4
Bi-M./Slow	18.1
Bi-M./Annealed	15.7
Lamellar	32.0

Fracture Toughness Tests

A clear tendency for the fracture toughness to increase with decreasing cooling rate can be seen in Table IV. The lowest fracture toughness was measured for the Bi-M./Fast condition (11.8 MPa $m^{1/2}$) and the highest value for Bi-M./Slow (18.1 MPa $m^{1/2}$). Table IV also shows that the toughness of Bi-M./Annealed (15.7 MPa $m^{1/2}$) is higher than for Bi-M./Fast (11.8 MPa $m^{1/2}$) and Bi-M./Medium (14.4 MPa $m^{1/2}$). The highest toughness of all microstructures investigated in this study was measured for the lamellar structure (32 MPa $m^{1/2}$).

To evaluate the parameters which influence the toughness, the crack front profiles (crack front perpendicular to the direction of the crack propagation) of the bi-modal and lamellar conditions were investigated. Since the crack front profiles of the bi-modal conditions appear very similar in the light-microscope, the Bi-M./Fast structure was choosen for further investigations. The crack front profile of the lamellar structure (Figure 6b) is very

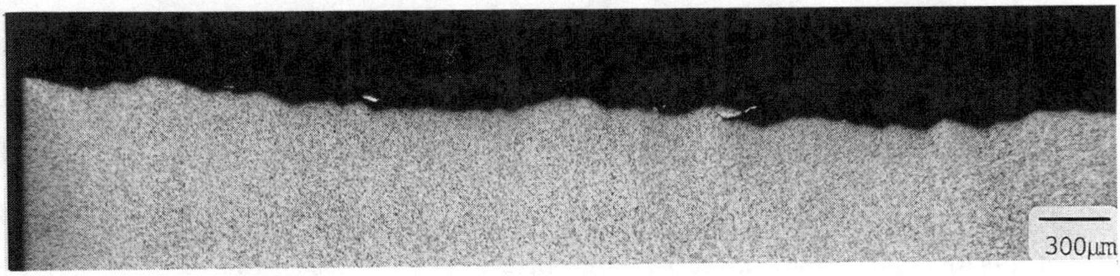

Figure 6a: Crack front profile of the Bi-M./Fast condition.

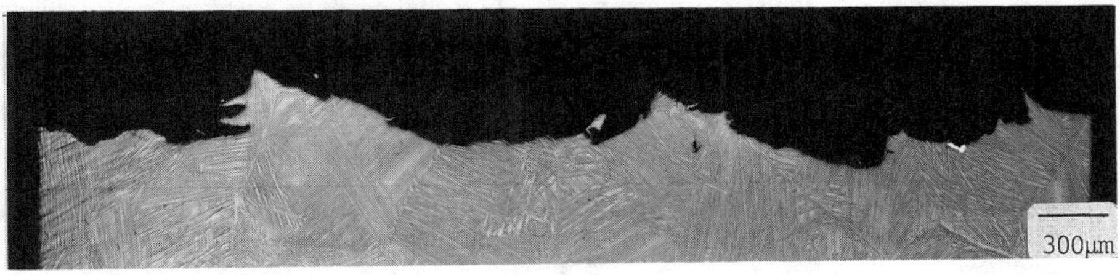

Figure 6b: Crack front profile of the lamellar condition.

a) Bi-M./Fast b) Lamellar

Figure 7: Fracture surfaces of the fracture toughness specimens.

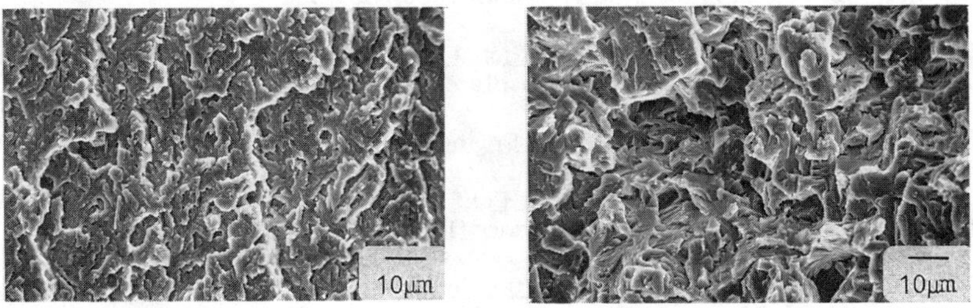

a) Bi-M./Fast b) Bi-M./Slow

Figure 8: Fracture surfaces of the fracture toughness specimens.

rough compared to that of the Bi-M./Fast condition (Figure 6a). Due to different crystallographic orientations between adjacent prior β grains in the lamellar structure, the crack is propagating locally on different planes. A large β grain size leads to correspondingly high ledges explaining the rough crack front profile (Figure 6b). The tearing of the ledges requires high energy to propagate the crack, resulting in a high fracture toughness.

Corresponding fractographs of the Bi-M./Fast and the lamellar condition are shown in Figures 7a and 7b. The results of the fracture surface examinations confirm the previous results of the crack front profile investigations. The fracture surface of Bi-M./Fast is very flat (Figure 7a), while the lamellar condition shows a rough fracture surface with large ledges (Figure 7b), explaining the large difference in fracture toughness.

The influence of the cooling rate on the fracture surface morphology of the bi-modal structures is shown in Figure 8 comparing Bi-M./Fast and Bi-M./Slow. Decreasing the cooling rate results in a rougher fracture surface. The increase in roughness is caused by the larger size of the secondary α_2 plates, deflecting the crack and increasing the fracture toughness. Even a low volume fraction of large secondary α_2 plates (Bi-M./Annealed) leads to an increase of the fracture toughness compared to the bi-modal conditions with a fine lamellar matrix.

Conclusions

1. Refining the lamellar matrix of the bi-modal structure results in a higher yield stress without significant influence on the ductility.
2. Refinement of the lamellar matrix also improves the creep resistance.
3. The rough crack front profile is the origin for the high fracture toughness of the lamellar condition.
4. Large secondary α_2 plates have a beneficial effect on the fracture toughness of the bi-modal conditions, even in low volume fractions.
5. The goal of this investigation to improve the fracture toughness of the bi-modal structures without major loss in strength, room temperature ductility, and creep resistance was found in a bi-modal structure with a second annealing treatment. The lamellar matrix of this structure is containing a combination of large and very small secondary α_2 plates.

Acknowledgment

Part of the work at the Technical University Hamburg-Harburg was supported by the Deutsche Forschungsgemeinschaft (DFG).

References

1. H.A. Lipsitt, in High-Temperature Ordered Intermetallic Alloys, vol.39, ed. C.C. Koch, C.T. Liu, N.S. Stoloff, Pittsburgh, PA., MRS (1985), 351-364

2. G. Lütjering et al., in Proc. of International Symposium on Intermetallic Compounds, JIMIS-6, ed. by O. Izumi, The Japan Institute of Metals (1991), 537-541

3. G. Proske et al., Materials Science and Engineering, to be published

4. C.H. Ward et al., in Proc. 6th World Conf. on Titanium, vol.II, eds. P. Lacombe, R. Tricot, G. Béranger, Les Ulis Cedex, France (1988), 1103-1108

5. J.C. Williams, G. Lütjering, Titanium '80, Science and Technology, Proc. of the Fourth International Conference on Titanium, vol.1, eds. H. Kimura, O. Izumi, Kyoto, Japan (1980), 671-681

6. W. Cho, A.W. Thompson, and J.C. Williams, Metall. Trans. A, 21A (1990), 641-651

Effects of microstructure on fatigue crack growth in a Ti_3Al based aluminide at ambient and elevated temperatures

R.J.T. Penton, M.T. Cope+ and P. Bowen,

School of Metallurgy and Materials/IRC in Materials for High Performance Materials,
University of Birmingham, Edgbaston, Birmingham, B15 2TT, U.K.
+Rolls Royce plc, P.O. Box 31, Derby, DE2 8BJ, U.K.

Abstract

Effects of microstructure on fatigue crack growth resistance have been considered for the Ti_3Al based alloy "Super α_2" both at room temperature and at the elevated temperature of 700°C. Microstructures comprising primary α_2 precipitates in a fully transformed β matrix exhibit increased crack growth rates with respect to a fully transformed β condition at room temperature. Marked effects of mean stress are also observed and since the primary α_2 precipitates cleave at room temperature it appears that such static events are responsible for the poorer crack growth resistance of the α_2 + transformed β condition. In contrast, at the test temperature of 700°C, this condition now exhibits superior crack growth resistance to the fully transformed β condition, and it has been deduced that the lower yield stress of the latter condition may be important in this respect.

Introduction

It is widely recognised that the titanium alloys currently used in aerospace applications have little further to offer in the way of increased elevated temperature capabilities. Thus it may be necessary to utilise new families of alloys in order to achieve a radical improvement in the operating temperature of the next generation of gas turbine engines. One such series of alloys is based on the intermetallic compound Ti_3Al. They represent one of the most developed of these innovative materials to date and include such alloys as Ti-25Al-10Nb-3Mo-1V and Ti-24Al-11Nb (atomic %)(1,2). It is critical in the design and production of actual components from these materials that their mechanical properties are understood. In this respect their fracture toughness(3,4,5) and resistance to fatigue crack growth(6,7,8) has been investigated.
This current paper addresses fatigue crack propagation at ambient and elevated temperatures, with an emphasis on the effects of heat treatment and the concomitant microstructure for a given processing route. Such studies should allow the development of improved, or more consistent, crack growth resistance through careful thermomechanical processing to produce optimised microstructures.

Experimental

Tests have been performed on "Super α_2" (Ti-25Al-10Nb-3Mo-1V atomic %) cut from an α_2 + β rolled plate processed by Timet Inc., USA. The study concentrated on two heat treatments. In both cases the specimens were heat-treated as oversized blanks. The first involved a solution treatment of 1 h at 1050°C, followed by air-cooling to room temperature. In the second blanks

Titanium '92
Science and Technology
Edited by F.H. Froes and I. Caplan
The Minerals, Metals & Materials Society, 1993

were solution treated at 1160°C followed once again by air cooling to room temperature. Ageing of both sets of test pieces at 850°C for 2h was followed by air-cooling to room temperature. For all specimens surface grinding to a depth of 2mm ensured the removal of any oxidation products formed during heat-treatment. Tensile properties are given in Table I, for both conditions at room temperature and 700°C.

Table I - Tensile properties of the titanium aluminide alloy Super α_2

Microstructural Condition	Temperature (°C)	0.1% Proof Stress (MPa)	UTS (MPa)	Elong. %
α_2+ transformed β Solution treated at 1050°C for 1 hour Aged at 850°C for 2 Hours	20	910	1064	1.7
	700	585	911	2.8
fully transformed β Solution treated at 1160°C for 1 hour Aged at 850°C for 2 Hours	20	750	957	0.5
	700	509	825	1.2

Fatigue crack propagation tests were carried out at room temperature on an Amsler Vibrophore electromagnetic resonant testing machine at a frequency of approximately 70 Hz in air, and on an ESH servo-hydraulic testing machine at 700°C, at a frequency of 20 Hz in air. Tests were carried out at two stress ratios, R = 0.1 and R = 0.5 at room temperature, (where R is the ratio of K_{min}/K_{max} and K_{min}, K_{max} are the minimum and maximum values of the applied stress intensity over the fatigue cycle) and at R = 0.1 only at elevated temperature. In all tests specimens were fatigued to catastrophic failure.

Optical micrographs were taken after mechanical polishing and etching in Krolls II reagent. Fracture surfaces were examined using an Hitachi field emission gun scanning electron microscope, operating at 0° tilt and 20 kV operating voltage.

Results

The microstructures after heat-treatment are shown in Figure 1 and 2. In the α_2 + transformed β condition it consists of primary α_2 precipitates (light regions) of size 5-20 μm and volume fraction approximately 20%, in a transformed β matrix with an average grain size of approximately 30 μm. The structure of the transformed β was unresolvable optically. The fully transformed β condition has a typical 'basketweave' microstructure and a grain size of approximately 1mm.

Fatigue crack growth resistance curves at room temperature are shown in Figure 3. There is a slight but significant effect of microstructural condition at room temperature: crack growth rates are increased for the α_2 +transformed β heat treated condition, particularly at high mean stress, where crack growth rates are faster by an order of magnitude, figure 3. Marked effects of mean stress are observed for both microstructural conditions. At the elevated test temperature of 700°C, figure 4, crack growth rates increase in general for both microstructural conditions. However, at this elevated temperature crack growth rates observed in the β heat treated material can be up to an order of magnitude higher than for the α_2 + transformed β heat treated condition. In both cases the dependence of crack growth rate, da/dN, on applied stress intensity, ΔK is reduced compared to that observed at room temperature. If a Paris-Erdogan law of the type,

$$da/dN = A\ \Delta K^m \ ..... \ (1)$$

is assumed, where A, m are materials constants, then values of A, m are given in Table II.

Table II - Constants derived from empiracal power law equations describing fatigue crack growth.

Microstructural Condition	Temperature (°C)	R ratio	A (mm/cycle)	m
α_2 + transformed β	20	0.1	1.3 E-13	5.1
	20	0.5	3.3 E-15	13.1
	700	0.1	1.53 E-08	4.0
fully transformed β	20	0.1	4.12 E-10	4.1
	20	0.5	4.2 E-14	9.9
	700	0.1	3.2 E-06	2.1

Fatigue surfaces at room temperature show quite distinct regions of flat transgranular cleavage, see Figure 5a, in the α_2 + transformed β heat treated condition. The average size of these cleavage facets is approximately 12 µm (with a range of sizes from 3-20 µm) and they are deduced therefore to be cleaved α_2 precipitates. Such areas are surrounded by more angular regions of irregular stepped growth, which is presumably the crack path through the transformed β matrix phase. The failure surface in the β heat treated condition is observed to be quite similar to that of the α_2 + transformed β heat treated condition without the presence of the larger scale cleavage facets, figure 5b. In contrast, at test temperatures of 700°C in air, fatigue surfaces tend to be flatter and exhibit no clear microstructural features, see Figures 6a and 6b.

Discussion

At room temperature marked effects of R ratio have been observed, figure 3 and table II. Closure could also result in a variation in growth rates with R ratio but it is difficult to see how it could result in such a large scale systematic variation over the whole of the Paris regime and in a situation of constant applied load range i.e. increasing ΔK. Therefore, this is considered to be unlikely to provide a complete explanation. More plausible perhaps is the influence of K_{max}, particularly because static modes of failure have been observed in the α_2 + transformed β microstructural condition. If crack growth rates are plotted versus K_{max}, figure 7, then the difference between the crack growth rates decrease and indeed for the α_2 + transformed β microstructural condition they are found to be similar. Thus the influence of Kmax in promoting static contributions to the process of fatigue crack growth can be supported. The mechanisms of fatigue crack growth in these materials have yet to be fully established. Static modes of cleavage are observed in the α_2 + transformed β microstructural condition of a size corresponding to the primary α_2 regions in the microstructure (figure 5a) together with smaller discrete steps. A similar fracture surface consisting solely of the discrete steps is observed in the β heat treated condition (figure 5b). Since the crack growth rates in the β heat treated condition are observed to be lower than those in the α_2 + transformed β material this is deduced to be a consequence of the static failure of the primary α_2 precipitates in the α_2 + transformed β material, thus accelerating crack growth. Although the mechanism is unclear the overall rates of crack growth are not inconsistent with the repeated failure of laths within the transformed β matrix of approximate size 0.05μm x 0.4μm, in agreement with previous work on other Ti_3Al alloys (6,7). These steps show no obvious signs of plasticity.

At the higher temperature of 700°C both conditions generally exhibit higher growth rates than at room temperature, figure 4. There is also a significant difference in fatigue crack growth behaviour between the two microstructural conditions under study at this temperature, with the β heat treated condition exhibiting crack growth rates of up to an order of magnitude faster than those observed in the α_2 + transformed β heat treated condition. This is in contrast to the effect of microstructure at room temperature, where the β heat treated condition shows improved fatigue crack growth resistance. However, careful examination of the fatigue surfaces has provided little clear evidence of differing mechanisms with no visible evidence of the influence of microstructure, figure 6. In other work such observations of decreased crack growth resistance at 700°C for β heat treated condition have also been made in vacuum (9) and hence it appears to be fundamental to the two microstructural conditions. From the tensile data, table I, it can be seen that the fully transformed β microstructural condition has a lower yield stress than the α_2 + transformed β condition. It is difficult to isolate effects of yield stress on the crack growth rates but in the Paris regime it can be argued that lower strength materials may undergo more rapid crack propagation. This is because the range of crack opening displacement over the fatigue cycle, $\Delta\delta$, is inversely proportional to the yield stress and as the crack growth rate da/dN is proportional to $\Delta\delta$, crack growth will be slower in this range for a material with a higher yield stress. This would, however, necessitate the adoption of a strain controlled mechanism of fatigue crack growth. Such an influence of yield stress would not therefore be expected at room temperature in the alloy under consideration here. Fatigue crack growth in both materials is transgranular, figure 6, but it is unclear why the microstructure of the α_2 + transformed β material should be more fatigue resistant then that of the fully transformed β material. In fact in traditional titanium alloys it is general the fully transformed β microstructures that are found to provide the best fatigue crack growth resistance as a consequence of crack tip deflection (10,11). Crack tip deflection is usually facilitated by the presence of regions of retained β material between the laths as observed in other Ti_3Al alloys (7). The materials under consideration here have been aged at 850°C and microstructural work currently being undertaken has given no evidence of a retained β phase to date (12). It is of interest to note in the fast fracture regions, the β heat treated material showed an intergranular ductile microvoid coalescence mode of failure while the α_2 + transformed β remained fully transgranular. This is discussed more fully elsewhere (13). It is clear, therefore, that fatigue crack growth behaviour at elevated temperature is a complex problem in these alloys and further work is still required to quantify the effects of microstructure on fatigue crack growth in these emerging alloys.

Conclusions

Effects of microstructure on fatigue crack growth resistance have been observed at both room temperature and at the test temperature of 700°C. Microstructures comprising primary α_2 precipitates in a fully transformed β matrix exhibit decreased crack growth rates at room temperature but increased resistance at 700°C with respect to a fully transformed β condition. At room temperature cleaved α_2 precipitates are observed and it appears that they are responsible for the poorer crack growth resistance of the α_2 + transformed β condition, seen particularly at high mean stress. Conversely, at 700°C, the reduced yield stress of the fully transformed β condition has been deduced to increase crack growth rates relative to the α_2 + transformed β condition.

Acknowledgements.

One of the authors (RJTP) has been supported during the course of this work by a SERC case studentship with Rolls Royce PLC, Derby.

References

1. H.A. Lipsitt "High temperature intermetallics", ed. C.C. Koch, et al.,MRS Symp. Proc. Vol. 39, Pittsburgh PA MRS 1985, pp. 351-364.

3. R.J.T. Penton, M.T. Cope and P. Bowen, Conf. Proc. ICM-6, Kyoto, Japan, July 28 - August 2, 1991, p 205 - 210, Pergamon, 1991.
4. K.S. Chan Met Trans Vol. 21A (1990), p. 2687-2699.
5. J.M. Larsen et al. in C.T. Liu., D.G. Pettifor, D.P. Pope, V.K. Sikka, J.A Stiegler, S.H. Wang and M. Yagamuchi eds. 'High temperature aluminides and intermetallics', TMS/ASM International, USA, 1990.
6. R.A. Chave, C.J. Beevers and P. Bowen, Euromat '91, Conf. Proc., eds. T.W. Clyne and P.J. Withers, Institute of Materials, London, p333, 1992.
7. P.B. Aswath. and S. Suresh, Mat. Sci. Eng., A114 (1989) L5-L10.
8. R.J.T. Penton, M.T. Cope, C.J. Beevers and P. Bowen, Conf. Proc. 7th World Titanium conference, San Diego, 29th June - 2nd July, 1992
9 R.J.T. Penton, PhD. Thesis, University of Birmingham, 1992
10 N. E. Paton, J.C. Williams, J. C. Chesnutt and A.W. Thompson, AGARD Proc. Conf., 185, National Technical Information Service, Springfield VA, 1976
11 H. Margolin, J.C. Williams, J.C. Chesnutt and G. Lueterjing, 4th World Titanium Conference, Kyoto, May 19th - 22nd, 1980
12 R. J. T. Penton and M. Strangwood, Unpublished work
13 R.J.T. Penton, M.T. Cope and P. Bowen, Conf. Proc. IRC 92 Processing Properties and Applications of Metallic and Ceramic Materials, Sept. 7 - 10, 1992

$133\mu m$

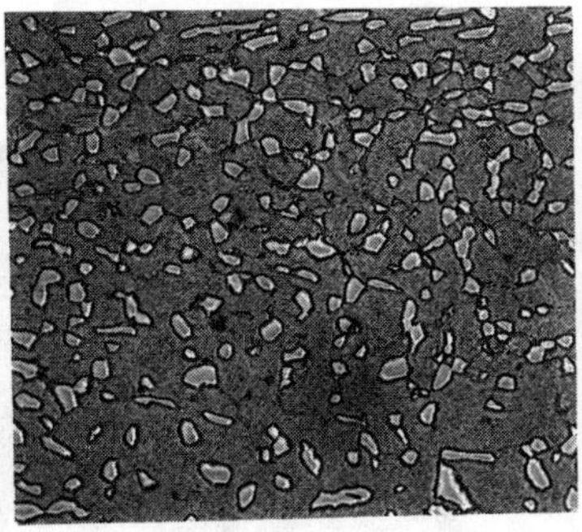

Figure 1 - As heat treated microstructure for the α_2 + transformed β condition.

Figure 2 - As heat treated microstructure for the fully transformed β condition.

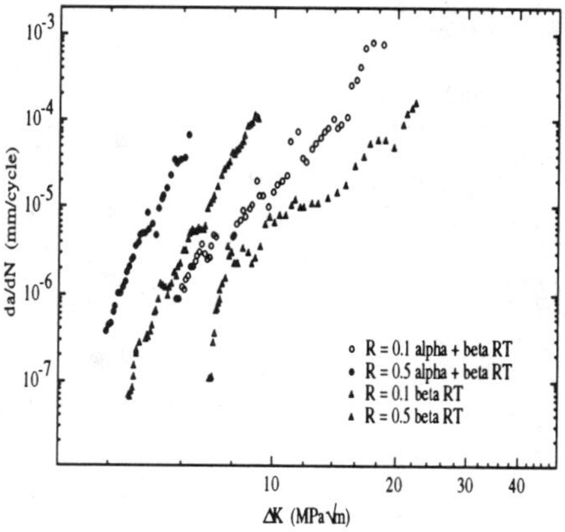

Figure 3 - Fatigue crack growth resistance curves at room temperature in air.

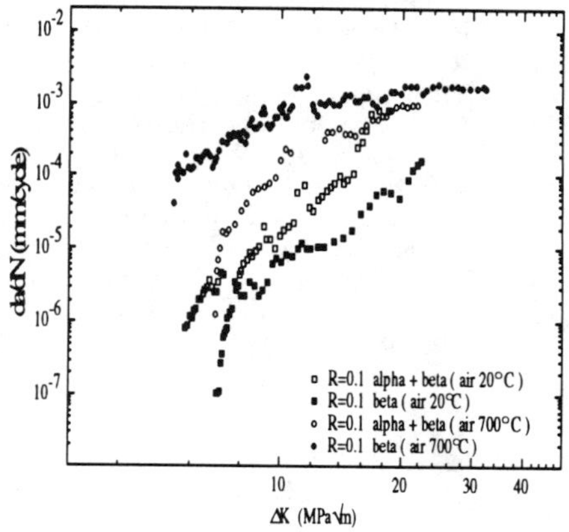

Figure 4 - Fatigue crack growth resistance curves at 700°C in air.

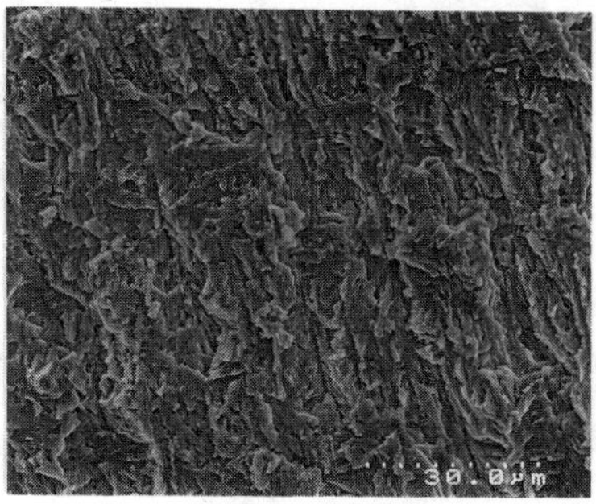

Figure 5 - Fatigue fracture surfaces after testing at R = 0.1 in air at room temperature, a) α_2+ transformed β condition, b) fully transformed β

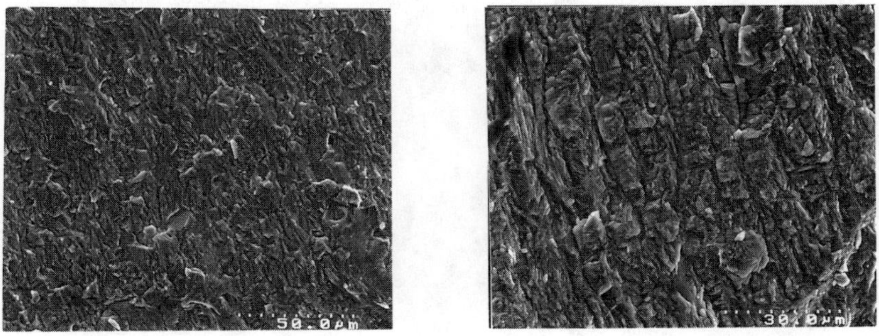

Figure 6 - Fatigue fracture surfaces after testing at R = 0.1 in air at 700°C, a) α_2+ transformed β condition, b) fully transformed β

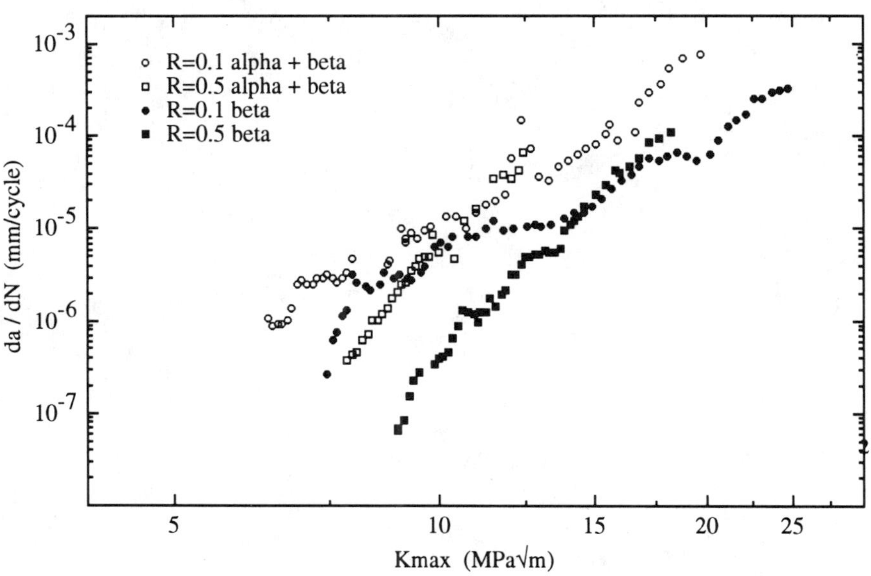

Figure 7 - Fatigue crack growth rate versus K_{max} at room temperature in air.

RECRYSTALLIZATION AND ITS EFFECT

ON THE FLOW BEHAVIOUR OF TiAl

N. K. Park, Y. T. Hyun, Y. T. Lee and A. K. Ghosh*

Korea Institute of Machinery and Metals, 66 sangnamdong,
Changwon, Kyungnam, Republic of Korea
* Dept. of Materials Science and Engineering, The Univ. of Michigan,
Ann-Arbor, MI 48109, USA

Abstract

This study is focused on the static and dynamic recrystallization of Ti-45Al and Ti-48Al-1Nb during heat treatment and compression test. Heat treatment was done near the eutectoid temperature, and compression tests were condcuted at 1300 K with strain rates ranging from 10^{-5} s^{-1} to 10^{-1} s^{-1}. X-ray diffraction, optical and TEM techniques were used to investigate the recrystallization behaviour.

Static and dynamic recrystallization result in the formation of fine grains, but their characteristics are different. The different behaviour of the recrystallization processes is reflected in step strain rate tests; i.e., the flow behaviour of two phase TiAl does not depend on heat treatment history, but on strain rate history.

Introduction

To fully utilize two phase TiAl developed for high temperature applications, the selection of a processing route is an important issue. Ingot metallurgy is generally accepted to be promising even with its own difficulties in process control for quality ingots. Limited work on the conversion of TiAl ingot has shown that it is relatively difficult to form into shape but dynamic recrystallization may be utilized for its conversion.

Dynamic recrystallization occurs near the grain boundaries and cross-twinning in two phase TiAl[1,2)], eventually leading to fine grains at large strains. TiAl grains can also be refined by static recrystallization. However, many features of recrystallization and their effects on flow behaviour await clarification to make the most use of it. This work is conducted to extend the understanding of the recrystallization behaviour in two phase TiAl. Alloys with different fraction of γ and α_2 are chosen to investigate the effect of the alloy chemistry and microstructure on the recrystallization bahaviour.

Alloys and Experimental

The material investigated in this study is two kinds of TiAl; i.e. Ti-45Al and Ti-48Al-1Nb. Microstructure of the as-received Ti-45Al is composed of lamellar and γ grains, having grain size of ~150 μm. The Ti-45Al ingot was homogenized at 1373 K/150 hrs, followed by hot isostatic pressing at 1100 K/5hrs. Ti-48Al-1Nb was cast/homogenized/forged, and the resulting microstructure consists of γ grains with less than 5% of lamellar phase. Average grain size of the as-received Ti-48Al-1Nb is about 15 μm.

To understand the static recrystallization behaviour in two phase TiAl, the alloys were heat-treated at several temperatures around 1400 K under vacuum atmosphere, and subsequently water-quenched. Compression tests were conducted under argon atmosphere

Titanium '92
Science and Technology
Edited by F.H. Froes and I. Caplan
The Minerals, Metals & Materials Society, 1993

with various strain rates ranging from 10^{-5} to 10^{-1} s^{-1}. The specimens were rapidly cooled to freeze the deformed microstructure and reduce any static recovery and/or recrystallization.

Results

Grains of two phase TiAl are known to be refined by heat treatment above the eutectoid temperature (~1400 K)[3]. The grain refinement by heat treatment would depend strongly on Al content in the two phase TiAl, since the equilibrium amount of α_2 (or α) phase formed by phase separation hinder the grain growth[4]. To understand the refinement of grain size during heat treatment, one set of experiment was conducted. The specimens were heat-treated at 1400 K for 24 hrs to obtain coarse γ grains, and followed by heat-treatment at 1540 K from 1 to 24 hrs. The change in microstructure of Ti-45Al and Ti-48Al-1Nb during heat-treatment is presented in Fig. 1 and 2, respectively. The coarse grains are refined by acicular type morphology which is running through the γ grains. The resulting microstructure is in contrast to that of the as-received alloys subjected directly to heat-treatment at 1540 K.

X-ray spectroscopy indicates that the composition of the acicular phase formed in Ti-48Al-1Nb is $Ti_{0.6}Al_{0.4}$, which is likely to be an off-stoichiometric Ti-deficient Ti_3Al. Antiphase domain boundaries are often observed in the α_2 phase. Titanium and aluminium atoms are substitutional, and the excess aluminium atoms occupy vacant titanium sites. Crystallography indicates that six orientation variants are possible between α_2 and γ , i.e. $\alpha_2\{0001\}//\gamma\{111\}$ and $\alpha_2<11\bar{2}0>//\gamma<1\bar{1}0>$[5].

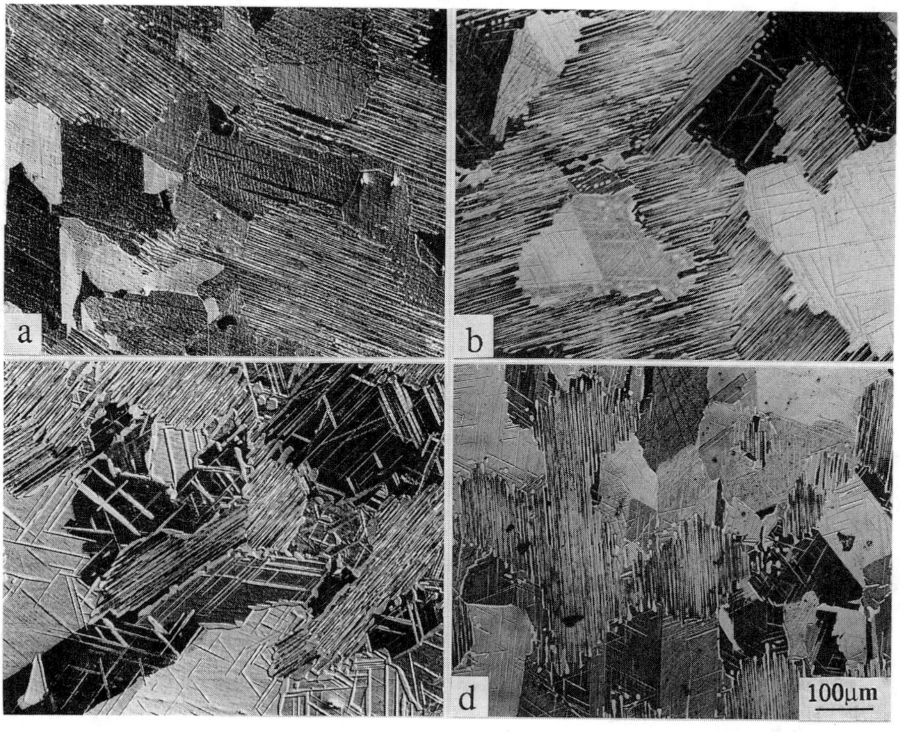

Fig.1 - Metallographs of heat-treated Ti-45Al
a) 1400K/24h/W.Q b) 1400K/24h/W.Q→ 1540K/1h/W.Q
c) 1400K/24h/W.Q→ 1540K/24h/W.Q d) 1540K/1h/W.Q

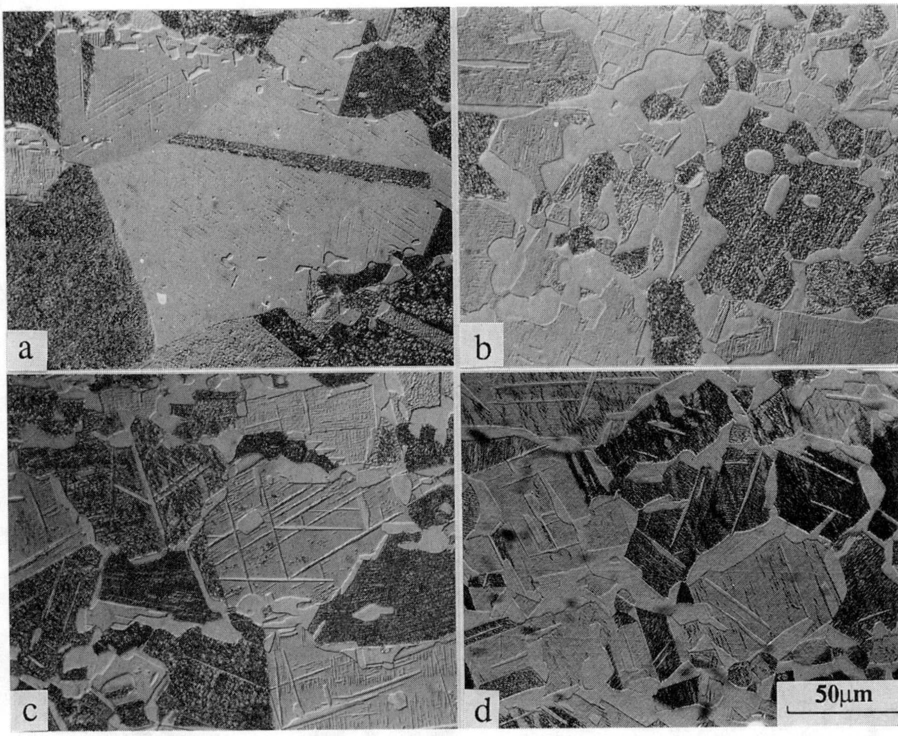

Fig. 2 - Metallographs of heat-treated Ti-48Al-1Nb
a) 1400K/24h b) 1540K/24h
c) 1400K/24h → 1540K/1h d) 1400K/24h → 1540K/24h

The flow behaviour of Ti-45Al and Ti-48Al-1Nb was observed at strain rates 10^{-2} s^{-1} and 10^{-4} s^{-1}. Fig. 3 shows that the flow curves of the alloys depend on the strain rate and alloy composition. The level of flow stress is higher in Ti-45Al than in Ti-48Al-1Nb. However, the flow curves show a stress peak regardless of the strain rate and composition, and reaches a saturation value at large strain. The decrease in the flow stress after the peak stress is due to dynamic recrystallization, as confirmed by optical micrographs. Fig. 4 shows the recrystallization of Ti-45Al. Similar behaviour is also noted in Ti-48Al-1Nb.

Several microstructural features are to be noted in the dynamic recrystallization of Ti-45Al and Ti-48Al-1Nb. It is shown in Fig. 4 that dynamic recrystallization of Ti-45Al occurs predominantly in γ grains compared to that in lamellar grains. The average grain size of the recrystallized Ti-45Al and Ti-48Al-1Nb increases with decreasing strain rate. Fig. 5 shows that recrystallization of Ti-48Al-1Nb leads to fine grains through formation of new grains near the primary lamellar grains. New grains are nucleated in such a way that α_2 and γ grains are formed at the end of the lamellar structure. Similar behaviour is not observed in the statically recrystallized TiAl.

Fig. 6 shows a bent lamellar structure formed by localized shear. Dynamic recrystallization occurs prominently in lamellar grains as well as in γ grains at 10^{-4} s^{-1}. It is to be noted that α_2 plates in the lamellar structure are gradually transformed into α_2 islands maintaining a small orientation difference < 10° with each other. Similar orientation relationship is also noted between two α_2 islands formed in neighbouring α_2 plates in the lamellar structure. This behaviour is often observed in the specimen deformed at 10^{-4} s^{-1} not in the specimen deformed at 10^{-2} s^{-1}.

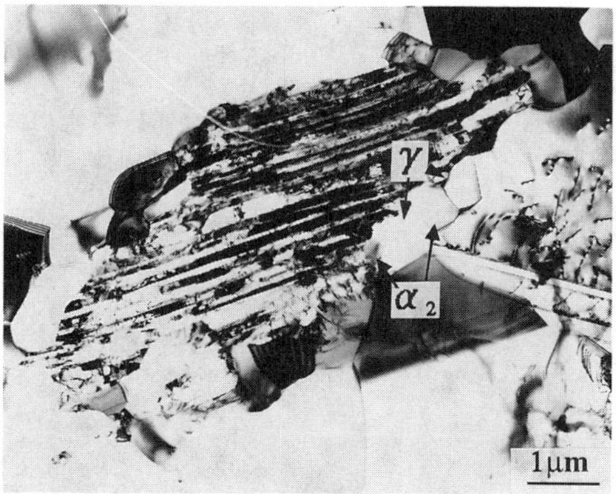

Fig. 5 - Formation of equiaxed γ and α_2 grains from the twinned lamellae. $T=1300K$; $\dot{\varepsilon} = 10^{-2}\ s^{-1}$; $\varepsilon = 1.4$

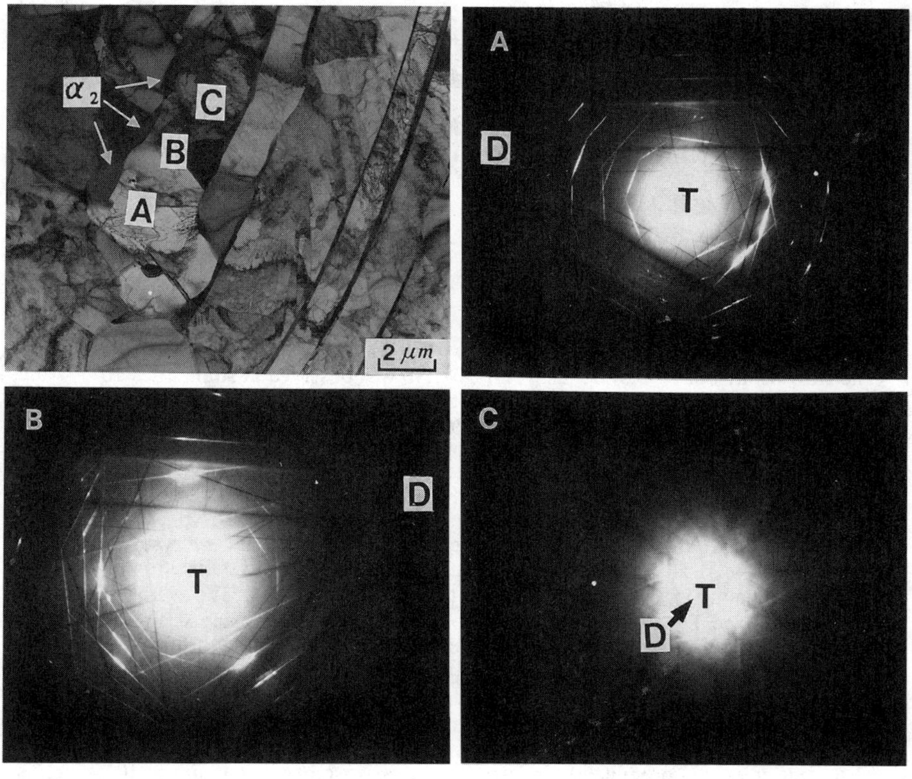

Fig. 6 - Orientation relationship between γ grains formed from lamellar plates in Ti-45Al. $T=1300K$; $\dot{\varepsilon} = 10^{-4}\ s^{-1}$; $\varepsilon = 0.9$ T and D in the microdiffraction pattern (A,B,C) representing Transmitted beam and D= $Z[0\ 1\ \bar{1}\ 0]$ respectively.

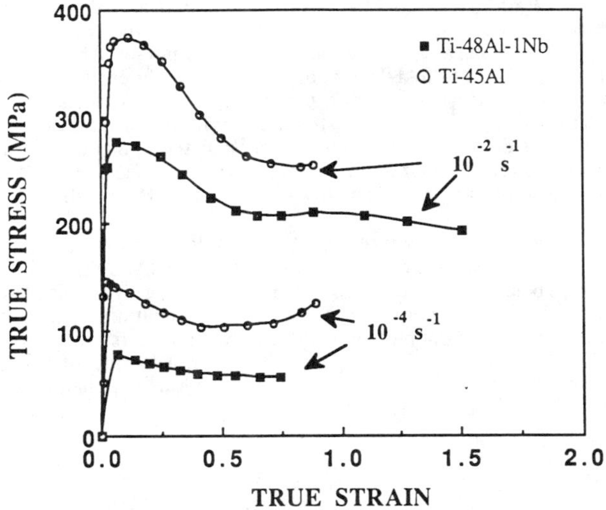

Fig. 3 - Flow curves of Ti-45Al and Ti-48Al-1Nb at 1300 K

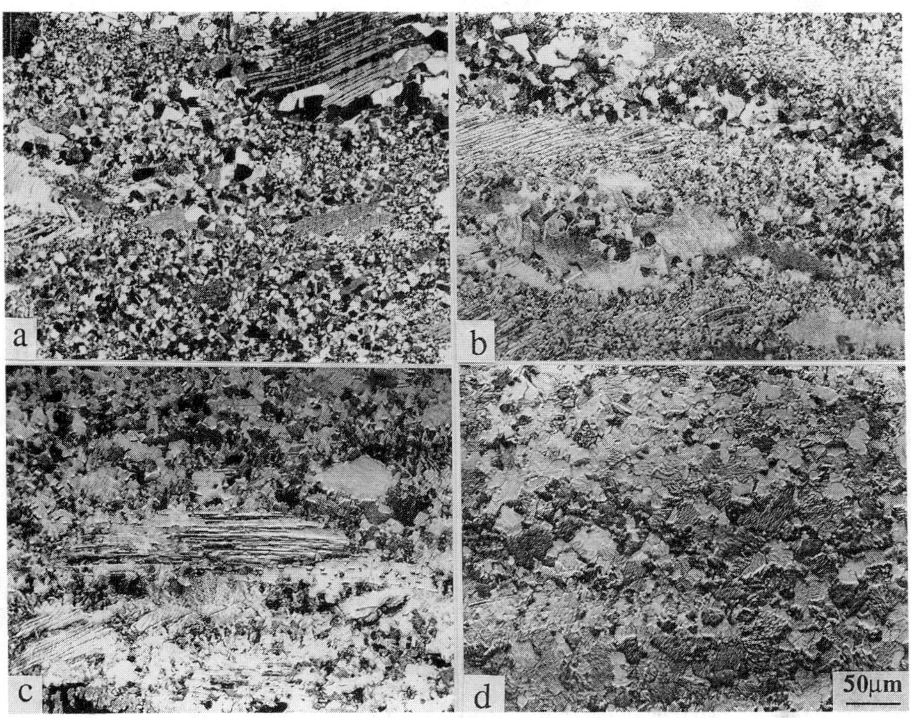

Fig. 4 - Optical micrographs of Ti-45Al deformed at different strain rates.
T = 1300 K, $\varepsilon = 0.9$
(a) $\dot{\varepsilon} = 10^{-1}\ s^{-1}$ (b) $\dot{\varepsilon} = 10^{-2}\ s^{-1}$ (c) $\dot{\varepsilon} = 10^{-3}\ s^{-1}$ (d) $\dot{\varepsilon} = 10^{-4}\ s^{-1}$

Since the strain rate sensitivity is an important parameter in metal forming, it is obtained by the step strain rate test between 10^{-5} s^{-1} and 10^{-2} s^{-1} with Ti-48Al-1Nb of different heat treatment history, and with Ti-48Al-1Nb and Ti-45Al of different strain rate history. To investigate heat treatment effect on the flow curve, step strain rate tests were conducted on the as-received Ti-48Al-1Nb and the heat-treated Ti-48Al-1Nb at 1540 K for 1hr. The heat-treated sample was composed of statically recrystallized grains of ~ 8 μm in diameter. To observe the strain rate history effect, the samples were strained at 10^{-2} s^{-1} up to the strain 0.5, followed by step strain rate tests from 10^{-5} s^{-1} to 10^{-2} s^{-1}. The results are summarized in Fig. 7.

The level of flow stress of the as-received Ti-48Al-1Nb and of the statically heat-treated Ti-48Al-1Nb is similar. However, the flow stress of the Ti-48Al-1Nb strained at 10^{-2} s^{-1} up to the strain 0.5 drops below that of the as-received or statically recrystallized specimens after a strain rate change to 10^{-5} s^{-1}, while it reaches a similar level at high strain rates. It is also shown that the flow stress of dynamically recrystallized Ti-45Al is quite comparable to that of Ti-48Al-1Nb at 10^{-5} s^{-1}, but it regains higher value at 10^{-2} s^{-1}. Accordingly, the strain rate sensitivity given by the slope in Fig. 7 is increased in effect due to dynamic recrystallization, and its effect is greater in Ti-45Al than in Ti-48Al-1Nb.

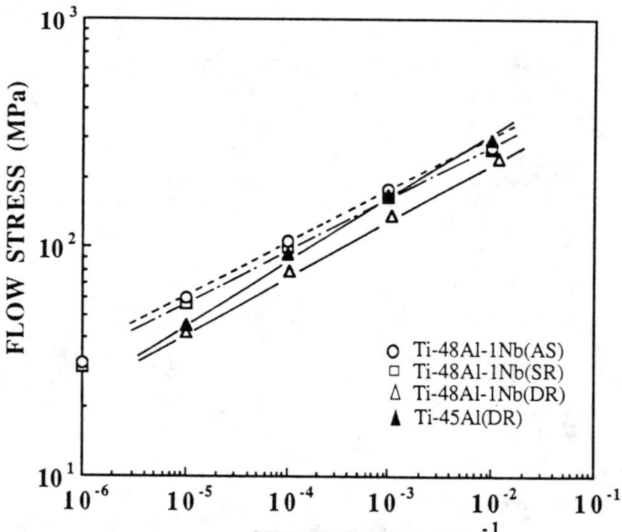

Fig. 7 - Variation of the flow stress with heat-treatment and deformation history. Step strain rate tests employed from low strain rates to high strain rates at 1300 K AS = As Received ; SR = Heat-treated at 1540 K/1hr ; DR = Deformed at $\dot{\varepsilon} = 10^{-2}$ s^{-1} up to $\varepsilon = 0.5$

Discussion

Recrystallization behaviour of two phase TiAl varies with heat treatment temperature and holding time. Fig. 1 and 2 show that the grain size refinement is significant in Ti-48Al-1Nb during heat treatment above the eutectoid temperature (~1400 K). The coarse γ grains are refined by acicular type phases during heat-treatment above the eutectoid temperature. The phases may have a transient form of structure and the recrystallization mechanism may be described as follows[4]; lamellar grains are formed by the precipitation of α grains in γ matrix, their subsequent growth to α_2 plates, and eventual formation of lamellar grains during cooling. The reason why the new α_2 phase has an acicular shape in coarse γ grains is not clear at present. However, it is likely that the diffusion rate of the constituents depend on the crystallographic orientation, resulting in the needle-like microstructure. The diffusion rate of the constituents in Ti-48Al-1Nb is not so high as to increase the grain size rapidly with an increase in holding time at 1540K[6].

In contrast to the static recrystallization, dynamic recrystallization results in a different type of morphology. Fig. 4 and 5 show that dynamic recrystallization occurs predominantly in the γ grains, adjoining the lamellar grain. The recrystallization behaviour in TiAl is rationalized by the deformation behaviour in γ grains. The deformation mode of twinned TiAl varies with the orientation of the loading axis and the characteristics of $\alpha_2 + \gamma$ lamellar boundaries[7,8]. The dynamic recrystallization of TiAl is also observed to be dependent on the mode and amount of deformation, and the restoration process appears firstly in the region where deformation is conducted by slip mode instead of twinning mode[9]. Twinned microstructure is then to be maintained during the recrystallization process, since the rearrangement and annihilation of dislocations occur firstly in γ grains neighboring the twinned lamellae.

Furthermore, geometrically necessary dislocations are likely to be generated in γ grains to meet the strain compatibility between the neighbouring lamellar and γ grains due to relatively large number of active slip systems in γ grains[10]. Dynamic recrystallization is possible in the γ region unless the dislocations are restored by recovery process. Accordingly, the ratio of lamellar to γ grains is expected to be an important factor in determining the deformation mode and recrystallization behaviour.

In the lamellar grains, most of the recrystallization process is confined to each lamellar boundaries at low strains, indicating that the recrystallization is effectively hindered by α_2 phase or phase boundaries between α_2 and γ. When the amount of deformation is large enough to induce bending of lamellar structure, the plate type α_2 phase is transformed to discrete α_2 islands. Even though α_2 grains or phase boundaries between α_2 and γ act effectively as an obstacle against dislocation motion at low strains, the change in α_2 morphology at high strains appears to be similar to that of conventional $\alpha+\beta$, titanium alloys. In this alloy, the lamellar phase breaks up and transforms to equiaxed grains either through subboundary formation in α plates or through localized shearing across the α plates during deformation followed by penetration of β, phase to complete the separation[11].

When Ti-48Al-1Nb is compressed at 10^{-2} s^{-1} to the strain 0.5, the microstructure is composed of finely recrystallized grains since γ, that is relatively easy to recrystallize, is the dominant phase. The grain boundaries formed by dynamic recrystallization are likely to have high angle boundaries and their contribution to deformation is high at low strain rates. Accordingly, when the strain rate is decreased from 10^{-2} s^{-1} to 10^{-5} s^{-1} at the strain 0.5, the flow stress would decrease below that of the as-received Ti-48Al-1Nb. Similar behaviour would be observed in Ti-45Al to a different extent, since the ratio of lamellar to γ grain is high and dynamic recrystallization is limited within the lamellar boundaries. Fig. 7 shows indeed that the flow stress of dynamically recrystallized Ti-45Al is comparable to that of Ti-48Al-1Nb at 10^{-5} s^{-1} even though the ratio of α_2 to γ in the former is higher than that in the latter, but soon the flow stress regains a high value at high strain rates due to the limitation of the dislocation activity in α_2 phase.

Conclusion

1) Static recrystallization and dynamic recrystallization are shown to result in fine grains in two phase TiAl, but the refined grains are different in morphology and characteristics.
2) Lamellar grains have an active role in dynamic recrystallization of the Ti-45Al and the Ti-48Al-1Nb. Alloy chemistry is an important factor in recrystallization since the ratio of lamellar to γ grains is very much affected.
3) The α_2 phase in the lamellar structure hinders dislocation motion and deters recrystallization in the lamellar structure, but it enhances dynamic recrystallization in the neighbouring γ grains.

References

1) C. R. Feng, D. J. Michel and C. R. Crowe, Proceed. Mat. Res. Soc. Symp., Vol. 133, MRS, (1989), 699-74.

2) O. A. Kaibyshev, Recrystallization '90. Proceed. Inter. Conf. on Recrystallization in Metallic Materials, ed. T. Chandra, (Warrendale, PA: TMS, 1990), 855-60.

3) Y. -W. Kim and F. H. Froes, Proceed. High Temperature Aluminides and Intermetallics Symposium, eds. S. H. Whang, et al., (1989), 465-92.

4) Y.-W. Kim, Microstructure/Property Relationships in Titanium Aluminides and Alloys, eds. Y-W. Kim and R. R. Boyer, (Warrendales, PA: TMS, 1990).

5) S. M. L. Sastry and H. A. Lipsitt, Titanium '80, eds. H. Kimura and O. Izumi, (Warrendale, PA: TMS-AIME, 1980), 1231-43.

6) N.K. Park, Unpublished work, 1990.

7) H. Inui, A. Nakamura, and M. Yamaguchi, Proceed. Proceed. Mat. Res. Soc. Symp., Vol. 213, 1991 MRS, 569-74.

8) H. Inui, A. Nakamura, M. H. Oh and M. Yamaguchi, Submitted to Phil. Mag., (1991).

9) M. H. Oh, H. Inui, A. Nakamura and M. Yamaguchi, Submitted to Acta Metall., (1991).

10) M. F. Ashby, Phil. Mag., 21, (1970), 399-424.

11) I. Weiss, F. H. Froes, D. Eylon and G. E. Welsch, Metall. Trans.,17A, (1986), 1935-47.

DEOXIDATION OF

TiAl INTERMETALLIC COMPOUNDS

T.H.Okabe, K.Fujiwara, T.Oishi and K.Ono

Department of Metallurgy, Kyoto University,
Yoshida Honmachi, Sakyo-ku, Kyoto 606, JAPAN

Abstract

Removal of oxygen in titanium aluminide (TiAl) by chemically active calcium-aluminum (Ca-Al) alloy was carried out around 1373K with the purpose of obtaining extra-low-oxygen TiAl.

The deoxidation experiments were preceded by an investigation of the phase equilibria of the system Ti-Al-Ca at 1273 and 1373K, and it was found that a Ca-Al alloy containing around 20 to 50mol%Ca would equilibrate with TiAl at 1273K. The compositions of the Ca-Al alloy deoxidant, which equilibrates with TiAl, and the experimental conditions suitable for the deoxidation were investigated.

In experiments in which Ti-Al samples were submerged in liquid Ca-Al alloys at 1373K, the surfaces of the samples had severely deteriorated and become nodular. When TiAl powders were mixed with CaO to prevent sintering, and the deoxidant was supplied in vapor form, powders which initially contained 510, 1100 and 4200 ppmO were deoxidized to about 160, 490 and 670 ppmO after deoxidation at 1373K for 86.4ks (1day). Among many conditions tested, the use of TiAl powders mixed with $CaCl_2$ was most effective for deoxidation at 1373K. $CaCl_2$ was used as a flux to facilitate the deoxidation by decreasing the activity of the deoxidation product CaO. In case that TiAl powders were mixed with $CaCl_2$ and reacted with Ca-Al vapor at 1373K for 86.4ks, the powders initially containing 510, 1100 and 4200 mass ppmO were deoxidized to a level of 62, 140 and 190 mass ppmO, respectively. No significant change in morphology of the particle after deoxidation was observed. The nitrogen concentrations in the powders remained constant, whereas calcium, which was present only in trace amounts initially, increased up to 160 mass ppm after deoxidation treatment.

Introduction

In recent years, there has been considerable interest in developing titanium aluminide (TiAl) as a structural material, and mechanical properties of TiAl have been intensively investigated. In particular, plasticity of TiAl has proven to be influenced by impurities such as oxygen.[(1)]

For laboratory use, the grade of TiAl is between 99.9% and 99.999% pure (excluding gaseous elements), where the main impurity is oxygen, at a level of at least several hundred mass ppm oxygen. This oxygen mainly originates from impurities in the titanium starting material which is produced by the Kroll process. Even though low oxygen-containing titanium has been successfully produced using other less-widely used methods, such as electrolytic refining or iodide refining methods, oxygen contamination of TiAl is inevitable during subsequent processing. Oxygen removal directly from TiAl to a level below 100 mass

Titanium '92
Science and Technology
Edited by F.H. Froes and I. Caplan
The Minerals, Metals & Materials Society, 1993

ppm seems to be very difficult, because TiAl has a strong affinity for oxygen. For this reason, titanium and aluminum are usually alloyed to make titanium aluminide in an oxygen-free atmosphere, taking great care to avoid oxygen pick up.

From a thermodynamic view point, external gettering used in solid state refining is one of the most promising methods for direct deoxidation of titanium and its alloys. Recently, the authors have developed several effective methods(2,3,4) for the deoxidation of titanium by using calcium-saturated halide flux. The titanium produced contained below 50 mass ppm oxygen. In the present study, one of these methods has been applied to the deoxidation of TiAl.

Deoxidation of TiAl by reaction with a chemically-active element, namely calcium, through surface contact has been examined in the temperature range 1273 to 1473K, to obtain low oxygen-containing TiAl. Calcium is thought to be the most effective deoxidation agent because it has an extremely strong affinity for oxygen. Furthermore, the deoxidant calcium and the reaction product CaO can be easily removed by acid treatment.

Since the detailed principle of deoxidation of titanium using calcium has been reported elsewhere (2), only a brief outline will be given here. Oxygen dissolved in TiAl can be removed by calcium according to the following reaction:

$$\text{O (in TiAl)} + \text{Ca (in Ca–Al)} = \text{CaO} \qquad (1)$$

Co-existence of calcium and the reaction product, CaO, fixes the equilibrium oxygen partial pressure, and hence, the amount of residual oxygen in TiAl is thermodynamically fixed. As there is no available information on activity of oxygen in TiAl, the deoxidation limit of TiAl cannot be estimated with accuracy; however, the activity coefficient of oxygen in TiAl is expected to be larger than that in pure beta-titanium considering the affinities of oxygen for titanium and aluminum and the solubilities of oxygen in titanium and aluminum. For this reason, the deoxidation limit of TiAl may be lower than that of titanium under same oxygen partial pressure.

Similar to the previous study(3), $CaCl_2$ was used as a flux to dissolve the deoxidation agent calcium and, in addition, to facilitate the deoxidation reaction by diluting the reaction product CaO, i.e., by decreasing the activity of the by-product CaO.

At first glance, the deoxidation of TiAl presents some difficulties. Pure calcium, which is effective in deoxidation of titanium, seems to be inadequate for the deoxidation of TiAl, because of the strong affinity between calcium and aluminum which results in Ca-Al alloy formation. For example, TiAl allowed to react with pure calcium or a Ca-Al alloy of lower aluminum activity than TiAl, will undergo aluminum depletion at the surface and form Ti_3Al at the interface. On the other hand, reaction between TiAl and Ca-Al alloy of higher aluminum activity than TiAl will result in diffusion of aluminum from Ca-Al alloy to TiAl and form aluminum-rich compounds ($TiAl_3$,$TiAl_2$) at the surface. Therefore, knowledge of the phase equilibria existing between TiAl and Ca-Al alloys is essential for selection of a suitable Ca-Al alloy deoxidant.

Prior to the deoxidation experiments, phase equilibria in the Ti-Al-Ca system at the reaction temperatures were investigated to determine suitable Ca-Al alloys in equilibrium with TiAl. Using this information of phase relation, the composition of the Ca-Al alloy deoxidant suitable for the preparation of extra-low-oxygen TiAl was decided.

Experimental

Phase Equilibria Between TiAl and Ca-Al Alloys

The laboratory investigation of the phase equilibria of the Ti-Al-Ca system was comprised of two types of experimental procedures - one was diffusion coupling of the two materials, and the other, isothermal annealing of metal powder mixtures of the Ti-Al-Ca system.

Diffusion couple experiments were conducted in order to identify Ca-Al alloys in equilibrium with TiAl. The small pieces of 99%-pure TiAl or powders(produced by plasma rotating electrode process) used are listed in Table I. The starting materials of the phase equilibrium investigation of the Ti-Al-Ca system at 1273K consisted of 99%-pure titanium powder, 99.5%-pure aluminum powder and 98%-pure calcium granules. The apparatus and experimental method used for these experiments were similar to those described previously.[5,6] Samples of various compositions were prepared by thorough mixing of starting materials to the desired ratios using an agate mortar and pestle. 10–gram sample mixtures were placed in molybdenum crucibles which were inserted into a stainless steel tube, and then sealed at both ends by welding in an argon gas atmosphere. The assembled tube was heated in an electric furnace to a temperature of 1273K. The holding time was between 73ks and 1Ms. Afterwards, the reaction tube was removed from the furnace and quenched in water. The samples obtained were subjected to phase analysis.

Table I TiAl samples used in this study.

Samples	Purity+	Oxygen (mass ppm)	Nitrogen (mass ppm)	Titanium (mass%)[mol%]	notes
TiAl powder*	3N	510	70	64.0 [50.1]	42~60 mesh
TiAl powder*	3N	1100	30	65.0 [51.2]	-60 mesh
TiAl powder*	3N	4200	50	65.6 [51.8]	-60 mesh
TiAl small pieces	3N	1000	150	63.5 [49.5]	~0.5g each

* : Produced by the plasma rotating electrode process.
+ : Excluding gaseous elements.

Deoxidation of TiAl by Ca–Al Alloys

Fig.1 shows the arrangement of the reaction tube used for the deoxidation of TiAl. Specifications and chemical analysis of the feed materials are listed in Table I. In some experiments, TiAl powder which had been thoroughly mixed with CaO powder, was contained in a titanium cup. In other experiments, TiAl in the form of powder and small pieces (about 20–40g) was placed on a titanium dish within a titanium cup, which was filled with about 20g of $CaCl_2$ powder. The anhydrous $CaCl_2$ and CaO used were reagent grade (99.9%) powders. These were dried for more than 200ks at 800K and for more than 100ks at 1300K, respectively, before use. Cups containing sample and flux were sealed in a stainless steel tube in addition to a mixture of calcium and aluminum granules, as shown in Fig.1. The total amount of calcium and aluminum was about 15g, and the initial composition of the Ca–Al alloy was varied from Ca–40mol%Al to Ca–60mol%Al, which was the composition range shown to be favorable for deoxidation of TiAl by the preceding phase equilibrium experiments. To avoid contamination by impurities present in calcium, especially nitrogen, calcium–aluminum alloy was isolated from the samples, and supplied to the samples in vapor form.

The assembled sealed tube was heated in an electric furnace to a temperature between 1273 and 1473K. The holding time was between 86.4ks and 260ks, after which the reaction tube was removed from the furnace and quenched in water.

After the deoxidation, the Ca–Al alloy, CaO and/or $CaCl_2$ in the titanium cup was removed by leaching with (1+1) acetic acid, and the remaining TiAl sample was carefully cleaned in warm aqueous HCl solution followed by water, alcohol and acetone rinses, and then allowed to dry under vacuum.

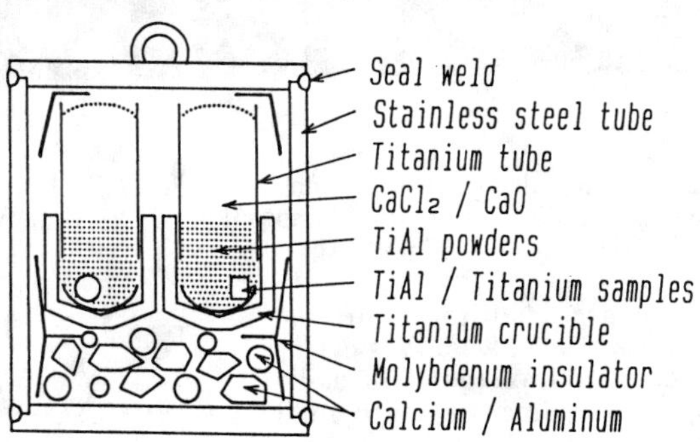

Fig.1 Schematic illustration of reaction tube used for deoxidation of TiAl.

Analysis

Oxygen and nitrogen analyses of samples were made using an inert gas fusion infrared absorption method (LECO TC–336 analyzer). About 0.1g of TiAl sample, mostly in powder form, was sealed in a tin capsule and wrapped in 1g of platinum foil for the purpose of oxygen and nitrogen extraction. In some analyses, 1g of nickel was used as the bath for oxygen extraction. To ascertain the accuracy of the analyzed values of oxygen, oxygen analysis of some samples was performed independently by a company.

Titanium and calcium concentrations of the sample were determined by a flameless atomic absorption spectrometric (AA) method.

Identification of the existing phases in the sample was carried out by powder X–ray diffractometry(XRD) using a Philips PW1700 system. X–ray microprobe analysis (XMA, Hitachi X–650) was done to obtain the compositional profiles.

Results and Discussion

Phase Diagram of the Ti–Al–Ca System at 1273K

Fig.2 shows some typical compositional and morphological changes of TiAl small pieces reacted with Ca–Al alloys of various compositions at 1273K for 86.4ks. The surfaces running vertically on the left hand side of the photographs mark the interfaces between TiAl sample and Ca–Al alloy which was removed by leaching. In the case of reaction between TiAl and pure calcium, as shown in Fig.2(a), aluminum near the surface of the sample reacted with liquid calcium to form Ca–Al alloy. As a result, aluminum depletion occurred and porous Ti_3Al phases formed in the vicinity of the surface. In the case of reaction with Ca–50mol%Al, as shown in Fig.2(b), no intermetallic compound other than TiAl is observed. Thus it is demonstrated that Ca–50mol%Al can equilibrate with TiAl at 1273K. It is worth noting that even though TiAl was in equilibrium with Ca–50mol%Al alloy, the originally planar surface of the TiAl sample became gouged, as can be seen in Fig.2(b). In the case of reaction between TiAl and Ca–67mol%Al, aluminum from the Ca–Al alloy diffused into TiAl to form aluminum–rich alloys such as $TiAl_2$ and $TiAl_3$, as shown in Fig.2(c).

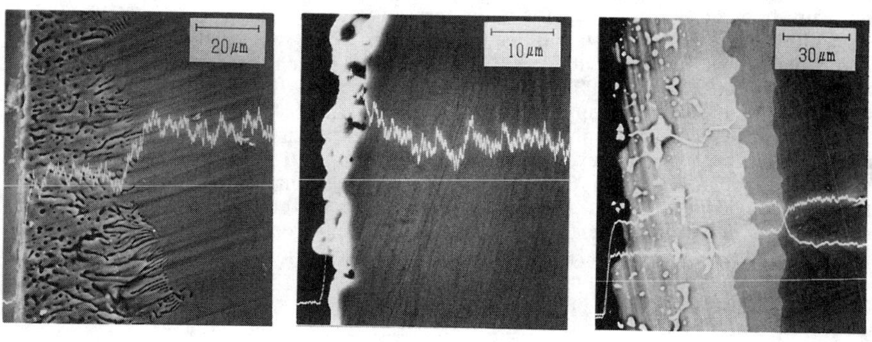

(a) (b) (c)

Fig.2 Scanning electron micrographs of sectioned TiAl small pieces reacted with various Ca–Al alloys at 1273K for 86.4ks. Composition of Ca–Al alloys were: (a) pure calcium, (b) Ca–50mol%Al and (c) Ca–67mol%Al.

Phase equilibria of the Ti–Al–Ca system were thus determined. In addition to the diffusion couple experiments, isothermal annealing of metal powder mixtures of various initial compositions was performed to verify the phase equilibrium results. Phases present were identified by XRD and XMA. It was confirmed that calcium does not dissolve in TiAl, as it was not detected by XMA. Results of this type of experiment for the Ti–Al–Ca system at 1273K, most of which were established in previous works[5,7], are compiled in Fig.3. The phase diagram in Fig.3 illustrates that

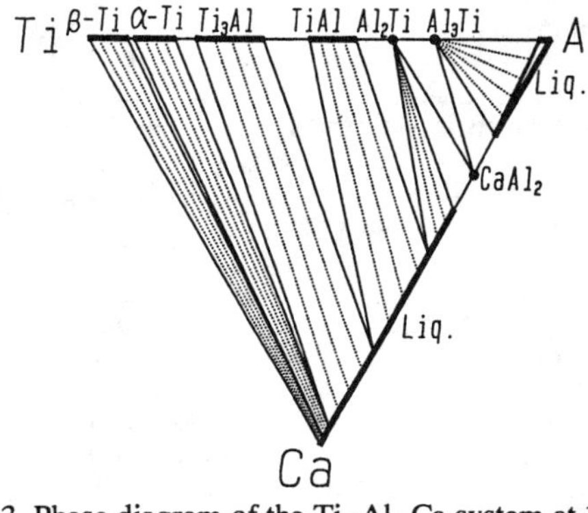

Fig.3 Phase diagram of the Ti–Al–Ca system at 1273K.

Ca–Al alloy of aluminum concentration between 20 and 50mol%, equilibrates with TiAl at 1273K. For oxygen extraction to a very low level, higher calcium concentration is preferred; however, high aluminum content alloys (about 50 mol%Al) were mainly employed as deoxidants in this study, to compensate for subsequent aluminum depletion of the alloy during reaction with metallic assemblies of the reaction tube.

Deoxidation of TiAl

Deoxidation of TiAl in Liquid Ca–Al Alloys The analyzed oxygen concentrations of the small pieces of TiAl, used in the diffusion couple experiments at 1273K, which reacted with Ca–50mol%Al, were in most cases about 1000 mass ppm, and did not vary much from the initial value. In some cases, there was an increase but with large scatter. The reaction time and the composition of the Ca–Al alloys did not appear to influence the final oxygen concentration of the TiAl small pieces. This is probably due to a low rate of oxygen diffusion in TiAl, and the increase in oxygen level may be due to oxygen pick up from impurities trapped on the rough surface.

When the reaction temperature was raised to 1373K, the TiAl small pieces (initial concentration of 1000 mass ppm oxygen) were, in some cases, deoxidized to a several hundred mass ppmO using Ca–50mol%Al and a reaction time of 86ks. Based on these results, the reaction temperature was raised to 1373K or 1473K with the purpose of shortening the reaction period. Moreover, TiAl samples in powder form (see Table I) were mainly used to diminish the problem of slow oxygen diffusion.

In the deoxidation experiments at 1373K and even more at 1473K, aluminum depletion by reaction with the stainless steel reaction tube was pronounced. The aluminum concentration in Ca–Al alloys decreased by 5 to 20 mol% at 1373K, and this decrease was more significant at 1473K. The initial composition of Ca–Al alloy adopted in subsequent deoxidation experiments was, therefore, Ca–60at%Al to compensate for aluminum depletion of the alloys.

A scanning electron micrograph of the TiAl powder used as starting material is shown in Fig.4 (a). Morphology typical of TiAl powder(Exp.No.33–liq) which had been reacted in liquid Ca–Al alloy at 1373K for 86.4ks and then acid leached to remove Ca–Al alloys, is displayed in Fig.4(b). The photographs in Fig.4 demonstrate a change in the surface of the TiAl spheres from planar to nodular, following reaction in liquid Ca–Al alloy. The concentration of aluminum in the TiAl powder remained unchanged, whereas the composition of the Ca–Al alloy co-existing with the powders of TiAl changed to Ca–53mol%Al as determined by XMA analysis. The average oxygen concentration of the TiAl powder which was initially 4200 mass ppm (refer to Fig.4) decreased to 870 mass ppm with a scatter of about 90 ppmO about the average analytical value. Measured oxygen concentrations, are listed in Table II (Exp.No.33–liq). Nitrogen concentrations of the deoxidized sample increased by 50 to 200 mass ppm, randomly. The large scatter in oxygen and nitrogen concentrations is probably due to impurities trapped on the rough particle surface.

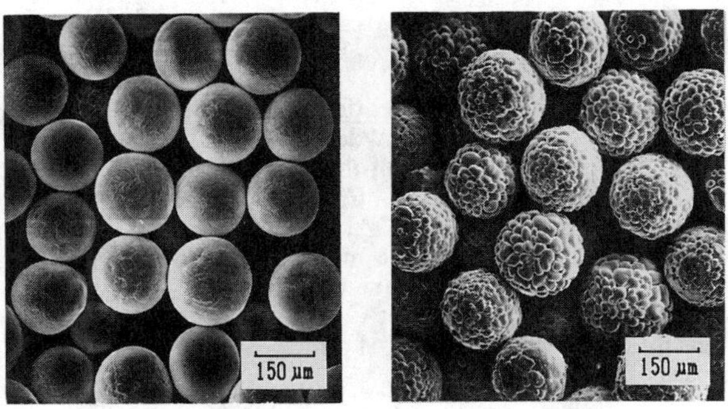

(a) (b)

Fig.4 Scanning electron micrographs of TiAl powders produced by the plasma rotating electrode process: (a) starting material, and (b): after deoxidation in liquid Ca–Al alloy at 1373K for 86.4ks. Oxygen concentrations of TiAl powders before and after deoxidation were 4200 and 870 mass ppm, respectively.

The method described above proved to be effective in deoxidizing TiAl, however, the resulting nodular surface led to trapped impurities.

Table II Analytical results of TiAl powders deoxidized by Ca-Al vapor at 1373K for 86.4ks using various fluxes.

Exp. No.	Flux	Oxygen concentration (mass ppm)		Titanium conc. (mass%)		Calcium conc. (mass ppm)
		Initial	After exp.	After exp. AA[1]	XMA	After exp. AA[1]
38-U	$CaCl_2$	510	66[2], 73[2], 33, 69, 70[3]	63.9	64.5	150
38-L	$CaCl_2$ (see Fig.6)	510	76[2], 87[2], 34, 48, 80[3]	63.9	64.0	160
39	CaO[4]	510	91[2], 80, 130[3]	62.1	61.1	220
39-ac	CaO[4]	510	85[2], 44	-	63.3	-
35-1	$CaCl_2$	1100	110, 160, 110[3], 190[3]	65.7, 65.3	63.2	250, 250
35-2	$CaCl_2$[5]	1100	190, 200, 200[3]	64.5	64.1	270
35-3	CaO	1100	430, 450, 600[3]	64.9	65.3, 66.6	870
40	$CaCl_2$	4200	210[2], 150, 180, 230[3]	62.9	63.3	460
33-M	CaO (see Fig.5)	4200	630[6], 660, 660[6], 660, 670, 700[6], 710	-	64.3, 63.7[6]	-
33-liq	Ca-Al liq. (see Fig.4(b))	4200	780, 820, 920, 960	-	61.4, 61.5	-

1 : Analyzed by flameless atomic absorption spectrometric method.
2 : Analyzed by inert gas fusion infrared absorption method using Pt bath.
3 : Analyzed by company A using Ni bath.
4 : $CaCl_2$ was added to CaO to facilitate reaction, and reacted for 260ks.
5 : 1g of aluminum granules were added to $CaCl_2$ flux before deoxidation.
6 : Samples were rinsed with acid twice.

Deoxidation of TiAl Embedded in CaO Powder by Ca–Al Vapor After deoxidation by Ca–Al vapor of TiAl powders mixed with CaO, a soft yellow cake was obtained. A scanning electron micrograph of the cross–sectioned cake which had been reacted at 1373K for 86.4ks, is shown in Fig.5(a). Characteristic X–ray images corresponding to Fig.5(a) show that the cake was a mixture of TiAl powder, CaO and Ca–Al alloy. The micrographs illustrate that TiAl and Ca–Al alloys are mutually insoluble. The composition of TiAl powder did not change after reaction, and the composition of Ca–Al liquid alloy remaining in the bottom of the reaction tube was about Ca–53mol%Al. The CaO and Ca–Al alloy in the cake could be easily removed from the TiAl powders by leaching in acid. Fine TiAl particles were obtained as shown in Fig.5(f).

In the case where TiAl powders were isolated from liquid Ca–Al alloy by using CaO powder, and deoxidation was effected by Ca–Al vapor, particle surface irregularity was considerably less than that of powders submerged in liquid Ca–Al alloy. The oxygen concentration of the isolated powders decreased from 4200 ppm to about 670 ppm for a reaction period of 86.4ks, and the amount of scatter in oxygen concentration among the analyzed samples was small, as shown by Exp.No.33–M in Table II.

As shown by Exp.No.35–3 in Table II, TiAl powders initially containing 1100 mass ppm oxygen were deoxidized to 400 – 600 ppmO, by the treatment at 1373K for 86.4ks. The bulk titanium concentration of the powders, measured by flameless atomic absorption, did not change during deoxidation, while the surface concentration of titanium in the powders increased by 1 to 2 mass%, as determined by XMA. The difference between the bulk and surface analytical values cannot be fully discussed at this stage because factors such as aluminum depletion at the surface by acid leaching, and uncertainty in XMA results due to surface conditions, have yet to be quantified.

From the results of 12 sets of different experimental conditions, a lower initial oxygen concentration of powder and a longer reaction time are to be desired in obtaining lower oxygen–containing TiAl powders. These results also indicate that the equilibrium oxygen concentration in TiAl was not reached under these experimental conditions. This is probably due to a low rate of oxygen diffusion in TiAl.

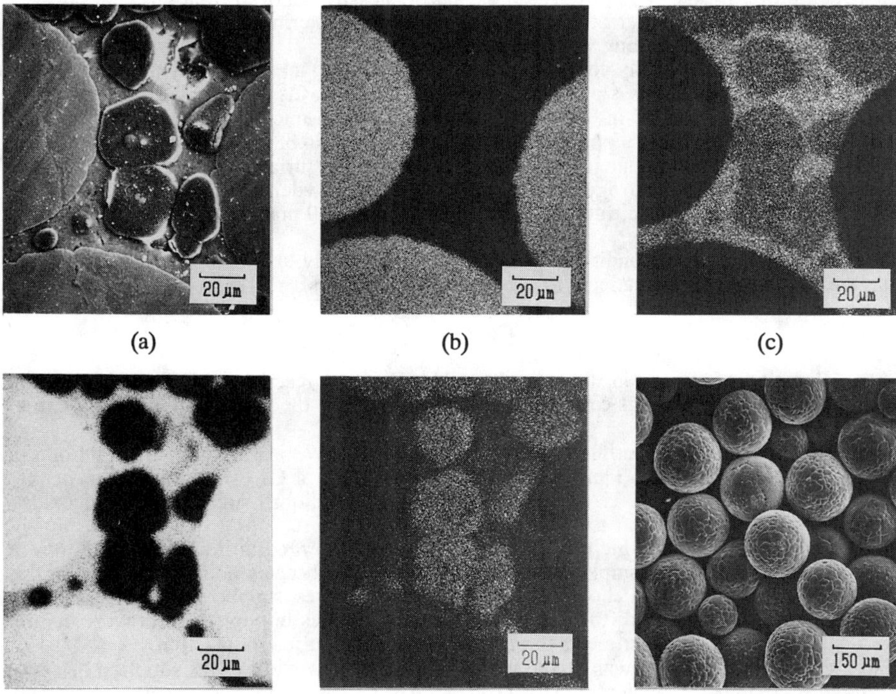

Fig.5 Scanning electron micrographs of TiAl powder – CaO mixture deoxidized by Ca–Al vapor, and corresponding elemental maps: (a) SEM image of mixture deoxidized at 1373K for 86.4ks. (b) titanium elemental map, (c) calcium, (d) aluminum, (e) oxygen, (f) SEM image of reacted TiAl following acid leaching to remove CaO and Ca–Al alloy. Average oxygen concentration decreased from 4200 mass ppmO to 670 mass ppmO, after reaction.

Nitrogen concentration in the samples was independent of the experimental conditions, and increased from the initial value by about 50 mass ppm. Calcium concentration of the TiAl powders increased to several hundred mass ppm after deoxidation. A reaction time longer than 260ks was found to be undesirable not only because of surface aggravation of the TiAl powders, which leads to inclusion of impurities, but also because of aluminum depletion of the Ca–Al alloy. In the following section, therefore, $CaCl_2$ was used as a flux to facilitate the deoxidation reaction.

Deoxidation of TiAl in $CaCl_2$ by Ca–Al Vapor

Following deoxidation, TiAl samples were found to be fused in $CaCl_2$ in the titanium cup. No metallic calcium from the Ca–Al alloy was detected on the surface of the fused salts, and no aluminum was present in the $CaCl_2$, as determined by XMA. This was the case even when 1g of small pieces of aluminum was added to the $CaCl_2$ flux before reaction (Exp.No.35–2 in Table II). TiAl particles with smooth surfaces were obtained

Fig.6 Scanning electron micrograph of TiAl powders deoxidized in $CaCl_2$ flux at 1373K for 86.4ks. TiAl powders, initially containing 510 mass ppmO, were deoxidized to about 70 ppm.

after reaction for 86.4ks at 1373K. The fused salt was easily removed from the TiAl powders by acid. A reaction period greater than 100ks resulted in sintering of the TiAl particles, and removal of the flux by leaching was time consuming.

TiAl powders initially containing 510, 1100 and 4200 mass ppm oxygen were deoxidized to 62, 140 and 190 mass ppm, respectively, by using $CaCl_2$ flux at 1373K, as shown by Exp.No.38–U, 35–1 and 40 in Table II. A scanning electron micrograph of deoxidized TiAl particles containing 66 mass ppmO as average value (Exp.No.38–L in Table II) is presented in Fig.6. No significant change in morphology of the particle surface following deoxidation is observed. Titanium and nitrogen concentrations of the powder were unchanged, whereas calcium, initially present in trace amounts, increased to 160 mass ppm by the deoxidation treatment.

Among the many conditions investigated in this study, the method using $CaCl_2$ flux was the most effective in obtaining low–oxygen TiAl powders.

Conclusions

Removal of oxygen in titanium aluminide (TiAl) by reaction with chemically active calcium–aluminum alloy was carried out around 1373K with the purpose of obtaining extra-low–oxygen TiAl.

The deoxidation experiments were preceded by an investigation of phase equilibria of the Ti–Al–Ca system at 1273 and 1373K. The compositions of Ca–Al alloys used as deoxidant, which equilibrate with TiAl, and the experimental conditions suitable for the deoxidation were examined.

In experiments in which Ti–Al samples were submerged in liquid Ca–Al alloys at 1373K, the surface of the samples severely deteriorated and became nodular. To prevent this surface deterioration, Ca–Al alloy deoxidant was, in most cases, supplied to the TiAl samples in vapor form. As oxygen diffusion in TiAl is slow, samples in powder form were mainly used in the deoxidation experiments to minimize the time required for reaction.

When TiAl powder was mixed with CaO, and the deoxidant was supplied in vapor form, powders initially containing 510, 1100 and 4200 ppmO were deoxidized to about 160, 490 and 670 ppmO, respectively, after deoxidation at 1373K for 86.4ks.

Among many conditions tested, utilization of TiAl powders mixed with $CaCl_2$ proved to be the most effective for deoxidation at 1373K. $CaCl_2$ was used as a flux to facilitate the deoxidation by decreasing the activity of the reaction product CaO. TiAl powders mixed with $CaCl_2$ and reacted with Ca–Al vapor at 1373K for 86.4ks, were deoxidized to levels of 62, 140 and 190 mass ppmO from initial concentrations of 510, 1100 and 4200 mass ppmO, respectively. No significant change in morphology of the particle after deoxidation was observed. The titanium and nitrogen concentrations of the powders remained unchanged, whereas calcium, initially present in trace amounts, increased to 160 mass ppm by the deoxidation treatment.

Acknowledgments

The authors wish to express their sincere appreciation to Mr.T.Unezeki, Mr.M.Maeno, and Mr.T.Ueki for their assistance in the experimental work of this study, and to Mr.T.Yamauchi (Osaka Titanium Co.Ltd.) for sample preparation and analyses. Last, but not least, the authors are very grateful to Dr.K.Yoshiki–Gravelsins for stimulating discussions and suggestions. Some analyses of titanium were done at Osaka Titanium Ltd.

References

1. T.Degawa, K.Kamata and Y.Nagashima: Proc. International Symposium on Intermetallic Compound (JIMIS–6), Sendai, Japan (1991), pp.1003–1008.
2. T.H.Okabe, R.O.Suzuki, T.Oishi and K.Ono: Testu–to–hagane, 77(1991),93–99.
3. T.H.Okabe, T.Oishi and K.Ono: J. Alloys and Compounds, (1992), in press.
4. N.Nakamura, T.Ueki, T.H.Okabe, T.Oishi and K.Ono: proceeding of JIM fall meeting, (1991), p.581.
5. R.O.Suzuki, T.Ueki, M.Ikezawa, T.H.Okabe, T.Oishi and K.Ono: Mater.Trans. JIM, 32 (1991) pp.272–277.
6. T.H.Okabe, K.Fujiwara, T.Oishi and K.Ono: Met. Trans. A, (1992) in press.
7. R.O.Suzuki, M.Ikezawa, T.H.Okabe, T.Oishi and K.Ono: Mater.Trans. JIM, 31 (1990) pp.61–68.
8. T.H.Okabe, R.O.Suzuki, T.Oishi and K.Ono: Materials Trans. JIM, 32,(1991) pp.485–488.

Effects of Environment on Fatigue Crack Growth in a Ti_3Al Based Aluminide

R.J.T. Penton, M.T. Cope* and P. Bowen.

School of Metallurgy and Materials / IRC in Materials for High Performance Applications, University of Birmingham, Edgbaston, Birmingham, B15 2TT, U.K.
*Rolls Royce plc, P.O. Box 31, Derby, DE2 8BJ, U.K.

Abstract

The influence of test temperature and environment on the fatigue crack growth resistance of a commercial Ti_3Al based alloy ("Super α_2") has been studied for a single microstructural condition. At the test temperature of 400°C, crack growth rates both in air and vacuum are similar and a reduction in the slope of the fatigue crack growth resistance curves are observed. Effects of environment are seen at the test temperature of 700°C. Crack growth rates observed in air are faster than those obtained in vacuum, particularly at low cyclic frequency of loading. Possible mechanisms of environmental interaction are discussed briefly.

Introduction

In recent years there has been increasing interest in the possibility of using intermetallic materials for high temperature applications (1,2). The most developed of these materials to date are those based on the titanium aluminide intermetallics TiAl and Ti_3Al. The Ti_3Al alloys such as Ti-25Al-10Nb-3Mo-1V and Ti-24Al-11Nb (atomic %) (3,4) have been the subject of wide-ranging research around the world. A variety of processing routes have been investigated and a number of trial components have been produced. However, the environment in which these materials are to serve is an exacting one and the resistance of these materials to environmental attack may be critical. Fatigue crack growth resistance can be strongly influenced by environment and therefore fatigue properties must be investigated with respect to the envisaged operating environment.
This current paper addresses, in particular, fatigue crack propagation at elevated temperatures. One major problem for α_2 based alloys is their relatively poor oxidation resistance at temperatures of 650°C and above (3,5). Therefore, their resistance to crack growth under both static and cyclic loading at such temperatures is of concern and requires careful study.

Experimental

Tests have been performed on "Super α_2" (Ti-25Al-10Nb-3Mo-1V atomic %) cut from a $\alpha_2 + \beta$ rolled plate processed by Timet Inc., USA. Specimens were heat-treated as oversized blanks. A solution treatment of 1 h at 1050°C was followed by air-cooling to room temperature. Subsequently, test pieces were aged at 850°C for 2 h followed by air-cooling to room temperature. Surface grinding to a depth of 2mm ensured the removal of any oxidation products formed during heat-treatment. Tensile properties and fracture toughness values are given in Table I, for this condition at ambient temperature and 700°C.

Titanium '92
Science and Technology
Edited by F.H. Froes and I. Caplan
The Minerals, Metals & Materials Society, 1993

Table I - Tensile properties of the titanium aluminide alloy Super α_2.

Microstructural Condition	Temperature (°C)	0.1% Proof Stress (MPa)	UTS (MPa)	Elong. %	Fracture toughness (MPa√m)
	20°C	910	1064	1.7	17
α_2+ transformed β	400°C	656	1041	8	---
	700°C	488	851	7	51

Fatigue crack propagation tests were carried out at room temperature on an Amsler Vibrophore electromagnetic resonant testing machine at a frequency of approximately 70 Hz, and on an ESH servo-hydraulic testing machine at test temperatures of 400 and 700°C and at a frequency of 20 Hz. Tests at 400 and 700°C were performed in both air and in vacuum (of better than 10^{-4} mbar). Tests were also carried out at two lower frequencies of 1 and 0.1Hz in air and at a frequency of 1Hz in vacuum, at the single test temperature of 700°C. All tests were carried out at a low stress ratio, R = 0.1, where R is the ratio of K_{min}/K_{max} and K_{min}, K_{max} are the minimum and maximum values of the applied stress intensity over the fatigue cycle. In all tests crack growth was monitored by the direct current potential drop method. All tests were performed under constant cyclic load conditions i.e. increasing ΔK with crack growth.

Optical micrographs were taken after mechanical polishing and etching in Krolls II reagent. Fracture surfaces were examined using an Hitachi field emission gun scanning electron microscope, operating at 0° tilt and 20 kV operating voltage.

Results

The microstructure after heat-treatment is shown in Figure 1. It consists of primary α_2 precipitates (light regions) of size 5-20 μm and volume fraction approximately 20%, in a transformed β matrix with a average grain size of approximately 30 μm. The structure of the transformed β was unresolvable optically.

Fatigue crack growth resistance curves are shown in Figure 2, 3 and 4 for tests performed at 20, 400 and 700°C, in air and in vacuum. It is noted that the increase in temperature from 20 to 400°C results in a reduction in the dependence of the crack growth rates on the applied stress intensity range. However, with a further increase in temperature to 700°C in air the dependence of crack growth rate on the applied stress intensity range increases once more. If a Paris-Erdogan law of the type

$$da/dN = A\ \Delta K^m \quad ..... (1)$$

is assumed, where A and m are materials constants, then values of A, m are given in Table II.

Table II - Constants derived from empirical power law equations describing crack growth.

Microstructural Condition	Temp. (°C)	Frequency (Hz)	Environment	A (mm/cycle)	m
α_2 + transformed β	20	>70	air	1.3 E-13	5.1
	400	20	air	1.75 E-07	2.3
	400	20	Vacuum	3.22 E-10	4.4
	700	20	air	1.53 E-08	4.0
	700	20	Vacuum	4.93 E-08	2.3
	700	1	air	8.77 E-08	4.3
	700	0.1	air	2.82 E-08	5.1
	700	1	Vacuum	5.26 E-08	2.8

Also shown in Table 2, are the materials constants obtained for fatigue crack growth at low frequencies both in air and vacuum at 700°C. Crack growth resistance curves obtained at 700°C at several test frequencies are shown in figure 4 and where it can be seen that crack growth rates measured under vacuum conditions are slower than those obtained in air, especially at low test frequencies.

Fatigue surfaces tend to be flat and featureless, see figure 5, for tests performed at both 400 and 700°C, in air and in vacuum. Thus crack growth does not appear to depend directly upon underlying microstructure.

Discussion

It may be deduced from the reduction in the Paris exponent, Table II, observed for tests performed in air as the test temperature is increased from 20 to 400°C, that this increase in temperature may lead to the introduction of a more plastic fatigue mechanism than the essentially brittle mechanism proposed at 20°C in such materials (6,7). It is observed that the yield stress undergoes a marked reduction at a test temperature of 400°C, Table I (8) and this is broadly in line with other work which shows an increase in dislocation activity at higher temperatures (9,10). Such an increase in the amount of slip would then allow the operation of traditional reversed plasticity fatigue crack growth mechanisms and hence a reduction in the Paris exponent would be expected with an increase in test temperature and this is indeed observed, Table II, if environmental attack is minimal, i.e. as in vacuum and at a test temperature of 400°C in air. One interesting and unusual observation is that fatigue crack growth resistance curves cross as the test temperature is raised from 20 to 400°C in air, figure 2. This implies that crack growth at a test temperature of 400°C and low ΔK, i.e. $\leq$ 11.5MPa$\sqrt{m}$, is actually faster than at room temperature. Without near threshold crack growth data it is unclear at this stage whether this relates to a higher crack threshold for the brittle mechanism of crack growth proposed at 20°C (6,7) or whether it is simply a natural consequence of a change in fatigue mechanism.

At the test temperature of 400°C the effect of environment on fatigue crack propagation rates is minimal with growth rates observed to be very similar both in air and in vacuum. This is in strong contrast with the effect observed at 700°C, where the growth rates in air are up to an order of magnitude higher than those measured in vacuum for a given value of ΔK. It is clear, therefore, that the environment has a strong effect on fatigue crack resistance in these alloys at temperatures of around 700°C. Previous studies (3,5,8) have shown that the oxidation properties of this material are a matter for some concern at these temperatures and this is again reflected here. At low frequencies, i.e. 1 Hz, in air the fatigue crack growth rates are observed to be far faster than those observed at 20 Hz. This effect could feasibly be the result of creep fatigue interactions and / or further oxidation attack. However, testing at 700°C in vacuum at 1Hz gives fatigue crack growth rates that are only slightly higher than those observed at 20 Hz in vacuum at this temperature. Hence it is clear that increased crack growth rates, da/dN, at the lower test frequency of 1Hz are a result of external environmental attack. Some evidence of an α_2 case has been observed after testing it is likely that the material around the crack tip has become saturated with oxygen during testing. Such material has been found to be more susceptible to crack growth under creep conditions (8). At lower test frequencies the material will plausibly have a longer time to become embrittled ahead of the crack tip before crack advance by mechanical fatigue. However, these individual crack growth increments through embrittled material may then extend further in one mechanical fatigue cycle and indeed crack growth rates with respect to time are actually observed to be closely similar at 1Hz and 20Hz in air at 700°C (11). In the present work as the cyclic frequency is reduced further to 0.1Hz, for tests at 700°C in air, no further increase in crack growth rate, da/dN, above that observed at 1Hz are seen, figure (5). This unusual observation appears to suggest that the embrittlement apparently reaches a limit with time prior to crack advance by mechanical fatigue. It is difficult to rationalise this result unless the rate of ingress of aggressive species (presumably oxygen) itself becomes self limiting. Oxide induced closure arguments are difficult to sustain over the entire range of ΔK applied and for tests performed at a constant applied cyclic load range. One mechanism might involve the build up of an oxide layer. In other work the oxide layers formed by this alloy have been shown to be more coherent than that formed by traditional titanium alloys (12) and hence suffer less spalling. The effect of this can be to severely limit the supply of oxygen available unless the oxide is disrupted mechanically. Hence, there are two competing effects: the effect of the oxygen which promotes the formation of less crack resistant material that accelerates crack growth rates and that of oxide layer build up which may limit the diffusion. Based on the work presented here, it would be predicted that at 700°C in air, further reductions in test frequency below 0.1Hz would not result in increased crack growth rates, da/dN.

Conclusions

Test temperature has a strong influence on the crack growth properties of Super α_2. With test temperature increase from 20 to 400°C, a marked reduction in the slope of the crack growth resistance curve is observed and crack growth rates are similar in both air and vacuum. With further in test temperature to 700°C, an influence of the environment is observed, with crack growth rates in air faster than those in vacuum. Such effects of environment are more pronounced at a low cyclic frequency of 1Hz, but a decrease in cyclic to 0.1Hz does not increase crack growth rates further.

Acknowledgements.

One of the authors (RJTP) has been supported during the course of this work by a SERC case studentship with Rolls Royce PLC, Derby.

References

1. G. Sauthoff, Z. Metallkde, 80, 1989, p337
2. R.L. Fleischer, D.M. Dimiduk and H.A. Lipsitt, Annu. Rev. Mater. Sci., 19, 1989, p.231
3. H.A. Lipsitt "High temperature intermetallics", ed. C.C. Koch, et al., MRS Symp. Proc. Vol. 39, Pittsburgh PA MRS 1985, pp. 351-364.
4. M.J. Blackburn and M.P. Smith , Tech. Report No. AFWAL-TR-82-4086.
5. Y.W. Kim and F.H. Froes, in S.H. Wang, C.T. Liu and D. Pope eds. 'High temperature aluminides and Intermetallics' , TMS Warrendale, PA (1990).
6. R.J.T. Penton, M.T. Cope and P. Bowen, Proc. Conf, 7th World Titanium Conference, San Diego, June 28 - July 2, 1992

7. R.A. Chave, C.J. Beevers and P. Bowen, Euromat '91, Conf. Proc., eds. T.W. Clyne and P.J. Withers, Institute of Materials, London, p333, 1992.
8. M.T. Cope, Rolls Royce internal report, MEG 034236, 1990.
9. S.M.L. Sastry and H.A. Lipsitt, Proc. Conf. 4th World Titanium Conference, eds. H. Kimura and O. Izumi, TMS - AIME, Warrendale PA, 1980, p.1231.
10. H.A. Lipsitt, D. Shectman and R.E. Schafrik, Met. Trans. A, 11A (1980), p. 1369.
11 R.J.T. Penton, PhD. Thesis, University of Birmingham, 1992
12. T.J. Johnson, M.H. Loretto and M.W. Kearns, Proc. Conf, 7th World Titanium Conference, San Diego, June 28 - July 2, 1992

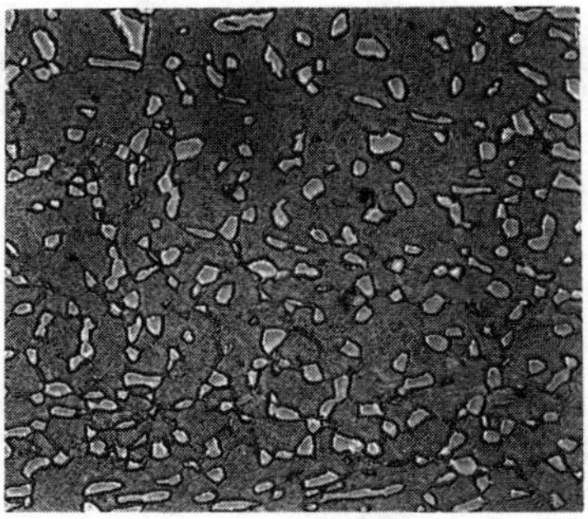

133μm

Figure 1 - Optical micrograph of the as heat treated α_2 + fully transformed β

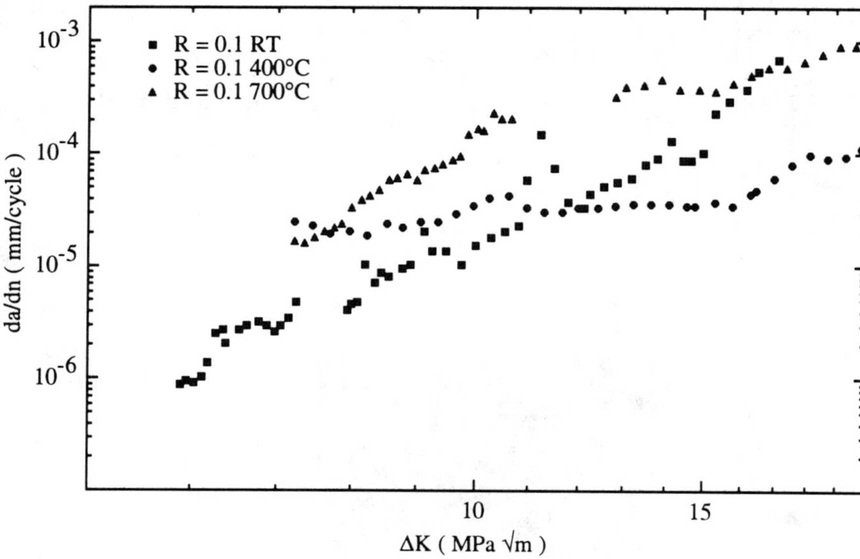

Figure 2 - Fatigue crack growth resistance curves at test temperatures of 20, 400 and 700°C in air.

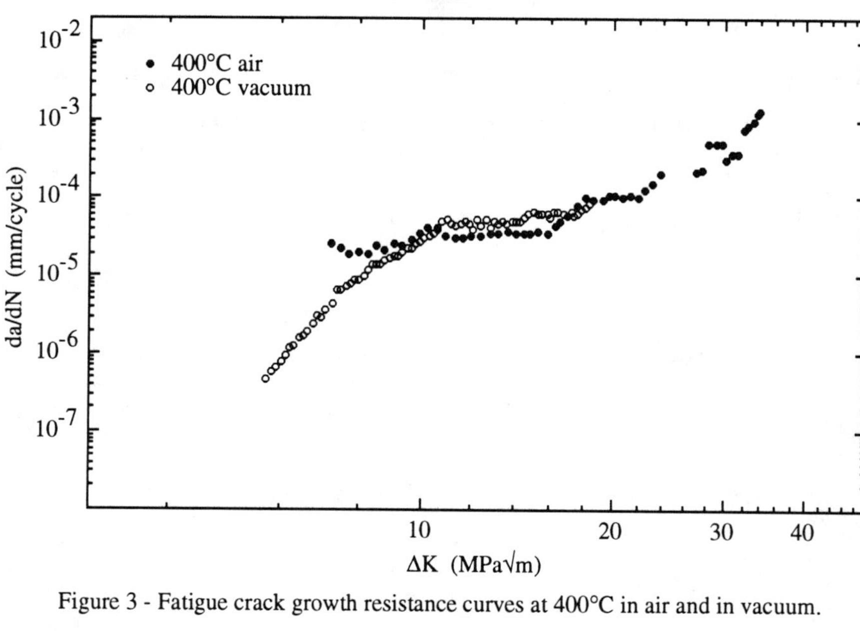

Figure 3 - Fatigue crack growth resistance curves at 400°C in air and in vacuum.

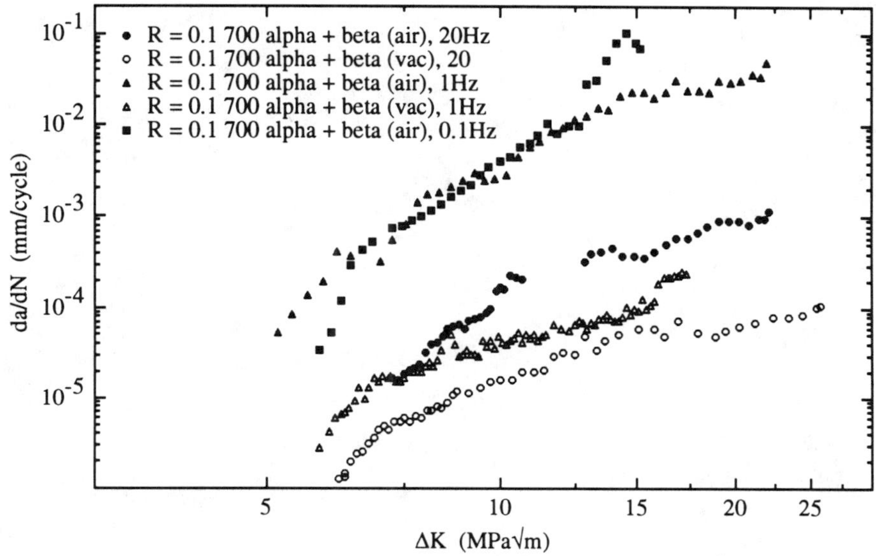

Figure 4 - Fatigue crack growth resistance curves at 700°C in air and in vacuum.

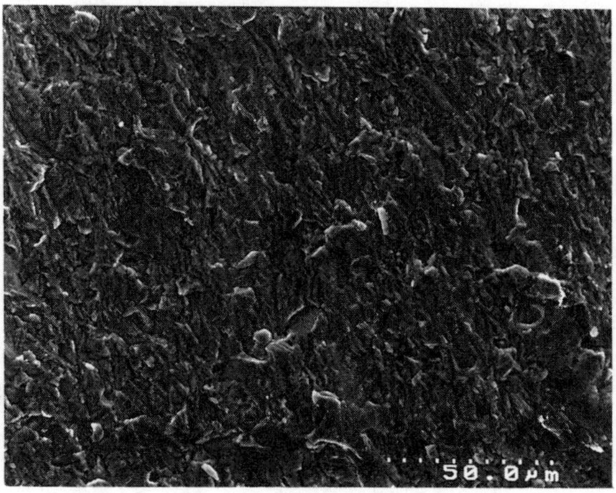

Figure 5 - The fracture surface after fatigue testing at R = 0.1 in air at 700°C.

CRACK GROWTH RESISTANCE OF ALPHA-TWO BASED TITANIUM ALUMINIDES UNDER CYCLIC LOADING AT AMBIENT AND ELEVATED TEMPERATURES

R.A.Chave, S.Kerry†, C.J.Beevers and P.Bowen

School of Metallurgy and Materials / IRC in Materials for High Performance Applications,
University of Birmingham, Edgbaston, Birmingham, B15 2TT, UK.
† Defence Research Agency, Aerospace Division, Royal Aerospace Establishment, Pyestock,
Hampshire, GU14 0LS, UK.

Abstract

The fatigue crack growth resistance of an experimental alloy based on an α_2 titanium aluminide (Ti_3Al), with and without the addition of molybdenum (Mo), has been assessed at ambient temperature and 600°C. The importance of monotonic modes of failure during cyclic loading at ambient temperature has been demonstrated and a mechanism of crack growth involving cleavage steps through α_2 laths followed by fatigue through the surrounding β phase has been proposed. The addition of Mo refines the microstructural scale significantly but does not appear to affect the fatigue crack growth rate. At 600°C fatigue proceeds predominantly by a classical reverse plastic flow mechanism, with striations evident on the fracture surface.

Introduction

Titanium aluminides are an important class of intermetallic compounds developed to satisfy the demands of the aerospace industry for low density materials with high temperature strength. To fulfil their potential as replacement materials for some of the super-alloy components in gas turbine engines, such materials must have roughly the same mechanical properties and oxidation resistance as nickel-based super-alloys but should possess considerable weight advantages over these more conventional engine materials [1].

The ordered intermetallic compounds Ti_3Al (α_2) and TiAl (γ) have been studied extensively in recent years. Both possess good high temperature strength and modulus retention, but their use is currently limited by poor toughness and low ductility at room temperature. Alloying Ti_3Al with niobium (Nb) results in a significant improvement in room temperature ductility and strength. The Nb stabilises the β phase and acts to increase non-basal slip activity [2]. Ti_3Al–Nb alloys form the basis of current research into α_2-based titanium aluminides. Two commercial alloys based on this system have been produced: Ti–24Al–11Nb, and Ti–25Al–10Nb–3V–1Mo ("Super α_2").

One factor which hinders the application of α_2-based titanium aluminides is their oxidation resistance. Although better than conventional titanium alloys, the α_2 titanium aluminides do not have the required oxidation resistance to challenge the supremacy of nickel-based super-alloys in the hotter parts of a gas turbine engine. Alloys based on the γ aluminide (TiAl) have much better high temperature potential, but a successful way of improving both very poor room temperature ductility and low fracture toughness has not yet emerged.

The research effort directed toward titanium aluminide development over the past two decades has made the incorporation of α_2-based alloys in some non-critical components a practical possibility. The processing of these alloys into thin foil has also produced a use as a

Titanium '92
Science and Technology
Edited by F.H. Froes and I. Caplan
The Minerals, Metals & Materials Society, 1993

low density matrix for intermetallic matrix composites, which offer further weight and performance advantages over conventional materials.

Experimental

The materials under investigation in this present study are based on the intermetallic compound Ti_3Al and form part of an alloy development programme being carried out at DRA Aerospace Division [3]. The base alloy composition for the development programme, Ti–23Al–11Nb–0·9Si (atomic %) has shown potential for reasonable room temperature elongation (up to 3%) and excellent creep rupture strength (in excess of 1000 hours at 625°C and 250MPa stress). Another alloy containing molybdenum (Mo), of composition Ti–23Al–9Nb–2Mo–0·9Si has also been investigated as part of this present study. The thermo-mechanical processing route employed for these alloys involves extrusion of the cast ingots at 1180°C (in the β phase field), followed by carefully controlled heat treatments to produce microstructures both with and without regions of allotriomorphic α_2. Tensile properties have been determined for all such heat treatments [4], but crack growth studies have concentrated initially on β-solution treatments. The base alloy composition has been studied in two heat treated conditions: (i) 1150°C for 1 hour followed by ageing at 625°C for 2 hours, (ii)1150°C for 1 hour followed by ageing at 800°C for 1 hour. The Mo containing alloy has been examined to date in the second of these two conditions only. These heat treatments result in a Widmanstatten 'basket-weave' type of microstructure. This microstructure provides optimised creep rupture strength at the expense of a small reduction in both tensile strength and elongation to failure [4].

Single edge notched (SEN) test specimens were prepared from $50x10x10mm^3$ blanks cut from the 50mm diameter extruded bar by electrical discharge machining. A notch of 2·5mm depth was put into one side using a 150μm thick diamond blade. Some specimens were mechanically polished subsequently on one or both faces and etched using Krolls reagent. Fatigue tesing at room temperature was carried out in three-point bending on an Amsler Vibrophore electro-magnetic resonance machine at a frequency of approximately 60Hz. A reliable pre-cracking routine was established with crack growth initiated and continued a short distance of about 200μm at a stress intensity range (ΔK) of $10MPa\sqrt{m}$, and at a R ratio of 0·1–0·2 (where R is the ratio of minimum to maximum stress intensity applied during the fatigue cycle). Once initiation had occurred the loads were reduced quickly by $\leq$10% and the pre-crack grown a further distance of 0·5–1mm, typically finishing at a ΔK value of $5–6MPa\sqrt{m}$. The crack was then grown under a constant load range so that ΔK increased as the crack length increased. High temperature fatigue crack growth tests were mostly conducted in air in a controlled environment chamber attached to an ESH servo-hydraulic machine operating at a frequency of 20Hz, although some testing was carried out using a furnace attached to a Vibrophore machine. Pre-cracks were grown at room temperature in the same manner as on the Vibrophore machines, and then the specimens were heated to 600°C before the crack was grown out at a constant load.

Monitoring of crack growth for all the room temperature tests and most of the elevated temperature tests was carried out using a direct current potential difference technique. This allows for continuous crack growth monitoring and gives an average through thickness crack length rather than relying on surface observations alone. For tests performed at high temperatures on the Amsler Vibrophore the close fitting funace did not allow the use of the potential difference technique. In this case an optical system [5] was used to provide in-situ observations of crack growth.

A detailed examination of the fracture surfaces was carried out on an ISI 100A scanning electron microscope (SEM), operating at 0° tilt, 20kV, and an Hitachi S2000 FEG SEM under similar operating conditions. In addition, some metallographic sections cut transverse to fracture surfaces were also examined.

Results

The optical and SEM micrographs in Figure 1 show the 'basket weave' microstructures of Ti–23Al–11Nb–0·9Si and Ti–23Al–9Nb–2Mo–0·9Si in the β heat treated conditions. Qualitative energy dispersive X-ray (EDX) analysis on the 625°C aged condition has shown a composition difference between the laths which compose the majority of the structure and the thin region surrounding each lath (the dark and light regions of Figure 1*b* respectively). This surrounding region is richer in niobium but has a lower aluminium content compared to the centre of the laths. Examination of foils in a transmission electron microscope (TEM) has shown this surrounding phase to be retained β. There is little discernible difference at an optical level between the microstructures for the base alloy in both ageing conditions. However the addition of Mo refines the microstructure extensively.

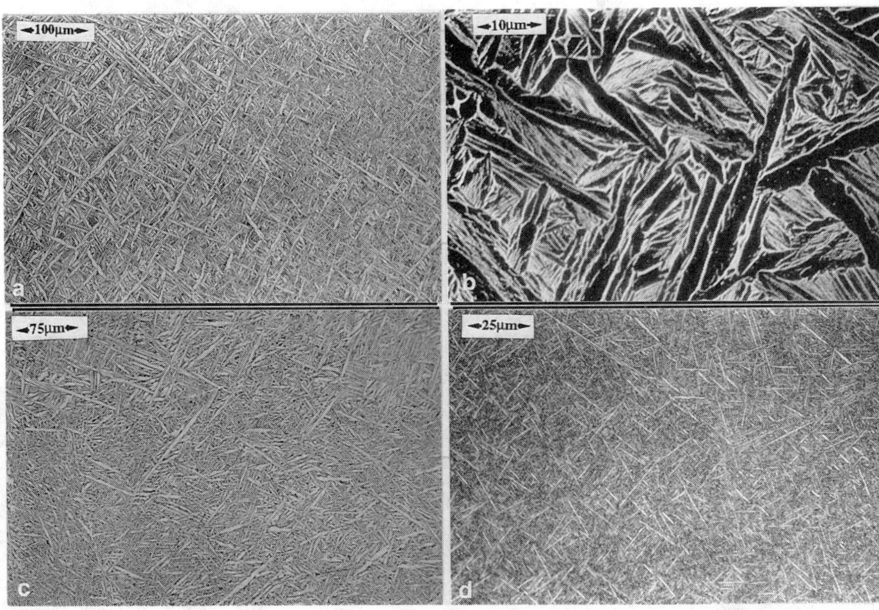

Figure 1 - Microstructures of *a), b)* base alloy in the 625°C aged condition, *c)* base alloy in the 800°C aged condition, *d)* Mo containing alloy in the 800°C aged condition.

The fatigue crack growth resistance results are shown in Figures 2 and 3. A comparison of ageing temperatures in the base alloy (Figure 2) shows that a marginal improvement in room temperature fatigue crack growth resistance is exhibited at the lower ageing temperature. At 600°C the improvement appears to be more substantial with the material aged at 625°C providing increased resistance to crack growth. The slope of the elevated temperature crack growth resistance is different markedly from that obtained at room temperature which is consistent with a change in mechanism of crack growth. The addition of Mo to the base alloy would appear from Figure 3 to have little effect on fatigue crack growth behaviour at either room temperature or at the test temperature of 600°C.

Fractographs of the room temperature fatigue surfaces of all three alloy conditions are shown in Figure 4. The scale of features of the Mo alloy fatigue surface are much finer than those of the others, consistent with the finer scale of the underlying microstructure. Some areas of the room temperature fatigue surfaces in the base alloy show periodic steps of sizes ranging from about 1-3μm (most clearly evident in Figure 4*a*, and also observed in studies on Super α_2 [6]), and which could be construed to correspond to individual increments of crack growth. Indeed, clear evidence of cleavage of the laths has now been obtained fractographically (Figure 4*b*). At elevated temperature the fatigue surfaces of the base alloy composition shows striations clearly (Figure 5). The Mo alloy does not appear to show such features, and the

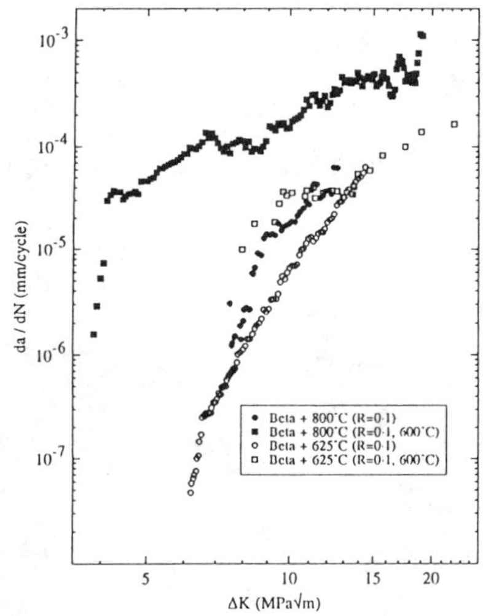

Figure 2 - Fatigue crack growth rates versus stress intensity range for the base alloy in two different ageing conditions tested at room temperature and at 600°C.

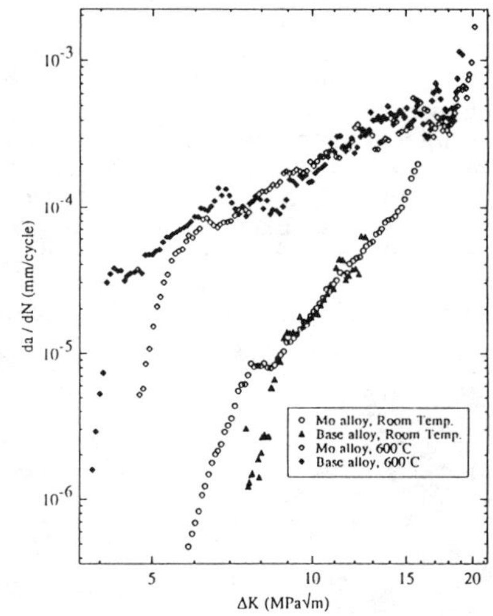

Figure 3 - Fatigue crack growth rates versus stress intensity range for the base alloy and Mo containing alloy both in the 800°C aged condition tested at room temperature and at 600°C.

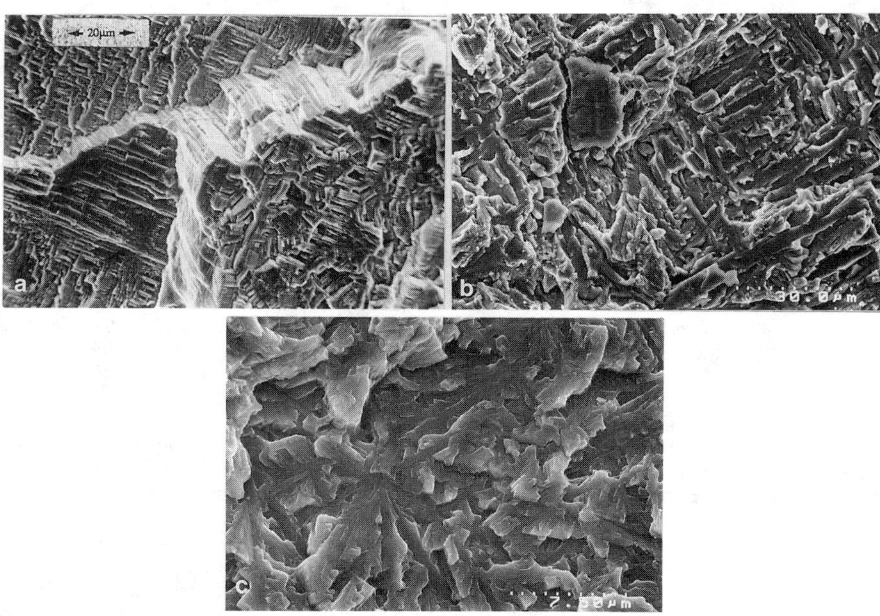

Figure 4: SEM fractographs of room temperature fatigue in *a)* base alloy, aged at 625°C, *b)* base alloy, aged at 800°C, *c)* Mo containing alloy, aged at 800°C.

fracture surface (a representative area of which is shown in Figure 5*b*) has a very diferent appearence from that of the base alloy. Measurement of the striation spacings at different levels of stress intensity in one of the base alloys shows very good agreement with observed crack growth rates (Figure 6).

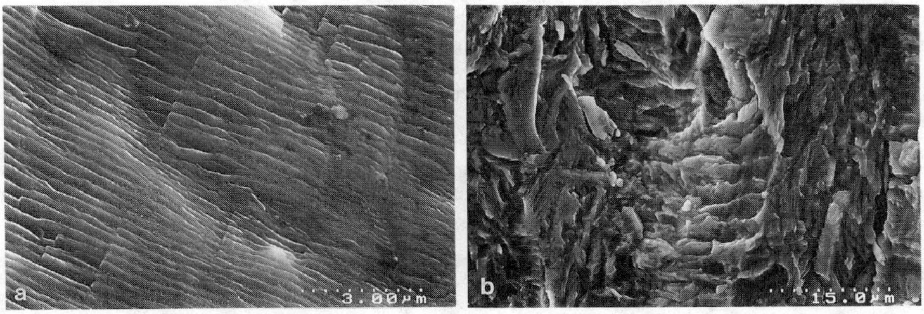

Figure 5: SEM fractographs of fatigue at 600°C in *a)* base alloy, *b)* Mo containing alloy.

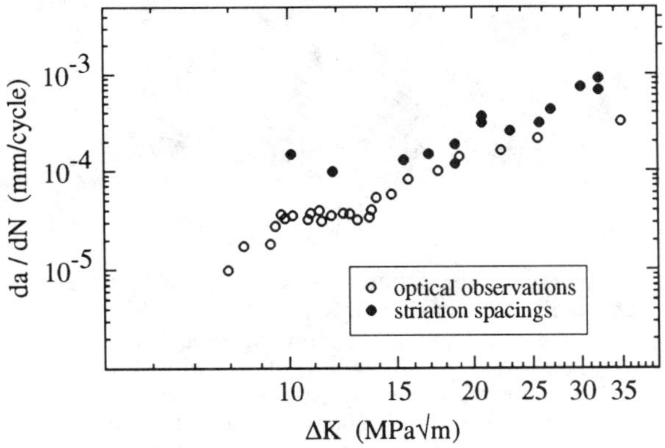

Figure 6: A comparison of fatigue crack growth rate versus stress intensity range for the base alloy aged at 625°C, measured by in-situ optical observations and striation spacings after failure.

Discussion

At room temperature, the brittle appearence of the fatigue surfaces for all three alloy conditions combined with the steep slopes of the fatigue crack growth resistance curves suggests that monotonic modes of failure may be important to the fatigue crack growth mechanism. Earlier work has shown that there is a significant effect of mean stress on the fatigue crack growth in these alloys [7] and this implies that the value of K_{max} (a monotonic parameter) plays an important role. The step-like growth patterns observed in places on the surface of base alloy specimens tested at room temperature is reflected in a sectioned fracture surface micrographs shown in Figure 7*a*, and it has been suggested that crack growth occurs by cleavage steps across the α_2 laths which are stopped by the intervening retained β phase [7]. Figure 7*b* also shows a secondary microcrack behind the main fatigue crack surface where it is possible to see a ductile β ligament still bridging the crack. Since the individual crack growth increments across α_2 laths are larger than the measured crack increment per cycle, da/dN, then it can plausibly be deduced that crack advance must involve fatigue through the surrounding β phase.

If a mechanism dominated by cleavage increments across the α_2 laths was operating in these materials then it would be expected that a finer lath size might retard the fatigue crack growth rate since the individual increments of crack growth would be smaller. The addition of Mo has the effect of significantly reducing the microstructural scale (Figure 1), and yet there is no difference in the observed crack growth rates between this alloy and the base alloy in the β heat treated condition. It seems likely that the precise character of the β-phase surrounding the laths could affect crack growth rates. As a β stabilising element, the Mo should be found preferentially in the β phase, and as such will be expected to influence the properties of this phase. Indeed, initial fractographic evidence suggests that cracks can propagate across individual laths in the Mo containing alloy in a more continuous fashion than in the base alloy, compare Figure 4*b* and *c*, but further work is required to confirm these preliminary observations. (Under fast fracture conditions the cleavage of groups or packets of laths is observed for the Mo containing alloy as distinct from the cleavage of individual laths for the base alloy[8])

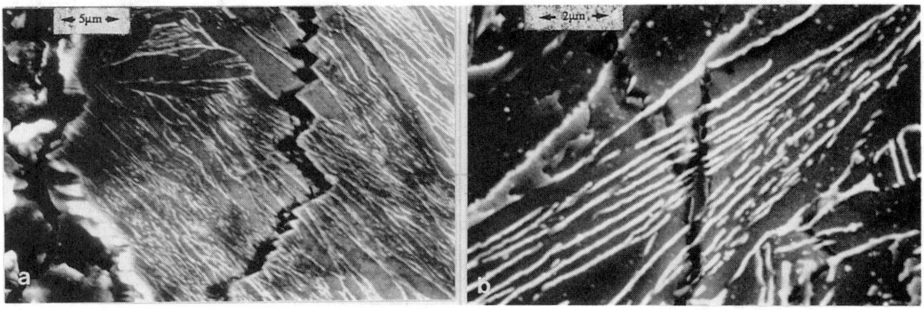

Figure 7: SEM micrographs of the sectioned fatigue surface of the base alloy aged at 625°C.

At elevated temperatures the dominant mechanism of failure in the base alloy would appear to be on of classical, reverse plastic flow, and which produces striations on the fatigue surface. Striations have been observed to a certain degree in Ti_3Al at 700°C [9], but the extent to which they are exhibited here suggests significant plasticity at the lower test temperature of 600°C. At this stage it remains an interesting observation only that striations are not seen for the Mo containing alloy under similar test conditions, and that the addition of Mo appears to change the fracture surface morphology. The origin of such changes is yet to be established.

Conclusions

1. There is little effect of ageing temperature on the fatigue crack growth rates at room temperature in the base alloy composition. Both α_2 lath width and the retained β phase are deduced to be important in the mechanism of crack growth. The addition of Mo to the base alloy significantly refines the lath width, but crack growth rates at room temperature are unaffected.

2. At the elevated temperature of 600°C a continuum reverse plasticity mechanism of crack growth dominates in the base alloy. Measurements of striation spacings are closely similar to the observed crack growth rate per cycle. The addition of Mo does not change the observed crack growth rates but does produce a change in fracture surface morphology and striations are no longer obvious on the fatigue fracture surface.

Acknowledgements

This work has been carried out with the support of the Defence Research Agency. One of the authors (RAC) is grateful for a SERC CASE award with DRA, Aerospace Division, Pyestock.

References

1. H. A. Lipsitt, High Temperature Ordered Intermetallic Alloys, C. C. Koch, C. T. Liu, and N. S. Stoloff, eds., (MRS, Pittsburgh, PA, 1985), 351-64.

2. S. M. L. Sastry and H. A. Lipsitt, Metall. Trans. A, 8A, (1977), 1543-52.

3. S.Kerry, Royal Aerosrace Establishment Technical Report 89049, 1989.

4. S. Kerry, Defence Research Agency, Aerospace Division, Private Communication, 1991.

5. J. Healy: PhD Thesis, University of Birmingham, 1990.

6. D. L. Davidson, J. B. Campbell, and R. A. Page, Metall Trans A, 22A, (1991), 377-91.

7. R. A. Chave, C. J. Beevers, and P. Bowen, Proceedings of the 2nd Eoropean Conference on Advanced Materials and Processes, T. W. Clyne and P. J. Withers Eds., (Instituite of Materials 1991), 333-9.

8. R. A. Chave, et al, (Paper presented at the Seventh World Conference on Titanium, 1992).

9. S. M. L. Sastry and H. A. Lipsitt, Acta Metall., 25, (1977), 1279-88.

MICROSTRUCTURE and PROPERTIES of FORGED PLASMA ARC MELTED and VACUUM ARC REMELTED TITANIUM ALUMINIDES

C. Huang,* Z. Chen,* T.A. Dean*+ and M.H. Loretto*

*IRC in Materials for High Performance Applications,
+School of Manufacturing and Mechanical Engineering,
The University of Birmingham, Edgbaston, B15 2TT, UK.

ABSTRACT

The properties of the Ti aluminides are critically dependent upon the thermomechanical processing route as well as upon the composition of the alloy and the aim of the present programme is to determine the influence of these parameters on the properties of plasma melted material. The Ti aluminide used in this investigation, which is part of a larger programme aimed at optimising the structure and properties of Ti aluminides, has a composition of Ti-25Al-10Nb-3V-1Mo (at.%). The microstructures of samples heat treated and forged using different thermomechanical processing routes have been assessed using optical microscopy and analytical transmission electron microscopy. The room temperature mechanical properties have been assessed using tensile tests and the properties interpreted in terms of the microstructure and processing route used to produce the microstructure.

INTRODUCTION

Ti_3Al-based titanium aluminides are potential aerospace engine materials because of superior creep resistance combined with low density [1-4]. However, these materials have low toughness and ductility at ambient temperature[2,4]. There have been considerable efforts toward improving room temperature ductility and toughness without sacrificing good mechanical properties at elevated temperatures for the last two decades. The second generation Ti_3Al based-titanium aluminide, super $\alpha2$ (Ti-25Al-10Nb-3V-1Mo (at.%)), has shown very promising mechanical properties at room temperature and at high temperatures. This alloy is analogous to α/β titanium alloys in terms of microstructure/property relationship.Therefore the mechanical properties of the alloy are strongly dependent on microstructure and hence on thermomechanical processing (TMP) route. Previous research workers have achieved significant microstructural variations such as: primary $\alpha2$ volume fraction, secondary $\alpha2$ plate morphology and the presence of β grains, by various TMP routes [8-10]. Quite a few studies have been undertaken to gain an understanding of the relationship between microstructure and mechanical properties at both room and elevated temperatures on various Ti_3Al-based intermetallic alloys made by vacuum arc remelting(VAR)[11-16]. Furthermore, it has been found that the oxygen level in these alloys has a significant effect on mechanical properties and high oxygen content usually results in low ductility at room temperature.[17] Therefore the ductility of these alloys can be significantly improved by reducing the oxygen level in these alloys during

Titanium '92
Science and Technology
Edited by F.H. Froes and I. Caplan
The Minerals, Metals & Materials Society, 1993

casting. Plasma arc melting(PAM) technique is a recently developed clean melting technique for advanced materials and super α2 with low oxygen may be easily obtained by PAM, provided the feedstock is low in oxygen.

In this paper the microstructures and properties of super α2 alloy made by VAR are compared with those of an alloy of the same composition made by PAM. The work is focused on the influence of oxygen content, thermomechanical processing and heat treatment on microstructure and mechanical properties with a view to optimising the properties of the alloy. The work forms part of a larger programme designed to gain a full understanding of the interaction of the processing parameters on properties and to identify any differences that might arise between VAR and PAM material.

EXPERIMENTAL

The chemical compositions of the VAR and the PAM super α2 and the feedstock prior to plasma melting are given in Table I. The PAM ingot was singly melted and was 100mm in diameter and 300mm in height, whereas the VAR ingot was significantly larger in diameter. The PAM and VAR ingots were electro-discharge machined into cylindrical forging billets of size in the range 36 to 46mm diameter, 16 to 54mm in height, as listed in Table II.

Table I. Chemical analysis of VAR and PAM super α2 and the feedstock compacts prior to PAM.(Ti: Balance)

Alloy	Composition(at%)				
	Al	Nb	V	Mo	O_2(wt)
VAR	25.1	9.7	3.2	1.0	1000ppm
PAM	25.1	9.8	3.0	0.94	650ppm
feed	25.2	10.3	3.0	1.0	600ppm

Table II. Thermomechanical Processing Parameters of VAR and PAM super α2*.

Alloy	Size (mm) (dia.x height)	Preheat. Temp.(°C)	Forging Field	Reduction (%)
VAR	36x16	1050	(α + β)	55
	36x31	1150	β	76
PAM	46x50	1050	(α + β)	70
	46x54	1150	β	71

*Note one forged sample for each forging condition was heat treated at 800°C and air cooled

The axes of the billets were in line with the axis of the ingot from which they were cut. The billets were coated first with a high temperature glaze and then wrapped with thermo-insulation ceramic cotton in order to limit oxidation and heat loss during thermomechanical processing. They were preheated before forging by placing them in an electric resistance muffle furnace at a temperature of either 1050°C or 1150°C for 1h. The preheated billets were upset forged on open dies using a Wilkins & Mitchell 200 ton single point crank press. The upset forged pancakes were then cooled in air to room temperature. The reductions were around 70% for most forgings and the strain rate was about $20s^{-1}$. The press dies were coated with a thick layer of colloidal graphite lubricant brushed on to their surfaces and were preheated up to 180°C.

The upset forged pancakes were sliced into 10x10mm cross section pieces by using electro-discharge machining and some of the pieces were further aged at 800°C for 2h then air cooled to room temperature. These heat treated and untreated specimens were then machined into tensile and three point bend-test specimens. The axis of the tensile specimens were orientated at right angles to the axis of the forging from which it was taken as shown in Fig. 1.

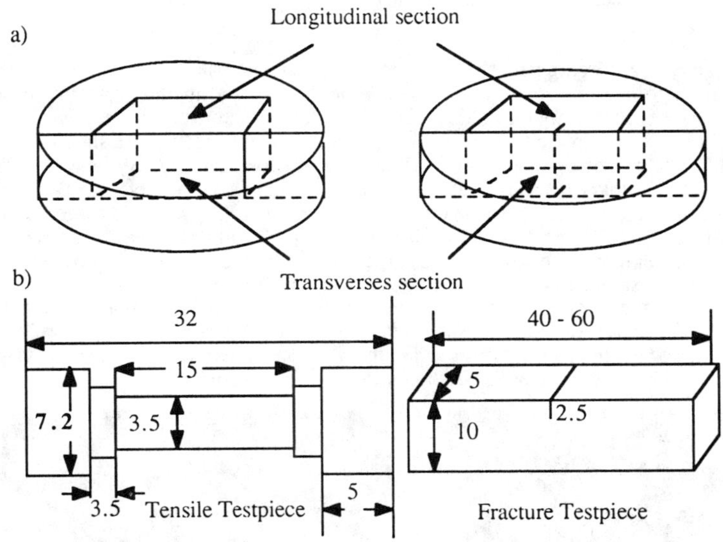

Fig.1 a) Schematic drawing of the positions of tensile and three points bending test specimens in a upset forging pancake, b) schematic drawing of tensile and three point bend test specimens.

The tensile specimens with 15mm gauge length and 3.5mm diameter were obtained by turning and polishing with Grade 800 carborundum paper. Tensile test was carried out at room temperature on a Hounsfield tensile testing machine under quasi-static conditions. As the stiffness of the test-pieces were high compared with that of the testing machine determination of Young's modulus from the autographic recording could not be achieved with accuracy. Therefore an electrical resistance strain-gauge was attached to each test-piece before testing and its output used to calculate the elastic modulus of the material. Three point bending specimens 5mm wide, 10mm high and 40 to 60mm long were machined with their axes at right angles to those of the forgings from which they were taken (Fig.1). Notches 0.25mm deep, at right angles to the forging direction, were machined by electro-discharge machining at the mid length of each test-piece. Test-pieces were pre-cracked using an Amsler Vibrophore fatigue testing machine (with R=0.1). Three point bend tests were carried out on a 10KN ESH servo-hydraulic mechanical test machine.

Microstructures of as cast, as forged and heat treated specimens were examined using optical and transmission electron microscopy. The phases in these specimens were identified by electron diffraction using a JEOL-4000FX transmission electron microscope.

The β transus temperatures of the PAM and VAR super $\alpha 2$ were determined metallographically as 1100° and 1130°C respectively. This observation is consistent with chemical analyses of these two alloys given in Table 1 which shows that the oxygen content in the PAM alloy is about 350ppm lower than that of the VAR alloy. It is well known that oxygen is a $\alpha 2$ phase stabiliser and thus a higher β transus temperature would be expected for the VAR material, because it has higher oxygen content and also the coarse primary $\alpha 2$ laths would result from the higher oxygen level in the alloy.

RESULTS

Typical basketweave microstructures of the as cast VAR and PAM super α2 are shown in Fig.2. It can be seen that the PAM super α2 has much finer microstructure than that of VAR super α2.

Fig.3 shows a typical forging pancake and a cross section macrostructure, which shows fairly uniform deformation across the specimen, though there is a small deformation dead zone at the surfaces contacting the cold forging dies. Microstructures of the VAR and PAM specimens forged in the (α2+β) and β fields at 1050 and 1150°C, are shown in Figs. 4 and 5. The microstructure of the VAR sample forged in the (α2+β) field (Fig.4a) consists of coarse primary α2 and a featureless matrix, which is mainly retained B2 (Fig.6a). In contrast, the microstructure of the PAM specimen, (Fig 4b) also forged in the (α2+β) field, consists of fine and more equiaxed primary α2 and basketweave matrix with no obvious retained B2 phase (Fig. 6b). The volume fractions of primary α2 in the VAR and the PAM samples were measured as 20% and 12% respectively. The microstructures of the β-forged PAM and VAR specimens are very similar and both have a fine basketweave microstructure, as shown in Figs.5a and 5c. Aging at 800°C for 2h either decomposed the retained B2 observed in the VAR specimens or the fine secondary α2 or orthorhombic laths coarsened, as shown in Figs.5b and 5d.

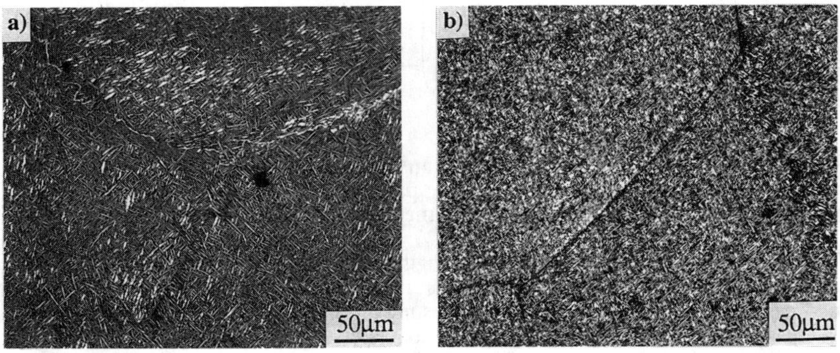

Fig.2 Optical microstructures of as cast a)VAR and b)PAM super α2 showing a finer basketweave microstructure in the PAM than in the VAR super α2.

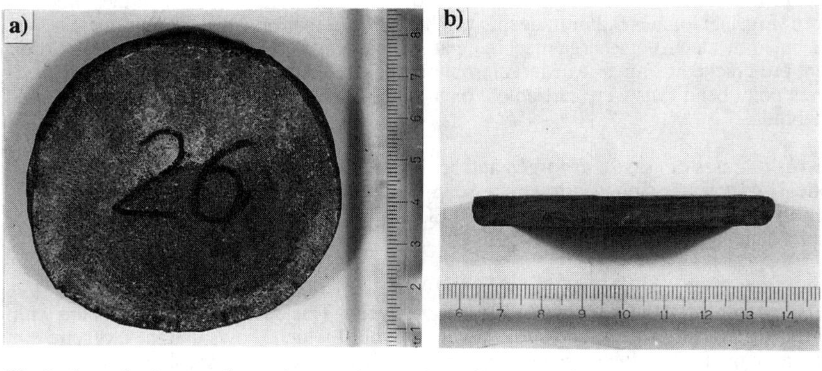

Fig.3 A typical upset forged pancake. a)plan view and b)macrostructure of cross section showing virtually uniform deformation across the specimen.

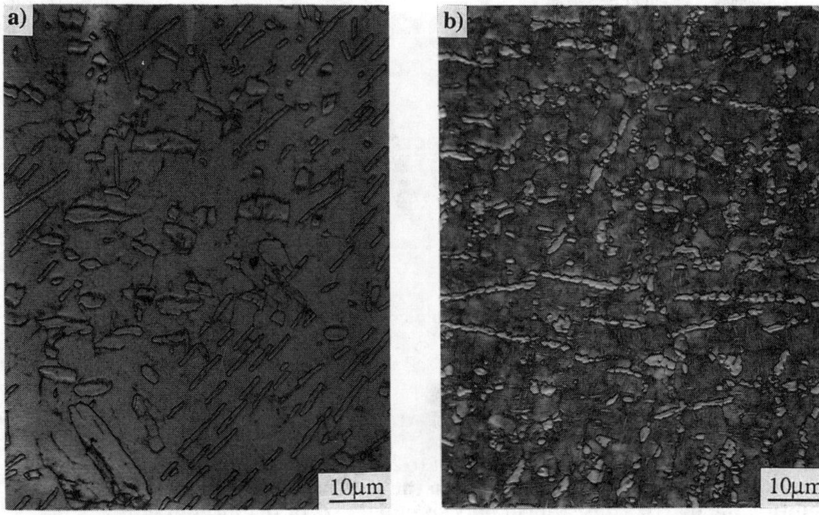

Fig.4 Optical micrographs of (α2+β)-forged a)VAR specimen showing coarse primary α2+featureless matrix and b)PAM specimen showing finer and more equiaxed primary α2+fine basketweave matrix microstructure.

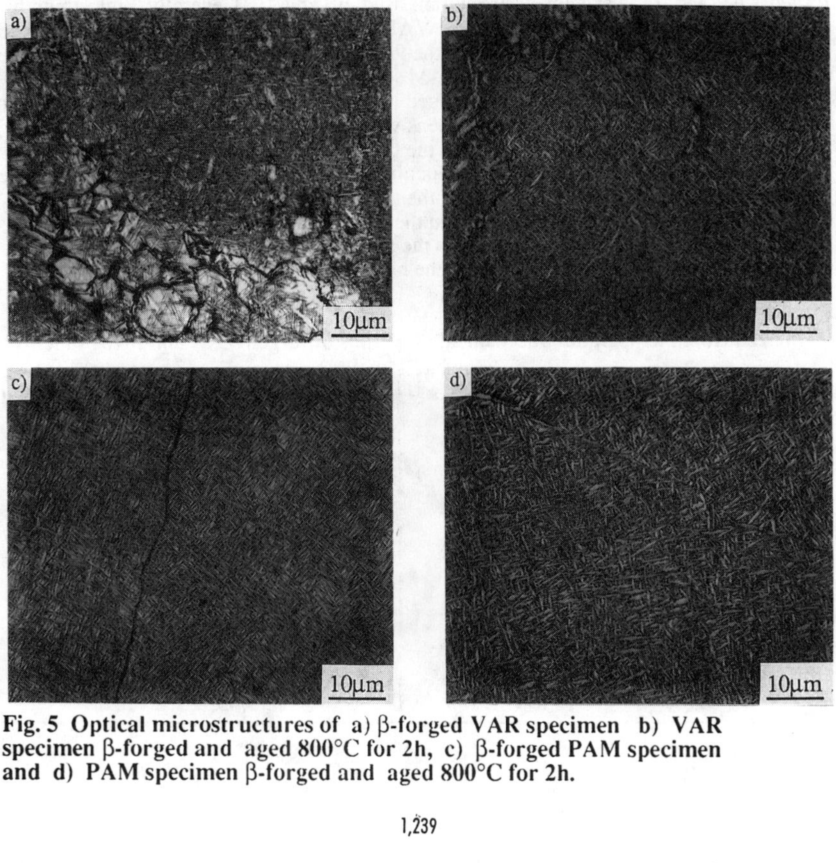

Fig. 5 Optical microstructures of a) β-forged VAR specimen b) VAR specimen β-forged and aged 800°C for 2h, c) β-forged PAM specimen and d) PAM specimen β-forged and aged 800°C for 2h.

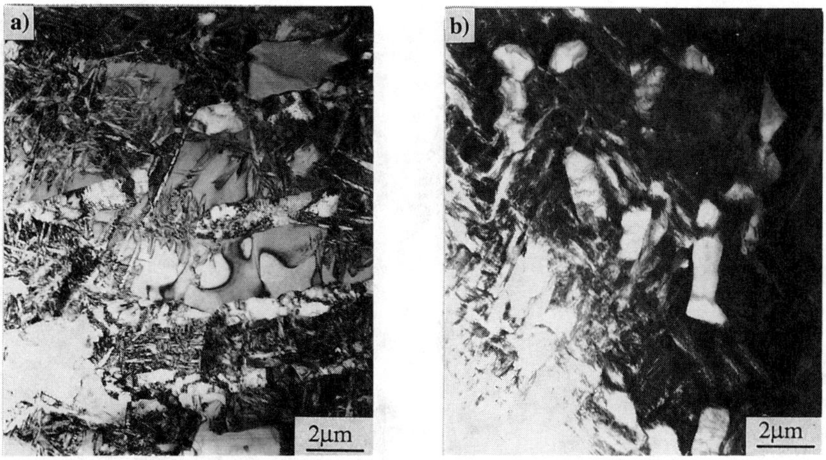

Fig.6 Many beam TEM micrographs of (α2+β)-forged samples a) VAR specimen showing retained B2 grains in the matrix,b) PAM specimen showing the matrix consisting of only secondary α2 or orthorhombic laths.

Room temperature mechanical properties such as strength, ductility and toughness, of the forged VAR and PAM specimens are summarised in Table III. Generally, high strength but low toughness were obtained from the VAR specimens and excellent toughness, good ductility and strength were obtained from the PAM specimens. The VAR specimens were much stronger and more brittle than the PAM specimens even when they have similar microstructures. In fact the VAR alloy forged in the β field has higher strength and lower elongation than the (α2+β) forged specimens whereas for the PAM alloy the (α2+β) forged specimens were a bit stronger than the β-forged specimens, but had a similar ductility. Furthermore it appears that the ductility of the β-forged specimens was improved dramatically by aging at 800°C for 2h and the toughness of the PAM specimens seems independent of the forging temperature and the aging treatment. For the VAR specimens, much higher toughness was obtained from the specimens forged in the β-field than the specimens forged in (α2+β)-field though the best toughness value is still significantly lower than that of the PAM specimens.

Table III. Room temperature mechanical properties of forged VAR and PAM super α2.

	$YS_{0.2}$(MPa)	UTS(MPa)	EL(%)	K_Q(MPa$\sqrt{m}$)
VAR				
(α + β)-forged	1144	1269	1.33	11
β-forged	1256	1378	1.00	16
β-forged + 2h at 800°C	1232	1388	0.95	---
PAM				
(α + β)-forged	1010	1170	2.55	20
	990	1196	1.6	21
(α+β)-forged+2h at 800°C	1009	1144	1.01	18
	960	1144	2.70	19
β-forged	909	1081	2.15	21
	855	1009	1.65	22
β-forged + 2h at 800°C	873	1123	3.30	20
	891	1123	3.45	20

DISCUSSION

Three different microstructures have been obtained from the ($\alpha2+\beta$) and β-forged PAM and VAR specimens in this work. These three microstructures are: (i) primary $\alpha2$ plus featureless matrix consisting of mainly retained B2; (ii) primary $\alpha2$ plus basketweave matrix and (iii) basketweave only. The first two microstructures were produced in the specimens forged in the ($\alpha2+\beta$) field and the last one obtained from the β-forged specimens.

The microstructure of primary $\alpha2$ plus retained B2 was expected from the specimens forged in the ($\alpha2+\beta$) field, because the niobium content of the β phase will be higher than the average concentration in the alloy and B2 would be retained to room temperature during air cooling. It was therefore surprising to find that the microstructure of the ($\alpha2+\beta$) forged PAM specimens consisted of primary $\alpha2$ and basketweave matrix. It is believed that the formation of the basketweave in the ($\alpha2+\beta$) forged PAM specimens may be associated with the following factors: the lower oxygen content; the higher total reduction and the use of larger forging pieces than in the VAR specimens forged in the same field. The lower oxygen content results in lower β transus temperature and thus a lower fraction of primary $\alpha2$ at a given temperature in the ($\alpha2+\beta$) field. Consequently the niobium concentration in the matrix of the PAM specimens would be lower than in the matrix of the VAR specimens forged at the same temperature. The B2 matrix of the PAM specimens will then be less stable than the matrix of the VAR specimens. Larger total reduction of the forged PAM specimens also gives more stored energy to the specimens than to the forged VAR specimens reduction, which again makes the matrix of the PAM specimens more unstable. The larger PAM specimen size leads to a slower cooling rate, which is also favourable to the matrix transformation. As a result, the different microstructures were obtained in the PAM and VAR specimens. It appears that the first two facts are the most important in the matrix transformation of the $\alpha2+\beta$ field forged PAM specimens.

Substantial variations of room temperature mechanical properties such as strength, elongation and toughness have been observed in the forged PAM and VAR specimens in this work, which may be interpreted by the different compositions of the VAR and PAM alloys and different microstructures produced by different TMP routes. It is very interesting to note that for the VAR alloy the β-forged specimens were stronger than the ($\alpha2+\beta$)-forged specimens and for the PAM alloy opposite is true. This is because the finer microstructures were produced in β-forged VAR specimens and in the ($\alpha2+\beta$)-forged PAM specimens than the ($\alpha2+\beta$)-forged VAR and the β-forged PAM specimens. This also indicates the importance of the composition of an alloy, oxygen in particular, and total reduction of forging pieces on the microstructure and final mechanical properties.

The significance of oxygen content in the alloys may be appreciated by the facts that the β-forged PAM specimens were much tougher and more ductile than the β-forged VAR specimens and the VAR samples were stronger, though they have very similar fine basketweave microstructure. The only obvious difference is the oxygen content in these two alloys. This confirms that oxygen content in the super $\alpha2$ alloys has a significant effect on the room temperature ductility and toughness and even a small difference in oxygen could substantially affect room temperature ductility and toughness. A low oxygen content improves ductility and particularly toughness with only a moderate drop in strength which is analogous to Ti alloys. However the true significance of oxygen in the super $\alpha2$ may not be accurately assessed until the exact forging conditions between VAR super $\alpha2$ with high oxygen level and PAM super $\alpha2$ is met, which is in our future research programme.

The presence of large amount of B2 phase also affects the ductility and toughness. Firstly the room temperature ductility was improved by the retained B2, since it is body centred cubic and it should be intrinsically more ductile than $\alpha2$ or orthorhombic lath. The improvement in ductility by the retained B2 over the secondary $\alpha2$ or orthorhombic laths was, however, found to be very limited as shown in Table III. Secondly, the toughness was significantly reduced by the presence of retained B2 because cracks propagate straight across the B2 grains without any deviation. It seems that the presence of small amount of primary $\alpha 2$ (12% in volume) has little effect on the room temperature toughness as long as the matrix consists of basketweave secondary $\alpha 2$ or orthorhombic laths.

Finally, the results shown in Table III illustrate an aspect that is not understood at this stage. The PAM material in the forged condition has 2% ductility and this increases to 3.5% on aging with no apparent change in strength. The structural changes responsible for this change in ductility have not been identified and further work is underway.

SUMMARY

Microstructures and room temperature mechanical properties of PAM and VAR super $\alpha 2$ are sensitive to TMP routes and oxygen content of the alloys. It appears that low oxygen level in the alloys may substantially increase the ductility and toughness with only a moderate drop in strength.

Room temperature ductility of super $\alpha 2$ could be only marginally improved by retained B2 but with big drop in room temperature toughness. It is therefore not desirable to improve ductility by introducing large quantity of retained B2 in super $\alpha 2$.

REFERENCES

1. M.G. Mendiratta and H.A. Lipsitt, J. Mater. Sci., 15(1980) 2985.
2. H.A. Lipsitt, R.E. Schafrik and D. Shechtman, Metall. Trans.A, 11(1980) 1369.
3. R.A. Perkins, K.T. Chiang and G.H. Meier, Scr. Metall., 21(1987) 1505.
4. Young-Won Kim, J. Met., 42,July (1989) 24
5. R.G. Rowe, High Temperature Aluminide and Intermetallics, Edited by S.H.Whang, C.T.Liu et. al.(The Minerals, Metals & Materials Society, 1990), 375
6. C.H. Ward, J.C. Williams, A.W. Thompson, D.G. Rosenthal and F.H. Fores, Proceedings of Sixth World Conference on Titanium, France, 1988, 1130
7. J.A. Peters and C. Bassi, Scr. Metall., 24(1990)915
8. C.H. Ward,J.C. Williams and A.W. Thompson, Scr. Metall., 24(1990), 617
9. C. Bassi and J.A. Peters, Scr. Metall., 24(1990), 1363
10. W.A. Baeslack III and T. Broderick, Scr. Metall., 24(1990), 319
11. D.A. Koss, D. Banerjee, D.A. Lukasak and A.K. Gogia, High Temperature Aluminide and Intermetallics, Edited by S.H.Whang, C.T.Liu et. al.(The Minerals, Metals & Materials Society, 1990), 175
12. D.A. Lukasak and D.A. Koss, Metall. Trans. A, 21(1990), 135
13. A.K. Gogia, D. Banerjee and T.K. Nandy, Metall. Trans. A, 21(1990), 609
14. D. Banerjee, A.K. Gogia and T.K. Nandy, Metall. Trans. A, 21(1990), 627
15. Wonsuk Cho, A.W. Thompson and J.C. Williams, Metall. Trans. A, 21(1990), 641
16. J.M.Larsen, K.A.Williams, S.J.Balsone and M.A.Stucke, High Temperature Aluminide and Intermetallics, Edited by S.H.Whang, C.T.Liu et. al.(The Minerals, Metals & Materials Society, 1990), 521.
17. J.M.Larsen, K.A.Williams,S.J.Balsoue and M.A.Stucke, High Temperature Aluminide and Intermetallics, Edited by S.H.Whang, C.T.Liu et. al.(The Minerals, Metals & Materials Society, 1990), 375

PHASE STABILITY IN Ti_3Al+X BASE (X=Nb, Mo, V, Ta, Fe) ALLOYS

M. Thomas, S. Naka, and T. Khan

Office national d'Etudes et de Recherches Aérospatiales,
29, Av. de la Division Leclerc, 92322 CEDEX, France

Abstract

In the present work, the relative stability of the β (or B2)-phase through the addition of various elements such as Nb, Mo, V, Ta and Fe in Ti_3Al was investigated. Both the variation of β-transus temperature along the tie line Ti_3Al-X_3Al or Ti_3Al-X and the α_2 / β (or B2) partitioning coefficient were used as criteria in order to compare the β-stabilising effectiveness of these elements. The stability of the B2 phase retained after quenching was examined during subsequent ageing treatments. When the amount of β-stabilising elements is not sufficient, the retained B2 phase decomposes into intermediate phases (ω, O). The decomposition sequence depends on both the species and the amount of each alloying element.

Introduction

Phase equilibria near the Ti-rich corner of the ternary Ti-Al-Nb alloy system has been progressively clarified during the last fifteen years [1-10]. The earliest investigations dealt with the detemination of the α_2+β (B2) / β (B2) transus in Ti_3Al-based alloys with increasing niobium content, from 5 to 17 at% Nb [4]. Such information is generally very useful for determining both the thermomechanical and heat treatment conditions required to optimize the microstructure. Structural parameters such as the size, the volume fraction and the morphology of the equilibrium phases comprising α_2 (DO_{19}) and β (or B2) can thus be suitably controlled.

For a successful development of new Ti_3Al (or Ti_2AlX) based alloys, the knowledge of the phase boundaries should be extended to Ti-Al-X (X= Mo, V, Ta, Fe etc.) alloy systems other than Ti-Al-Nb. In particular, the variation of transus temperature along the tie line Ti_3Al-X_3Al (or Ti_3Al-X) as well as the α_2 /β (or B2) partitioning coefficient can be regarded as guidelines, not only for microstructural but also for compositional optimization. The present work was therefore aimed at establishing an equivalence in β-stabilising effectiveness between those alloying elements.

Although the above parameters (transus temperature and partitioning coefficient) enable us to evaluate the β-stabilising effectiveness, they are not sufficient to predict the atomic concentration which allows the alloy to remain single-phase B2 during ageing subsequent to quenching. There is no clear evidence that the β-stabilising equivalence provided by the above thermodynamic criteria will necessarily match with the equivalence deduced from a kinetic criteria, i.e. the minimum atomic concentration needed to preserve a retained B2 phase during ageing subsequent to quenching. In this investigation, the stability of the B2 phase du-

Titanium '92
Science and Technology
Edited by F.H. Froes and I. Caplan
The Minerals, Metals & Materials Society, 1993

ring ageing was therefore studied with increasing amount of β-stabilising elements. In ternary alloys in which the B2 phase is not sufficiently stabilised, its decomposition gives rise to a number of transition phases. Various sequences of decomposition encountered during thermal exposure in the low temperature range (450-650°C) are presented. Finally, the srong hardening effect of these phases is briefly mentioned.

Experimental procedure

Small ingots were prepared by vacuum arc melting using appropriate amounts of Ti (99.5%), Al (99.95%), Nb (99.9%), Mo (99.8%), V (99.7%), Ta (99.8%) and Fe (99.9%). A good homogeneity was ensured by remelting the buttons several times. The compositions of the studied alloys are listed in Table 1. The β-alloying concentration typically varies from 6 to 30 at% depending on the atomic species.

The α_2 + β (B2) / β (B2) transus temperature was determined by optical microscopy after isothermal heat treatments followed by water quenching. Quantitative image analysis was used in order to determine the volume fractions of different phases.

Compression tests were performed on some as-cast alloys in order to obtain the temperature dependence of yield stress (between 20 and 950°C). These tests were conducted under a vacuum of 10^{-5} Torr at a strain rate of 10^{-4} sec^{-1}.

Crystal structure of the transition phases which formed during ageing was examined by Transmission Electron Microscopy (TEM). For this purpose, specimens were mechanically sliced down to 150μm. Electropolishing was then performed in a cooled mixture (at 0°C) of sulfuric acid and methanol for Nb, Mo and Fe containing alloys and in a cooled mixture (at -40°C) of perchloric acid, butyl-cellosolve and methanol for V containing alloys. On the other hand, thinning of Ta containing alloys was performed by ionic polishing. A JEOL 200CX microscope equiped with a double tilt goniometer was used to obtain dark field images and corresponding diffraction patterns.

Table 1 Chemical compositions of the alloys studied in this work (at.%).

Alloy	Al	Nb	Mo	V	Ta	Fe	Ti
N°1	21.7	11.5	-	-	-	-	balance
N°2	19.9	17.7	-	-	-	-	balance
N°3	25.0	25.0	-	-	-	-	balance
N°4	24.0	29.5	-	-	-	-	balance
N°5	26.3	-	3.5	-	-	-	balance
N°6	22.6	-	5.5	-	-	-	balance
N°7	24.9	-	6.9	-	-	-	balance
N°8	20.0	-	10.9	-	-	-	balance
N°9	25.0	-	25.0	-	-	-	balance
N°10	25.1	-	-	8.6	-	-	balance
N°11	22.4	-	-	9.9	-	-	balance
N°12	25.2	-	-	11.0	-	-	balance
N°13	24.9	-	-	13.4	-	-	balance
N°14	22.6	-	-	-	12.0	-	balance
N°15	20.1	-	-	-	18.1	-	balance
N°16	16.7	-	-	-	25.6	-	balance
N°17	24.0	-	-	-	-	6.2	balance
N°18	22.5	-	-	-	-	12.2	balance
N°19	21.0	-	-	-	-	18.0	balance
N°20	24.2	4.0	3.9	4.2	-	-	balance
N°21	23.5	6.7	1.0	4.9	-	-	balance
Super α_2	24.9	9.6	0.9	3.0			balance

Results

1. Transus temperature variation

The variation of transus temperature as a function of increasing content of various elements is an indication of their β-stabilising effectiveness. Fig.1 and 2 show such a variation, evaluated from the experimental results, along the tie lines Ti_3Al-X and Ti_3Al-X_3Al respectively. In these figures, the steeper the slope, the higher the effectiveness of the refractory element for stabilising the β (or B2) phase. By using data from Fig.1, the following equivalence in β-sta-

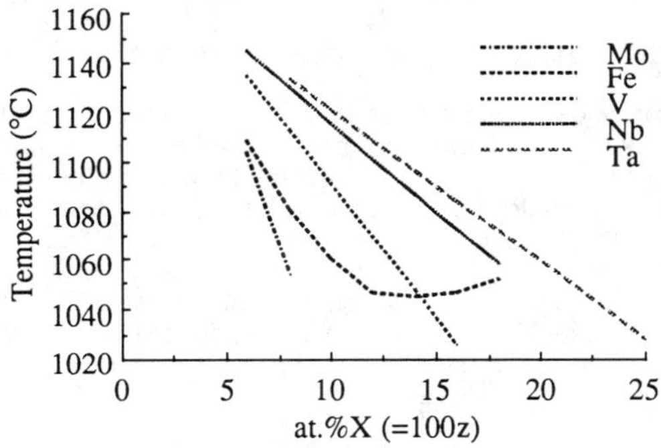

Fig.1 β transus curves along the tie line Ti_3Al-X (X=Nb, V, Mo, Fe, Ta). The lower concentration side of these curves is drawn with extrapolation from the experimental data.

bilising effectiveness can be obtained in the case of $(Ti_3Al)_{1-z}X_z$ (z=0 to 1) alloys (for the sake of clarity, the niobium atomic percent is normalized to 25 at.%) :

$$11.5\%Mo \Leftrightarrow 13.5\%Fe \Leftrightarrow 18\%V \Leftrightarrow 25\%Nb \Leftrightarrow 26.5\%Ta \text{ (at.\%)} \qquad (1)$$

Due to the fact that the transus temperature decreases, at first, with increasing Fe content and then increases, a behaviour which is very different from that of other elements, the above equivalence for Fe is only valid up to a concentration of about 10%. The reason for this singularity will be discussed later.

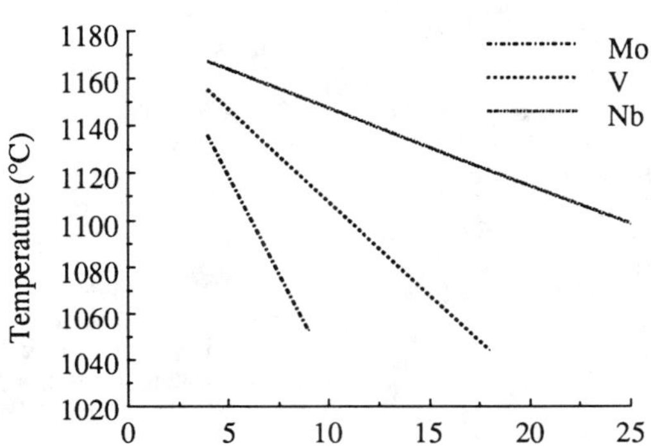

Fig.2 β transus curves along the tie line Ti_3Al-X_3Al (X=Nb, V, Mo). The lower concentration side of these curves is drawn with extrapolation from the experimental data.

Experimental data from Fig.2 provide the following euivalence in β-stabilising effectiveness in the case of $(Ti_{1-z}X_z)_3Al$ (z=0 to 1) alloys :

7%Mo <=> 10.5%V <=> 25%Nb(at.%) (2)

Since most of the useful Ti_3Al based alloys contain slightly less than 25at.%Al, a convenient equivalence in β-stabilising effectiveness between the three major elements (Nb, V and Mo) may be obtained by taking roughly the average values of (1) and (2), such as :

9%Mo <=> 14.5%V <=> 25%Nb(at%) (3)

2. β-transus approach curve (Temperature dependence of volume fraction of the β phase)

Fig.3 shows the temperature dependence of volume fraction of the β (or B2) phase determined experimentally on four alloys. This figure also comprises the case for the Ti-24Al-11Al alloy examined in our previous work [11]. It is clear that the four alloys investigated in the present work (including the Super α_2 alloy) are more β-stabilised than the Ti-24Al-11Al alloy. As illustrated in this figure, the temperature dependence of volume fraction of the β (or B2) phase for the Super α_2 alloy (Ti-25Al-10Nb-3V-1Mo) is quite similar to that for the Ti-23,5Al-6,7Nb-4.9V-1Mo alloy, in spite of some small differences. The same situation exists when one compares the Ti-24.9Al-13.4V and the Ti-24.2Al-4Nb-4.2V-3.9Mo alloys. In other words, we can reasonably consider that there are following equivalences in β-stabilising effectiveness:

10%Nb + 3%V + 1%Mo <=> 6,7%Nb + 4.9%V + 1%Mo (4)

13.4%V <=> 4%Nb + 4.2%V + 3.9%Mo (5)

The equivalence (4) gives us 3.3%Nb <=> 1.9%V. Combining this with the equivalence (5), we can get another relationship : 12%Nb <=> 4%Mo.

This results in a ratio of approximately 3 for Mo/Nb and 5/3 for V/Nb and leads to the following equivalence in β-stabilising effectiveness :

8.3%Mo <=> 15%V <=> 25%Nb(at%) (6)

It is worth noting that the above relationship is very close to (3) which was obtained from the transus temperature determination. Note also that two β-approach curves corresponding to the above two groups of alloys were injected in Fig.3.

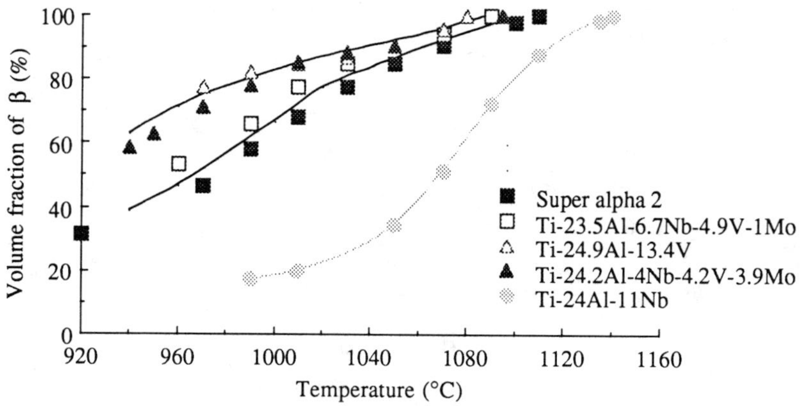

Fig.3 Volume fraction of the β-phase as a function of temperature.

Finally, based on these equivalences (3) and (6), two alloys Ti-25Al-4Nb-4V-4Mo and Ti-25Al-13,5V are considered having approximately 22% of Nb equivalent while the Super α_2 and Ti-23,5Al-6,7Nb-5V-1Mo alloys contain only 18% of Nb equivalent.

3. Decomposition of the B2 phase

Depending on the nature and the amount of β alloying additions, the decomposition of the B2 phase, retained after quenching, was examined during ageing in the intermediate temperature range (450-650°C). The precipitates resulting from this decomposition are responsible for a considerable strengthening of the alloys, as illustrated by the yield stress versus temperature curves (Fig.4). Compression test specimens were in this case prepared after quenching. Indeed, the variation of yield stress showed a yield stress peak around 550°C in the Ti-25Al-25Nb, Ti-24.9Al-13.4V, Ti-23.5Al-6Mo and Ti-24.2Al-4Nb-4.2V-3.9Mo alloys. Such an anomaly is due to the strenghtening effect resulting from a second phase precipitation, as illustrated by TEM observations. Note that the Ti-25Al-4Nb-4V-4Mo alloy exhibits a less pronounced strenghtening compared with the other three alloys. This is probably because of the lower metastability of this alloy. From these experiments, the following equivalence can be tentatively proposed for the minimum atomic concentration needed to preserve a retained B2 phase during ageing subsequent to quenching :

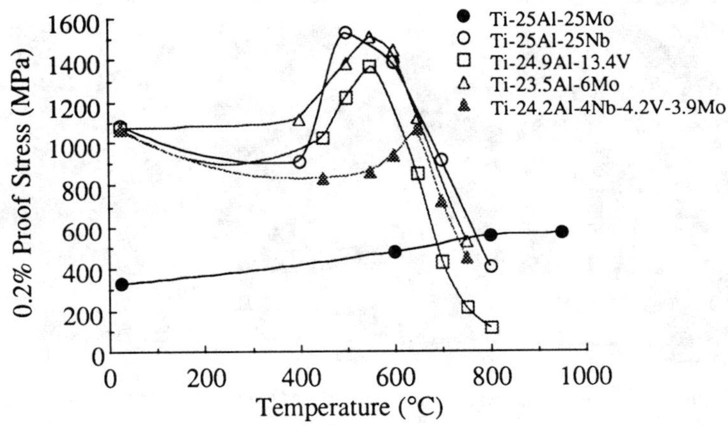

Fig.4 Temperature dependence of yield stress.

6%Mo <=> 13.5%V <=> 25%Nb (8)

The Ti-25Al-25Mo alloy exhibits, on the contrary, a totally different temperature dependence of yield stress. Although this alloy shows a slight and gradual increase in yield stress over the entire temperature range, no precipitation is observed and the alloy remains single phased B2. Work is presently in progress in order to elucidate the reason for the gradual increase in yield stress with temperature.

TEM micrographs in dark field condition exhibit a homogeneous dispersion of very fine precipitates in the aged Ti-24.9Al-13.4V, Ti-25Al-25Nb, Ti-23.5Al-6Mo, Ti-24.2Al-4Nb-4.2V-3.9Mo and Ti-22Al-12Fe alloys (Fig.5). The appearance of the precipitates around the peak yield stress temperature clearly indicates that these second-phase precipitations are responsible for the yield stress anomalies. The selected area diffraction patterns (SADP) from the $<110>_{B2}$ zone axis of the different alloys (Fig.6) reveal that different transition phases occur depending on the nature of the β-stabilising element.

In the case of V and Fe containing alloys, the observed precipitation is related to an ω-type

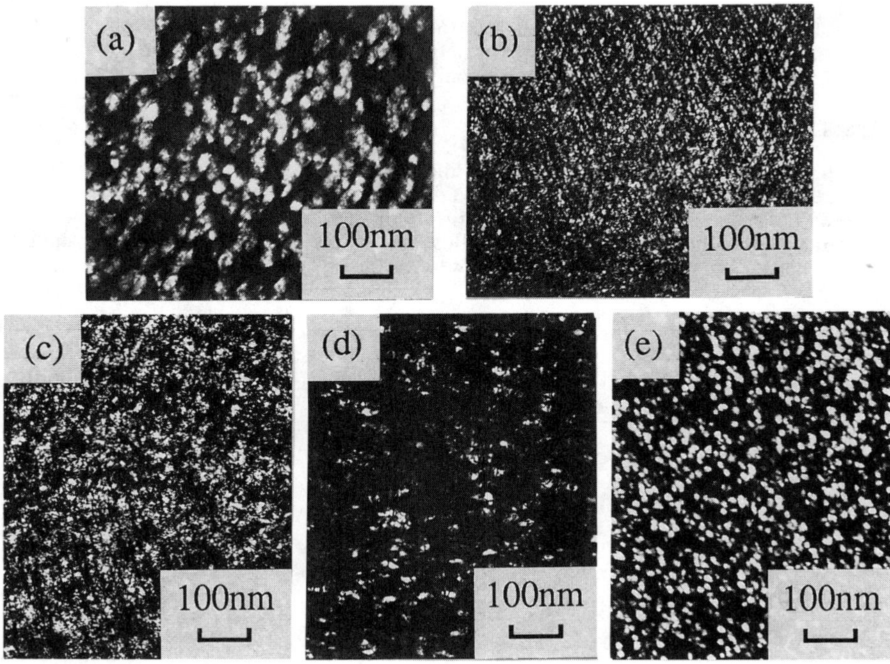

Fig.5 Fine precipitates observed after ageing at 550°C : (a) Ti-24.9Al-13.4V, (b) Ti-25Al-25Nb, (c) Ti-23.5Al-6Mo, (d) Ti-24.2Al-4Nb-4.2V-3.9Mo and (e) Ti-22Al-12Fe.

precipitation process, like the one described by Bendersky et al. in Ti-Al-Nb alloys [6,7]. The orientation relationship, i.e. $(0001)_{\omega}$ // $\{111\}_{B2}$ and $\langle 1\bar{1}00\rangle_{\omega}$ // $\langle 1\bar{1}0\rangle_{B2}$, is indicative of four rotational variants with their c-axis parallel to the four $\langle 1\bar{1}1\rangle_{B2}$ zone axes. The presence of extra-spots with respect to the disordered ω phase, i.e., at the 1/3 111 $_{B2}$, 1/3 112 $_{B2}$ and

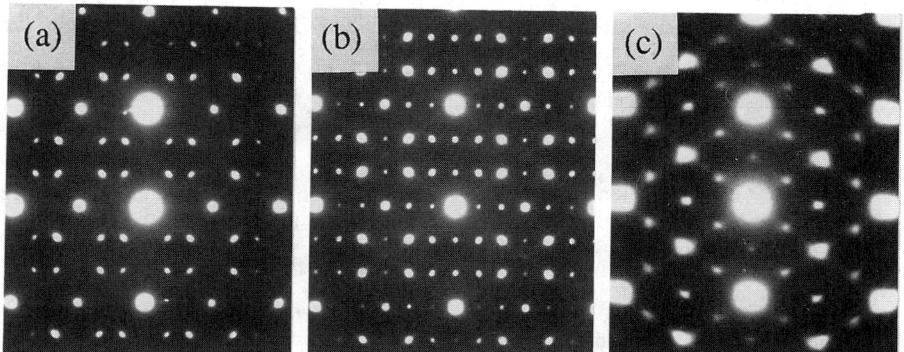

Fig.6 SADP from the $\langle 110\rangle_{B2}$ zone axis : (a) ω ($B8_2$) in Ti-20.5Al-6Fe, (b) ω ($D8_8$ or Ga_4Ti_5 type) in Ti-24.9Al-13.5V and (c) O in Ti-25Al-25Nb. (Ageing temperature : 550°C)

1/3 110 $_{B2}$ positions reveals that a further ordering sequence takes place. A variety of ordered ω phases ($B8_2$ type, $D8_8$ type or Ga_4Ti_5 type) could be observed depending upon the aging treatments as well as the nature of the β-stabilising element. The formation of these ω phases involves both displacive and short range diffusive mechanisms, and the lattice instability along $<111>_{B2}$ is at the origin of this displacive movement.

In the case of Nb and Mo containing alloys, SADP reveals that the extra spots at the 1/2 110_{B2} and at the 1/4 112_{B2} are not equivalent, which reflect the two-fold symmetry of an orthorhombic cell. According to the orientation relationship, i.e. $<110>_{B2}$ // $[001]_O$ and $\{1\bar{1}2\}_{B2}$ // $\{110\}_O$, three $<1\bar{1}0>_O$ variants are distributed in the $<111>_{B2}$ zone axis. From a crystallographic point of view, the orthorhombic (O) phase is an ordered form of the α" phase encountered in conventional disordered titanium alloys [8]. As is the case for the ω formation, this transformation involves both displacive and short range diffusive mechanisms. A martensitic shearing of alternate $\{1\bar{1}0\}_{B2}$ planes due to the lattice instability along $<110>_{B2}$ takes place with a displacement vector equal to a3/2[110]. The chemical rearrangements consist in an ordering of niobium atoms with respect to the DO_{19} lattice.

The fact that Nb, Mo containing alloys promote O-type precipitates whereas V, Fe containing alloys promote ω-type precipitation indicates a strong influence of the atomic species on the transformation sequence, but the origin of such a difference is still not clear.

Discussion

Although a substantial additional experimental data are still required to grasp an overall view of phase aquilibria in Ti_3Al +X alloy systems, the results of the present study already constitute useful information for future alloy development of Ti_3Al (or Ti_2AlX) based alloys. In the metallurgy of conventional titanium alloys, the information about the β-stabilising effectiveness of various alloying elements is well documented in the literature [12]. Such information is scarce, at least in the open literature, for Ti_3Al-based alloys. It should be emphasized that the knowledge about the variation of β transus temperature with the amount of alloying elements as well as the temperature dependence of volume fraction of the β (or B2) phase is essential for developing new alloys.

Among the β-stabilising elements investigated in this work, three of them (Nb, V and Mo) were well characterized and an equivalence of β-stabilising effectiveness was proposed such as 1 : 1.67 (=5/3) : 3 ≈ %Nb : %V : %Mo. Work is in hand to get more information about Ta, another promising refractory element. As for Fe, its behaviour for β stabilisation was found to be very different. In view of the fact that in the binary Ti-Fe system, the β phase decomposes through a eutectoid reaction into α and TiFe (B2), contrary to the other elements (Nb, V, Mo and Ta) which are β-isomorphous with Ti and that in the ternary Ti-Al-Fe system, this B2 phase extends in solid solution even up to compositions around Ti_2AlFe [13], it is plausible that there is a eutectoid type reaction along the tie line Ti_3Al-Ti_2AlFe. This may lead to the increase in the β-transus temperature at concentrations above 10%Fe (see Fig.1). A fairly strong interaction between Ti, Fe and Al may also be responsible for the decrease of β-stabilising effectiveness of Fe; Fe was found to be less β-stabilising than Mo, while it is quite the opposite in binary Ti-X systems.

Concerning the decomposition of the B2 phase during ageing, it should be emphasized that under specific heat treatment conditions (for example, after heat treatment just below the transus temperature, leading to the formation of the primary α_2 phase), the metastability of the B2 matrix is changed in such a manner that the ω phases could appear in Nb or Mo containing alloys and the O phase in V or Fe containing alloys. a more detailed understanding of these phases is important because of the beneficial effect on ductility and strength of the O phase discovered recently [14].

Conclusions

In the present investigation, the β-stabilising effectiveness of various alloying elements (Nb, Mo, V, Ta and Fe) in Ti_3Al based alloys was evaluated by investigating the variation of β transus temperature with the amount of alloying elements as well as the temperature dependence of volume fraction of the β (or B2) phase. Such information seems to be essential for the future development of new Ti_3Al (or Ti_2AlX) based alloys. For three of those elements, the following equivalence coud be deduced : 1 : 1.67 (=5/3) : 3 ≈ %Nb : %V : %Mo.

When the amount of β-stabilising elements is not sufficient, the retained B2 phase decomposes giving rise to the formation of intermediate phases (ω, O). The decomposition sequence depends on both the species and the amount of each alloying element.

Acknowledgements

This work was conducted with a financial support of the French Ministry of Defence (DRET). The authors are grateful to A. Bachelier-Locq and P. Thévenin for their technical assistance.

References

1. S.M.L. Sastry and H.A. Lipsitt, "Ordering Transformations and Mechanical Properties of Ti_3Al and Ti_3Al-Nb Alloys", Met. Trans. A, 8A (1977) 1543.
2. M.G. Mendiratta and H.A. Lipsitt, "Steady-State Creep Behaviour of Ti_3Al-Base Intermetallics", J. Mater. Sci., 15 (1980) 2985.
3. M.J. Blackburn and M.P. Smith, US Patent N° 4,292,077, 1981.
4. R. Strychor, J.C. Williams and W.A. Soffa, "Phase Transformations and Modulated Microstructures in Ti-Al-Nb alloys", Metall. Trans. A, 19A (1988) 225.
5. D. Banerjee, T.K. Nandy and A.K. Gogia, "Site Occupation in the Ordered Beta Phase of Ternary Ti-Al-Nb Alloys", Scripta Metall., 21 (1987) 597.
6. L.A. Bendersky, W.J. Boettinger, B.P. Burton, F.S. Biancaniello and C.B. Shoemaker, "The Formation of Ordered ω-Related Phases in Alloys of Composition Ti_4Al_3Nb", Acta Metall., 38 (1990) 931.
7. L.A. Bendersky, B.P. Burton, W.J. Boettinger and F.S. Biancaniello, "Ordered ω-Derivatives in a Ti-37.5Al-20Nb at.% Alloy", Scripta Metall., 24 (1990) 1541.
8. L.A. Bendersky, W.J. Boettinger and A. Roytburd, "Coherent Precipitates in the B.C.C./Orthorhombic Two-Phase Field of the Ti-Al-Nb System", Acta Metall. et Mater., 39 (1991) 1959.
9. K. Muraleedharan, A.K. Gogia, T.K. Nandy, D. Banerjee and S. Lele, "Transformations in a Ti-24Al-15Nb Alloy - Part-I : Phase Equilibria and Microstructure", to be published in Acta Metall. Mater..
10. K. Muraleedharan, T.K. Nandy, D. Banerjee and S. Lele, "Transformations in a Ti-24Al-15Nb Alloy - Part-II : A Composition Invariant $\beta_o \Rightarrow$ O Transformation", to be published in Acta Metall. Mater..
11. M. Thomas, S. Naka and T. Khan, "Stability of the B2 Phase in Ternary Ti-Al-X Alloys (X=Nb, Mo, V)", paper presented at the 2nd European Conf. on "Advanced Materials and Processes - EuroMat 91", Cambridge, United Kingdom, July, 1991.
12. U. Zwicker, Titan und Titanlegierungen, Berlin, Springer Verlag, 1974.
13. A. Seibold, "Phasengleichgewichte in den ternären Systemen Ti-Fe-O und Ti-Fe-Al", Z. Metallkde., 72 (1981) 712.
14. M. Thomas, S. Naka, M. Marty, W.G. Smarsly and T. Khan, "Mechanical Behaviour of Extruded I/M and P/M Super α_2 Alloys, in this conference.

CREEP OF Ti-24Al-11Nb

Diane E. Albert and Anthony W. Thompson

Department of Materials Engineering
Carnegie Mellon University, Pittsburgh, PA 15213 USA

Abstract

The stress and temperature dependences of the steady state creep rate (SSCR) of various microstructures of Ti-24Al-11Nb (a/o) have been determined. SSCR *vs.* stress and strain *vs.* time curves were determined in order to investigate the effect of microstructure on creep resistance. SSCR *vs.* temperature curves were determined to find Q_a, the apparent creep activation energy. Experimentally determined values of Q_a were found to be between 106 kJ/mol and 156 kJ/mol; the equiaxed α_2 phase type microstructure exhibited the smallest Q_a at each test stress. Steady state creep rate *vs.* stress tests have shown that microstructures with a coarse α_2 lath size in general display improved creep resistance over fine-lathed structures; however, specimens with the equiaxed α_2 phase microstructure exhibit much greater creep rates at all stresses and temperatures, implying that dislocation slip length, proportional to α_2 phase dimension, does not always control creep in these materials and that texture may adversely affect creep resistance.

Introduction

Titanium-base materials are attractive for use in the aerospace industry because their strength to density ratio is quite high. The development of Ti_3Al-based titanium aluminides, such as Ti-24Al-11Nb and Ti-25Al-10Nb-3V-1Mo, have increased the temperature capability of conventional titanium alloys. At high temperatures, the creep strength of structural parts becomes important. However, at this time the relationship between microstructural features and the creep behavior of α_2-based titanium aluminides is still not fully understood, in spite of several studies reported in the literature **[1-6]**.

In order for the titanium aluminides to become useful engineering materials, more detailed information on elevated-temperature creep must be available. Therefore, the objective of this study was to determine the stress dependence and the temperature dependence of the steady-state creep rate of different microstructures of Ti-24Al-11Nb.

Titanium '92
Science and Technology
Edited by F.H. Froes and I. Caplan
The Minerals, Metals & Materials Society, 1993

Experimental Procedure

A. *Microstructures*

The Ti-24Al-11Nb alloy used in this research was supplied as 1.27 cm thick cross-rolled plates. The plates had been hot rolled, annealed at 1038°C for one hour, then air cooled. The chemical analysis of the alloy is shown in Table I.

Table I. Chemical Composition of Ti-24Al-11Nb

Al	Nb	Fe	O	N	C	H	Ti
14.0	20.6	0.08	0.054	-	0.01	4.8x10-5	bal(wt%)
24.6	10.5	0.068	0.161	-	0.039	0.002	bal (at%)

Transverse and longitudinal sections of the as-received microstructure of the cross-rolled Ti-24Al-11Nb are shown in Figure 1. The transverse direction of the plate is parallel to the thin dimension of the plate. Parallel to one rolling direction, designated RD1 (rolling direction 1), elongation of α_2 phase is clearly evident. There is evidence of inhomogeneous deformation through the thickness of the plate. Since the plate was cross rolled, elongation of the α_2 phase is also evident at an angle of 90° from RD1 in the plane of the plate. This direction is designated RD2, or rolling direction 2. RD1 and RD2 are in the sheet plane. There is less deformation, as indicated by density of elongated α_2 phase, in this direction. Elongation of α_2 phase is not evident in the transverse direction (TD). Evidence of deformation texture in the as-received cross-rolled plate is displayed by the non-random distribution of poles in the (0002) pole figure procurred from a thin slice of material cut perpendicular to the RD1 direction, shown in Figure 2. Maximum intensity is 7.5 times random, indicating a moderate texture. Thus, basal poles tend to be nearly parallel to RD1 in the plane of the plate. It can be assumed that basal poles, to a lesser extent, are also aligned near RD2. In an α_2+ß microstructure, the ß {110} peak overlaps the α_2 (0002) reflection [7]; however, in this microstructure an insignificant amount of ß phase is present and all information can be assumed to correspond to α_2 phase.

To produce other microstructures, the AR material was solution heat treated either in the ß range (1200°C) or the α_2+ß phase region (1000°C) in air in a ceramic tube resistance furnace with temperature control of ± 2°C. All heat treated blanks were subsequently air cooled with a cooling rate of 10°C/second. To produce α_2+ß solutionized microstructures, the material was first ß solutionized for one hour to return the microstructure to single-phase ß, then immediately transferred to a furnace at 1000°C and held for either 15 minutes or 30 minutes, then air cooled. The resulting microstructures are shown in Figure 3a and 3b and are designated 1000AC-15 and 1000AC-30, respectively. In the 1000AC-15 specimens, the α_2 laths are small and are arranged in a basketweave morphology. A small amount of globular α_2 is also present, most likely formed during α_2+ß solutionizing. The α_2 lath width in the 1000AC-30 microstructure is significantly larger than in the 1000AC-15 microstructure and also has a basketweave morphology. A microstructure with a very fine basketweave morphology was produced by solutionizing at 1200°C for one hour with a subsequent air cool, as seen in Figure 3c. This microstructure was designated 1200AC. The as-received microstructure is again seen in Figure 3d. Prior-ß grain sizes and α_2 lath widths as measured in all of these microstructures are listed in Table II. The optical micrographs indicated a somewhat inhomogeneous distribution morphology and distribution of the α_2 phase and thus precise determination of grain size and lath width was difficult. However, the width of the α_2 laths in 1000AC-30 and the diameter of the equiaxed α_2 phase in the as-received plate are significantly larger than the α_2 phase in other microstructures. Three microstructures, 1200AC, 1000AC-30 and 1000AC-15, are characterized by a large prior ß grain size of nearly 1 mm. No grain boundaries were discernable in the cross-rolled plate as-received microstructure. No appreciable volume fraction of the colony structure, such as is found in Ti-25Al-10Nb-3V-1Mo cooled very slowly from the α_2+ß phase field [8], was found in any heat treated specimens derived from this alloy.

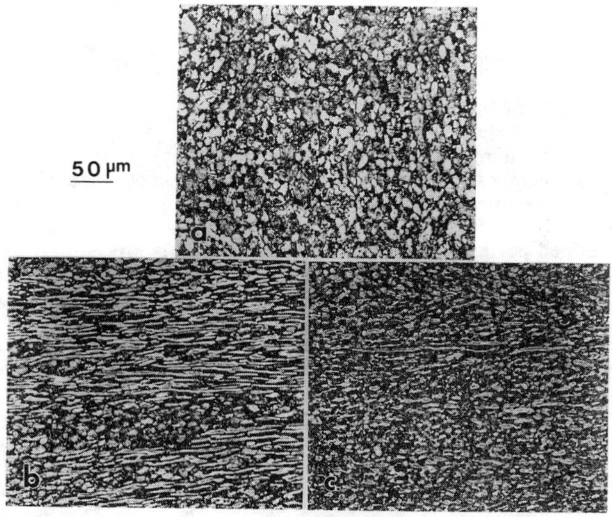

Figure 1. As-received microstructure of Ti-24Al-11Nb. **a)** Transverse direction, **b)** longitudinal (RD1) direction, and **c)** longitudinal (RD2) direction.

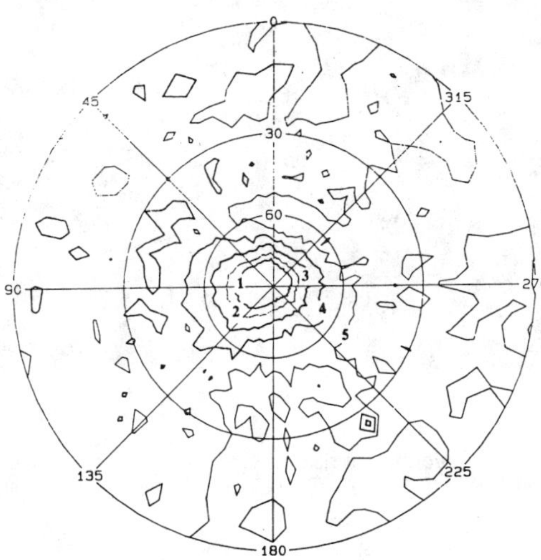

1=highest intensity
RD1 out of paper

Figure 2. The (0001) pole figure obtained from the as-received plate.

Table II. Size of α_2 laths and prior-ß grains in Ti-24Al-11Nb.

Microstructure	Prior-ß grain size (mm)	α_2 lath width (µm)
1200AC	0.709	2.0
1000AC-30	0.836	5.2
1000AC-15	0.817	2.3
AR	Not discernible	7.0

B. *Experimental Methods*

Constant temperature creep tests [9] were conducted at 650, 760, and 870°C in air, using single lever creep machines with a load ratio of 20:1. A "stress decrement" procedure was used where first a predetermined load, less than the yield stress of the material, was applied to the specimen. After sufficient time had elapsed for the steady state regime to be reached, a predetermined portion of the load was removed. This procedure was repeated a number of times until the steady state creep rate's dependence on stress became evident. Constant load creep tests were conducted at stresses of 100 MPa and 31.4 MPa to determine the apparent activation energy of creep; a similar "temperature decrement" procedure was used. Creep strain was measured using a high-temperature extensometer in unison with a super linear variable capacitor (SLVC) transducer. A strain resolution of 5×10^{-6} was achieved.

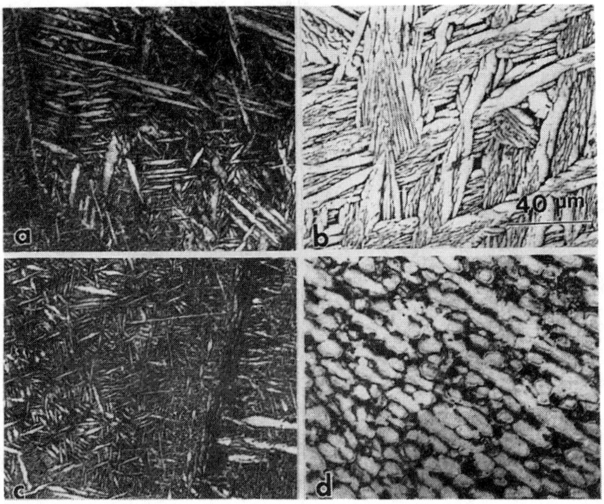

Figure 3. Microstructures of heat treated cross rolled plate, designated **a)** 1000AC-15, **b)** 1000AC-30, **c)** 1200AC, and **d)** AR; all at the same magnification.

Results

A. *Creep Strain as a Function of Time*

The rate of elongation through the primary creep range and into the steady state creep range of various microstructures derived from the cross-rolled plate is shown in Figures 4a,b, and c. Each plot represents test data collected under conditions of constant stress and temperature. In all cases, the primary creep regime was very small in terms of absolute time elapsed and total strain accumulation. At all three test temperatures, the 1000AC-30 microstructure was most creep resistant while the 1200AC and 1000AC-15 microstructures had very similar, but faster creep rates. The as-received microstructure crept at a significantly greater rate than all other structures.

B. *Steady State Creep Rate as a Function of Stress*

In Figures 5a-c, the SSCR as a function of stress at 650°C, 760°C, and 870°C, respectively, is shown for multiple specimens representing all four microstructures considered in this work. Test stresses ranged from a high of 312 MPa to a low of 31.4 MPa. The as-received material, possessing an equiaxed α_2 microstructure, had the highest creep rate at all stresses and, in general, the microstructures with a smaller α_2 lath size (1200AC and 1000AC-15) crept faster than the material with a larger lath size, *i.e.*, the 1000AC-30 material. The inferiority of the creep resistance of the as-received microstructure was reduced at high stresses. The intermediate creep rate of the 1200AC and 1000AC-15 materials was quite similar. At all temperatures, the distinctive result is the inferior creep resistance of the as-received material.

C. *Steady State Creep Rate as a Function of Temperature*

Figures 6a and 6b show the effect of temperature on SSCR for three microstructures. The slope of log SSCR *vs.* reciprocal temperature (1/T) is Q_a, the apparent activation energy of creep. Figure 6a shows Q_a for three microstructures, AR, 1000AC-15, and 1200AC, crept at the constant stress of 31.4 MPa. Figure 6b shows values of Q_a obtained when the same three microstructures were crept at 100 MPa. The as-received microstructure has the smallest Q_a at both stresses, while the two lath-type microstructures have larger creep activation energies. As stress level increases, activation energy decreases.

Discussion

The term "plate" has been used to describe the morphology of the α_2 phase seen in titanium aluminide alloys, but it is proposed that a more appropriate term to use to describe this phase is "lath," with dimensions $a<b\ll c$. In this research and other work [10], the equiaxed α_2 phase was seen with a diameter equal to the α_2 lath width, not the long dimensions of the laths. Considering results from the various creep tests performed in this research, the most obvious conclusion is that the as-received plate with an equiaxed α_2 microstructure has the poorest creep resistance by at least an order of magnitude at all stresses and temperatures. Indeed, as seen in Figure 7, this microstructure's poor creep resistance is highlighted when compared to creep test results obtained by other researchers on titanium aluminides [1,3-5].

The slip length, corresponding to the lath width or equiaxed α_2 phase diameter, has been suggested to be an important factor determining creep rate in titanium alloys [11,12]. Dislocations glide on slip planes until stopped by a barrier such as ß film-α_2 lath interfaces or α_2/α_2 interfaces. Pile-ups form and plastic deformation is effectively halted, as has been observed in the creep at low stresses of Ti-24Al-11Nb [5]. The stress level during creep testing is usually below the yield strength of a material so the chance for a slip band to break a barrier is quite small and slip length thus limits the extent of creep deformation. Mobile dislocations are confined to individual α_2 laths.

As lath dimensions increase, slip length increases and the end result after creep deformation should be a greater strain or creep rate. Thus, fine-grained materials or fine-lathed microstructures should have superior creep resistance, contrary to the evidence found in this work which suggests that the coarse-lathed 1000AC-30 microstructure has a creep rate superior to that of the fine-lathed 1200AC and 1000AC-15 microstructures. The data in Table II, which lists the diameter of the equiaxed α_2 phase and the α_2 lath widths found in microstructures used in this research, do not suggest a correlation between increase in lath size and increase in creep rate. In addition, the as-received equiaxed α_2 phase diameters are not significantly larger than the average width of the coarse α_2 laths. Therefore, phase dimension-slip length considerations do not appear to be sufficient explanations of the creep test results. The degree of lath coarseness in Ti-24Al-11Nb basketweave specimens may not be as important as the α_2 phase morphology and deformation texture present in this alloy because of the type of slip system which is commonly observed to operate in the α_2 phase.

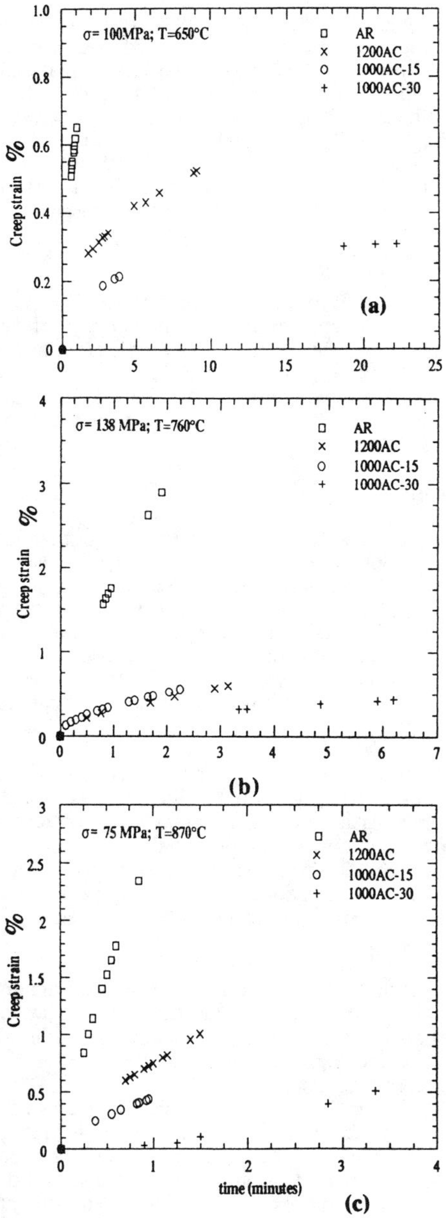

Figure 4. Creep strain *vs.* time at **a)** 650°C, **b)** 760°C, and **c)** 870°C.

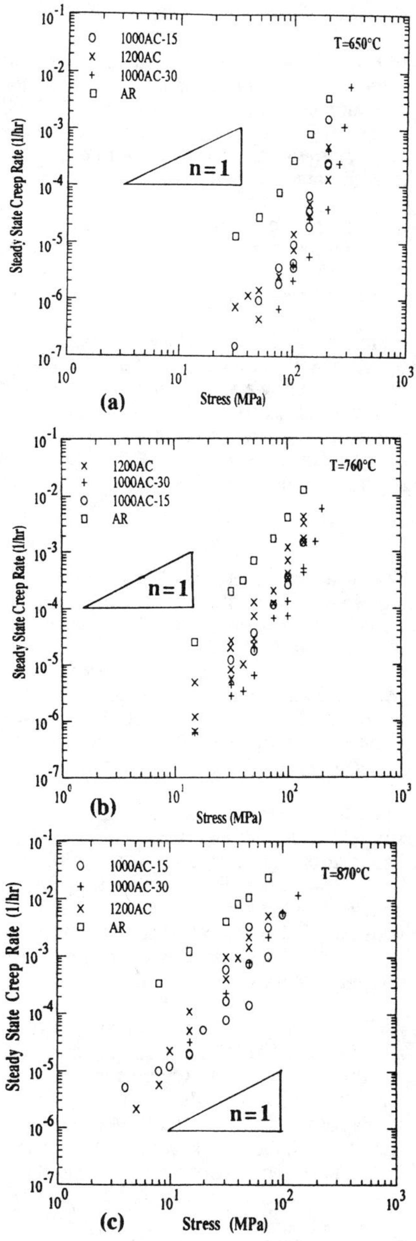

Figure 5. SSCR *vs.* stress at **a)** 650°C, **b)** 760°C, and **c)** 870°C.

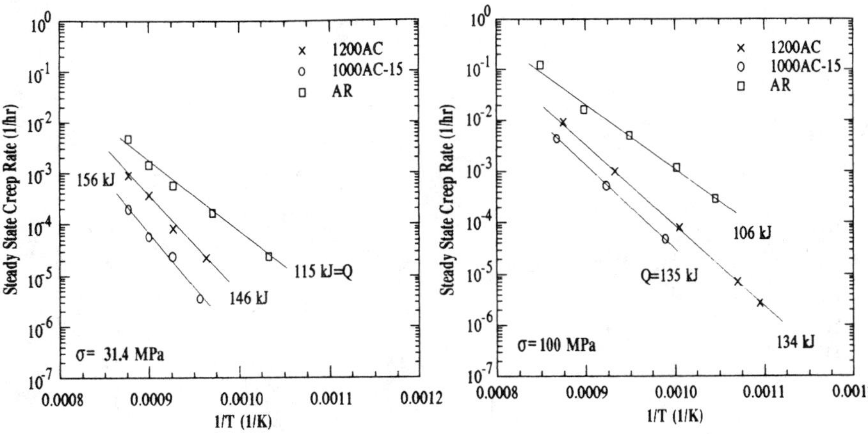

(a) (b)

Figure 6. SSCR *vs.* 1/T at **a)** 31.4 MPa and **b)** 100 MPa.

Orientation of the α_2 crystal lattice to the tensile axis may be as significant as dimensional aspects. In a basketweave microstructure, the laths' orientations are related if they nucleate and grow in the same ß grain. However, the crystal lattices within individual laths might be expected to have a variety of orientations to the creep specimen load axis during a creep test. The colony microstructure in titanium alloys has been suggested to possess a longer effective slip length because of parallel orientation of α or α_2 lath crystal lattices in each packet. Individual dislocations are not able to move from one lath to the next, but the addition of the individual slip lengths within each lath leads to a longer effective slip length. It has been reported that when there is lath alignment resulting in a colony structure in conventional titanium alloys, creep rate increases [13,14]. However, ß-solutionized, slow-cooled Ti-25Al-10Nb-3V-1Mo with a colony-type structure was shown to have creep resistance superior to a basketweave microstructure at all stresses at 760°C [8]. At 650°C, the colony microstructure had the best creep resistance only at "relatively low" stresses, which were actually less than 300 MPa and greater than 150 MPa. These observations, as well as the results of this work, lead to the conclusion that factors other than slip length, such as diffusion rates through the ß phase, are important in describing the creep of these alloys.

The equiaxed α_2 phase presents approximately the same diameter when viewed or stressed from any direction, although some elongation of equiaxed α_2 is present, as seen in Figure 1. The evidence of deformation texture manifested in the (0001) pole figure obtained from the as-received plate, Figure 3, suggests that there is a preferred orientation of basal planes perpendicular or nearly perpendicular to the tensile axis. A possible result of the preferred orientation is the enhanced alignment of crystal lattices from one equiaxed region to another. Thus, it may be possible for slip bands to have effective lengths longer than the diameter of the equiaxed phase. In other words, when dislocations reach the equiaxed α_2 phase boundary, they will most likely not be able to slip into an adjacent area of equiaxed α_2, but it will probably be easier to nucleate dislocations in this adjacent area if there is alignment of slip planes. Also, if there are dislocation sources in adjacent areas of α_2 phase, a reasonable assumption, this will lead to a longer effective slip length, as seen schematically in Figure 8. The direction of elongation of the α_2 phase due to cross-rolling is parallel to the creep load axis of the creep specimen, as a result of the method of machining. Therefore, the basal planes should be oriented at right angles to the long dimension of this phase. Elongation of the equiaxed α_2 phase therefore should not contribute to lengthening of slip distance.

The equiaxed morphology in the cross-rolled plate is present due to deformation of the material, and consequently not seen in undeformed material. Strain introduced into the material with subsequent annealing in the α_2+ß range has caused recrystallization of the material. The structure which results after deformation and annealing consists of a two-phase microstructure

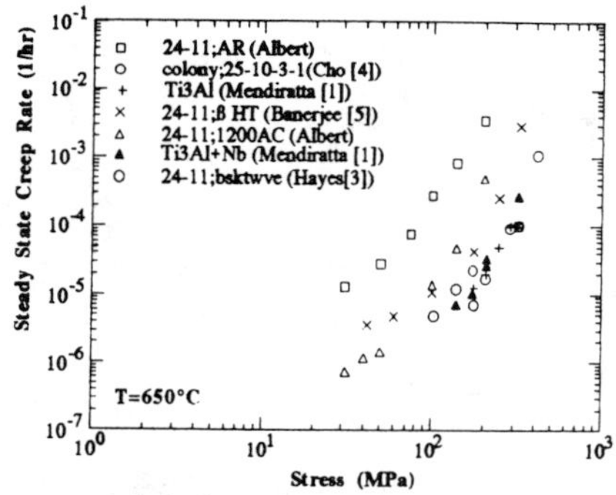

Figure 7. SSCR *vs.* stress at 650°C for various titanium aluminides.

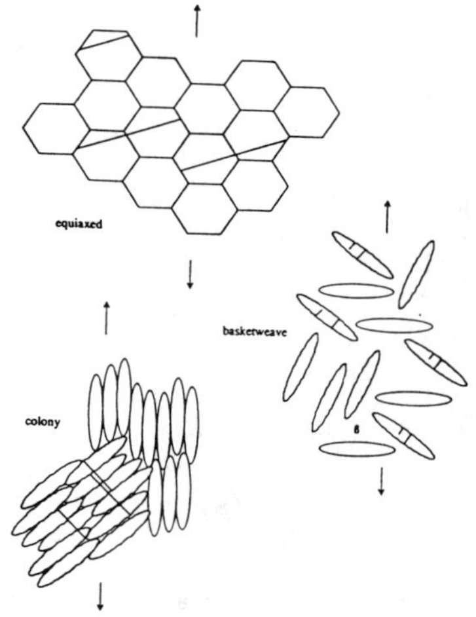

Figure 8. Slip length of three possible α_2 morphologies.

References

1. M. Mendiratta and H.A. Lipsitt: *J. Mat. Sci.*, 1990,Vol. 15, pp. 2985-2990.
2. D.E. Albert and A.W. Thompson: in *Microstructure/Property Relationships in Titanium Aluminides and Alloys*, Y.-W. Kim and R.R. Boyer, eds., TMS, Warrendale, PA, 1991, pp.399-406.
3. R.W. Hayes: *Scripta Met.*, 1989, Vol. 23, pp. 1931-1936.
4. W. Cho, A.W. Thompson and J.C. Williams: *Metall Trans A*, 1990, Vol. 21,pp. 641-51.
5. R.S. Mishra and D. Banerjee: *Mater.Sci. Eng.*, 1990, Vol. A130, pp. 151-164.
6. W.O. Soboyejo, R.J. Lederich, and D.S. Schwartz: in *Microstructure/Property Relationships in Titanium Aluminides and Alloys*, Y.-W. Kim and R.R. Boyer, eds., TMS, Warrendale, PA, 1991, pp. 407-422.
7. E139: in *Annual Book of ASTM Standards*, ASTM, Philadelphia, PA, 1979.
8. D.B. Knorr and N.S. Stoloff: *Materials Sci. and Eng.*, 1990, Vol. A123, pp. 81-87.
9. *Metals Handbook*, Vol. 9, American Society for Metals, Metals Park, OH, 1985.
10. F.C. Dary: PhD Thesis Overview, Carnegie Mellon University, 1991.
11. M.A. Imam and C.M. Gilmore: *Metall. Trans. A*, 1979, Vol. 10, pp. 419-425.
12. S.J. Ashton and L.H. Chambers: in *The Science, Technology and Application of Titanium*, R.I. Jaffee and N.E. Promisel, eds., Pergamon Press, Edinburgh, 1968, pp. 879-889.
13. R.T. Chen, J.W.H. Miller and J.E.A. Starke: in *Titanium Science and Technology*, Vol. 4, G. Lütjering, U. Zwicker and W. Bunk, eds., Deutsche Gesellschaft für Metallkunde e.V., Germany, 1984, pp. 2363-2370.
14. W.H.Miller, R.T. Chen and E.A. Starke: *Metall Trans A*, 1987, Vol. 18, pp. 1451-68.
15. D. Banerjee, A.K.Gogia and T.K.Nandi: *Metall Trans A*, 1990, Vol. 21, pp. 627-639.
16. T. Furuhara and H.I. Aaronson: *Acta Met.*, in press.
17. J.C. Williams and A.W. Thompson: in *Titanium and Titanium Alloys*, Vol. 1, J.C. Williams and A.F. Belov, eds., Plenum Press, New York, NY, 1982, pp. 467-496.
18. H.J. Lee and H.I Aaronson: *J. Mater. Sci.*, 1988, Vol. 23, pp. 150-160.

PHASE EQUILIBRIA IN Ti-Al-Nb ALLOYS NEAR Ti_2NbAl

R. G. Rowe*, D. Banerjee†, K. Muraleedharan†, M. Larsen*,
E.L. Hall*, D.G. Konitzer++ and A.P. Woodfield++

*GE Corporate Research and Development, Schenectady, NY 12301 USA
†Defence Metallurgical Research Laboratory, Hyderabad -500 258, INDIA
++GE Aircraft Engines, 1 Neuman Way, Cincinnati, OH 45215 USA

Abstract

The titanium aluminide Ti_2NbAl (O phase) has been shown to have potential as an elevated temperature structural material, but the mechanical properties of O phase-containing alloys are strongly dependent upon microstructure and phase constitution. Good understanding of phase relations in this system are needed for alloy design. Isothermally aged and water quenched Ti-Al-Nb samples were examined by analytical electron microscopy to determine the identity and composition of constituent phases. Isothermal phase relations between the O (ordered orthorhombic,Cmcm), α_2 ($D0_{19}$) and β_o (B2) phases were determined at 900°C.

Introduction

Recent alloy development to improve both creep resistance and room temperature fracture toughness of Ti_3Al-base alloys has led to the identification of a new class of ordered orthorhombic Ti_2NbAl-base alloys [1-3]. Studies of the mechanical properties of these alloys have shown that they have higher strength, specific strength and fracture toughness than alloys based upon Ti_3Al [3-7]. More active slip systems have been observed in the ordered Ti_2NbAl phase than in Ti_3Al at both room temperature and 650°C [8, 9].

The attractive properties of these alloys have driven a need to understand phase relations and transformations so that the effects of processing and heat treatment can be interpreted on the basis of known phase relations and compositions. This paper is an attempt to establish an equilibrium ternary isothermal phase section for the Ti-Al-Nb system at the important processing temperature of 900°C. A preliminary isothermal section at 900°C was published in reference 10 using alloy A to establish the ternary O+β_o+α_2 phase field. The present study presents results on 24 additional compositions.

Experimental Technique

Alloys were triple non-consumable arc melted to assure homogeneity. Two series were prepared. One series of alloys, M through Z, was drop cast into 12.5 mm diameter copper chill molds and homogenized at 1400°C for 24 hours. These alloys were used to determine the extent of the single phase B2 field at 900°C. The second series of alloys, C through L, was drop cast into 30 mm diameter chill molds then hot die forged or extruded. Alloy A was consumable VAR double melted followed by forging and hot rolling. Alloy B was drop cast and heat treated (there was no extrusion). The second series was homogenized

Titanium '92
Science and Technology
Edited by F.H. Froes and I. Caplan
The Minerals, Metals & Materials Society, 1993

at 1200 or 1250°C and used to establish tie-line compositions between the β_o, O and α_2 phase fields.

Samples were abrasively cleaned and etched with a nitric-hydrofluoric solution to remove surface oxide prior to encapsulation. Alloys C-L were wrapped in Ta foil and welded into CP titanium tubes in an argon glovebox. The titanium tubes were then encapsulated with a piece of yttrium metal in a quartz tube filled with high purity argon gas. 12.5 mm diameter samples of alloys M-Z were wrapped in Ta foil and directly encapsulated in quartz tubes in high purity argon. Encapsulated samples were placed in a furnace after homogenization in the high temperature single phase field and aged at 900°C. After the heat treatment at 900°C, the capsules were broken under water to quench the samples.

Samples of alloys I, J, K and A were water quenched from the single phase B2 or beta phase field for use as bulk composition standards. They were analyzed by electron microprobe (EPMA) using elemental standards. The average composition of remaining alloys was then determined using X-ray fluorescence (XRF) and EPMA of homogenized beta quenched samples using alloy I as a standard. There was good agreement between samples analyzed using both XRF and EPMA. The results of titanium, aluminum and niobium analysis are tabulated in Table 1.

Gas fusion analysis was used to determine the interstitial content of some of the alloys. Oxygen contents ranged from 500 to 1000 wppm. Carbon contents were less than 275 wppm. It should be noted that ultra low interstitial starting materials were not used because results which corresponded to commercial purity materials were desired for alloy design and interpretation. The oxygen contents of alloys U, V and W were analyzed after heat treatment and found to be 1170, 900 and 610 wppm, respectively. This verified that oxygen pickup during heat treatment was not a significant factor. Nonetheless, individual phase composition analyses were made from regions well away from the surface of the samples.

Table 1. Alloy Compositions and analysis technique[1].

Alloy	Composition, at.%	Alloy	Composition, at.%
C	Ti-19.9Al-20.2Nb ••	M	Ti-5.3Al-30.6Nb †
D	Ti-19.5Al-29.6Nb ••	N	Ti-6.1Al-35.0Nb †
E	Ti-24.7Al-29.8Nb •	O	Ti-6.1Al-52.0Nb †
F	Ti-28.5Al-24.1Nb•	P	Ti-8.0Al-37.6Nb †
G	Ti-20.7Al-31.9Nb •	Q	Ti-10.1Al-34.6Nb †
H	Ti-23.2Al-17.1Nb ••	R	Ti-9.9Al-39.2Nb †
I	Ti-26.4Al-23.3Nb †	S	Ti-10.4Al-30.3Nb †
J	Ti-25.8Al-21.0Nb †	T	Ti-9.3Al-44.7Nb †
K	Ti-21.8Al-25.4Nb †	U	Ti-14.0Al-30.7Nb †
L	Ti-22.1Al-22.0Nb •	V	Ti-14.0Al-35.0Nb †
A	Ti-25.2Al-15.7Nb	W	Ti-14.2Al-39.1Nb †
B	Ti-30.8Al-12.5Nb	X	Ti-19.1Al-26.0Nb †
Z	Ti-30.3Al-25.7Nb †	Y	Ti-19.9Al-21.0Nb †

The compositions of individual phases in samples which were isothermally aged at 900°C were established by EPMA or analytical electron microscopy

1 • - Wet chemical analysis
•• - Electron microprobe using Alloy I as a standard
† - X-ray Fluorescence using Alloy I as a standard

(AEM) using appropriate correction factors for absorbtion as described by Muraleedharan, et.al. [10, 11].

Experimental Results

A series of 25 alloys was annealed for times ranging from 24 to 600 hours at 900°C. The heat treatments are tabulated in Table 2. The number of phases in each sample was determined by optical microscopy, backscattered electron microscopy or AEM to establish single, two phase and three-phase regions of the isothermal phase diagram. The crystal structures of the single phase α_2, β_o, and O fields were determined by selected area electron diffraction (SAD) and convergent beam electron diffraction (CBED) of two and three phase samples. The results were plotted on a partial 900°C isothermal Ti-Al-Nb section between Ti, Ti-60Al and Ti-60Nb, Figure 1. The positions of the β_o (B2), O (Cmcm) and α_2 ($D0_{19}$) phase boundaries were based upon the compositions of single phase alloys plus analysis of the volume fractions of the lath and matrix phases in two phase samples near the phase boundaries.

Table 2. Heat treatment and number of phases present after annealing at 900°C.

Alloy	Heat Treatment	Number of Phases
J	1250°C/100 hr/CC + 900°C/600 hr/WQ[1]	1
I	1250°C/100 hr/CC + 900°C/600 hr/WQ	1
F	1250°C/24 hr/CC + 900°C/600 hr/WQ	1
M	1400°C/24 hr/CC + 900°C/600 hr/WQ	1
N	1400°C/24 hr/CC + 900°C/600 hr/WQ	1
O	1400°C/24 hr/CC + 900°C/600 hr/WQ	1
P	1400°C/24 hr/CC + 900°C/600 hr/WQ	1
Q	1400°C/24 hr/CC + 900°C/600 hr/WQ	1
R	1400°C/24 hr/CC + 900°C/600 hr/WQ	1
S	1400°C/24 hr/CC + 900°C/600 hr/WQ	1
T	1400°C/24 hr/CC + 900°C/600 hr/WQ	1
U	1400°C/24 hr/CC + 900°C/600 hr/WQ	2
V	1400°C/24 hr/CC + 900°C/600 hr/WQ	2
W	1400°C/24 hr/CC + 900°C/600 hr/WQ	2
X	1400°C/24 hr/CC + 900°C/600 hr/WQ	2
Y	1400°C/24 hr/CC + 900°C/600 hr/WQ	2
C	1300°C/24 hr/WQ + 900°C/500 hr/WQ	2
D	1300°C/24 hr/WQ + 900°C/500 hr/WQ	2
E	1250°C/100 hr/CC + 900°C/600 hr/WQ	2
G	1250°C/100 hr/CC + 900°C/600 hr/WQ	2
H	1250°C/100 hr/CC + 900°C/600 hr/WQ	2
K	1250°C/100 hr/CC + 900°C/600 hr/WQ	2
L	1250°C/100 hr/CC + 900°C/600 hr/WQ	2
B	1060°C/4 hr/0.1°C/sec + 900°C/16 hr/WQ	2
A	1020°C/6 hr/WQ + 900°C/24 hr/WQ	3
Z	1400°C/24 hr/CC + 900°C/600 hr/WQ	3

1 CC: Encapsulated sample pulled into a water-coooled cooling chamber under Ar gas.
WQ: Water quenched, breaking the quartz capsule.

Only the phase fields which defined the two phase O+β_o and α_2+β_o regions and the three phase O+β_o+α_2 region have been represented in Figure 1. The positions of ternary alpha, omega and gamma phases were not determined. Binary Ti-Nb alloys have a disordered bcc structure at 900°C, but the compositions along the β_o solvus line all had ordered B2 ordering [12]. This indicated a β/β_o boundary which was left undetermined.

The microstructure of the alloy J (Ti-25.8Al-21.0Nb) is shown in Figure 2(a). It was single phase ordered orthorhombic (O) as determined by CBED, Figure 2(b). The symmetry of the crystal structure was consistent with Cmcm [1].

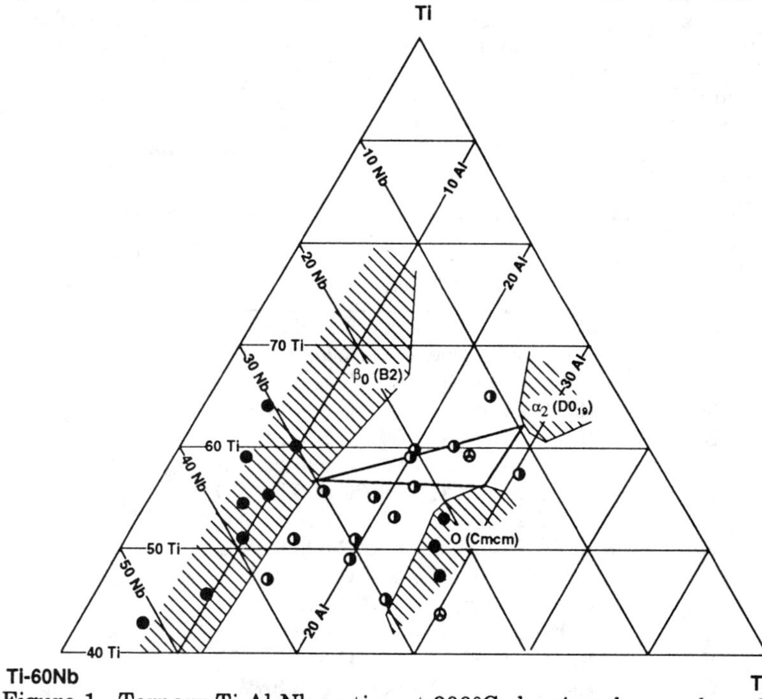

Figure 1. Ternary Ti-Al-Nb section at 900°C showing the number of phases present in annealed samples. The solid points were single phase at 900°C, half-filled points had two phases and open points divided into three segments contained three phases.

Two phase alloys near the O phase field boundary had little β_o matrix phase and a high O phase lath density. An example is alloy K (Ti-21.8Al-25.4Nb), Figure 3. The crystal structures of both the matrix and lath phases of alloy K were identified by CBED. It consisted of O laths in a β_o (B2) matrix. Lower aluminum alloys whose composition was near the β_o phase field boundary had low O phase lath volume fractions. An example was alloy U (Ti-14.0Al-30.7Nb), Figure 3. The highest density of O laths in alloy U was observed at grain boundaries.

Tie-line determinations: The tie lines between the β_o, O and α_2 phases were determined in samples taken from the two and three phase fields between these phases at 900°C. Quantitative AEM analyses were made from both interior lath and matrix phases as well as grain boundary phases using alloys I and A as composition standards. The alloy I standard was also used to determine calibration factors for EPMA and AEM analyses of the composition of individual phases in 900°C heat treated samples.

Continuous O phase plus an O-depleted zone occurred at grain boundaries of most samples. The grain boundary O phase had the same composition as O phase laths in grain interiors. EPMA and AEM analyses agreed. Tie line compositions determined by AEM and EPMA are shown in Figure 5. Tie lines were drawn from a center point denoting the average analyzed composition of the alloy. The tie-line of alloy E was shown dashed because chemical analysis of that alloy was by wet chemical analysis, not XRF or EPMA. Tie line compositions reported by Kestner-Weykamp, et.al [13] which were determined by annealing the alloy Ti-24Al-11Nb for 24 hrs. at 900°C were included.

Figure 2(a). Transmission electron micrograph of alloy J (Ti-25.8Al-21.0Nb) annealed 600 hours at 900°C.

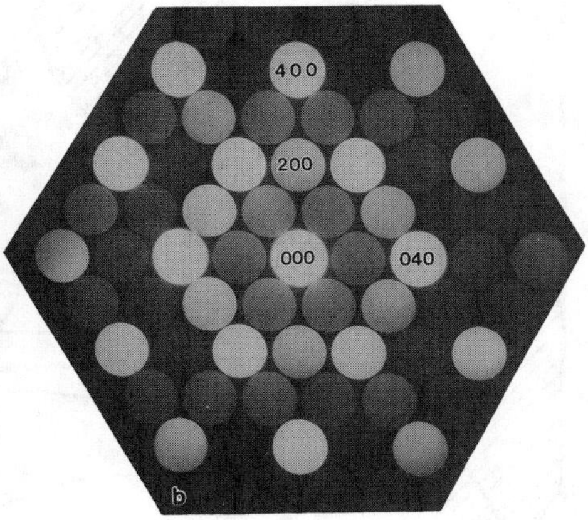

Figure 2(b). Convergent beam electron diffraction pattern of the basal zone of the O phase observed in alloy J.

Figure 3. Transmission electron micrograph of alloy K (Ti-21.8Al-25.4Nb) annealed 600 hrs at 900°C.

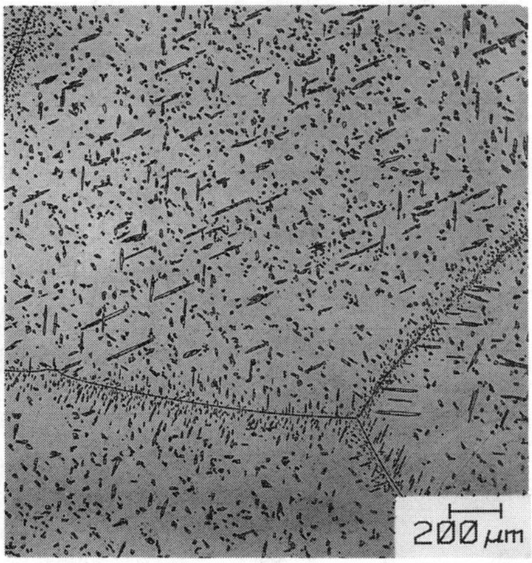

Figure 4 Optical micrograph of alloy U (Ti-14.0Al-30.7Nb) which was annealed 600 hours at 900°C.

The identities of the phases in Figure 5 are denoted by the symbols at the ends of the tie lines. Filled circles denote O phase, cross-hatched circles denote β_o phase and shaded circles denote α_2 phase. Phase identifications were by CBED and

SAD. Phase identity was consistent with the phase boundaries defined in Figure 1. Tie lines determined only by electron microprobe are shown terminated by open circles.

Tie lines were not straight lines through the average composition. Tie-line β_o phase aluminum contents did not agree with the phase boundaries determined from single phase samples and the estimates of the phase boundary position based upon the volume fractions of lath phase in two phase alloys near the β_o phase boundary. In contrast, there was good agreement between O phase tie line compositions and the phase boundary defined by single phase O samples.

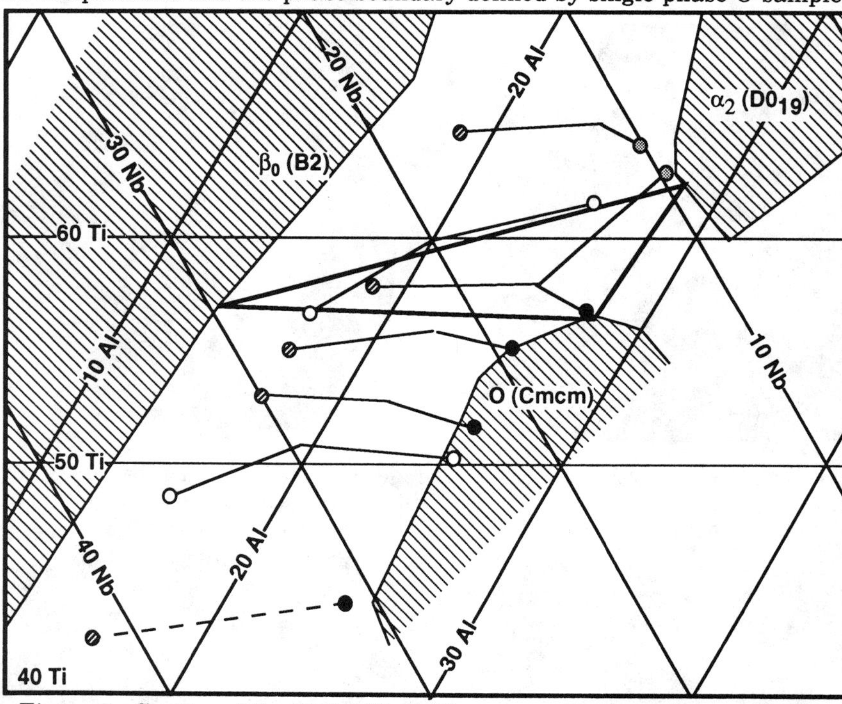

Figure 5. Section of the 900°C Ti-Al-Nb ternary diagram showing tie line compositions determined by EPMA and AEM. The tie line results are superimposed with phase boundaries from Figure 1.

Table 3. Tie-line compositions for alloys J and K at 1000°C.

	Average Composition	Heat Treatment	O Phase at.%	β_o Phase at.%
J	Ti-25.8Al-21.0Nb	1250°C/100 hr + 900°C/600 hr/WQ	Ti-27.2Al-19.6Nb	Ti-25.0Al-23.0Nb
K	Ti-21.8Al-25.4Nb	1250°C/100 hr + 900°C/600 hr/WQ	Ti-25.5Al-15.7Nb	Ti-21.8Al-25.3Nb

At 1000°C, alloys J and K both lay in the two phase β_o+O phase field. The tie-line compositions between β_o and O for alloys J and K, which were given the same homogenization treatment and aged for 600 hours at 1000°C, are shown in Table

3. The important observation from this 1000°C data is that the aluminum content of the β_0 phase of both alloys is close to the average composition of each alloy. This suggests that it was difficult to achieve an equilibrium β_0 aluminum content in 600 hours, even at an annealing temperature of 1000°C.

Discussion of Results

There were several indications that the β_0 composition of the two and three phase samples annealed 600 hours at 900°C had not fully equilibrated. The first was that the aluminum composition of the β_0 tie-line compositions was 3 to 4 at.% higher than the composition determined by interpolation between alloys straddling the phase boundary. It was also apparent that tie lines between O and β_0 phase compositions did not pass through the average composition of the alloy. Third, there were indications that even at 1000°C, equilibrium aluminum content in the β_0 phase was hard to achieve over large compositional separations between the average alloy composition and the β_0 phase boundary.

We hypothesize that the reason for the lack of equilibration of the aluminum content in the β_0 phase is a low diffusivity of Al in the β_0 phase. Anti-phase domains (APD's) were not observed in the β_0 phase (B2) after quenching, and from this, we inferred that the β_0 phase was ordered at 900°C. If the β_0 phase had ordered upon quenching one would expect to see fine APD's [13, 14].

The O phase compositions that were determined from the tie-lines agreed with O phase compositions of single phase annealed samples I and J. This suggests that the compositions defined by the tie lines of alloys E, D, K, and L indicate the O phase boundary. It shows that the Nb solubility range of the O phase is large, approximately 15 at.%. The lowest Al content for the O phase was approximately 25 at.% over this Nb range, and the highest Al content was less than 30 at.%. This lower range of solubility for aluminum relative to niobium may be due to the site occupation of Al and Nb in the O phase structure. Al atoms have been shown to exclusively occupy one set of sites, while Nb atoms are preferred for one set of Ti,Nb subsites but occupy a fraction of the other subsites [1, 2]. It would thus appear that variation in the ratio of niobium to titanium atoms may be accomodated by adjustment of the occupation fraction of niobium between these two subsites.

Some samples had a thick grain boundary O phase or a higher lath density near grain boundaries than in grain interiors. The composition of the grain boundary O phase was the same as the composition of thick grain interior laths, however, suggesting that nucleation of O phase rather than diffusion kinetics led to this variation in O phase volume fraction and growth mechanism.

Conclusions

- It has been established that at 900°C, an ordered orthorhombic phase with Cmcm symmetry lies between 25 and 28 at.% Al for Nb contents of 15 to 30 at.%.
- The β_0 phase had an ordered B2 (CsCl structure) crystal structure for compositions near the aluminum solubility limit for this phase (approximately 13-15 at.% Al).
- Single and two phase samples which had average compositions near the β_0 phase boundary composition were used to define the β_0 boundary. Compositional shifts for the β_0 phase were not large in these alloys so that precipitation of O phase indicated the equilibrium boundary position.
- Disagreement between the β_0 phase boundary compositions and β_0 tie-line compositions suggested that equilibration of the β_0 phase was not achieved in 600 hours at 900°C over the 10 to 12 at.% Al composition range between the β_0 and O phases.

References

1. D. Banerjee, *et al.*, "A New Ordered Orthorhombic Phase in a Ti_3Al-Nb Alloy", Acta Metall., 36, 871-882, (1989).
2. B. Mozer, *et al.*, "Neutron Powder Diffraction Study of the Orthorhombic Ti_2AlNb Phase", Scripta Met. et Mater., 24, 2363-2368, (1990).
3. R.G. Rowe, "The Mechanical Properties of Titanium Aluminides Near Ti-25Al-25Nb". in Microstructure/Property Relationships in Titanium Alloys and Titanium Aluminides. Edited by Y. Kim and R.R. Boyer, TMS, AIME, Warrendale, OH, 1990, 387-398
4. R.G. Rowe, "Mechanical Properties of Ternary and Quaternary Ti_2NbAl-based Titanium Aluminide Alloys". in Proceedings, Seventh World Conference on Titanium, San Diego, CA., TMS-AIME, Warrendale, PA, 1992.
5. R.G. Rowe, "Tri-Titanium Aluminide Alloys Containing at Least Eighteen Atom Percent Niobium", U.S. Patent # 5,032,357, Jul. 16, 1991.
6. R.G. Rowe, "Advanced Ti_3Al-base Alloy Property Comparison". in Synthesis, Processing and Modelling of Advanced Materials. Edited by F.H. Froes and T. Khan, ASM International, Paris, France, 1991, To be published
7. R.G. Rowe, *et al.*, "Tensile and Creep Behavior of Ordered Orthorhombic Ti2AlNb-Based Alloys", Mat. Res. Soc. Symp. Proc, 213, 703-708, (1991).
8. D. Banerjee and R.G. Rowe, "Deformation of the Orthorhombic Phase in Ti-Al-Nb Alloys". in Proceedings, Seventh World Conference on Titanium, San Diego, CA. TMS-AIME, Warrendale, PA, 1992.
9. D. Banerjee, R.G. Rowe, and E.L. Hall, "Deformation of the Orthorhombic Phase in Ti-Al-Nb Alloys", Mat. Res. Soc. Symp. Proc, 213, 285-290, (1991).
10. K. Muraleedharan, *et al.*, "Transformations in a Ti-24Al-15Nb Alloy, Part I. Phase Equilibria and Microstructure", Metall. Trans. A, 23A, 401-415, (1992).
11. K. Muraleedharan and D. Banerjee, "Alloy Partitioning in Ti-24Al-11Nb by Analytical Electron Microscopy", Metall. Trans. A, 20A, 1139-1142, (1989).
12. J.L. Murray, Phase Diagrams of Binary Titanium Alloys, ASM International 1987, Metals Park, OH.
13. H.T. Kestner-Weykamp, *et al.*, Scripta Met., 23, 1697-1702 (1989).
14. R. Strychor and J.C. Williams, "Phase Transformations in Ti-Al-Nb Alloys". in Proceedings of the International Conference on Solid-Solid Phase Transformations. Edited by H.I. Aaronson, *et al.*, TMS-AIME, Warrendale, PA, 1982, 249-253

Acknowledgements

The author would like to acknowledge the assistance of Ms. R. L. Casey and Mr. R. J. Zabala in all aspects of the experimental work. X-ray fluorescence analyses were done at GE CR&D by Ms. C.J. Hayden and EPMA analyses were performed by Mr. L.A. Peluso. This work was supported at GE Corporate R&D by GE Aircraft Engines under the Independent Research and Development Program and by the Defence Metallurgical Research Laboratory, Hyderabad, India.

STRUCTURE-PROPERTY RELATIONSHIPS OF INVESTMENT CAST GAMMA TITANIUM ALUMINIDES

D.A. Wheeler, B. London, and D.E. Larsen, Jr.

Howmet Corporation
Operhall Research Center
1500 South Warner Street
Whitehall, MI 49461

Abstract

Investment cast gamma titanium aluminides are candidate materials for use within advanced aircraft engine designs due to their low density and elevated temperature capabilities. To better define the role of these materials for such applications, the relationship between microstructure and mechanical properties have been characterized for several cast gamma alloy compositions. In this paper, the relative effects of equiaxed gamma and lamellar (alpha2/gamma) constituents on room and elevated temperature tensile properties, fracture toughness, and creep behavior are discussed. It is shown that the presence of lamellar colonies increases the fracture toughness and creep resistance of cast gamma alloys, while maximum tensile ductilities are associated with the equiaxed gamma phase. Therefore, a duplex microstructure containing relatively equal fractions of equiaxed gamma and lamellar constituents results in the best overall balance of mechanical properties.

Introduction

Monolithic gamma titanium aluminides are candidate materials for the replacement of nickel-based superalloys in selected aircraft engine components due to their low density, high modulus and strength at elevated temperatures, and good environmental resistance (1). However, their poor room temperature ductility and impact resistance have traditionally limited their widespread use within advanced engine designs. In recent years, the development of new alloy compositions has created considerable interest in gamma alloys for selected applications, such as low pressure turbine blades and structural components (2). Yet, the successful manufacture of these components by wrought processing methods is rather difficult due to the relatively low ductility of these materials at moderate hot working temperatures. By contrast, investment casting technology offers several advantages for the

Titanium '92
Science and Technology
Edited by F.H. Froes and I. Caplan
The Minerals, Metals & Materials Society, 1993

processing of gamma alloys; namely, the elimination of complex thermal processing sequences and the ability to manufacture near-net shape components (3-5). Unfortunately, the vast majority of mechanical property measurements on gamma alloys have been performed on wrought processed material. Therefore, an effort was undertaken to expand our knowledge of the relationships between the microstructures and mechanical properties of common gamma alloys in the investment cast product form. This paper will summarize these efforts, with particular attention paid to the effects of microstructural constituents on the tensile, fracture toughness, and creep properties of these materials.

Microstructural Development in Cast Gamma Alloys

Preliminary efforts targeted the identification of baseline heat treatments to produce a full range of microstructures in common two-phase gamma alloy compositions. Prior to heat treatment, all cast material is subjected to a hot isostatic pressing (HIP) cycle at 1260°C/172 MPa for 4 hours to close any microporosity retained from the casting operation. To tailor the relative fractions of equiaxed gamma phase and lamellar (alpha2/gamma) constituent, elevated temperature heat treatments in the alpha+gamma phase field are utilized. These investigations resulted in the selection of the following heat treatments for further mechanical property evaluation:

Identification	Heat Treatment	Microstructure
A	1300°C/20h/GFC	Equiaxed γ
C	1350°C/0.5h/GFC	Lamellar($\alpha 2/\gamma$)
F	1350°C/0.5h/1300°C/10h/GFC	Lamellar + γ
	(GFC=gas fan cooling at 1°C/s)	

Typical microstructures produced by these heat treatments are shown in Figure 1 for Ti-48Al-2Nb-2Cr (at.%), providing evidence that cast gamma alloys can be readily manipulated to produce a wide variety of microstructures. Of particular importance is the relative size of the cast, HIP'ed and heat treated microstructures, which are generally coarser than typical wrought-processed material. The microstructures shown in Figure 1 and the mechanical properties cited below were taken from the center portion of 15.8 mm diameter cast test bars. As expected, the microstructural feature size of cast gamma titanium aluminides is known to scale directly with the cast section size (6). Therefore, the resulting mechanical properties of cast gamma alloys can be somewhat influenced by the section size of the casting being evaluated and should be interpreted accordingly.

<u>Heat Treat 'A'</u>
1300°C/20hrs/GFC

<u>Heat Treat 'C'</u>
1350°C/0.5hr/GFC

<u>Heat Treat 'F'</u>
1350°C/0.5hr/GFC to
1300°C/10hrs/GFC

Figure 1: Characteristic microstructures of cast, HIP'ed and heat treated Ti-48Al-2Nb-2Cr (at.%)

Mechanical Properties of Cast Gamma Alloys

Tensile Properties

Room and elevated temperature tensile properties of several gamma alloy compositions are given in Table I as a function of initial microstructure. These data indicate that the tensile ductility of cast gamma is maximized by the presence of the equiaxed gamma phase. This result is rationalized by the observation of reduced transfer of slip and twinning dislocations across alpha2 laths within the lamellar colonies when compared to the dislocation activity within the equiaxed gamma grains (7). However, the overall strength and ductility of the cast alloys are approximately 40-50% lower than material processed by wrought or powder metallurgy techniques (8,9). This is likely due to the generally increased grain size of the cast product form relative to typical wrought-processed material.

Table I - Room and Elevated Temperature Tensile Properties of Cast Gamma Titanium Aluminide Alloys

Alloy Composition (at%)	Test Temperature (°C)	Heat Treatment	Yield Strength (MPa)	Ultimate Strength (MPa)	Plastic Elongation (%)
Ti-48Al-2V	20	A	338	475	1.6
		C	357	414	0.5
		F	303	386	0.9
Ti-48Al-2Nb-2Cr	20	A	324	443	2.3
		C	336	432	1.5
		F	292	397	1.6
	760	A	268	391	58.0
		F	301	426	57.2
Ti-48Al-2Mn-2Nb	20	A	276	422	2.2
		C	328	391	0.9
		F	277	406	2.0
	760	A	258	384	40.7
		F	278	409	36.7

Fracture Toughness

The effect of microstructure on the room temperature short rod fracture toughness values of several gamma alloys is shown in Figure 2. In general, the presence of lamellar colonies increases the fracture toughness of cast gamma alloys to levels approaching, if not exceeding, that measured for wrought-processed material (8). The measured increase in toughness is attributed to the formation of shear ligaments due to significant crack deflection and branching during the fracture of

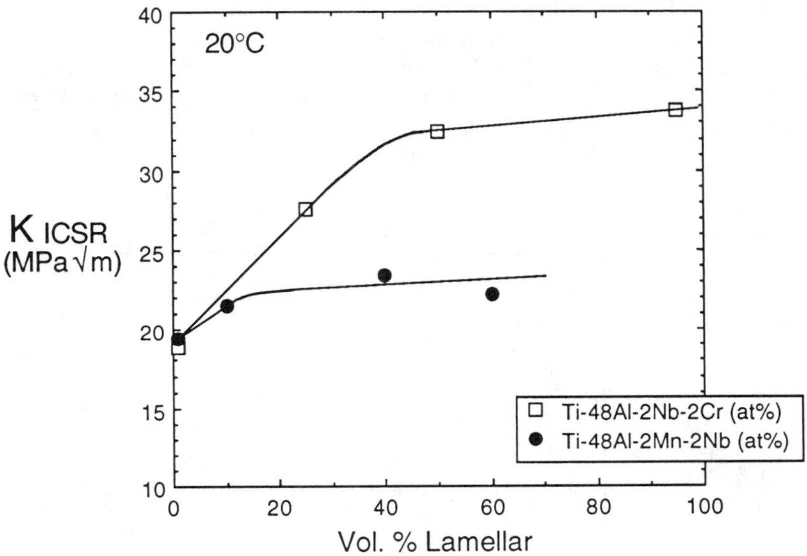

Figure 2: Effect of lamellar content on the room temperature short rod fracture toughness of cast gamma

Figure 3: Fracture surface of Ti-48Al-2Nb-2Cr (at.%) room temperature K_{ICSR} specimen, HT F (K_{ICSR} = 32 MPa√m)

the lamellar colonies (10). Figure 3 provides support for this premise, as the low energy cleavage failure of equiaxed gamma grains is accompanied by the relatively high energy fibrous tearing of adjacent lamellar colonies. It is important to note, however, that a minimal presence of lamellar constituent can produce a significant increase in measured fracture toughness. Therefore, a mixture of equiaxed gamma and lamellar constituents would result in satisfactory tensile ductility without a great sacrifice in room temperature fracture toughness.

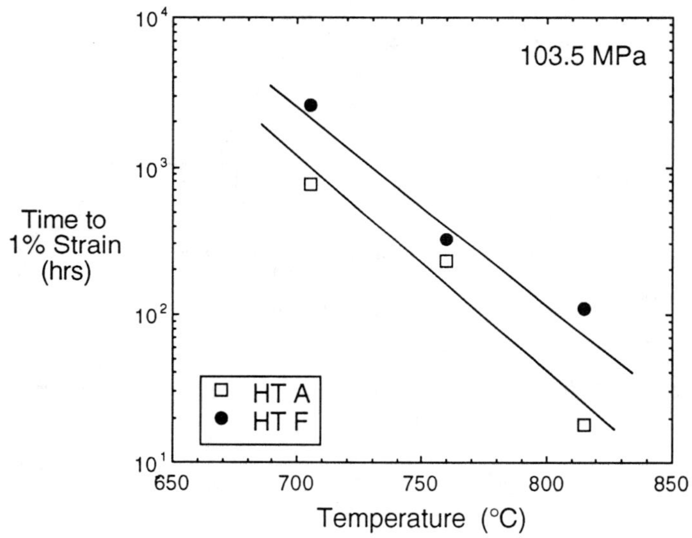

Figure 4: Effect of microstructure on the creep resistance of cast Ti-48Al-2Nb-2Cr (at.%)

<u>Elevated Temperature Creep Behavior</u>

The effect of microstructure on the elevated temperature creep behavior of Ti-48Al-2Nb-2Cr (at.%) is shown in Figure 4. The presence of lamellar colonies increases the overall resistance of the material to creep deformation, presumably due to the difficulty in moving dislocations through the lamellar laths. A similar result is obtained when performing more detailed steady state creep experiments as a function of temperature and stress level, as shown in Figure 5. It is interesting to note the change in stress exponent, n, from n=3 to n=7 at high temperature and stress levels in the duplex microstructure, indicating a change in controlling creep mechanism from diffusional processes at low stress levels to dislocation processes at higher stress levels (11). By comparison, these data indicate that the creep resistance of cast gamma alloys is comparable to that measured for the wrought product form (7).

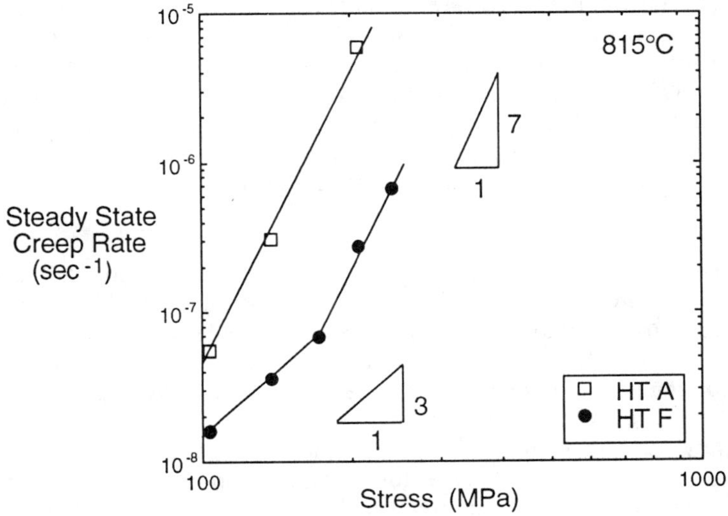

Figure 5: Effect of microstructure on the power-law creep behavior of cast Ti-48Al-2Nb-2Cr (at.%)

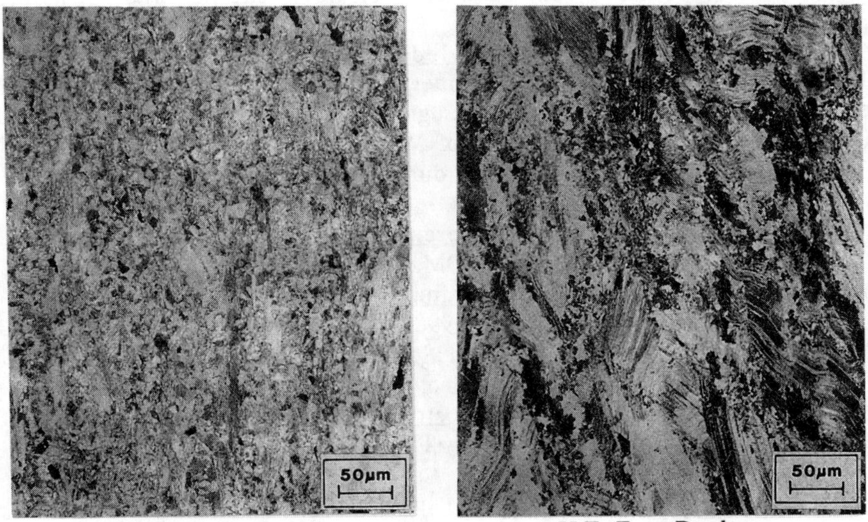

H.T. A - Equiaxed γ H.T. F - Duplex

Figure 6: Dynamic recrystallization in the uniform gage section of Ti-48Al-2Nb-2Cr (at.%) creep specimens deformed to 10% plastic strain at 815°C

In Figure 6, longitudinal sections of specimens tested to failure at 815°C reveal significant amounts of dynamic recrystallization. As expected, the extent of recrystallization is decreased in microstructures containing lamellar constituent. Similar metallographic examinations of specimens tested at 760°C did not reveal recrystallization. This result suggests that a critical temperature exists below which the microstructure of cast gamma alloys is stable during long-term elevated temperature exposures. Of course, since dynamic recrystallization is a function of both temperature and stored strain energy, the critical temperature for recrystallization would be somewhat higher than indicated here for material experiencing lower overall plastic strains.

Conclusions

It has been demonstrated that cast gamma titanium aluminides can be heat treated to produce a wide range of microstructures. The presence of equiaxed gamma phase strongly increases tensile ductility, while the presence of lamellar colonies produces greater room temperature fracture toughness and elevated temperature creep resistance. As a result, the best overall combination of mechanical properties is attained by a mixture of equiaxed gamma and lamellar constituents. While cast gamma alloys are shown to have lower tensile properties than wrought-processed material, their fracture toughness and elevated temperature creep properties are comparable. These properties, coupled with its demonstrated near-net shape capabilities, make investment casting a viable processing method for the production of gamma titanium aluminide components for advanced aircraft engine and structural applications.

Acknowledgments

This work was supported by Howmet Corporation internal research and development funds. The authors gratefully acknowledge the helpful discussions with Prof. H.A. Lipsitt of Wright State University and Prof. T. Bieler of Michigan State University during the course of this work.

References

1) Y-W. Kim and D.M. Dimiduk, JOM, vol. 8, (1991), pg. 40
2) J.M. Larsen, etal., High Temperature Aluminides and Intermetallics, TMS, S.H. Whang, etal., eds., (1990), pg. 521
3) D.E. Larsen, Jr., etal., Materials Science and Engineering, vol. A144, (1991), pg. 45
4) B. London and T.J. Kelly, Microstructure/Property Relationships in Titanium Aluminides and Alloys, TMS, Y-W. Kim and R.R. Boyer, eds., (1991), pg. 285
5) S. Sen and D.M. Stefanescu, JOM, vol. 5, (1991), pg. 30
6) D.A. Wheeler, unpublished research, Howmet Corporation, 1992
7) D.S. Shih, etal., Microstructure/Property Relationships in Titanium Aluminides and Alloys, TMS, Y-W. Kim and R.R. Boyer, eds., (1991), pg. 135
8) S-C. Huang, etal., Intermetallic Compounds-Structure and Mechanical Properties, JIMIS-6, O. Izumi, (1991), pg. 363
9) Y-W. Kim, JOM, Vol. 7, (1989), pg. 24
10) K.S. Chan and Y-W. Kim, Met. Trans. A, vol. 23, (1992), pg. 1663
11) D.A. Wheeler, etal., Scripta Met., vol. 26, (1992), pg. 939

THE EFFECT OF PROCESSING ON THE MICROSTRUCTURE AND TENSILE PROPERTIES OF A γ-TiAl BASED ALLOY

G.E. Fuchs

General Electric Company
P.O. Box 1072
Schenectady, NY 12301-1072

Abstract

A great deal of effort has been devoted to the ductilization of γ-TiAl based alloys. Compositional and microstructural control, and alloying additions have been the primary methods used to increase the low temperature ductility of TiAl-based alloys. However, very little has been done to characterize the effect of processing on ductile γ-TiAl alloys. This study examines the effect of processing on the microstructure and properties of cast, wrought and powder metallurgy TiAl-based alloys with a similar nominal composition. The alloy selected for evaluation was Ti-48Al-2Nb-2Cr (at%). After processing, all of the samples were heat treated in the $\alpha + \gamma$ phase field. The microstructures were characterized by optical, scanning electron and transmission electron microscopy. The tensile properties were evaluated in the temperature range 25°-1000°C in air. The powder processed alloys exhibited the highest strength and ductility at test temperatures up to 850°C. All of the alloys exhibited similar tensile properties at 1000°C. Examination of the fracture surfaces indicated that the tensile failure occurred by predominently transgranular cleavage at temperatures below 700°C. However, at temperatures above 850°C, significant plastic deformation was observed, particularly in the powder metallurgy samples. The inter-relationships of processing-microstructure-mechanical properties of Ti-48Al-2Nb-2Cr (at%) will be discussed.

Introduction

Due to their low density, high strength, and oxidation resistance, TiAl-based alloys are candidate materials for elevated temperature applications [1]. The vast majority of the work reported in the open literature is based on cast and wrought ingot metallurgy (I/M) materials. However, cast TiAl-based alloys are difficult to hot work due to the high strength of the material and the inhomogeneous microstructure observed in the cast condition. In order to homogenize the microstructure, high temperature solution anneals are required. After these solution anneals, the segregated cast structure is reduced or eliminated. However, the grain size is often very large (>1mm). High temperature (>1100°C), isothermal processing is then necessary to break down this microstructure. It should be noted that powder metallugy (P/M) processing is often utilized in other alloy systems in order to increase the homogeneity, refine

Titanium '92
Science and Technology
Edited by F.H. Froes and I. Caplan
The Minerals, Metals & Materials Society, 1993

the microstructure, improve mechanical properties and increase fabricability. Similar benefits from P/M processing also might be expected in TiAl-based alloys.

The purpose of this study is to evaluate the microstructure and tensile properties of I/M and P/M processed near gamma (α_2/γ) alloys with similar nominal compositions - Ti-48Al-2Nb-2Cr (at%). Samples were prepared by several I/M and P/M techiques, heat treated to similar microstructures and tensile tested. The fractured samples were examined by SEM and TEM.

Experimental Procedures

Materials with a nominal composition of Ti-48Al-2Nb-2Cr (at%) were prepared by both I/M and P/M processing techniques (Table 1). The I/M materials were vacuum induction skull melted at Duriron Company, Dayton, Ohio. Alloy TA-40B was investment cast into 1.27 cm diameter x 15.24 cm cylinders and was HIPed (1260°C/172MPa/4hrs) in order to eliminate any casting porosity. A second alloy (TA-5) was cast into a 7 cm diameter x 91 cm cylindrical graphite ingot mold and HIPed (1175°C/103MPa/4hrs). One portion of the ingot was utilized to make the Plasma Rotating Electrode Process (PREP) powder, as discussed below. The remaining material was cut in half and homogenized at two different temperatures prior to isothermal forging. Samples designated TA-5A were homogenized at 1400°C for 5 hours and TA-5D samples were homogenized at 1200°C for 48 hours. The 1200°C heat treatment (TA-5D) resulted in a fine grained, duplex material. However, the variations in grain size observed in this sample (TA-5D) indicated that some of the segregation and cast structure remained after this heat treatment. The 1400°C heat treatment eliminated any evidence of the cast structure; however, the grain size of this material had increased significantly. The materials were then isothermally forged at Wyman Gordon Co., North Grafton, MA (1175°C/0.1min^{-1}/75%).

The P/M samples were produced by Plasma Rotating Electrode Process (PREP) and gas atomization (GA) techniques. A PREP electrode was prepared from the TA-5 ingot. The electrode was atomized by Nuclear Metals Inc. (NMI), Concord, MA. The -35 mesh (<500μm) powder was canned and HIPed (1230°C/103MPa/4hrs) to full density (TA-16A). In addition, a heat of powder was produced by gas atomization (GA) at Crucible Compaction Metals (CCM), Oakdale, PA. Approximately half of the -35 mesh (<500μm) yield was canned and HIPed (1230°C/103MPa/4hrs) to full density (TA-11A). The remainder of the powder was canned and direct extruded (1290°C/16:1, TA-43A).

All of the samples, except the cast (TA-40B) material, were given a 1300°C/2hrs/Furnace Cool (FC) to 900°C/6hrs/FC heat treatment. The TA-40B samples were given a 1300°C/6hrs/FC to 900°C/6hrs/FC heat treatment.

Metallographic samples were examined by optical metallography, scanning electron microscopy (SEM) and transmission electron microscopy (TEM). Threaded tensile samples with a 3.81mm diameter x 1.27 cm gage section were tested in air in the temperature range 25°-1000°C at an initial strain rate of 0.1/minute. The fractured samples were then examined by SEM and TEM in order to determine the fracture mechanisms.

Table 1
Composition (in at%) of Ti-48Al-2Nb-2Cr Alloys

Alloy ID	Ti	Al	Nb	Cr	O_2	H_2	N_2
TA-5A TA-5D	Bal	48.59	1.91	2.03	0.158	0.058	0.025
TA-11A TA-43A	Bal	48.25	2.00	1.78	0.243	0.122	0.015
TA-16A	Bal	48.32	2.01	2.21	0.162	0.062	0.023
TA-40B	Bal	46.50	1.94	2.14	0.160	0.064	0.029

Table 2
Grain Size of Heat Treated Ti-48Al-2Nb-2Cr samples.
Heat treatment:
TA-5A,D, TA-11A, TA-16A, TA-43A - 1300°C/2hrs/FC→900°C/6hrs/FC
TA-40B - 1300°C/6hrs/FC→900°C/6hrs/FC

Alloy ID	Processing	Grain size (μm)	Comments
TA-5A	I/M + Forge	34	Equiaxed γ grains, fine α_2 grains
TA-5D	I/M + Forge	41	Equiaxed γ grains, fine α_2 grains
TA-11A	GA + HIP	43	Equiaxed γ grains, fine α_2 grains
TA-16A	PREP + HIP	40	Equiaxed γ grains, fine α_2 grains
TA-40B	Inv. Cast	120	Coarse Duplex micrsotructure (30μm γ grains, $\geq$300μm α_2/γ lamellar grains)
TA-43A	GA + Extrude	24	Fine equiaxed γ grains, fine α_2 grains

Results and Discussion

Microstructures

In all cases, the microstructure of each alloy was comprised of both the γ-TiAl phase and the α_2-Ti_3Al phase (Figure 1 and Table 2). The heat treatment in the $\alpha + \gamma$ phase field was intended to produce a duplex microstructure consisting of a mixture of single phase γ grains and lamellar α_2/γ grains. The single phase γ grains were observed in all of the alloys. However, the α_2 phase was not always present in the form of lamellar α_2/γ grains. The heat treated cast + forged I/M alloys, TA-5A and TA-5D, exhibited a two phase $\gamma + \alpha_2$ microstructure, with the α_2 present in the microstructure as small single phase grains. This type of equiaxed duplex microstructure has been reported [2] in near gamma alloys heat treated close to the γ-solvus or heat treated with direct cooling from the solution heat treatment temperature to the aging temperature [10]. The material homogenized at 1400°C (TA-5A) exhibited a finer grain size (34μm) than the material heat treated at 1200°C (TA-5D, 41μm). The reason for this difference in grain size due to homogenization treatment is not clear and additional work is required. The P/M alloys also exhibited an equiaxed duplex microstructure. The grain size of the HIPed P/M alloys were similar to the cast + forged I/M materials. The grain sizes of the HIPed P/M materials were very similar for both the GA (TA-11A, 43μm) and the PREP (TA-16A, 40μm) alloys. The extruded GA alloy (TA-43A) exhibited the finest grain size. The investment cast materials exhibited the largest grain size of any of the materials evaluated. In addition, the solidification microstructure was still present in TA-40B after the 6 hour exposure at 1300°C. The average grain size was approximately 120μm; however, the equiaxed γ and lamellar α_2/γ duplex microstructure had a bi-modal grain size. The single phase γ grains were quite small in size ($\approx$30μm); whereas, the lamellar grain size was approximately 300μm.

Tensile Properties

The tensile properties of the alloys were determined at 25°, 600°, 700°, 850° and 1000°C in air. All of the alloys, except the I/M + forged alloys, possessed at least 2% plastic strain at all test temperatures (Table 3 and Figure 2). This level of ductility at room temperature is in agreement with published data for I/M Ti-48Al-2Nb-2Cr alloys [2,3]. However, the ductility of the P/M alloys and the investment cast alloy are significantly higher than the cast + forged I/M alloys (TA-5A and TA-5D). The increased ductility observed in the investment cast alloy may be due to the low Al content (Table 1) of the alloy or the difference in the microstructures (equiaxed duplex versus lamellar duplex). The increased homogeneity and more uniform microstructures may be responsible for the greater ductility exhibited by the P/M alloys. The strength of the P/M alloys was significantly higher than the I/M alloys at test temperatures up to 850°C (Table 3 and Figure 2). In particular, the extruded GA alloy (TA-43A) exhibited the highest strength of all of the materials tested. The HIPed PREP (TA-16A) and GA (TA-11A) material exhibited superior strength and ductility in comparison to the cast + forged (TA-5A and TA-5D) samples. The higher strengths and ductilities observed in the P/M alloys probably resulted from the reduced grain size and increased homogeniety of the P/M materials. Typically, increasing the oxygen content in TiAl-based alloys results in increased strength and decreased ductility [4-5]. However, the P/M alloys, with higher oxygen contents, exhibited higher strengths and greater ductility than the I/M alloys. Additional work is required to determine the effect of oxygen content on the properties of TiAl-based alloys. All of the alloys exhibited similar tensile properties at 1000°C. All of the alloys tested also exhibited a slight increase in the yield strength with increasing temperatures (Figure 2). This temperature effect has been observed in large grained single phase γ-TiAl alloys [7-9], but has not been reported in fine grained, near gamma alloys.

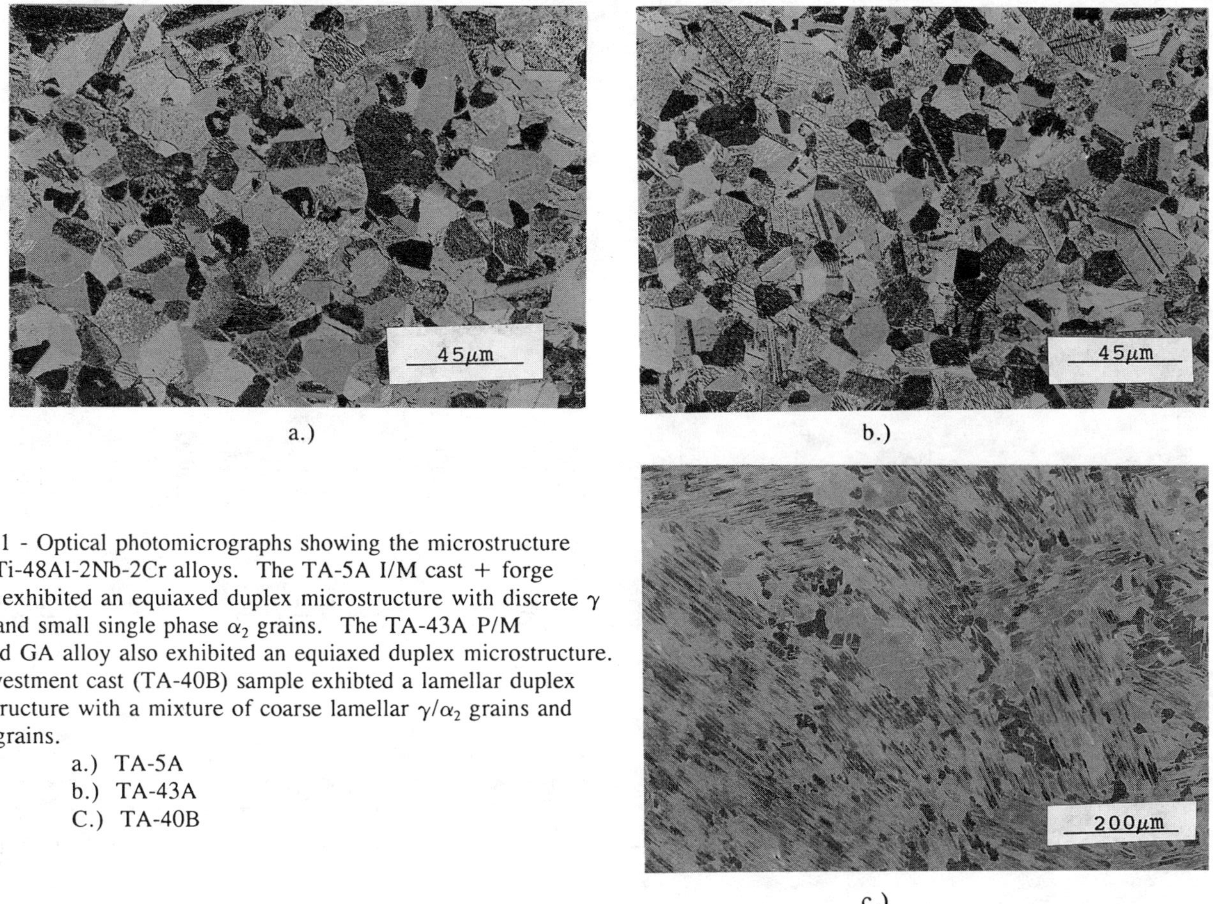

Figure 1 - Optical photomicrographs showing the microstructure of the Ti-48Al-2Nb-2Cr alloys. The TA-5A I/M cast + forge sample exhibited an equiaxed duplex microstructure with discrete γ grains and small single phase α_2 grains. The TA-43A P/M extruded GA alloy also exhibited an equiaxed duplex microstructure. The investment cast (TA-40B) sample exhibted a lamellar duplex microstructure with a mixture of coarse lamellar γ/α_2 grains and fine γ grains.

a.) TA-5A
b.) TA-43A
C.) TA-40B

Table 3

Tensile Properties of Ti-48Al-2Nb-2Cr Alloys (YS, UTS units: MPa)

Room Temperature

Alloy ID	0.2% YS	UTS	Elongation	RA
TA-5A	377	443	1.9%	3%
TA-5D	348	362	0.8%	--
TA-11A	419	489	3.2%	--
TA-16A	359	443	2.0%	--
TA-40B	401	521	3.6%	3%
TA-43A	510	597	2.9%	2%

600°C

Alloy ID	0.2% YS	UTS	Elongation	RA
TA-5A	334	429	2.1%	2%
TA-5D	312	471	4.6%	4.4%
TA-11A	334	432	3.4%	2%
TA-16A	Not Tested At This Temperature			
TA-40B	319	514	5.8%	4.4%
TA-43A	414	568	3.7%	3%

700°C

Alloy ID	0.2% YS	UTS	Elongation	RA
TA-5A	345	455	3.7%	--
TA-5D	321	450	4.0%	2.6%
TA-11A	348	501	6.3%	6%
TA-16A	316	452	3.8%	5%
TA-40B	331	523	5.8%	6.9%
TA-43A	421	581	5.2%	3%

850°C

Alloy ID	0.2% YS	UTS	Elongation	RA
TA-5A	341	414	4.5%	3%
TA-5D	320	383	3.0%	4.7%
TA-11A	349	426	60%	44.0%
TA-16A	314	421	33.9%	25.0%
TA-40B	337	441	67.0%	62.1%
TA-43A	389	482	41.0%	42.0%

1000°C

Alloy ID	0.2% YS	UTS	Elongation	RA
TA-5A	224	231	100.6%	86.9%
TA-5D	219	222	128.6%	84.1%
TA-11A	229	234	80.0%	68.2%
TA-16A	213	226	106.0%	59.1%
TA-40B	231	252	56.5%	62.9%
TA-43A	235	241	81.0%	75.3%

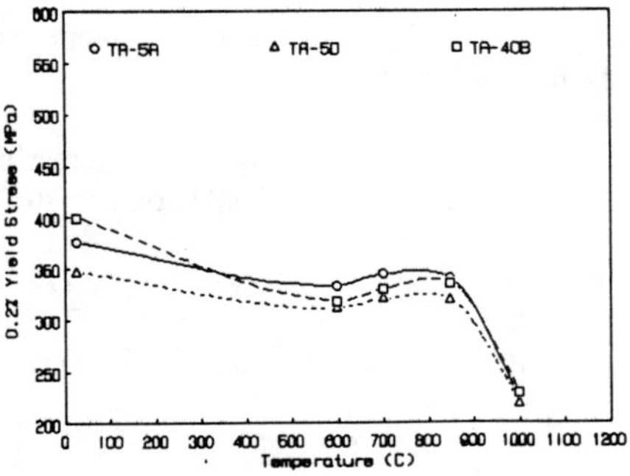

a.)

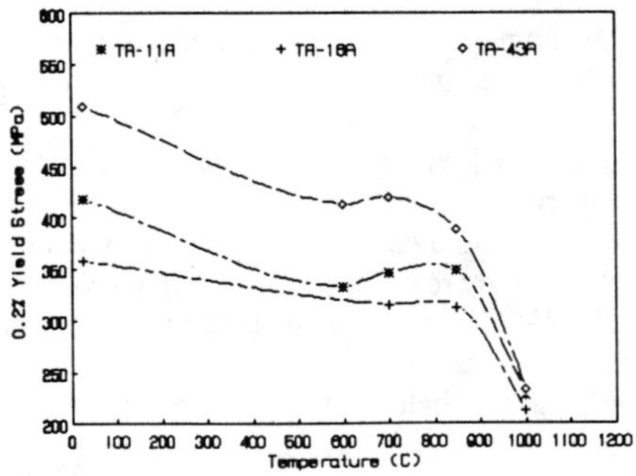

b.)

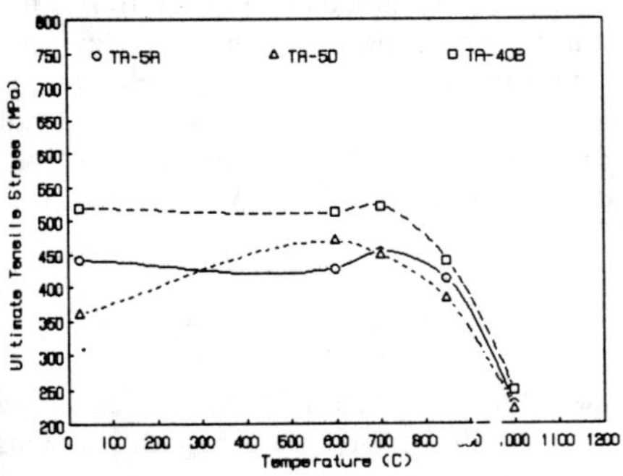

c.)

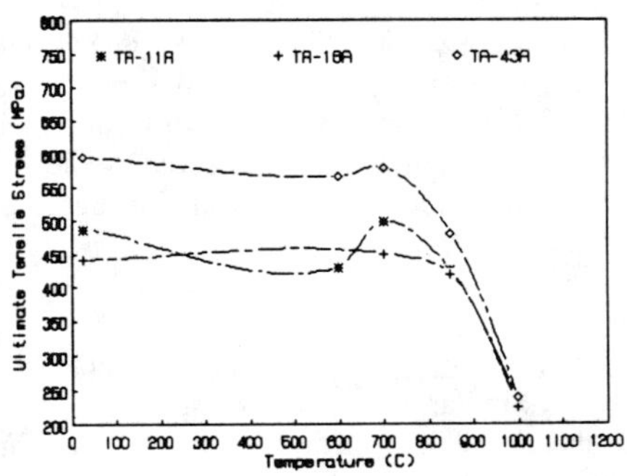

d.)

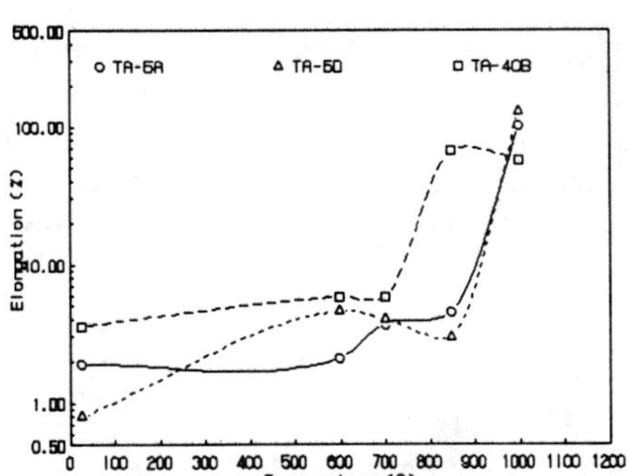

e.)

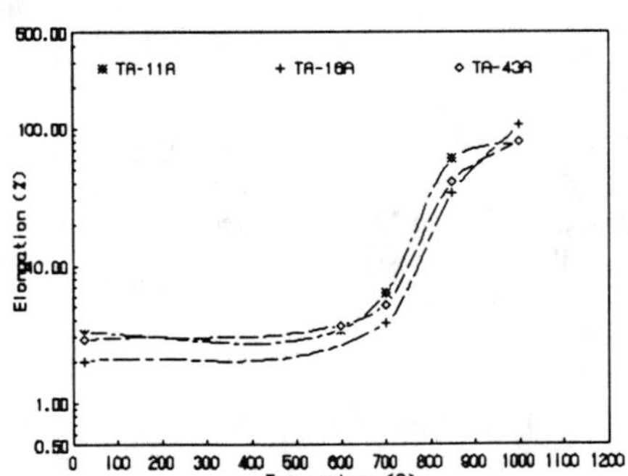

f.)

Figure 2 - The effect of temperature on the tensile properties of the Ti-48Al-2Nb-2Cr alloys.

a.) I/M alloys - yield stress
b.) P/M alloys - yield stress
c.) I/M alloys - UTS
d.) P/M alloys - UTS
e.) I/M alloys - elongation
f.) P/M alloys - elongation

The anomalous yield strength behavior in these alloys may be due to the high Al content and, therefore, large volume fraction of γ-phase present in the microstructure.

TiAl-based alloys typically exhibit a rapid increase in ductility with increasing temperature. This "ductile-brittle transition temperature" is dependent on the Al content, strain rate, interstitial content, and heat treatment (and therefore microstructure); however, most alloys exhibit an increase in ductility at approximately 750°C [4-6]. The cast + forged alloys, TA-5A and TA-5D, exhibited this type of increase in ductilty at temperatures in excess of 850°C. The P/M and the investment cast alloy, however, exhibited a rapid increase in ductility at temperatures below 750°C.

Fractography

SEM analysis of the fracture surfaces indicated that the fracture mechanisms did not change with processing techniques utilized. All of the materials tested at temperature less than approximately 850°C, exhibited a mixed transgranular/intergranular fracture. Some evidence of interlamellar separation was observed in some of the samples. At high test temperatures, the failure mechanism became almost entirely intergranular. However, in many cases, ductile creep-type cavities were observed, indicating a high degree of plastic deformation. These observations are consistent with fractographic evaluation of binary TiAl alloys reported in the literature [2-6].

TEM Evaluation of Deformation Mechanisms

Similar to reported investigations of TiAl-based alloys, all of the Ti-48Al-2Nb-2Cr alloys in this study deformed primarily by twinning and $1/2<110]\{111\}$-type ordinary dislocations. The dislocation and twinning activity was observed to increase with increasing test temperatures. This trend is in agreement with the observed increase in ductility with increasing temperatures.

Conclusions

Powder metallurgy processing of Ti-48Al-2Nb-2Cr alloys resulted in a signficant increase in ducility and strength at all test temperatures. This improvement in properties is probably due to the refinement and homogenization of the microstructure inherent in powder metallurgy processing.

References

1.) H.A. Lipsitt, Proc. Met. Res. Soc., 33, p. 351 (1984).
2.) S.C. Huang and D.S. Shih, Proc. Microstructure/Property Relationships in Ti-Aluminides and Alloys, Eds. Y.W. Kim and R.R. Boyer, TMS, (1991), pp. 105-122.
3.) D.S. Shih, S.C. Huang, G.S. Skarr, H. Jang and J.C. Chesnutt, Ibid., pp. 135-148.
4.) Y.W. Kim, Ibid., pp. 91-103.
5.) Y.W. Kim, JOM, 41, (1989), pp. 24-30.
6.) Y.W. Kim, JOM, 43, (1991), pp. 40-47.
7.) T. Kawabata, T. Kanai and O. Izumi, Acta Met, 33, (1985), pp. 1355-1366.
8.) S.C. Huang, Scripta Met, 22, (1991), pp. 1885-1888.
9.) S.C. Huang and E.L. Hall, Scripta Met, 25, (1991), pp. 1805-1809.
10.) Y.W. Kim, Acta Met, 40, (1992), 1121-1134.

EFFECTS OF MICROSTRUCTURE IN HYDROGEN EMBRITTLEMENT OF TITANIUM ALUMINIDE ALLOYS

Anthony W. Thompson

Dept. of Materials Science and Engineering
Carnegie Mellon University
Pittsburgh, PA 15213

Abstract

Hydrogen is now known to have major embrittling effects on titanium aluminide alloys. It has now been shown that large amounts of hydrogen are readily absorbed at high temperature in these alloys. On cooling to room temperature, virtually all this hydrogen is precipitated as a hydride phase or phases. This hydride phase affects mechanical properties in ways similar to what is observed in other hydride-forming materials, although effects per unit volume fraction of hydride are not particularly severe. Extensive data exist for alloys based on both α_2 and γ phases. It is particularly evident that microstructure plays an important role in controlling the severity of hydrogen effects in both kinds of alloys. Processes which occur on the scale of the metallurgical microstructure, which are called micromechanisms, are affected by hydrogen for both deformation and fracture. A preliminary description of these micromechanisms is presented and matched to experimental observations on titanium aluminides and literature analyses.

Introduction

It is becoming recognized that hydrogen can significantly affect alloys based on titanium aluminides, either the DO_{19} Ti_3Al phase (α_2) or the $L1_0$ TiAl phase (γ), which is of concern because of developmental interest in these alloys [1,2]. Considerable study of their physical metallurgy in recent years has established that alloying additions which stabilize the BCC β phase in α_2 alloys, or the α_2 phase in γ alloys, are a major benefit to low-temperature ductility and toughness [2-6]. These alloying additions are frequently the classic β-stabilizing elements known for Ti alloys, such as Nb, V, and Mo. For example, one of the best-studied α_2 alloys has the composition Ti-24 Al-11 Nb (at. pct.), which is designated Ti-24-11 below. It also appears that mechanical properties exhibit a marked sensitivity to microstructure, as is true for many structural alloys.

There is both fundamental and applied (for example, relating to the National Aerospace Plane or NASP) interest in hydrogen effects in these alloys [7]. Substantial amounts of hydrogen dissolve in titanium aluminides [8-12], and this dissolution increases at elevated temperature. If equilibrium is maintained during cooling to room temperature, all or nearly all this hydrogen precipitates as a hydride phase or phases [12]. The reversibility of this dissolution permits calculation of thermodynamic parameters of the hydride [12]. The presence of hydrides gives rise to extensive and

Titanium '92
Science and Technology
Edited by F.H. Froes and I. Caplan
The Minerals, Metals & Materials Society, 1993

largely deleterious effects on mechanical properties, similar to what has been observed in other hydride-forming alloys [13-15]. These effects, which are of concern for NASP applications, are beginning to be documented for titanium aluminide alloys, as is summarized here. The mechanism(s) and details of these effects, however, are poorly understood to date, and needs for additional work to answer such questions are included in the present paper.

Hydrides

Titanium and its alloys have long been known to interact strongly with hydrogen, a behavior which has been summarized in previous reviews [13-16]. The Ti-H binary diagram is the basis for interpretation of alloy effects. In that diagram, the principal hydride phase, which has the fluorite structure with Ti atoms on an FCC lattice and hydrogen randomly arranged in tetrahedral interstices, was originally designated γ. When it was recognized [17,18] that there exist at least two other tetragonal hydrides, as is also true in the Zr-H system, it was suggested [17] that the FCC hydride of titanium be named δ instead of γ. In addition, it was proposed that the low-hydrogen hydride of FCT structure, with $c/a > 1$, be called γ, and the high-hydrogen hydride, also FCT structure but with $c/a < 1$, be called ε (formerly called δ). With these changes, the phase names in the Ti-H diagram become entirely analogous to the phase nomenclature of hydrides in the Zr-H system. The literature on titanium hydrides of the last several years [17-20] has largely accepted this new nomenclature, and it is used in the present paper. The phase diagram then looks like Figure 1, with the γ phase regarded [20] as metastable.

The hydride which is most commonly observed, δ, has a range of stoichiometry. If the hydride compound is designated TiH_x, evidence indicates [13,17,20] that x varies from about 1.5 to 1.94. A number of other transition metal hydrides also exhibit ratios of hydrogen to metal which are sub-integral, i.e. with $x < 2$ [13,15], probably a consequence of the hydrogen vacancies thought to stabilize these phases throughout their stoichiometry range [13,15,21]. The FCC lattice parameter increases within this range of x values from 0.440 to 0.446 nm. At low temperature and high hydrogen contents, as x approaches 2, the δ hydride evidently transforms to the tetragonal ε [15,20]. At higher temperature, however, both tetragonal hydrides tend to transform to the cubic δ phase, e.g. above 200°C [20]. These tetragonal phases have usually been designated FCT in the literature on Ti and Zr hydrides, although BCT (identical to FCT except for unit cell size) is the normally preferred nomenclature for this structure. Use of "FCT" emphasizes the close relation to the FCC phase, which is less evident from the BCT description.

Rudman, Reilly and Wiswall [22] first extended these observations to titanium aluminides, with their determination that a hydride of lattice parameter 0.435 nm forms in Ti_3Al and that it has the same fluorite structure as the δ hydride of pure Ti. Rudman, et al. also reported that the hydride above 200°C seemed to contain no aluminum, that is, that Al was rejected from the TiH_x hydrides, a somewhat surprising observation in light of the expected mobility of aluminum at so low a temperature. Previous studies of the role of aluminum in hydride behavior in Ti alloys [23-25] have shown that solubility, hydride habit plane, and other characteristics are a function of Al content. More work on this aspect of hydrides in aluminides may be important, from both structural and compositional perspectives.

The first published report on hydrogen dissolving in an alloy based on the α_2 aluminide appears to have been for Ti-24-11 [8]. Subsequently, there have been a number of published reports [11,12,26-32] and reviews [9,10,33,34] on hydrogen effects in Ti-24-11 and other α_2 aluminide alloys. The solvus for hydrogen in Ti-24-11 has been determined [12] and is shown in Fig. 2. Note that the solvus is displaced to higher temperatures, compared to the Ti-H diagram in Fig. 1, corresponding to a reduced relative solubility for hydrogen. The solubility difference could be due to the Nb content of the α_2 alloy, or to structural differences between HCP Ti and the DO_{19} structure of Ti_3Al. However, the partial molar enthalpy for formation of the hydride, ΔH_δ, has been measured for Ti-24-11 to be -141 kJ/mol H_2, which is similar to the range of values reported for δ hydride in pure Ti, which is -125 to -142 kJ/mol H_2 [15,35].

In Ti-Al binary alloys, it has been shown that aluminum additions change the hydride habit plane from prism to basal [25]. For α_2 alloys, a limited number of results on hydride crystal structure and habit planes have been presented to date, without a definite pattern of behavior emerging. Shih, *et al.* [28] reported that the hydride habit plane changed from basal in Ti_3Al, to prism with Nb additions, which is opposite to the alloying observations for Al in Ti alloys [25]. There is clear evidence for both the δ and ε hydride in these alloys, but the observations of tetragonality do not yet seem consistent. One published study [32] reported that the ε hydride in Ti-24-11 exhibited almost the exact c/a ratio found for this phase in pure Ti [36]. To date, stoichiometry ranges of the presumed TiH_x hydrides in α_2 alloys have not been determined, nor has there been any clear indication whether Al, Nb or other elements participate in the hydride. It would be no surprise if Nb is involved, due to its strong interaction with hydrogen, and at low temperatures, Al might be part of the structure simply for kinetic reasons, so a stoichiometry of the form $(Ti,Al,Nb)H_x$ is possible. It could also be speculated that equilibrium or metastable hydrides other than δ may form in these alloys, particularly in high-Nb alloys such as those based on Ti_2AlNb [37], although none have yet been reported. It would also be useful to have systematic investigations of changes in structure, in the c/a ratio, or in other variables as functions of x or temperature.

It must be recognized in hydrogen studies of α_2 alloys that hydrogen is a potent stabilizer of the beta phase in these aluminides. Comparison of Figs. 1 and 2 suggests that this must be primarily a kinetic effect, since the eutectoid temperature evidently is elevated in Fig. 2, and depression of the β transus, is modest, consistent with measurements that 30 at. % H in Ti-24-11 reduced T_β less than 50°C. [38]. Profound microstructural changes, relative to hydrogen-free material, are possible in the presence of hydrogen, presumably due to transformation kinetics, and have been observed [39]. This means that otherwise convenient combinations of heat treatment and hydrogen charging into a single thermal cycle must be used with care. However, it should also be pointed out that the refined microstructural scale which results from beta stabilization [39] can provide significant improvement in subsequent hydrogen-free mechanical properties.

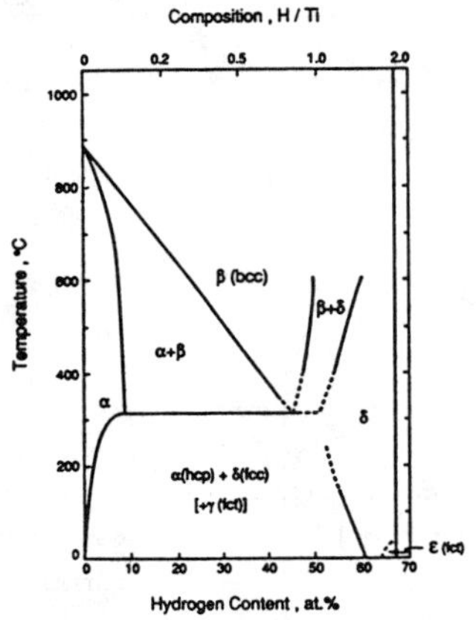

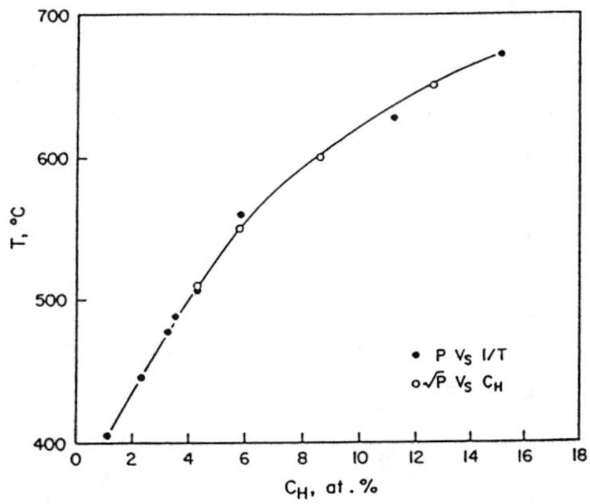

Fig. 1 (Left). Phase diagram for titanium-hydrogen system, after refs. 17 and 20, using revised nomenclature as suggested by Numakura and Koiwa [17]. Both the γ and ε phases are of fct structure, but γ has c/a > 1 and ε has c/a < 1.

Fig. 2 (Right). Terminal solubility of hydrogen in the α_2 phase in equilibrium with the hydride phase, i.e. solvus for hydride in α_2, for Ti-24-11. From ref. 12.

There is less information for TiAl-based alloys. Early reports [34,40-42] were that hydrogen was difficult to dissolve in this compound, or that hydrides did not form. It is now clear that these reports merely reflect an absence of hydrogen pickup. When conditions, particularly cleanliness of specimen surfaces, of the Sieverts apparatus itself, and of the charging hydrogen gas, are suitable for hydrogen uptake, the γ phase absorbs appreciable amounts of hydrogen (though less than an α_2 aluminide under the same conditions [8,12,43,44]). The hydrogen solvus has not yet been determined for γ, but the total uptake of hydrogen, i.e. the sum of hydrogen in solution and hydrogen precipitated as hydrides, has been measured [11]. It shows that, like the α_2 alloys, the γ phase does precipitate copious hydrides on cooling [10,11,44-46]. The δ hydride characteristic of the Ti-H system, Fig. 1, has been identified, as has a new hexagonal hydride [47], and there may be other hydride phases as well [45]. To an even greater degree than the α_2 aluminides, the γ aluminides are in an early stage of metallurgical understanding [4-6], and considerable work remains to be done in both development and understanding of behavior in these alloys. The effects of hydrogen are certainly among the topics which will repay careful study.

Mechanical Behavior

A general understanding is emerging for the mechanical behavior of titanium aluminide alloys [2-6,39,48-54]. Knowledge of hydrogen effects on this behavior is much more limited [10,11,30,45,53,55]. Hydrides have a relatively modest strengthening effect, and participate in a pronounced strain rate effect, much like what has been observed in Ti alloys containing hydrides [16,56]. Tensile properties of a single microstructure of Ti-24-11, as a function of hydrogen content [30,53], are shown in Fig. 3. There are not many comaprable data for γ alloys, but Fig. 4 gives an example for single-phase γ. Yield strength is far smaller and hydrogen strengthening is also less.

Measurements have also been made for fracture toughness and crack growth resistance or R-curve behavior for Ti-24-11 with hydrogen, as reported in more detail elsewhere [30,53,57]. Fig. 5 shows that the local true fracture stress σ_F, the plane strain fracture toughness K_{Ic}, and the threshold K for sustained-load cracking [38,57], called K_{IH}, all have a similar dependence on hydrogen content as do the uniform and fracture strain measures of ductility depicted in Fig. 3.

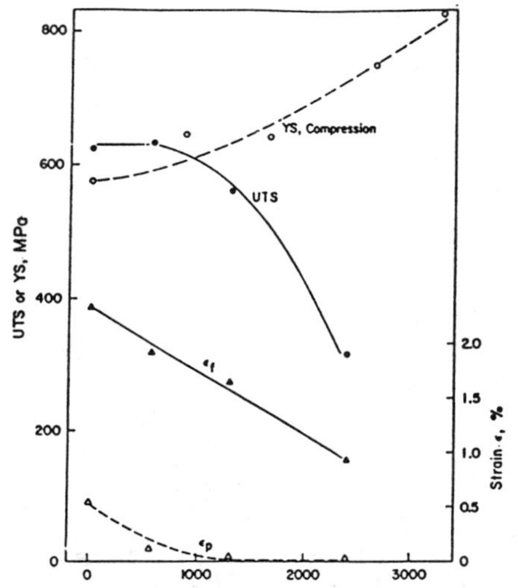

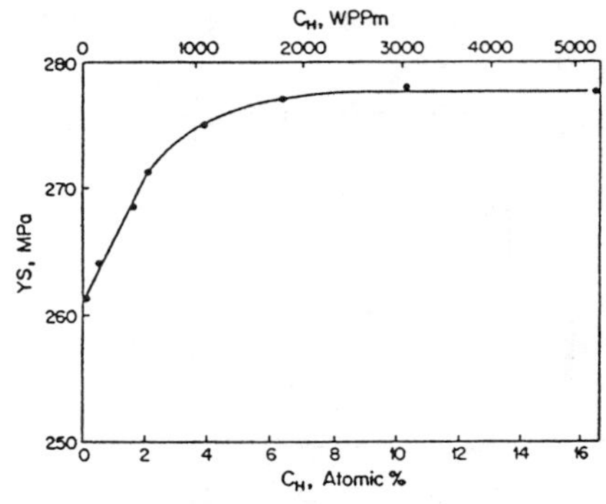

Fig. 3 (Left). Variation of compressive yield strength (YS), failure stress (either UTS, ultimate tensile strength, or fracture stress), and ductilities (uniform true strain ε_p and true fracture strain ε_f) with hydrogen concentration, largely hydride. Each point averages 2 to 5 tests. Ref. 53.

Fig. 4 (Above). Yield strength dependence on hydrogen content in Ti-50 Al (single phase). Ref. 45.

Fracture Micromechanisms

As the basic mechanical properties become established, it becomes possible to explore micromechanisms of behavior, i.e. mechanisms taking place on the scale of the metallurgical microstructure. Work on micromechanisms of fracture is just beginning in these alloys [58]. The fractographic appearance of both α_2 and γ fracture surfaces tends to resemble classical cleavage, and that term has been extensively used in the literature on these alloys. It has been pointed out, however, that some care in use of this term may be appropriate [45,53,58]. It still has not been shown that the fractures obey all the criteria for classical cleavage [59]. This is of particular concern for fracture facets in single-phase γ alloys, where fracture facets are commonly observed to be curved or conchoidal [6,45,54], or of primary α_2 particles with moderate to large dimensions in alloys in α_2 alloys.

One approach to discussion of this problem is to assume that fracture in aluminide alloys is in fact a cleavage process. That assumption permits comparison of data on α_2 alumindes to cleavage theories [53,58], in which the appropriate mechanism to compare would be the Smith model for homogeneous cleavage [60], that is, cleavage which is not nucleated by a second phase, such as a grain boundary carbide in iron.

In such situations, it has been shown [61,62] that the controlling factor in fracture is the effective energy, γ_{eff}. The value which has been experimentally determined [53] for γ_{eff} in fracture of Ti-24-11 which was not hydrogen charged is 27 J/m^2, at least an order of magnitude greater than the expected value of the true surface energy of the lattice, which is a γ_s value of about 0.9 J/m^2 for Ti_3Al [53]. Such an "effective" energy value is broadly consistent with results on iron [61,62], where measured values of γ_{eff} are about 14 J/m^2 and the known value of γ_s is about 2 J/m^2. It should be reiterated that this γ_{eff} corresponds to crack propagation [62] and thus is not directly related to the total fracture energy $\gamma = 2\gamma_s + \gamma_{irr} + \gamma_p$, which in Ti-24-11 has a value [53] of about 3000 J/m^2.

The alternative to the foregoing estimates from a cleavage model would be a "critical distance" approach, as first postulated by Ritchie, Knott and Rice [63]. This approach uses the measured ratio of σ_F to the yield strength YS to determine the distance l^* ahead of the crack tip at which that stress ratio must be maintained to propagate fracture [61,63,64]. For the Ti-24-11 on which measurements were made [53], $\sigma_F / YS = 2.37$, and the calculation [53] then yields a value for l^* of about 16 μm. This estimated l^* value, equalling about three α_2 lath widths, is broadly consistent with the microstructural dimensions found in other determinations [63,64] of l^*. It would be of interest to examine microcracking in α_2 laths to permit application of a statistical model [65] to Ti-24-11.

The "effective energy" and "critical distance" calculations just described are for hydrogen-free

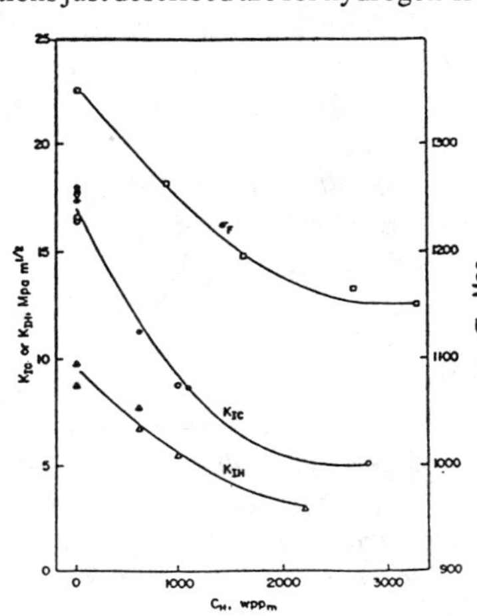

Fig. 5. Dependence of fracture toughness K_{Ic} and local true fracture stress in notched bars, σ_F, on amount of hydrogen (essentially all precipitated as hydride). From ref. 53.

α_2 material. In the presence of hydrogen, a variety of effects are possible [57]. As the hydride phase does not itself appear to act as a nucleus for cracks in this material [30,53], it may be essential to refine understanding of the controlling factors [53] in a matrix strengthened with hydride laths. It appears in such a case [46] that the major fracture parameter affected by hydrogen is the slip resistance τ_f, with some change also expected in γ_{eff}. These ideas are consistent with hydrogen assistance of local plasticity [14,66], which would require a fracture model with a strain component.

If, however, it should prove to be the case that cleavage of hydride laths stimulates cleavage of the α_2 matrix, as appears to be the case in many hydride-forming metals [13-16], then the "particle cracking" version of Smith's model [67] would require comparison to the hydride results. This is likely to be a challenge, because the stoichiometry and other aspects of hydrides in these aluminides remain unclear, not to mention hydride elastic properties or such parameters as surface energy, which are needed if quantitative comparison of theory [67] to experiment is to be performed. Additional metallurgical study, of phase transformations [68] and microstructural dependence of mechanical properties of aluminides would also pay dividends. Whether or not hydrides nucleate cleavage, it appears that considerable work remains to be done to elucidate the processes of hydrogen fracture in titanium aluminide alloys.

Acknowledgements

I am grateful for discussions with W.-Y. Chu on hydrogen effects in titanium aluminides, and with J.C. Williams, H.A. Lipsitt and R. Strychor on aluminide metallurgy. I particularly appreciate the extensive experimental work performed by Prof. Chu during his research visit to Carnegie Mellon University. Preparation of this manuscript has been supported by the David Jackson Fund for Scholarly Endeavors and by the Alcoa Center for Engineered Materials at Carnegie Mellon, and has benefited from the warm hospitality of the Materials Department at the University of California, Santa Barbara, during my sabbatical visit.

References

1. H.A. Lipsitt, in *Ordered Intermetallic Alloys*, MRS Symp. Vol. 39, ed. C.C. Koch, C.T. Liu and N.S. Stoloff (Pittburgh: Mater. Res. Soc., 1985), 351-364.
2. R. L. Fleischer, D.M. Dimiduk and H.A. Lipsitt, *Ann. Rev. Mater. Sci.*, 19 (1989), 231-263.
3. C.H. Ward, J.C. Williams, A.W. Thompson, D.G. Rosenthal, and F.H. Froes, in *Proc. 6th World Conf. on Titanium*, Vol. 2, ed. P. Lacombe, R. Tricot and G. Beranger (Paris: Les Editions de Physique, 1989), 1103-1108.
4. D.A. Koss, D. Banerjee, D.A. Lukasak and A.K. Gogia, in *High Temperature Aluminides and Intermetallics*, ed. S.H. Whang, C.T. Liu, D.P Pope and J.O. Stiegler (Warrendale, PA: TMS-AIME, 1990), 175-196.
5. J. M. Larsen, K. A. Williams, S. J. Balsone and M. A. Stucke, in *High Temperature Aluminides and Intermetallics*, ed. S.H. Whang, C.T. Liu, D.P Pope and J.O. Stiegler (Warrendale, PA: TMS-AIME, 1990), 521-556.
6. Y.-W. Kim, in *Microstructure/Property Relationships in Titanium Alloys and Titanium Aluminides*, ed. Y.-W. Kim and R.R. Boyer (Warrendale, PA: TMS-AIME, 1991), 91-103.
7. H.G. Nelson, *SAMPE Quarterly*, 20, No. 1 (1988), 20-23.
8. A.W. Thompson, W.-Y. Chu and J.C. Williams, in *Summary Proceedings of the 2nd Workshop on Hydrogen-Materials Interaction*, NASP Joint Program Office Workshop Pub. 1004, ed. H.G. Nelson (Moffett Field, CA: NASA-Ames, 1988), 133-135.

9. N.S. Stoloff, M. Shea and A. Castagna, in *Environmental Effects on Advanced Materials*, ed. R.H. Jones and R.E. Ricker (Warrendale, PA: TMS-AIME, 1991), 3-19.
10. A.W. Thompson, in *Environmental Effects on Advanced Materials*, ed. R.H. Jones and R.E. Ricker (Warrendale, PA: TMS-AIME, 1991), 21-33.
11. W.-Y. Chu and A.W. Thompson, *Scripta Metall. Mater.*, 25 (1991), 2133-38.
12. W.-Y. Chu and A.W. Thompson, *Acta Metall. Mater.*, 40 (1992), 455-462.
13. G.G. Libowitz, *J. Nucl. Mater.*, 2 (1960), 1-22.
14. H.K. Birnbaum, in *Hydrogen Effects on Material Behavior*, ed. N.R. Moody and A.W. Thompson (Warrendale, PA: TMS-AIME, 1990), 639-658.
15. W.M. Mueller, in *Metal Hydrides*, ed. W.M. Mueller, J.P. Blackledge and G.G. Libowitz (New York: Academic Press, 1968), 337-383.
16. N.E. Paton and J.C. Williams, in *Hydrogen in Metals*, ed. I.M. Bernstein and A.W. Thompson (Metals Park, OH: Amer. Soc. Metals, 1974), 409-431.
17. H. Numakura and M. Koiwa, *Acta Metall.*, 32 (1984), 1799-1807.
18. O.T. Woo, G.C. Weatherly, C.E. Coleman and R.W. Gilbert, *Acta Metall.*, 33 (1985), 1897-1906.
19. D.S. Shih, I.M. Robertson and H.K. Birnbaum, *Acta Metall.*, 36 (1988), 111-124.
20. A. San-Martin and F.D. Manchester, in *Phase Diagrams of Binary Titanium Alloys*, ed. J.L. Murray (Metals Park, OH: ASM Int., 1987), 123-135.
21. P.E. Irving and C.J. Beevers, *J. Mater. Sci.*, 7 (1972), 23-30.
22. P.S. Rudman, J.J. Reilly and R.H. Wiswall, *J. Less-Common Metals*, 58 (1978), 231-240.
23. J.D. Boyd, *Trans. ASM*, 62 (1969), 977-988.
24. N.E. Paton, B.S. Hickman and D.H. Leslie, *Metall. Trans.*, 2 (1971), 2791-2796.
25. N.E. Paton and R.A. Spurling, *Metall. Trans A*, 7A (1976), 1769-1774.
26. D.E. Matejczyk and R.P. Jewett, in *Summary Proceedings of the 2nd Workshop on Hydrogen-Materials Interaction*, NASP Joint Program Office Workshop Pub. 1004, ed. H.G. Nelson (Moffett Field, CA: NASA-Ames, 1988), 137-139.
27. E. Manor and D. Eliezer, *Scripta Metall.*, 23 (1989), 129-134.
28. D.S. Shih, G.K. Scarr and G.E. Wasielewski, *Scripta Metall.*, 23 (1989), 973-978.
29. E. Manor and D. Eliezer, *Scripta Metall.*, 23 (1989), 1313-1318.
30. W.-Y. Chu, A.W. Thompson, and J.C. Williams, in *Hydrogen Effects on Material Behavior*, ed. N.R. Moody and A.W. Thompson (Warrendale, PA: TMS-AIME, 1990), 543-553.
31. D. Eliezer, E. Manor and F.H. Froes, in *Hydrogen Effects on Material Behavior*, ed. N.R. Moody and A.W. Thompson (Warrendale, PA: TMS-AIME, 1990), 523-532.
32. M. Gao, J.B. Boodey and R.P. Wei, *Scripta Metall. Mater.*, 24 (1990), 2135-2138.
33. N.S. Stoloff, in *Hydrogen Effects on Material Behavior*, ed. N.R. Moody and A.W. Thompson (Warrendale, PA: TMS-AIME, 1990), 483-495.
34. D. Eliezer, F.H. Froes and C. Suryanarayana, *J. Metals*,43, No. 3 (1991), 59-62.
35. R.M. Haag and F.J. Shipko, *J. Am. Chem. Soc.*, 78 (1956), 5155-5159.
36. L.D. Jaffe, *Trans. AIME*, 206 (1956), 861.
37. R.G. Rowe, in *Microstructure/Property Relationships in Titanium Alloys and Titanium Aluminides*, ed. Y.-W. Kim and R.R. Boyer (Warrendale, PA: TMS-AIME, 1991), 387-398.
38. W.-Y. Chu and A.W. Thompson, unpublished research, Carnegie Mellon University, 1990.
39. W.-Y. Chu and A.W. Thompson, *Metall. Trans. A*, 22A (1991), 71-81.
40. L. Christodoulou and J.A. Clarke, in *Hydrogen Effects on Material Behavior*, ed. N.R. Moody and A.W. Thompson (Warrendale, PA: TMS-AIME, 1990), 515-520.
41. L.G. Fritzemeier and M.A. Jacinto, in *Hydrogen Effects on Material Behavior*, ed. N.R. Moody and A.W. Thompson (Warrendale, PA: TMS-AIME, 1990), 533-542.

42. D.E. Ulmer and D.E. Matejcyzk, in *Summary Proceedings of the 3rd Workshop on Hydrogen-Materials Interaction*, NASP Joint Program Office Workshop Pub. 1007, ed. H.G. Nelson (Moffett Field, CA: NASA-Ames, 1990), 201-203.
43. B.S. Majumdar, H.J. Cialone and J.H. Holbrook, in *Summary Proceedings of the 3rd Workshop on Hydrogen-Materials Interaction*, NASP Joint Program Office Workshop Pub. 1007, ed. H.G. Nelson (Moffett Field, CA: NASA-Ames, 1990), 73-87.
44. D. Legzdina, I.M. Robertson and H.K. Birnbaum, *J. Mater. Res.*, 6 (1991), 1230-1237.
45. A.W. Thompson and W.-Y. Chu, in *Microstructure/Property Relationships in Titanium Alloys and Titanium Aluminides*, ed. Y.-W. Kim and R.R. Boyer (Warrendale, PA: TMS-AIME, 1991), 165-177.
46. A.W. Thompson, *Mater. Sci. Eng.*, A153 (1992), 578-583.
47. D.E. Matejczyk and C.G. Rhodes, *Scripta Metall. Mater.*, 24 (1990), 1369-1373.
48. S.A. Court, J.P.A. Löfvander, M.H. Loretto and H.L. Fraser, *Phil. Mag.*, 61 (1990), 109-139.
49. S.A. Court, V.K. Vasudevan and H.L. Fraser, *Phil. Mag.*, 61 (1990), 144-158.
50. W.-Y. Chu and A.W. Thompson, *Scripta Metall. Mater.*, 25 (1991), 641-644.
51. E.L. Hall and S.-C. Huang, in *Microstructure/Property Relationships in Titanium Alloys and Titanium Aluminides*, ed. Y.-W. Kim and R.R. Boyer (Warrendale, PA: TMS-AIME, 1991), 47-64.
52. S.-C. Huang and D.S. Shih, in *Microstructure/Property Relationships in Titanium Alloys and Titanium Aluminides*, ed. Y.-W. Kim and R.R. Boyer (Warrendale, PA: TMS-AIME, 1991), 105-122.
53. W.-Y. Chu and A.W. Thompson, *Metall. Trans. A*, 23A (1992), 1299-1312.
54. K.S. Chan, *J. Metals*, 44, No. 5 (1992), 30-38.
55. A.W. Thompson and W.-Y. Chu, in *Summary Proceedings of the 3rd Workshop on Hydrogen-Materials Interaction*, NASP Joint Program Office Workshop Pub. 1007, ed. H.G. Nelson (Moffett Field, CA: NASA-Ames, 1990), 125-130.
56. N.R. Moody and J.E. Costa, in *Microstructure/Property Relationships in Titanium Alloys and Titanium Aluminides*, ed. Y.-W. Kim and R.R. Boyer (Warrendale, PA: TMS-AIME, 1991), 587-604.
57. W.-Y. Chu and A.W. Thompson, in *High Performance Composites for the 1990's*, ed. S.K. Das, C.P. Ballard and F. Marikar (Warrendale, PA: TMS-AIME, 1991), 143-157.
58. A.W. Thompson, "Micromechanisms of Fracture in Titanium Aluminide Alloys," this conf.
59. C.D. Beachem and R.M.N. Pelloux, in *Fracture Toughness Testing and Its Applications*, STP 381 (Philadelphia: A.S.T.M., 1965), 210-244.
60. E. Smith, *Acta Metall.*, 14 (1966), 985-989 and 991-996.
61. J.F. Knott, *Fundamentals of Fracture Mechanics*, Ch. 7 and 8, Butterworths, London, 1973, pp. 176-233.
62. A.W. Thompson and J.F. Knott, *Metall. Trans. A*, in press (scheduled for Dec. 1992).
63. R.O. Ritchie, J.F. Knott and J.R. Rice, *J. Mech. Phys. Solids*, 21 (1973), 395-410.
64. R.O. Ritchie and A.W. Thompson, *Metall. Trans. A*, 16A (1985), 232-248.
65. T. Lin, A.G. Evans and R.O. Ritchie, *Acta Metall.*, 34 (1986), 2205-2216.
66. A.W. Thompson and I.M. Bernstein, in *Advances in Corrosion Science and Technology*, Vol. 7, ed. M.G. Fontana and R.W. Staehle (New York: Plenum Press, 1980), 53-175.
67. E. Smith and J.T. Barnby, *Metal Sci. J.*, 1 (1967), 56-64.
68. R. Strychor, J.C. Williams and W.A. Soffa, *Metall. Trans. A*, 19A (1988), 225-234.

A RE-EXAMINATION OF THE Ti-Al EQUILIBRIUM DIAGRAM

IN THE RANGE 33-54 A/O ALUMINUM

Andi Bartz, Lisa Rothenflue, Mohammad Saqib,
Ralph Omlor, and Harry A. Lipsitt

Department of Mechanical and Materials Engineering
Wright State University, Dayton, Ohio 45435

Abstract

The central region of the Ti-Al equilibrium diagram has been thoroughly reexamined. The invariant is established as an eutectoid near 1115°C. The eutectoid composition is 39.6 a/o Al and the endpoints of the invariant are at 38.4 and 48.4 a/o Al, respectively. The alpha plus alpha2 region is established by TEM showing adjacent ordered and disordered areas. The high temperature gamma solvus has a lower slope to 1350°C than previously reported. The low temperature gamma solvus extends to greater than 50 a/o Al below 1000°C. The alpha plus beta field has been located at both 1200 and 1300°C.

This research was sponsored jointly by General Electric, Aircraft Engines, Cincinnati, Ohio and by the Howmet Corporation, Whitehall, Michigan.

Titanium '92
Science and Technology
Edited by F.H. Froes and I. Caplan
The Minerals, Metals & Materials Society, 1993

Introduction

Although the Ti-Al equilibrium diagram has been investigated repeatedly, the positions of several phase boundaries near the center of the diagram remain in question (1). Recent studies have shown that the high temperature phase in equilibrium with gamma is alpha, not beta (2-6). However, these studies have not revealed the exact position of the alpha plus beta field that must exist above about 1200°C and in the range 35-43 a/o Al (2-6). The nature of the invariant near 1125°C, eutectoid or peritectoid, also needs to be determined unequivocally (7-13). The "eutectoid" composition is subject to considerable variation and the compositions of the phases in equilibrium at the invariant are also open to discussion (7-13). Finally, the phase boundary positions at temperatures both above and below the invariant need to be established (1-13).

It is clear from the above that an exhaustive study of the central portion of this critical equilibrium diagram is needed. The resulting diagram would provide an unequivocal reference system within which much of the current alloy development results could be considered. This paper is the result of an intensive study of this important binary system using samples equilibrated between 1350 and 800°C for times as long as 1.75×10^7 seconds (29 weeks). The equilibrated microstructures have been examined by a variety of techniques (optical microscopy, bright and dark field TEM, electron diffraction, back scattered SEM, and electron probe microanalysis) and the studies pursued until all the data have agreed. The specimens were treated to produce sufficiently large individual phase regions so that unambiguous electron probe microanalytical data could be obtained. The nature and temperature of the invariant were investigated mainly by dark field TEM examination of quenched samples equilibrated in this range. The presence or absence of antiphase domain boundaries in the alpha2 phase in the specimens indicated whether the alloy had been disordered or ordered at temperature, respectively.

Experimental Procedure

Materials for this study were furnished through the courtesy of several sources. Dr. S.C. Huang, (GE, CRD, Schenectady, NY.) furnished button heads from broken tensile samples of nominally 48 and 52 a/o Al. Several alloys (nominally 35, 37, 39, 41, 43, 46, and 50 a/o Al) were received through the courtesy of Dr. D.M. Dimiduk (Wright Laboratories, Materials Laboratory, Wright-Patterson AFB, Ohio) as 150 gram finger ingots. The as-received materials were sealed in evacuated quartz tubes filled with Ti machining chips and homogenized for one week at 1000°C. The homogenized materials were cut into 3.8 cm long pieces and the pieces were forged in vacuum at 1000°C to a height reduction of approximately 80% (0.75 cm). Smaller pieces, approximately 0.75 cm cubes, and the button heads were double wrapped in Ta foil and sealed in evacuated quartz tubes filled with Ti machining chips. These smaller pieces were heat treated according to a preset schedule and water quenched.

The heat treatment schedule was developed by first achieving equilibrium at 1300°C. Samples exposed for various times were studied optically and by electron probe microanalysis until equilibrium had been achieved. This required between 5 and 6 hours at 1300°C. The activation energy for the steady state creep of Ti_3Al, 2.1×10^5 J/mol, was used to calculate a diffusivity at 1300°C, assuming that D_0 equalled unity (14). Next, the experimentally determined time of 6 hours and the calculated diffusivity were used to determine a necessary diffusion distance. This diffusion distance and the calculated diffusivity at each temperature were then used to determine the necessary diffusion time at the other temperatures. Table I shows the results of the calculations. The heat treatment for every specimen began with an equilibration treatment at 1300°C. After the required time at 1300°C, the furnace temperature was reduced to 1200°C and the specimen subjected to the full equilibration treatment at that temperature. This process was repeated at each temperature so that the complete treatment for each specimen to be quenched and examined from a given temperature included a cascade of equilibration treatments at 100°C increments at every higher temperature. It should be noted that the actual time used at 800°C was one half the calculated time. It was believed that this procedure was acceptable because the 800°C samples had previously been equilibrated at 900°C and above and because the composition change required at 800°C was small.

Table I Equilibration Calculations

Temperature °C	Diffusivity cm^2/s	Diffusion Distance cm^2	Diffusion Time seconds	Actual Time seconds
1350	1.74×10^{-7}	-	1.31×10^{4}	8.64×10^{4}
1300	1.06×10^{-7}	2.29×10^{-3}	2.16×10^{4}	8.64×10^{4}
1200	3.57×10^{-8}	"	6.40×10^{4}	8.64×10^{4}
1100	1.02×10^{-8}	"	2.24×10^{5}	2.59×10^{5}
1000	2.41×10^{-9}	"	9.49×10^{5}	9.50×10^{5}
900	4.44×10^{-10}	"	5.15×10^{6}	4.84×10^{6}
800	5.97×10^{-11}	"	3.83×10^{7}	1.75×10^{7}

The equilibrated and quenched samples were sectioned through the middle, mounted, polished, etched, and examined by optical microscopy. A band of affected material, mainly due to Al loss, existed around the outside of every specimen. If the band width were only a small part of the specimen thickness it was assumed that the interior microstructure was unaffected by exposure to the Ta foil, quartz, Ti chips, and the environment of the capsule. This was later verified by oxygen analyses performed through the courtesy of Dr. Blair London, Howmet Corporation, Whitehall, Michigan. The oxygen analyses are presented in Table II. The mounted specimens were repolished and examined by electron probe microanalysis (EPMA). Backscattered electron images of the microstructure were recorded in the process. The EPMA was carried out through the courtesy of Mr. J.C. Chesnutt, GE, Evendale, Ohio and Dr. D.M. Dimiduk, Wright Laboratories, Materials Laboratory, W-PAFB, Ohio. In some cases, specimens equilibrated at the same temperature were examined by more than one facility and the individual analyses are presented. The EPMA data are presented in Table III.

Table II Oxygen Analyses After Various Heat Treatments

Composition a/o	Temperature °C	Initial Oxygen wppm	Final Oxygen wppm
46 Al	1300	-	636
50 Al	1200	-	501
52 Al	1300	-	862
46 Al	1350	752	873
46 Al	1300	752	768
46 Al	1200	752	770
46 Al	1200	980	1185

Note: The times for the above heat treatments are given in Table I above.

Table III Electron Probe Microanalytical Data

Temperature ^0C	Composition a/o	Source	Alpha a/o	Alpha2 a/o	Beta a/o	Gamma a/o
1350	52	GE	48.6	-	-	54.0
1350	52	GE	47.7	-	-	53.0
1350	46	AF	45.9	-	-	-
1300	48	GE	45.4	-	-	52.3
1300	48	GE	44.8	-	-	51.7
1300	48	AF	44.3	-	-	50.6
1300	41	AF	41.2	-	-	-
1300	39	AF	39.22	-	38.35	-
1200	50	GE	40.3	-	-	50.9
1200	37	AF	37.14	-	-	-
1200	35	AF	34.74	-	33.93	-
1140	41	AF	39.8	-	-	48.3
1140	39	AF	39.6	-	-	-
1110	41	GE	-	38.4	-	48.4
1000	50	AF	-	34.61	-	50.42
800	50	AF	-	32.75	-	50.53
800	52	AF	-	-	-	51.04

The temperature of the invariant was determined in the TEM by imaging samples equilibrated for 2.59 x 10^5 seconds at and quenched from 1140, 1125, and 1110^{0}C. These treatments followed standard equilibration treatments at 1300 and 1200^{0}C. TEM foils were prepared by standard techniques from 0.25 mm thick slices cut from the quenched samples. The slices were mechanically polished to a thickness of 0.1-0.2 mm using a cloth wheel and an abrasive suspension. Then, cores were removed from the slice using a coring tool and diamond paste as an abrasive. Next, the 3 mm diameter cores were mechanically polished and finally jet polished in a solution of 10 ml $HClO_4$, 150 ml butyl cellusolve, and 250 ml methanol at -40^{0}C, using 12 milliamps and 30 volts. The foils were examined in a JEOL 100CX transmission electron microscope operating at 100 Kv. Grains of alpha2 were identified by the characteristic electron diffraction patterns of the DO_{19} structure. Next, a grain identified as alpha2 was imaged in centered dark field using a low index superlattice spot to illuminate the specimen. When antiphase domain boundaries (APB's) were found to be present, it was clear that the alpha phase had ordered on quenching and had been disordered at the equilibration temperature. When APB's were not found to be present in the equilibrated and quenched samples it was clear that the alpha2 phase had been ordered at the equilibration temperature.

The nature of the invariant was also determined in the TEM by the dark field examination of specimens of 35 and 37 a/o Al equilibrated for 2.59 x 10^5 seconds at and quenched from 1125, 1140, 1160, and 1180^{0}C. As above, these treatments followed standard equilibration treatments at 1300 and 1200^{0}C. Such samples were found to be either fully ordered alpha2 at temperature or were composed of areas that had been ordered, together with areas that had been disordered at the equilibration temperature. In addition, areas disordered at temperature could be differentiated from areas ordered at temperature by EDS examination in the TEM. In this way the Al content of the individual areas could be determined, although the data are neither as accurate nor as precise as are the EPMA data cited above. Nevertheless, there was an Al concentration difference measured between areas ordered at temperature and areas disordered at the equilibration temperature. The TEM data are presented in Table IV.

Table IV TEM Data on APB Presence and EDS Chemistries

Temperature °C	Composition a/o	APB's	Alpha a/o	Alpha2 a/o
1180	35	Y/N	36.4	35.5
1160	37	Y/N	37.5	36.3
1160	35	N	-	-
1140	39	Y	-	-
1140	37	N	-	-
1140	35	N	-	-
1125	39	Y	-	-
1125	35	N	-	-
1110	46	N	-	-
1110	43	N	-	-
1110	41	N	-	-
1110	39	N	-	-

Results and Discussion

The data from Table III are plotted in Figure 1. In addition, Figure 1 contains a few additional points from samples which were optically seen to be single phase and are so indicated. Finally, the TEM data from Table IV have been used to determine the temperature of the invariant and to indicate the approximate position of the alpha + alpha2 region on the diagram. In considering this equilibrium phase diagram and its relationship to other recent work, it is most convenient to consider the present data one area at a time.

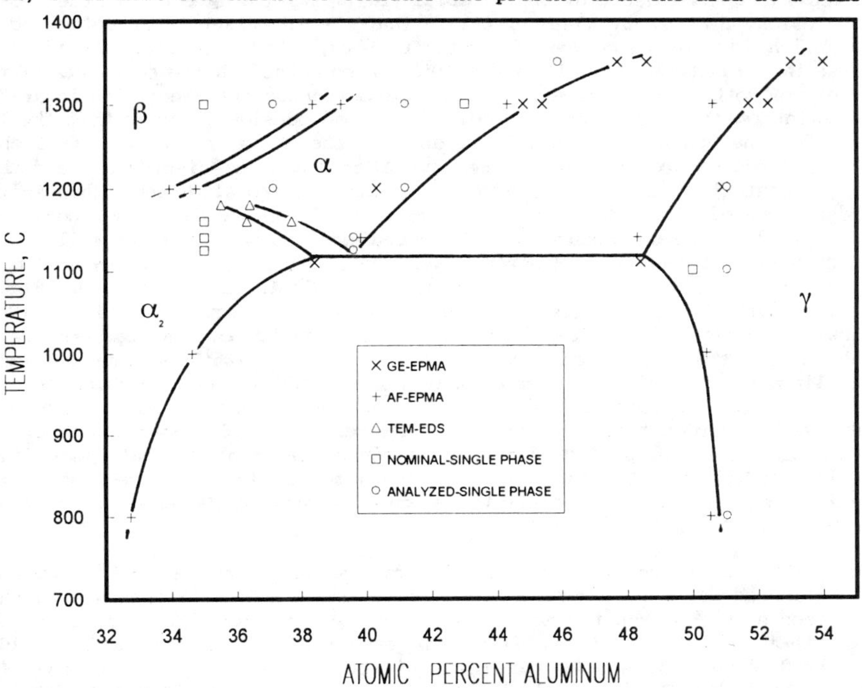

Figure 1 - The central region of the Ti-Al equilibrium diagram drawn from the present data.

The alpha transus shown here is plotted from ten points in the range 1125-1350°C. Tentative alpha transus lines were also shown in references 4, 5, and 6 and have been compared by Mishurda and Perepezko (1). The agreement between these data is quite good between 1125 and 1300°C, however, the present data indicate a lower slope at 1350°C. The present lower slope at 1350°C agrees with recent X-ray data (4). The transus line drawn by Mishurda and Perepezko followed their DTA data. Since it is expected that DTA data will yield transus temperatures that depend on the rate of heating, the agreement shown is really very good.

The present gamma solvus has a markedy lower slope than does that presented by Mishurda and Perepezko (1). Both the present diagram and that of Mishurda and Perepezko indicate the minimum Al composition of gamma to be about 48.5 a/o Al at the eutectoid. Their gamma solvus passes through 51 a/o Al at 1350°C whereas the present solvus line is at 53 a/o Al. The present slope is drawn using seven EPMA determined equilibrium gamma phase compositions. Thus, it should be more representitive of equilibrium than data from either DTA or high temperature X-ray studies where only limited time is allowed at each temperature and Al loss from the specimen surface is a real possibility. In addition, this placement is augmented by the optical examination of the equilibrated materials which shows that both a nominal 52 a/o Al (measured as 51.04 a/o Al) alloy equilibrated at 1300°C and a nominal 50 a/o Al alloy equilibrated at 1200°C contain only a small amount of alpha phase (<2% and all at the grain boundary triple points). It should also be noted that the present gamma solvus is similar to that drawn by Shull and Cline (13).

The present data clearly establish the invariant as an eutectoid under equilibrium conditions. This conclusion agrees with the previous DTA data of Mishurda, et al, (5) and of Shull, et al (10). It also agrees with the recent conclusion of Jones and Kaufman (12). That the invariant is an eutectoid is established by consideration of several factors, taken together. The first of these is the EPMA data at 1140 and 1110°C, which establish the endpoints of the reaction isotherm and indicate that the alpha transus intersects the invariant at an intermediate Al concentration. The second consideration is that the TEM data for the 35% Al alloy at 1180°C and for the 37% Al alloy at 1160°C show that two phases are in equilibrium, one ordered and one disordered, and that the Al content of the disordered phase is about 1 a/o Al greater than in the ordered phase. (Note that the Al contents of the alpha and alpha2 phases as determined by the unstandardized EDS measurement do not properly fall on the diagram here presented. However, even given the measurement error, it is clear that the alpha phase is richer in Al than is the alpha2 phase.) Figure 2 is a dark field micrograph of the 35 % Al alloy equilibrated at 1180°C showing the ordered and disordered phases in equilibrium. The temperature of the invariant is established as being 1110° < T < 1125°C because specimens equilibrated at the higher temperature contain APB's upon quenching, wheras specimens equilibrated at 1110°C are clear of APB's. Again, this is in agreement with the recent data of Jones and Kaufman (12). The composition of the eutectoid is established as 39.6 a/o Al since specimens of that composition could be equilibrated at 1125° as single phase materials. The same composition equilibrated at 1110°C precipitates a small amount of gamma at a few grain boundary triple points.

It is also clear from the data that the gamma solvus should extend to compositions in excess of 50 a/o Al at low temperatures. This is contrary to what has been shown on even the most recent assessed diagram(15), in spite of the fact that evidence to this effect has been available for some time (16). Figure 3 shows a 50 a/o Al specimen equilibrated for 1.75 x 10^7 seconds (29 weeks) at 800°C. The material clearly contains two phases. EPMA data for this specimen shows that the gamma phase composition at 800°C is 50.5 a/o Al. Thus,

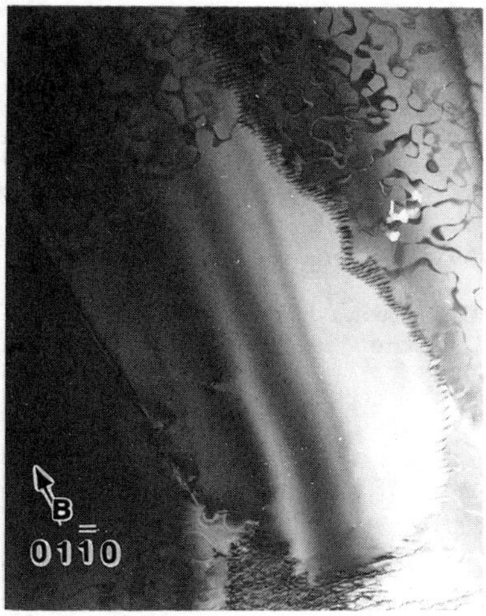

10K 1.0 μm

Figure 2 - A 35 a/o Al alloy equilibrated at 1180°C, showing disordered alpha in equilibrium with ordered alpha2.

200x 100 μm

Figure 3 - A 50 a/o Al alloy equilibrated for 29 weeks at 800°C, showing alpha2 in equilibrium with gamma.

it is clear that at low temperatures the stoichiometric composition of TiAl should be shown to lie in the two phase alpha2 plus gamma region. At temperatures below the invariant the present data show the alpha2 solvus to be steeply sloped. These data are presented without comment since the EPMA of the low temperature alpha2 phase was not very satisfactory. The total analysis did not add up to 100% and, as such, the data are open to some question.

Finally, the alpha plus beta field has been located at both 1200° and 1300°C. Samples quenched from this field exhibited microstructures composed of grains of alpha2 and grains of transformed beta. This microstructural analysis has also been verified by TEM and EPMA.

Conclusions

The central region of the Ti-Al equilibrium diagram has been carefully reassessed by studying various compositions which were equilibrated for long times in the range 800-1300°C. The specimens have been examined by several techniques which allow a number of conclusions to be drawn. The equilibrium alpha transus agrees with that previously drawn from DTA data to 1300°C, but shows a lower slope at 1350°C. The high temperature equilibrium gamma solvus has a markedly lower slope up to 1350°C than has previously been reported. The invariant is clearly established as an eutectoid with the eutectoid temperature established as 1110°C $<$ T $<$ 1125°C. The eutectoid composition is 39.6 a/o Al. The endpoints of the eutectoid isotherm are at 38.4 and 48.4 a/o Al. The alpha plus alpha2 region is established by dark field TEM photomicrographs and EDS data showing ordered alpha2 in equilibrium with disordered alpha. The disordered alpha is about 1 a/o richer in Al than is the ordered alpha2. The low temperature gamma solvus extends to greater than 50 a/o Al at temperatures lower than 1000°C. Thus, contrary to what many authors have drawn, the stoichiometric composition, TiAl, exists in a two phase region at low temperatures. The alpha plus beta field has been located at both 1200 and 1300°C.

References

1. J.C. Mishurda and J.H. Perepezko, "Phase Equilibria in Ti-Al Alloys", Microstructure and Property Relationshipsin Ti Alloys and Titanium Aluminides, ed. Y-W Kim and R.R. Boyer (TMS, Warrendale, Pa., 1991), 3-30.

2. J.J. Valencia, C. McCullough, C.G. Levi, and R. Mehrabian, "Microstructure Evolution During Conventional and Rapid Solidification of a Ti-50 a/o Al Alloy", Scripta Met., 21 (1987), 1341-1346.

3. C. McCullough, J.J. Valencia, H. Mateos, C.G. Levi, R. Mehrabian, and K.A. Rhyne, "The High Temperature Alpha Field in the Titanium-Aluminum Phase Diagram", Scripta Met., 22 (1988), 1131-1136.

4. C. McCullough, J.J. Valencia, C.G. Levi, and R. Mehrabian, "Phase Equilibria and Solidification in Ti-Al Alloys", Acta Met., 37 (1989), 1321-1336.

5. J.C. Mishurda, J.C. Lin, Y.A. Chang, and J.H. Perepezko, "Titanium-Aluminide Alloys Between the Compositions Ti_3Al and TiAl", High Temperature Ordered Intermetallic Alloys III, 133 (1989), 57-62.

6. S.C. Huang and P.A. Siemers, "Characterization of the High Temperature Phase Fields Near Stocichiometric TiAl",Met Trans., 20A (1989), 1899-1906.

7. Hisaoki Sasano, Tokuzo Tsujimoto, and Hirozo Kimura, "On the Quenched Structure and Aging Process of Ti-8~15 wt% Al Alloy", Titanium Science and Technology, ed. R.I. Jaffee and H.M. Burte, 3 (1973), 1635-1647.

8. F.J. van Loo and G.D. Rieck, "Diffusion in the Titanium-Aluminum System-II. Interdiffusion in the Composition Range Between 25 and 100 at.% Ti", Acta Met., 21 (1973), 73-84.

9. Ken-ichiro Ouchi, Yoshiaki Iijima, and Ken-ichi Hirano, "Interdiffusion in Ti-Al System", Titanium'80 Science and Technology, ed. H. Kimura and O. Izumi, 1 (1980), 559-568.

10. R.D. Shull, A.J. McAlister, and R.C. Reno, "Phase Equilibria in the Titanium-Aluminum System", Titanium Science and Technology, ed. G. Lutjering, U. Zwicker, and W. Bunk, 3 (1985), 1459-1466.

11. S.A. Jones, R.D. Shull, A.J. McAlister, and M.J. Kaufman, "Microstructural Studies of Ti-Al Alloys in the Vicinity of the "Eutectoid" Reaction", Scripta Met., 22 (1988), 1235-1240.

12. Steven A. Jones and Michael J. Kaufman, "Phase Equilibria and Transforma tions in Intermediate Titanium-Aluminum Alloys", Submitted to Acta Metallurgica et Materialia, September 1991.

13. Robert D. Shull and James P. Cline, "High Temperature X-ray Diffractometry of Ti-Al Alloys", High Temp. Sci., 28 (1989), 95-117.

14. M.G. Mendiratta and H.A. Lipsitt, "Steady-State Creep Behaviour of Ti_3Al-Base Intermetallics", J. Mat. Sci., 15 (1980), 2985-2990.

15. J.L. Murray, "The Al-Ti Phase Diagram", Binary Alloy Phase Diagrams, ASM International, Second Edition, 1990, 225-227. J.L. Murray, Phase Diagrams of Binary Tianium Alloys, ed. J.L. Murray, ASM International, 1987, 12-24.

16. D. Shechtman, M.J. Blackburn, and H.A. Lipsitt, "The Plastic Deformation of TiAl", Met. Trans., 5 (1974), 1373-1381.

MICROSTRUCTURE AND MECHANICAL PROPERTIES OF Ti-Al-(CrSi) ALLOYS

G. Frommeyer, Th. Kremser, and W. Wunderlich

Max-Planck-Institut für Eisenforschung GmbH, Düsseldorf/Germany

Abstract

Alloying elements such as Cr, Mn, V, Nb and Si have a strong influence on the mechanical properties as well as on the microstructure of two-phase intermetallic TiAl/Ti_3Al alloys. Considerable improvement in room temperature tensile ductility was achieved by additions of Cr and Si. The deformation mechanisms have been studied by TEM and APFIM investigations. Formation of deformation twins by glide of $<121]$-partial dislocations, which are initiated by misfit dislocations at the α_2/γ interfaces is the main deformation mechanism in the TiAl phase. The lamellar arrangement of the Ti_3Al phase with a volume fraction of about 10 or 15 % plays an important role in strengthening the TiAl/Ti_3Al alloy.

Titanium '92
Science and Technology
Edited by F.H. Froes and I. Caplan
The Minerals, Metals & Materials Society, 1993

Introduction

TiAl base alloys show great potential applications as structural light weight materials at elevated temperatures. The microstructural related strength properties, the ductility, and the fracture toughness at room and higher temperatures are very important for the use of titanium aluminide as an aerospace and motor component material.

Several investigations on mechanical properties [1-3] showed that the largest fracture strain can be achieved in the two phase alloys containing 48 at% aluminium. Maximum elongations to failure of about 2 to 3 % have been observed in tensile tests. The plastic strain is larger than that of the single phase materials Ti_3Al or TiAl. The stoichiometry seems to have a great influence on the mechanical properties. However, the deformation mechanisms have not been fully understood.

The partly revised phase diagram, recently published [4-6], shows that alloys within the composition range from 44 to 48 at% Al consist at high temperatures of pure α-Ti which transforms eutectoidally at 1125 °C into the α_2-Ti_3Al and γ-TiAl phase. TEM observations revealed that the coexisting phases TiAl and Ti_3Al are arranged in a lamellar microstructure [7,8].

The plastic deformation mechanisms of TiAl alloys have been studied thoroughly [9-15]. Several dislocation configurations and dissociations were observed in the TiAl phase [12,13,16].
The main slip systems in compression tests were found to be a/2<110], a/2<112], and a<011] on {111}. Another slip system a<001]{110} was observed after compression above 600 °C [12]. The special brackets notation of Hug et al. [11-13] was used to characterize the tetragonally distorted $L1_0$ structure.

The objective of the present paper is to describe and discuss the enhanced ductility of two phase TiAl/Ti_3Al alloys, modified by additions of chromium and silicon. The results of the tensile tests as well as the fracture toughness measurements were completed by TEM and FIM investigations. The important aspects of this investigation will be discussed on the basis of a deformation model.

Experimental Procedure

TiAl base alloys of different compositions were produced in 50 kg ingots by skull melting technique. The chemical analysis is listed up in Table I. The specimens for the tensile tests (dimensions: 15 x 3.5 x 5 mm^3) and for K_{IC} measurements (6 x 4 x 50 mm, with a 2 mm notch) were cut by spark erosion.
Tensile tests were performed from room temperature up to 1000 °C, applying a strain rate of $\dot{\epsilon} = 1.6 \cdot 10^{-3}$ 1/s. At higher temperatures the experiments were carried out under argon gas atmosphere to protect the samples from oxidation.

The fracture toughness of the investigated TiAl/Ti_3Al alloys were quantitatively determined by measuring the temperature dependent K_{IC} values in four-point bending tests [17]. The microstructure of the alloys were studied in greater detail by optical microscopy, scanning electron microscopy (SEM), transmission electron microscopy (TEM), and field ion microscopy (FIM).

Table I. Chemical analysis of the investigated TiAl/Ti_3Al alloys.

Alloy	Ti [at%]	Al [at%]	Cr [at%]	Si [at%]	O_2 [ppm]	N_2 [ppm]
Ti56Al44	56	44	-	-	<400	<500
Ti52Al48	52	48	-	-	<400	<500
TiAl46Cr1Si0.2	51.8	46	1	0.19	<400	<500

For TEM observations small discs with 3 mm diameter and 500 μm thickness were cut by spark erosion from the plastically deformed region as well as from the grip section of the tensile test specimens. Polishing was carried out by using Struers electrolyte A3 (600 ml methanol, 360 ml butylcellosolve, 60 ml perchloric acid) at 21 V and -25 °C. The TEM Philips EM420 was equipped with $\pm 60°/30°$ double tilt specimen holder. The lamellar Ti_3Al and TiAl phases were clearly distinguished from each other either by applying the quantitative EDX system or SAD diffraction.

Results and Discussions

Mechanical Properties

The mechanical properties of the TiAl/Ti_3Al alloys were investigated in detail. The flow stress, tensile strength and elongation to failure as a function of the temperature are shown in Fig. 1a-c. The yield stress curves (Fig. 1a) exhibit a slight increase from room temperature up to 600°C or 800°C. The chromium and silicon containing alloy reveals a maximum yield stress of 400 MPa. The Ti56Al44 alloy behaved brittly at temperatures up to 600°C. Above 800°C the yield stresses of all investigated TiAl alloys decreased monotonically and reached 150 MPa at 1000°C, respectively.

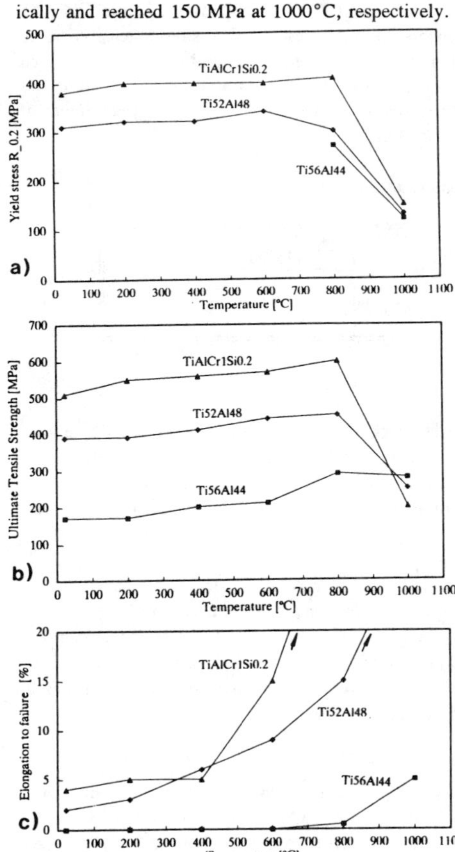

Fig. 1a-c. Mechanical properties of the three examined TiAl base alloys as a function of temperature (a) yield stress, (b) ultimate tensile strength, and (c) elongation to failure.

a)

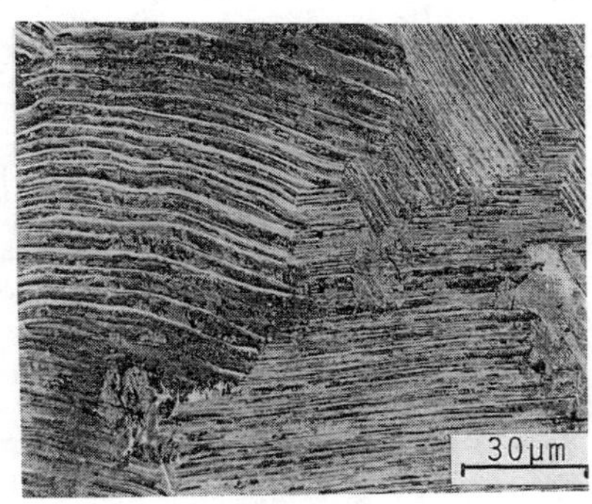

b)

Fig. 2a-b. Optical micrographs of the Ti52Al48 alloy after tensile deformation (a) at room temperature and (b) at 800 °C.

The ultimate tensile strength shows a similar temperature dependence (Fig. 1b). For the Ti56Al44 alloy the strength values increase slightly from 170 MPa at room temperature to 290 at 800 °C. The Ti52Al48 alloy reveals a maximum strength value of 450 MPa at 800 °C, and for the TiAl46Cr1Si0.2 material 600 MPa at 600 °C was achieved. Above 800 °C the ultimate tensile strength of all alloys decrease to about 250 MPa or 200 MPa.

Maximum elongations to failure of $\varepsilon_{pl} \approx 4$ % were measured for the TiAl46Cr1Si0.2 alloy at room temperature in tension tests. At elevated temperatures between 200 and 400 °C elongations of 5 % were determined. At 800 °C $\varepsilon_{ep} = 35$ % and at 1000 °C a fracture strain of $\varepsilon_{pl} \approx 100$ % was determined. In comparison to this a TiAl46Cr3 alloy was tested. It showed even a slightly higher elongation to failure than TiAl46Cr1Si0.2. Therefore it is proposed that Cr is an effective alloying element which improves the ductility in the lower temperature regime. For the Ti52Al48 alloy, lower elongations to failure (30 % at 1000 °C) were measured. However, the Ti56Al44 shows a rather brittle behaviour without any measurable fracture strain below 800 °C.

The temperature dependent fracture toughness in K_{IC} values are presented in Table II. The Ti56Al44 alloy reveals K_{IC} values of 15 and 21 MPa$\sqrt{}$m at room temperature and 400 °C. For the more ductile Ti52Al48 alloy, which exhibits a certain plastic strain in room temperature tensile tests, critical stress intensity factors of 25 MPa$\sqrt{}$m at room temperature and 29 MPa$\sqrt{}$m at 500 °C were measured. For the modified TiAl46Cr1Si0.2 alloy a superior fracture toughness of more than 33 MPa$\sqrt{}$m at room temperature and 500 °C were obtained. These values are beyond the upper limit of the test conditions for ductile samples of small sizes.

Table II. K_{IC} values [MPa$\sqrt{}$m] at room temperature and 500 °C.

	Ti56Al44	Ti52Al48	TiAl46Cr1Si0.2
RT	15	20	>33
500 °C	21	29	>35

Microstructure

The optical micrographs in Fig. 2 represent the lamellar microstructure of the as cast Ti52Al48 alloy, deformed at room temperature to $\varepsilon_{pl} = 2$ % (Fig. 2a) and strained at 800 °C to $\varepsilon_{pl} = 15$ % (Fig. 2b). After low tensile deformation at room temperature the microstructure remains unchanged (Fig. 2a) compared to that of the as cast condition. Large tensile deformation of 30 % at 800 °C (Fig. 2b) produces bent TiAl and Ti_3Al lamellae, exhibiting a wavy shape.

The TEM bright field images reveal more details of the lath type microstructure. In Fig. 3a, b the image conditions are adjusted in such a way that the Ti_3Al phase appears dark and the TiAl phase bright. The Ti_3Al lamellae are oriented mostly parallel to each other in one direction inside a grain. The orientation relationship between the Ti_3Al phase and the TiAl matrix is $(111)_\gamma \| (0001)_{\alpha2}$, $<110>_\gamma \| <1210>_{\alpha2}$ [18-20]. This is confirmed by high resolution FIM investigations, shown in Fig. 4. The FIM image reveals that the phase boundaries are atomically sharp. The $\alpha_2 \| \gamma$ interfaces are represented exactly by parallel aligned $(0001)_{\alpha2} \| (111)_\gamma$ planes [21].

The TEM images show the deformed state of the constituent phases. It is obvious that the microstructural features differ slightly among the three investigated TiAl alloys. In Ti56Al44 (Fig. 3a) the Ti_3Al lamellae are distributed homogeneously within the TiAl matrix. The lamellae thickness of about 200 nm is rather uniform. The TiAl phase reveals twin boundaries parallel to the Ti_3Al lamellae. No evidence for plastic deformation, e.g. dislocations or deformation twins, was found in either phase when deformed at room temperature or at 600 °C. The largest volume fraction of the Ti_3Al phase of 25 to 35 vol% was determined for the alloy with relatively high Ti concentration of 56 at%. The Ti_3Al phase is known to be harder and less ductile than the TiAl matrix. The larger amount of the Ti_3Al phase is the main reason for the poor plasticity of Ti56Al44.

A coarser lamellar microstructure is present in the Ti52Al48 alloy (Fig. 3b). The thickness of the Ti_3Al lamellae is three times larger than that of the other alloys (see e.g. Table III). Most of the Ti_3Al lamellae end up at grain boundaries. Discontinuous Ti_3Al lamellae were not observed. In some cases the ends of the

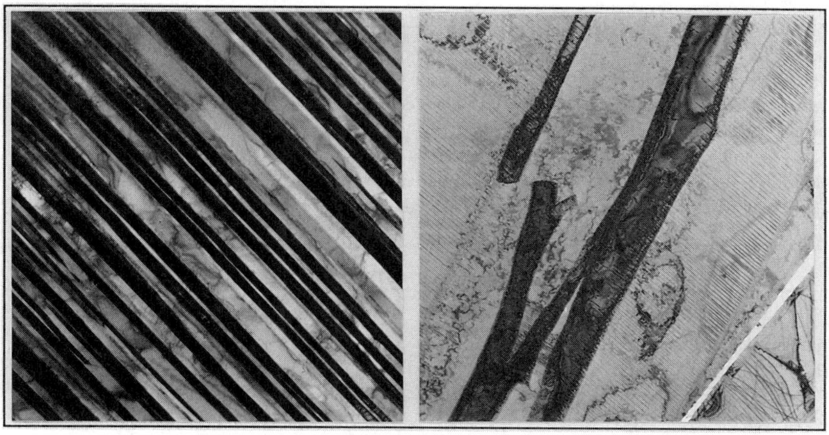

Fig. 3a-b. TEM micrographs of the deformed sections of tensile test samples of the alloy (a) Ti56Al44, and (b) TiAl46Cr1Si0.2.

Ti_3Al lamellae taper off. Fig. 5 shows the sharp tips at the end of two multiply sheared Ti_3Al lamellae in the Ti52Al48 alloy. The Ti_3Al lamellae are partly sheared by TiAl deformation twins coming from TiAl if a critical thickness of 1 μm is not overstepped. Below this critical thickness the deformation twins penetrate the Ti_3Al phase. This mechanism will generate deformation twins in Ti_3Al. Consequently enhanced plasticity of the alloy occurs. In contrast to this no deformation twins have been observed in single phase bulk specimens of Ti_3Al [3]. Thicker Ti_3Al lamellae shear off neither by deformation twinning nor by dislocation motion. Ti_3Al seems to act like a hard skeleton within the more ductile TiAl matrix.

Dislocations in Ti_3Al lamellae were only observed at deformation temperatures above 800°C. Hence, the Ti_3Al phase behaves brittly at lower temperatures.

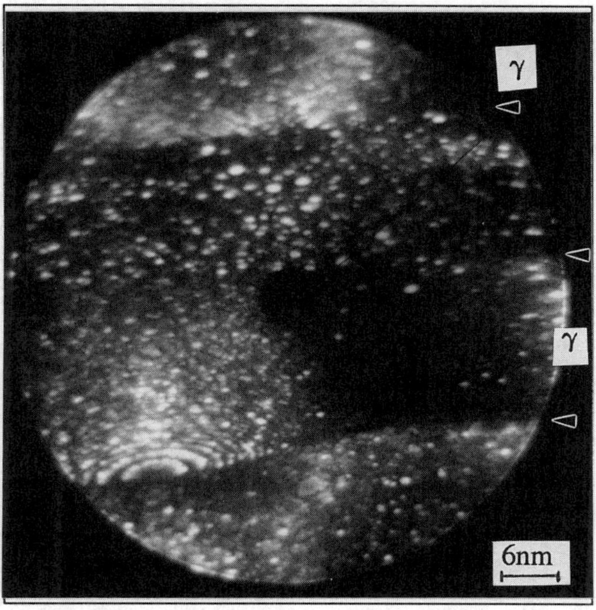

Fig. 4. FIM micrograph showing the α_2/γ interface on an atomic scale.

After annealing at 1000 °C antiphase boundaries appear in the Ti_3Al phase. This indicates that small ordered domains are present within the Ti_3Al lamellae.

The TiAl/Ti_3Al interfaces in the deformed Ti52Al48 sample are rather wavy and non planar. In the deformed tensile test specimens the TiAl matrix contains numerous dislocations and deformation twins. The deformation twins can be clearly distinguished from primary twins, formed during phase transformation from α-(TiAl) into γ-TiAl by their narrow spacings of 5 to 100 nm instead of 1 to 2 μm. From the edges of the deformation twins dislocations were emitted into the surrounding TiAl matrix. Twins as well as Ti_3Al lamellae are acting as dislocation sources. The highest density of deformation twins were found in specimens deformed at medium test temperatures (600 °C).

Table III. Thickness of Ti_3Al lamellae, interlamellar spacings, TiAl twin thickness, and volume fraction of the Ti_3Al phase for the three examined TiAl alloys.

Alloy		Ti56Al44	Ti52Al48	TiAl46Cr1Si0.2
Mean thickness of Ti_3Al lamellae	[nm]	200	500..2000	200
Interlamellar spacing of the Ti_3Al constituent	[nm]	500...1000	2000	500
Thickness of TiAl twins in as cast alloys	[nm]	100...500	100	100
After tensile deformation	[nm]	-	5...100	5...10
Volume fraction of the Ti_3Al phase	[vol%]	25...35	10...15	15...20

Fig. 5. TEM micrograph of a Ti_3Al lamellae (dark) sheared by deformation twins in the TiAl matrix (bright) at 600 °C.

Room temperature deformation, however, leads to finely spaced dislocation networks. These are localized in deformation bands.

The TEM image in Fig. 3b displays the microstructural features of the TiAl46Cr1Si0.2 alloy. The Ti_3Al lamellae exhibit a mean thickness of about 200 nm and a separation distance of 500 nm, respectively. The Ti_3Al fraction of 15 to 20 vol% is significantly higher than that in the Ti52Al48 alloy. This difference is surprising because the Al content in both alloys is almost the same. EDX analysis revealed that the solubility of Cr in the Ti_3Al phase is significantly higher than in the TiAl phase.

Cr addition stabilizes the Ti_3Al phase. The deformed specimen contains uniformly distributed deformation twins with a thickness of about 5 to 10 nm. Twin formation have been studied by field ion microscopy in greater detail. The FIM image in Fig. 6 shows that the twin boundary is extremely smooth and does not exceed 2 - 3 atomic layers in thickness.

The deformation twins change their orientations if they interact with primary twin boundaries. Lipsitt et al. [10] found cross twinning also in tensile specimen of TiAl, deformed at 700 °C. Thinner Ti_3Al lamellae of thickness dimensions much less than 1 μm were not sheared by deformation twins in contrast to those of the Ti52Al48 alloy. The high density of deformation twins causes the improved ductility of this alloy. The tensile test samples deformed at 600 °C reveal a larger amount of twins than the specimens deformed at room temperature. The latter ones showed mainly stacking faults. Dislocations were observed in both specimens.

Selected areas of the deformed specimen of the TiAl46Cr1Si0.2 alloy have been studied by TEM in greater detail. Fig. 7a represents a dark field and Fig. 7b shows a weak beam image with the marked $\vec{g} = \bar{1}11$.

To perform dislocation analysis, 12 different $\bar{g}$ vectors were chosen for imaging. The results showed that mostly the a/2 [110] and a/2 [$1\bar{1}0$] dislocation are dominant, some a/2 [101] and a/2 [$01\bar{1}$] dislocations have also been observed. The preferential slip planes were determined to be ($11\bar{1}$) and ($\bar{1}11$), respectively.

For understanding the important role of the deformation twins for the enhanced plasticity of this alloy special attention was focused on this subject. A model shown in the schematic drawing of Fig. 8 illustrates the generation of deformation twins in detail.

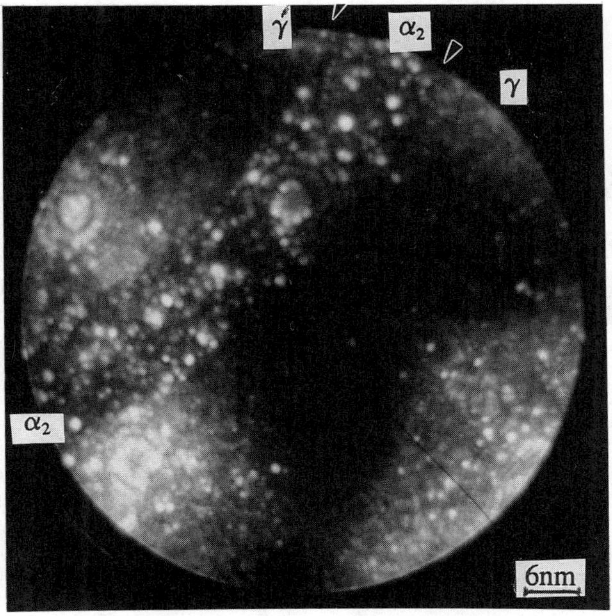

Fig. 6. FIM image displays the true γ/γ_{true} relation in the TiAl phase.

Low angle grain boundaries act as obstacles against the propagation of deformation twins. Some twins have been passed through, and most of them are locked. In front of the penetrating twins dislocations appeared. These are clearly visible in Fig. 7a by using the image condition $\bar{g}$ = ($\bar{1}11$) because then the ($11\bar{1}$) twin boundary is inclined steeply.

The dislocation analysis showed that a/n [$\bar{1}21$] partial dislocations are present on the ($11\bar{1}$) glide plane which is identical to the twin plane. a/6 [$\bar{1}21$] partial dislocation cannot be distinguished from similar a/3 and a/2 [$\bar{1}21$] dislocations without image computer simulation. However, Hug [13] has shown that dissociated a/6 <121] dislocations are preferentially present in deformed TiAl. The distance between the a/n [$\bar{1}21$] dislocations was determined to about 20 to 40 nm, respectively. The more the dislocations have penetrated into the small angle grain boundary the thicker are the twins observed in the left grain (Fig. 6). This leads to the conclusion that the ($11\bar{1}$) deformation twins are formed by the movement of the [$\bar{1}21$] partial dislocations.

The presence of the Ti_3Al phase with an optimum volume fraction of about 15 vol% in a TiAl matrix is necessary to improve the mechanical properties, especially the ductility of TiAl alloys. An important reason is also the high affinity of the Ti_3Al phase to interstitially dissolved elements such as N, O, C. It was found that Ti_3Al exhibits a preferential solubility especially for oxygen compared to TiAl [8,22,23]. Consequently the oxygen will be gettered by the Ti_3Al lamellae and small amounts of impurities will increase the low temperature plasticity by dislocation glide [4].

In addition, shearing of the Ti_3Al lamellae with thickness less than 1 μm was observed only in the alloy without Cr. This implies that chromium strengthens and stabilizes the Ti_3Al phase in the modified alloy. Chromium and silicon have another positive effect in Ti_3Al/TiAl alloys. The Ti_3Al lamellae are distributed homogeneously with finer spacings within the TiAl matrix. In the Cr and Si free Ti52Al48 alloy thicker lamellae with wider spacing were observed. The presence of small amounts of silicon supports the nucleation of the Ti_3Al phase which results in a finer lamellar microstructure. This improves the mechanical properties especially the yield strength as well as the fracture toughness and the ductility in general.

Fig. 7a-b. Dislocation structure in a deformed TiAl46Cr1Si0.2 tensile test sample. (a) dark field image, (b) weak beam images taken with γ = <111] vectors.

The Ti_3Al phase is stronger and less ductile than TiAl. Larger volume fractions of Ti_3Al in Ti56Al44 cause the poor ductility in comparison to the Ti52Al48 alloy. The TEM results clearly showed that the α_2-Ti_3Al phase acts like a hard scaffold. Dislocations in the Ti_3Al were detected only after tensile deformation above 800 °C, when Ti_3Al lamellae have been distorted. The strong decrease of the yield stress and ultimate tensile stress above 800 °C might be caused by weakening effects of Ti_3Al. Its lath type of morphology deflects a propagating crack and will spread the crack tip, which results in higher K_{IC} values.

The TiAl matrix was also found to be slightly modified by Cr and Si. TEM investigations clearly showed the presence of <121] dislocations in front of the deformation twins. For pure fcc metals with low stacking fault energy the model of twin formation by the movement of a/6 <112>. Shockley partial dislocations is well established [24]. For TiAl with an ordered $L1_0$ superlattice structure with a slight deviation from the fcc symmetry (c/a= 1.018) a similar mechanism can be assumed (Fig. 8). A moving a/6 <121] partial dislocation on a (111) glide plane generates an intrinsic stacking fault and one lattice plane is effectively removed. Consequently the stacking sequence is disturbed. Twinning occurs if one partial dislocation moves on every neighbouring plane. The glide planes of the partial dislocations are parallel to the twin planes. For the ordered $L1_0$ lattice the mechanism described takes place in the same way except that two types of atoms (Ti and Al) are present, marked in Fig. 8 by capital and small letters. After passing six a/6 <121] partial dislocations a small twin has been formed without disturbing the ordering of the lattice. A stepwise motion of the Shockley partials leads to an enlargement of smaller twins. The mechanism of twin formation and the enlargement by moving of the Shockley partial dislocations is in agreement with the TEM results (Fig. 7a-b). Hug and Veyssiere [25] found a similar mechanism in Mn doped TiAl alloys. They proposed that a/2 <112] superdislocations are forming twins by splitting into a/6 [112] dislocations. These are moving to the upper neighbouring glide plane either by climb or cross slip. Only a Frank-Read source is necessary to activate this process. It should be mentioned that besides this deformation twinning the so called structural twinning occurs. After peritectic solidification the formed α-Ti(Al) solid solution changes its lattice structure from hcp to fct ($L1_0$) structure with decreasing temperature. The structural twins are of larger size and have been observed in undeformed specimens.

Deformation twins and antiphase boundaries are mainly present in the Cr and Si containing alloy. For fcc alloys it is well known that deformation twins occur if the stacking fault energy is decreased. A low stacking fault energy increases the distance between partial dislocations and allows them to move more easily. The presence of deformation twins leads to the conclusion that chromium lowers the stacking fault energy in TiAl. Therefore an enhanced low temperature ductility of the Cr containing TiAl alloys by deformation twinning occurs. Similar results were obtained for binary TiAl alloys by Hall et al. [23]. Hanamura et al. [26] found enhanced twin formation during room temperature deformation of TiAl-2 % Mn.

The effect of alloying elements on the plasticity of TiAl is based on dislocation dissociation by solute atoms. Hug et al. [11-13] found favoured dissociations of <110] into <121] dislocations in pure TiAl. Hahn et al. [16] observed preferential dissociation in TiAl by additions of Nb. Hanamura et al. [26] assumed stabilization of <121] dislocations in the presence of Mn.

Chromium may stabilize the <121] partial dislocations. Plastic deformation might be easier and deformation twinning can be explained in the sense by increasing the mobility of partial dislocations.

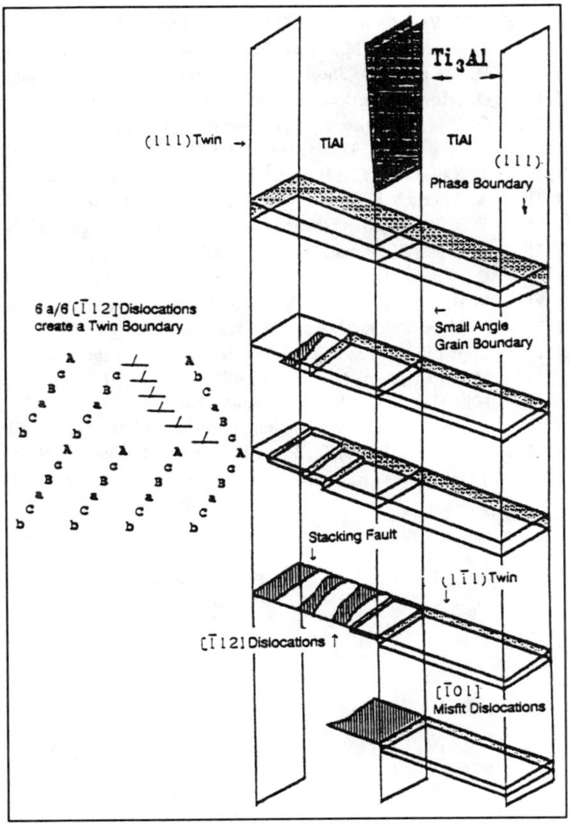

Fig. 8. Schematic drawing and model for generation and propagation of twins by movement of six a/6 [121] partial dislocation in an ordered $L1_0$ lattice.

Conclusions

Additions of small amounts of Cr and Si to binary Ti52Al48 alloys improve the strength properties as well as the room temperature tensile ductility and the fracture toughness. The enhanced ductility is caused mainly by glide of <121] partial dislocations and subsequent twin formation in the TiAl phase. Furthermore additions of Cr and Si stabilize the Ti_3Al phase and increase its strength.

For optimizing the microstructure and the mechanical properties thin Ti_3Al lamellae should be homogeneously distributed throughout the TiAl matrix. The Ti_3Al volume fraction does not exceed 20 vol%, otherwise the brittle character of Ti_3Al prevails.

References

[1] Young Wim Kim, J. of Metals, July (1989) 24-30

[2] E. J. Hall, Shyh-Chin Huang, J. Mater. Res. 4 (1989) 595-602

[3] H.A. Lipsitt, D. Shechtman, R.E. Schafrik, Metall. Trans. 11 (1980) 1369-1375

[4] J,.J. Valencia, C. McCullough, C.G. Levi, R. Mehrabian, Scripta Metall. 21 (1987) 1341-1346

[5] J.C. Mishurda, J.C. Lin, Y.A. Chang, J.H. Perzepzko, Mat. Res. Soc. Proc. 133 (1989) 57-62

[6] S.C. Huang, E.L. Hall, Mat. Res. Soc. Proc. 133 (1989)

[7] J.L. Murray, in Binary Alloy Phase Diagrams, T.B. Massalski (ed.), ASM, Metals Park, OH (1986) 173-176

[8] V.K. Vasudevan, M.A. Stuke, S.A. Court, H.L. Fraser, Phil. Mag. Letters 59 (6) (1989) 299-307
[9] D. Shechtman, M.J. Blackburn, H.A. Lipsitt, Metall. Trans. 5 (1974) 1373-1381
[10] H.A. Lipsitt, D. Shechtman, R.E. Schafrik, Metall. Trans. 6 (1975) 1991-1996
[11] G. Hug, A. Loiseau, A. Lasalmonie, Phil. Mag. 54 (1986) 47-65
[12] G. Hug, A. Loiseau, P. Veyssiere, Phil. Mag. 57 (1988) 499-523
[13] G. Hug, Ph.D. Thesis, Université de Paris Sud, Orsay (1989)
[14] T. Kawabata, Y. Takezono, T. Kanai, O. Izumi, Acta Metall. 36 (1988) 963-975
[15] A. Court, V.K. Vasudevan, H.L. Fraser, Phil. Mag. A 61 (1990) 141-158
[16] Y.D. Hahn, S.H. Whang, Scripta Metall. 24 (1990) 139-144
[17] S. Reuss, H. Vehoff, Scripta Metall. 24 (1990) 1021-1026
[18] C.R. Feng, D.J. Michel, C.R. Crowe, Phil. Mag. Letters 61 (3) (1990) 95-100
[19] M.H. Yoo, C.L. Fu, J.K. Lee, Mat. Res. Soc. Symp. Proc. 133 (1989) 11189-196
[20] S. Chen, Ph.D. Thesis, TU Hamburg-Harburg (1988)
[21] Z.G. Liu, G. Frommeyer, and M. Kreuss, Scripta Metall. 25 (1991) 1205-1210
[22] M.J. Kaufmann, D.G. Konitzer, R.D. Shull, H.L. Fraser, Scripta Metall. 20 (1986) 103ff
[23] E.L. Hall, S.C. Huang, J. Mater. Res. 4 (1989) 595-602
[24] F.R.N. Nabarro, Dislocation in Solids, Vol. 2, North-Holland Pub., Amsterdam (1979)
[25] G. Hug, P. Veyssiere, Proc. Int. Conf. TEM and Plasticity, Halzhau/Dresden (1989)
[26] T. Hanamura, R. Uemori, M. Tanino, J. Mater. Res. 3 (1988) 656-664

THE CHEMISTRY and STRUCTURE of PLASMA-MELTED γ-BASED TITANIUM ALUMINIDES

R. V. Ramanujan, X. D. Zhang, I. P. Jones, J. M. Young and J. W. Sears

IRC in Materials for High Performance Applications, The University of Birmingham, Edgbaston B15 2TT, U. K.

ABSTRACT

Plasma melting has been used to produce ingots of nominal compositions Ti-48Al-2Mn-2Nb, Ti-48Al and Ti-52Al (all compositions in atomic percent).The feedstock was either compacts of the elements or vacuum arc remelted material. Cylindrical ingots of diameter 100 mm and 150 mm and lengths up to 750 mm have been produced and their macrostructure and microstructure have been examined.The results show that close control of the chemistry can be achieved both from top to bottom and from centre to edge of the ingot. There is however, a significant variation in the macrostructure from top to bottom of an ingot. The microstructures of Ti-48Al and Ti-48Al-2Mn-2Nb consist of a lamellar structure of γ and α_2 and some equiaxed γ grains whereas the microstructure of Ti-52Al consists of a dendritic morphology with aluminium-rich segregate and dendrites composed of γ and α_2. We have found, through this and related studies, that the macrostructure causes the deformation behaviour and mechanical properties to be highly anisotropic. Heat treatment before thermomechanical processing is, therefore, needed to ensure uniform mechanical properties.

INTRODUCTION and EXPERIMENTS

Aggressive high temperature working environments require superclean alloys of rigorously controlled chemical composition and microstructure. For many years vacuum arc remelting (VAR) has been the established process for producing high integrity titanium alloys, but alternative processes based on cold hearth refining have recently been introduced(1-6).

The cold hearth refining process involves melting feed material into a water cooled hearth where it initially forms a skull. Subsequently, liquid metal flows over the skull via a lip into a water cooled crucible with a retractable base to form a continuous ingot. Advantages of the cold hearth process include elimination of refractory elements (and hence removal of the risk of non-metallic particle pickup) through the ability to remove both heavy and light inclusions by density separation in the hearth.

The Retech (Retech Inc., Ukiah, CA) plasma arc furnace (2) uses plasma as the heat energy source for the melting process which takes place under a positive pressure of argon. Such conditions prevent oxidation and loss of high vapour pressure alloying elments such as aluminium and chromium. In addition, the process is tolerant of

Titanium '92
Science and Technology
Edited by F.H. Froes and I. Caplan
The Minerals, Metals & Materials Society, 1993

outgassing and hence can be used to melt sponge.

The plasma arc furnace has been used to produce cylindrical γ-based titanium aluminide ingots with a diameter of 100 mm and 150 mm and lengths up to 750 mm. The nominal compositions were Ti-48Al-2Mn-2Nb, Ti-48Al and Ti-52Al (all compositions in atomic percent). Both vacuum arc remelted (VAR) material and compacts of the elements were used as feedstock. The process parameters used, the chemistry and the structure of the ingots obtained after plasma arc melting are described in this paper. Our results on the macrostructure can be used to rationalise the highly anisotropic deformation and mechanical behaviour observed in related studies (7,8) of as-cast material.

RESULTS

CHEMISTRY OF THE INGOTS

Table 1 shows the chemical analysis performed at various locations in the ingot. The composition of the substitutional elements was determined by wet chemical analysis, the levels of oxygen and nitrogen by a inert gas fusion technique and the level of hydrogen by a vaccuum extraction technique. The oxygen level in the feedstock varied from one melt to another in the range 570 ppm to 780 ppm. For example, the actual composition of the VAR feedstock used in melt 20 is Ti-49.7Al-2.5Mn-2.1Nb with an oxygen content of 600 ppm. From the results in Table 1 it is clear that the composition of the ingot can be held to close tolerances both from top to bottom and from centre to edge. These results are consistent with those obtained using quantitative microprobe analysis (Fig. 1) on samples from melt 27 (Ti-48Al-2Mn-2Nb).

STRUCTURE of Ti-48Al-2Mn-2Nb

Figs 2-4 show the transverse sections of the plasma melted ingot (melt 38), 50mm from the top of the ingot (referred to as the top section), 300 mm from the top (middle) and 25 mm from the bottom (called the bottom section), respectively. The transverse section of the top region consists of a chill zone (roughly 4 mm) at the outer edge of the ingot, long columnar grains (20 mm) extending towards the centre of the ingot and a central region consisting of apparently equiaxed grains. The columnar grains are not radial but are curved (Fig.2) as a result of the three dimensional nature of the solidification process and, to a lesser extent, electromagnetic stirring.

The transverse section of the middle of the ingot (Fig.3) is similar to the top section of the ingot with a smaller chill zone and a smaller "equiaxed" region.The transverse section of the bottom region (Fig.4) shows a small chill zone (1mm), a reduced area fraction of columnar grains and a larger fraction of "equiaxed" grains in the central region of the ingot.

The microstructure of this alloy consists of lamellar colonies and some apparently equiaxed γ grains(9). The variation in the microstructure can be explained in terms of the change in cooling rate from centre to edge in an ingot. The slower cooling in the central region of an ingot produces more γ grains.

STRUCTURE of Ti-48Al

The transverse sections were prepared at similar locations or areas in the ingot as in the case of Ti-48Al-2Mn-2Nb. The top section is similar to that of that of Ti-48Al-2Mn-2Nb with a smaller chill zone. The longitudinal section (Fig.5) is seen to consist of two sets of columnar grains. One set of grains is larger (approx. 2 mm in width) at the edge of the ingot and are thinner as they approach the centre of the ingot. These grains curve upwards (towards the top of the ingot) as they approach the central region. The second set of grains consists of columnar grains of constant width that run parallel to the length of the ingot. It is the cross-section of these grains that form the apparently equiaxed structure in the central region of transverse sections. The microstructure of the longitudinal section (Fig.6) is seen to consist of long lamellar colonies of different orientation.

The transverse section of the middle of the ingot is similar to the top section but it has a higher area fraction of equiaxed grains. Correspondingly, the longitudinal section contains a larger area fraction of columnar grains which are parallel to the length of the ingot. The bottom region (Fig. 7) consists of a small chill zone, a small columnar region and a large (roughly 35 mm) apparently equiaxed structure in the central region. Again, there is a corresponding increase, in the longitudinal section, of columnar grains which are parallel to the length of the ingot.

STRUCTURE of Ti-52Al

Macrostructures were prepared from the same locations in the ingots as Ti-48Al-2Mn-2Nb and Ti-48Al. The macrostructure of the top section (Fig.8) shows a columnar structure from the edge to the central region. This columnar structure consists of dendrites surrounded by aluminium-rich segregate. Quantitaive electron probe microanalysis showed that the average composition of the dendrites is Ti-50.2Al while that of the segregate is Ti-54.4Al.The central region is also dendritic but consists of cross sections of dendrites which have formed by solidification along the longitudinal direction. The dendritic nature of the strucuture can be seen from Fig. 9 which shows an aluminium-rich interdendritic region and aluminium-lean dendrites consisting of alternate γ and $\alpha 2$ plates. This microstructure is, therefore, very different from that of the as-cast Ti-48Al (Fig 6.). The middle section is similar to the top section. However, the bottom section shows a much larger area fraction of dendrites formed from the growth of dendrites in the longitudinal direction and a correspondingly small growth of dendrites in the radial direction. Towards the edge of the ingot roughly circular bands, with a thickness of approximately a millimetre, are found which etch differently from the surrounding regions. These areas, observed in all three sections, have been previously called "tree rings" (10).In titanium aluminides this effect has been attributed to minor compositional fluctuations(10). Crystal growth experiments (11) conducted on Ti-52Al also show a periodicity in the structure (and presumably the composition) along the length of the specimen.

DISCUSSION

For plasma arc melting γ-based titanium aluminide ingots the feedstock can be either elemental feedstock or VAR material. The surface quality of the ingots is generally smooth with occasional laps(3). The results of the chemical analyses show that close tolerances of the composition of the ingot can be achieved.

The macrostructure of the ingots can be understood in the following manner. For the three ingots studied, the interplay between the heat extraction from the bottom of the ingot and that from the side walls appears to determine the macrostructure. In all three sections (top, middle and bottom) the chill zone is the usual result of the cooling by the side walls. In the top and middle section of the ingots the heat extraction is mainly from the side walls. The direction of the growth of the grains is along the direction of maximum heat extraction (maximum thermal gradient). Consider a longitudinal section of the ingot during the melting operation. The directions of maximum heat extraction are given by the perpendiculars erected to the tangent drawn at each point of the melt pool/solid interface. An elliptical melt pool therefore gives rise to the curved columnar grains which point towards the top of the ingot in the longitudinal section. The transverse section is essentially a cross-sectional view of these grains.

The bottom sections show the effect of the heat extraction from the withdrawal crucible. This additional heat extraction causes a number of columnar grains to grow in the central region longitudinally from the withdrawal crucible. However, there is still substantial heat extraction from the side walls and curved columnar grains are still present, although to a lesser extent than in the top and middle sections. The main effect of electromagnetic stirring seems to be the creation of additional thermal gradients which cause the solidifying grains to "twist" as they solidify. This effect is particularly evident in Ti-48Al-2Mn-2Nb (melt 38).

The macrostructure plays an important role in several aspects of the mechanical behaviour of the as-cast plasma arc melted material. Our results on forging (7) of as-cast titanium aluminides show that a cylindrical billet of circular cross-section has an elliptical cross-section after forging. This behaviour is a direct result of the variation in flow stress as a function of the local orientation of the lamellae relative to the forging direction (12).James et al (8) have found that the fracture toughness of as-cast Ti-48Al specimens oriented in the transverse direction is more than 20 MPa√m while it is less than 10 MPa√m in the longitudinal direction. Even more dramatic is the change in the fatigue crack growth rate measured on Ti-48Al at 800°C in vacuum (8). In the transverse direction a value of 5.27×10^{-12} mm/cycle is obtained while a value of 1.51×10^{-31}mm/cycle is obtained in the longitudinal direction. From the present results on the macrostructure these differences in mechanical behaviour can be explained in terms of the difference in the orientation of the lamellae from the transverse to the longitudinal direction.

By appropriate heat treatment these effects of the macrostructure can be altered significantly. An example is shown in Fig. 10 where heat treatment for 140 hours at 1200°C converts the lamellar structure into equiaxed γ grains. When heat-treated cylindrical billets are forged, circular cross-sections remain circular after forging. Another method is to heat treat the alloy at 1200°C to produce equiaxed γ grains, followed by heat treatment in the α phase field followed by slow cooling.

CONCLUSIONS

1. Plasma arc melting of Ti-48Al-2Mn-2Nb, Ti-48Al and Ti-52Al has been carried out using both elemental and VAR feedstock. 100 mm diameter ingots 750 mm long have been produced and our results on the chemistry of the plasma melted alloys shows that close control can be achieved both from top to bottom and from centre to edge of the ingot.

2. The macrostructure of the ingots varies with position in the ingot. The transverse sections show that a chill zone exists at the edge, columnar grains growing from the edge to the centre and a central equiaxed region. The longitudinal sections show a chill zone and two types of long columnar grains. One set of grains start from the outer edge and curve upwards as they approach the centre becoming thinner as they grow. The second set of grains are of constant width and run parallel to the length of the ingot. This second set is seen as equiaxed grains in the transverse sections. There is an increase in the proportion of the second set of grains from the top to bottom of the ingot. This increase can be explained by the relative magnitude and direction of heat extraction from the side walls and the bottom withdrawal crucible.

3. The microstructures of Ti-48Al and Ti-48Al-2Mn-2Nb consist of a lamellar structure of γ and $\alpha 2$ and some equiaxed γ grains. The columnar grains of the macrostructure affect the microstructure by dividing lamellar colonies of different orientation. The microstructure of Ti-52Al consists of a dendritic morphology with aluminium rich segregate and dendrites composed of γ and α_2.

4. Our results show that characterisation of the macrostructure enables us to rationalise the highly anisotropic mechanical and deformation behaviour of as-cast plasma arc melted ingots.

ACKNOWLEDGEMENTS

Permission to quote the unpublished work of colleagues, discussions and experimental help of many colleagues in the IRC is gratefully acknowledged, as is financial support of the SERC through the IRC.

REFERENCES

1 J. W. Sears, JOM, 42 March (1990), 17-21.

2. J. W. Sears, Advances in Powder Metallurgy-1991, vol. 6, (Metal Powder Industries Federation, Princeton, New Jersey, 1991), 335-346.

3. J. W. Sears, Proc. of this Conf. (1992).

4. F. H. Froes, C. Suryanarayana and D. Eliezer, ISIJ Intl.,31(1991), 1235-1248.

5. S. Sen and D. M. Stefanescu, JOM, May (1991), 30-34.

6. P. Wightman and E. Hengsberger, Metals and Materials, 7 (7) (1991), 676-682.

7. X. D. Zhang, R. V. Ramanujan, T. A. Dean, R. Dewes, M. H. Jacobs and M. L. H. Wise, Proc. of this Conf. (1992).

8. A. W. James et al, Proc. of this Conf (1992).

9. R. V. Ramanujan et al, Proc. of IRC '92, University of Birmingham, U. K.(1992).

10.R. Boyer, Metals Handbook, Ninth Edition, vol. 9, (ASM, Metals Park, 1985), 465.

11.R. V. Ramanujan, manuscript in preparation (1992).

12. M. Yamaguchi, ISIJ Intl., 31 (1991), 1127-1133.

TABLE 1 CHEMICAL ANALYSIS OF THE PLASMA ARC MELTED INGOTS

Melt No.	Nominal Composition		Al (a/o)	Nb (a/o)	Mn (a/o)	O_2 ppm	N_2 ppm	H_2 ppm	C w/o
20	Ti-48Al-2Mn-2Nb	T	50.1	2.12	1.69	650	40	10	.015
27	Ti-48Al-2Mn-2Nb	T, C	49.7	2.1	2.1	650	30	10	0.025
		T, mr	49.8	2.1	2.0	650	35	<10	0.020
		T, E	50.1	2.1	1.88	700	40	<10	0.015
36	Ti-48Al	T	48.8			750	20	10	0.010
		M	49.5			650	20	10	0.015
38	Ti-48Al-2Mn-2Nb	T	50.4	2.14	1.96	750	40	10	0.010
		M	49.7	2.06	1.95	700	35	10	0.010
		B	50.1	2.09	1.90	850	40	10	0.030
42	Ti-52Al	T	53.5			600	10	10	0.015
		M	53.8			600	15	<10	0.015
		B	53.1			750	20	10	0.015

T= 50mm from top of ingot, M= 300 mm from top of ingot, B= 25 mm from the bottom of the ingot;
C=centre of the ingot, mr=midradius of the ingot, E= edge of ingot.

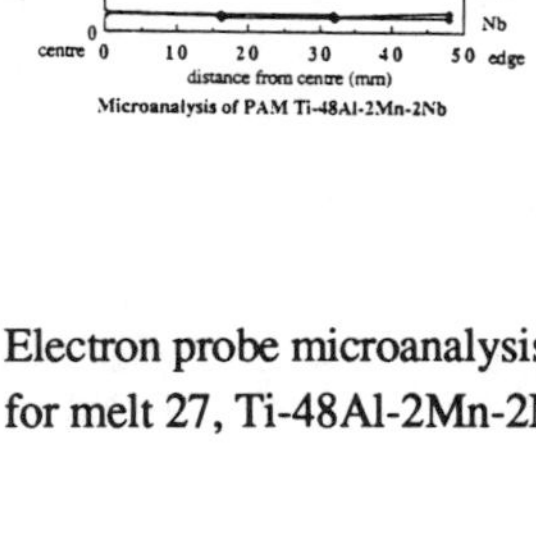

Fig. 1 Electron probe microanalysis results for melt 27, Ti-48Al-2Mn-2Nb

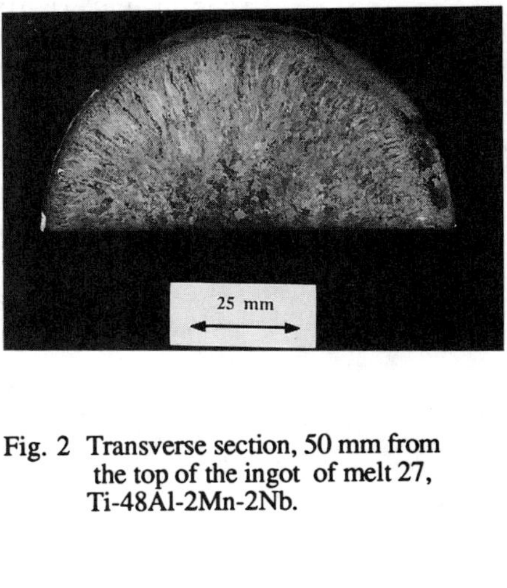

Fig. 2 Transverse section, 50 mm from the top of the ingot of melt 27, Ti-48Al-2Mn-2Nb.

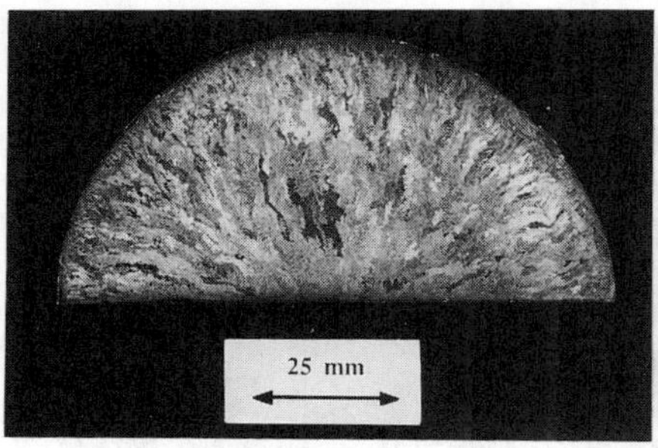

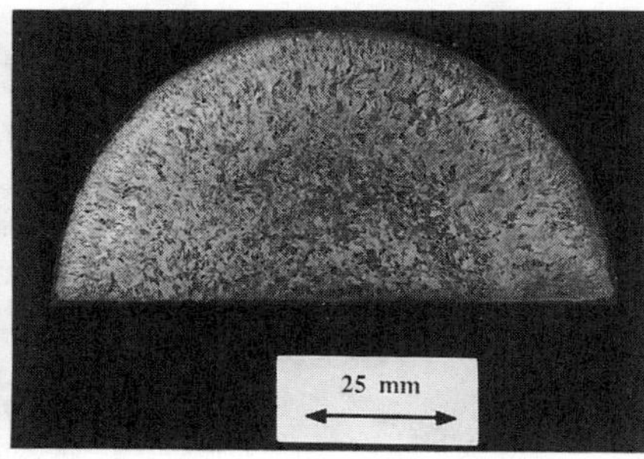

Fig. 3 Transverse section of Ti-48Al-2Mn-2Nb from the middle of the ingot.

Fig. 4 Transverse section of Ti-48Al-2Mn-2Nb 25 mm from the bottom of the ingot.

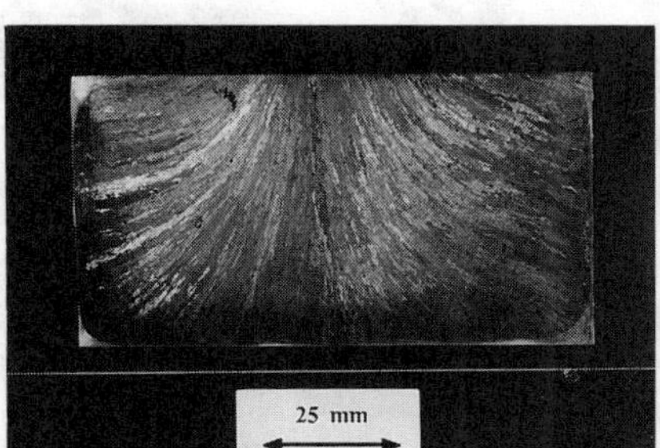

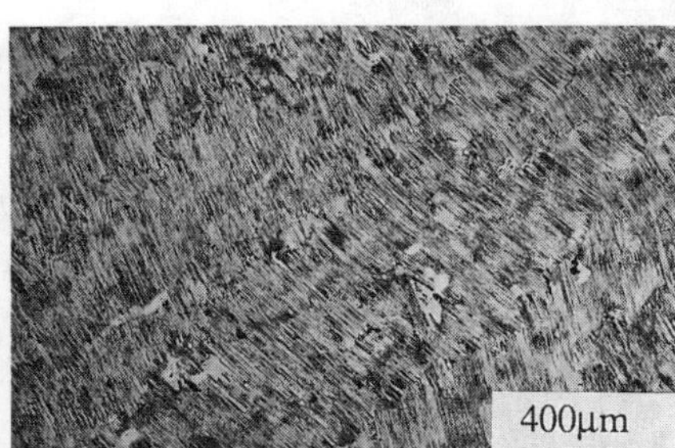

Fig. 5 Longitudinal section of Ti-48Al-2Mn-2Nb 50 mm from the top of the ingot showing columnar grains.

Fig. 6 Microstructure of as-cast Ti-48Al showing columnar grains containing lamellar colonies.

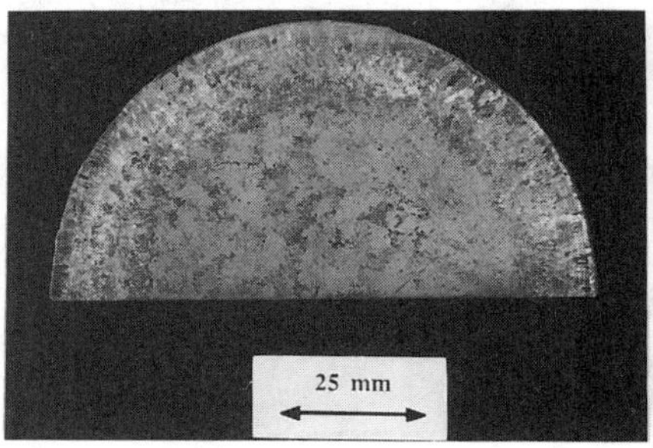

Fig. 7 Transverse section of Ti-48Al, 25 mm from the bottom of the ingot showing a large area fraction of apparently equiaxed grains.

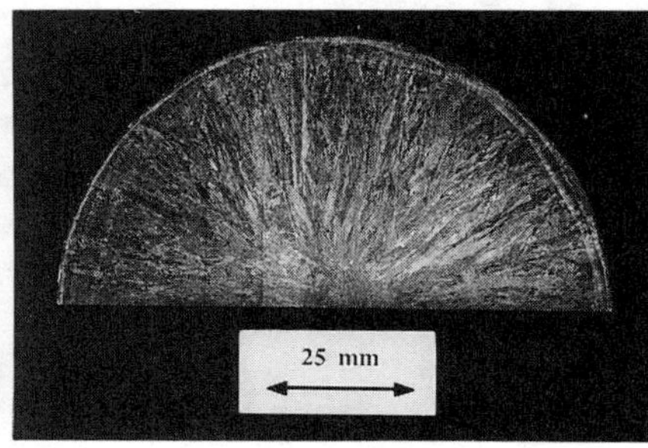

Fig. 8 Transverse section of Ti-52Al, 50 mm from the top of the ingot. Long columnar grains from the edge to the centre are evident.

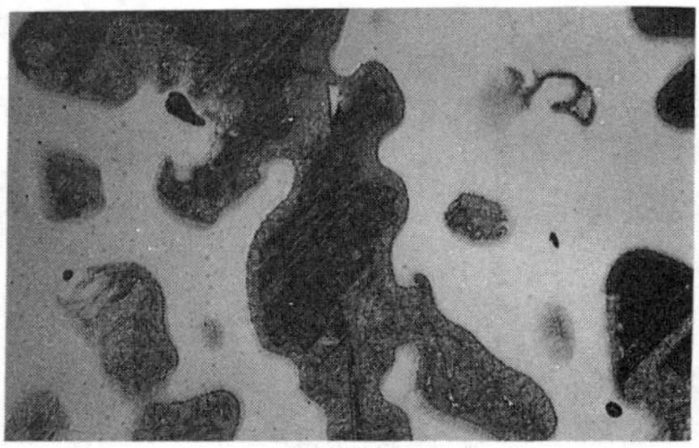

Fig. 9 Microstructure of as-cast Ti-52Al, showing aluminium-rich segregate and dendrites containing α_2 and γ.

Fig. 10 Microstructure of Ti-48Al-2Mn-2Nb heat treated at 1200°C for 140 h. followed by furnace cooling, showing equiaxed γ grains with a small amount of α_2.

FORGING, FORMING, JOINING, AND CASTING

Titanium - a Review of Current

Forming and Fabrication Techniques

J W Brooks, P J Bridges and D Stephen

Inco Engineered Products Ltd (IEPL)

28-30 Derby Road, Melbourne, Derbyshire, England DE7 1FE

Abstract

Titanium alloys are unique in terms of the wide range of forming and fabrication methods which can be used on them. This paper describes the modern methods of making shapes and components from titanium, and outlines the potential and limitation of these techniques.

Topics covered include isothermal forging, casting and superplastic forming/diffusion bonding.

Introduction

The increased use of titanium alloys in the engineering industries has given rise to a number of advanced forming technologies. The processes discussed in this paper, isothermal forging, casting and superplastic forming are used to produce a wide range of components and some of these are discussed in the context of the manufacturing route.

Isothermal Forging

Many of the alloys used to make gas turbine components have characteristics which require special consideration in relation to their processing.

Firstly, by the nature of the engineering requirement placed on them, these materials retain high strength at elevated temperatures even up to the normal forging range. This obviously has implications with respect to the size and cost of the forging equipment used. Secondly there is usually a stringent requirement for a particular distribution of microstructure and properties and a corresponding requirement for consistency between successive components and production batches. Thirdly, turbine materials are comparatively expensive because of the alloying elements used and the complex melting route needed to ensure a clean, homogeneous forging billet.

All these considerations have led to the development of isothermal forging. This section describes the process, the equipment, and the techniques used to produce turbine discs by this method.

Characteristics of Isothermal Forging

There are three main features of isothermal forging which distinguish it from the conventional forging process.

Titanium '92
Science and Technology
Edited by F.H. Froes and I. Caplan
The Minerals, Metals & Materials Society, 1993

(1) The dies and work-piece are maintained at the same temperature throughout the forging operation.

(2) The forging operation is carried out at a relatively slow, controlled, strain rate, generally in the range 10^{-5} to 10^{-1} per second.

(3) The forging operation is done in a vacuum or an inert atmosphere.

These features produce several benefits. With the dies and the work-piece at the same temperature there is no die-chilling effect. This facilitates the generation of a uniform microstructure across the section of the work-piece.

The slow strain rate used enables advantage to be taken of the corresponding reduced flow stress. Figure 1 illustrates this effect with reference to IMI 834. Reducing the strain rate by a factor of ten reduces the flow stress by up to 50%. The implication of this is that, for an equivalent work piece, press loads are much less in isothermal forging than in conventional forging. Thus the amount of superstructure is less and there is no environmental pollution in terms of noise and vibration. The slow action is also much more controllable and easier to record than in conventional processes.

In the development of specific microstructures in titanium based materials there is often a requirement to operate within a certain strain rate range during hot working. This is necessary to give the right balance between microstructure and properties.

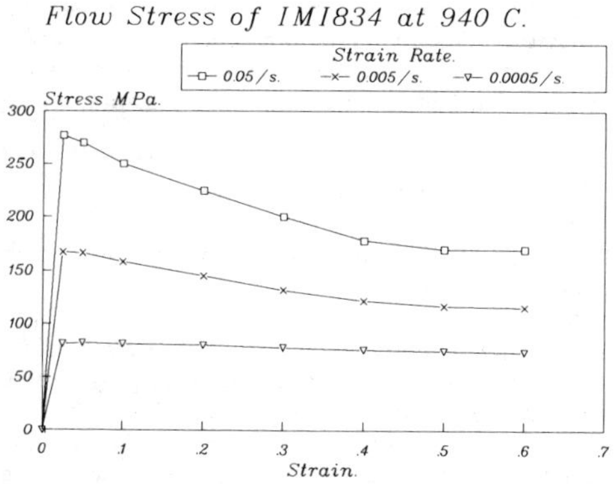

Figure 1 - The effect of strain rate on flow stress

The control of ram speed in isothermal forging is infinitely variable and can be pre-set to follow a specific strain rate programme. Thus, for example, a constant compressional strain rate along the axis of a cylindrical billet necessitates a corresponding decrease in the ram speed as the axial length decreases.

A further benefit of a slow strain rate is that there is no adiabatic heating in the work-piece. Adiabatic heating would, in a faster process, be generated in a non-uniform manner and could thereby generate non-uniform structures.

The production rates from an isothermal press are generally lower than those from a conventional press. However this disadvantage is offset by the fact that isothermal forging is a one-shot operation and there is no need for a series of shaping stages with the associated reheating and die changes.

The slow squeezing operation also enables pieces to be forged to near net shape. This can give significant savings on material cost in the forging and on subsequent machining operations.

The use of an inert atmosphere is for two reasons. The first is to prevent oxidation of the work piece during forging. This would normally be aggravated by the extended times that the piece is in the preheat furnace or in the press. However both these operations are done in an oxygen-free atmosphere and this results in a very clean,

Figure 2 - View of the IEP isothermal press

Figure 3 - As forged near net shape turbine disc

almost shiny, surface on the forging. The second reason for using this environment for forging relates to the fact that, for forging operations above about 1000°C, it is necessary to use a molybdenum based material for the dies. Only this has sufficient strength for the stresses imposed. However at these temperatures molybdenum oxidises very quickly and hence an inert atmosphere is essential.

The IEP Airfoils Isothermal Press

IEP Airfoils installed their isothermal press in 1986. It has a maximum load capability of 3200 tonnes and, at the start, had a relatively simple air-lock device for introducing the forging billets into the dies. Modifications to this system have involved building a robot-serviced pre-heating chamber in which the billets are heated before introduction into the dies. The pre-heating chamber contains the same inert atmosphere as the forge and, after forging, the piece is removed, by the robot arm, and stored in the chamber. A view of the plant is shown Figure 2.

The forging operation is viewed through a window at the front of the press and the whole operation is monitored and recorded using appropriate load and movement sensors. In this way it is possible to ensure that the process is consistent from piece to piece and from batch to batch. Temperature control in the dies is maintained by appropriate sensors and control of the induction heating and it is possible to maintain the temperature to within ±5°C throughout the dies and work-piece.

The maximum diameter of forging currently possible is 800mm and the maximum weight which the robot can handle is 230kg. Figure 3 illustrates the ability to produce complex near net shape components.

Modelling of the Isothermal Forging Process

The intrinsic cost both of the material being forged and of the molybdenum dies means that any failure during the process could be very expensive. Therefore mathematical modelling techniques are used to predict:-

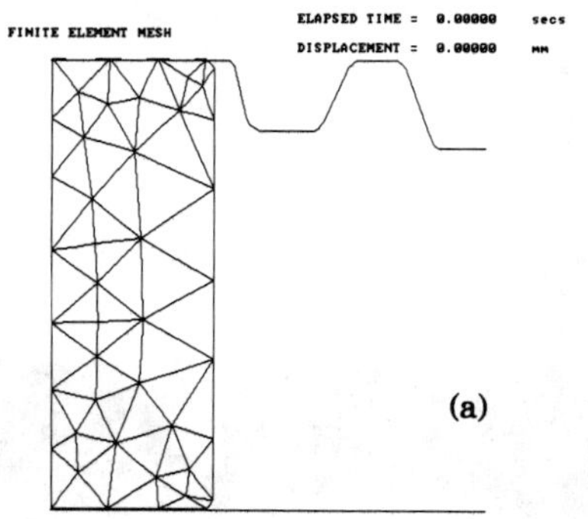

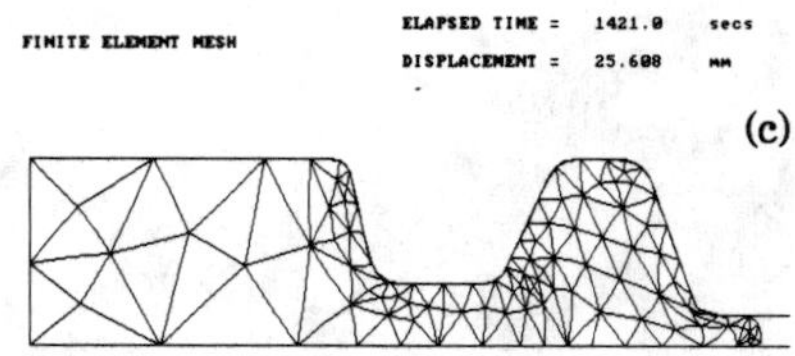

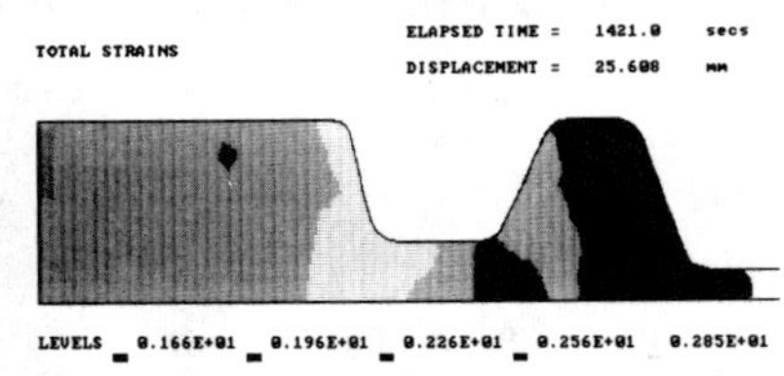

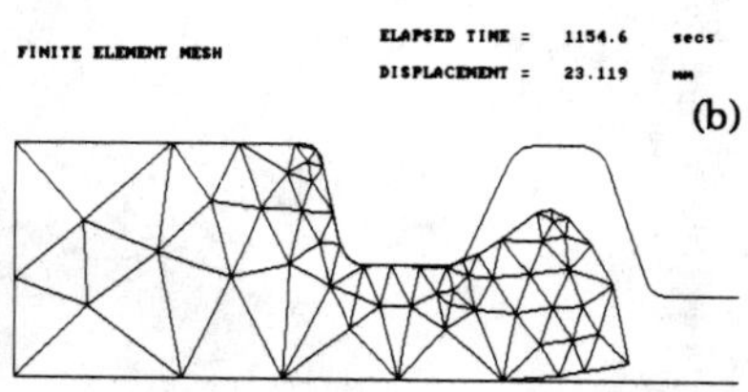

(d) Figure 4 - Typical finite element analysis output

(a) - (c) Meshes, (d) Total strain

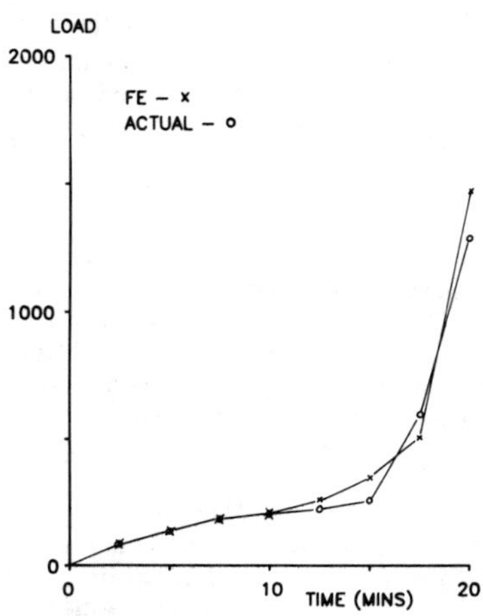

Figure 5 - Comparison between the actual press load and the finite element simulation of a titanium alloy disc forging

(a) the maximum load on the press during a forging operation

(b) the local stresses both in work-piece and die

(c) the strain rate regime and total strain in each part of the forging and, hence, the microstructure.

The model used is one developed in-house and is based on finite element methods(1). Because of the proliferation of relatively fast and cheap computers these methods have been applied to a number of engineering problems in recent years. The present system is based on the assumption that the material being forged behaves in an elasto-viscoplastic manner.

The forging piece and dies are presented graphically with the piece divided into triangular segments (Figure 4) as the dies are moved together the mesh is forced out sideways as shown. The relationship between each node in the mesh depends on various properties of the material and on the amount of friction between work-piece and die. Therefore, prior to the modelling, a significant amount of small scale testing has to be done in order that the deformation behaviour can be described adequately. In addition various assumptions have to be made about the constitutive relationships used to describe the physical processes being simulated.

Once this background information is generated for any one alloy it is relatively easy to predict total loads on the 3,200T press for a particular geometry of forging. The success of this technique can be illustrated by reference to Figure 5 which shows actual and predicted press loads for a titanium alloy IMI 834 disc forging.

A more challenging task is that of predicting microstructural development. The effective flow stress is often defined in material models as a power function of effective strain and strain rate. However this description is inadequate for many materials. In particular, dynamic recrystallisation can take place during hot deformation and this results in flow softening which is not described by the latter approach (2,4). As a consequence the constitutive relationship utilised incorporates a variable which follows the microstructural evolution that takes place during hot working. The primary use of this state variable is to allow accurate prediction of flow stress. However it is possible, given the necessary microstructural information and an understanding of the deformation process, to correlate the structural development with the parameter.

Potential and Limitations of Isothermal Forging

Process costs generally restrict use of this forming technique to high value items. The bulk of production has been gas turbine discs but there is scope for forging non-axisymetric components such as large airfoils. The advantage of the isothermal route is the consistent achievement of accurate shape with no residual stress.

A related application is for the forging of intermetallic materials which are very brittle below about 1000°C and for which conventional forging is not applicable. The reason for hot working is that the properties of many of these intermetallics are thereby enhanced.

Casting

Casting of titanium is now a well establish route for the manufacture of components both for commercial and aerospace applications. In the case of the latter, confidence in titanium castings is such that casting factors approaching unity are now commonplace at the design stage.

Casting Techniques

There are three major problems in the production of titanium castings.

(a) High melting point of the alloys, generally above 1700°C.
(b) Low fluidity of the metal at pouring temperatures.
(c) High reactivity with almost every gas, liquid or solid at temperatures above 500°C.

The usual method of melting titanium for the casting process is to use consumable electrode, arc-melting in a vacuum. The melting crucible is a double walled copper container with water cooling, which causes a solid skull of titanium to be formed on the inner surface. This skull and the absence of atmosphere protects the molten metal from contamination.

Once the titanium electrode has been consumed there is no more heat input to the system and the metal must be poured into the moulds as quickly as possible before further solidification in the crucible occurs. The batch weight of the IEP furnace in Belgium (SETTAS) is 1 tonne of metal and a centrifugal casting arrangement is used to move this metal into the moulds. The melting and casting equipment is shown diagrammatically in Figure 6. The mould assembly has a diameter of up to 3 metres and, when rotating, generates a centrifugal force of up to 60G at the periphery. This force ensures fast movement of the metal into the moulds and also provides extra pressure for feeding during solidification.

Moulding Methods

The two principle methods for making moulds are:

(a) Ceramic shells made by the lost wax process - investment casting.
(b) Precision sand and graphite - rammed sand.

The choice between the two depends on the complexity of the cast piece, the required dimensional tolerances and surface finish, and economic considerations.

The dimensional tolerances which are obtainable from the two processes are given in Table I. The investment casting technique can be used to make complex shapes with a surface finish of 3μ. However the cost of the wax pattern dies is high and therefore this method is better suited to long runs. It is also amenable to the use of robotics in the shell making process. Typical examples of castings made by the lost wax process are shown in Figure 7.

TABLE I
COMPARISON OF MOULDING METHODS

Method	Pattern	Moulding Material	Surface Finish (μm)	Diameter Tolerance (mm)	Thickness Tolerance (mm)
Precision Sand	Wood, Resin	Special Sand	6	750 +/-2	2.5 +/-0.6
Lost Wax	Metal + Wax	Ceramic	3	750 +/-1	2.5 +/-0.3

The precision sand method is generally used for simpler shapes and the surface finish is slightly inferior to that obtained on investment castings, typically about 6μ. However the tooling used to shape the sand/graphite mixture is relatively inexpensive and the technique can be used to make large pieces. Typical examples of castings made by the precision sand method are shown in Figure 8. This technique is a standard production route for aerospace parts and, with the flexibility offered, it is a preferred route during the development phase before the design is fixed.

Comparison of mechanical properties obtained from castings made by the two moulding methods are given in Table II. Both routes give material which meets the relevant ASTM specification.

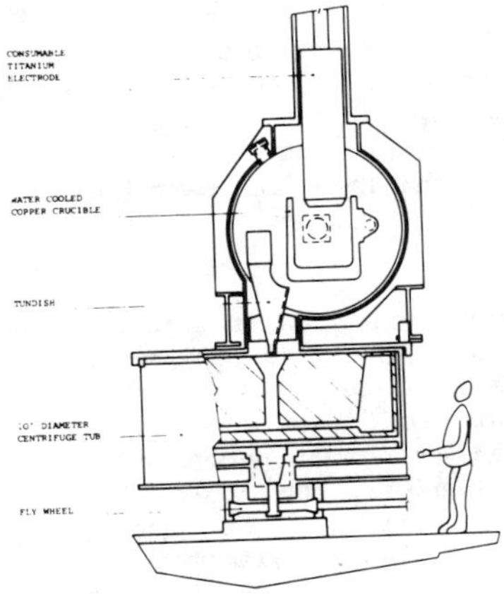

Figure 6 - Schematic diagram of a centrifugal casting furnace

Figure 7 - Typical titanium investment castings

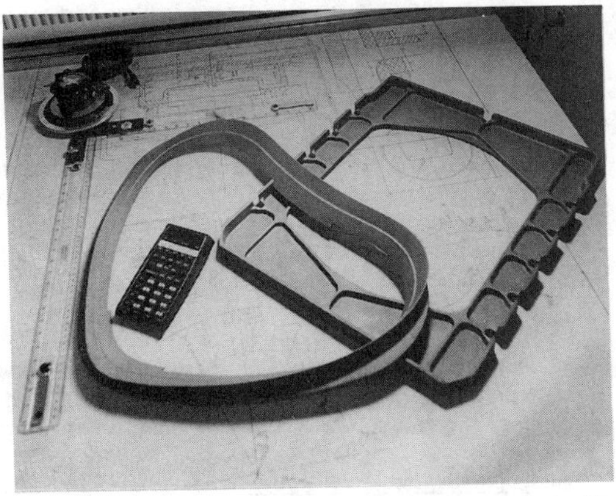

Figure 8 - Examples of titanium precision sand castings

TABLE II
MEHCANICAL PROPERTIES OF CAST Ti 6/4

(-summary of all casts 1986-91 with Oxygen Content in range 0.16-0.24% after HIP and anneal)

	0.2% YS (MPa)	TS (MPa)	el (%)
Sand Castings	891±31	1021±26	10.3±1.8
Investment Castings	871±22	984±21	9.9±1.7
ASTM B 367-83 Grade C5	825 (min)	875 (min)	6 (min)

Casting Quality

There are several other factors which have to be considered when aiming for a sound titanium casting. The first is shrinkage control to which is applied the normal rules calculating the size and position of risers and in-gates. In the precision sand process there is the possibility of using various moulding mixtures to give a progressive chilling effect but this is not so easy with investment casting shells. There is an extra complication imposed by the centrifugal casting process. This gives an added compressive force in the radial direction of the chamber. This also leads on to the second consideration which is mould distortion. All the usual precautions need to be taken during mould making, dewaxing, drying and firing to controi and monitor mould distortion and shrinkage. In addition, because of the extra hydrostatic and gravity forces during casting, particular care has to be taken in mounting and supporting the moulds in the chamber. There is also a need to ensure that the mould assembly is balanced in relation to the axis of rotation, both before, during and after filling, otherwise mechanical instability will result.

The third consideration in casting quality relates to surface contamination. Although titanium is cast under vacuum and in a mould made of an appropriate non-reactive material, it is quite difficult to avoid any contamination of the casting surface by oxygen, carbonaceous material or first coat oxides picked up by the molten metal from the mould surface at the time of pouring. The resulting layer of contamination, known as alpha-case, is very hard and can range in thickness from 0.6mm down to zero. The alpha-case, because it is hard and can easily initiate surface cracks, has to be removed and this is usually done by chemical milling in mixed acids. Obviously to achieve good dimensional tolerances and good surface finish in the casting, the alpha-case thickness and its removal need to be controlled and monitored. Removal of the alpha-case has the added benefit that it generally further improves the surface finish.

Hot Isostatic Pressing

The final factor to be considered in relation to quality is internal microporosity which is inevitably present in any casting. Although this is minimised by appropriate foundry engineering and the imposition of centrifugal force there is still a need, particularly in pieces for aerospace, to take further steps to ensure the absence of internal cavities. Hot isostatic pressing (HIP) is very effective in producing full densification of titanium and its alloys.

Trials have been done in order to define the effect of HIP on porosity level. Results are shown in the quantative metallography histogram in Figure 9. For this latter technique six micrographs from each sample, at a magnification of x50, were analysed in order to record a statistically significant number of pores.

Opportunities and Limitations

The maximum size of titanium castings is limited by the furnace capacity the largest of which is reported to be 3 tonnes (in CIS). The complexity is limited only by the skill of the tool designer and manufacturer, and the willingness of the customer to pay. Large and complex engine components are now appearing as titanium castings (Fig 10). In many of these, thin walls are called for and, although these cannot be cast directly, they can be sculptured using chemical milling.

Superplastic Forming and Diffusion Bonding

Twenty five years of development have brought the processes of superplastic forming and diffusion bonding to a state of maturity. These processes provide the designer with the opportunity to design components which are both cost and weight efficient. However, to achieve optimum performance, the designer needs to have an in-depth understanding of the freedoms and limitations provided by these processes.

A tradition has grown up over the years which limits the use of titanium to-

a)areas which demand its high temperature strength.

b)highly loaded, fatigue sensitive fittings.

This situation has arisen due, in the main, to the perceived high cost of material and the associated conventional manufacture of titanium. Designers need to be reminded

(a) (b)

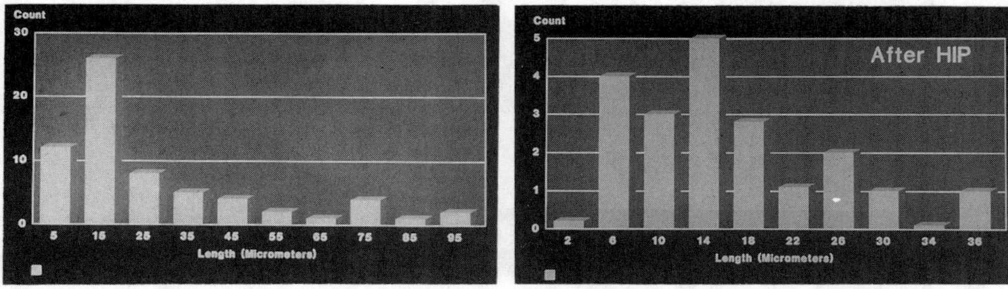

Figure 9 - Quantitative metallographic measurement of
pore size and distribution in:-
(a) as - cast titanium
(b) cast and HIP titanium

Figure 10 - Titanium 6-4
compressor casing

that the specific properties of titanium and it's corrosion and fatigue resistance make it an attractive structural material even for room temperature use. With the development of the combined processes of SPF/DB and their cost saving potential, there is now a growing body of evidence that titanium can compete with aluminium fabricated components in both cost and weight. In the light of this evidence, therefore, designers need to review their traditional views of this material.

Production Equipment

Early experimentation was conducted using primitive press/heated platen systems and in many cases using "hard back" bolted up tooling systems. Today, however, purpose built heated platen presses of a type similar to that shown in Figure 11. have become the mainstay of SPF and SPF/DB manufacture.

Variations on this basic theme have now been produced in, for example, the "Shuttle" platen press developed by Aerospatials, the Gantry press system recently installed by McDonnell/Douglas and the "C-Frame" restraint system developed by Grumman.

Alongside the press development there has been automation of the welding processes for cutting and stitching together of the preforms, and of the subsequent non-destructive testing.

Design and Manufacturing Aspects

All of the SPF/DB structural types which are presented schematically in Figures 12-14 have specific structural and manufacturing aspects which should be understood by the designer.

Figure 11 - SPF/DB platten press

Figure 15 - A 2 sheet SPF/DB part

Figure 16 - A 4 sheet SPF/DB part

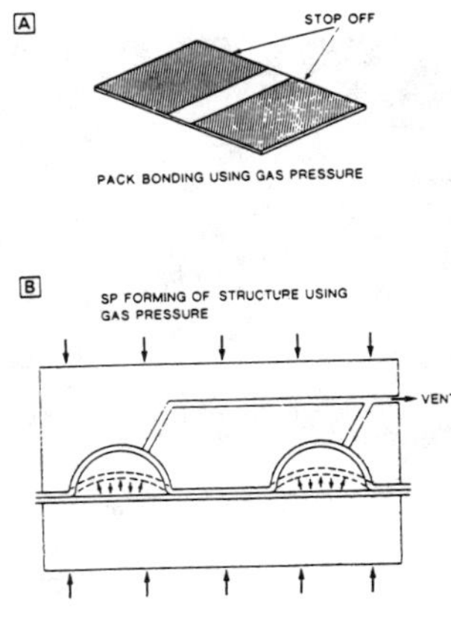

Figure 12 - 2 sheet structure

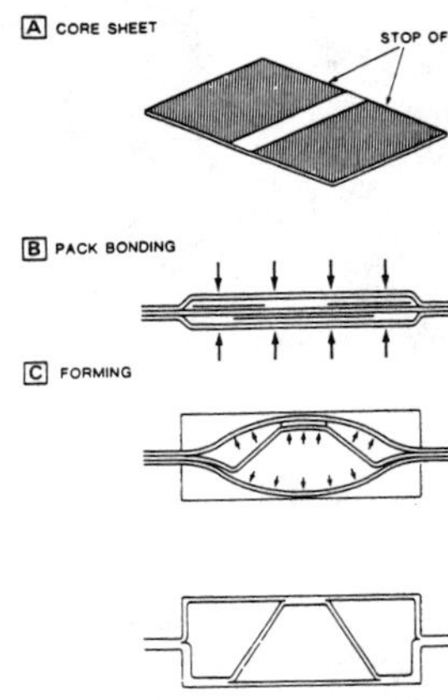

Figure 13 - 3 sheet structure

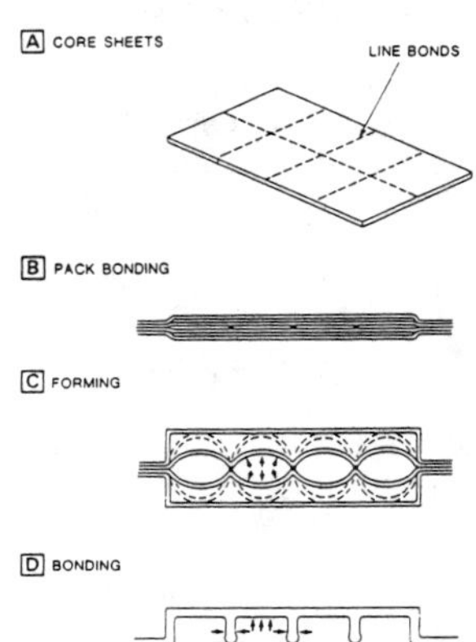

Figure 14 - 4 sheet structure

2-Sheet Structure The two sheet structure of the type illustrated in Figure 12 represents the SPF/DB equivalent of a common conventional form of construction and therefore, with the exception of the use of DB as a means of joining the two sheets, the structural advantages and limitations should already be well understood. Although the formed sheet will be subject to thinning, generally the shapes demanded in such structures do not present difficulties for the prediction of the thickness distributions.

In effecting the DB of such structures, two methods of manufacture are possible:-

a) pre bonding, either singly or in a pack, using gas pressure and delineating the bond areas using a bond-inhibitor (stop-off) then subsequently expanding in the form tool as a separate operation.

b) bonding by mechanical means in the tool prior to forming, this is usually a continuous but sequential operation.

The former route requires good alignment between the bonded areas of the blank and the corresponding features in the tool to ensure a consistent and optimum transition between the swaged section and the bonded area.

Although the latter route will have a bond area which is defined and effected by the pressure generated by the tooling features, the quality of the bond is totally dependent upon the matching of tool faces, including the flatness and combined thickness of the component blank in the area to be bonded(5).

3-Sheet Structure Unlike the two sheet SPF/DB form, which is controlled directly by the tool in terms of its final structural definition, the three sheet structural form is largely controlled by the definition of the starting blank. The tool in this case, only provides a restraint to the blank around its periphery and in addition provides the envelope shape of the finished component. This, therefore, places responsibility on the designer to define both the starting blank and the finished component and to accomplish this it is necessary for the designer to model and understand the forming process. For simple structures, such as that illustrated in Figure 14, the modelling is relatively easy but for the more complex stiffening patterns such as sine wave or discontinuous stiffening, the establishment of a finite element technique as a basis of an SPF CAD system is highly desirable and will greatly reduce the iterative trial and error development which is currently a feature of SPF/DB technology.

The fact that the tooling can, to a large degree, be divorced from the component structural definition, provides the designer in turn with freedom to change the structural definition without affecting the tool. This is comforting to the designer bearing in mind the cost and lead time of provisioning tools.

One feature of the three sheet structure, is the fact that the skin thicknesses used need to be significantly greater than the core thickness to ensure skin stability during forming and to overcome external waviness on the finished component.

A wide variety of core stiffening patterns are possible with the three sheet structure. These include straight, sine wave, and discontinuous features. Orthogonal stiffening is possible for a three sheet structure but requires a considerable increase in the blank complexity.

4-Sheet Structure The four sheet structure illustrated in Figure 16 results in cellular stiffening which has the virtue of having design and stressing characteristics which, although not identical in detail, are familiar to the designer in their conventionally fabricated equivalent.

As for the three sheet structure, the designer has the responsibility of defining the details of the starting blank to achieve his finished component. Again this cannot be achieved without an in depth understanding of the manufacturing process and this has been greatly assisted by the development of a finite element CAD analysis system.

Opportunities and Limitations

The plan view area of an SPF/DB structure is limited by the size of the press. The depth of the structure can be accommodated by cell design which allows for the metal thinning. This latter aspect can be alleviated either by using thicker starting material or by selective metal removal on the blanks to leave re-inforcement where needed on the blown piece. There is a limit to the increase metal thickness useable because in sheet above about 4mm thick it is not easy to achieve superplastic properties. The main advantages of the SPF/DB process are economic, due to higher material utilisation and simple starting blank forms combined with significant weight savings.

Conclusion

The advances in the production methods for titanium alloy components discussed in this paper have provided engineers with the ability to specify parts with good mechanical properties, high integrity and complex forms which can be manufactured economically.

References

(1) P.S.Bate: AGARD Conf. Proc. 426 (1987) 22.

(2) J.V.Bee, A.R.Jones, P.R.Howell: J. Mat. Sci. 15 (1980) 337.

(3) A.Y.Kandiel, J-P.A.Immarigeon, W.Wallace, M.C.de Malherbe: Met. Sci. 14 (1980) 493.

(4) R.G.Menzies, J.W.Edington, G.J.Davies: J. Mat. Sci. 15 (1981) 217.

(5) M.Ohsumi, M.Shimizu, A.Takahashi, T.Tsuzuku, Mitsubishi Heavy Industries - Technical Review, 21 (1) (1984) 42.

MICROSTRUCTURE AND PROPERTY MODIFICATION OF CAST ALPHA-2 TITANIUM ALLOYS BY THERMOCHEMICAL PROCESSING WITH HYDROGEN

L. S. Apgar[a], C. F. Yolton[b], M. Saqib[c], I. Weiss[c], and D. Eylon[a]

[a]Graduate Materials Engineering, University of Dayton, 300 College Park, Dayton, OH 45469-0240, USA

[b]Crucible Research Center, Pittsburgh, PA 15230, USA

[c]Department of Mechanical and Materials Engineering, Wright State University, Dayton, OH 45435, USA

ABSTRACT

Thermochemical processing (TCP) with hydrogen of $\alpha+\beta$ titanium alloys has been established as an effective method to modify the microstructure and enhance properties, such as tensile and fatigue strength. This process was also successful in substantially refining the microstructure of α_2 titanium based alloys of both powder metallurgy and wrought products. The objectives of the present work were to refine the microstructure and improve the mechanical properties of two *cast α_2 based titanium alloys* (Ti-24Al-11Nb at% and Ti-25Al-10Nb-3V-1Mo at%) by temporary alloying with hydrogen. Hot isostatically pressed coupons were hydrogenated by soaking in hydrogen at a positive pressure at temperatures ranging from 540 to 980°C. Various hydrogenation times at either isothermal or multiple-step temperatures were employed on the cast coupons. Under optimum conditions, 0.36 wt% of hydrogen could be absorbed into 6 mm cast sections of Ti-25-10-3-1. Control experiments with fine powder have shown that up to 0.9 wt% of hydrogen could be absorbed in Ti-25-10-3-1. Hydrogen was removed by vacuum annealing at 650°C for 48 hrs to a level below 50 ppm. As a result of the limited solubility of hydrogen in α_2 alloys, only TCP schemes involving either high temperature hydrogenation or multiple-step high temperature hydrogenation were successful in modifying the microstructure of Ti-25-10-3-1. As a result, TCP using a two-step or four-step temperature hydrogenation resulted in improvements in RT tensile ductility. The Ti-24-11 alloy did not respond well to any of the TCP treatments used in this study.

INTRODUCTION

Optimization of properties by microstructural modification of near net-shapes (NNS), such as castings, is limited to processes which do not involve material working. One unique process which is currently used on conventional cast titanium alloys, such as Ti-6Al-4V, is *thermochemical processing* (TCP), *i. e.* temporary alloying with hydrogen [1-9]. TCP has been used successfully to refine the microstructure, improve the mechanical properties, and enhance the workability of conventional Ti and Ti_3Al based alloys [1-5,10-13].

TCP involves the use of hydrogen as a temporary alloying element which diffuses into the material at elevated temperatures. This process is the result of a high hydrogen solubility in Ti and its alloys [3 wt% (60 at%) @ 595°C in α-Ti] which decreases with increasing temperature [14,15] followed by hydrogen removal at intermediate temperatures to levels below 100ppm by vacuum annealing [15]. The hydrogenation stage results in the formation of new phases and increased lattice strains which enhances both microstructural modification and workability. In addition, hydrogen stabilizes the β-phase (BCC) at lower temperatures, so that a larger volume

---

Work was sponsored by EMTEC-Edison Materials Technology Center, Kettering, OH 45420.

Titanium '92
Science and Technology
Edited by F.H. Froes and I. Caplan
The Minerals, Metals & Materials Society, 1993

fraction of β-phase is present at the processing temperature. As a result, TCP leads to unique processing capabilities for NNS which can not be achieved by conventional thermal processing methods.

The objectives of the present work were to investigate the possibilities of microstructural modification and property improvement of cast + HIP Ti-24Al-11Nb (at%) and Ti-25Al-10Nb-3V-1Mo (at%) alloys by using several different TCP treatments.

EXPERIMENTAL PROCEDURES

The TCP technique used in this study involved hydrogenation of investment cast coupons of the Ti_3Al (α_2) alloys Ti-24Al-11Nb[1] and Ti-25Al-10Nb-3V-1Mo[1] followed by dehydrogenation by vacuum annealing (VA). The starting hydrogen content for both materials was below 10ppm. Prior to hydrogenation, the as-cast material was hot isostatically pressed (HIP) at 1175°C/275MPa/6hr and then machined by EDM followed by surface grinding into 6mm x 6mm x 45mm rectangular bars representing a two-dimensional hydrogen diffusion case. Prior to hydrogenation, the specimens were cleaned and weighed to the nearest milligram and then wrapped in stainless steel (SS) and CP-Ti foil to getter any residual oxygen during the hydrogenation process. The specimens were hydrogenated by Crucible Research Corporation [16] either isothermally or by a multi-step temperature method using various soaking times/temperature listed in Table 1.

Table 1: Hydrogenation Cycles for both Alloys and all Specimens Studied.

ID No.	Hydrogenation Cycles Temperature [°C] / Time [hr]
1	540/20
2	650/20
3	760/20
4	870/20
5	980/20
6	980/2 + 870/4 + 760/6 + 650/8
7	540/48
8	650/48
9	980/10 + 650/10
10	870/10 + 760/10

Hydrogenation was accomplished in the constant temperature zone of a 304 SS vacuum tube furnace pre-evacuated to 5μm Hg at room temperature and then heated up to the hydrogenation temperature. Hydrogen gas (with < 500ppm oxygen) was admitted into the chamber and maintained at 0.2atm positive pressure for a period of time (Table 1). At the end of the cycle the gas valve was turned off. The furnace was either shut down or cooled to the next temperature where the hydrogenation cycle was repeated [16]. At the completion of the hydrogenation cycle the specimens were weighed to the nearest milligram to determine the weight gain after hydrogenation. Annealed Ti-25-10-3-1 gas atomized powder was placed in the hydrogenation chamber with the specimens. This was done to determine the hydrogen saturation limit which represents the conditions at the material surface since the powder particle has a high surface to volume ratio. **Dehydrogenation** was conducted by vacuum annealing at 650°C/48hr. Control samples, containing approximately 10ppm of hydrogen, were heat treated in a vacuum furnace at the same time and temperatures (Table 1) used for hydrogenation (2, 6, and 9 only) and dehydrogenation.

Specimens were examined by optical microscopy and scanning electron microscopy (SEM). Vicker's microhardness testing was conducted on polished cross-sections of several hydrogenated specimens. Hydrogen gas analysis was performed on the specimens following both hydrogenation and dehydrogenation using a Leco RH404 hydrogen analyzer with an argon carrier (traceable to NBS Standard 3352b) [17]. Uniaxial tensile testing was conducted on selected conditions (control and TCP'd specimens). Tensile testing was conducted at room temperature at a strain rate of 0.13 min^{-1} to 0.2% strain followed by a crosshead rate of 1.3 mm/min to failure. More details of this work are available in reference 18.

RESULTS

HYDROGENATED MATERIAL

The **starting microstructure** of the casting HIP'd at 1175°C are shown in Figs. 1a and b for both alloys. The macrostructure of **hydrogenated** coupons after various cycles (Table 1) are shown in Figs. 2a and b.

---

1 All compositions are reported in atomic percent (at%).

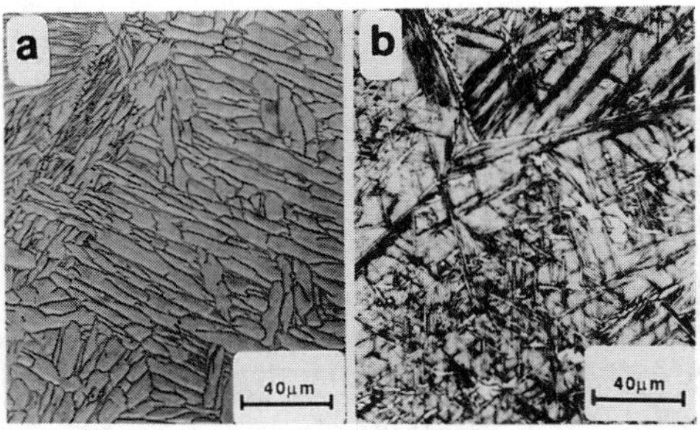

Figure 1: Microstructure of the Cast+HIP starting material; (a) Ti-24-11 and (b) Ti-25-10-3-1.

Figure 2: All hydrogenated specimens; (a) Ti-24-11 and (b) Ti-25-10-3-1 (Table 1).

Specimens hydrogenated at conditions 4, 5, 6, and 9 (Table 1) were unintentionally thermally etched during the hydrogenation cycle for both alloys (Fig. 2). It should be noted that the control specimens heat treated at these same conditions were not thermally etched. However, control specimens heat treated slightly above the β-transus (β_t) temperature were thermally etched. The surface topographies of specimens hydrogenated at cycle 6 are shown in Fig. 3 for both alloys. Weight gain as a result of hydrogenation, for all conditions, are illustrated graphically in Fig. 4. During cycle 5 the specimens lost surface material due to thermal etching. In addition, weight gain of Ti-24-11 specimens hydrogenated at conditions 9 and 10 could not be determined since a substantial amount of hydrogenwas absorbed into the material to cause a phenomenon known as surface decrepitation to occur (Fig. 5). Bulk hydrogen concentrations were limited only to Ti-25-10-3-1 material since hydrogen charging did not produce any significant microstructural changes in the Ti-24-11 material. Bulk hydrogen concentrations and powder hydrogen contents (represents the hydrogen concentration at the cast surface) are plotted in Fig. 6. Microhardness evaluation as a function of depth below the surface into the hydrogenated specimens was performed on polished cross-sections. It did not reveal any significant changes as would be expected from an alloy with high interstitial content. Averaged microhardness values for several hydrogenation conditions are plotted in Fig. 7.

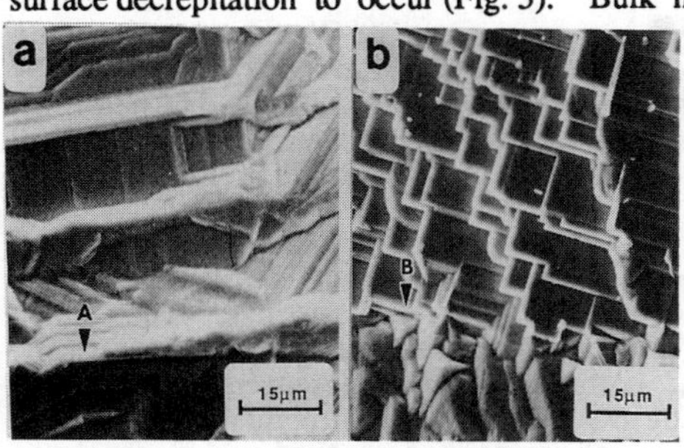

Figure 3: SEM image of thermally etched surfaces shown in Fig. 2 hydrogenated at condition 6; (a) Ti-24-11 and (b) Ti-25-10-3-1.

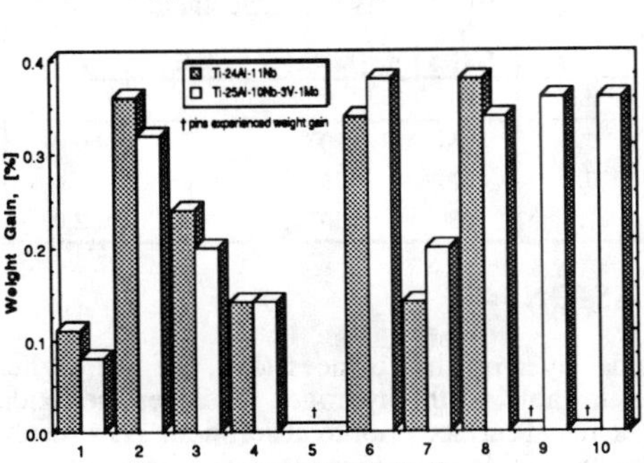

Figure 4: Weight gain as a result of hydrogenation for both alloys. See Table 1 for hydrogenation conditions.

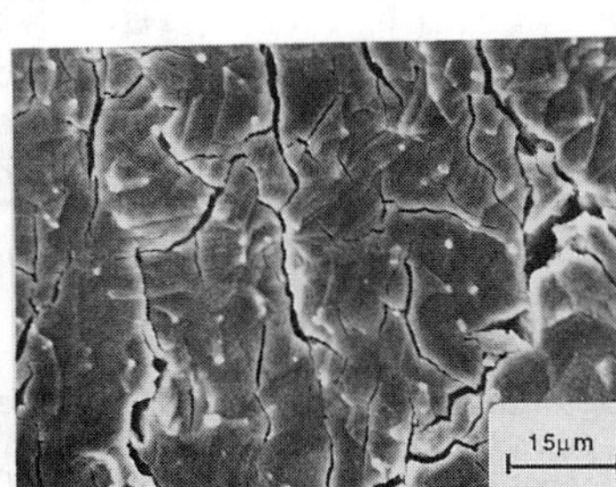

Figure 5: SEM of Ti-24-11 coupon hydrogenated at condition 9 showing surface decrepitation.

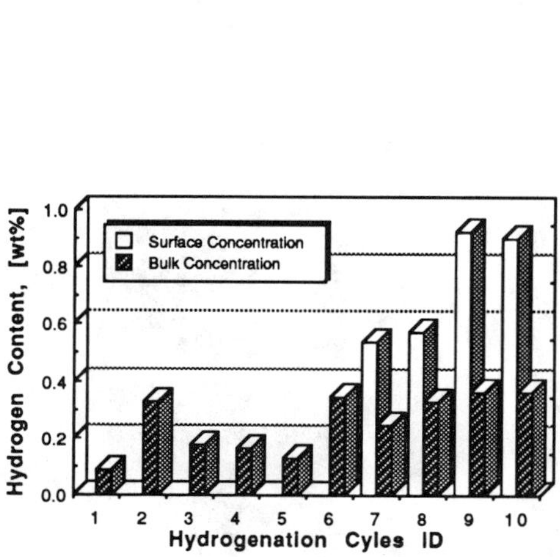

Figure 6: Bulk and surface hydrogen concentrations for the Ti-25-10-3-1 for different hydrogenation conditions (Table 1).

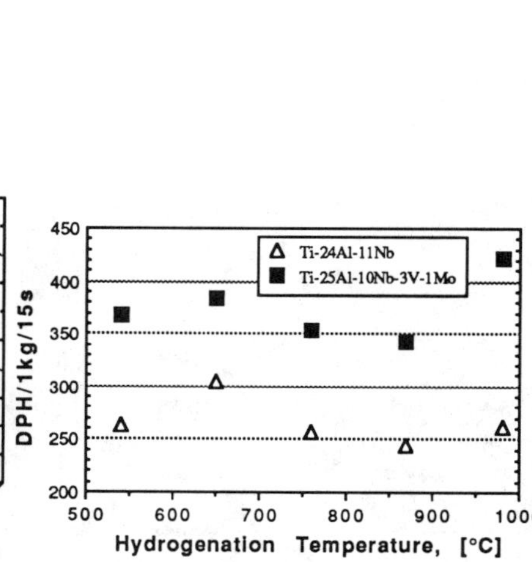

Figure 7: Variation in the average microhardness for isothermal hydrogenation conditions (20hr).

Microstructures of the hydrogenated conditions are shown in Figs. 8 and 9 for the Ti-24-11 and Ti-25-10-3-1, respectively. Fine, unconnected short cracks were observed in the microstructure of the Ti-24-11 material hydrogenated at conditions 2, 8, 9, and 10 (Fig. 8c). Ti-24-11 specimens hydrogenated at condition 6, and Ti-25-10-3-1 specimens hydrogenated at conditions 4, 5, 6, and 9 (Figs. 9b, c, d, and e, respectively) exhibited a different microstructures near the surface and at the center. Photomicrographs shown in Figs. 8 and 9 are taken from the center section of the hydrogenated specimens, except for Fig 9c.

DEHYDROGENATED MATERIAL

Microstructures of dehydrogenated Ti-25-10-3-1 after VA at 650°C/48hr and of control specimens heat treated at the same conditions but, with no hydrogen, are shown in Fig. 10. The hydrogen content after VA for these three specimens was below 50 ppm. It should be noted that the microstructures of the control samples heat treated at the hydrogenation conditions were unchanged after VA. Therefore, the microstructures are only shown once (Figs. 10b, d, and f).

MECHANICAL PROPERTIES

Results of mechanical testing conducted on three Ti-25-10-3-1 coupons are indicated in Table 2. These conditions contained both the largest concentration of hydrogen after hydrogenation and the highest degree of microstructural change as compared to the cast+HIP microstructures (Fig. 1a). Results are also compared to typical cast + HIP material with no subsequent heat treatment [18]. Since the Ti-24-11 material did not result in significant microstructural refinement as illustrated by comparing Figs. 1a and 8; mechanical testing was not performed on this material.

Table 2: Room Temperature Tensile Properties of TCP'd and Control Ti-25-10-3-1 Specimens.

TCP or HT Conditions	Control Specimens				TCP'd Specimens			
	YS [MPa]	UTS [MPa]	%E	%RA	YS [MPa]	UTS [MPa]	%E	%RA
As-HIP	841	945	0.7	0.5				
2 + VA 650°C/48hr	787	910	0.7	0.7	787	907	0.7	0.4
6 + VA 650°C/48hr	611	698	0.8	0.4	627	754	1.0	2.1
9 + VA 650°C/48hr	736	798	0.6	1.7	630	807	1.4	2.0

DISCUSSION

HYDROGENATION

Hydrogenation below 650°C produced an oxide layer on the surfaces (Fig. 2). At higher hydrogenation temperatures the oxide becomes unstable, or the hydrogen dissolves the oxide layer, permitting hydrogen to be adsorbed onto the metal surface prior to absorption. As a result, hydrogenation temperatures below 650°C should be avoided. Specimens hydrogenated above 870°C exhibited varying degrees of thermal etching on the surfaces (Figs. 2 and 3). At 980°C, both alloys exhibited extensive thermal etching in well defined geometric patterns (Fig. 3). The Ti-24-11 coupons exhibited a lesser degree of thermal etching (Fig. 3a) in comparison to the Ti-25-10-3-1 (Fig. 3b). The Ti-24-11 alloy displayed a hexagonal-shaped patterns (A in Fig. 3a) along with cubic-shaped patterns indicating that the hydrogenation was conducted high in the α_2+β phase field. However, the Ti-25-10-3-1 material exhibited only a cubic shaped pattern (B in Fig. 3b) indicating that hydrogenation could of occurred in the β-phase field. It was concluded that 980°C hydrogenation of Ti-25-10-3-1 lowered the β_t by approximately 100°C. However, such observations should be qualified by the fact that the surface chemistry may be different than that of the bulk as a result of a higher hydrogen concentration at the surface and possible surface loss of aluminum.

Hydrogenation using multiple-step temperatures (cycles 6, 9, 10) resulted in high levels of hydrogen for both alloys (Figs. 4 and 6). This method takes advantage of both high hydrogen diffusivity at high temperatures combined with high hydrogen solubility at low temperatures. Hydrogenation at 540°C/20hr exhibited the lowest concentration of hydrogen (Figs. 4 and 6) as a result of the development of a stable oxide layer during hydrogenation.

Ti-24-11 specimens hydrogenated at conditions 9 and 10, exhibited decrepitation at the corners of the pins, (Figs. 2 and 3). This is expected in compounds where the lattice expansion due to hydride formation is so large that it cannot be accommodated by the associated deformation of the material which results in cracking and flaking. As a consequence, the effective diffusion lengths are shortened and new clean surfaces for hydrogen adsorption are created spontaneously. Therefore, hydrogen concentrations are increased quickly to the saturation limit when decrepitation occurs.

No change of microhardness as a function of depth was observed in both alloys. A change in the *average hardness* occurred for different hydrogen cycles for both Ti-24-11 and Ti-25-10-3-1 (Fig. 7). The Ti-25-10-3-1 coupons exhibited a higher hardness value for all six hydrogenation cycles than the Ti-24-11coupons; a reflection of the higher strength of this alloy [18]. It should be noted that the effect of process temperature on the average microhardness was the same for both alloys (Fig. 7). This trend matches the trend of hydrogen content for isothermal hydrogenation conditions 1-5 shown in Fig. 6, except for the 980°C/20hr hydrogenation cycle. Therefore, the average microhardness values show a trend with hydrogen content which could be related to the amount of hydrides in the hydrogenated material.

HYDROGENATED MICROSTRUCTURES

Hydrogenation of Ti-24-11 did not result in significant microstructural changes compared to the starting material (Figs. 1a and 8). Hydrogenation at all conditions resulted in coarsening of the primary α_2 lathes, increasing the grain boundary α_2 width, and increasing the amount of retained discontinuous β-phase. As the hydrogenation temperature increases the amount of retained β-phase increases as shown in Figs. 8b and c. These changes are most likely the result of thermal exposure rather than introducing hydrogen into the material. However, hydrogenation results in increasing the amounts of retained β-phase over the equilibrium concentration of transformed β. Fine discontinuous cracks observed in several hydrogenation conditions are only present in the primary α_2-phase and appears to terminate or bypass the β-phase (Fig 8c). This is consistent with the fact that hydrides will form in the α_2-phase but not in the β-phase. It should be noted that, hydrogenation did not significantly modify the microstructure of the castings in comparison to work done on Ti-24-11 powders where significant refinement and morphological changes occurred as a result of using a multiple-step hydrogenation method during TCP [10-11].

Figure 8: Microstructures of hydrogenated Ti-24-11. Conditions (a) 2, (b) 5, and (c) 9 (Table 1).

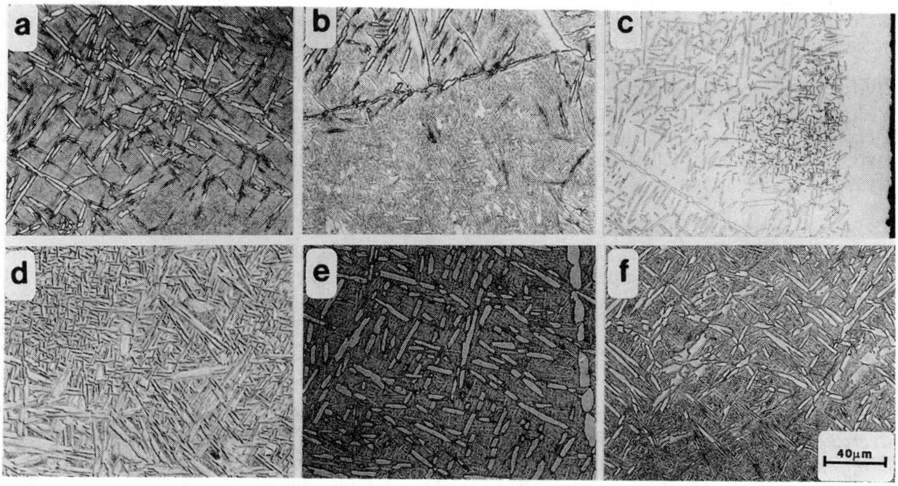

Figure 9: Microstructure of hydrogenated Ti-25-10-3-1. Conditions (a) 2, (b) 4, (c) 5, (d) 6, (e) 9, and (f) 10 (Table 2).

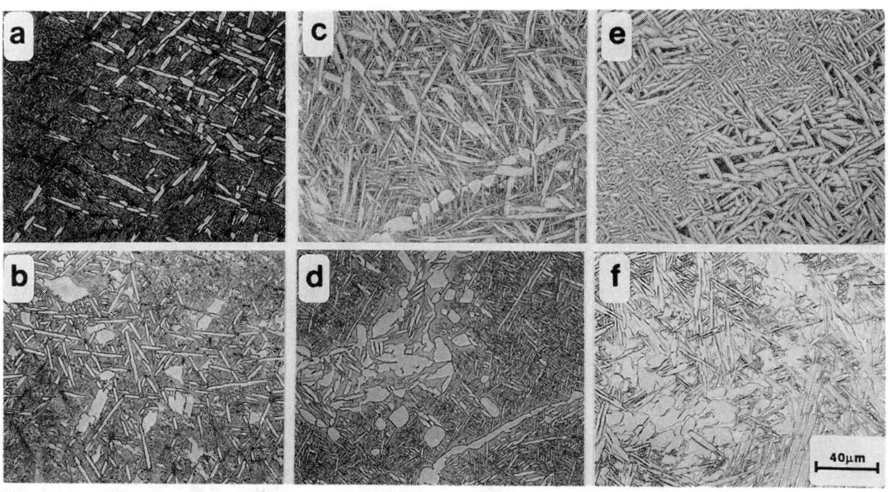

Figure 10: TCP'd microstructure of Ti-25-10-3-1 material hydrogenated at conditions (a) 2, (c) 9, and (e) 6 after dehydrogenation at 650°C/48hr. Control microstructures for the same time/temperature exposure conditions; (b), (d), and (f), respectively.

In the case of Ti-25-10-3-1, microstructural modification and refinement were obtained after hydrogenation (Fig 9). However, microstructural changes are probably due to both temperature exposure and hydrogenation. Comparison to control samples (Fig. 10b) reveals that the microstructures of isothermal hydrogenated samples charged between 650-870°C are similar. This indicates that microstructural modification is due to heat treatment in the α_2+β phase field rather than hydrogenation. The multiple-step temperature methods did reveal microstructural changes and some refinement as a result of hydrogenation. The hydrogenated microstructures contain two major microstructural elements. The first is a large primary α_2 lamellae probably the result of coarsening of the original cast+HIP α_2 plates. The primary α_2 lamellae are unevenly distributed throughout the material and arranged in a cell structure. The cell walls may contain up to 40 vol% of the α_2 phase and the cell interior close to 0 vol% of the large primary α_2 plates. The second microstructural unit is an ultra-fine high-aspect ratio α_2 lathes present in the matrix and in the cell interiors. These lathes can typically be resolved at very high magnification or by TEM. It should be noted that these fine lathes are not present in the baseline specimens heat treated using multiple-step temperatures. Therefore, hydrogenation using a multiple-step temperature method (Figs. 9d, e, and f) resulted in refinement of the α_2-phase. Present in all microstructures are large primary α_2 lathes which have been broken. Comparison to control sample reveals that this is due to thermal exposure in the 650-870°C range rather than hydrogenation. However, this type of feature is not present in control samples heat treated by a multiple-step temperature method indicating that hydrogenation resulted in breaking up the large α_2 lathes. Furthermore, it appears that the growth of the α_2 lathes as a result of thermal exposure were restricted due to the presence of hydrogen or hydrides in the material when compared to control samples. The uniform microstructure of condition 5 (Fig. 9c) is a result of hydrogenation high in the α_2+β phase field. The formation of large α_2 lathes are the result of thermal exposure, while the fine needle like morphology is the result of hydrogenation (Fig 9c).

DEHYDROGENATED (TCP) MICROSTRUCTURES AND MECHANICAL PROPERTIES

Vacuum annealing of the Ti-24-11 material did not result in any microstructural changes from the as-hydrogenated microstructure. However, vacuum annealing of the Ti-25-10-3-1 material resulted in further development of the second microstructural unit, the ultra fine α_2 plates, shown in Figs. 10a, c, and e as a result of aging. These ultra fine α_2 plates are uniformly distributed between the coarse primary α_2 lamellae. Thermo-chemical processing using multiple-step temperature hydrogenation methods + VA at 650°C/48hr, condition 6 (Fig 10e), condition 9 (Fig. 10c) and condition 10 resulted in the highest degree of microstructural refinement of α_2 between the coarse primary α_2 lamellae (compare to Figs. 10d and f, respectively). TCP using a 650°C/20hr hydrogenation cycle (No. 2) did not result in any improvement in the mechanical properties as compared to control specimens heat-treated at the same conditions (Table 2). However, TCP using a multiple-step temperature hydrogenation cycle resulted in improvement in ductility over control specimens (Table 2). TCP using a 980°C/10hr + 650°C/10hr hydrogenation cycle (No. 9) resulted in over a two fold increase in room temperature tensile ductility. Tensile results after TCP cycle 6 and 9 (Table 2) are superior than any thermal processing results obtained in a previous study [18] in terms of room temperature ductility.

SUMMARY AND CONCLUSIONS

Cast+HIP Ti-24-11 and Ti-25-10-3-1 coupons were *thermo-chemically processed* (TCP) using hydrogen at various time/temperature conditions. Two types of hydrogenation methods were used: isothermal hydrogen charging and multiple-step temperature hydrogen charging.

1. Low temperature (540°C) hydrogenation should be avoided since a stable oxide layer is formed. In addition, high temperature hydrogenation should be avoided for the Ti-24-11 material since surface cracking and flaking occurs.

2. Up to 0.35wt% of hydrogen was absorbed into both Cast+HIP titanium aluminide alloys.

3. The beta transus temperature for hydrogenated Ti-25-10-3-1 was reduced to below 980°C.

4. TCP treatments used in this investigation were unsuccessful in modifying or refining the microstructure of the Ti-24-11 castings.

5. Microstructural changes and refinement of the primary α_2-phase was obtained for TCP'd Ti-25-10-3-1 castings. TCP using high-temperature/multiple-step hydrogenation treatments resulted in the highest degree of microstructural changes and refinement.

6. Improvement in ductility was achieved by TCP using a multiple-step temperature hydrogenation treatment. A two fold increase in ductility was achieved using a 980°C/10hr + 650°C/10hr hydrogenation treatment + VA at 650°C/48hr. The TCP tensile results are superior to all thermally treated cast coupons tested in a previous study.

ACKNOWLEDGMENTS

This work was funded under EMTEC program entitled Property Improvement of Titanium Aluminide Castings. The authors wish to thank EMTEC for their administrative and financial support. Special thanks to the Duriron Company for providing castings and Crucible Research Company for hydrogenation and VA. The USAF, Wright Laboratory, Materials Directorate, Material Development Branch is acknowledged for continuing support of this project.

REFERENCES

1. R. G. Vogt, F. H. Froes, D. Eylon, and L. Levin, "Thermo-Chemical Treatment (TCT) of Titanium Alloy Net Shapes," in ***Titanium Net Shape Technologies***, ed. by F. H. Froes and D. Eylon, TMS, Warrendale, PA, 1984, pp. 145-153.
2. J. Stinson, R. G. Vogt, and D. Eylon, "Improvement of HCF Properties in Ti-6Al-4V Castings via Thermochemical Treatment," in ***High Integrity Castings***, ed. by C. V. White, D. Eylon, and F. H. Froes, ASM-International, Metals Park, OH, 1988, pp. 3-8.
3. D. Eylon, F. H., Froes and W. J. Barice, "Effects of Treatments on Mechanical Properties of Ti Alloy Castings," ***SAMPE Quarterly***, January, 1989, pp. 42-45.
4. C. F. Yolton, D. Eylon, and F. H. Froes, "Microstructure Modification of Titanium Alloys by Temporary Alloying with Hydrogen," in ***Sixth World Conference on Titanium Proceedings***, ed. by P. Lacomb, R. Tricot, and G. Beranger, Les Editions de Physique, Cedex, France, 1989, pp. 1641-1646.
5. D. Eylon, C. F. Yolton, and F. H. Froes, "Property Enhancement of Titanium Alloys by Microstructure Modification," in ***Sixth World Conference on Titanium Proceedings***, ed. by P. Lacomb, R. Tricot, and G. Beranger, Les Editions de Physique, Cedex, France, 1989, pp. 1523-1528.
6. F. H. Froes, and D. Eylon, U. S. Patent 4,808,250 February, 28, 1989.
7. D. Eylon, F. H. Froes, and C. F. Yolton, U.S. Patent 4,820,360, April 11, 1989.
8. D. Eylon, and F. H. Froes, U.S. Patent 4,832,760, May 23, 1989.
9. R. J. Smickley and L. E. Dardi, U. S. Patent 4,505,764, March 19, 1985.
10. L. S. Steele et al, "Effect of Temporary Alloying with Hydrogen on the Compactability, and Microstructure of Titanium Aluminides Powders," in ***Sixth World Conference on Titanium Proceedings***, ed. by P. Lacomb, R. Tricot, and G. Beranger, Les Editions de Physique, Cedex, France, 1989, pp. pp. 1009-1014.
11. L. S. Steele, D. Eylon, and F. H. Froes, "Microstructure Control of Titanium Aluminide Powder Compacts by Thermo-Chemical Treatment," in ***Advances in Powder Metallurgy***, Vol. 3, ed. by T. G. Gasbarre and W. F. Jandeska, MPIF/APMI, Princeton, NJ, 1989, pp. 509-523.
12. W-Y. Chu and A. W. Thompson, "Effect of Hydrogen as a Temporary β stabilizer on Microstructure and Brittle Fracture Behavior in a Titanium Aluminide Alloy," ***Met. Trans. A***, Vol. 22A, 1991, pp. 71-81.
13. W. Cho, A. W. Thompson, and J. C. Williams, "Creep Behavior of Ti-25Al-10Nb-3V-1Mo," ***Met. Trans. A***, Vol. 21A, 1990, pp. 641-651.
14. A. San-Martin and F. D. Manchester, ***Phase Diagrams of Binary Titanium Alloys***, ed. by J. Murray, ASM International, Metals Park, OH, 1987, pp. 123-135.
15. W. M. Albrecht and M. W. Mallett, "Hydrogen Solubility and Removal for Titanium and Titanium Alloys," ***Trans. of TMS of AIME***, April, 1958, pp. 204-210.
16. Private communication with Fred Yolton, Crucible Research Corp., 1992.
17. Private communication from Analytical Services Inc., WRDC/MLSA, 1992.
18. D. Eylon et al, ***Property Improvement of Titanium Aluminide Casting***, EMTEC/CT-8/TR-90-8 Final Report, EMTEC, Kettering, OH, 1991.

˙THERMOMECHANICAL β-PROCESSING OF HIGH STRENGTH MEDIUM TEMPERATURE TI-ALLOYS : COMPARAISON OF TI17 AND β-CEZ

P.E. MOSSER *, N. MARNIER *, Y. HONNORAT **

* SNECMA, Materials and Processes laboratory, BP 48
92234 GENNEVILLIERS – FRANCE
** SNECMA, Materials and Processes laboratory, BP 91
91003 EVRY – FRANCE

ABSTRACT

When designing the two LP and HP compressor of a jet engine, there is a need for an alloy with optimized properties in the medium temperature range. Good candidates are the β metastable alloys. β-CEZ, a new alloy developed by CEZUS and SNECMA, belongs to this category as well as Ti17 to which it is compared.

A basic approach of process metallurgy shows the deformation mechanisms during forging and the kinetics of α precipitation. Heat treatment optimization leads to select a possible range for solutioning temperature, by which adjustments of room temperature strength and creep resistance are possible. Feasibility of large ingots and forgings is demonstrated.

One β-CEZ pancake and one Ti17 pancake were forged using the same process and extensively characterized, including low cycle fatigue and crack propagation. The comparaison reveals an overall advantage of β-CEZ over Ti17.

Titanium '92
Science and Technology
Edited by F.H. Froes and I. Caplan
The Minerals, Metals & Materials Society, 1993

1. INTRODUCTION

When designing the LP and HP compressor of a jet engine, there is a need for an alloy with optimized properties in the medium temperature range (300-450°C). At this temperature, creep resistance of Ti6Al4V is insufficient and the tensile properties of α-alloys are too low. Moreover in the search for reducing engine weight, Ti6Al4V, despite its low density, could also be replaced by an alloy with better strength at room temperature. Good candidates for this use are the β-metastable alloys. β-forged Ti17, used in the SNECMA M88 engine for instance, already offers a good compromise between room temperature properties and creep resistance at 400°C. β-CEZ is a β metastable alloy developed by CEZUS and SNECMA, which has been designed for better characteristics than Ti17.

The first characterization [1] conducted on small parts demonstrated an excellent property level. A new characterization was then undertaken to determine the alloy's properties for large parts. It started with a basic approach to process metallurgy definition. The heat treatment conditions were reoptimized within the thermomechanical processing window covered by the process for a disc. A quenchability study proved the predictability of static properties versus quenching rate. Lastly a forge shop processed pancake was extensively characterized, including LCF and crack propagation testing, and compared with a similarly processed Ti17 pancake.

2. BASIC METALLURGY OF β-CEZ

The nominal composition of β-CEZ is given on Table 1. β-transus lies between 880 and 900°C.

elements	Al	Mo	Cr	Fe	Sn	Zr
%	5.	4.	2.	1.	2.	4.

Table 1 : Nominal Composition of β-CEZ

In order to focus the basic metallurgical studies, a generic process had to be determined. As β-CEZ is derived from Ti17, a similar process could be applied (Table 2). Furthermore, an α/β blocking was a shaping operation without strong consequences on mechanical properties. Consequently, attention was focused on the β-finishing and the heat treatment operations and thus a processing window covering the hot die forging of large discs was explored.

blocking	α/β forging
finishing	β-hot die forging + controled cooling
solutioning	controlled quench rate
aging	

Table 2 : Generic Process of β-CEZ to be Optimized

The work performed was a collaboration of French universities and industrial companies in order to fully understand the microstructure evolution occuring during the forge and heat treatment process. The main results, following the process, are the subject of this paper.

The α solutioning curve shows the typical behaviour of β-metastable alloys [2]: the α content change with temperature is not very rapid (Fig. 1). α precipitation kinetics was also investigated [3]. The TTT diagram for β-CEZ shows a metastable domain for β equivalent to that of Ti17 (Fig. 2). The β-grain size before final forging is easy to control due to the low β-transus and no occurence of abnormal growth has been detected [4].

Flow stress was measured [5], it is characterized by an almost constant stress during deformation. The influence of temperature and strain rate on the level stress is shown in Fig. 3.

Deformation mechanisms were also investigated; it was found that the new grains are formed by progressive misorientation of the subgrains produced by dynamic recovery [6]. It was also shown by SNECMA that the β-phase remains metastable during deformation even for temperature as low as 800°C. It was observed that only a small amount of α precipitated at the grain boundaries during deformation. This observation enhances forgeability.

Another interesting feature of the precipitation phenomena in β-CEZ is the sluggish occurence of ω phase. It has been detected only for solutioning temperatures higher than 830°C, which will be later determined to be the upper limit for optimum solutioning temperature and only after very rapid quenching, not expected during a normal forging TMP.[7].

The basic metallurgy of β-CEZ is now well known and reveals a rather confortable behaviour.

3. HEAT TREATMENT OPTIMIZATION

Experimental observation

The evaluation of thermal treatments was performed on small scale pancakes and based on tensile strength and fracture toughness at 20°C, and creep at 400°C.

The material came from a 530mm diameter ingot, the composition of which is given on Table 3.

elements	Al	Mo	Cr	Fe	Sn	Zr
top	4.98	4.05	2.17	1.04	2.02	4.19
bottom	4.82	3.94	2.10	0.98	1.99	4.00

Table 3 : Composition of the Material Used (Ingot n°200487)

Pancakes were hot-die forged on a 150t hydraulic press. The test specimen coupons were finish-machined after heat-treatment.

Local deformation conditions were known through FORGE2 process modeling and measurements (flow-lines, post-forging cooling rates). They appeared to be homogeneous in the pancake.

Test matrix

Only one deformation condition was tested :

- heating temp. 920°C/die temp. 860°C
- deformation $\dot{\epsilon} = 10^{-2}s^{-1}$ / $\epsilon = 0.7$
- post forge quench 80°C/min.

Heat treatment conditions were evaluated according to the following test matrix :

- Solutioning: 830°C/4h – 800°C/4h
- Quench: 120°C/min
- Age: 570°C – 600°C and 630°C/8h

The solutioning temperature was choosen in order to ensure high strength or high creep resistance respectively. The relatively low quenching rate is representative of the center section of a water quenched 120mm thick pancake.

Results

At low magnification, grains and subgrains appear surrounded by α needles somewhat thicker than intragranular precipitates. Observed at high magnification, the intragranular α consists of entangled needles (Fig. 4). The higher the solutioning temperature, the thicker and sparser the needles. Not shown here is the secondary α precipitated between primary needles and revealed only at higher magnification. These microstructures are representative of solutioning with rapid quenching.

A higher solutioning temperature results in a high yield strength (Fig. 5). This effect is enhanced when the quenching rate is rapid. For both solutioning temperatures, the influence of aging on tensile stress is classical: higher aging temperature, the lower the strength. Ductility remains constant, although toughness increases with aging temperature. Contrary to most results on other alloys, the toughness/yield stress relationship appears to be dependent of solutioning temperature (Fig. 6).

Two possible solutioning treatments are considered : 830°C/4h to promote a high tensile strength or 800°C/4h to emphasize creep properties.

A remarkable behaviour for β-CEZ is the stability of creep resistance with aging temperature. On the other hand, the creep resistance is increased when the solutioning temperature is lowered from 830°C to 800°C (Fig. 7). This behaviour is opposite to that of most α/β and α alloys, for which the creep resistance increases when the solutioning temperature is raised. This is consistent with the fact that creep strength is mainly provided by large α needles. The same effect of increasing the large α needles content is reached while lowering the solution temperature for β processed alloys, while raising this temperature solutionises globular α for α/β processed alloys. It also appears that the creep results are much more scattered when the aging temperature is low.

Thus, a 600°C/8h age is optimum for creep properties.

4. PROCESSING AND CHARACTERIZATION OF LARGE PANCAKES

Two objectives were assigned to this study: demonstrate the feasability of large β-CEZ ingots and the forgeability and treatability of large pancakes.

The first goal was achieved by delivery of a 250 mm dia. billet transformed from a 660 mm dia. ingot. The billets proved to be chemically homegeneous (Table 4) and without microsegregation. The microstructure was an homogeneous equiaxed α/β type. No beta-fleck was detected after a beta-transus –30°C solutioning.

elements	Al	Mo	Cr	Fe	Sn	Zr
top	4.72	3.98	2.11	0.92	1.95	3.86
middle top	4.71	3.97	2.12	0.91	1.93	3.86
middle bottom	4.73	3.98	2.11	0.93	1.95	3.84
bottom	4.84	4.08	2.20	0.97	1.96	3.91

Table 4 : Chemical Homogeneity of the Ingot n°200554

Seven mults were forged into pancakes (dia. 430mm, h. ≈ 120mm) on a hot die press. The applied process consisted of an α/β preforming and a β-finishing (50% height reduction) followed by a water quench. Then the center of the disc was removed for a better quench efficiency at the bore. The choosen heat treatment was solution 830°C/4h/WQ + age 600°C/8h. Two of these pancakes were utilized to investigate the quenchability of β-CEZ.

The first one was thermocoupled before the solutioning and water quench and a quench modeling was performed at the same time. The correlation between the measured and calculated cooling rate curves was good. The second pancake was cut-up to measure the evolution of mechanical characteristics with increasing depth. Two characteristics were focused on: tensile properties at room temperature and creep resistance at 400°C. The cut-up plan and the cooling rate iso-contours are shown at the same scale at Fig. 8.

Results

Fig. 9 shows the evolution of YS and A% with cooling rate superimposed with the limits determined using the different sources of subscale experiments. The agreement between the results obtained on large and small pancakes is good and allows derivation of the quenchability curve after a 830°C-solutioning. The ductility is generally acceptable, except at the surface (cooling-rate over 1300°C/min); this was demonstrated to be generated by an intermediate ω precipitation resulting, after the 600°C treatment, in a very fine secondary alpha population which embrittles the alloy [5].

On the other hand, the creep results obtained on the large pancakes are at the upper limit of the life reached on small specimen (Fig. 10). This favorable creep behaviour remains to be explained.

Thus, the feasability of large discs taken from large billets is now demonstrated. Their properties are expected to be predictable from small size experiments. In particular, the effect of cooling rate has been quantified, allowing for a good prediction of massive part properties.

5. CHARACTERIZATION OF A MEDIUM SIZE PANCAKE – COMPARAISON WITH TI17

A medium size pancake (dia. 270mm / h.60mm) was hot die forged, heat treated and cut up in order to evaluate the complete set of in-service characteristics. For comparaison purposes, a Ti17 pancake of same size was also forged using the same process on the same equipment. The β-CEZ pancake has been applied a 800°C/4h/WQ solutioning treatment and a 600°C/8h aging. The heat treatment for the Ti17 pancake is a 800°C/4h/WQ solutioning and a 635°C/8h aging. Thus, except for the aging temperature, the characteristics of both pancakes can be compared on a constant process basis.

For any testing conditions no effect of the test specimen localization has been detected, so that the material can be considered as homogeneous.

The comparison between both alloys is given for the different characteristics on the following figures :

Fig. 11 : Shows a 50 to 80 MPa advantage for β-CEZ for UTS and YS between RT and 450°C. The elongation to rupture though slightly lower is still at an acceptable level (> 7%).

Fig. 12 : Creep stress to reach 0.2 % plastic elongation is plotted on a Larson-Miller diagram for β-CEZ and Ti17. This shows an advantage of 100MPa for β-CEZ to reach 0.2% elongation in 100 hours at 400°C.

Fig. 13 : Shows LCF life at 300 and 400°C. With an equivalent Young's modulus, at 5 10^3 cycles, an advantage in pseudo-stress is demonstrated over Ti17 : at 300°C +200MPa
at 400°C +50MPa.

Fig. 14 : The crack growth behaviour of both alloys is equivalent. The crack growth rate doesn't vary for test temperature comprised between 20°C and 450°C, including 90 s hold time experiments.

6. CONCLUSION

1. Based on pancake forgings, in the medium temperature range up to 450°C, β-CEZ provides better characteristics than Ti17.

2. Although β-CEZ is a new alloy, its metallurgical behaviour during thermomechanical processing has been widely investigated.

3. At several laboratory and industrial scales, quench modeling has proved capable of predicting the tensile and creep properties of a thick pancake.

4. The risk for the introduction of β-CEZ in an industrial process is then reduced.

7. ACKNOWLEDGMENTS

The authors acknowledge the sponsorship of the STPA and DRET. Our thanks are due to CEZUS for the provision of the β-CEZ alloy used in this investigation and for the support of many experiments made at their Research Center, Ugine.

REFERENCES

[1] B. PRANDI , "β-CEZ a new titanium alloy", (Paper presented at the TDA meeting, Orlando, 3-10/10/90)

[2] C. ANGELIER , J. BECHET , O. LACAGNE , "Etude des transformations structurales de l'alliage de titane β-CEZ : Ti-5Al-2Sn-4Zr-4Mo-2Cr-1Fe" , (Rapport de synthèse du GS Titane , CNAM : Laboratoire de métallurgie , 11/12/90)

[3] B. CORMIER , "Contribution à l'étude des transformations structurales dans les alliages de titane Ti-15-3-3-3 et β-CEZ", (Mémoire présenté en vue d'obtenir le diplôme d'ingénieur CNAM en métallurgie)

[4] N. COME , D. DELANNOY , E. GAUTHIER , A. SIMON , (Rapport de synthèse du GS Titane , Ecole des Mines de Nancy : LSG2M , 11/12/90)

[5] D. DAJNO , "Rhéologie globale et structurale des alliages de titane TA6V et β-CEZ dans les domaines α+β et β" , (Thèse de l'Ecole des Mines de Saint Etienne)

[6] F. MONTHEILLET , D. DAJNO , N. COME , E. GAUTHIER , A. SIMON P. AUDRERIE , A.M. CHAZE , Ch. LEVAILLANT , "Hot deformation of a new high strengh titanium alloy", (Paper presented at the 7th World Conference on Titanium, San Diego, 29/07-2/08/92)

[7] A HENRI , A. VASSEL , "Relation microstructure – propriétés d'emploi de l'alliage de titane β-CEZ", (Technical report n°21/3578M, ONERA : Direction des Matériaux)

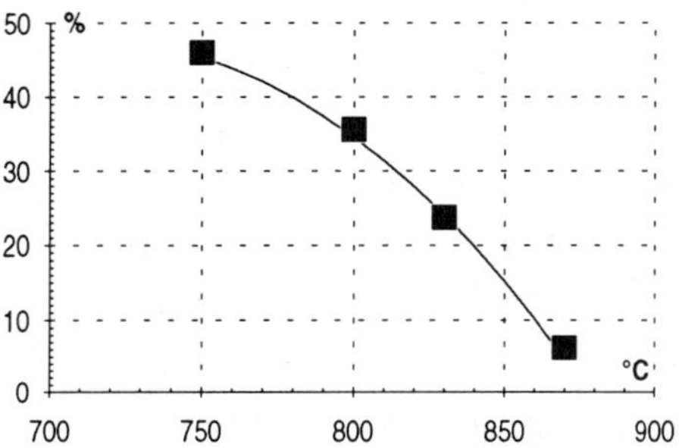

Fig.1: α-content vs. Solutioning Temperature

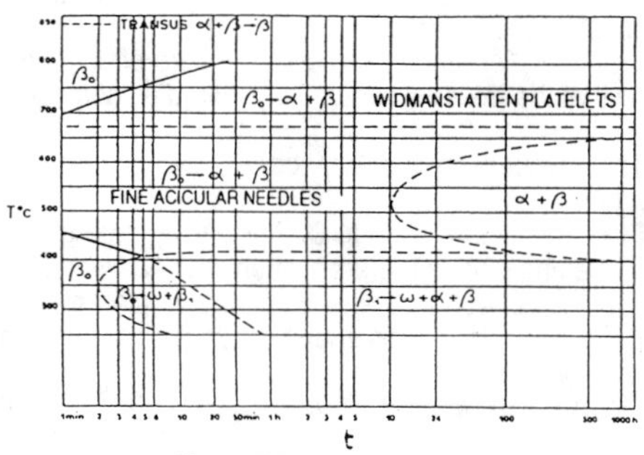

Fig. 2: T.T.T. Diagram : β-CEZ

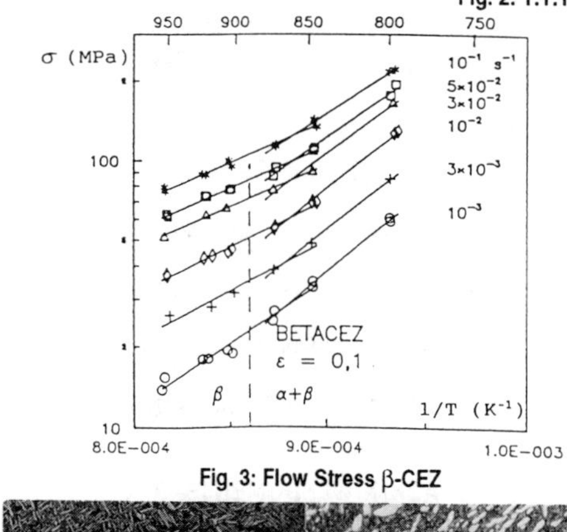

Fig. 3: Flow Stress β-CEZ

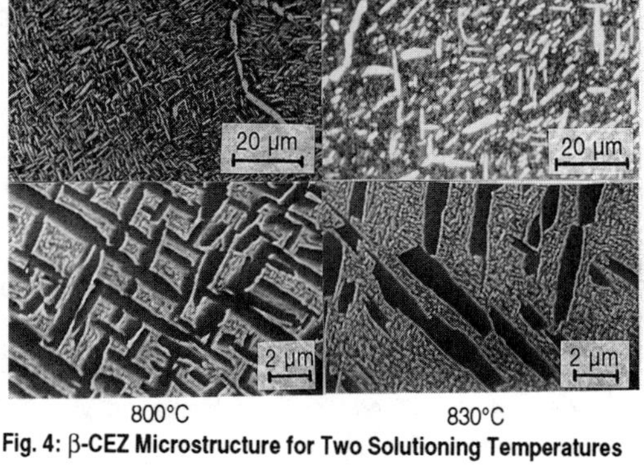

Fig. 4: β-CEZ Microstructure for Two Solutioning Temperatures

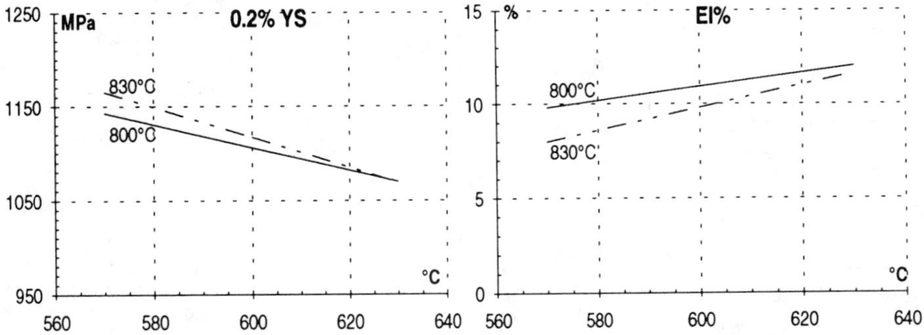

Fig. 5: RT Properties vs. Aging Temperature

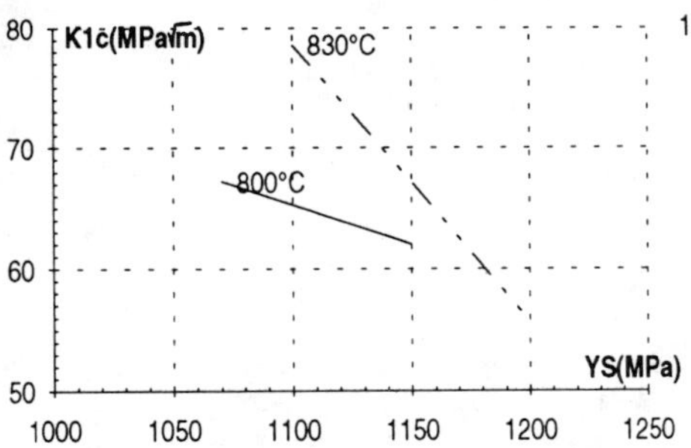

Fig.6: Toughness vs. 20°C Yield Stress

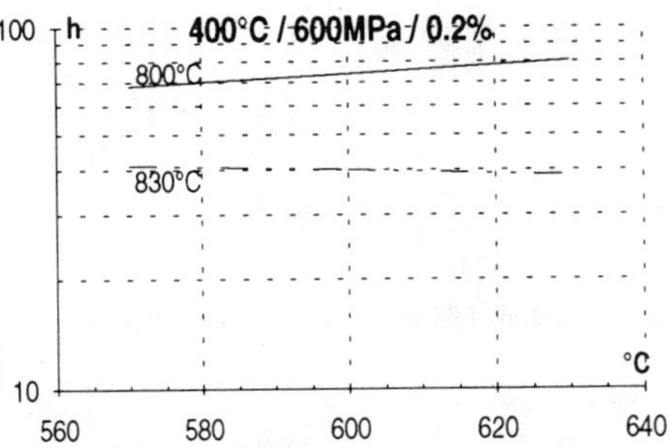

Fig.7: Creep Resistance vs. Aging Temperature

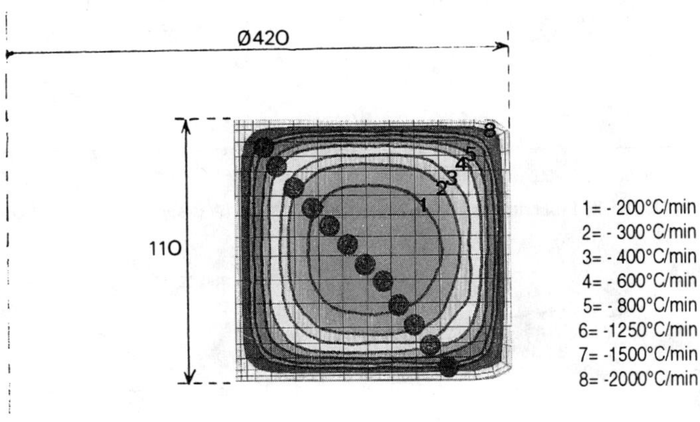

Fig. 8: Cut-Up Plan of the 20000t Pancake Superposed With Iso-Cooling Rates

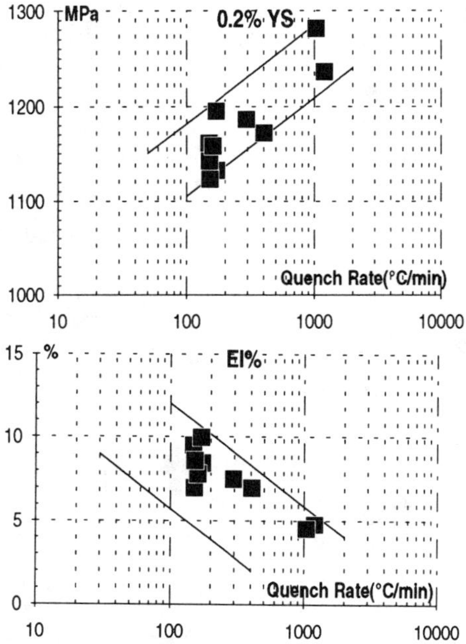

Fig.9: Evolution of RT Properties with Cooling Rate

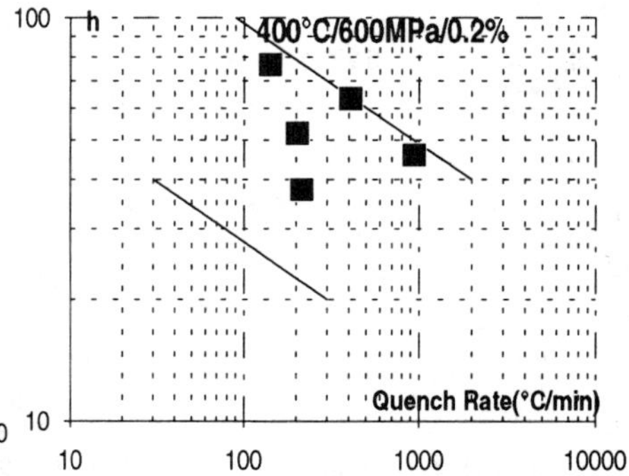

Fig.10: Evolution of Creep Properties with Cooling Rate

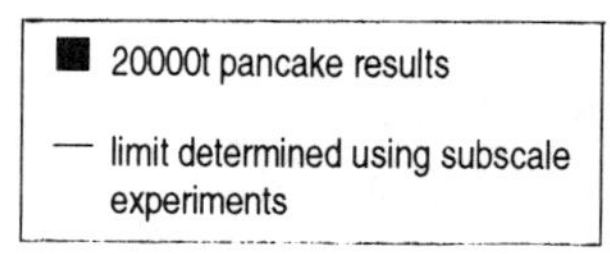

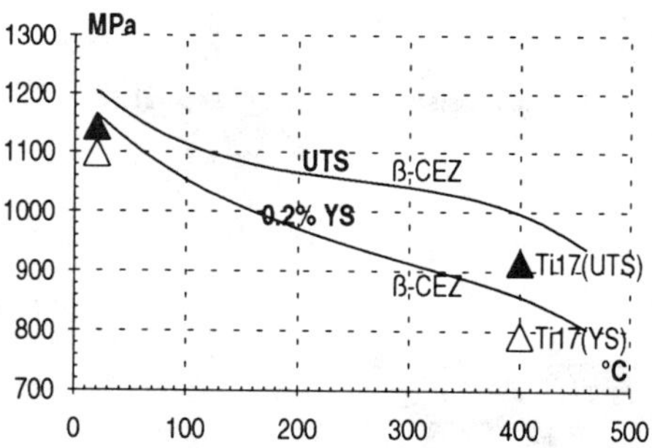

Fig.11: Tensile Properties vs. Test Temperature

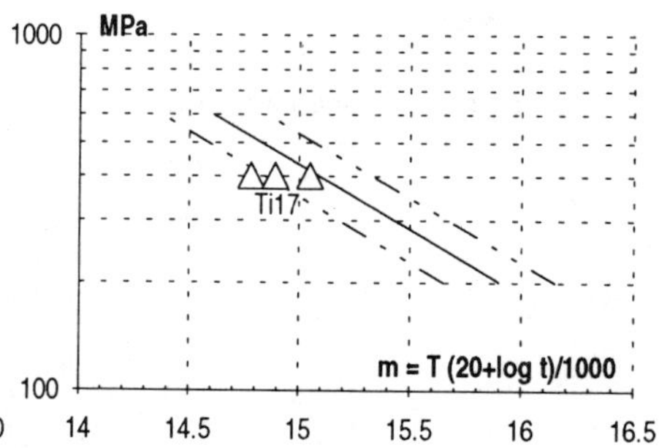

Fig.12: Creep Properties vs. Larson Miller Parameter

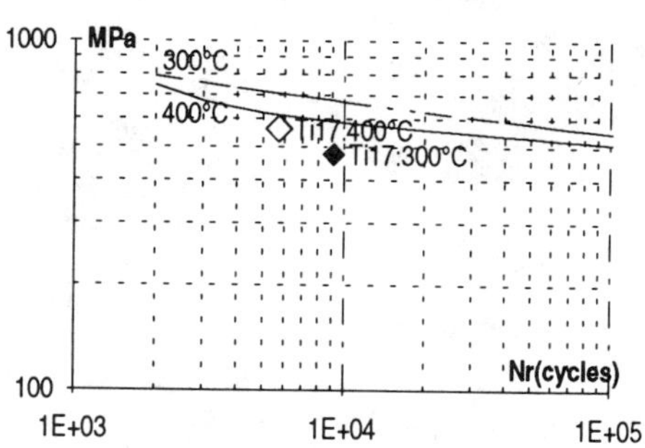

Fig.13: LCF Life (R_ε=0)

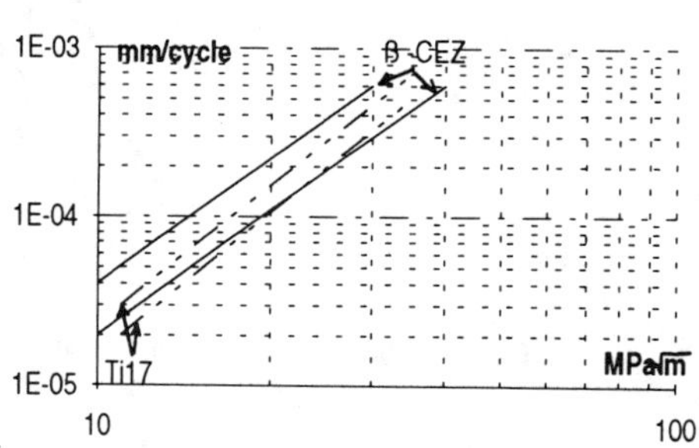

Fig. 14: Crack Propagation Rate Between 20°C and 400°C (scatter band for all testing temperatures)

HOT DEFORMATION OF THE HIGH STRENGTH *BETACEZ* TITANIUM ALLOY

F.Montheillet[1], D.Dajno[1], N.Côme[2], E.Gautier[2], A.Simon[2],
P.Audrerie[3], A.-M.Chaze[3] and Ch.Levaillant[3]

[1]Ecole des Mines de Saint-Etienne, Département Matériaux,
158 cours Fauriel, 42023 Saint-Etienne cedex 2, France
[2]Ecole des Mines de Nancy, LSG2M, Parc de Saurupt,
54042 Nancy cedex, France
[3]Ecole des Mines de Paris, CEMEF, Sophia Antipolis,
06560 Valbonne, France
(Groupement Scientifique Titane-Traitements Thermomécaniques)

Abstract

The mechanical behavior during hot working of the new ß metastable titanium alloy *Betacez* was studied by tension, compression and torsion tests. The flow rule and dependence on strain, strain rate and temperature of the associated rheological parameters were determined. The microstructural evolutions were investigated at the grain scale by optical and scanning electron microscopy. It is shown that a continuous recrystallization process takes place in the ß phase leading to a constant flow stress and producing a grain refinement. In the two phase field, the evolution of the morphological texture provides a reasonable explanation to the strain softening observed in compression.

Introduction

A new titanium alloy, hereafter designated *Betacez*, with high tensile strength and excellent creep resistance up to 450 °C, has been recently developed by the CEZUS Company for jet engine compressor disk applications [1]. The chemical composition, *viz.* 5% Al, 4% Zr, 4% Mo, 2% Sn, 2% Cr and 1% Fe can be rationalized by aluminium and molybdenum equivalent contents of about 6% and 10%, respectively, which locates this alloy close to the Ti17 standard [2]. *Betacez* thus belongs to the ß metastable type of titanium alloys for which the ß phase can be kept at room temperature by rapid cooling. The ß transus is about 890 °C.

This paper aims to describe the hot deformation behavior of this alloy and the associated microstructural evolutions at a grain scale in both the ß single phase and α+ß two phase temperature ranges. Uniaxial tension (strains lower than 0.3) and compression as well as torsion (strains larger than 1) experiments were carried out between 800 and 950 °C at strain rates from 10^{-3} to 1 s^{-1}. Experimental details are given in references 3-5.

Titanium '92
Science and Technology
Edited by F.H. Froes and I. Caplan
The Minerals, Metals & Materials Society, 1993

Stress-Strain Curves

Initial Yield Point

The mechanical response of the *Betacez* alloy submitted to hot compression tests is illustrated in Figs 1a and b. It is apparent that all stress-strain curves exhibit a sharp initial peak, for a total strain lower than 0.05, in both the single and two phase domains. This also holds for the torsion (Figs 2a and 2b) and tension tests. When the elastic part of strain is substracted, the peak appears to be an initial *yield point*. As in the case of mild steels at lower temperatures, it increases with strain rate and the time of annealing heat treatment before deformation (1 h in the present case). Although electron microscopic investigations do not provide any microstructural evidence, this yield point can be reasonably attributed to the anchoring of dislocations by solute atoms. For instance, the 850 ppm oxygen content of the alloy is likely to be enough to explain such a phenomenon. Since it has been shown that iron increases the diffusion rate in titanium alloys [6], this would agree with the fact that the yield point is more pronounced in *Betacez* than in other titanium alloys such as Ti6-4 and Ti6-2-4-2.

Strain Softening Behavior in Compression and Torsion Tests

In the $\alpha+\beta$ field (Fig.1a), the *compression* flow stress decreases until a steady state takes place. Although this softening can be partly attributed to deformation heating at large strain rates, it will be shown below that it can be correlated with the evolution of the α phase aspect ratio during straining. In the β single phase domain (Fig.1b), softening is much lower and would probably vanish without any thermal effect.

Fig.2a shows the evolution of flow stress in *torsion* test [7]. Here adiabatic heating has been taken into account. It appears that the corrected stress remains almost constant after the initial yield point. It should be noted that Figs 1a and 2a are relative to an initially equiaxed microstructure. On the other hand, Fig.2b shows that when the initial microstructure is lamellar, a strong softening occurs, which clearly again cannot be attributed to deformation heating only. Such a decrease of flow stress associated with needle-like or lamellar microstructures has been frequently reported in titanium alloys [8].

Temperature and Strain Rate Dependence

The classical ln σ-ln $\dot{\epsilon}$ and ln σ-1/T diagrams show that the flow rule of the *Betacez* alloy can be expressed to a first approximation in the simple form

$$\sigma = k\ \dot{\epsilon}^m \exp(mQ/RT)$$

The average strain rate sensitivity exponent m is about 0.3, that is significantly higher than for Ti6-4 in the same strain rate range. Since a large amount of experimental data were available, the dependence of m on strain, strain rate and temperature (i.e. *second order* variations of σ) were determined. It was shown that m is an increasing function of temperature, except at low strain rates in the two phase range, which can be attributed to the occurrence of some grain boundary sliding (superplastic behavior), and at large strain rates in the ß domain. m is also a decreasing function of strain rate, which is in agreement with the above result according to the equivalence between strain rate and temperature. Finally, it was found that it decreases with strain. Although the same trend has been observed for Ti6-4 [3], it has not received any physical explanation to date.

The average apparent activation energy is found to be about 200 kJ/mol from compression tests performed in a high temperature domain including the stable ß phase range together with the higher part of the two phase range (850 °C<θ< 950 °C), Fig.3. However, when experiments carried out on the only (stable or metastable) ß structure are used (750 °C<θ< 920 °C), a somewhat lower value of about 160 kJ/mol is determined [4]. In the lower part of the α+ß domain, Q becomes larger, with an average value of about 360 kJ/mol. Thus the 200 kJ/mol value is likely to be influenced by the deformation of the α phase in the two phase domain, so that the value of 160 kJ/mol for the ß phase can be considered to be more reliable. It should be noted that it is very close to the activation energy for self diffusion in ß titanium, i.e. 153 kJ/mol [7], which can be related to the mechanism of dynamic recovery. In the two phase domain, however, the physical interpretation of Q is not straightforward, since the volume fraction of the phases varies with temperature [9].

Microstructural Evolutions

The initial and deformed microstructures were investigated by optical and scanning electron microscopy in order to provide some physical or micro-mechanical interpretations to the above stress-strain curves at the scale of grains.

Initial Microstructures

The heating schedule and time of annealing heat treatment before deformation must be carefully controlled to get significant stress-strain curves since they determine the initial microstructure of the test. In the ß range, the grain size is about 100 μm after 5 min holding time at 920 °C and grows up to 400 μm after 4 h [4]. In order to be close to the industrial hot working conditions, most of the experiments were carried out after 1 h holding time at the test temperature, leading for instance to an average grain size of 300 μm at 920 °C. In the two phase domain, the evolution is less significant, although a slight size decrease and spheroidization of the α grains were observed [10]. After 1 h holding time, the α grain size was typically 4 μm with an aspect ratio of 1.3.

Deformation of the ß Phase

Due to the metastability of the ß structure, it was possible to clearly characterize the strained microstructures after water quenching. Figs 4a and b show that the grain boundaries become corrugated during tensile straining, while subgrains progressively develop. At large strains (ϵ>1) obtained in compression (Fig.4c) or in torsion, new 'recrystallized' grains of small size (≈ 10 μm) are observed near the initial grain boundaries and then fill entirely the initial flattened (or elongated) grains.

The same deformation process has already been observed in other titanium alloys, such as Ti10-2-3 [11] and other metals (aluminium, α iron, ferritic steels) or minerals (NaCl, KCl, olivine), always characterized by a high stacking fault energy, *i.e.* a high dislocation mobility [12]. The observed new grains are likely to be formed by the *progressive* misorientation of the subgrains produced by dynamic recovery. This 'continuous recrystallization' mechanism is more efficient in the neighborhood of the initial grain boundaries where there is a larger number of active slip systems. It should be emphasized that it differs from the mechanism of 'dynamic recrystallization' commonly observed in low stacking fault energy metals and alloys such as copper or austenitic steels, since it does not involve any nucleation or growth of the new 'recrystallized' grains. Accordingly, whatever the temperature and strain rate conditions, the stress-strain curves do not exhibit the classical 'single peak' or 'multiple peak' shape associated with dynamic recrystallization. Another interpretation for the hot deformation of high stacking fault energy metals at very large strains has been recently proposed [13], in which the 'new' grain boundaries are nothing else than the initial elongated and fragmented ones ('geometric dynamic recrystallization'). However the latter does not seem to apply here, where the investigated strains do not exceed $\epsilon \approx 2$. From a practical point of view, it is worthy to note that deformation in the ß range is likely to produce a significant refinement of the microstructure.

Deformation in the α+ß Domain

For the investigated temperature and strain rate ranges, it does not seem that any dynamic recrystallization occur in the α phase, which is in agreement with previous work on Ti6-4 [3]. To a first approximation, the volume fraction of α phase decreases linearly from about 0.35 to zero between 800 and 890 °C [14]. Since the latter is considerably harder than the ß structure, the α grains can be considered to behave like hard inclusions within a softer matrix. This is illustrated by Fig.5 which shows that the strain of the α phase is 0.3 the overall strain between 800 and 840°C. Thus it can be thought that **strain is not large enough to start dynamic recrystallization** in the α phase. Another effect of the two phase structure is that strain concentrations must arise in the softer ß matrix in the neighborhood of the α grains, in order to accommodate their low deformation (and rotation in torsion), which is likely to favour the continuous recrystallization process within the ß phase.

The differences between torsion and compression stress-strain curves reported above can be related to the evolution of the α grain shape : in compression, since the strain direction coincides with the long axis of the grains, the aspect ratio *decreases* during the test. On the other hand in torsion, the latter *increases* (Figs 6a and b). Theoretical models have been developed to predict the rheological behavior of two phase aggregates [15]. Whatever the precise assumption used for the homogeneization procedure, they show that the strain rate within the hard phase (which is always less than the overall strain rate), is *larger* when the grains are elongated or flattened and *lower* when they are globular. Therefore, the flow stress is higher in the first case, which agrees with the strain softening observed in compression (in particular for a lamellar initial microstructure) and would predict some strain hardening in torsion, as long as there is no fracture of the elongated grains. In the latter case, however, it is also possible that the correction for adiabatic heating was somewhat overestimated.

In conclusion, the rheological behavior and the associated evolutions of microstructure in the *Betacez* and Ti6-4 alloys are quite comparable : in particular, the ß phase appears to give rise to a process of 'continuous recrystallization', leading to a small ß grain size. The rheological parameters m and Q are very similar. Finally, the shape of the stress-strain curves in the two phase range can be interpreted in terms of the evolution of the material 'morphological texture'.

References

1. B.Prandi et al., "ß-Cez, a High Performance Titanium Alloy for Aerospace Engines", International Conference on Titanium 88, ed. P.Lacombe, R.Tricot and G.Béranger (Paris, Société Fançaise de Métallurgie, 1988), 811-818.
2. Y.Combres and B.Champin, "Titanium Alloys Processing : State of Art and Perspectives", Matériaux et Techniques, 1991, no. 5-6:1-11.
3. D.Dajno, "Rhéologie globale et structurale des alliages de titane TA6V et *Betacez* dans les domaines α+ß et ß" (Thèse, Ecole des Mines de Saint-Etienne, 1991).
4. N.Côme, "Déformation à chaud et évolutions microstructurales des alliages de titane TA6V et ß-Cez dans le domaine ß" (Thèse, Ecole des Mines de Nancy, 1991).
5. D.Dajno and F.Montheillet, "Rhéologie globale et locale des alliages de titane TA6V et *Betacez*", in press.
6. M.L.Meier and A.K.Mukherjee, "Consequences of Using the Average Diffusivity in Correlating the Enhanced Superplasticity of Iron Modified Ti-6Al-4V", Scripta Metall. Mater. 25 (1991), 1471-1476.
7. Y.Combres and B.Champin, "Rhéologie du ß-Cez et transformation des lingots", in press.
8. J.-G.Malcor and F.Montheillet, "Loi de comportement de l'alliage de titane Ti-6Al-4V dans le domaine du forgeage à chaud", Mém. Etud. Sci. Rev. Métall. 80 (1983), 89-103.
9. F.Montheillet, "Rhéologie des alliages de titane : méthodologie, progrès et problèmes", in press.
10. P.Audrerie et al. (G.S. Titane Report, june 1991).
11. I.Weiss and F.H.Froes, "The 'Processing Window' for the near beta Ti-10V-2Fe-3Al Alloy", International Conference on Titanium 84, ed. G.Lütjering, U.Zwicker and W.Bunk (Oberursel, Deutsche Gesellschaft für Metallkunde, 1984), 499-506.

12. F.Montheillet, "Comportement mécanique et structural des matériaux à forte énergie de défaut d'empilement sous grande déformation à chaud", 24ème Colloque de Métallurgie 'Les traitements thermomécaniques' (Saclay, INSTN, 1981), 57-70.
13. J.K.Solberg et al., "Influence of Ultra-High Strains at Elevated Temperatures on the Microstructure of Aluminium, Part I", Phil. Mag. A60 (1989), 447-471.
14. J.Bechet, C.Angelier, B.Cormier, O.Lacagne and M.Lhermitte, "Phase Transformations in a Near β-Ti Alloy : Microstructure and Mechanical Properties" (This Conference).
15. Ph.Vernusse and F.Montheillet, "Etude théorique de la rhéologie à chaud d'alliages à deux phases ductiles", Mém. Etud. Sci. Rev. Métall. 89 (1992), 83-91.

Acknowledgments. The authors gratefully acknowledge the CNRS (Centre National de la Recherche Scientifique), the DRET (Direction des Recherches Etudes et Techniques) and the CEZUS Company for the support of this research.

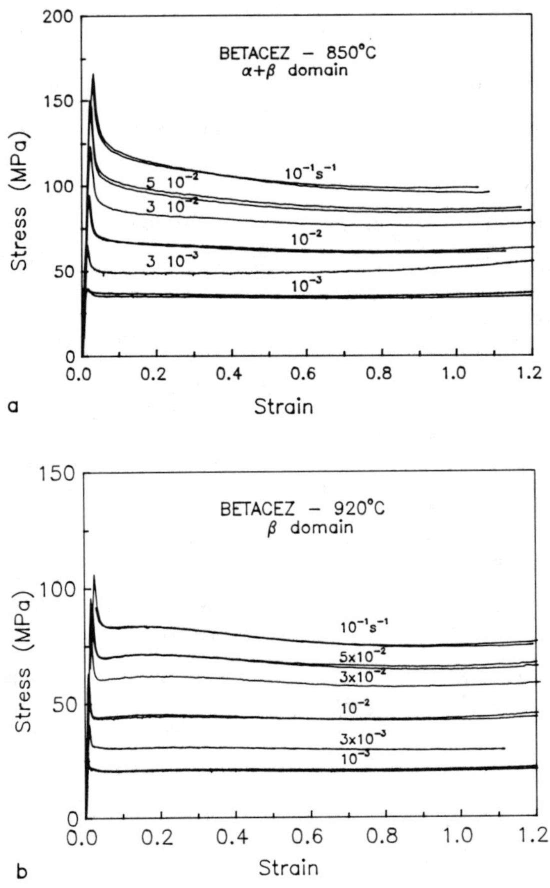

Figure 1 : Compression stress-strain curves of the *Betacez* alloy in the two phase (a) and single phase (b) domains.

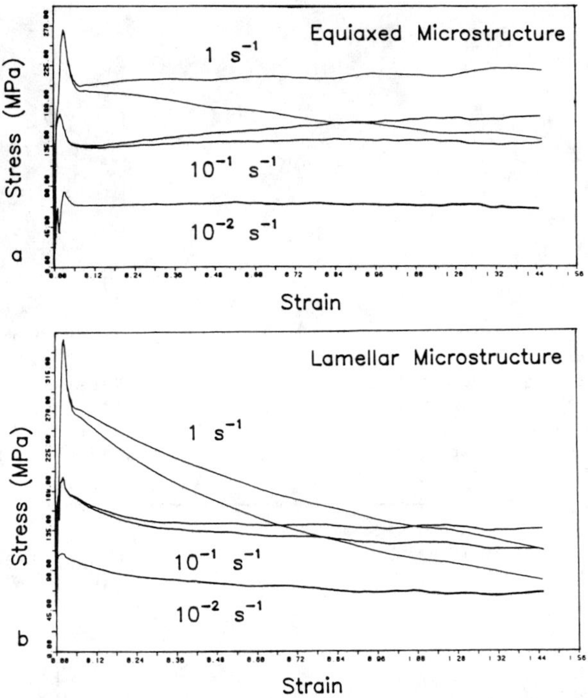

Figure 2 : Torsional stress-strain curves of the *Betacez* alloy in the two phase domain (Θ = 800 °C), for an initially equiaxed (a) or lamellar (b) microstructure [7].

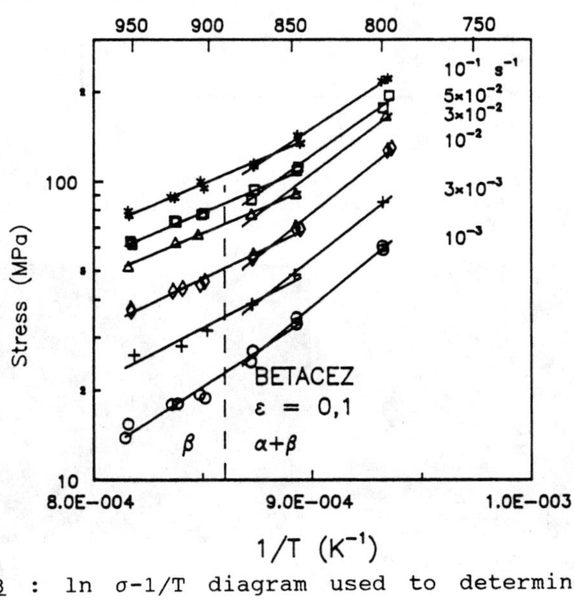

Figure 3 : ln σ-1/T diagram used to determine the apparent activation energy Q.

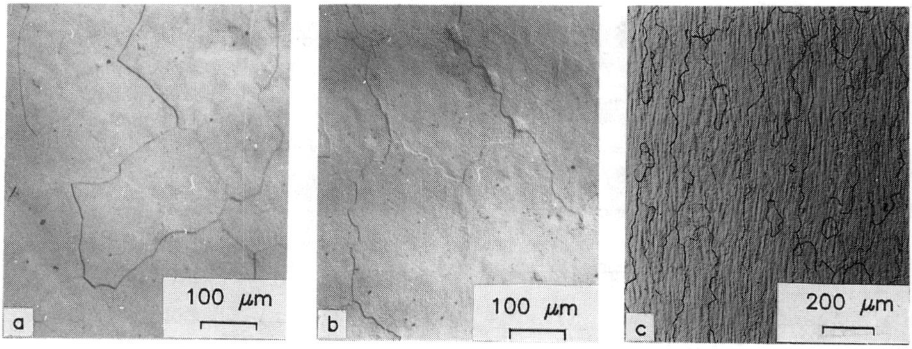

Figure 4 : Microstructure of quenched *Betacez* alloy after straining in the β domain. (a) and (b) : tension along vertical axis, $\dot{\epsilon} = 10^{-2}$ s^{-1}, $\Theta = 920$ °C , $\epsilon \approx 0.1$ and 0.26, respectively ; (c) compression along horizontal axis, $\dot{\epsilon} = 10^{-3}$ s^{-1}, $\Theta = 905$ °C, $\epsilon \approx 1.2$.

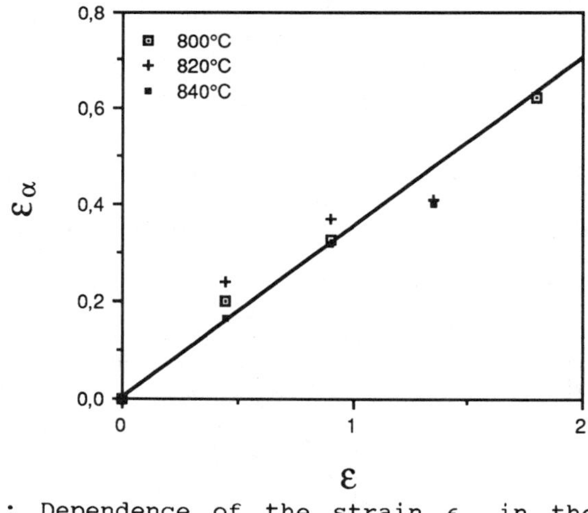

Figure 5 : Dependence of the strain ϵ_α in the α phase with respect to the overall strain ϵ.

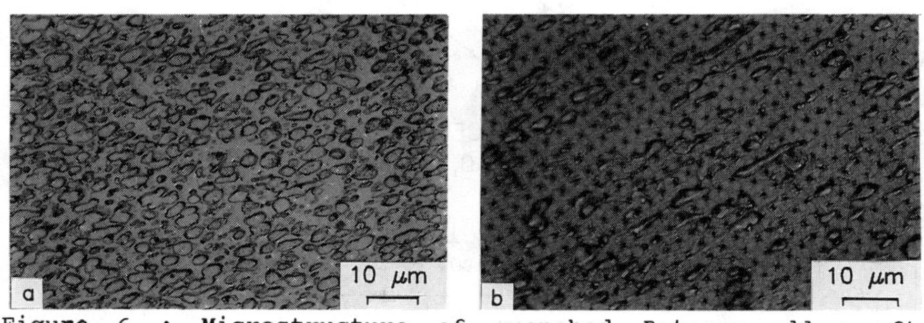

Figure 6 : Microstructure of quenched *Betacez* alloy after torsional straining (horizontal axis) in the two phase domain. $\dot{\epsilon} = 10^{-1}$ s^{-1}, $\epsilon = 1$. (a) $\Theta = 800$ °C ; (b) $\Theta = 850$ °C.

MANUFACTURE AND PROPERTIES OF HEAVY OPEN-DIE FORGINGS FROM

LOW - ALLOY Ti-0.3Mo-0.8Ni FOR CHEMICAL PROCESSING EQUIPMENT

D. Fischer, K. Hülse

Deutsche Titan GmbH, 4300 Essen/GERMANY

Abstract

The low-alloyed titanium grade Ti-0.3Mo-0.8Ni, ASTM/ASME Grade 12, exhibits improved corrosion resistance, higher strength and permits higher design loads than unalloyed grades. This combination of properties has led to an increased economic interest in this alloy.

Thick-walled heavy open-die forgings cannot be considered as standard components. Nevertheless, they must comply with the standard material requirements in order to benefit from the advantages of the Grade 12 alloy.

By setting an optimum composition of the vacuum-melted ingot in conjunction with suitable forming and heat treatment parameters it is shown that the desired characteristics can be reached with a high rate of repeatability with properties nearly equivalent to those of thin-walled forgings.

The review of the properties of low-alloy Grade 12 titanium parts is followed by a comparisation with the characteristics of common unalloyed titanium grades at room temperature and elevated temperatures.

Introduction

Titanium has long been appreciated as a corrosion-resistant structural material and has proven its value in chemical processing equipment over several decades.

An outstanding combination of beneficial properties such as:

- excellent corrosion resistance
- low specific weight
- high strength
- good workability

and the availability of titanium in all standard semi-finished forms, as specified in national and international standards, have brought this material into universal use and given rise to vast manufacturing, processing and application know-how.

Titanium '92
Science and Technology
Edited by F.H. Froes and I. Caplan
The Minerals, Metals & Materials Society, 1993

The low-alloy titanium grade Ti-0.3Mo-0.8Ni, ASTM/ASM Grade 12, which was first introduced around 1974 (1), was originally developed as a corrosion-resistant low-cost alternative to palladium-alloyed titanium grades with special regard to crevice corrosion behaviour (2).

It is known from the literature (3, 4) that the corrosion resistance of Titanium Grade 12 varies with its microstructure, and hence with the heat treatment applied. While under many conditions Ti-0.3Mo-0.8Ni is nearly equivalent to Grade 7 or Grade 11 in terms of its corrosion resistance, it is superior in terms of strength (Table I). This combination of properties has led to an increased economic interest in this alloy wherever elevated corrosion, pressure and temperature loads are encountered.

ASTM/ASME B 381	O max	N max	C max	H max	Fe max	Mo	Ni	Pd	residuals max each	residuals max total	Ti	DTG TIKRUTAN
Grade F-2	0.25	0.03	0.10	0.0150	0.30	–	–	–	0.1	0.4	remainder	RT 15
Grade F-3	0.35	0.05	0.10	0.0150	0.30	–	–	–	0.1	0.4	remainder	RT 18
Grade F-7	0.25	0.03	0.10	0.0150	0.30	–	–	0.12 / 0.25	0.1	0.4	remainder	RT 15 Pd
Grade F-12	0.25	0.03	0.08	0.0150	0.30	0.2 / 0.4	0.6 / 0.9	–	0.05	0.3	remainder	LT 27

ASTM/ASME B-381	tensile strength min N/mm²	tensile strength min ksi	yield strength min N/mm²	yield strength min ksi	elongation (4D) min %	reduction of area min %	DTG TIKRUTAN
Grade F-2	345	50	275	40	20	30	RT 15
Grade F-3	450	65	380	55	18	30	RT 18
Grade F-7	345	50	275	40	20	30	RT 15 Pd
Grade F-12	483	70	345	50	18	25	LT 27

Table I
ASTM/ASME B 381
Requirements
Chemical composition
(weight %)
Mechanical Properties

Like commercially pure titanium grades, Grade 12 can be welded without problems. The standard filler metal is bare wire of the same composition as the base material. Instead of Grade 12 filler wire it is also possible to use a welding wire alloyed with palladium.Titanium is also readily amenable to site welding(5), which is for example of great importance in offshore applications.

Figure 1
Tubular heat exchanger with solid tube plates from TIKRUTAN LT 27 / ASTM B 381 grade F-12 (courtesy of OLMI S.p.A. Italy)

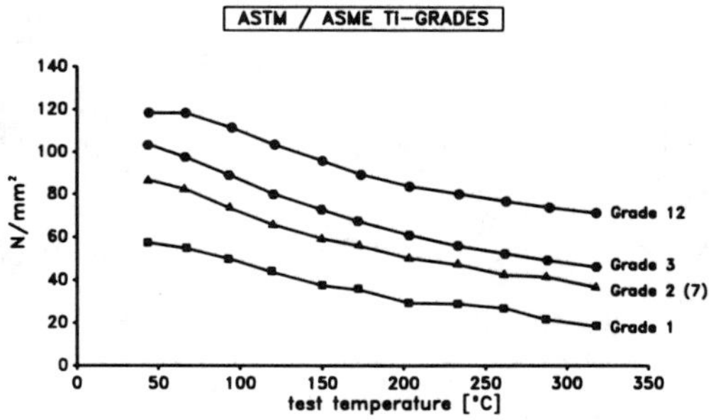

Figure 2
Maximum design allowables in tension for ASME Code-qualified titanium grades (6)

The present paper mainly deals by way of example with tube plates for tubular heat exchangers for North Sea offshore use, and with industrial-type centrifugal machinery such as decanting centrifuges.
The heat exchangers measured about 1500mm dia.x 6800mm length; their shells consisted of titanium-lined steel (Figure 1).
Grade 12 titanium has been specified on the basis of ASME design values (Figure 2).
The centrifuge is used in a coal preparation plant to remove salt crystals from hot washing water containing very high chloride concentrations (approx. 160,000 ppm). The application required resistance to both corrosion and wear with expected service life of around 35 years.
The housing of around 700 mm dia. x 1500 mm length was shape-forged as were the remaining components (Figure 3, Figure 4).

Figure 3
Casing, shape forged,
weight approx.750 kg,
TIKRUTAN LT 27 /
ASTM B 381 grade F-12

Figure 4
Flange shafts, shape forged,
weights approx.120/160 kg,
TIKRUTAN LT 27 /
ASTM B 381 grade F-12

Manufacture of the Forgings

In order to manufacture forgings of the aforementioned size and weight from Grade 12 titanium, it is necessary to take both metallurgical and forming aspects into account.
In the absence of any previously known or published external experience, economic production methods had to be developed, beginning with ingot melting, in order to ensure the required reproducibility in terms of shape and properties.

The main requirements were:

- achievement of the required mechanical properties
- shape accuracy and freedom from defects
- uniform microstructure
- practically positioned integral specimen material
- minimum risk of failure.

The starting ingots were double melted in a vacuum arc furnace using consumable electrodes, with both alloying elements and oxygen content being closely controlled.

The forging operation was carried out on a hydraulic 17-MN double-stand underfloor press with tool quick-change system.

The open-die forging press is served by an integrated manipulator operating on rails as well as by a freely movable material feeding and tool manipulator.
The ingots were heated in bogie hearth furnaces and rotary hearth furnaces fired with natural gas.

The forging steps used to manufacture the tube plates are shown in Figure 5. This illustration shows two different temperature ranges (7).

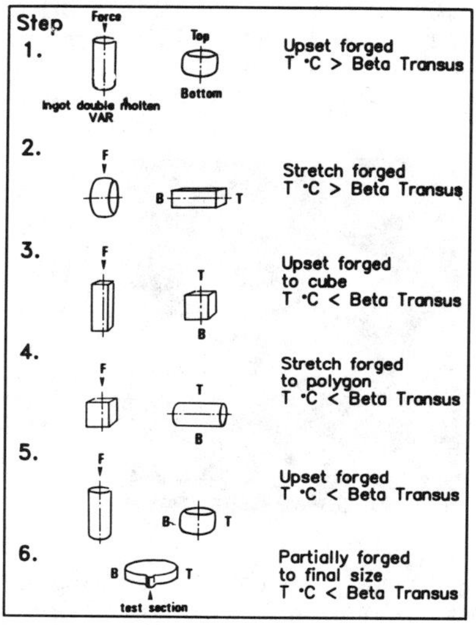

Figure 5
Forging route of heavy tube plates from Titanium on forging press 17 MN

At temperatures above the beta transus temperature, the as-cast structure was destroyed and the structure homogenized. At the same time, any cavities were closed by upsetting operations.
Refining of the microstructure to produce the required mechanical properties and final shaping were carried out at temperatures below the beta transus temperature.

The centrifuge forgings were produced according to the same specification. All parts were then annealed at approx. 650 °C/-AC to optimize the material properties.
Figure 6 shows the finished tube plates after an ultrasonic inspection. Figure 7 shows a shape-forged housing component after heat treatment and machining.

Description and Evaluation of Test Results

The properties of all products presented in this paper - tube plates, forgings, and for comparison plate and sheet material - were evaluated on the basis of the requirements of ASTM/ASME B-381 for Grade 12 titanium material.

The target values for chemical composition and mechanical properties are shown in Table I. To allow comparison with other titanium materials used in chemical process equipment, Table I additionally shows the values specified for the unalloyed grades F-2, F-3 and the Pd-alloyed grade F-7.

Figure 6
Tube plates forged, annealed, machined, ultrasonically tested

Figure 7
Casing, shape forged, machined, weight approx. 160 kg, TIKRUTAN LT 27 / ASTM B 381 grade F-12

The actual concentrations of the two metallic alloying elements nickel (Ni) and molybdenum (Mo), and the nonmetallic element oxygen (O), which exerts a key influence on properties, are shown in Table II for each heat analysed.

The results of the chemical analysis show that all values for the aforementioned three elements as well as for the remaining specific elements are within the range specified in ASTM B-381. The iron concentrations, which are of importance for the corrosion resistance of the material, remained at an entirely uncritical level below 0.10 % by weight in all heats analysed.

In the manufacture of large forgings, where the material volume comprises virtually the entire heat, the uniformity of all properties is critically dependent on the chemical homogeneity of the ingot. Hardness measurements have been found to be a highly reliable method of verifying this chemical homogeneity on the individual ingot. Figure 8 shows the surface hardness measured in ingot longitudinal direction for each of the heats. For heats no. 1, no. 2 and no. 3, which were assigned to the manufacture of the tube plates and forgings, the differences in HB 30 hardness values amount to at most 4 units on an individual ingot and lie within a range of 8 units HB 30 for all three ingots. These values are sufficient to permit the manufacture of heavy forgings displaying homogeneous and reproducible properties.

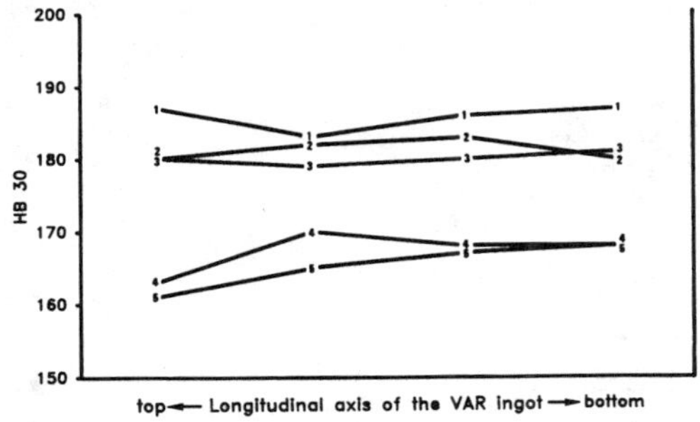

Figure 8
Ingot hardness of TIKRUTAN LT 27 / ASTM B 381 grade F-12

The same observations can be made for the homogeneity of heats no. 4 and 5, which exhibited somewhat reduced but still sufficiently uniform average hardness values and were therefore assigned to the production of small-to-medium sized forgings as well as sheet/plate material.

The results of mechanical tests of all forgings made from heats no. 1 to 5 are listed in Table II.

Table II
Mechanical properties and chemical composition of various products from TIKRUTAN LT 27 / ASTM B 381 grade F-12

DTG TIKRUTAN LT 27				Product type dimension / weight	Tensile strength min. 483 N/mm²	Yield strength min. 345 N/mm²	Elongation (4D) min. 18%	Reduction of area min. 25%
Melt No.	Comp. (wt.-%) Ni	Mo	O					
				Tube plates				
1	0,66	0,28	0,17	Ø1570 x 322 2840 kg	537	426	21	32
2	0,78	0,32	0,16	Ø1570 x 322 2840 kg	512	392	21	35
3	0,71	0,32	0,15	Ø1130 x 125 547 kg	490	367	27	38
				Ø1130 x 135 622 kg	494	388	26	41
				Forgings				
3	0,71	0,32	0,15	416 kg	555	452	24	42
				119 kg	529	425	22	36
4	0,72	0,32	0,15	138 kg	578	479	23	38
				42 kg	516	379	23	40
5	0,74	0,30	0,14	208 kg	524	391	22	41
				128 kg	522	407	19	42
				Plate / sheet				
4	0,72	0,32	0,15	50 mm	521	400	24	41
				6 mm	562	457	21 (L_0 = 50mm)	–

All individual values are above the required minimum thresholds by a sufficient margin. This applies both to strength characteristics and to ductility.

It will be noted that in spite of the size differences between the products, as shown in Table II, the mechanical properties lie within a relatively narrow band. This can only be achieved by close control of the composition and hardness of the ingot material.

Figure 9 shows a graphic representation of the averages and scatter ranges of all yield strength values as measured at room temperature. For comparison, the diagram also gives the corresponding yield strength averages and scatter ranges for similar forgings made of Grade F-2 and Grade F-3 material.

A number of representative values measured at temperatures of 150 °C, 200 °C and 250 °C for Grade F-3 and Grade F-12, and at 120 °C for Grade F-2, illustrates the temperature dependence of the mechanical properties.

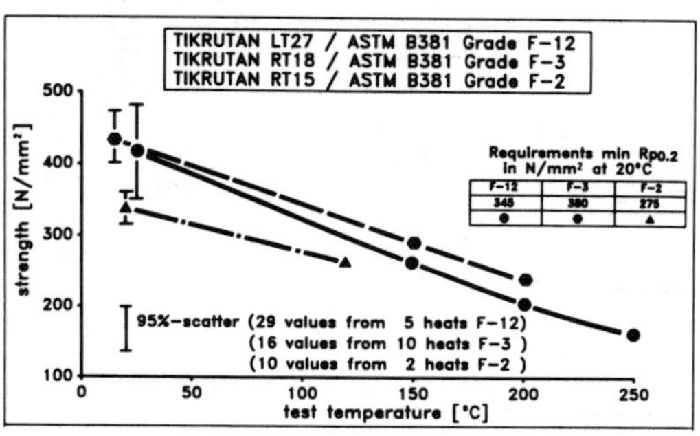

Figure 9
Yield strength 0.2% of investigated material

These results show that the mechanical properties of Grade F-3 and Grade F-12 are nearly identical at room temperature, and that the two materials are clearly similar in terms of the temperature-dependence of their yield point.

Figure 10 shows the microstructures of two large tube plates made of F-3/RT 18 and F-12/LT 27, respectively. The basic difference between these two materials is that F-3/RT 18 exhibits a single-phase equiaxed structure while F-12/LT 27 shows a dual-phased (alpha / Ti_2Ni), finely globular to finely lamellar structure.

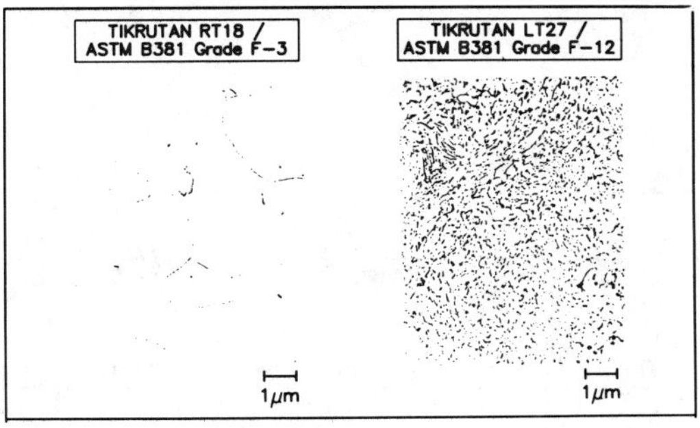

Figure 10
Microstructure of forged tube plates

Despite these differences, which are due to metallurgical influences, it is possible to achieve similar mechanical properties provided the manufacturing conditions are adequately selected. However, differences in structure and grain size can become visible in ultrasonic inspection with the result of better signal-to-noise ratio for Grade 12/LT 27 titanium.

Conclusions

This paper deals with the composition, mechanical properties, corrosion resistance and forming characteristics of the titanium alloy Ti-0.3Mo-0.8Ni, ASTM/ASME Grade 12.

A comparison was made with heavy thick-walled open-die forgings made from the unalloyed titanium materials Grade 2 and Grade 3 and with smaller forgings, sheet and plate made of Grade 12.

The results show that the main objectives set were reliably and reproducibly achieved through use of appropriate measures in raw material selection, ingot melting, forming and heat treatment.
This concerns in particular:

- achieving specified mechanical properties irrespective of wall thickness
- shape accuracy and freedom from internal and external defects
- uniform microstructure over the entire cross section
- optimal testability of all properties discussed
- cost-efficient manufacture with zero defects

In summary it can be stated that open-die-forged thick-walled non-standard components such as tube plates for heat exchangers or centrifuge housings with weights of up to approx. 3000 kg and wall thicknesses in excess of 300 mm can be manufactured in accordance with the standard specifications.
A comparison of Grade 12 and Grade 3 shows that in terms of mechanical properties at room temperature and elevated temperature they are almost equivalent.

Where more stringent demands are made on corrosion resistance Grade 12 should be preferred, also in view of the higher long-term creep resistance as expressed in the design characteristics given in the ASME Boilder and Pressure Vessel Code.

References

1. Ward W. Minkler, Titanium for Chemical Processing Equipment (Metal Progress, February 1978, American Society for Metals, 1978), 27-31

2. R.W. Schutz, Titanium Alloy Crevice Corrosion: Influencing Factors and Methods of Prevention (Cannes, France, Sixth World Conference on Titanium, Société Francaise de Métallurgie, Proceedings - part IV, 1988), 1917 - 1922

3. Y. Wu et al, Effect of Heat Treatment on the Corrosion Behaviour of Ti-0.3Mo-0.8Ni, (Cannes, France, Sixth World Conference on Titanium, Société Francaise de Métallurgie, Proceedings - part IV, 1988), 1717 - 1721

4. J.A. Hall et al, The Relationships of Structure and Corrosion Behaviour of Ti-0.3Mo-0.8Ni (Ti Code 12), (Munich, FRG, Proceedings of the Fifth International Conference on Titanium, Deutsche Gesellschaft für Metallkunde, Volume 4, 1984), 2603 - 2610

5. A. Broström, Welding of Titanium, (Trondheim, Norway, International Conference on Titanium in Practical Application, 1990)

6. ASME Boiler and Pressure Vessel Code, Maximum Allowable Stress Values in Tension for Nonferrous Metals, (Section VIII, Division 1, 1983 Edition), Table UNF-23.4

7. D. Fischer, K. Hülse, Manufacture and Properties of Heavy Solid Titanium Tube Plates for Highly Stressed Tubular Heat Exchangers (Terni/Spoleto, Italy, 11th International Forgemasteds Meeting, 1991), X.8 1-11

A STUDY ON THE TRANS-BETA HOT DIE FORGING

OF Ti-10V-2Fe-3Al ALLOY

Wang Jinyou and Ye Bin

Institute of Aeronautical Materials
Beijing 100095, China

Abstract

The alloy Ti-10V-2Fe-3Al(Ti-10-2-3) alloy has lower hot deformation temperature and is very suitable for trans-beta hot die forging. A systematical study on the relationship between processing parameters and mechanical properties has been done and optimum conditions for trans-beta hot die forging has been determined. The stock temperature of 820 ^{o}C and the die temperature of 700-750 ^{o}C with the thickness reduction of about 50 % should be used in order to achieve the best combination of mechanical properties.

Introduction

The alloy Ti-10V-2Fe-3Al(Ti-10-2-3) is a high strength and high toughness near-beta titanium alloy developed to satisfy the demands of damage-tolerant design. As a result of high content of beta stabilizing elements, the alloy has lower beta transus temperature and excellent forging capability. The typical forging temperatures range from 780 to 820 ^{o}C.

The lower hot deformation temperature makes it very suitable for hot die forging so as to produce near-net-shape forgings. The comparison of forging stress for Ti-10-2-3 and the widely used Ti-6Al-4V alloy has shown that at the same die temperature the Ti-10-2-3 is more easily deformable. When the die temperature is increased from 300 ^{o}C to 800 ^{o}C the forging stress is decreased 2-3 times.[1]

The hot die forging process is characterized by die temperature higher than those in conventional forging, but lower than those in isothermal forging. Typical die temperatures in hot die forging are 100 to 200 ^{o}C lower than the temperature of the stocks. If the stocks were heated to the temperature above beta transus, the temperature of the stocks will be decreased gradually to the temperature of the dies, in this case there will be essentially a trans-beta hot die forging process.

Our work is aimed at a systematical study of the relationship between processing parameters and mechanical properties under different conditions of hot die forging. The purpose of this paper is to determine the optimum trans-beta hot die forging conditions in order to satisfy the mechanical property requirements of AMS 4986.

Titanium '92
Science and Technology
Edited by F.H. Froes and I. Caplan
The Minerals, Metals & Materials Society, 1993

Material and Experimental

Preparation of the Stocks

All stocks used were prepared from a 250Kg ingot triple melted in a vaccum arc furnace with consumable electrodes. The chemical composition of the ingot is listed in Table I. The beta transus temperature of the material is about 800 ^{o}C. The finished forging of the stocks is carried out in the alpha+beta field with deformation of about 25 %.

Table I: Chemical composition of the ingot used

Elemets	Al	V	Fe	O	N	H	C
wt.%	3.08	9.95	2.07	0.079	0.016	0.001	0.03

Experimental Procedures

All experiments were accomplished on a 315 tons hydraulic press by upseting the stocks with different heights to the pancakes of similar thickness. The stocks were heated in a separate electric furnace and the dies were heated with an induction coil. The upseting time was about 60-90 seconds, the corresponding strain rates were about $10^{-2}s^{-1}$.

The work was divided into two phases. During the first phase, the influences of stock temperature and cooling method after upseting on mechanical properties of hot die forged pancakes were studied. During the second phase, the influences of die temperature and thickness reductions was investigated.

All pancakes were solution treated at 760 ^{o}C for 2 hours followed by water quenching and aging at 520 ^{o}C for 8 hours. The tensile specimens were cut in the cord direction and the fracture toughness specimens in the C-R direction of the pancakes. Tensile specimens 5 mm in diameter and 25 mm in calibrated length were used. All fracture toughness values were obtained from compact tension specimens with B=15 mm and W=30 mm.

Results and Discussion

The Influences of Stock Temperature and Cooling Method

It can be seen from Table II that under the condition of heating the dies to 800 ^{o}C the temperature of stocks above or below beta transus did not display obvious influence on the tensile properties and fracture toughness. During conventional forging, the deformation at temperatures above beta transus may lead to lower ductility and fracture toughness. However, in the process of hot die forging one may adopt higher heating temperature for stocks without worrying about unfavourable influence of heating above beta transus.

It is worth to point out that an attractive combination of strength, ductility and fracture toughness was obtained and the technical requirements of AMS 4986 are fully satisfied (820 ^{o}C/75 %/WQ did not meet K_{IC}=60 MPa$\sqrt{m}$, but UTS much higher than 1105 MPa). At relatively higher strength

Table II: Room Temperature Experimental Results of First Phase

Stock Temp.	Reduction	Cooling Method	Direction	UTS MPa	YS MPa	Elong. %	RA %	K_{IC} MPa$\sqrt{m}$
820 °C	50 %	Water	R	1237	1171	11.2	38.5	
			C	1239	1166	10.0	38.8	
			C-R					60.3
780 °C	50 %	Water	R	1200	1130	9.5	38.5	
			C	1217	1150	8.7	36.0	
			C-R					61.3
820 °C	75 %	Water	R	1210	1151	9.8	34.5	
			C	1215	1137	8.7	32.6	
			C-R					50.8
820 °C	50 %	air	R	1209	1135	9.9	35.1	
			C	1212	1134	9.0	34.5	
			C-R					63.0
Technical Requirements of AMS 4986 , min			C	1105	1035	6.0	10.0	
			C-R					60.0

Note: The dies were heated to 800 °C for all experiments.

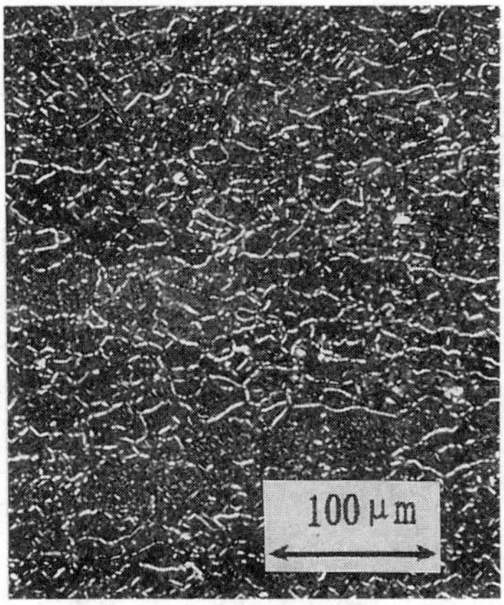

(a) Stock temperature: 820 °C
Die temperature: 800 °C
Thickness reduction: 50 %
Cooling method: water

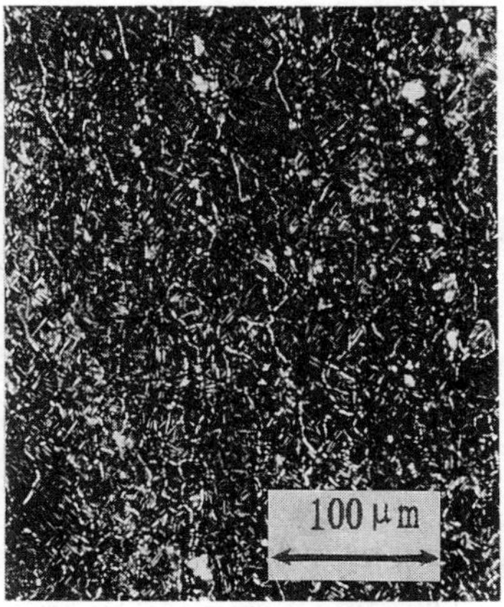

(b) Stock temperature: 820 °C
Die temperature: 800 °C
Thickness reduction: 50 %
Cooling method: air

Figure 1 Microstructures of trans-beta hot die forged pancakes after (a) water and (b) air cooling level of 1210 MPa the elongation is 9.0 % and the fracture toughness reaches 60 MPa$\sqrt{m}$. Such a good result is attributed to dynamic recrystallization occurred during hot die forging at the temperatures near to beta transus. Metallographic observation showed the dynamic recrystallization resulted in formation of small beta grains (see Figure 1), the grain size after dynamic recrystallization is about 25 um.

For hot die forged pancankes with thickness of no more than 30 mm, the air cooling could improve the fracture toughness. Comparing Figure 1(a) and 1(b) we can see the particles of primary alpha phase after air cooling have larger size and aspect ratio than those after water cooling. For this reason the fracture toughness of pancakes after air cooling is better than that after water cooling.

<u>The Influence of Thickness Reduction</u>

The experimental results of the second phase are listed in Table III. Figure 2 illustrates the influence of thickness reduction on tensile properties. It is well evident that the pancakes upseted with 50 % reduction have the best combination of tensile strength and ductility. There is a maximum on the dependence curves of tensile strength on thickness reduction at 50 %. The reduction of area increased significantly for

Table III: Room Temperature Experimental Results of Second Phase

Die Temp.	Reduction	Direction	UTS MPa	YS MPa	Elong. %	RA %
800 ^oC	75 %	C	1154	1123	12.6	42.0
800 ^oC	50 %	C	1194	1141	10.3	41.0
800 ^oC	25 %	C	1175	1129	9.7	34.0
700 ^oC	75 %	C	1178	1124	13.0	38.6
700 ^oC	50 %	C	1213	1153	10.6	40.2
700 ^oC	25 %	C	1195	1137	9.4	30.3
600 ^oC	75 %	C	1143	1097	14.0	36.0
600 ^oC	50 %	C	1148	1116	13.7	36.1
600 ^oC	25 %	C	1144	1091	9.8	29.9

Note: 1) The stocks were heated to 820 ^oC for all experiments.
2) After upseting air cooling was used for all experiments.

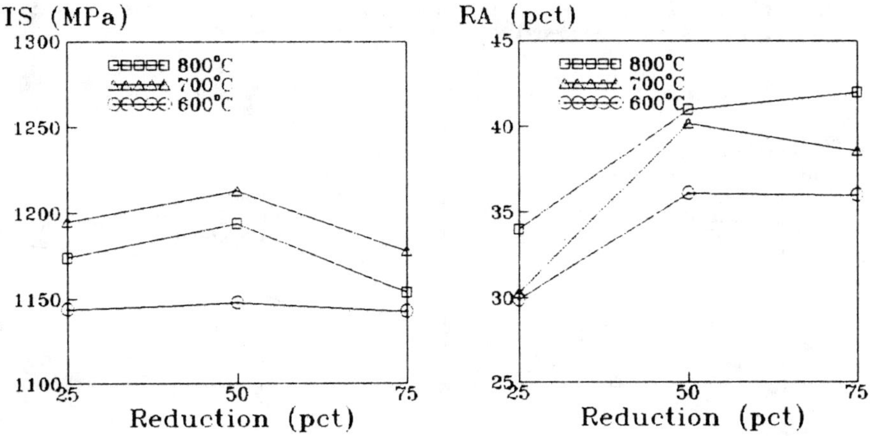

Figure 2 Dependence curves of tensile strength and ductility on the thickness reduction amount

thickness reduction up to 50 % and beyond 50 % it has been kept at nearly the same level with further raising of thickness reduction.

When the thickness reduction is 25 %, the deformation is not enough to break up the coarse prior beta grains and to induce dynamic recrystallization of beta grains with the result that lower strength and ductility are obtained. But when the thickness reduction is 75 %, fine beta grains are obtained as a result of complete dynamic recrystallization, therefore the tensile strength is slightly decreased. Generally speaking, the fine grain structure has relatively lower fracture toughness. For this reason, we deem it neccessary to adopt the thickness reduction of about 50 % in the process of trans-beta hot die forging of Ti-10-2-3 alloy in order to achieve the best combination of mechanical properties.

The Influence of Die Temperature

The dependence curves of tensile strength and ductility on the die temperature are shown in the Figure 3. When die temperature is raised from 600 °C to 700 °C, both the tensile strength and ductility are increased. If the die temperature is further raised to 800 °C, the tensile strength is decreased and the ductilily is slightly increased. Therefore, in the process of trans-beta hot die forging the die temperature within the range of 700-750 °C could guarantee the best combination of mechanical properties.

From Figure 4 it can be seen that when the die temperature is 600 °C, the microstructure of trans-beta hot die forged pancakes is similar to that of conventional forgings. But when the die temperature raised to 700 °C, there are obvious dynamically recrystallized grains with more homogeneous grain size than those obtained at die temperature of 800 °C. In order to take advantage of the beneficial effect of dynamic recrystallization, the optimum die temperature should be no less than 700 °C.

When compared with isothermal forging, the lowering of die temperature allows wider selection of die materials. Therefore, the trans-beta hot die forging of the Ti-10-2-3 alloy is especially attractive because of availability of relatively inexpensive alloys for die materials.

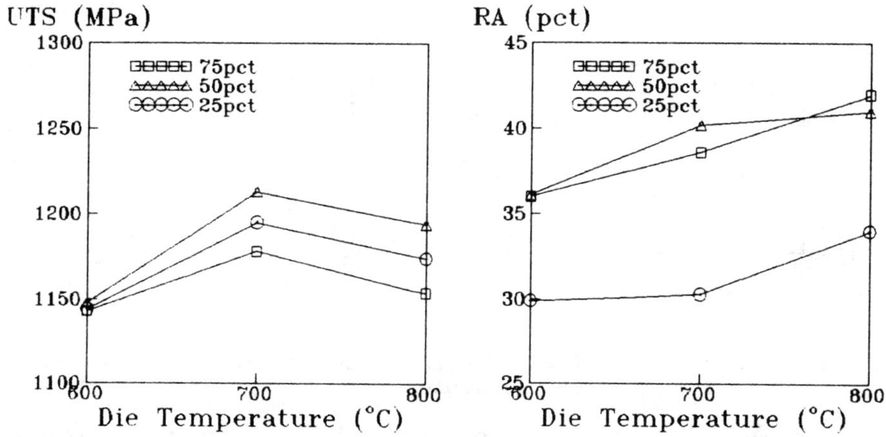

Figure 3 Dependence curves of tensile strength and ductility on the die temperature

Conclusions

1) During hot die forging process, heating the stocks to temperatures above or below beta transus has no obvious influence on the tensile property and fracture toughness.

2) For hot die forged pancakes with thickness of no more than 30 mm, substitution of air cooling for water cooling could improve the fracture toughness when keeping the strength at the same level.

3) In order to produce trans-beta hot die forgings with best combination of mechanical properties, the neccessary thickness reduction is about 50 % and the optimum die temperature is within the range of 700-750 °C.

References

1. G.W.Kuhlman,R.Pishko,J.R.Kahrs & J.W.Nelson, Beta Titanium Alloys in the 80's edited by R.R.Boyer and H.W.Rosenberg (New York, A Publication of The Metallurgical Society of AIME, March 8, 1983), 255-280.

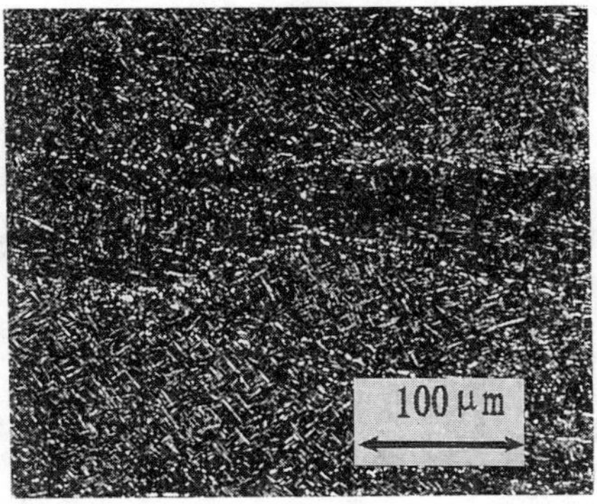

(a) Stock temperature: 820 ^{o}C
Die temperature: 600 ^{o}C
Thickness reduction: 50 %
Cooling method: air

(b) Stock temperature: 820 ^{o}C
Die temperature: 700 ^{o}C
Thickness reduction: 50 %
Cooling method: air

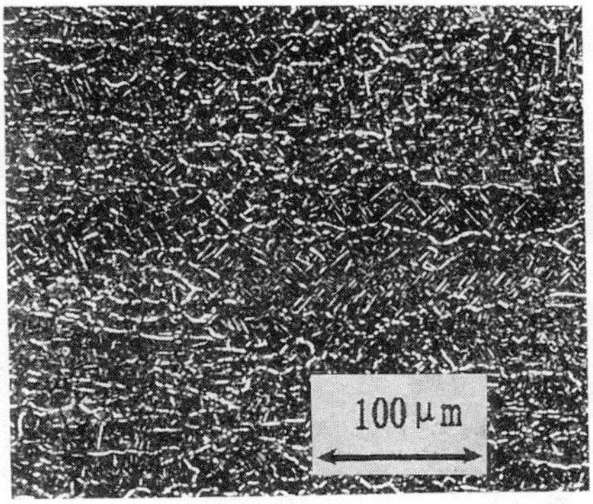

(c) Stock temperature: 820 ^{o}C
Die temperature: 800 ^{o}C
Thickness reduction: 50 %
Cooling method: air

Figure 4 Microstructures of trans-beta hot die forged pancakes under the condition of different die temperatures

CHARACTERIZATION OF Ti-6-22-22S: A HIGH STRENGTH ALPHA-BETA TITANIUM ALLOY FOR FRACTURE CRITICAL APPLICATIONS

G. W. Kuhlman*, K.A. Rohde* and A. K. Chakrabarti#

* Alcoa Forging Division, Cleveland, Ohio USA
# Alcoa Laboratories, Alcoa Center, Pa. USA

Abstract

Ti-6Al-2Sn-2Zr-2Mo-2Cr + Si (Ti-6-22-22S) is a high strength, silicon modified, $\alpha + \beta$ titanium alloy originally developed for elevated temperature gas turbine engine applications. When $\alpha + \beta$ processed, this highly beta stabilized alloy has an elastic modulus 10% higher than other $\alpha + \beta$ alloys. The alloy is now a material of choice for advanced, high performance airframe structures. Through judiciously designed thermomechanical processing (TMP) for forgings, superior combinations of room and elevated temperature strength, fracture toughness, high and low cycle fatigue, creep and fatigue crack growth resistance, the latter being the most critical property for the current intended application, have been achieved. TMP's exploit controlled β processing techniques to meet microstructural and stringent mechanical property objectives. Presented is a complete characterization of Ti-6-22-22S first and second tier properties achieved with optimal TMP sequences. Mechanical property achievements are rationalized in terms of key microstructure-processing relationships.

Introduction

Alloy Ti-6Al-2Sn-2Zr-2Mo-2Cr + Si (Ti-6-22-22S) was developed in the 1970's by RMI Titanium Co. as a deep hardenable, elevated temperature $\alpha + \beta$ alloy for thick section engine forgings. [1,2] While not exploited commercially, the alloy demonstrated fracture toughness, fatigue crack growth resistance, and fatigue properties and superior, by up to 10%, elastic modulus. Composition limits for Ti-6-22-22S are outlined in Table 1. Cr and Mo in the alloy improve strength and toughness. Small

Table 1: Ti-6Al-2Sn-2Zr-2Mo-2Cr + Si CHEMICAL COMPOSITION LIMITS (Weight Percent)

	Al	Sn	Zr	Mo	Cr	Si	Fe	O	C	N	H
Min.	5.25	1.75	1.75	1.75	1.75	0.20					
Max	6.25	2.25	2.25	2.25	2.25	0.27	0.15	0.13	0.04	0.03	125 ppm

amounts of Si in combination with Mo improve strength without affecting ductility and Si additions are known to enhance creep. Sn and Zr contribute to solid solution strengthening. Oxygen content is kept low to contribute strength without loss of toughness. [3]

Titanium '92
Science and Technology
Edited by F.H. Froes and I. Caplan
The Minerals, Metals & Materials Society, 1993

Typical physical properties for the alloy are presented in Table 2.

Table 2: Ti-6Al-2Sn-2Zr-2Mo-2Cr + Si PHYSICAL PROPERTIES

Density:	4.65 g/cm^3 (0.165 lbs/in^3)	Beta Transus:	960°C (1760°F)
Young's Modulus (Tensile):		Poisson's Ratio:	0.33
Conventionally Processed:	122 x 10^3 MPa	(17.7 x 10^3 Ksi)	
Beta Processed:	115 x 10^3 MPa	(16.8 x 10^3 Ksi)	
Bulk Modulus:	117 x 10^3 MPa (17.0 x 10^3 Ksi)	Shear Modulus:	46 x 10^3 MPa (6.7 x 10^3 Ksi)
Coefficient of Thermal Expansion:	9.18 x 10^{-6} m/m/°C (5.1 x 10^{-6} in/in/°F)		

Fracture mechanics design criteria are now institutionalized in design of most critical airframe components, particularly military aircraft. For advanced, high performance airframes, Ti-6-22-22S has been confirmed to be a superior alloy for such applications that demand high strength in combination with excellent fracture toughness and in particular, superior fatigue crack growth resistance (FCGR) to meet stringent safe life requirements. Ti-6-22-22S, with appropriate TMP, offers a significant weight savings advantage, 15 to 20% or more, over incumbent alloys such as Ti-6Al-4V ELI recrystallized anneal (Ti-6-4 ELI RA) and beta annealed (Ti-6-4 ELI BA) exploited in forgings in a wide range commercial and military airframes since the 1970's. [4]

It is well recognized that a transformed β microstructure in α + β alloys, consisting of Widmanstatten and/or colony α, optimizes fracture related properties such as FCGR. [4] Enhancement of FCGR is rationalized by the bifurcation and deflection of growing cracks due to the size of the Widmanstatten α platelets and/or α colonies. The colony α structure, usually achieved with subtransus working followed by β heat treatment (e.g. multi-step solution treatment and anneal, referred to simply as β anneal) is recognized in alloys such as Ti-6Al-4V as providing the highest FCGR. [4] The transformed Widmanstatten α structure is achieved with β or through transus deformation followed by subtransus heat treatment. Thus, for the application of Ti-6-22-22S requiring superior FCGR, both major β processing techniques were considered. Target minimum mechanical properties for successful application of Ti-6-22-22S were yield strength 930 to 1000 MPa, ultimate strength 1035 to 1140 MPa, elongation 6%, reduction in area 15%, fracture toughness 77 to 88 MPa$\sqrt{m}$, and FCGR on a par with Ti-6-4 ELI BA, e.g. at ΔK of 11 MPa$\sqrt{m}$, a da/dN of 8 to 9 x 10^{-9} m/cycle.

<u>TMP Rationale for Ti-6-22-22S</u>

The strong interaction between TMP and microstructure and microstructure and mechanical properties in titanium alloys is well known. [5] Optimization of FCGR properties from prior work [4,6] established that a lamellar or acicular α morphology, with or without a small amount of primary α, may develop the desired combination of high strength, high toughness and high FCGR. Further, a transformed β microstructure consisting of Widmanstatten and/or colony α, optimizes fracture related properties such as FCGR. [4,6] Ti-6-4 ELI BA illustrates this TMP and is currently accepted, and commercially exploited, as having the lowest rate of fatigue crack growth (e.g. highest FCGR) of all commercially used α + β alloys; however, high FCGR is achieved with a sacrifice in strength, high cycle fatigue and ductility properties.

From this basis with similar α + β alloys, two major β processing TMP's were designed: TMP (1) β forging followed by α + β heat treatment and TMP (2) α + β forging followed by β heat treatment. TMP (1), where final forging processes are conducted in a through-the-transus manner, was postulated to provide a Widmanstatten α structure essentially free of grain boundary α. However, supratransus forging processes (at β_t + 25 to 35°C for Ti-6-22-22S) for α + β alloys have found limited commercial success due difficulties in process control and product variation. [6] In TMP (1) final subtransus heat treatment is accomplished at β_t - 27°C (e.g. ≈ 925°C), 1 hour, fan cool, plus age at 537°C, 8 hours, air cool. [1]

Considerable commercial experience is available for TMP (2) for Ti-6-4, thus extension to Ti-6-22-22S was logical. [6] However, Ti-6-22-22S is more highly β stabilized, and therefore, within TMP (2) several process variants have been studied including two-step and three-step (triplex) β heat treatments; post β heat treatment cooling rates (e.g. fast - oil quench and slow - air or fan cool); and in triplex heat treatments, temperature of the second step (e.g. β_t - 27 to 50°C) and cooling rate (air/fan cool). Previous work [6] on β heat treated Ti-6Al-4V and highly β stabilized alloys Ti-6Al-6V-2Sn and Ti-6Al-2Sn-4Zr-6Mo suggested that an intermediate subtransus treatment (termed stabilization anneal), following the initial β solution treatment, successfully modified the size of the lamellar or acicular α platelets in α colonies and achieved a better balance in strength, ductility and fracture-related properties. Early evaluations of two-step β heat treatments suggested inferiority to triplex βHT processes, and thus two-step β heat treatments are not discussed further. Subsequently, discussed below, studies of the final aging step have been conducted. Table 4 presents optimum triplex β heat treatment (now termed βSTSA for beta solution treated, stabilized and aged) confirmed from studies; used in TMP (2) to fabricate all recent commercial scale Ti-6-22-22S forgings; and further employed to fully characterize the alloy as reported herein.

Microstructures achieved with TMP (1), termed βF, and the two cooling rate variations (oil quench, fan cool) in TMP (2), βHT, are presented in Figure 1. TMP (1) is chara-

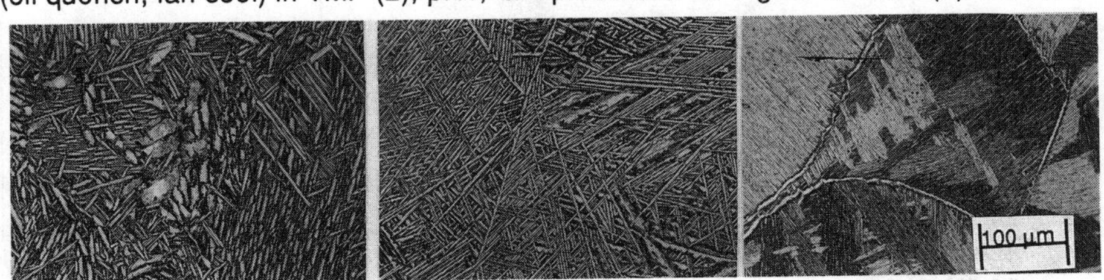

β Forge, α+β STA | α+β Forge, Triplex β H.T., Fan Cool | α+β Forge, Triplex β H.T., Oil Quench

Figure 1: Microstructures of Ti-6-22-22S With Selected TMP Variants.

cterized by transformed, fattened lamellar α, with some evidence of rounded and bent primary α, and little or no grain boundary α. For TMP (2), microstructures for the triplex heat treatment with slow (fan cool) and fast (oil quench), designated βHT, are presented as well. The post-solution treatment cooling rate requirements for Ti-6-22-

22S have been shown [1,2] to constrain property achievements and to restrict applicable section sizes. Fan cooling from β heat treatment provides a coarser acicular α colony structure with more grain boundary α than does the faster oil quench. Property achievements with both quench variations are discussed below.

First and Second Tier Mechanical Property Characterization of Ti-6-22-22S Forgings

A. Physical Properties: Table 2 summarizes key physical properties for the alloy form current investigations and Ref. 1, 2. Density of Ti-6-22-22S is 2.5% greater than Ti-6-4. Conventionally processed Ti-6-22-22S may have tensile modulus as much as 10% higher than other α + β alloys; however, with β processing TMP's reported here, precision modulus determinations suggest lower tensile modulus, but it may still be as much as 6% higher than competitive α + β alloys.

B. Room and Elevated Temperature Tensile Properties and Fracture Toughness:

Table 3 summarizes room temperature tensile and fracture toughness properties achieved in several Ti-6-22-22S forging shapes with the βF and triplex βHT processes. Pancake forgings were used to develop the TMP and a web-rib die forging to

Table 3: TYPICAL TENSILE MECHANICAL PROPERTIES OF Ti-6-22-22S FORGINGS

Forging Type	TMP History Forge	H.T.	Dir	Sect Thk in (mm)	Yld Str Ksi (MPa)	Ult Str Ksi (MPa)	El (%)	RA (%)	K1c (Kq*) Ksi√in (MPa√m)
Pancake Forging	β	α+β	L	0.5 (13)	146 (1007)	163 (1123)	10	17	74 (81) T - L
(Low Oxygen)		(Fan Cool)	T		142 (979)	157 (1082)	9	15	
	α+β	β	L	0.5 (13)	141 (972)	159 (1096)	10	16	78 (86) T - L
		(Fan Cool)	T		137 (945)	155 (1069)	8.5	15	
Pancake Forging	β	α+β	L	1.5 (38)	138 (951)	154 (1062)	12	17	79 (87)* T - L
(Std. Oxygen)		(Fan Cool)	T		137 (945)	152 (1049)	13	18	
	α+β	β	L	1.5 (38)	154 (1062)	176 (1213)	7.5	12	63 (69) T - L
		(Oil Quench	T		152 (1048)	169 (1165)	6.5	11	
Closed-Die	β	α+β	L	3 (75)	146 (1006)	163 (1123)	12	21	71 (78) S - L
Forging		(Fan Cool)	L	1 (25)	144 (979)	161 (1110)	11	19	
(Std. Oxygen)			T	1 (25)	148 (1028)	160 (1103)	12.5	24	
			T	0.5 (13)	147 (1013)	165 (1138)	17	26	
	α+β	β	L	3 (75)	138 (951)	154 (1062)	10	17	74 (81) S - L
		(Fan Cool)	L	1 (25)	141 (972)	159 (1096)	19	24	
			T	1 (25)	137 (945)	154 (1062)	9	16	
			T	0.5 (13)	144 (993)	159 (1096)	14	18	

confirm scalability of TMP and provide second tier characterization data. For initial TMP, the metal employed had either standard oxygen content of 0.11% or a low oxygen content, 0.08%. From these data, similar tensile and toughness properties are achieved with both major TMP's. The triplex βHT with a fan cool, preferred due to process control and product uniformity advantages, clearly provides properties meeting targets. Thus, this process was selected for scale-up and production prototype forging fabrication to develop design allowables for Ti-6-22-22S - βSTSA. Triplex βSTSA with an oil quench provides higher strengths and lower toughness than a fan cool, if such tradeoffs are feasible in applications for the alloy. In subsequent production prototype activities with large Ti-6-22-22S forgings, the triplex βSTSA, fan cool process has been explored further and optimized. It has been found that proximity to the transus and cooling rate in the second step of the triplex βSTSA TMP can be manipulated to adjust strength and toughness. While the βF TMP (1) provides attractive property combinations, this TMP was not selected for scale-up due to uniformity and process control problems on the large forgings contemplated for commercial capture. Table 4 presents guaranteed minimum mechanical properties for the two cooling rate versions of the triplex βHT (βSTSA) commercially proceeding.

Table 4: MINIMUM MECHANICAL PROPERTIES Ti-6-22-22S FORGINGS to 75 mm (3 in)

I. TMP: α + β Forge Plus Triplex βSTSA (slow cooling rate):
YS: MPa (Ksi) 931 (135); UTS: MPa (Ksi) 1034 (150); El: (%) 6; RA : (%) 10
Fracture Toughness: MPa√m(Ksi√in) 77 (70)
Recommended Heat Treatment:
βt + 27°C, Fan Cool plus βt - 27 to 50°C, Air Cool, plus 537°C, 8 Hrs., Air Cool

II. TMP: α + β Forge Plus Triplex βSTSA (fast cooling rate):
YS: MPa (Ksi) 1000 (145); UTS: MPa (Ksi) 1069 (155); El: (%) 6; RA: (%) 10
Fracture Toughness: MPa√m (Ksi√in) 66 (60)
Recommended Heat Treatment:
βt + 27°C, Oil Quench plus βt - 27 to 50°C, Air Cool, plus 537°C, 8 Hrs., Air Cool

Elevated temperature properties for Ti-6-22-22S forgings representing the triplex βSTSA TMP (but representative of other TMP variations studied due to similarity in room temperature properties) are presented in Figure 2. The alloy retains nearly 70% of it room temperature yield strength at temperatures up to 480°C.

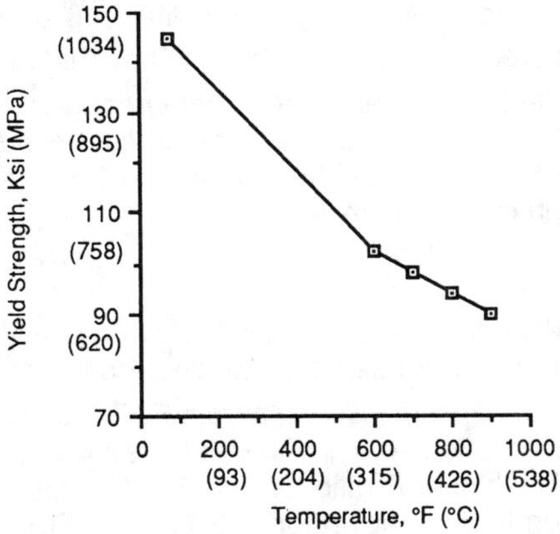

Figure 2 - Elevated Temperature Tensile Properties Ti-6-22-22S, 100 Hrs., Test at Temperature,Triplex βSTSA.

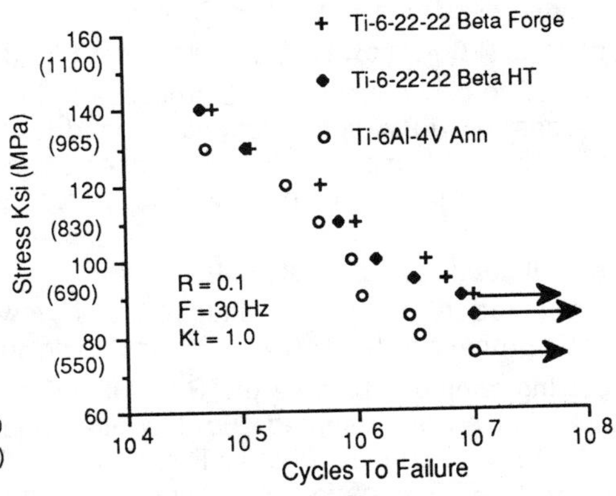

Figure 3 - Smooth High Cycle Fatigue Properties, Ti-6-22-22S With βF and βHT TMP's.

C. Smooth and Notched High Cycle Fatigue Properties:

Figures 3 and 4 present, respectively, smooth and notched high cycle fatigue data obtained from Ti-6-22-22S closed-die forgings processed with both the βF and triplex βSTSA (fan cool) TMP's. For comparison, smooth and notched fatigue data for Ti-6Al-4V standard composition, α + β forged and annealed (Ref.7) are included. The strength properties of the isothermally forged and annealed Ti-6-4 were 937 to 972 MPa. From Ref. 4, 6, smooth and notched fatigue performance of Ti-6-4 ELI BA are expected to be inferior, for it is accepted [4, 5, 6] that a transformed structure in Ti-

6-4 degrades initiation related properties, e.g. high cycle fatigue.

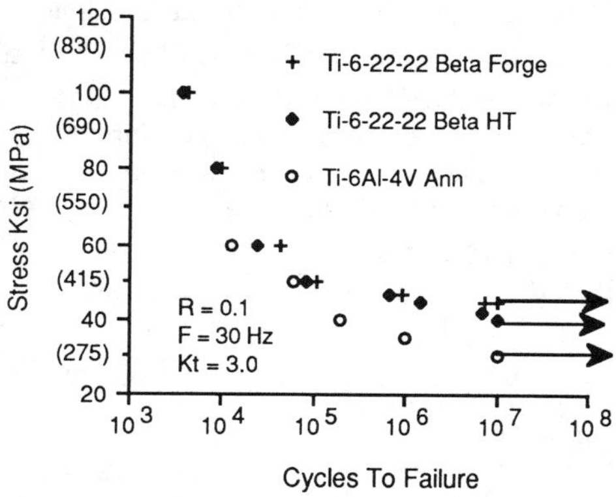

Figure 4: Notched High Cycle Fatigue Properties, Ti-6-22-22S, With βF and βHT TMP's.

From the data presented, it appears that the fatigue performance of Ti-6-22-22S with selected TMP is excellent and superior to that reported for Ti-6-4 isothermally forged and annealed. The smooth high cycle fatigue advantage may be rationalized on the basis of higher strengths. The improved notched performance may be attributed to the acicular structures achieved with the successful βHT TMP that reduces fatigue crack propagation rates that may be rate controlling. From these data, it appears there may be a a slight fatigue advantage for the βF TMP over the βHT TMP.

D. Strain Controlled Low Cycle Fatigue:

Strain-controlled low cycle fatigue (LCF) behavior of the triplex βSTSA, fan cool TMP has been compared with β forged and annealed Ti-6Al-4V from other work. [8,9] From limited test data, Ti-6-22-22S was found to have superior LCF life to β processed Ti-6-4; however, it is likely that the β processed Ti-6-22-22S will have inferior LCF to conventional, α + β processed Ti-6Al-4V, especially when compared to processes for Ti-6-4 specifically developed [8] to enhance strain and stress controlled LCF lives. It appears that the higher strength of Ti-6-22-22S enhances LCF in a manner parallel to that observed above for high cycle fatigue lives.

E. Fatigue Crack Propagation:

Presented in Figure 5 are fatigue crack growth rate (FCG) data for the two major β TMP processes for Ti-6-22-22S, including the fast (oil quench) and slow (fan cool) cooling rates used for the βSTSA TMP. For comparison, FCG data from Ref. 4 for Ti-6-4 ELI BA are presented. For the latter data, R ratio is 0.01, whereas tests for Ti-6-22-22S were conducted at an R ratio of 0.1. The higher R ratio for Ti-6-22-22S would increase FCG rates for this alloy in comparison to that shown for Ti-6-4. From these data, it is evident that FCG of β processed Ti-6-22-22S is excellent. The FCG of triplex βSTSA, fan cool approaches that reported for Ti-6-4ELI BA. However, the FCG of the βF TMP appears to be faster than triplex βSTSA, fan cool but slower than triplex βSTSA, oil quench. The FCGR of Ti-6-22-22S as a function of the two major TMP process variants and cooling rates within βHT TMP's may be rationalized on the basis of the microstructural differences achieved (Figure 1). The βSTSA, fan cool TMP achieves a colony, Widmansatten α structure with significant grain boundary α shown by others [4,6] to provide the greatest resistance to crack propagation. The βSTSA, oil quench TMP provides a much finer acicular α with significantly less grain boundary α and much higher strength and lower fracture toughness. The

FCGR of βF TMP (1) is intermediate between the other processes due to a structure that is a combination of fattened Widmanstatten α platelets and colony α but essentially free of grain boundary α. Optimal FCGR of Ti-6-22-22S is achieved with the βSTSA, fan cool TMP. Characterization data suggest a minor compromise in toughness and fatigue properties in comparison to the βF TMP. FCGR is the most critical property, thus βSTSA fan cool was selected to commercialize. Production prototype forgings processed with variations in second-step temperatures have demonstrated FCGR properties identical to those shown, suggesting that the βSTSA, fan cool TMP for Ti-6-22-22S is robust and scalable to large parts.

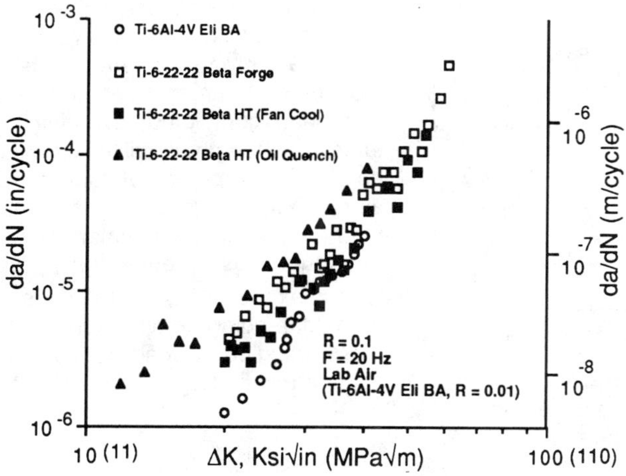

Figure 5: Fatigue Crack Growth Rate Properties of Ti-6-22-22S Processed With βF and βHT TMP's.

F. Creep Properties:

Table 5 summarizes creep properties measured for the βSTSA, fan cool TMP. It is evident that creep performance is on a par with other beta processed α + β alloys.

Table 5: CREEP PROPERTIES Ti-6Al-2Sn-2Zr-2Mo-2Cr + Si βSTSA Forgings

Temperature (C)	Stress (MPa)	Time (Hrs) 0.1%	0.2%	Rupture
370	415	225	687	687
425	345	190	606	606
480	240	69	251	251

G. Producibility and Commercial Capture:

Through the course of TMP investigations and developmental forging activities, forging producibility, product consistency and scalability issues were a major consideration. Extensive production experience exists with Ti-6Al-4V ELI and Std. Comp. BA. This forging product has been demonstrated [6] on a statistical basis to be the most consistent and uniform Ti-6-4 forging product currently produced. The triplex βSTSA, fan cool TMP for Ti-6-22-22S is identical in design, and thus the forging product in this alloy is expected to demonstrate the same consistency found for Ti-6-4 ELI BA. Design allowables forging fabrication to prepare for commercialization has presented the opportunity to explore critical aspects of the triplex βSTSA process: (1) the temperature and cooling rate in the second step of the heat treatment and (2) the temperature of the final aging process. Second step H.T. process studies on closed-die forgings evaluated temperatures from as high as β_t minus 27°C to as low β_t minus 50°C and with slow (air cool) and fast (fan cool) cooling rates. This work [9] has suggested that strength - toughness combinations, may be adjusted should that be necessary, with such process modifications; however, critical FCGR properties are unaffected. Most recently, an aging process study has been completed, examining the effect on strength, ductility, toughness and FCGR properties of aging temperatures in the range of 425 to 620°C. Figure 6 presents a summary of fractture toughness vs. aging temperature for triplex βSTSA process. It is evident that aging temperatures above 565°C must be avoided. FCGR tests confirm this finding.

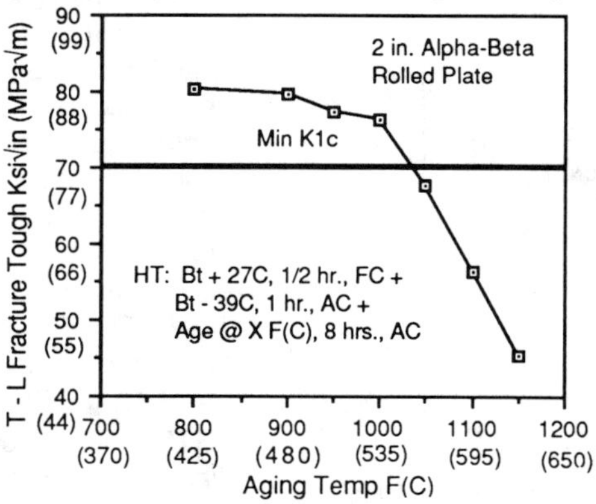

Figure 6: Fracture Toughness versus Aging Temperature, Ti-6-22-22S

Conclusions:

(1) Ti-6-22-22S may be successfully beta processed to achieve superior FCGR in combination with excellent strength, fracture toughness, and ductility properties.

(2) The βSTSA, fan cool TMP offers the best combination of mechanical properties on the basis of optimum FCGR and demonstrates superior smooth and notched fatigue properties, excellent LCF properties and excellent creep performance.

(3) Increasing the cooling rate in the first βHT step (via oil quench) or in the second $\alpha + \beta$ step degrades FCGR and/or toughness properties. The βF TMP offers slightly better toughness and fatigue but results in greater forging product variation.

(4) The final aging temperature for Ti-6-22-22S - βSTSA must be kept below 565°C to assure achieving target fracture toughness and FCGR properties.

References

1. P. A. Russo, et al., Development of a 900°F Titanium Alloy, U.S. Air Force Materials Laboratory, AFSC, Wright Aeronautical Center, Ohio, Technical Report AFML-TR-70-125, 1970 July.

2. L. J. Barto, et al., Deep Hardenable Titanium Alloy, U.S. Air Force Materials Laboratory, AFSC, Wright Aeronautical Center, Ohio, Technical Report AFML-TR-73-122, 1973 May.

3. S. R. Seagle, et al., "Development of a Deep Hardenable Alloy For Intermediate Temperature Applications," RMI Company, Niles, Ohio, TMS Proceedings, 1973 March.

4. J. C. Chesnutt, A. W. Thompson, J. C. Williams, Influence of Metallurgical Factors On The Fatigue Crack Growth Rate in Alpha-Beta Titanium Alloys, U.S. Air Force Materials Laboratory, Wright Aeronautical Center, Ohio, Technical Report AFML-TR-78-68, 1978 May.

5. Titanium and Its Alloys, M. J. Donachie, Editor, ASM International, Materials Park, Ohio, 1982.

6. G. W. Kuhlman, F. R. Billman, "Selecting Process Options For High-Fracture Toughness Titanium Airframe Forgings," Metal Progress, Vol. III, No. 3, 1977 March, pp. 38 - 49.

7. A. J. Vasquez, A. F. Hayes, Isothermal Forging of Reliable Structural Forgings, U.S. Air Force Materials Laboratory, Wright Aeronautical Center, Ohio, Report AFML-TR-74-123, 1974 June.

8. G. W. Kuhlman, A. K. Chakrabarti, S. R. Seagle, "Ti-6Al-4V Forgings With Enhanced Fatigue Characteristics For Engine and Airframe Applications," Titanium Products and Applications 1990, Titanium Development Association, Vol. 1, 1990, pp. 111 - 120.

9. G. W. Kuhlman and A. K. Chakrabarti, "Characterization of Ti-6-22-22S: A High Strength Alpha-Beta Titanium Alloy," EUROMAT 91, FEMS, Cambridge, England, in Publication.

HIGH TEMPERATURE FLOW LOCALIZATION IN COARSE GRAIN WIDMANSTÄTTEN TI-6Al-4V

H. J. Rack and A. Wang[1]
Materials Science and Engineering Program
Department of Mechanical Engineering, Clemson University,
Clemson, South Carolina 29634-0921

ABSTRACT

The compressive elevated temperature deformation behavior of Widmanstätten $\alpha+\beta$ Ti-6Al-4V has been examined in the temperature range 800 to 965°C between strain rates of 10^{-4} and 10^{0} s^{-1}. Flow localization due to adiabatic heating and microstructural instability was observed at high temperatures/high strain rates and low temperatures/low strain rates, respectively. In both instances the severity of flow localization can be quantified in terms of the flow localization parameter α. Metallographic observations demonstrated that microstructurally derived localization was associated with the breakup and spheroidization of α platelets situated either at the prior β grain boundaries or within Widmanstätten colonies.

INTRODUCTION

Various investigators[1-5] have reported that the incidence of flow localization, compressive plastic instability and unstable bulging in $\alpha+\beta$ titanium alloys is critically dependent upon alloy microstructure. For example, transformed Widmanstätten $\alpha+\beta$ microstructures have been found, in comparison with equiaxed $\alpha+\beta$ microstructures, to promote localization. While it has been suggested that this microstructurally dependent behavior is related to deformation assisted breakup and spheroidization of the Widmanstätten structure[6,7], a detailed interrelationship between macroscopic instability, i.e., flow localization, and microscopic instabilty is still lacking. The present investigation has therefore examined the effects of temperature, strain rate and strain, on macroscopic and microstructural stability in Widmanstätten $\alpha+\beta$ Ti-6Al-4V. The objective was to reveal the microstructural origins of flow localization and to correlate the degree of flow localization with microstructural stability.

MATERIAL AND EXPERIMENTAL PROCEDURES

Coarse grained β processed Ti-6Al-4V (5.86 Al, 4.01 V, 0.28 Fe, 0.02 C, 0.04 N, 0.112 O) was investigated.[2] The initial microstructure consisted of a Widmanstätten colony structure and grain boundary α, Figure 1, the average prior β grain size and the mean Widmanstätten interlamellar spacing being 1.1 mm and 1.8 μm, respectively.

---

[1] Professor of Mechanical Engineering and Metallurgy, Clemson University, Clemson, SC, and formerly Research Associate, Clemson University, currently Senior Research Engineer, Howmedica Division, Pfizer Hospital Products Group, Inc., Rutherford, NJ, respectively.

[2] The β transus of this alloy, as determined by calorimetric differential thermal analysis, was 973°C.

Titanium '92
Science and Technology
Edited by F.H. Froes and I. Caplan
The Minerals, Metals & Materials Society, 1993

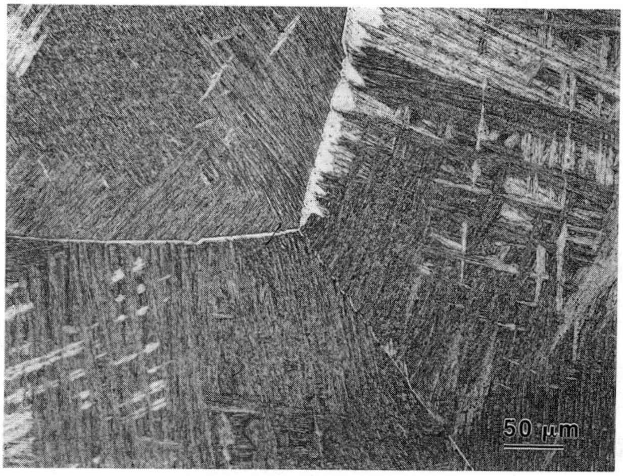

Figure 1. Microstructure of Ti-6Al-4V alloy.

Elevated temperature characterization of this alloy utilized both ring and cylindrical compression samples. The former were 25.4 mm OD, 12.7 mm ID and 8.458 mm height (6:3:2) and were included for simultaneous determination of friction factors and flow stress behavior. Compression samples were 20.3 mm in diameter and 30.5 mm in height and served to further define the stress-strain behavior, frictional correction of the compressive stress-strain curves being accomplished utilizing procedures described previously[8].

Hot compression of the ring and cylindrical specimens was conducted within the temperature range 800 to 965°C, i.e., below the β transus, in a ultra-high purity argon atmosphere (O_2 content less than 10^{-7} ppm) at three different constant strain rates (10^{-4}, 10^{-2}, 10^{0} s^{-1}). The specimens were heated to temperature in 20 to 30 minutes and soaked at temperature for 5 minutes prior to testing, with boron nitride serving as the lubricant between the specimen and platens. All specimens were furnace cooled at approximately 20 °C min^{-1} to room temperature after deformation.

RESULTS

Stress-Strain Curves

Representative true stress-true strain curves are shown in Figure 2. In each instance the true stress increases to a maximum and then decreases, the flow stress tending at higher strains towards a saturation value. The extent of this flow softening was a function of both strain rate and temperature. For example, at 800°C the extent of softening increased with decreasing strain rate, while at 965°C it increased with increasing strain rate. Similarly, at a strain rate of 10^{-4} s^{-1} the extent of flow softening decreased with increasing temperature, while at 10^{0} s^{-1} the extent of softening increased with increasing temperature. Finally, at the highest temperature examined, i.e., 965°C, periodical oscillations in the true stress-true strain curves were observed, the amplitude of these oscillations tending to be a maximum at intermediate strain rates.

The effect of temperature on the maximum flow stress and the flow stress at a true strain of 0.30 are shown in Figure 3. Both measurements of the flow stress were sensitive to temperature at lower temperatures, this sensitivity decreasing as the β transus was approached. If a power law relationship between flow stress and strain rate, i.e., $\sigma = k\dot{\epsilon}^{m}$, is assumed, the rate sensitivity parameter m can be obtained from a log-log plot of flow stress vs. strain rate. This determination, Table 1, suggests that the strain rate sensitivity of Widmanstätten Ti-6Al-4V is generally lower than that observed in equiaxed microstructures[9]. Moreover the strain rate sensitivity was a function of both temperature and strain rate, generally decreasing with increasing strain rate at each temperature investigated.

Macroscopic Observations

Macroscopic observations indicated that the dimensional changes of the ring samples after compression were often non-uniform. Both external and internal barreling was observed, the degree of barreling varying with temperature and strain rate. A measure of the specimen non-uniformity was obtained by normalizing the areal difference between the maximum and minimum areas following deformation, i.e.,

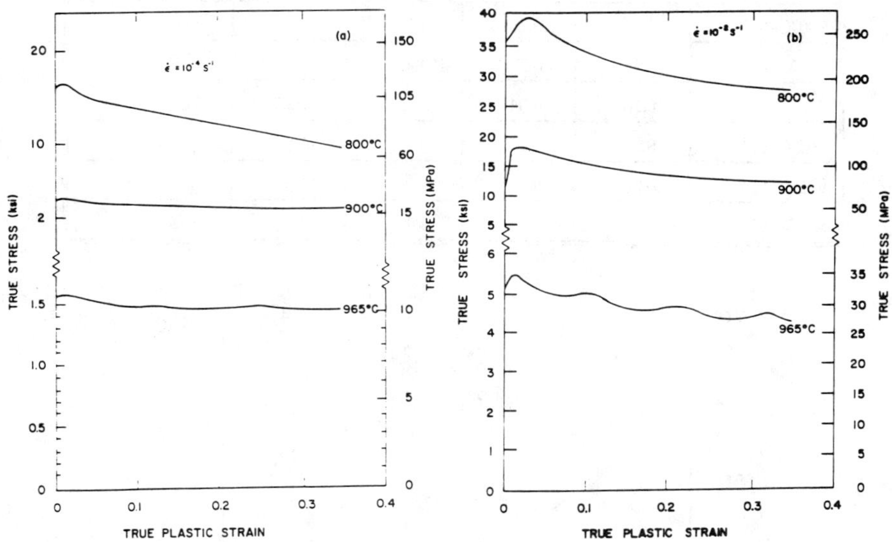

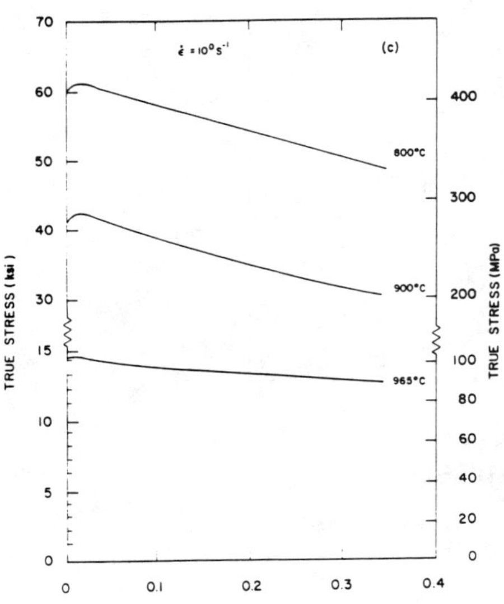

Figure 2. True stress-true strain curves at a strain rate of (a) 10^{-4} s^{-1}, (b) 10^{-2} s^{-1} and (c) 10^{0} s^{-1}

Table 1. Flow Uniformity and Localization

Temp./Rate	λ	m	γ'	α	Localization
800°C-10^{-4}	0.34	0.22	5.0	18.2	yes
800°C-10^{-2}	0.22	0.15	2.65	11	yes
800°C-10^{0}	0.05	0.1	0.98	-0.2	no
900°C-10^{-4}	0.33	0.28	4.50	12.5	yes
900°C-10^{-2}	0.31	0.27	4.40	15.5	yes
900°C-10^{0}	0.16	0.04	2.25	31.25	yes
965°C-10^{-4}	0.23	0.26	4.25	12.5	yes
965°C-10^{-2}	0.30	0.3	5.81	16.0	yes
965°C-10^{0}	0.20	0.05	1.10	2.0	yes

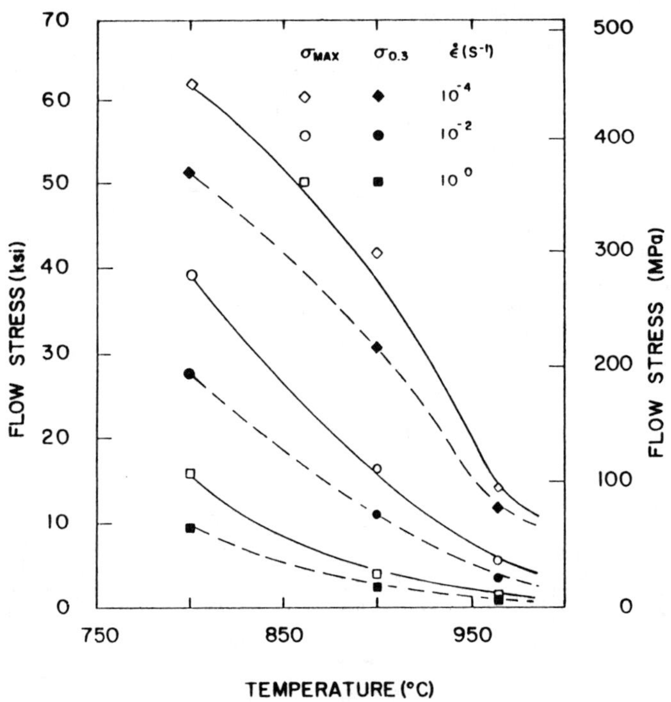

Figure 3. Flow stress as a function of temperature and strain rate.

$$\lambda = \frac{A_{max} - A_{min}}{A_o} \qquad (1)$$

with λ denoting the degree of non-uniformity introduced during deformation of the ring samples, and A_{max}, A_{min} and A_o representing respectively the maximum, minimum and original cross-sectional areas. Application of this procedure, Table 1, shows that, at 800° and 900°C the degree of non-uniformity decreased with increasing strain rate, while at 965°C the degree of non-uniformity was a maximum at intermediate strain rates.

Microstructural Observations

Microstructural observations within regions where non-uniform deformation occurred revealed various degrees of grain boundary and lamellar α spheroidization, Figure 4. At 800°C and low strain rates(10^{-4} and 10^{-2} s^{-1}) breakup and spheroidization occurred preferentially within the grain boundary α region, particularly at grain boundary triple points, Figure 4a. Increasing the deformation temperature to 900°C promoted spheroidization of both grain boundary and matrix lamellar α, Figure 4b. The latter occurred initially through unstable kinking of the α lamellae within those colonies which were favorably oriented in the direction of the maximum shear stress, i.e., at approximately 45° to the compression axis. When the deformation temperature was increased to 965°C, spheroidization occurred within all colonies, although its extent varied substantially from grain to grain and from colony to colony, Figure 4c.

DISCUSSION

Macroscopic Aspects of Flow Localization

Jonas et al.[10] and Dadras and Thomas[11] have suggested that the onset of compressive plastic instability will occur when:

$$\alpha = -\frac{\delta \ln \dot{\epsilon}}{\delta \epsilon} = \frac{\gamma' - 1}{m} > 0 \qquad (2)$$

$$\gamma' > 1 \qquad (3)$$

where α is the rate of strain rate localization subsequent to the onset of instability, ϵ the true plastic strain, $\dot{\epsilon}$ the strain rate, m the strain-rate sensitivity and γ' is related to the strain-strain-rate parameter, γ,

$$\gamma' = \gamma + \left(\frac{\eta\phi}{\rho c}\right) - \left(\frac{\partial \ln \sigma}{\partial \epsilon}\right)_{\dot{\epsilon},T} + \left(\frac{\eta\phi}{\rho c}\right) \qquad (4)$$

where η is the efficiency of deformation heating, ρ, the density, c, the specific heat and $\phi = \partial\sigma/\partial T$.

Numerical evaluation showed excellent agreement between the prediction of flow localization, i.e., $\alpha > 0$, and the observed non-uniformity, Table 1.

Microscopic Aspects of Flow Localization

A prominent characteristic of the hot-deformation behavior of the Ti-6Al-4V alloy under the compression coditions examined was extensive flow localization at relatively low temperatures and low strain rates. This phenomenon cannot be ascribed to adiabatic heating. For example, flow localization associated with adiabatic heating is expected, at a given deformation temperature, to occur more readily at higher strain rates. Indeed flow localization at 965°C - 10° sec^{-1} appears to be promoted by adiabatic heating.

In contrast, flow localization at lower temperatures and rates is related to microstructural instabilities. Under these conditions metallographic observations revealed that the onset of flow localization was always associated with breakup and subsequent spheroidization of the α phase either at grain boundaries or within those Widmanstätten colonies having a preferred orientation for shear deformation. Enhanced spheroidization of lamellar α during deformation in Ti-6Al-4V is expected to involve: (i) breakup of the α platelets by deformation, and (ii) diffusion-assisted coalescence of the broken lamellae. While the microstructures of deformed specimens showed some degree of breakup and spheroidization of both grain boundary and lamellar α, thermally exposed specimens exhibited only a limited degree of spheroidization below 965°C, Figure 5. Therefore, at low temperatures and low strain rates, thermal activation by itself is not

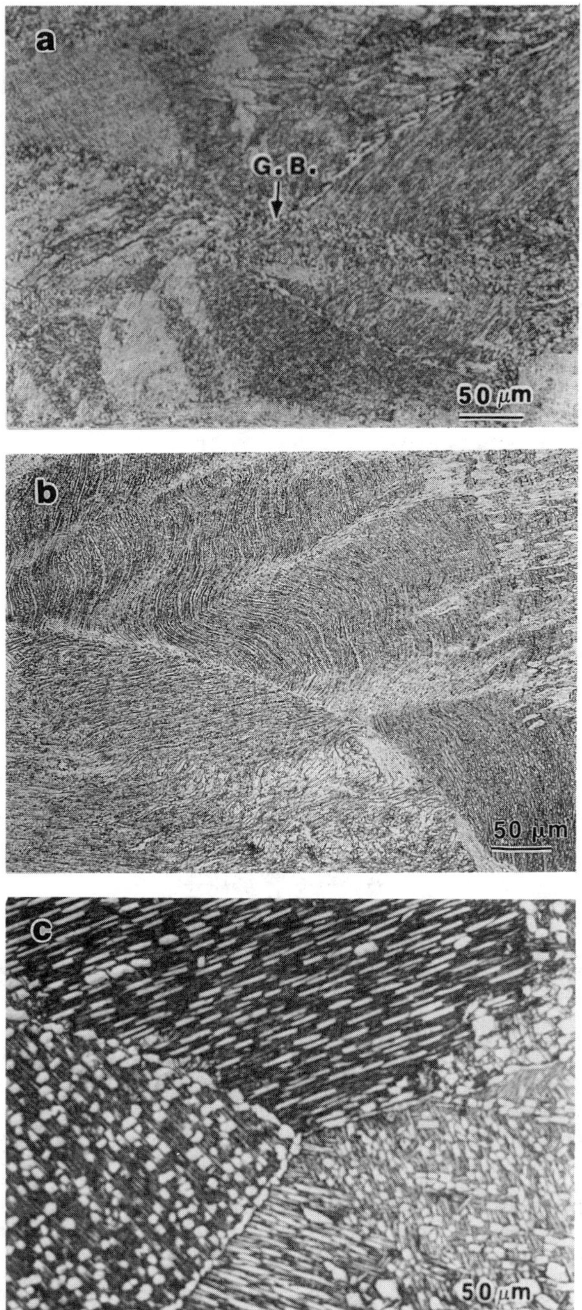

Figure 4. Optical micrographs of sample after $\epsilon = 0.3$ at (a) 800°C/10^{-4} s^{-1}, (b) 900°C/10^{-2} s^{-1}, and (c) 965°C/$10^{-4}s^{-1}$.

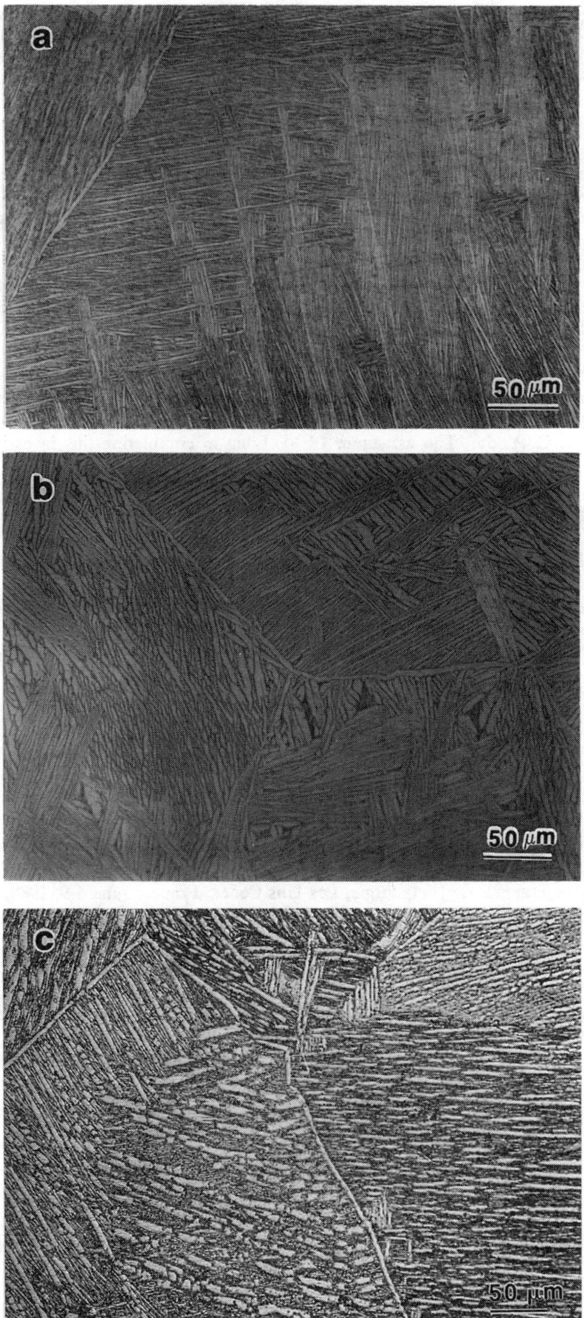

Figure 5. Optical micrographs of samples exposed for 1 hr at (a) 800°C, (b) 900°C and (c) 965°C.

sufficient to cause spheroidization of the α phase. However, at the highest temperature examined, e.g., 965°C, partial spheroidization of α lamellae occurred prior to deformation, therefore flow localization at this temperature under low strain rates is expected to be, and indeed was, significantly reduced when compared with specimens deformed at lower temperatures, see Table 1.

CONCLUSIONS

The hot deformation characteristics of Widmanstätten $\alpha+\beta$ Ti-6Al-4V has been examined under compression conditions in the temperature range 800 to 965°C between strain rates of 10^0 to 10^{-4} s^{-1} utilizing both ring and conventional cylindrical samples. Flow localization arising from either adiabatic heating or microstructural instabilty was observed under various conditions; the severity of this macroscopic flow localization being quantified in terms of the flow localization parameter α. Metallographic observations also revealed that flow localization associated with microstructural instability was initiated by breakup and spheroidization of α platelets situated either at the prior β grain boundaries or within Widmanstätten colonies.

ACKNOWLEDGEMENTS

The authors would like to acknowledge receipt of the experimental material from RMI, Inc., Niles, Ohio. This effort was funded by DARPA thru ONR Contract No.0014-89-J-3166 supervised by Mr. W. Barker and monitored by Dr. G. Yoder. The assistance of M. Long in completing this investigation is gratefully acknowledged.

REFERENCES

1. S.L. Semiatin, G.D. Lahoti and T. Altan, Process Modeling: Fundamentals and Applications to Metals, T. Altan, H. Burte, H. Gegel and A. Male, eds., ASM, Metals Park, OH, 1980, p. 387.
2. S.L. Semiatin and G.D. Lahoti,"Deformation and Unstable Flow in Hot Forging of Ti-6Al-2Sn-4Zr-2Mo-0.1Si", Metall. Trans., 12A (1981) 1705-1717.
3. P. Dadras and J. Thomas, Jr.,"Characterization and Modeling for Forging Deformation of Ti-6Al-2Sn-4Zr-2Mo-0.1Si", Metall. Trans., 12A (1981) 1867-1876.
4. S. L. Semiatin, J. F. Thomas, Jr., and P.Dadras, "Processing-Microstructure Relationships for Ti-6Al-2Sn-4Zr-2Mo-0.1Si", Metall. Trans., 14A (1983) 1867-2374.
5. S. L. Semiatin and J. J. Jonas, Formability and Workability of Metals, ASM, Metals Park, Ohio, 1984, p. 55.
6. I. Weiss, F.H. Froes, D. Eylon and G.E. Welsch, "Modification of Alpha Morpholgy in Ti-6Al-4V by Thermomechanical Processing", Metall. Trans., 17A (1986) 1935-1947.
7. G. Welsch, I. Weiss, D. Eylon and F.H. Froes, "Shear Deformation and Breakup of Lamellar Morpholgy in Ti-6Al-2Sn-4Zr-2Mo Alloy", Sixth World Conf. on Titanium, Part III, P. Lacombe, R. Tricot, G. Berangen, eds., Societe Francais de Metallurgie, Les Ulis Cedex, France, June 6-9, 1988, p. 1289.
8. J. R. Douglas and T. Altan, "Flow Stress Determination for Metals at Forging Rates and Temperatures", Trans. ASME, 97B (1974) 66-76.
9. V. Seetharaman, L.Boothe and C. M. Lombard, "Compressive Deformation Behavior of a Ti-6Al-4V Alloy at High Temperatures and Strain Rates", Microstructure/Property Relationships in Titanium Aluminides and Alloys, Y-W. Kim and R. R. Boyer, eds., The Minerals, Metals and Materials Society, Warrendale, PA, 1991, p.605.
10. J.J. Jonas, R.A. Holt and G.E. Coleman, "Plastic Instability in Tension and Compression", Acta Met., 24 (1976) 911-918.
11. P. Dadras and J.F. Thomas, Jr., "Compressive Plastic Instability and Flow Localization in Ti-6242", Res Mechanica Letters, 1 (1981) 97-103.

COMPUTER SIMULATION OF CHANGES IN TEMPERATURE AND STRESSES IN TITANIUM ALLOY FORGINGS DURING HEAT TREATMENT

M. Ya. Brun*, V. M. Olshansky**, G. Z. Malkin**,
A. N. Lozhko**, L. A. Bykova*
* All-Russia Institute of Light Alloys, Moscow, Russian Federation
** Dnepropetrosk Metallurgical Institute, Ukraine

Abstract

One of the most important problems of improvement of technology of heat treatment of $\alpha+\beta$ titanium alloy semiproducts is the optimization of conditions of their cooling from solid solution temperature. These conditions determine, to a great extent, the level of mechanical properties and the value and distribution of residual stresses. Such optimization is especially complicated for forgings, which have different thickness in different cross-sections.

The present study focused on development of procedure of computer aided calculation of change in temperature field through the bulk of titanium alloy disk forgings during their cooling, estimation of state stresses after cooling. The above procedure allows simulation of various cooling conditions for forgings and choice of the most optimum conditions.

Introduction

Cooling rate from solid solution temperature has a significant effect on mechanical properties of $\alpha+\beta$ titanium alloy semiproducts. The control of this rate within certain limits (0.3-1.0 °C/s for VT3-1 and VT25u alloys; 1.0-5.0 °C/s for VT6, VT8 and VT9 alloys ensures the best combination of short-time strength, low-cyclic fatigue and high-cycle fatigue, as well as tensile ductility and toughness [1-3].

The development of the necessary cooling rates for semiproducts of different cross-sections requires application of different cooling media and cooling conditions (such as continuous, step-by-step, etc.). Application of special measures ensuring sufficient uniformity of cooling of different thicknesses could be necessary during heat treatment of complex forgings with different thicknesses in different scross-sections. Such differences in cooling rates in different zones of a forging can result not only in substantial inhomogeneity of mechanical properties, but also in a high level of volumetric residual stresses. The unfavourable stress distribution is im-

Titanium '92
Science and Technology
Edited by F.H. Froes and I. Caplan
The Minerals, Metals & Materials Society, 1993

permissible because of the possibility of degradation of component performances, risk of distortion and even failure of initial billets during machining.

The most effective current method for determination of optimum cooling conditions is a mathematical simulation of changes in temperature gradient, level of residual stresses and their distribution in forgings depending on cooling conditions. The present study was performed on the basis of titanium alloy disk forgings. It consisted of development of mathematical models for description of temperature and state of stresses of such forgings and evaluation of such models for optimization of forging cooling.

1. Development of Mathematical Models of Description of Temperature and State of Stresses of Titanium Alloy Disk Forgings During Cooling

Cooling of bodies is determined by conditions of their convecting and conductive heat transfers. Accordingly, control of cooling processes changes the conditions of heat transfer on the body surfaces with the surraundings. Such a control depends on conditions of internal heat transfer.

Analytical solutions describing convective and conductive heat transfers are known [4]. However, these solutions, as a rule, do not take into account temperature dependence of the main thermophysical constants which reduces their accuracy. In addition, they are applied only to simple-shape bodies and cannot be used for thermomechanical calculations of complex forgings.

Heat conduction equation in a cylindrical coordinate systems, with allowance for changes in heat condition and heat capacity of a material with change in temperature, was used for discribing the heat transfer processes. Boundary conditions of the third type with allowance for radiant and convective heat transfers and arbitrary initial conditions were used.

The solution of the above equation allowed development of mathematical model for calculation of temperature fields developed during cooling of forgings. The model was developed to be used on personal computers. The service advantages of these programmes are: 1) simplicity of the description of forging geometry; the description is performed by introducing of boundary point coordinates (the forging surface between these points is taken as a rectilinear one); 2) the possibility of arbitrary control of density of arrangement of calculation grid nodes; 3) tabulated representation of heat capacity and heat conduction depending on temperature; 4) arbitrary distribution of boundary conditions on the forging surface; 5) the possibility of repeated calculations of temperature fileds on the same grid; 6) the possibility of visualization of calculation grid and temperature fields.

To describe stressed conditions of forgings, a system of three differential equations concerning static equilibrium conditions, generalized Hook's law and plane-stressed state was used [5]. The model of quasielastic loading was used for solution of these equations. In this model, parameters, such as Young's modulus and Poisson's ratio, usually taken as constants in elastic region, were considered as variables, which were dependent on stress level at each point on the basis of the strain curve $\sigma - \varepsilon$ for the appropriate alloy.

In addition, unloading theorem was used to take into account the history of loading and to form distribution of residual stresses. According to this theorem, the strain curve changes at the given point depending on the obtained value of plastic strains.

Mathematical model and programme of calculation of stresses arising in disk forgings due to nonuniformity of their cooling have been developed as a result of the solution of the equations.

The performed theoretical investigations allowed to develop a universal algorithm of search of optimum cooling technique of the forgings. Such technique satisfied the desired requirements for cooling rate, nonuniformity of temperature field, value and distribution of volumetric residual stresses to maximally possible degree. For evaluation of an optimum combination of these factors, a special functional "C" can be used. This functional includes values of root-mean-square deviations of the calculated values of these factors from the desired or allowable ones. Thus, the problem of the search consists in finding of cooling conditions ensuring minimum value of the functional.

Evaluation of adequacy of the created model by comparison between calculated data and experimental ones in real forgings that relative error of determination of temperature and stresses does not exceed 12 and 25 % respectively (Fig. 1, 2). Such accuracy is quite acceptable for engineering calculations, which permits practical application of this model.

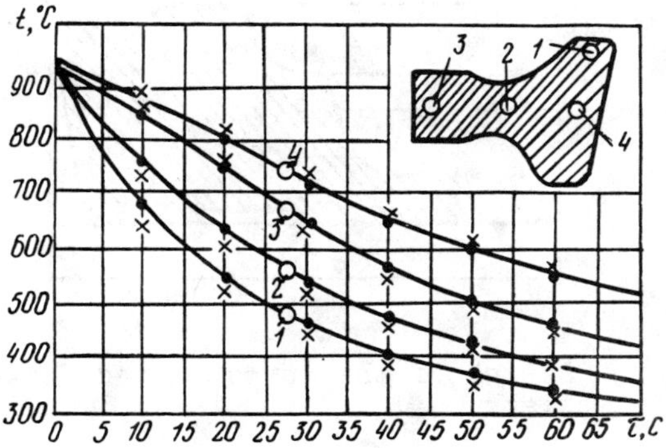

Figure 1 - Curves of forging cooling:
o - experimental data;
x - calculated data.

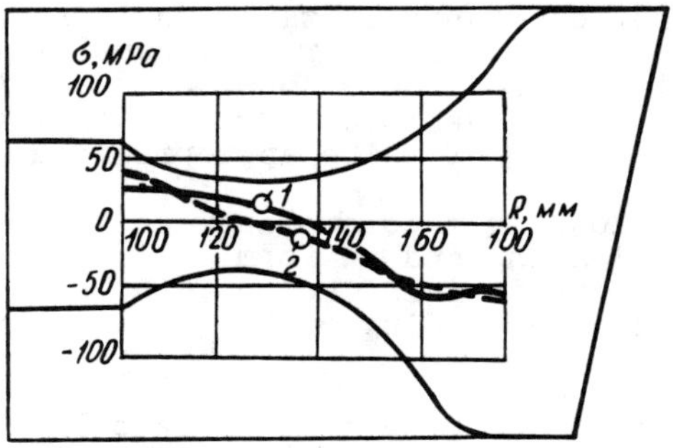

Figure 2 - Curves of tangential residual stresses along the forging radius:
1 - experimental data;
x - calculated data.

2. Investigation of Temperature Fields and Stresses in Forgings

The procedure of the search of the optimum cooling conditions for forgings is called iterative one (i.e. procedure gradually approaching the desired results). It consists of a determination, according to the developed algorithm, of the required distribution of cooling media differing in heat transfer coefficient α, over the surface of forgings.

Figs 3-7 show the results of such search with reference to one of the investigated forged disks. The task to be achieved was improvement of mechanical properties of the disk by regulation of cooling rate within the limits of 1.0-2.0 °C/s (in temperature range from 950 down to 600 °C) with maintaining of cooling rate gradient through the bulk of the disk, as well as maximum level of tensile stresses.

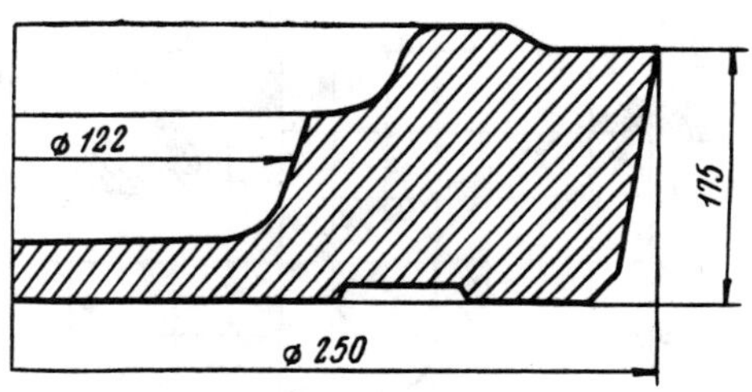

Figure 3 - Sketch of the investigated forged disk.

Six iterative calculations were made in total at boundary conditions appropriate to change in heat transfer coefficient from 20 to 2,000 $W/m^2 \cdot K$ (from air to water). Fig. 4 shows the best distribution of media over various surfaces of cooling.

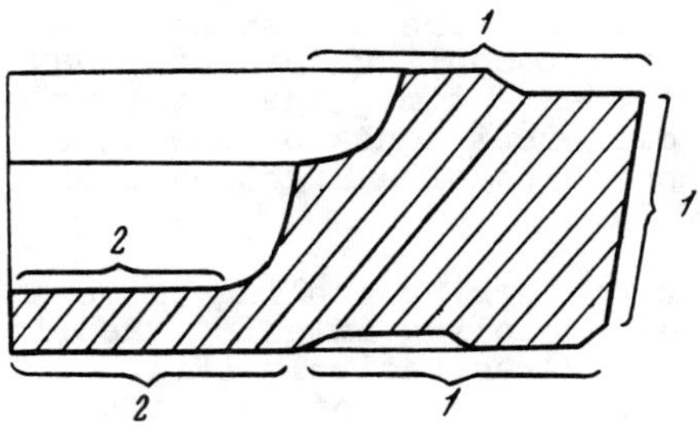

Figure 4 - Optimum distribution of boundary conditions on the forging surface:
1 - jet cooling by water-air mixture;
2 - the same, but with application of heat insulation.

By way of example, Figs 5 a and b show temperature fields in the air-cooled forging in 6 and 20 min after the beginning of cooling. The analysis shows that cooling of the center of the forging rim, especially at the initial period of time, is substantially slower than that of the web.

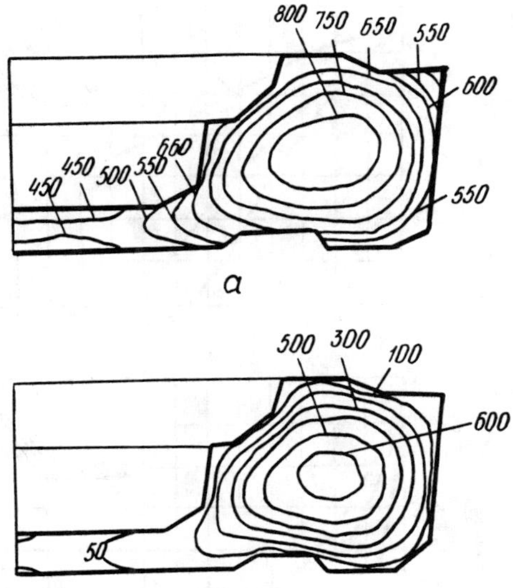

Figure 5 - Temperature fields in the forged disk in (a) 6 and (b) 20 min after the beginning of air cooling.

Quantitative differences in cooling character of these zones at various moments of time are shown in Fig. 6 and Table 1. One can see that air cooling does not ensure the desired cooling rate. At the same time, the existing differences in cooling rates of different zones of the given forging do not lead to the generation of a high level of residual **tensile** stresses (Fig. 7).

Intensification of forging cooling process by oil quenching allows increase in cooling rates of the rim and of the web up to 0.6 and 5.3 °C/s respectively (Fig. 6, Table 1). However, it results also in increase of nonuniformity of cooling and rise in tensile stresses (Fig. 6, Table 1, Fig. 7b).

Of the six evaluated variants, jet cooling by water-air mixture ensures the best cooling conditions provided that heat insulation of the web was made (Fig. 4, 6, 7, Table 1). Cooling rate at all the points of the forging is within the limits from 0.7 to 1.7 °C/s, which is close to the preset range. Maximum temperature gradient between different zones of the forging and through their thickness is 210 °C and the value of volumetric residual stresses does not exceed 200 MPa (after aging - 60 MPa). The above values are comparable with those obtained after air cooling. Accordingly, the value of the optimization functional "C" at such cooling is $0.2 \cdot 10^4$ whereas at air cooling and oil cooling it is equal to $0.8 \cdot 10^4$ and $1.5 \cdot 10^4$ respectively.

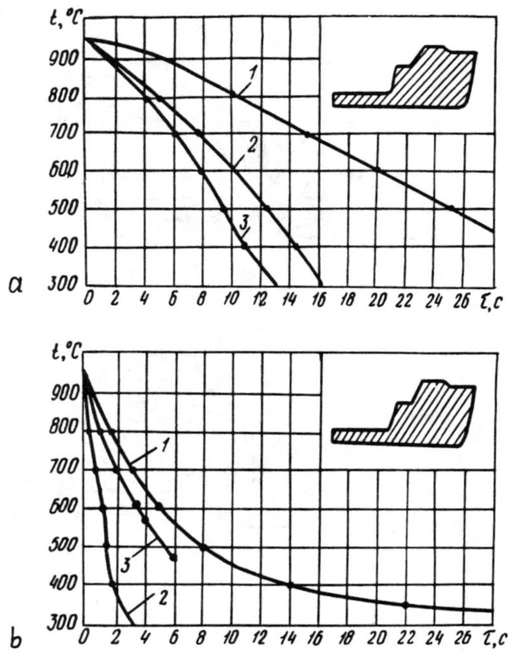

Figure 6 - Temperature change in the center of: (a) the rim and (b) in the web during air cooling (1), oil cooling (2) and water-air mixture cooling (3).

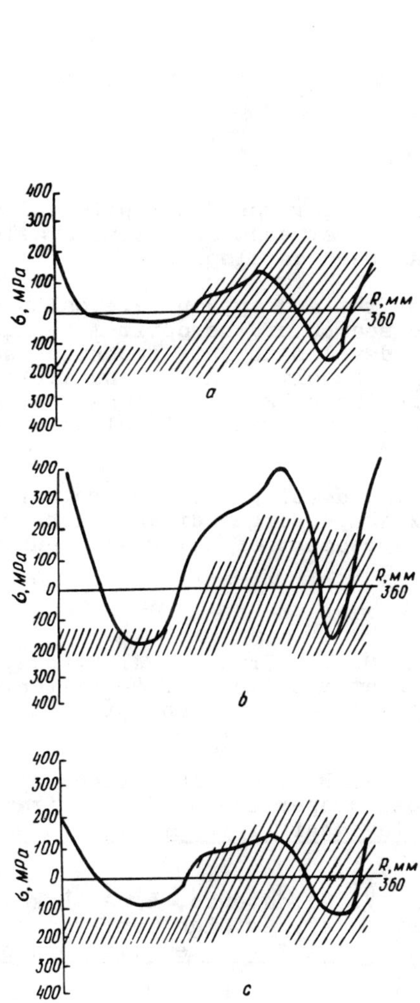

Figure 7 - Curves of tangential stresses in the forged disk after air cooling (a), oil cooling (b) and water-air mixture cooling (c).

Table 1 Cooling Rates of the Forging in Different Cooling Media, °C/s*

Temperature range, °C	α, W/m²·K ≈20 air	≈500 oil	≈1000 water-air mixture
950-850	0.2/1.4	0.4/5.0	0.6/1.8
850-750	0.3/1.4	0.6/5.1	0.7/1.7
750-650	0.3/1.3	0.8/4.9	0.9/1.6
950-650	0.3/1.4	0.6/5.3	0.7/1.7

* In the center of the rim/in the web.

Conclusions

1. Mathematical model for the description of temperature and the state of residual stresss of titanium alloy disk forgings during cooling has been developed.

2. Universal algorithm and appropriate calculating programs for the automatic search of the optimum technoque of forging cooling have been developed. This technique satisfies the desired requirements for cooling rate, nonuniformity of temperature field through the bulk of forgings, value and uniformity of distribution of volumetric residual stresses to maximum degree.

3. Evaluation of the developed procedure for prediction of cooling of complex shape forged disks has been carried out and the satisfactory agreement of calculated and experimental data has been shown.

References

1. N. F. Anoshkin, M. Ya. Brun, L. A. Bykova, "On conditions of strengthening heat treatment of $\alpha+\beta$ titanium alloys" (Proc. of the 3rd Intern. Conf. on Titanium, Moscow, VILS, 1978), 589-598.

2. M. Ya. Brun, L. A. Bykova, "The effect of cooling conditions on mechanical properties of VT3-1 titanium alloy at heat treatment", Metallovedenie i termicheskaya obrabotka metallov, (12) (1985), 36-38.

3. M. Ya. Brun, I. V. Soldatenko, L. A. Bykova, "Structure and mechanical properties of new high temperature titanium alloy VT25u", Metallovedenie i termicheskaya obrabotka metallov, (1) (1992), 29-31.

4. A. A. Sernov, Engineering methods of calculation of heat transfer in complex shape bodies (Moscow, 1977), 80.

5. I. A. Birger, R. R. Mavlutov, Strength of Materials (Moscow: Metallurgiya, Nauka, 1986), 560.

LASER FORMING OF NEAR SHAPES

F. G. Arcella

The Johns Hopkins University
Applied Physics Laboratory
Aeronautics Department
Laurel, MD 20723

Abstract

A new method of forming near shapes directly from metal powders is described and experimental structures are shown. Sequential laser fusion of thin, laminar metal powder layers to form integral, 100% dense structures has been demonstrated with titanium, titanium elemental alloys, and other metals. An 18KW, CW CO_2 high energy laser was used to form titanium cylinders and model impellers, and stellite nozzles for an incinerator. The process is very advantageous for hard to cast, forge, or form materials such as stellite, titanium, titanium gamma aluminides, or composites. Near shape structures of these and other materials can be formed directly from metal powders and elemental mixtures without using molds or dies. The size of the laser cast part produced by this new process could extend to several feet in dimension and to several hundred pounds; and would be independent of the process. Microscopic examination of laser cast titanium and titanium alloys show the product to be 100% dense, and alloys formed from elemental powder precursors to be homogeneous in composition. Projected economics of forming laser cast parts from titanium precursor powders by the direct downloading of shape parameters from CAD systems to CNC tables targeted by the laser metalworking beam are reviewed.

Summary

A method of casting metals into near shapes by a new process has been evaluated.[1] Incremental layers of stellite, stainless steel, or titanium powder have been sequentially fused by a high power CO_2 laser into near shapes. Special powder handling methods have resulted in the production of high purity titanium and titanium alloy castings from as-received Hunter-Kroll sponge fines. Shaped casting geometries can be achieved by computer controlled movements of a target metal powder bed beneath a laser beam. The powder not fused by the beam acts as a "sand mold" to retain the beam-melted powder until it solidifies. As a result, elemental and alloy near shape castings can be made without molds or dies by a completely computer driven process (Figure 1). Significant technical and cost advantages are projected for the fabrication of titanium (alloy) marine and aerospace structures by this method.

Titanium '92
Science and Technology
Edited by F.H. Froes and I. Caplan
The Minerals, Metals & Materials Society, 1993

Initial work in titanium was very promising. If successfully developed, this process for forming titanium (alloy) near shapes could reduce the cost of selected marine and defense materials structurals significantly.

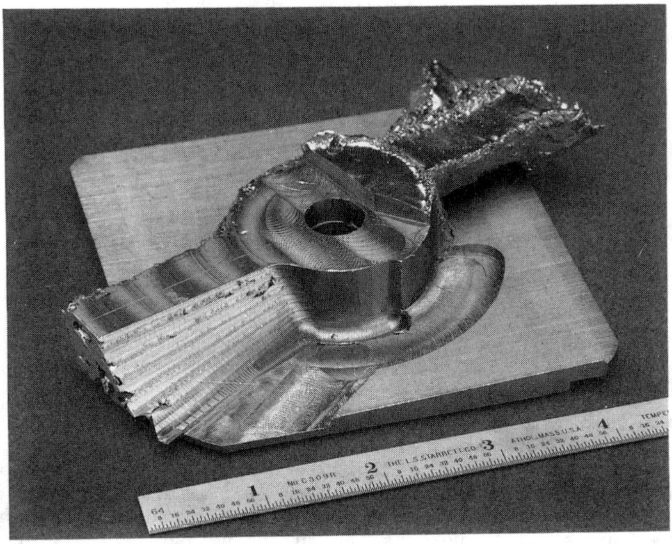

Figure 1 - Laser cast titanium impeller structure formed from as-received Hunter-Kroll sponge fines. One vane has been rough machined by a CNC milling machine.

Introduction

The moldless casting technology described here can be applied to many metal powders. However, this paper will focus upon its application and development with titanium. Although titanium and its alloys are important, critical defense-related materials, extensive use of titanium in defense materials systems has not occurred because of technology and cost reduction impediments.[2] A reduction in the cost or calendar delivery time and the elimination of technological barriers in processing (shaping) of titanium structures would be significant achievements and would open opportunities for the metal.

Titanium is produced in Kroll reactors by the reduction of $TiCl_4$ by either sodium or magnesium.[3-5] The resulting sponge must be consolidated, double or triple vacuum consumable arc melted to drive off the residual chlorine and sodium (Hunter) or magnesium and to blend in alloy ingredients. The ingot thus produced is then processed by one of the following methods:

- Forged and machined to a desired shape with continual heat treatment between forging events (a series of blocker and final dies) and extensive machining after final forming.
- Melted and vacuum cast into rammed graphite molds (prefired and dried) and hot isostatic pressure weld (HIP) consolidated to remove casting porosity. Final machining is again necessary.
- Melted and vacuum die cast into fired ceramic molds where the mold leaves a case and case removal etchants have left hydrogen embrittlements.[6]
- Formed into controlled particle size powder by the rotating electrode process (REP). Titanium parts are formed by HIP'ing the powder in shaped

envelopes in high pressure, high temperature autoclaves.
- Rolled to plates or slabs. The desired part is machined by discarding 80 to 90% of the material.

All of the casting processes require mold designs and fabrication, firing, etc., which can delay delivery of the desired casting for three to six months. Since the molds are seldom reusable, the cost of the mold must be added to that of the casting. HIP processing of the casting to remove casting porosities also adds to the cost and the delivery time. Forgings must proceed through a series of blocker dies (incremental shapes) with intermediate heat treatments to remove work hardness, and the final forging requires rather extensive machining to arrive at the final net product shape (fly-to-buy ratios are down to one pound for each 6 to 8 pounds made). Some forging schedules extend 12 and 16 months. Autoclave HIP'ing of REP powder gives a good near net shape product, but the elaborate practices to achieve the proper powder result in its costing from $60 to $600 per pound. The forming of large plate sections for hull constructions is impractical due to casting size limitations and the absence of large domestic mills to work the plates down and curve them. There is a world oversupply of titanium materials producers and a depressed business posture for the finishers, yet no one has been able to reduce costs any further. The base line cost for titanium materials still inhibits its use and deployment.[7]

In response to this serious problem, a new, lower cost process to manufacture near shape metallic (titanium) structures from precursor materials has been studied.[1] The process overcomes the technical, size, and schedule constraints of present casting and forging operations, and forms near shape structures at a very significant cost advantage over other forming methods.

Technical Approach

The new near shape forming method described here has grown from efforts in laser welding at the Science & Technology Center of the Westinghouse Electric Corporation and was performed on an 18 KW CO_2 AVCO high power laser operating on full scale production equipment in five metalworking cells.

New Technology

The new forming technology developed is very simple. Fine metal powder (sieved Hunter-Kroll sponge fines) is placed in a flat bed mounted on X-Y slides. A high power CO_2 laser beam is focused vertically down onto the powder in the bed, and the bed is caused to move by computer control of the X-Y slides. The motion of the bed causes the laser beam to melt the metal powder in the lithography of one cross sectional plane of the desired part. The unmelted powder surrounding the beam melted powder acts as a retaining structure and prevents the melted powder from flowing out of the desired lithographic shape. The melted powder solidifies in the desired shape, and a second layer of powder is added over the entire bed surface. The bed moves again by computer control of the X-Y slides and the next cross section of the desired part is melted by the laser beam. This melted layer is offset from the solidified layer cast in the previous step, but sufficient layer-to-layer registry is retained that it fuses also to the fused metal sublayer. Another layer of powder is placed over the bed, and the process is repeated sequentially until the desired part is incrementally constructed, layer by layer, to the final desired near net shape. Figure 1 illustrates the results of the buildup process.

Although others have proposed and have performed the fabrication of net shape structures by stick and arc weld methods,[8] and by laser/powder processes[9-

[11]), this is the first effort to perform, and file for patent on net shape forming by a laser in a variable surface metal powder bed. This process presents several unique opportunities for working with titanium alloys. This process is the innovative subject matter of U.S. Patent 4,818,562 which is assigned to the Westinghouse Electric Corporation.[1]

Titanium Forming

The process can be used with a metal powder to form near shapes in one step by computer driven means. The process appears ideally suited for near shaping of titanium. The normal Kroll or Hunter-Kroll routes to titanium parts and their problems have been described earlier. These routes prevent the further cost reduction of titanium marine or aerospace parts. Several technical features of titanium and its net shaping by the laser process contribute significantly toward the success of laser casting:

- Cast titanium (and newly developed casting alloys) has nearly the same fatigue and strength properties as forged and HIP'ed titanium.[12-16] This means that the near shape (alloy) structures made by this laser forming process should, with heat treatment, have parity in properties with traditionally formed titanium product.
- The laser process uses a patented, inert gas protected chamber at oxygen levels of 10 to 15 ppm. There is evidence that this small contamination of the titanium product will not detract from its strength or fatigue properties, but may in fact enhance these properties.[17]
- Experimental evidence indicates that the surface tension of liquid titanium is very low; thus it lies very flat when fused and after solidified as opposed to that of stainless steel. This means that vertical registry of sequentially fused layers will be facilitated, and lateral surfaces will be more uniform.
- This laser process is not limited to the size of vacuum casting chambers, HIP furnace volume, or vacuum ladle capacity. Thus, the laser optics may be mounted on a gantry above a large titanium powder bed and the powder fused directly into hatch rings, bulkhead plates, supporting ribbing, etc. Size does not appear to be a limitation of this laser casting process.
- Most important, it may be possible to skip all of the post-Kroll sponge processing steps, and form the titanium (alloy) near shaped part directly from sieved sponge fines in one step.

These technical, economic, and process features of the laser process for titanium metal structural formation should contribute toward its success. Notably, the thermophysical properties of titanium work for its success in this process.

Accomplishments

A feasibility demonstration of the process was conducted. Work was performed at the Westinghouse Science & Technology Center. For these tests the 18 KW CO_2 laser was used. Sieved (-100 mesh) Hunter-Kroll titanium sponge fines were obtained from commercial sources. Small aliquots of the as-received powder were placed on titanium substrates in an open top argon purged (sweep gas) chamber and melted with a stationary laser beam at beam energies of 6 to 8 kW. Below this energy, the melted powder did not wet the titanium substrate. The roughness of the powder absorbed most of the laser beam energy. The resulting product was a bright, flat titanium deposit wetted onto the titanium substrate. Notable in these tests were the following facts:

1. The titanium powder absorbed most of the laser energy (for good efficiency).

2. The melted titanium has low surface tension and lies and solidifies very flat on the substrate. Thus, additional powder layers could be added and fused evenly without losses or irregularities.
3. Contamination (oxygen and nitrogen) was controllable by a small purge of sweep gas (argon) in the patented target chamber, which was open at the top. The solidified titanium deposit was bright and lustrous.
4. The deposits were 100% dense with no voids or porosities. Density is very important when one is considering making castings of this material in one step directly from the Hunter-Kroll fines.

Most important, when chemical analyses were performed on the solidified metal product, residual sodium was not found and residual chlorine was between 50 and 150 ppm. The product of the laser fusion of as-received Hunter-Kroll titanium sponge fines was therefore high purity titanium metal with a chlorine content that made the product weldable. Table 1 shows typical chemistries before and after laser processing, and the ability of the process to drive off the chlorine and sodium impurities.

Table I The Laser Fusion Process in the Gas Sweep Target Chamber Purifies the Titanium of Chlorine and Sodium Contaminants

Element	As-Received Kroll Fines	Laser Cast Titanium Product
Cl	1500 ppm	50-150 ppm
Na	1500	0
Si	2000	800
Fe	900	900
*O	1100	1300
*N	100	100

*Leco; - 2% - NBS traceable.

Titanium powder was placed in the argon sweep purged chamber mounted on an x-y table. Tests were performed to determine beam energy, beam optic manipulation frequency, and linear travel speed to fuse a linear strip of titanium. Proper settings were determined to fuse a flat deposit of titanium 0.5-in. wide by any length. Bright, lustrous deposits of pure titanium were made in one-layer (0.030-in.) and four-layer (0.120-in.) thickness. Analysis of the product showed the following:

1. Residual sodium was 0 ppm and chlorine 50 to 100 ppm.
2. Bulk oxygen remained below 1300 ppm, and nitrogen remained unchanged at 100 ppm.
3. There was no casting porosity. Almost all of the chlorine and sodium were evolved.
4. The metallurgical structure was acicular alpha with a weak outline of the previous beta structure.
5. There was no residual demarcation to denote where the layers had formed or were fused one to another. Interestingly, the casting of one layer could be the heat treatment for the previous layer.
6. The lateral unfused powder acted to retain the melted titanium until it solidified, and resulted in a sharp replication (layer by layer) of the desired shape.

The previous test was repeated with one change: 60-40 commercial aluminum-vanadium powder was blended 10-90 with the as-received Hunter-Kroll sponge fines. The laser parameters used in the previous test were employed, and a flat, linear deposit of Ti-6Al-4V alloy 0.5-in. wide by 0.040-in. thick was made in situ in one step. SEM analysis showed that the in situ manufactured alloy was homogeneous in composition and free of porosity. Its sodium content was 0 ppm and its chlorine was 50 to 150 ppm. Alloy powder could thus be blended with the as-received Hunter-Kroll fines, and titanium alloys produced and near shape laser cast in one step.

The as-received Hunter-Kroll titanium sponge fines were next placed onto a rotary table mounted inside the sweep gas target chamber that was on an x-y table. The laser energy levels, target travel speeds, powder thicknesses, laser optic actions, etc., determined in the prior tests were employed to produce the following structures:

1. A strictly rotary motion of the powder target beneath the laser beam permitted the sequential layer casting of a titanium cylinder 3.0-in. OD by 1-in. high with a 1-in. center hole. Vertical registry of the cylindrical layers could be controlled with less than 1/32-in. offset.
2. An off-set linear motion was added after each rotary movement as above, and a hub with two canted impeller vanes (2.5-in. radial extension from the hub) was cast (Figure 1).

These accomplishments set the technical and feasibility foundation for a new method of processing titanium and its alloys. The following sections describe advantages it offers over current practice.

Technical Advantages

The main technical advantage of this process is that it is a completely flexible, computer controlled process by which metal shapes and part geometries may be fabricated (cast) by software coding of the motion of the end adapters for the laser fixture and the robotic assist devices, i.e., cast without molds or dies. The part may be designed on an auto-CAD computer system and downloaded to the process control computer, which calculates the incremental planes taken through the desired part geometry and controls the laser lithography and planar registry required to form the part by the incremental fusion process described. For modern foundries or die forging shops where minimum lot sizes of 500 or 1000 pieces are required to cover the cost of the die and or mold, or for rapid prototyping for lots of one or two, without the cost and time delays to design and machine molds and/or dies, this process shows merit.

Improvements in practice over prior art include the following unique features:

- Structures are formed to near shape without molds or dies.
- Structures can be formed by downloading CAD/CAE generated x-y motions for each planar section of the desired part to the casting bed, making the process extremely flexible, instantly changeable (product design), amenable to small lot sizes (lots of "one"), and the product is automatically formed (minimal skilled labor content).
- The compositional makeup of the part being formed may be varied throughout its dimension to optimize desired part operational requirements.
- The inert fludizing gas acts to sweep away all evolved gases and vaporous weld (fusion) products to clean the weld (fusion) zone and produce a bright, lustrous metal product.
- Only the near shape volume of material of the desired part is consumed in the process, thus conserving the remaining powder material in the bed for reuse later (100% raw material utilization).
- The same powder bed may be used for the sequential production of a series of different geometrically shaped parts without any physical changes except for the software that controls the CNC table that provides the x-y motion for the bed.
- The sweep gas may be varied during the fabrication process to impart

certain hardness properties to the product. For instance, periodically adding nitrogen or methane to the inert gas for ferrous or superalloy part fabrication could impart bands of varying hardness to the structure.
- Heating the powder bed with buried, metallic sheath cartridge heaters will result in a hot powder feed to the weld fusion zone, which could improve the metallurgical properties of the resulting fused shape form.

Since laser welds have been successfully performed using loose powders fed by a tube into the beam, or on loose powders laid before the beam,[9,10] the probabilities for success of this process using the powder casting bed are high. The potential of the process for the metals fabricating industry is good. Potential product lines include near shape castings, forging preforms, repair of fractured structures, and the coating and glazing of dissimilar metal structures. Similar powder forming processes are now being reported.[18]

Cost Advantages

Cash flow analyses of the new forming process for titanium and titanium alloy structures as standalone profit centers were made. Projected product costs have been compared with those of forged and HIP'd product using cost curves obtained from public reports.[19,20,21] Titanium alloy structurals for aerospace application formed by this process typically would cost around 30 to 40% of the marginal cost of the same product formed by forging and HIP'ing. Autoclave costs and starting material costs are the high contributing factors for HIP'ing, while excessive machining and heat treatment costs are the factors for forging.[19,21] The new laser process described here promises significant cost reductions for titanium structurals for several reasons:
- It starts with low cost Kroll fines as opposed to $60 to $600 a pound REP powder for HIP'ing, or $10.50 a pound and up ingot for forging.
- It purifies, alloys, and near (net?) shapes the titanium sponge fines to final product in one step.
- It does NOT require molds, dies, vacuum, or pressure autoclave equipments.
- Since it is totally software driven, prototype lot sizes of "one" can be made.
- Unused titanium powders (Kroll fines) are not consumed, do not need to be reprocessed, and can be recycled immediately.

This new laser forming process could be developed to contribute significantly toward increasing opportunities and markets for titanium.

Acknowledgements

The author appreciates and acknowledges the contribution of R. A. Miller and the support of G. Bruck, R. Obrokta, and A. Mandel at the Westinghouse Science and Technology Center, helpful discussions with G. G. Lessmann, and encouragement from H. Bomberger and M. Goldberg.

References

1. Casting Shapes with Laser/Fusible Powder, US Patent #4,818,562, F. G. Arcella and G. G. Lessmann, April 4, 1989.

2. W. W. Minkler, Near Net Shape Falls Short as Titanium Cost Cutter, Frank Haflich, American Metal Market/Metalworking News, p. 19, Nov. 10, 1986.

3. H. Bomberger, F. H. Froes, P. H. Monton, Titanium - A Historical Perspective, Titanium Technology, Titanium Development Associates, p. 3, 1985.

4. Titanium 80: Science and Technology, Proc. Fourth International Conf. on Titanium, Vol. 1 to 4, Kyoto, Japan, May 19-22, 1980.

5. Titanium, Science and Technology, Proc. Fifth International Conf. on Titanium, Vol. 1 to 4, Munich, September 10-14, 1984.

6. C. Liesner, Titanium Investment Casting - Production, Properties, Applications, ibid, Vol. 1, p. 157.

7. Titanium: Past, Present, and Future, National Materials Advisory Board, January 1983, PB83-171132.

8. R. R. Irving, Shape Welding: A New Concept in Fabrication, Iron Age, November 23, 1981, p. 111.

9. C. O. Brown, et al., Method for Fabricating Articles by Sequential Layer Deposition, U.S. Patent No. 4,323,756, April 6, 1982.

10. D. B. Snow, et al., Rapid Solidification Processing of Superalloys Using High Power Lasers, Superalloys, 1980, by ASM, pp. 189-203.

11. T. C. Peng, et al., Exploratory Study of Laser Processing of Titanium Alloys, McDonnell Douglas, ONR Contract N00014-76-C-0626, 1981.

12. F. H. Froes, et al., Developments in Titanium Metallurgy, J. of Metals, Vol. 32, No. 2, February 1980, pp. 47-54.

13. F. H. Froes and D. Eylon, Powder Metallurgy of Titanium Alloys, Titanium, Science and Technology, Proc. Fifth International Conf. on Titanium, Munich, September 10-14, 1984, Vol. 1, p. 267.

14. R. H. Colloy, Engineer's Guide to Titanium Castings, ibid, Vol. 1, p. 151.

15. Structural Ti Castings to Fly in Helicopters, Metal Progress, June 1986, pp. 13-16.

16. New Titanium Alloys Offer High Performance, Precision Metal, October 1986, pp. 13-16.

17. R. I. Jaffee, An Overview on Titanium Development and Application, Proc. Fourth International Conf. on Titanium, Vol. 1 to 4, Kyoto, Japan, May 19-22, 1980, p. 53.

18. P. J. Dvorak, Chipless Machining Solves Modern Cutting Problems, Machine Design, pp. 131-134, December 10, 1987.

19. R. H. Witt and A. L. Ferreri, Titanium Near Shape Components for Demanding Air Frame Applications, Grumman, Proc. 17th National SAMPE Technical Conf., October 22-24, 1985, p. 296.

20. E. J. Kubel, Jr., Titanium Alloy Technology Upgrade, Metal Progress, pp. 41-51, June 1986.

21. F. H. Froes, H. B. Bomberger, D. Eylon, and R. G. Rowe, Potential of Titanium Powder Metallurgy, First International SAMPE Metals Conference, Cherry Hill, N.J., August 1987.

SUPERPLASTIC FORMING OF INDUSTRIAL PARTS:

A NUMERICAL SIMULATION.

Christine CORSINI *, Michel BELLET**, Jérome PORA***

* TRANSVALOR, Les Espaces Delta, B.P. 037, 06901 Sophia Antipolis Cedex, France

** CEMEF, Ecole des Mines de Paris, B.P. 207, 06904 Sophia Antipolis, France

*** AEROSPATIALE, 12 rue Pasteur, B.P. 76, 92152 Suresnes Cedex, France

Abstract

A finite element model for superplastic forming, SUPFORM3, is presented. The mechanical formulation is based on membrane approximation. The model includes contact and friction with dies of complex geometry. A special attention is devoted to the calculation of the "optimum pressure cycle". The advantages of using a domain-specific code has been amply illustrated by applying SUPFORM3 to industrial problems. Since this model has been already tested in a great number of cases, the aim of this paper is to demonstrate the capability of the code to deal with real industrial problems.

Introduction

Superplastic sheet forming of titanium alloys has been developed for the last twenty years in the aerospace industry. When associated with diffusion bonding (SPF/DB), this technique allows the manufacturing of near net shape complex structural components in a single forming operation [1]. This process takes advantages of the excellent superplastic properties of the widespread structural Ti-6Al-4V alloy [2]: in a certain range of temperature and strain rate, the plastic deformation is relatively stable and it is possible to produce large deformations under low flow stress. It is to be noticed that the single sheet process is now more and more applied to alumunium superplastic alloys and that active research is in course concerning the diffusion bonding of Al alloys [3].
The basic forming operation is sheet bulging by a one-sided gas pressure in a shaped mould. Although the technology seems conceptually simple, the influence of the process parameters (pressure cycle, lubrication conditions, die geometry...) on the final thickness profile is far from well established. A number of theoretical analysis, based on iterative techniques, were developed to predict thickness variations in the forming of specific bidimensional shape only [4-6]. However, the previous analysis are not generally applicable to the 2D (and a fortiori 3D) shaped components.
Now, there is a great need from aerospace industry for a better control of superplastic forming operations, especially for complex parts. Consequently, research has been very active in this area during the recent years [7-10], with intensive use of the finite element method which is more general and flexible than previous models [11]. Six years ago, the CEMEF laboratory (Centre for material forming processes in France) has developed a 3D finite element model for superplastic forming called SUPFORM3 [7-8].
The purpose of the present paper is to illustrate the capabilities of SUPFORM3. A general presentation is summurized and then industrial applications of SUPFORM3 are presented. For more detail on the scientific background the interested reader can refer first to the appendix and secondly to the bibliography.

Titanium '92
Science and Technology
Edited by F.H. Froes and I. Caplan
The Minerals, Metals & Materials Society, 1993

General presentation

The modelling of an industrial situation is done with SUPFORM3 using a group of modules which carry out the finite element analysis, pre and post processing functions. The general mechanical frame is a plane stress membrane formulation for an isotropic strain-hardening material. The finite element method is used, associated with an incremental viscoplastic flow rule obtained by the implicit time integration of the constitutive relations over the increment. The general solving procedure is a Newton Raphson algorithm. The element types that can be used in SUPFORM3 are 3-nodes triangles or 4-nodes quadrangles. The model includes contact and friction with dies of complex geometry, and pressure optimization in order to remain in the superplastic strain-rate range. SUPFORM3 can treat single sheet forming problems only, while a 2D model SUPFORM2 exists and is able to simulate the forming of diffusion bonded components.
The simulation stands in three steps: first the data generation (surfacic mesh of the die issued from commercial CAD system, design of the finite element mesh of the sheet with an interactive graphic pre-processor, input of the rheological and tribological data), secondly the computation itself, and then the results exploitation with graphic post-processing. The color graphic post-process enables the user to visualize the deformed mesh and the iso-lines of each computed result: equivalent strain rate, equivalent strain, stress, thickness, contact zone ....

Applications

Three dimensional SPF/DB component

This investigation is related to the design of a test part made of Ti-6-4 alloy (figure 1). It is a 4-sheet component with an internal vertical rib, which is formed by two core sheets, and divides this circular part in two. The deformation of internal sheets only has been studied, as it is the most critical. By reason of symmetry, only a upper quarter of the system has been considered.

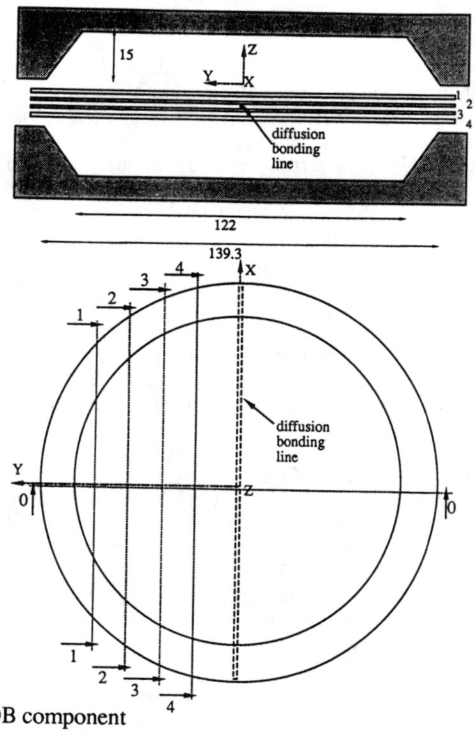

Figure 1: 4-sheet SPF/DB component

The finite element mesh is made of 1027 nodes and 1944 triangular elements. The friction conditions are the following: as the studied sheet is to contact the external sheet without any lubrication (to allow diffusion bonding), the contact is considered sticking, except in the rib region. To form the vertical rib, the internal sheet comes into contact with itself and this can be simulated by a sliding contact on the vertical fictitious wall. The initial thickness of the core sheets is 0.74 mm.

The study proceeded in three steps. The 3D code was run first, in order to compute the optimum pressure curve, for a given value 3.10^{-4} s^{-1} of the maximum equivalent strain rate, and taking into account an upper limit for the forming pressure (figure 2). In a second phase, the part was formed experimentally with a pressure curve fitting as well as possible the previous one. Finally, the code was run again with the pressure curve recorded during the experiment.

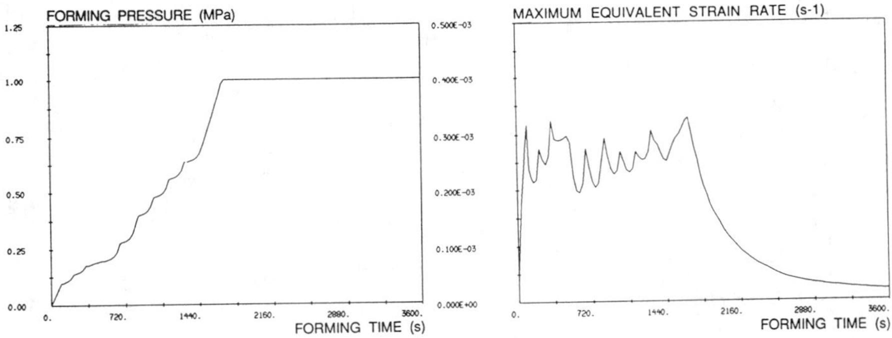

Figure 2: Optimum pressure curve and strain rate regulation

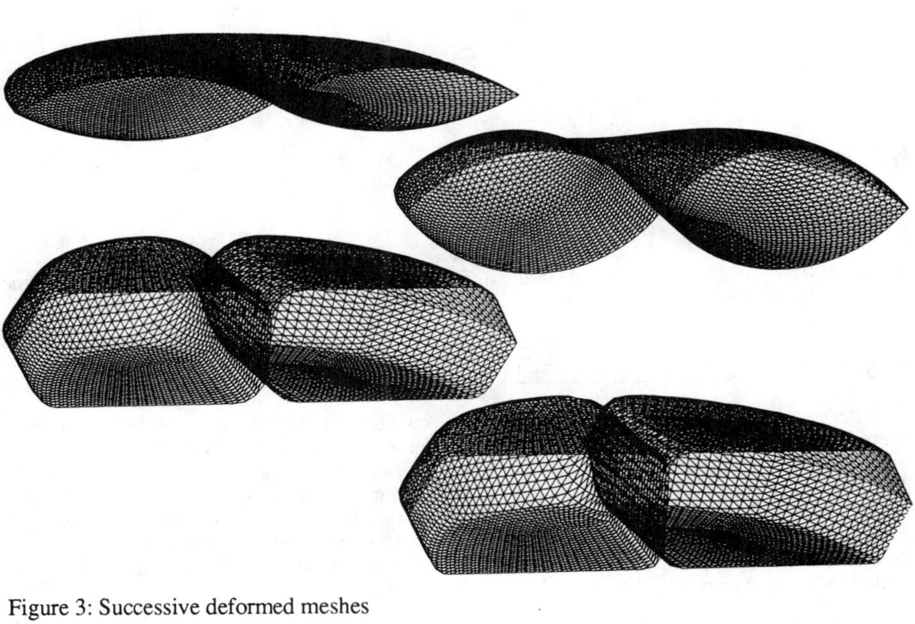

Figure 3: Successive deformed meshes

The comparisons between the computation and the experiment have been done at two different forming times 1500 and 2500s. The figure 3 shows the successive computed deformed meshes which are in rather good agreement with the experiment. However, at the end, of the forming, the filling of the corners is slower with the model. This is probably due to a coarse mesh in these regions, and it illustrates clearly the need for a remeshing operation. Concerning the computation cost, 54 increments were necessary to get the final configuration, resulting in about 1 hour CPU time on a vectorial computer FPS 64/60.

As regards the thickness distribution, several cross sections of the part have been analysed (figure 1) and generally, the agreement with the experiment is quite good (figure 4). It is worth to notice that the hypothesis of the sliding contact against the fictitious wall has been confirmed by experimental results: for a given x coordinate, the thickness has been found uniform throughout the rib.

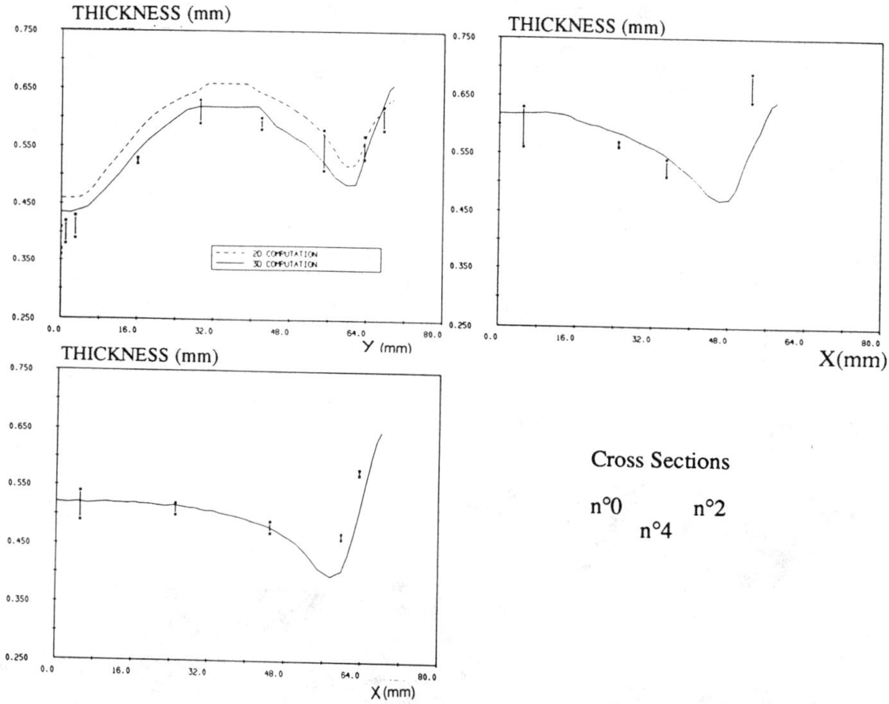

Figure 4: Computed thickness profile in the different cross sections. Vertical segments are for the experimental results

<u>Simulation of an "helicopter V fin"</u>

In order to evaluate the capability of SUPFORM3 on aluminium alloys also, the following trials were carried out. The part represented the border of an "helicopter V fin" made in 8090 superplastic aluminium alloy as shown on figure 5. The thickness of the sheet was 2.5mm. The forming step was a classic sheet bulging by a one sided gas pressure in a shaped mould. The part was divided in three pieces for the simulation operation: the body, the cross and the wings. For each piece SUPFORM3 gave the following results: a pressure curve for a given value 3.10-4 s-1 of the equivalent strain rate, a Z axis deflection curve (for a given section) as a function of time and a thickness distribution diagram.

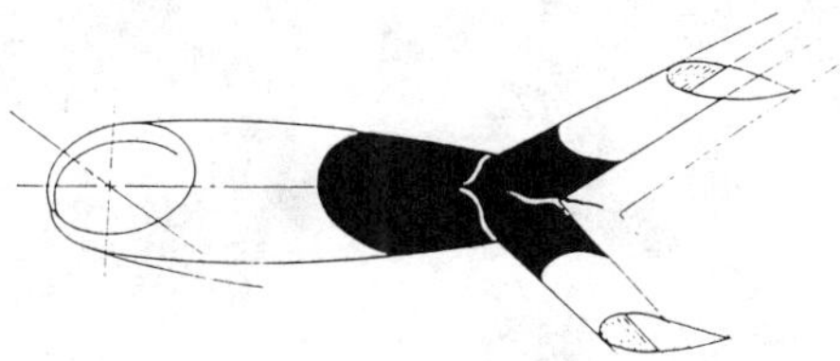

Figure 5: Shematic description of the part.

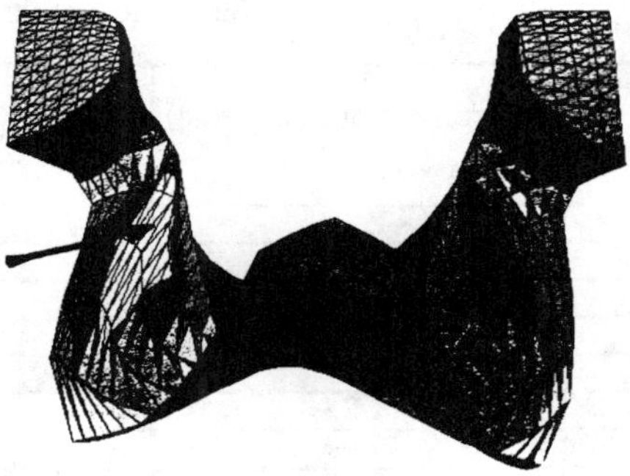

Figure 6: Thickness distribution of the cross (maximum=1.9, minimum=0.76).

The combination of the 3 pressure curves gave the pressure cycle used to realize the part. The deflection curves showed that the body was formed first, then came the cross and at last the wings. The thickness distribution diagrams were very useful. They permited to evaluate the equivalent strain rate everywhere in the part, showing that the maximum strain rate had to be obtained in the wings and was over the performances of the material (over 1000%). The thickness distribution diagram of the cross is presented on figure 6. It shows that a critical area in the part is located at the basis of the rigth wing, near the joining edge with the cross. This is due to an important re-entrant shape.

Figure 7: Picture of the real part.

Real part . Simulation results.

Z deflection (mm)	Experiment thickness (mm)	SUPFORM3 results (mm)
0	2.30	2.28
-16	2.00	2.05
-32	1.85	1.83
-54	1.64	1.61
-86	1.41	1.40
-114	1.25	1.17
-152	0.99	0.94
-210	0.70	0.72
-292	0.50	0.50

Figure 8: Comparison between the simulation and the experiment in the body area.

The part was formed to check the good reliability of the simulation. This part is shown on figure 7. Two failures of the material were observed on the part. One was located on the cross at the basis of the rigth wing and the other one was on the extreme area of the left wing where the equivalent strain rate was 1070%.
The part was cut in the body area to check the simulated thickness distribution. The results are presented on figure 8. The agreement with the experiment is excellent. The error of the simulation is about 3%. This is a remakable for an industrial application.
In order to check the reproducibility of the experiment an other part was produced using the same material and the same forming cycle. The results were confirmed.
An other simulation of the part was performed with an other superplastic aluminium alloy: 7475. Due to different rheological data, the pressure curve and the thickness distribution diagram were different. An other forming trial was done and once again a very good corelation was obtained between simulation and experiment.
The following step would be to optimize the thickness distribution of the part using a sheet with different tickness areas. SUPFORM3 is able to deal with this kind of problem.

Conclusion

Now SUPFORM3 is able to solve a great number of industrial problems. The examples previously presented have brievly proven some of the capabilities of the code. The results are promising but in order to increase the efficiency of the code, further developments should focused on automatic remeshing.
SUPFORM3 is implemented on different kind of computer and specially on the new powerfull workstation where a realistic simulation (more than 10 000 triangular elements) can be performed for less than one night of CPU time.

Appendix

Flow rule

Fine grain size and high temperature are the two basic requirements for titanium alloys superplasticity. The main characteristic of the superplastic behavoiur is the high sensitivity of the flow stress to the rate of strain, which ensures the deformation stability. This phenomenon occurs in a certain range of low strain rates; the flow stress has then low rule (deduced from tension tests) is of the following usual form:

$$\sigma_{eq} = k\, \dot{\bar{\varepsilon}}^{m} \qquad (1)$$

with

$$k = k\,(\bar{\varepsilon}) \qquad (2)$$

$$m = m\,(\bar{\varepsilon}) \qquad (3)$$

where σ_{eq} is the equivalent Von Mises stress, $\dot{\bar{\varepsilon}}$ is the equivalent strain rate, $\bar{\varepsilon}$ is the equivalent strain and m the strain rate sensitivity index. In the present models, the sheet material is assumed to be purely viscoplastic with an associated flow rule of type (1). Relations (2) and (3) can be given by analytical formulae or point by point tables.

Mechanical formulation

Considering the thickness of the superplastically formed sheet, with respect to their radii of curvature, bending effects are assumed to be negligible in front of expansion effects and membrane equations can be used to describe the deformation and the equilibrium of the deformed sheet. The deformed sheet is then assumed to be a geometric surface and material points are identified by two convected material coordinates (θ^1 and θ^2) which are initialized as the cartesian coordinates in the undeformed plane state. For each material point, the position vector **X** and the associated local thickness depend on θ^i. A covariant tangential vector basis ($\mathbf{g}_1,\mathbf{g}_2,\mathbf{g}_3$) is then defined : $\mathbf{g}_1$ and $\mathbf{g}_2$ are tangent to θ^1 and θ^2 lines, and $\mathbf{g}_3$ is the local unit normal vector.

$$\mathbf{g}_1 = \frac{\partial \mathbf{X}}{\partial \theta^1} \qquad \mathbf{g}_2 = \frac{\partial \mathbf{X}}{\partial \theta^2}$$

The principle of virtual work is used to write the equilibrium of the deformed sheet : for any arbitrary velocity field **v***, we have :

$$\int_{\Omega} \sigma^{ij}\, \varepsilon^{*}_{ij}\, h\, ds - \int_{\Omega_p} (-p)\, \mathbf{g}_3.\mathbf{v}^{*}\, ds - \int_{\Omega_c} \mathbf{t}.\mathbf{v}^{*}\, ds = 0 \qquad (4)$$

where σ^{ij} are the contravariant components of the stress tensor (plane stress : i,j =1,2), ε^{*}_{ij} are the covariant components of the virtual strain rate tensor connected with **v***, **t** are the friction stresses applied to the sheet regions Ω_c contacting the mold, and p is the fluid pressure applied to the domain Ω_p of the deformed surface Ω.

Numerical resolution

Time discretization

The model is of incremental type : at each time increment [t,t+Δt], starting with a balanced deformed configuration Ω at time t, the problem lies in the calculation of the unknown equilibrated state Ω' at t+Δt. By application of (4) at t+Δt, and denoting "prime" the mechanical variables at t+Δt, we should have, for any virtual velocity field **v*** :

$$\int_{\Omega'} \sigma'^{ij}\, \varepsilon^{*}_{ij}\, h' ds' + \int_{\Omega'_p} p' \mathbf{g}'_3.\mathbf{v}^{*}\, ds' - \int_{\Omega'_c} \mathbf{t}'.\mathbf{v}^{*}\, ds' = 0 \qquad (5)$$

This equation may be solved in terms of incremental displacements **U** between Ω and Ω', during Δt, provided that we can calculate **σ',t'** and h' as functions of **U**. Those relations are exposed hereunder and the solution procedure further.

Incremental flow rule

The incremental strain tensor $\Delta\varepsilon$ is function of the covariant derivatives of the displacement field **U**. The relation $\sigma' = \sigma'(\sigma, \Delta\varepsilon)$ is then based on a semi-implicit time integration of the strain rate, during Δt [13].

Contact

The die surface is discretized in triangular elements (3D model) or in linear elements (2D model), so that it is possible to evaluate sheet points relative positions easily, at each time increment. On the surface, different regions with different contact conditions can be defined. In the contacting regions, contact can be sticking or, if not the tangential friction stress **t** is given by a Coulomb model. p denoting the applied forming pressure, μ the friction coefficient, we have :

$$\mathbf{t} = -\mu\, p\, (1 / \|\mathbf{U}\|)\, \mathbf{U} \qquad (6)$$

Thickness calculation

The new local thickness h' is deduced from the equation of volume conservation. The material is here considered as incompressible: the elastic contribution to the thickness variation is neglected. Taking into account that the elementary surface ds is given by $ds^2 = \det(\mathbf{g})\, d\theta^1\, d\theta^2$, where **g** is the metric tensor, this yields :

$$h' = h \sqrt{\det(\mathbf{g}) / \det(\mathbf{g}')} \qquad (7)$$

<u>Spatial discretization by the finite element method</u>

The membrane surface Ω is meshed with two-dimensional finite elements (linear triangles or linear quadrangles, 3D model) or one-dimensional finite elements (2D model). The finite element implementation is given in details in [8] and [12]. The resolution of the discretized equilibrium equation (5) at every time step leads to the resolution of a non-linear system, the unknown variable of which is the vector **U** of the nodal displacement components :

$$\mathbf{R}(\mathbf{U}) = \mathbf{0} \qquad (8)$$

This set of equations is solved iteratively by the Newton-Raphson method : the tangent stiffness used is the symmetric part of the exact tangent stiffness $\partial\mathbf{R}/\partial\mathbf{U}$.

Computation of the optimum pressure curve

To avoid necking and to optimize the forming time are the two basic ideas for such a computation. This is illustrated by the following figure.

An automatic pressure computation has been worked out to regulate the maximum strain rate in the sheet around the upper bound of the superplastic domain: most of the sheet is then deformed in superplastic conditions and the forming time is minimum.

At the beginnning of each time increment, the pressure correction is computed in function of the previous differences between the aimed and the maximum value of the strain rate.

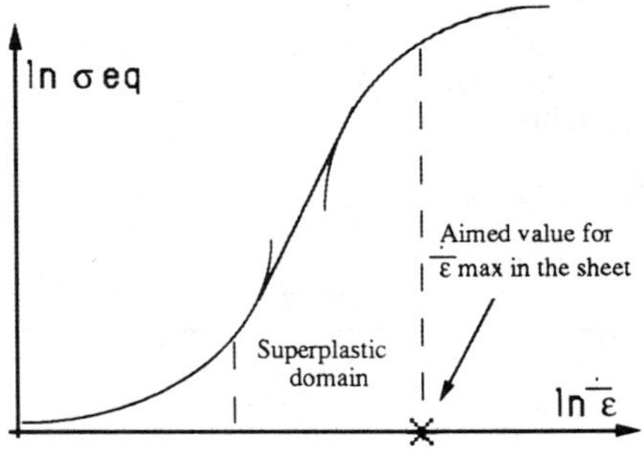

Figure 1: Typical behaviour law of a superplastic alloy and basic idea of the automatic pressure computation

References

1. D. Stephen,"Designing for Superplastic Alloys", NATO-AGARD Lecture Series n°154 Superplasticity, (1987),7-1
2. A.K. Ghosh and C.H. Hamilton,"Mechanical Behavior and Hardening Characteristics of a Superplastic Ti-6Al-4V alloys", Met. Trans., 10A, (1979), 699
3. P.G. Partridge, "Diffusion Bonding of Metals", NATO-AGARD Lecture Series n°154 Superplasticity, (1987),5-1
4. G.C. Cornfield and R.H. Johnson, "The Forming of Superplasticity Sheet Metal", Int. J. Mech. Sci., 12,(1970), 479
5. A.K. Ghosh and C.H. Hamilton,"Superplastic Forming of a Long Rectangular Box Section", ASM Conf. on Process Modelling, (Thousand Oaks, CA, 1980),303
6. S. Tang,"Analysis of Superplastic Forming of Titanium Sheet into Shallow Hat Section", AIAA/ASME 19th Structures, Structural Dynamics of Materials Conf.,(1980),256
7. M.Bellet and J.L. Chenot,"Numerical Modelling of 3D Superplastic Forming of Titanium Sheet",Proc. Int. Conf. on Titanium Products and Applications, (San Fransico, CA ,2,1986),1175
8. M.Bellet ,E. Massoni and J.L. Chenot,"Numerical Simulation of Thin Sheet Forming Processes by the Finite Element Method", Eng. Comp., (vol7,1990),21
9. J.S. Doltinis, J. Luginsland and S. Nölting,"Some Developments in the Numerical Simulation of Metal Forming Processes",Eng. Comp.,(4 ,1987),266
10. J.Bonet and R.D. Wood,"Solutions Procedures for Finite Element Analysis of Superplastic Forming of Thin Sheet",(ibid) 927
11. K.N. Shah and J.M. Story,"SPF Model Applications for Aluminium",J. of Mat. Shap. Tech., (vol7, n°3,1989)
12. M.Bellet,"Modélisation numérique du Formage Superplastique de Tôles", PhD dissertation (in french), Ecole des Mines de Paris, (1988)
13. M. Bellet, E. Massoni and J.L.Chenot,"A Viscoplastic Formulation for 3D Analysis of Thin Sheet Metal Forming",Proc. Int. Conf. on Computational Plasticity, (Barcelona,SP, Pineridge Press, 1987), 917

SUPERPLASTICITY AND POST-FORMING ROOM TEMPERATURE TENSILE PROPERTIES OF Ti-2.5% Cu ALLOY

M.A. Hassan and C. Hammond

School of Materials, The University of Leeds,
Leeds LS2 9JT, UK.

Abstract

The superplastic deformation properties of the alloy Ti-2.5% Cu have been assessed in the temperature range 775-890°C and for three different microstructural conditions, i.e. α + Ti_2Cu, α+ß and ß. The influence of uniaxial superplastic deformation followed by an ageing treatment process on the room temperature tensile properties has also been investigated. The results show that the alloy deformed superplastically when tested in the α+ß phase field in which the superplastically deformed structure consisted of α grains surrounded by ß grain boundary network, but it showed no superplasticity when tested in either α + Ti_2Cu or ß phase fields. The strain hardening which was observed in the α+ß phase field (the superplastic field) is associated with grain growth and/or the redistribution of the phases during the deformation process.

The results of the room temperature tensile properties also showed that significant tensile strength reductions occur after superplastic deformation as compared to the annealed condition after the same ageing treatments. These reductions were due to factors such as grain growth and to changes in the distribution of the eutectoid phase.

Introduction

The alloy Ti-2.5% Cu is a formable sheet alloy with better high temperature characteristics than commercially pure titanium [1]. It is also the only commercial titanium alloy which is heat treated to exploit the precipitation hardening effects of the Ti_2Cu eutectoid phase, the eutectoid reaction in the Ti-Cu system being considerably less sluggish than in other eutectoid titanium alloys.

This alloy provides the opportunity of (a) investigating its (superplastic) deformation properties in temperature ranges corresponding to the α + Ti_2Cu, α+ß and ß phase fields, (b) the effects on the deformation behaviour in changing the deformation temperature from one phase field to another and (c) the effects of the (superplastic) deformation and the elevated temperature exposure on the subsequent ageing behaviour and room temperature tensile properties in relation to the annealed material.

Titanium '92
Science and Technology
Edited by F.H. Froes and I. Caplan
The Minerals, Metals & Materials Society, 1993

Alloy and Experimental Procedure

The alloy Ti-2.5% Cu was supplied as rolled sheet 1.5 mm thick manufactured by IMI Titanium Ltd., Birmingham, UK, from which tensile test specimens with a gauge length of 20 mm and width of 5.0 mm were machined. The high temperature tensile tests were carried out on an Instron tensile testing machine on which was mounted a three-zone tube furnace accurate to $\pm$ 2°C. Oxidation of the specimens was kept to a minimum by diffusing high purity argon up through the furnace.

Prior to testing, the specimens were solution treated in the furnace at the testing temperatures for 1 hr. prior to the commencement of straining. The testing procedure involved prestraining the specimens to an initial strain (15-20%) at a cross-head speed corresponding to a high strain rate of about 2 x 10^{-3} s^{-1}. During this stage a transient work-hardening peak was observed followed by a constant flow stress. Prestraining therefore is important in that it establishes a microstructural condition in which transient work hardening is not observed and which would otherwise confuse the subsequent stress-strain rate - strain data. The cross-head speed was reduced to give a low strain rate (about 4.0 x 10^{-5} s^{-1}) and the load again allowed to stabilise. The cross-head speed was then increased in stages to the highest strain rates (about 4.0 x 10^{-3} s^{-1}); the load being allowed to stabilise between each change. Then the cross-head speed was reduced in stages to give the lowest strain rates. The sequence of this cycling was repeated at increasing strain levels, and enables an assessment to be made of the effects of strain and time as well as strain rate on the flow stresses. The flow stresses and the corresponding strain rates were evaluated assuming a uniform elongation and a constant volume. A curve fitting computer program was used to draw a smooth curve through the data points and the strain rate sensitivity (m) values were obtained from the slopes between adjacent points [2].

The room temperature properties for both the superplastically deformed and annealed specimens in the temperature range 775-960°C were evaluated after being subjected to a duplex ageing treatment consisting of 24 hrs at 400°C plus 8 hrs at 475°C [1]. This ageing treatment is recommended for Ti-2.5% Cu since it gives a consistent and reliable ageing response and reduces the total ageing time to an acceptable duration. Following the ageing treatment, the deformed and the annealed specimens were tested in tension at room temperature on an Instron tensile testing machine employing a strain rate of 4 x 10^{-5} s^{-1} [3].

Microscopical observations were carried out on the superplastically deformed and annealed specimens (gauge length and the grip sections of the test pieces) in order to analyse the microstructures developed and to help establish the mechanisms operating during the deformation process.

Results

The superplastic deformation properties of the alloy were evaluated as described above in the temperature range 775-890°C, and the results of these tests are summarised in Table 1, which indicates that the strain rate sensitivity (m) values are greater than 0.5 when the alloy is tested in the temperature range 810-850°C (i.e. in the $\alpha+\beta$ phase field), whereas the m values are less than 0.4 when the alloy is tested at 775°C (α + Ti_2Cu phase field) and 890°C (β phase field).

Table I. Maximum m-values observed in the temperature range 775-890°C for the alloy Ti-2.5% Cu.

Phase Field	Temp. °C	Max. m-value	Strain-rate range ϵ / s^{-1}	Elong. %
α + Ti_2Cu	775	0.31	1.1 x 10^{-5} - 7.1 x 10^{-4}	-
α + ß	810	0.53	3.6 x 10^{-5} - 3.2 x 10^{-3}	140
α + ß	830	0.50	2.6 x 10^{-5} - 2.9 x 10^{-3}	180
α + ß	838	0.59	2.7 x 10^{-5} - 2.9 x 10^{-3}	176
α + ß	850	0.71	3.1 x 10^{-5} - 2.8 x 10^{-3}	175
ß	890	0.28	2.5 x 10^{-5} - 2.7 x 10^{-3}	-

Figure 1 shows the stress-strain rate and strain rate sensitivity data (up to total strains of about 140%) for the test carried out at 810°C. This shows that the flow stresses increase with strain, but the effect is most marked at low levels of strain (compare the data for cycles 1 and 2 with those for 4 and 5). This indicates an approach to an equilibrium microstructural state in which the flow stresses are strain independent.

The microstructures of the undeformed (annealed) and superplastically deformed alloy at 810°C are compared in Figs. 2(a) and 2(b). These clearly show enhanced α grain growth in the superplastically deformed alloy and a redistribution of the ß phase around the α grains to become, in effect, a continuous matrix phase. In addition to constant temperature tests (Table 1), strain rate cycling tests were also carried out during which the temperature was changed so as to straddle the (α + Ti_2Cu) and (α + ß) phase fields. Fig. 3 shows the results on increasing the deformation temperature from 790°C (cycles 1, 2 and 3) to 805°C (cycles 4 and 5). It clearly shows the decrease in flow stresses and increase in m values in the small temperature between the (α + ti_2Cu) and (α + ß) phase fields.

Fig. 4 shows the changes in microstructure associated with the data shown in Fig. 3. Fig. 4(a) shows the (α + Ti_2Cu) microstructure of the alloy annealed at 790°C; Fig. 4(b) shows that on raising the annealing temperature from 790°C to 805°C, the ß phase nucleates from the Ti_2Cu phase and grows preferentially at the α grain boundaries. Fig. 4(c) shows the microstructure of the alloy superplastically deformed at this temperature. The intragranular ß phase particles have dissolved to be replaced by further grain boundary ß phase, which provides the optimum microstructural condition for superplastic deformation in (α + ß) titanium alloys.

The room temperature tensile properties for the alloy superplastically deformed in the α+ß phase field, quenched and duplex aged, are compared with those of the alloy annealed for the corresponding time and temperature in the α+ß phase field, quenched and duplex aged are summarised in Table 2. The results clearly indicate reductions in strength and ductility for the superplastically deformed material.

The results of the transmission electron microscopy of the deformed and annealed specimens in the α + ß phase field are shown in Figure 5. Figure 5(a) shows the formation of a coarse lamellar eutectoid product on the grain boundary for the specimen deformed at 830°C. On the other hand a fine lamellar eutectoid has formed on the grain boundary of the specimen annealed at 830°C (Fig. 5(b)). The distribution of the Ti_2Cu ageing precipitate is also different.

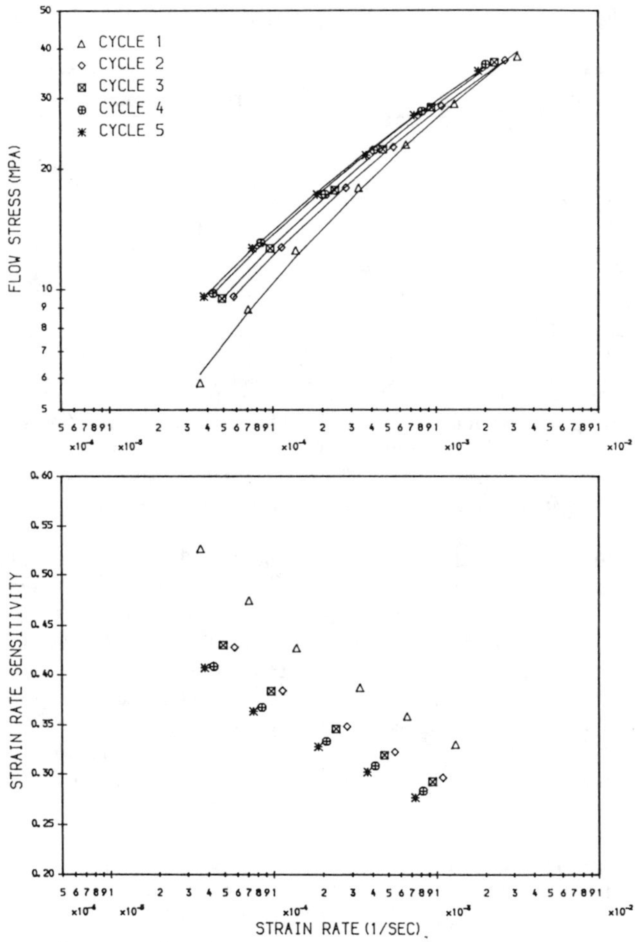

Fig. 1. Stress-strain rate and strain rate sensitivity (m-value) data for the alloy deformed to increasing levels of strain (cycles 1-5, to 140% strain) in the $\alpha+\beta$ phase field (810°C).

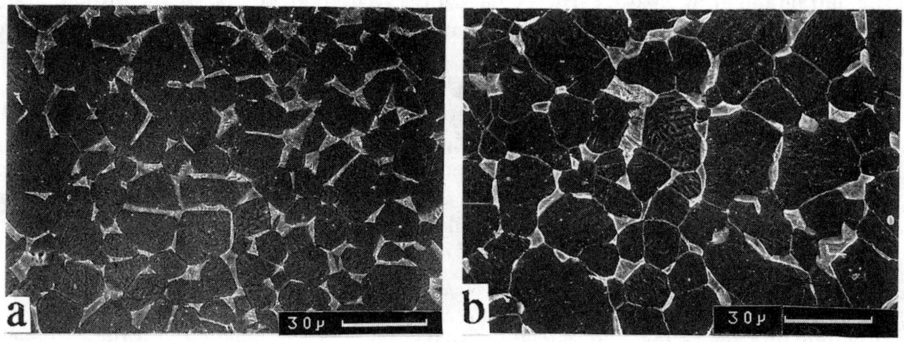

Fig. 2. Microstructures of the alloy (a) undeformed (annealed) and (b) superplastically deformed at 810°C to a total strain of 140% showing grain coarsening of the α phase and a redistribution of the β phase around the α grain boundaries.

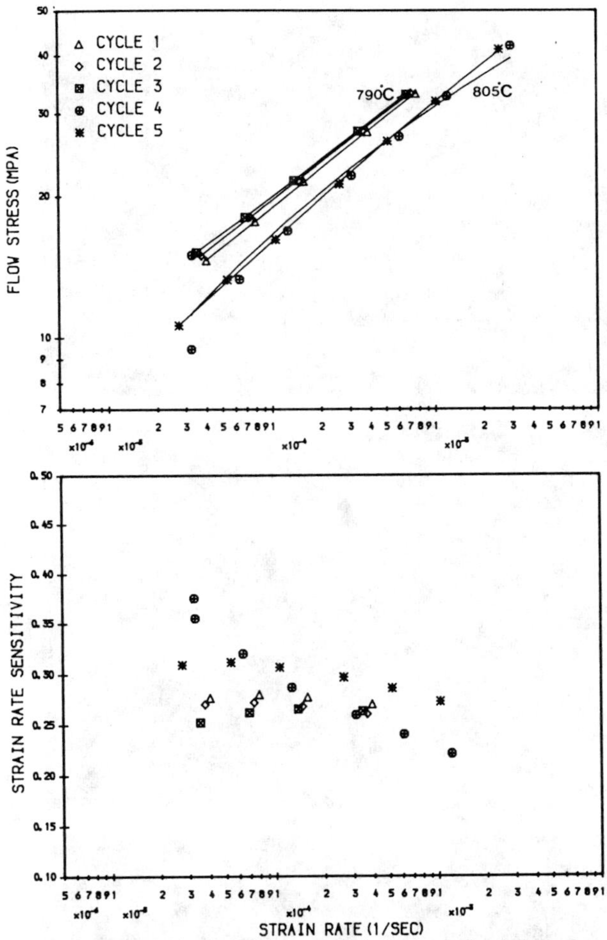

Fig. 3. Stress-strain rate and strain rate sensitivity (m-value) data for the alloy deformed at 790°C (α + Ti_2Cu phase field, cycles 1, 2 and 3) and 805°C ($\alpha+\beta$ phase field, cycles 4 and 5). The flow stresses decrease, and m values increase in passing from the (α + Ti_2Cu) to the ($\alpha+\beta$) phase fields.

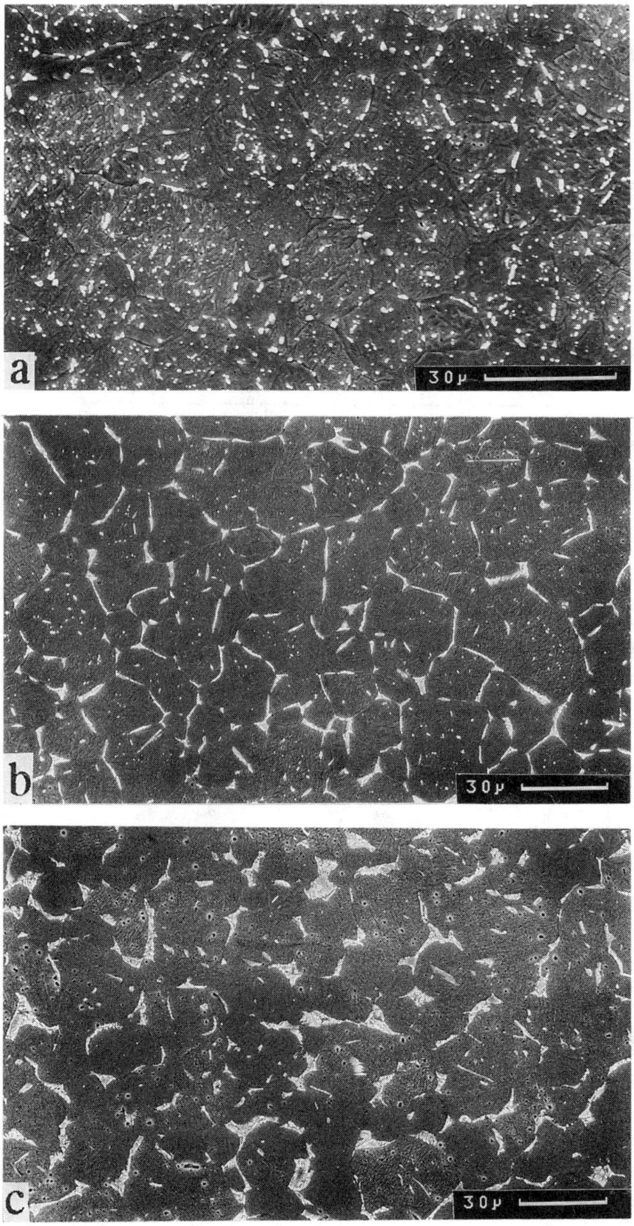

Fig. 4. (a) (α + Ti_2Cu) microstructure of the alloy annealed at 790°C, (b) the (α+β) microstructure of the alloy annealed at 790°C and 805°C and (c) the (α+β) microstructure of the alloy superplastically deformed at 790°C and 805°C. The sequence of micrographs shows the replacement of the intragranular Ti_2Cu phase (a) by intragranular and intergranular (grain boundary) β phase, (b) and (in the superplastically deformed alloy) almost complete redistribution of the β phase to the grain boundaries (c).

Table 2. Comparison of the room temperature tensile properties of the alloy superplastically deformed or annealed in the temperature range 775°C-960°C and duplex aged at 400°C for 24 hr + 475°C for 8 hr.

Alloy condition	Temp. / °C	Tensile strength / MPa	Yield strength / MPa	Reduction in Area / %	Elong. / %	α grain size / μm
Superplastically deformed and duplex aged.	775	450	633	12.9	14.8	-
	810	615	683	3.5	3.6	15.0
	830	506	586	5.2	5.4	17.0
	850	620	659	3.1	3.2	-
	960	646	650	3.5	3.6	-
Annealed condition.	775	469	699	19.6	24.5	-
	810	673	965	19.9	24.9	14.0
	830	-	803	17.0	20.4	15.0
	850	630	877	15.5	18.4	-
	960	750	882	11.0	12.3	-

There is a precipitate free zone near the grain boundary in the case of the deformed specimen (Fig. 5(a)) which was not seen to occur with the annealed specimen (Fig. 5(b)).

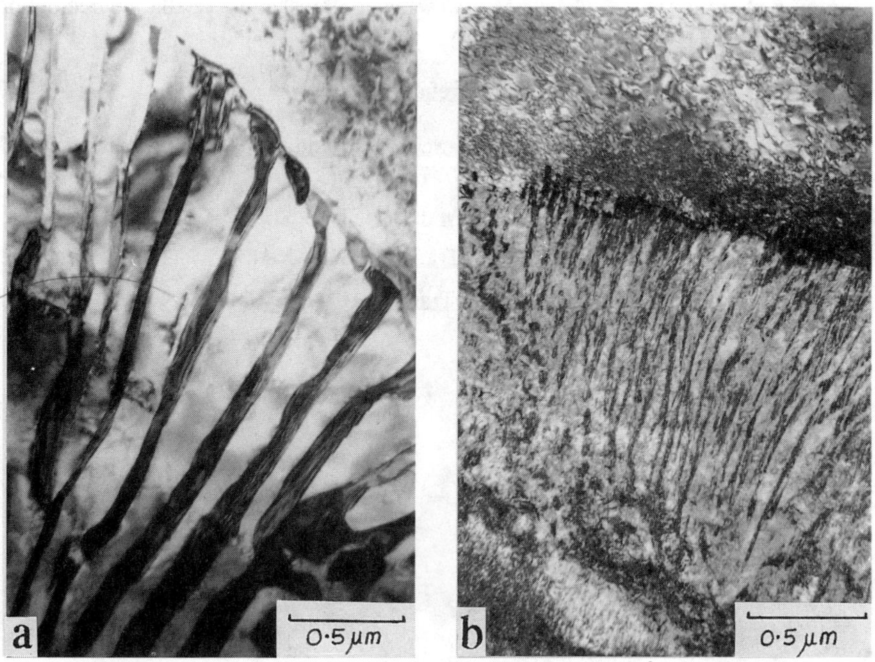

Fig. 5. TEM micrographs of the alloy (a) deformed at 830°C, quenched and aged, showing transformation of the intergranular ß phase to a coarse lamellar eutectoid product; (b) annealed at 830°C, quenched and aged, showing transformation of ß to a fine lamellar eutectoid product.

Discussion

The results show that the alloy Ti-2.5% Cu exhibits superplasticity in the α + ß phase field, and that the increase in flow stresses with the strain (Fig. 1) is due to the α grain growth which stabilises as ß becomes the continuous grain boundary phase (Fig. 2). The poor superplastic response of the alloy in the α + Ti_2Cu and ß phase fields may be due to the replacement of the ß phase by the undeformable eutectoid phase when it is tested in the α + Ti_2Cu field, and to the ß grain growth when it is tested in the ß field.

The decrease in the flow stresses and the increase in m-values upon changing the test temperature from the α + Ti_2Cu field (i.e. 790°C) to the α + ß field (i.e. 805°C) in Fig. 3 is associated with the changes in the microstructure. When the temperature is changed from 790°C to 805°C, the Ti_2Cu particles dissolve across the eutectoid line into the ß phase, which is situated at the α grain boundaries, and this results in optimum microstructural condition for superplasticity (Fig. 4(c)). The reduction in the room temperature properties in the superplastically deformed specimens as compared with the annealed one may be due to the increase in the α grain size (Table 2). Also the formation of a coarse lamellar eutectoid product in the deformed specimens (Fig. 5(a)) as compared with a fine lamellar product (Fig. 5(b)) in the annealed one might explain the results of the reduction in the room temperature properties of the deformed specimens. The formation of the precipitate free zone (PFZ) in the deformed specimens at the α/ß boundaries (Fig. 5(b)) may lead to grain boundary separation [4]. The lower ductility obtained in the deformed specimens is explained by the fact that on cooling the continuous ß phase in the α + ß field becomes a continuous grain boundary eutectoid of α + Ti_2Cu and this results in lower ductility.

References

1. P.A. Blenkinsop and R.E. Goosey, The Sci. Tech. and App. of Ti, ed. Jaffee and Promisel (The Metall. Soc. of AIME), 1970), 783.

2. R. Kent, Ma Jmin and C. Hammond, J. Mat. Sci. 21 (1986) 475.

3. G. Lutjering and S. Weissmann, Metall. Trans. 1 (1970) 1641.

4. A.K. Ghosh and C.H. Hamilton, Metall. Trans. A, 13A (1982) 733.

Superplastic Behavior of Ti-6Al-4V and ß-Enhanced Iron-Modified Ti-6Al-4V

M.L.Meier and A.K.Mukherjee[1]

Universität Erlangen-Nürnberg
Institut für Werkstoffwissenschaften, Lehrstuhl I
Martensstaße 5, 8520 Erlangen, Germany

[1]Department of Mechanical, Aeronautical and Mechanical Engineering, University of California at Davis
Davis, CA 95616, USA

Abstract

The superplastic behavior of Ti-6Al-4V and iron-modified Ti-6Al-4V has been characterized in terms of: 1. a weighted average diffusivity for the ß phase which emphasizes the faster diffusing species and takes into account alloy element partitioning, 2. a d^2 dependence on the α grain size and 3. a dependence on the volume fraction of the α phase which can be expressed as $(1-f_\alpha)^q$. This behavior is in general agreement with Chen's model except that q is approximately a factor of two low. This is probably due to grain boundary sliding which is in addition to the dislocation creep of the softer phase which Chen's model assumes. Also, deformation which is dominated by either the α phase ($q<0$) or the ß phase ($q>0$) has been observed.

Titanium '92
Science and Technology
Edited by F.H. Froes and I. Caplan
The Minerals, Metals & Materials Society, 1993

Introduction

The industrial and commercial importance of Ti-6Al-4V is well known. Is well established in aerospace, marine, chemical and other industries and accounts for approximately half of the titanium produced. There is also an ever increasing interest in the superplastic properties of titanium alloys due to the savings in weight and cost and the potential for improved designs. In this respect Ti-6Al-4V is in effect a standard to which other superplastic titanium alloys are compared and efforts to exceed this standard include developing completely new alloys, finding new ways to process the existing alloy, or to make minor changes in the composition of Ti-6Al-4V. Regarding altering the composition of Ti-6Al-4V, two strategies have been tried: adding a fine dispersion of inert oxides which would stabilize grain growth and modifying specifically the properties of the ß phase to lower forming temperature and increase the forming rate. The latter involves additions of 1-3 percent Ni, Co, Fe and/or other high-diffusivity ß-stabilizers.

Aside from the potential for improved superplastic properties, these ß-modified alloys offer a unique opportunity to more fully understand the superplastic behavior of the base Ti-6Al-4V alloy, for although Ti-6Al-4V is a commercial success, its relatively complicated and variable (at superplastic temperatures) microstructure have made it difficult to determine the fundamental rate controlling mechanism(s) responsible for its renowned superplastic properties. Other than the tendency for concurrent grain growth, the basic problem seems to be in the fact that it is a two phase alloy in which the properties of each phase are supposedly quite different and where the phase proportion varies significantly, particularly in the superplastic temperature regime. A series of ß-modified Ti-6Al-4V alloys can provide different phase proportions at any given temperature. Also, since the ß-modifying elements, which are eutectoid formers, generally have very low solubilities in the ß phase the composition (and properties) of the α phase will be relatively fixed in all cases. As a result it should be easier to characterize the base alloy and a series of ß-modified alloys than the base alloy alone. This is because it allows one to differentiate between the temperature dependence of phase proportion and composition and the temperature dependence of the kinetics of the deformation mechanism.

In this paper we describe the characterization of the superplasticity of Ti-6Al-4V and iron-modified Ti-6Al-4V in terms of the Mukherjee, Bird, Dorn [1] equation:

$$\frac{\dot{\epsilon}kT}{DGb} = A\left(\frac{\sigma}{G}\right)^n\left(\frac{b}{d}\right)^p \qquad (1)$$

(n is the stress exponent, p is the grain size sensitivity, d is the grain size, b is the magnitude of the Burger's vector, G is the shear modulus and D is the diffusivity). The dependence of the stress exponent n on the volume fraction of the ß phase f_β is investigated and an estimate of D is made using published data on the tracer diffusivities, equilibrium phase diagrams and on our experimental measurements of f_β. The remaining dependence of the deformation behavior on temperature is then analyzed in terms of the mechanics of deformation of two-phase materials. The grain size dependence is not investigated.

Procedure

Material

Fine grained 1.6 mm sheet Ti-6Al-4V was supplied by Northrop Aerospace (Los Angeles, CA). The microstructure of the as received material was elongated but after annealing at 925°C for 2 hours, the microstructure consisted of approximately 50% ß phase and equiaxed α grains. Five Ti-6Al-4V alloys containing controlled amounts of iron were supplied by Timet (Henderson, Nevada). The concentration of iron ranged from 0.115 to 3.04 weight percent. The microstructure of the five alloys consisted of different amounts of the ß phase but the α grain size ranged only from 3 to 6 μm. The dependence of f_β on temperature and concentration of iron was determined and have been reported elsewhere [2].

Equipment

All testing was conducted using an automated MTS system which had been outfitted with a radiant heating furnace capable of rapid temperature changes and a retort which allowed testing to be done in an atmosphere of purified argon. The testing software and servo-hydraulic system was capable of providing up to 20 true strain rates in a single test with total strains as high as 2.0. Specimens used for testing had dimensions of 25.4 mm in length and 6.3 mm wide and 1.6 mm thick in the gage section.

Stepped-Strain Rate Tests

A series of stepped-strain rate tests was conducted to determine the value of the stress exponent for each alloy. Each test consisted of an initial 20% true strain at a strain rate of 2×10^{-4} s^{-1} after which the strain rates were changed incrementally, at intervals of 0.05 true strain. After reaching the maximum strain rate of 1×10^{-3} s^{-1} the strain rate was decremented until the original strain rate was obtained. The total strain on any specimen was only 0.54. This range of strain rates was chosen because it was expected, based on our previous results for Ti-6Al-4V, that region II behavior would be observed for all alloys. Three temperatures were investigated: 775°C, 825°C and 875°C. It was in this temperature range that, for all alloys, f_β was relatively insensitive to temperature but was different for each alloy.

Stepped-Temperature Tests

A series of stepped-temperature tests was conducted at a strain rate of 2×10^{-4} s^{-1} and temperatures between 695 and 879°C. Each test consisted of an initial 20% true strain at 810°C followed by an abrupt increase in temperature of 10-15°C every 0.03 strain. After reaching a temperature of 879°C (or slightly less, depending on the alloy) the temperature was decremented by 10-15°C until the temperature reached 695°C or, judging from the stress-strain curve, the specimen appeared to have started to neck.

Results and Discussion

Strain Rate Dependence

For all six alloys the value of n was consistently close to 2. Note that while the volume fraction of phases is very different for each alloy the value of n was essentially constant. This showed that, in the present temperature range, n is not effected by f_β.

Temperature Dependence

Figure 1 shows an Arrhenius-type plot of the results of the stepped-temperature tests. The six lines are curved and approximately parallel, shifting to higher temperatures as the concentration of iron decreases. These results, however, require two corrections (or compensations) before one can begin to estimate the activation energy. The first is the effect of the small differences in the initial grain size and the expected variation in grain size with temperature. This was done by assuming $p=2$ and that small variations of d with temperature are such that the number of grains is constant but their size varies to produce the required value in f_β:

$$d = d_r \sqrt[3]{\frac{f}{f_r}} \qquad (2)$$

where f_r and d_r are measured, reference values of the volume fraction and grain size of the same phase. In this the α phase is considered primarily because it is practically impossible to resolve the ß/ß grain boundaries. (Note that this adds an additional $f(T)^{-p/3}$ term to equation 1.)

The second correction concerns the expected variation of D as a result of alloy composition and alloy element partitioning. This was done by estimating the composition of the ß phase at different temperatures and replacing D_o in the diffusivity term D in equation 1 with the equivalent term from the average diffusivity for the ß phase:

$$D_\beta^{AV} = \sum_i^j x_i D_{o_i} \exp\left(\frac{-Q_i}{kT}\right) \qquad (3)$$

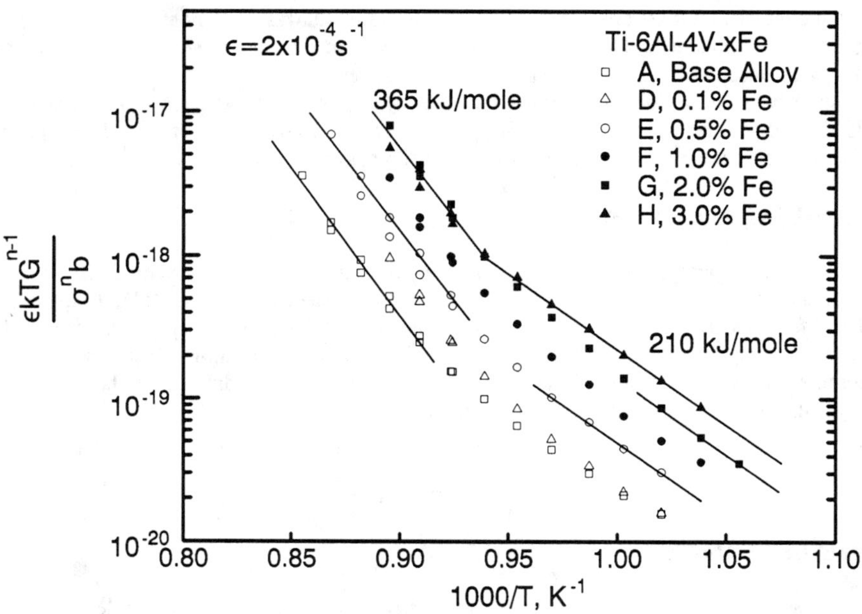

Figure 1. Results of the stepped-temperature tests plotted in an Arrhenius fashion.

where x_i is the concentration of the ith of j elements. (See reference 2 for details.) As a result of these two corrections (figure 2) the lines for all alloys at temperatures below 780°C have condensed to one line whose slope indicates that the activation energy is 210 kJ/mole, a value typical for the superplasticity of titanium alloys [3]. At higher temperatures the line for the base alloy does not deviate but those for the other alloys branch upwards with a slope suggesting the activation energy is 485 kJ/mole.

While the correlation among the different alloys is excellent, both of these activation energies are too high to associate with a conventional diffusion controlled mechanism. Note, however, that the lower value happens to be 1.6 times the activation energy expected from D_β^{AV} and that this is equal to the sum of the activation energies for lattice and grain boundary diffusion, Q_L^{AV} and Q_{gb}^{AV}, respectively. Therefore we split this into two parts, the activation energy for lattice diffusion stays with the diffusivity term in equation 1, which is in agreement with Suery and Baudelet's model [4], and the remaining Q_{gb}^{AV} represents the temperature dependence of the A-parameter. At this point this seems to be the only way to deal with such high activation energies. What we are basically proposing is that there appear to be two mechanisms operating and that they interact in such a way that the observed deformation behavior is the product of these two processes (the activation energies are additive) which is different from the series/parallel approach to accounting for the simultaneous contributions of multiple deformation processes. Following the conventional way on thinking, the most likely explanation is that these processes involve the deformation mechanism itself plus strain enhanced microstructural evolution.

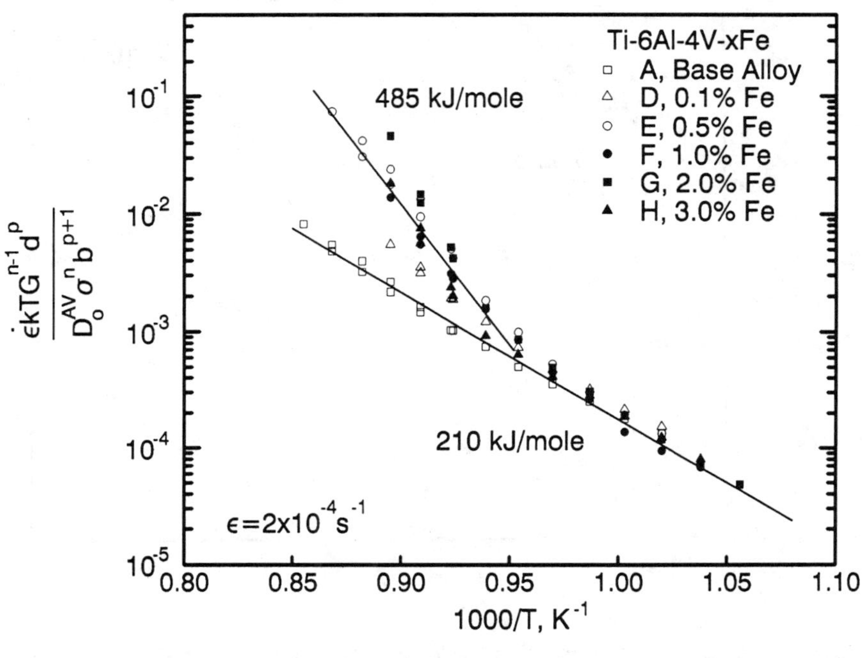

Figure 2. Results of stepped-temperature tests corrected for minor variations in grain size, differences in alloy composition and alloy element partitioning.

Volume Fraction Dependence

As stated in the previous section we find that the A-parameter has its own temperature dependence given by:

$$A = A_o \exp\left(\frac{-Q_{gb}^{AV}}{kT}\right) \quad (4)$$

Figure 3 shows that the pre-exponential term in the A-parameter, A_o, follows the $(1-f_\alpha)^q$ relation suggested by Chen [5] except in this case the values of q are low by a factor of nearly 2. Chen's model says that when $n=2$, the values of q should be -3.1 or 3 depending on which phase is considered the matrix. A value close to 4 (3.785) was found to fit Suery and Baudelet's data on α/ß brass [4] and this tolerable discrepancy was attributed to the nonspherical shape of the grains and the resulting suppression of rheological flow around sharp corners, etc. By the same token lower values should be expected for equiaxed grains especially if there is a significant contribution from grain and interphase boundary sliding. Chen's model considers flow in the matrix phase and not in the grain boundaries.

Figure 3 also shows the transition from a behavior typical of soft particle is a hard matrix ($q<0$) to hard particles in a soft matrix ($q>0$) which Chen's model also describes. (For Ti-6Al-4V-xFe α is the hard phase and ß is the soft phase.) This transition, however, is observed only for the alloys where f_β can fall below approximately 0.25, roughly the void space in a

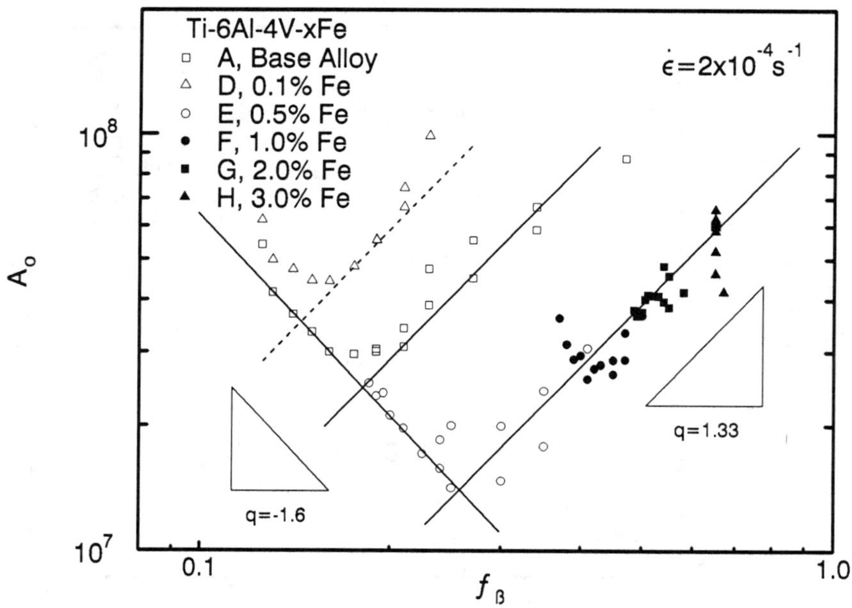

Figure 3. Pre-exponential of the A-parameter as a function of volume fraction of the ß phase.

close-packed arrangement of spheres. When this is the case, lowering the concentration of iron shifts this transition to lower f_β. This transition can be seen as the intersection of two lines, one where $q<0$ (β in a matrix of α) and the other where $q>0$ (α in a matrix of β). The first line probably represents deformation through the α grains as well as β-assisted sliding between these grains, consistent with the TEM observations of Gurewitz [6] for Ti-6Al-4V. The second line probably represents deformation which takes place entirely in the β phase (especially regions adjacent to the α grains) but modified by the presence of the harder α phase.

The pre-exponential term in the A-parameter can now be written as:

$$A_o = A_f\,(1-f_\alpha)^{q} \tag{5}$$

For the case where $q<0$ A_f has a unique value which one can think of as a "coefficient of friction" between sliding α grains, a structural constant for the α phase, or both. For the other case where $q>0$ A_f has the same meaning but in this case A_f decreases with increasing concentration of iron until 0.5% iron and then is constant. This suggests a saturation-type behavior which is in addition to the rheological effects described above. This might be the behavior described by Chen [5] where the non-steady state nature of diffusion across sliding interphase boundaries leads to the saturation of these boundaries with the fast diffusing species. If this is the case then iron has the effect of lowering the "coefficient of friction" or at least increasing its importance in A_f when $q>0$. It also suggests that this is the strain-dependent process which results in high value of the activation energy.

Conclusions

1. The superplastic behavior of Ti-6Al-4V and five iron-modified Ti-6Al-4V alloys can be described using a single function based on the Mukherjee, Bird, Dorn equation:

$$\frac{\dot{\epsilon}kT}{DGb} = A\left(\frac{\sigma}{G}\right)^{n}\left(\frac{b}{d}\right)^{p} \tag{1}$$

   where $n=2$, $p=2$, d refers to the α grain size, D is replaced by the weighted average diffusivity for the β phase which includes the effect of alloy element partitioning:

$$D_\beta^{AV} = \sum_i^j x_i D_{o_i} \exp\left(\frac{-Q_i}{kT}\right) \tag{3}$$

   where x_i is the concentration of the ith of j elements and the A-parameter which is a function of temperature and phase proportion:

$$A = A_f\,(1-f_\alpha)^{q} \exp\left(\frac{-Q_{gb}^{AV}}{kT}\right) \tag{6}$$

2. Excellent agreement between experimental and computed stresses has been obtained, figure 4.

3. The results also suggest that the high values of the activation energy (often reported for the superplasticity of titanium alloys) may be due to an interphase boundary sliding related saturation of the interphase boundaries with the faster diffusing elements, particularly iron.

Acknowledgements

This work was supported by grant number 91-NC-251 from AFOSR (Alan Rosenstein) and Lawrence Livermore National Laboratory (Don Lesuer and Ed Dalder).

References

1. A.K.Mukherjee, J.E.Bird and J.E.Dorn, Transactions of ASM, vol.62, pp.155-179, (1969).
2. M.L.Meier and A.K.Mukherjee, Scripta Metallurgica et Materialia, vol.25, pp.1471-1476, (1991).
3. N.E. Paton, C.H.Hamilton, Titanium Science and Technology, eds. G.Lütjering, U.Zwicker and W.Bunk, Deutsche Gesellschaft Für Metallkunde E.V., pp. 649-672, (1984).
4. M.Suery and B.Baudelet, Philosophical Magazine, vol.40A, pp.41-64, (1980).
5. I-W. Chen, Superplasticity, International Conference on Superplasticity, eds. B.Baudelet and M.Suery, Éditions du Centre National de la Recherche Scientific, Paris, pp.5.1-5.20, (1985).
6. G.Gurewitz, Ph.D. dissertation, University of California, Davis, CA, (1984).
7. B.Hidalgo, Ph.D. dissertation, University of California, Davis, CA, (1986).

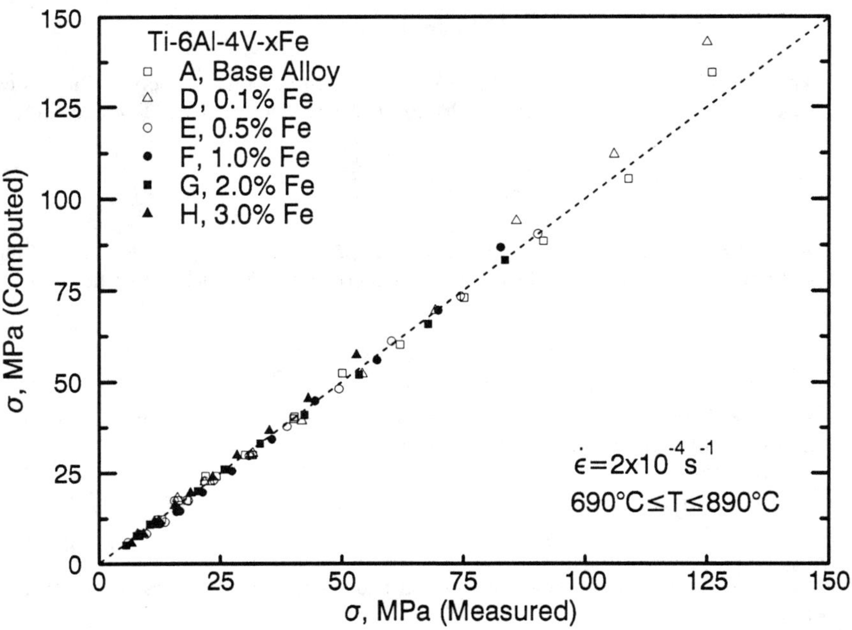

Figure 4. Comparison of stresses measured from the stepped-temperature tests and stresses computed using equations 1, 3 and 6.

SEAMLESS THIN-WALLED TiAl6V4 TUBES MANUFACTURED BY FLOW-FORMING PROCESS

M. Koch*, H. Sibum*, H. Jelenak**, E. Plänker**

* Deutsche Titan GmbH, Essen/Germany
** Metall-Spezialrohr GmbH, Schwerte/Germany

Abstract

The three-roller flow forming process was applied to the (alpha+beta)-titanium alloy TiAl6V4. This process forms thick-walled tube blanks into thin-walled, close-tolerance rotationally symmetrical workpieces (tubes, tapered reducers, etc.). As well as permitting high reductions the process offers advantages such as homogeneous stress distribution and the ability to hold close tolerances on diameter, wall thickness and straightness. Thanks to the smooth surface finish obtained subsequent machining is not necessary. The mechanical properties of manufactured TiAl6V4 tubes are presented.

Introduction

The manufacture of titanium tubes has been an established technology for many years. Processes commonly used include extrusion, rotary piercing and pilger rolling for seamless tubes and rolling and welding for welded tubes. Most tubes are made from commercially pure titanium or low-alloy titanium alloys. They are mainly used in the chemical industry in applications utilizing titanium's resistance to corrosion and oxidation.

In addition to these applications, an increasing number of uses are emerging which require extremely homogeneous characteristics over the entire tube cross section.

One of titanium's major advantages is its high specific strength, i.e. the ratio of density to strength. To exploit this a manufacturing process is required which can ensure rotationally symmetrical homogeneity, very close tolerances and very light wall thicknesses. Aerospace applications in particular require high strength tubes for structural components, calling for the use of high-alloy titanium materials too. The following report presents and discusses ways of manufacturing seamless thin-walled tubes and their properties.

Titanium '92
Science and Technology
Edited by F.H. Froes and I. Caplan
The Minerals, Metals & Materials Society, 1993

Flow forming

The three-roller flow forming process has proved particularly advantageous for special materials because of the high material costs involved. Materials that have been worked by the process include heat-treatable steels, corrosion-resistant, high-tensile, high-temperature and heat-resistant steels, special alloys based on nickel, cobalt, tungsten and molybdenum, and nonferrous metals such as aluminium and aluminium alloys, commercially pure titanium and various copper materials.

The starting material for flow-forming (1) is thick-walled tube blanks. These are formed over a rotating mandrel into thin-walled rotationally symmetrical parts (Fig. 1). The wall thickness reduction is effected by three forming rollers arranged symmetrically around a mandrel. Their axis of rotation is parallel to the mandrel axis and they are moved or fed against the blanks. The contact pressure of the rollers plastically forms the material, causing it to flow. The length of the blank increases while its inside diameter remains constant. The blank becomes a thin-walled tube.

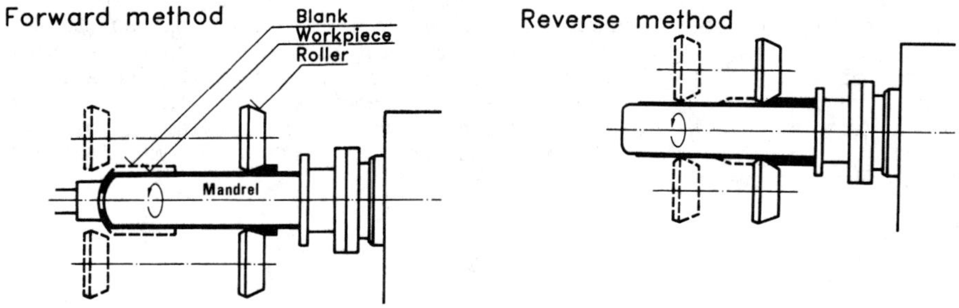

Fig. 1: Flow forming (schematic)

The amount of reduction is determined by the infeed of the rollers and can be split over several passes. Percentage reductions of up to 90 % can be achieved depending on material. Forming speed is controlled via the feed rate of the longitudinal axis of the machine.

As shown in Fig. 1 there are two methods of flow forming - forward and reverse. Reverse flow forming can be used to produce very long cylindrical parts thanks to the fact that the material flows in the opposite direction to the feed direction of the rollers.

The following parameters have an effect on the production and quality of flow-formed components:

- radial and axial offset of the rollers
- roller approach angle
- working radius and smoothing angle of the rollers
- roller feed
- mandrel speed
- cooling of rollers, mandrel and workpiece
- lubrication of rollers and mandrel
- roller to mandrel diameter ratio.

The key advantages of flow forming which led to it being chosen for the manufacture of titanium tubes are as follows:

- heavy reduction in wall thickness
- high accuracy of diameter, wall thickness and straightness
- symmetrical stress distribution
- smooth surface finish
- ready-to-instal components

Elevated-temperature flow forming

In contrast to unalloyed and low-alloy titanium materials and the beta-alloys, the (alpha+beta) alloys show limited formability at room temperature. The high ratio of yield point to tensile strength of usually over 90 % and the low amount of elongation without necking restrict plastic formability to a very narrow, technically hardly useable range. At higher temperatures both the yield point ratio and the ratio of modulus of elasticity to yield point change, allowing components of (alpha+beta) alloys to be (hot) formed (2, 3). Hot forming takes place above the recrystallization temperature, with the temperature range being limited by microstructure requirements, resistance to forming and susceptibility to cracking.

For the manufacture of ausformed very high strength steel tubes a flow forming machine was developed which makes it possible to carry out forming in a narrow elevated temperature range (4). Fig. 2 shows such a flow forming machine.

Test procedure

The forming tests covered the production of flow-formed tubes of two different sizes. The tubes of the first test series have the finish dimensions Dm. 102/88 x 300 mm, those of the second series the finish dimensions Dm. 50/46 x 750 mm. The starting materials were hot-rolled tubes (1st test series) and drilled forged bars (2nd test series) made of titanium alloy TiAl6V4 (DTG trade name TIKRUTAN LT 31). The chemical composition of the starting material is shown in Table I.

Fig. 2: Detail of a flow forming machine showing mandrel, workpiece and rollers

Table I Chemical composition

Table I: Chemical composition of starting material need for tube manufacture (max. %)

	Al	V	Fe	O	C	N
1st test series	6.4	4.1	0.14	0.16	0.02	0.01
2nd test series	6.5	4.0	0.14	0.16	0.02	0.01

The blanks were prepared by machining. The inside diameter was turned to the finish size and in the case of the hot-rolled blank the rolled surface was removed from the outside diameter. The blanks were pre-heated, pushed onto the pre-heated mandrel and formed according to a predefined progressive forming sequence (number of passes, radial and axial offset of rollers). After forming, the finished tubes were cooled in air. The effective overall percentage reduction in both test series was more than 70 %.

The oxide surface layer caused by hot forming can be removed by the pickling treatment usual for titanium materials.

Test results and discussion

The first test series was primarily intended to optimize flow forming parameters and determine some of the properties of flow-formed tubes made of titanium alloy TiAl6V4. The starting material was tubes rolled in the beta-range. The beta-(alpha+beta) mixed structure of the starting material is shown in Fig. 3.1.
The turned blanks were flow formed to final dimensions in one pass.
The structure of the finished tube in longitudinal direction is shown in Fig. 3.2. The worked structure displays pronounced grain stretching, indicating the presence of work hardening in the tubes, which is confirmed by tensile tests on longitudinal specimens.

Table II shows a comparison of the mechanical properties of the starting material and the finished tubes.

The danger that exists when hot forming titanium materials of hardening or embrittlement as a result of high hydrogen pickup can be excluded. The H_2 contents given in Table II make this clear.

Measurement of the surface roughness gave a value of R_a = 1.6 µm.

3.1: Starting material

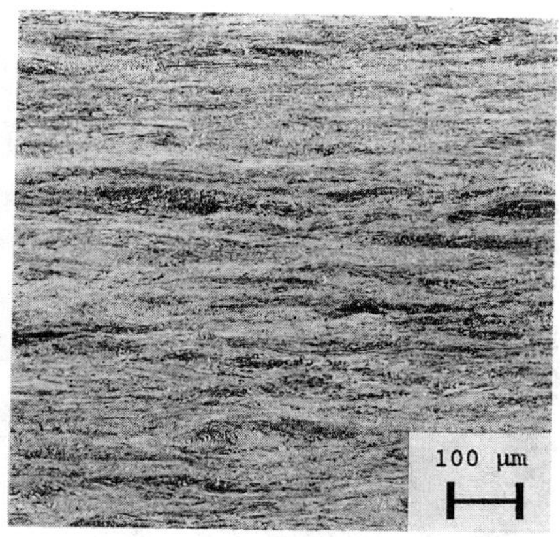

3.2: Tubes, unannealed

Fig. 3: Micrographs (longitudinal) of starting material and finished tubes - 1st test series

More extensive tests were carried out on the tubes in the second test series. The manufacture of these tubes was based on an actual aerospace application. The following dimensional and shape tolerances had to be observed:

Surface roughness:	$R_a \leq 2.0$ µm
Diameter and length:	+/- 0.1 mm
Roundness:	+/- 0.1 mm
Wall thickness:	+/- 0.05 mm
Straightness:	$\leq$ 1.0 mm/m

The second test series was carried out on forged bars which were drilled to make the blanks. The structure of the starting material is shown in Fig. 4.1. This structure is a globular (alpha+beta) standard structure conforming to standard series ETTC 2 (5). As with the first test series the blanks were formed into tubes with a wall thickness of 2 mm in one pass.
The microstructure in the longitudinal direction of the tubes is shown in Fig. 4.2. A comparison of Figs 4.1 and 4.2 shows that despite the difference in starting materials the tube microstructures of the two test series show no significant differences. In both cases pronounced stretching of the microstructure, due to the process employed, is visible. The

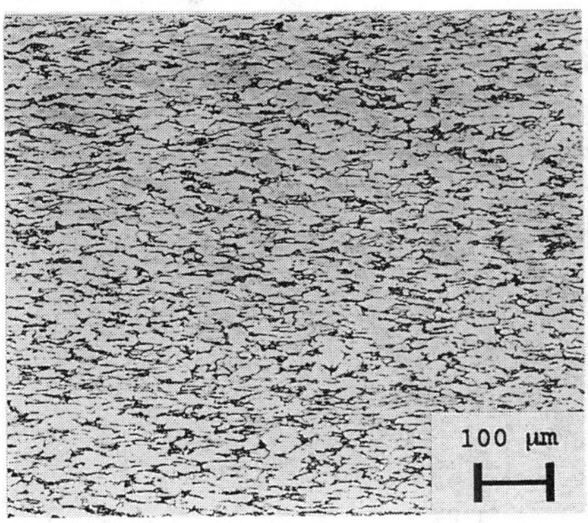

4.1: Starting material

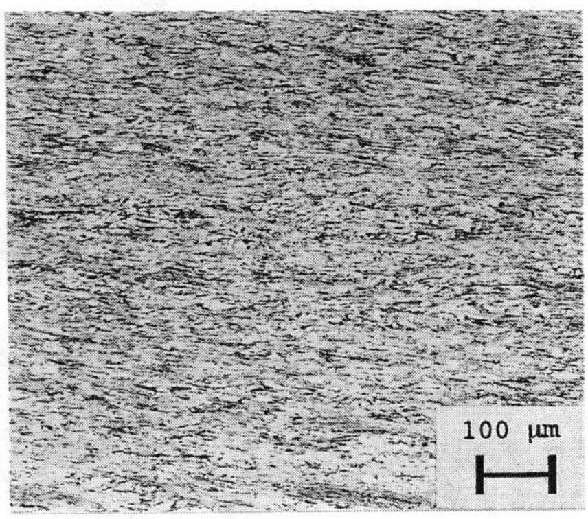

4.2: Tube, unannealed

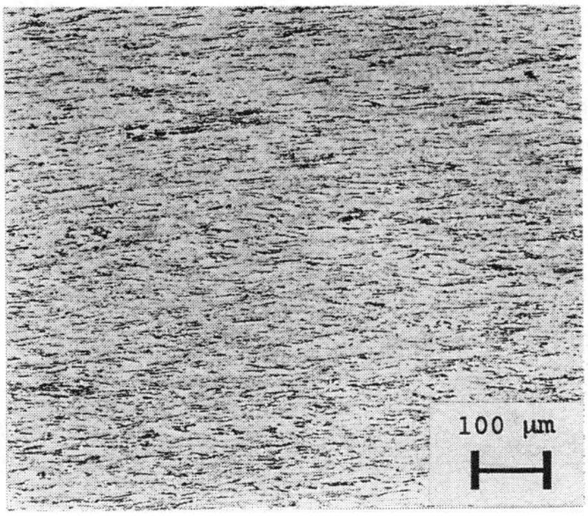

4.3: Tube, annealed

Fig. 4: Micrographs (longitudinal) of starting material and finished tube - 2nd test series

strength values of the tubes of the second test series as determined in tensile tests are also listed in Table II. Here, an increased level of work hardening can be observed. As in the first test series, an increased hydrogen content was not measured in the tubes.
To relieve the work hardening some tubes were heat treated under conditions normal for the titanium alloy TiAl6V4 (700 °C/1h/air). A comparison of the microstructures in longitudinal direction between unannealed, work-hardened tubes (Fig. 4.2) and annealed tubes (Fig. 4.3) shows no significant difference. This shows that this annealing treatment has no effect on the microstructure caused by forming.

If a comparison is made between the mechanical properties of the unannealed and the annealed tubes in the longitudinal direction (Table II) it is noticeable that whereas the tensile strength and yield point of the annealed tubes decrease as expected, the elongation at fracture by contrast remains almost constant. One explanation for this at first sight contradictory behaviour is that the heat treatment cancels out to a large extent the work hardening, i.e. the limitation on dislocation movements, and thus leads to a reduction in the strength values. In the case of elongation at fracture, which is a measure of the plastic formability of the tubes, work hardening has only a small effect compared with the microstructure. The pronounced stretching of the grains in longitudinal direction therefore seems to have a positive effect on attainable elongation at fracture.

Table II Mechanical properties

Table II: Mechanical properties and hydrogen contents of starting material and tubes

		Rp0.2 (N/mm²)	Rm (N/mm²)	A5 (%)	A50 (%)	H_2 (ppm)
1st test series	starting materials	881	949		12.5	50
	tubes	994	1206	10		52, 53
2nd test series	starting materials	927	986		12.0	40
	tubes unannealed	1050	1124	13.4		88, 91
		1064	1137	12.6		88, 91
	tubes annealed	925	970	13.2		
		910	956	13.6		

Table III Surface roughness

Table III: Surface roughness of a tube (2nd test series)

	OD		ID	
Ra (μm)	0.79	0.65	1.10	1.21
Rz (μm)	3.56	3.27	5.63	6.60
Rmax. (μm)	4.44	4.20	7.52	8.08

The structure of the unannealed tubes in transverse direction (Fig. 5), in contrast to that in longitudinal direction, exhibits a very fine globular pattern. Under radial loads this can be expected to result in a marked difference in plastic formability between unannealed and annealed material. This is confirmed by the results of tube flattening tests (DIN 510136) in which the work hardening of the tubes is very noticeable. Whereas the compressing of an unannealed tube gave a value of 6 mm for the reduction in diameter, the annealed sample gave a value of 18 mm. The unannealed tube sample broke at this change of length without incipient cracks having been macroscopically visible, whereas the annealed sample showed macroscopically visible cracks when the stated change in length was reached and the test was broken off.

The flattening tests show that the plastic formability of the unannealed tube in circumferential direction is very much lower than that of the annealed tube. The unannealled sample broke at a force of 7.5 kN, whereas the test on the annealed sample was aborted at a force level of 8.5 kN. In both cases the rate of loading was 4 mm/min.

These test results show that for many applications of flow-formed TiAl6V4 tubes in which a minimum ductility value is specified annealing of the tubes is recommendable. Tests will have to be carried out to clarify whether by changing the process parameters the tendency towards brittle fracture in peripheral direction can be reduced.

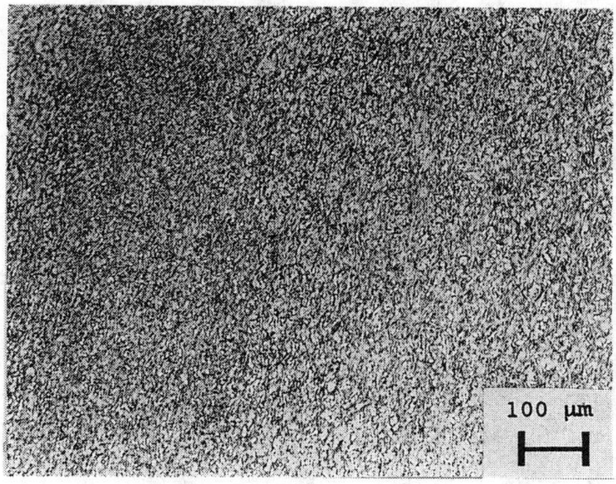

Fig. 5: Micrograph (transverse) of a finished, annealed tube
2nd test series

When final annealing is carried out it must be ensured that the reduction in residual stresses does not have a negative effect on dimensional and shape accuracy. Because of the smooth surfaces produced by flow forming annealing should be carried out in a vacuum or a controlled atmosphere.

Measurement of the surface roughness on the inside and outside diameters of a tube from the second test series gave the results listed in Table III. A surface crack inspection by the dye penetrant method showed no incipient cracks. Fig. 6 shows tubes from both test series.

Summary

The present report describes a process with which for the first time it was possible to manufacture seamless thin-walled tubes of titanium alloy TiAl6V4 meeting close dimensional and shape tolerances. Tubes were produced in the finish dimensions Dm. 92/88 x 300 mm and Dm. 50/46 x 750 mm. The more than 70 % reduction made corresponds to the percent reduction usual for other materials formed by this process.

The measured mechanical properties show that there are differences in plastic formability between the longitudinal and transverse directions of the tubes. With regard to possible applications in which specific requirements are placed on the ductility of the tubes in circumferential direction, further tests are necessary.

Possible applications of flow-formed tubes or other rotationally symmetrical components of the titanium alloy TiAl6V4 are in the aerospace, offshore and auto industries and in special areas of general engineering in which the emphasis is on a homogeneous strength distribution.

Fig. 6: Tubes from the two test series

References

1. R. Kopp, H. Gaile
   Das-Rollen-Fließdrücken
   Dornier-Post 4/72, S. 29 - 32

2. H. Bühler, H.-W. Wagner
   Umformeigenschaften von Titan und Titanlegierungen
   Bänder, Bleche, Rohre 6 (1965) Teil 1: H. 11 S. 625 - 630, 667
   Teil 2: H. 12 S. 677 - 684

3. U. Zwicker
   Titan und Titanlegierungen
   Springer Verlag 1979

4. D. Frodl, E. Plänker, K. Vetter
   Austenitformgehärtete, höchstfeste Rohre
   Stahl und Eisen 101 (1981), H. 12, S. 75 - 80

5. ETTC 2, Publication ETTC, Technical Committee of European Titanium Producers, Essen, Birmingham, Ugine, Spring 1979

Effect Of Interfacial Microvoid Geometry During Diffusion Bonding of Ti 6-4 Alloy

H.A. Nied
GE Corporate Research and Development
Schenectady, N.Y

ABSTRACT

Earlier diffusion bonding models published in the open literature were developed by assuming simplified geometries and loading conditions without microvoid interaction in order to obtain closed form solutions. A finite element formulation is described which provides a more versatile tool for studying microvoid collapse during diffusion bonding. A power law strain rate sensitive constitutive equation was used to model the time dependent deformation of Ti6-4. The FE method provided an effective method for determining the effect of different initial geometries, void interaction and non-uniform temperature distributions. The paper describes the difference in the behavior by showing the deformation at various times during the collapse. This paper focuses attention on the effect of initial void geometry.

INTRODUCTION

Modeling Background

The solid-state elevated temperature diffusion bonding process for metals is considered to consist essentially of three stages of interfacial asperity or void collapse during the joining process. These stages of diffusion bonding were identified experimentally and described by King and Owczarski in Reference (1). Stage 1 occurs during the plastic deformation of the surface asperities and waviness when external loads or pressures are exerted on the components. During stage 2, the external pressures have been found to play a minor role on interfacial microvoid collapse, since surface tension becomes the dominant driving force during this period. The final stage 3 of the bonding process consists of grain boundary diffusion and migration. Earlier published research has shown that most of the bonding time occurs during the combined time for stages 1 and 2 to produce microvoid closure.

Since most of the time for bonding is consumed during stages 1 and 2, a process model for predicting either the degree of void closure or the time for total collapse is very appealing. Process models have been reported in the literature based on using simple closed form solutions for various geometries. References (2) through (11) describe many of the models proposed for simulating the bonding mechanisms and predicting the time of microvoid collapse.

Hamilton (2) is considered to have developed the first quantitative diffusion bonding model. His model consisted of a series of long ridges to represent the contacting asperities.

Titanium '92
Science and Technology
Edited by F.H. Froes and I. Caplan
The Minerals, Metals & Materials Society, 1993

The application of temperature and pressure produced a collapse of the serrated ridges so that the interface voids were represented as the collapse of a diamond shaped cavity. Hamilton's experiments revealed that during the diffusion bonding process, titanium can dissolve minor amounts of absorbed gases and thin oxides and diffuse them away from the bonding surface. Sufficient pressure must be applied initially to overcome surface asperities since bondline porosity produces bond strengths less than parent metal.

Garmong et al. (3) developed a closed form solution which accounted for periodic surface waviness and microvoids to predict time for complete diffusion bonding. The authors refined Hamilton's ridge model and also included the diffusion mechanism. Their models included the flattening of the long-wave surface uneveness and short wave surface asperities by plastic flow and void reduction by both plastic flow and vacancy diffusion. The initial geometry of surface roughness was modeled as plane strain ridges with the final stages of bonding modeled as spherical microvoids. It was assumed that void spacing was such that their interaction did not need to be considered. In addition, Garmong's studies and experiments revealed that the pressure needed to overcome long-wavelength asperities can dominate the bonding process. Also, microvoid closure is a combination of both plastic deformation and diffusion.

In a series of papers, Derby and Wallach developed extended diffusion bond models in References (4), (5), (6) and (9), which accounted for the many diffusion bonding mechanisms. In (4) the authors presented a computer model that determines interface void collapse by considering plastic deformation, power law creep, Coble creep and Herring-Nabarro creep based on plane strain assumptions. In (5) they made extensive use of sintering theory in their model and included seven possible diffusion mechanisms that were incorporated into an analytic expression. The geometry of the interfacial voids was an array of triangular ridges. In (6) they extended their model published in 1982 by including vapour phase mass transfer. A plane strain 2D gometry was still retained. In (9) they provided a description and discussion of the many models appearing in the literature. They also discussed the future of DB modelling. In all of these models, void interaction was not considered.

Pilling et al. in (7) considered the void distribution at the bonding interface as an array of circular cavities. The creep collapse was based on the yield stress of the material rather than on the strain rate dependent flow properties. The diffusion relations used by Chen and Argon were incorporated into the model. The important consideration of grain boundary diffusion was included in the model as a series of layers parallel to the bond interface. As the bonding proceeds, the voids become smaller retaining the circular shape, and the number of active grain boundaries that intersect the void diminish. The interaction of the voids was not included in this model. In (8), the authors extended their model previously published to include grain boundary diffusion. Void interaction and the diffusion of surface oxides or contaminates were not considered. In (10) Pilling introduces a different model geometry that was developed to predict time to attain full interfacial contact during DB under isostatic stress and isothermal temperature conditions. The voids at the interface are modeled as a composite circular cylinder whose axis is perpendicular to the bond interface. Both axial pressure and transverse outer pressure are applied to the model. Grain boundary diffusion is accounted for in radial planes. Void interaction was also not considered. Pilling's analyses and tests had revealed that bonding is primarily due to superplastic flow of material into interfacial voids. In the superplastic regime, he found that temperature and pressure have little effect on kinetics of bonding. Surface roughness has a substantial effect on bonding time. Long-range surface waviness plays as significant role as short wave roughness. He determined that isostatic modeling underestimates the time required to attain full interfacial contact.

Finally, in Reference (11), Hill and Wallach provided a comprehensive summary and description of the various diffusion bonding models. In that paper the authors described their latest theoretical model for diffusion bonding which is based on the isotropic coalescence of an elliptic microvoid. The authors introduced an elliptic geometry to represent interface voids considered more representative of void geometries observed. Grain size and grain boundary diffusion were also included. Void interaction was considered by accounting for plastic deformation between voids to account for the collapse of surface asperities. A power law creep mechanism was applied in the model for bonding during stage I.

Present Modeling Approach

Based on this modeling survey, a finite element formulation of the diffusion bonding process was chosen that could account for surface waviness, interfacial voids and their interaction, non-isothermal conditions, advanced material models and diffusion transport mechanisms in the future.

The initial FE code was developed which could model any shape microvoid to study the effect of interfacial microvoid geometry on time until total collapse was attained. The computer code developed had the capability for analyzing both planar and axisymmetric microvoids. This finite element formulation provides a realistic representation of the combined stage 1 and 2 microvoid coalescence. A viscoplastic strain rate-dependent material model provides the mechanism for the stage 1 coalescence until the microvoid geometry is such that the surface tension dominates because of the surface energy during stage 2.

For the initial study, three different planar microvoid geometries were studied which could be identified with surface imperfections or machine tool marks. Attention was focused on v-shaped, circular and elliptic interfacial voids. Figure 1 shows a comparison of the different models that were investigated.

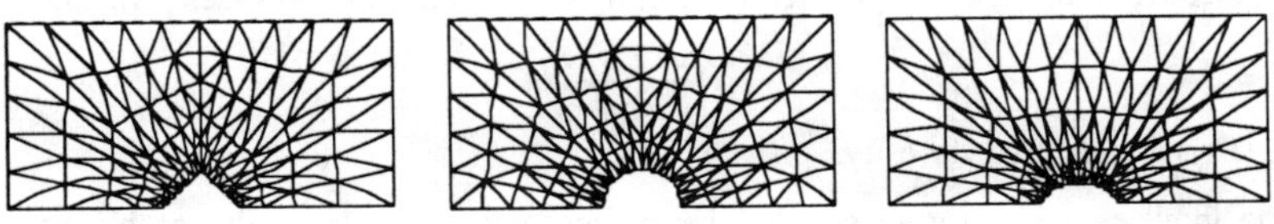

Figure 1. Microgeometry of interfacial cavities.

GOVERNING EQUATIONS

The temperature distribution is described by the classical transient heat conduction equation:

$$\frac{\partial}{\partial X_i}\left(k\frac{\partial\theta}{\partial X_i}\right)+\dot{Q}=\rho C_p\frac{\partial\theta}{\partial t}$$

where θ is the temperature, $\dot{Q}$ is the internal heat generation term, k is the thermal conductivity, ρ is the density, C_p is the specific heat and X_i is the spacial coordinates written in indicial notation.

The mechanical deformation of a viscoplastic solid can be determined by computing the material velocity distribution. Once the velocity field is known, the material displacement can be obtained by time integration. The point of formulating the mechanical behavior in terms of the velocity instead of displacements is a tremendous simplification and yet provides all the necessary description for large local deformation. Accordingly, the flow and deformation of the solid subjected to mechanical loading are described by the conservation of momentum and mass equations. When inertia and body forces are neglected, the equilibrium equation can be written as

$$\frac{\partial \sigma_{ij}}{\partial X_i} = 0$$

where σ_{ij} is the total stress tensor composed of the deviatoric and hydrostatic stress tensors.

The deformation rate tensor d_{ij} is related to the material velocity gradients by

$$d_{ij} = \frac{1}{2}\left(v_{i,j} + v_{j,i}\right)$$

The material incompressiblity assumption requires that the trace of the deformation rate tensor vanish, accordingly

$$d_{ii} = v_{i,i} = 0$$

In this formulation, the mechanical constitutive equation assumes that the elastic material response is small compared to the large plastic deformation. Consequently, a power law viscoplastic material model was chosen to represent Ti6-4 for the flow stress dependency on deformation rate. This relationship can be written in terms of the second invariant of the deviatoric stress tensor as

$$\sigma'_{II} = Ce^{\frac{Q_a}{R\theta}} d_{II}{}^{m}$$

where Q_a is the activation energy, R the universal gas constant, θ the absolute temperature and m the strain rate sensitivity factor. The components of the deviatoric stress tensor σ'_{ij} can be obtained from the deformation rate tensor using the associated Levi-Mises flow rule:

$$\sigma'_{ij} = 2\mu^{e} d_{ij}$$

where μ^{e} is an effective viscosity.

The above equations form a coupled thermomechanical boundary value problem with the appropriate thermal and mechanical boundary conditions.

FINITE ELEMENT FORMULATION

A finite element program was developed to investigate microvoid deformation based on a formulation developed by Dawson Reference (12). Six noded isoparametric triangular elements were used which provide the capability to handle highly non-linear local deformation gradients. The six-node trianglular FE is used to determine the velocity and temperatures at the corner and midpoint nodes. The hydrostatic stress is determined at the corner nodes by using the Langrangian multiplier method.

The finite element solution for the transient heat conduction equation was obtained using Galerkin's method of weighted residuals formed by the governing equation and a set of weighting functions N_j. When Green's theorem is applied, the relationship for the residual is determined by

$$R_\theta = \int_V \left[\frac{\partial N_j}{\partial X_i} k \frac{\partial \theta}{\partial X_i} + \dot{Q} N_j - N_j \rho C_p \frac{\partial \theta}{\partial t}\right] dV - \int_S \bar{q} N_j dS$$

where $\bar{q}$ is the applied heat flux. This equation can be reduced to a matrix equation for each element by using an approximating interpolation function for the temperature and applying the Crank-Nicholson method.

The finite element approximation of the equilibrium equation for the determination of the velocity field is based on the variational statement for virtual work. The functional J is written as

$$\delta J = \int_V \sigma'_{ij} \delta d_{ij} dV - \delta \int_V p d_{ii} dV - \int_S T_i \delta v_i dS$$

where δ denotes variation of function or variable, p is a Lagrangian multiplier having the physical meaning of pressure, T_i the applied tractions. The velocity and pressure variables are approximated by using interpolation functions.

The stiffness matrices for the formulation are obtained by quadrature, and the governing equations are finally reduced to a set of matrix equations that are solved using the frontal technique to obtain the nodal velocities and pressures. The Newton-Raphson method was used to obtain the nonlinear solution since the stiffness matrix depends on the effective viscosity, which is a function of the deformation rates.

Finally a quasi-Lagrangian reference frame was used in the formulation. An Euler integration scheme was used for each time step to update the mesh in time. The updated mesh was then used as the reference frame for the next time step.

PROBLEM STATEMENT

Figure 1 shows the v-notch, circular and elliptic cavities geometries investigated. Because of the symmetry of the model, only half of the geometry needed to be modeled. The width of the model could be adjusted to account for void interaction with the appropriate boundary conditions imposed. Time varying pressure loading can be applied on external or internal surfaces as well as isothermal or non-isothermal temperature fields depending on the process parameters used.

Figure 2 is a graph showing the titanium alloy Ti6-4 flow stress data used in the analyses. The power law strain rate relationship material parameters are also identified in Figure 2 for the material model:

$$\sigma_{eff} = K e^{\frac{Q_a}{R\theta}} \dot{\varepsilon}_{eff}^{m}$$

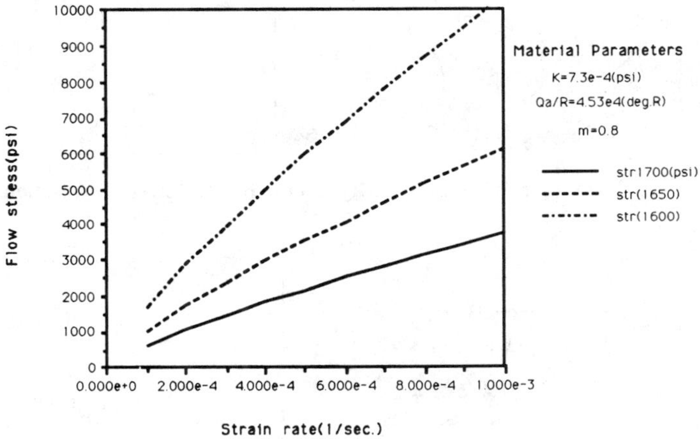

Figure 2. Ti6-4 Flow stress vs. strain rate data.

The cases studied were for an applied processing pressure of 300 psi (2.07 MPa) and temperature of 1700°F (927°C). A surface energy γ value of 5.7×10^{-3} in.-lb/lb (1×10^{-4} Joule/cm^2) was used in this analysis.

RESULTS

Figure 3 shows a typical cavity deformation and the associated equivalent stress field surrounding the cavity. The existence of a void at the bondline acts as a local stress concentration. From this result it can be noted that even though the pressure loading was only in the Y-direction, the coalescence of the circular void is circular. This indicates that the assumption originally made by Hamilton, Garmong and Pilling—that there is a symmetrical deformation of the flaw—is valid. Figure 4 shows the completely collapsed geometry for the elliptic microvoid and the associated effective stress contours at the bond interface. A comparison of the behavior for these three types of interfacial microgeometries for bonding titanium sheet alloy Ti 6Al-4V subjected to 300 psi at 1700 °F was conducted. There is a pronounced difference in the behavior, between the V-shaped and the other two geometries, as shown in Figure 5 by the displacement of the center high point in the void as a function of the time. Circular and elliptic voids tend to take the same time to collapse during the bonding process. This occurs in the final phase of coalescence of elliptic voids which become circular because of the increased surface tension forces at the region of highest curvature. If elliptic voids have a large aspect ratio, they are unstable and will collapse to form smaller microvoids. In the latter stage of diffusion bonding, pressure plays a minor role in the coalescence process compared to the surface tension and diffusional transport. Figure 6 shows the rate of microvoid collapse with time for an elliptic microgeometry. The results in that case show that any increase in time would have very little effect on further coalescence.

Our studies have shown that the finite element method (FEM) can be effectively used to study:

(1) The effect of various interfacial geometries,

(2) The significance of flow stress properties,

(3) The optimization of bonding process by using both time-dependent pressure and temperature profiles.

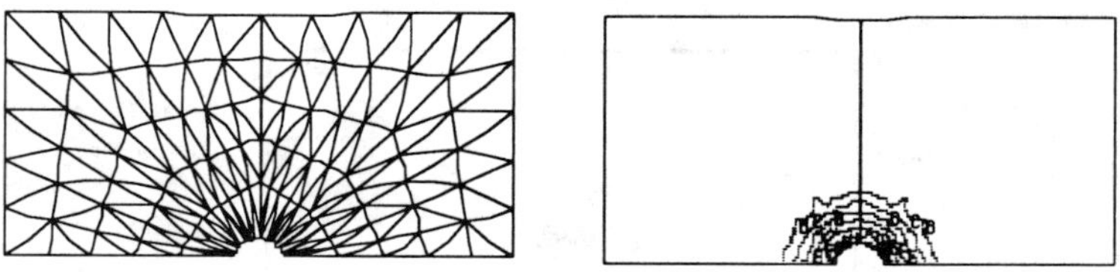

Figure 3. Circular cavity deformation and stress distribution.

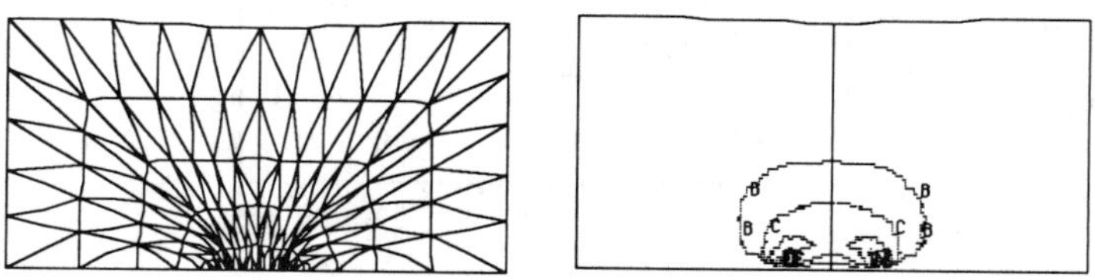

Figure 4. Elliptic cavity deformation and stress distribution.

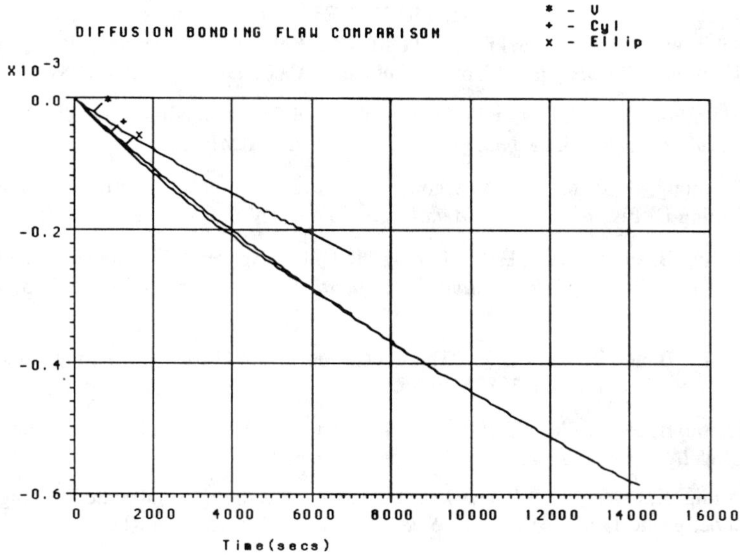

Figure 5. Comparison of cavity collapse as function of time.

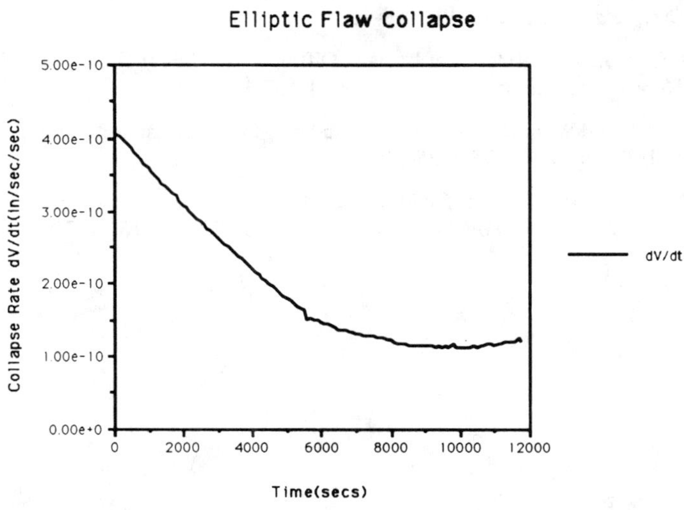

Figure 6. Surface cavity collapse rate as a function of time.

REFERENCES

(1) King, W.H. and Owczarski,W.A.; "Diffusion Welding of Comercially Pure Titanium," *Welding Jour.*, Vol. 46, Publisher AWS, 1967, pp. 289s- 298s.

(2) Hamilton, C.H.; "Pressure Requirements for Diffusion Bonding Titanium," *Titanium Science & Technology*, Vol. 1, Publisher AIME, 1973.

(3) Garmong, G.; Paton, N.E. & Argon, A.S.; "Attainment of full Interfacial Contact During Diffusion Bonding," *Metallurgical Trans.*, Vol. 6A, (Jan-Jun) 1975.

(4) Derby, B. and Wallach, E.R.; "Joining Methods in Space: A Theoretical Model for Diffusion Bonding," *International Astronautical Federation.*, IAF-79-ST-15, (Sept. 1979).

(5) Derby, B. and Wallach, E.R.; "Theoretical Model For Diffusion Bonding," *Metal Science.*, Vol. 16, (Jan. 1982), 49-56.

(6) Derby, B. and Wallach, E.R.; "Diffusion Bonding: Development of Theoretical Model," *Metal Science,* Vol. 18, (Sept. 1984), 427-431.

(7) Pilling, J., Livesey, D.W., Hawkyard, J.B. and N. Ridley; "Solid State Bonding in Superplastic Ti-6Al-4V," *Metal Science,* Vol. 18, (March 1984), 117-122.

(8) Pilling, J. and N. Ridley; "Solid State Bonding of Superplastic AA7475," *Materials Science and Tech.*, Vol. 3, (May 1987), 353-359.

(9) Wallach, E.R. and Derby, B.; "Modeling of Diffusion Bonding," *Proc. of Intl. Conf. Cranfield,* Vol. 1, (July 1987), 9-16.

(10) Pilling, J.; "The Kinetics of Isostatic Diffusion Bonding in Superplastic Materials," *Materials Science and Eng.,* 100 (1988), 137-144.

(11) Hill, A. and Wallach, E.R.; "Modelling Solid-State Diffusion Bonding," *Acta Metall.*, Vol. 37, No. 9, (1989), 2425-2437.

(12) Dawson, P.R.; "Viscoplastic Modeling of Upset welding Using the Finite Element Method," *Process Modeling Tools*, Proc. ASM 1980, pp. 151-172.

SOME ASPECTS OF FORMATION OF JOINTS OF VARIOUS CONFIGURATIONS DURING DIFFUSION BONDING OF TWO-PHASE TITANIUM ALLOYS

N. F. Anoshkin, A. A. Gelman, V. M. Pavlov

All-Russia Institute of Light Alloys
Moscow, Russian Federation

Abstract

This study has covered some aspects of the production of butt- and T-joints in two-phase titanium alloys via vacuum diffusion bonding with regard to effects of "deformation-time" parameters of the process, alloy composition and structure of bonded components.

In butt bonding under triaxial stress conditions and relatively short holding times (≤0.5 h), the degree of macroplastic deformation has the principal effect on bond formation and mechanical properties, but the effect of alloy composition and structure of bonded components on bond quality is negligible. Regarding bond times, bond quality depends on the combined effect of plastic deformation and time of its accumulation, as well as on alloy composition. The change from butt to T-joints produces an effect on the strained metal condition at the interface. It has been shown that, due to low degree of deformation, realized at the interface the production of the sufficient T-joint can be obtained by a noticeable increase in bonding times.

A defect morphology diagram has been plotted on "degree of deformation-time" coordinates and fields of formation of microvoids-free joints have been defined.

The present paper has focused on some aspects of the formation of joints in two-phase titanium alloys via vacuum diffusion bonding, and specifically on the effects of deformation-time parameters of the process, alloy composition, as well as on configuration and structure of components bonded on bond quality.

The investigation was carried out on structural alloys including VT6 (Ti-6Al-4V), VT3-1 (Ti-6Al-2.5Mo-2Cr-0.5Fe-0.3Si), VT22 (Ti-5Al-5Mo-5V-1Fe-1Cr), which differed in β-phase content, polymorphous transformation temperatures ($T_{p.t.}$) and resistance to deformation at comparable $\alpha+\beta$ field bonding tempera-

Titanium '92
Science and Technology
Edited by F.H. Froes and I. Caplan
The Minerals, Metals & Materials Society, 1993

tures ($T_{bond.} = T_{p.t.}$ - 20 to 30 °C). The effect of alloy structure and composition on bond quality was studied on butt-joints of VT6 alloys bonded under triaxial-stress conditions. The types of microstructures under study are as follows: fiber and coarse-grained, lamellar. Formation of T-joints was studied under plane-strain deformation. Characteristic features of the as-deformed condition at the interface after bonding were evaluated by Moire strip method according to a specific procedure [1]. Bonding temperatures are as follows: 950 °C for VT3-1, 920 °C for VT6, 850 °C for VT22 (i.e., 20-30 °C below $T_{p.t.}$ of every alloy).

Relationships between the joint quality, macroplastic deformation degree (ε), and bonding times(t) have been studied. In butt bonding, the deformation degree was defined by the relative change in the total specimen height. In T-bonding, it was defined by the relative change in rib height. This can be used in practice as quality control parameters for an adaptive system. Ranges of deformation degree (ε) and bonding times (t) were 3 to 16 % and 0.3 to 3 h, respectively. Completeness of joint formation was evaluated by joint impact toughness-to-base metal impact toughness ratio ($\overline{KCV}$). This ratio is one of the most sensitive to defects in the contact surface. The average degree of deformation of the whole specimen (Table 1, column 1) and of every bonding element (Table 1, column 2, 3) was evaluated for joints produced from components with dissimilar structure. In this case $\overline{KCV}$ was defined by joint impact toughness-to-weaker base metal impact toughness ratio. Microdefect morphology at the interface was studied according to a previously developed procedure [2].

Titanium alloy joints are characterized by two major groups of microvoids, namely slot-type defects formed at the early stages of joint formation and pinhole porosity spaced in packets or scattered porosity [2, 3]. Fig. 1 represents a "ε -t" diagram of defect morphology for a VT6 fiber structure alloy. The diagram region with pinhole porosity covers the whole range of defects of this type (from porosity packets to single scattered porosity). The line of transition from slot-type microvoids to pinhole porosity for bonding times of more than 1 h is shown conventionally as the range of deformation degree below 3 % and was not studied. The diagram shows that an increase in the degree of macroplastic deformation for bonding times up to 1 h is accompanied by a regular change in defect type from slot-type defects to pinhole porosity, which practically disappear completely at ε =10 to 15 %, depending on the process duration. In this case, a change in bonding times in the range from 0.1 to 0.5 h does not affect the location of regions with various types of defects. For longer bonding times (t=3 h), the stage of slot-type defect fragmentation resulting in pinhole porosity formation has enough time to be completed at the deformation degree under the low limit in question ($\varepsilon \leq 3$ %). In joints formed at this bonding times in the range of ε =3 to 6 % only defects in the form of pinhole porosity dissappearing with $\varepsilon > 6$ % are observed.

According to observed variations in defect morphology with bonding times within 1 h, the $\overline{KCV}$ (ε) relationship has two specific areas (Fig. 2) expressed especially clearly at relatively short holding times ($t \leq 0.5$ h). The most pronounced increase in relative impact toughness is observed at the initial stages of joint formation with deformation degree giving rise to fragmentation of slot-type defects and a decrease in their surface fraction. Deformation degree resulting in a change of the $\overline{KCV}$ (ε) relationship seems to agree with the stage of transition from packets to scattered pinhole porosity. It should be noted that a change in bonding times by 0.1 to 0.5 h without a significant effect on the morphology and amount of defects in the joining zone (Fig. 1) does not affect the $\overline{KCV}$ (ε) relationship. In this case, joint quality is determined by the degree of macroplastic deformation only.

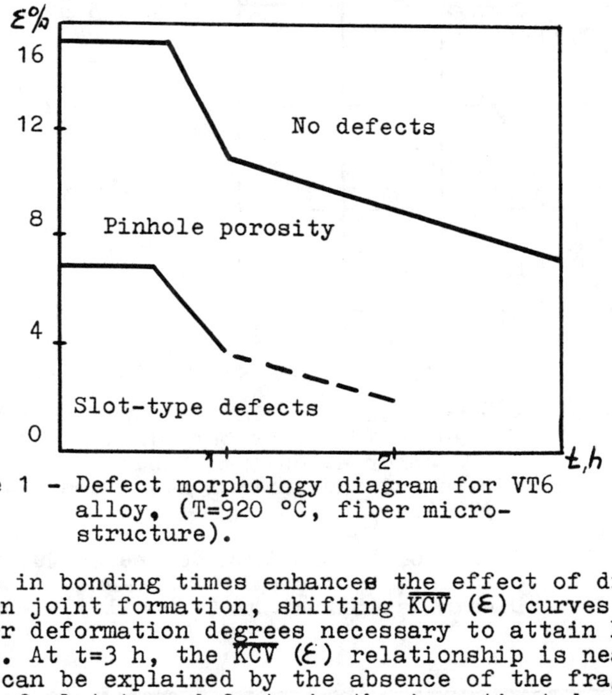

Figure 1 - Defect morphology diagram for VT6 alloy, (T=920 °C, fiber microstructure).

An increase in bonding times enhances the effect of diffusion processes on joint formation, shifting $\overline{KCV}$ (ε) curves left towards lesser deformation degrees necessary to attain high quality joints. At t=3 h, the $\overline{KCV}$ (ε) relationship is nearly linear. This can be explained by the absence of the fragmentation stage of slot-type defects in the investigated range of deformation degrees (Fig. 2).

At the same temperature and relatively short bonding times ($t \leq 0.5$ h) a change in structure has a minimal effect on the $\overline{KCV}$ (ε) relationship. Joint quality depends on the degree of macroplastic deformation of metal (Fig. 2). Under these bonding conditions, the certain favourable effect of the fiber structure manifests itself, mainly, in a noticeable decrease in pressure necessary to provide deformation required. Deformation and diffusion processes proceeding in finer-grain component have the major effect on the formation of joints bonded of elements with dissimilar structure (fiber+coarse grain, lamellar). It can be seen from comparison of the data presented in Table 1 and in Fig. 2. Under similar bonding conditions, equal $\overline{KCV}$ values of joints bonded from elements with dissimilar structure and with fiber structure are achieved at similar "ε" values of the element with fiber structure in a dissimilar joint and of the specimen with the same structure of both elements.

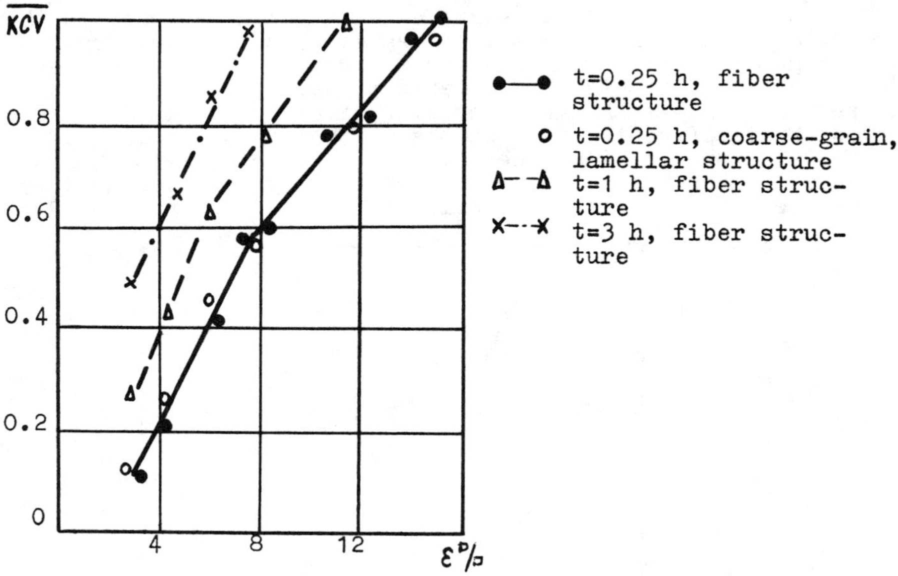

Figure 2 - Dependence of relative impact toughness of VT6 alloy butt joints upon the deformation-time bonding parameters and the initial structure (T= 920 °C).

TABLE 1 Effect of Deformation Degree on Quality of Joints Produced from Elements with Dissimilar Structure (VT6 alloy, T=920°C, t=0.25 h)

Deformation degree, %			Impact toughness, J/cm^2			
Average	Element with fiber structure	Element with lamellar structure	Joint	Base metal: Fiber structure	Base metal: Coarse-grain lamellar structure	$\overline{KCV}$
2	3.5	1	10	63	74	0.16
3	5	1	15	56	68	0.25
7	13	1	50	59	72	0.85

A change in the composition of two-phase alloys diffusion bonded at relatively short holding times does not affect the $\overline{KCV}$ (ε) relationship and maintains a principle importance of macroplastic deformation in joint formation. With longer bonding times an increase in quantity of β-phase in the alloy resulting in lowering of $T_{p.t.}$ and bonding temperature in the two-phase region, shifting the $\overline{KCV}$ (ε) relationship curve right towards the greater deformation degrees assures satisfactory joints (Fig. 3).

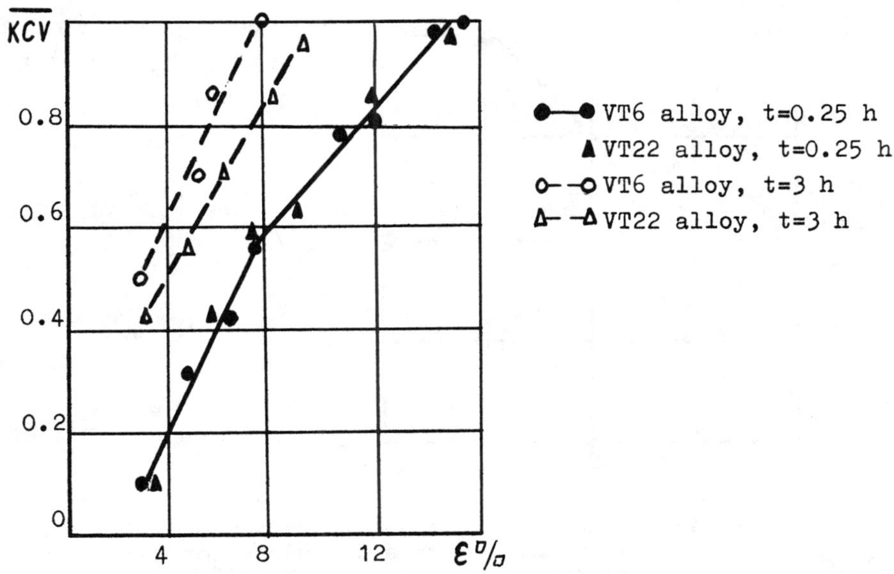

Figure 3 - Dependence of relative impact toughness of butt joints upon the deformation-time parameters of bonding and the alloy composition (fiber structure).

The nature of the $\overline{KCV}$ (ε) relationships in T-joint differs from that in butt joints. Analysis of the deformed state of metal at the interface by a Moire method [1] has shown that in T-joining of VT6 alloy under plane-strain deformation correlation between the degree of rib deformation in height and the nature of metal plastic flow in bonding zone is observed in a narrow range of deformation degree only ($\varepsilon \leq 7$ %). An increase in deformation degree up to 15 to 16 % does not change noticeably deformation parameters at the interface and hence does not improve joint quality under relatively short bonding times (t=0.25 h) (Fig. 4). It can be achieved only due to longer bonding time and more complete diffusion processes.

The mechanisms determined for joint formation gave rise to development of some processes for manufacture of intricate-shaped diffusion-bonded billets of panel-, blade- and monodisk-types in investigated titanium alloys (Fig. 5, 6). Table 2 exemplifies serviceability of VT6 alloy bonded joints.

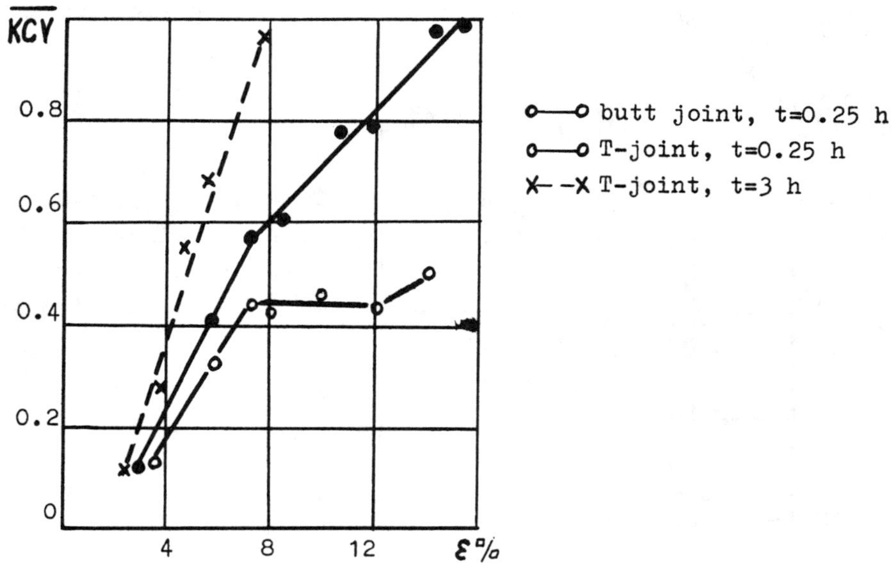

Figure 4 - Effect of joint configuration on deformation degree dependence of relative impact toughness (VT6 alloy, T=920 °C, fiber structure).

Figure 5 - Diffusion bonding vacuum installation. P=2.46 MN.

a.

b.

Figure 6 - Panel (a) and blade (b) billets produced by diffusion bonding.

TABLE 2 Mechanical Properties of Bonding Joints of VT6 Alloy Panels

Characteristic	Value	Weld-to-base metal ratio
UTS, MPa	910	1.0
El., %	14	1.0
RA, %	30	1.0
K_{1c}, MPa $\sqrt{M}$	84	0.92
Stress-rupture strength, MPa (450 °C, 100 h)	450	1.0
$\overline{KCV}$, J/sq. sm	44	0.8
LCF, σ =500 MPa, f=40 Hz	56700-145100	1.0

Conclusions

1. In butt bonding under triaxial stress conditions and relatively short bonding times ($t \leq 0.5$ h), the degree of macroplastic deformation has a determinant effect on joint formation and mechanical properties. Bonding times, the alloy structure and composition have negligible effects. At longer bonding times, joint quality depends both on plastic deformation and the time of its accumulation, as well as on the alloy composition.

2. A defect morphology diagram has been plotted on "degree of deformation-time" coordinates. Using this diagram one can define bonding conditions to produce high-quality joints in terms of impact toughness characteristic being one of the most sensitive to bonding defects.

3. In changing from butt to T-joints, the variation of configuration and stress-strain condition affects the nature of metal plastic flow at the interface and the $\overline{KCV}$ ($\mathcal{E}$) relationship. In T-bonding under plain-strain deformation the value of accumulated deformation at the interface has a slight dependence upon the degree of rib deformation in height. It is evidently insufficient for perfect joint formation under short bonding times. The strength level of the base metal can be achieved by an increase in bonding times only. This favours more complete diffusion processes.

References

1. L. M. Puchkova, A. A. Gelman, V. M. Pavlov, "Analysis of the Deformed Condition Nature in Diffusion Bonding of Standard Elements of VT6ch Alloy Ribbed Structures", Svarochnoye proizvodstvo, 11 (1991), 13-15.
2. A. A. Gelman, et al. "Effect of Defects on Impact Toughness of VT6 Alloy Bonded Joints", Svar. proiz., 10 (1985), 16.
3. G. Garmong, N. E. Paton, A. S. Argon, "Attainment of Full Interfacial Contact During Diffusion Bonding", Metallurgical Transactions 5A (6) (1977), 1269-1279.

Interface Characteristics of Solid-Phase Welds Between Ti-6Al-2Sn-4Zr-2Mo-0.1Si and Ti-14Al-21Nb Titanium Aluminide

M.C. Juhas,[1] W.A. Baeslack III,[2] H.L. Fraser,[1] P. Threadgill,[3]
D. Phillips,[4] and T.F. Broderick[5]

[1]Dept. of Materials Science and Eng., The Ohio State University, Columbus, Ohio USA
[2]Dept. of Welding Engineering, The Ohio State University, Columbus, Ohio USA
[3]The Welding Institute, Abington Hall, Abington, Cambridge, England
[4]G.E. Aircraft Engines, Cincinnati, Ohio USA
[5]Materials Directorate, WPAFB, Dayton, Ohio USA

Abstract

Microstructures of dissimilar-alloy linear-friction and diffusion welds between Ti-14Al-21Nb titanium aluminide and Ti-6Al-2Sn-4Zr-2Mo-0.1Si were evaluated using optical, scanning electron, and transmission electron microscopy. The linear-friction weld interface was characterized by localized mechanical mixing, negligible alloying element interdiffusion, and a fine, dynamically recrystallized prior beta grain structure. Beta decomposition occurred in the Ti-6242 to fine colonies of alpha and alpha-prime martensite, and in the Ti-14Al-21Nb to alpha-two prime and beta. In contrast, the diffusion weld exhibited substantial interdiffusion of Nb into the Ti-6242 which locally suppressed the beta transus and promoted subsequent beta decomposition to a coarse colony alpha + transformed beta structure.

Introduction

The increasing demand for advanced lightweight materials with elevated-temperature properties and oxidation resistance superior to that of conventional titanium alloys has prompted the development of a new class of materials based on the Ti_3Al (alpha-two, DO_{19} structure) intermetallic. These alloys potentially exhibit a relatively low density, high elastic modulus, and high creep resistance compared to conventional near-alpha titanium alloys. Although the binary compound exhibits unacceptably low room-temperature ductility and toughness, improvements in these properties have been achieved through the addition of the beta-stabilizing elements Nb, Mo, and Ta (1,2). The equilibrium microstructure of these alloys correspondingly is comprised of both alpha-two and beta (BCC) phases. Ti-14Al-21Nb (wt%) is a first-generation Ti_3Al-Nb type alpha-two Ti aluminide which has been studied in detail. Regarded principally as a model system, thermomechanical processing and heat treatment information determined for Ti-14Al-21Nb has provided an important data base for the further development of alpha-two titanium aluminides.

The effective integration of alpha-two titanium aluminides into gas turbine engines will require their joining both to themselves and to conventional elevated-temperature Ti alloys. The application of solid-phase welding processes (which precludes melting and solidification experienced during fusion welding) offers a strong potential for producing high-integrity joints between dissimilar alloys. The mechanical properties of these welds are dependent principally on the integrity and microstructural characteristics of the interface region between the dissimilar

Titanium '92
Science and Technology
Edited by F.H. Froes and I. Caplan
The Minerals, Metals & Materials Society, 1993

materials. In the present study, detailed microstructural analysis was performed to characterize the interface region of solid-phase diffusion and linear-friction weldments produced between Ti-14Al-21Nb (hereafter referred to as Ti-14-21) and the conventional, near-alpha titanium alloy Ti-6Al-2Sn-4Zr-2Mo-0.1Si (hereafter referred to as Ti-6242).

Experimental

The beta-forged Ti-14-21 microstructure was characterized by coarse, equiaxed prior beta grains ~ 1-3 mm in diameter and transformed beta in the form of Widmanstätten alpha-two platelets bounded by thin ribs of retained beta-phase, Figure 1a. The Ti-6242 originated as an alpha/beta processed and heat-treated forging and exhibited a microstructure of equiaxed alpha grains with a smaller amount of coarse, transformed beta in the form of fine alpha platelets bounded by retained beta, Figure 1b.

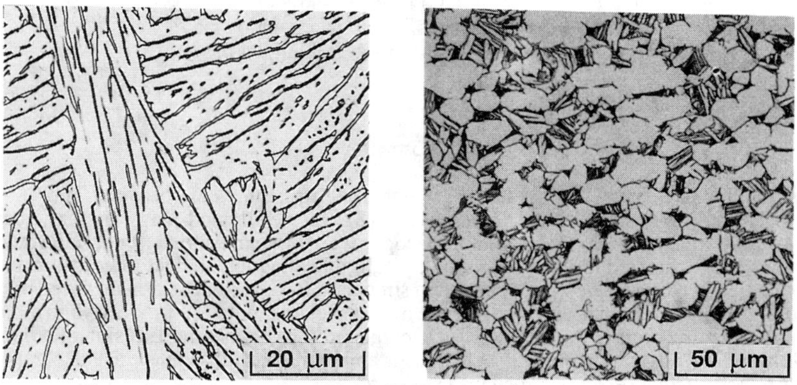

Figure 1: (a) As-received Ti-14-21 showing Widmanstätten alpha-two platelets bounded by retained beta; (b)As-received Ti-6242 showing equiaxed alpha surrounded by fine alpha platelets and retained beta.

The diffusion weld was produced between blocks 50 mm x 50 mm x 25 mm in a vacuum hot press at 970°C for 2 hours under 3.45 MPa (500 psi) pressure and furnace cooled. Linear-friction welds were produced between blocks 30 mm x 15 mm x 50 mm (length) using a weld frequency of 25 Hz, amplitude of 2 mm, friction and forge forces of 50 kN, and a burnoff (upset) of 4 mm.

Specimens were prepared for metallographic examination using standard polishing techniques and etched with Kroll's reagent (3 ml HF + 6 ml HNO_3 + 100 ml H_2O). Conventional light and scanning electron microscopy (SEM) were used to evaluate the metallographic samples. Thin foils for transmission electron microscope (TEM) examination were slurry drilled from various regions of both solid phase weld samples. The desired area of each sample was selectively thinned by dimpling to a final thickness of approximately 100 microns. The samples were then jet electropolished to perforation in a solution of 10% H_2SO_4 in methanol at a temperature of -20°C and a current of ~30 mA (10V). Ion milling was used as a final step to obtain additional thin area as needed.

Electron probe microanalysis (EPMA) was performed on metallographic samples to obtain compositional information across the interfaces of both solid phase welds using a *JEOL 730* electron microprobe. Conventional TEM was performed on a *JEOL 200CX* equipped with an energy dispersive spectrometer (EDS) and a *JEOL JSM 840F SEM* was employed to obtain secondary and backscattered electron images (BEI) of metallographic samples.

Results and Discussion

Linear-friction Weld

A photograph of the linear-friction weld and a macrograph of a metallographic cross-section through the weld zone (WZ) appear in Figures 2a and 2b, respectively. It can be seen in Figure 2a that the combination of frictional heating and forge force produced flash principally in the Ti-6242 that semi-continuously skirted the joint. Figure 2b reveals a slight curvature to the weld interface concave into the Ti-14-21, and a decrease in the width of the weld heat and deformation zone (HDZ) from the center to the outer periphery.

Figure 2: (a) Linear-friction weld between Ti-14-21 and Ti-6242;
(b) Macrograph showing weld cross-section.

Optical micrographs of the linear-friction weld interface are shown at low magnification in Figures 3a and 3b. The HDZ extends approximately 700-800 microns from the bond line into each base metal. Figure 3b reveals a metallurgically sound but rather irregular interface with a substantial amount of highly-localized mechanical mixing between the two alloys. Dynamically recrystallized beta grains, approximately 15-20 microns in diameter and darkly outlined, are clearly observed in the Ti-14-21. Transformation products in the HDZ of the Ti-14-21 are ultra-fine platelets of acicular alpha-two and, most probably, alpha-two prime martensite (3,4) surrounded by beta phase which may be ordered, as shown in Figure 4a. On the Ti-6242 side of the weld interface, a fine, dynamically recrystallized beta grain structure is observed in the HDZ. The transformed-beta microstructure appears as platelets of alpha (HCP) as well as alpha-prime martensite, as shown in Figure 4b. Microhardness values of approximately 375 KHN (200g load) in this region are relatively high for colony structure alpha (5). This is due to the small beta grain size which promoted a fine colony platelet size.

EPMA data for the friction weld are shown in Figures 5a and 5b. The most striking feature in the composition profile is the virtual absence of interdiffusion across the interface. In Figure 5b, the anomalous drop in the levels of Nb and Al on the Ti-14-21 side of the interface reveals mechanical mixing as shown in the micrograph in Figure 3b. The apparent detection of limited interdiffusion in Figure 5a is probably due to beam overlap effects as the probe traversed the interface. It should be noted that the thermal excursion into the beta phase field, albeit short in duration, promoted local compositional homogeneity on both sides of the interface region.

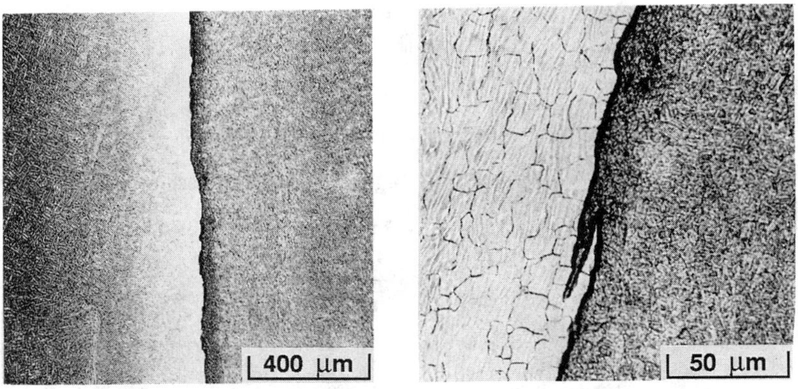

Figure 3: Optical micrographs of linear-friction weld: (a) Low magnification of interface region; (b) Dynamically recrystallized beta grains in both Ti-14-21 and Ti-6242.

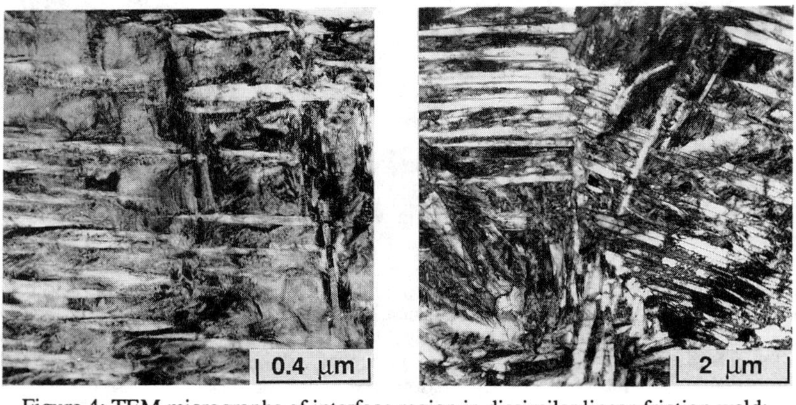

Figure 4: TEM micrographs of interface region in dissimilar linear-friction weld: (a) Ti-14-21; (b) Ti-6242.

Diffusion Bond

The microstructure of the diffusion weld interface region, Figure 6a, reveals the absence of macroscopic deformation. In this weld, the two blocks of material were heated and held at the constant temperature, producing a compositional gradient at the interface. The welding temperature of 970°C was well below the beta transus of Ti-14-21, leaving the microstructure essentially unaffected. In contrast, the thermal excursion was only 20° below the Ti-6242 beta transus, producing a decrease in the relative amount of equiaxed alpha in the base metal as well as its coarsening. Figure 6b is a low magnification micrograph of the weld interface illustrating a narrow diffusion zone (~50 microns) in the Ti-6242 that is devoid of the equiaxed alpha grains, revealing only fine alpha platelets <1 micron wide. Selective etching of the Ti-6242 by a very light etching of the entire interface revealed preferential attack on the Ti-14-21 side only at the alpha-two grain boundaries in immediate contact with the interface. This suggests a limited amount of Mo enrichment principally in the beta phase on the Ti-14-21 side of the boundary.

A backscattered-electron SEM image of the diffusion weld interface is shown in Figure 7. Several features are noteworthy in this micrograph. Beta phase was continuous across the bond line, however, the change in contrast between alpha and alpha-two suggests that the transformed-beta remained essentially distinct across the bond line in the Ti-6242 and Ti-14-21, respectively. Several beta ribs in the Ti-6242 reveal ultra-fine acicular alpha platelets (~100 - 200 nm) that transformed on cooling. These are shown in the TEM micrograph in Figure 8. It should be noted in Figure 8 that selected alpha/beta interfaces near the bond line in the Ti-6242 display an irregular, scalloped shape indicating local instability.

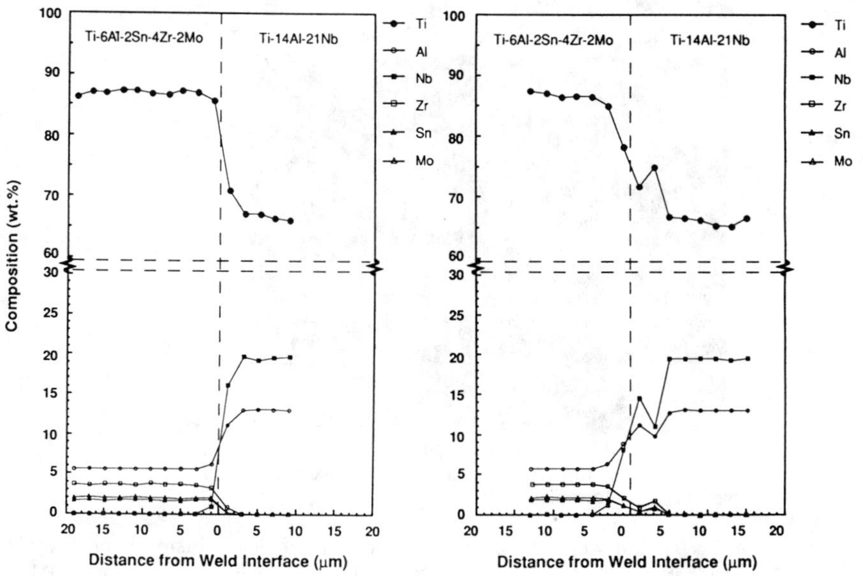

5a 5b

Figure 5: Composition profile across: (a) smooth friction weld interface region; (b) ragged friction weld interface region (see Figure 3b).

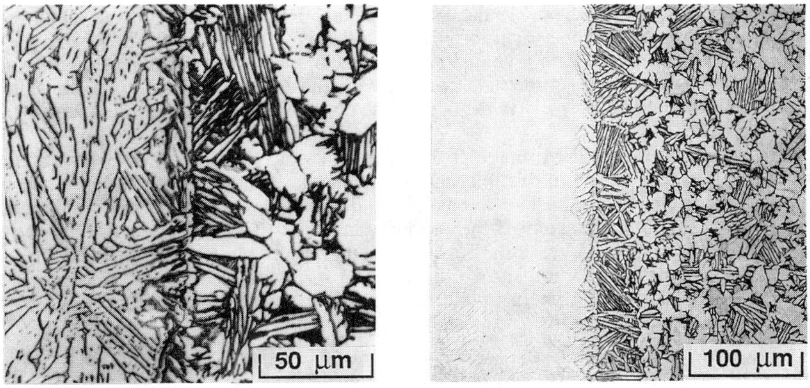

Figure 6: Micrographs of diffusion weld: Ti-14-21 on left.

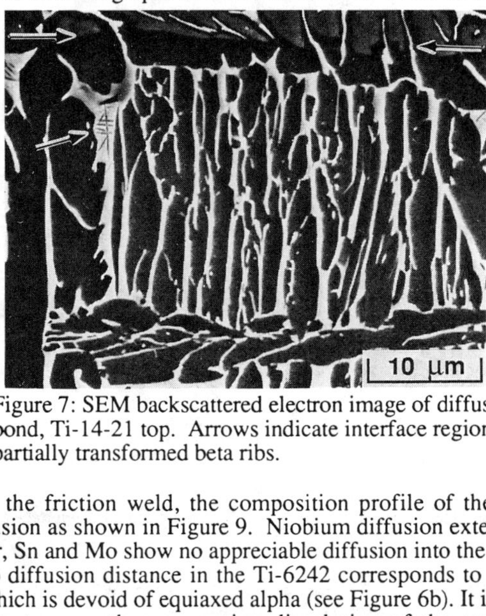

Figure 7: SEM backscattered electron image of diffusion bond, Ti-14-21 top. Arrows indicate interface region and partially transformed beta ribs.

In contrast to the friction weld, the composition profile of the diffusion bond reveals extensive interdiffusion as shown in Figure 9. Niobium diffusion extends >60 microns into the Ti-6242 whereas Zr, Sn and Mo show no appreciable diffusion into the Ti-14-21. It is interesting to note that the Nb diffusion distance in the Ti-6242 corresponds to the width of the band of transformed beta which is devoid of equiaxed alpha (see Figure 6b). It is apparent that Nb locally suppressed the beta transus, thus promoting dissolution of the equiaxed alpha grains and subsequent beta decomposition which resulted in a plate-like transformation product. It should also be noted that the composition near the bond line is quite inhomogeneous compared with the friction weld. These compositional fluctuations correspond to partitioning of Mo and Nb (beta stabilizers) and Al (alpha stabilizer) to the beta, and alpha or alpha-two phases, respectively.

At welding temperatures, the equilibrium Nb concentration in the beta phase dropped to about one-half the concentration at room temperature. Excess elemental Nb then became available to diffuse into the neighboring alpha-two grains and eventually across the interface into the Ti-6242 to eliminate the concentration gradient. Although the cooling rate was relatively slow, especially compared with the friction weld, it is reasonable to assume that, upon cooling, Nb *locally* partitioned into the beta phase on both sides of the interface. The microstructural evidence in Figure 6 clearly supports this argument. The narrow band of transformed beta that is devoid of equiaxed alpha in the Ti-6242 illustrates the Nb diffusion distance across the interface. In this region, only compositionally-induced beta remained stable during welding. The retained beta was susceptible to local instability as seen in Figure 7. Arrows point to beta ribs which locally transformed to very fine, acicular alpha.

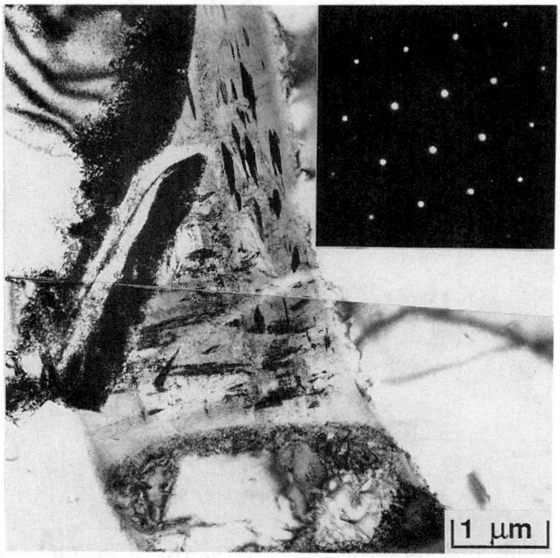

Figure 8: TEM micrograph of Ti-6242 side of diffusion bond showing ultra-fine alpha platelets within beta ribs

Summary

In this study, two solid-phase welding processes were employed to produce dissimilar joints between Ti-14-21 and Ti-6242. The rapid thermal cycle and extensive deformation experienced by the linear-friction weld resulted in dynamic recrystallization of beta phase in both the Ti-14-21 and Ti-6242 with negligible interdiffusion between the two materials. The rapid cooling promoted martensitic beta decomposition in both alloys and the ordering of retained beta phase in Ti-14-21. The processing temperature of the diffusion weld promoted appreciable Nb interdiffusion into the Ti-6242, resulting in local suppression of the beta transus temperature. This was manifested in a region near the interface that was characterized by transformed beta in the form of alpha platelets and was devoid of equiaxed alpha. The beta phase in this region was relatively unstable as evidenced by ultra-fine acicular alpha platelets that transformed within the beta ribs on weld cooling.

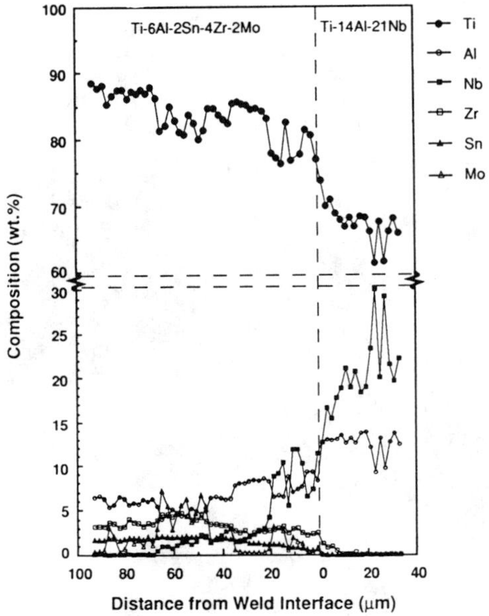

Figure 9: Compositional profile across diffusion bond

Acknowledgements

The authors wish to thank the Edison Welding Institute and the Center for Materials Research at The Ohio State University for their support of this work. Appreciation is further expressed to Mr. David Nicholas at TWI for production of the linear-friction weld. We are grateful to Mr. Scott Apt of UES, WPAFB for specimen preparation and Mr. Jeff Moser of The Ohio State University for the SEM analysis. Finally, we thank Mr. Cameron Begg of UES, WPAFB for the electron probe microanalysis.

References

(1) M.J. Blackburn and M.P. Smith, "Research to Conduct an Exploratory Experimental and Analytical Investigation of Alloys", AFWAL-TR-81-4046, Interim Tech. Rept., May 1977 - Dec. 1978, Contract #F33615-75-C-1167, June, 1981.

(2) M.J. Blackburn and M.P. Smith, "Titanium Alloys of the Ti_3Al Type", U.S. Patent #4,292,077, Sept., 29, 1981.

(3) H.T. Weykamp, K.R. Baker, D.M. Paxton, and M.J. Kaufman, Scripta Met., 24, (1990), 445.

(4) M.J. Cieslak, T.J. Headley, and W.A. Baeslack III, "Effect of Thermal Processing on the Microstructure of Ti-26Al-11Nb: Applications to Fusion Welding", Metall. Trans., 21A, (1990), 1273.

(5) W.A. Baeslack III and F.D. Mullins, Metall. Trans., 15A, (1984), 1948.

PLASMA ARC WELDING OF TA6V TITANIUM ALLOY

C. BOUCHER, F. MESSAGER
Institut de Soudure
32, Bd de la Chapelle -75882 PARIS CEDEX 18

F. GAILLARD, J.L. HEUZE
Service Technique des Constructions et Armes Navales,
Groupe Matériaux et Structures Navals
8, Bd Victor - 75732 PARIS CEDEX 15

ABSTRACT

This paper presents the results of an exhaustive study of the plasma arc welding of TA6V titanium alloy :
- welding conditions,
- maximum penetration as a function of the welding position, with or without GTA dressing,
- fit-up tolerances,
- effect of various factors such as the type of plasma gas, flared plasma flow, etc...

High quality straight welds were obtained without filler metal on plates up to 18 mm thick welded in the flat position with GTA dressing and up to 14 mm thick welded in the vertical. and overhead positions.
Assessment of the mechanical properties of a butt weld in standard TA6V alloy, 16 mm thick, welded without filler metal : UTS = 1000 MPa, YS = 855 MPa, elongation = 5%, KV = 36 J (Charpy at - 20°C), K_Q = 65 MPa Vm (CT15 at 20°C).

1. INTRODUCTION

Titanium alloys require welding processes capable of ensuring protection of the weld and its adjacent zones from the atmosphere. This requirement is met with various welding processes : welding in a vacuum (electron beam welding) and welding under inert gas shield: GTAW, GMAW, and plasma arc welding.
Plasma arc welding has been the subject of many studies and its attractiveness for the production of very high quality welds in one run on thin parts (generally up to 8-10 mm) is well established (1) (2) (3).
In the case of titanium alloys, the technological benefits inherent in the plasma welding process are supplemented by the capability to efficiently protect the weld and its adjacent zones through the use of a trailing shield. Furthermore, the physical-chemical properties of the material promote its weldability ; as a matter of fact, the performances of the plasma process benefit by the low thermal conductivity and low density of titanium and its alloys which permit the weldable thickness ranges to be markedly increased with the key-hole technique. Moreover, in multipass welds, the process offers many advantages for the deposition of the root bead in thick material. It is precisely this application which was the subject of an exhaustive research work including feasibility studies in different configurations : welding from outside or inside.

Titanium '92
Science and Technology
Edited by F.H. Froes and I. Caplan
The Minerals, Metals & Materials Society, 1993

2. EXPERIMENTAL PROCEDURE

2.1. Welding equipment

- SAF NERTINOX TH 500 generator with a highly drooping dynamic characteristic optimized by a self-regulation of the energy as a function of the arc length. This characteristic permits the weld sagging to be delayed and facilitates the welding operation.
- Type SP6 plasma/GTAW torch with a permissible peak current of 400 A.
  The electrodes used for this investigation were 2% thoriated tungsten electrodes in a diameter of 4.8mm. The nozzles were of the convergent-divergent type (1).
- Microprocessor-piloted control cabinet including a weld-cycle programmer, a box for piloting the main parameters (voltage, current, travel speed, wire feed rate) during welding, and a programmer for the plasma gas flow rate (from O to 10 liters/min). The crater-filling parameters were programmable.

2.2. Protection of the welds

Root of weld : protection was ensured by a backflow of argon ensured by a water-cooled copper device machined with a relief ; the argon flow rate was 30 liters/min.

Face of weld : a trailing shield incorporated in the plasma torch was produced with a design taking account of the heat radiated by the plasma. Argon was diffused through a porous plate in sintered bronze and the whole system was water-cooled. The trailing shield was positioned on the surface of the workpiece ; tightness was improved by means of thin skirts made of brass (see Figure 1 at the end of text). This equipment was used in all positions.

2.3. Joint preparation

Two types of welded specimens were produced, depending on the welding procedure :
- first pass at the root of joint : U-groove with sloping sides, with root face and sufficient gap for torch penetration. For this one-side welding configuration, the torch size was reduced.
- first pass on the opposite side : this configuration was more conventional for plasma arc welding. U-groove with sloping sides, without root face, was used.

Test specimens in a length of 500 mm were welded.

2.4. Alloy studied

The study was carried out by using rolled and forged flats in standard TA6V alloy to ASTM B 265, supplied by CEZUS and RMI. A chemical composition is listed in Table I.

Table I - Chemical analysis in the weld metal

		%C	%Al	%V	%N	%O
	BASE METAL	0.010	6.1	3.83	0.016	0.17
Back-side welding + GTA dressing	Constricted	0.009	6.1	3.88	0.025	0.17
	Flared	0.007	6.1	3.90	0.020	0.15
Face-side welding	Constricted	0.006	6.1	3.89	0.016	0.19

3. RESULTS AND COMMENTS

3.1. Generals on the plasma arc welding of TA6V

Plasma arc welding is an automatic welding process wherein fusion is obtained by melting the groove faces without any filler metal being necessary. In this technique, an arc maintained between a tungsten electrode and the workpiece is forced through a constricting (water cooled) nozzle, and a very hot plasma is obtained. For heavy gauges (over 3 mm) the welding current and the plasma gas flow rate are such that the plasma penetrates completely through the workpiece (keyhole technique). A double flow of gas protects the weld. However, in the case of titanium, this protection is insufficient and the additional protection by a trailing shield is absolutely necessary.

3.2. **Feasibility study**

The tests were conducted with a view to finding the maximum weldable thickness, with a GTA dressing being applied systematically (see below). The contemplated configurations of assemblies corresponded to straight welds produced with run-on and run-off tabs ; no particular investigation of the crater filler system was made.

Two configuration of the welding torch were employed : with the normal configuration, the (transferred) plasma is constricted a result of the forced passage of the plasma gas through the converging-diverging torch. A hot and stiff plasma is then obtained. However, because of the torch design, the plasma current was limited to 280 A. It was possible, by slightly advancing the tungsten electrode, to increase the current to 400 A (maximum delivered by the generator) without any erosion of the nozzle. This change, which produced a flared plasma flow, did not permit any increase in penetration, but notably increased (by 40%) the welding speed, for instance from 1.7 mm/sec to 2.3 mm/sec for a thickness of 14 mm.

Tests were performed with pure argon and with argon-helium mixtures (between 30 and 70% helium). The addition of helium caused the plasma to expand and hence the volume of molten metal to increase, with the inherent risk of burn-through. Finally, only pure argon was used as the plasma gas.

The main parameters for the production of sound welds were as follows.

Current : to avoid any damage to the equipment used, it was limited to 280 A. With a flared plasma, the current could be increased to 400 A.
Voltage : it determines the arc length during welding. In our tests, it was adjusted in such a way that stand-off distances of the order of 4 mm were obtained (i.e. 25V).
Flow rate gas : the flow rate ranged from 7 to 10 liters/min. It was adjusted to the current. The higher the flow rate, the stronger was the blowing effect, with a tendency toward weld sagging and deeper undercutting.
The argon flow rate in the trail shield was 30 liters/min.
Travel speed it depended on the welding energy and material thickness. For a thickness of 16 mm, it ranged from 1.3 to 2.2 mm/sec with constricted plasma, and from 2 to 2.7 mm/sec with flared plasma. Below the lower limits of these ranges, the weld pool tended to collapse ; at speeds exceeding the maximum values, a consistent penetration was no longer ensured.
The maximum thickness weldable without undercut was 12 mm (14 mm in vertical welding). Beyond this thickness, a GTA dressing was necessary. The maximum weldable thickness in keyhole welding was 18 mm. However, penetration became less regular. The optimum welding conditions with a 16 mm root face are shown in Table II.

Table II - Welding conditions for plasma butt welding in the flat position

	Constricted	Flared
Intensity (A)	260	350
Voltage (V)	25	24
Welding speed (mm/sec.)	1.3	2.2
Gaz flow rate (l/min)	8	10

3.2.1. Vertical welding

Vertical upward

Welding tests on progressive thicknesses showed that up to 12 mm the execution did not pose any particular problem. At about 14 mm, it was impossible to control the backward collapse of the weld pool which occurred in an erratic way. is welding technique was not applicable to thicknesses in excess of 12-13 mm.

Vertical downward

The best results obtained in the plasma welding of TA6V were recorded in the vertical downward position. The maximum weldable thickness was 14 mm ; up to this thickness no undercut was observed and GTA dressing was not necessary.
Contrary to flat position welding, the plasma gas flow rate was here a prime factor. Penetration remained very regular whatever the parameters adopted.
This technique, which would even permit the elimination of GTA dressing, should be recommended, even as a substitute for the conventional flat position welding.

3.2.2. Welding in the horizontal position

The tests performed on thicknesses ranging from 8 to 16 mm showed that this position was very suitable for the plasma arc welding of titanium. However, this position produced an asymmetry of the bead surface which however remained acceptable. Undercutting was comparable to that produced in the flat position.
The settings were those adopted for the welding of the same thicknesses in the flat position. On the other hand, since no weld pool collapse was to be feared, this position allowed higher settings of the welding energy to be adopted.
The maximum thickness still weldable at reasonable speeds was about 16 mm.

3.2.3. Welding in the overhead position

On 14 mm thick material, welding could be performed in the overhead position without any difficulty. The operating conditions were very flexible and the molten pool was very efficiently supported, even at low travel speeds.
However, in the overhead position, weld initiation was difficult and was accompanied by the formation of a marked bulge of weld metal. Consequently, welding in the overhead position should be started on a run-on tab (to be removed afterwards) or at a conical hole in which the excess metal could sink.
In overhead position welding with the weld inclined at 45 and 60 degrees, good results were obtained on 12 mm thick material.

3.2.4. Determination of the fit-up tolerances

For assemblies made in one pass like the first pass of a multipass weld, the selected process should be capable of accommodating fit-up defects that form unavoidably.
The tolerance limits were assessed for the following defects :

- gap between 0 and 3 mm,
- high-low between 0 and 3 mm,
- fit-up angle (0 and 5 deg.)
- combined gap + high-low (from 0+0 to 3+3 mm)

These limits were established for a thickness of 16 mm in flat position welding and 14 mm in vertical-down welding. In the overhead position (thickness :

14mm) an exhaustive investigation was underway ; the first results showed that the permissible tolerances were at least equal to those in the flat position.
The flared plasma permitted a better accommodation.

3.2.4.1 - Joint gap : Between 0 and 1 mm, penetration was improved without any alteration of the bead appearance. Beyond 1 mm, the bead collapsed (depression on the face side, excessive root penetration). With a flared plasma, the collapse was less marked.Welding with the addition of a filler metal improved the weld quality. However, deeper undercutting formed on the face-side.

3.2.4.2 - High-low : On closed joints, the tolerance of the plasma process was wide. Of course, the appearance of the assembly was not satisfactory when high-low was important but the weld was very sound and the bond with the groove faces was excellent.

3.2.4.3 - Fit-up angle : Such a defect was easily accommodated by the plasma process. With an angle of 5 degrees, corresponding to a gap of 1 mm on the wider open side (thickness of 16mm), the weld appearance was still quite good.

3.2.4.4 - Combined high-low + gap : Up to a gap of 1 mm, high-low was the most susceptible defect ; however, it did not cause important appearance defects when it was not greater than 2mm.

3.3 Characterization of the welds

The characterization of the welds was performed by using three coupons produced by back-side welding, respectively with a constricted plasma, with a flared plasma + GTA dressing, and with a constricted plasma on the face-side, without GTA dressing. The three coupons were subjected to radiographic and ultrasonic inspections. No volumic defect was detected by radiographic inspection. No planar defect was detected by ultrasonic inspection.

3.4. Structure

Figure 2 shows two macrographs of a weld produced in the flat position with a constricted plasma, respectively, before GTA dressing and after GTA dressing. The penetration contour was regular, and the welds were perfectly sound. The GTA dressing produced a hardly visible reinforcement on the reverse side.
A grain coarsening was observed in the area where GTA dressing was applied (lower FIGURE 2).
The flared plasma did not change the structure ; at most, a slight increase in HAZ and wel metal sizes was observed. The microstructure of the beads was essentially fine transformed beta, but a thin border of alpha phase was observed at the grain boundaries. No notable difference existed between the GTA-dressed zone and the plasma-melted zone, or between the two welds made with a constricted plasma and a flared plasma.

3.5 Mechanical properties of the joints

The following mechanical tests were used for characterizing the joints :
- hardness tests (Table III),
- tensile tests on round specimens taken from the weld metal in the direction of welding (Table IV) and on rectangular specimens transverse to the weld (Table V),
- KCV impact tests at -20°C (Table VI) and
- toughness tests at room temperature (Table VII).

Table III - HV 10

Base metal	305 - 325
HAZ	310 - 330
Plasma bead	320 - 330
Plasma bead + GTA dressing	320 - 340

It appears from the table that the hardness values of the HAZ and the weld metal zone were comparable, evidencing that the weld metal was efficiently protected.

Table IV - Tensile test in the weld metal

Round specimen Weld metal (*)		YS 0.2% (N/mm2)	UTS (N/mm2)	Elong. %	R of A %
Back-side welding +	constricted	880	990	5	11
GTA dressing	flared	890	990	5	9
Face-side welding	constricted	855	1000	5	10

(*) mean value of two specimens

No difference was observed between the constricted and the flared plasma. Elongation was low (5%), albeit homogeneous. This low value should be related to the high oxygen content of the base metal (1600 ppm) which led to high values for the welded joint .

Table V - Tensile test transverse to the weld metal

Prismatic specimen transverse to the weld		UTS (N/mm2)	Failure
Back-side welding +	constricted	965	in base metal
GTA dressing	flared	950	in base metal
Face-side welding	constricted	980	in base metal

(*) mean value of two specimens

Table VI - Charpy V specimen - Test at - 20°C

		KCV (*) (J/cm2)
Base metal		28
Back-side weld	constricted plasma pass	43
Back-side weld	constricted plasma pass	45
+ GTA dressing	flared plasma pass	40
Face-side weld	constricted plasma pass	48

(*) average of three specimens notch parallel to the welding direction.

As compared to the base metal, a marked increase in the impact strength of the weld metal was observed, due to the fine transformed beta structure. The GTA dressing did not reduce the impact strength of the weld metal. No significant difference between face-side weld and back-side weld was noted.

The weld metal toughness was assessed at room temperature by using standard CT15 specimens, with a notch parallel to the welding direction and perpendicular to the surface of the welded coupon.

Table VII - Toughness in the weld metal - test at + 20°C

		K_Q (MPa Vm)
Back-side weld	constricted plasma pass	63-68-72
GTA dressing	flared plasma pass	63-67-70

As was the case for impact strength, the K_Q values for the weld metal were rather medium. This was attributable to the high oxygen content in the weld as a result of the remelting of a base metal containing 1600 ppm of oxygen.

4 - CONCLUSION

Plasma arc welding appears as a valuable process for joining titanium and its alloys, inasmuch as the physical-chemical properties of these materials enhance the performances of the process in terms of maximum penetration and ease of position welding.
Quite deep penetrations could be obtained on TA6V alloy (over 16mm in the flat position) provided a GTA dressing was made (as a matter of fact, beyond 12mm, undercutting formed on the face-side of the welds). Such a defect was more marked in flat position welding. GTA dressing was made with the same equipment, by advancing the electrode tip in such a manner that a "soft plasma" was obtained.
In a less-common application consisting in the execution of the first pass of a multipass weld, the plasma arc welding process offers decisive advantages :

- high quality penetration (appearance and regularity)
- wide fit-up tolerances in any position
- possibility to obtain very thick weld displaying a high resistance to shrinkage during the filling runs
- lower costs and improved productivity.

The TA6V weldments displayed excellent quality, without any porosity cracks, or lack of fusion. Radiographic inspection is well adapted.
The mechanical properties were quite satisfactory : strength and toughness levels were at least as high as those obtained with other fusion welding processes on alloys with comparable interstitial contents.

REFERENCES

1 - **P. DEMARS, J.P. SCHULTZ, F. CUNY** - "Les nouvelles possibilités du procédé plasma en soudage", Soudage et techniques connexes, (04-05) (1973), 211-230.

2 - **D. LAVIGNE , P. VAN DER HAVE M. MAKSYMOVICZ** - "Principe et applications du soudage automatique par plasma d'arc", Soudage et techniques connexes, (07-08) (1986), 221.

3 - **E. CRAIG** - "The plasma Arc process - A. Review", Welding Journal, (02) (1988), 19-25.

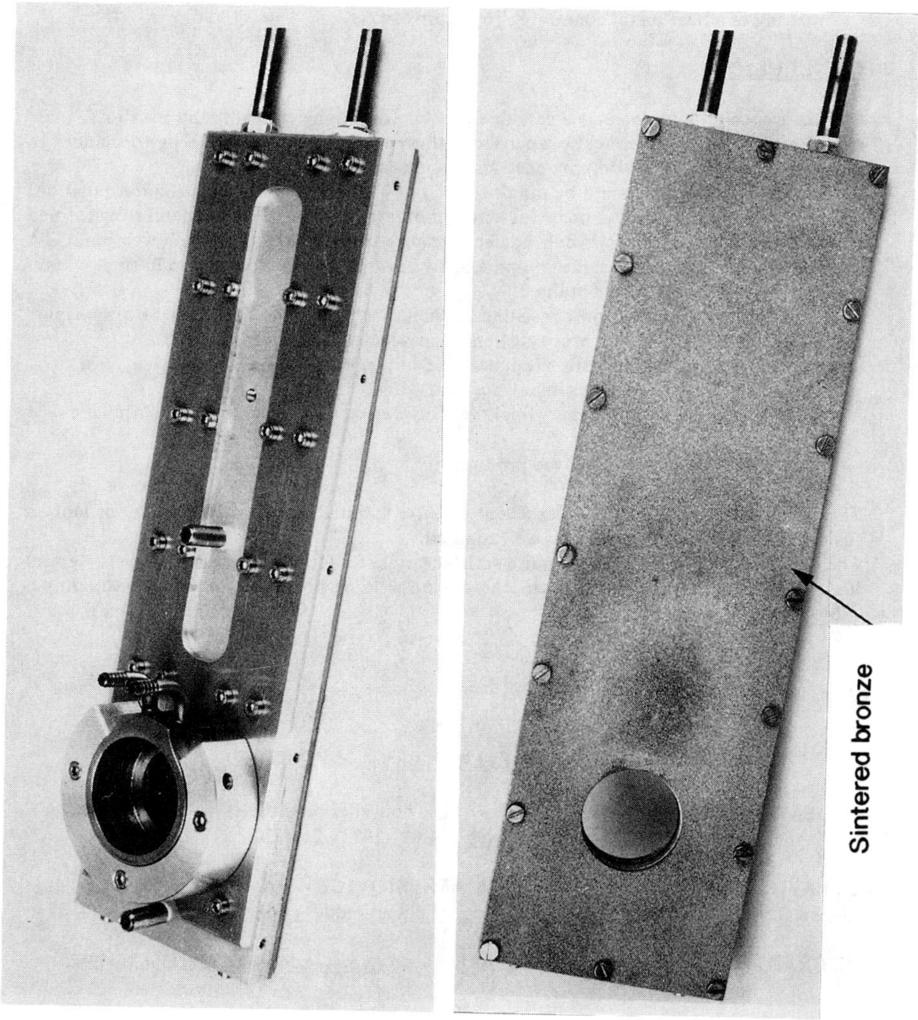

FIGURE 1 - TRAILING SHIELD WITH ARGON DIFFUSER AND WATER COOLING

POSTWELD HEAT TREATMENT OF GTA WELDS IN A HIGH-STRENGTH,

METASTABLE-BETA TITANIUM ALLOY - BETA-C™

W. A. Baeslack III,* P. S. Liu,* D. P. Barbis,** J. R. Schley,** J. R. Wood**

*Department of Welding Engineering
The Ohio State University
Columbus, OH 43210 USA

** RMI Titanium Company
Niles, OH 44446 USA

Abstract

The effects of postweld heat treatment on the microstructure, mechanical properties and fracture characteristics of autogenous gas tungsten-arc (GTA) welds in sheet material of the high-strength, metastable-beta titanium alloy Beta-C™ (Ti-3wt%Al-8wt%V-6wt%Cr-4wt%Mo-4wt%Zr) were investigated. Microstructure characterization revealed a non-uniform aging response in the base metal which resulted from the presence of unrecrystallized regions in the prior solution heat-treated microstructure. Although the weld heat-affected zone (HAZ) exhibited a homogeneous aging response, microsegregation residual from weld solidification promoted the nonuniform precipitation of alpha phase in the fusion zone. An increase in postweld heat treatment temperature resulted in a coarsening of alpha precipitates within the base metal and weld zone beta grains and the growth of coarser and more continuous alpha phase at beta grain boundaries. Consistent with microstructural observations, a higher postweld heat treatment temperature promoted a decrease in base metal and weld zone hardness and strength and an increase in ductility. Postweld heat treatment at 482°C/24hr provided base and weld metal tensile strengths exceeding 1400 MPa, but ductilities of only 6.5 and 2.5%, respectively. Heat treatment at 593°C/8hr resulted in moderate base and weld metal tensile strengths of 1104 and 1120 MPa, respectively, and comparatively high respective tensile ductilities of 15 and 8 percent. Fractography revealed transgranular fracture in the as-welded and low-temperature aged fusion zones. Heat treatment at higher temperatures promoted a mixture of transgranular and intergranular fracture, which was attributed to an increase in alpha phase at beta grain boundaries.

Introduction

Beta-C™ is a metastable-beta titanium alloy which can be thermo-mechanically processed and heat treated to provide an excellent combination of yield and tensile strength, ductility, fracture toughness and fatigue strength. The alloy contains the beta-isomorphous elements Mo and V and the sluggish beta-eutectoid element Cr at levels which depress the beta transus temperature to approximately 750°C and promote retention of the high-temperature beta phase down to room temperature on air cooling. Solution heat treatment above the beta transus temperature (787-870°C) followed by aging at temperatures approximately 150-250°C below the beta transus temperature promote the precipitation of a fine distribution of alpha phase in the beta phase matrix.

Previous weldability studies on metastable-beta titanium alloys (1-4), including a study on the GTA welding of Beta-C™ pipe (4), have shown the fusion and near-HAZ regions of fusion welds to be characterized by low-strength, retained-beta microstructures in the as-welded condition. Although postweld aging at various temperatures can provide a wide range of weld zone strength levels, and high joint efficiencies, the coarse beta-grain size in these regions markedly

Titanium '92
Science and Technology
Edited by F.H. Froes and I. Caplan
The Minerals, Metals & Materials Society, 1993

reduces their ductility relative to the base metal. The present study systematically investigated the postweld heat treatment response of autogenous GTA welds in Beta-C™ sheet. The effect of postweld heat treatment temperature on the weld structure, mechanical properties and fracture behavior was determined and interrelationships were developed.

Experimental

Beta-C™ sheet utilized in this study was provided by RMI Titanium Co. is the solution heat-treated condition (788 °C AC). The chemical composition of the sheet was Ti-3.4wt%Al-7.8wt%V-6.1wt%Cr-4.1wt%M-3.9wt%Zr-0.094wt%O-0.02wt%C-0.016wt%N-0.03wt%Si.

Full-penetration, square-butt welds were produced between coupons of the Beta-C™ sheet 1.35 mm in thickness using the automatic GTA welding process. Welding parameters were: 68.5 amperes DCSP, 8.5 volts, 4.2 mm/s travel rate, 1.56 mm thoriated-tungsten electrode, 12 l/min torch, trailing and backing shielding gas flow rates. Following radiographic inspection, postweld heat treatments were performed on small coupons at 482, 538 and 593°C for 8, 16 and 24 hours. Based on the microstructure, hardness and bend ductility of these first iteration specimens, and desired ultimate tensile strength levels of approximately 1033, 1206 and 1378 MPa, a second set of heat treatments was performed at 482 and 538°C/24 hr and 593°C/8hr for tensile property evaluation and fractographic examination. Following heat treatment, coupons were bead blasted and pickled in an HF + HNO_3 + H_20 solution prior to machining into bend and tensile specimens.

The as-welded and postweld heat-treated coupons were sectioned, metallographically prepared and etched using Kroll's reagent for analysis with light microscopy. Compositional variations in the weld zone were evaluated using electron-probe microanalysis (Cameca SX-50 electron microprobe, 15 kV, 20 mA, one micron beam size). Hardness traverses were produced across welded specimens using Diamond Pyramid hardness testing with a load of 1000 grams. Three-point guided bend testing was performed on base metal and longitudinal-weld oriented specimens 75 mm in length and 25.4 mm width. During testing, the top surface of the weld was placed in tension and the bending strain (elastic + plastic) was determined at which cracking initiated. Tensile testing was performed on longitudinal-weld (essentially all fusion zone) and transverse-weld oriented specimens having a 25.4 mm gage section length and a cross-section of 6.25 x 1.27 mm (weld surface machined smooth). Following mechanical testing, transverse-weld oriented tensile specimens were metallographically prepared and examined to identify the fracture location. Since fracture of longitudinal-weld oriented bend specimens was found to initiate in the weld fusion zone, SEM fractography was performed exclusively on longitudinal-weld oriented tensile specimens.

Results and Discussion

Microstructure Analysis

Solution heat treatment of the base metal sheet at 788°C/30min.AC resulted in a microstructure comprised principally of recrystallized beta grains, with a small proportion of incompletely recrystallized grains. For all postweld heat treatments, these unrecrystallized grains provided nucleation sites for more uniform, denser alpha precipitation versus the surrounding regions. The solution heat treated microstructures also contained a network of fine particles (revealed as etch pits) which are believed to be titanium silicides, $(Ti,Zr)_5Si_3$.

Base metal specimens postweld heat treated at 482°C/8hr were characterized by an incomplete, nonuniform precipitation reaction resulting in appreciable regions of retained-beta phase. As shown in Figures 1A and B, heat treatment at this temperature for 16 and 24 hours promoted the formation of fine, locally homogeneous alpha precipitation with negligible evidence of grain boundary alpha phase formation. Note that the dark-etching regions, which corresponded to unrecrystallized regions in the solutionized base metal, exhibited a denser and finer alpha precipitate structure. Postweld heat treatment at 538°C/8 hours also resulted in a low alpha precipitate density. Aging for 16 and 24 hours (Fig. 1C), however, resulted in a uniform density of precipitates somewhat coarser than those observed in Fig. 1B. Base metals aged at 593°C appeared essentially identical for all heat treatment times. As shown in Figure 1D, the alpha platelets were much coarser than in specimens aged at lower temperatures and, although less apparent, the regions of finer alpha precipitation remained. An increase in postweld heat treatment temperature also resulted in increased intergranular alpha formation.

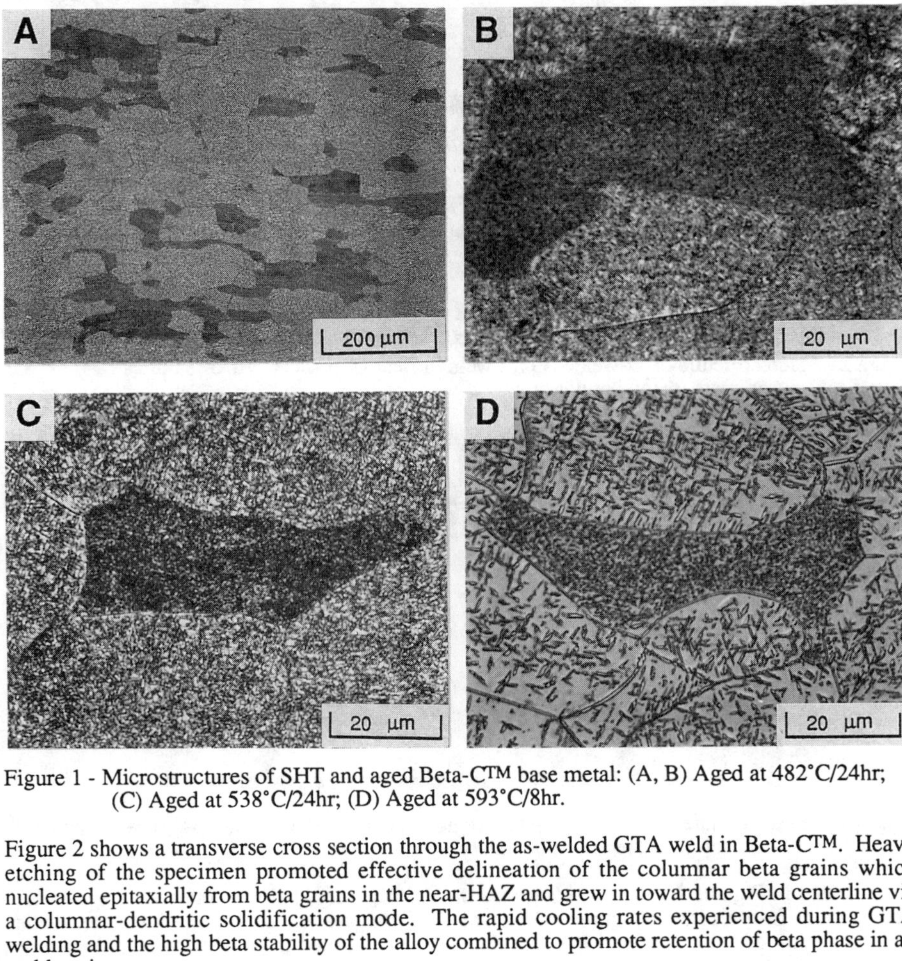

Figure 1 - Microstructures of SHT and aged Beta-C™ base metal: (A, B) Aged at 482°C/24hr; (C) Aged at 538°C/24hr; (D) Aged at 593°C/8hr.

Figure 2 shows a transverse cross section through the as-welded GTA weld in Beta-C™. Heavy etching of the specimen promoted effective delineation of the columnar beta grains which nucleated epitaxially from beta grains in the near-HAZ and grew in toward the weld centerline via a columnar-dendritic solidification mode. The rapid cooling rates experienced during GTA welding and the high beta stability of the alloy combined to promote retention of beta phase in all weld regions.

Figure 3A shows the fusion boundary region in the as-welded condition, and reveals a distinct partially-melted region (PMR) directly adjacent to the fusion zone. This region is characterized by two different beta grain structures. "Liquated" beta grain boundaries are delineated by diffuse-appearing regions of dark contrast and originate from compositional differences between liquid originally present at these boundaries and the adjacent intragranular regions. Superimposed upon the liquated grain boundary is a "migrated" beta grain boundary structure delineated by grain boundaries which exhibit sharp contrast and which are continuous with grain boundaries observed in the near-HAZ. These migrated grain boundaries are "true" high angle beta grain boundaries which exist at room temperature. Electron-probe microanalysis across the liquated grain boundaries revealed a slight decrease in Mo and V (less than 1 wt%) and an increase in Cr (approximately 1 wt%). These results are consistent with the effect of these elements in suppressing the alloy solidus temperature. Note that these compositional variations were likely appreciably greater at solidification temperatures and prior to solid-state cooling. Figure 3B shows the fusion boundary region after postweld heat treatment at 482°C/24hr, and indicates an influence of segregation in the PMR on subsequent alpha precipitation.

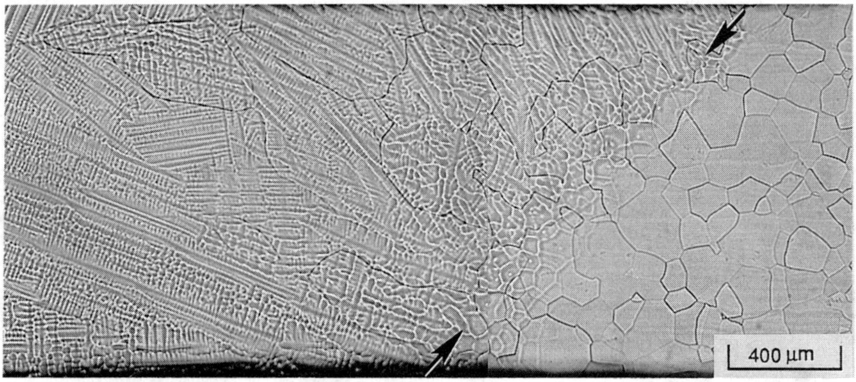

Figure 2 - Macrostructure of as-welded GTA weld in Beta-C™ sheet. Arrows indicate approximate fusion boundary.

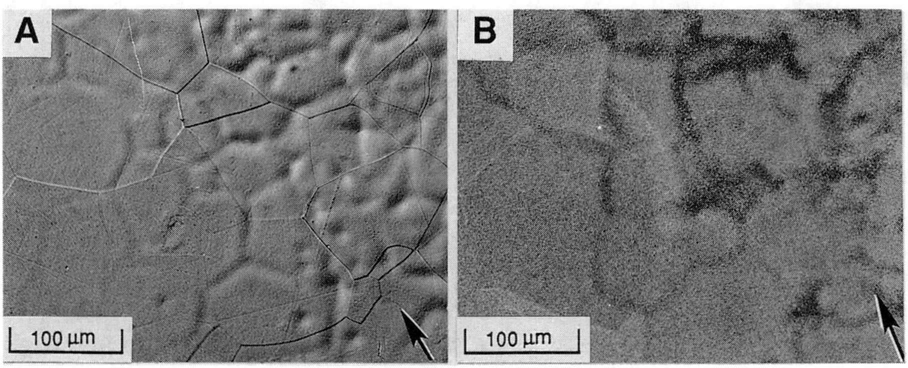

Figure 3 - Microstructures of fusion boundary region in GTA welded Beta-C™ sheet. Arrows indicate approximate fusion boundary.

As indicated in Figures 2 and 3A, beta grains in the PMR provided substrates for the epitaxial nucleation of fusion zone beta grains. As in the PMR, large-angle beta grains which existed at the completion of solidification have straightened and migrated from their original locations. Electron-probe microanalysis across the fusion zone showed an increase in Cr and a decrease in Mo and V in the dendrite interstices at levels comparable to those observed at liquated grain boundaries. These observations are consistent with the equilibrium partition ratios of these elements in titanium.

Alpha precipitation in the weld regions was generally finer than observed in the recrystallized base metal regions for identical heat treatment conditions. Light micrographs in Figures 4A and B show the influence of microsegregation in the weld fusion zone on the alpha precipitation response, with the dendrite interstices exhibiting a different etching response than the relatively uniform dendrite cores. Aging at higher temperatures promoted a more uniform etching response, a coarser alpha precipitate size, and the formation of increased grain boundary alpha phase (Figures 4 C - F). As shown in Figures 4E and F, aging at 593°C/8hr resulted in nearly homogeneous precipitation of coarse alpha precipitates and the growth of continuous grain boundary alpha phase.

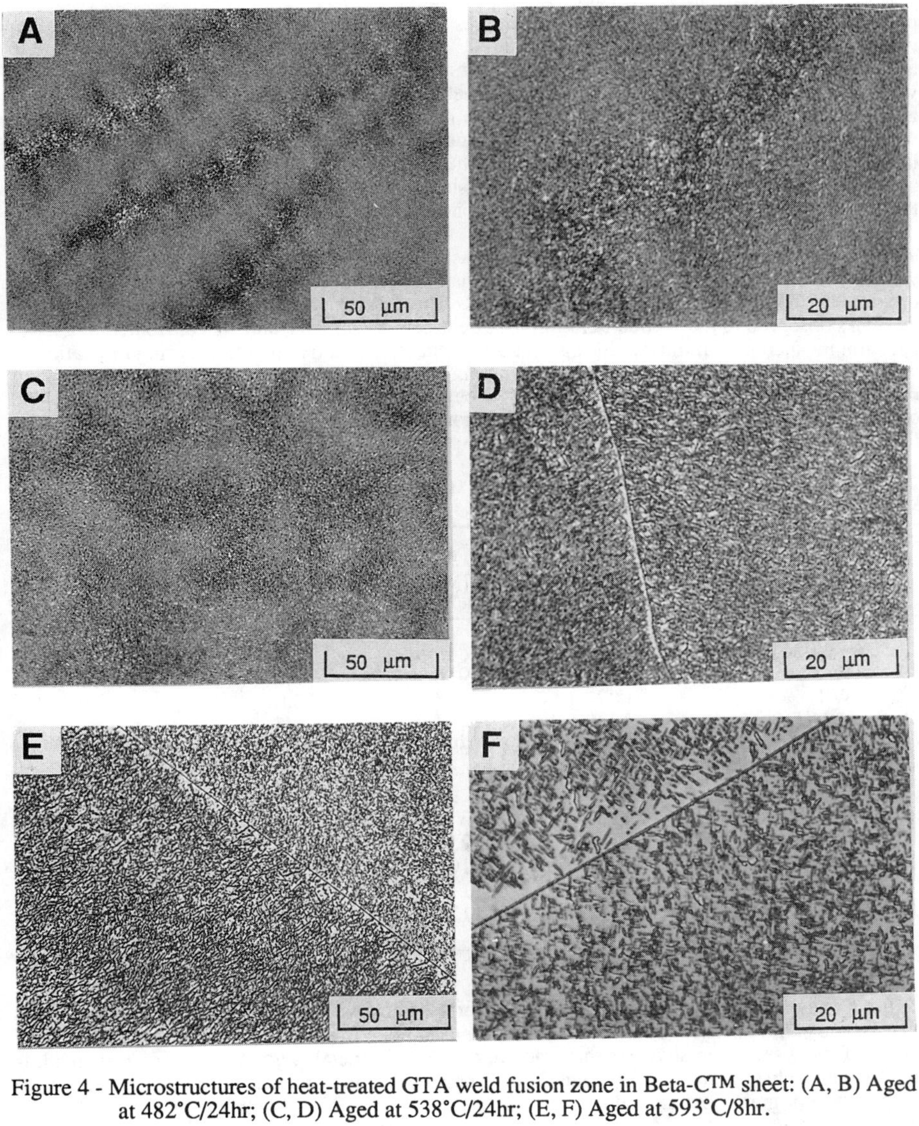

Figure 4 - Microstructures of heat-treated GTA weld fusion zone in Beta-C™ sheet: (A, B) Aged at 482°C/24hr; (C, D) Aged at 538°C/24hr; (E, F) Aged at 593°C/8hr.

Mechanical Properties

Consistent with microstructure analysis, hardnesses of base metals heat treated at 482 and 538°C/8hr were relatively low (DPH 343 and 342, respectively) due to incomplete precipitation. The finer, denser alpha distributions present in the dark-etching regions of these specimens were significantly harder (up to DPH 60 greater) than the light-etching matrix regions. Except for these two heat treatments, hardness was not influenced by heat treatment time.

The weld fusion and HAZ regions age hardened much more rapidly than the base metal, apparently due to the higher dislocation structure induced by thermal strains generated on weld cooling. Hardness and three-point bend testing showed heat treatment at 482°C for 16 and 24 hours to promote high base and weld metal hardnesses (about 390 and 415 DPH, respectively) but marginal weld zone ductility of approximately 3%. An increase in heat treatment temperature to 538°C for heat treatment times of 16 and 24 hours correspondingly promoted a softening of these weld regions (350 and 370 for base and weld metal, respectively), and an improvement in weld zone ductility to about 7.4%. At the maximum heat treatment temperature of 593°C, base and weld metal hardnesses decreased to about 320 and 330, respectively, and the weld zone ductility increased to 9%. Hardness data obtained from these preliminary heat treatments were utilized in conjunction with previously developed relationships between base metal hardness and ultimate tensile strength (5) to select the three aforementioned heat treatments.

Table I summarizes tensile property data for base metal, longitudinal-weld-oriented and transverse-weld-oriented specimens. Tensile strengths obtained for the three heat treatments were approximately 75-100 MPa higher than anticipated. This discrepancy correlated to a slightly higher hardness observed in the second iteration heat treatment specimens, and was attributed to a modification in specimen fixturing during heat treatment (to preclude distortion during relaxation of weld residual stresses) which slowed heating and cooling rates. As in the first iteration specimens, weld zone ductilities were consistently below those of the base metal. Transverse-weld oriented specimens failed exclusively in the fusion zone at a strength level comparable to that of the base metal, indicating joint efficiencies of nearly 100%.

Table I: Properties of Beta - C™ Base Metal and GTA Welds*

Heat Treatment/ Weld Region	YS (MPa)	UTS (MPa)	Tensile Elong. (%)	Failure Location
SHT/BM	854.4	888.8	10.0	--
AW-T	910.9	910.9	9.0	FZ
AW-L	877.1	888.1	13.0	--
482 °C/24hr/BM	1349.8	1454.0	6.5	--
482 °C/24hr/WM/L	1299.6	1382.4	2.5	
482 °C24hr/WM/T	1344.2	1446.2	3.0	FZ
538 °C/24hr/BM	1176.8	1298.1	9.0	--
538 °C/24hr/WM/L	1160.4	1217.6	3.5	
538 °C/24hr/WM/T	1167.9	1253.3	3.5	FZ
593 °C/8hr/BM	994.9	1103.8	15.0	--
593 °C/8hr/WM/L	1040.4	1121.1	8.0	
593 °C/8hr/WM/T	1051.4	1121.7	7.5	FZ

* Average of two specimens/condition.

Figure 5 shows a plot of UTS versus hardness for base and weld metal specimens, and indicates that for the same hardness level the weld metal exhibits a lower tensile strength than the base metal. This observation suggests that fracture of the weld metal is governed by factors in addition to matrix characteristics of the transformed-beta microstructure, i.e., prior-beta grain size and the presence of grain boundary alpha.

Fractography

Fracture of the as-welded, longitudinal-weld oriented tensile and bend specimens occurred transgranularly by a void nucleation, growth and coalescence mechanism commonly observed in single-phase beta titanium (Figs. 6A and 6B). Figures 6C and 6D show the fracture surface of a longitudinal-weld oriented tensile specimen postweld aged at 482°C/24hr. Fracture occurred transgranularly, with large facets visible which correspond to underlying columnar fusion zone beta grains. Despite the macroscopic appearance and low tensile ductility, on a microscopic scale fracture occurred by a ductile-appearing mechanism. Similar fracture characteristics have been reported in Ti-8Mo-8V-2Fe-3Al (2) and Ti-15V-3Al-3Cr-3Sn GTA welds (3) postweld aged at temperatures below approximately 538°C. Fusion zone fracture of welds postweld aged at

538°C/24 hours also occurred in a predominantly transgranular manner, but did exhibit evidence of occasional intergranular fracture. The occurrence of intergranular fracture was attributed to the presence of increased grain boundary alpha. Figures 6E and 6F show the fracture surface of a weld fusion zone heat treated at 593°C/8 hr. Consistent with the increased thickness and continuity of alpha phase at prior-beta grain boundaries in this microstructure, a greater proportion of intergranular fracture was observed.

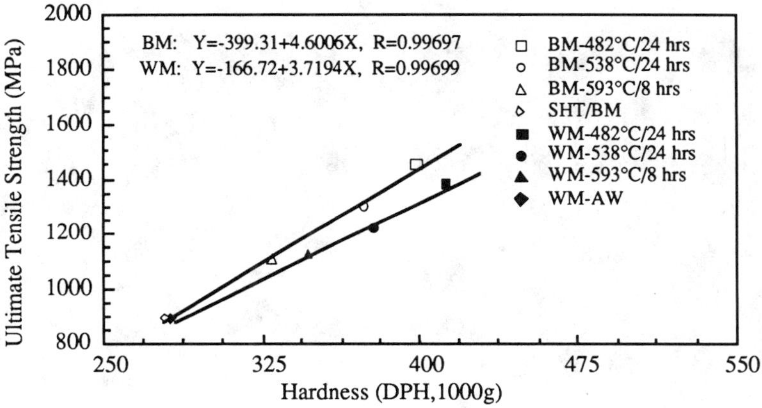

Figure 5 - Plot of ultimate tensile strength versus hardness for heat-treated base and weld metals in Beta-C™ sheet.

It is of significance to note that despite the increased proportion of grain boundary alpha phase and intergranular fracture with an increase in postweld aging temperature, tensile and bend ductility of the base and weld metal specimens increased. This result indicates the importance of additional factors in determining macroscopic ductility, including microstructural and compositional characteristics of the interphase boundary between the grain boundary alpha and beta phase (which may influence slip behavior and void formation which leads to intergranular crack initiation) and deformation characteristics of the two-phase grain interior.

Summary

Relationships between postweld heat treatment temperature and time and resulting base metal and weld zone microstructures, hardness and tensile properties, and weld zone fracture behavior have been determined for GTA welds in Beta-C™ sheet.. Heat treatment at 482°C promoted high hardness and strength in the base and weld metals, but limited weld zone ductility. The observation of transgranular fracture in the weld zone of these specimens indicated that low ductility resulted from the fine alpha distribution, rather than from the presence of grain boundary alpha. The utilization of higher postweld heat treatment temperatures promoted coarser alpha phase, lower strength microstructures, but improved ductility. Fractographic analysis showed that although higher temperature heat treatments promoted greater grain boundary alpha precipitation and higher proportions of grain boundary fracture, ductility of the weld zones increased with aging temperature.

References

1. J. F. Rudy, F. A. Crossley and H. Schwartzbart, "Heat Treatment and Welding of 13%V-11%Cr-3%Al Titanium Base Sheet Alloy," Welding Journal, Research Supplement, 40 (1961) 447s-458s.
2. M. A. Greenfield and C. M. Pierce, "Postweld Aging of a Metastable-Beta Titanium Alloy," Welding Journal, Research Supplement, 52 (1973) 524s-527s.

3. D. W. Becker and W. A. Baeslack III, "Property-Microstructure Relationships in Metastable-Beta Titanium Alloy Weldments," Welding Journal, Research Supplement, 59 (1980) 85s-92s.
4. D. E. Thomas, "Structure and Properties of a Welded Beta Titanium Alloy,"Proceedings of Sixth World Conference on Titanium, ed. P. Lacombe, R. Tricot, G. Beranger (Les Ulis Cedex, France, esles editions de physique, 1988), 1421-1427.
5. Unpublished Research, RMI Co., Niles, Ohio.

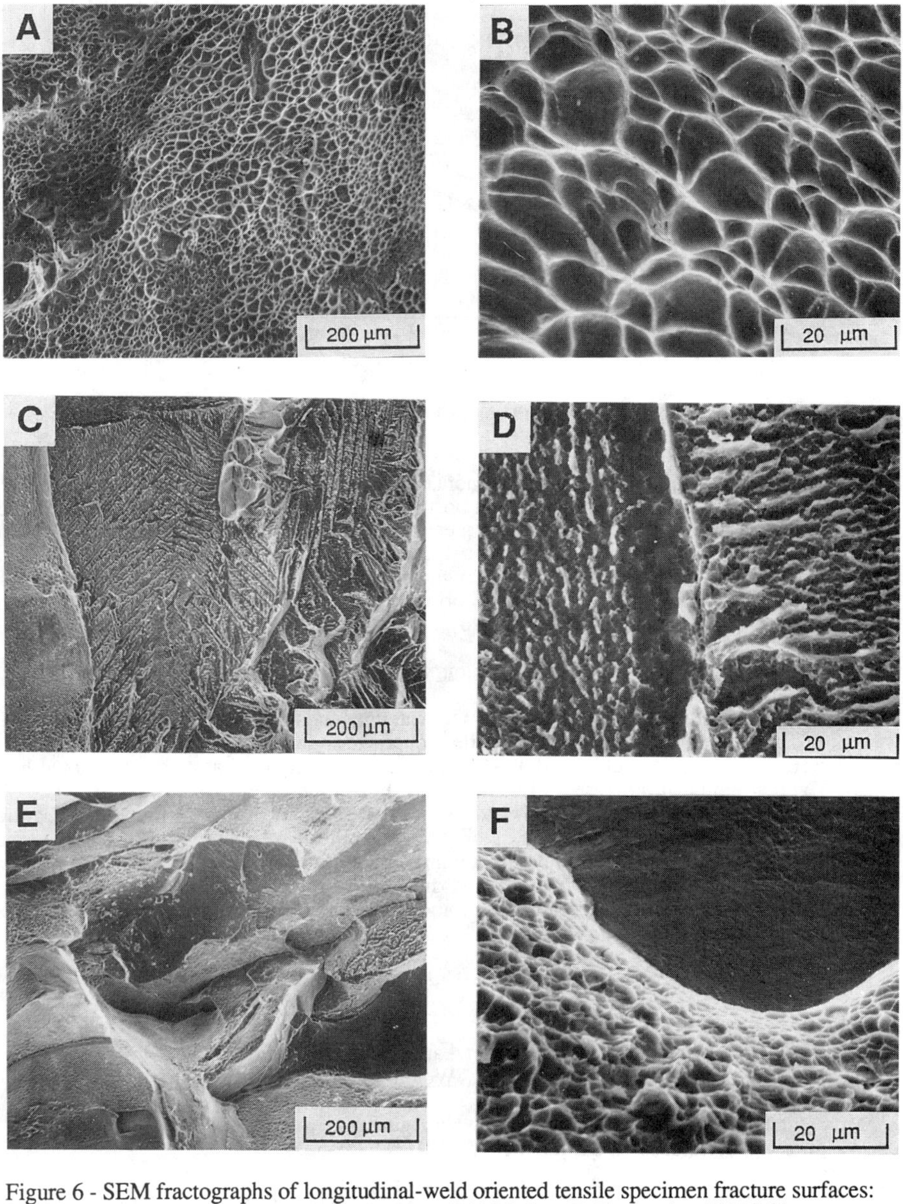

Figure 6 - SEM fractographs of longitudinal-weld oriented tensile specimen fracture surfaces: (A, B) As-Welded; (C, D) Aged at 482°C/24hr; (E, F) Aged at 593°C/8hr.

LASER WELDING OF AN OXIDATION RESISTANT,

METASTABLE-BETA TITANIUM ALLOY - BETA-21S

P. S. Liu,* K. H. Hou,* W. A. Baeslack III* and J. Hurley**

*Department of Welding Engineering
The Ohio State University
Columbus, OH 43210 USA

** Edison Welding Institute,
Columbus, OH 43212 USA

Abstract

Laser welding was utilized to generate high-integrity welds in 1.5 mm thick Beta-21S sheet. Welds were characterized by narrow fusion and heat-affected zones and a relatively fine beta grain size. Postweld heat treatment at 482°C/20hr promoted extremely fine alpha precipitation in the base and weld metals and correspondingly high tensile strengths (exceeding 1400 MPa), but marginal ductility. Heat treatment at 593°C/4hr resulted in appreciably coarser alpha precipitation both intragranularly and at beta grain boundaries. This microstructure was associated with a decreased tensile strength (1027 MPa) but a significant improvement in ductility. An intermediate combination of strength and ductility was achieved using a duplex heat treatment of 482°C/20hr + 593°C/30min. Fractographic analysis of the welded specimens indicated a transition from transgranular to predominantly intergranular fusion zone fracture with an increase in heat treatment temperature, which was attributed to the presence of appreciable alpha phase at beta grain boundaries.

Introduction

Beta-21S is a new, metastable-beta titanium alloy (nominal composition: Ti-15wt%Mo-2.7wt%Nb-3wt%Al-0.2wt%Si) which exhibits superior oxidation resistance and mechanical properties at elevated temperatures and excellent corrosion and hydrogen resistance (1,2). These properties, combined with excellent cold formability and age hardening characteristics, have made Beta-21S a strong candidate for applications in metal-matrix composites, "warm" aerospace structures (i.e., from 204-427°C) and downhole oil service.

As a structural alloy, fabricability by fusion welding represents an important consideration in determining potential alloy applications. Although metastable-beta titanium alloys are generally considered to be very weldable (3,4), the presence of coarse, columnar beta grains in the fusion zone of conventional gas tungsten-arc (GTA) welds can promote the degradation of weld zone ductility relative to the base metal, particularly in high-strength conditions. The high energy density and low net energy input utilized during laser welding generates narrow weld fusion and HAZ's characterized by a fine beta grain structure (5), and correspondingly the potential for enhanced weld zone ductility versus arc welds. The objective of the present study was to evaluate the laser welding characteristics of Beta-21S sheet and determine the influence of postweld aging on the weld structure, mechanical properties and fracture characteristics.

Experimental Procedure

Beta-21S sheet evaluated in this study was provided by TIMET Co. in the solution heat-treated condition (843°C/30min.AC). The 1.5 mm thick sheet was laser cut into coupons 150 mm x 50

Titanium '92
Science and Technology
Edited by F.H. Froes and I. Caplan
The Minerals, Metals & Materials Society, 1993

mm with the coupon length oriented parallel to the sheet rolling direction. Prior to welding the coupons were degreased in acetone.

Full-penetration, bead-on-plate laser welds were produced in the coupons using a GE Fanuc C 3000 CO_2 laser equipped with a 190 mm plano-convex ZnSe lens. A laser power of 3000 watts and beam traversing rate of 95 mm/s were utilized. Welding was performed in a helium-purged collapsible chamber in order to prevent atmospheric contamination.

Following welding, coupons were sectioned and heat treated at temperatures of 482°C and 593°C for 2, 4, 8, 16 and 24 hours. Based on Diamond Pyramid Hardness testing (1000 g load) and heat treatment information developed previously by TIMET Co. (1,2), three postweld heat treatments were selected for more detailed characterization: 1) 482°C/20hr AC; 2) 593°C/4hr AC; 3) a duplex heat treatment of 482°C/20hr + 593°C/30min.

Characterization of the as-welded and postweld heat treated welds included: 1) light and analytical-electron microscopy; 2) DPH hardness testing; 3) three-point bend testing of longitudinal and transverse-weld oriented specimens; 4) tensile testing of transverse-weld oriented specimens (12.7 mm gage length, 1.5 mm x 3.2 mm gage cross-section); 5) fractographic examination using scanning-electron microscopy.

Results and Discussion

Base and Weld Metal Age Hardening Response

Figure 1 shows a plot of DPH hardness versus aging time for the Beta-21S base metal and laser weld fusion zone. As shown, the as-welded fusion zone exhibited a hardness slightly higher than that of the solution heat-treated base metal. Since both microstructures were comprised of retained-beta phase, this difference was attributed principally to the presence of a fine solidification subgrain structure in the weld fusion zone. The base metal and fusion zone aging curves for heat treatment at 482°C showed the fusion zone hardnesses to be slightly greater than those of the base metal, with peak hardness achieved in 16 hours and the occurrence of slight overaging at 24 hours. Base metal and fusion zone specimens aged at 593°C reached peak hardness at four and two hours, respectively, and experienced noticeable overaging at longer aging times. Hardness data obtained in the present investigation generally paralleled tensile strength data developed by TIMET Co. (1), with a principal difference being more significant overaging at 593°C. Based on the results of this study, and in order to allow a direct comparison with previous work by TIMET Co. (1), the aforementioned single-cycle heat treatments were used in the present investigation.

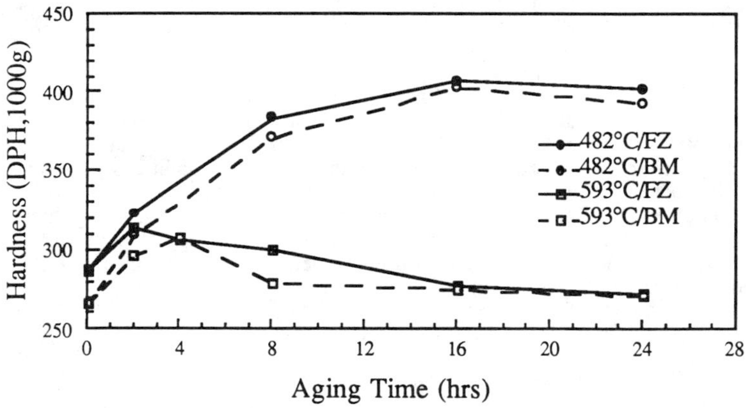

Figure 1 - DPH hardness of Beta-21S base metal and laser weld fusion zone versus postweld heat treatment (aging) time.

Microstructure Analysis

As shown in Figure 2A, the unaffected base metal microstructure was comprised predominantly of nearly-equiaxed, recrystallized beta grains approximately 100 microns in diameter and occasional unrecrystallized regions elongated in the rolling direction. Although not apparent at low magnification, distinct solute bands and the presence of fine silicides were visible in the solution heat-treated base metal microstructure.

The weld fusion zone was characterized by a "crown" which experienced 3-D (conduction-limited) heat flow on the top surface, a narrow region at the center of the sheet which experienced 2-D heat flow, and a smaller crown on the bottom surface. Examination of the fusion boundary region indicated a narrow HAZ and the epitaxial nucleation and growth of columnar beta grains in the fusion zone directly from beta grains in the HAZ. Figure 2B shows a transition from a cellular growth mode near the fusion boundary (light etching region of fusion zone in Figure 2A) to a cellular-dendritic growth mode nearer to the weld centerline. Solidification of the weld crowns occurred completely via a cellular-dendritic mode.

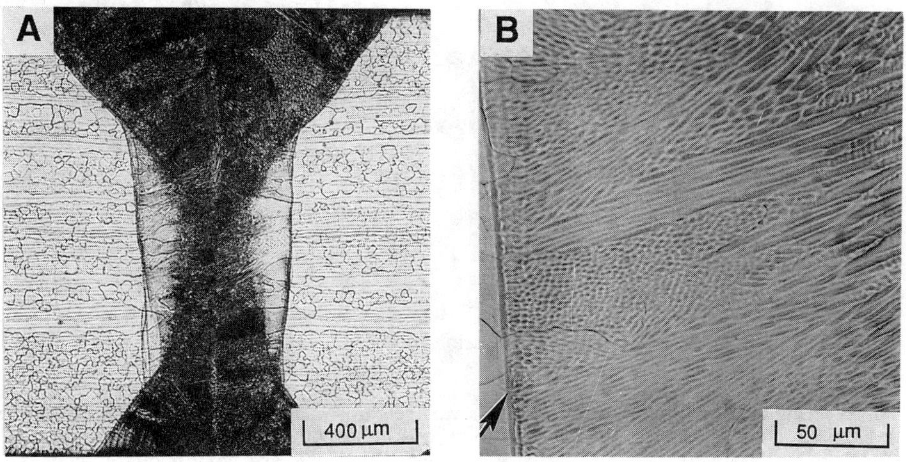

Figure 2 - Laser weld in Beta-21S sheet. Arrow in (B) indicates fusion boundary.

Figure 3 shows light (A,C) and TEM bright-field (B,D) micrographs of the base metal and fusion zone regions of the laser weld heat-treated at 482°C/20hr. Although nearly unresolvable with light microscopy, TEM analysis revealed a transformed-beta microstructure comprised of fine alpha platelets which had nucleated both intragranularly and from beta grain boundaries. The alpha precipitate structure in the weld zone was likewise difficult to delineate using light microscopy, although an effect of weld solidification segregation on alpha precipitation was apparent in Figure 3C. TEM analysis of this microstructure (Fig. 3D) revealed the precipitation of extremely fine alpha platelets somewhat finer than those observed in the base metal. Although a direct correlation was not made, variations in the coarseness of these platelets in different regions of the thin foils may have been related to the contrast differences apparent in Figure 3C and associated with compositional variations in the fusion zone residual from microsegregation experienced during weld solidification. It is important to note that only occasional evidence of extremely thin grain boundary alpha was observed in the low-temperature aged base metal and fusion zone microstructures.

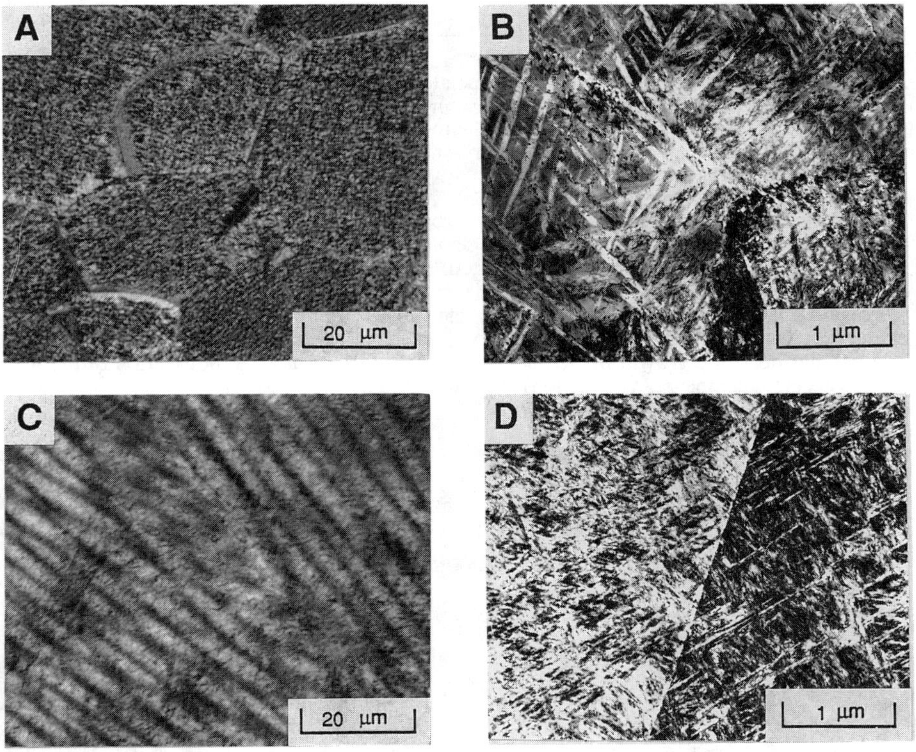

Figure 3 - Base metal (A,B) and laser weld fusion zone (C,D) in Beta-21S postweld heat treated at 482°C/20hr.

Alpha precipitation during the 593°C/4hr postweld heat treatment resulted in beta decomposition to an appreciably coarser alpha platelet structure, with alpha phase nucleating both intragranularly and from beta grain boundaries (Figs. 4A-C). The presence of coarse, continuous alpha phase at beta grain boundaries was visible optically and more clearly resolved using electron microscopy. The lower hardness and strength of this microstructure relative to the base metal heat treated at lower temperatures was consistent with the coarse alpha platelet size and large regions of retained beta phase. Light microscopy of the laser weld fusion zone (Fig. 4D) indicated a finer alpha precipitate size and an influence of solidification-induced segregation on the alpha morphology. TEM analysis confirmed the finer alpha platelet size (Fig. 4E), although significant differences in alpha platelet coarseness within the fusion zone were not apparent. Although not resolvable in Figure 4D, Figure 4F clearly reveals the presence of coarse, continuous alpha phase at a beta grain boundaries.

Figures 5A-D show light and TEM bright-field micrographs of the duplex heat-treated laser weld. In general, the base metal and fusion zone regions appeared comparable to those of the low-temperature heat-treated specimen, with the short temperature excursion at 593°C promoting a slight coarsening of intragranular and intergranular alpha phase. This minimal coarsening was consistent with a hardness only slightly below that of weld aged at 482°C. The original objective in utilizing the duplex heat treatment was to promote the nucleation of fine, intragranular alpha phase at low temperatures and its coarsening at a higher temperature, versus aging at an intermediate temperature which promotes nucleation and growth of both intragranular and grain boundary alpha. Although the microstructure did exhibit a higher ductility as compared to the single-cycle, low-temperature aged specimen, this improvement was relatively marginal.

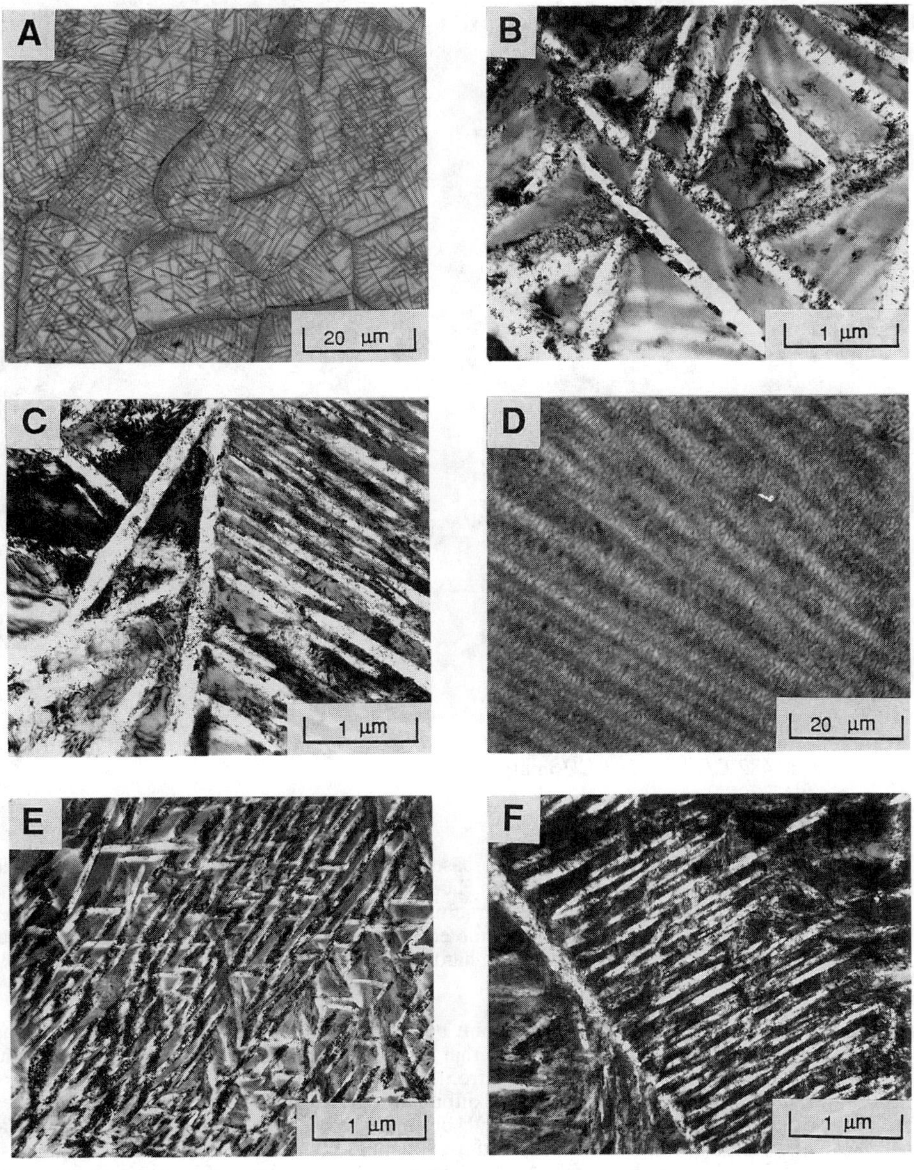

Figure 4 - Base metal (A-C) and laser weld fusion zone (D-F) in Beta-21S postweld heat treated at 593°C/4hr.

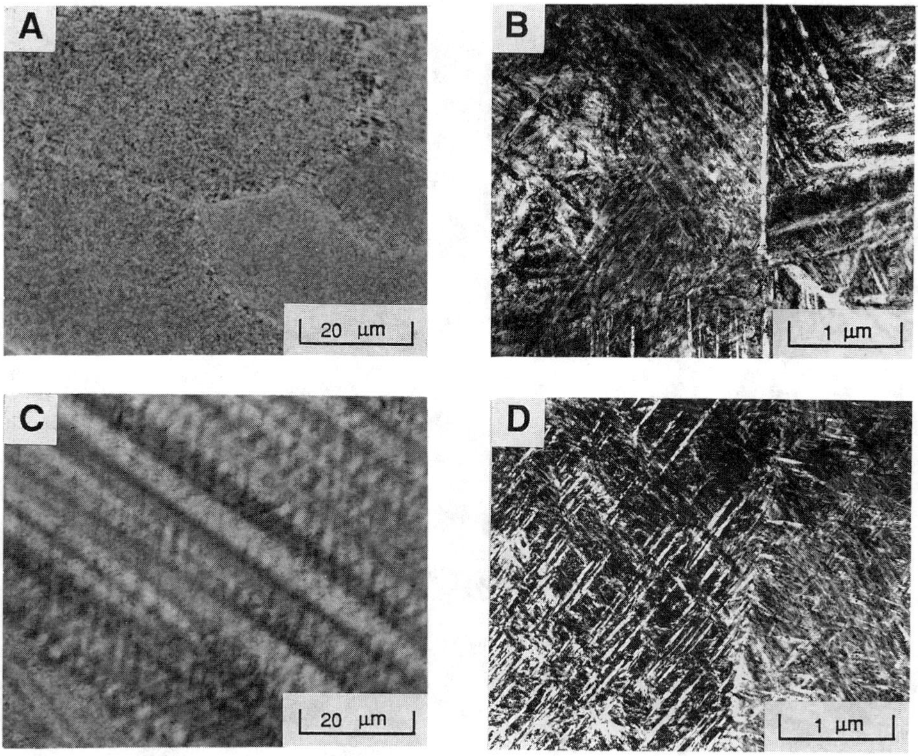

Figure 5 - Base metal (A,B) and laser weld fusion zone (C,D) in Beta-21S postweld heat treated at 482°C/20hr + 593°C/30min.

Mechanical Properties

Figure 6 shows hardness traverses from the base metal to the weld centerline for the postweld heat-treated Beta-21S laser weld specimens. Hardness values were relatively constant across the weld zone and were consistent with microstructural characteristics described above (i.e., a finer intragranular alpha precipitate structure promoted a higher hardness). For the single-cycle heat treated specimens, the weld HAZ exhibited a hardness slightly higher than that of the comparable fusion zone and base metal microstructures.

Table I summarizes the results of three-point bend and tensile testing for the as-welded and postweld aged Beta-21S laser welds. Note that all transverse-weld oriented tensile specimens failed in the unaffected base metal remote from the weld zone, indicating joint efficiencies of 100%. The retained-beta microstructures exhibited by the solution heat-treated and as-welded laser welds promoted relatively low strength but high tensile elongation. Postweld aging at 482°C/20hr resulted in a high base metal and weld yield and tensile strengths and moderate base metal tensile ductility (approximately 4-5%). Three-point bend testing of longitudinal-weld oriented weld specimens, however, indicated a low fusion zone bend ductility of 2% (elastic + plastic strain at crack initiation). Application of the high temperature postweld heat treatment resulted in moderate strength but promoted high base metal tensile ductility (approximately 10%) and good weld zone bend ductility (5.7%). The results of DPH hardness analysis and three-point bend testing indicated that the duplex heat treatment provided strength and ductility levels intermediate to those of the low and high-temperature single-cycle heat treatments.

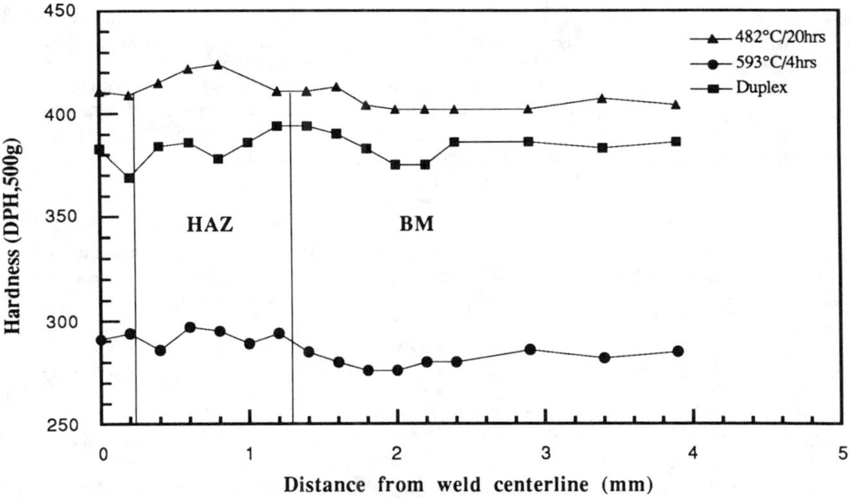

Figure 6 - DPH hardness versus distance from weld centerline for postweld heat-treated laser welds in Beta-21S.

Table I: Mechanical Properties of Beta-21S Laser Welds

Condition/ Orientation	YS (MPa)	UTS (MPa)	Tensile Elong. (%)	Bend Ductility (%)
AW/TVS	620.1	861.3	18.7	--
482 °C/20hr BM*	1364.2	--	--	4.6
-- --/20hr TVS	1405.6	1467.6	6.1	3.8
-- --/20hr Long	--	--	--	2.0
593 °C/4hr BM*	1019.7	1102.4	8.5	10.0
-- --/4hr TVS	964.6	1033.5	12.5	11.0
-- --/4hr Long	--	--	--	5.7
Duplex**/BM	--	--	--	6.4
-- --/Long	--	--	--	3.2

* TIMET Co. Data, Average of duplicate L and T oriented tests (1,2)
** Duplex heat treatment: 482°C/20hr + 593°C/30 min.

Fracture Analysis

Figure 7 shows fracture surfaces of longitudinal-weld oriented bend specimens in the as-welded and postweld heat treated conditions. In the as-welded condition (Figs. 7A and 7B), fracture surfaces were characterized by a ductile, transgranular appearance, indicating fracture by a void nucleation, growth and coalescence mechanism. Fracture of the specimens heat treated at 482°C/20hr (Figs. 7C and 7D) and those given the duplex heat treatment also occurred transgranularly in a manner which macroscopically exhibited a cleavage-type appearance. Despite the relatively low bend ductility of these specimens, examination of the fracture surface at increased magnification indicated a microscopically ductile appearance on the facet surfaces (Fig. 7D). A similar behavior has been observed in the weld fusion zone of high-strength alpha-beta titanium alloys (6). As shown in Figure 7E, heat treatment at 593°C/4hours promoted a transition from transgranular to predominantly intergranular fracture. Examination of the intergranular fracture surfaces at increased magnification revealed a relatively smooth surface, although some evidence of microscopic ductility was observed (Fig. 7F). A similar transition in weld zone fracture morphology with an increase in postweld heat treatment temperature has been observed in the metastable-beta titanium alloys Ti-8V-7Cr-3Al-4Sn-1Zr (wt%) (3) and Ti-8Mo-8V-2Fe-3Al (wt%) (4), and was attributed to the presence of coarse, continuous grain boundary alpha in the microstructure of welds postweld aged at high temperature. It is of interest to note that this transition in fracture path was associated with an increase in tensile ductility. This observation indicates that the extent of plastic deformation which occurs intragranularly within the beta grains prior to grain boundary fracture is important in determining macroscopic ductility, and that the presence of grain boundary alpha does not necessarily indicate a degradation in ductility.

Conclusions

1. High-integrity welds exhibiting a fine beta grain structure were produced in Beta-21S sheet using the CO_2 laser welding process. The retention of beta phase in the weld fusion and heat-affected zones during rapid weld cooling promoted moderate weld strength and high ductility in the as-welded condition.
2. The precipitation of fine alpha phase during postweld aging at 482°C/20 hours resulted in high base metal and weld zone tensile strengths (>1467 MPa) but relatively low ductility in the weld zone (2%). An appreciably coarser alpha precipitate structure produced during heat treatment at 593°C/4hr promoted a moderate weld tensile strength (>1033 MPa) and a weld zone ductility of 5.7%. The application of a duplex postweld heat treatment of 482°C/20hr + 593°C/30min. resulted in intermediate strength and ductility levels.
3. A transition in fracture mode from transgranular in the as-welded and low-temperature postweld heat treated conditions to predominantly intergranular fracture for the high-temperature postweld heat treated condition was attributed to the presence of appreciable grain boundary alpha phase at beta grain boundaries in the welds heat treated at higher temperature.

References

1. P. J. Bania and W. M. Parris, "Beta-21S: A High-Temperature Metastable-Beta Titanium Alloy," Titanium 90 - Products and Applications, (Dayton, OH: Titanium Development Association, 1990), 784.

2. P. J. Bania and W. M. Parris, Presentation at 119th TMS Annual Meeting, Anaheim, CA (1990).

3. D. W. Becker and W. A. Baeslack III, "Property-Microstructure relationships in Metastable-Beta Titanium Alloy Weldments," Welding Journal, Research Supplement, 59 (1980) 85s.

4. M. A. Greenfield and C. M. Pierce, "Postweld Aging of a Metastable-Beta Titanium Alloy," Welding Journal, Research Supplement, 52 (1973) 524s.

5. W. A. Baeslack III and C. M. Banas, "A Comparative Evaluation of Laser and Gas Tungsten-Arc Welds in High-Temperature Titanium Alloys," Welding Journal, Research Supplement, 60 (1981) 121s.

6. W. A. Baeslack III and D. W. Becker, "Fusion Zone Fracture Behavior of Weldment in Alpha-Beta Titanium Alloys," Metallurgical Transactions, 10A (1979), 1803.

Acknowledgement

The authors express appreciation to Dr. Paul Bania of TIMET Co., Henderson, Nevada, for providing the Beta-21S sheet material.

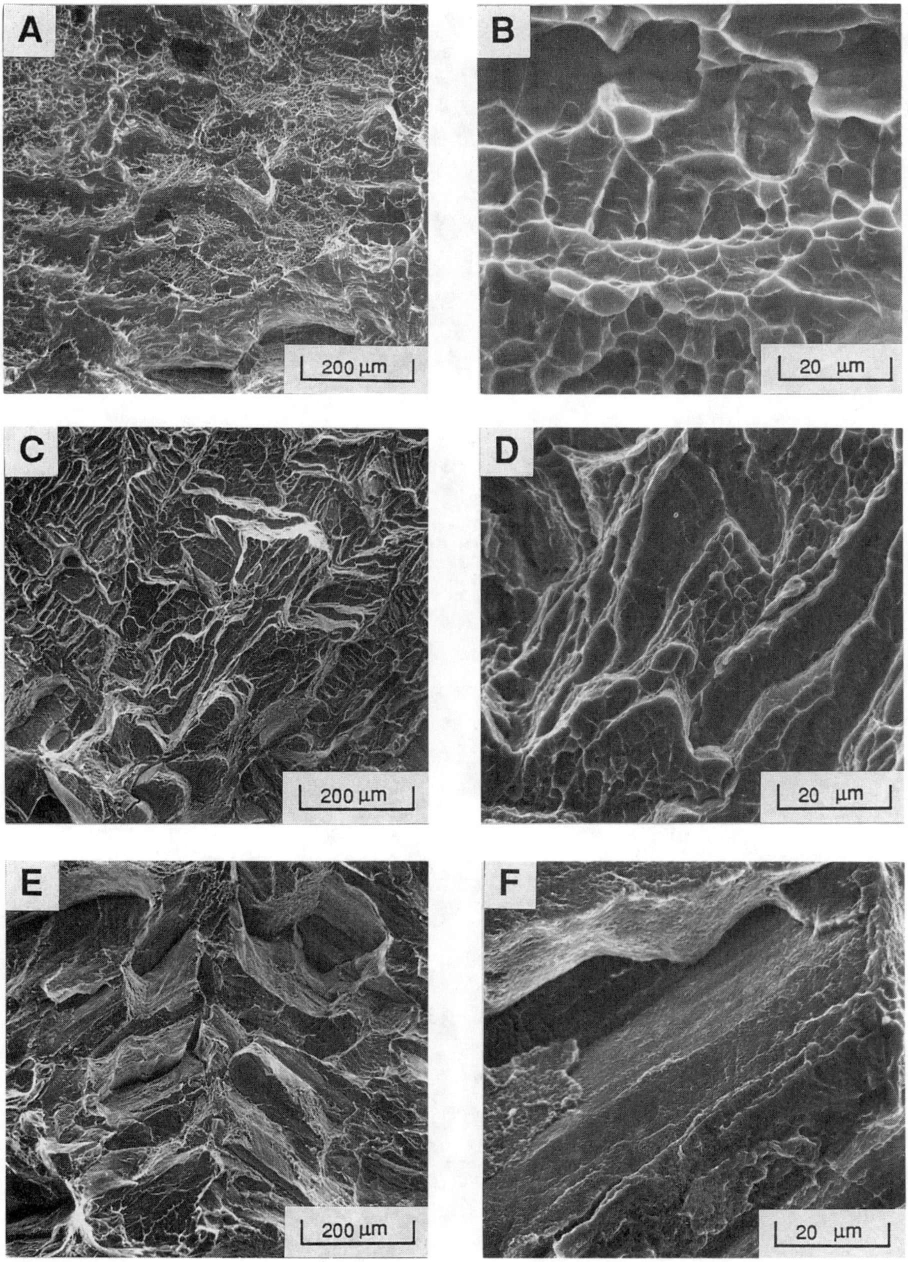

Figure 7 - SEM fractographs of longitudinal-weld oriented three-point bend specimens: (A,B) As-welded; (C,D) Aged at 482°C/20hr; (E,F) Aged at 593°C/4hr.

NEAR-NET SHAPE PROCESSING OF GAMMA AND XD™ GAMMA

TITANIUM ALUMINIDES BY INVESTMENT CASTING

D.E. Larsen, Jr., D.A. Wheeler and B. London

Howmet Corporation
Operhall Research Center
1500 South Warner Street
Whitehall, MI 49461

Abstract

Gamma and XD gamma titanium aluminides could potentially replace superalloys and other high temperature materials in aerospace and automotive applications. Their low density, high temperature properties and oxidation resistance make them attractive materials for advanced engine designs and aerospace systems. However, the near-net shape processing of these advanced materials is critical to their implementation. The investment casting process has been demonstrated for the manufacture of gamma alloy near-net shape components using only minor modifications to conventional titanium alloy melting and casting practices. The resulting components respond favorably to heat treatment, producing a wide range of microstructures and corresponding mechanical properties. Examples of complex cast components, microstructures and mechanical properties, demonstrating the potential of these materials for high temperature applications, are presented.

Introduction

Titanium alloys, both cast and wrought, have been accepted materials for use in aerospace and other applications for over three decades.[1] However, conventional titanium alloys do not have useful strength or oxidation resistance above approximately 600°C. In addition, under certain service conditions, conventional titanium alloys can ignite resulting in a titanium fire which can result in severe component damage.[2]

Titanium aluminide alloys based on the gamma (TiAl) system have low density, high temperature strength, oxidation resistance, good modulus retention, and are titanium fire resistant. Gamma alloys can perform to temperatures that can exceed 815°C at approximately half the density of current superalloys. However, these alloys suffer from low room temperature ductility, fracture toughness, and impact resistance. Typically, gamma alloys contain 45-48 at.% aluminum and alloying elements such as vanadium, niobium, tantalum, tungsten, chromium, silicon and manganese.[2,3] These alloys are actually two-phase containing a small amount of alpha-two (Ti_3Al). XD gamma alloys contain a fine dispersion of TiB_2 particles which promote grain refinement resulting in strength increases over the base alloy.[4,5,6]

Titanium '92
Science and Technology
Edited by F.H. Froes and I. Caplan
The Minerals, Metals & Materials Society, 1993

A large portion of gamma titanium aluminide research has been performed using powder metallurgy and/or thermomechanical processing (TMP) routes.[2,3] However, this type of processing is difficult since it requires temperatures in the 1000 - 1300°C range and an inert atmosphere to avoid interstitial contamination. These high processing temperatures require complex canning or pack configurations and the use of expensive die or containment materials. Gamma alloys are machinable but not to the degree of conventional titanium alloys. Therefore, the extensive machining needed to produce an actual component from a powder metallurgy or wrought product is expensive. The manufacture of near-net shape or net shape gamma titanium aluminide components by investment casting represents a large potential cost savings. These cost savings are obtained through decreased machining and use of existing conventional titanium production facilities.[7,8] Additionally, complex shapes with hollow passages can potentially be manufactured that would be difficult or impossible by powder metallurgy or TMP approaches.

Investment Cast Processing

The preferred melting method for ingot manufacture of gamma alloys is vacuum arc remelting (VAR). Gamma alloys are very sensitive to tramp element contamination.[2] The absence of ceramics in the cold hearth VAR process produces ingots with less than 500 ppm of oxygen and low levels of other detrimental elements. Careful selection of alloying elements that contain minimal oxygen and other contaminants also is critical for the manufacture of high quality gamma alloys. Gamma alloys have a lower melting point than several of the alloying elements used in their formulation. Therefore, a minimum of two melts for casting ingot and three melts for forging ingot is necessary to ensure proper solution of alloying elements.

The VAR process also is used to melt gamma alloys for component casting. One disadvantage of the VAR casting process is that the method produces a low superheat in the molten metal prior to casting. This lowers the metal's ability to flow into components with thin sections under 2.0 mm thick. Therefore, centrifugal casting is used to fill thin sections by accelerating the metal into the mold at forces that can exceed 25g.

Ceramic investment casting molds used to cast gamma alloys are similar to those used for conventional titanium alloys. The high aluminum content in gamma alloys lowers their reactivity making them less reactive with the mold system. Therefore, reaction layers (e.g., alpha-case) found in conventional titanium alloys can be reduced or avoided in gamma alloys by using a low reactivity Mono-Shell® molding system.[7] In the as-cast condition, gamma alloys are brittle requiring innovative techniques to avoid component cracking following casting. Lowering the strength of the ceramic mold system to less than that of the component can help avoid component cracking. Gating used to feed metal into the component can be used as a reinforcement or to relieve stresses by allowing it to break during cooling of the casting.[9] All of these factors represent minor changes to conventional titanium casting and contribute to successful manufacture of gamma alloy components.

Solidification shrinkage and gas porosity found in conventional titanium castings also are prevalent in gamma alloy castings. Both of these types of porosity can be eliminated by hot isostatic pressing (HIP). We have performed HIP studies that show complete closure and bonding of porosity using a processing condition of 1260°C/172 MPa/4h.

Component Applications

Many different component geometries have been identified for potential cast gamma alloy application. These include both aerospace and automotive applications. Rotating components have the highest payoff in terms of weight savings. The low density of gamma alloys reduces the self-imposed loading and also reduces the strength requirements (resulting in further weight reduction) of components that secure the rotating component. For example, reducing the weight of an aircraft engine low pressure (LP) turbine blade by 30-50% could allow weight reductions in the turbine disk, turbine shaft, and bearing support housings. This "trickle down" effect can result in impressive overall efficiency increases and weight reductions. Examples of rotating gamma alloy applications include low pressure turbine blades, high pressure compressor vanes, rotating seals, rotating diffusers, and turbocharger wheels.[7,8,9] Figure 1 shows examples of an investment cast aircraft engine LP turbine blade, and Figure 2 shows an industrial gas turbine bucket. Both components were cast near-net shape and passed X-ray inspection and fluorescent penetrant inspection (FPI) quality standards equivalent to superalloy components.

Figure 1 - XD gamma LP turbine blade.

Reduced weight and increased performance benefits also can be gained in structural nonrotating components.[8,9] These benefits are a result of the low density/high modulus and low coefficient of thermal expansion characteristics of gamma alloys. These factors can greatly reduce stresses and/or reduce running clearances in the gamma components and other surrounding structures. Applications of these types of components include compressor cases, compressor vanes, turbine cases, diffuser housings, missile fins, missile structures, gun blast diffusers, structural brackets, and automotive exhaust valves.

Figure 3 shows an example of investment cast XD gamma missile fins produced for General Dynamics as part of a Navy development program. Casting these components in the XD gamma alloy resulted in reduced weight and increased high temperature performance compared to conventional materials. These components were subjected to the same flight qualification tests as the conventional materials they are intended to replace. The larger component was tested under static load, heat/load, and deployment conditions. In heat/load testing, the component passed test load requirements of 2365 kg at temperatures ranging from 600 - 815°C. Deployment tests required the component to be erected into flight position at room temperature by a spring loaded actuator in 75 milliseconds. The fin passed deployment testing surviving an impact load of 54 N-m at the termination of the deployment sequence. The smaller fin also passed similar flight qualification tests.

Figure 2 - Cast gamma industrial gas turbine bucket.

Figure 3 - Cast XD gamma missile fins. (Naval Surface Warfare Center Contract #N60921-89-CA155).

The complex configuration of the F-18 gun blast diffuser was cast for McDonnell Douglas under a Navy contract (Figure 4). This component houses a gattling gun that fires at 3600 rounds per minute. The current stainless steel component needs to be coated to survive the hot particle erosion imparted by shell propellent materials exiting through the diffuser. Preliminary tests performed by McDonnell Douglas have shown that the hot erosion resistance of the XD gamma alloy significantly exceeds that of the uncoated stainless steel material. The full gun blast diffuser assembly requires welding a 50 mm diameter XD gamma tube with 6 mm thick walls to the back of the casting. McDonnell Douglas has demonstrated that the electron beam welding process used in production of the stainless steel component can produce crack free, mechanically sound welds in the cast XD gamma alloy. Moreover, the use of a gamma alloy will reduce the weight of the component by 50% allowing more weight in fuel, avionics, or weapons systems to be installed on the aircraft.

Figure 4 - Cast XD gamma F-18 gun blast diffuser. (Produced under Naval Air Development Center Contract #N62269-90-C-0260).

Automotive exhaust valves also can benefit from cast gamma technology (Figure 5). The potential yearly market for gamma valves on a single engine type could exceed 2 million components. Replacing steel valves with lighter gamma valves lowers the spring tension necessary to retain the valve during operation. This results in lower friction in the valve train allowing fuel economy to be increased by as much as 0.3 kpl. The benefit of casting gamma valves near-net shape allowed the component in Figure 5 to be machined in less than two minutes.

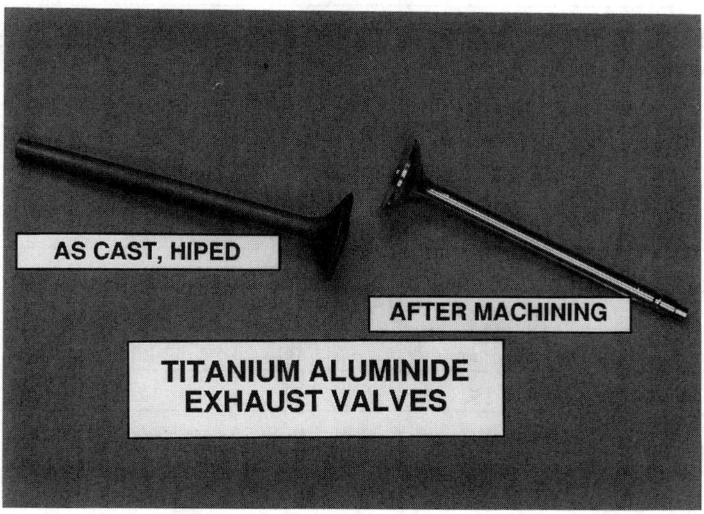

Figure 5 - Cast gamma automotive exhaust valves.
(Machining and Photograph Courtesy Ford Motor Company)

Microstructure and Mechanical Properties

Heat treatment is a key element in component casting, allowing the properties of the component to be tailored to a specific application. The gamma and alpha-two phases found in current cast gamma alloys can be manipulated through thermal treatment to form three basic types of microstructure: lamellar, lamellar + equiaxed gamma (duplex), and equiaxed gamma.[7,8] Typically, a lamellar structure increases high temperature strength and fracture toughness. The equiaxed gamma structure increases room and elevated grain refinement produced in XD alloys increases microstructural uniformity and strength over the base alloy composition. However, any amount of TiB_2 addition results in reduced room temperature ductility.

Table I lists a summary of cast gamma, cast XD gamma and superalloy properties.[7,8,10,11] The summary is based on a wide range of alloys from all three systems. Alloy composition has a large effect on the strength and ductility of gamma alloys. Cast gamma alloys retain their strength to approximately 815°C. Above this temperature, only a limited number of alloys have usable strength, ductility levels, and oxidation resistance. Gamma alloys also have a brittle to ductile transition temperature which occurs for most of the alloys at approximately 650°C. Above this temperature, the alloys have considerable ductility and behave more like conventional metallic materials.

Table I Properties of Cast Gamma, Cast XD Gamma, and Cast Superalloys

Property		Gamma TiAl Based	XD Gamma TiAl Based	Superalloys
Density (g/cm^3)		4.0	4.0	8.3
RT Modulus (GPa)		160-175	160-190	200-220
RT Yield Strength (MPa)		276-510	415-621	690-1035
RT Ductility (%)		1-2.5%	0.5-1.5%	4-10
650°C Yield Strength (MPa)		265-500	405-610	690-1035
650°C Ductility (%)		2-10%	1-5%	4-10
815°C Yield Strength (MPa)		207-414	310-450	590-710
815°C Ductility (%)		15-40%	15-30%	4-14%
RT K_{IC} (MPa $\sqrt{n}$)		15-33	10-18	50-120
Creep Life (h) 815°C/138 MPa	Time to 1%	6-50	10-80	-
	Time to Rupture	90-200h	90-200h	-
Creep Life (h) 760°C/132 MPa	Time to 1%		>120h	-
	Time to Rupture		>600h	-
760°C HCF 10^7 Cycles R = 0.1 (MPa)		~276	~414	-
760°C LCF 10^5 Cycles R = 0.1 ($\Delta\epsilon p$)		-	0.6	-

Conclusions

Gamma alloys with their unique combination of low density and high temperature strength appear to be key materials for advanced systems applications. Investment casting offers a cost effective way to produce complex near-net shape components using conventional titanium alloy processing methods and equipment. Cast gamma microstructures can be manipulated through heat treatment and/or application of the XD process. Through alloy development, current cast gamma materials have reasonable room temperature elevated temperature properties. Further alloy development will, hopefully, further improve these properties and lead to the acceptance of gamma alloys as engineering materials.

References

1. H.B. Bomberger et al., Titanium Technology, Dayton, OH, TDA, 1985, pp. 3-17.
2. Y-W. Kim, JOM, vol. 7 (1989), pp. 24-30.
3. Y-W. Kim and D.M. Dimiduk, JOM, vol. 8, (1991), pp. 40-47.
4. L. Christodoulou et al., High Temperature/High Performance Composites, ed. F.D. Lemkey et al., Pittsburgh, PA, MRS 1988, pp. 29-34.
5. D.E. Larsen et al., Intermetallic Matrix Composites, ed. D.L. Anton et al., MRS 1990, vol. 194, pp. 285-292.
6. D.E. Larsen, Microstructure/Property Relationships in Titanium Aluminides and Alloys, ed. Y-W. Kim and R.R. Boyer, TMS, 1991, pp. 345-352.
7. D.E. Larsen et al., Materials Science and Engineering, vol. A144, 1991, pp. 45-49.
8. B. London and T.J. Kelly, Microstructure/Property Relationships in Titanium Aluminides and Alloys, ed. Y-W. Kim and R.R. Boyer, 1991, pp. 285-295.
9. D.E. Larsen et al., International Conference on High Temperature Aluminides and Intermetallics, ASM, San Diego, CA, September 16, 1991.
10. D.A. Wheeler et al, Structure-Property Relationships of Investment Cast Gamma Titanium Aluminides, Seventh World Conference on Titanium Proceedings, ed. F.H. Froes, TMS, 1992.
11. INCO, High Temperature/High Strength Nickel-Base Alloys, International Nickel, Saddle Brook, NJ 07662.

Acknowledgements

This work was supported by Howmet Corporation internal research and development funds, customer funds, and U.S. government funds. Howmet acknowledges the support of Dr. Charles Blackmon at the Naval Surface Warfare Center and Ed Neu at the Naval Air Development Center. Special thanks to Martin Marietta Corporation, General Dynamics, and McDonnell-Douglas Aircraft Company for their continuing support.

RECENT ADVANCES IN TITANIUM INVESTMENT CASTING

M.S. Cianci, R.G. Vogt, and G.N. Colvin

Howmet Corporation
Titanium Casting Division
1600 South Warner Street
Whitehall, MI 49461

Abstract

The technology utilized to investment cast titanium components has rapidly advanced during the past several years to accommodate the need for near net shape airframe and jet engine components. Improvements in processing have been made in areas as diverse as: molding, casting, and fixturing which have improved component dimensional and metallurgical integrity. Processes have been developed to produce castings in nontraditional TiAl and Beta titanium alloys. In addition, in an attempt to reduce overall manufacturing cycle times, technology to rapidly prototype engine hardware has been developed.

Introduction

Increasing competitiveness in the aerospace industry has resulted in new market opportunities and added expectations on the suppliers of titanium alloy investment castings. The need to reduce manufacturing costs and lead times as well as reduce airframe and engine weight, thereby improving specific fuel consumption, has resulted in significant changes in the titanium investment casting industry such as: Increased component complexity and size coupled with reduced costs and manufacturing cycle times, decreased prototype cycle times, and development of processes to cast advanced alloy systems.

The manufacturing processes required to provide production hardware with the above characteristics is currently being developed at a rapid pace. Advanced manufacturing technologies currently being implemented include: Ceramic systems with improved chemical and dimensional stability, thermal forming, and casting parameters tailored for advanced alloys.[1][2] The introduction of these new technologies has helped meet aerospace industry needs and keep the titanium casting industry competitive with other sources of titanium hardware.

In general, there are three major steps inherent in all investment cast processes: (1) mold manufacture, (2) casting, and (3) postcast processing and inspection. Investment molds are manufactured by coating an injected wax replica pattern with multiple layers of ceramic slurries. These ceramic slurries are dried on the wax pattern and the wax subsequently removed by firing. Titanium alloys normally are melted using vacuum arc remelting (VAR) consumable electrodes into water cooled copper crucibles. The molten titanium is then poured into the ceramic mold and centrifugal casting techniques may be used to assist fill. Because VAR produces minimal superheat in the molten pool, centerline shrinkage porosity is common in the cast part. These defects, however, are collapsed and diffusion bonded by hot isostatic pressing (HIP).

The depressions resulting from HIP closure subsequently must be eliminated by weld repair in an inert atmosphere using tungsten arc welding methods. All conventional titanium alloys are very readily weldable.

Titanium '92
Science and Technology
Edited by F.H. Froes and I. Caplan
The Minerals, Metals & Materials Society, 1993

During casting, titanium reacts with the ceramic mold to form an oxygen enriched layer, or "alpha case," at the mold-titanium interface. This reaction layer, which is minimized by using ceramic oxides in the mold system, is subsequently removed via chemical milling.

Discussion

Increased Customer Expectations

In an attempt to reduce overall manufacturing costs, aircraft engine and airframe manufacturers have turned to net and near net shape titanium investment castings in order to minimize machining, improve raw material utilization, and reduce the amount of weldments. For example, studies made on a series of sixteen parts on a commercial aircraft indicated an average cost of $749 for each of 200 units machined from forgings. The average cost of investment casting the same production quantity was $227.[3]

In order to satisfy the demand for highly complex castings several enabling technologies recently have been developed: (1) improvements in chemical and dimensional stability of ceramic systems, (2) increased melting and component handling capacity, and (3) the ability to thermally form components.

Development of an ultralow reactivity mold facecoat has improved casting surface finish, enabled fill-out of very thin section thicknesses and reduced the amount of alpha case formation. The ability to fill thinner sections will enable parts to be cast without the same degree of overstock traditionally required, thereby reducing the amount of chemical milling and improving dimensional control.

Continuous improvements made in ceramic cores and investment technology have enabled components with complex internal passages to be cast. For example, outer fan struts for a large turbine engine are hollow and have wall thicknesses of 0.140 cm (0.055") to minimize component weight. The hollow cavities are formed with several unique modifications to the standard mold building process. Smaller passages can be produced using injected ceramic cores. Passages down to 0.318 cm (0.125") diameter and up to 15.2 cm (6") long can be formed using this approach.

Figure 1. Fan Frame Case Investment Cast in Ti-6Al-4V.

Manufacturing capacity has been increased allowing up to 320 kg (700 lb) components to be processed. In several instances this has enabled a number of smaller individual castings to be combined into one very large casting. An example of this is a fan frame case in Figure 1. The fan frame case, which was originally made as 10 strut castings and one inner ring casting, will reduce machining and welding costs and, therefore, result in significant cost savings to the customer.

In order to maintain the extremely tight dimensional tolerances now being specified, hot forming techniques are now often employed. At temperatures exceeding 760°C (1400°F), conventional alloys such as Ti-6Al-4V can easily be formed to shape. Castings with very thin walls such as 0.102 cm (0.040") to 0.152 cm (0.060") can be formed at temperatures down to 583°C (1100°F).

Driving the technology that has enabled the production of complex casting geometries is the need to improve foundry efficiencies and cycle times. Reductions in casting scrap and repair operations have been made possible by: (1) lowering process variation with statistical process control and factory automation, (2) identifying key input variables using design of experiments, and (3) reorganizing and empowering the work force to facilitate continuous process improvement. Some casting companies have reached complete cellular manufacturing levels. Here, cross-functional teams work together to solve problems and eliminate inefficiencies in order to reduce cost and cycle times.

Decreasing Prototype Cycle Times

Aerospace manufacturers are increasingly demanding reductions in the time required to produce usable titanium hardware from blueprint designs. Traditionally in investment casting, once a component design is complete the next step requires fabrication of a wax injection die. Depending on the component geometry, die fabrication can take up to 40 weeks to complete. Recently advanced prototyping techniques have been used to produce wax or plastic replica patterns prior to actual fabrication of permanent wax injection tooling.[4] These patterns allow the design engineer to visualize their concept and quickly verify or improve their design. The foundry engineer can produce ceramic investment molds and preliminary castings from the wax patterns. Proposed gating schemes can be evaluated and castings can be provided for mechanical property verification and microstructural characterization.

Prototyping technologies range from simple wax mockup techniques (hand fabrication) to more complex technologies for producing temporary injection tooling such as laser stereolithography. Selection of prototyping method depends on purpose of the project, geometric complexity of the component, availability of electronic dimensional data, and number of components to be made. As an example, wax mockup techniques were used to produce prototype heat shields for new commercial aircraft, Figure 2. Typically this part would have required 20 weeks from design to first article; however, these components were taken from design and fully processed castings in less than eight weeks.

A temporary wax injection tool may be specified if, for instance, several dozen patterns are required and if electronic dimensional data is available. Stereolithography is a recently developed technique which can be used to develop a precise master pattern directly from CAD/CAM data.[5] Commercially available stereolithography machines use an ultraviolet laser to trace a thin cross section of the target object onto the surface of a vat of liquid polymer, selectively hardening the photopolymer. Complex geometry components such as airfoils can be produced in as little as eight hours to twelve hours with linear tolerances of ± 0.008 cm (0.003").

If hard patterns are available, such as from stereolithography, then there are a variety of methods and materials which can be used to produce temporary wax injection tooling. These include REN (resin epoxy filled with glass beads), polyurethane, and common silicon rubber. These materials may be poured around the master pattern in a cope/drag fashion. A metal injection die can be made by arc spray coating a master pattern with aluminum.[5][6] Spray metal dies can be dimensionally stable for up to 1000 injections and may be produced in one week.

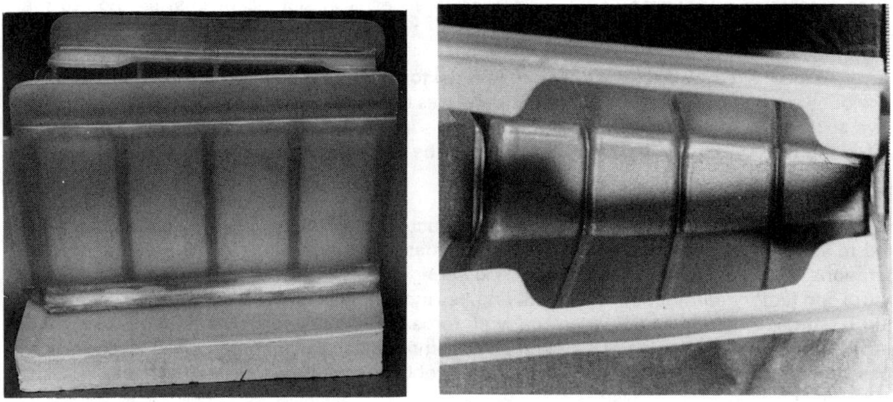

A) B)

Figure 2. Aerospace Heat Shield A) Prototyped in Wax and B) Investment Cast in Ti-6Al-2Sn-4Zr-2Mo.

Advanced Alloy Systems

In response to demands for higher performance systems, aerospace industry casting users are evaluating new classes of titanium alloys to replace current Ni or Fe based components. Material attributes such as improved high temperature properties, lower density, burn resistance, and high strength at intermediate temperatures are necessary to satisfy industry demand. Aircraft engine companies are most interested in the improved high temperature alloys such as gamma titanium aluminides, Ti-1100, or IMI-834 while airframe companies favor high strength beta alloys such as Ti-15V-3Al-3Cr-3Sn.[7][8] Other new titanium alloys include Pratt & Whitney's proprietary "Alloy C", which offers good high temperature strength and burn resistance and Timet's Beta 21S which is a high strength alloy with improved oxidation resistance.

Gamma Titanium Aluminides

Gamma titanium aluminide alloys are melted using VAR and ISR (Induction Skull Remelting) methods and, therefore, exhibit many of the same casting characteristics as do conventional titanium alloys. Gamma titanium aluminides are also prone to surface connected microporosity and cracking, but developments in the ceramic molding and casting areas have enabled complex geometries to be cast, Figure 3. Because of TiAl's low as-cast ductility gating has to be designed to accommodate thermal contraction differences between the mold and the alloy during cool down.

Over the past five years much experience has been gained in casting GE Aircraft Engine's gamma titanium aluminide alloy, Ti-48Al-2Cr-2Nb. Work has focused on producing turbine blades and nozzles but have included structural components as well. Over 80 molds of Ti-48Al-2Cr-2Nb have been cast on a production scale using greater than 4500 kg (10,000 lbs) of ingot.

Gamma titanium aluminide alloys reinforced via XD™ (exothermic dispersion strengthened) processing have also been receiving attention from titanium users.[9][10] XD™ is a Martin Marietta Laboratories' process for incorporating titanium diboride particulate in alloy systems.

XD™ additions make titanium aluminides more difficult to VAR melt but often improves castability due to the inoculation effect of the titanium diboride. In 1989, under a DARPA DICE contract, Howmet produced 100 overstocked CF6-80C2 Stage 5 low pressure turbine blades (LPT) in Ti-48Al-2V + 7 vol% TiB^2.[11] The blades were cast overstocked and subsequently electrochemically machined and conventionally ground to size. In 1992 net shape GE LPT blades and Rolls-Royce LPT blades were produced using Ti-48Al-2Cr-2Nb + 2vol%TiB2. Both blade designs were metallurgically sound.

Figure 3. Examples of Investment Cast Gamma TiAl Components Including a Missile Fin, Compressor Case, Low Pressure Turbine Blade, Actuator Housing, and High Pressure Turbine Blades.

Beta Titanium Alloys

The ability to heat treat beta titanium alloys to high strength levels make them an attractive substitute for precipitation hardened stainless steels. Beta alloy's processing behavior is different than that of Ti-6Al-4V and requires modified techniques. For example, decreased fluidity necessitates more gating and overstock for fill while tendency toward intergranular attack and hydrogen pickup during chemical milling dictates nonstandard milling solutions. A comparison of the processing behavior of several cast titanium alloys is shown in Table I.

Table I - Comparison of Processability of Various Titanium Alloys

Alloy	Ti-6Al-4V	Ti-6-2-4-2Si	Ti-1100	IMI-834	Beta-C	Ti-15-3
Fluidity	100	80	50	60	70	70
Melting	Basis	Outgasses	Similar	Similar	Outgasses	Outgasses
Alpha Case	Basis	Same	Same	Same	Same	Same
Grain Size	Basis	Fine	Slightly Larger	Slightly Larger	Finer	Finer
Chemical Milling	Basis	Same	TBD	TBD	IGA	H_2 Pickup
HIP Response	Good	Good	Good	Good	Good	Good
Weldability	Basis	Same	OK	OK	OK	OK

Table II - Comparison of Typical Tensile Properties of Investment Cast and HIP'ed Titanium Alloys

Alloy	Temp (°F)	UTS (ksi)	0.2% YS (ksi)	El (%)	RA (%)
Ti-6Al-4V [1]	70	135.4	120.8	11.4	18.8
Ti-6Al-2Sn-4Zr-2Mo + Si [2]	70	144.0	128.3	9.0	13.1
Ti-6Al-2Sn-4Zr-2Mo + Si [2]	1000	92.2	72.2	17.5	33.1
Ti-1100 [2]	70	135.0	121.0	8.2	14.5
Ti-15V-3Al-3Cr-3Sn [3]	70	180.0	163.2	8.7	14.0
Beta-C [4]	70	193.4	181.8	6.8	9.8
Alloy C [1]	70	145.1	139.2	13.8	24.8
Alloy C [1]	1000	120.8	96.9	13.0	27.0
17-4 pH Stainless Steel [2]	70	160.0	145.0	15.0	55.0

1. Annealed
2. Solution treated and aged
3. Aged only
4. Solution treated and duplex aged

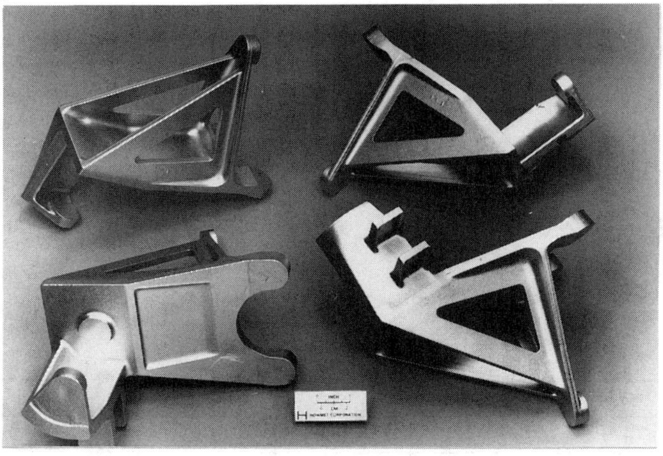

Figure 4. Investment Cast Ti-3Al-8V-6Cr-4Mo-4Zr Components for Airframe Applications.

Weight reduction is a must for modern aerospace systems but it cannot come at great sacrifices to mechanical properties. Table II shows a comparison of a number of cast titanium alloys versus two typical precipitation hardened stainless steels. In the case of Ti-15V-3Al-3Cr-3Sn nearly equivalent properties can be achieved with a 37% weight savings. For certain beta alloys the HIP'ing operation acts as a solution treatment, and the castings can be directly aged to full strength eliminating costly heat treatment steps. A number of configurations have been cast in Ti-15V-3Al-3Cr-3Sn for airframe applications, Figure 4. Pratt & Whitney's Alloy C maintains good strength up to 537°C (1000°F) which, combined with its burn resistance, could replace some Ni-based superalloys.

Summary

The recent demands placed on the aerospace industry have resulted in an increased interest in titanium investment castings and pressure for less expensive, more rapidly delivered components. The investment casting industry has responded by developing technologies such as ultrastable ceramics, stereolithography wax pattern making, and processing for advanced alloy systems. High performance aerospace systems require the attributes investment cast titanium has to offer: design flexibility and complexity, improved economics of net shape processing, and alternate alloys to replace higher density materials on a one-to-one basis. The benefits of investment casting have long been realized for producing gas turbine engine hardware and airframe applications are rapidly developing.

Acknowledgments

The authors would like to thank Bill Armstrong, Larry Bednarz, John Klepeisz, and Keith Ramthun for providing information for this manuscript.

References

1. J.K. Thorne, W.J. Barice, "Advances in Titanium Alloy Casting Technology", 2nd International SAMPE Metals Conference, August 2-4, 1988.

2. R.L. Saha, T.K. Nandy, "On the Evaluation of Stability of Rare Earth Oxides as Face Coats for Investment Casting of Titanium", Metallurgical Transactions B, Volume 21B, June 1990.

3. D.L. Squier, "The Outlook for Titanium Castings", International Titanium Conference, New York, NY, December 4, 1989.

4. J. Ingalls, unpublished work, 1991.

5. "Metal Spray Toolmaking Gains Investment Cast Foothold", Netcast, June 1991.

6. D. Hartung, "Metal Spraying for Prototype Molds", 39th AFS Northwest Regional Conference, August 1990.

7. M.A. Daeubler, D. Helm, D.F. Neal, "Application of IMI 834 in Aeroengines-A Collaborative IMI/MTU Programme", 1990 International Conference on Titanium Products and Applications, October 1990.

8. R. Boyer, "New Titanium Applications on the Boeing 777 Airplane", JOM, Vol 44, Number 5, May 1992.

9. D.E. Larsen, L. Christodoulou, "Investment Cast Processing of XD™ Near-Gamma Titanium Aluminides", Materials Science and Engineering, Vol 144, 1991.

10. D.E. Larsen, "Effects of XD™ TiB_2 Volume Fraction on the Microstructure and Mechanical Properties of a Cast XD™ Near Gamma Alloy", Microstructure/Property Relationships in Titanium Aluminides and Alloys, TMS, Y-W. Kim, R. Boyer Ed., 1991.

11. G. Colvin, M. Cianci, B. Kleyn, L. VanderLeest, "Manufacture of XD™ Gamma Titanium Aluminide Airfoils, "Microstructure/Property Relationships in Titanium Aluminides and Alloys, TMS, Y-W. Kim, R. Boyer Ed., 1991.

EFFECTS OF HIP AND HEAT TREATMENT ON THE MICROSTRUCTURE AND ROOM TEMPERATURE PROPERTIES OF NEAR-γ TITANIUM ALUMINIDE CASTINGS

P.E. Jones [a], M. Saqib [b], T.J. Kelly [c], W. J. Porter [a], I. Weiss [b], and D. Eylon [a]

[a] Graduate Materials Engineering, University of Dayton
300 College Park, Dayton, OH 45469-0240, USA

[b] Department of Mechanical and Materials Engineering
Wright State University, Dayton, OH 45435, USA

[c] General Electric Aircraft Engines, Mail Drop M89
PO Box 156301, Cincinnati, OH 45215-6301, USA

ABSTRACT

Casting is one of the most promising routes to produce complex shapes from titanium aluminide alloys. The objective of this work was to determine the effect of hot isostatic pressing (HIP) process parameters on the microstructure and room temperature tensile properties of Ti-48Al-2Nb-2Cr (a%) investment castings following a stabilization heat treatment (ST). The microstructures were characterized in the as-cast, as-HIP, and ST conditions by optical and scanning electron microscopies as well as x-ray diffraction. HIP conditions ranged from 1800 to 2300°F (980 to 1260°C) at 20 to 40 ksi (140 to 275 MPa) for 4 to 8 hours. The results show significant effects of HIP conditions on the microstructure, phase morphologies, and volume fraction of the phases present. The effects of the subsequent ST at 2375°F (1300°C) for 20 hours on the microstructure and tensile properties were studied for selected HIP conditions. Tensile strengths as high as 70 ksi (480 MPa) were achieved. Additionally, elongations as high as 2.1% could be consistently obtained from a wide range of HIP conditions. No significant effect of HIP temperature on room temperature tensile properties was observed after ST provided pore closure was achieved.

INTRODUCTION

Commercial introduction of alloys based on the intermetallic compound γ-TiAl has been hampered by difficulties in fabricating components by ingot metallurgy routes. Near-net-shape (NNS) processes may become a key factor in the successful introduction of these alloys. One NNS process being developed for these alloys is investment casting. The shape-making capability of this technique has been demonstrated for several γ-TiAl alloys [1-5]. Because of the demanding applications envisioned for these alloys and their relatively poor crack growth resistance and toughness, hot isostatic pressing (HIP) is required to close gas and shrinkage pores. Most castings made thus far have been HIP'd at 2300°F (1260°C)/25 ksi (170 MPa)/4 hr [1,6,7]. The high temperature selected increases the cost of the process, as well as the potential for distortion of the NNS components and surface depletion of Al during HIP. This study examined the effects of the HIP temperature, pressure, and time on the microstructural development and room temperature tensile properties of stabilization heat treated (ST) Ti-48Al-2Nb-2Cr (a%) investment castings.

Work sponsored by EMTEC-Edison Materials Technology Center, Kettering, OH 45240, USA.

EXPERIMENTAL PROCEDURE

MATERIAL

Investment cast test bars with an aim chemistry of Ti-48Al-2Nb-2Cr (a%) were used in this study. The ingot and casting compositions are shown in Table I. The material was vacuum arc remelted and cast as nominally 0.5"(13mm) φ x 2.72"(70mm) long oversized tensile bars. The as-cast microstructure and typical centerline microshrinkage are shown in Fig. 1. The microshrinkage was visible on radiographs prior to HIP. The alpha transus was estimated to be 2440°F (1340°C).

Table I: Ingot and test bar compositions.

Form	Al (a%)	Nb (a%)	Cr (a%)	O (ppmw)	Fe (ppmw)	Other (ppmw)
Ingot	48.02	1.79	1.96	710	820	N=20, H=26, C=8
Cast Bar	47.3	2.0	2.0	880	670	N=30, H=N/A, C=187, Si=880

HOT ISOSTATIC PRESSING (HIP)

HIP was performed in argon over the range of temperatures, pressures, and times listed in Table II with a cooling rate from the HIP temperature of about 180°F (100°C)/min. The grain sizes of the equiaxed γ and lamellar (prior α) grains were measured by the comparison method of ASTM E112. Volume fractions of equiaxed γ and lamellar (prior α) components were determined by point counting. Both the transformed and untransformed lamellar constituents were counted as lamellar grains [8].

STABILIZATION HEAT TREATMENT (ST)

A stabilization heat treatment (ST) at 2375°F (1300°C) for 20 hr in flowing argon was performed on those test bars which had been HIP'd at 25 ksi (170 MPa) for 4 hr (Table II). The specimens HIP'd at 1800°F (980°C) were cooled from 2375 (1300) to 1300°F (705°C) at a rate of 150°F (85°C)/min, while the remaining specimens were cooled through this range at 120°F (70°C)/min. The cooling rate was decreased to 20°F (10°C)/min through the ductile-brittle transition temperature range (from 1300 (705) to 200°F (95°C)), resulting in an average cooling rate of 33°F (18°C)/min. The grain sizes and volume fractions of the constituents were determined per ASTM E112 and point counting as described above.

Figure 1: A transverse section of an as-cast bar showing the columnar prior α grains which have transformed into α_2+γ lamellae. Equiaxed γ grains were found along the prior α grain boundaries. The microshrinkage evident in these micrographs was typical of the bars before HIP. (Unetched, polarized light.)

TENSILE TESTING

Tensile specimens were machined from the ST bars to 0.130"(3mm) φ x 0.65"(17mm) long gage sections. The bars were longitudinally polished to 16AA prior to testing at room temperature in air. Load-strain curves were recorded through failure using a load cell and an extensometer. The strain rate was 0.005/min through 0.2% yield. After yielding the cross head rate was increased to 0.03"(0.76mm)/min to failure. The average properties obtained from two bars per condition, including the plastic elongation measured both conventionally by micrometer and from the extensometer, are reported in Table IV. Samples with low elongation were investigated by SEM fractography and metallography to detect insufficient pore closure during HIP, since unhealed porosity cannot always be detected by post-HIP radiography.

RESULTS AND DISCUSSION

AS-CAST AND AS-HIP MICROSTRUCTURES

The as-cast microstructure (Fig. 1) was predominantly columnar prior α grains with intergranular equiaxed γ. The prior α grains transformed into lamellar α_2+ γ, resulting in the mostly lamellar microstructure typical of as-cast near-γ alloys. These lamellar regions were discontinuous, with the α_2 lamellae terminating where the local composition favored γ phase.

The volume fraction of equiaxed γ was compared to that predicted by the Ti-Al equilibrium diagram [9] with the α transus decreased 55°F (30°C) to account for the 2%Cr in this alloy [10]. After HIP at 1800°F (980°C), the as-cast macrostructure (Fig. 1) was retained, but small γ grains were nucleated within the columnar prior α grains (Table II and Fig. 2). The volume fraction of equiaxed γ increased with increasing HIP temperature and time (Fig. 3), approaching equilibrium only after 8 hour HIP cycles.

The equiaxed γ grain size changed during HIP at higher temperatures. Coarsening was noted in 4 hr at temperatures above 2150°F (1175°C) and in 8 hr at temperatures above 1800°F (980°C) (Fig. 4). However, the equiaxed γ grain size decreased when the HIP temperature increased from 2200 (1205) to 2300°F (1260°C). Similarly, the prior α grain size was unchanged up to 2150°F (1090°C) for 4 hr, but doubled between 2150 (1175) and 2300°F (1260°C). These observations indicate that thermally driven grain growth during HIP was competing with constitutionally driven phase transformations [11]. Few α_2 plates were seen in the equiaxed γ grains after HIP (Fig. 2).

Table II: HIP cycle matrix and as-HIP microstructure characteristics. The cooling rate was ~180°F (100°C)/min. for all cycles. The α transus was estimated to be 2440°F (1340°C).[a]

Temperature (°F)	(°C)	Pressure (ksi)	(MPa)	Time (hr)	%γ	GS$^\gamma$ (µm)	%L	GS$^\alpha$ (µm)
As Cast		N/A		N/A	9	22	91	Columnar
1800	980	20	140	8	10	22	90	65
1800	980	25	170	4	10	32	90	65
2000	1090	20	140	8	81	90	19	45
2000	1090	25	170	4	26	45	74	65
2150	1175	25	170	4	21	45	79	65
2200	1205	0 (control)		6	85	90	15	65
2200	1205	20	140	8	86	125	14	65
2200	1205	40	275	6	82	125	18	125
2300	1260	25	170	4	24	90	76	125

a %γ = vol. fraction equiaxed gamma, GS$^\gamma$ = Avg. grain size of equiaxed gamma, %L = vol. fraction of lamellar (prior α) grains, GS$^\alpha$ = Avg. grain size of prior alpha grains.

The effect of HIP pressure on microstructural development was examined on samples HIP'd at 2200°F (1205°C). Increasing the HIP pressure, from 0 to 20 (140) and 40 ksi (275 MPa), had no reproducible effect on the microstructure based on optical microscopy (Table II). However, shifts in the X-ray diffraction peaks of the α_2 and γ phases suggested a change in the lattice parameters of the two phases due to HIP pressure. The lattice parameters measured as a function of pressure are shown in Table III. The unit cell volume of the α_2 increased by about 2% while the unit cell volume of the γ decreased by about 2%. The c/a ratio of of the α_2 decreased slightly, while the c/a ratio for γ remained essentially constant. Further experiments are required to determine the reproducibility of this effect.

Table III: Effect of HIP pressure on the lattice parameters of α_2 and γ for specimens HIP'd at 2200°F (1205°C).

Pressure		Time	α_2		γ		α_2		γ	
(ksi)	(MPa)	(hr)	c (Å)	a (Å)	c (Å)	a (Å)	c/a	Vol. $(Å)^3$	c/a	Vol. $(Å)^3$
0	0	6	5.75	4.63	4.06	3.99	1.24	320	1.02	64.6
20	140	8	5.72	4.68	4.06	3.99	1.22	325	1.02	64.6
40	275	6	5.72	4.68	4.05	3.96	1.22	325	1.02	63.5

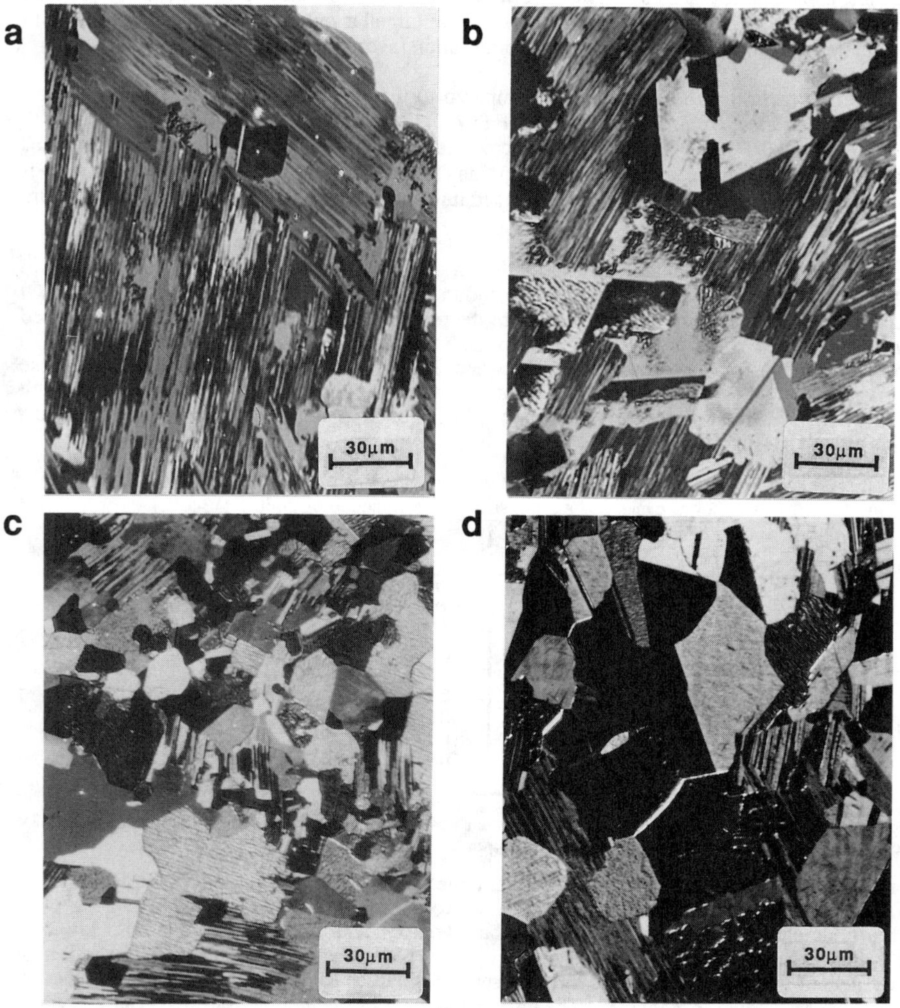

Figure 2: As-HIP microstructures. a) 1800°F(980°C)/25ksi(170MPa)/4 hr; b) 2000°F(1090°C)/25ksi (170MPa)/4 hr; c) 2150°F(1175°C)/25ksi(170MPa)/4 hr; and d) 2300°F(1260°C)/25ksi(170MPa)/4 hr. (Unetched, polarized light.)

AS-ST MICROSTRUCTURES

Bars from four different HIP temperatures were ST at 2375°F (1300°C) for 20 hr to stabilize the microstructure. All ST bars had been HIP'd at 25ksi (170MPa) for 4 hr. The ST was selected to produce a duplex microstructure with equilibrium phase fractions of about 80% prior α and 20% equiaxed γ based on the modified Ti-Al equilibrium diagram [9,10]. During cooling from 2375°F (1300°C), α_2 laths formed within the equiaxed γ grains to allow the γ to reach the equilibrium composition at lower temperatures.

Bars HIP'd at 1800°F (980°C) retained a nearly lamellar structure through the ST which was similar to the as-cast and as-HIP conditions (Fig. 5). All bars HIP'd at higher temperatures developed the desired duplex structure during ST (Fig. 5 and Table IV). The average γ and α grain sizes coarsened to about 75μm after the ST, except for the bars which had been HIP'd at 2300°F (1260°C). The prior α grain size of the material HIP'd at 2300°F (1260°C) decreased by a factor of two during ST, while the average equiaxed γ grain size decreased slightly. This grain refinement may result from constitutionally driven phase transformation to achieve compositional equilibrium of the α and γ phases [11].

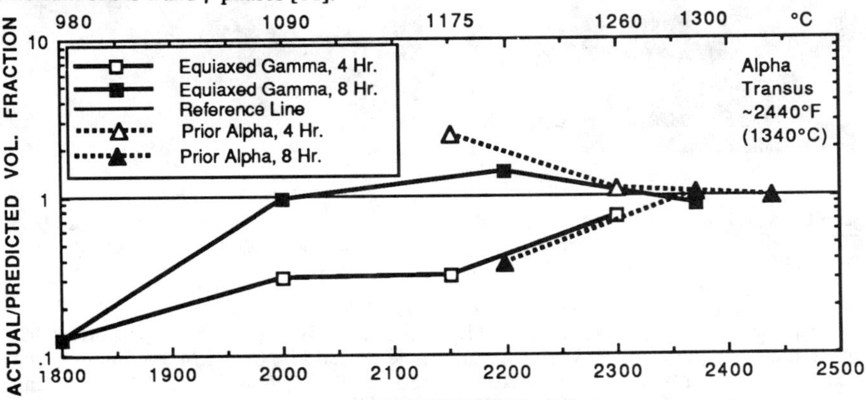

Figure 3: The volume fractions of equiaxed γ and prior α approached those predicted by the Ti-Al equilibrium diagram as the HIP or ST time and temperature were increased.

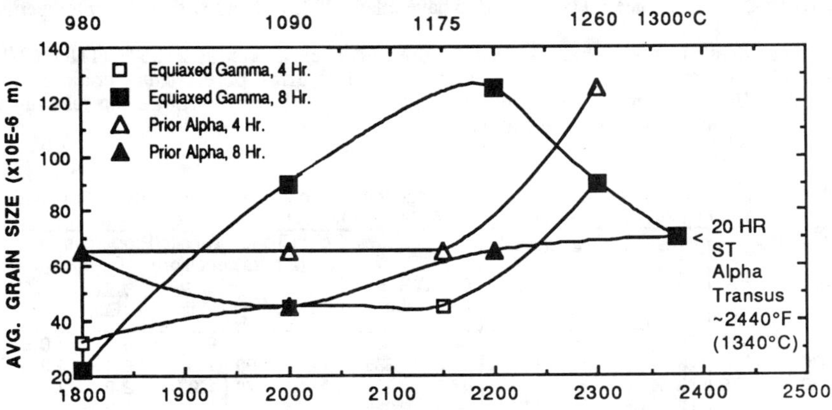

Figure 4: The effects of time and temperature on the average grain size of equiaxed γ and prior α suggest that thermally driven coarsening competes with phase transformations. Grain refinement occurred during the 2375°F (1300°C)/20hr ST.

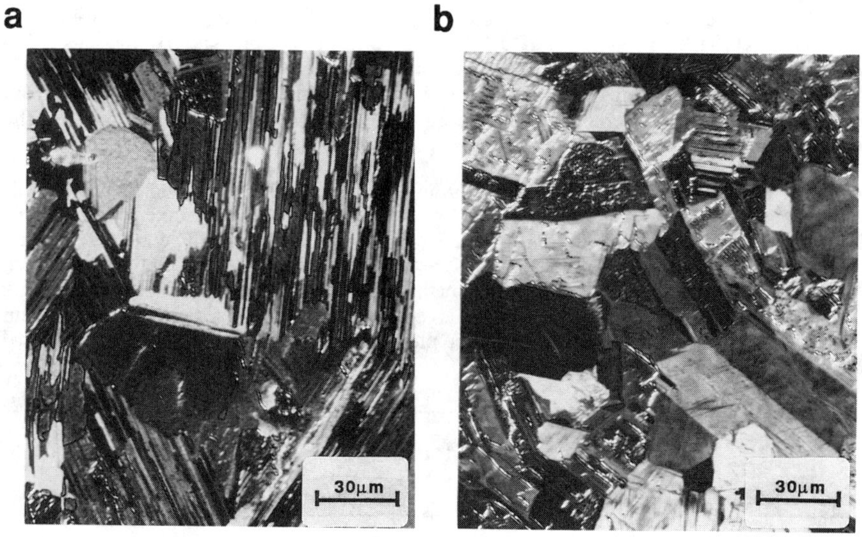

Figure 5: As-ST microstructures: a) HIP at 1800°F (980°C)/25ksi (170MPa)/4 hr followed by ST at 2375°F (1300°C)/20 hr; b) HIP at 2300°F (1260°C)/25ksi (170MPa)/4 hr followed by ST at 2375°F(1300°C)/20 hr. (Unetched, polarized light.)

ROOM TEMPERATURE TENSILE PROPERTIES

The tensile data from the ST bars highlight the insensitivity of these properties to HIP temperature when pore closure was achieved (Table IV and Fig. 6). SEM fractography showed unhealed porosity only in specimens HIP'd at 1800°F (980°C)/25ksi (170MPa)/4 hr (Fig. 7). Pore closure, accomplished by HIP at 2000°F (1090°C) and above, was associated with good tensile elongation and strength. The room temperature tensile strengths and elongations in this study were comparable to those reported by London and Kelly [1]. The low data scatter was quite remarkable for a material considered to have relatively low ductility at ambient temperature.

Since these data did not include lower pressure HIP cycles, such as 2165°F (1185°C)/15ksi (104MPa), no assessment of the role of pressure in assuring pore closure could be made. Certainly, HIP at a lower pressure than that used in this study would require careful evaluation before being adopted for aerospace components.

Table IV: Microstructure characteristics and average tensile properties of bars stabilized at 2375°F (1300°C)/20hr after HIP at 25ksi (170MPa) for 4 hr over a range of temperatures.

HIP Temp.		As-ST Microstructure [a]				Avg. Room Temp. Tensile Properties (2 bars/condition)						
(°F)	(°C)	%γ	GSγ (μm)	%L	GSα (μm)	0.2%YS (ksi)	(MPa)	UTS (ksi)	(MPa)	%EL[b] (Mic.)	%EL[c] (Ext.)	%RA
1800	980	10	32	90	75	57.6	400	64.1	440	0.5	0.5	0.5
2000	1090	13	65	87	75	48.7	335	69.2	480	1.9	2.1	2.0
2150	1175	18	75	82	75	47.6	330	67.9	470	1.9	2.2	1.8
2300	1260	12	65	88	65	47.2	325	67.3	465	2.0	2.2	2.3

[a] Same as Table I, [b] Plastic Elongation measured by micrometer, [c] Plastic Elongation measured from the load-strain curves.

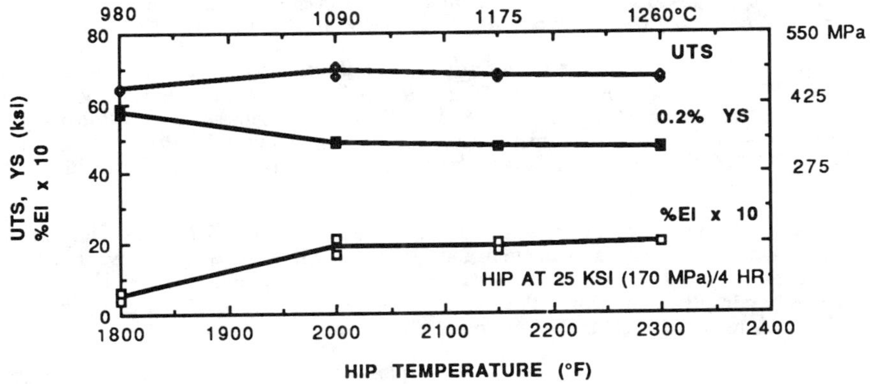

Figure 6: Effect of HIP temperature on the 70°F (23°C) tensile properties of Ti-48Al-2Nb-2Cr (a%).

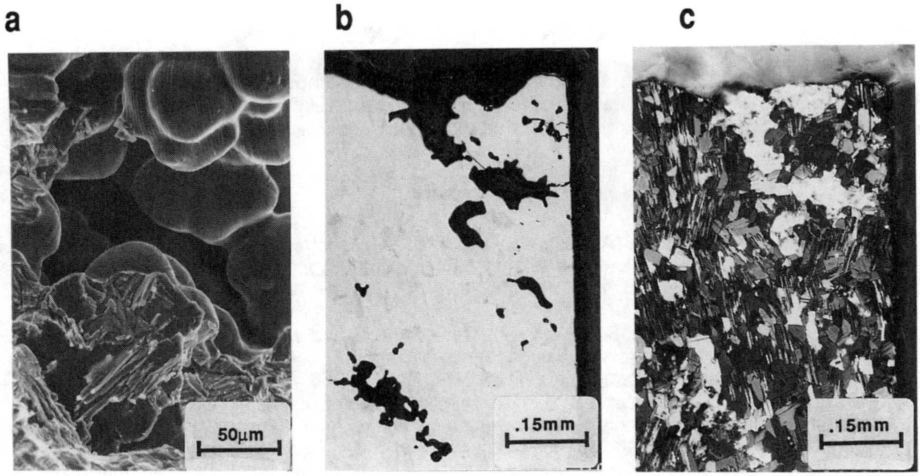

Figure 7: HIP at 1800°F (980°C)/25ksi (170MPa)/4 hr did not heal all casting porosity. a) SEM fractograph showing porosity, delamination, and cleavage on a tensile bar. b) Section through the porosity shown in (a). (Bright field optical micrograph) c) Same area as (b) (Unetched, polarized light).

SUMMARY AND CONCLUSIONS

The effects of HIP parameters and ST at 2375°F (1300°C) for 20 hr on the microstructures and tensile properties of cast Ti-48Al-2Nb-2Cr (a%) were studied. It was observed that:

1. The predominantly lamellar as-cast microstructure transformed to a duplex microstructure during thermal processing, except for specimens HIP'd at 1800°F (980°C).

2. Metastable phase fractions were developed during 4 hr HIP cycles below 2300°F (1260°C). Longer HIP cycles of 8 hr at temperatures above 1800°F (980°C) approached the predicted equilibrium γ phase fraction, but HIP at 2300°F was required to achieve nearly equilibrium phase fractions of both α and γ.

3. ST at 2375°F (1300°C)/20 hr created duplex microstructures with equilibrium volume fractions of α and γ, with the exception of specimens HIP'd at 1800°F (980°C).

4. Duplex microstructures developed during 4 hr HIP at 2000°F (1090°C) and 2150°F (1175°C) coarsened slowly during ST in the α+γ field. Grain refinement took place during ST for material HIP'd at 2300°F (1260°C) for 4 hr. This suggests that thermally driven coarsening dominates microstructural evolution between 2000°F (1090°C) and 2300°F (1260°C), but constitutionally driven phase transformation competes with grain growth at higher temperatures.

5. The effects of HIP temperature on room temperature tensile properties were small when casting pore closure was achieved and HIP was followed by ST above the HIP temperature. This conclusion suggests that lower temperature, lower cost HIP cycles could be considered for near-γ alloy castings. Further work is required to assure casting pore closure in components HIP'd at lower temperatures and pressures.

6. Average ultimate tensile strengths of 68 ksi (470 MPa) and plastic elongations of 2% were consistently achieved in material HIP'd at and above 2000°F (1090°C) under 25ksi (170MPa) pressure for 4 hr followed by ST at 2375°F (1300°C) for 20 hr.

ACKNOWLEDGMENTS

The support of the Edison Materials Technology Center, its members, General Electric Aircraft Engines, and the Wright Laboratory of Wright Patterson Air Force Base made this study possible. Their contributions, both tangible and intangible, are greatly appreciated.

REFERENCES

1. B. London and T. J. Kelly, "A Preliminary Study of the Microstructure-Property Relationships in Cast Gamma Titanium Aluminide Alloys", Microstructure/Property Relationships in Titanium Aluminides and Alloys, ed. Y-W. Kim and R. R. Boyer, TMS, Warrendale, PA, 1991, pp. 285-296.
2. D. E. Larsen, M.L. Adams, S. L. Kampe, L. Christodoulou, and J. D. Bryant, "Influence of Matrix Phase Morphology on Fracture Toughness in a Discontinuously Reinforced XD™ Titanium Aluminide Composite", Scripta Met., V. 24, 1990, pp. 851-856.
3. D. E. Larsen, D. Wheeler, and B. London, "Applications of Investment Cast Gamma and XD™ Gamma Titanium Aluminide Alloys", presented at the International Conference on High-Temperature Aluminides and Intermetallics, San Diego, CA, Sept. 16, 1991.
4. Y-W. Kim and D. M. Dimiduk, "Progress in the Understanding of Gamma Titanium Aluminides", JOM, V. 43, No. 8, August, 1991, pp. 40-47.
5. G. Colvin, M. Cianci, B. Kleyn, and L. VanderLeest, "Manufacture of XD™ Gamma Titanium Aluminide Airfoils via Investment Casting and Machining", ref 1, pp. 361-369.
6. S.L. Kampe, P. Sadler, D. E. Larsen, and L. Christodoulou, "Microstructure Morphology Effects on Fracture Toughness of a Cast Near-Gamma Titanium Aluminide Alloy", ref 1, pp. 313-322.
7. D. E. Larsen, "Effects of XD™ TiB_2 Volume Fraction on the Microstructure and Mechanical Properties of a Cast XD™ Near-Gamma Alloy", ref 1, pp. 345-352.
8. Y. Zheng, L. Zhao, and K. Tangri, "Microstructure Evolution During Heat Treatment of a Cr-Bearing Ti3Al+TiAl Alloy", Scripta Met., V. 26, 1992, pp. 219-224.
9. A. Bartz, L. Rothenflue, M. Saqib, R. Omlor, and H. A. Lipsitt, "A Re-examination of the Ti-Al Equilibrium Diagram in the Range 33-54 A/O Aluminum" to be published in these proceedings, 2/13/92.
10. S-C. Huang and E.L. Hall, "Characterization of the effect of Cr additions to TiAl-base alloys", MRS Vol. 186: Alloy Phase Stability and Design, G. M. Stocks, D. P. Pope, and A. S. Giamei eds., MRS, Pittsburgh, PA., 1991, pp. 381-386.
11. M. Mendiratta, private communication, 3/5/92.

EFFECTS OF MICROSTRUCTURE ON THE MECHANICAL PROPERTIES OF Ti-15V-3Cr-3Al-3Sn CASTINGS

W. J. Porter[a], R. R. Boyer[b], and D. Eylon[a]

[a] University of Dayton, Graduate Materials Engineering,
300 College Park Dr., Dayton, OH 45469-0240, USA

[b]Boeing Commercial Airplane Group
P. O. Box 3707, M/S 73-43, Seattle, WA 98124, USA

ABSTRACT

The relationships between microstructure, room temperature tensile strength and fatigue for investment castings of the high strength beta alloy, Ti-15V-3Cr-3Al-3Sn (wt%) (Ti-15-3), were studied. Following hot isostatic pressing (HIP) to close casting-related porosity, various heat treating and aging schemes were used to obtain high strength and high ductility conditions. Static and dynamic mechanical testing yielded tensile strengths ranging from 160 to 170 ksi (1120 to 1190 MPa) and fatigue runout stress levels from 120 to 140 ksi (840 to 980 MPa). In most cases, fatigue crack initiation was found to be associated with the presence of grain boundary alpha phase, with residual casting microporosity playing a lesser role. Thin, continuous, and planar grain boundary alpha phase oriented at about 45° to the applied stress was responsible for the majority of fatigue initiation events. Grain boundary alpha thickness appears to play a role in the formation of tensile voids. Material with thicker grain boundary alpha phase resulted in tensile fracture predominantly along prior beta grains and was associated with lower tensile elongation. The role of grain boundary alpha phase in tensile failure and fatigue crack initiation is discussed, as well as possible methods for controlling its presence.

INTRODUCTION

The lightweight, high strength, and corrosion resistant nature of titanium and titanium alloys has made it a popular candidate material for various engineering applications. However, the inherent cost of titanium requires investigation of more competitive manufacturing methods. Near-net-shape manufacturing processes, such as investment casting and powder metallurgy, were introduced to address the cost issue and are commonly used in the aerospace community. Investment casting is currently the most widely accepted and fastest growing titanium net-shape manufacturing process [1].

Although the majority of current titanium investment castings have involved the $\alpha+\beta$ alloy Ti-6Al-4V, significant work has also been done in the area of beta titanium castings [1]. Beta alloy castings have received much attention due to their excellent combination of strength, ductility and toughness. With appropriate processing, tensile strengths as high as 180 ksi (1260 MPa) with 9% elongation and runout fatigue strengths as high as 140 ksi (980 MPa) have been obtained for various beta alloys [1-3]. This study examined investment castings of the beta alloy, Ti-15V-3Cr-3Al-3Sn (wt%) (Ti-15-3) obtained from different sources. It evaluated various post-casting hot isostatic pressing (HIP) and heat treatment routes in terms of microstructure and the resulting mechanical properties. Special attention was given to room temperature tensile and fatigue properties.

Parts of this work were funded by Boeing Commercial Airplane Group

Titanium '92
Science and Technology
Edited by F.H. Froes and I. Caplan
The Minerals, Metals & Materials Society, 1993

EXPERIMENTAL PROCEDURES

MATERIAL

The chemical compositions of some of the Ti-15-3 castings used in this study are shown in Table I. The castings came from a number of suppliers and in a wide array of shapes. Some of these cast shapes included oversized blanks with 0.5" (12.5 mm) gage diameter x 3" (75 mm) length, rectangular test bars from 1" x 1" (25 x 25 mm) to 1" x 6" (25 x 150 mm), test plates from 0.15 to 1" (4 to 25 mm) in thickness and step castings increasing in thickness from .125"- .25"- .5"- 1", respectively. More details on this work are available elsewhere [4].

Table I: Compositions of Ti-15V-3Cr-3Al-3Sn Castings Used in This Work

	Element (wt %)									
Supplier	V	Cr	Al	Sn	Fe	O	N	H	C	Ti
A	15.6	3.0	3.0	3.1	0.11	0.125	0.013	0.003	0.009	Bal.
B	14.0	1.8	3.1	2.8	0.10	0.117	0.008	0.013	NA	Bal.
C	14.3	2.6	2.9	2.6	0.10	0.110	0.011	0.002	NA	Bal.
D	14.9	2.9	3.2	2.5	0.12	0.108	0.008	0.006	NA	Bal.

PROCESSING

The various HIP and heat treatment schedules used are shown in Table II. The beta-transus temperature for Ti-15-3 was determined to be 1400±15°F (760±8°C) [5]. Condition 1 and 2 specimens were provided by supplier A (Table I). Condition 3 specimens were provided by suppliers B, C, and D.

Table II: HIP and Heat Treatments Used for Ti-15V-3Cr-3Al-3Sn Castings

Condition	HIP Cycle	Solution Treatment	Aging
1	1650°F/15ksi/2hr (900°C/105MPa/2hr)	1750°F/1hr/GFC[a] (955°C/1hr/GFC)	975°F/12hr/AC[b] (525°C/12hr/AC)
2	1750°F/15ksi/2hr (955°C/105MPa/2hr)	1750°F/1hr/GFC[c] (955°C/1hr/GFC)	975°F/8hr/AC (525°C/8hr/AC)
3	1650 or 1750°F/15ksi/2hr (900 or 955°C/105MPa/2hr)	-- none --	975°F/12hr (525°C/12hr)

[a]GFC - Gas Fan Cooling 150°F(85°C)/min [b]AC - Air Cool 50°F(30°C)/min [c]GFC - 350°F(195°C)/min

MECHANICAL TESTING

Specimens HIP'd and heat treated according to Conditions 1 and 2 (Table II) were taken from cylindrical casting blanks (Table III). They were machined into a unified tensile and fatigue specimen geometry with a 1.25" (32 mm) gage length, 0.2" (5 mm) gage diameter, and 0.518" (13.2 mm) x 20 threaded grip. Tensile testing was conducted at room temperature with a strain rate of 0.005 min^{-1} through 0.2% yield, followed by a crosshead speed of 0.05 in (1.3 mm)/min to failure. Multiple bars from each condition were tested (Table III). Condition 3 tensile specimens, which were HIP'd and direct aged, were machined from the oversized cast shapes (Table III).

High cycle fatigue (HCF) (axial load) testing was done at room temperature in laboratory air using a triangular load-time waveform, with frequencies ranging from 20-30 Hz. An R ratio ($R = \sigma_{min}/\sigma_{max}$) of 0.1 (tension-tension mode testing) was chosen to preserve fracture surfaces for fractographic analysis. The smooth specimens for the limited fatigue testing of the direct aged material (Condition 3) were taken from 0.5" (13 mm) diameter cast round bars provided by supplier C.

RESULTS

MICROSTRUCTURE

Representative microstructures of Conditions 1, 2, and 3 (Table II) are shown in Figs. 1a, b and c, respectively.

TENSILE PROPERTIES

Room temperature tensile and fatigue properties of Conditions 1, 2 and 3 are listed in Table III. The grain boundary alpha (GBα) thickness of the tensile specimens was measured only for Conditions A1, A2, and C3 (0.5" plate).

Typical tensile fracture surfaces for Conditions 1 and 2 are shown in Fig. 2. These are also representative of Condition 3 fracture surfaces.

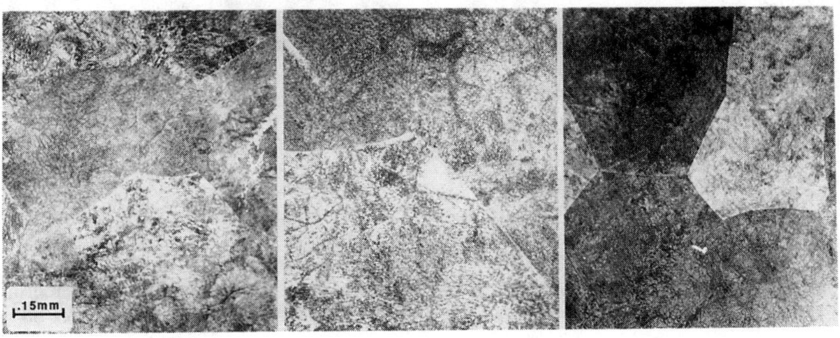

a **b** **c**

Figure 1: Optical micrographs of Ti-15V-3Cr-3Al-3Sn castings: (a) Condition 1, (b) Condition 2, and (c) Condition 3 as defined in Table II. Original magnification is 100X.

Table III: Mechanical Properties of Condition 1, 2, and 3 Ti-15V-3Cr-3Al-3Sn Castings

Supplier	Cond.	Cast Geometry	UTS ksi (MPa)	.2%YS ksi (MPa)	El. (%)	R.A. (%)	No. of Tests	Fatigue Ratio[a]	Avg. GBα Thickness (nm)
A	1	.5" dia x 3" length bars	181 (1265)	169 (1180)	3.7	9.3	3	0.69	60±20
	2	.5" dia x 3" length bars	174 (1215)	159 (1110)	6.6	15.0	3	0.70	40±15
B	3	1" dia. bars	174 (1215)	165 (1155)	4.5	4.5	2	NA	
C	3	over size test bars	181 (1265)	164 (1145)	6.3	9.7	3	0.78	
	3	.375x1x6" test bars	182 (1270)	170 (1190)	7.5	NA	2	NA	
	3	.375x4x6" plate	180 (1260)	167 (1170)	6.3	12.8	4	NA	
	3	0.5" plate	170 (1190)	164 (1145)	2.0	NA	2	NA	200±25
		<u>Step casting</u>							
D	3	.125" step	181 (1265)	172 (1200)	4.5	5.8	2	NA	
	3	.25" step	176 (1230)	168 (1175)	4.0	4.9	2	NA	
	3	.5" step	174 (1215)	167 (1165)	3.0	3.5	2	NA	
	3	1" step	162 (1130)	158 (1105)	3.0	7.0	2	NA	

[a]fatigue ratio = runout fatigue strength/ UTS

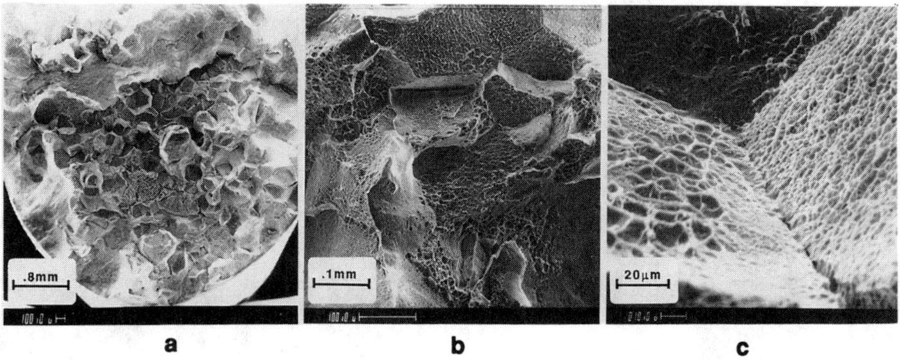

Figure 2: Tensile fracture surfaces of: (a) Condition 1, (b) Condition 2 specimens, and (c) high magnification of Condition 1 highlighting ductile dimpling along GBα fracture surface.

FATIGUE PROPERTIES

A comparison of the high cycle fatigue test data (S-N curves) for smooth fatigue specimens from all three HIP and heat treatment conditions is plotted in Fig. 3.

Condition 1 and 2 fatigue specimens exhibited two types of fatigue crack initiation sites. Examples of the *grain boundary alpha* and *micropore* related initiation sites are shown in Fig. 4.

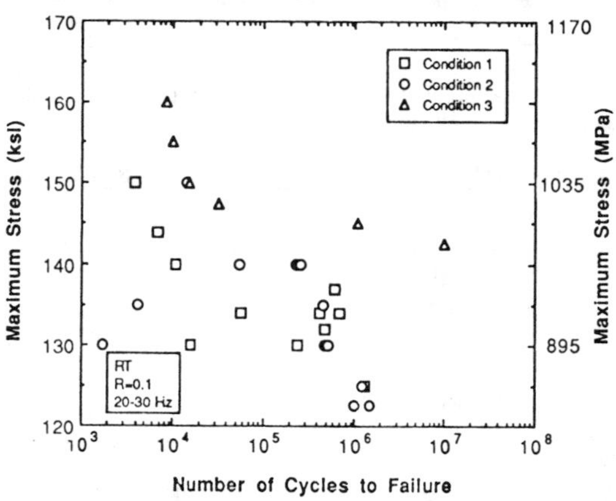

Figure 3: Comparison of high cycle fatigue data for all three HIP+heat treatment conditions.

DISCUSSION

MICROSTRUCTURE

While the room temperature tensile and high cycle fatigue test results indicate that the mechanical properties of Ti-15-3 castings were influenced by changes in the HIP parameters and the subsequent heat treatments, the optical microstructures for the three conditions (Fig. 1) were not significantly different. Each condition resulted in microstructures exhibiting fine alpha precipitates in a recrystallized beta matrix. A small difference in the amount of alpha precipitate between the solution treated Conditions 1 and 2 was noticeable, with Condition 1 having more. This was due to the longer aging time of 12 hours for Condition 1 versus 8 hours for Condition 2.

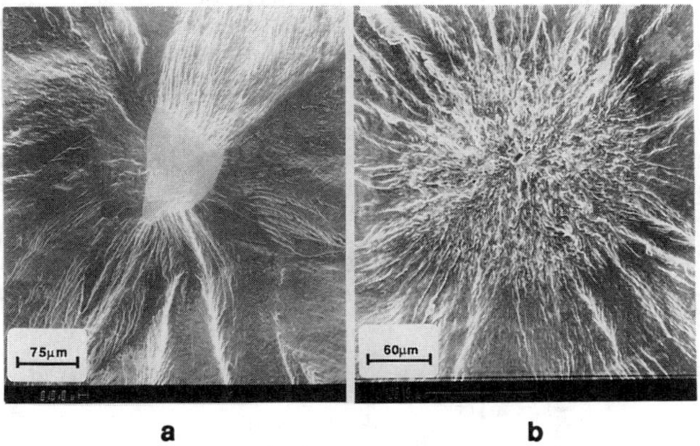

Figure 4: Fatigue crack initiation sites for Conditions 1 and 2: (a) grain boundary alpha at 45° to tensile direction, and (b) micropore on a plane oriented 90° to tensile direction.

One of the most prominent features of the microstructure was the continuous GBα phase on the prior beta grain boundaries. GBα is common in beta solution treated, solute rich, beta titanium alloys such as the Ti-15-3 alloy used in this study [1-3,6,7]. The formation of GBα occurs during cooling through the $\alpha+\beta$ phase field. Condition 2 exhibited somewhat thinner GBα phase than that found in Condition 1 (40 vs 60 nm, Table III) due to differences in the cooling rates. The faster cooling rate for Condition 2, 350°F(195°C)/min, as opposed to 150°F(85°C)/min for Condition 1, was employed in an attempt to minimize the thickness of the GBα. Ideally, this should lead to improved mechanical properties, such as tensile elongation and fatigue strengths. In Condition 3, the HIP cycle acted as a solution treatment with a very slow cooling rate, leading to formation of the thickest GBα observed in this study (200 nm, Table III). The 975°F(525°C)/12hr direct aging of Condition 3 following HIP, resulted in a microstructure similar to Condition 1, which was also aged for 12 hours at 975°F(525°C) following solution treatment.

TENSILE PROPERTIES

Although the differences in the microstructures of each condition were subtle, their effect on mechanical properties was readily apparent. While both Conditions 1 and 2 were beta solution-treated, the longer aging times for Condition 1 (which led to a larger amount of α-phase precipitation) resulted in higher yield and tensile strengths when compared with Condition 2 (Table III). The short α/β interfaces formed upon precipitation of the alpha in the beta structure acted as effective obstacles to dislocation motion and crack propagation and strengthened the material accordingly [3]. It was also interesting to note the differences in the tensile properties obtained using the direct age (Condition 3). In general, the specimens with the thicker cross sections exhibited lower ductility. This was most notable in the step castings where the section thickness varied from 0.125 to 1" (3 to 25 mm) (Table III). The lower ductility was attributed to the slower cooling rates encountered with thick cross sections and the subsequent formation of large beta grains and thicker GBα [2].

In an attempt to correlate the loss of tensile elongation to the existence of thicker GBα, the thickness of this phase was measured in three microstructural conditions (Table III). These selected conditions represented very low, intermediate, and high tensile elongations [C3 (0.5" plate), A1 and A2, respectively]. The RT tensile elongation was plotted against the GBα thickness in Fig. 5. It should be noted that this study was based on limited data and therefore may only provide a general guideline.

As would be expected, the higher strength of Condition 1 was associated with a corresponding drop in ductility in spite of the very dimpled nature of the fracture surface (Fig. 2c). The drop in ductility was related to the strength differences between the high strength $\alpha+\beta$ structure material and the relatively soft GBα regions. The presence of flat and continuous GBα lead to long, relatively soft zones which preferentially deformed and eventually fractured intergranularly during straining (Fig. 2)[2,3]. This resulted in large *localized plastic strains* which were distributed over the relatively small volume of the GBα material [3] (Fig. 2c). Due to the planar nature of the GBα-phase and the corresponding long slip paths found in the β-solution treated and aged (β-STA) material, high stress concentrations and local strains were developed at the grain boundary triple points. The inherent strength of the finely precipitated α+β structure enhanced the strain localization in the relatively soft GBα. In these triple point regions, fracture occurred at low *macroscopic strains* even though high *local strains* were accommodated at the GBα as evidenced by the highly dimpled facets (Fig. 2c) [3].

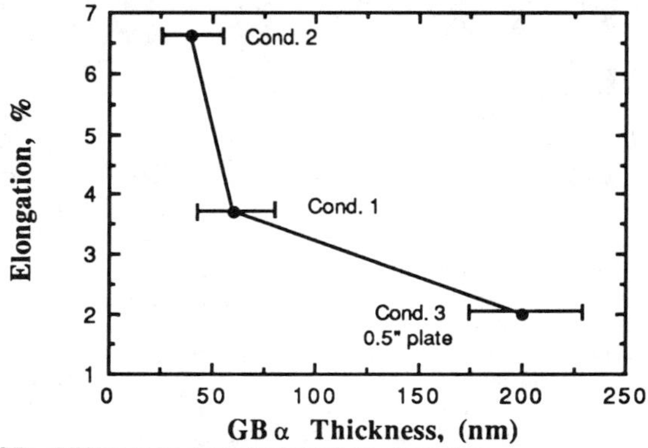

Figure 5: Influence of GBα thickness on room temperature elongation.

A number of factors have been identified to reduce tensile crack nucleation and gain ductility in β-STA beta alloys . These include minimizing the thickness and length of the relatively soft GBα zones and reducing the strength differences between the GBα material and the α+β structure. Both issues were addressed by the changes made in the heat treatments from Condition 1 to Condition 2. Condition 2 material was cooled at a faster rate to reduce the thickness of the GBα material. Also, the shorter aging times used in Condition 2 resulted in a lower strength α+β structure by reducing the amount of fine alpha phase precipitation. Therefore, lower strength differences between the GBα and the α+β structure resulted in higher tensile elongation.

The relatively small reduction in tensile strength from Condition 1 to 2 was associated with a twofold increase in elongation (Table III). It was also associated with a change in fracture mode from dimpled *intergranular* fracture (Condition 1, Fig.2a) to predominantly dimpled *transgranular* fracture (Condition 2, Fig. 2b). The high strength of Condition 1 localized the deformation in the relatively soft GBα leading to a highly ductile but localized fracture zone along the prior beta grains at the GBα phase. On the other hand, the lower strength Condition 2 resulted in more plasticity in the α+β matrix structure, leading to more macrodeformation throughout the sample and hence the higher tensile elongation.

It is also possible to correlate the room temperature elongation to the GBα thickness since a thick GBα will increase the strain localization into the lower strength zone. As can be seen from Table III, the lowest elongation (Condition 3- 0.5" plate) was associated with a 200 nm average GBα thickness, while the highest elongation (Condition 2) displayed only a 40 nm average thickness as illustrated in Fig. 5. Similar findings have been reported for α+β alloys where tensile voids in the GBα zone were shown to grow at faster rates as GBα thickness increased [8].

FATIGUE PROPERTIES

Condition 2 displayed slightly better fatigue ratio (Table III) than Condition 1. In both conditions the fatigue cracks initiated internally in most of the specimens tested (Fig. 4). This was unusual since fatigue crack initiation is typically a surface-related phenomena.

The most common initiation site in both conditions was along the GBα phase. A feature common to the GBα-initiated cracks was long, planar and continuous GBα inclined about 45° to the tensile direction (Fig. 4a), coinciding with the angle of maximum shear stress. Initiation sites with similar characteristics have been reported for Ti-6Al-4V [9] and for the beta alloy, Ti-8Mo-8V-2Fe-3Al [10].

Thick, planar regions of GBα oriented near 45° to the direction of applied stress nearly always accounted for loss of fatigue life in Ti-6Al-4V [9], which agrees with the results of this work. While Condition 2 samples had lower UTS than Condition 1 samples [174 (1215) vs 181 ksi (1265MPa)], the fatigue ratio ($S = \sigma_f / \sigma_{UTS}$) was higher for Condition 2 than Condition 1. The higher fatigue ratio indicated a lower propensity to nucleate cracks at GBα due to the relatively lower α+β matrix strength. This may also be the result of the thinner GBα in Condition 2 versus the thicker GBα in Condition 1.

About 15% of initiation sites were associated with residual casting micropores. The fact that micropores were not readily evident in any of the tensile specimens examined is an indication that there were probably only very few present due to HIP. Further, most of these pores were 10 to 15 μm in diameter (Fig. 4b). The microporosity responsible for fatigue crack initiation was always found on GBα fracture facets. This fact links pore initiation to GBα initiation. Micropore-initiated fatigue cracks do not necessarily start at GBα oriented 45° to the stress direction since the pore stress risers were sufficient to initiate cracks at other orientations. It was very intriguing that *all* samples associated with *micropore initiation* exhibited *above average* fatigue lives. It is possible that the samples exhibiting micropore-initiated failure had correspondingly thin or non-planar regions of GBα, so that initiation shifted to the next weakest link: the areas of stress concentration surrounding the micropores.

The fatigue ratios for all conditions, Condition 1- S = 0.69, Condition 2- S = 0.70, and Condition 3 -S = 0.78 (Table III) are similar to those found in another beta alloy, Beta C, and are generally superior to cast+HIP Ti-6Al-4V. Ti-6Al-4V typically range from 0.65 to 0.71 [1].

The high strength and fatigue properties, in addition to the low densities, of the beta titanium alloys make them an attractive option to replace precipitation-hardened steel castings such as 17-7 PH. However, additional work is needed to reduce property scatter, particularly low fatigue lives and low tensile elongation. Controlling the formation of GBα phase will be required to meet these challenges and increase the aerospace applications of the β alloys.

SUMMARY AND CONCLUSIONS

Investment castings of the high strength beta titanium alloy, Ti-15V-3Cr-3Al-3Sn (Ti-15-3), were studied to determine the tensile and fatigue behavior and the mechanisms leading to tensile and fatigue crack initiation. Three conditions were examined by applying different hot isostatic pressing (HIP), solution treating and aging parameters.

1. The three process conditions provided relatively high ultimate strengths in the range of 160 (1120) to 180 ksi (1260 MPa).

2. The lower strength Condition 2 material showed an almost twofold increase in tensile elongation over Condition 1 material. The tensile fracture surface of the Condition 1 showed a highly faceted, ductile, intergranular mode of failure while the Condition 2 fracture surface was somewhat faceted and exhibited a higher percentage of transgranular fracture.

3. The large strength mismatch between the finely precipitated α+β structure and the grain boundary alpha (GBα) phase in Condition 1 resulted in localized ductile intergranular tensile fracture surfaces yet with low *macrodeformation.* The smaller strength mismatch in Condition 2 allowed for more transgranular deformation and therefore higher elongation.

4. Limited data correlated the tensile elongation to the GBα thickness. Condition 3 with the thickest GBα phase displayed the lowest elongation while Condition 2 with the thinnest GBα phase had the highest tensile elongation.

5. Fatigue cracks initiated at GBα phase for Conditions 1 and 2 and was sometimes associated with occasional residual casting microporosity.

6. GBα fatigue crack initiation was due to the presence of a thin, continuous, planar GBα phase oriented at about 45° to the direction of the applied stresses. GBα initiation resulted in premature fatigue failure for each condition.

7. Fatigue cracks that initiated at micropores were observed only in specimens exhibiting **longer than average** fatigue lives. These were always associated with GBα regions normal to the tensile direction.

8. The combination of high tensile strength, fatigue strength, and low density will make castings of the beta titanium alloy, Ti-15-3, strong candidates for replacing castings of precipitation-hardened steels (PH) in demanding aerospace and industrial applications.

REFERENCES

1. D. Eylon, W. J. Barice, R. R. Boyer, L. S. Steele, and F. H. Froes, "Casting of High Strength Beta Titanium Alloys," Sixth World Conference on Titanium, Part II, P. Lacombe, R. Tricot, and G. Beranger, eds., Les Editions Physique, Les Ulis Cedex, France, 1989, pp. 655-660.

2. R.R. Boyer, W.J. Porter, and D. Eylon, "Microstructure/Properties Relationships in Ti-15V-3Cr-3Al-3Sn High Strength Castings,"Microstructure/Property Relationships in Titanium Aluminides and Alloys, Y-W. Kim and R. R. Boyer, eds., TMS, Warrendale, PA, pp. 511-520.

3. G. T. Terlinde, T. W. Duerig, and J. C. Williams, "Microstructure, Tensile Deformation and Fracture in Aged Ti-10V-2Fe-3Al", Metallurgical Transactions A, Vol. 14A, October 1983, pp. 2101-2115.

4. W. J. Porter, Crack Initiation in Beta Titanium Castings, Master's Thesis, Graduate Materials Engineering, University of Dayton, June, 1990.

5. P. J. Bania, G. A. Lenning, and J. A. Hall, "Development and Properties of Ti-15V-3Cr-3Sn-3Al (Ti-15-3)," Beta Titanium Alloys in the 1980's, R. R. Boyer and H. W. Rosenberg, eds., TMS-AIME, Warrendale, PA, 1983, pp. 209-229.

6. T.W. Duerig and J. C. Williams, "Overview: Microstructure and Properties of Beta Titanium Alloys," Beta Titanium Alloys in the 1980's, R. R. Boyer and H. W. Rosenberg, eds., TMS-AIME, Warrendale, PA, 1983, pp. 19-67.

7. S. Ankem and S. R. Seagle, "Heat Treatment of Metastable Beta Titanium Alloys," Beta Titanium Alloys in the 1980's, R. R. Boyer and H. W. Rosenberg, eds., TMS-AIME, Warrendale, PA, 1983, pp. 107-126.

8. M. A. Greenfield and H. Margolin, "The Mechanism of Void Formation, Void Growth, and Tensile Fracture in an Alloy Consisting of Two Ductile Phases," Metallurgical Transactions, Vol. 3, October 1972, 2649-2659.

9. D. Eylon, "Fatigue Crack Initiation in Hot Isostatically Pressed Ti-6Al-4V Castings," Journal of Materials Science, Vol. 14, 1979, pp. 1914-1922.

10. R. Chait and T. S. Desisto, "The Influence of Grain Size on the High Cycle Fatigue Crack Initiation of a Metastable Beta Ti Alloy," Metallurgical Transactions A, Vol. 8A, June 1977, pp. 1017-1020.

A NEW CALCIA TECHNOLOGY FOR INVESTMENT CASTING OF TITANIUM

ALLOYS WITH THE LOST WAX PROCESS

Serge PRIGENT and Jean DEBUIGNE

Laboratoire de Métallurgie et Physico-Chimie des Matériaux
URA CNRS 1495
I.N.S.A. 20, avenue des Buttes de Coësmes
F 35043 RENNES Cédex FRANCE

Abstract

The investment casting process of titanium and titanium alloys, usually generates a significant pollution of the superficial layers by mould/metal reaction. After an outline of the physical and chemical properties which bring about this contamination, we show the detrimental effects of Alumina, Zirconia and Magnesia on Calcia ceramics. An original dip slurry composed of pure Calcia, enables us to produce ceramic moulds through the lost wax casting process. An experimental melting device, simulating the casting conditions of titanium is described. The reaction zone of the specimens is studied by metallography, microhardness and EPMA. The " α-case" is never observed, even at high temperature.

Introduction

Following the lost wax process, the precision casting of titanium alloys, invariably engenders superficial contamination. Titanium's high affinity for light atoms such as B,C,N,O..., which are the basic compounds of the ceramics, is the starting-point of mould/metal reaction. The migration of these atoms, which are located in the interstitial sites of the metal, modifies the microstructure and the properties of the outer layers of the metal, and gives rise in particular to an increase of its hardness.

The reaction of the mould materials with titanium has been widely studied. The published results are as follows :

- Silica reacts strongly with the metal even in a cold mould process. Authors have noticed rutile formation at the mould/metal interface [1].
- Magnesia, Alumina, Zircon and Zirconia strongly contaminate the titanium. An "α-case" layer grows quickly with an increase in oxygen concentration. Suboxide can be observed when the holding time is long enough [2,3,4].
- Rare earth oxide studies give prominence to the thermochemical inertia of the yttrium oxide facing Titanium [5]. However its manufacture on an industrial scale is restricted by its prohibitive cost.

Titanium '92
Science and Technology
Edited by F.H. Froes and I. Caplan
The Minerals, Metals & Materials Society, 1993

• The reaction of the Calcium Oxide is limited, but its inertia decreases with an increase of the content of SiO_2 and Fe_2O_3 contents appearing in industrial Calcia [6].

Some authors highlight the ceramic contamination by the binders employed in precision casting, which modify the chemical composition of the ceramic layers. In particular, Ethyl Silicate and Colloïdal Silica, increase the silica content, whose high reactivity with molten titanium causes irreversible damage by hot mould process.

The ceramic properties giving rise to the mould/metal reaction must be as follow :

• The formation free enthalpy of the ceramic oxide has to be as low as possible.
• The solubility of the ceramic basic metal, in titanium, has to be as low as possible.
• The melting point of the ceramic has to be higher than the temperature of the superheated titanium before casting.

Calcia would seem to be a very promising material for the inner wall of the moulds in view of its inertia facing titanium. However, the high hydration ratio and the poor sintering capability of the Calcia force study of the properties of the Calcia ceramics with different contents of Alumina, Zirconia and Magnesia.

The reactivity of Molten Titanium

Through a qualitative study of the reactivity of the molten titanium has been conducted, keeping a molten metal sample for one minute into contact with various cold ceramics. These compounds belong to the binary systems: Alumina/Calcia, Calcia/Zirconia and Calcia/Magnesia. The results are as follows :

Molten titanium reacts strongly with Zirconed and Alumined Calcia. These results are in complete accord with published studies on thermodynamic equilibrium [4,6]. The formation free enthalpy of the Zirconia and Alumina oxides is below that of TiO_2 but is fairly close to that of TiO. It is important to report that the migration of the oxygen located in the interstitial sites and the migration of the metal located in the substitutional sites is issuing from the reduction of the ceramic oxide, giving rise to a transformation of the properties and of the microstructure of the titanium. (The dissolution of zirconium and aluminium in titanium has been shown by EPMA). Furthermore, some molten phases of the Alumina/Calcia diagram appear below the temperature of the superheated titanium before casting, which can destroy the mould.

The formation free enthalpy of the Calcia/Magnesia phases are below those of titanium suboxides. Furthermore the solubility of calcium and magnesium is quite low, as we are unable to detect these atoms by EPMA. These properties seem to be the starting-point of the slow kinetics of the reaction of the ceramics facing molten titanium.

Calcias applied to the inner wall of the casting moulds have to be free from Zirconia and Alumina, for the reactivity of these compounds with titanium is much too high. Since the sintering properties of the ceramics and the hydration ratio are not improved with Magnesia. An original dip slurry of pure Calcia was developed which allows for the production of ceramic moulds through the lost wax process.

Experimental procedure

The production of the ceramic crucibles from a mother sample made of wax (∅15mm, h25mm) is as follows :

- Soaking in the dip slurry (the slurry is composed of a liquid of special composition, a binder and an electrofused Calcia [size : 200 mesh]),
- Stuccoing (the size of the Calcia sand is 200-500 μm for the first coat, and 500-1000 μm for the others),
- Drying/setting (the setting time is nearly 2 hours for the first and second layer and overnight for the third),

Figure 1 : Calcia crucibles

- Flash dewaxing in a vertical furnace heated at 1000°C, follows the building of the shell. A presintering of the calcia ceramic is employed during this sequence. The crucible is then sintered in a muffle furnace, for 2 hours at 1450°C. Figure 1 shows some crucibles produced by the process detailed above.

The melting device drawn in Figure 2 can simulate a titanium casting test in the laboratory and check certain experimental conditions. The operating processes are as follows :

- Degassing of the ceramic in primary vacuum at high temperature.
- Cleaning the chamber with purified Argon.
- Ceramic preheating (1200° C).
- Titanium superheating (1700-1800°C) crucible pouring (=3g Ti).
- Isothermal holding at high temperature (1200-1400°C, 10 minutes).
- Free cooling (15 minutes).

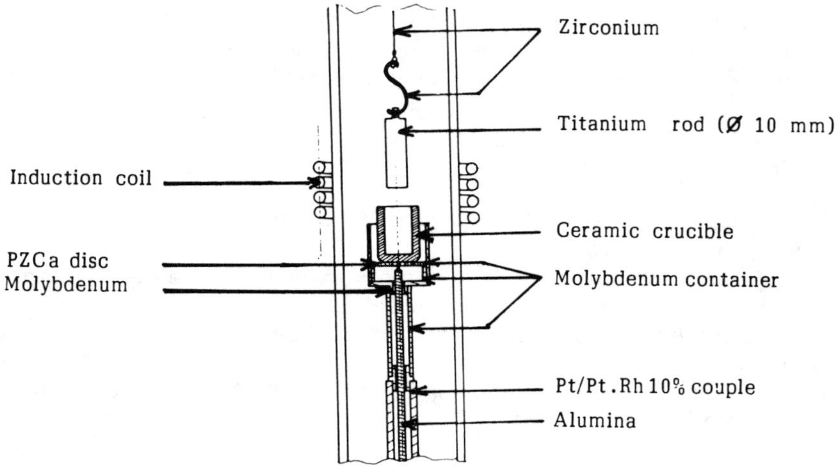

Figure 2 : Melting device for titanium samples

After reaction, titanium samples are cut perpendicularly to the reaction layer, with a diamond saw, and embedded into a thermosetting resin. They are then polished with SiC abrasive paper (240-2400 mesh), diamond paste (0,5 μm) and finally OPS (Struers). The metallographical observation is conducted on a PME OLYMPUS Microscope, a microhardness profile is taken, under 8g applied load for 20 seconds. The Ca, O, Ti concentrations are recorded by EPMA, (CAMEBAX Sx50).

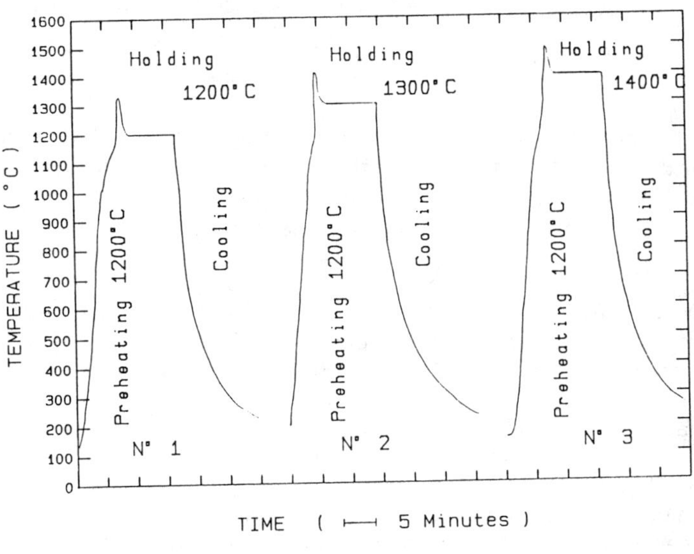

Figure 3 : Thermal exposure curves

Three different tests have been conducted. Figure 3 draws the thermal exposure curves of the samples, recorded with a Pt/Pt.Rh.10% thermocouple. This process is comprised of the crucible preheating, titanium casting, isothermal holding for 10 minutes at 1200, 1300 and 1400°C and overall cooling.

Results

Three different grades of electrofused Calcia, whose chemical composition is shown in Table I, has been studied. The building process of the ceramic shell does not modify this composition.

Oxides	Concentration in wt%		
CaO	98 min.	99 min.	99.4 min.
MgO	0.5 to 2	0.6	0.4
SiO_2	1	0.2	0.08
Al_2O_3	0.3	0.05	0.04
Fe_2O_3	0.25	0.05	0.04

Table I - Composition of industrial electrofused Calcias

Figure 4 : Microstructure of the external layer of titanium

The cast titanium samples have the same metallographical structure (Figure 4). titanium, alloyed with a low oxygen content, has a poor polarising power, without chemical attack. Nevertheless, the acicular structure of the mould/metal interfacial zone of titanium was observed, which is quite different from the structure of the internal zone. This probably results from the action of the oxygen on the formation and growth of the widmanstatten needles, caused by the $\beta \rightarrow \alpha$ transformation during the cooling of the metal. However the difference between the concentrations of oxygen located in the internal zone and the external zone remains difficult to quantify. The depth of penetration of the needles, which have a narrow basis, can extend to almost 1 mm. The metallographical observation of the titanium cast in a Calcia crucible results in no distinction either of the different cooling treatments or of the different grades of the electrofused Calcia used in the crucibles.

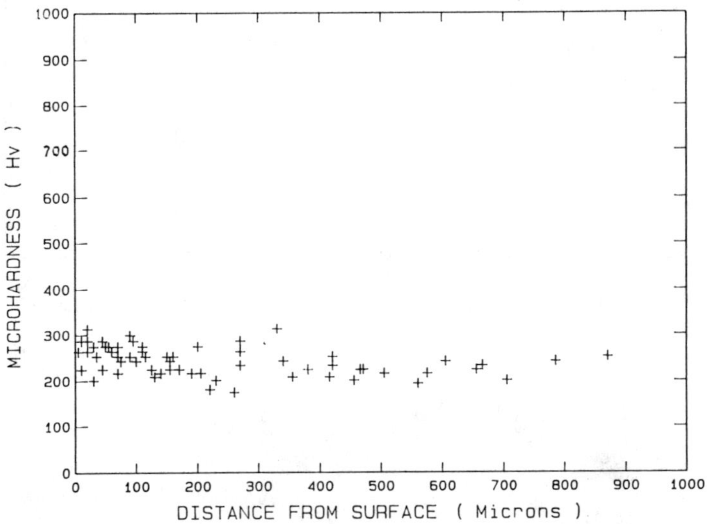

Figure 5 : Microhardness profile of cast titanium

Microhardness of titanium is strongly influenced by oxygen in solid solution. It is immediately clear (Figure 5) that the variation in hardness is much too low to be able to quantify the metal reaction with oxygen. The average of hardness data is 215Hv for the initial metal and 250 Hv in the external layer after reaction. These results correspond well with the metallographical observations and confirm a fairly low reaction of titanium with oxygen.

The concentration of Ca, O, Ti by EPMA were recorded. The data on oxygen contents are not accurate due to :

- An apparent increase in oxygen concentration when the polishing of the titanium samples was carried out by oxides (Al_2O_3, OPS). The level of this increase fluctuates according to the location on the samples, and remains difficult to estimate.

• The natural oxidation of titanium, in the air, provides a small layer of oxide equivalent to a layer of TiO_2 which is about 60 to 80Å wide. This property of titanium increases the oxygen concentration to about 1 at %.

• The precision of oxygen determination is limited to 1500 ppm even on standards.

Thus analysis show relative variations in concentration. However a high level of background noise prevents any conclusions being made because of the low variation of the oxygen content in titanium, even near the surface, after reaction with the Calcia crucibles.

The diffusion of calcium in titanium is very limited. Its concentration is below 0.1 at %, which is the minimum amount measurable by EPMA, even a few microns from the surface.

Calcia behavior at very high temperature

This simulation of casting titanium into pure Calcia crucibles shows that the oxidation of solid titanium, facing pure Calcium oxide, is restricted, even at high temperature (1400°C). It seems useful, therefore, to study the reaction between molten titanium and hot Calcia crucibles. Thus, in the melting device, a sample of titanium (5g) was placed inside a crucible and heated them together up to 1700°C. No reaction due to temperature increase was observed until 1660°C. Above this temperature, an important release of vapor was noted, with a bubbling of the metal in the crucible. The failure of the Pr/Pt.Rh.10% couple at the melting point of the titanium prevented a precise recording of the temperature of molten titanium.

Molten titanium reduces calcium oxides, then oxygen diffuses into the liquid metal, thus increasing the hardness of the metal after cooling. Holding times of 1,2 and 3 minutes in the molten state have shown an increase of about 50 Hv for each additional minute of holding time.

The solubility of calcium in molten titanium being low, these atoms are in free state at temperatures above their boiling point (=1440°C). This explains why the sample releases vapor which crystallises on the inner wall of the chamber. Calcium hydroxide has been identified by X-Ray analysis of chamber deposits. The reaction between calcium and moistured air is well known, and its end product is calcium hydroxide.

It is important to note that while titanium is in a molten state and flows with a convectional movement, the kinetics of crucible reactions is no longer limited by diffusional reactions. Unlike the oxidation of titanium in the solid state, the crucible reaction is almost entirely interfacial, with a rapid elimination of the reacting products : oxygen by dissolution into the molten titanium and calcium by vaporisation.

Conclusions

(1) Contamination of titanium by oxygen cast in Calcia crucibles can not be quantified by traditional methods of investigation such as metallography, microhardness and EPMA examination.

(2) This study enabled differentiation of neither the different cooling treatments nor the different grades of the electrofused Calcia, because the increase in the oxygen content of the metal was much too low, even at high temperature.

(3) The lack of reaction of pure Calcia facing titanium is derived from the low value of the formation free enthalpy of the Calcia, and from the low solubility of the calcium in both liquid and solid titanium.

(4) Molten titanium reacts strongly with hot Calcia crucibles when the holding time is long enough. The boiling point of calcium (1440°C) seems to be a detrimental to use of Calcia for titanium-melting crucibles.

References

1. R.L Saha et al., "Evaluation of the reactivity of titanium with mould materials during casting," Bull Mater Sci, 12 (5) (1989), 481-493.

2. R.L. Saha et al., "Metal-Mold reactions during casting of titanium in Zircon sand molds", AFS Transactions, 98, (1990), 253-260.

3. D. Mukherji, R.L. Saha, and C.R. Chakravorty, "Titanium casting in Zircon sand mould", Trans of the Ind Ins of Metals", 38(6) (1985), 465-471.

4. A.K. Misra, "Reaction of Ti and Ti-Al alloys with Alumina", Metal transA, 22 (3) (1991), 715-721.

5. R.L Saha et al. , "On the evaluation of stability of rare earth oxides as face coats for investment casting of titanium", Metal Trans B, 21(6) (1990) 559-566.

6. T. Degawa, K. Kamata, and Y. Yoneda, "Melting and precision casting of pure titanium using Calcia", 6th world conf on Ti, vol. II, France 1988, 707-713.

SUPERPLASTIC DEFORMATION BEHAVIOUR OF

Ti-4Al-4Mo-2Sn-0.5Si (IMI 550)

L.B. Duffy* and N. Ridley†

†Materials Science Centre
University of Manchester/UMIST
Grosvenor Street, Manchester M1 7HS, England

*Magnesium Elektron Limited
Regal House, London Road
Twickenham TW1 3QA, England

Abstract

The superplastic behaviour during uniaxial tensile deformation of microduplex IMI 550 sheet material has been examined as a function of strain, strain rate and deformation temperature in the range 810-940°C. Changes in the strain rate sensitivity of flow stress, m, as a function of strain, strain rate and temperature have been related to microstructural changes. Elongation to failure tests showed that microduplex IMI 550 has an excellent potential for SPF with tensile elongations > 1500% being obtained at the optimum deformation temperature of 900°C, when the microstructure contained approximately equal proportions of the α and β phases. Above 900°C, tensile elongations decreased and strain hardening increased due to rapid grain growth.

Introduction

It is well established that microduplex Ti-6%Al-4%V can exhibit excellent superplastic formability at temperatures in the range 850-940°C.(1-5). However, there is a continuing requirement for higher strength superplastic titanium alloys and this has led to an interest in a number of alternative materials including the heat treatable alloy, Ti-4%Al-4%Mo-2%Sn-0.5%Si (IMI 550)(6). This alloy has been shown to exhibit creep and elevated temperature tensile properties superior to those of Ti-6%Al-4%V(7). Published work on SPF of IMI 550 is relatively limited although McDarmaid(8) and Hammond et al(9) have reported substantial tensile elongations for hot rolled bar material.

The alloy has also been produced in the form of sheet with a microduplex structure and good superplastic formability has been reported for a limited range of conditions during equibiaxial bulging(10). It has also been shown that any loss of tensile properties during SPF can be restored by heat treatment(10,11). In the present work the SP behaviour of the sheet material in uniaxial tension has been examined as a function of strain, strain rate and deformation temperature, and the microstructural stability has also been investigated. Comparisons are made with properties/behaviour reported for Ti-6Al-4V.

Titanium '92
Science and Technology
Edited by F.H. Froes and I. Caplan
The Minerals, Metals & Materials Society, 1993

<u>**Experimental**</u>

The titanium alloy was supplied in the form of cross-rolled 2.1 mm thick sheet which had been manufactured by IMI Titanium under contract from the Royal Aircraft Establishment, Farnborough. The sheet had a predominantly basal texture and showed tensile properties in the plane of the sheet which were reasonably isotropic (10). The wt% composition obtained by chemical analysis was: Al - 4.10, Mo - 3.74, Sn - 1.96, Si - 0.35, C - 0.04, Fe - 0.025, O - 0.0019, Ti - bal.

Tensile specimens with a 10 mm gauge length and 5 mm gauge width were machined from the sheet. Deformation was carried out in an argon atmosphere in a three-zone split furnace attached to an Instron tensile testing machine, interfaced with a computer. Metallographic observations were made on specimens etched in Krolls reagent (1% vol. HF, 12% vol. HNO_3 aqueous solution). Mean linear intercepts were measured for as-received, heat treated and superplastically deformed specimens in the longitudinal (d_A), long transverse (d_B) and short transverse (d_C) directions. The mean grain size, d_M, was the average of these three measurements. For the as-received sheet, d_M = 2.1 μm. Point counting was used to measure the phase proportions of samples water quenched from various holding temperatures. The effect of temperature on β-phase is shown in Table I.

TABLE I: Variation of β-phase proportion with temperature for IMI 550

Temperature °C	**810**	**840**	**870**	**900**	**930**
β-phase Vol %	37	40	44	50	62

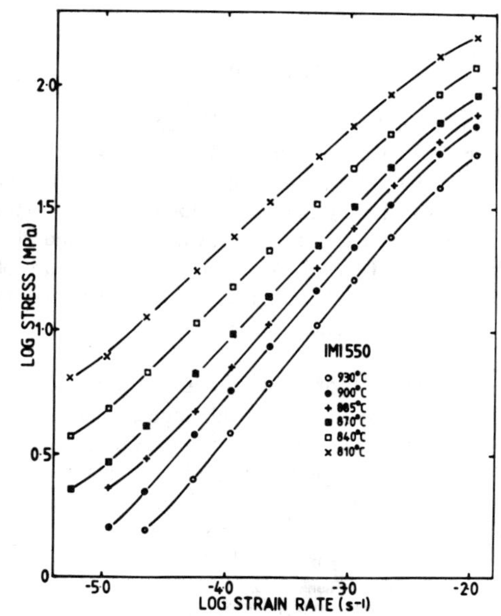

Figure 1: Logarithmic plots of stress versus strain rate for IMI 550

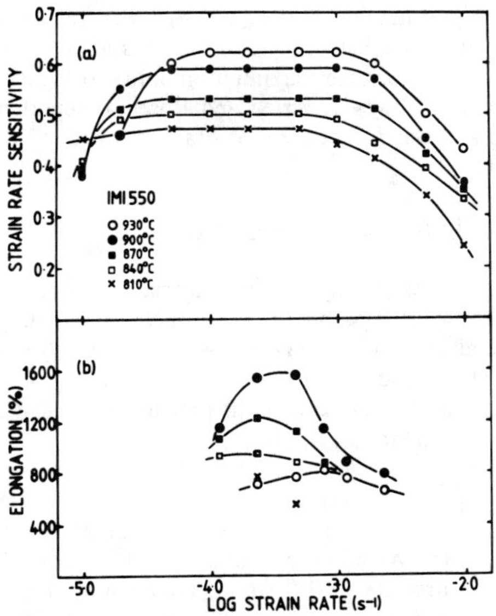

Figure 2: Variation of (a) strain rate sensitivity and (b) elongation to failure, with strain rate and temperature

Results and Discussion

Stress-Strain Rate Behaviour

Figure 1 shows the log stress-log strain rate curves for the as-received IMI 550 sheet deformed in the range 810-930°C and these exhibit many of the characteristics typical of superplastic materials. A sigmoidal relationship exists between stress and strain rate and there is a clearly defined linear region at intermediate strain rates (Region II) where optimum superplastic behaviour is normally observed. As the deformation temperature is raised, the flow stress decreases at all strain rates and the maximum slope of the curve, m, increases from 0.47 at 810°C to 0.62 at 930°C (Fig. 2). The curves in Fig. 2 show a broad relatively flat top end and the strain rate of maximum 'm' moves to higher strain rates as the temperature is increased. Each curve in Figure 1 is the mean of two curves obtained from increasing and decreasing strain rate cycles.

At each temperature in the range 810-870°C, the stress-strain rate curves3 for both cycles almost superimpose on each other, and m values are therefore virtually identical. However, at temperatures of 900°C and 930°C, the increasing strain rate cycle gives clearly higher m values than the decreasing cycle due to the increased extent of grain growth during testing. This is supported by grain size measurements which show that enhanced grain growth at the higher deformation temperatures gives rise to a hysteresis effect between the two cycles. Similar observations have been reported by Hammond et al (9) for deformation of microduplex IMI 550 bar material.

The m values measured by McDarmaid(8) on IMI 550 bar material of 0.48 at 880°C and 0.65 at 928°C are similar to those obtained in the present work, but are somewhat lower than those often quoted for other α/β titanium alloys.(1-3, 5). Several studies of superplasticity in Ti-6%Al-4%V alloy have shown that an increase deformation temperature up to 950°C leads to an increase in m and displaces the maximum m value to higher strain rates(1, 5, 12). In contrast to these studies, however, Arieli and Rosen(4) reported that the m value for Ti-6%Al-4%V was 0.5 and was independent of strain rate within Region II, and temperature in the range 700-950°C.

Effect of Strain, Strain Rate and Temperature.

Strain rate sensitivity values for superplastic materials are usually measured over a range of tensile elongations from 30-70% but this is insufficient to fully characterise superplastic behaviour. A material may exhibit high m values after small strains but may fail after say 200% strain due to microstructural changes such as grain growth. It is the instantaneous m value that determines the resistance to neck propagation and it is, therefore, important to know how m changes with strain.

Figure 3 shows the effect of superplastic strain on the strain rate sensitivity of IMI 550 of 900°C. The general trend is a decrease in m with strain although the extent of this decrease depends on the strain rate. At strain rates of 2.3 x 10^{-4} s^{-1} and 7.8 x 10^{-4} s^{-1}, m remains fairly constant for the first 100-150% strain, after which it decreases. However, at the slowest strain rate of 7.8 x 10^{-5} s^{-1}, m falls continuously after a very small strain. This decrease is attributable to the occurrence of grain growth during deformation which is most pronounced at the lowest strain rate. Measurements showed that the mean grain size, d_M, increased with increasing superplastic strain at all strain rates although the rate of grain growth was strongly dependent on strain rate. At the fastest strain rate of 7.8 x 10^{-4} s^{-1}, d_M increased by 90% after 500% elongation whereas as 7.8 x 10^{-5} s^{-1}, d_M increased by 215%. This is due to increased deformation time experienced at the lower strain rates which gives increased opportunity for grain growth.

Data from Figure 3 and grain size measurements have been used to plot strain rate sensitivity against instantaneous mean grain size, d_M, in Figure 4. This shows that m decreases most rapidly with increasing d_M at the highest strain rate of 7.8 x 10^{-4} s^{-1} because this strain rate lies at the high strain rate end of Region II. Since an increase in grain size generally causes

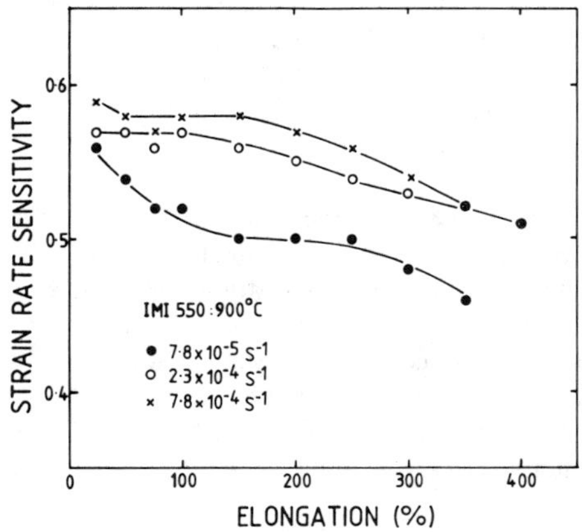

Figure 3: Effect of strain and strain rate on 'm' value at 900°C

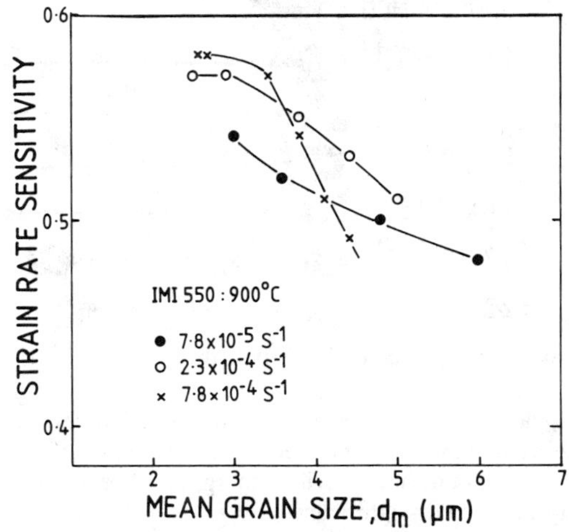

Figure 4: Variation of m value with the instantaneous grain size and strain rate at 900°C.

Region II to be displaced to lower strain rates, grain growth during the test causes the deformation mode to shift to the relatively non-superplastic Region III giving a marked decrease in m. The relatively gradual decrease in m with d_M observed at the lower strain deformation mechanism. Further measurements showed that m decreases more rapidly with strain as the deformation temperature is raised.

Several authors have examined the variation of m with strain in Ti-6%Al-4%V over the temperature range 760-950°C(1, 13, 14). Most of the tests showed that m decreased with increasing strain and this was attributed to coarsening of the microstructure during deformation. At 760°C, however, Furushiro et al(14) reported that m was either strain independent or increased up to ~ 100% elongation. Microstructural examination revealed that these trends coincided with either the grain size remaining constant during deformation or grain refinement occurring during the early stages of a test. The authors concluded that the m value was directly related to the grain size which is in broad agreement with results from the present work.

Constant strain rate tests.

The results of elongation to failure tests carried out at a range of strain rates at temperatures between 810°C and 930°C are shown in Figure 2b. The alloy clearly has considerable superplastic deformation potential and tensile elongations in excess of 1,500% can be achieved. Figure 2b also shows that, for a given strain rate, the elongation to failure increases with increasing deformation temperature up to a maximum at 900°C. As the deformation temperature is increased, the strain rate at which the maximum elongation occurs also increases. Figure 5 shows the profiles of fractured IMI 550 specimens deformed at several strain rates at 900°C.

The largest elongation of 1,565% is considerably higher than values reported for Ti-6%Al-4%V(3, 5, 14). To some extent, this may be due to the extremely fine microstructure of the IMI 550 alloy which has a mean grain size of only 2.1 μm in the as-received condition. In contrast, Paton and Hamilton(3) and Cope(5) reported maximum elongations for Ti-6/4 of

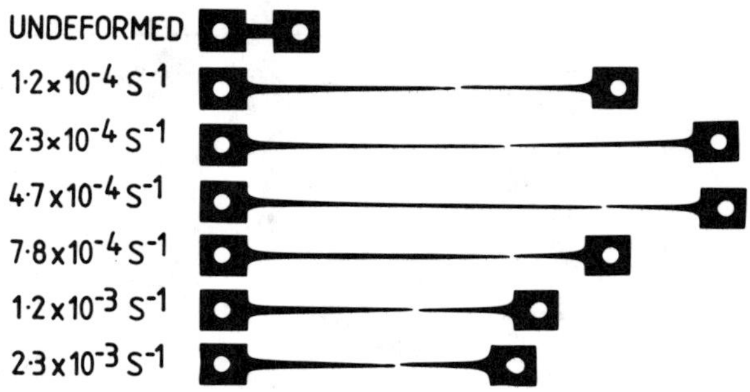

Figure 5. Fracture profiles of IMI 550 pulled to failure at 900°C at various strain rates

675% for 3.7 μm material and 1250% for 5.3 μm material, respectively. Clearly, the maximum attainable elongation does not necessarily increase with decreasing grain size since other features such as grain size distribution, grain aspect ratio, phase proportions and specimen geometry are also of significance(3).

Although the increase in elongation to failure over the temperature range 810-900°C is consistent with the increasing m values over the same range, elongations are smaller at 930°C than at 900°C despite the apparently higher m value at the highest temperature. This is due to the increased amount of grain growth at 930°C which leads to a steadily decreasing m during deformation. Figure 2 shows that, at temperatures of 840°C and 870°C, maximum elongations are obtained at strain rates at the centre of Region II but, at 900°C and 930°C, the highest elongations are found to occur near to the high strain rate end of Region II. This is observed because, as the deformation temperature is raised, a higher strain rate is necessary to give the maximum elongation since grain growth reduces m at lower strain rates. Boyer and Magnusson(15) have reported similar behaviour for Ti-6%Al-4%V.

Paton and Hamilton(16) examined the relationship between both elongation to failure and strain rate sensitivity and the β-phase proportion using data for α/β titanium alloys from the literature. Tensile elongations were highest at a β-phase content of 20-40% whereas maximum m values occurred when 40-60% of beta phase was present. The difference in location of these peaks was attributed to grain growth during deformation since higher diffusivities in the β-phase lead to increased coarsening rates as the β-phase proportion is raised. Cope et al(17) also reported that, at a strain rate of $2.3 \times 10^{-4}\ s^{-1}$, maximum elongations for Ti-6%Al-4%V and Ti-6%Al-2%Sn-4%Zr-2%Mo were obtained when the β-phase proportions were 40% and 42%, respectively.

Figure 6 is a plot of elongation at a strain rate of 1.4% min^{-1} ($2.3 \times 10^{-4}\ s^{-1}$), for a range of temperatures, against the β-phase proportion for IMI 550 and also includes the data of Cope et al(16). For IMI 550, maximum elongations are observed at 900°C at all strain rates for a β-phase proportion of 50%. Further, m values are highest at 930°C where 62% of beta-phase is present. Hence, the optimum microstructural condition for IMI 550 appears to be different from that for other α/β titanium alloys. This may be related to the presence of 4%Mo, which has a much lower diffusivity in β-Ti than Ti, or other beta phase stabilisers (i.e. V, Fe, Co, Ni)(9,18). The presence of 4%Mo may cause diffusional accommodation of grain boundary sliding to be more difficult at lower deformation temperatures where optimum superplastic behaviour would normally be observed. Hence, optimum behaviour is pushed to a higher deformation temperature where a larger beta-phase proportion is present. Figure 7 shows the true stress versus true strain for a strain rate of 1.4% min^{-1} (2.3 x 10^{-}

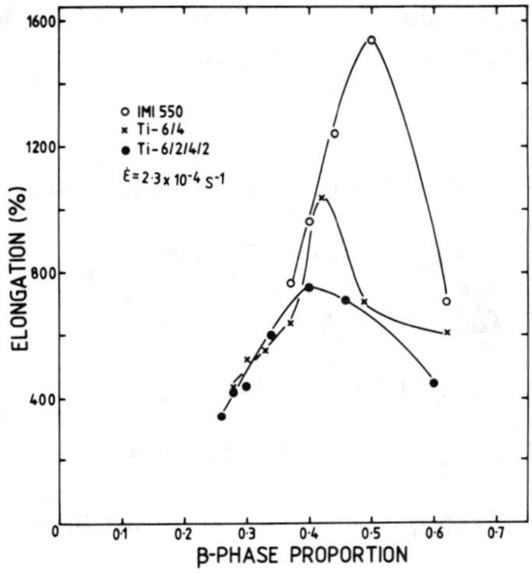

Figure 6: Effect of β-phase proportion on elongation to failure.

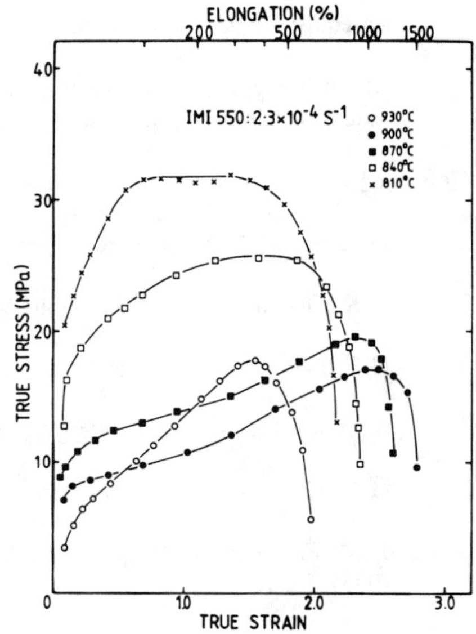

Figure 7: True stress versus true strain for IMI 550

4 s^{-1}) at temperatures in the range 810-930°C. At 870°C and 900°C continuous strain hardening up to approximately 1,000% elongation due to grain growth followed by a decrease in stress due to neck propagation. The sample deformed at 930°C, however exhibits rapid strain hardening and after 200% elongation, the flow stress is greater than that at 870°C. This is due to the marked grain growth at 930°C. Samples deformed at the lower temperatures show a gradual increase in flow stress up to 100-150% elongation after which a steady state is reached. The flow stress then remains relatively constant until plastic instability leads to failure. The attainment of a steady state flow stress at 810°C and 840°C reflects the relatively stable grain sizes at these temperatures.

Conclusions

1. In the temperature range 810-930°C, IMI 550 alloy exhibited m values of 0.47-0.62 which are characteristic of superplastic materials. Values of m increased with increasing temperature.

2. At temperatures of 870-930°C, strain rate sensitivity was strain dependent and, in general, an increase in superplastic strain led to a decrease in m. This effect was most pronounced at the highest temperature (930°C) due to higher diffusivity and hence faster grain growth, and at lower strain rates due to increased deformation times. The strain rate sensitivity is related to the instantaneous mean grain size, d_M.

3. In microduplex sheet form, IMI 550 alloy has considerable potential for superplastic forming and tensile elongations in excess of 1,500% can be achieved under optimum deformation conditions. Tensile elongation increased with increasing temperature up to 900°C but then decreased markedly at 930°C due to increased grain growth. Optimum elongations were obtained at 900°C where the microstructure consisted of

an equi-volume fraction of α and β phases.

4. Strain hardening occurs during superplastic deformation due to grain growth. The extent of hardening increases with increasing deformation temperature.

References

1. D. Lee and W.A. Backofen, Trans. Met. Soc. A.I.M.E., 239 (1967), 1034-1040.

2. A.K. Ghosh and C.H. Hamilton, Metall. Trans. A., 10A (1979), 699-706.

3. N.E. Paton and C.H. Hamilton, Metall. Trans. A., 10A, (1979), 241-250.

4. A. Arieli and A. Rosen, Metall. Trans. A., 8A, (1977), 1591-1596.

5. M.T. Cope and N. Ridley, Mater. Sci. and Technology, 21 (1986), 140-145.

6. T.W. Farthing in Metallurgy of Light Alloys, (London, Institut. of Metallurgists, 1983), 9-31.

7. R.M. Duncan and R.T.J. Hubbard, in Titanium Science and Technology, Vol. I ed. R.I. Jaffee and R.M. Burke, (New York, NY, Plenum Press, 1972), 91-103.

8. D.S. McDarmaid, RAE Technological Report 82106, (1982).

9. J. Ma, R. Kent and C. Hammond, J. Mater. Sci., 21(1986), 475-487.

10. L.B. Duffy, J.B. Hawkyard and N. Ridley, Mater. Sci. and Technology, 4(1988), 707-712.

11. C.D. Inglebrecht, Designing with Titanium, (London, Inst. of Metals, 1986), 252-260.

12. S.M.L. Sastry, R.J. Lederich, T.L. Mackay and W.R. Kerr, J. Metals, 35 (1983), 48-53.

13. A. Arieli and A.K. Mukherjee, Mater. Sci. Eng., 43 (1980), 47-54.

14. N. Furushiro, H. Ishibashi, S. Shimoyama and S. Hori, Proc. 4th Int. Conf. on Titanium, (1980), 993-1000.

15. R.R. Boyer and J.E. Magnusson, Metall. Trans. A., 10A (1979), 1191-1193.

16. N.E. Paton and C.H. Hamilton, Titanium Science and Technology, Vol. 2, ed. G. Lütjering, U. Zwicker and W. Bunk (Oberusel, FRG, DGM, 1985).

17. M.T. Cope, D.R. Evetts and N. Ridley, J. Mater. Sci., 21 (1986), 4003-4008.

18. J.A. Wert and N.E. Paton, Metall. Trans. A., 10A (1979), 1191-1193.

SUPERPLASTIC DEFORMATION PROPERTIES OF AS-RECEIVED, ROLLED AND BETA TRANSFORMED Ti-2% Al-2% Mn ALLOY

M.A. Hassan and C. Hammond

School of Materials,
The University of Leeds,
Leeds LS2 9JT, UK.

Abstract

The superplastic deformation properties of the alloy Ti-2% Al-2%Mn in three different initial microstructural conditions have been investigated in the temperature range 813°C - 920°C by the technique of strain rate cycling.

The as-received microstructure consisted of equiaxed α and ß grains with a non-uniform α grain size distribution in the range 5 μm - 22 μm. The optimum superplastic properties were obtained at 860°C corresponding to a ß phase proportion of 30%. Above this temperature substantial grain coarsening occurred.

The alloy was (a) hot rolled in the α+ß field in order to develop a more uniform equiaxed grain structure (of 12 μm grain size), and (b) solution treated and slack quenched from the ß field in order to give a transformed ß structure (of α platelets 40 μm in length and 1.5 μ thick). During superplastic deformation at 860°C and 830°C, grain coarsening, leading to higher flow stresses and lower m-values, occurred in (a) the equiaxed α+ß structure, but that grain refinement, leading to lower flow stresses and higher m-values occurred in (b), the transformed ß structure. The results indicate that, on strain-rate cycling to increasing levels of strain, the flow stresses, m-values and grain sizes for the initially widely different microstructures approach common values.

Introduction

Ti-2% Al-2% Mn is an (α+ß) titanium alloy with similar transformation characteristics to Ti-6% Al-4% V, i.e. the ß transus temperatures and phase proportions in the (α+ß) phase field at the same temperatures are similar. The object of the research programme was to assess the superplastic deformation properties of this alloy in three different initial microstructural conditions: equiaxed α+ß with a wide α grain size distribution (as-received condition); equiaxed α+ß with a narrow α grain size distribution (as-received and α+ß hot rolled condition), and acicular α (ß transformed condition). These microstructures correspond to those which may be encountered in practice, in particular the ß transformed condition is characteristic of that in the weld and heat affected zone of weld microstructures.

Titanium '92
Science and Technology
Edited by F.H. Froes and I. Caplan
The Minerals, Metals & Materials Society, 1993

Alloy and Experimental Procedure

The alloy was supplied as a 19.0 mm diameter bar manufactured by IMI Titanium Ltd., Birmingham, UK, from which tensile test specimens with a gauge length of 20 mm and a diameter of 6.4 mm were machined. In order to obtain a finer grain structure with a smaller grain size distribution, a piece 20 mm long was hot rolled in the $\alpha+\beta$ field to a 3.0 mm thick plate from which flat tensile specimens with a gauge length of 20 mm and width of 5.0 mm were machined. The coarse ß transformed structure was obtained by heat treating specimens machined from the plate material at a temperature of 1000°C (in the ß field) for 2 hrs followed by a water quench.

The high temperature tensile tests were carried out on an Instron tensile testing machine on which was mounted a three-zone tube furnace with a 300 mm uniform temperature zone accurate to ± 2°C. Oxidation of the specimens was reduced by diffusing high purity argon up through the furnace.

The specimens were solution treated at the testing temperatures for 1 hr and were initially strained prior to each cross-head speed cycling test to about 20% at a cross-head speed corresponding to a high strain rate of $1.67 \times 10^{-3}\ s^{-1}$. During this initial pre-straining stage a transient work hardening peak was observed followed by a constant flow stress. This procedure is important in that it establishes a microstructural condition in which transient work hardening is not observed and which would otherwise confuse the subsequent stress-strain rate-strain data. The cross-head speed was then reduced to give a lowest strain rate of $4.17 \times 10^{-5}\ s^{-1}$ and the load allowed to stabilise. Then the cross-head speed was increased in stages to the highest strain rate (about $4.2 \times 10^{-3}\ s^{-1}$), the load being allowed to stabilise between each change. The cross-head speed was then reduced to the lowest strain rate. This sequence of cross-head speed cycles was repeated at increasing strain levels and enables the effects of strain and time on the flow stress, the evolution of the superplastic deformation properties, and the effects of changing microstructure to be determined. The experimental load/cross-head speed/time data were computed and plotted as log σ versus log ϵ, assuming a uniform extension and a constant volume. A curve fitting computer program was used to draw a smooth curve through the data points and to find strain rate sensitivity (m) values from the slopes between adjacent points(1). Specimens for light and electron microscopy were taken from both the gauge length and (deformed) and grip (undeformed) sections of the specimens and prepared in the usual way in order to provide microstructural information on the mechanisms which are operating during the deformation process.

Results

The Microstructure of the Alloy Ti-2% Al-2% Mn

The microstructure of the as-received bar material consists of an equiaxed α phase with a non-uniform grain size between 5.0 μm and 22.0 μm and a ß phase which is distributed around the α grain boundaries and within the α grains as shown in Figure 1a. The microstructure of the alloy after hot rolling may be seen in Figure 1b, which shows again a matrix phase (α grains) surrounded by the ß phase with a more uniform equiaxed grain size of 12 μm as compared to the as-received material. It also shows that during the rolling process the ß phase has been redistributed at the α grain boundaries and no longer occurs within the α grains.

The microstructure of the ß transformed structure is shown in Fig. 1c and consists of a coarse acicular structure. These different initial starting microstructures enable a characterisation to be made of the flow behaviour of the alloy as a function of the microstructure.

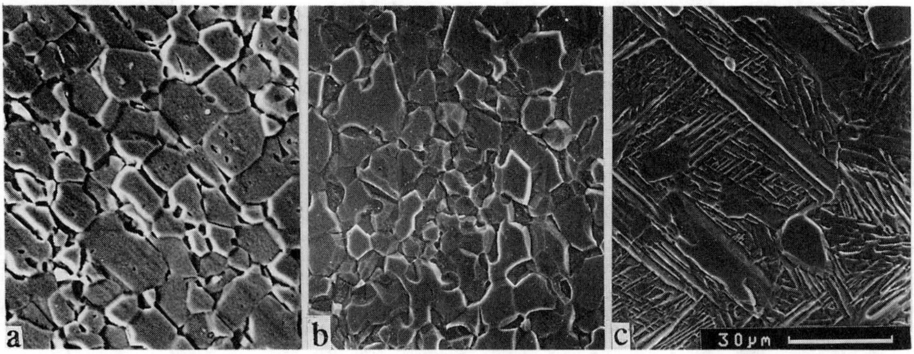

Fig. 1 - Microstructures of (a) the as-received alloy showing non-uniform α grain size, (b) the as-received and hot rolled alloy showing uniform equiaxed α grain size, and (c) the ß transformed microstructure showing coarse acicular α platelets.

The Stress-Strain Rate-Strain Behaviour

The superplastic deformation properties of the alloy in the as-received condition were evaluated in the temperature range 813-920°C in the α+ß phase field, in order to study the effects of increasing the ß phase volume fractions on the flow behaviour. The results are summarised in Table I, which indicates that increases in the test temperature correspond to decreases in the flow stresses and increases in the stress sensitivity (m) values, and that the alloy is superplastic (m-values greater than 0.4) over a wide range of temperatures and strain rates. The results suggest that the optimum deformation temperature for superplasticity in terms of maximum elongation (420%, m-value > 0.4,) occurs at 860°C, corresponding to a microstructural condition at which the volume fraction of the phases is 70% α / 30% ß.

The stress-strain rate-strain behaviour of the alloy in the as-received condition (Fig. 1a) at this temperature is shown in Fig. 2. These data clearly show that the flow stresses increase with strain in the early stages of deformation and then become approximately strain-independent (compare, for example, the flow stresses for cycle 1 and with those for cycles 4 and 5). The m-values correspondingly decrease with strain (Fig 2).

The microstructures of the grip (undeformed and annealed at 860°C) and gauge length (deformed to a strain of 420%) sections are shown in Figs. 3a and 3b respectively. Whereas the microstructure in the grip section is substantially unchanged (compare Fig. 1a with Fig. 3a), that in the gauge length section shows a coarser and more uniform α grain structure with the ß phase solely distributed around the α grain boundaries. Clearly, the increase in the flow stresses with strain is a consequence of α grain growth.

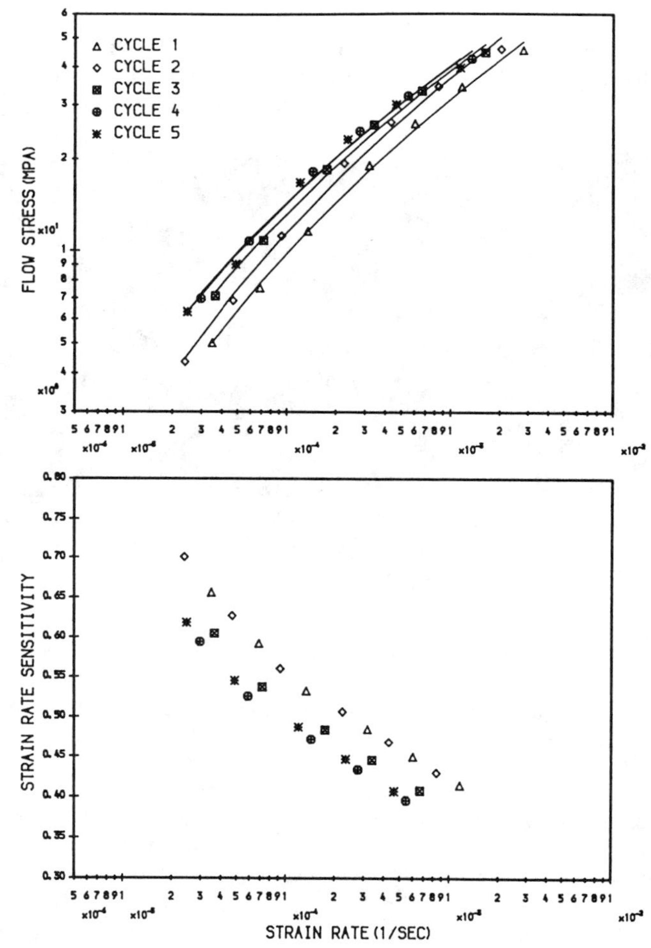

Fig. 2 Stress-strain rate and strain rate sensitivity data for the as-received alloy at 860°C showing increased in flow stress and decreases in strain rate sensitivity with increasing strain (strain range 420%).

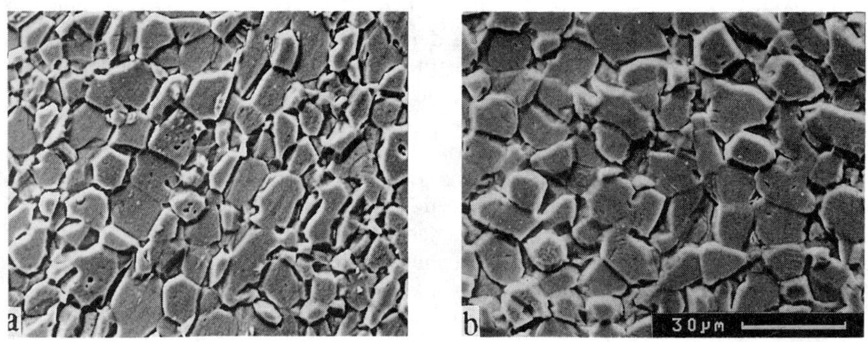

Fig. 3 Comparison of (a) the grip (undeformed) and (b) the gauge length (deformed) microstructures for the as-received alloy deformed at 860°C to a strain of 420%.

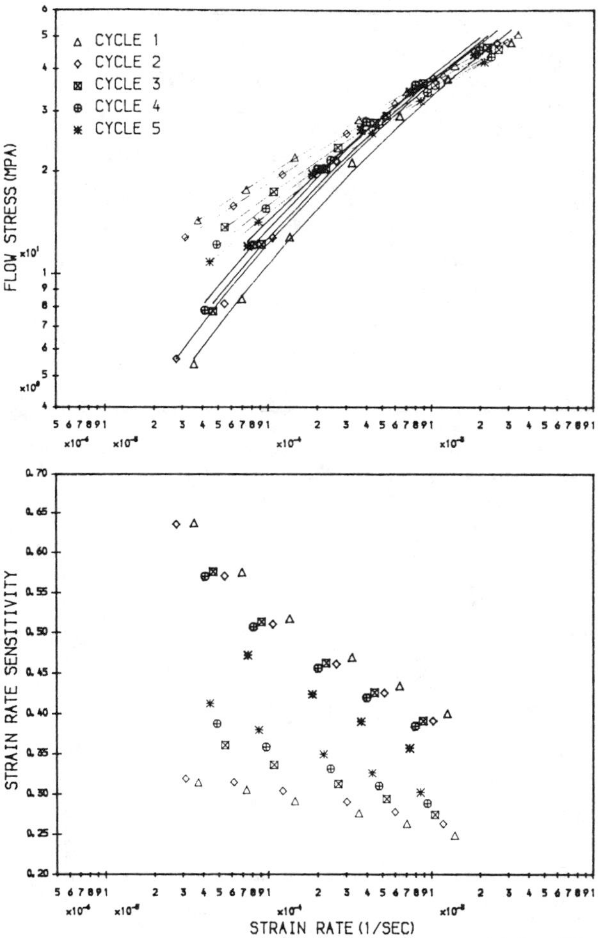

Fig. 4 Comparison of stress-strain rate and strain rate sensitivity data for the hot rolled (heavy print) and ß transformed alloy (light print) at 860°C showing the convergence of the flow stresses and m-values at increasing levels of strain.

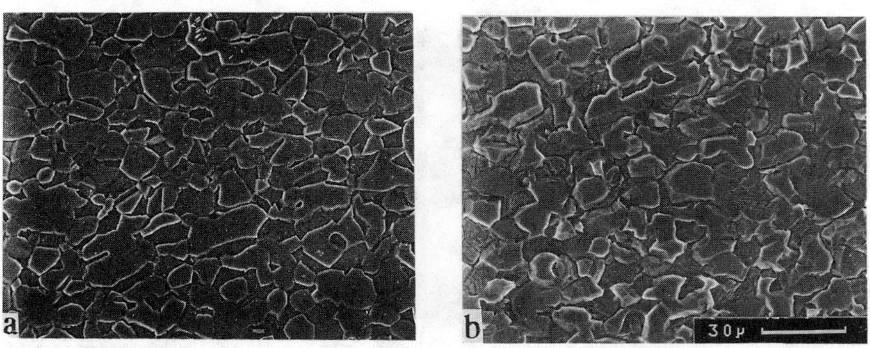

Fig. 6 Microstructures of the hot rolled alloy (a) undeformed grip region and (b) superplastically deformed at 860°C.

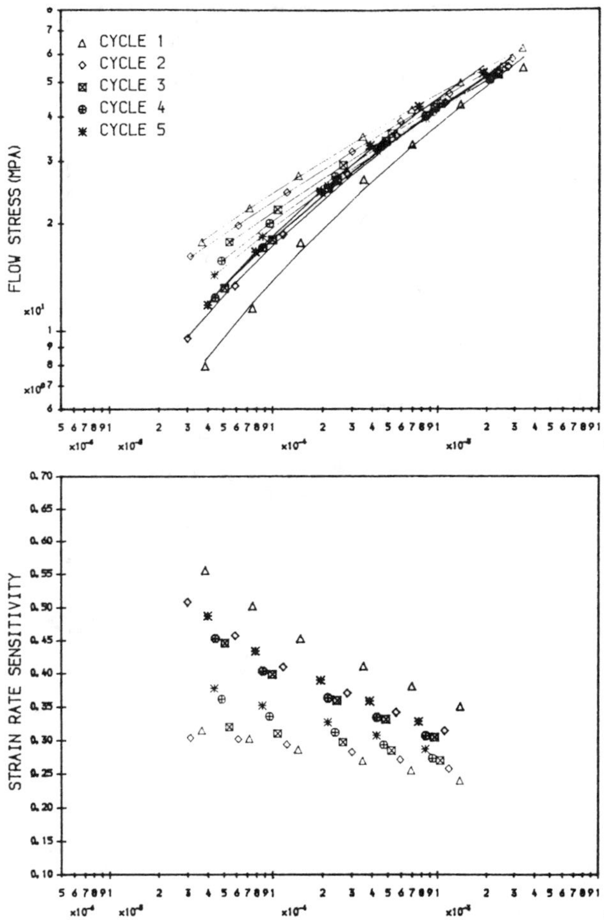

Fig. 5 Comparison of stress-strain rate and strain rate sensitivity data for the hot rolled alloy (heavy print) and ß transformed alloy (light print) at 830°C showing the convergence of the flow stresses and m-values at increasing levels of strain.

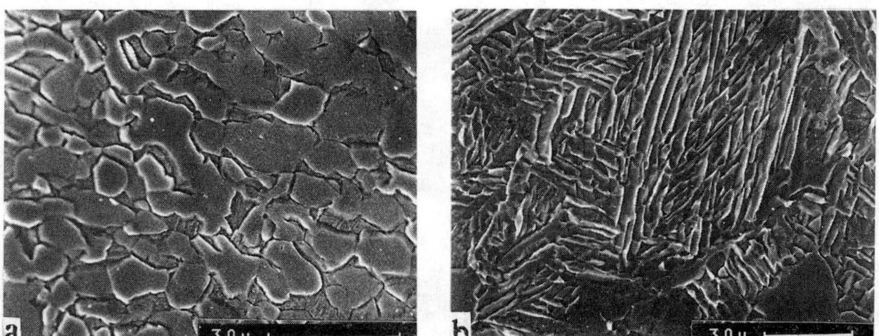

Fig. 7 Microstructures of the ß transformed alloy (a) superplastically deformed at 860°C and (b) the undeformed grip region.

Table I

The m-values and flow stresses at a strain rate of $3 \times 10^{-4}\ s^{-1}$, the elongations and volume fractions for the as-received alloy in the temperature range 813°C - 920°C.

Temp. °C	Strain-rate range $\epsilon\ s^{-1}$	m-value at $\epsilon = 3 \times 10^{-4}\ s^{-1}$	Flow stress σ MPa at $\epsilon = 3 \times 10^{-4}\ s^{-1}$	Elong. %	Vol. fr. of α/ß phases %
813	1.8×10^{-5} - 2.9×10^{-3}	0.41	25	407	74/26
860	2.1×10^{-5} - 2.8×10^{-3}	0.48	18	420	70/30
880	2.5×10^{-5} - 2.9×10^{-3}	0.50	17	304	62/38
900	2.6×10^{-5} - 2.9×10^{-3}	0.54	13	240	50/50
920	2.6×10^{-5} - 3.0×10^{-3}	0.66	10	255	40/60

The stress-strain rate-strain behaviour of the alloy in the initial hot-rolled condition (Fig. 1b) and ß-transformed condition (Fig. 1c) are compared in Figs. 4 and 5 which show the data at 860°C and 830°C respectively. These show that, whereas the flow stresses for the fine-grained equiaxed (hot-rolled) alloy *increase* with strain in the same manner as those for the as-received alloy (Fig. 2), those for the coarser acicular (ß transformed) structure *decrease* with strain; furthermore the flow stresses and m-values appear to approach common, strain-independent values at the higher levels of strain (compare the data for cycles 5 in Figs. 4 and 5).

The corresponding microstructures of the grip (undeformed) and gauge length (deformed) sections for the alloy deformed at 860°C are shown in Figs. 6 and 7. Again, it can be seen that whereas the microstructures in the grip sections are substantially unchanged (compare Fig. 6a with Fig. 1b and Fig. 7a with Fig. 1c), those in the gauge length sections show coarsening of the initial equiaxed microstructure (Fig. 6b) and the development of an equiaxed microstructure from the initial acicular microstructure (Fig. 7b). In short, the deformation process is such as to develop similar and strain-independent microstructures.

The transmission electron microscopy of the alloy is complicated by the occurrence of the ß → martensitic α transformation on rapid cooling from the deformation temperature. This makes a comparison of the initial and deformed microstructures difficult. However, the results clearly show the development of a "mantle" of the ß phase around the α grain boundaries (Fig. 8a) and the development of sub-grain networks within the α phase (Fig. 8b).

Discussion

The results show that the alloy Ti-2%Al-2% Mn exhibits superplastic deformation properties in the (α+ß) field and that these properties are microstructure dependent. Furthermore, the strain rate cycling deformation process appears to be such as to establish an equilibrium microstructure consisting of equiaxed α grains surrounded by a mantle of

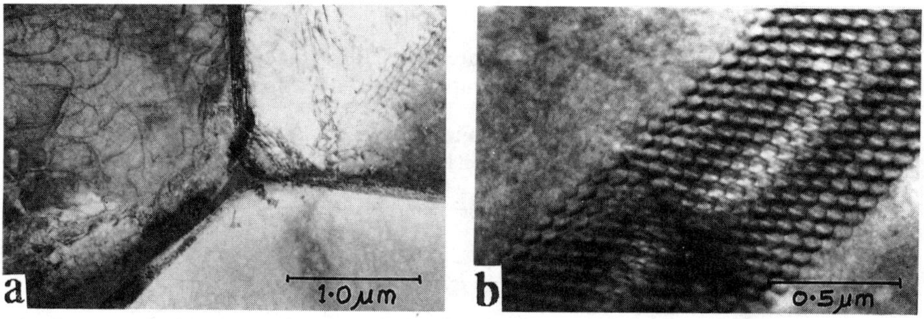

Fig. 8 - (a) Development of a "mantle" of β phase at a triple point between α grains and (b) development of dislocation sub-grain networks within the α phase (deformation conditions as for Fig. 7(b)).

the β phase, the equilibrium α grain size and α/β volume fraction being dependent upon the testing temperature. Grain coarsening occurs in initially fine-grained or non-uniform grain size microstructures (giving rise to increases in flow stress) and grain refinement and the development of equiaxed microstructures occurs in initially acicular microstructures (giving rise to decreases in flow stress).

The second case is of particular interest for two reasons; it indicates the occurrence of dynamic recrystallisation processes[2] which require that the deformation is carried out, in part, in regimes in which dislocation generation and climb processes dominate. The process of strain rate cycling in the present work is such that the strain rates extend from Region I (diffusional creep processes dominate) through Region II (optimum region for superplasticity) to Region III (dislocation generation and climb processes dominate). The boundaries between these regions are not fixed; as the initial acicular microstructure is broken up the transition from Region II to Region III occurs at increasing strain rates. The results are also of great practical significance since they show that superplastic properties can be established as a result of the deformation in initially non-superplastic microstructures such as those characteristic of welds. The work of Cope[3] on the alloy Ti-6Al-4V in which tensile specimens containing welds exhibiting transformed β microstructure similar to the β transformed structures in the present work, has shown that superplasticity may be developed in such a structure during constant strain rate tests in which the strain rate corresponds to Region III for the initial coarse microstructure but to Region II in the refined microstructure.

The mechanisms of superplastic deformation in Region II in this alloy are not established. However, the results clearly show the redistribution with strain of the β phase, forming a mantle around the α grains and thereby facilitating the processes of grain rotation and sliding[4,5,6].

References

1. J. Ma, R. Kent and C. Hammond, J. Mat. Sci. 21 (1986) 475.
2. T. Sheppard et al., Met. Sci. 17 (1983) 481.
3. M.T. Cope, Ph.D. Thesis, University of Manchester/UMIST, UK (1984).
4. B. Hidalgo-Prada and A.K. Mukherjee, Scr. Metall. 19 (1985) 1235.
5. G. Gurewitz, Diss-Abstr-Int., 44, (April 1984) p. 3168-B.
6. R.C. Gifkins, J. Mater. Sci. 13 (1978) 1926.

Hot Bar Rolling of Ti-6Al-4V in a Continuous Mill

Mikio Moriga* , Yoichi Takahashi* , and Mikio Shigenaga**

* Mechanical Engineering Research Laboratory,

KOBE STEEL,LTD., 1-5-5, Takatsukadai Nishi-Ku,

Kobe, Hyogo, 651-22 JAPAN

** Kobe Works, KOBE STEEL,LTD., 2, Nadahamahigashicho

Nada-ku, Kobe, Hyogo, 657 JAPAN

Abstract

To obtain high quality $\alpha+\beta$ processed Ti-6Al-4V bar and rod, a heat-build-up phenomenon during hot rolling is to be suppressed as much as possible. In typical commercial mills for bar and rod, rolling temperature changes are calculated to predict this phenomenon and gain optimum rolling conditions in titanium alloy rolling, by using empirical mathematical models (developed by the authors,) for carbon steel bar and rod to estimate lateral spread, forward slip, rolling force, and rolling torque. These models are found to be applicable to rolling of titanium alloy Ti-6Al-4V. The flow stress equation adopted in Ti-6Al-4V rolling temperature simulation was obtained by compression tests and rolling tests.

Introduction

In hot rolling materials with high flow stress and low heat conductivity such as titanium alloys, the temperature at the core of the rolled materials rises as the number of passes increases, depending on rolling conditions (e.g. delivery temp., rolling speed, and so on). This is called heat-build-up phenomenon. Once this phenomenon occurs, not only are grains coarsened but also phase trasformation occurs, degrading toughness and other impotant mechanical properties.

Before quantitatively predicting the heat-buid-up, flow stress of titanium alloy (Ti-6Al-4V) was investigated by a compression test and model mill test. Taking account of these test data, an equation of the flow stress was obtained. Lateral spread is a typical example of deformation characetristics in hot rolling, and thus the spread of titaninum alloy was compared with that of comercially pure titanium (C.P.T.) and that of carbon steel. Then, numerical simulations of rolling temperature changes were carried out , using empirical mathematical models[1],which have been developed by the authors for carbon steel bar and rod to estimate lateral spread, rolling load, etc. As a result, the optimum rolling conditions of typical mills(for example Bloom Mill, Continuous Mill, Block Mill, and so on) are clarified and sound $\alpha+\beta$ titanium alloy bar and rod were obtained.

1. Flow Sress of Titanium Alloys

Little research has been done on flow stress of titanium alloys, and if any, data published are primarily concerned with flow stress at low strain rates. Data that detemine rolling conditions are not readily available[2-4]. Therefore, in order to simulate the strain and strain rate conditions during rolling in actual operation, compression tests were carried out. Moreover, rolling tests were performed to calucuate rollling flow stress.

1.1 Compression Test

Compression test at working temperatures of 850,900, and 950 ℃ and at constant strain rates of 2 and 10 per second were carried out to simulate rolling in actual operation. The size of the

Titanium '92
Science and Technology
Edited by F.H. Froes and I. Caplan
The Minerals, Metals & Materials Society, 1993

specimen used was 8 mm in diameter and 12 mm in height. A carbon sheet was used when the specimen was brought in contact with tools to avoid the effects of friction.

Fig. 1 shows examples of the stress-strain relationship for tianium alloy,Ti-6Al-4V. The figure indicates that as strain increaces, the specimen appears to work-soften. The apparent work-softening is more marked as the strain rate increases. It is probale that specimen temperature rises are due to adiabatic change during the test and as deformation is increased.

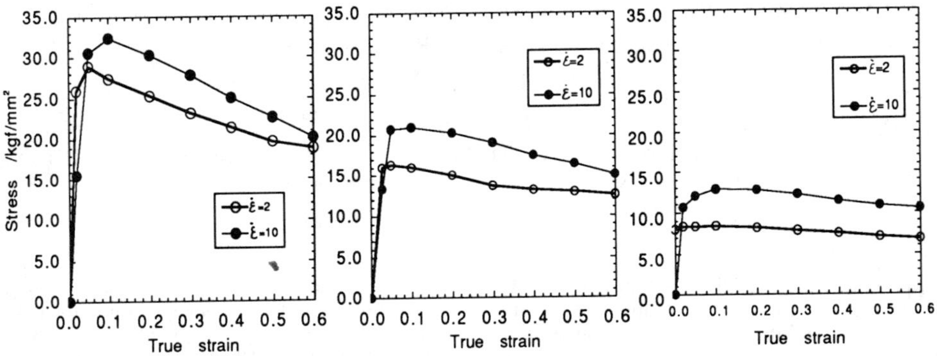

a) 850 ℃ b) 900 ℃ c) 950 ℃

Fig. 1 Relationship between true stress and true strain by a compression test (strain rate;2 and 10 /sec.,Temp.;850, 900, and 950 ℃).

<u>1.2 Rolling Test</u>

A plate 6 mm thick and 36 mm wide was rolled in a two high mill (220 mm diamter) at reduction ratio of 0.2 and 0.35, at strain rates of 2, 10, and 50 /sec., and at various rolling temperatures (760-900 C) to determine the rolling flow stress experimentally. Based on the equation of R.B. Sims[5] and arranged according to rolling temperature, the results are shown in Fig. 2. The figure indicates that increased rolling temperature has the greatest effect, and flow stress does not depend significantly on the reduction ratio. It is observed that when the strain rate is 50 /sec. and the greater the reduction ratio, the smaller are apparent flow stress and back-calculated flow stress. This is explained by the fact that the rolled specimen is to subjected to adabatic effects and is work-softened since it is deformed at high speed.

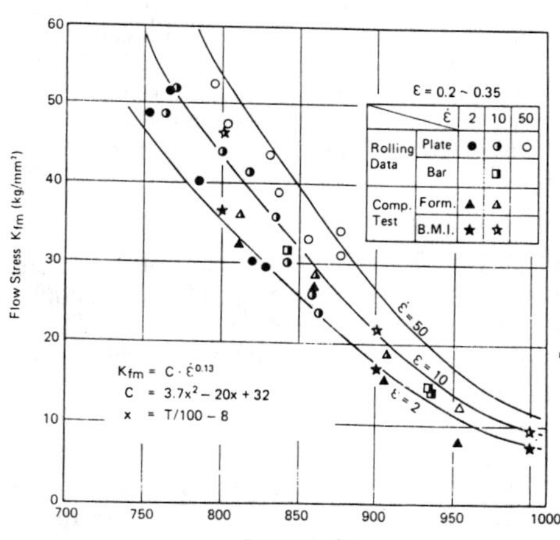

Fig. 2 Flow stress of Ti-6Al-4V at various rolling temperatures.

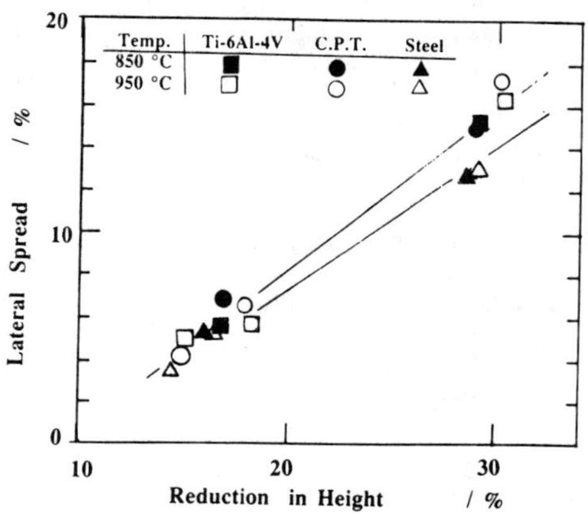

Fig.3 Relation between spread and reduction in Ti-6Al-4V, C.P.T., and carbon steel (C%=0.18) rolling (specimen;6.2mm in hight , 8.6mm in width, and roll;220mm in dia.)

In this way, to predict rolling of titanium alloys with an empirical experimetal equation, it becomes important to be able to estimate this apparent work-softening. In this paper, the empirical flow stress equation is obtained below by empirically adopting a strain rate of 10 /sec. and equation (1).

$$Kfm = C\,\dot{\varepsilon}^{0.13} \qquad (1)$$

$$C=3.7x^2-20x+32,\ x=T/100-8$$

where T is rolling temperature (°C) , and $\dot{\varepsilon}$ is strain rate.

2. Deformation of Titanium Alloy(Ti-6Al-4V) during Hot Rolling:

Titanium alloys have less slip planes and are strongly anistropic caused by the hcp structure. Therefore, flat hot rolling of titanium alloy, commercially pure tianium(C.P.T.), and carbon steel (carbon contetes 0.18%) was studied at 850 and 950 ℃ on a test mill (the size of specimen:6.2 mm in height and 8.6 mm in width, roll diamter 220mm). In Fig. 3 the lateral spreads are arranged by the reduction ratio. At a reduction ratio of around 15%, no difference is observed due to differences in materials. A difference is obseveved at a reduction ratio of around 30%, but about 10% at most, suggesting the successful use of the pass schedule of cabon steel. That is titanium alloy can be rolled with the passs schedule designed for carbon steel with roll gap varied[6].

3. Prediction of Temperature Changes during Rolling Titanium Alloy

Deformation and load during rolling is predicted by performing the above referenced empirical model[1]. Work-heat generation is evaluated by the flow stress of Eq.(1). Radiation of heat and heat conductivity of rolled material may be taken into account and calculated by using Finite Difference Method[7]. Below, examples of calculated results are shown in various types of rolling mills.

3.1 Reverse Rolling

In mills used for the blooming process and in obsolete rolling mills, reverse rolling is carried out, in which several grooved passes are porvided in one roll. Fig. 4 shows the layout of a typical reverse rolling mill comprised of one roughing and one finishing roll each. The roughing roll is 500 mmm in diameter and 1500mm long at the roll barrel, and the finshing roll is 300 mm in diametr and 800 mm long at the roll barrel. The roll peripheral speeds are 2.1 m/sec. and 3.0m/sec. respectively. the billet size of titanium alloy is 85 mm square and is rolled into a round bar of 20 mm in diameter.

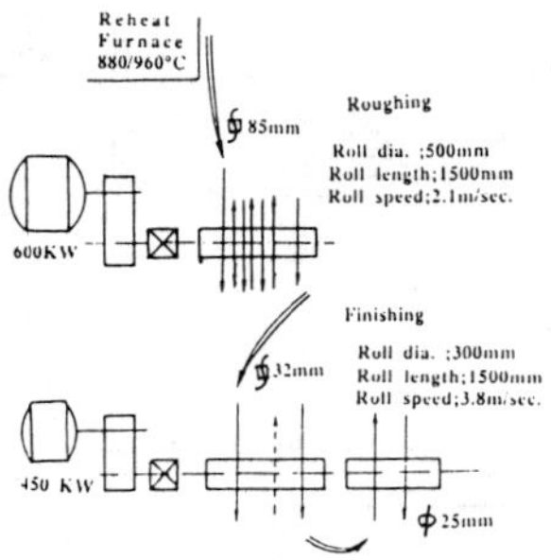

Fig. 4 A typical layout of reverse mill for Ti alloys.

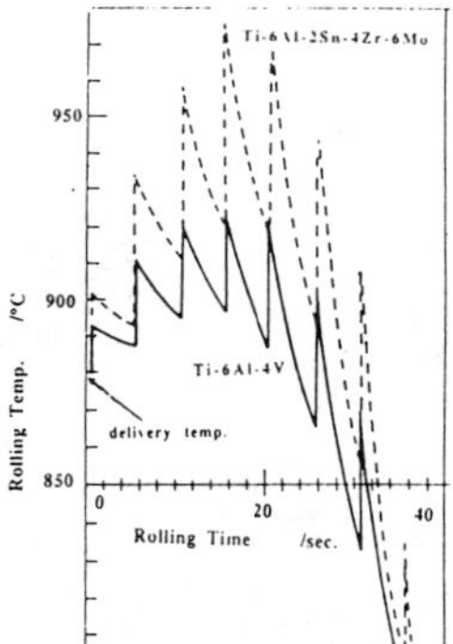

Fig. 5.1 Calculated core temp. of titanium alloys in the roughing roll.

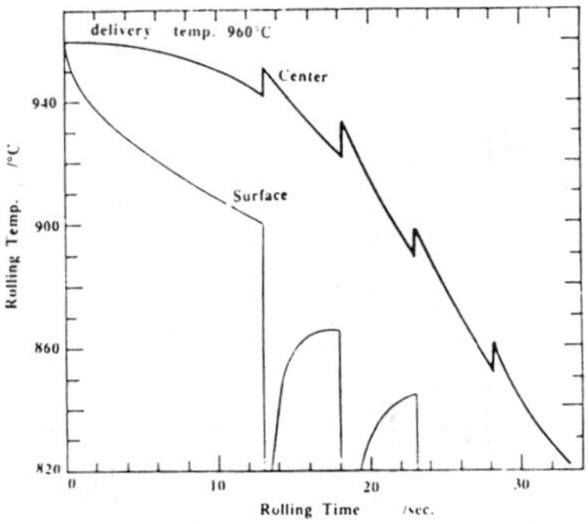

Fig. 5.2 Calculated rolling temp. of Ti-6Al-2Sn-4Zr-6Mo in the finishing roll.

Fig. 5.1-5.2 show calculated temperture changes at a typical pass schedule, a total of 12 passes consisting of 8 passes from 85 mm square to 32 mm square in a roughing roll and 4 passes from 32 mm square to 25 mm round in a finishing roll. Fig. 5.1 shows the temperature changes at the core of Ti-6Al-4V and Ti-6Al-2Sn-4Zr-6Mo at the delivery temp. of 880 ℃ in the roughing roll. Fig. 5.2 shows the changes of the core temp. and the surface temp. of Ti-6Al-2Sn-4Zr-6Mo at the delivery temp. of 960 ℃ in the finshing roll.

In the roughing roll(Fig. 5.1), the core temperature of the rolled titanium tends to decrease after reaching a peak from pass 4 to pass 5. But in the finishing roll(Fig. 5.2), heat -build-up caused by working heat is instantaneously recorded but as a whole it tends to decrease rapidly. It is conceivable that in reverse rolling, rolling take place at nearly constant intervals, and therefore, relatively spanning, the surface area of rolled material is increased by rolling at the later rolling stages. When titanium alloy is reverse-rolled, special attention must be given to heat-build-up at the first stage of rolling in which the cross section is large, and rolling must be carried out with the effects of phase transformation on material properties taken into account.

Fig. 6 shows the predicted results of temperature changes of rolled material when 350mm square Ti-6Al-4V is rolled into 170mm round in a high productivity iron blooming mill, and its derivery temperature is varied from 900, 920, 940, and 960 °C. The total number of passes is 21. The figure indicates that surface temperature rapidly decreases between #1 and #2 bloomer, where the rolled titanium billet is being transported, but it slowly rises when rolling at #2 bloomer begins. From this result, titanium stock was bloomed with reheat furnace delivery temperature controlled to 920-940 ℃ in order to obtain a large round bar of 170 mm in diameter with sound structure. Fig. 7 shows microphotographs of macro- and microstructures when titanium is air-cooled after rolling at the delivery temp. of 930±10 °C. The figures shows satisfactory structures free from structural change due to heat-build-up.

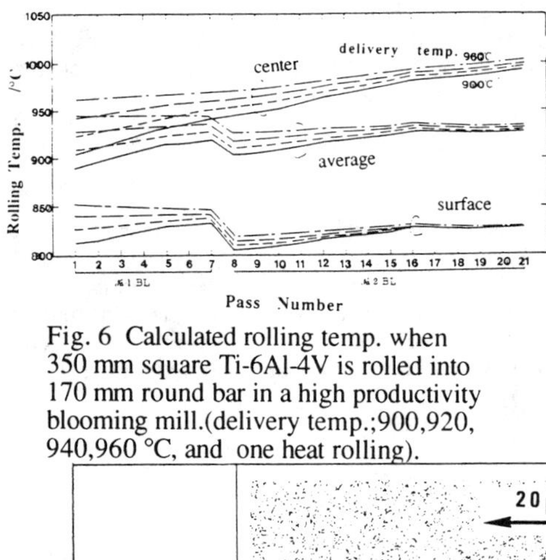

Fig. 6 Calculated rolling temp. when 350 mm square Ti-6Al-4V is rolled into 170 mm round bar in a high productivity blooming mill.(delivery temp.;900,920, 940,960 °C, and one heat rolling).

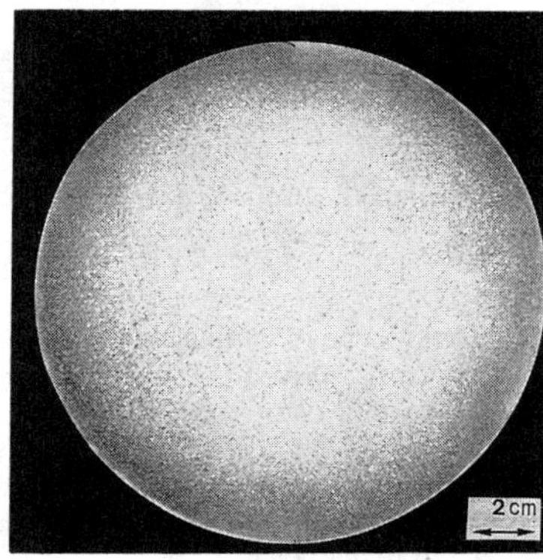

a) macrostructures

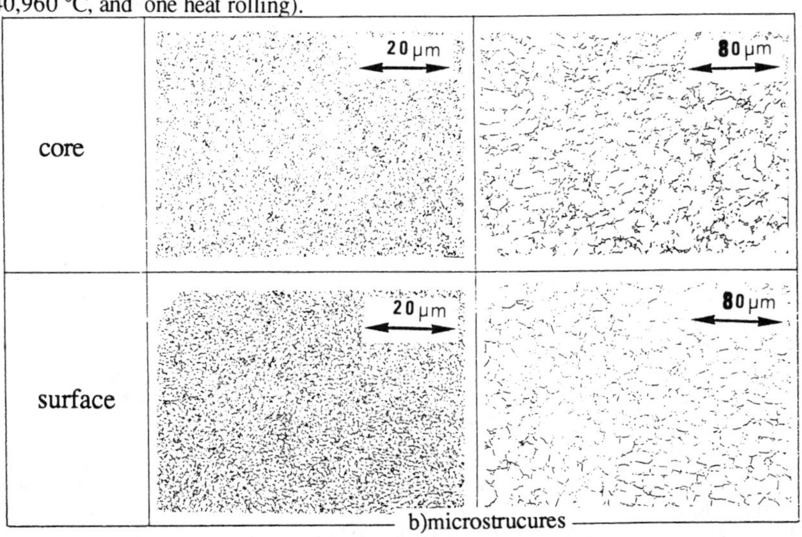

b)microstrucures

Fig.7 Macro- and microstructures at the delivery temp. of 930±10 ℃ when 350mm square Ti-6Al-4V is rolled into 170 mm round bar in a high producty blooming mill (two heat Rolling).

3.2 Repeater Rolling

In mills specialized for rolling special alloys, repeater rolling is generally adopted. Fig. 8 shows an example of a typical plant. Repeaters are frequently located between trains, such as roughing, intermediate, and finishing trains, and have become popular for moitoring variation in tension between stands during rolling. For example, rolled material left longer than neccesary at the repeater indicates faulty speed control in the trains. It also indicates that compressive force is being genarated between trains and the material is about to be misrolled. Because the distance of this repeater may be as long as 20 meters, the temperature of the rolled material is drastically lowerd at the zone.

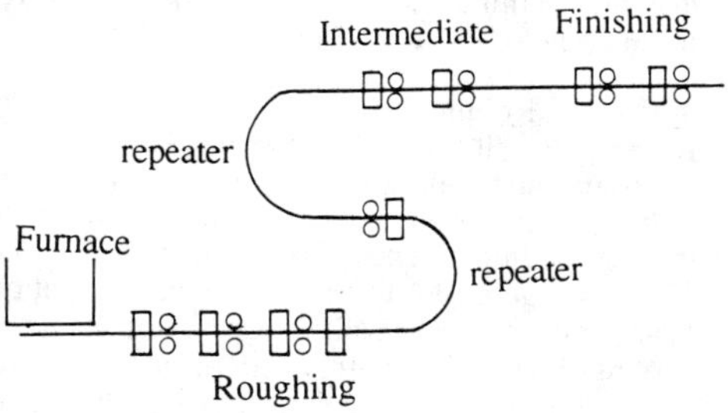

Fig.8 A typical repeater mill layout .

This is desirable for rolling materials such as titanium alloy, which does not permit coarsening of grains or phase transformation caused by heat-build-up. Fig. 9 shows calculated results and measured values of change in rolling temperature and rolling consumption energy when Ti-6Al-4V is rolled from 115mm square to 26 mm round bar by 12 passes in the plant shown in Fig. 8. In Fig. 9, which shows the temperature history, the loaction of repeater is shown cross-hatched. It indicates that the presence of the repeater may restrict heat-build-up.

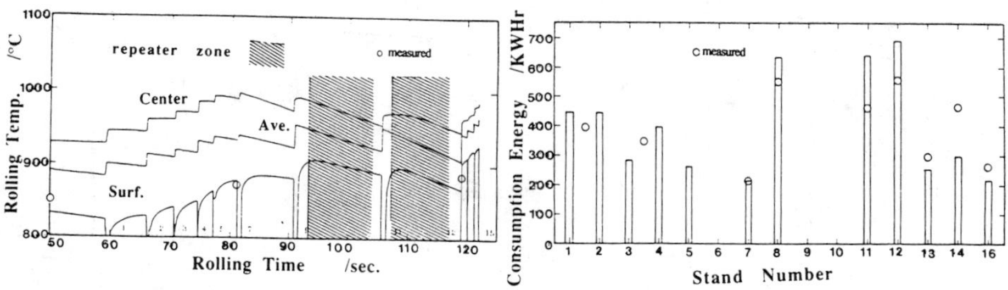

Fig. 9 Rolling temp. and rolling consumption energy in a repeater mill.
(ϕ115mm - ϕ26mm, delivery temp.;930 °C, finishing speed;7.5m/sec.).

3.3 Continuous Rolling

In a plant which produces carbon steel in a large quantities, no repeater exists, the pass line is straight and stands are located at short intervals, giving the first priority to productivity. Fig.10 shows calculated result of temp. changes when Ti-6Al-4V is rolled from 155 mm square to 70 mm round bar at the same rolling speed as that of carbon steel and 50% lower than that speed, and at delivery temperature of 930 °C.

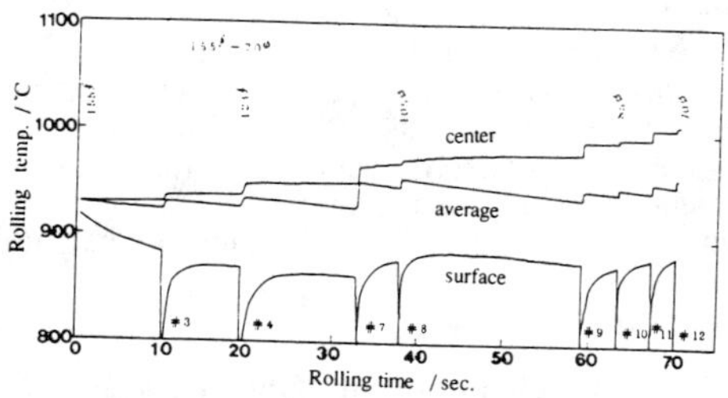

a) Normal speed rolling.

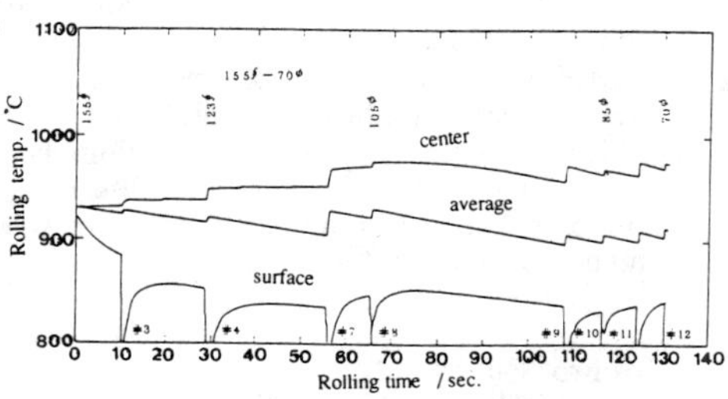

b) Half of normal speed rolling

Fig. 10 Effect of roll speed on the rolling temp. in a continuous mill
(ϕ155mm - ϕ70mm, delivery temp.;930 °C) .

The figure clearly shows that lowering the rolling speed can restrict heat-build-up. Data suggest that with continuous mills there is no alternative but to roll at lower boundary-limit speed of the rolling mill to obtain a rolled, titanium alloy with a sound structure.

3.4 Block Rolling

Recently,to roll 5 mm diameter fine steel wire, high-productivity rolling (block rolling) is carried out by tension-rolling using a common drive. Block rolling positively provides tension of about 20% of rolling flow stress instead of conventional zero tension betweens stands, and produces the least additional shear deformation. It is shown that with little shearing work, block rolling of Ti-6Al-4V generates a relatively small amount of heat casused by rolling work, resulting in small heat-build-up during rolling. On the other hand, the process produces a problem of insufficient cooling time due to narrow stand intervals. Block rolling of titanium alloy was carried out to reduce 10mm round into 6.4 mm round wire using 4 passes. Calculations were also carried out at the same time to determine the changes in rolling teperature. Fig. 11 shows calculated results of finishing speed at 10 m/sec. and 15 m/sec. at the delivery temperature 820 °C. Fig. 12 shows microstructure of rolled cross sections. From the figure, the rolled titanium passess the β structure at 15 m/sec. And that it is possible roll the α+β region at rolling speeds up to 12.5 m/sec..

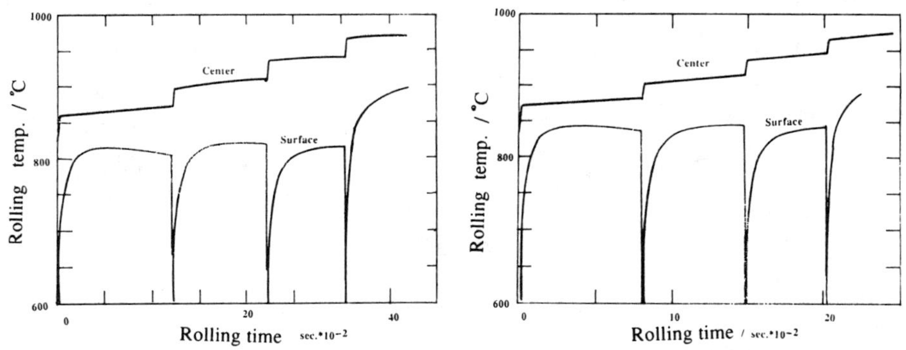

a) Finshing rolling speed;10 m/sec. b) Finshing rolling speed;15m/sec.

Fig. 11 Effect of rolling speed on calculated rolling temp. in a block mill at the delivery temp. 820 ℃ on Ti-6Al-4V(10 mm - 6.4 mm under 4 passes).

Summary

Rolling conditonsof Ti-6Al-4V have been determined by experiments and calculations to obtain a structure with equiaxed α+β grains with good mechanical porperties. As a result, the following has been clarified.:

• In a reverse rolling mill, such as a blooming mill, heat-build-up tends to occur at the stage with a large mass of rolled material. Therefore, rolling from large size ingots should be avoided as much as possible or ingots should be returned to the furnace in the course of hot rolling.

• In repeater rolling, since heat-build-up is restricted at the repeater, placing attention on the delivery temperature of rolled titanium alloy enables rolling of products with sound structure.

•In high-productivity continuous rolling, heat-build-up inevitably occurs. The rolling speed should be lowered to the lower limit of the equipment for rolling titanium alloys.

• In block rolling which imparts tension between stands, it is possible to roll titanium wire with sound equiaxed,α+β grain.

Reference

1.Yoshiro Saito et al., "A New Method for Calculating Deformation and Force Parameters in Steel Rod Rolling and Its Apllication to Roll Pass Design,"Sosei to Kakou," 24(273)(1983),1070-1077.

2.Sadahiro Yamamoto and Chiaki Ouchi,"Deformation Resistance in Plate Rolling of Commercially Pure Titanium and Ti-6Al-4V Alloy,"Tetsu to Hagane,72(6)(1986),53-60.

3.Hirowo G. Suzuki et al.,"Hot Workability of Ti-6Al-4V Alloys,"Tetsu to Hagane, " 72(6)(1986),53-60.
4.Battele Memorial Institute,Columbus Lab.,Topical Report,Review of Flow Stress Data,No16(1982/11)
5.R.B.Sims,"The Calculation of Roll Force and Torque in Hot-Rolling Mills,"Proc. Inst. Mech. Eng.,166(1954),75-81.
6.Yoshiro Yamaguchi et al.,"Mill Set-up Calculation and Roll Gap Adjustment in Bar rolling,"Kobe Steel Engineering Report,35(2),(1985),36-40.
7.Yoshiro Yamaguchi et al.,"Temperature Model in Bar Rolling,"Kobe Steel Engineering Report,35(2),(1985),32-35.

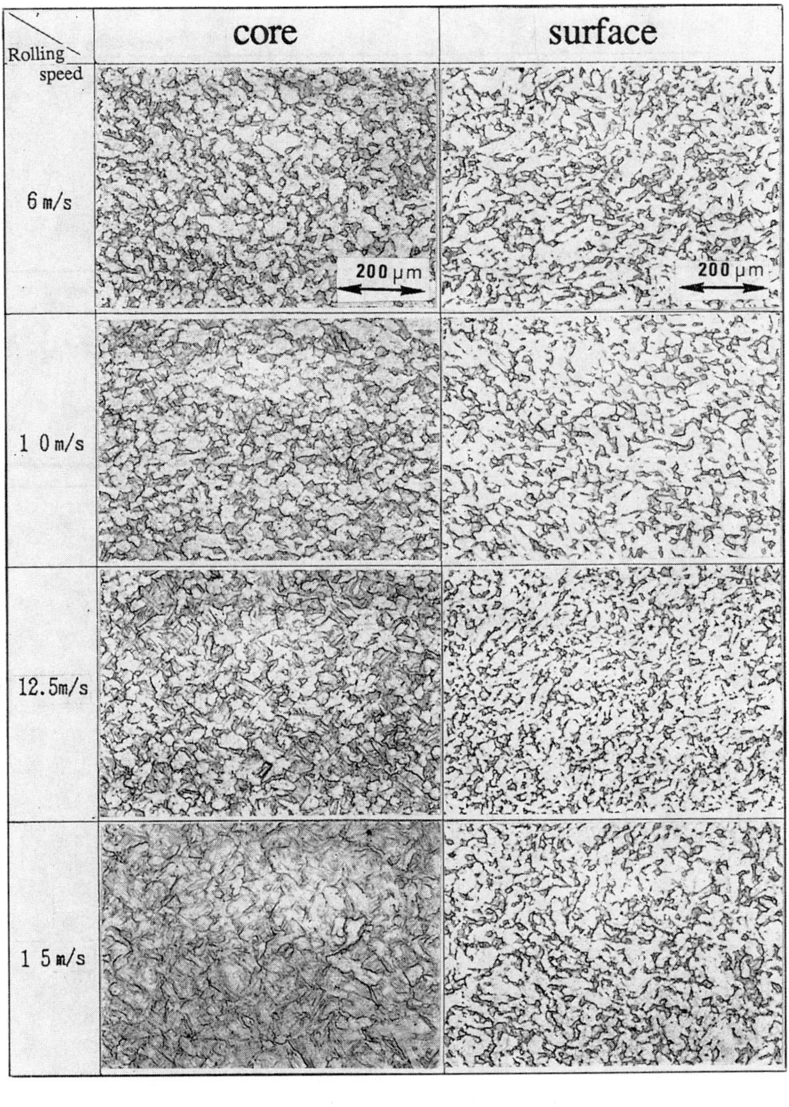

Fig. 12 Microstructures of Ti-6Al-4V rolling in a block mill (pass schedule: ϕ10mm - ϕ6.4mm under 4 passes,finishing speed; 8,10,12.5, and 15 m/sec.).

STRENGTHENING AND TOUGHENING OF ALPHA-BETA TITANIUM ALLOYS BY THERMOMECHANICAL PROCESSING

Minoru Okada

Sumitomo Metal Industries, Ltd.
1-8, Fuso-cho, Amagasaki City, 660, Japan

Abstract

A new thermomechanical process (TMP) has been developed for alpha-beta titanium alloys, which consists of beta working with high reduction level, rapid cooling after beta deformation and annealing at the temperature where the volume fraction of alpha and beta become almost equal. The material processed by this thermomechanical process showed high strength and high fracture toughness. This new thermomechanical process was applied for the extruded shape of Ti-6Al-6V-2Sn alloy. Developed extruded shapes show higher fatigue strength with the optimum TMP.

1. Introduction

It is well known that alpha-beta titanium alloys have acicular microstructure in the condition processed or heat treated in beta region, and they have equiaxed microstructure in the condition processed and heat treated in alpha-beta region[1]. In general, equiaxed microstructure processed and heat treated in alpha-beta region has been utilized as the microstructure of aerospace materials, because the material with equi-axed microstructure shows higher ductility and higher fatigue strength compared with those of acicular microstructure[2][3]. The workability of alpha-beta titanium alloy is superior in beta region compared with alpha-beta region, so more cost savings will be developed if beta processing is utilized. It has been reported that materials with acicular microstructure show higher fracture toughness and creep strength than those with equiaxed microstructure[4][5], so it is also expected that the materials with good combination of strength, ductility and fracture toughness will be developed by the application of acicular microstructure by processing in beta region.

In this paper, we will discuss new process with alpha-beta alloys with high levels of reduction in beta region, and the possibility to produce the material with a good combination of strength, ductility and fracture toughness by this new beta processing thermomechanical process. This new process consists of beta deformation followed by annealing at a high temperature in alpha-beta region. The material thermomechanically treated with this process has a fine bi-modal microstructure, which displays a good combination of strength, ductility and fracture toughness.

We also discuss the properties of the extruded shapes of Ti-6Al-6V-2Sn alloy, which has been manufactured by this new, beta processed thermomechanical process.

Titanium '92
Science and Technology
Edited by F.H. Froes and I. Caplan
The Minerals, Metals & Materials Society, 1993

2. Materials and Method

Table 1 shows the chemical compositions of the materials used. Material A is Ti-6Al-4V and was used to check the effect of cooling rate. Material B is Ti-6Al-4V EL1 and was used to check the effect of reduction level in plate rolling. Material C is Ti-6Al-6V-2Sn, which was used for extruded shapes.

Table 1 Chemical Compositions. (wt.%)

	Al	V	Sn	Fe	Cu	O	H	N	C	Ti
A	6.5	4.3	—	0.17	—	0.18	0.002	0.01	0.01	bal.
B	5.9	4.1	—	0.19	—	0.10	0.005	<0.01	0.01	bal.
C	5.6	5.3	2.0	0.53	0.51	0.17	0.002	0.01	0.01	bal.

Figure 1 shows the proposed thermomechanical process, which consists of beta rolling, rapid cooling after rolling and alpha-beta annealing used for Ti-6Al-4V. Figure 2 shows the extrusion processes tested with Ti-6Al-6V-2Sn.

Microstructures and mechanical properties were examined on the materials thermomechanically processed. Fracture toughness tests were performed using the specimens specified in ASTM E399 (t=12.5mm), and fatigue tests were performed using rotating beam fatigue tests.

Three types of shapes were extruded by thermomechanical processes shown in Fig. 2 and were compared for microstructures, mechanical properties, fatigue properties and fracture toughnesses.

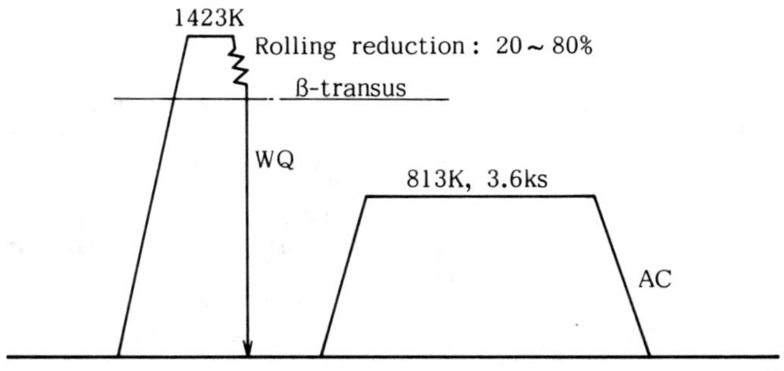

Fig. 1 Thermomechanical Process for Alpha-beta Alloy Ti-6Al-4V

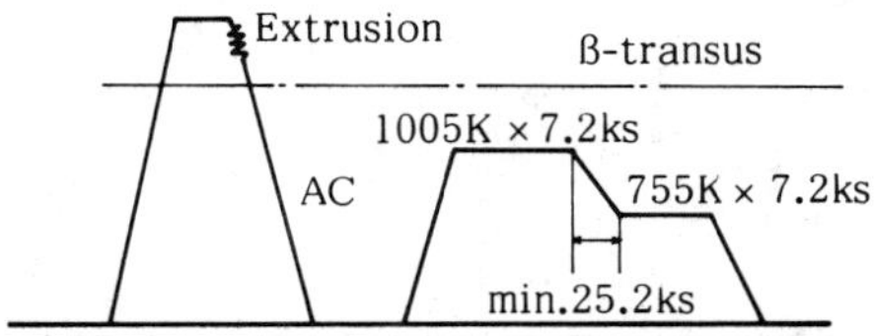

(a) ß Extruded, Air Cooled and Two-step Annealed

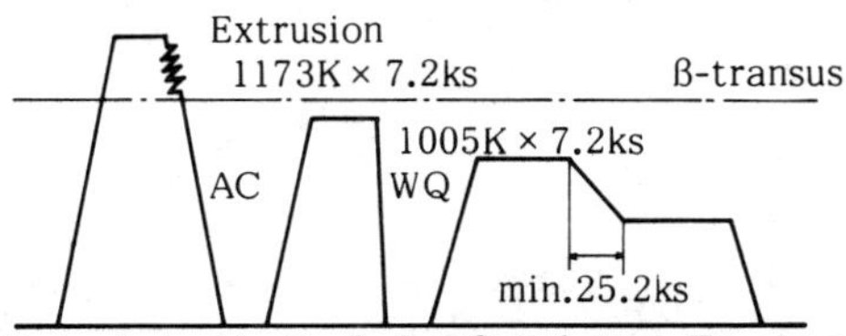

(b) ß Extruded, Air Cooled σ+ß Solution Treatment, Quenched and Two-step Annealed

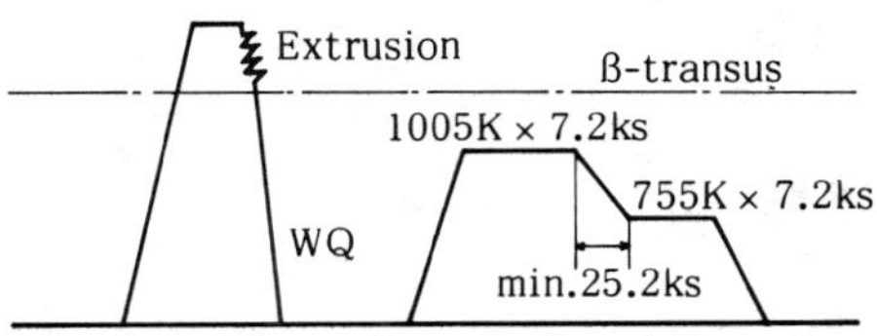

(c) ß Extruded, Direct Quenched and Two-step Annealed

Fig. 2 Thermomechanical Processes for Ti-6Al-6V-2Sn Extruded Shapes.

3. Results and Discussions

3.1 The effect of thermomechanical processing conditions on the microstructure and mechanical properties

The reason for brittleness in alpha-beta alloys that were beta processed are deemed to be the grain growth and the large alpha precipitation on grain boundaries of beta grains. We have intend to develop new thermomechanical process consisting of large reductions in beta deformation and rapid cooling after beta processing, to prevent the grain growth of beta grains and prevent the precipitation of excessive grain boundary alpha. The most important factors in this process were deemed to be cooling rates after beta reduction and the level of reductions in beta working. Figure 3 shows the change of microstructures with cooling rates. In the cooling rate of 0.05K/s or 1K/s, it was observed that large alpha plates precipitated on grain boundaries of beta grains and large acicular alpha plates precipitated in beta grains. On the other hand, it was observed in the rapid cooling rate 10K/s that no grain boundary alpha plates precipitated and acicular alpha of small width and length precipitated in beta grains. (Fig.3(a)-(c)) After anneal at 1193K, which is 40K below beta-transus, acicular alpha plates remained in the cooling rates of 0.05K/s and 1K/s. On the other hand, a half of acicular alpha plates changed their morphology to an equi-axed structure, and the rest became fine acicular alpha with the fast cooling rate of 10K/s. It became bi-modal microstructure. (Fig. 3(d)-(f)) It was clear that a bi-modal microstructure could be easily produced by a high temperature alpha-beta anneal for the samples cooled with a rapid cooling rate. Figure 4 shows the mechanical properties after anneal with the change of cooling rate after beta working. The reduction in beta processing was 70% constant in this test. The strength is constant in spite of various cooling rates, but the ductility increased with the increase of cooling

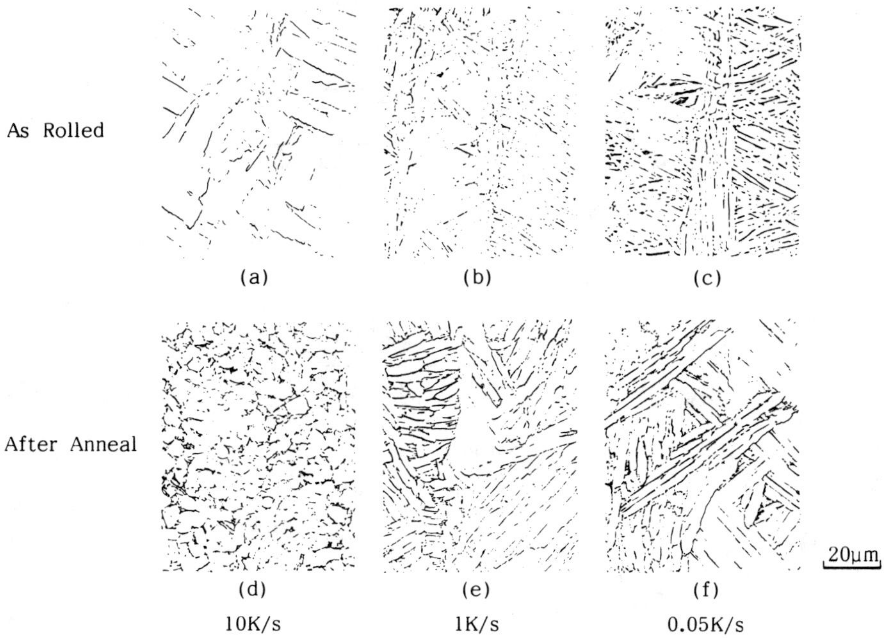

Fig. 3 The Effect of the Cooling Rate after ß-Rolling on the Microstructures. (Ti-6Al-4V, ß-Rolling Ratio : 70%, Annealing : 1193K, 3.6ks, AC) (a) (b) (c) : as rolled, (d) (e) (f) : after annealed.

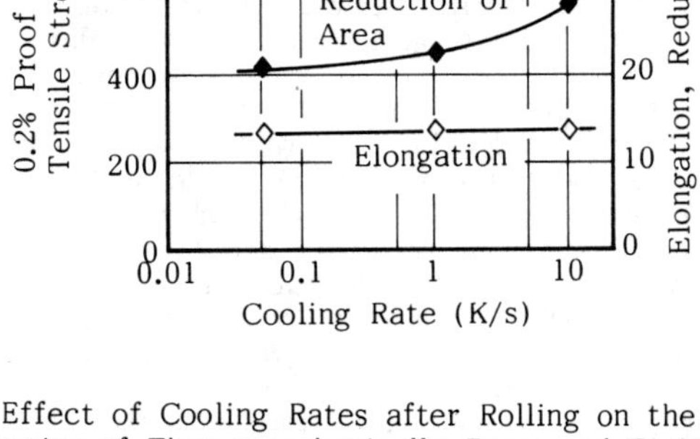

Fig. 4 The Effect of Cooling Rates after Rolling on the Mechanical Properties of Thermomechanically Processed Ti-6Al-4V. (Rolling Reduction : 70%, Anneal : 1193K, 3.6ks, AC)

rate. At the cooling rate over 5K/sec, the reduction of area shows over 25% with high beta reductions (70%), which approaches those of equi-axed microstructures.

Figure 5 shows the mechanical properties with the change of reductions in beta processing. The cooling rate was constant, water cooling in this test. The ductility increased with the increase of reduction levels in beta processing. Over 70% beta reductions, high ductility was attained as high as those of equi-axed microstructure with rapid cooling rates.

From the results shown in Fig. 4 and Fig. 5, it is concluded that this new TMP must have over 70% beta reduction and cooling rates over 5K/s to obtain ductility equal to equi-axed microstructure. Figure 6 shows the comparison of our data of the mechanical properties between thermomechanically processed plate samples and conventionally alpha-beta processed forged samples. Thermomechanically processed samples show much higher fracture toughnesses than the samples of conventional alpha-beta forging processes. The reason for higher fracture toughness is deemed to be the formation of fine bi-modal microstructure in the new thermomechanical process.

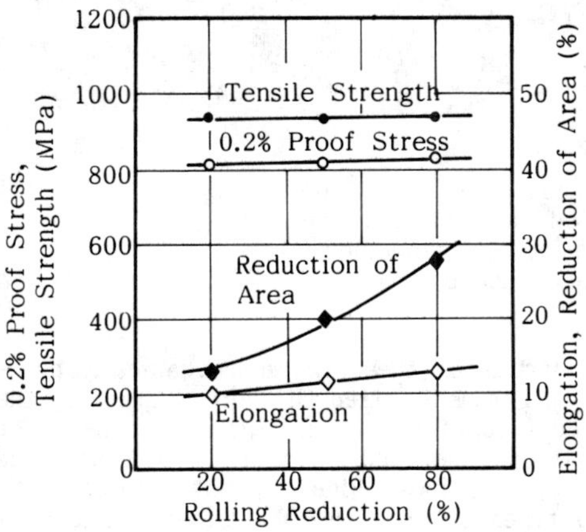

Fig. 5 The Effect of Rolling Reduction in Thermomechanical Process on the Mechanical Properties after Annealing.
(water cooling after ß rolling)
(Anneal : 1193K, 3.6ks, AC)
(Ti-6Al-4V)

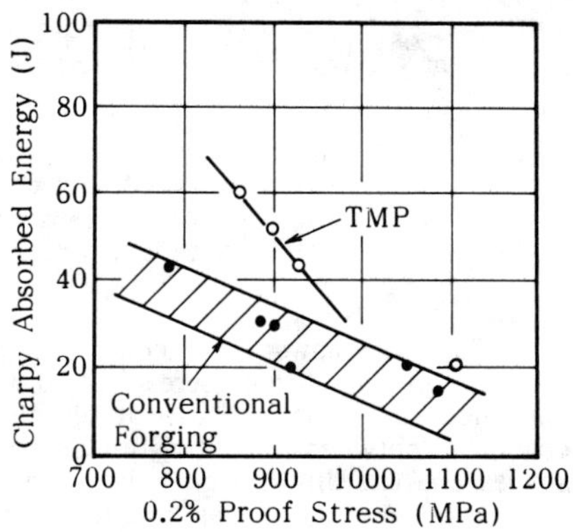

Fig. 6 Comparison of Charpy Absorbed Energy and 0.2% Proof Stress between Forgings and Thermomechanically Processed Ti-6Al-4V.

3.2 The development of the new thermomechanical process for extruded titanium shapes

The new thermomechanical process for alpha-beta titanium alloys has been investigated in production of extruded shapes of Ti-6Al-6V-2Sn with high strength, high ductility and high fracture toughness for airframe structures. Three thermomechanical processes shown in Fig. 2 were investigated for strengthening and toughening. The first process consisted of air cooling and annealing. The second process consisted of solution treatment, quench and annealing after beta extrusion and air cool. The third process was water quench after beta extrusion and annealing.

Figure 7 shows the microstructures of extruded Ti-6Al-6V-2Sn shapes thermomechanically processed. The microstructure of the air cooled shape consisted of grain boundary alpha and the acicular alpha-beta microstructure (Fig. 7-a). Growth of grain boundary alpha and acicular alpha was observed after solution treatment and annealing (Fig. 7-b). On the other hand, very fine and uniform microstructure was observed in the shape directly quenched and annealed, with no grain boundary alpha precipitates and no large acicular alpha plates (Fig. 7-c).

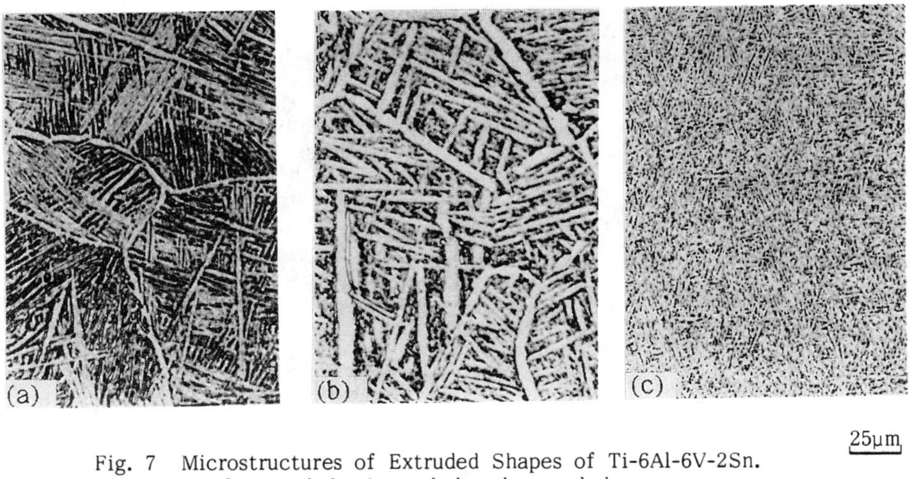

Fig. 7 Microstructures of Extruded Shapes of Ti-6Al-6V-2Sn.
(a) β extruded, air cooled and annealed
(b) β extruded, α+β quenched and annealed
(c) β extruded, direct quenched and annealed

Table 2 shows the mechanical properties of these thermomechanically processed shapes. Proof stress of 0.2% and tensile strength were higher in the quenched shapes than those of air cooled shape. Significant differences were not observed between the tensile properties of the solution treated and annealed shapes and the directly quenched and annealed shapes. Figure 8 shows the comparison of fatigue properties of these two types of processes for titanium alloy extruded shapes. The fatigue strength of the shape directly quenched and annealed was as high as 590 MPa, and the fatigue retio (fatigue strength/tensile strength) was high value of 0.54. On the other hand, the fatigue strength of solution treated and annealed shape was 510 MPa, and the fatigue ratio was 0.47. This high fatigue strength of the directly quenched shape was deemed to be due to the very uniform and fine microstructure (Fig. 7-c).

Table 2 Mechanical Properties of Ti-6Al-6V-2Sn Extruded Shapes after various Thermomechanical Processings.

	YS (MPa)	TS (MPa)	El (%)	RA (%)
β Extruded, Air Cooled and Annealed	919	1051	19.2	37.0
β Extruded, (α+β) Quenched and Annealed	979	1079	18.6	38.2
β Extruded, Direct Quenched and Annealed	986	1089	18.1	39.1

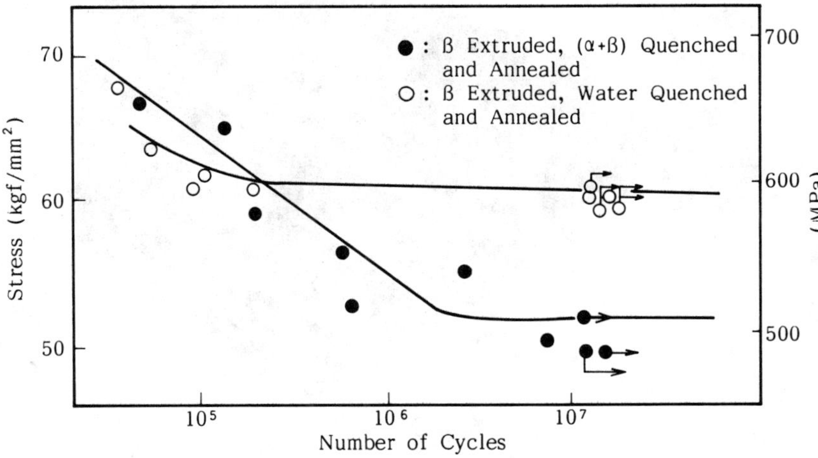

Fig. 8 S-N Curves of Ti-6Al-6V-2Sn, Extruded Shapes.

Figure 9 shows the fracture toughness value of the Ti-6Al-6V-2Sn shape directly quenched and annealed, compared with those of alpha-beta processed Ti-6Al-4V and Ti-6Al-6V-2Sn plate and forgings from reported literatures. It is observed that the fracture toughness of thermomechanically processed extruded shape showed rather high value in spite of high strength. Figure 10 shows the aspect of an extruded shapes processed using this new thermomechanical process.

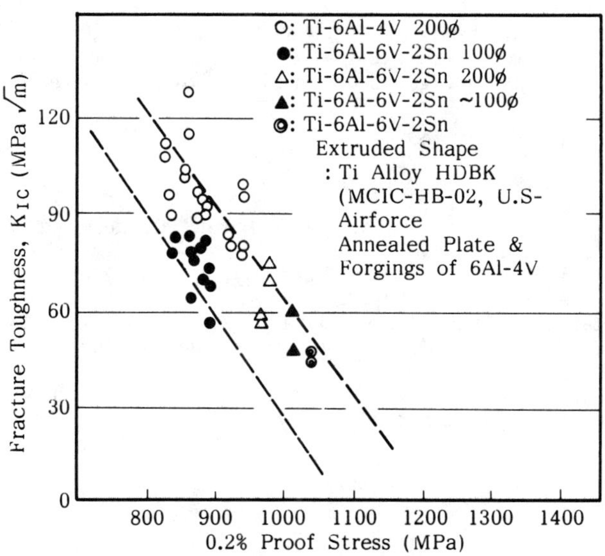

Fig. 9 The Relation between 0.2% Proof Stresses and K_{IC} Values of α+ß Ti Alloys.

Fig. 10 Aspect of High-strength Extruded Shape of Ti-6Al-6V-2Sn.

4. Conclusion

(1) The new thermomechanical process has been developed for alpha-beta titanium alloys, which consist of high reduction rate over 70% in beta region, rapid cooling over 5K/s after beta working and bi-modal annealing. The microstructure after this thermomechanical process consisted of fine beta grains and fine acicular alpha in beta grains.
(2) The shapes of fine bi-modal microstructure processed in this new thermomechanical process showed as high strength and ductility as those of the materials with equi-axed microstructure.
(3) High strength extruded shapes of Ti-6Al-6V-2Sn has been developed by this thermomechanical process. The developed extruded shapes showed uniform fine microstructure and higher fatigue strength than those of conventional processed.

References

1) H. Kusamichi, Y. Murakami, K. Kimura and O. Izumi, Titamium and its Applications (1983), p.42 (Nikkan Kogyo)
2) S.R. Seagle and L.J. Barlo, Metals Eng. Quarterly, 8, No.3 (1968), p.1
3) R.R. Boyer, J. of Metals, 32(1980), p.61
4) M. Katcher, Metals Eng. Quarterly, 8, No.3(1968), p.19
5) G.H. Heitman, J.E. Coyne and R.P. Galipeau, Metals Eng. Quarterly, 8, No.3(1968), p.15
6) M. Peters, A. Gysler and G. Lutzering, Titanium '80 Science and Technology, ed. by H. Kimura and O. Izumi, (1980, p.1777, AIME

HIGH STRENGTH $\alpha+\beta$ AND β-TITANIUM ALLOYS FOR FASTENERS AND TECHNOLOGY OF THEIR PRODUCTION

I. S. Polkin, I. A. Vorobyev, V. A. Sheenkov,
V. V. Dolgov
All-Russia Institute of Light Alloys, Moscow
SPA "Normal", Nijny Novgorod,
Russian Federation

Essay

At present $\alpha+\beta$-titanium alloys with the strength limit of up to 1000 MPa are widely used for production of fasteners. The modern level of aircraft-and-space material conditions that the requirements to the materials, the members of constructions are going to be made of, must be much more stringent for the sake of increasing its safety and reliability in operation. For successful solution of this problem one should use new fastening fixtures having high operating characteristics which can be provided on the basis of utilization of new materials and optimization of technology of their production.

In this work possibility of making bolts of high-strength $\alpha+\beta$ and β-titanium alloys characterized by different degree of strength has been investigated and estimation based on comparison of their operational characteristics with those of the bolts made of the production-type titanium alloy VT16 has been carried out.

The technology of production of bolts of the alloys under investigation in a hardened-by-heat-treatment state has been developed. Perspective of utilization of fasteners characterized by the strength limit of more than 1270 MPa made of β-titanium alloys has been shown.

Titanium '92
Science and Technology
Edited by F.H. Froes and I. Caplan
The Minerals, Metals & Materials Society, 1993

Methods

The inquires have been carried out with the use of high-strength $\alpha+\beta$-titanium alloys of martensite class VT16, VT23 and beta-titanium alloys VT19 and VT35.

Chemical composition of the alloys is tabulated in Table I.

Table I

Alloy trade mark	Content of alloying elements, wt.%								β-phase stabilisation factor, K_β	Temperature of polymorphous conversion, T_{nn}, °C
	Ti	Al	Mo	V	Cr	Fe	Sn	Zr		
VT16	Base	3.30	4.70	4.30	-	-	-	-	0.70	870
VT23	Base	4.17	2.28	4.65	0.97	0.51	-	-	0.78	900
VT19	Base	2.70	6.00	3.20	5.00	-	-	1.45	1.52	780
VT35	Base	3.10	1.70	13.60	3.00	-	3.10	1.00	1.47	740

360 mm dia. ingots of VT16 and VT23 alloys have been made according to the quantity production technology on the basis of the double remelting method, whereas ingots of VT19 and VT35 alloys have been smelted in a skull furnace with the use of an expendable electrode for the first remelting, the next remelting has been carried out in a vacuum electric arc firnace.

The ingots have been hammered to become the blanks of 130x850 mm dia in size and rolled on a small-section mill into 16 mm dia rods.

6 and 10 mm dia rods of VT16 alloy have been made in accordance with the quantity production technology.

Tension tests have been carried out on round samples with the working member of 5 mm dia, whereas jumping-up tests have been carried out on cylindric samples with $H=2d_o$. Degree of deformation at which the sample is damaged in the course of upsetting has been calculated on the basis of relative change of the sample's height:

$$\varepsilon_h = \frac{H_o - H_k}{H_o} \cdot 100\ \%,$$

where H_o and H_k are heights of the sample prior and after upsetting (mm).

Carrying out of Experiments

Calculation of a conventional stabilization factor of beta-phase (K_β) has proved that VT16 and VT23 alloys under investigation having $K_\beta=0.70$ and 0.78 correspondingly are classified as belonging to the martensite class, whereas VT35 and VT19 alloys with $K_\beta=1.47$ and 1.52 correspondingly are classified as belonging to class β-titanium alloys (see Table I).

Construction of the curves of beginning of desintegration of metastable β-phase in VT35 and VT19 alloys has proved that these alloys may be hardened by 100 % of beta-phase being allowed to cool in the open, the latter makes it possible to provide high plasticity properties of them in the hardened state (Fig. 1).

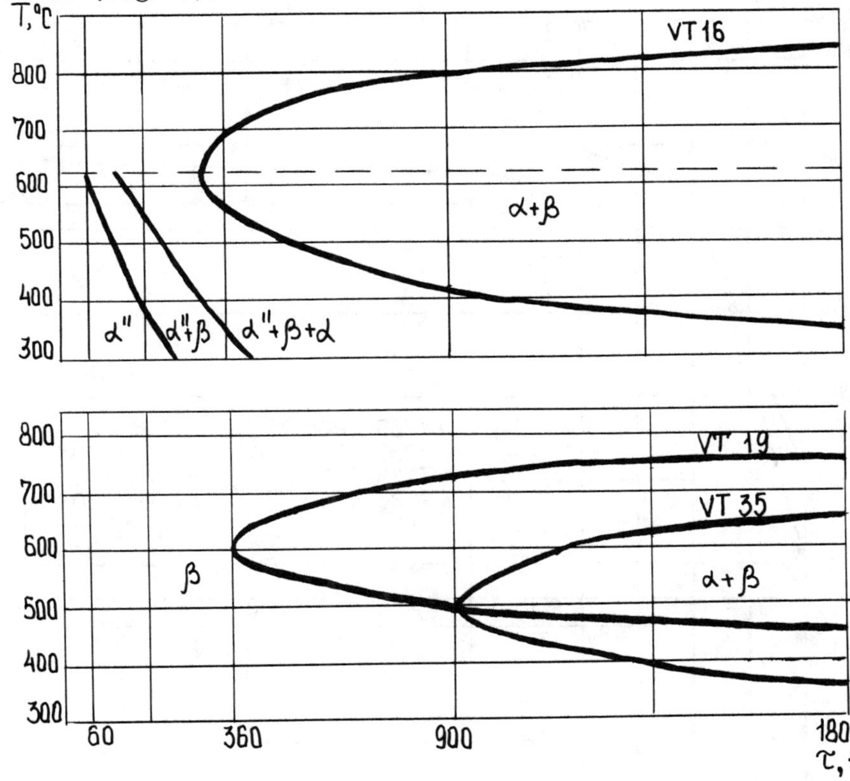

Figure 1 - The curves of beginning of desintegration of solid solution of alloys hardened from β-field: a - VT16, b - VT19, VT35.

As distinct from beta-alloys martensite phase α'' is fixed in VT16 and VT23 alloys from β-field, thet leads at the successive ageing to worsening of plasticity properties of these alloys in a hardened-by-heat treatment state. Therefore in contrast to beta-alloys hardening of $\alpha+\beta$-titanium alloys should be performed from two-phase field.

In Fig. 2 the diagrams of change of the ultimate strength and relative contraction of VT19, VT35 and VT23 alloys depending on holding time at various temperatures of ageing are presented. Hardening of VT19 and VT35 alloys prior to ageing has been carried out from β-field (T=800 °C), whereas hardening of VT23 alloy - from $\alpha+\beta$-field (T=750 °C).

These experiments have proved that beta-alloys having strength higher by 50-100 MPa than that possessed by VT23 $\alpha+\beta$-

alloy are characterized by much higher degree of plasticity thus providing relative compaction which is higher by 10-15 % than that of VT23 alloy. The same degree of plasticity (ψ=15-18 %) which is characteristic of VT23 alloy at strength of 1200-1300 MPa is obtained on VT35 β-alloy at ultimate strength of ~1400 MPa.

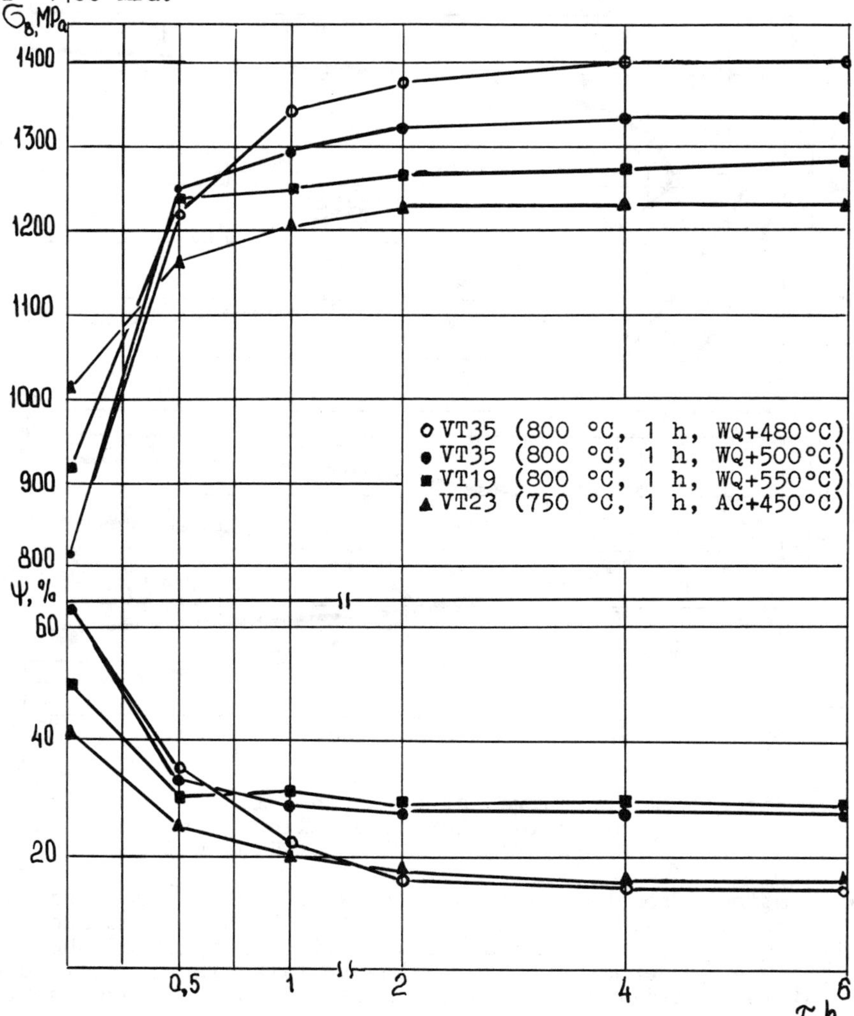

Figure 2 - Effect exerted by time of ageing on the ultimate strength and relative contraction of alloys: ● - VT35, o - VT35, ■ - VT19, ▲ - VT23.

Taking into account the obtained data it was of interest to evaluate nature of changes of mechanical properties of the rods made of β-alloys depending on the test temperature.

In this work there has been considered effect exerted by the test temperature on mechanical properties of the rods made of VT19 and VT35 alloys. The rods made of VT35 alloy have been subjected on thermal treatment for 1150 and 1330 MPa strength, whicle the rods made of VT19 alloy - for 1100-1220 MPa and to tension test at various temperatures ranging from 196 up to 450 °C.

As it is clear from the data presented in Fig. 3 both alloys have high plasticity properties within the whole investigated range of temperatures up to the test temperature of 196 °C, at which relative compaction of VT19 alloy is at the level of 18-20 %, and that of VT35 alloy - 20-22 %.

At the same time the alloys under inquiry have rather high values of short-time atrength at the elevated temperatures, for example during testing at a temperature of 450 °C both alloys have the ultimate strength exceeding 800 MPa.

In order to determine possibility of use of the alloys for making bolts by means of cold heading there have been carried out estimation of adaptability to manufacture of the alloys under inquiry. The alloys adaptability to manufacture has been estimated basing upon the results of the upsetting test at a temperature of 20 °C of cylinder-shape samples of height $H=2d_o$, where d_o is the sample's diameter.

In Table 2 the results of the upsetting test of the samples of the alloys under investigation after various conditions of thermal treatment are tabulated.

Table 2

Alloys to be investigated	VT16	VT23	VT19		VT35	
	annealing	annealing	annealing	hardening from $\alpha+\beta$-field	annealing	hardening from β-field
Ultimate strength, σ_B, MPa	880	920	900	960	890	800
Deformation degree during upsetting, ε_h, %	75	60	70	60	70	85

As it is clear from the results of the test tabulated in Table 2 the ultimate strength of the alloys is at the level of the corresponding characteristics of the production-type VT16 alloy in the annealed state with the exception of VT35 alloy in the hardened state the ultimate strength of which is lower by 80 MPa.

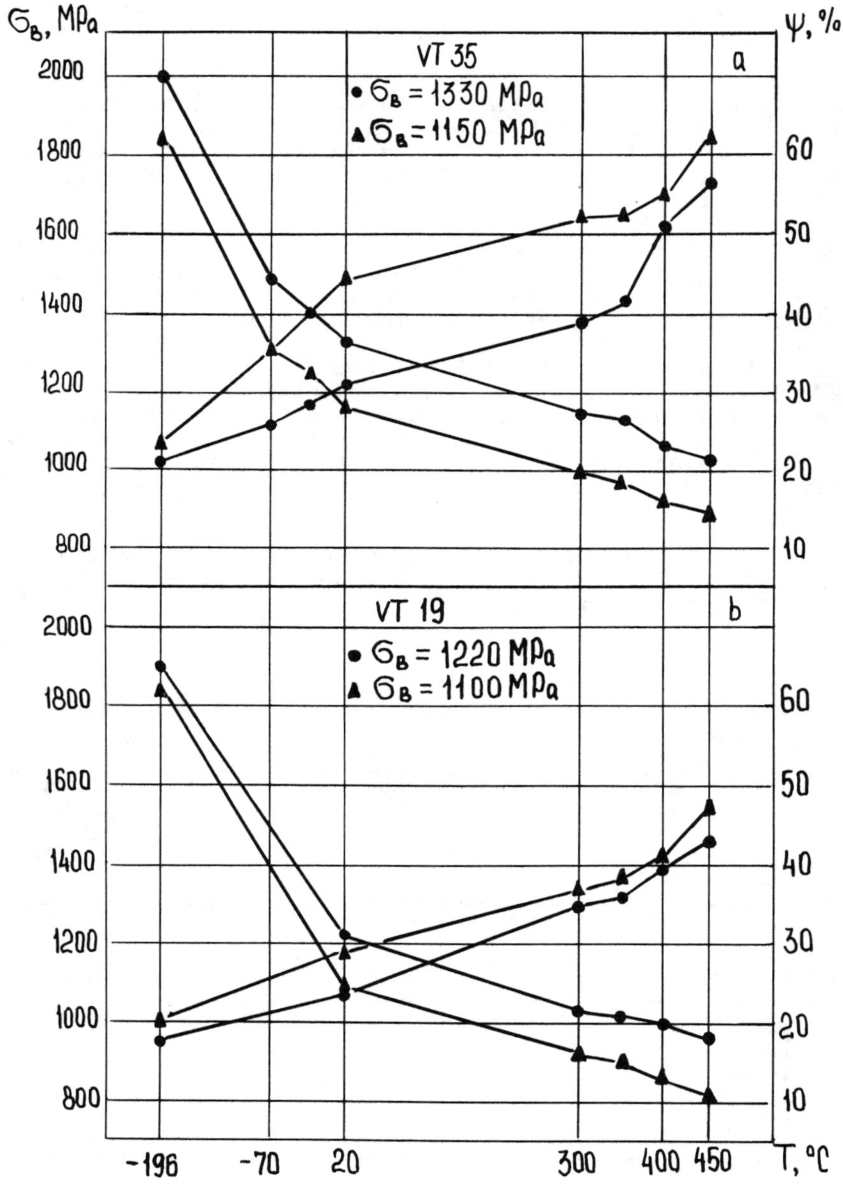

Figure 3 - Effect exerted by test temperature on mechanical properties of rods: a - VT35, b - VT19.

It should be noted that as far as the process plasticity is conserned VT35 alloy in the hardened state has much higher plasticity as compared with the production-type VT16 alloy, whereas VT23, VT19 and VT35 alloys in the annealed state are inferior to VT16 alloy.

In Fig. 4 microstructure (x200) of VT35 alloy is shown prior to and after upsetting with a degree of ε_h=85 %, as well as after upsetting and ageing at a temperature of 500 °C for 6 hours.

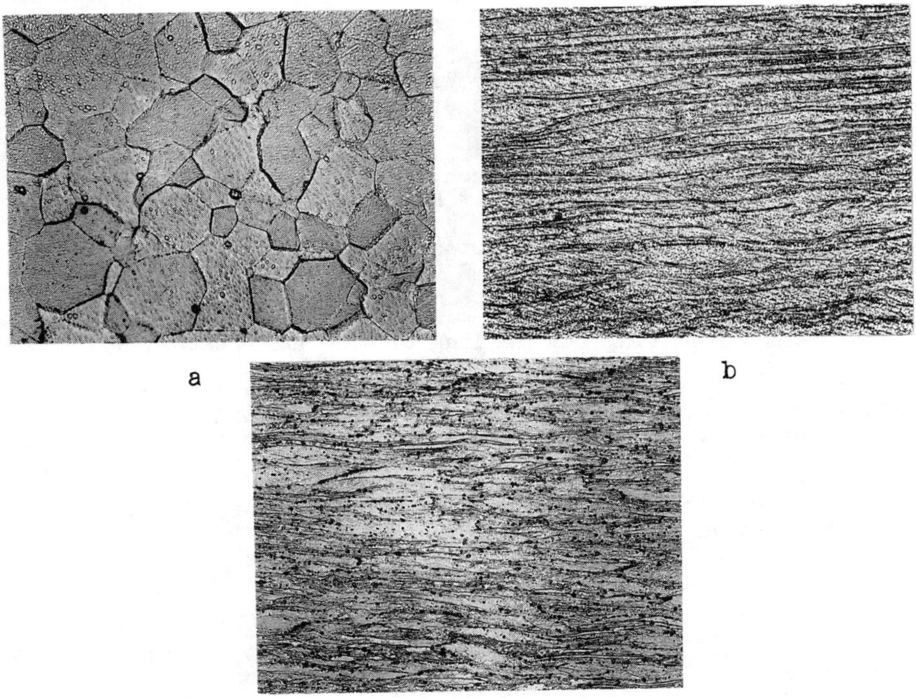

a b c

Figure 4 - Microstructure (x200) of VT35 alloy: a - initial state, b - upsetting with ε_h=85 %, c - upsetting+ageing 500 °C, 6 hours, cooling in the open.

In the course of the work we have carried out testing of making bolts of VT19 and VT23 alloys by means of cold heading. The heading has been carried out on the tool and at the transitions of the process of manufacture.

In Table 3 the results of cold heading of bolts are tabulated.

As the data tabulated in Table 3 show 8 % of the bolts made of VT19 alloy and 20 % of the bolts made of VT23 alloy had cracks on their heads located at an angle of∠45° that corresponds to the results of upsetting of these alloys.

Table 3

Alloy	Size of bolt	Number of produced bolts, pcs	% of bolts having cracks
VT19	M6	25	8
VT35	M6	25	0
VT23	M6	50	20
VT16	M6	50	0

Making of bolts under hardened-by-heat treatment condition has been carried out in accordance with the process of manufacture which included the following operations:

- hot heading of bolt head;
- reduction of body for successive thread rolling;
- strengthening thermal treatment;
- grinding of radius under bolt head and diameter of plane part;
- spinning of radius under bolt head;
- thread rolling;
- LUM control.

Conditions of heating for heading of the bolt head are: for β-alloys - 740±10 °C and for $\alpha+\beta$-alloys - 780±10 °C with holding at the heating temperature for 10-12 s. The bolts have been subjected to strengthening thermal treatment with the aim of obtaining different degree of strength and comparative evaluation of their operating characteristics has been carried out.

The results of inquiries into mechanical properties and operating characteristics are tabulated in Table 4.

Conclusion

1. Possibility of use of titanium alloys of various classes for making fasteners has been investigated. It was proved that beta-titanium alloys have much higher technological plasticity as compared with $\alpha+\beta$-titanium alloys.

2. It was determined that the use of high-strength $\alpha+\beta$ and β-titanium alloys characterized by the ultimate strength higher than 1200 MPa makes it possible to improve operating characteristics of bolts. The bolts made of beta-alloys are less sensitive to skewing during tension test as compared with the bolts made of $\alpha+\beta$-titanium alloys provided that they have the same level of strength.

3. Perspective of the use of beta-alloys for making fasteners characterized by the ultimate strength of up to 1400 MPa was shown. It is advisable to carry out additional inquiry to specify operating characteristics of the bolts made of beta-titanium alloys characterized by the ultimate strength of 1200-1400 MPa.

TABLE 4. - PROPERTIES OF BOLTS UNDER CONDITIONS OF HARDENING BY HEAT TREATMENT

Alloy	Mechanical properties		Bolts size	Properties of bolts				
	σ_B, MPa	ψ, %		P, kg	P_1, kg	P_2, kg	τ_{cp}, MPa	LCF** $N \cdot 10^3$ cycles
VT16	1050	60	M6	2230-2410	-	1810-2320	630	18-66
	1200	40	M10	6640-7780	4800-7100	5440-5840	750	-
VT23	1270	20	M6	2500-2780	2460-2620	2000-2420	860	18-33
VT19	1050	54	M6	2300-2390	2260-2320	2160-2310	770	9-95
VT35	1270	35	M8	4480-4530	-	-	860	12-15*
	1400	18	M8	4280-4300	-	-	930	16-17*

Notes: P - tension test of bolts;
P_1 - tension test of bolts with skewing of 8° under the head;
P_2 - tension test of bolts with skewing of 8° under the nut.

* Values of LCF of VT35 alloy have been determined on plain corset samples at $\sigma_{max}=0.75\sigma_B$

** Low cycle fatigue

STRUCTURE AND MECHANICAL PROPERTIES OF VT22 ($\alpha+\beta$) HIGH STRENGTH TITANIUM ALLOY SEMIPRODUCTS

I. S. Polkin, V. L. Rodionov, A. N. Stroshkov,
T. V. Ishunkina, V. G. Kudryashov, O. S. Korobov
All-Russia Institute of Light Alloys,
Moscow, Russian Federation

Abstract

VT22, a high strength Ti-5Al-4.5Mo-4.5V-1Cr-1Fe titanium alloy, is widely used in production of rolled and forged rods, drop forgings, press forgings, extruded tubes, and plates.

The main advantage of this alloy is its wide range of mechanical properties (UTS, El, RA) and characteristics of serviceability (KCT, K_{1c}, FCGR, LCF) through thickness up to 250 mm owing to high hardenability.

The globular microstructure, or microstructure approximated to such, obtained during deformation is the necessary condition for insuring high strength, ductility and serviceability. Formation of such microstructure is complicated by the difficulties of insuring sufficient deformation in $\alpha+\beta$ field because of the low temperature of polymorphic transformation, 860 °C. The principle schemes for formation of necessary structures and optimum conditions of heat treatment are given in this paper. Changes in characteristics of serviceability at various levels of strength (1050-1400 MPa) are also shown. The data obtained indicate that VT22 alloy is very promising for application in highly stressed structures.

1. Phase transformation and structure changes in VT22 alloy

Aging at which improved strength can be obtained due to decomposition of β-solid solution is used as one of the main methods of hardening $\alpha+\beta$ titanium alloys. Attainment of high strength is attributed to two factors: quantity of the age-hardenable β-phase and quantity of strengthening phase formed at aging. Relationships between these two factors are determined by alloying and heat treatment conditions.

Generalized phase diagram [1] for two-phase ($\alpha+\beta$) alloys (Fig. 1) shows that in the alloys with concentration of alloying elements less than critical level (C_k), β-phase is not

Titanium '92
Science and Technology
Edited by F.H. Froes and I. Caplan
The Minerals, Metals & Materials Society, 1993

fixed by quenching and at quenching martensite phases α' (with hexagonal crystal lattice) or α'' (with ortorhombic lattice) are formed. In alloys with higher concentration of alloying elements than critical (C_k), quantity of α-phase formed at aging (N_α cm) is maximum for quenched alloys with concentration close to C_k. The quantity of α-phase decreases with the increases of β-solid solution concentration (Fig. 1b). Then, maximum age-hardening can be attained in alloys where concentration of alloying elements is close to C_k.

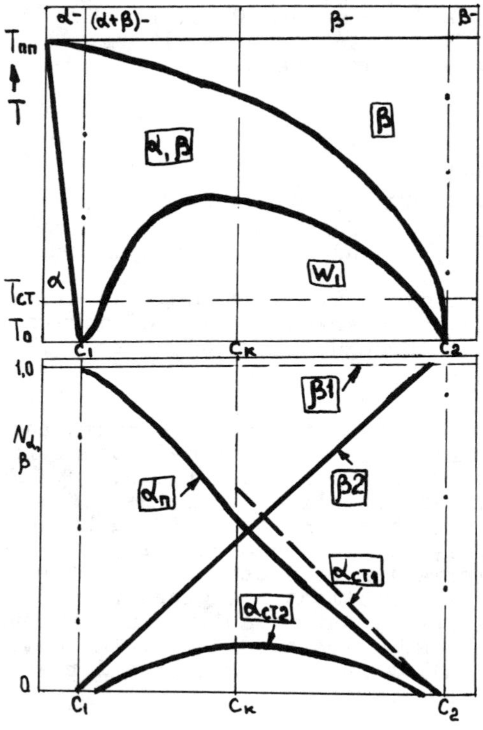

Figure 1 - Generalized phase diagram of two-phase titanium alloys and change in phase composition during aging:
a - phase diagram;
b - change of amount of α and β -phases ($N_{\alpha,\beta}$) during aging after quenching ($\alpha_{CT}1$) and after slow cooling (α_n, $\alpha_{CT}2$, $\beta 2$).
where: α_n - primary α-phase; α_{CT}, α-phase precipitating during aging; $\beta 1$, β-phase fixed by quenching; $\beta 2$ - residual phase

In cooling at slower rates than those of quenching, partial $\beta \rightarrow \alpha$ transformation resulting in formation of primary α-phase occurs. Change in quantity of this phase ($N_{\alpha n}$) with concentration is shown in Fig. 1b. Partial transformation during cooling of alloys results in formation of certain α- and β-phase relationships, schematically shown by the curve (W_1) in Fig. 1a. At aging temperature, further decomposition of β-solid solution with precipitation of disperse α-phase occurs in the alloys with changed quantity of α and β phases after cooling. Schematic representation of change in alloy composition during such treatment is shown in Fig. 1b. It can be seen, that the quantity of α-phase precipitated at aging is also near maximum for alloys with concentration of alloying elements close to C_k.

VT22 alloy is classified among the alloys in which concentration of alloying elements in near critical but above it. One of the advantages of this alloys is its high hardenability, which allows one to prevent decomposition of solid solution during cooling after deformation or heat treatment. This offers ample control of structure and properties and, as a result, allows one to use this alloy as a high strength material for production of a wide range of products for aerospace.

On the other hand, the alloy's chemical composition, close to critical C_k, results in rather complex character of phase transformations during aging. Fig. 2 shows phase transformation diagram for this alloy aged from the state obtained by quenching from β-field. It can be seen that depending on the age temperature, ω, α'' and α-phases can precipitate. It is shown that fields of precipitation of phases of various types overlap forming more complex phase states. Hardness isoline network traced on the diagram shows close agreement of change in properties with change in phase states.

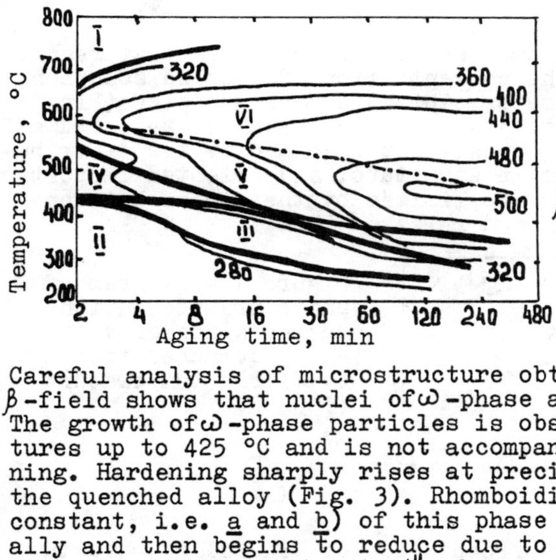

Figure 2 - Diagram of phase transformation during aging of VT22 alloy quenched from field temperature (900 °C): I - β, II- $\beta+\omega$, III - $\beta+\omega+\alpha''$, IV - $\beta+\alpha''$, V - $\beta+\alpha+\alpha''$, VI - $\beta+\alpha$.

Careful analysis of microstructure obtained by quenching from β-field shows that nuclei of ω-phase are present in the alloy. The growth of ω-phase particles is observed at aging temperatures up to 425 °C and is not accompanied by pronounced hardening. Hardening sharply rises at precipitation of α''-phase in the quenched alloy (Fig. 3). Rhomboidity (ratio of lattice constant, i.e. a and b) of this phase remains constant initially and then begins to reduce due to relaxation of internal stresses and transition to α''-phase and to α' (α) phase. This latter process is accompanied by the improvement in hardness and strength.

The oriented character of precipitation of α''-particles, in accordance with Burgers orientation relation [2,3], texture and character of particle distribution in the structure; and preset character of distribution of ω-particles create oriented microstresses and can lead to spontaneous cracking [ ] of the alloy in aging. Such cracking can be also observed, in isolated cases, during heating of quenched, coarse-grain ingot to higher aging temperatures when ω-phase particles are formed at the first stages of heating. Thus, a treatment including quenching from β-field + aging is not always suitable for VT22 products as it can cause cracking.

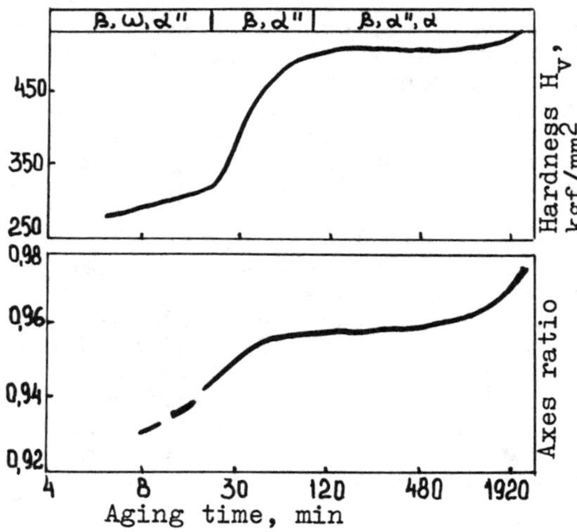

Figure 3 - Change in hardness and rhomboidity of lattice in VT22 alloy quenched from β-field and aged at 400 °C (ratio of rhomboidity: a and b axes ratio; at $b=a\sqrt{3}$) rhombic lattice becomes hexagonal).

Studies have shown that lowering of quenching temperatures to the α+β-field changes, the concentration of β-solid solution and does not cause ω-phase precipitation during aging (Fig. 4). In this connection the changes in structure character of distribution of α'' (α)-particles are not governed by the distribution of ω-phase particles in structure, and oriented microstresses arising during aging are limited by the fields between primary α-phase particles and relax on them.

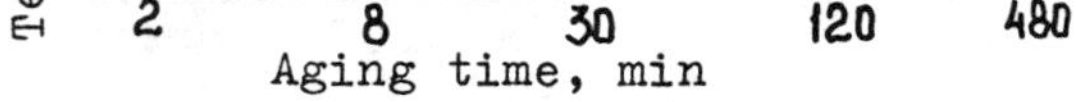

Figure 4 - Diagram of phase transformation during aging of VT22 alloy quenched from α+β field temperature (750 °C).

Fig. 5 shows schematic representation of VT22 alloy structures forming during the main types of heat treatments. Bimodal structure is optimum and allows combination of high strength and high ductility. This structure results from isothermal holding in α+β-field near the temperature of polymorphic transformation ($T_β$), slow cooling to quenching temperature and subsequent one- or two step aging.

Isothermal holding close to $T_β$ after deformation allows refining of β-phase structure due to recrystallization and polygonization. The subsequent cooling results in precipitation of α-phase particles of optimum size along β-grain (subgrain) boundaries. This permits a reserve of ductility and obtaining an amount of β-phase hardened by the subsequent aging, which is necessary for attainment of desired strength level and to avoid ω-phase precipitation due to change in concentration of β-solid solution.

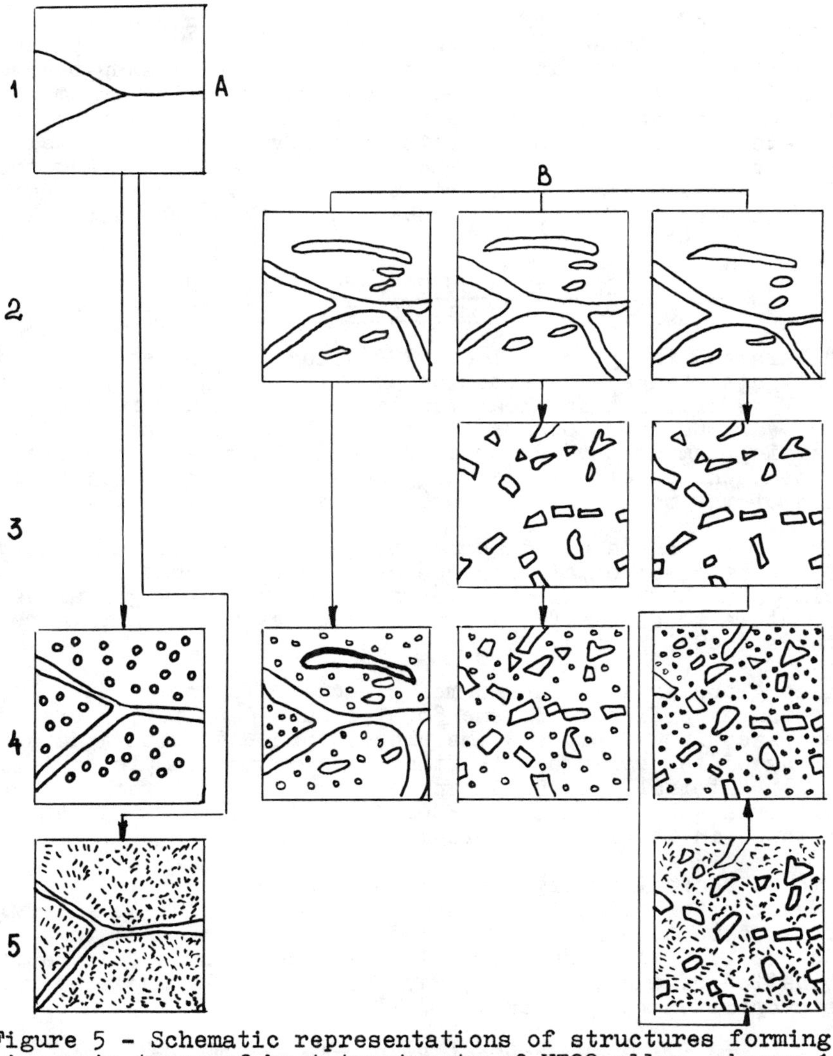

Figure 5 - Schematic representations of structures forming during main types of heat treatments of VT22 alloy, where: 1 - deformation in β-field, quenching from β-field; 2 - deformation in α+β field, isothermal holding in α+β field close to $T_β$; formation of β-phase structure due to recrystallization and polygonization, formation of primary α-phase structure; 3 - slowed cooling to quenching temperature; formation of primary α-phase structure, change in concentration of β-solid solution, establishment of necessary proportions of α and β-phases; 4 - aging at intermediate temperatures for attainment of necessary strength level due to α-phase precipitation; 5 - low temperature aging for uniform distribution of α-particles during aging, spontaneous cracking can occur in case of aging after quenching from β-field.

The results of studies of one and two step aging conditions have shown that the desired degree of hardening, while preserving sufficient ductility, can be obtained during step-by-step aging, due to the higher degree of β-solid solution decomposition. In this case, the scheme includes low-temperature aging + high-temperature aging is preferable for structures obtained after such treatment are more homogeneous due to the fact that at low-temperature aging distribution of precipitating α-particles is more uniform and is inherited by the second (high temperature) step of aging.

2. Change of Structure After Hot Deformation of VT22 Alloy

To understand the potential of VT22 alloy, it is necessary to investigate change and formation of structure as the joint effect of deformation and heat treatment. During deformation, a set of various types of structures can be formed in the alloy by varying temperature-deformation parameters. Fig. 6 shows selected schemes of deformation processing which result in formation of the most typical structures for VT22 alloy.

According to Scheme 6a, deformation is carried out in β-field, with the major deformation process being the second step. As a result, nonrecrystallized grain structure having high values of fracture toughness (K_{1c}) and fatigue crack growth rate resistance (FCGR) formed.

A mixed globular and transformed structure is formed during deformation according to Scheme 6b. The main deformation process takes place in β-field and the process proceeds with gradual lowering of temperature with final deformation completed in $\alpha+\beta$ field. Material with such type of structure has rather high ductility level, but K_{1c} and FCGR value are slightly lower than those in Scheme 6a material.

Scheme 6c shows that after preliminary deformation in β-field, the major deformation is carried out in $\alpha+\beta$ field. As a result, globular structure is formed. Studies have shown that this structure is optimum for high strength and ensures favourable combination of ductility (El, RA), fatigue (LCF) and good rupture properties (K_{1c}, FCGR).

3. Mechanical Properties of Thick VT22 Alloy Semiproducts

As VT22 alloy is used with an ultimate tensile strength of 1150-1200 MPa, its rupture properties are of great importance. Figs 7 and 8 show change in fracture toughness, fatigue crack growth rate and low cycle fatigue depending on ultimate tensile strength for various types of structures.

Scheme 6a material has a higher level of K_{1c} and FCGR properties in comparison to globular structure material (Scheme 6c) within the whole range of strength. The plot shows that optimum values of strength for VT22 alloy fall in the range of 1200-1250 MPa (Fig. 7), to achieve desired combination of K_{1c} and FCGR.

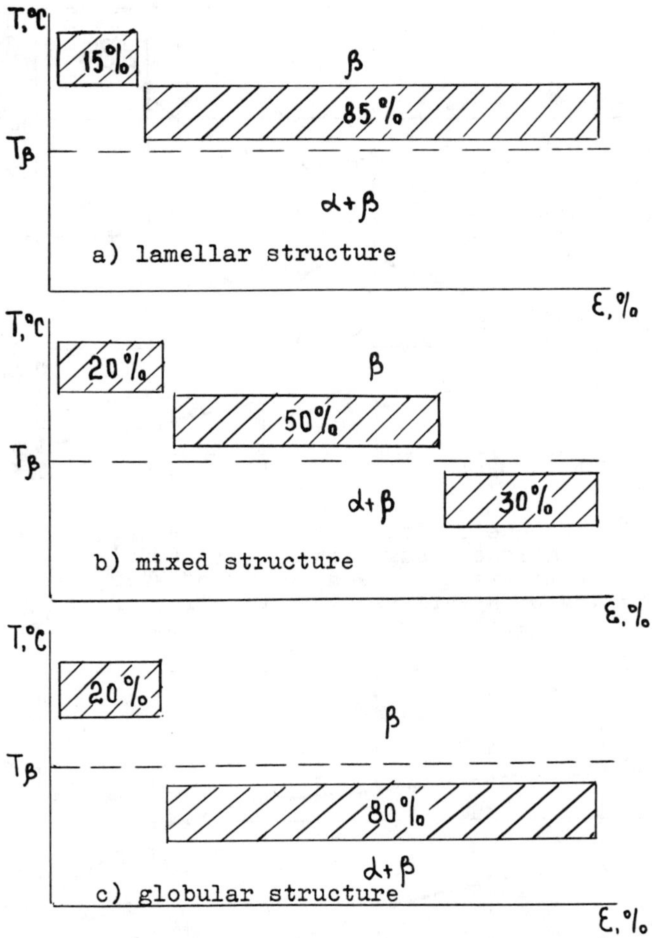

σ_B, MPa	ψ, %	K_{Ic}, MPa$\sqrt{m}$
1200	16	77-87
1200	20	65-71
1200	25	53-62

Figure 6 - Schemes of deformation used for production of semiproducts in VT22 alloy with various types of structures and strength levels.

Fig. 8 demonstrates advantages of the globular structure, material with a high level of LCF, LFR up to 1250-1300 MPa. Sharp decrease in LCF values are observed at higher level of strength.

A problem of formation of a homogeneous structure throughout the thickness of semiproducts may arise because of the fact that in formation of globular structure, the main deformation process proceeds in $\alpha + \beta$ field at rather low temperatures, because of relatively low temperature of polymorphic transformation (860 °C) of VT22 alloy.

Thoroughly developed conditions of deformation and heat treatment allow production of material with sufficient homogeneity of structure and distribution of properties throughout the bulk of semiproducts.

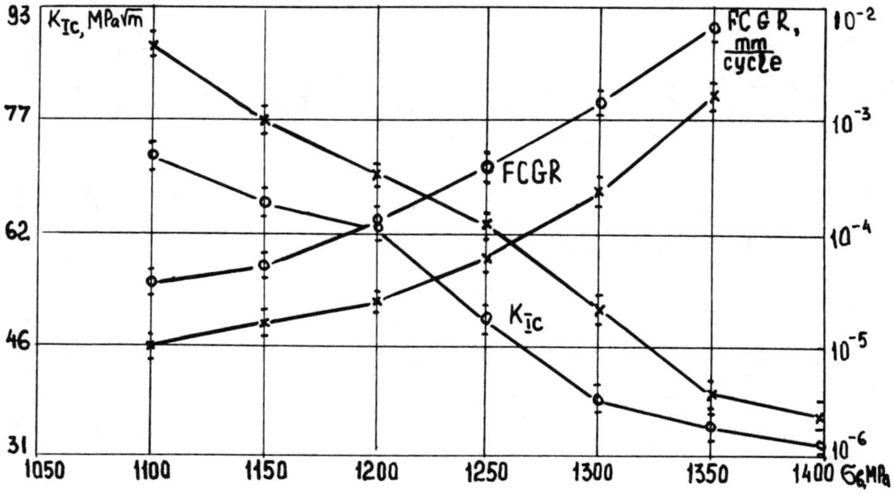

Figure 7 - Dependence of fracture toughness (K_{1c}) and fatigue crack growth rate (FCGR) on ultimate tensile strength of VT22 alloy with various types of structures: x - lamellar structure; o - globular structure.

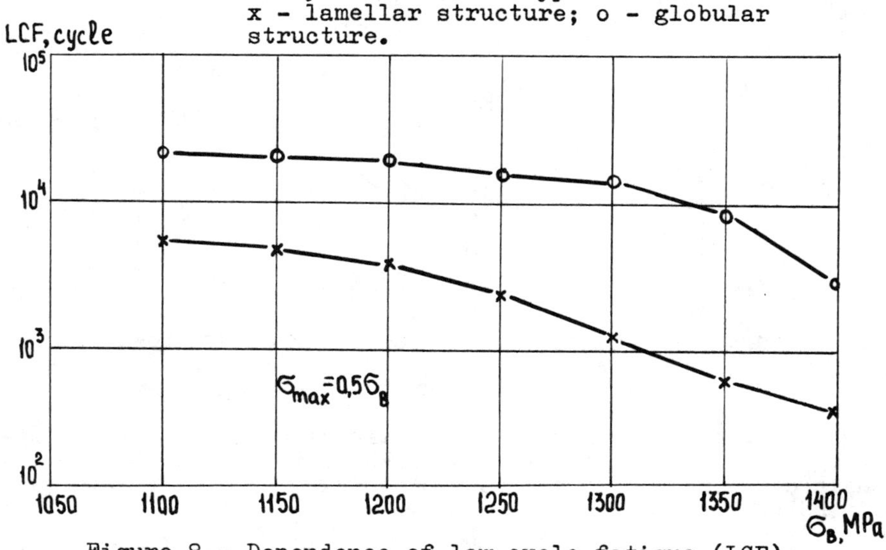

Figure 8 - Dependence of low cycle fatigue (LCF) on ultimate tensile strength of VT22 alloy, where: o - globular structure (Scheme 6c); x - lamellar structure (Scheme 6a).

Fig. 9 shows change in strength and fracture toughness through the thickness of semiproduct at two levels of initial strength. The degradation of strength as a function of thickness is related to the amount of metastable β-phase at various cooling rates at different points of the cross-section. During decomposition of this phase, microstructure differing in amount and morphology of α-phase may then make varying contribution to strengthening of the material.

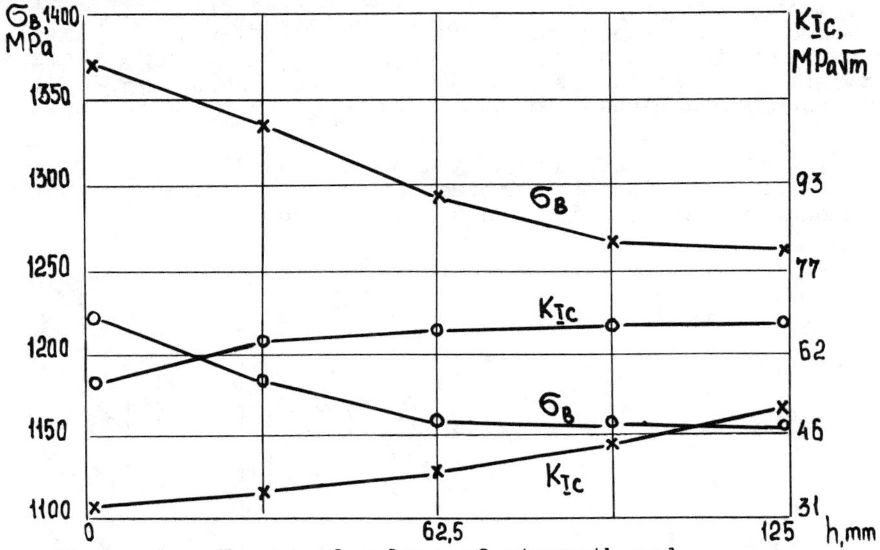

Figure 9 - Change of values of strength and fracture toughness through the thickness of VT22 alloy semiproduct.

Fig. 10 shows comparative data on fracture toughness of several titanium alloys treated to obtain strength above 1200 MPa with various values of β-phase stabilization coefficient (K_{β}) which is used to characterize the degree of alloying of these materials. This plot corroborates the merit of VT22 alloy in terms of its rupture strength in a high-strength state. This property determines, to a considerable extent, the application of the alloy in highly stressed structures.

Conclusion

1. The main mechanisms of phase transformations in highly alloyed titanium alloys have been discussed. It has been found that these alloys with chemical composition close to a critical level have the largest potential to achieve combination of high mechanical properties.

2. Decomposition of β-phase in these alloys can be accompanied by precipitation of metastable α and ω phases which have unfavourable effect on ductility properties.

Possible approaches to heat treatment which avoid formation of embrittling phases during aging of commercial have been demonstrated.

3. Thermomechanical treatment conditions for VT22 alloy semiproduct have been developed. These conditions permit forma-

tion of tailored structure ensuring optimum combination of mechanical and special properties.

4. Mechanical properties and rupture properties of thick semiproducts treated so as to attain different strength levels have been investigated. The possibility of attaing a ultimate tensile strength above 1200 MPa in the centre of cross section (~250 mm) has been shown.

5. Investigation of fatigue and rupture strength properties has shown satisfactory combination of values of all properties through the whole thickness of semiproducts.

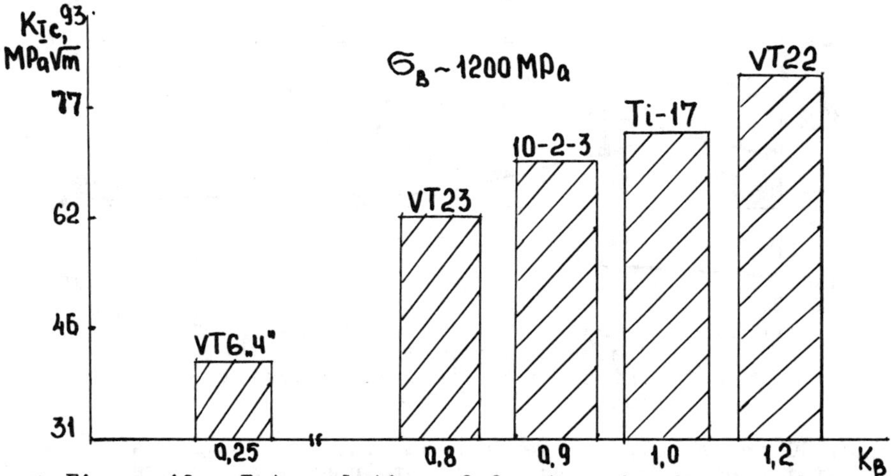

Figure 10 - Interrelation of fracture toughness and β-phase stabilization coefficient for titanium alloys treated so as to attain strength above 1200 MPa.

References

1. I. S. Polkin, V. A. Kolachev, "Phase Transformation, Thermal and Thermomechanical Processing of Titanium Alloys", Memoires et Etudes Scientifique Revue de Metallurgie, (October 1989), 659-665.

2. U. Zwikker, "Titanium and its Alloys", (Moscow, Metallurgiya, 1979), 511.

3. I. B. Gorynin, A. M. Parshin, S. S. Ushakov, I. I. Yarmolovich, "Embrittlement and "Spontaneous" Cracking of Quenched + Titanium Alloy During Aging", Titanium Physical Metallurgy and Technology. Vol. 2 (Moscow, VILS, 1976).

EFFECTS OF STRAIN PATH CHANGES ON WORK HARDENING IN TITANIUM SHEET

W.T. Roberts and R.A. Chamberlain*

School of Metallurgy and Materials and IRC in Materials for High Performance Applications
University of Birmingham
*now at Luxfer UK Ltd., Nottingham, England

Abstract

Tensile tests involving a 90° change in direction of applied stress after selected levels of prestrain were carried out on commercial purity titanium of grade IMI 125. Stress-strain and work hardening data were compared with the behaviour of monotonically stretched samples. At similar total strain levels, work hardening rates in prestrained samples were higher than in the monotonically stretched samples, leading to an increase in uniform stretching. The behaviour is similar to that observed in face-centred cubic alloys with a low stacking fault energy.

Introduction

The effects of strain history or strain path on the mechanical behaviour of sheet metals with cubic crystal structures have been extensively studied. A list of relevant publications is given in reference (1). The practical stimulus for such studies arises from the fact that complex strain path changes often occur in real sheet metal forming operations. In the laboratory, simulative tests involving sudden changes in straining can be made and, in principle, the changes in mechanical response can be measured and related to microstructural features.

Accurate measurements of work hardening rate are conveniently made in uniaxial tests, and therefore tests involving uniaxial prestrain followed by uniaxial restraining in a different direction offer the possibility of observing in detail the transient changes in work hardening behaviour usually associated with a change in strain path.

It has been observed in cubic metals that the transient changes can be described either as a lowered initial flow stress with an increased hardening rate (Type 1) or an increased initial flow stress with a reduced hardening rate (Type 2) (2). Type 2 behaviour has been associated with metals of high stacking fault energy (SFE) which contain well-developed dislocation cell structures after deformation. When the applied stress direction is changed, these cell structures tend to be destroyed leading to a reduction in hardening rate. Continued deformation in the second mode results in the generation of a new cell structure and a recovery in the hardening rate, Fig. 1(a). Thus type 2 behaviour can result in a drastic reduction in uniform elongation in the restrain test following prestrain above a certain level. Type 1 behaviour has been observed in face centred cubic materials with low SFE. Such materials are less susceptible to transient reductions in work hardening rate than the high SFE materials, Fig. 1(b).

Titanium '92
Science and Technology
Edited by F.H. Froes and I. Caplan
The Minerals, Metals & Materials Society, 1993

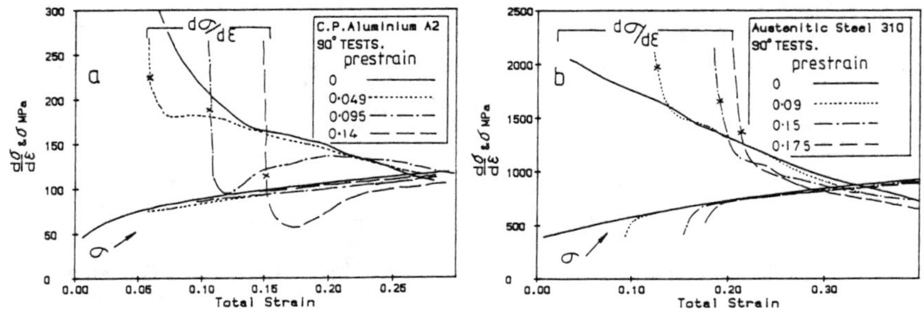

Figure 1 - Work hardening rates following 90° strain path changes in uniaxial tension (a) aluminium, representative of high SFE materials, (b) austenitic stainless steel, representative of low SFE material (1).

Another factor which may contribute to changes in work hardening rate after a change in strain path is related to the internal stress system in different regions of the specimen. Although the largest effects are observed in two-phase alloys (2), the behaviour of certain single phase materials with restricted slip may be strongly influenced by residual stresses. Also, in alloys containing sufficient solute concentration, dynamic strain ageing can inhibit the dislocation rearrangement process characteristic of type 2 behaviour, and thus reduce the transient fall in work hardening rate.

These observations on cubic metals provide a useful background for a study of strain path changes in a metal with hexagonal close packed structure. However, it should be borne in mind that differences in deformation modes will almost certainly be reflected in differences in response to strain path changes. The most obvious difference is the propensity of some hcp materials to exhibit extensive amounts of deformation twinning. The tendency for hcp titanium to form deformation twins depends strongly on the chemical composition of the alloy and its grain size (3,4). Published information about the influence of strain path changes in titanium is very sparse (5,6). Both of these studies were mainly concerned with the effects of strain history on forming limit diagrams. They involved combinations of uniaxial and biaxial stretching, and no attempt was made to measure changes in work hardening behaviour. The present investigation describes a preliminary attempt to address this aspect.

Material and Experimental Procedure

The material used in this study was commercially pure titanium sheet of grade IMI 125. This is an intermediate grade typically containing a relatively high oxygen content of about 0.15 wt.%. The sheet was received in the mill-annealed condition at a nominal thickness of 1.00 mm. The average grain diameter was about 20 μm, and the material had the normal type of texture developed in commercially processed annealed titanium sheet with basal plane poles tilted away from the sheet normal towards the transverse direction, Fig. 2. The ODF representation indicates in addition that there is a spread between $\langle 10\bar{1}0 \rangle$ and $\langle 11\bar{2}0 \rangle$ directions aligned with the rolling direction. As a result of the formation of this texture, the sheet exhibits a degree of planar anisotropy, and uniaxial tensile properties measured parallel to the rolling direction (RD) and at 90° to RD show some distinctive differences.

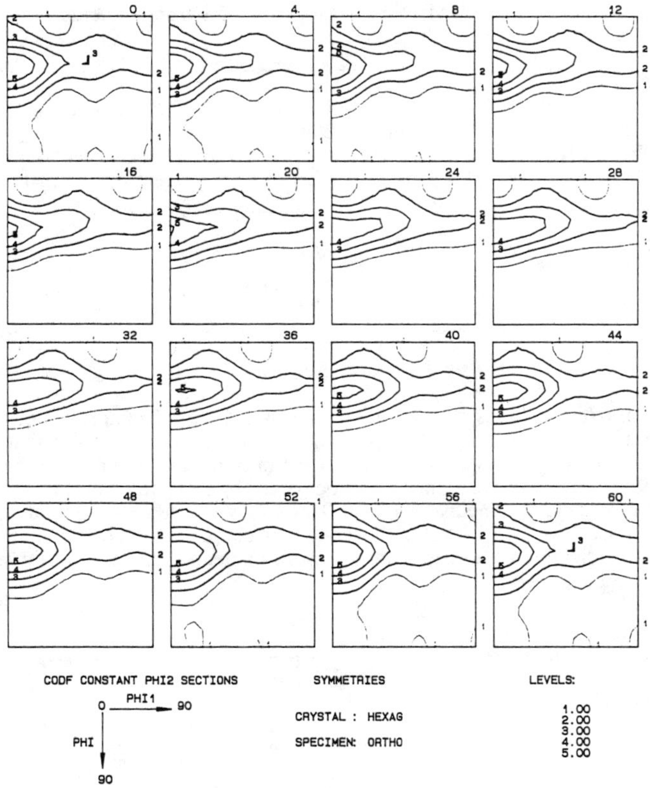

Figure 2 - Crystallographic texture of the as-received material. Constant ϕ_2 sections.

The mechanical properties of the as-received sheet are listed in Table I. For the two-stage tests, large tensile test pieces with a parallel gauge length of 350 mm and a width of 112 mm were machined with the tensile axis either at 0° or at 90° to RD. These were prestrained to preselected strain levels up to maximum values corresponding approximately to the values of uniform elongation listed in Table I. Small tensile test pieces with the tension axis at 90° to the axis of the large specimens were cut out using an electro discharge machine and these small specimens were subsequently extended. The parallel gauge length of these test pieces was 35 mm and elongations were measured on a 25 mm gauge length.

From the load-extension signals, a computer programme gave numerical and graphical outputs of true stress, σ, true strain, ε, and the relationship between $d\sigma/d\varepsilon$ and ε. For further processing of the data, the selected total stain range was divided into 50 intervals with small intervals in the small strain range and larger interval sizes at large strains. The average slope of the σ/ε relationship within each of these intervals was derived from a regression coefficient calculated from the distribution of σ and ε values collected within each of the strain intervals. With the strain rates used in the tests, each of the calculations of $d\sigma/d\varepsilon$ was based on several hundred datum points. Finally, an optional smoothing routine can be applied before the derived relationships are plotted. This involves a repeatable five-point least squares fitting procedure. All the $d\sigma/d\varepsilon$ relationships shown in this work have been smoothed in this way. Using these procedures excellent reproducibility is obtained in repeat tests.

Table I Mechanical Properties of Starting Material

Direction of Test Relative to RD	0.5% PS MPa	UTS MPa	Uniform Elongation	Total Elongation on 25 mm percent	Total Elongation on 50 mm percent	Plastic Strain Ratio r
0	397	468	0.120	37	27	1.49
90	424	469	0.075	44	32	2.86

Results

True stress-strain curves and dσ/dε strain curves for a selected range of prestrains for 0-90° tests and 90-0° tests are shown in Figs 3 and 4 respectively. The shapes of the σ–ε curves in two-stage tests are distinctly different in the two figures. In the 0-90 tests, a high initial rate of work hardening is accompanied by an increase in flow stress relative to the monotonic curve. The two-stage 90-0 tests show a much lower initial flow stress and a gradual rise towards the level of the monotonic curve. In neither case does there appear to be any indication of transient drop in work hardening rate as is observed with most common cubic metals.

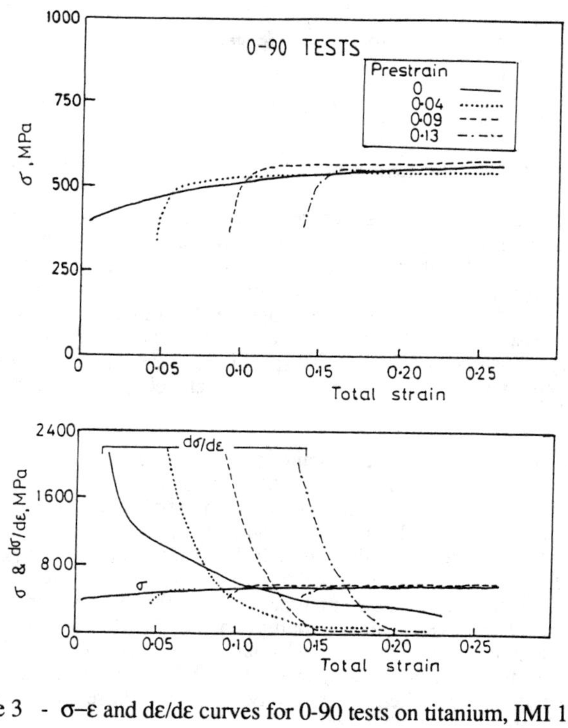

Figure 3 - σ–ε and dε/dε curves for 0-90 tests on titanium, IMI 125 grade.

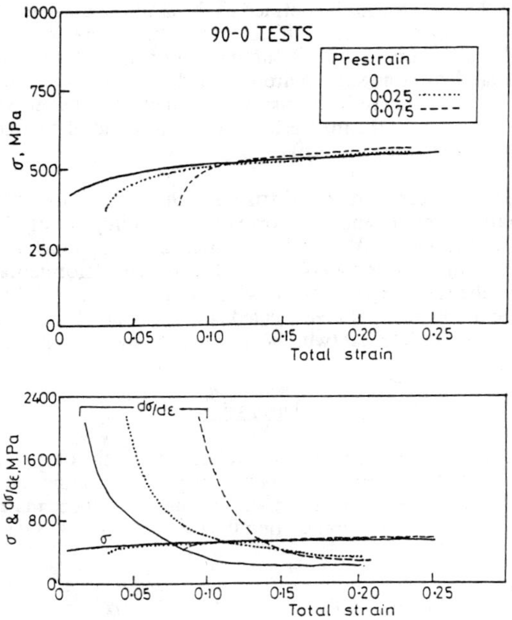

Figure 4 - σ–ε and dε/dε curves for 90-0 tests on titanium, IMI 125 grade.

Table II Results of Strain Path Change Tests

Prestrain	Residual Strain	Plastic Strain Ratio r	Residual Total Elongation on 25 mm %
		0-90 Tests	
0 (Monotonic)	0.075	2.9	37
0.041	0.052	2.7	32
0.044	0.048	3.2	34
0.056	0.052	2.4	34
0.056	0.049	3.3	31
0.087	0.035	-	28
0.093	0.033	-	26
0.127	0.036	2.8	34
0.137	0.036	3.8	34
		90-0 Tests	
0 (Monotonic)	0.120	1.5	44
0.023	0.087	1.8	33
0.027	0.085	1.4	35
0.027	0.092	1.7	33
0.028	0.083	1.8	33
0.036	0.088	1.4	36
0.076	0.059	-	32
0.079	0.059	-	28

Measurements of the uniform strain, defined as the strain rate when $d\varepsilon/d\varepsilon$ is equal to σ, are given in Table 2. Also quoted in this table are the values of the plastic strain ratio, (r), and the total elongation. A simple summation of the prestrain and residual strain values indicates that there is no sudden drop in residual ductility above a certain prestrain level - a feature that is observed with many cubic metals. A better way of representing the results, taking account of the measured r values, is to plot the values on a strain path diagram as shown in Figs.5(a) and 5(b).

In order to make more meaningful comparisons with the behaviour of cubic metals, it is necessary to establish the extent of deformation twinning during deformation of the titanium used in the present work. Tracings of optical micrographs of samples deformed at the limits of uniform extension are shown in Fig. 6. The deformation twins are shaded in black and from the relative sparsity of twins, it can be assumed that the predominant deformation mode in the material investigated is crystallographic slip. This is consistent with earlier observations (3,4) that twinning is inhibited by a fine grain size and by a high interstitial content.

Discussion

Although strain path changes are commonly referred to by the change in direction of the applied stress (e.g., 0-90), a more fundamental way of describing the path changes can be defined using the components of the strain tensor for sheet materials. The angle between two normalised vectors representing the two strain states is then given by (7):

$$\cos\phi = \frac{(1+\cos^2\alpha)(1+pq) - (1+\sin^2\alpha)(p+q)}{2\sqrt{1-p+p^2}\sqrt{1-q+q^2}}$$

in which α is the angle between two sets of tensile axes in the sheet plane and p and q are measures of anisotropy:

$$p = \frac{r_1}{1+r_1} \qquad q = \frac{r_2}{1+r_2}$$

r_1 and r_2 being the plastic strain ratios in the two test directions respectively.

For isotropic materials, p = q = 0.5 and for a 90° change in tensile direction in the plane of the steel, the angle ϕ is equal to 120°. However for high values of p and q, the angle is greater. When $r_1 = 1.5$ and $r_2 = 3.0$, as in the present material, ϕ is 143°, which is approaching a complete reversal of the strain vector.

Under such circumstances, the influence of directional internal stresses may be of greater significance. Long-range directional stresses in softer regions combine with the new applied stress to cause plastic deformation to start in the softer regions at a relatively low applied stress (1). In single phase fcc materials, the effect of internal stresses which enhance work hardening rates in the second stage of straining is more pronounced in materials in which cross slip is more difficult, i.e., materials with a low SFE. Results obtained with 70/30 brass and a stable fcc stainless steel support this view.(Fig. 1(b)).

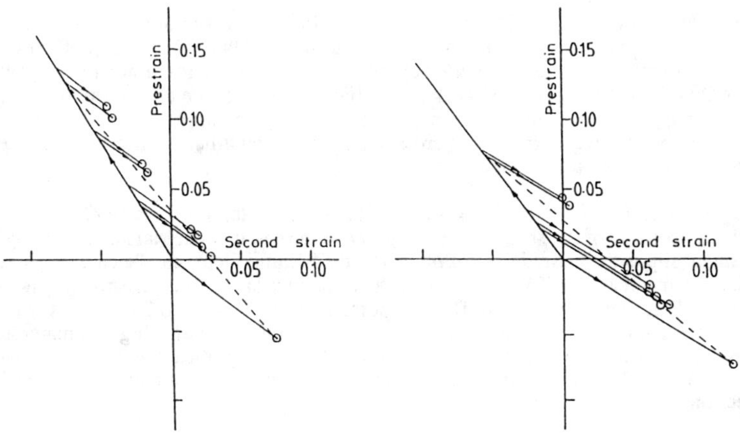

Figure 5 - Surface strains after extensions in two perpendicular directions. The slopes of the full lines are derived from the measured r values.

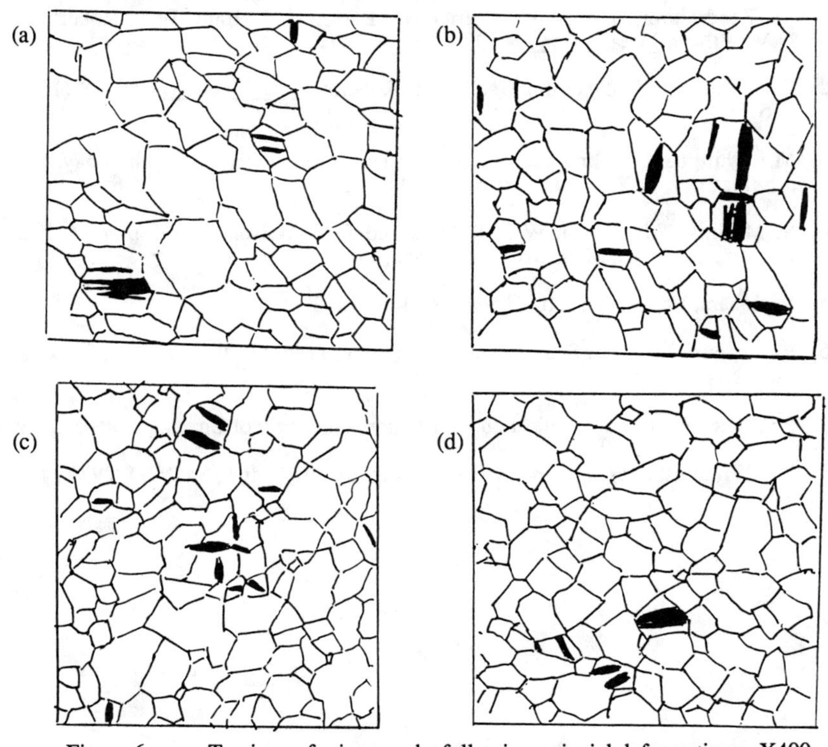

Figure 6 - Tracings of micrographs following uniaxial deformations. X400
(a) and (b) Monotonic tests. (a) 0°, $\varepsilon = 0.12$, (b) 90°, $\varepsilon = 0.075$,
(c) 0-90 test. ε_1 (0°) = 0.057, ε_2 (90°) = 0.107,
(d) 90-0 test. ε_1 (90°) = 0.028, ε_2 (0°) = 0.11

It has been shown (8) by TEM observations that high levels of interstitial oxygen in commercially pure titanium tend to produce planar slip arrays after strains of a few percent, compared with more tangled dislocation arrays in the low-interstitial alloys. Thus the relatively high rate of work hardening and the absence of a sharp transient reduction in hardening rate observed in the second stage of straining in IMI 125 may be related to the apparent difficulty of cross slip in high interstitial titanium. Further tests on low-interstitial titanium sheet are planned.

The pronounced rise in flow stress observed in the 0-90 tests is probably due to the planar anisotropy of the sheet. With the given texture in the material and assuming that prismatic slip is the main deformation mode, the Schmid factor can be calculated. For uniaxial tension at 0° to RD, the Schmid factor on two of the three prismatic systems is 0.43, while for tension at 90° to RD, the corresponding value is 0.29 (3). It is reasonable to assume that following the relatively small strains involved in testing this materials, the change in texture during tensile deformation will not be significant and thus the flow stress to cause deformation in the 90° direction is higher than that required in the 0° direction.

References

1. M. Zandrahimi, S. Platias, D. Price, D. Barrett, P.S. Bate, W.T. Roberts and D.V. Wilson, Metall. Trans. A, 20A (1989), 2471.

2. D.V. Wilson, M. Zandrahimi and W.T. Roberts, Acta Metall.Mater., 38 (1990), 215.

3. M.J. Philippe, C. Esling and B. Hocheid, Titanium Science and Technology, Proc. 5th Int. Conf. on Titanium, Germany, 1984), 1805.

4. S. Mullins, T.K. Sinha, W.T. Roberts and D.V. Wilson, Developments in the Drawing of Metals, Metal Society, London, 1983.

5. S. Kohara, M. Katsuta and H. Aoki, ICM 4, Vol. 2

6. R. Arrieux, Titanium Science and Technology, Proc. 6th Int. Conf. on Titanium, France, 1988, 1391.

7. P.S. Bate, private communication with author, University of Birmingham.

8. J.C. Williams, A.W. Sommer and P.P. Tung, Metall. Trans., 3 (1972), 2979.

Manufacturing of Large Turbine Blades of Titanium Alloy

Takao Satoh, Shinya Ishigai, Toshio Matsumoto

Engineering Dept., Steel Casting and Forging
Plant, Takasago, Kobe Steel, Japan

Abstract

The 1000mm(40 inch) large Ti－6Al－4V blade was forged with 392MN•m counter blow hammer. The forging process is designed by computer simulation techiniques which are based on primary analytical slab method and rigid viscoplastic finite element method. As the simulation, three steps of forging, flattening, twisting, finishing, has less unbalanced force and has uniform material flow than direct forging.The titanium blade made by three steps of forging method is uniformly deformed and has good material properties.

Introduction

The most effective way to improve the heat efficiency of a steam turbine for power generation is the use of long blades at the low－pressure final stage [1]. The use of long blades, however, raise a problem when the conventional 12Cr material is used because its strength is questionable. Thus, titanium alloy, being noted as a substitute for 12Cr material because of its characteristics of light－weight,high－strength,and corrosion－resistance, has been a candidate for long blades. But,in the production of large titanium blades, a new problem was arising that a titanium has quite a high deformation resistance in the $\alpha + \beta$ zone , making it difficult to provide thin－walled blades especially with a 392MN •m counter blow hammer. To solve this problem. A technique to provide a near－net shape through a forging process using three steps of flattening, twisting, and finishing was developed. This paper describes these results of mass production system.

1. Manufacturing Process of Large Titanium Blades

The manufacturing process of titanium alloy (Ti－6Al－4V) blades is as follows.Ingots are forged in the β zone, rolled in the $\alpha + \beta$ zone, and undergo β process. Then, forged preforms are prepared in the $\alpha + \beta$ zone again and die－forged into the desired shape with a counter blow hammer. After forging, blades are heat－treated,inspected, and shipped.

2. Outline of Forging Design

Fig.1 shows the flow diagram in the blade forging design. The forging design is done automatically by CAD system, where the coordinates of machining geometry are input. They are developed to forging and flattening geometry. These geometries are used for forging balance simulation and metal flow simulation. The forging balance simulation ,which is based on slab method, determines whether deformation is permitted. It also determines the geometrical positional relationship of the die cavity. The limitation of forging is based on the concept that forging is possible when the plastic deformation energy of anvil due to

Titanium '92
Science and Technology
Edited by F.H. Froes and I. Caplan
The Minerals, Metals & Materials Society, 1993

forging load is not greater than the energy given by the hammer [2]. The metal flow simulation, which is based on rigid viscoplastic finite element method (NASKA[3]), predict the under filling and strain distribution, and then decide the preform geometry.

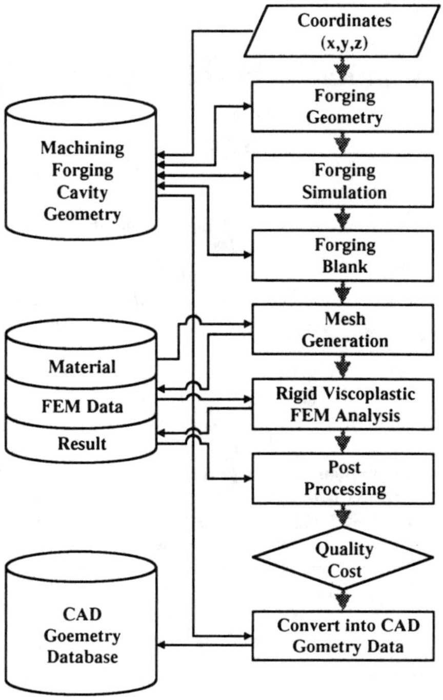

Fig.1 Design flow for blade forging.

2.1. Forging Balance Simulation

The blade is twisted in the longitudinal direction. So at the forging, the dies and hammer receive thrust Fx, Fy and moment Mx, My, Mz as shown in Fig.2. As a result, it is difficult to provide a precise forging. It is thus necessary to minimize the thrust Fx, Fy, and moment Mx, My acting on the forging machine by optimizing the angle in longitudinal direction of die cavity. If the design is made in this way, only the moment that rotates the dies in the horizontal direction will remain as an unbalanced force. This unbalanced force is equal in the upper and lower dies, so it can be absorbed by the counter lock machined on the cavity to minimize the damage given to forging machine. Although a three - dimensional analysis is need to obtain exactly these conditions, it is easy to solve them as plain strain problem when the metal flow in the longitudinal direction is small as in this case of the turbine blade. In the forging balance simulation, the blade is divided in the longitudinal direction as shown in Fig.3, then each section is divided into rectangular modules, and the slab method is applied to each module to add up the forging load [4]. The fundamental equation to obtain the stress distribution, the forging load, and the thrust are shown below [5].

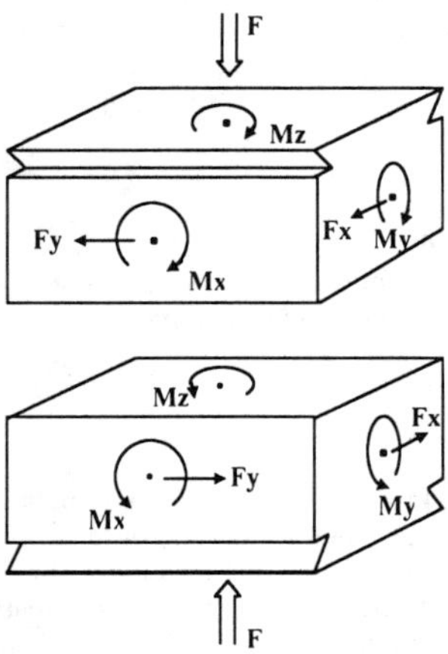

Fig.2 Moment and thrust effect on dies during forging.

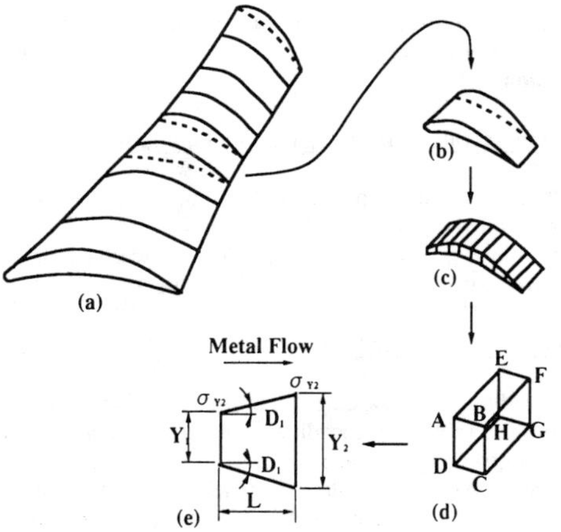

Fig.3 Determination of module for slab method.

1)Stress distribution and forging load

when $D_1+D_2=0$

$$\sigma_{y1}=(A_2/A_1)\ln(Y_2/Y_1)+\sigma_{y2} \quad (1)$$

$$P_y=(A_2/A_1)[Y_1(\ln Y_1-1)-Y_2(\ln Y_2-1)]+[\sigma_{y2}+(A_2/A_1)\ln Y_2]L \quad (2)$$

when $D_1+D_2\neq 0$

$$\sigma_{y1}=(A_2/Y_1)L+\sigma_{y2} \quad (3)$$

$$P_y=(A_2/2Y_1)L^2+\sigma_{y2}L \quad (4)$$

where

$A_1=\tan(D_1)+\tan(D_2)$

$A_2=-2\sigma A_1/\sqrt{3}+\tau[2+\tan^2(D_1)+\tan^2(D_2)]$

σ_{y1}, σ_{y2}: Stress at the inlet, stress at the outlet

σ : flow stress

τ : Shear stress

D_1,D_2 : Inclination of upper and lower module

L : Length of module

2)Thrust

$$P_{tu}=P_y\tan(D_1)-\tau[\tan^2(D_1)+1]L \quad (5)$$

$$P_{tl}=P_y\tan(D_2)-\tau[\tan^2(D_2)+1]L \quad (6)$$

where

P_{tu}, P_{tl} : Thrusts applied to the Upper and Lower module

The force which forges the blade can be obtained by applying above equations (1) to (6) to each module. At this time, the neutral surface is obtained and, based on the resulting stress distribution, the forging load and thrust are obtained.

2.2. Prediction Method of Preform Geometry

It is important to prepare a suitable preform to obtain more precise forgings. So as to obtain suitable preform, the reverse step method, which is based on slab method was developed. Fig.4 shows the concept of the reverse step method. This method is as follows.

(1)predict final neutral surface by adopting a slab method to the finishing geometry.

(2)calculate the areas of Sr0,Sl0 which are divided by neutral surface.

(3)set back die to the upper direction by Δ H

(4)calculate the material edges according to the condition of area Sr0=Sr1 and Sl0=Sl1.

(5)Predict the neutral surface again by adopting slab method to region ER1－EL1.

(6)get neutral surface continuously by repeating from (2) to (5).

Neutral surface (i.e. preform setting location) can be easily predicted by this method because of need for less computer power to calculate by the slab method.

3.1. Prediction of Unbalanced Forces Due to Forging

In order to clarify the capability of precision

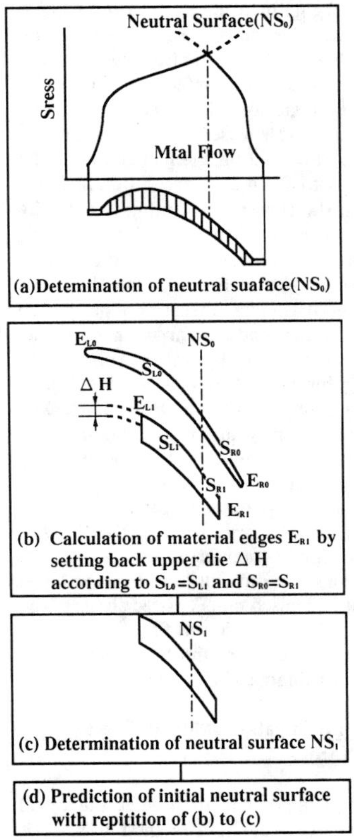

Fig.4 Prediction Method of nuetral surface

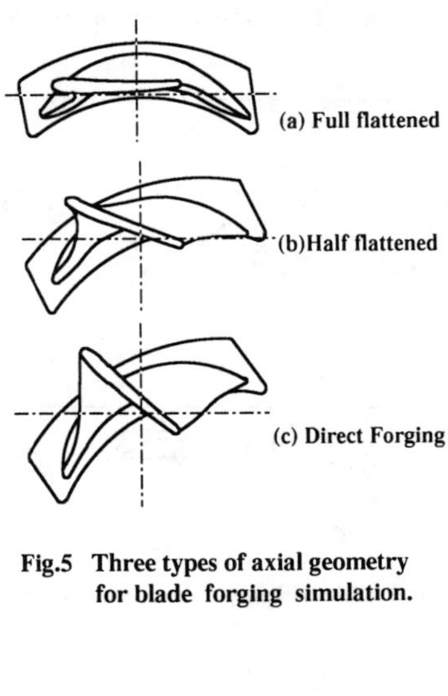

Fig.5 Three types of axial geometry for blade forging simulation.

forging, flattening and direct forging were compared. Axial view of flattening and direct forging of blade are shown in Fig.5. Flattened geometry is obtained by correcting the finish or twisted geometry.

Fig.6 shows the effect of the cavity rotation angle on the thrust on direct forging. In this case, the thrust can be minimized by setting the inclination of the die cavity to about 36 degrees. But, in direct forging, the thrust distribution is not fully satisfied because of the large thrust on both ends of longitudinal direction of the blade is elevated as shown in Fig.7. On the other hand, in flattening, thrust is almost zero along the longitudinal direction of blade. Fig.8 shows the horizontal moment due to forging for three types of forging, which are flattening,twisting and finishing. The moment becomes lower and lower when twist angle is reduced, then it becomes almost zero when fully flattened.

The above analysis revealed that, for a blade having a large twist angle, the dies receive a great amount of moment, thrust, which makes it difficult to reduce the blade thickness. Thus, a non-twisted shape of the same sectional profile was first forged, then the twist and the finish forging were applied to obtain a thin blade.

3.2. Investigation of Metal Flow by NASKA

It is impossible to predict the metal flow by using the slab method because it considers only finishing geometry of forging. So, metal flow of 4 typical sections of two types of blades were predicted by the FEM code NASKA. The preform location is

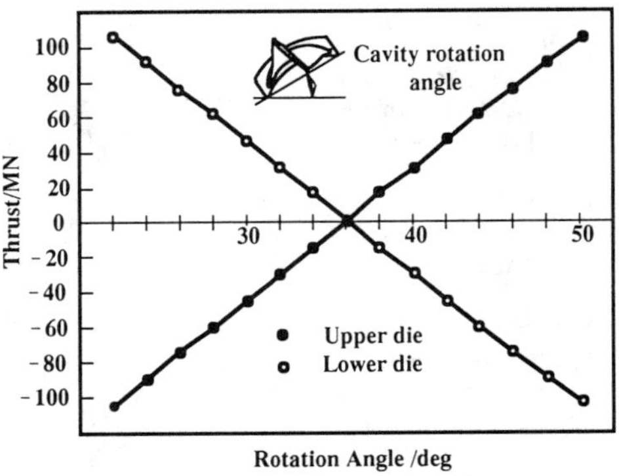

Fig.6 Effect of cavity rotation angle on thrust for direct forging.(simulation)

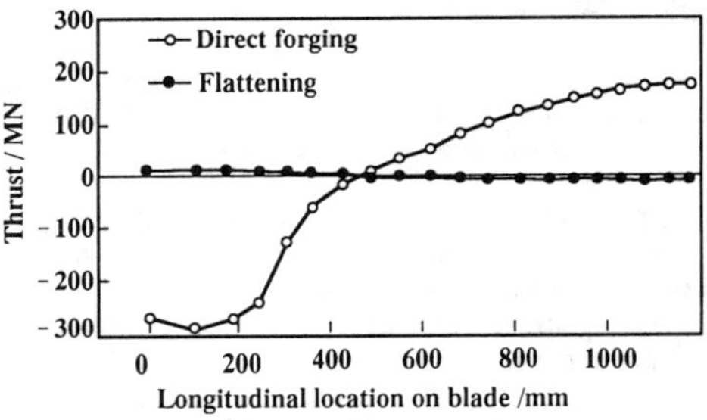

Fig.7 Comparison of thrust distributions for flattening and direct forging.

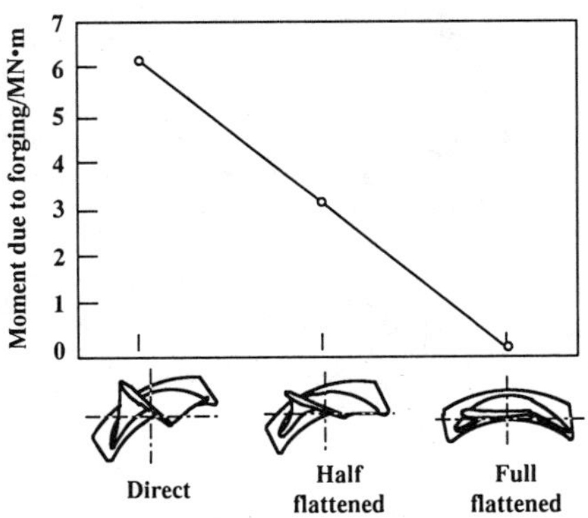

Fig.8 Efferct of forging geometry on moment

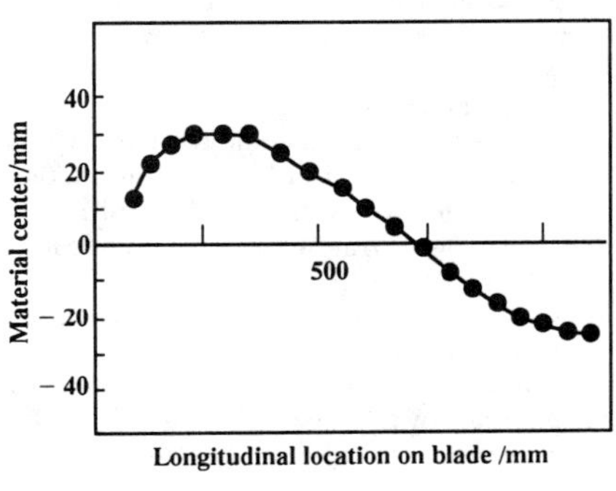

Fig.9 Calculated material center distribution for flattening.

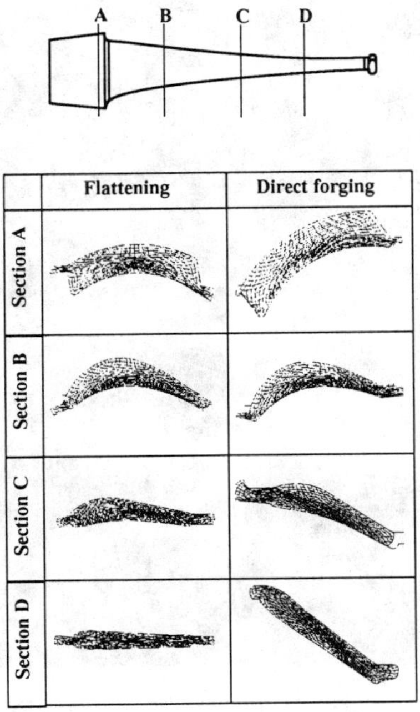

Fig.10 Metal flow of each section for flattening and direct forging simulated by NASKA

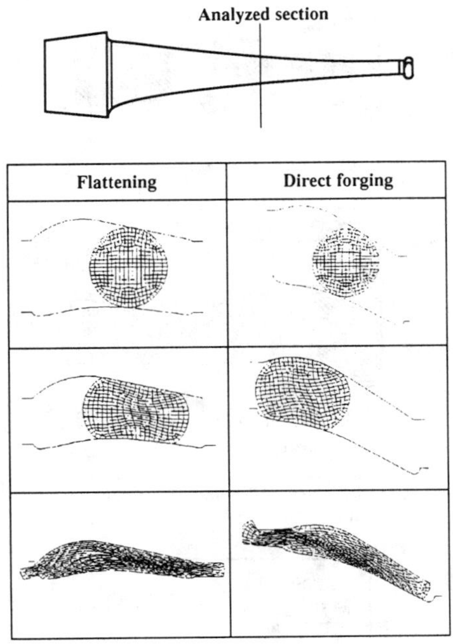

Fig.11 Metal flow for flattening and direct forging simulated by NASKA

based on the reverse step method described before, which is shown in Fig.9. Simulation conditions are as follows.

Flow stress $\sigma = 1.0 \times \varepsilon^{0.5}$mPa
Shear factor m = 0.3

The simulation results are shown Fig.10. In flattening process, the metal can be filled almost uniformly all of 4 sections. But in direct forging, metal cannot be filled uniformly on section C. Fig.11 shows filling process of section C for two types of forgings. In direct forging, material moves to the upper left of the cavity, as the result the cavity of lower right in the die remains underfilled until the end of forging.On the other hand, in flattening, material moves uniformly in both directions. To compare with analysis to the real forging, the three - step forging and direct forging were forged as a test. Results are shown in Fig.12. In direct forging, underfill appears around the analyzed section C as predicted by FEM simulation but does not appear in flattening of three steps forging. As the result, it is clarify that the three - step forging is better to obtain more precise shape for this type of blade. Photo 1 shows the finished titanium blade forged with the three - step forging processes, which is precise and has uniform machining tolerance.

4. Mechanical Properties of Forged Products

Fig.13 shows the relationship between the tensile strength at room temperature and the portions of the 1000mm (40 inch) turbine blade manufactured and toughness throughout the full length of the blade. The tensile strength tends to be higher toward the tip of the blade, because the air foil is cooled faster than the root. However, the variations are within several percent and cause no problem. Fig.14 shows the fatigue strength at 10^7 cycles is as high as 519 MPa.

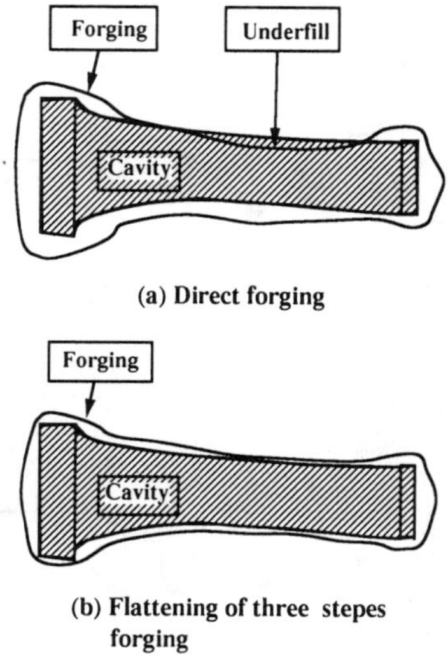

Fig.12 Effect of flattening on reducing underfill

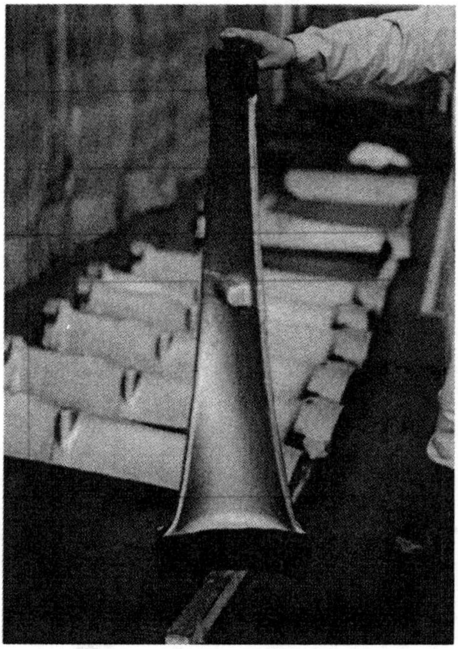

Photo.1 40-inch Ti-6Al-4V blade forging produced by three steps of forging.

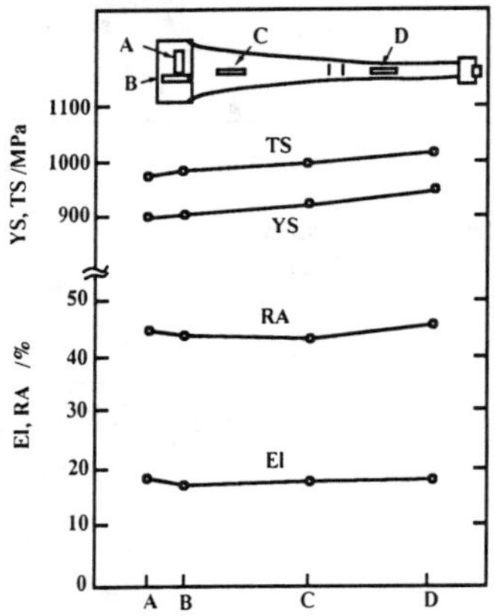

Fig.13 Mnufacturing properties of 40 inch large Ti-6Al-4V blade.

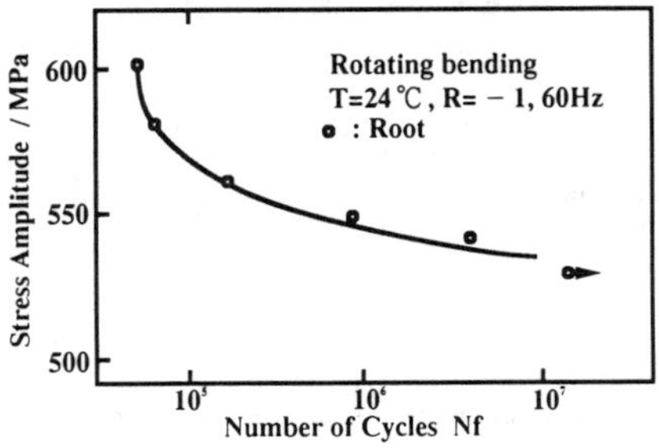

Fig.14 High Cycle fatigue strength of 40 inch Ti-6Al-4V blade

5. Conclusion

With the processes of flattening, twisting, and finishing developed using the elementary slab analytical simulation technique and the finite element method, it has been made possible, by the use of a 392 MN•m counter blow hammer, to forge a 1000mm (40 inch) large titanium alloy turbine blade which previously had been impossible to produce. It has been verified that the titanium alloy blade manufactured with this process has good material properties.

References

1)R.Araki et al.:Thermal and Nuclear Power,39, (1988)6, p.15
2)C.W.Frame:North American Metalwork Res. Conf., (1976), p.94
3)S.Toyosima et al.:Proc. of 39th Japanese Joint Conf. for Tech. of Plasticity,(1988), p.347
4)N.Akgerman et al.: Trans. ASME J.of Eng.for Ind., 98(1978),A(1976), p.290
5)T.Altan et al.: Trans. ASME J.of Eng.for Ind ,93(1971), p.477

EFFECT OF PROCESSING ON TENSILE AND CREEP PROPERTIES OF TWO CAST α_2 TITANIUM ALUMINIDE ALLOYS

D. Eylon[a], L. S. Apgar[a], D. J. Evans[c], M. Saqib[b], W. J. Porter[a], and I. Weiss[b]

[a]University of Dayton
Graduate Materials Engineering, Dayton, OH 45469-0240, USA

[b]Wright State University
Mechanical and Materials Engineering, Dayton, OH 45435, USA

[c]U.S. Air Force, Wright Laboratories
Materials Directorate, WL/MLLM, Wright-Patterson AFB, OH 45433-6533, USA

ABSTRACT

Post casting treatments of the ordered α_2 alloys Ti-24Al-11Nb[a%] (Ti-24-11) and Ti-25Al-10Nb-3V-1Mo[a%] (Ti-25-10-3-1) were studied and compared. The alloys were hot isostatic pressed (HIP) at a range of 1900 to 2150°F(1040-1175°C), followed by solution treatment (ST) at 1900 to 2125°F(1040-1165°C) range, and cooled after ST at rates from 50(30) to 360°F(200°C)/min. Complete casting pore closure was accomplished at all HIP conditions. Mechanical test data showed no HIP temperature effect on tensile or creep results. However, the higher ST temperature produced better properties for both alloys. Maximum tensile results were obtained in the Ti-25-10-3-1 alloy, using ST cooling rate of 130°F(70°C)/min. The Ti-25-10-3-1 had finer α_2 grain structure in the as-cast, cast+HIP, and ST conditions as well as substantially superior tensile strength and creep resistance at all test temperatures when compared with the Ti-24-11 alloy. While the high temperature tensile and the creep strengths of Ti-25-10-3-1 are at levels exceeding conventional high temperature titanium alloys, the room temperature elongation of both cast alloys is still at an unacceptable level of below 1%.

INTRODUCTION

Titanium net-shape castings are rapidly gaining acceptance in the aerospace, chemical, and general engineering industries [1-3]. Most of the cast alloy volume is in Ti-6Al-4V and Ti-6Al-2V-4Sn-2Mo [2]. The hot isostatic pressing response of cast Ti-6Al-4V is well documented [4], and the post-casting heat treatment range has also been studied and reported in detail [5,6]. By practicing alloy control, investment casting techniques and post-casting HIP+heat treatments, today it is possible to produce parts with properties as good as premium forgings [1-6]. Another group of ingot metallurgy (IM) alloys currently being explored for net-shape castings is the metastable beta alloys [7-9] which can have tensile strength of 200ksi(1360MPa) and fatigue runout strength in excess of 140ksi(950MPa) [7].

Net-shape casting technology for intermetallic alloys, based on the ordered α_2 (Ti_3Al) or γ (TiAl) phases [10], presents a challenging opportunity. On one hand, the difficulty in machining brittle alloys favours net-shape technologies. On the other hand, the coarse cast structure and occasional casting defects can be very detrimental to such brittle materials. Many Japanese companies are currently experimenting with production of low cost investment cast titanium aluminide automotive components, such as turbocharger impellers [11] and exhaust valves. The objectives of this work were to study the effects of HIP and ST on microstructure and mechanical properties of two α_2 based alloys: Ti-24Al-11Nb and Ti-25Al-10Nb-3V-1Mo.

Work was sponsored by EMTEC - Edison Material Technology Center, Kettering, Ohio 45420

Titanium '92
Science and Technology
Edited by F.H. Froes and I. Caplan
The Minerals, Metals & Materials Society, 1993

EXPERIMENTAL PROCEDURES

MATERIAL

The alloys Ti-24Al-11Nb[a%] (Ti-14Al-21Nb w%), and Ti-25Al-10Nb-3V-1Mo[a%] (Ti-14Al-20Nb-3V-2Mo w%), were melted in a 50lbs(23kg) capacity *induction skull melting* furnace by the Duriron Company, Inc. [12] into 0.75in(20mm) dia. x 5in(125mm) long test bars. Chemical analysis and estimated beta transus temperatures are listed in Table I.

Table I: Chemical Analysis and Estimated Beta Transus Temperature of the Cast Alloys Studied

ELEMENT	Al	Nb	V	Mo	Fe	Si	C	O	H	N	Estim. Beta Transus
	wt%	wt%	wt%	wt%	wt%	wt%	wt%	ppm	ppm	ppm	°F (°C)
Ti-24-11	14.70	21.40	-	-	0.07	0.08	0.043	770	10	53	2070-2100 (1130-1155)
Ti-25-10-3-1	14.93	19.47	3.38	2.06	<0.10	0.05	0.019	640	10	87	1980-2020 (1080-1105)

PROCESSING

Since earlier in the program, the Ti-25-10-3-1 showed better properties, more process conditions were experimented on this alloy. **HIP study** was carried out following casting (Table II). **Solution treatment study** was done on HIP'd bars at a range of temperatures and cooling rates (Table II) in a controlled cooling rate recirculating argon furnace. The 2125°F(1165°C) ST was above beta transus for both alloys. All other ST temperatures were at or below the transus. All bars were aged at 1400°F(760°C)/1hr. More details on the alloys, casting process, HIP and ST conditions are available in Reference 13.

Table II: List of HIP, ST Temperatures, ST Cooling Rates and Creep Test Conditions Used

HIP Cond., [°F(°C)/ksi]:	1700(925)/38;	1800(980)/38;	1900(1040)/38;	2150(1175)/15;	2150(1175)/40.
ST Temp., [°F(°C)]+ 1400°F(760°C)/1hr:	1900 (1040);	1950 (1065);	2025 (1105);	2125 (1165).	
ST Cooling Rates, [°F(°C)/min]:	50 (30);	130 (70);	150 (85);	360 (200).	
Tensile Test Temp., [°F(°C)]:	75 (25);	1050 (565);	1200 (650);	1400 (760).	
Creep Cond., [°F(°C)/ksi(MPa)]:	1050(565)/35(240);	1200(650)/20(135);	1200(650)/55(375);	1400(760)/10(70).	

Characterization of both alloys in the as-cast, cast+HIP and HIP+ST conditions was performed by standard metallographic techniques. Selected samples were characterized by TEM and XRD.

MECHANICAL TESTING

Unified tensile/creep samples, with a gage section 0.2in(5mm) dia. x 1.25in(32mm) long, were machined from the treated bars. **Tensile testing** was done at a strain rate of 0.005min^{-1} up to 0.2% strain followed by a crosshead rate of 0.002mm/min to failure. Duplicate tensile tests were performed for each alloy condition. **Creep testing** was done on Arcweld frames in accordance with ASTM E139. Only a few samples were tested to rupture. Most samples were unloaded between 200 to 2,300 hrs. All test conditions are listed in Table II. Creep at 1200°F(650°C)/55ksi(375MPa), with rupture time over 50 hrs, was established by some aerospace manufacturers as a "bench-mark" acceptance criterion for α_2 alloys.

RESULTS

MICROSTRUCTURE

The *as-cast* and *cast+HIP* microstructures of both alloys are shown in Figs. 1 and 2 respectively, for both alloys. Optical and TEM micrographes are shown in Figs. 3a through d for the 2150°F(1175°C) HIP+2125°F(1165°C) at 130°F(70°C)/min ST condition of both alloys.

Figure 1: As-cast microstructure (400X) of: (a) Ti-24-11, and (b) Ti-25-10-3-1.

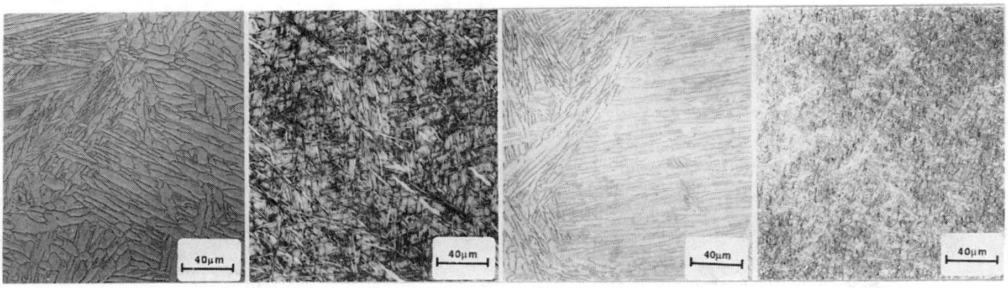

Figure 2: Cast+HIP microstructures (400X) after 2150°F(1175°C)/40ksi(280MPa) HIP: (a) Ti-24-11, and (b) Ti-25-10-3-1, and after 1900°F(1040°C)/38ksi(265MPa) HIP: (c) Ti-24-11, and (d) Ti-25-10-3-1.

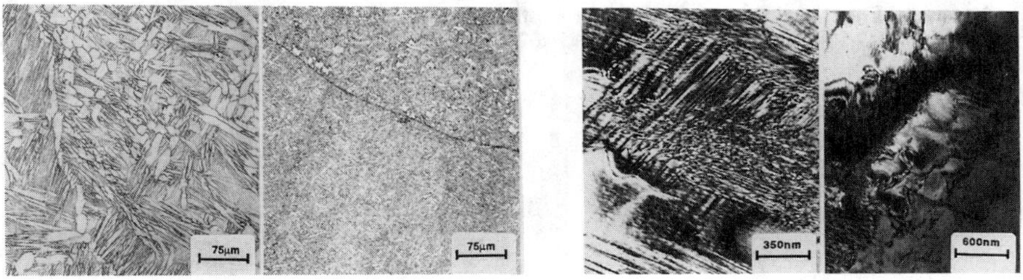

Figure 3: HIP (2150°F/15ksi/2hr)+ST (2125°F, 130°F/min) microstructures of: (a) Ti-24-11, 400X, (b) Ti-25-10-3-1, 400X, (c) Ti-24-11 (TEM micrograph), 46kX, and (d) Ti-25-10-3-1 (TEM micrograph) 26kX.

PHASE IDENTIFICATION

Phase identification of both alloys in the as-cast, Cast+HIP, and HIP+STA conditions was studied by X-ray diffraction. Volume fractions of the phases present were determined by using the Miller Method [14], and are listed in Table III. B2 refers to the ordered beta phase.

Table III: XRD Determination of Phases Volume Fraction Under Different Processing Conditions

HIP Conditions, [°F/ksi(°C/MPa)]	Ti-25-10-3-1	Ti-24-11
1. As-cast	12% B2[a] + α_2 + orthor.	<5% B2[a] + α_2 + orthor.
2. 2150/15(1175/105)	15% B2[a] + α_2 + orthor.	12% B2[a] + α_2 + orthor.
3. 1900/38(1040/260)	25% B2 + α_2	12% B2[a] + α_2 + orthor.
4. 1800/38(980/260)	45% B2 + α_2	12% B2[a] + α_2 + orthor.
5. 2150/15(1175/105) + 2125°F(1165°C) STA	<5% B2[a] + α_2 + orthor.	<5% B2[a] + α_2 + orthor.

[a] Ignoring the contribution of the orthorhombic phase to the $(02\bar{2}1)\,\alpha_2 + (221)$ orth.-peak

TENSILE PROPERTIES

The effects of test temperature, HIP temperature, ST temperature, ST cooling rate on room and elevated temperature tensile test results of both alloys are summarized in Figs. 4, 5, 6, 7 and 8 respectively. *These results are the average of two tests for each condition.*

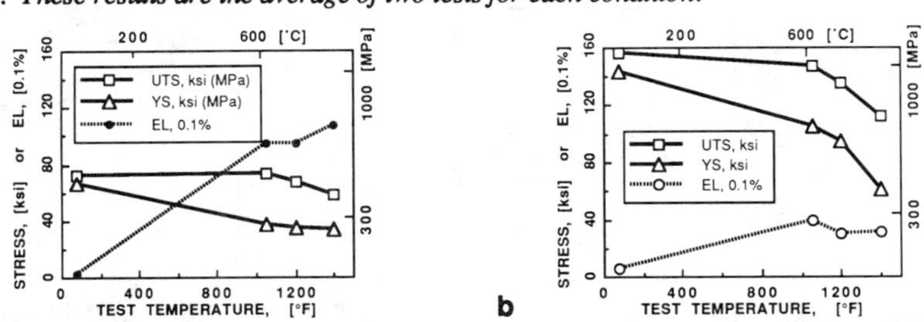

Figure 4: Comparison of tensile properties of: (a) Ti-24-11 and (b) Ti-25-10-3-1 at RT to 1400°F (760°C) range for Beta HIP'd and Beta ST conditions. EL scale is 10X magnified to highlight variations.

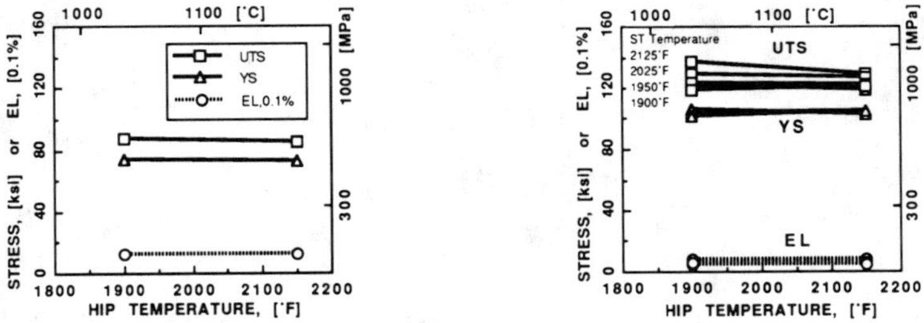

a b

Figure 5: Effect of HIP temp. on RT tensile properties of: (a) Ti-24-11 ST at 2125°F @350°F/min., and (b) Ti-25-10-3-1 ST at 1900, 1950 and 2025°F at 50°F/min. and 2125°F at 130°F/min. (EL 10X.)

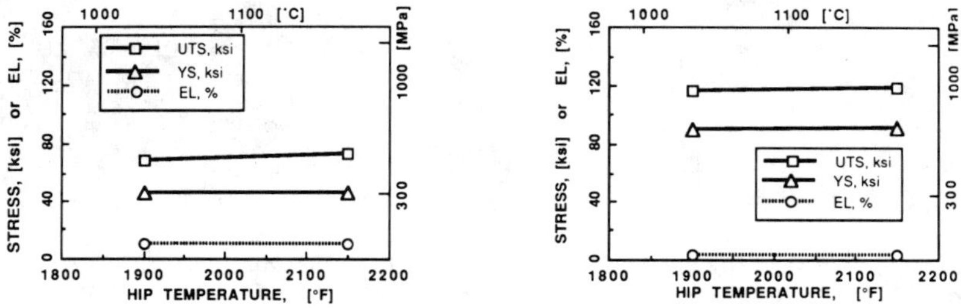

a b

Figure 6: Comparison of the effect of HIP temperature on 1200°F(650°C) tensile properties of 2125°F (1165°C)[360°F(200°C)/min] solution treated: (a) Ti-24-11, and (b) Ti-25-10-3-1. (EL scale is 1X.)

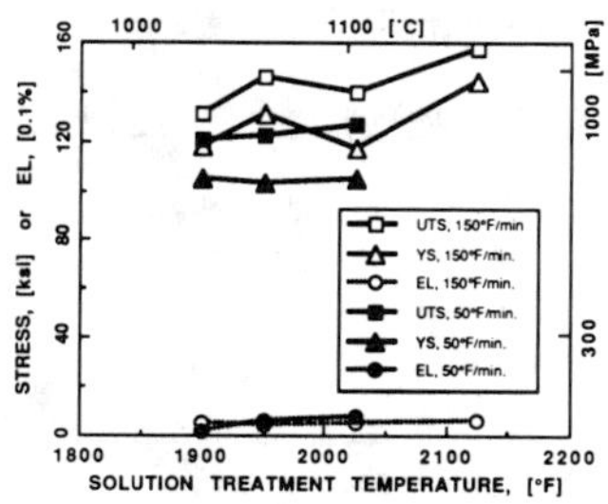

Figure 7: Effect of ST temperature followed by 50(30) or 150°F(70°C)/min cooling rate on RT tensile properties of Ti-25-10-3-1. (EL scale is 10X.)

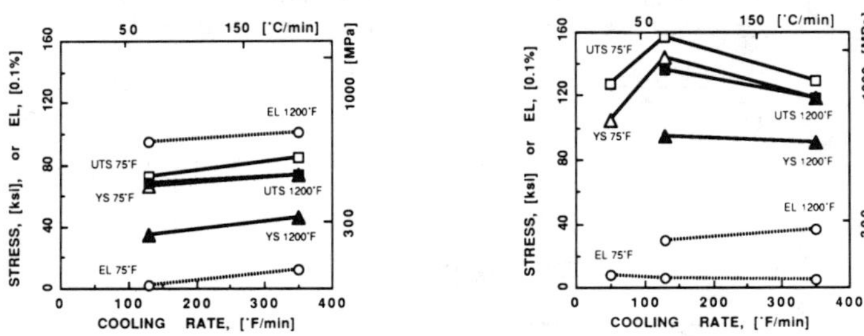

a b

Figure 8: Effect of cooling rate from the Beta ST temperature on RT and 1200°F(650°C) tensile properties of Beta HIP'd: (a) Ti-24-11, and (b) Ti-25-10-3-1. (EL scale is 10X.)

CREEP PROPERTIES

Typical creep curves of the two alloys at 1200°F(650°C)/55ksi(375MPa) following 2125°F (1165°C) beta ST are compared in Fig. 9. The creep stress vs. 0.2% creep strain Larson-Miller Parameter (LMP) is plotted in Fig. 10 for both alloys. The LMP is calculated by the following equations:

$$K\ (LMP) = (460 + T°F) \times (20 + \log t[hrs]) \times 10^{-3}$$

or:

$$K\ (LMP) = (492 + 1.8T°C) \times (20 + \log t[hrs]) \times 10^{-3}.$$

The effects of HIP temperature, ST temperature, and ST cooling rate on the creep strength of Ti-25-10-3-1 under 1200°F(650°C)/55ksi(375MPa) creep conditions are plotted in Figs. 11, 12, and 13, respectively. The effects of creep exposure on high temperature *post-creep tensile properties* are shown in Figs. 14a and b for both alloys.

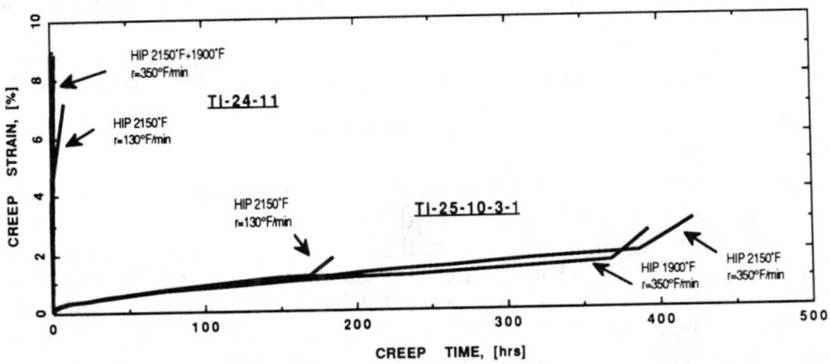

Figure 9: Creep curves of both alloys at 1200°F(650°C)/55ksi(375MPa) following 2125°F(1165°C) beta ST.

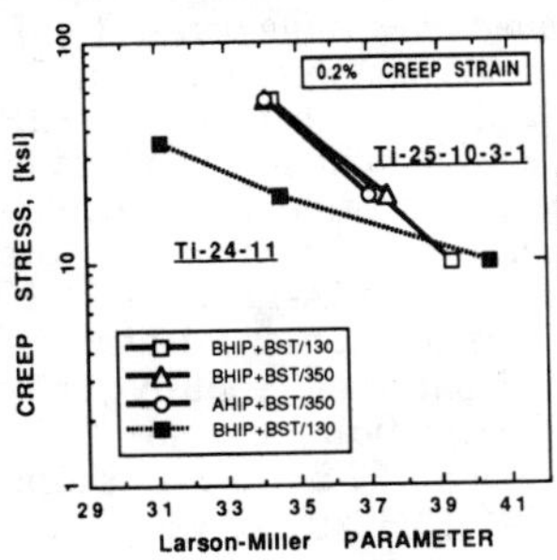

Figure 10: Stress vs. 0.2% creep deformation LMP for both alloys.

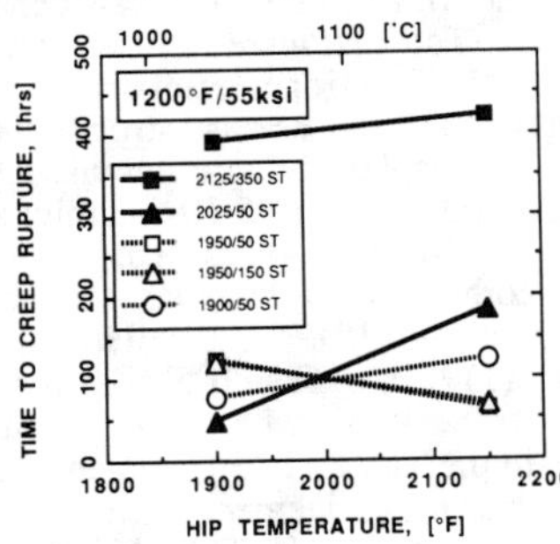

Figure 11: Effect of HIP temp. on rupt. life of Ti-25-10-3-1 at 1200°F(650°C)/55ksi(375MPa).

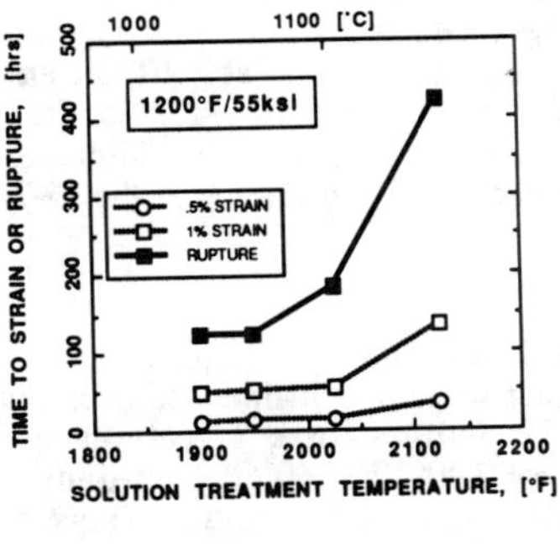

Figure 12: Effect of ST temp. at 150°F/min. on Ti-25-10-3-1 1200°F(650°C)/55ksi(375MPa) creep.

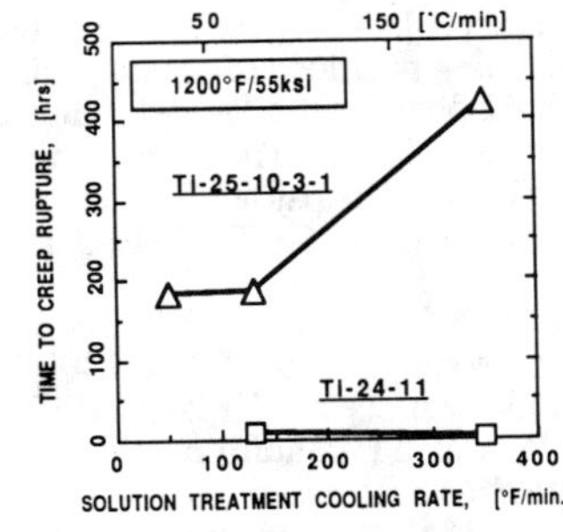

Figure 13: ST cooling rate effect on rupture time for both alloys (Beta HIP'd and Beta ST).

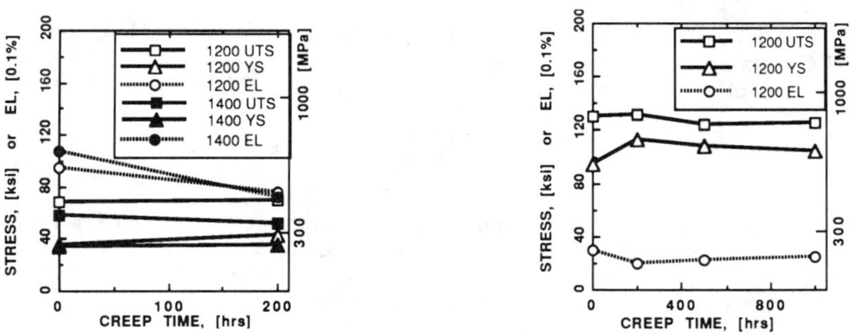

a b

Figure 14: Effect of creep exposure time on post creep tensile of: (a) Ti-24-11 and (b) Ti-25-10-3-1.

DISCUSSION

MICROSTRUCTURE

The **as-cast microstructure** of both alloys (Figs.1a and b) consists of *coarse prior β grains,* the result of the solidification process in the β phase field, which contains a fine α_2 *plate structure*. The Ti-24-11, α_2 plates are on the order of 50 μm long x 5 μm thick in a *colony arrangement*. The Ti-25-10-3-1 plates are finer and on the order of 20 μm long x 1 μm thick in a *basketweave arrangement*. In addition, while the Ti-25-10-3-1 alloy displays a thin continuous grain boundary α_2 phase, the Ti-24-11 alloy displays a much thicker, but non-continuous, grain boundary α_2 phase. The microstructure of the Ti-24-11 is *coarser* than that of the Ti-25-10-3-1 in *all process conditions*: in the as-cast (compare Figs. 1a and b), in the cast+HIP (Figs. 2a and b; and c and d), and in the HIP+ST conditions (Figs. 3a and b). *All HIP parameters used in this program (Table II) were sufficient to close all non-surface connected casting pores.* The Ti-24-11 **cast+HIP microstructure** and the **HIP+ST microstructure** were not significantly different than the as-cast structure (compare Figs.1a, 2a and 3a). However, the HIP+ST condition displayed some low aspect ratio α_2 plates and the grain boundary α_2 phase is less pronounced. The Ti-25-10-3-1 microstructure of all the process conditions was essentially the same as shown in Figs. 1b, 2b and 3b. Fine scale phase separation in α_2, forming α_2+orthorhombic (O) lamellar structure, was observed in the HIP+ST Ti-24-11 and Ti-25-10-3-1 (Figs. 3c and 3d). A smaller quantity of the O phase was also observed after HIP at 2150°F(1175°C)/15ksi(105MPa) for both alloys. In the **HIP parameters study** (Table II), lower temperatures with higher pressures were applied with no significant refinement of the α_2 phase in the Ti-24-11 (Figs. 2a and 2c). However, in the case of the Ti-25-10-3-1, substantial microstructure refinement was achieved during the 1900°F(1040°C)/38ksi(265MPa) HIP when compared to the 2150°F(1175°C) /40ksi(280MPa) HIP (Figs. 2b and d).

Volume Fraction of the Phases

XRD analysis showed that both alloys in as-cast, cast+HIP, and HIP+STA conditions contain varying amounts of three phases α_2, B2, and O characterized earlier by Kaufman [15] and Banerjee [16]. Volume fraction of the O phase in the alloys could not be determined because of overlapping positions of the major O peaks with α_2 peaks. As noted from Table III, the amount of the B2 phase in the Ti-24-11 alloy is consistently less than in the Ti-25-10-3-1 under the same process conditions. Transformation of β phase during cooling was completed to a higher extent in the Ti-24-11 because of an order of magnitude higher rate of transformation in the Ti-24-11 when compared to the Ti-25-10-3-1 alloy [17].

MECHANICAL PROPERTIES

Tensile strength from RT to 1400°F(760°C) of HIP+ST Ti-25-10-3-1 is almost *twice as high as* similarly processed Ti-24-11 (Figs. 4a and b). This higher strength is maintained over a range of HIP and ST conditions (Figs. 5, 6 and 8). No **HIP temperature effect on tensile properties** was observed for both alloys in the range of 1900°F(1040°C) (sub-transus) to 2150°F(1175°C) (supra-transus) when followed by a wide range of ST conditions (Figs. 5 and 6). This indicates that a wide range of HIP conditions can be applied as long as pore closure is accomplished. *All HIP conditions applied in this study (Table II) resulted in full pore closure.* The **effect of ST temperature on tensile properties** was studied only on the Ti-25-10-3-1. While the 1900°F(1040°C) ST produced low RT tensile strength and elongation, the 2150°F(1175°C) ST resulted in the highest values of 157ksi(1080MPa) strength and 0.6% EL. The **ST cooling rate effect** showed a slight increase in strength and elongation with increasing cooling rate for the Ti-24-11(Fig. 8a). The Ti-25-10-3-1 presented maximum tensile strength at 130°F(70°C)/min (Fig. 8b). *The higher strength of the Ti-25-10-3-1 is attributed to the finer α_2 plate structure and to the solid solution strengthening of the V and the Mo.*

Creep testing clearly showed the *Ti-25-10-3-1* to be a *superior alloy*. Processing at 2150°F HIP+2125°F(1165°C) ST at 360°F(200°C)/min, produced very high creep rupture life [422hrs at 1200°F(650°C)/55ksi(375MPa)], which exceeded values typically obtained from IM material (Fig.9). The LMP plot in Fig. 10 shows that the Ti-24-11 is better only at the very low stress of 10ksi(70MPa) and very high temperature of 1400°F(760°C). **HIP temperature effect** was studied only on Ti-25-10-3-1. HIP at 2150°F(1175°C) significantly improved rupture life when compared to 1900°F(1040°C) HIP only for material ST at 2025°F(1105°C) (Fig.11). All other ST conditions showed moderate or no benefit for the supra-transus HIP temperature. The most significant effect found in the creep study was the **effect of ST temperature on creep rupture life** of Ti-25-10-3-1. The *supra-transus ST significantly increased the creep resistance*, with 2150°F(1175°C) ST showing the best results (Fig. 12). The **effect of ST cooling rate** was studied for both alloys. While the Ti-24-11 did not show any cooling rate effect, the *Ti-25-10-3-1* showed *marked improvement in creep strength* only *at 360°F(200°C)/min.* **Post creep** high temperature **tensile** testing after 1200°F(650°C) and 1400°F(760°C) exposure surprisingly showed almost *no degradation* in HT strength and elongation for both alloys (Figs. 14a and b). Since the Ti-25-10-3-1 had finer lamellar structure than the Ti-24-11, the improvement in the creep strength could only be attributed to the solid solution strengthening of the V and the Mo. However, it should be noted that since both alloys were crept at the same stress and temperature ranges, the Ti-24-11 was loaded above its HT yield strength (Fig. 4a), and therefore did not present true creep but rather stress-rupture conditions. This explains why this alloy displayed superior creep strength only at the very low stress level of 10ksi(70MPa) (Fig.10) which provided true creep conditions. However, this also reflects lower creep loading capability for the Ti-24-11when compared to the Ti-25-10-3-1 alloy.

SUMMARY AND CONCLUSIONS

The effect of post-casting processing on microstructure and mechanical properties of two α_2 alloys Ti-24-11 and Ti-25-10-3-1 was investigated. It was found that:

1. Proper casting pore closure was obtained for both alloys under HIP conditions ranging from 2150°F(1175°C)/15ksi(105MPa) to 1700°F(925°C)/38 ksi(265MPa).
2. Cast Ti-25-10-3-1 alloy developed finer α_2 lamellar structure than the Ti-24-11 alloy after *all processing stages,* namely: as-cast, cast+HIP, and ST conditions.
3. Substantial microstructural refinement was observed in the Ti-25-10-3-1 HIP'd at 1900°F(1040°C)/38ksi(265MPa).
4. The tensile strength of Ti-25-10-3-1 alloy was significantly higher than similarly processed Ti-24-11 castings. This is attributed to the finer α_2 plate structure and to the solid solution strengthening of the V and the Mo alloy additions.
5. The creep strength of Ti-25-10-3-1 was much higher than the Ti-24-11 alloy in all process conditions. It was attributed to V and Mo solid solution strengthening.
6. Higher HIP temperature did not significantly affect tensile or creep behavior.
7. Supra-transus ST temperatures markedly improved creep strength of the Ti-25-10-3-1 alloy.
8. In the Ti-25-10-3-1 alloy, the maximum RT tensile strength was achieved at 130°F(70°C)/min ST cooling rate; maximum RT elongation was achieved 360°F/min(200°C/min) and the maximum creep strength at 360°F(200°C)/min.
9. No degradation of tensile properties was observed after creep at 1200(650) and 1400°F(760°C) in both alloys.

ACKNOWLEDGEMENTS

The authors wish to thank EMTEC for the encouragement and the financial support, with special thanks to Dr. Frank Moore, Mr. Percy Gros, and Dr. Larry Midolo. Mrs. Cheryl Seitz is acknowledged for her assistance in manuscript preparation. The Duriron Company and especially Shelby Scott, Scott Reed, and Mike Jones, are gratefully acknowledged for their continuing support of this program. Bill Barice of PCC is acknowledged for his assistance in heat treating. The USAF Wright Laboratory, Materials Directorate at WPAFB (WL/MLLM) is acknowledged for supporting this project.

REFERENCES

1 D. Eylon, F.H. Froes, and R.W. Gardiner, "Developments in Titanium Alloy Casting Technology," Journal of Metals, 35 (2) (Feb. 1983), 35-47.

2 J.R. Newman, D. Eylon, and J.K. Thorne, "Titanium and Titanium Alloys," A chapter in Metals Handbook, Ninth Edition, Volume 15, Castings, (Metals Park, OH: ASM, 1988), 824-835.

3 D. Eylon and F.H. Froes, "Titanium Casting - A Review," Titanium Net Shape Technologies, ed. F.H. Fores and D. Eylon (Warrendale, PA: TMS-AIME, 1984), 155-178.

4 D. Eylon, F.H. Froes, and L. Levin, "Effect of Hot Isostatic Pressing and Heat Treatment on Fatigue Properties of Ti-6Al-4V Castings," Titanium, Science and Technology, Vol. 1, ed. G. Lutjering, U. Zwicker, and W. Bunk (Oberursel, Germany: DGM, 1985), 179-186.

5 D. Eylon, F.H. Froes, and W.J. Barice, "Effects of Treatments on Mechanical Properties of Ti Alloy Castings," Space Age Metals Technology, Vol. 2, ed. F.H. Froes and R.A. Cull (Covina, CA: SAMPE, 1988), 28-36.

6 D. Eylon, F.H. Froes, and W.J. Barice, "Property Improvement of Titanium Alloy Castings by Microstructure Modification," Proceedings of the First International Conference on the Metallurgy and Materials Science of Tungsten, Titanium, Rare Earths and Antimony, Vol. II, ed. F. Chongyue, L. Jianchun, and L. Songren (Beijing, China: International Academic Publishers, 1988), 824-833.

7 D. Eylon, W.J. Barice, R.R. Boyer, L.S. Steele, and F.H. Froes, "Casting of High Strength Beta Titanium Alloys," Sixth World Conference on Titanium, Part II, ed. P. Lacombe, R. Tricot, and G. Beranger (Paris: Les Editions de Physique, 1989), 655-660.

8 D. Eylon, W.J. Barice, and F.H. Froes, "Beta Titanium Alloy Castings for Demanding Fatigue Applications," Technology for Premium Quality Castings, ed. E.M. Dunn and D.R. Durham (Warrendale, PA: TMS, 1988), 29-39.

9 D. Eylon, W.J. Barice, R.R. Boyer, L.S. Steele, and F.H. Froes, "Improved Properties in Beta Titanium Alloy Castings," High Integrity Castings, ed. C.V. White, D. Eylon, and F.H. Froes (Metals Park, OH: ASM International, 1988), 17-22.

10 H.A. Lipsitt, "Titanium Aluminides - An Overview," High Temperature Ordered Intermetallic Alloys, Vol. 39, ed. C.C. Koch, C.T. Liu, and N.S. Stoloff (Pittsburgh, PA: MRS, 1985), 351-364.

11 Y. Nishiyama, T. Miyashita, T. Nakamura, H. Hino, S. Isobe, and T. Noda, "Development of a Titanium Aluminide Turbocharger Rotor," Proceedings of the 1987 Tokyo International Gas Turbine Congress, Vol. III (Tokyo, Japan, 1988), 264-270.

12 D.J. Chronister, S.W. Scott, D. R. Stickle, D. Eylon, and F.H. Froes, "Induction Skull Melting of Titanium and Other Reactive Alloys," Journal of Metals, 38(9) (1986), 51-54.

13 D. Eylon at al. "Property Improvement of Titanium Aluminide Castings" EMTEC/CT-8/TR-90-8 report, (1991), 1-291. (EMTEC, Edison Materials Technology Center, Kettering, Ohio 45420).

14 R.L. Miller, "A Rapid X-Ray Method for the Determination of Retained Austenite," Trans. ASM, 57 (1964), 892.

15 M.J. Kaufman, "Phase Relations in the Ti_3Al+Nb System," AFWAL-TR-88-4113 Report, , (1988).

16 D. Banerjee, A.K. Gogia, T.K. Nandi, and V.A. Joshi, "A New Orthorhombic Phase in Ti_3Al-Nb Alloy," Acta Met., 36(4) (1988), 871-882.

17 W.A. Baeslack III and T.F. Broderick, "Effect of Cooling Rate on the Structure and Hardness of Ti-25-10-3-1 Titanium Aluminide," Scripta Met., 24(2) (1990), 319-324.

EFFECT OF CONTAMINATION BY AIR ON THE TIG WELDING OF COMMERCIALLY PURE TITANIUM SHEET

M F Gittos and M H Scott

TWI
Abington Hall
Abington
Cambridge CB1 6AL
UK

Abstract

Autogenous TIG welds were made in 1.5mm thick commercially pure Ti sheet using Ar shielding gas which was progressively contaminated with O or air. The welds were examined visually for surface colour and by an ESCA technique. The welds were analysed for O and N pick-up. Tensile and hardness tests carried out on selected welds. Surface discoloration was found to be primarily controlled by conditions under the trailing shield rather than the torch nozzle and to be a poor indicator of the level of contamination, except when this was gross. Contamination of the electrode tip appeared to be at least as sensitive.

Introduction

For Ti, reaction with O and N begins to be significant at approximately 500°C (1) and any Ti at or above this temperature must be protected. This includes the weld metal, both when it is molten and after it has solidified, and the parent metal adjacent to the weld. If this protection is inadequate, severe embrittlement in the weld region can occur (2).

For small components, one solution is to use chambers flow or vacuum purged with inert gas (3,4). Gas shielding, additional to that provided by the torch or gun, enables TIG or MIG welding to be used (4-6). The face of the weld is protected by a trailing shield attached to the torch, the back of the weld by gas channelled in a backing bar or by carefully positioned foil etc. (5,6). However, it is impossible to guarantee no contamination and the question then arises as to how contamination may be detected. For test pieces, destructive testing can be used but for production components, the only non-destructive test for judging the success of the shielding of a weld has been its colour. The colour is caused by interference within an oxide film (7) and Table 1 gives the sequence usually quoted for Ti (8), together with a commonly accepted interpretaton of its meaning.

It has always been appreciated that the use of colour as a criterion was not soundly based (4) since it necessarily assumes that information about contamination on the surface (which is what the colour basically

Titanium '92
Science and Technology
Edited by F.H. Froes and I. Caplan
The Minerals, Metals & Materials Society, 1993

gives) is an indication of contamination in the bulk. There has been relatively little information on the relationship between colour and mechanical properties, or pick-up of O and N. Some work on this aspect has been carried out using a welding chamber (9), but open air welding with auxiliary shields is more widely used. A programme of work has therefore been carried out to investigate these aspects in open air welding. This paper deals with surface effects and pick-up of O and N and their influence on tensile properties and hardness.

Table 1 Rule of thumb classification for acceptability of Ti welds, based on colour (8)

Colour of weld zone	Interpretation
Silver	Correct shielding, satisfactory
Light straw	Slight contamination, but acceptable
Dark straw	Slight contamination, but acceptable
Light blue	Heavier contamination, may be acceptable depending on service
Dark blue	Heavy contamination, unlikely to be acceptable
Grey blue	Very heavy contamination, unacceptable
Grey	Very heavy contamination, unacceptable
White (loose deposit)	Very heavy contamination, unacceptable

Method of Investigation

Material

Commercially pure Ti sheet to ASTM B265-76 was selected. A sheet thickness of 1.5mm was chosen as being suitable for autogenous welding, thus avoiding the complication of adding filler metal.

Welding

Mechanised TIG welding was adopted since accurate control of the welding speed was essential to ensure a consistent rate of contamination. All welds were made using a trailing shield, extending 70mm behind the torch. The edges which were to be welded were degreased in acetone immediately before being positioned in a restraining jig over a grooved backing bar which was plugged at each end and filled with Ar flowing at the rate of 2.5 litres/min. The groove was flow purged for about 5 min before welding. Welding grade Ar (99.996% purity) and the contaminating gas (air or an Ar/O_2 mixture) were fed at a pressure of 2 bar through ball flow meters and mixed in a tank containing stainless steel wire wool to act as a diffuser. Flow was maintained for 5 min to allow the mixture to reach equilibrium before welding. Where it was required to feed contaminated gas through the torch but uncontaminated Ar through the trailing shield, the gases were separated by interposing a curtain of asbestos cloth between the nozzle and the trailing shield.

Weld evaluation

The O and N contents of the welds were measured on full thickness specimens taken from the welds after they had been surface ground to remove any oxide films. The measurements were made in a TC30 Leco Nitrogen Oxygen Determinator. The H contents were determined by vacuum hot extraction.

Flat tensile specimens (20mm gauge length, 4.75 x 1.5mm cross section) were machined from the weld parallel to the welding direction and parent metal specimens of similar dimensions were also tested. The 0.2% proof stress was estimated from the load-time trace. All specimens were tested at room temperature without machining the surface, since the weld beads were almost flush with the surfaces.

In order to establish the nature of the surface film, the surfaces of selected welds were subjected to surface analysis by an ESCA technique in which Al X-radiation incident on the specimen produces photoelectrons with energies characteristic of the elements present. Due to the small escape depth of the photoelectrons (<10nm) only the surface of the specimen is examined. The analysis of the energy distribution of the emitted photoelectrons also yields information about the chemical state of the elements and semi-quantitative analysis of the surface can be produced.

Results

In initial trials a common gas supply was used for both the torch nozzle and the trailing shield. Using welding grade Ar, bright silvery welds were produced. Indeed, the surface of the heat-affected zone was also

Table 2 Details of trial welds, made with $Ar-O_2$ shielding gas mixture

Weld number	Shielding gas: Torch	Shielding gas: Trailing	O_2% in gas mixture	Colour	Analysis of weld wt%: O	Analysis of weld wt%: N
1	Ar	Ar	0	1st order silver	0.16	0.002
2	$Ar-O_2$	$Ar-O_2$	0.03	1st order pale straw	0.16	0.002
3	$Ar-O_2$	$Ar-O_2$	0.06	1st order golden-straw	0.17	0.001
4	$Ar-O_2$	$Ar-O_2$	0.09	1st order reddish mauve	0.17	<0.001
5	$Ar-O_2$	$Ar-O_2$	0.12	1st order blue	0.17	0.001
6	$Ar-O_2$	$Ar-O_2$	0.14	1st order blue	0.17	0.001
7	$Ar-O_2$	$Ar-O_2$	0.29	2nd order greenish blue	0.17	0.001
8	$Ar-O_2$	$Ar-O_2$	0.42	high order pink green	0.17	0.002
9	$Ar-O_2$	Ar	0.17	1st order silver	0.20	0.002
Parent metal	-	-	-	-	0.14	0.002

brighter than the relatively dull appearance of the unaffected sheet, presumably because surface contamination is dissolved by the hot metal and not replaced when the quality of the shielding is good. When air was used as the contaminating gas, colours representing a high level of surface contamination were produced at the lowest practical flowmeter settings. In order to achieve the required sensitivity it was necessary to mix Ar with Ar-O_2 mixtures (1-5%O_2). A range of welds was produced with surfaces varying in colour from silver to grey, Table 2. For weld No.9, the trailing shield was fed with Ar and the torch with Ar-0.17%O_2. Such a mixture fed to both trailing shield and torch would have given a blue colour (cf No.6), but weld No.9 was in fact silver.

The technique of feeding contaminated gas to the torch only was used for all subsequent welds and, in this way, it was found possible to use air as the contaminant without excessive surface oxidation of the weld. This was desirable because air is the most likely contaminant in practice. The details of these welds are given in Table 3. They varied in colour from silver to straw-mauve, except for the grossly contaminated welds 18 and 19 which were grey.

Table 3 Details of welds, made with Ar-air shielding gas mixtures

Weld number	% air in torch shielding gas	Colour	Analysis of weld, wt%		H, ppm
			O	N	
10	0	1st order silver	0.15	0.001	17
11	0	1st order silver	0.15	0.003	14
12	0.3	1st order dull silver	0.15	0.021	16
13	0.3	1st order dull silver	0.17	0.029	18
14	0.9	1st order pale straw	0.17	0.045	15
15	0.9	1st order pale straw	0.17	0.076	16
16	1.3	1st order straw-mauve	0.21	0.12	14
17	1.3	1st order golden straw	0.20	0.14	14
18	3.3	High order grey	0.26	0.20	14
19	8.8	High order dark grey	0.26	0.20	13
Parent metal	-	-	0.14	0.002	13

Observations were also made of the appearance of the electrode: as the concentration of contaminant in the shielding gas increased so did the whiskering on the electrode tip, Fig.1. Each electrode was ground before welding and photographed after being used for 330mm of weld (107sec arc time).

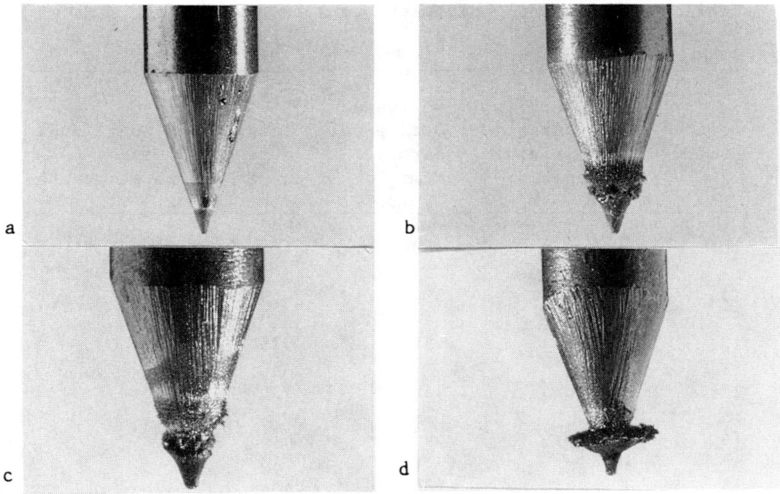

Fig.1 Tungsten electrode tips after 330mm welds: (a) in welding grade argon (b)-(d) contamination with air. (a) Weld 11; (b) Weld 12.03% air; (c) Weld 14,0.9% air; (d) Weld 16,1.3% air.

The wide scan ESCA spectra for welds made in Ar as well as mixtures of Ar with O_2 and air showed intense photoelectron peaks for Ti, O and C. The weld surfaces had relatively low Ti (13 at %) with differing levels of O and carbonaceous contamination: the latter is typical of surfaces analysed without ion-beam cleaning in the instrument. The wide scan spectrum for weld 18 (contaminated with air) was different from the others recorded in that some W was present on the surface, in the form of WO_3: this originated from oxidation of the electrode. Narrow scan spectra showed that Ti was present in a chemically bound as TiO_2. No evidence was found for the formation of any other oxide states. The O peak from weld 18 (air contaminated) is similar to those observed in Ar and Ar-O_2 mixtures. The narrow scan spectra also indicate low binding energy photoelectron peaks of low intensity in the cases of welds 1 and 9. These small peaks correspond to Ti in metallic form. They suggest that the oxide layer was very thin (<2.0nm), since photoelectrons from the underlying metal were able to penetrate the oxide and be detected.

The results of O and N analyses on welds made with O as the contaminant are given in Table 2 and with air as the contaminant in Table 3: the results of H analyses on the latter welds are also given in Table 3.

Vickers hardness tests (at a 10kg load) were made on a traverse across the welds at mid section. There was a marked variability in the hardness of the parent metal, HV10 165-195, and between adjacent readings in the weld zone, although hardness did increase with contamination in the weld metal.

The results of the longitudinal weld metal tests are given in Table 4.

Discussion

A number of different Ti oxides have been reported but oxidation of Ti in air or O usually results in the formation of TiO_2 (1). This was supported by ESCA with TiO_2 being detected on welds made in both the O and air mixtures. Since the technique derives information from within a few atom layers of the surface only, it is possible that layers of different

Table 4 Longitudinal weld metal tensile tests

Weld number	%air in torch shielding gas	Estimated 0.2% proof stress N/mm^2	Tensile strength N/mm^2	Elongation % 20mm gauge length
10	0	363	466	18
11	0	394	459	28
12	0.3	451	525	18
13	0.3	468	538	15
14	0.9	547	616	10
15	0.9	539	588	5
16	1.3	617	657	5
17	1.3	637	653	5
Parent metal	-	339	447	35

oxide states existed beneath TiO_2 on the surface. However, the detection of the Ti metal peak on the welds having thin oxide films suggests that in these cases, at least, no other oxide was present. Therefore, it would appear that differences in coloration are due only to differences in oxide film thickness.

To assist the interpretation of colours, a piece of Ti was heated on one edge in air. The sequence observed is that produced by the interference effect of a gradually thickening thin film on the surface of the metal (7). At its thinnest the film is invisible and on thickening it passes through yellow, reddish mauve, blue and finally again becomes invisible. This second invisible region is the so-called 'silvery-hiatus'. Increasing film thickness yields second order colours with green present at the end of the sequence instead of silver. After about four orders of colours, the oxide becomes opaque grey and then starts to flake off. The higher order colours form a characteristic pink-green mixture which was observed on the welds of highest surface contamination (7 and 8). This sequence is basically that observed on welds 1-8 (Table 5), and is at some variance with the generally accepted sequence given in Table 1. In particular,

1. a reddish mauve colour was observed between straw and blue
2. pale blue appeared after dark blue and
3. pale blue merged into invisibility at the silvery hiatus.

In practice contamination is unlikely to occur within the precise limits required to produce no visible interference colour. However, the phenomenon was observed in setting up conditions and the importance of the effect may lie in the fact that compared to the intense yellow, reddish mauve and blue of lower oxide thicknesses, the pale blue or yellow observed close to the critical thickness appear relatively innocuous. Thus thicknesses of oxide which are close to, as well as in the range of the silvery hiatus, may cause a false judgement of the state of oxidation.

Moreover, Table 5 shows that, except in the case of the weld (No.1) made with no deliberate contamination, there is no direct relationship between the degree of O pick-up (which itself varied very little with $\%O_2$ in the shielding gas) and weld colour. In contrast, Table 6 shows that welds 10-17 varied very little in colour, but significantly in contamination (mainly by N). In addition, weld 9 (Table 5) proved that given contaminated gas through the torch, but not the trailing shield, a silver weld could be obtained. (The ESCA examination confirmed that this silver was not due to the silvery hiatus, but to very thin oxide.)

Thus, a silver coloured weld can indicate three things:

i. that the shielding gas has been satisfactory

ii. that contamination has occurred, but the contaminants have dissolved in the metal at high temperature. (filler metal may also introduce contaminants e.g. from an oxidised tip).

iii. that the oxide thickness is such that it is in the silvery hiatus.

Therefore, it must be concluded that the colour of a weld is an indifferent guide to the degree of contamination, unless high order colours are present. It could also be concluded that even a highly coloured weld is not necessarily contaminated to an unacceptable degree, but this will only be true in certain very limited circumstances and it would be highly injudicious to make any such assumption in judging a weld.

Examination of the electrode tips provides an alternative method of judging contamination. It has the advantage over discoloration in that it occurs when the metal reaches the highest temperatures and is most susceptible to embrittlement. The disadvantage is that the observation must be made at the time of welding and not on subsequent inspection. Although the extent of growth of the contamination on the electrode will be time as well as O and/or N concentration dependent, its formation can be observed during welding and it was sufficiently sensitive to occur with the lowest level of air contamination used, weld 12. It is uncertain whether this behaviour is representative of other situations e.g. when air entrainment occurs.

The analyses on the welds made with air as the contaminant showed that the pick-up of H during welding was very low, the highest figure being only 5ppm on weld 13. The absolute increase in N was higher than that for O, from 0.002 to 0.20% for the former, from 0.14 to 0.26% for the latter in the worst contaminated welds. Taking the parent metal specification figures as a guide, weld 13 represents the maximum permissible contamination.

The hardness traverses showed, as expected, higher hardness in welds than parent metal: the change was more marked as the contamination level increased. Hardness has been suggested as a means of checking contamination, but the results show the difficulty of applying this: the parent metal hardness varied and the scatter in the weld metal readings makes interpretation difficult.

The longitudinal weld metal tensile tests show that both the proof and tensile strengths increased with contamination, while the elongation decreased. Taking the parent metal specification proof stress levels, weld 12 is just outside specification and weld 13 well out. Comparison between the O and N analyses (Table 6) and the corresponding longitudinal weld metal tensile tests (Table 8) shows that N has a greater effect on tensile properties than does O.

Conclusions

1. Discoloration of the weld was a function of surface film thickness and this was primarily controlled by the conditions under the trailing shield rather than the torch nozzle.

2. The surface film was found to consist of TiO_2, whether the weld was made under Ar, Ar contaminated with O or Ar contaminated with air.

3. There was no good correlation between surface colour and level of O and N pick-up, except at gross levels of contamination.

4. The generally accepted colour sequence was found to be wrong. In particular, a light blue colour indicated greater oxide thickness than dark blue.

5. Contamination of the electrode tip occurred at all levels of air contamination used and was readily observable during welding. As a guide to the probable level of contamination of the weld, it appeared more promising than the use of colour.

6. The proof stress and tensile strength of all the weld metals were greater than those of the parent metal, increasing as the levels of O and N increased, the latter being more significant in this respect.

7. The marked scatter in hardness measurements on both parent and weld metal rendered very uncertain the interpretation of hardness as a guide to contamination at those levels where it most needs to be known.

References

1. Kubaschewski O and Hopkins B E, Oxidation of metals and alloys, (Butterworths 1962).

2. Martin D C, "Effect of carbon oxygen and nitrogen on welds in titanium", Weld J. 32 (3) (1953), 139s-154s.

3. Barrett J C , Lane J R and Huber R W, "Effect of atmospheric contaminants on arc welds in titanium," Weld. J. 32 (6), 1953, 283s-291s.

4. Faulkner G C and Voldrich C B, "Interpretive Report on Welding Titanium and Titanium Alloys," Welding Research Council Bulletin, (56), Dec.1959.

5. Scott M H, "Arc Welding Titanium," Weld Inst. Res. Bull., 20 (6) (1979), 163-168.

6. Woolcock A and Ruck R J, "Argon shielding techniques for TIG welding Ti," Met. Con., (May 1980), 219-221.

7. Evans U R, Oxidation and corrosion of metals (Edward Arnold, London, 1960).

8. British Standard Code of Practice. CP3003: Part 9: 1970. Lining of vessels and equipment for chemical processes, Part 9, Titanium.

9. Morgan-Warren E J, "Atmosphere control criteria for welding titanium," (Conf. Proc. Advances in welding processes, Harrogate, May 1978), 343-349.

An Investigation of the Characteristics of Bonding Strength

in Titanium Clad Steel

Takao Kawanami, Shyuichi Shirasuna, Shigenori Shirogane, and Akio Segawa

Kanazawa Institute of Technology
Nonoichimachi, Kanazawa-City, 921 Japan

Abstract

The interfacial strength has been studied for test specimens made by hot-roll, explosive weld, H.I.P.-bonding processes. An investigation of the microstructure, distribution of hardness, inter-diffusion layer and precipitates, and the results of fracture toughness testing shows that explosive welded and hot-rolled clad were much stable to H.I.P. clad in bonding strength. a) From the results of fracture toughness testing, the values of fracture toughness K_c of explosive-welded titanium clad steel are larger than those of the hot-rolled method. b) Crack tip opening displacement, δ, of the explosive-welded titanium clad steel is more easily affected by the working condition of pre-cracks than in the case of the hot-rolled method. c) The existence of intermetallic compounds such as TiC, βTi, FeTi and so forth was confirmed at the bonding inter-face of the hot-rolled titanium clad steel.

Introduction

Titanium clad steel, being used in various fields, is a composite material of high function having both titanium's superior corrosion resistance and heat resistance, and steel's superior strength, economy, processability, heat conductance and weldability. Its high corrosion resistance has recently attracted public attention, and because of this quality it is being considered for applications in such areas as the petrochemical industry, atomic power generation, seawater desalination plants and offshore structures. Its great demand in such a field connected with seawater will still be expected.[3] The conventional and most important evaluations for the interface bonding strength in the titanium clad steel have been by tests for tensile strength, bending and shear strength. The ram tensile strength test may reportedly be added to these. Among them, the results of the shear strength test may vary relatively greatly depending on the production accuracy of test samples and thinly-made materials. Since such clad materials are manufactured by bonding different types of metals,there is a risk of defects or cracks at a bonding interface. A fracture would be subject to the state of stress distribution in the vicinity of the tips of cracks. Therefore, an evaluation method for the bonding strength at the bonding interface using stress intensity factors would be valuable. In this report, an investigation of microstructure observation with optical microscopy, hardness distribution in the vicinity of the bonding interface, and inter-diffusion with EPMA liner analysis,is described using test specimens of the titanium clad steel processed by respective hot rolling, explosive welding and HIP (Hot Isostatic Press)

Titanium '92
Science and Technology
Edited by F.H. Froes and I. Caplan
The Minerals, Metals & Materials Society, 1993

bonding methods. Further, the strength characteristics of said specimens have been evaluated through a K_{IC} fracture toughness test.

Experimental Methods

Test Specimens

Chemical compositions of ultralow-carbon steels, the test materials prepared by the respective hot rolling, the explosive welding and HIP bonding methods, are shown in Table I, chemical compositions and mechanical properties of titanium in Table II, and the manufacturing methods and their test material specifications in Fig. I. The hardness distribution, the microstructure observation and the inter-diffusion were investigated using specimens cut out in proper size from steels with carbon contents of 0.005%C, 0.01%C, and 0.03%C.

Table I Chemical Composition of Base Materials.[3)]

Materials		Chemical Composition (Wt.%)							
	Aim.C	C	Si	Mn	P	S	Al	O	N
Ultra-low Carbon Steel	0.005	0.0045	0.01	0.01	0.001	0.0017	0.001	0.0057	0.0020
	0.010	0.012	0.01	0.01	0.001	0.0018	0.001	0.0066	0.0020
	0.030	0.030	0.01	0.01	0.002	0.0013	0.001	0.0060	0.0029

Table II Chemical Composition and Mechanical Properties of Clad Materials.

Materials	Chemical Composition $\times 10^{-3}$ (wt.%)					Mechanical Propaties			
Titanium JIS H 4600	H	O	N	Fe	Ti	TS kgf/mm^2	YS	El %	HV
TP28H t= 3.2mm	5	47	40	29	Bal.	34	19	46	126
TP35H t=20.0mm	8	95	40	46	Bal.	39	31	32	146
TP35H t= 4.0mm	8	103	50	69	Bal.	43	32	40	154

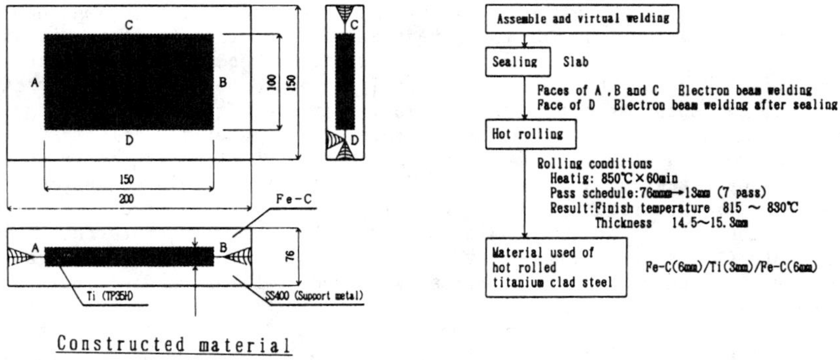

Hot Rolling Method[1)]

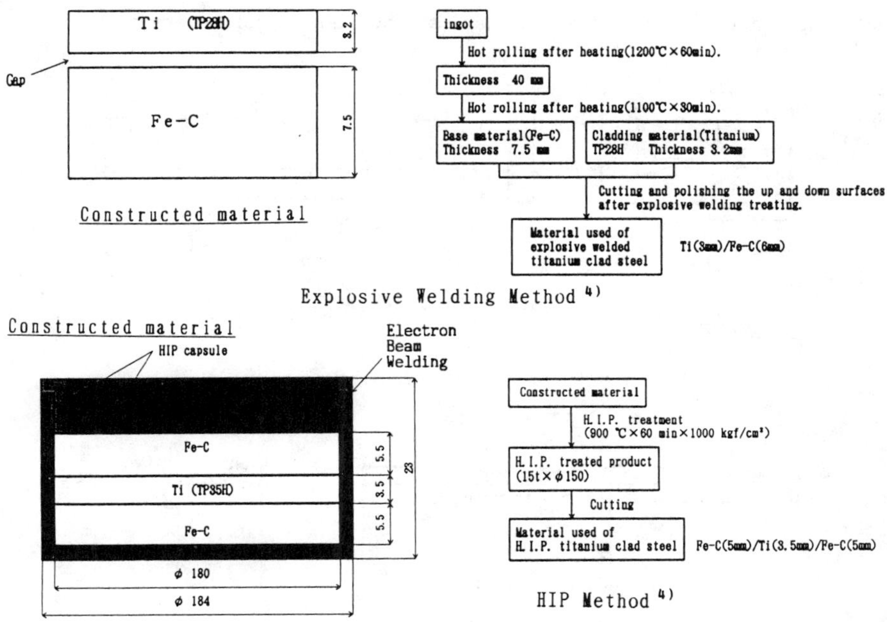

Fig. I Manufacturing process and bonding conditions of material used of clad steel.

Hardness Distribution Test

The hardness distribution test was carried out according to JIS Z 2251 with using a micro-Vickers hardness testing machine which is fit for JIS B 7725.

Microstructure Observation Test

Specimens were first hand-polished with emery papers of #400 through #1200 and buffed with chromium (Ⅲ) oxide for a mirror finish, and then corroded. Table Ⅲ shows corroding conditions for the steel and titanium.

Table Ⅲ Etching Condition

Material under test	Etching solution	Etching time
Fe Side	2% Nital	20 Sec
Ti Side	1% HF Water Solution	140 Sec

Inter-diffusion State Test

The state of inter-diffusion was evaluated by EPMA (Electron Probe Micro Analyzer), the analytic conditions of which are shown in Table Ⅳ.

Table Ⅳ EPMA line scanning condition.

Analysis element	Ti, Fe
Acceration voltage	25 kV
Feed velocity (Test piece)	100 μm/min

Fracture Toughness Test

In the fracture toughness test, specimens for CT(Compact Type) and three point bending were applied. The size and shape of each test specimen are as shown in Fig. II.

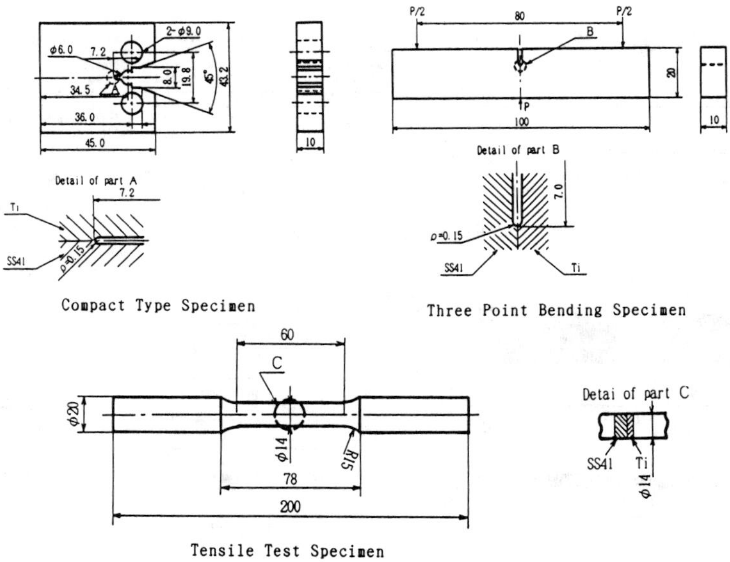

Fig. II The size and shape of each test specimen.[5]

The fracture toughness test was carried out under conditions in accordance with both ASTM E 399 and BS 5762 which are the standards for test of COD(Crack Opening Displacement). As it proved difficult to introduce pre-cracks to the specimens through repeated loads, a scratch of 0.2mm was made along the interface with a handsaw. A schematic diagram of this is shown in Fig. III. The amount of COD in the fracture toughness test was determined with a clip gauge for the test specimen, which was set up with a knife-edge worked along the line of the load, while with a separate type knife-edge for the three point bending test specimen, as shown in Fig. IV, since knife-edge portion was not prepared in the latter case. Further, the specimen's fracture surface after carrying out the fracture toughness test was investigated by X-ray diffractometer to identify materials existing on the surface of the specimen.

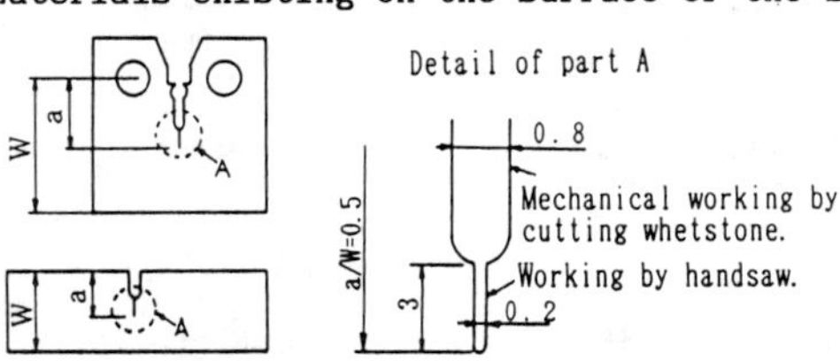

Fig. III Detail of pre-crackiing method.

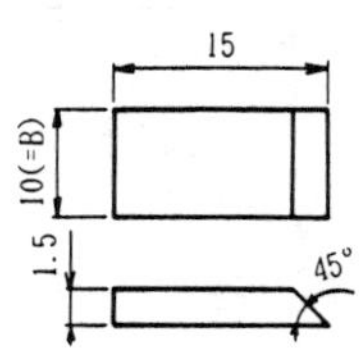

Fig. IV Attachable knife-edge.

Results and Considerations

Microstructure Observations

Photos of the specimen's microstructure in the vicinity of the bonding interface are shown in Fig. V.

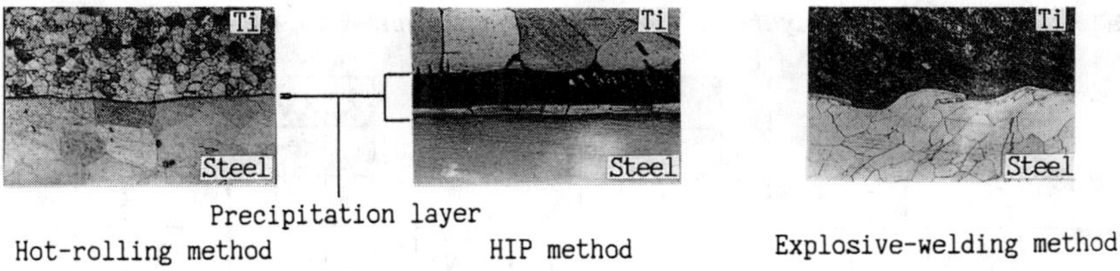

Fig. V Microstructure of Ti/Fe bonding interface. (×100)[5)]

From the figure, a precipitation layer is observation at the bonding interface in each specimen processed both by the hot rolling method and HIP method, but not in the specimen from the explosive welding method which, on the other hand, presents a flexuous appearance with its actual bonding area being large compared with the other two methods. In the case of the HIP method, it was observed that a thicker precipitation layer was generated at the bonding interface.

Hardness Distribution

Fig. VI shows the investigation results of hardness distribution in the area adjacent to the bonding interface in specimens processed by each method.

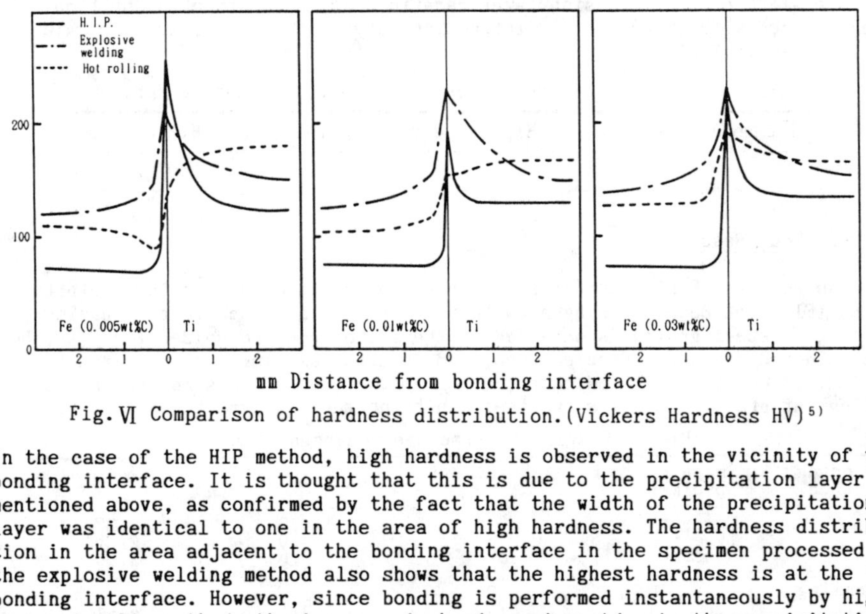

Fig. VI Comparison of hardness distribution. (Vickers Hardness HV)[5)]

In the case of the HIP method, high hardness is observed in the vicinity of the bonding interface. It is thought that this is due to the precipitation layer mentioned above, as confirmed by the fact that the width of the precipitation layer was identical to one in the area of high hardness. The hardness distribution in the area adjacent to the bonding interface in the specimen processed by the explosive welding method also shows that the highest hardness is at the bonding interface. However, since bonding is performed instantaneously by high energy in this method, the increase in hardness is notdue to the precipitation layer generated by inter-diffusion in the area adjacent to the bonding interface, but is thought to be mainly due to work-hardening by plastic strain generated during bonding in the area adjacent to the bonding interface. On the other hand, in the case of the hot rolling method, such a significant increase in hardness at the bonding interface was not observed, unlike the other two method of explosive welding and HIP.

Inter-diffusion

The results of EPMA linear analysis are shown in Fig. VII. From the figure, the width of the diffusion layer in the specimen processed by HIP method is widest.

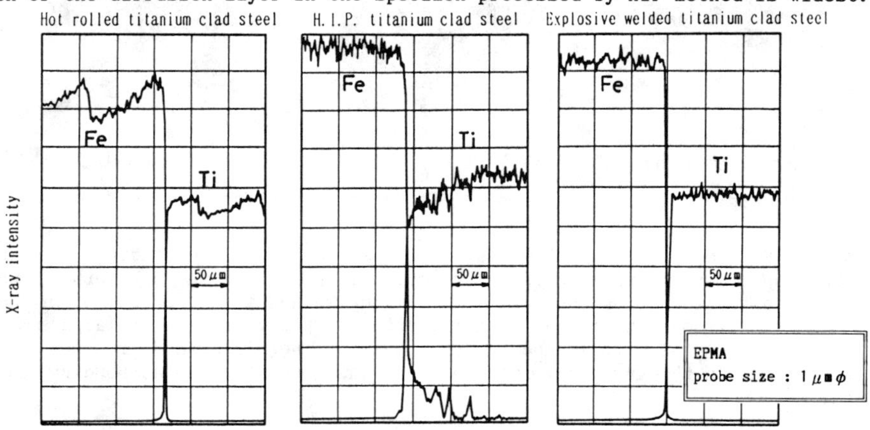

Fig. VII Results of EPMA line-scanning across the bonding interface.[5)]

A very narrow diffusion layer is also observed in the specimen processed by the hot rolling method, but is not in the case of the explosive welding method. With the width of the diffusion layer redefined as the width at a position being exposed with 5% Fe and Ti diffraction intensities, the resulting values are as shown in Table V.

Table V Width of diffusion layer with each bonding method.

Hot rolling method	Explosive welding method	HIP method
5 μm	2 μm	50 μm

Fracture Toughness

From the results of microstructure observation in the area from the vicinity of the EB (Electron Beam) welding area to the bonding interface in the specimen, a refining of steel grains is observed in the vicinity of EB welded zone, which is considered the heat-affected area, and it measures 1mm in width. From the results, it can be concluded that the heat-affected area does not extend to the bonding interface. Table VI shows the results of the fracture toughness test.

Table VI Results of fracture toughness test.

Bonding Method	Shape of Specimen	Break Load, kgf	K_C MPa · $m^{1/2}$	$\delta_C \times 10^{-6}$, m
Hot Rolling Method	Compact Type Specimen	291	*14.0	2.37
		273	*14.0	2.36
		267	*13.6	2.56
		215	*10.6	1.11
	Bend Specimen	224	*18.5	4.47
		235	*17.4	4.88
		172	*12.3	2.49
		133	*11.0	2.11
Explosive Welding Method	Compact Type Specimen	813	46.4	137
		925	45.8	27.7
		850	43.2	23.6
		660	31.1	12.6
	Bend Specimen	476	46.2	102
		544	40.8	67.8
		460	39.4	293
		491	34.6	78.9

* K_{IC}

Those values markedwith * are the plane stress fracture toughness (K_{IC}) as these met the qualifications of plane stress conditions. Since the evaluation of bonding interface strength by the stress intensity factor K is not considered pertinent to such material as explosive-welded specimen in which an area of plasticity extends over broad area, although on the other hand it may be an apposite evaluation for the case of the hot-rolled material in which the plasticity area generated at the crack tips in specimen is fairly narrow, a strength evaluation with a crack tip opening displacement of δ which is not restricted by the expansion of the area of plasticity, is shown together in Table VI.

From the results, relatively high δ values are observed in the case of the explosive welding method. These are considered, from our observation on the fracture surface of the specimen on its titanium side, to be due to steel more thickly adhering to the vicinity of pre-crack's tips than in other resulting in cracks on the steel side in the clad material in the early stages of fracture. In addition, δ values of the explosive-welded specimen vary widelycompared with the hot-rolled specimen. In the latter case, there is high probability, even with a variability of the pre-crack tips, of the fact[2] that the cracks originate at and spread from the precipitation layer which is considered as a fracture inducing factor. In the explosive-welded specimen, onthe other hand, there is scarcely any precipitation layer and, in addition, a processing condition at the tips of pre-cracks exerts influence more easily, resulting in δ values that vary greatly. Fig. VIII shows the results of X-ray diffraction analysis on the fracture surface obtained after the fracture toughness test. On both fracture surfaces on the titanium side and the steel side of the hot-rolled specimen, a small amount of TiC, βTi and FeTi are observed. This indicates that the fracture originated at the bonding interface. On the other hand, there is more steel on the titanium side of the fracture surface of the explosive-welded specimen, which indicates that the fracture likely did not start from the bonding interface.

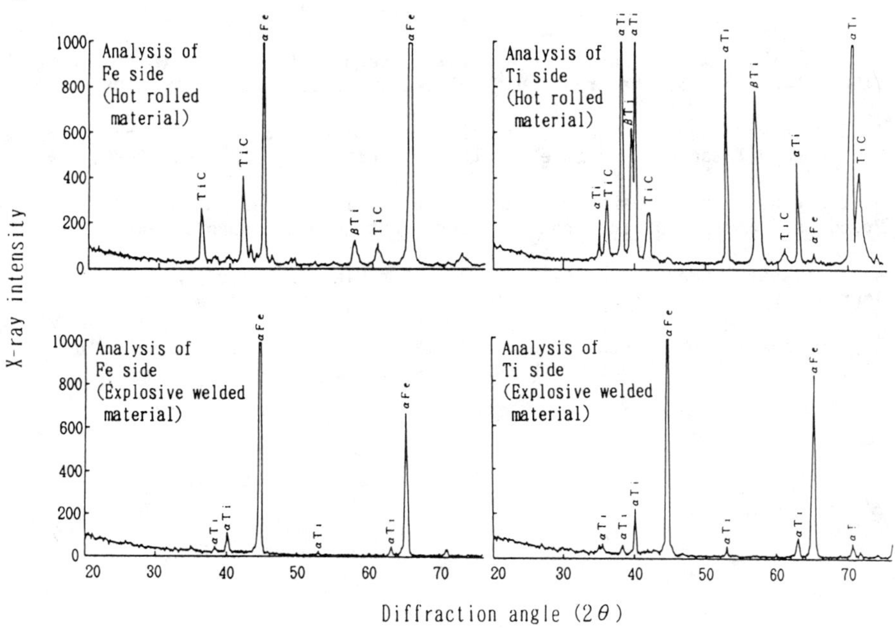

Fig. VIII Results of X-ray diffraction on fracture appearance.[5]

Conclusions

From the above results, the following have been determined.

(1) A broadest diffusion layer is obtained by the HIP method, followed by the hot rolling method, and then the explosive welding method.

(2) The straightest bonding interface is obtained by the HIP method, while the bonding interface in the explosive-welded specimen is flexuous.

(3) A heat-affected area in EB welding does not extend to the bonding interface, resulting in a very narrow width.

(4) From the results of fracture toughness test, the values of fracture toughness K_{IC} of explosive-welded titanium clad steel are larger than those of the hot-rolled method.

(5) Crack tips opening displacement, δ, of the explosive-welded titanium clad steel is more easily affected by the working condition of pre-cracks than in the case of the hot-rolled method.

(6) In the case of hot-rolled titanium clad steel, fractures originate at and spread out from the bonding interface.

(7) In the case of the explosive-welded titanium clad steel, appreciable Fe is detected on the fracture surface on the titanium side. It is presumed from this and the related conclusions in (2) and (3) that fractures in this material do not simply start from its bonding interface.

(8) The existence of the intermetallic compounds such as TiC, βTi, FeTi and so forth, was confirmed at the bonding interface of the hot-rolled titanium clad steel.

Ackowledgments

The authors are very grateful to Dr. Fumio Kurosawa of Nippon Steel Corporation for his support in the fracture surface analysis.

References

1. S.Yoshiwara, T.Kawanami, and K.Suzuki, "Bonding Condition of Titanium Clad Steels," Journal of The Iron and Steel Institute of Japan ,72-6(1986),137.

2. S.Yoshiwara and T.Kawanami, "Bonding Conditions and Material Properties of Roll-Bonded Titanium-Clad steel," 6th World conf. on Titanium, (1988)

3. T.kawanami and S.Yoshiwara, "Manufacturing Techniques for Rolled Clad Products," Journal of The Iron and Steel Institute of Japan ,74-4(1988)

4. T.Kawanami, "Manufacturing Techniques for Clad Products" Journal of The Japan Society For Technology of Plasticity ,32-760(1991)

5. T.Kawanami, S.Shirasuna, and S.Shirogane, "Study on Bonding Interface of Titanium Clad Steel," 42th JSTP Autumn conf. (1991)

EFFECT OF DIAMOND WHEEL GRAIN SIZE ON THE GRINDABILITY OF Ti-6Al-4V

Nobuo Kumagai*, Kiyoshi Kamei*, and Yukio Tanaka**

* Department of Material Science and Engineering
** Department of Mechanical Engineering
Faculty of Engineering, Kansai University, Suita, 564, Japan.

Abstract

In order to further improve the effect of the grain size of a diamond wheel upon the grindability of Ti-6Al-4V alloy, surface plunge grinding experiments, while injecting high-pressure grinding fluid at grinding points, have been carried out. The main results are as follows: (1) As the grain size decreases, the grinding ratio decreases, and specific grinding force increases. There exists an optimum grain size at which the finished surface roughness is minimized. (2) Higher injecting pressure of the grinding fluid results in improved grindability.

1 Introduction

Titanium alloys are difficult to grind, and a method of grinding these materials efficiently and to a high quality is desired. The authors have conducted experimental studies to improve on the grinding method of titanium alloys. The wheel low-peripheral speed grinding, conventionally known as the low stress grinding method(1), has the drawbacks of large grinding force and poor grinding efficiency. For this reason, a method to grind the alloy at a wheel peripheral speed of 32 m/s while jet-injecting high-pressure grinding fluid at grinding points has been proposed, and wheels of various grains have been used to perform grinding experiments in order to determine the grinding characteristics. It was found that diamond wheels provide the best grinding characteristics for titanium alloys, followed by the cubic boron nitride(CBN) wheels(2). However, diamond wheel grains come in various types affecting the grinding characteristics. Using resinoid bonded diamond wheels SDC80R100B having a friability rating of 25, 40 and 80 %, Ti-6Al-4V alloy was ground in experiments under a pressure of 2 MPa. These experiments revealed that a high grade wheel using a diamond wheel grain with relatively great toughness (friability rating of 40 %) is the best. Using this wheel, the titanium alloy was ground while injecting fluid at 2 MPa, making possible a more efficient grinding with the consequent grinding characteristics. It was also revealed that with this grinding method, the performance of the grinding fluid is heavily dependent on the injection pressure (jet pressure), and that the higher the pressure, the better the grinding characteristics(3).

Since the grain size and type of bonding material of a diamond wheel affect grinding characteristics as well as the roughness of the finished ground

Titanium '92
Science and Technology
Edited by F.H. Froes and I. Caplan
The Minerals, Metals & Materials Society, 1993

surface and surface integrity, they are significant factors in selecting wheels. The present report discusses the results of the experiments carried out to determine the influence of the grain size and bonding material of a diamond wheel upon the grindability of Ti-6Al-4V alloy.

2. Experimental equipment and methods

Grinding experimental conditions and a workpiece are indicated in Table I. The same rolled plate of Ti-6Al-4V alloy as used in the previous report(3) was employed. After the workpiece was machined into a geometry as shown in Figure 1, it was subjected to a vacuum annealing of 7.2 ks at 973 K to remove the residual stresses which occurred during previous machining. When mounting the work on a grinding experiment device (Figure 1), the collar of the work was pressed and fixed so that the stress occurring during the fixation would not affect the ground surface. The same grinding experiment equipment was employed as had been used in the previous report, but a crystal piezo-electric dynamometer was used for the measurement of grinding force. The test wheel and the wheel dressing method are shown in Table II. As wheels, resinoid bonded diamond wheel (called "SDC wheel") and vitrified bonded diamond wheel (called "SD wheel") of different grain sizes were employed. Both wheels use a bond with great retention power of wheel grains. In some experiments, GC80K8V wheel was used for comparison. Grinding experiments were conducted with the wet surface plunge grinding method. Grinding characteristics were evaluated in terms of stock removal, wheel wear, grinding force, ground surface roughness, and grinding layer of the finished surface and residual stress. In order to determine wheel wear mechanisms, the measurement of the profile of the wheel surface and the microscopic observation of grain cutting edge (by the replica method) were performed. Residual stress was measured with an X-ray stress measurement apparatus (MSF-2M type, Rigaku corporation). The work which had been subjected to the stress measurement was cut at an angle of 10°

Table I Workpiece and experimental conditions

Workpiece		Ti-6Al-4V(KS130AV, Kobe Steel Co. Ltd.), σ_B=932 MPa, HV330, Annealed (973K, 7.2 ks), L60xW6 mm,
Type of operation		Surface plunge grinding, up cut
Peripheral wheel speed	V	32 m/s
Table speed (Work speed)	v	0.1 (0.123 ~ 0.045 *) m/s
Wheel depth of cut	t	5 ~ 150 (20 ~ 150 *) μm/pass
Theoretical material removal rate, Z_{th}**		0.5 ~ 6.68 (2.45 ~ 6.68 *) $mm^3/mm{\cdot}s$
Grinding fluid		JIS K 2241, W1-1 (Yushiroken GC) x10
	Jet pressure P_j	0.1 ~ 2.0 MPa

* Theoretical grain depth of cut g_{th}= 0.96 μm (Constant)

**

$$Z_{th}=\frac{W_{th}}{B' \cdot T}=\frac{B \cdot t \cdot v}{B'}, \qquad (mm^3/mm{\cdot}s)$$

W_{th}= Volume of metal removed (theoretical value), (mm^3).
T= Time, (s).
B= Width of workpiece, (mm).
B'= Width of wheel, (mm).
= B (Plunge grinding).
t = Setting wheel depth of cut, (mm).
v = Table speed (workpiece speed), (mm/s).

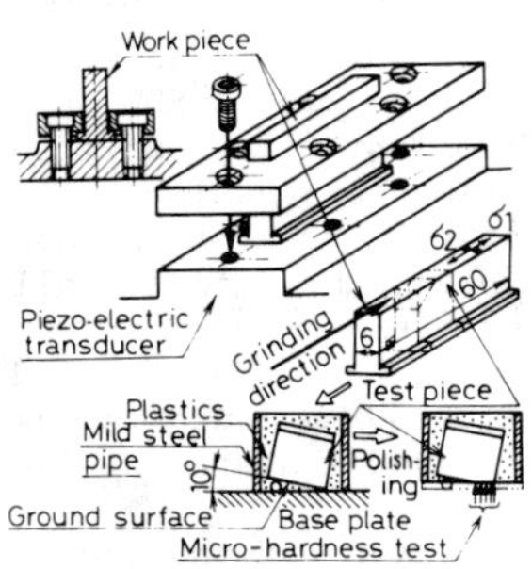

Figure 1 – Shape of workpiece, and measuring methods of grinding residual stress and grinding layer.

Table II Grinding wheels and dressing

Grinding wheel	Bond	Friability of grain (%)	Dressing for grinding wheel	Dressing wheel (ϕ80x25xϕ12.5mm)	Dressing stone (□20mm stick)
SDC80R100B, (A)	MT	40			
SDC80R100B, (B)	MT	80		C60K7V	WA220G7V
SDC80R100B, (C)	MT	25			
SDC140R100B, (H)	MT	about 60	BDD		WA400G7V
SDC230R100B, (H)	MT	about 60			WA600G7V
SDC325R100B, (I)	MT	about 30		C80K7V	
SD100P75V, (E)	BT	about 30			———
SD170S75V, (E)	MT	about 30			WA220G7V
CBN 80R100B, (G)	MT	about 40	BDD	C60K7V	WA220G7V
GC80K8V *	BT		SDD	———	———

Grinding wheel size and edge type: ϕ175x10(*13)x31.75mm. 1A1.
Note: MT: Matrix type, BT: Bridge type,
BDD: Brake dressing device, SDD: Single diamond dresser,

and the resulting section was polished carefully. Thereafter, the hardness of the section was determined with a Micro Vickers hardness tester.

3. Results and considerations

3.1 Grinding wheel wear

The wear of grinding wheel is the most important factor, since it affects the other grinding characteristics. Figures 2(a) and (b) show the "cutting point spacing" and "Abott's bearing curve". These figures were plotted by measuring the wheel surface profile after dressing and in steady-state grinding conditions. The horizontal axis of Figure 2(a) and the longitudinal axis of Figure 2(b) indicate the depth from the most protruding tip of the cutting edge. Grinding wears a wheel, so the tip position of the cutting edge recedes slightly. In these figures, the position is plotted with regard to the tip position of the cutting edge. These figures show that for the already dressed wheel, the cutting point spacing and the pores (chip pockets) on the wheel surface decrease with decreasing grain size (increasing mesh number). Under the steady state grinding conditions, for the SDC140R wheel, the cutting point spacing and the pores on the wheel surface decrease more than after the wheel has been dressed. On the other hand, for the SDC325R wheel, the cutting point spacing and the pores have increased. This indicates that as the grain size decreases, there is an increase in the fracture and breaking down of the grains, or that there occurs a deformation of the wheel surface configuration.

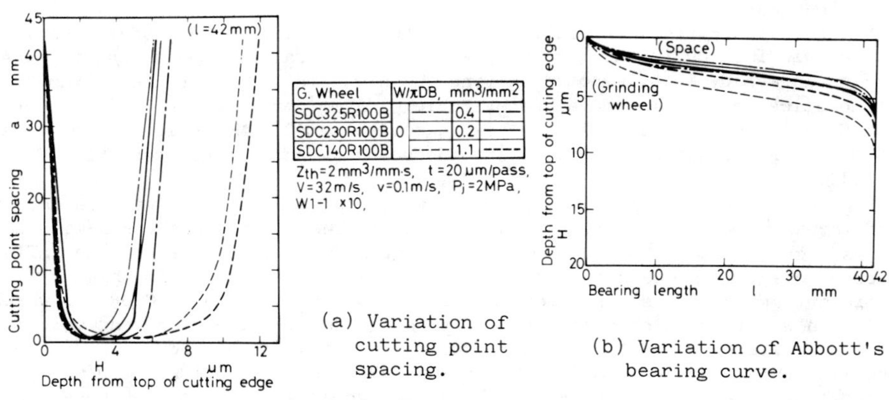

(a) Variation of cutting point spacing.

(b) Variation of Abbott's bearing curve.

Figure 2 — The wheel surface profiles after dressing and in steady state grinding conditions.

No loading of a diamond wheel occurred when the wheel was ground at a grinding fluid jet pressure of Pj=2 MPa. Figure 3 shows the effect of grain size upon grinding ratio G [(volume of metal removed)/(volume of wheel worn), called "G ratio"] . In all cases, immediately after the initiation of grinding, there was greater wear on the wheel resulting in a lower G ratio. With increasing stock removal, however, the G ratio increases gradually to bring the wheel into steady-state grinding conditions, causing the wheel to exhibit its original grinding performance. As is evident from figure 3, as the grain size decreases, the G ratio decreases markedly. This is because as the grain size decreases, the total quantity of heat generated during the grinding operation is decreased only slightly even with decreased grain size. Also, as the grain size decreases, the pores on the wheel surface become shallower [Figure 2(b)], and it becomes harder for the grinding fluid to enter the grinding point. Matrix-type wheels are particularly susceptible to the effects of heat. As

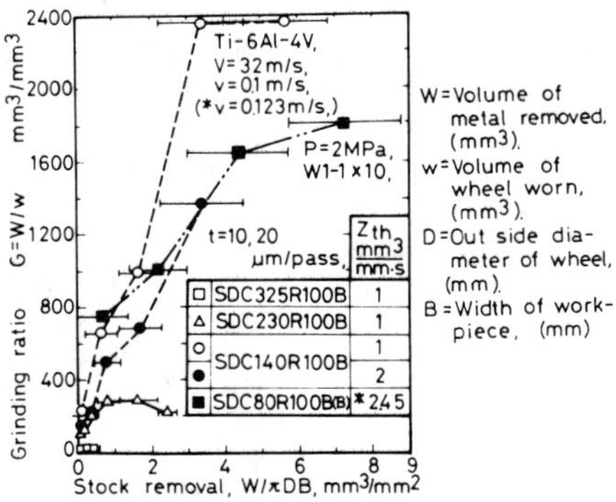

Figure 3 - Variation of grinding ratio depending upon stock removal for the different grinding wheels.

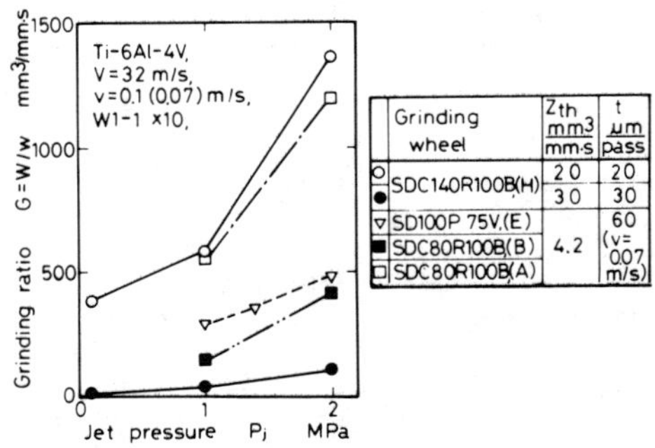

Figure 4 - Variation of grinding ratio depending upon jet pressure and theoretical material removal rate for the different grinding wheels.

has been discussed above, for fine grain wheels, grinding heat cannot be cooled sufficiently, so the wear on the wheel increases. The increased wear is also responsible for the fact that the bond on these types of wheel has less power to retain grains and that they are sensible to the effect of grinding heat.

Figure 4 indicates the relationships between G ratio and Pj under steady-state grinding conditions. For both wheels, the G ratio increases vastly when Pj is increased. This is noted in fine grain wheels . For instance, when Zth (theoretical material removal rate, Table I)=2 $mm^3/mm \cdot s$ for an SDC140R wheel, the G ratio is about 3.5 times as large as that at 0.1 MPa which can be obtained when Pj is set at 2 MPa. As Zth increases, G ratio decreases (SDC140R wheel).

Figure 5 shows the relationships between the peak G ratio of various wheels and actual material removal rate Z′(Pj=2 MPa). The G ratio decreases witn decreased grain size, and the greater the figure for Z′, the greater the influence becomes. This means that rough grain wheels are suitable for highly efficient grinding. As discussed in the previous report, the SDC80R100B wheel (A) is a wheel in which comparatively tough grains are employed, and which offers greater G ratio than a wheel the grains of which are not tough (B) or one with very tough grains (C). The SD100P75V wheel is a bridge-type wheel using tough grains and a vitrified bond with great grain retention power (high wettability). This wheel is found to show performance close to that of the SDC80R100B wheel (C). In general, the bridge-type vitrified bonded wheel has features that allow comparatively easy dressing. Accordingly it is of use for operations at factory sites. In the case of a CBN wheel, as with the diamond wheel, its comparatively tough grains make it suited to the grinding of titanium alloy[2] . The Gratio of the CBN80R100B wheel using fine crystal grains is markedly inferior to that of the SDC80R100B wheel.

3.2 Grinding force

Figure 6 indicates the relationships between the grinding force and wheel grains under steady-state grinding conditions. With decreasing grain sizes, grinding force increases slightly, and Ft/Fn decreases. The SDC80R100B wheel offers approx. one-half of the percentage of grains than the GC80K8V wheel. The Ft/B and ks of the SDC wheel are smaller than those of the GC wheel, while Fn/B is greater. For GC wheels, grain fractures and fallings frequently occur, and consequently the cutting edge shape is sharp and the Fn/B is small,

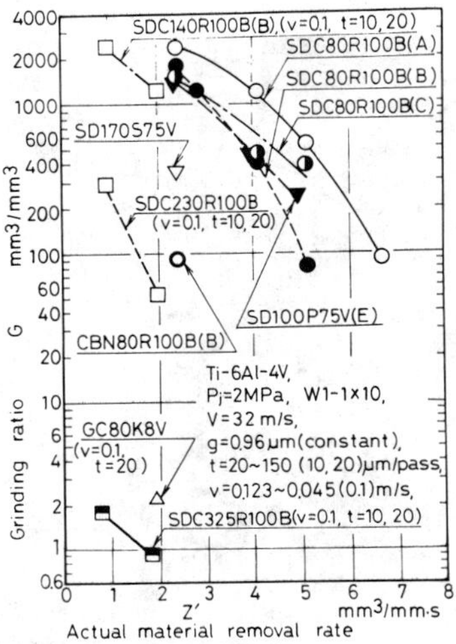

Figure 5 - Effects of various wheels and actual material removal rate upon grinding ratio.

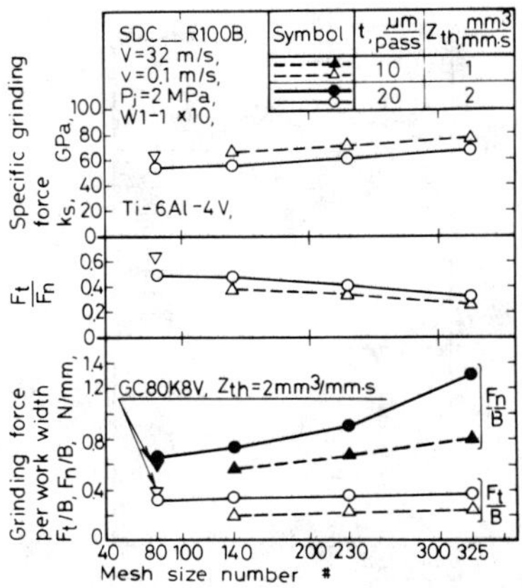

Figure 6 - Effect of grain size upon grinding forces.

Ft: Tangential force
Fn: Normal force

Ft/Fn being a great value. As Zth increases, Ft/B and Ft/Fn increase, and ks decreases. This is qualitatively the same tendency as for the conventional grinding of steel.

3.3 Roughness of finished surface

Figure 7 indicates the changes with stock removal in the roughness of the finished surface when the grinding is done with SDC wheels of different grain sizes. Roughness is controlled by the wear of the wheel. As has been described above, the SDC325R wheel shows heavy wear, and becomes rough due to the irregular cutting edges. For this reason, this type of wheel should be used under grinding conditions where Zth is much lower. The SDC230R wheel is not so rough immediately after the initiation of grinding, but rapidly becomes rougher with the slight increase in stock removal. Accordingly, the wheel must be used at shorter intervals of dressing. In the case of the SDC140R wheel, roughness is small and stable. This means that when titanium alloy is ground, since the wear of the wheel increases tremendously if the grain size is decreased, roughness is not reduced simply by using a wheel with small grain size. To make the finished surface less rough without any sacrifice of grinding efficiency, it would be advisable to use a wheel with a rough slightly grain size and perform dressing so that the cutting point spacing is made as small as possible.

3.4 Finished surface integrity

The finished surface integrity is a very important consideration. In these experiments, when an SDC wheel was used for grinding under an injection fluid of Pj=2 MPa, there occurred no grinding burn mark or grinding crack. Figure 8 depicts an example of micro-structures of the section (inclined section) of the work ground under Zth=2 $mm^3/mm \cdot s$, and Pj=2 MPa, using an SDC140R wheel. The structures have become slightly finer in the vicinity of the grinding surface, but there is no smeared metal on the ground surface. Figure 9 illustrates an example of hardness distribution on the final griund surface. Work hardness can be found near the ground surface due to plastic deformation, but

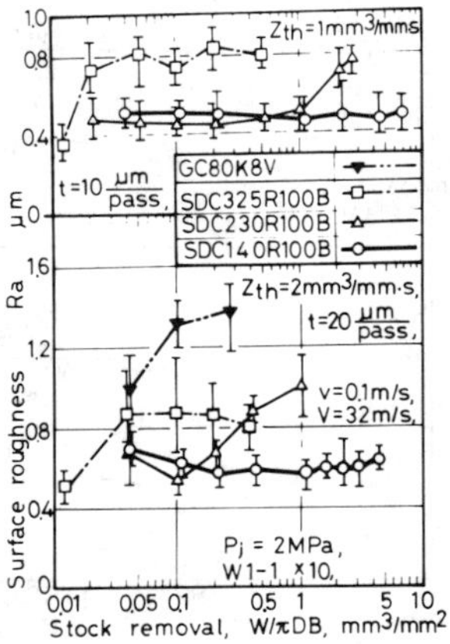

Figure 7 - Variation of ground surface roughness depending upon stock removal for the different grinding wheels.

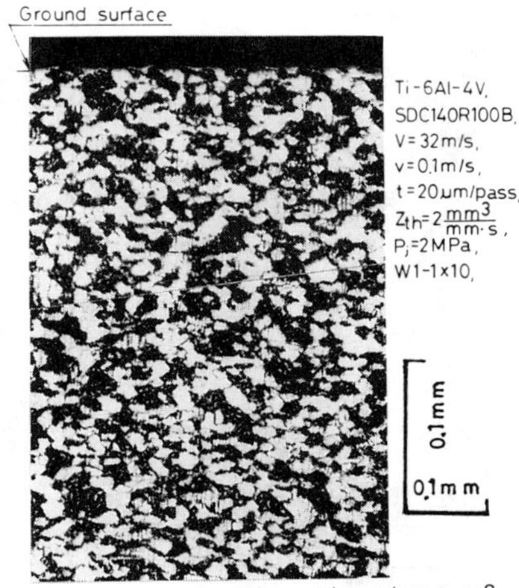

Figure 8 - Microstructure of workpiece(inclined section).

the ground surface itself is the hardest point. The depth of grinding layer and the hardness of grinding surface increase with increased Zth. It was also found that there are fewer grinding layers for the SDC140R wheel than for the GC80K wheel.

E.K. Henriksen[4] states that residual grinding stress consists of (1) tensile stress due to grinding heat, (2) tensile stress due to the cutting action of grains and (3) compressive stress due to the burnishing action of grains, and that a stress synthesizing these three stresses remains as residual stress inside the material.

Figures 10 and 11 show examples of distribution of residual stress on the finished surface ground under Zth=2 $mm^3/mm \cdot s$. σ_1 is a stress in the grinding direction, and σ_2 is a stress in a direction perpendicular to the grinding direction. Figure 10 illustrates a comparison of residual stresses under Pj= 0.1 MPa. For the finished surface ground with the SDC wheel, the stress on the surface becomes a compression stress, and the peak stress (C) on the compression side appears slightly inside the surface. Of these compressive stresses in the vicinity of the surface, σ_2 is larger than σ_1. By contrast, for the finished surface ground with the GC80K8V wheel, the surface stress and the peak stress (T) on the tension side become a tensile stress, and the

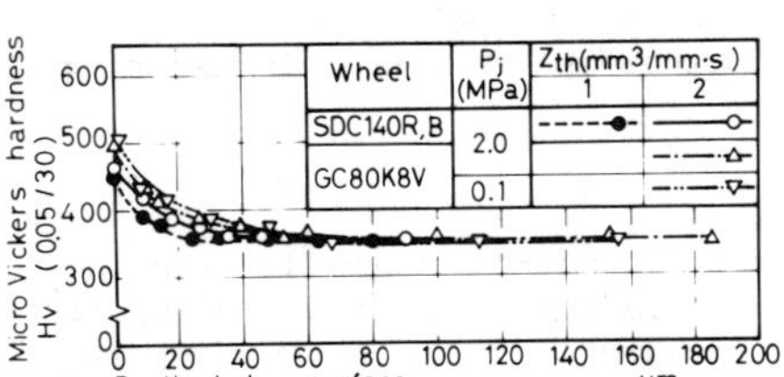

Figure 9 — The hardness distribution of a section of workpiece.

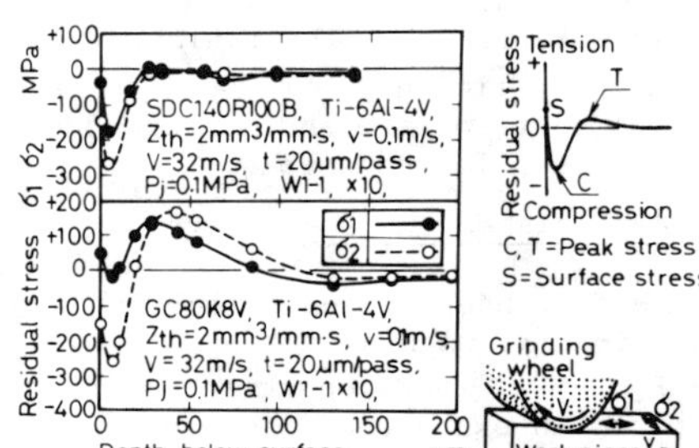

Figure 10 - Variation of residual stress distribution for the different grinding wheels (low pressure coolant).

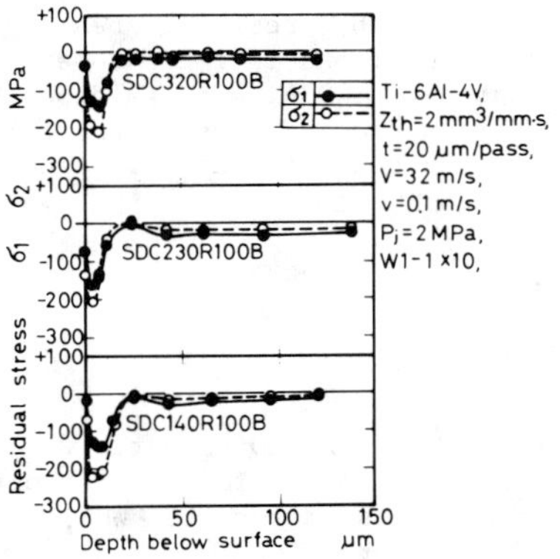

Figure 11 – Effect of grain size upon residual stress.

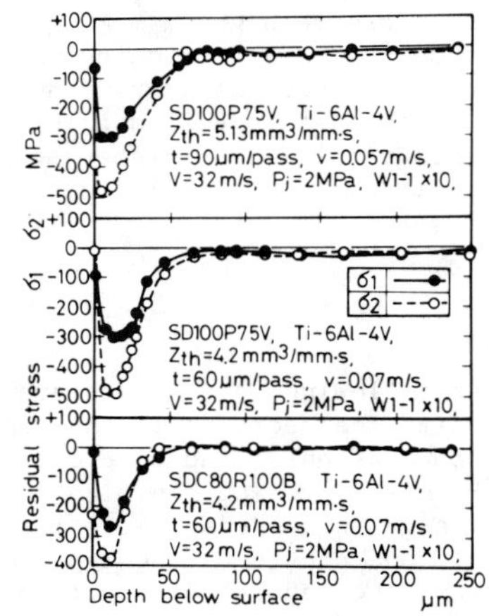

Figure 12 - Effect of grinding wheel and Zth upon residual stress.

value exceeds 150 MPa. This tensile stress decreases with increasing Pj, and becomes a tensile residual stress of approx. 100 MPa even at Pj=2 MPa (not shown). The penetration depth of residual stress is far greater than in the case of the SDC140R wheel. The results are typical of both wheels. Figure 11 illustrates the effect of the grain size of a diamond wheel when the grinding is done under Pj=2 MPa. The distribution patterns of residual stress are of compressive type(5) . As is evident from the figure depicting the SDC140R wheel, no tensile stress is generated by increasing Pj, and the peak stress and the penetration depth of the stress are smaller than in the foregoing cases. As has been discussed above, the presence of compressive residual stress in the vicinity of the grinding surface is desirable from a viewpoint of fatigue strength. The penetration depth and amount of residual stress decreases slightly as the grain size becomes smaller, but as far as the present experiments are concerned, the effect of grain size on residual stress is negligible.

To ensure highly efficient grinding without any sacrifice of finished surface integrity, it is effective to use a wheel the grain sizes of which are a little coarse. Figure 12 indicates an example of distribution of residual stress on a finished surface ground under Pj=2 MPa, using the SD100P75V wheel. The figure shows that even with a highly efficient grinding under Zth=5.13 mm^3/mm·s (t=90 μm/pass, v=0.057 m/s), the residual stress on the grinding surface becomes of compressive type, and that grinding heat is sufficiently cooled. Virtually the same result can be obtained for the SDC80R100B. A finished surface of such superior integrity can be obtained because, as in the previous report(3), the mean temperatures at the grinding point of the grains (above room temperature) is 100 K or below, and because the surface temperature of the workpiece is cooled to the original temperature(2) as soon as the grains have passed, decreasing the amount of heat which flows into the workpiece and the tensile residual stress caused by the grinding heat. For the diamond wheel, grinding force is small, and tensile stress due to cutting force is limited. In addition, diamond grains have a low friction coefficient, subject to burnishing action, and generating a large compressive stress on the grinding surface. For these reasons, the distribution pattern of grinding residual stress becomes of compressive type. The peak stress on the compression side attains a value as high as apporox. 500 MPa. If a grinding under the same Zth is carried out with the same wheel under Pj=0.1MPa, grinding temperatures become higher, making grinding impossible by greatly increasing the wear on the wheel. Grinding under the above conditions is impossible with a fine grain wheel if a grinding fluid is injected at Pj=2 MPa. As described above, the jet injection grinding method with a diamond wheel permits low stress grinding at a higher efficiency than the conventional low speed grinding method.

4. Conclusion

The following conclusions are derived:

(1) With the decrease in the grain size of a wheel, grinding ratio is decreased remarkably, and specific grinding force increases. There exist optimum grain sizes which minimize the roughness of finished surface. Wheels of fine grain are unsuitable for highly efficient grinding.

(2) If one considers the types of diamond grains and the grain retention power of the bond, the vitrified bonded diamond wheel can offer almost the same grinding characteristics as the resinoid bonded diamond wheel can.

(3) If the pressure on the grinding fluid is increased, the grinding ratio increases, resulting in marked improvements in finished surface roughness and surface integrity.

(4) When a diamond wheel is employed to perform grinding under fluid at high pressure, it is possible to give highly efficient grinding without causing tensile residual stress in the vicinity of the finished surface, and thus more efficient low stress grinding than normal becomes possible.

Acknowledgements

The authors wish to thank Kobe Steel, Ltd., Osaka Diamond Industry, Ltd., FSK, Ltd. and Yushiro Chemical Industry, Ltd. for their kind cooperation in this research. Thanks are also due to Mr. Hiroshi Tabuchi, Mr. Hidehisa Nishibu, Mr. Ikuhiro Takeyoshi and Mr. Takashi Hokazono, graduates of Kansai University, for their kind help in the present experiments.

References

(1). Guy Bellows, ed., Low Stress Grinding for Quality Production(No.MDC 78-103)(Cincinnati, Ohio: Machinability Data Center, Metcut Research Associates inc.,1978).

(2). Nobuo Kumagai and Kiyoshi Kamei, "Grinding of Titanium with Jet Infusion of Grinding Fluid", Titanium Science and Technology (5th. Int. Conf. on Titanium, (2) (1984), 1015.

(3). Nobuo Kumagai, Kiyoshi Kamei, and Seiichi Inoue, "Grinding of Titanium with Jet Infusion of Grinding Fluid (2nd report)" Proceedings of 6th. World Conf. on Titanium, (IV) (1988), 1841.

(4). E. K. Henriksen, Discussion on by H. R. Letner, and H. J. Snyder: "Grinding and Lapping Stress in Manganese Oil-Hardening Tool Steel", Trans. ASME, 75, (1953), 881.

(5). Kenjiro Okamura, Toshikatsu Nakajima, Youichi Inoue, and Yoshiyuki Uno, "Surface Integrity in Transitional Cutting Process (2nd report) - Residual Stress in Plunge Grinding - " J. Japan Society Precision Engineering, 39, 8, (1973), 794. (in Japanese)

ISOSTATIC DIFFUSION BONDING OF SUPERPLASTIC

Ti-6Al-4V AND SUPER ALPHA-2 INTERMETALLIC ALLOY

N. Ridley*, M.F. Islam* and J. Pilling†

*Materials Science Centre
University of Manchester/UMIST
Grosvenor Street, Manchester M1 7HS, England

†Department of Metallurgical and Materials Engineering
Michigan Technological University
Houghton, Michigan, U.S.A.

Abstract

Studies of isostatic diffusion bonding have been made for microduplex Ti-6Al-4V and Super Alpha-2 sheet materials. The conditions of temperature, pressure and time required to produce a sound bond have been examined for surfaces which were either chemically cleaned or fine ground. Bond integrity in Ti-6Al-4V was assessed using optical metallography, lap shear strength measurements and fractography, while bonds in Super Alpha-2 were examined metallographically. Experimental bonding times were compared with those predicted by a theoretical model for isostatic bonding, and it was found that they showed reasonable agreement within the limits of reliability of the material property data. For Ti-6Al-4V sound bonds could be readily obtained at temperatures close to those used for SPF. However, for Super Alpha-2 it was clear that the production of sound bonds at temperatures near to those associated with optimum superplasticity, 960-980°C, would require higher pressures and/or longer times than those used for Ti-6Al-4V.

Introduction

Diffusion bonding (DB) combined with superplastic forming (SPF) is now a well established route for the production of complex structures from α/β titanium alloys. The procedure can result in both cost and weight savings compared with conventional forming techniques.
Several models of DB have been proposed to predict the time required to form a sound bond between surfaces of varying roughnesses(1-7). The models show, for the most part, reasonable agreement with the limited experimental data available to test them. However, these models assume that bonding occurs in plane strain, whereas there is an increasing commercial interest in the DB of superplastic materials under conditions of isostatic compression. It has been predicted that isostatic DB will lead to a reduction in bonding times compared with plane strain conditions[(8)]. However, there appears to be no published experimental information on isostatic bonding. In the present work, studies have been made of the times to produce sound bonds in superplastic Ti-6Al-4V and Super Alpha-2 sheet materials under isostatic conditions. Experimentally determined bonding times have been compared with those predicted by theoretical modelling.

Titanium '92
Science and Technology
Edited by F.H. Froes and I. Caplan
The Minerals, Metals & Materials Society, 1993

Experimental

Materials

The Ti-6Al-4V alloy was received in the form of superplastic grade sheet of 1.6 mm thickness, with a mean grain size (mean linear intercept) of ~ 6μm. The Super Alpha-2 alloy, which was based on the composition: Ti-25Al-10Nb-3V-1Mo, was obtained as 3 mm rolled sheet. The β-transus temperature was determined as 1080-1085°C, and the as-received sheet had an $\alpha_2+\beta$ microstructure with a mean linear intercept of ~ 3μm. Both materials had been manufactured by Timet.

Diffusion Bonding

For the preparation of diffusion bonding couples, blanks measuring 88 mm x 22 mm were cut from the sheet materials. For the Ti-6Al-4V, bonding was investigated for two surface conditions: (i) as-received and chemically cleaned (see Ref. 9) and (ii) ground to P60 grade silicon carbide followed by ultrasonic cleaning in acetone. For the Super Alpha-2, surfaces were ground to a P1200 grade silicon carbide grit finish before ultrasonic cleaning. Pairs of blanks were electron beam welded around their peripheries in an evacuated system to form a DB couple. The Ti-6Al-4V specimens were subjected to various isostatic pressures up to 2.1 MPa (300 psi) at temperatures in the range 907°C - 940°C (1180 - 1213K) for times of up to 4 hours. For Super Alpha-2 the temperatures examined were from 950°C - 1050°C (1223-1323K), with pressures up to 2.8 MPa (500 psi) and times up to 4 hours.

For Ti-6Al-4V alloy, bond quality was assessed on the basis of optical metallography, lap shear strength measurements and scanning electron metallography of the shear fracture surfaces, while for the Super Alpha-2 only optical metallography has so far been used. Sections for optical examination were etched in 2% HF and 10% HNO_3 by volume, in water. Lap shear test pieces with an overlap of 1.5t, where t is sheet thickness, were machined from the bonded samples, and tested in compression. A sound bond was one which reached a shear strength of 95% of the parent metal tested after a comparable thermal cycle, showed no metallographic evidence of the bond line and exhibited a fully ductile shear fracture.

Comparison of Experiment with Theoretical Prediction

Isostatic Bonding Model

In diffusion bonding, the application of an external pressure will cause the points of contact between two surfaces to collapse until the cross-sectional area of contact can support the applied pressure. This leads to the creation of a planar array of interfacial voids. In the isostatic DB model of Pilling(8) it was assumed that the voids had a cylindrical shape with their axes aligned perpendicular to the plane of the bond. However, in recent work on the isostatic DB of Ti-6Al-4V, a pyramidal/ellipsoidal void morphology was adopted, following metallographic observations of the shape of collapsing voids(9).

In the model, the time to attain full interfacial contact is calculated by integrating the rate of reduction of area fraction of the voids between the start of bonding and the completion of bonding. The reduction in area fraction is considered to be the sum of two independent processes involving time dependent plastic collapse, and diffusion from the bond interface, and also from grain boundaries intersecting the void, into the void. The magnitude of each process will vary with stress and temperature. In duplex alloys, the material properties such as shear modulus and diffusion coefficient will depend on the proportions of phases present. From measurements of the volume fractions of phases at different temperatures, the properties of the alloys can be calculated from those of the constituent phases using a rule of mixtures.

(a) (b)

Figure 1. Microstructure of bonds formed in Ti-6Al-4V at 940°C and 1 MPa for chemically cleaned surfaces (a) 20 mins, (b) 45 mins.

Measurement of Material Parameters

The material parameters required to test the model include the α/β phase proportions at various bonding temperatures, the flow stress-strain rate behaviour, the average grain size as a function of bonding time and temperature. Surface roughness is also required.

Phase proportions were determined by point counting of specimens which were metallographically prepared after water quenching, following a 30 minute hold at various temperatures. The data is given in Tables I and II for the two materials being examined. For the determination of stress-strain rate relationships, tensile specimens of 10 mm gauge length and 5 mm gauge width were held at various temperatures in a furnace attached to a computer controlled Instron testing machine. Velocity jump tests were performed at cross-head speeds corresponding to strain rates in the range 10^{-2}-$10^{-5}s^{-1}$. From the resulting record of load versus time, true stress-true strain rate relationships were calculated. For the Super Alpha-2 alloy additional data was taken from the work of Mukherjee(10), Ghosh(11) and co-workers, and Collins(12). The effect of test temperature on the flow stress was removed by plotting modulus compensated stress, $\hat{\sigma}(=\sigma/G(T))$ and a temperature compensated strain rate $\hat{\epsilon}(=\dot{\epsilon}/D(T))$. From the compensated data plot the stress exponents and pre-exponential terms in the constitutive equations for flow were obtained and are listed in Tables I and II.

The variation of mean grain size with time at various bonding temperatures was determined metallographically for water quenched specimens. The average grain size expressed as a mean linear intercept is given in Tables I and II. Surface roughness, obtained from Talysurf measurements, was (i) as-received and chemically cleaned, R_a 0.6μm, and wavelength λ = 10.5μm, (ii) ground surface finish P60, R_a = 0.8μm and λ = 14.8μm, (iii) ground surface finish P1200, R_a = 0.1μm and λ = 10.1μm. Other material parameters such as diffusivity,

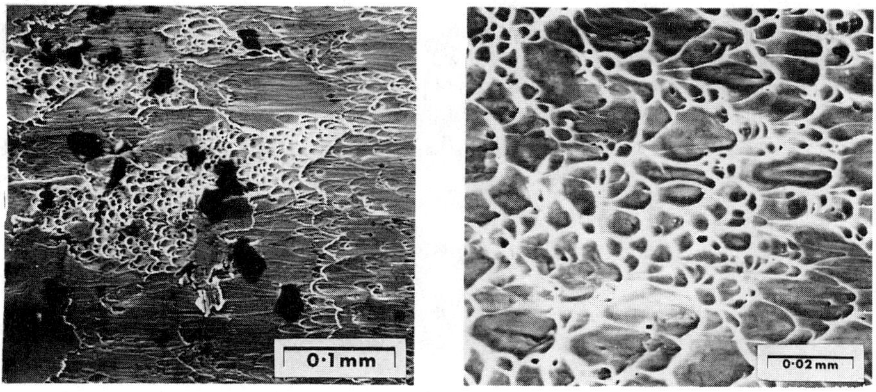

(a) (b)

Figure 2. Shear fracture surfaces for Ti-6Al-4V bonded at 940°C for (a) 2.1MPa, 5 mins, (b) 1MPa, 45 mins. Chemically cleaned surface.

shear modulus and its temperature dependence, atomic volume were obtained from the literature, given in Tables I and II. Where the information required for Super Alpha-2 was not available, the values used were estimated from the corresponding values for Ti-6Al-4V.

Ti-6Al-4V : Results and Predictions

The conditions to produce sound bonds have been examined at three temperatures. Figure 1 shows the bonded regions of specimens with a chemically cleaned surface, bonded at 940°C (1213K) under a pressure of 1 MPa(144 psi) for 20 minutes and 45 minutes, respectively. The bond interface is clearly visible in Fig. 1(a) but not in 1(b). Lap shear test pieces showed a systematic change in the appearance of the fracture surface as the bond quality improved. When the strength was low the fracture surface was almost planar, with evidence of the original surface clearly apparent (Fig. 2a). With increasing time and pressure the failure mode became more ductile, with high strength bonds showing 100% ductile tearing (Fig. 2b). The metallographic observations, both optical and SEM, were combined with the conditions for which the lap shear strength ratios were greater than 0.95, to give the conditions required to form sound bonds at temperatures of 907°C (1180K), 927°C (1200K) and 940°C (1213K), for chemically cleaned surfaces, and at 940°C for a P6O ground surface finish (Figure 3a-d).

The predicted bonding times, based on the assumption that the collapsing voids have an ellipsoidal rather than a cylindrical geometry, show essentially the same variation with pressure as the experimental bonding times. As can be seen in Fig. 3a-d the model tends to overestimate the actual bonding time although the agreement is good for the chemically cleaned surfaces at lower bonding pressures. Hence, at the higher pressures/shorter times either the calculated rate of mass transfer is too small and/or the volume of material to be transferred to the interfacial voids is too great. Overall, however, the agreement is satisfactory within the limits of uncertainty of the materials property data. It is interesting to note that at the temperatures examined 907- 940°C, the experimental studies showed that it was possible to get sound bonds in times of 15-30 minutes at a pressure of 2.1 MPa (300 psi) for as-received sheet which had been chemically cleaned.

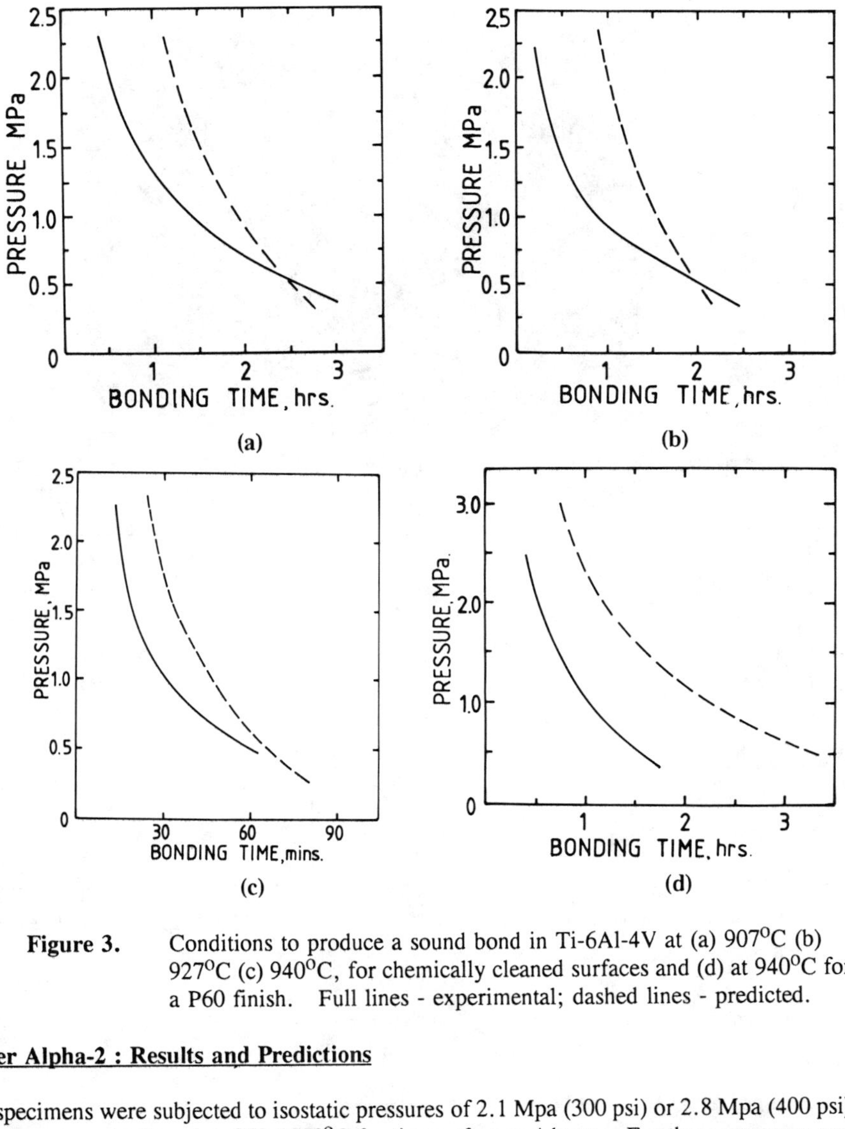

Figure 3. Conditions to produce a sound bond in Ti-6Al-4V at (a) 907°C (b) 927°C (c) 940°C, for chemically cleaned surfaces and (d) at 940°C for a P60 finish. Full lines - experimental; dashed lines - predicted.

Super Alpha-2 : Results and Predictions

DB specimens were subjected to isostatic pressures of 2.1 Mpa (300 psi) or 2.8 Mpa (400 psi) at temperatures in the range 950-1050°C for times of up to 4 hours. For these pressures and times, sound bonds were not obtained at temperatures of 1000°C or below (950°C, 970°C). However, metallographically sound bonds were obtained at both 1030°C and 1050°C. At 1030°C a pressure of 2.8 MPa applied for 4 hours was sufficient to remove the bond line while at 1050°C a pressure of 2.1 MPa developed a sound bond after both 2 and 4 hours. At 1050°C the alloy contains about 80% β-phase (Fig. 4). For the temperature range of interest (950°C - 1050°C) the calculated bonding times varied by a factor of two at constant pressure and by more than an order of magnitude at constant temperature for the pressure range 0.35 - 3.5 MPa. However, given the high flow stresses of Super Alpha-2 compared to Ti-6Al-4V for a similar microstructure it is easy to understand why sound bonds were

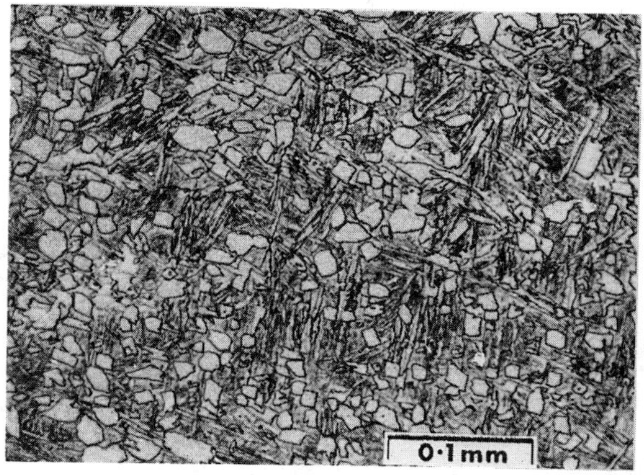

Figure 4. Microstructure of bond formed in Super Alpha-2 at 1050°C, 2.1 MPa 2 hours.

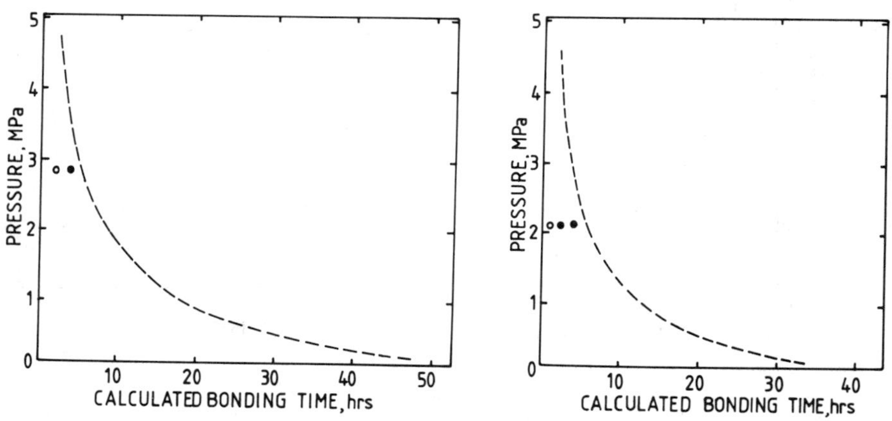

(a) (b)

Figure 5. Predicted bonding time versus pressure at (a) 1030°C (b) 1050°C. Experimental data: 0 - poor bond; ● - sound bond.

not developed at 1000°C and below for the times and pressures investigated(11).

At 1000°C bonding times of 15 hours at 2.1 MPa and 10 hours at 2.8 MPa are predicted. Clearly, even longer times and/or higher pressures would be required to bond at the optimum SPF temperatures of 960- 980°C. The variation of predicted bonding time with pressure at both 1030 and 1050°C is shown in Fig. 5a and 5b. The predicted times of 4.1 hours at 1050°C/2.1 MPa and 5.6 hours at 1030°C/2.8 MPa compare favourably with experiment.

Summary

The times required to produce sound isostatic diffusion bonds for Ti-6Al-4V sheet material have been determined experimentally for temperatures of 907-940°C at pressures in the range

0.5 - 2.1 MPa for two surface finishes. For as-received sheet material subjected to a commercial chemical cleaning procedure it was observed that sound bonds could be obtained for times of 15 - 30 minutes over the temperature range examined for an isostatic pressure of 2.1 MPa (300 psi). The experimental data shows reasonable agreement with bonding times predicted theoretically, within the limits of uncertainty of the materials property data used in the calculations. The main differences between experiment and prediction occur at higher bonding pressures, where the experimental times are shorter than those predicted.
For Super Alpha-2, bonding was investigated for the temperature range 950-1050°C using pressures up to 2.8 MPa (400 psi). For the conditions examined a sound bond was not obtained at 1000°C or below. However, sound bonds were obtained at 1030°C in 4 hours at 2.8 MPa, and at 1050°C in 2 hours at 2.1 MPa. These bonding times were in reasonable agreement with those predicted. For the lower temperatures examined, ≤ 1000°C, the model predicted that higher pressures/longer times would be required for bonding, than those investigated. At 1000°C a bonding time of 15 hours at 2.1 MPa, and 10 hours at 2.8 MPa is predicted. At the optimum SPF temperature range of 960-980°C measured for the material, even longer times/higher pressures would be required.

Acknowledgements

The authors are grateful to SERC for funding to purchase an ABB Q1H-9 Mini Hipper.

References

1. C.H. Hamilton, in Titanium Science and Technology, ed. R.I. Jaffee and R.M. Burte, (New York, N.Y. Plenum, 1973), 625-647.

2. G. Garmong, N.E. Paton and A.S. Argon, Metall. Trans. A., 6 (1975), 1269-1279.

3. B. Derby and E.R. Wallach, Metal Sci., 16 (1982), 49-56.

4. B. Derby and E.R. Wallach, Metal Sci., 18 (1984), 427-431.

5. J. Pilling, D.W. Livesey, J.B. Hawkyard and N. Ridley, Metal Sci., 18 (1984) 117-122.

6. Z.X. Guo and N. Ridley, Mater. Sci. Technol., 3 (1987) 945-954.

7. A.D. Hill and E.R. Wallach, Acta Metall., 37 (1989) 2425-

8. J. Pilling, Mater. Sci. Eng., 100 (1988) 137-144.

9. M.T. Salehi, J. Pilling, N. Ridley and D.L. Hamilton, Mater. Sci. Eng., A150 (1992)1-6.

10. H. Yang, P. Jin and A.K. Mukherjee, Scripta Metall., 25 (1991) 1223-1228.

11. A.K. Ghosh and C-H. Cheng, in Superplasticity in Advanced Materials, ed. S. Hori, M. Tokizane and N. Furushiro, (Osaka, Japan, J.S.R.S. 1991), 299-310.

12. E.W. Collings, The Physical Metallurgy of Titanium Alloys, (ASM, Metals Park, OH 1984), 116-117.

Table I Material properties for α/β Ti-6Al-4V

Property	α-Phase	β-phase
Melting point T_m(K)	1933	1933
Atomic Volume (m^3)	1.76×10^{-29}	1.81×10^{-29}
Shear Modulus (MPa)	$4.36 \times 10^4[1-0.5(T-300)/T_m]$	$2.04 \times 10^4[1-1.2(T-300)/T_m]$
Grain boundary diffusion ($m^3\ s^{-1}$)	$3.6 \times 10^{-16} \exp(-97{,}000/RT)$	$5.4 \times 10^{-17} \exp(-150{,}000/RT)$
Surface energy (Jm^{-2})	1	1

$$\text{Strain rate}(s^{-1})\ (\sigma \text{ in MPa}) \text{ for SP flow} = \frac{1.3 \times 10^4}{T} \exp\left(-\frac{150{,}000}{RT}\right)\sigma^{1.83}$$

$$\text{Strain rate } (s^{-1})\ (\sigma \text{ in MPa}) \text{ for power-law creep} = \frac{6.6 \times 10^{-1}}{T} \exp\left(-\frac{150{,}000}{RT}\right)\sigma^{4.3}$$

Temperature	907°C	927°C	940°C
Volume fraction β-phase	0.47	0.55	0.60
Mean grain size (μm)	7.9	11.3	14.0

Table II Material properties for Super Alpha-2

Property	α_2-Phase	β-phase
Melting point T_m(K)	2223	2223
Shear Modulus (GPa)	58	35

$$\text{Strain rate } (s^{-1})\ (\sigma \text{ in MPa}) \text{ for SP flow} = \frac{2 \times 10^9}{T} \exp\left(-\frac{308{,}000}{RT}\right)\sigma^{2.18}$$

Temperature	1000°C	1030°C	1050°C
Volume fraction β-phase	0.60	0.70	0.80
Mean grain size	5 μm	5 μm	5 μm

PHYSICAL AND MECHANICAL PROPERTIES

MECHANICAL PROPERTIES OF TITANIUM ALLOYS

G. Lütjering, J. Albrecht, A. Gysler

Technische Universität Hamburg-Harburg
2100 Hamburg 90, FRG

Abstract

For $\alpha + \beta$ and near-α alloys some important issues concerning fatigue crack nucleation, fatigue crack propagation of self-initiated microcracks, fracture toughness tests on small cracks, and creep are pointed out in this paper. These mechanical properties are discussed in connection with critical microstructural parameters such as lamellar versus bi-modal microstructures, cooling rate, texture, and intermediate annealing. For Super Alpha 2 the importance of cooling rate and final heat treatment for balancing the mechanical properties is pointed out.

Introduction

This paper tries to point out some of the possibilities to optimize or balance the mechanical properties of titanium alloys through microstructural control. The discussion will concentrate on $\alpha + \beta$ alloys, high temperature near α alloys, and Ti_3Al based intermetallics. In each of these alloy systems the microstructure can be varied from fully lamellar over bi-modal (equiaxed primary α or α_2 in a matrix of lamellar grains) to fully equiaxed. With respect to the most important mechanical properties (yield stress, tensile ductility, fatigue strength, fatigue crack propagation resistance, fracture toughness, creep resistance) it is easily possible to optimize the microstructure for individual properties, for example a bi-modal structure for high yield stress, tensile ductility, fatigue strength, or a fully lamellar structure for high fatigue crack propagation resistance, fracture toughness, creep resistance, but difficulties arise when trying for example to optimize for high fatigue strength combined with high creep resistance (1).

In dealing with some of these problems it is important to realize by which step of the thermomechanical treatment (TMT) or production route the most important microstructural parameters are adjusted. Taking the bi-modal microstructure of $\alpha + \beta$ and near α alloys the TMT can be subdivided into five important steps:

- Cooling rate from the β phase field prior to final $\alpha + \beta$ deformation (including heating rate to deformation temperature) limiting the attainable primary α (α_p) size.
- $\alpha + \beta$ deformation (degree, temperature, mode) determining the crystallographic texture of the hexagonal α phase.
- $\alpha + \beta$ recrystallization (temperature, time) adjusting α_p volume fraction and α_p size, both together determining β grain size.
- Cooling rate from recrystallization treatment determining width and length of α lamellae.
- Final heat treatment (temperature, time) determining degree of precipitation (Ti_3Al, Ti_5Si_3) and final volume fraction of β phase.

For the fully lamellar microstructure only the last two steps are applicable (cooling rate

Titanium '92
Science and Technology
Edited by F.H. Froes and I. Caplan
The Minerals, Metals & Materials Society, 1993

in this case from the β recrystallization treatment and final heat treatment). In the section on Ti_3Al based intermetallics the major differences for these materials with respect to the above outlined TMT will be pointed out.

From the mechanical properties mentioned above, yield stress and tensile ductility will not be discussed in this paper.

Fatigue Crack Nucleation

Many applications of titanium alloys require a high HCF strength. A popular example in recent years is the effort of the Electric Power Research Institute (Dr. R.I. Jaffee) to maximize the HCF strength of Ti-6Al-4V for application as material for large steam turbine blades (2). To optimize the resistance against fatigue crack nucleation the bi-modal microstructure was taken as the basic structure of the Ti-6Al-4V material containing 0.2 % oxygen. The problem with the bi-modal microstructure is that both parts of the structure, the primary α and the lamellar grains, have to be optimized against fatigue crack nucleation. The primary α was optimized in two steps: First, the volume fraction of primary α was adjusted to about 35 % to separate the individual primary α grains. Second, the size of the primary α was decreased as far as possible. This decreased also the β grain size. It turned out that below a primary α size of about 15-20 μm no noticable further improvement in HCF strength was obtained (2,3). A careful examination of the fatigue crack nucleation sites showed that for these microstructures the cracks nucleated within the lamellar grains (Figure 1). A further improvement of the HCF strength can then be obtained only by optimizing the lamellar grains. The easiest way is to increase the cooling rate from the bi-modal recrystallization temperature (3). With increasing cooling rate the width of the α lamellae is decreased as shown in Figure 2 by comparing air-cooling to water-quenching. Due to the reduction in lamellae width the resistance against crack nucleation is increased raising the HCF strength. The second possibility to increase the HCF strength would be to induce a strong crystallographic texture in the blading material (3). Figure 3 shows such a strong and for comparison a weaker crystallographic texture (0002 pole figures) derived from different rolling procedures. The improvement of HCF strength by these two methods is shown in Figure 4 (cooling rate) and Figure 5 (texture). In production, water-quenching of the forged blades has to be done in a fixture to assure geometrical stability. The texture method has not been used successfully up to now because this would require a completely new blade forging practice.

Fatigue Crack Propagation

For improving LCF strength crack propagation has to be taken into account in addition to crack nucleation. From testing fatigue crack propagation separately by conventional fracture mechanics methods, for example on CT-specimens, it is well known that fully lamellar microstructures exhibit slower crack growth rates than bi-modal microstructures. Surprisingly, when testing LCF strength on non-precracked round specimens bi-modal structures are always superior to lamellar structures. An example (4) is shown in Figure 6 for the IMI 834 alloy comparing a lamellar microstructure with two bi-modal microstructures having the same α_p volume fraction of about 13 % but different α_p sizes (Bi-M.1: 40 μm, Bi-M.2: 20 μm). The cooling rate from the recrystallization temperature was identical for the three different microstructures tested. Separating crack nucleation from propagation and measuring the crack propagation rates of these self-initiated small fatigue cracks the lamellar microstructure exhibited faster crack growth rates than the bi-modal microstructures (Figure 7) (4). In agreement with (5), no measurable differences in propagation rates were found for the two different bi-modal microstructures tested (Bi-M.1 and Bi-M.2). The difference in fatigue crack growth rates between small cracks and large through-cracks for the lamellar microstructure was attributed to the development of a serrated crack front profile with increasing crack front length, hindering crack growth (6). This behavior of the fully lamellar microstructure can be visualized best by ascribing to each crack size a separate fatigue crack propagation curve (Figure 8) (7) which also explains the often observed stress dependence for microcrack propagation. Since for the fine bi-modal microstructure the crack front remains very smooth even for large cracks, not much difference is measured in this case with increasing crack front length or between small and large cracks.

Fracture Toughness

A logical consequence of the observed dependence of fatigue crack propagation rate on crack front length for the fully lamellar microstructures is that the onset of unstable crack propagation (i.e. fracture toughness) also might depend on crack front length. Two microstructures of the IMI 834 alloy, a coarse lamellar and a bi-modal microstructure with a primary α volume fraction of about 13 %, were compared in fracture toughness testing. In one case conventional pre-cracked specimens with a width of about 6 mm and in the other case small surface fatigue cracks with a crack front length of about 1 mm in round specimens with a diameter of about 3 mm were tested. The measured fracture toughness values are shown in Figure 9 and the corresponding crack front profiles in Figure 10 (small cracks) and Figure 11 (wide cracks) with the crack propagation direction being normal to the plane of the micrographs. It can be seen from Figure 9 that the fracture toughness values for the small cracks are similar for both microstructures with a value of about 20 MPa $\cdot$ $m^{1/2}$ and perhaps with the tendency that the coarse lamellar even has a lower toughness value. A lower toughness value would agree with the lower ductility of the lamellar as compared to the bi-modal microstructure (tensile elongation 4.5 % and 10.5 % respectively). In contrast, the conventional fracture toughness specimens with a width of about 6 mm exhibited the well-known dependence of fracture toughness on microstructure, the higher value of about 50 MPa $\cdot$ $m^{1/2}$ for the lamellar microstructure as compared to about 35 MPa $\cdot$ $m^{1/2}$ for the bi-modal microstructure (Figure 9, wide cracks). The corresponding crack front profiles demonstrate clearly that for the wide cracks a tremendous difference in roughness exists between the bi-modal and the lamellar microstructures (Figure 11), whereas for the small cracks the difference in roughness is much less pronounced (Figure 10). The observed results suggest that the common viewpoint of a superior fracture toughness behavior of lamellar microstructures as compared to fine bi-modal or equiaxed microstructures might not be correct for highly stressed parts with relatively small critical crack sizes.

High Temperature Properties

For most high temperature applications of titanium alloys creep properties as well as fatigue properties play a role. The above outlined relationships between microstructure and fatigue properties at room temperature can be taken as a first approximation also for high temperature fatigue. It was pointed out that for an optimized bi-modal microstructure fatigue crack nucleation occurs in the lamellar grains and that in this case the easiest way to increase the fatigue properties further would be an increase in cooling rate from the bi-modal recrystallization temperature increasing the resistance against fatigue crack nucleation. This improvement is also obtained for fatigue at high temperature, an example is shown in Figure 12, in which the LCF strength for 10^4 cycles is plotted as a function of cooling rate for the Ti-6242 alloy with a bi-modal microstructure tested at 450° C (8). It should be mentioned that the yield stress at high temperature showed qualitatively the same dependence on cooling rate pointing out that dislocation motion (crack nucleation) is affected. Unfortunately, this improvement in fatigue strength with increasing cooling rate cannot be utilized if at the same time a high resistance against creep deformation is needed. This is shown in Figure 13 in which the creep resistance (plastic strain after 100 hours) is plotted as a function of cooling rate for the same material condition as in Figure 12 (8). It can be seen that for the high cooling rates for which the improvement in fatigue properties was observed, the creep resistance drastically declined (9). The first thought to explain this behavior was the increased dislocation density associated with the martensitic transformation at high cooling rates. More detailed investigations performed on the IMI 834 (10) alloy proved that this explanation was wrong. Increasing the dislocation density artificially by cold rolling increased the plastic creep strain drastically but over the whole range of cooling rates studied, the original dependence of creep resistance on cooling rate remained unchanged. Also an intermediate annealing treatment of 2h 830°C before the final aging treatment of the IMI 834 (2h 700°C) removing most of the dislocations did not change to any noticable degree the high plastic creep strain of the fast cooled condition for the fully lamellar structure as well as for the bi-modal structure with 15 % primary α (Figure 14) (10). One microstructural feature which did not change during the 2h 830°C intermediate annealing treatment was the crystallographic misorientation between neighbouring lamellae. Whereas for the slow cooled conditions, in which the α lamellae grow by a diffusion controlled process into the β grains, the orientation difference between neighbouring lamellae is very small (Figure 15a), the misorientation is much larger for the fast cooled condition, in which a martensitic transformation occurs (Figure 15b) (10). If interface sliding also plays a role in the creep deformation and assuming that the interface sliding is much more pronounced for boundaries between lamellae with large misorientations then the observed dependence of creep strain on cooling rate can be summarized as follows: In the regime of slow cooling rates (perfect lamellae boundaries, no boundary sliding) dislocation creep dominates and decreasing the lamellae width by increasing the cooling rate decreases the creep strain. In the regime of fast cooling rates beyond the maximum in creep resistance, the tendency for martensitic transformation (misoriented lamellae boundaries) increases with increasing cooling rate and, assuming that boundary sliding dominates, the creep strain increases drastically.

The observed result that the intermediate annealing of 2h 830°C improved the creep resistance of the bi-modal microstructure of the IMI 834 alloy in the slow-cooled regime appreciably (Figure 14) might be due to the redistribution of aluminum from the primary α into the α lamellae increasing subsequent age-hardening by coherent Ti_3Al particles during the final heat treatment of 2h 700°C. This improvement can only be utilized for alloys containing silicon and, as a consequence, Ti_5Si_3 particles at lamellae boundaries hindering the coarsening of the lamellae. For example, intermediate annealing of the Ti-6Al-4V alloy decreased the creep resistance in this slow-cooled regime drastically by coarsening of the lamellae (10).

Super Alpha 2

From the Ti_3Al based alloys, the Super Alpha 2 alloy is the most advanced and will be taken in this paper therefore as example for these types of materials. The microstructure of the Super Alpha 2 alloy (Ti-14Al-20Nb-3.2V-2Mo, in wt.%) can also be varied by TMT from lamellar over bi-modal to equiaxed (Figure 16) (11). Comparing the bi-modal microstructure of the Super Alpha 2 (Figure 16c) with the commercial bi-modal structure of IMI 834 (Figure 16d) it can be seen that the microstructure of the Super Alpha 2 is so much finer that the lamellar grains are only visible by TEM (for example Figure 17c). The important difference to conventional α + β and near α alloys is that upon fast cooling from the recrystallization temperature the β phase does not transform martensitically but is retained (Figure 17a). Decreasing cooling rate increases the volume fraction of diffusion controlled grown α_2 lamellae (Figure 17b) until at a cooling rate of about 100°C/min the process of α_2 lamellae growth is nearly completed during cooling (Figure 17c). This means that for fast and intermediate cooling rates the last step in the TMT, the final heat treatment, determines to a large extent the details of the microstructure within the former β grains. For the Super Alpha 2 alloy it is therefore much more important to optimize the final heat treatment for the bi-modal microstructures as it is for conventional α + β and near α alloys. The mechanical properties at 600° C of the microstructures shown in Figure 16 are compared in Figure 18 (HCF) and Figure 19 (creep resistance). It can be seen that the properties of the Super Alpha 2 alloy with such an optimized bi-modal microstructure are much better than those of the IMI 834 alloy. Including the room temperature ductility and fracture toughness in the problem of balancing the mechanical properties, this bi-modal structure of the Super Alpha 2 shows sufficient room temperature tensile elongation (5 %) but the fracture toughness is fairly low with about 12 MPa $m^{1/2}$. The result from conventional fracture toughness testing that the fully lamellar microstructure shows superior toughness values of about 30 MPa $m^{1/2}$ (Figure 20, wide cracks) should not lead to the direction of optimizing the fully lamellar microstructure for application purposes, since this superior toughness of the fully lamellar condition is totally absent for small crack sizes (Figure 20, small cracks). The very low toughness of the lamellar microstructure for small cracks is in agreement with the poor ductility of this structure (tensile elongation near zero). The crack front profiles corresponding to the toughness tests of Figure 20 are shown in Figure 21 (small cracks) and Figure 22 (wide cracks), the crack propagation direction being again normal to the plane of the micrographs. As pointed out already in detail for the IMI 834 alloy (see Figures 9-11) the serrated crack front profile of the lamellar microstructure developing with increasing crack front length (Figure 22b) seems to be the reason for the high toughness of conventional pre-cracked specimens.

Taking the approach of improving the fracture toughness by modifying the lamellar part of the bi-modal microstructure the following observation was helpful: With decreasing cooling rate from the bi-modal recrystallization temperature and concomitant coarsening of the lamellar portion of the microstructure (Figure 23a-c) the fracture toughness increased from $12\ \text{MPa} \cdot \text{m}^{1/2}$ to $18\ \text{MPa} \cdot \text{m}^{1/2}$ (12). Unfortunately, all other important mechanical properties measured (yield stress, tensile elongation, creep resistance, HCF strength) decreased at the same time. By modifying the final heat treatment it was possible to create within the lamellar regions of the bi-modal microstructure a structure containing only a few large lamellae, the rest being very fine lamellae (Figure 23d). This modified microstructure ("annealed" condition) exhibited a fracture toughness of about $16\ \text{MPa} \cdot \text{m}^{1/2}$ keeping all the other important mechanical properties mentioned above at the high level. These results suggest that further research effort along this route may increase the fracture toughness of Super Alpha 2 to an acceptable level.

Conclusions

Instead of summarizing the results presented in this paper, a few areas are listed below in which additional research work seems to be necessary:

- Fatigue crack propagation of microcracks.
- Fracture toughness on naturally nucleated small cracks.
- Optimization of TMT for a combination of high fatigue strength at elevated temperature and creep resistance.
- Optimization of heat treatment of Super Alpha 2.

References

1. D.F. Neal, in Titanium, Science and Technology, DGM (1985), 2419-2424.

2. H. Puschnik et al., "Optimization of Service Properties for Titanium Forgings with Bi-modal Microstructure", these proceedings.

3. L. Wagner, G. Lütjering, and R.I. Jaffee, in Microstructure/Property Relationships in Titanium Aluminides and Alloys, TMS (1991), 521-531.

4. G. Lütjering, in BDLI-Werkstofftag '91, BDLI (1992), 139-148.

5. A.L. Dowson, A.C. Hollis, and C.J. Beevers, Int. J. Fatigue, 14 (1992), 261-270.

6. G. Lütjering, A. Gysler, and L. Wagner, in Sixth World Conference on Titanium, Les Editions de Physique (1988), 71-80.

7. U. Tacke-Messing, L. Wagner, and G. Lütjering, in Gefüge und Bruch, Gebr. Borntraeger (1990), 199-209.

8. S. Saal et al., Z. Metallkde., 90 (1990), 535-540.

9. W. Cho et al., in Sixth World Conference on Titanium, Les Editions de Physique (1988), 187-192.

10. C. Andres, A. Gysler, and G. Lütjering, "Correlation Between Microstructure and Creep Behavior of the High-Temperature Ti-Alloy IMI 834", these proceedings.

11. G. Lütjering et al., in Proc. 6th Conf. on Intermetallic Compounds - Structure and Properties, The Japan Institute of Metals (1991), 537-541.

12. G. Proske et al., "On the Influence of the Phase Morphology on the Mechanical Properties of the Two-Phase Intermetallic Compound Super Alpha 2", these proceedings.

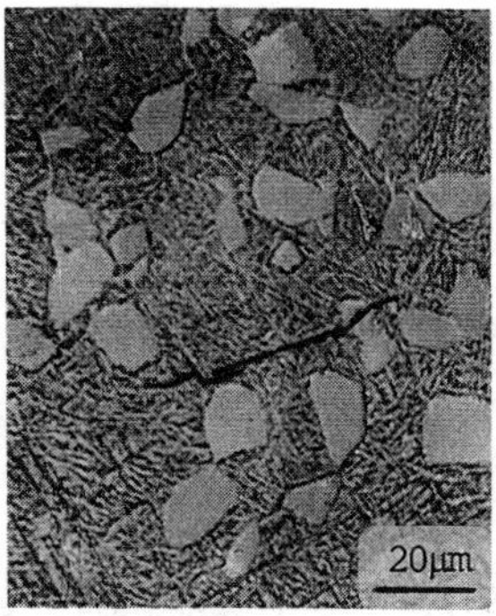

Figure 1: Ti-6Al-4V, (LM), crack nucleation.

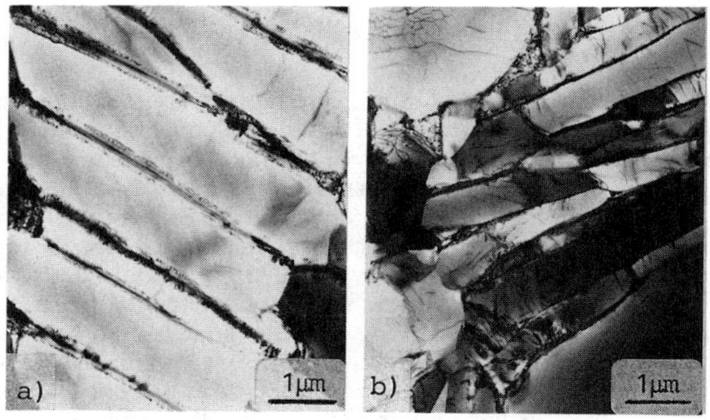

Figure 2: Ti-6Al-4V, (TEM), a) AC, b) WQ.

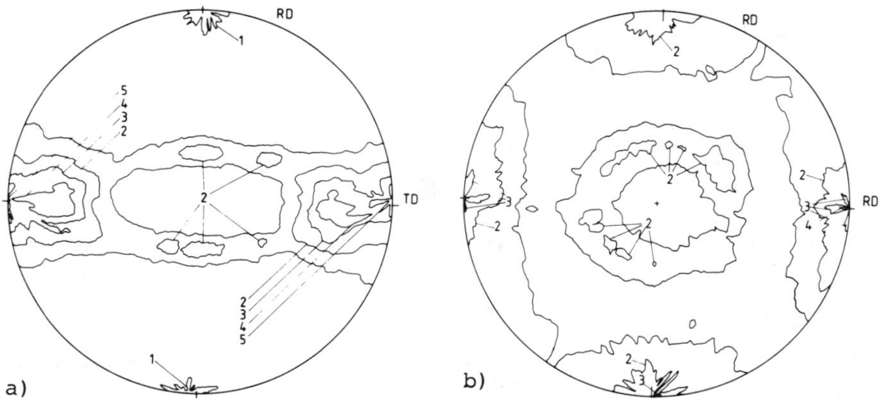

Figure 3: Ti-6Al-4V, (0002 pole figures), a) UR 875°C, b) CR 940°C.

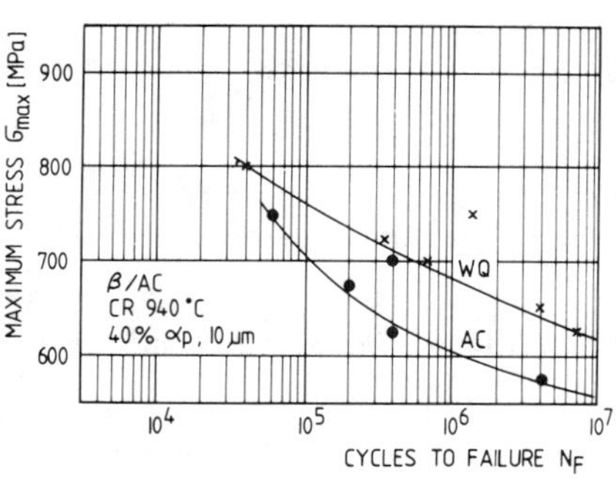

Figure 4: Ti-6Al-4V, HCF, (R = 0.1).

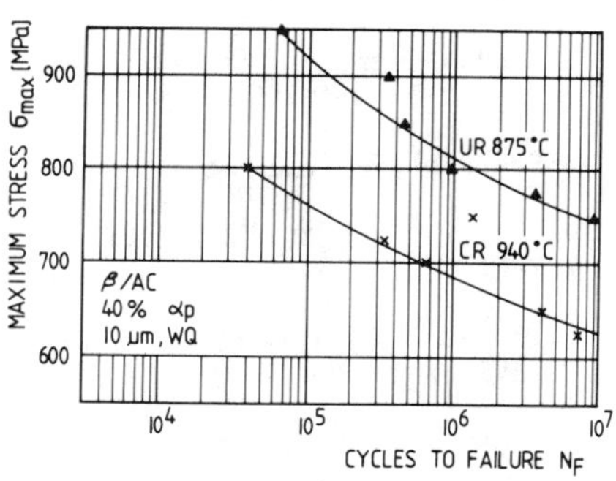

Figure 5: Ti-6Al-4V, HCF, (R = 0.1).

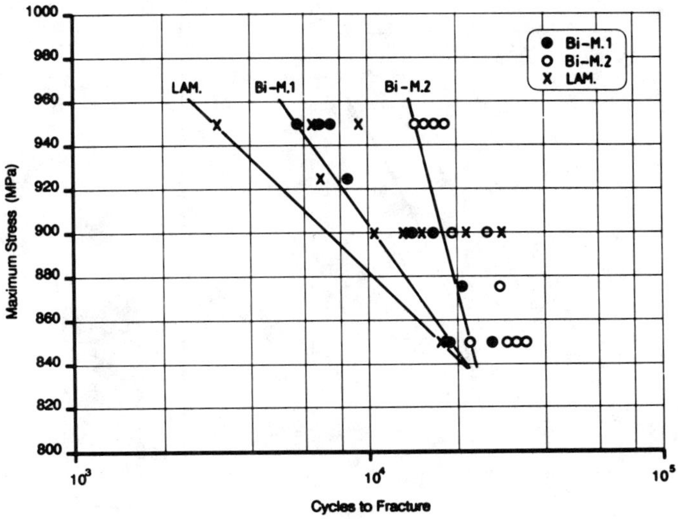

Figure 6: IMI 834, LCF, (R = 0.1).

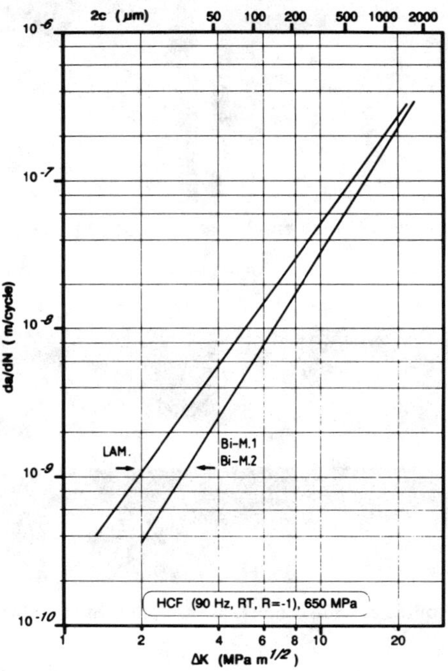

Figure 7: IMI 834, propagation rates of microcracks.

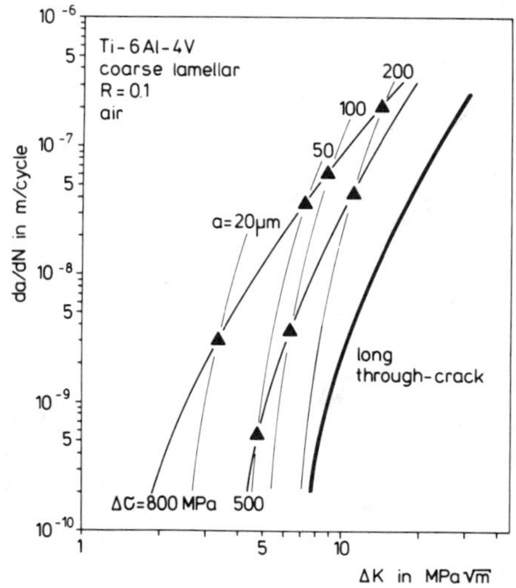

Figure 8: Ti-6Al-4V, propagation rates of microcracks, (schematically).

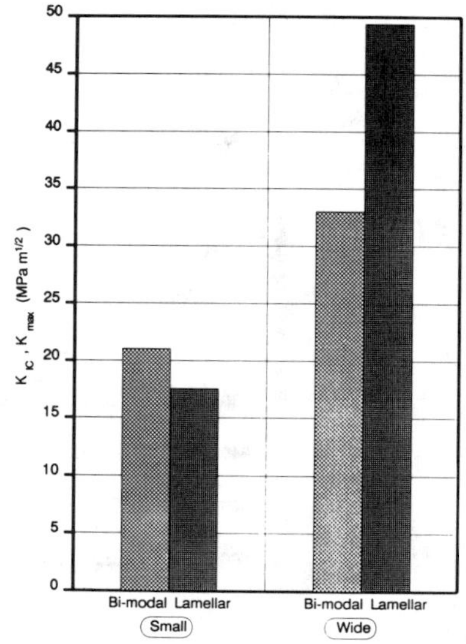

Figure 9: IMI 834, fracture toughness values.

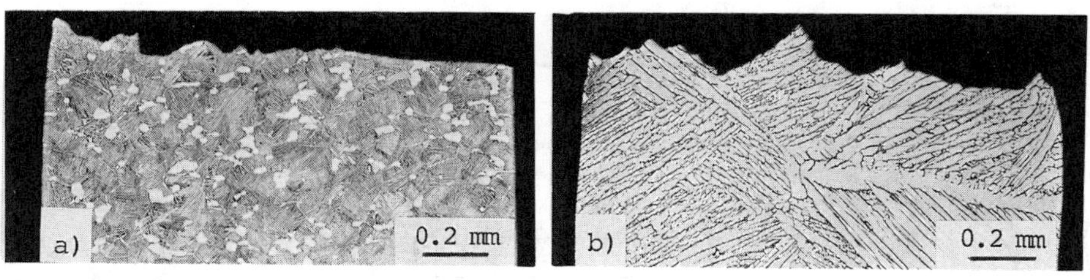

Figure 10: IMI 834, profiles of small cracks, a) bi-modal, b) lamellar.

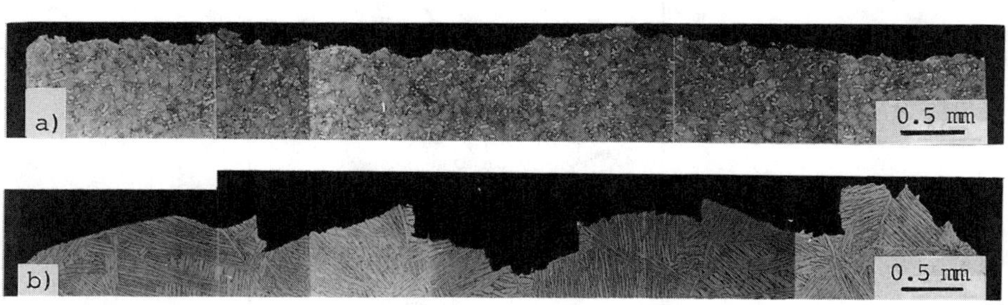

Figure 11: IMI 834, profiles of wide cracks, a) bi-modal, b) lamellar.

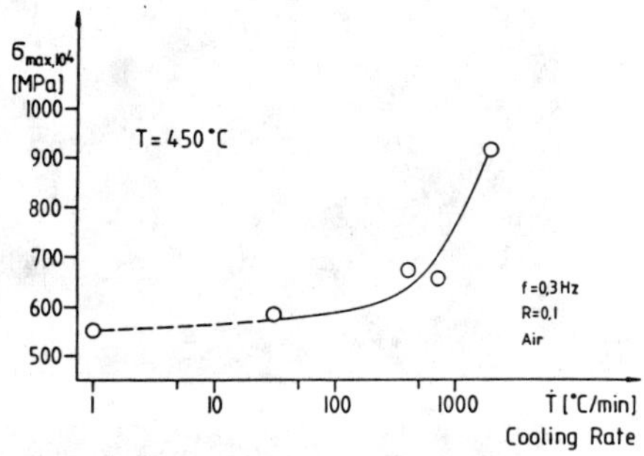

Figure 12: Ti-6242, influence of cooling rate on LCF ($\sigma_{max}10^4$).

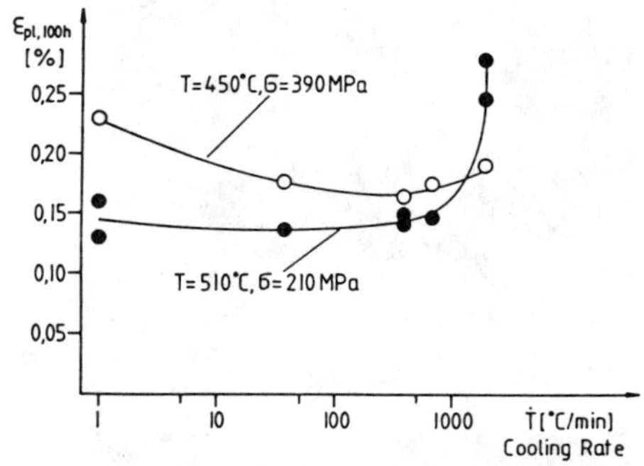

Figure 13: Ti-6242, influence of cooling rate on creep strain (ε_{pl} 100h).

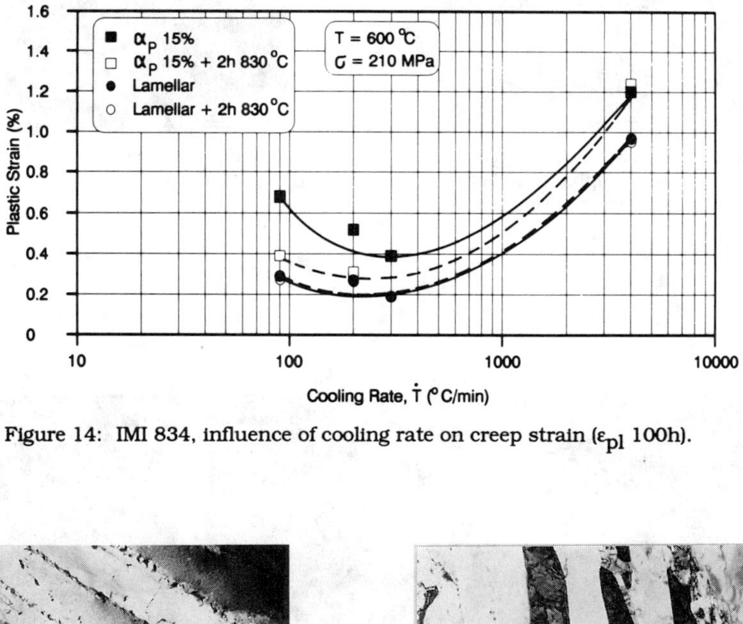

Figure 14: IMI 834, influence of cooling rate on creep strain (ε_{pl} 100h).

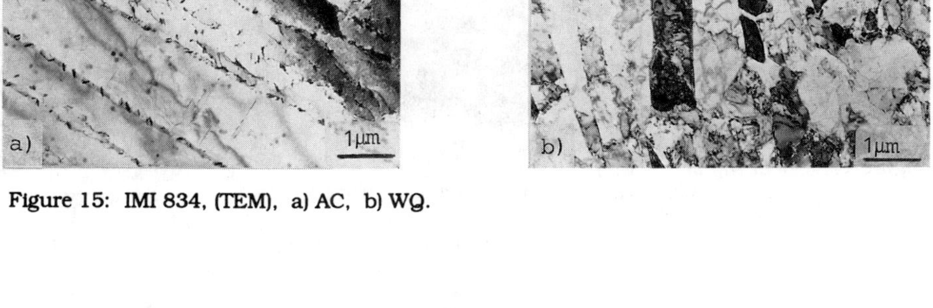

Figure 15: IMI 834, (TEM), a) AC, b) WQ.

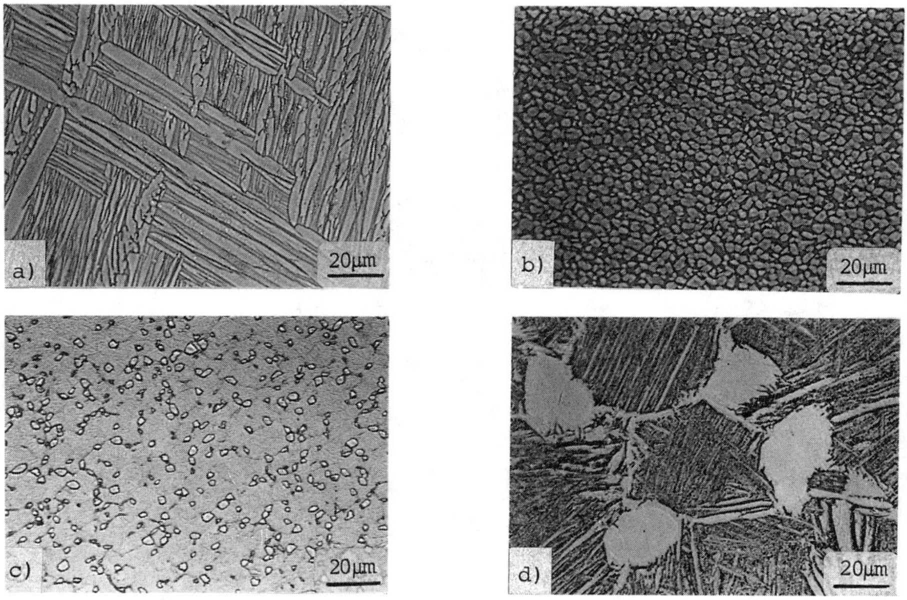

Figure 16: Microstructures, (LM).
Super Alpha 2: a) lamellar, b) equiaxed, c) bi-modal.
IMI 834: d) bi-modal.

Figure 17: Super Alpha 2, (TEM).
Bi-modal microstructures after cooling with different $\dot{T}$ from the recrystallization temperature.
a) $\dot{T} = 1100^{o}C/min$, b) $\dot{T} = 450^{o}C/min$, c) $\dot{T} = 100^{o}C/min$.

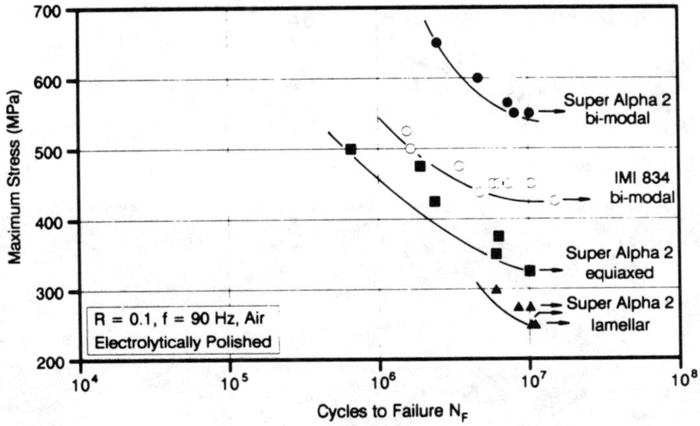

Figure 18: HCF of microstructures of Figure 16, (600°C).

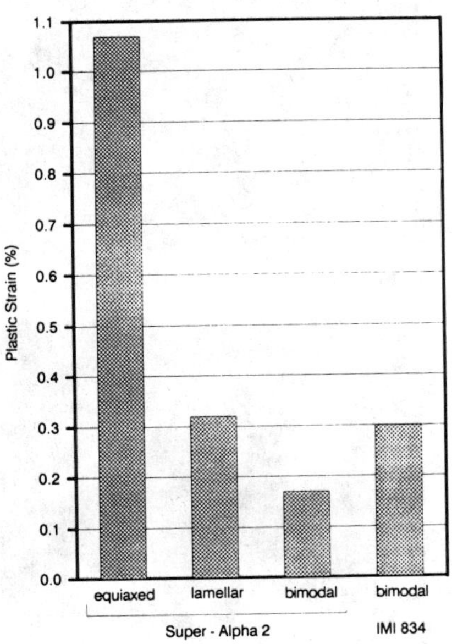

Figure 19: Creep resistance of microstructures of Figure 16, (600°C, 210 MPa, ε_{pl} 100h).

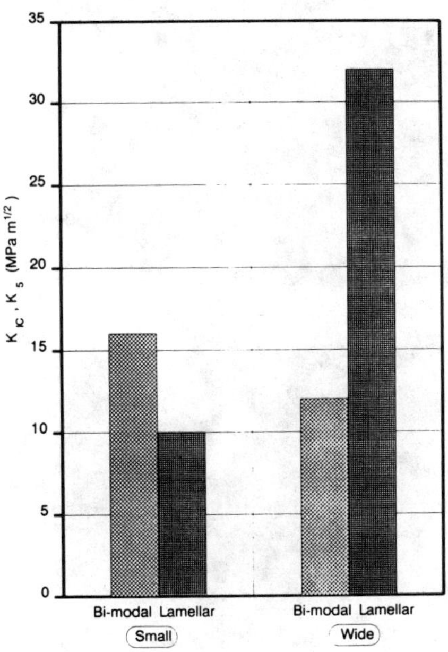

Figure 20: Super Alpha 2, fracture toughness values.

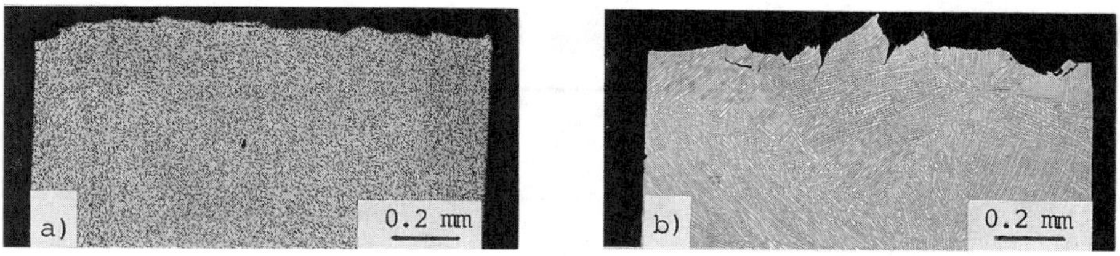

Figure 21: Super Alpha 2, profiles of small cracks, (LM), a) bi-modal, b) lamellar.

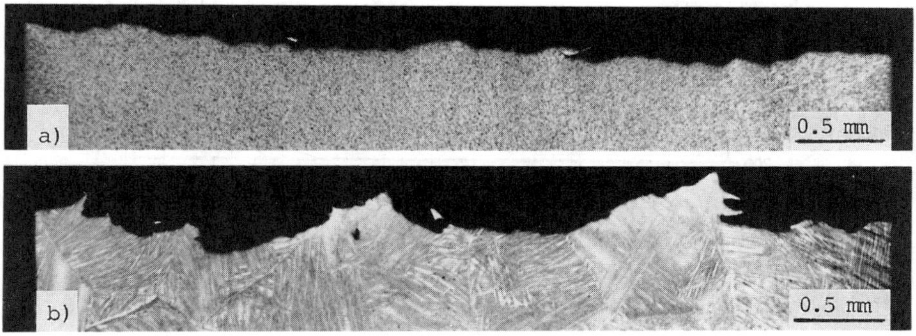

Figure 22: Super Alpha 2, profiles of wide cracks, (LM), a) bi-modal, b) lamellar.

a) Fast, K_{Ic} = 11.8 MPa $\sqrt{m}$

b) Medium, K_{Ic} = 14.4 MPa $\sqrt{m}$

c) Slow, K_{Ic} = 18.1 MPa $\sqrt{m}$

d) "Annealed", K_{Ic} = 15.7 MPa $\sqrt{m}$

Figure 23: Super Alpha 2, (TEM).
Microstructures and fracture toughness values of different bi-modal conditions:
a) - c) different cooling rates, d) "annealed" condition.

THE PROPERTIES OF TI-6AL-2SN-2ZR-2MO-2CR SHEET

R.R. Boyer and A.E. Caddey

Boeing Commercial Airplane Group
P.O. Box 3707, MS 73-44
Seattle, WA 98124

Abstract

Ti-6Al-2Sn-2Zr-2Mo-2Cr+Si (Ti-6-22-22) is an alpha-beta Titanium alloy developed by RMI in the early 1970's as a deep hardenable material. This alloy will be used for forgings and plate on the Lockheed/Boeing/General Dynamics Advanced Tactical Fighter. Recent studies initiated by the authors indicate the alloy also offers potential advantages for sheet applications.

Ti-6-22-22 can provide high strengths, in excess of 1500 MPa, and exhibits excellent superplastic forming (SPF) characteristics. Superplastic forming has been accomplished at temperatures as low as 800°C. The strength of the as-formed parts is a function of the forming temperature. The effects of post SPF heat treatments on mechanical properties were also studied.

The mechanical properties of sheet were studied as a function of solution treatment temperature and aging temperature. This study resulted in tensile strengths ranging from about 1035 to 1520 MPa. Fracture toughness was measured as a function of heat treatment (strength) to provide further guidance toward selection of heat treatments with optimal design properties.

Optical, scanning electron microscopy and thin foil scanning transmission electron microscopy examinations were used to analyze the microstructures and provide a correlation between heat treatment, microstructure and properties.

Introduction

Ti-6Al-2Sn-2Zr-2Mo-2Cr+Si (Ti-6-22-22) was developed by RMI in the early 1970's as a deep hardenable, moderate temperature alloy. It was studied by several aerospace companies but was never implemented into production. Recent studies by Boeing and Lockheed indicated that this alloy had damage tolerance characteristics similar to that of ß-annealed extra low interstitial Ti-6Al-4V with a significant strength advantage. This ultimately resulted in Ti-6-22-22 becoming the baseline Titanium plate and forging alloy on the Lockheed/Boeing/General Dynamics F-22 advanced fighter aircraft. While producing plate and forged block for evaluation, wheer was also rolled. This sheet was found to have excellent superplastic forming characterisitcs and was capable of being heat treated to high strengths. These attributes led to this study.

The purposes of this investigation were twofold:

(1) to determine the mechanical properties and fracture toughness of Ti-6-22-22 sheet using a range of solution treatment and aging temperatures to optimize the heat treatment procedure with respect to the design properties, and

(2) to characterize the superplastic forming behavior, the mechanical properties after SPF, and the effect of post SPF heat treatments on the mechanical properties of the formed alloy.

Titanium '92
Science and Technology
Edited by F.H. Froes and I. Caplan
The Minerals, Metals & Materials Society, 1993

Experimental Procedures

Most of the investigations were conducted using 1.27 mm thick Ti-6-22-22 sheet, although 1.14 and 0.635 mm sheet were tested in some cases. These sheets were all from the same heat and had the following chemistry:

Table I. Chemical Composition of Ti-6-22-22

Element	C	N	O	Fe	Al	Sn	Zr	Cr	Mo	Si
Weight %	.01	.007	.102	.05	5.64	1.90	1.91	1.98	1.86	.218

The transus temperature was determined by extrapolation of the volume percent of the a phase in the microstructure as a function of the solution treatment temperature. A beta transus of 950°C was obtained. This value compares favorably with the 957°C reported by RMI on the basis of their differential thermal analysis results.

For the purposes of heat treatment optimization, solution treatment temperatures in the range of 815 to 925°C and aging temperatures in the range of 480 to 675°C were employed. Solution treatment times were always 30 minutes, followed by an air cool, and aging times were 8 hours.

Tensile testing was accomplished in accordance with ASTM E8 and fracture toughness of the sheet was measured by using ASTM E561 procedures using 229 mm wide center cracked panels fabricated from the 1.27 mm thick sheet.

Superplastic forming trials were conducted at temperatures in the range 800 to 900°C using pressures between 3.45 and 5.2 MPa. Four different part configurations were produced, each using different forming parameters. Post forming aging was conducted at temperatures in the range of 510 to 565°C.

Results and Discussion

Heat Treat Study

Samples were solution treated between 815 and 925°C. Solution treatment temperature had a very pronounced effect on both strength and ductility as can be seen in Figure 1. Tensile testing indicated that a minimum yield strength occurred at approximately 845°C, beyond which the yield strength increased rapidly as the solution treatment temperature was increased. A transmission electron microscopy (TEM) study of the microstructure indicated that at solution treatment temperatures up to 845°C most of the "soft" ß is retained on cooling which would reduce the yield strength. However, as the solution treatment temperature was increased above 845°C, the ß transformed martensitically upon cooling. The extent of the martensitic transformation increases as the solution treatment temperature is increased, resulting in the markedly increased yield strength and ductility loss. Microstructures of the two solution treatment temperature extremes, 815 and 925°C, supporting this premise, are presented in Figure 2. A similar behavior, i.e., a strength minimum as a function of solution treatment temperature, has been reported in Ti-6Al-4V by Hall and Hammond (1). Bliss (2) reported a minimum yield strength for Ti-6-22-22 (from the same heat of material studied here) to be at a higher temperature, about 870°C. The strength minimums are attributed to the retention of the maximum amount of ß, which would reduce the yield strength.

The material solution treated at 815°C had a highly polyganized dislocation substructure. The higher solution treatment temperatures resulted in a more equiaxed, dislocation free structure. Both the a and ß grain sizes increased as the solution treatment temperature was increased as would be expected.

Specimens solution treated at 870, 900 and 925°C were aged over the range from 480 to 675°C. (In all cases the aging time was 8 hours.) The yield strength as a function of aging temperature is presented in Figure 3. For a given aging temperature, the strength increases with increasing solution treatment temperature as expected. The aging response of the high solution treatment temperature was different than that of the two lower solution treatment temperatures. There was scatter in the data, but the yield strength of the material solution treated at 925°C was virtually independent of the aging temperature up to 565°C, while the material solution treated at 870 and 900°C showed a strength maximum at an intermediate aging temperature.

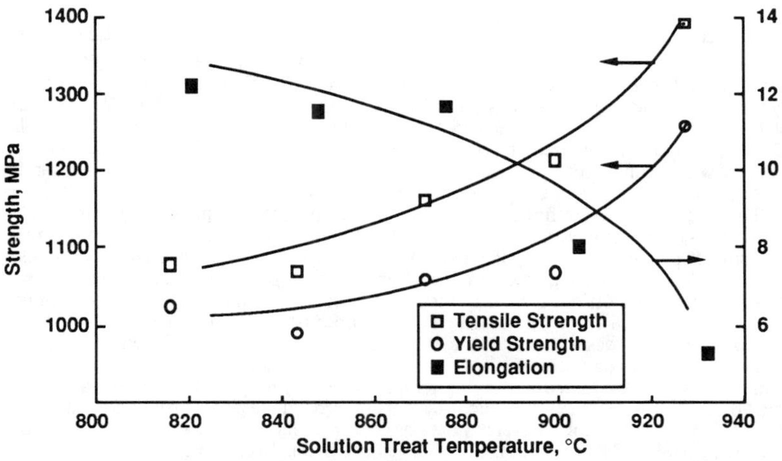

Figure 1. Strength and Ductility as a Function of Solution Treat Temperature. Solution Treat Time is 30 Minutes With an Air Cool

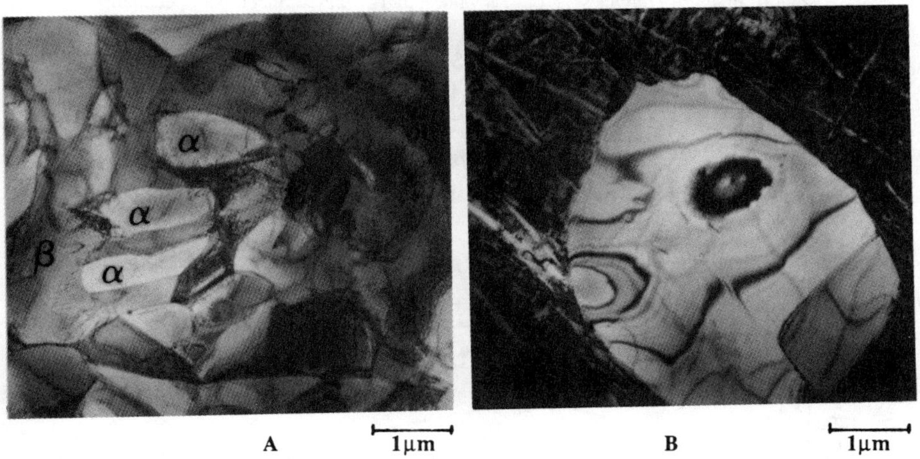

Figure 2. TEM Micrographs of Ti-6-22-22 Solution Treated at A) 815°C and B) 925°C

The independence of the yield strength on aging temperature for the high solution treatment temperature is not fully understood. The ß has almost fully transformed to martensite on cooling from the solution treatment temperature. The strength is about 140 MPa higher than the solution treated material, so some aging is obviously occurring. The aging temperatures studied were apparently not high enough to overage this structure. The ß phase resulting from the lower temperature solution treatments transformed to martensite to a lesser extent. Some of the observed aging response is the result of the decomposition of the retained ß. The 900°C solution treatment had a larger amount of martensite than the 870°C solution treated material, accounting for its higher strength (see Figures 1 and 2). In addition, the material solution treated at 900C would have retained ß-phase that is leaner in stabilizers than that formed at the lower solution treatment temperature (870°C) making it decompose more readily. This would account for the peak strength occurring at a lower aging temperature for the 900°C solution treatment. The decrease in yield strength at higher aging temperatures is an over-aging effect.

Fracture toughness was measured for 480, 565 and 675°C aging treatments subsequent to a 900°C solution treatment to determine the effect of strength on toughness. The results are shown in Table II. Contrary to expectations, the highest toughness was not associated with the lowest strength. The

$K_{apparent}$[1] value corresponding to the highest strength condition, which was obtained at the lowest aging temperature (480°C), was about 35% higher than that for the higher aging temperatures (565 and 675°C). This data is difficult to rationalize from two viewpoints: (1) The 480 and 565°C aging resulted in essentially the same strength, but toughness for the higher temperature age was about 35% lower than that of the lower aging temperature, and (2) The 675°C age resulted in a strength drop of about 10% relative to the 480°C aged material, but this strength decrease was accompanied by the same 35% toughness drop. In titanium alloys, when aging does not affect the strength one would not expect it to affect the fracture toughness, and when aging causes a decrease in strength it is normally accompanied by an increase in toughness. The results were re-confirmed by running the experiment a second time. The data in Table II are the averages of both experiments. It was speculated that this behavior may be related to ordering of the alpha or to a silicide precipitation during the higher temperature aging treatments. Thin foil samples were examined using a TEM in an attempt to determine if either of these possibilities could be the cause of this anomaly. No evidence of ordering or differences in silicide precipitates for the different aging temperatures were observed from the TEM investigation. Polished samples were also studied in the electron microprobe in the backscatter and Cr and Si X-ray modes; again, no differences were observed. Kuhlman (3) observed the same behavior in Ti-6-22-22 forgings as a function of aging temperature, i.e., a drop in toughness with an increase in aging temperature. In his case the drop in toughness was accompanied by a reduction in ductility, which was not observed in the present study. Silicides would not seem to be a likely cause for the discrepant data as their presence would also result in a significant reduction in ductility (4,5). The cause of the 35% difference in toughness for the two higher temperature aging conditions is still under investigation. The precipitation of a Cr-eutectoid intermetallic compound is another possibility, but no evidence to support this hypothesis was observed.

Table II. Fracture Toughness, $K_{apparent}$ values, of Ti-6-22-22 Sheet. All Specimens Were Solution Treated at 900°C for 30 Minutes; Aging Times Were 8 Hours.

Temperature °C	UTS MPa	TYS MPa	Elongation %	Toughness, MPa $\sqrt{m}$
480	1274	1162	11.7	165
565	1242	1172	11.2	109
675	1138	1078	10.4	112

Note: The tensile properties are the average of 6 values each and the toughness values are the averages of 2 tests each.

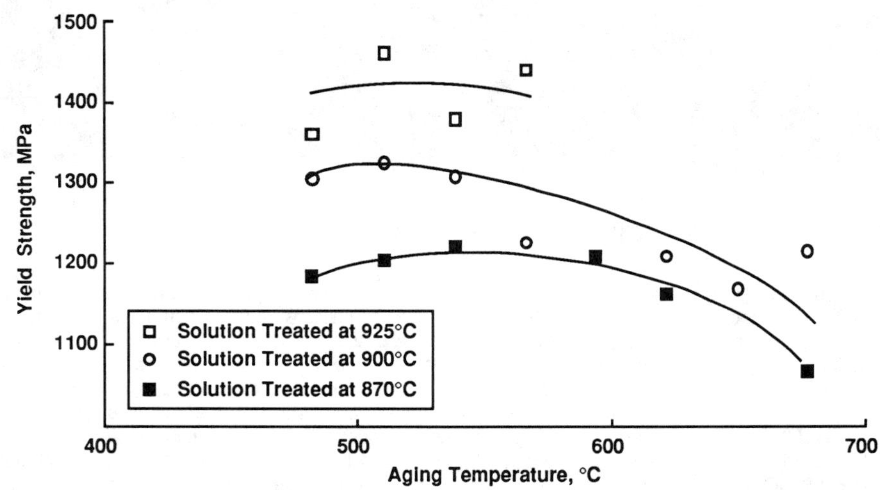

Figure 3. Yield Strength as a Function of Aging Temperature

[1] $K_{apparent}$ is calculated similarly to K_c, the plane stress fracture toughness, except that the initial crack length rather than the final unstable crack length is used. This permits relating a fracture toughness value to the inspectable crack length.

SPF Study

Sheet material, 1.27 mm in thickness, was supplied to Jet Die, Inc. for a cursory study of the superplastic characteristics of this alloy. Four different configurations were formed, each using different forming parameters, Figure 4. A comparison of the SPF parts formed at the three temperatures revealed the greatest thinning with the lowest forming temperature (about 0.88 thickness strain). It should be pointed out that no effort was made to optimize the strain rates at a given temperature.The off-the-press tensile properties, Table III, were dependant upon the forming temperature, as would be expected. Based on the results of the heat treat study, the material formed at 900°C would be expected to have a higher strength than that formed at 885°C, Figure 1. In fact, the material formed at 900°C had a lower strength which may be ascribed to the higher thickness strain in the material formed at 900°C (about .3 for the former versus .08 for the latter) and the consequent increase in the processing time. The 885°C forming cycle was 40 minutes long, as opposed to 80 minutes required for the 900°C forming cycle. This resulted in greater grain coarsening for the latter, possibly contributing to the lower strength for the higher temperature forming cycle (the 900°C formed material had a final grain size of about 6.7μm while the material formed at 885°C had an approximate grain size of 5μm).

Figure 4. Configurations Superplastically Formed from the Ti-6-22-22 Alloy. Compliments of Jet Die

Table III. Tensile Properties of SPF Ti-6-22-22 As-Formed

Forming Temperature °C	UTS MPa	TYS MPa	Elongation %
900	1210*	1072	3.4
885	1236**	1051	8.0
800	1024***	949	10.0

* Avg. of 6 tests
** Avg. of 3 tests
*** Avg. of 5 tests

The engineering thickness strains obtained from all the parts ranged from 0.04 to 0.88. The strengths obtained off the press compared favorably with those obtained during the solution treatment study for corresponding solution treatment temperatures.

Post SPF Heat Treatments

This study was undertaken to determine (1) if the material superplastically formed at the lower temperature could be aged to higher strength (with retention of good ductility), and (2) if the strength of the material formed at a higher temperature could be reduced so as to improve the ductility.

Aging the material formed at 800°C at temperatures in the range of 510 to 565°C provided a substantial strength increase with retention of good ductility, Table IV.

Aging of the material formed at the higher temperature, 885°C, also resulted in substantial strengthening with only a modest loss in ductility. Based on the fracture toughness test results, aging at 510 or 540°C for 8 hours would be recommended if the part were to be used at elevated temperature, in order to impart a stabilization treatment. (Bliss (2) indicated significant aging of solution treated material with thermal exposure at temperatures in the range of 370 to 480°C.) Higher temperature ages would not be recommended due to a significant loss in fracture toughness.

Table IV. The Effect of Post SPF Aging of Ti-6-22-22

	Aging Temperature °C	UTS MPa	TYS MPa	Elongation %
Formed at 800°C				
	510	1247	1110	11.0
	540	1197	1069	9.2
	565	1216	1111	10.0
	No Age	1024	949	10.0
Formed at 885°C				
	510	1349	1271	7.5
	565	1431	1282	6.0
	No Age	1236	1051	8.0

Note: Aging time was 8 hours

Conclusions

The Ti-6Al-2Sn-2Zr-2Mo-2Cr+Si sheet alloy is capable of achieving very high strengths with good ductility and fracture toughness. It exhibits excellent superplastic forming capabilities, which could provide a high strength SPF part. A post SPF age would be recommended for high temperature applications to provide added thermal stability, which also provides an improvement in strength with minimal or no loss in ductility.

Further work is needed to understand the metallurgy of the alloy. The higher temperature aging treatments, at temperatures above 480°C, result in approximately a 35% reduction in the fracture toughness, which may or may not be accompanied by a reduction of strength. In addition, the aging kinetics for the different solution treatments is not understood. Studies are continuing to analyze these phenomena.

ACKNOWLEDGEMENTS

The authors would like to extend their appreciation to Jet Die for their efforts in the superplastic forming of the parts and to Accurate Metallurgical Services for their efforts in providing data in an expeditious manner. We would also like the thank Dr. A. Eikum for performing the thin foil analysis and his helpful comments and Dr. H. Narayanan for his critical review of the manuscript.

REFERENCES

1. I.W. Hall and C. Hammond, Titanium and Titanium Alloys, ed. J.C. Williams and A.F. Belov (Plenum Press, 1982) 601.
2. R.C. Bliss, "Evaluation of Ti-6Al-2Sn-2Zr-2Cr-2Mo-.23Si Sheet", Proceedings of this Conference.
3. G.W. Kuhlman, K.A. Rohde and A.K. Chakrabarti, "Characerization of Ti-6-22- 22S: A High Strength Alpha/Beta Titanium Alloy for Fracture Critical Applications", Proceedings of this Conference.
4. P.A. Russo, S.R. Seagle and H.B. Bomberger, "Development of a 900F Titanium Alloy (Report AFML-TR-70-125, July 1970).
5. C. Ramachandra and V. Singh, "Effect of silicides on tensile properties and fracture of alloy Ti-6Al-5Zr-0.5Mo-0.25Si from 300 to 823K," Journal of Materials Science, 25 (1988), 835-841.

TENSION CREEP OF A SINGLE PHASE γ-TiAl ALLOY

R.W. Hayes
Metals Technology, Inc.
Northridge, CA 91324
and
P. L. Martin
Rockwell Science Center
Thousand Oaks, CA 91360

Abstract

The creep deformation behavior of a single phase gamma TiAl alloy: Ti-53Al-1Nb (a/o) in uniaxial tension has been studied within the temperature range 760 to 900°C and initial stress levels ranging from 32 to 345 MPa. Two distinct regions of creep deformation behavior are identified. Depending on the temperature and stress level, these regions correspond to diffusional creep controlled by grain boundary diffusion (Coble creep) and power law creep controlled by dislocation motion. In addition, at the lowest test temperature (760°C), a transition to power law breakdown from the diffusional creep region was observed. The mechanisms are identified by analysis of the stress and temperature dependence of the minimum strain rates. The regimes of creep deformation and of power law breakdown are summarized in the construction of a preliminary deformation mechanism map.

Introduction

The gamma (γ) TiAl alloys based upon the tetragonal $L1_0$ crystal structure are potential light weight high temperature alloys for use in advanced aircraft engines and structures. Based on present knowledge of the behavior of γ-TiAl alloys, the intended use temperature ranges from about 700°C to 900°C. In the past decade, significant advancement in the understanding of microstructure property correlations have been made; much of this work is summarized in the review article by Kim (1). Because of the limited ductility at room temperature, the primary focus of fundamental research on γ-TiAl has been the understanding of the deformation behavior at room temperature. In a recent series of papers, Huang and Hall (2-4) discuss the influence of composition and microstructure on the deformation behavior of binary and ternary γ-TiAl alloys in bending and in uniaxial tension. Compositions and microstructures which lead to enhanced room temperature plasticity have been identified along with the corresponding dislocation types.

In this paper we will focus our attention on the high temperature creep deformation behavior of the single phase γ-TiAl alloy Ti-53Al-1Nb. The single phase γ composition has been selected because this phase accounts for the major volume of the near-γ two phase ($\gamma + \alpha_2$) alloys currently under development. From a microstructural viewpoint, the single phase γ alloys are simpler than the multi-phase alloys of engineering significance. This allows for an easier

Titanium '92
Science and Technology
Edited by F.H. Froes and I. Caplan
The Minerals, Metals & Materials Society, 1993

correlation between properties without the intervening complexity of the concurrent deformation activity taking place at the γ-α_2 lath interfaces. The major goal of this study is to correlate the tension creep behavior with the mechanisms associated with different deformation regimes.

Experimental

The material used in this program was melted and isothermally forged into a 235 mm diameter by 13 mm thick pancake shape at Timet Corporation, Henderson, Nevada. The chemical composition is presented in Table I. Specimen blanks, arranged radially in the pancake forging, were machined prior to a recrystallization heat treatment at 1300°C for 1 hour. Following the heat treatment, creep specimens of a buttonhead configuration were fabricated by low stress crush grinding. These specimens had a gage length of approximately 30 mm and a gage diameter of 5 mm. Tests were conducted using uniaxial tension on constant load lever arm creep machines in air over the temperature range from 760 to 900°C at constant initial applied stress levels ranging from 32 to 345 MPa. Creep strain was measured with a dual dial gage averaging extensometer attached to the shoulders of the specimen. The strain resolution was $\pm 10^{-5}$. Most tests were conducted to failure.

Table I. Chemical composition (atomic percent for all elements).

Ti	Al	Nb	Fe	O	N	H	C
bal.	52.9	0.91	0.048	0.153	0.003	0.022	0.054

Results

Microstructure

The optical microstructure of the Ti-53Al-1Nb following the 1300°C recrystallization heat treatment is presented in Figure 1. This heat treatment resulted in an equiaxed recrystallized grain size of about 260 µm. Figure 2 shows a TEM micrograph of the initial microstructure. No evidence for the existence of the Ti_2AlN particles observed by other investigators (5,6) could be found due to the low nitrogen content of the forging. These particles, when present, interact strongly with dislocations and thus exert a strong influence on creep deformation behavior (6,7).

Creep Deformation

Table II is a summary of the creep deformation conditions and the resultant minimum strain rate, approximate time to tertiary creep, and rupture time for each test. The temperature dependence of Young's modulus for γ-TiAl was taken from the work of Schafrik (8). Figure 3 shows a log-log plot of minimum strain rate versus stress normalized by the (temperature dependent) modulus. At 760°C, between (log) modulus normalized stress levels of -3.32 and -2.73, the apparent stress exponent was calculated to be 1. At modulus normalized stress levels exceeding -2.73, a transition in stress exponent from 1 to 21 occurred. At 832°C, a slope equal to 1 was measured within a modulus normalized stress range of -3.48 to -3.32. Above -3.32, the stress exponent increased to a value of 6. At 900°C, over the entire stress level, a slope equal to 6 was observed. According to conventional creep deformation analysis, the transition in apparent stress exponent from 1 to 6 can be taken as a transition from creep controlled by a diffusional process ($n \approx 1$) to a dislocation controlled creep deformation process ($n \approx 6$).

An Arhennius plot of log minimum strain rate vs 1/T at a modulus compensated stress of -3.32 is presented in Figure 4. These data suggest that the creep activation energy at this stress

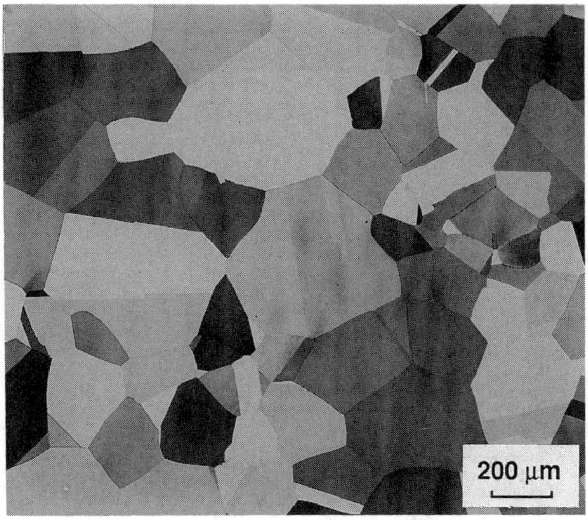

Figure 1. Optical microstructure of single phase gamma Ti-53Al-1Nb following 1300°C/1 hour recrystallization heat treatment. Grain size determined by line intercept is on the order of 260μm.

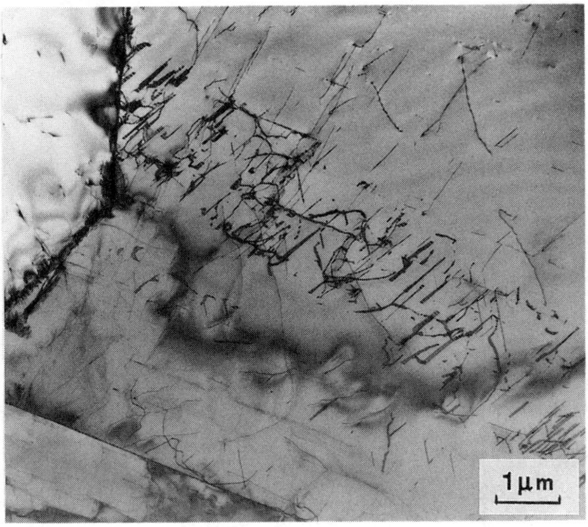

Figure 2. TEM micrograph of Ti-53Al-1Nb following 1300°C/1 hour recrystallization. Note low initial dislocation density and lack of Ti_2AlN particles.

level is not constant over the temperature range investigated in this study. Figure 4 shows a low temperature region (760°C to 796°C) having an activation energy of approximately 190 kJ/mole and a high temperature region (832°C to 900°C) having a value of 560 kJ/mole.

Table II. Summary of Ti-53Al-1Nb Creep Data

T (°C)	σ (MPa)	$\log(\sigma/E)$	$\varepsilon_{min.}$ (per hr.)	t_t (hrs.)	t_R (hrs.)
760	69.4	-3.32	$2.2X10^{-6}$	250.0	2294.5
760	103.4	-3.14	$3.9X10^{-6}$	180.0	586.8
760	206.8	-2.85	$1.0X10^{-5}$	20.0	68.4
760	275.7	-2.73	$1.4X10^{-5}$	9.0	24.8
760	344.7	-2.62	$1.0X10^{-3}$	1.3	2.45
796	69.4	-3.32	$4.1X10^{-6}$	10.0	
832	48.2	-3.48	$6.9X10^{-6}$	20.0	866.1
832	69.4	-3.32	$1.3X10^{-5}$	16.0	209.1
832	103.4	-3.15	$8.1X10^{-5}$	11.0	45.3
832	137.8	-3.03	$6.0X10^{-4}$	4.5	17.5
871	69.4	-3.32	$2.0X10^{-4}$	9.0	
900	31.7	-3.65	$2.5X10^{-6}$	60.0	377.9
900	48.3	-3.45	$1.7X10^{-4}$	12.0	91.0
900	69.4	-3.31	$5.9X10^{-4}$	6.0	26.4
900	103.4	-3.15	$5.2X10^{-3}$	1.6	4.8

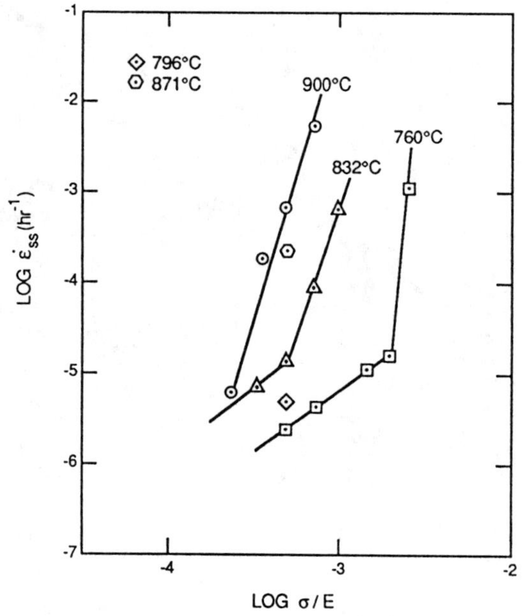

Figure 3. Minimum strain rate plotted versus modulus compensated stress, the slope of which is the stress exponent.

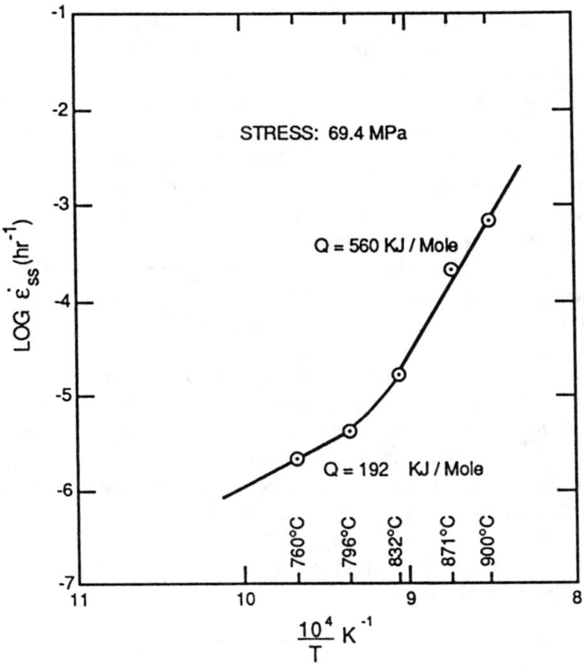

Figure 4. Arrhenius plot of minimum strain rate versus 1/T at modulus compensated stress of -3.3 (applied stress of 69.4 MPa).

Discussion

The data presented in Figures 3 and 4 indicate that two regions of creep deformation behavior exist in the Ti-53Al-1Nb alloy within the stress and temperature regime of the present study. Recently, Rosler et al (9) have reported an activation energy for volume diffusion in TiAl of 330 kJ/mole (this value is actually taken from compression creep experiments reported in reference 15). The activation energy for grain boundary diffusion, Q_b, in γ-TiAl has been postulated by Ashby as $Q_b \approx 0.611 Q_v$, where Q_v is the lattice self diffusion activation energy (10). This agrees well with results previously reported by Langdon and Vastava (11). Using these criteria, Oikawa's data (15) provides the an activation energy for self diffusion for TiAl of approximately 200 kJ/mole. This calculation is in excellent agreement with the low temperature activation energy reported in this study of 190 kJ/mole (Figure 4). The high temperature creep activation energy of 560 kJ/mole is considerably higher than the reported value for self diffusion (9).

The tests corresponding to a modulus compensated stress exponent of 1 correspond exactly to the temperature regime having an activation energy of 190 kJ/mole. Using conventional creep analysis, we would suggest that the stress and temperature dependence taken together indicate that Coble creep, where grain boundary diffusion is the rate limiting process, is the dominant mechanism in this regime. The constitutive equation for the minimum strain rate during Coble creep can be written in the form (12):

$$\varepsilon_{min} = \frac{8\, w\, D_{g.b.}\, \Omega\, \sigma}{d^3\, k\, T} \qquad (1)$$

where w $D_{g.b.}$ is the product of the grain boundary thickness and grain boundary diffusivity, Ω is the atomic volume, d is the grain diameter, σ is the applied creep stress, k is Boltzman's constant, and T is the absolute temperature. As indicated in Eq.1, the activation energy for Coble creep is that for grain boundary diffusion and shows a d^{-3} grain size dependence. A grain size dependence of the minimum strain rate was not determined in this study. However, prior work (15) has failed to show a strong grain size dependence in compression creep of single phase TiAl. Careful examination of the test parameters will show that this prior work was conducted at higher stresses and temperatures with consequently higher minimum strain rates. These authors were working in the normal dislocation controlled creep regime in agreement with our data discussed below. Thus, our results in the low temperature region of creep deformation (760°C to 796°C) are consistent with the correct stress and temperature dependence for Coble creep. Determining the grain size dependence in this regime is necessary to prove this contention.

The increase in stress exponent at 760°C from 1 to 21 as the modulus compensated stress exceeds -2.73 has several possible sources. One possible rationale for stress dependencies of this magnitude can be found in systems strengthened by small scale incoherent particles (13). However, no particle was observed in the Ti-53Al-1Nb. Another explanation is a direct transition from diffusional creep controlled by grain boundary diffusion to power law breakdown where dislocation glide or twinning become the dominant deformation processes. Direct transition from diffusional flow to dislocation glide has been observed in some gamma prime strengthened superalloys (14). Once the stress exceeds a critical level, it becomes high enough to activate glide or twinning dislocation sources.

Having established that Coble creep is the most likely process controlling creep at temperatures below 800°C, the high temperature region must be rationalized. Referring to Figure 3, the stress exponent determined for creep in the high temperature region is approximately 6. This value agrees with the stress exponent associated with dislocation climb controlled creep in pure metals and simple alloy systems. A typical power law representation for the minimum strain rate in the high temperature region is written:

$$\dot{\varepsilon}_{min} = A\,(\sigma/E)^n \exp(-Q/RT) \qquad (2)$$

where A is a structure dependent constant, σ is the applied stress, E is the unrelaxed dynamic Young's modulus, n is the modulus corrected stress exponent, Q is the creep activation energy, and R and T have their usual meaning. Equation 2 predicts that the activation energy should be equal to that for lattice self diffusion. The 560 kJ/mole activation energy for high temperature creep of Ti-53Al-1Nb (in the region corresponding to n=6) is significantly higher than the self diffusion value reported (9,15). Hence, the stress dependence of the minimum strain rate is in accord with Eq.2 but the temperature dependence is not. In compression creep, Oikawa (15) measured the activation energy for creep in Ti-53.4Al and found a value in excess of 600 kJ/mole at low strain which decreased with increasing strain reaching a plateau of approximately 350 kJ/mole at strains of 20%. The higher strain region, corresponding to 350 kJ/mole, is within the tertiary creep regime for these alloys in tension. In the low strain region there appears to be reasonable agreement between our data and Oikawa's compression data. From a qualitative standpoint, this would tend to indicate that the activation energy for creep controlled by dislocation motion at least in the low strain region is higher than that for self diffusion in γ-TiAl. However, this is clearly not the case in near-γ two phase TiAl alloys where several investigators have measured creep activation energies in the power law regime which are in the range of 330 to 370 kJ/mole (6, 16-18).

At present, a quantitative explanation for the high activation energy in the high temperature region is difficult. Creep activation energies, as high as 2-3 times the self diffusion value of the matrix, have been reported for particle strengthened alloys (13). As in the case of the high stress exponents, the high activation energies are rationalized in terms of dislocation-particle interactions. Incorporation of a back stress or threshold stress term into Eq. 2 allows the determination of the true activation energy for creep and results in a decrease of the initial activation energy value to the value for self diffusion of the matrix. This approach is not applicable in the present case as no such particle was found. In addition, the stress dependence is rational. A more fundamental difference between the near-γ two phase alloys and the single phase γ alloy of this study must be used to explain the observed differences in the creep activation energies. One such difference may be the relative abundance of dislocation sources available to support deformation. It has been shown that the near-γ alloys have a relatively high dislocation density (2). Some number of these dislocations may be available to support creep deformation at the onset of the test. The many γ-α_2 interfaces and their abundance of potentially mobile dislocations are not present in single phase alloys. Using an argument of a single process controlling creep deformation, the high activation energy of 560 kJ/mole may be that necessary to create glide or twinning dislocations in the $L1_0$ structure. The above argument is speculative; however, it could also apply to the low strain region in compression and thereby account for the correlation between the low strain activation energy measured by Oikawa (15) and that measured in the current study. By measurement of the temperature dependence of the time in tertiary creep (defined as the rupture time minus the time to start tertiary), the current data gives a temperature dependence (activation energy) of 354 kJ/mole. The tertiary creep region in tension corresponds to a higher strain and thus might be considered as the equivalent of Oikawa's plateau. As the strain increases, a greater abundance of mobile dislocations would be expected, leading to creep deformation being limited by normal climb process with an activation energy more nearly equal to that for self diffusion.

As a summary of the present results, the two regions of creep deformation and the power law breakdown are plotted as a preliminary deformation mechanism map in Figure 5. The deformation field boundary separating the Coble creep region from the power law region was not defined rigorously by equating the rate equations for the two mechanisms. We have qualitatively defined this boundary based on experimental observation of rate limiting mechanisms and corresponding microstructures.

Conclusions

Based on uniaxial tension creep experiments on Ti-53Al-1Nb, the following conclusions can be drawn:

1. The low temperature creep consists of a stress exponent of 1 and an activation energy of 190 kJ/mole probably due to grain boundary diffusion in γ-TiAl.
2. The second region, at higher temperatures, consists of a stress exponent of 6 and an activation energy of 560 kJ/mole. The high activation energy is suggested to represent dislocation creep with the rate controlling process being that of dislocation generation.
3. The activation energy decreases to 354 kJ/mole in the tertiary creep regime corresponding to high strain deformation. This suggests that an increase in the mobile dislocation density gives rise to creep controlled by a normal climb process.
4. A preliminary deformation mechanism map showing the proposed regions of creep deformation and power law breakdown has been constructed based on experimental creep data and microstructural observations.

References

1. Y.W. Kim, JOM, 41, 1989, pp 24-30.
2. S.C. Huang and E.L. Hall, Met Trans A, 22A, 1991, pp 427-439.
3. S.C. Huang and E.L. Hall, Acta Metall. Mater., 39, 1991, pp 1053-1060.

4. S.C. Huang and E.L. Hall, Met. Trans A, 22A, 1991, pp 2619-2627.
5. M.J. Kaufman, D.G. Konitzer, R.D. Shull and H.L. Frazer, Scr. Metall., 20, 1986, pp 103-108.
6. P.L. Martin, D.H.Carter, R.M. Aiken Sr., R.M. Aiken Jr., and L. Christodoulou, Creep and Fracture of Engineering Materials and Structures, B. Wilshire and R.W. Evans, eds., The Institute of Metals, London, 1990, pp 265-275.
7. A. Loiseau and A. Lasalmonie, Materals Sci. and Eng., 67, 1984, pp163-168.
8. R.E. Schrafrik, Met Trans A, 8A, 1977, pp 1003-1006.
9. J. Rosler, G. Bao and A.G. Evans, Acta Metall. Mater., 39, 1991, pp2733-2738.
10. M.F. Ashby, in Operating Manual for HIP 487, Cambridge Univ., 1987.
11. T.G. Langdon and R.B. Vastava, Mechanical Testing for Deformation Model Development, R.W. Rohde and J.C. Swearenger eds., ASTM, 1980, pp 445-451.
12. J.K. Gregory, J.C. Gibeling and W.D. Nix, Met. Trans A, 16A, 1985, pp 777-787.
13. V.C. Nardone, D.E. Matejczyk and J.K. Tien, Acta Metall., 32, 1984, pp 1509-1517.
14. M.F. Ashby, Proceedings, Third International Conference on Strength of Metals and Alloys, vol 2, Cambridge, England, 1973, p 8.
15. H. Oikawa, High Temperature Aluminides and Intermetallics, S.H. Whang, C.T. Liu, D.P. Pope and J.O. Stiegler, eds., TMS, 1990, pp 353-373.
16. P.L. Martin, M.G. Mendiratta and H.A. Lipsitt, Met. Trans A, 14A, 1983, pp 2170-2174.
17. S.L. Kample, J.D. Bryant and L. Christodoulou, Met. Trans A, 22A, 1991, pp 447-454.
18. R.W. Hayes and B. London, Creep Deformation of a Cast Near Gamma TiAl Alloy Ti-48Al-1Nb, 1991, submitted to Acta Metall. Mater.

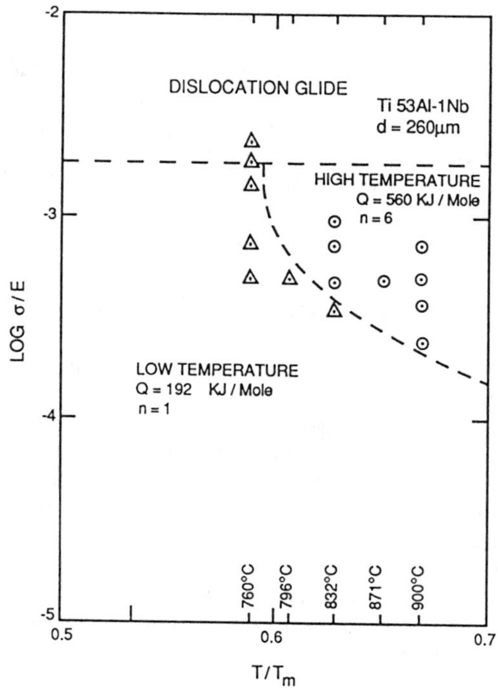

Figure 5. Preliminary deformation mechanism map for tension deformation in Ti-53Al-1Nb. Regions dominated by power law dislocation creep, grain boundary diffusion (Coble) creep, and plasticity (power law breakdown) are identified.

STRAIN-RATE AND TEMPERATURE EFFECTS ON THE DEFORMATION AND CRACK INITIATION BEHAVIOR OF Ti-48Al-1V

L. S. Harbison* , D. A. Koss, and R. J. Bourcier+

* Department of Materials Science and Engineering
Penn State University, University Park, PA 16802.

Abstract

The TiAl-based alloy Ti-48Al-1V has been deformed in compression to monitor both the stress-strain response and microcracking behavior over a range of strain rates from 10^{-4}/s to 10^{2}/s and temperatures from 77 to 1073K. The susceptibility of the alloy to strain-induced microcracks is found to be very sensitive to strain rate. At strain rates of 10^{-4}/s and 10^{-1}/s, only an occasional, isolated microcrack was observed after compressive straining, even at 77K where very planar slip dominated. In contrast, deformation at 10^{2}/s induced large populations of short microcracks ($\leq$ 10μm long), presumably due to local strain incompatibilities. The microcracks appeared to concentrate in certain grains within the microstructure and were prevalent even to test temperatures of 623K, despite more uniform slip. Associated with the microcracking at 10^{2}/s was a decreased rate of strain hardening when compared to that at 10^{-4}/s or 10^{-2}/s.

The stress-strain behavior during compression at strains up to 0.10 is also presented for a wide range of temperatures from 77 to 1073K. Notable among these data are (1) the presence of serrated yielding and negative strain-rate hardening at 623K for strain rates between 10^{-4} and 10^{-1}/s and (2) an unexplained decrease in strain hardening observed for tests performed at 10^{2}/s..

Introduction

As is evidenced in this conference, there is a large amount of interest in the mechanical behavior of intermetallic alloys based on the compound TiAl. Previous studies of the deformation behavior of TiAl alloys have examined in detail factors such as alloy content, effects of temperature, influence of processing treatments, deformation substructure, and dislocation as well as twinning mechanisms (see ref's 1-3 for reviews). However, the combined effects of strain rate and temperature, especially with regard to deformation behavior at high strain rates, has only been recently examined in a study by Gray [4]. That research uses Ti-48Al-1V as a model TiAl system and examines the stress-strain response and associated deformation-induced substructure from strain rates of 10^{-3}/s to 7.5×10^{3}/s over the temperature range 298 to 973K. The results indicate deformation to occur by a mixture of slip and twinning with increasing test temperature and/or strain rate increasing the incidence of twinning. However, only minor influences of temperature and strain rate were detected in the work hardening behavior, although a significant strain-rate hardening component was reported [4].

* Currently: Babcock and Wilcox Nuclear Technologies, P. O. Box 10935, Lynchburg, VA 24506.

+ Metallurgy Department, Sandia National Laboratories, Albuquerque, NM 87185.

Titanium '92
Science and Technology
Edited by F.H. Froes and I. Caplan
The Minerals, Metals & Materials Society, 1993

The purpose of the present study is to complement that of Gray by examining the same Ti-48Al-1V alloy, heat treated in an identical manner. In this study, experimental observations examining the influence of temperature and strain-rate on the mode of slip and the occurrence of microcracking are presented for the temperature range 77 to 973K and strain rates 10^{-4}/s to 10^{2}/s.

Experimental Procedures

As in the Gray study [4], the Ti-48Al-1V was donated by TIMET and had a composition in w/o of Ti-34.3Al-1.45V. There was also a considerable level of carbon (0.120 w/o); in addition 0.075 0 and 0.008 N (w/o) were present. The alloy, hereafter referred to as Ti-48-1, was supplied in the form of an upset forging which was roughly 5 cm thick after forging at ≈1645K and subsequently at 1450K. After specimen preparation, a final heat treatment of 1173K for six hours in a vacuum followed by a furnace cool was performed.

All tests were performed in compression, and two specimen geometries were used . For all of the tests at strain rates of 10^{-4}/s and 10^{-1}/s, rectangular specimens (12.7x6.35 x6.35 mm) were used. One face of each specimen was mechanically polished through 0.05 μm alumina for microscopy purposes, while the remaining three sides were ground to 600 grit. End faces were ground parallel to within ±0.013 mm. Some of the high strain rate (10^{2}/s) tests were also performed using cylindrical specimens (12.7x6.35 mm diameter); stress-strain results at high strain rates from the cylindrical and rectangular specimens were identical.

Compression tests were performed in a screw-driven load frame at initial strain rates of 10^{-4}/s and 10^{-1}/s, while the high strain-rate tests at 10^{2}/s were performed using a compression cage fixture mounted in a servo-hydraulic load frame [5]. The validity of the high rate stress-strain data was confirmed by testing annealed 304L stainless steel and duplicating data from similar high strain-rate tests performed elsewhere [6]. Tests at 773 and 1073K were performed in an atmosphere of ultra high purity argon (99.999% pure).

Results and Discussion

Microstructure

Fig. 1 shows the microstructure of the Ti-48-1 alloy used in this study to consist of about 80% large (≈30μm) equiaxed gamma-phase grains combined with regions comprised of elongated grains, about 30 μm long, which are a mixture of gamma and alpha-two phases in the form of alternating lamallae. Roughly 5 to 10 vol. pct. of the microstructure is the alpha-two phase. Previously, Gray has shown by TEM of the undeformed microstructure that a high density of fine precipitates (≤0.05 μm) exist; he concluded that they probably are Ti_2AlC, heterogeneously nucleated on dislocations within the gamma phase as a result of the high carbon content (1200 wt.ppm) of the present alloy [4].

Stress-Strain Response at Low Strain Rates

When tested in compression, the Ti-48-1 specimens deform plastically with only an occasional microcrack (<10 μm long) present on the surface. Given the absence of specimen buckling and minimal barreling (<2%), Fig. 2 should represent an accurate depiction of the quasi-static true stress-true strain response of this Ti-48-1 alloy over a wide range of temperatures for strains up to 0.10. As observed in other gamma Ti alloys [7] Fig. 2 shows that the yield stress is essentially athermal between 448 and 1073K. Similar to the behavior of Ti-48Al [8] there is also a moderate increase in yield strength with decreasing temperature from 448 to 77K.

In addition to increasing the yield strength at low temperatures, Fig. 2 shows that decreasing test temperature below room temperature influences work hardening. Specifically, the work hardening rate ($d\sigma/d\varepsilon$) is large, especially at 77K where for $0.04 \leq \varepsilon \leq 0.10$, $d\sigma/d\varepsilon = 5650$ MPa compared to $d\sigma/d\varepsilon = 3570$ MPa at room temperature. The latter value compares to a value of 3790 MPa observed by Gray at room temperature at strain rates of 10^{-3}/s [4]. Assuming a Taylor factor of 3.1 for TiAl, we note that in terms of shear stress τ, shear strain γ, and shear modulus G of 70 GPa [9], the strain hardening of Ti-48-1 in the absence of recovery at 77K is $d\tau/d\gamma \cong$

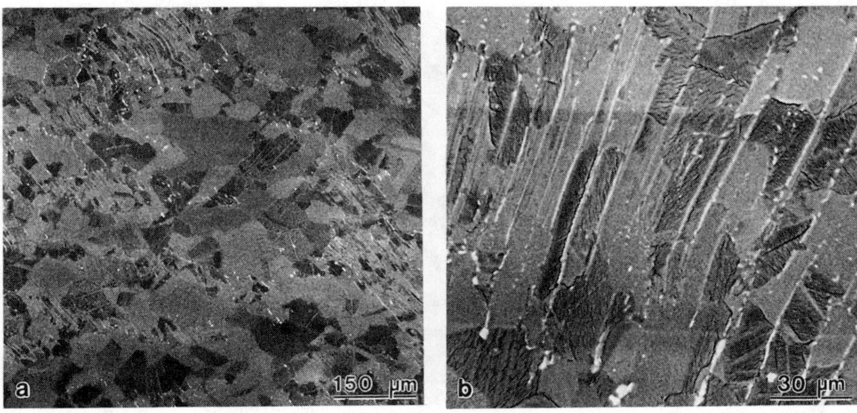

Figure 1 - Optical (a) and scanning electron micrographs (b) of the Ti-48Al-1V alloy. Arrows denote lath regions.

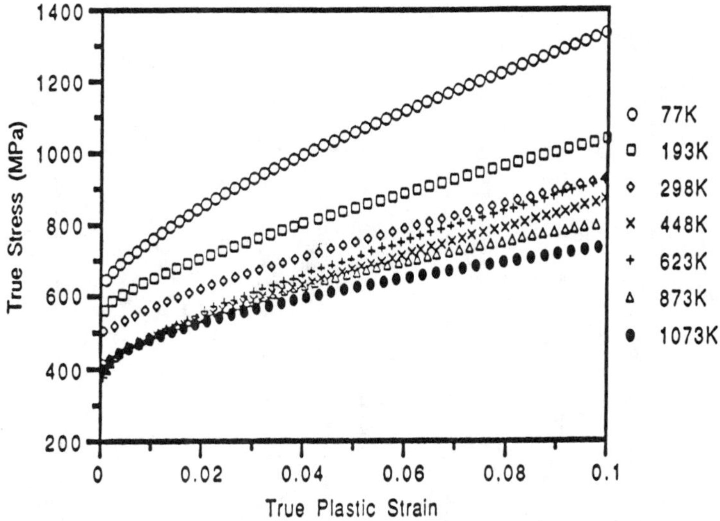

Figure 2 - True stress-true strain responses of Ti-48Al-1V as measured in compression.

G/120. This compares to $d\tau/d\gamma \cong G/120$, which is typically observed for Stage II strain hardening in fcc single crystals and the early stages of strain hardening in polycrystalline fcc alloys [10]. This increase in work hardening is not surprising in view of the character of slip in TiAl, its low stacking fault energy, and the presence of twinning, as well as slip as a mode of deformation [2,4].

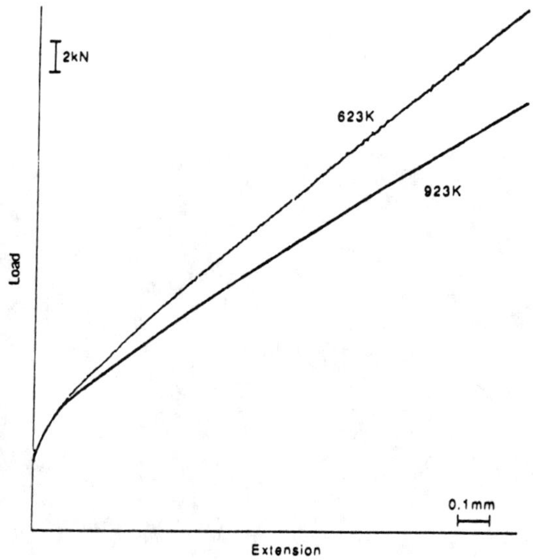

Figure 3 - Load-extension data showing jerky flow Ti-48Al-1V at a 10^{-4}/s and 623K.

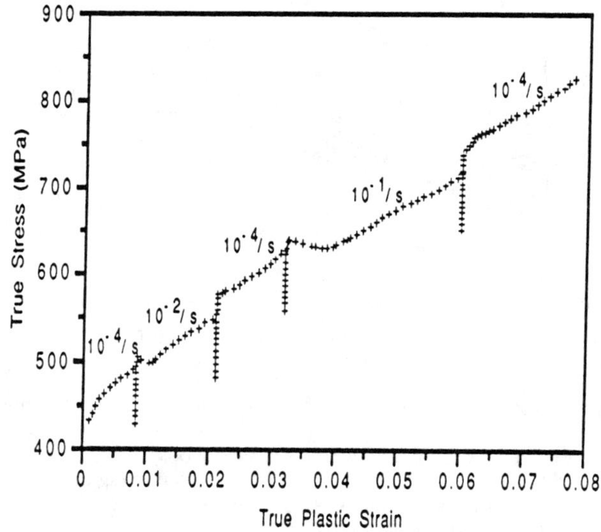

Figure 4 - Strain-rate change responses at 623K showing strain-rate softening.

Perhaps the most significant observation regarding the work hardening behavior above room temperature is the presence of serrated yielding at 623K and 10^{-4}/s. This is manifested by (a) jerky flow, see Fig. 3, (b) negative strain-rate hardening during jump tests between 10^{-4}/s and 10^{-1}/s, see Fig. 4, and (c) an increase in strain hardening at 623K compared to either lower or higher test temperatures, see Fig. 2. These three effects occur only at 623K and 10^{-4}/s and are characteristic of the Portevin-Le Chatelier mechanism in which mobile solute atoms, probably interstitial oxygen, carbon or nitrogen, are involved in a dynamic strain-aging process involving a pinning, unpinning, and repinning sequence with mobile dislocations. While serrated yielding (or jerky flow) has been observed in alpha as well as beta Ti alloys [11-13], we are not aware of any report of its occurrence in TiAl-based alloys. Interestingly, at comparable strain rates, the dynamic strain aging in single phase alpha or beta Ti alloys occurs in the vicinity of 600 to 760 K [11-13] or comparable to the 623K condition for this TiAl alloy. It is also noteworthy that in alpha Ti the dynamic strain aging causes a significant increase in work hardening as is observed in our TiAl alloy in Fig. 2.

High Strain Rate Behavior and Microcracking

When compared to the stress-strain response at 10^{-4}/s and 10^{-1}/s, our tests show work hardening to decrease at 10^{2}/s as shown in Fig. 5. This result differs with the observations of Gray who observed no significant differences between work hardening of the same Ti-48-1 alloy at 10^{-3}/s, 10^{-1}/s, and 3.5 to 7.5×10^{3}/s [4]. We are not able to explain this apparent inconsistency. Adiabatic shear banding could account for a decreased strain hardening response. While this is the case in alpha-beta Ti alloys [14-17] as well as Ti_3Al-based alloys [18], it was not observed in our Ti-48-1 specimens during the high strain-rate testing.

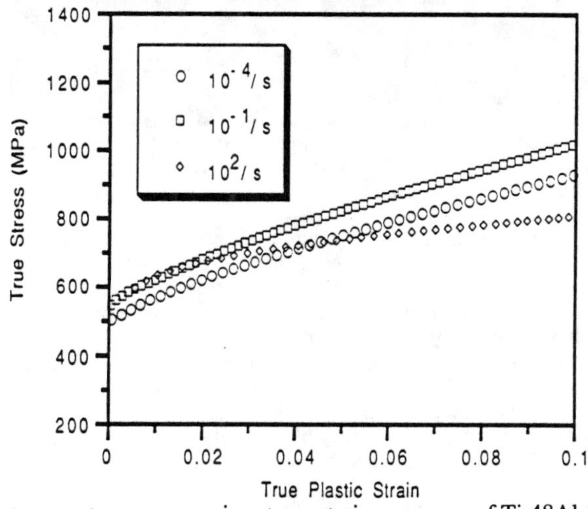

Figure 5 - Room temperature compressive stress-strain responses of Ti-48Al-1V at strain rates of 10^{-4}/s, 10^{-1}/s, and 10^{2}/s.

We do, however, observe strain-rate-induced differences in deformation characteristics and especially surface microcracking. As shown in Fig. 6a&b, deformation to strains of 0.10 at 10^{-4}/s (as well as 10^{-1}/s) result in extensive surface relief caused by either coarse, planar slip or twinning within the gamma phase. Similar deformation traces were observed in quasi-static deformation of TiAl single crystals at 77K [19]. Unlike Ti_3Al alloys [18], only a few isolated microcracks are seen, despite the large compressive strains and the small tensile ductilities present in these materials. In contrast, deformation at 10^{2}/s causes locally high densities of short microcracks within certain gamma-phase grains and near grain boundaries; see Fig. 6c and d. The microcracking only occurs at high strain rates of 10^{2}/s and is observed at the test temperatures of 298 to 623K. Furthermore, the microcracks originated despite a tendency for more uniform deformation; Fig. 6a and b compared to Fig. 6c and d. Cross-section microscopy indicates that such cracking is a surface effect; no microcracks are initiated below the surface were detected. Furthermore it should be noted that the specimens containing the microcracks after 10^2/s deformation were prepared in an identical manner to those tested at lower strain rates.

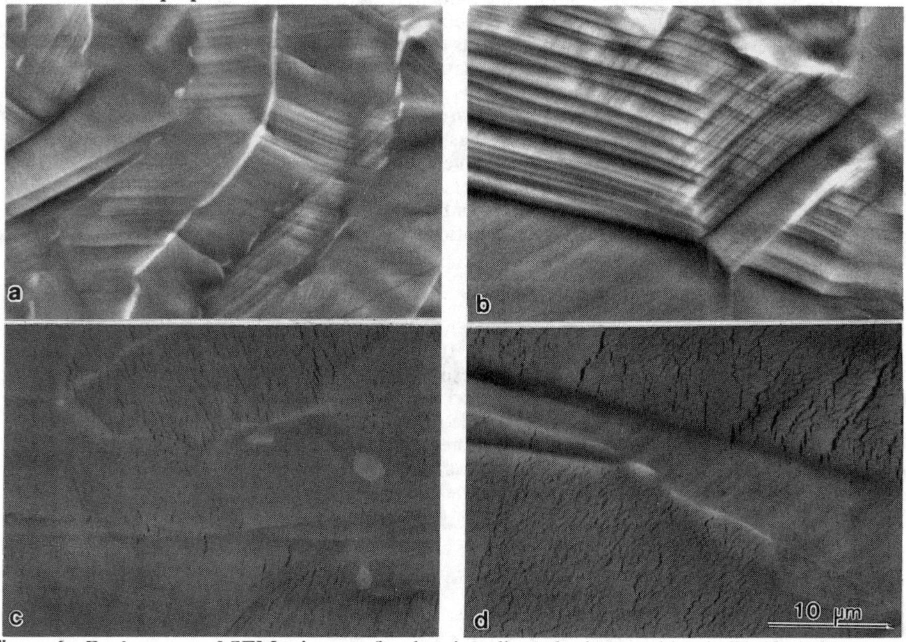

Figure 6 - Back-scattered SEM micrographs showing slip and microcracking in Ti-48Al-1V at 0.10 compressive strain at (a) 10^{-4}/s and 298K (b) 10^{-4}/s and 623K, (c) 10^2/s and 298K, and (d) 10^2/s and 623K.

We do not know why the high strain-rate deformation causes microcrack formation when no such cracking is observed at either 10^{-4}/s or 10^{-1}/s, even at 77K where the flow stress is quite high. The cracks occur within the gamma phase and their orientation, which is parallel to the stress axis, orients them normal to the maximum tensile strain axis (the transverse direction across the specimen). As a result, they are very likely a result of tensile stresses which arise locally within the gamma grains due to strain incompatibilities within the microstructure. This suggests a change in deformation mode at high strain rates such that strain incompatibilities increase. Gray observes an increase in the incidence of twinning at high strain rates [4]. However, deformation twinning usually accompanies increased ductility of TiAl alloys at elevated temperatures[2]; thus, we might expect improved crack initiation resistance at high strain rates and not a decreased resistance.

Finally, we note that the presence of microcracks induced at high strain rates suggest a dependence of such a material to loading history. For example, a component experiencing an impact loading would be very likely to exhibit poor high cycle fatigue resistance as a result of the short cracks, initiated during impact loading, and subsequently growing under cyclic loading. Similar sensitivity to environmental embrittlement might be envisioned.

Summary

The most significant observation of this study is the occurrence of locally large populations of microcracks ($\leq$ 10μm long) which result from compressive deformation at 10^2/s but not at 10^{-4}/s or 10^{-1}/s. Such crack initiation occurs over a range of test temperatures from 298 to 623K despite the presence of slip which appears more uniform than at lower strain rates. This suggests that components made of TiAl alloys may be susceptible to premature failure under certain conditions, such as high cycle fatigue, if the component has suffered pre-existing damage due to high strain-rate loading. Finally, we also note the presence of serrated yielding or jerky flow at 623K and a strain rate of 10^{-4}/s.

Acknowledgments

The experimental assistance of D. Schmale, P. Hatch, and D. Grant is gratefully acknowledged. We also wish to thank Dr. G. T. Gray for helpful discussions. This research was supported by the U.S. Department of Energy under Contract No. DE-AC04-76-DP00789.

References

1. H. A. Lipsitt in High-Temperature Ordered Intermetallic Alloys, vol. 39, C. C. Koch, C. T. Liu, and N. S. Stoloff (eds.), (Pittsburgh, PA: Materials Research Society, 1985) 352.

2. E. L. Hall and S-C. Huang in Microstructure/Property Relations in Titanium Aluminides and Alloys, Y-W. Kim and R. B. Boyer (eds.), (Warrendale, PA: TMS, 1991) 47.

3. Y-W. Kim in ref. 2, pp. 91-104; see also in High-Temperature Ordered Intermetallic Alloys IV, vol. 213 (Pittsburgh, PA: Materials Research Society, 1991) 777.

4. G. T. Gray III in Microstructure/Property Relations in Titanium Aluminides and Alloys, Y-W. Kim and R. B. Boyer (eds.), (Warrendale, PA: TMS, 1991) 263.

5. L. S. Harbison, M. S. Thesis, The Pennsylvania State University, 1991.

6. M. G. Stout and P. S. Follansbee, Trans. ASME, 108 (1986) 344-353.

7. J. M. Larsen, K. A. Williams, S. J. Balsone, and M. A. Stucke in High Temperature Aluminides and Intermetallics, S. H. Whang, C. T. Liu, D. P. Pope, and J. O. Stiegler, eds., (Warrendale, PA: TMS, 1990) 521.

8. S. C. Huang and E. L. Hall in High-Temperature Ordered Intermetallic Alloys III, vol. 133, C. T. Liu, A. J. Taub, N. S. Stoloff, and C. C. Koch, eds. (Pittsburgh, PA: MRS, 1989) 373.

9. R. E. Schatrik, Metall. Trans. A, 8A (1977), 1003.

10. W. D. Nix, J. C. Gibeling, and D. A. Hughes, Metall. Trans. A, 16A (1985) 2215.

11. R. Zeyfang and H. Conrad, Acta Metall. 19(1971) 985.

12. A. Garde, A. T. Santhanam, and R. E. Reed-Hill, Acta Metall. 20 (1972) 215.

13. G. Hari Narayan and T. F. Archbold, Metall Trans., 2(1971) 1264.

14. R. F. Recht, J. Appl. Mech. Trans. ASME, 31 (1964) 189.

15. R. S. Culver in Metallurgical Effects at High Strain Rates, R. W. Rohde, B. M. Butcher, J. R. Holland, and C. H. Karnes, eds., (Plenum Press, New York 1973) p. 519.

16. S. P. Timothy and I. M. Hutchings, Acta Metall 33 (1985) 667.

17. H. Grebe, H. Pak, and M. A. Myers, Metall. Trans. A, 17A (1985) 761.

18. L. S. Harbison, R. J. Bourcier, and D. A. Koss in Microstructure/Property Relationships in Titanium Aluminides and Alloys, Y-W. Kim and R. R. Boyer eds. (Warrendale, PA: TMS, 1991) 437.

19. T. Kawabata, T. Abumiya, T. Kanai, and O. Izumi, Acta Metall, 38 (1990) 1381.

MECHANICAL BEHAVIOUR OF EXTRUDED I/M AND P/M SUPER α_2 ALLOYS

M. Thomas [1], S. Naka [1], M. Marty [1], W.G. Smarsly [2] and T. Khan [1]

1 : ONERA, 29, Av. de la Division Leclerc, 92322 Châtillon CEDEX France
2 : MTU, Postfach 50 06 40, 8000 Munich 50, Germany

Abstract

This investigation deals with the improvement of mechanical properties of the Super α_2 alloy through extrusion. With this processing route, a strength increment of more than 50% was observed, without any loss in ductility at room temperature, over the properties obtained through the isothermal forging route. Such an improvement results from the formation of an orthorhombic (O) phase, probably induced by a fairly strong plastic strain expected during extrusion; the transformation kinetics giving rise to this O phase seems to be accelerated. Optimization between strength and ductility can be achieved by suitable ageing treatments.

Introduction

The investigations conducted up to now on the two-phase α_2 (Ti_3Al) + β alloys have shown that these materials are of great technological interest, mainly because of their strength potential at temperatures in excess of 600°C. One of the most advanced alloys in this category is the so-called Super α_2 alloy (Ti-25Al-10Nb-3V-1Mo at.%) developed by TIMET [1]. This alloy is now being evaluated for various engineering applications at temperatures up to 650°C. However, the mechanical properties of this alloy depend very strongly on the microstructure resulting from the thermomechanical processing. A wide variety of microstructures can be produced through these treatments. The most frequently applied processing route is isothermal forging. The initial idea of the present study was to evaluate the effect of extrusion (another useful processing route) on the microstructure and the mechanical properties.

Alloying Ti_3Al with higher amounts of refractory elements (up to 25at.%) provides a substantial increase in tensile strength. The introduction of β-stabilizing elements such as niobium is also known to retain the high temperature β structure in its ordered form B2. Rowe [2] discussed the possibility of preserving high strength in the Ti_2AlNb (Ti-25Al-25Nb) alloy up to 815°C arrising from an additional transformation sequence B2 $\Rightarrow$ O. This orthorhombic phase results from the distortion of the ordered hexagonal α_2 (DO_{19}) structure due to the ordering of niobium atoms. According to Banerjee et al. [3], the O phase provides a substantial ductility enhancement due to the increase in the number of slip systems, compared to that of the α_2 phase. Keeping in mind these advantages, O containing alloys are now being studied in order to develop alloys of a new generation possessing higher high temperature strength with a higher room temperature ductility.

Specific alloying concentrations and atomic species lead to the formation of the O phase and

Titanium '92
Science and Technology
Edited by F.H. Froes and I. Caplan
The Minerals, Metals & Materials Society, 1993

the field of its existence in the phase diagramme seems to be located around Ti_2AlNb, although many controversies still exist about the nature of this phase. As shown in the present paper, some thermomechanical and heat treatments can also promote the O phase formation, even in alloys containing much less than 25at.% of β-stabilizing elements (viz. far from Ti_2AlNb) such as the Super α_2 alloy. In this alloy, the isothermal forging route was not found efficient to trigger off the O phase formation. Only a specific combination of certain conditions (strain, temperature and cooling rate) as encountered, for example, in extruded materials are favorable for developing the satisfactory microstructure comprising the O phase. In this paper, attempts were made to establish a relationship between the microstructure and tensile properties of extruded Super α_2 alloy. Since the O phase can undergo further decomposition into extremely fine scale α_2, O and B2 phases, a necessary requirement is to carefully control the stabilization or the decomposition of this phase in the temperature range of applications. The influence of the starting material, either I/M (Ingot Metallurgy) or P/M (Powder Metallurgy) processed, is also discussed.

Experimental procedure

Two starting materials were used in this work. The first one corresponds to a fairly high purity Super α_2 ingot produced by consumable electrode melting (I/M), whereas the second one corresponds to gas atomized Super α_2 powder (P/M). These alloys were received from separate sources and they slightly differ in composition. Chemical analyses revealed small differences in Al, V and interstitial contents between I/M and P/M products, as shown in Table 1. Consistently, the measured transus temperature was different, i.e., 1110°C and 1145°C in I/M and P/M Super α_2 respectively. It is worth noting that the difference in interstitial content (410ppm O_2 in I/M against 1170ppm O_2 in P/M) can yield a shift of 35°C, i.e., 5°C per 100ppm O_2.

Extrusions were performed at temperatures between 1000°C and 1080°C at a strain rate of 0.37m/s. Mechanical properties were determined on the materials extruded either at 1025°C or at 1050°C. The extrusion ratio (18:1) was chosen to obtain 9mm diameter rods. Rods were then cut for metallographic examinations and tensile tests.

Table 1 Chemical analysis of the two starting materials (wt.%).

Alloy	Al	Nb	Mo	V	Fe	C	O (ppm)	N	H
I/M Super α_2	14.1	18.8	1.86	3.24	0.09	0.023	410	0.006	0.002
P/M Super α_2	14.5	18.8	1.95	2.84	0.05	0.016	1170	0.018	0.001

Subsequent to the extrusion, a number of heat treatments was performed including some solution heat treatments and low temperature aging treatments. The solution heat treatments were carried out just below the transus temperature followed by furnace cooling. It is recalled that if these treatments are applied to the forged Super α_2, they produce the conventional duplex microstructure consisting of about 30 vol.% of primary globular α_2 phase and the Widmanstätten lamellae. Thus, the extrusion route can be directly compared to the forging route for which the mechanical properties of the duplex microstructure are well characterized. On the other hand, low temperature aging treatments at 700 or 800°C were performed in order to stabilize the as-extruded structure.

The presence of various phases (α_2, O and B2) was identified by transmission electron microscopy (TEM). In addition, backscattered scanning electron imaging was used to obtain some information on the amount of niobium contained in these phases.

Results

1. As-extruded condition

In the as-extruded condition, optical micrograph typically exhibits primary globulae dispersed

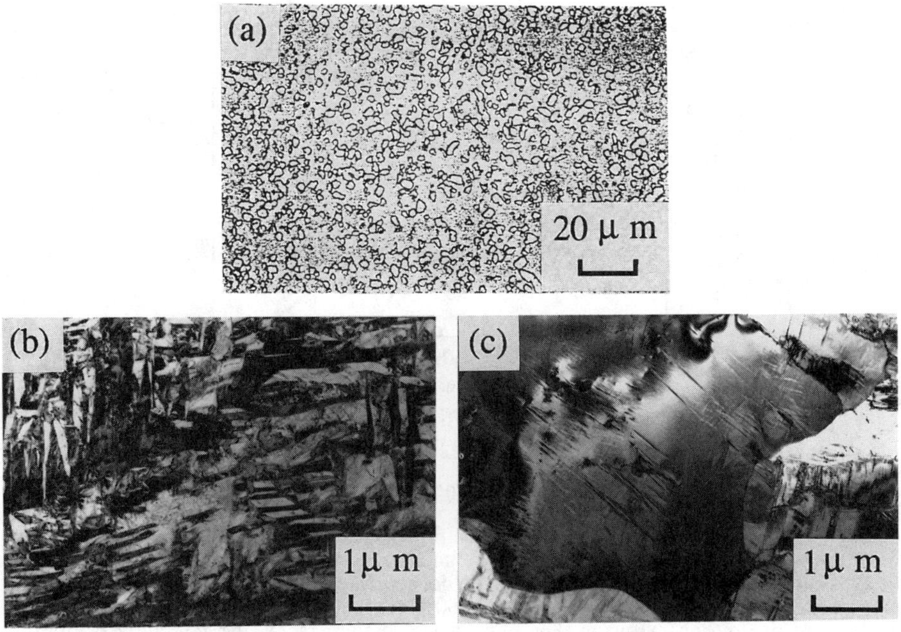

Fig.1 (a) Optical micrograph of the as-extruded Super α_2,
(b) Very fine lamellae and (c) monolithic grains observed by TEM.

in the apparently monolithic and polycristalline matrix (Fig.1a). At a high magnification, TEM reveals that the matrix is in fact often composed of very fine lamellae (Fig.1b). Monolithic regions are also present, although in much lower proportion (Fig.1c). The selected area diffraction patterns (SADP) in Fig.2a and b show the orthorhombic symmetry of the lamellae and of the monolithic regions respectively. Primary globulae also exhibit the orthorhombic symmetry but the lattice distortion with respect to that of the hexagonal α_2 phase appears to be of a lower magnitude than that observed in the O matrix (Fig.2c).

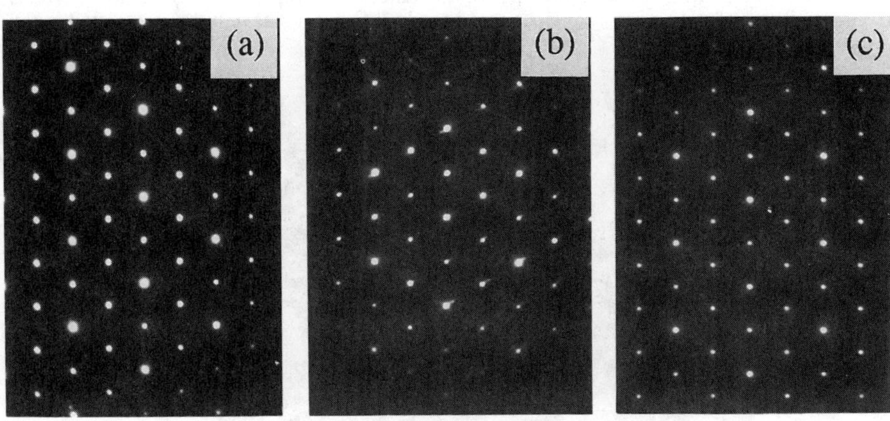

Fig.2 Selected area diffraction patterns : (a) in lamellae,
(b) in monolithic grains and (c) in primary globulae.

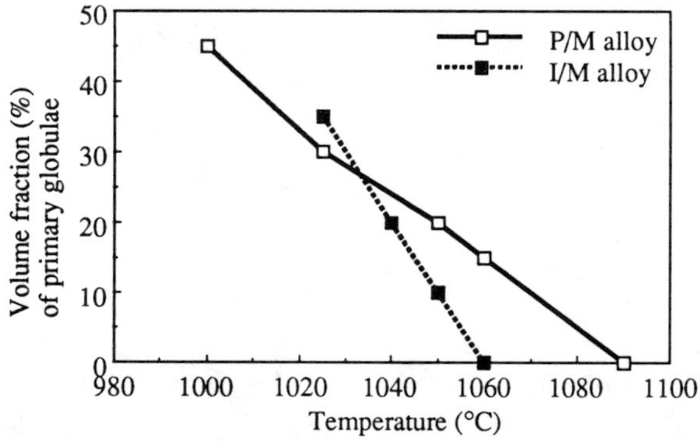

Fig.3 Variation of the volume fraction of primary globulae as a function of extrusion temperature in I/M and P/M alloys.

Different extrusion conditions were then examined as a function of the extrusion temperature. The volume fraction of primary globulae increases with decreasing extrusion temperatures, although in different manners for I/M and P/M products (Fig.3). It should be noted here that the temperatures corresponding to the zero volume fraction of primary globulae (1090°C for the P/M alloy and 1060°C for the I/M alloy) do not coincide with the above mentioned transus temperatures (1145°C for the P/M alloy and 1100°C for the I/M alloy) because of an adiabatic temperature rise during extrusion. For a given temperature range, the I/M alloy exhibits a larger variation in globulae volume fraction. In the case of extrusion at 1050°C, the I/M alloy is characterized by a lower volume fraction compared to that in the P/M alloy. At 1025°C, optical micrographs reveal that the volume fraction is similar in both for the products.

When examining the morphology of the matrix, it seems that two types of the B2 ⇒ O transformations take place. In some areas, the B2 matrix is transformed "massively" into an orthorhombic (O) phase, possibly through a recristallization process. In other areas, the B2 matrix is decomposed into fine tangled O lamellae. The reason why two different transformation mechanisms simultaneously occur is not clear. The extrusion parameters such as temperature, deformation rate and reduction in cross section which can modify the driving force for such transformation mechanisms are now under investigation. Compositional effects are also expected to play a key role since the "massive" O phase has only been detected in the

Fig.4 Duplex microstructure observed after solution annealing.

P/M alloy.

2. Solution annealed condition

When solution heat treatments on the extruded products were applied just below the transus temperature followed by furnace cooling, the observed microstructure was identical to that observed in the forged products with the same solutioning heat treatment, i.e. 30 vol.% of primary α_2 together with a fine dispersion of Widmanstätten lamellae which results from the decomposition of the matrix B2 $\Rightarrow$ B2 + α_2; this microstructure is here designated as the duplex microstructure (Fig.4).

This duplex microstructure exhibits a slightly higher mechanical strength than that reported in the literature (1) for the forging route (Table 2). This table also shows that for the same extrusion temperature (1050°C), the I/M alloy possesses better properties at 20 and 650°C than the P/M alloy. If the I/M alloy is extruded at lower temperature (1025°C), the refinement of primary globulae leads to an even higher yield stress.

Table 2 Results of the tensile tests (after extrusion + solution annealing).

Alloy	Test Temp. (°C)	0.2%TYS (MPa)	UTS (MPa)	EL (%)
Super α_2 forged	20	790	1000	2
P/M extruded at 1050°C	20	890	950	0.5
I/M extruded at 1050°C	20	925	1055	1.2
I/M extruded at 1025°C	20	1055	1135	0.7
Super α_2 forged	650	585	790	10
P/M extruded at 1050°C	650	595	755	2.6
I/M extruded at 1050°C	650	660	845	5.5

It should be noted that in the I/M alloy, surface cracking occurs during tensile tests at 650°C, while this was not observed in the P/M alloy. In fact, there is no clear explanation for the absence of surface cracking in the P/M alloy or for the relatively low ductility observed at 650°C in both I/M and P/M alloys. It should be stressed that in the case of I/M alloy the ductility is much higher.

3. Aged condition

It must be pointed out that ageing was performed in the present study directly after extrusion.

In a preliminary investigation, we examined the response of the O phase to a number of heat treatments performed below the transus temperature. Ageing at 800°C for 116 hours preserves the very fine extruded microstructure. No microstructural changes are detected with respect to the as-extruded condition. On the other hand, ageing at 900°C for 24 hours leads to an extensive coalescence of the primary globulae. As a result, the matrix is strongly confined to interglobular areas. Back-scattered scanning electron imaging (Fig.5a and 5b) exhibits a totally different contrast depending on the ageing temperature (800 or 900°C). At 800°C, the concentration of Nb is apparently less in the globulae than in the matrix. On the other hand, the higher ageing treatment caused a change in composition by extensive diffusion operating between the primary globulae and the matrix. The globulae preserve their prior niobium concentration whereas the periphery is slightly richer in niobium. The remaining matrix appears to be very rich in niobium.

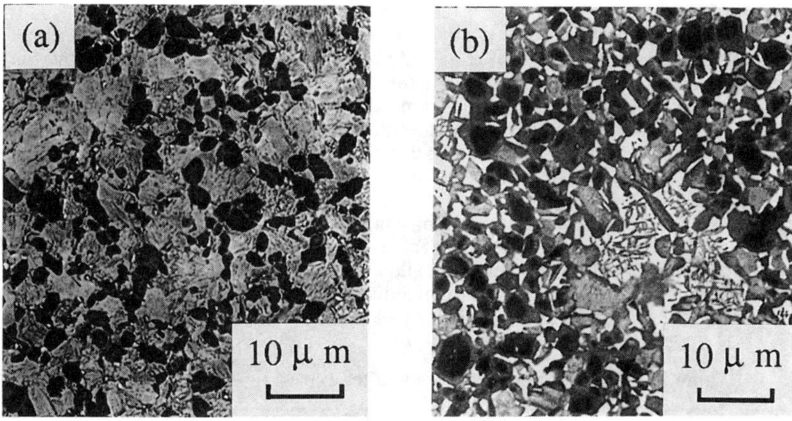

Fig.5 Back-scattered scanning electron micrographs of the aged P/M alloy : aged (a) at 800°C and (b) at 900°C.

As described by Muraleedharan et al. [4,5], a number of reactions are expected to occur in the composition range around the tie-line between Ti_3Al and Ti_2AlNb of the ternary Ti-Al-Nb phase diagramme; these reactions take place involving α_2, O and B2 phases. Unfortunately, the lack of phase diagramme data for complex alloy systems (Ti-Al-Nb-V-Mo) with several atomic species such as the Super α_2 alloy does not enable us to define the phase boundary temperatures in an alloy such as Super α_2. Indeed, the above microstructural observations in the extruded Super α_2 alloy are indicative of a more complex situation. This point will be discussed later.

Since the microstructure becomes unstable at temperatures beyond 800°C, the ageing prior to mechanical tests was performed at 800°C. The results of the tensile tests (Table 3) show a significant enhancement of the mechanical properties, compared to the case of a conventional duplex microstructure. This improvement is presumably related to the presence of a large volume fraction of O phase. Here again, for a given extrusion temperature (1025 or 1050°C), I/M alloy is characterized by a higher yield stress and ultimate tensile strength. TEM micrographs reveal that the I/M microstructure consists of a much higher volume fraction of secondary lamellae. As previously mentioned, monolithic grains were nor present in I/M

Table 3 Results of the tensile tests (after extrusion + ageing).

Alloy	Test Temp. (°C)	0.2%TYS (MPa)	UTS (MPa)	EL (%)
P/M extruded at 1050°C	20	1020	1190	2.6
P/M extruded at 1025°C	20	1235	1295	0.7
I/M extruded at 1050°C	20	1275	1370	0.8
I/M extruded at 1025°C	20	1435	1580	1.8
P/M extruded at 1050°C	650	750	790	0.4
P/M extruded at 1025°C	650	795	755	0.4
I/M extruded at 1050°C	650	875	845	4.3

alloy. For both alloys, a further increase in strength is provided by lowering the extrusion temperature from 1050 to 1025°C, which allows a higher volume fraction of the retained fine primary globulae.

Discussion

The present study has shown that the O phase induced through extrusion has a beneficial effect on the mechanical properties of the Super α_2 alloy. Considering the fact that the field of existence of the O phase is around the composition, Ti_2AlNb, it is a little surprising to observe this phase in alloys such as the super α_2 lean in β-stabilising elements. However, it should be remembered that vanadium shows a β-stabilising power about 5/3 times stronger and molybdenum about three times stronger than that of niobium [6]; the composition of the Super α_2 alloy ($Ti_{61}Al_{25}Nb_{10}V_3Mo_1$) therefore corresponds to $Ti_{57}Al_{25}Nb_{18}$. From this simple β-stabilising criterion, the O phase should be easily observed in this alloy, since Banerjee and co-workers [4,5] found this phase in an alloy less rich in Nb such as $Ti_{61}Al_{24}Nb_{15}$. The issue seems, however, to be more complicated, since the O phase can only be found after severe thermomechanical treatments like extrusion but not after forging. These results indicate that, in fact, two factors may play a role in the formation of this phase : chemistry and thermomechanical treatments. As far as the chemistry is concerned, the formation of the O phase is affected by the nature of the β stabilising elements. Another important factor for the O phase formation is the plastic strain which is extremely high in an extrusion than in a forging. Moreover, a higher cooling rate expected subsequent to extrusion may prevent recovery effects. Therefore, it can be assumed that a high plastic strain produced through extrusion accelerates the kinetics of the transformation giving rise to this O phase. In other words, the field of existence of this phase is extended toward the lean side of β-stabilising elements at the expense of the α_2 phase.

It is interesting to recall here that similar improvements in strength have already been reported in the Super α_2 alloy owing to a thermomechanical treatment consisting of a six step swaging at 950°C [7]. At such a low temperature, the formation of the "massive" O phase is responsible for the sharp increase (10%) in ductility at room temperature. A very short-time solution treatment followed by water quenching and aging results in the subsequent decomposition of the matrix into extremely fine α_2, O and B2 particles according to a peritectoïd reaction. This decomposition leads to a sharp increase in yield strength while a moderate ductility is still preserved. From an industrial viewpoint, it is difficult to apply the above processing sequence because it involves a very short solution treatment followed by water quenching in order to optimize the tensile properties. Alternatively, in this investigation, extrusion at higher temperature enables us to obtain a material which has already decomposed into fine O+ B2 phases without any solution heat treatment.

Finally, the issue of surface cracking encountered in the I/M Super α_2 alloy remains a serious industrial problem, and a more detailed understanding of its origin is required. For example, work is now in progress to determine the strain at which this cracking appears. The absence of such a cracking in the P/M alloy is also not clear; shift of the brittle-ductile transition temperature toward high temperature side (beyond 650°C) due to its high impurity (oxygen) content might be an explanation.

Conclusions

In the extruded and aged condition, both the I/M and P/M Super α_2 alloys exhibit, in the temperature range 20-650°C, a strength level 50% higher than the data reported in the literature for the forging route. Microstructural investigations lead to the conclusion that the decomposition of the B2 matrix giving rise to extremely fine lamellae is responsible for the increase in strength. Concurrently, the orthorhombic phase, identified by selected area diffraction, contributes in preserving some ductility at room temperature. At 650°C, however, the I/M alloy was prone to surface cracking contrary to the case of the P/M alloy.

One promising feature is that the fine lamellar orthorhombic microstructure which develops an encouraging room temperature ductility also shows attractive high temperature strength.

Acknowledgements

This work was conducted with a financial support of the French Ministry of Defence (STPA). The authors are grateful to J.L. Raviart for his technical assistance.

References

1. M.J. Blackburn and M.P. Smith, US Patent N° 4,716,020, 1987.
2. R.G. Rowe, "Recent Developments in Ti-Al-Nb Titanium Aluminide Alloys" in Proc. ASM/TMS-AIME Symp. on "High Temperature Aluminides and Intermetallics", Indianapolis, TMS, 1989, 375.
3. D. Banerjee, R.G. Rowe and E.L. Hall, "Deformation of the Orthorhombic Phase in Ti-Al-Nb alloys", in Proc. Mat. Res. Soc. Symp. on "High-Temperature Ordered Intermetallic Alloys", Boston, MRS, 1991, Vol. 213, 285.
4. K. Muraleedharan, A.K. Gogia, T.K. Nandy, D. Banerjee and S. Lele, "Transformations in a Ti-24Al-15Nb Alloy - Part-I : Phase Equilibria and Microstructure", to be published in Acta Metall. Mater..
5. K. Muraleedharan, T.K. Nandy, D. Banerjee and S. Lele, "Transformations in a Ti-24Al-15Nb Alloy - Part-II : A Composition Invariant $\beta_0 \Rightarrow$ O Transformation", to be published in Acta Metall. Mater..
6. M. Thomas, S. Naka and T. Khan, "Phase stability in Ti_3Al-based Alloys", in this conference.
7. J. Kumpfert, K.J. Grundhoff, H. Schurmann, Y.T. Lee and M. Peters, "Thermomechanical Treatment to Achieve Ultra-High Strength in Ti_3Al Based Alloys", paper presented at the 2nd European Conf. on "Advanced Materials and Processes - EuroMat 91", Cambridge, United Kingdom, July, 1991.

THE DEFORMATION AND FRACTURE BEHAVIORS OF A Ti-25Al-10Nb-3V-10Mo ALLOY AT DIFFERENT TEMPERATURES

Fusheng Sun, Chenggong Li, and Chunxiao Cao

Beijing Institute of Aeronautical Materials

Beijing 100095, P.R.China

Abstract

The tensile deformation and fracture behaviors of a Ti-25Al-10Nb-3V-1Mo (at%) alloy have been investigated over the temperature range of 25-980°C in either air or vacuum. The tensile strength and ductility are influenced not only by the deformation structures (such as dislocation, twinning and dislocation loop) corresponding to the temperature but also by the environmental factors (such as oxidation). Both the strength and the ductility of samples deformed at the elevated temperature (700°C) in vacuum are superior to that in air. The fracture modes of samples deformed at the elevated temperature (700°C) are also affected by oxidation and possibly hydrogen embrittlement.

Introduction

The titanium aluminide alloys such as Ti-25Al-10Nb-3V-1Mo (at%) have been exploited as new promising commercial materials because of their potential for use in high-temperature structural application[1]. The mechanical properties, such as tension, creep, fatigue, and toughness, were extensively evaluate[2-5]. So was the influence of microstructure on property examined[6]. The temperature dependence and the environmental dependence of deformation and fracture behaviors of titanium aluminides have been distinctly observed. However, it is noted that there is little data published on the influence of temperature and environment on the tensile deformation and fracture of Ti-25Al-10Nb-3V-1Mo alloy. This investigation was aimed at understanding the change of tensile strength and ductility with temperature in air or vacuum over the temperature range of 25-980°C , and elucidating the deformation and fracture mechanism of Ti-25Al-10Nb-3V-1Mo alloy.

Experimental

The material used in the present study was a Ti-25Al-10Nb-3V-1Mo (at%) alloy with analyzed chemical composition (in weight percent) of 14.4 Al, 19.3 Nb, 3.5 V, 2.2 Mo, 0.10 O, 0.018 N, and the remainder titanium, which was prepared by consumable electric arc melting. After $\alpha_2+\beta$ region forging and subsequently $\alpha_2+\beta$ region heat treating, round samples with 25mm gauge length (5mm in diameter) were machined. Tensile tests were performed over the temperature range of 25-980°C in air. Some samples (with 15mm gauge length) were tested at 700°C in either air or vacuum (1×10^{-3} torr) so as to examine the environmental dependence of tensile deformation. The deformation

Titanium '92
Science and Technology
Edited by F.H. Froes and I. Caplan
The Minerals, Metals & Materials Society, 1993

structures were investigated by transmission electron microscope by g b invisibility criterion. The fracture surfaces were examined by scanning electron microscope.

Results

Microstructure

The typical microstructure resulted from $\alpha_2+\beta$ region forging and $\alpha_2+\beta$ region heat treating is demonstrated in Figures 1, which consists of about 30-40 percent volume fraction of primary alpha-2 phase (black blocks in Figure.1a) and transformed, beta phase (white or grey blocks in Figure.1a). The primary alpha-2 is equiaxed, while the transformed beta phase is characterized as lenticular secondary alpha-2 with retained beta film present around (Fig.1b).

Tensile Behavior in Air

Figure 2 shows the change of strength and ductility with tensile test temperature over a

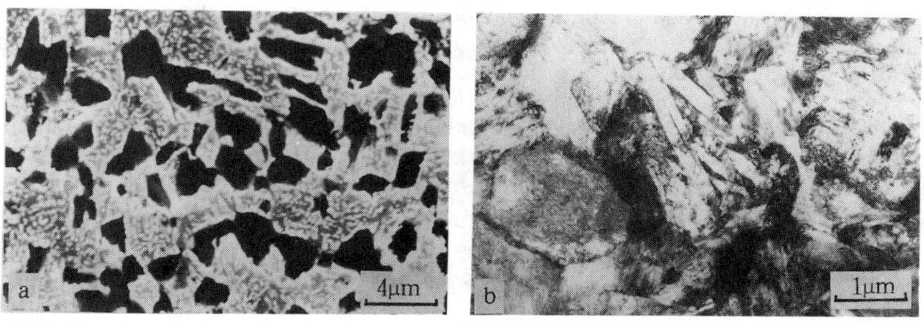

Figure 1-Typical microstructure examined by (a) SEM and (b) TEM.

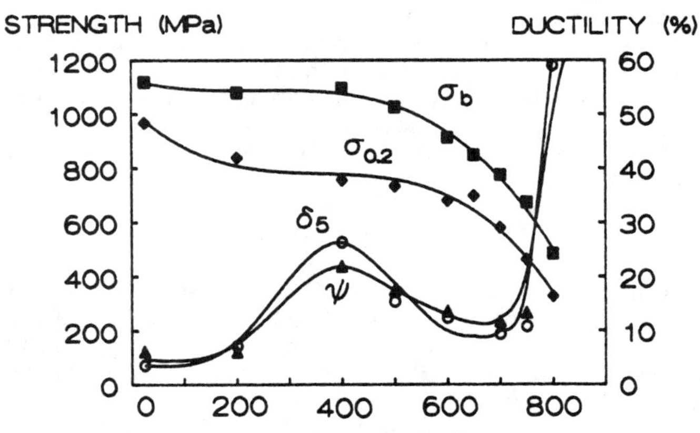

Figure 2-Influence of temperature on the tensile properties over the temperature range of 25-800°C in air.

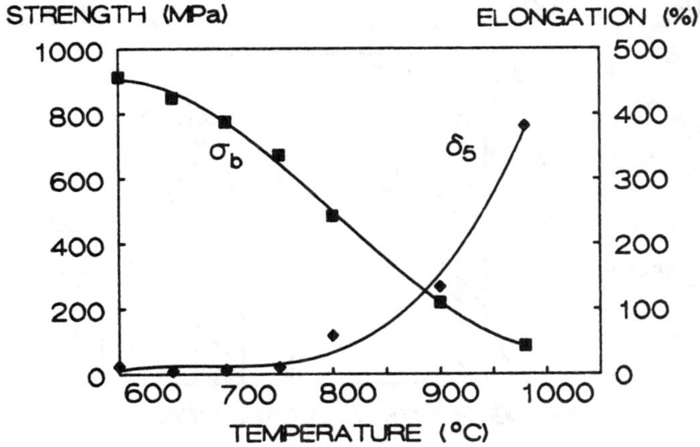

Figure 3-Influence of temperature on the tensile properties over the temperature range of 600-980°C in air.

temperature range of 25-800°C in air. As can be seen, both the ultimate tensile strength and the yield strength decrease gradually with the increase of temperature below 700°C, but decrease significantly with the increase of temperature up to 700-800°C. It is strikingly noted that the response of ductility to test temperature is complex. The ductility first increases remarkably with the increase of temperature in the temperature range of 200-400°C, and subsequently decreases with the increase of temperature at 400-750°C. Therefore, a peak of tensile ductility is noticed at 400°C . With the further increase of temperature up to 750°C , an abrupt enhance of ductility is observed in the present study. Furthermore, an excellent superplasticity is recognized in Ti-25Al-10Nb-3V-1Mo (at%) alloy at a strain rate of 1×10^{-3} sec^{-1} as shown in Figure 3, which its total elongation is more than 380% at 980°C.

Tensile Behavior in Vacuum

The influence of environment on the tensile deformation behavior at 700°C is illustrated in Fig.4 by constructing bar chart. Every three samples were strained at 700°C in either air or vacuum so as to evaluate the strength and ductility with or without the effect of oxygen and hydrogen. Both the ultimate tensile strength and ductility of samples deformed in vacuum are superior to that in air. It is interesting to note that the tensile elongation is enhanced more than one-fold at 700°C in vacuum, showing a very good ductility.

Discussion

Deformation Structures

A systematic investigation has been carried out to explore the deformation structures of a Ti-25Al-10Nb-3V-1Mo alloy under different temperature tension ranging from 25-980°C . The experiment result indicates that the majority of dislocations in alpha-2 phase are $\vec{a}$ type (Figure 5a) with only a few $\vec{c}+\vec{a}/2$ type dislocations present (Figure 5b) at room temperature, which the $\vec{c}+\vec{a}/2$ type dislocations are characteristic of dislocation pairs with corrugated appearance. The $1/6<11\bar{2}0>\{10\bar{1}0\}$ and $1/6<11\bar{2}0>\{0001\}$, as well as occasional $1/6<11\bar{2}6>\{20\bar{2}1\}$ slip systems play a significant role in contributing

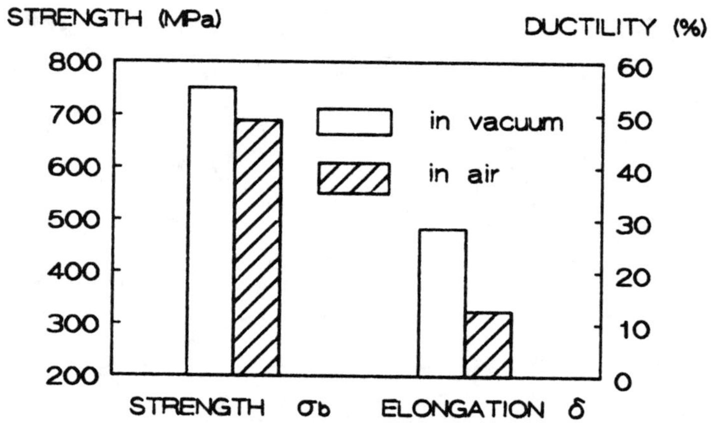

Figure 4-A bar chart showing the comparison of tensile properties at 700°C in air with that in vacuum.

to the onset of ductility at lower temperatures (25-200°C). However, the ductility is limited due to the insufficient presence of dislocation systems at lower temperatures.

In addition to $\vec{a}$ type dislocation slip systems, which is similar to that at room temperature, the extensive presence of deformation twinning in alpha-2 grains (Figure 5c) is identified to contribute to the remarkable increase of ductility at 400°C . The twin plane is determined to be $\{20\bar{2}1\}$. But the non-basal slip of $\vec{c}+\vec{a}/2$ type dislocation is more rarely observed in alpha-2 phase at 400°C.

Figures 5d-5f demonstrate that the ductile deformation of alpha-2 phase at higher temperature (700°C) occurs primarily by $\vec{a}$ type dislocation slip systems on prismatic, basal and pyramidal planes, such as slip systems $<11\bar{2}0>\{1\bar{1}00\}$, $<11\bar{2}0>\{0001\}$ and $<11\bar{2}0>\{20\bar{2}1\}$.The $\vec{a}$ type dislocation pairs are often observed on pyramidal planes in alpha-2 grains (Figure 5e), but hardly can $\vec{c}+\vec{a}/2$ type dislocation pairs be noticed in this case. In addition, a very few of isolated $\vec{c}$ type deformation dislocations are found at elevated-temperatures (Figure 5f).

With the further increase of temperature to 900-980°C, it exhibits excellent superplastic property. The superplastic model for grain boundary sliding with the accommodation processes such as diffusion, dislocation slip, dislocation loop (Figure 5g) and stacking fault (Figure 5h) accommodation processes provides a good representation of the superplastic deformation of a Ti-25Al-10Nb-3V-1Mo alloy at 980°C with a strain rate of $1 \times 10^{-3}\ s^{-1}$. The majority of dislocation loops are identified as shearing ones with $b=1/6[10\bar{1}0]$.

Fractography

Close examination of fracture surfaces by scanning electron microscope indicates that transgranular cleavage appears to be the predominant mode of fracture at either room-temperature or 200°C (Figure 6a). The cleavage-like appearance is lessened with the increase of temperature up to 400°C as exhibited in Figure 6b. However, the fracture behavior at the elevated-temperature (700°C) becomes more complex. Ductile fracture mode absolutely predominates at 700°C in vacuum, characteristic of deep ductile dimples with occurrence of necking (Figures 6c and 6d). In contrast, a propensity for cleavage

Figure 5-TEM micrographs showing the deformation structures in alpha-2 phase at different temperatures. (a) 25°C, $\vec{a}$ type dislocations, (b) 25°C, dark field, $\vec{c}+\vec{a}/2$ type dislocations, (c) 400°C, twinning, (d) 700°C, $\vec{a}$ type dislocation network, (e) 700°C, $\vec{a}$ type dislocation pairs, (f) 700°C, isolated $\vec{c}$ type deformation dislocations, (g) 980°C, dislocation loops, (h) 980°C, stacking faults.

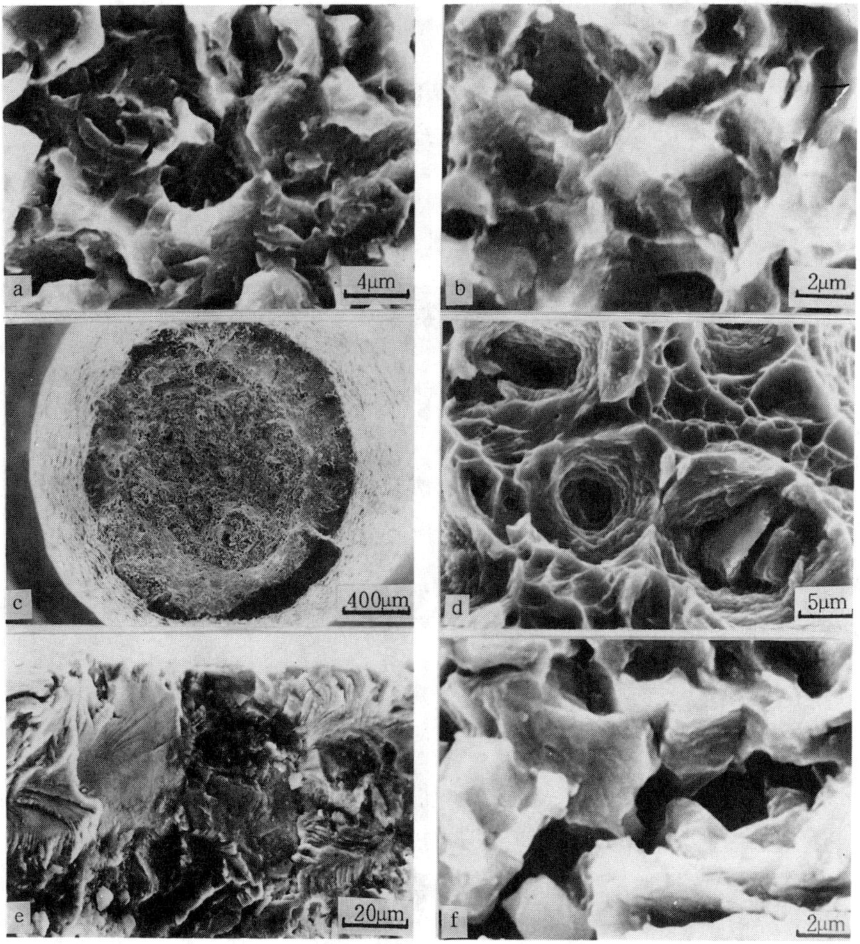

Figure 6-SEM micrographs showing the fracture surfaces at different temperatures. 25°C, in air, (b) 400°C ,in air,(c)-(d) 700°C, in vacuum, (e)-(f) 700°C, in air.

fracture in crack initiation region (Figure 6e) and intergranular fracture in crack growth region (Figure 6f) becomes more pronounced at 700°C in air.In addition, the remarkable influence of oxidation layer on fracture behavior is recognized.The cleavage facets associated with oxidation region can evidently observed in Figure 6e. The presence of oxidation or oxygen-riched layer will promote the microcrack initiation due to the distinct uncompatible deformation between the oxidation layer and the matrix. While the interganular fracture is more likely attributed to hydrogen embrittlement due to disbonding in grain boundary. Thus, it deteriorates the mechanical properties such as ductility of Ti-25Al-10Nb-3V-1Mo alloy at elevated temperatures in air. However, a direct evidence of hydrogen embrittlement is badly required.

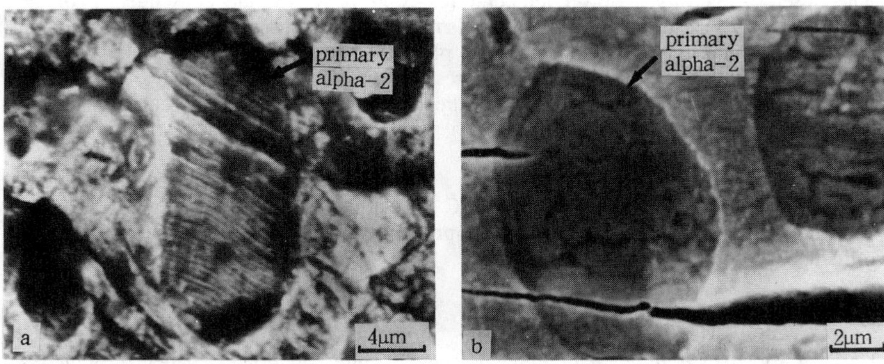

Figure 7-The free surface appearance of tensile specimen after deformation at 700°C. (a) in vacuum, (b) in air.

Slip Line Pattern

Figures 7 demonstrated the free surface appearance of tensile samples after tensile fracture at 700°C in either vacuum (Figure 7a) or air (Figure 7b). Because the free surfaces of samples had been polished and slightly chemically etched before deformation, slip bands and microcracks could be easily discerned in this case. As can be seen, the equiaxed primary alpha-2 grain is distinctly elongated after deformation, which suggests that the alpha-2 phase become more ductile at elevated temperature in vacuum.Moreover, the slip line pattern on primary alpha-2 grain can be clearly seen in Figure 7a. It is noteworthy that some slip lines are curved, indicative of the possible presence of cross-slip of dislocation or other slip activity. The ductile deformation of alpha-2 phase may lie in the fact that the cross-slip activity and the abundant presence of $\tilde{a}$ type dislocation slip systems such as $<11\bar{2}0>\{20\bar{2}1\}$ with occasionally isolated $\tilde{c}$ type dislocation present would introduce an effective ductile deformation region around the microcrack tip. Thus, the cleavage cracking is curbed.

In contrast, a direct evidence of the influence of oxidation on fracture behavior at 700°C in air was shown in Figure 7b, indicating that the microcracks readily occurred on the free surface of specimen. The microcrack easily propagated in the oxidation layer, no matter what the primary alpha-2 grain or transformed beta phase was. Hardly can be the slip lines discerned on the primary alpha-2 phase in this case.

Conclusions

1. Deformed over the temperature range of 25-980°C in air, both the ultimate tensile strength and yield strength of Ti-25Al-10Nb-3V-1Mo alloy decrease with the increase of temperature. The ductility first increases remarkably with the increase of temperature at 200-400°C , and subsequently decrease with the increase of temperature at 400-750°C. However, with the further increase of temperature up to 900-980°C, it exhibits an excellent superplasticity at a strain rate of 1×10^{-3} sec^{-1}.

2. Either the strength or the ductility of samples deformed at 700°C in vacuum is superior to that in air. The ductility is enhanced more than one-fold at the elevated-temperature in vacuum.

3. The onset of ductility at lower temperatures (25-200°C) is the consequence of the activation of $\vec{a}$ type dislocation slip on prismatic and basal planes, as well as occasion $\vec{c}+\vec{a}/2$ type dislocation slip on pyramidal planes. The significant increase of ductility at 400°C is attributed to the extensive presence of deformation twinning. While the oxidation and possibly hydrogen embrittlement deteriorate the ductility at the elevated -temperature (700°C) in air.

4. Generally, cleavage fracture is lessened with the increase of temperature. The ductile fracture absolutely predominates at 700°C in vacuum. But a propensity for cleavage and intergranular fracture becomes more pronounced at 700°C in air, indicative of the influence of environment.

References

1. Harry A.Lipsitt, "Titanium Aluminides-Future Turbine Materials" (paper presented at Conf. on Advance High Temperature Alloy, USA, 1985), 157-164.

2. A.K.Gogia, D.Banerjee, and T.K.Nandy, "Structure, Tensile Deformation, and Fracture of a Ti_3Al-Nb Alloy", Metall. Trans., 21A (1990), 609-625.

3. R.S.Mishra, and D.Banerjee, "Microstructure and Steady State Creep in Ti-24Al-11Nb", Mater. Sci. and Eng. A130 (1990), 151-164.

4. D.L.Davidson, J.B.Campbell, and P.A.Page, "The Initiation and Growth of Fatigue Cracks in a Titanium Aluminide Alloy", Metal. Trans., 22A (1991), 377-391.

5. K.S.Chan, "Fracture and Toughening Mechanisms in an α_2 Titanium Aluminide Alloy", Metal. Trans., 21A (1990), 2687-2699.

6. Wonsuk Cho, Anthony W.Thompson, and James C.Williams, "Creep Behavior of Ti-25Al-10Nb-3V-1Mo", Metal. Trans., 21A (1990), 641-651.

Improvement of Ductility and Strength of

Ti-25Al-10Nb-3V-1Mo by TMT

J. Kumpfert[*/**], C.H. Ward[**], Y.T. Lee[***], M. Peters[*]

[*] DLR - German Aerospace Research Establishment
Institute for Materials Research, Cologne, Germany

[**] Wright Laboratory
Materials Directorate, Wright-Patterson AFB, OH, USA

[***] KIMM - Korea Institute of Machinery and Metals
Department of Materials Research, Changwon, Korea

Abstract

In order to optimize the ductility and strength of the Ti_3Al-based alloy Ti-25Al-10Nb-3V-1Mo (Super Alpha 2), various thermomechanical treatments were performed. These included different extents of mechanical deformation by swaging, solution heat treatment temperatures, cooling rates and aging time and temperatures.

After swaging in the $\alpha_2+\beta$ field, solution heat treatments at temperatures well below the ß-transus led to a more than three-fold increase in ductility without loss in yield and tensile strength when compared to the as-received material. In order to generate high strength, solution heat treatments were performed just below the β-transus temperature. After optimizing the deformation degree and the aging treatment, yield strength and ultimate tensile strength values close to 1500 MPa and 1800 MPa, respectively, were achieved. These high strength values were also maintained at elevated temperatures, e.g. UTS at 650^{0}C was still beyond 1200 MPa.

Introduction

Conventional high temperature Ti-based alloys, like the near alpha type alloys Ti-1100 and IMI 834, have reached a temperature barrier set by their creep and oxidation resistance below 600^{0}C. The primary needs for advanced aerospace systems are both, higher service temperature and specific strength at the same time. Currently, the materials that appear most promising for these demanding needs are alloys based on titanium aluminides [1,2].

In this investigation different thermomechanical treatments were applied to the Ti_3AL based alloy Ti-25Al-10Nb-3V-1Mo (at.%) to optimize the microstructure with respect to ductility and strength. The influence of the different parameters during thermomechanical processing on microstructure were investigated and related to the mechanical properties. The extremely high strength values, close to 1800 MPa (UTS), in this alloy with nearly no loss of ductility compared to the as-received material, result from a very fine microstructure after an optimized thermomechanical process. The density corrected strength values based on these results represent

Titanium '92
Science and Technology
Edited by F.H. Froes and I. Caplan
The Minerals, Metals & Materials Society, 1993

the top values among the high temperature alloys like TiAl based alloys, Ni based alloys and near alpha alloys in the temperature range up to 650°C.

Experimental Methods

The material was supplied by Timet and RMI as 254 mm/286 mm diameter bar material with the chemical composition Ti-23.0Al-10.8Nb-2.32V-0.85Mo (at.%) and Ti-23.3Al-10.3Nb- 2.45V-0.90Mo, respectively; The β-transus temperature (T_β) was determined to be 1070°C (Timet) and 1100°C (RMI). The oxygen content for both alloys was about 700 ppm.

Swaging was carried out in 9 steps at 950 °C to impart a total deformation varying from 76% and 93%. The swaged bars were then air cooled. Afterwards, the bars were solution heat treated (SHT) for 15 minutes at temperatures between 800 °C and 1110 °C, then cooled to room temperature at different cooling rates. The cooling schemes were as follows: furnace cooling (FC), Air cooling (AC), Oil quenching (OQ) and water quenching (WQ) with average cooling rates of 0.02 K/s, 9 K/s, 80 K/s and 120 K/s, respectively, when measured between the solution heat treatment (SHT) temperature and 600 °C. Finally, the specimens were annealed at 650 °C and 700 °C for different times. Vicker's hardness was measured after the solution heat treatment as a function of the annealing time.

Tensile tests were performed at RT and at elevated temperatures up to 700°C in air on cylindrical specimens with a gauge length of 30 mm and a diameter of 6 mm at a strain rate of $5.6 \cdot 10^{-4}\ s^{-1}$. Specimens were soaked at test temperatures for 15 minutes before testing.

Results and Discussion

Microstructures

Fig. 2 shows the microstructures that were obtained after TMT at solution heat treatment (SHT) temperatures T_β-120 °C, T_β-20 °C and T_β+20 °C and cooling rates between 120 K/s (WQ) and 0.02 K/s (FC). Figures 2(c), 2(e) show microstructures which were additionally aged at 700 °C/4h. Before aging, the matrix in almost all cases remained single phase B2 after WQ [3]. The intermediate cooling rates, OQ (80 K/s) and AC (9 K/s), led to the same type of microstructure as seen by optical microscopy as that obtained after WQ. After a SHT temperature 120 °C below T_β, the variation of the cooling rate caused only a little change in the equiaxed microstructure (Fig. 2(a), 2(b)). The high volume fraction of primary α_2 (~50%) in these microstructures is finely distributed with a fine grain size of about 4 to 8 μm.

Higher SHT temperatures reduced the primary α_2 volume fraction (Fig. 2(c)), and cooling rates down to 9 K/s (AC) enabled a high volume fraction of supersaturated retained B2 [3]. Slower cooling rates led to the formation of coarse secondary α_2 laths by heterogeneous nucleation and growth (Fig. 2(d)). Compared to the furnace cooled β-processed material (Fig. 2(f)), however, the sub-transus microstructure is markedly finer (Fig. 2(d)) and shows a 'basketweave' structure instead of the coarse lamellar structure with large aligned regions. Thus, the effects of a small volume fraction of primary α_2 become apparent here. First, primary α_2 pins the recrystallized β/B2 grains and second, reduces the extent of sideplate formation.

Mechanical Properties

Solution heat treatment temperature

Fig. 3 shows the results of the room temperature tensile tests for Ti-25-10-3-1 depending on the SHT temperature at a constant cooling rate and aging treatment. Specimens were swaged at T_β-120 °C to 76% reduction, solution heat treated at temperatures ranging from 800 °C to 1110 °C, then quenched (120 K/s) and aged for 4 hours at 700 °C. Yield and ultimate tensile strengths increased with increasing

SHT temperature and reached maximum values just at or below the β-transus temperature after UTS dropped drastically. Elongation values dropped with increasing SHT temperatures from about 9 to 10% at temperatures up to 950 ^{0}C to essentially zero, once the transus was passed. Hardness values show very similar behavior to the ultimate tensile strength except that the maximum hardness is reached at 1090 ^{0}C, i.e. in the β-phase field. The higher hardness can be attributed to a higher driving force for precipitation during aging, leading to a finer microstructure in the prior β grains. However, these grains are very large after β processing (Fig. 2(e)), leading to large plastic incompatibilities across the prior β grains and thus very low ductility (~0%) [4]. After a SHT temperature of T_β - 20^{0}C and quenching, the retained β had a much lower grain size due to finely distributed primary α_2 particles, which hindered the grain growth. Thus, this microstructure (Fig. 2(c)) has a reduced dislocation pileup length resulting in smaller incompatibility strains at the grain boundaries, which leads to higher fracture elongation. Due to this increased ductility, the high ultimate tensile strength potential of the fine matrix can be realized. SHT temperatures far below the β-transus increase the primary α_2 volume fraction (Fig. 2(a)) and decrease the volume fraction of secondary α_2 in the matrix. Additionally, the distribution and morphology of the β phase is much different. Thereby the high fracture elongation of this microstructure (Fig. 2(a)) can be explained by a uniformely distributed ductile matrix, which is of sufficient microstructural dimensions to arrest cracks nucleated in the primary α_2 (Fig. 1(a)).

Cooling Rate

The extremely high room temperature yield and ultimate tensile strength for the fine bi-modal microstructure (Fig. 2(c), 3; SHT at 1050^{0}C) results from a marked age hardening effect (Fig. 4). For sub-transus SHT temperatures (T_β - 20^{0}C) the age hardening is not influenced by variation of the cooling rate between 120 K/s (WQ) and 9 K/s (AC) (Fig. 4(b)). However, FC leads to a coarse 'basketweave' and stable microstructures (Fig. 2(f)) with a lower hardness. It is also apparent that the very fine microstructure formed after aging in the high cooling rate specimens are metastable at 700^{0}C. A stable microstructure is reached after 100h of aging, resulting in the same hardness level as after FC. For super-transus SHT temperatures (T_β + 20^{0}C) a marked age hardening effect is also observed at medium cooling rates (OQ, AQ; Fig. 4(b)).The much larger grain size (~ 400 μm) of the β processed material (Fig. 2(e)) seems to increase the stability of this microstructure due to a lower driving force for recrystallisation [5]. The low hardness level of the stable microstructure is not reached before 500h of aging at 700^{0}C.

However, water quenching of the β-processed specimens leads to a lower age hardening response. Different phase transformations and the occurrence of the ω phase might be the reason for this behavior [6,7]. More investigations need to be done to understand this hardening response of the β-processed specimens.

SHT temperatures 120 ^{0}C below T_β (Fig. 4(a) also have an age hardening response, but less pronounced, as a result of the lower volume fraction of remaining matrix.

Phase Transformations

For an age hardening effect, the content of β-stabilizing elements is important and thus not readily observed in Ti-24-11, if only cooling rates below 120 K/s are considered [3]. The higher content of β-stabilizing elements in Ti-25-10-3-1 slows down transformation kinetics, allowing increased processing flexibility and microstructural control [8]. The major advantage of the more highly β stabilized Ti-25-10-3-1 alloy is that cooling rates down to 9 K/s (AC) produce a large volume fraction of supersaturated B2. Subsequent aging treatments lead to O phase/α_2 precipitation at a very high driving force resulting in a fine microstructure [3,9,11].

Deformation Degree

Increasing the deformation degree from 76% to 93% improves the room temperature yield and ultimate tensile strength by about 5%, i.e. from 1400/1640 MPa to 1460/1725 MPa. In spite of these high strength values, fracture elongation was in the range of 3% due to the fine microstructure after sub-transus SHT (T_β - 20^{0}C) (Fig. 2(c)). First investigations on the crystallographic orientation showed that a pronounced <10$\bar{1}$1> fibre texture is formed during thermomechanical processing, which might contribute to these high strength levels. Further work is in progress to get more insight into the influence of texture on strength.

Stability of the Microstructures

Fig. 5 shows the hardness values of the fine bi-modal α_2+β processed microstructure (Fig. 2(c)) and the coarse β processed structure (Fig. 2(e)). All specimens were first aged at 700^{0}C/6h and afterwards further aged at lower temperatures ranging from 550^{0}C to 650^{0}C. As already observed, the coarse β processed material is more stable than the fine bi-modal structure (Fig. 4(b), 4(c)). The hardness level of the stable microstructure is reached in the bi-modal structure after an aging time of about 500h. However, at temperatures up to 600^{0}C this high strength microstructure seems to be fairly stable (Fig. 5).

Mechanical Properties at Elevated Temperatures

In Fig. 6 the specific ultimate tensile strength of Ti-25-10-3-1 is compared with the most common Ni base alloys, the latest near alpha alloys (IMI 834, Ti-1100), the new γ-TiAl alloys represented by Ti-48Al-2Cr-2Nb and the as received material (AR). The cross hatched region contains all strength values found for Ti-25-10-3-1 in the literature. In the temperature range up to 650^{0}C Ti-25-10-3-1 possesses the highest strength potential, after optimizing the microstructure by TMT as shown in this investigation. However, the major drawback of Ti-25-10-3-1 is its behavior in a oxidizing environment. Altough the oxidation rates of this alloy is improved compared to conventional alloys, it is much more sensitive to an environmentally embrittled surface due to its lower capability for damage tolerance (Fig. 1(b)) [10]. Fig. 1(b) shows the typical tensile failure behavior of Ti-25-10-3-1 at elevated temperatures due to circumferencial surface cracking. Fracture propagates across the specimen normal to the loading direction. Thus, a protective coating would probably need to be found for service temperatures above 600^{0}C.

Conclusions

1. Microstructural optimization enables strength values in Ti-25-10-3-1 which are among the highest reported for titanium base alloys.
2. A marked age hardening effect can be observed in Ti-25-10-3-1 after cooling rates from solution heat treatment down to at least 9K/s. Therefore microstructural control and processing flexibility is ehanced in this alloy as compared to Ti-24-11.
3. A pronounced <10$\bar{1}$1> fibre texture is formed during thermomechanical treatment.
4. The fine bi-modal microstructure with the high strength values and reasonable ductility seems to be fairly stable up to 600^{0}C.
5. The equiaxed microstructure with the high volume fraction of primary α_2 (T_β-120^{0}C) enables fracture elongations up to 10% at room temperature.
6. Testing at elevated temperatures in air leads to predominant surface cracking in Ti-25-10-3-1.

Acknowledgements

The authors highly appreciate the experimental support given by M. Alperth, J. Brien, J. Eschweiler, K. Klaukien, H. Schurmann, W.-D. Zimmermann.

References

[1] H.A. Lippsitt, High-Temperature Ordered Intermetallic Alloys, Vol. 39, ed. C.C. Koch, C.T. Liu, N.S. Stoloff (Pittsburgh, PA: MRS 1985) 351-364.

[2] D.M. Dimiduk, D.B. Miracle, Y-W. Kim, M.G. Mendiratta, "Recent Progress on Intermetallic Alloys for Advanced Aerospace Systems," ISIJ International, 32 (1991) 1223-1234.

[3] J. Kumpfert, K.-J. Grundhoff, H. Schurmann, Y.T. Lee, C.H. Ward, M. Peters, EUROMAT 91 - The 2nd European Conference on Advanced Materials and Processes, (London, England: The Institute of Metals, 1992), in print.

[4] D.L. Lukasak, D.A. Koss, "The Flow and Fracture of a Ti_3Al-Nb Alloy", Met. Trans. A, (21A) (January 1990), 135-143.

[5] E. Hornbogen, "Combined Reactions", Met. Trans. A, (10A) (August 1979), 947-972.

[6] L.A. Bendersky, W.J. Boettinger, B.P. Burton, F.S. Biancaniello, "The Formation of Ordered ω-Related Phases in Alloys of Composition Ti_4Al_3Nb", Acta Met., (38) (1990), 931-943.

[7] T.F. Broderick, private communication, Wright Laboratory, June 1992.

[8] W.A. Baeslack III, T.F. Broderick, "Effect of Cooling Rate on the Structure of a Ti-26Al-10Nb-3V-1Mo Titanium Aluminide", Scripta Metall., (24) (1990), 319-324.

[9] C.H. Ward, Microstructural Evolution and its Effect on the Fracture Behavior of Ti_3Al Intermetallics" (Accepted for publication in Int. Met. Rev., 1992).

[10] C.H. Ward, S.J. Balsone, Microstructure/Property Relationships in Titanium Aluminides and Alloys, ed. Y-W. Kim, R.R. Boyer (Warrendale, PA: TMS, 1990), 373-386.

[11] H.T. Weykamp, D.R. Baker, D.M. Paxton, M.J. Kaufman, "Continuous Cooling Transformations in Ti_3Al + Nb Alloys", Scripta Metall., (24) (1990), 445-450.

Figures

(a) 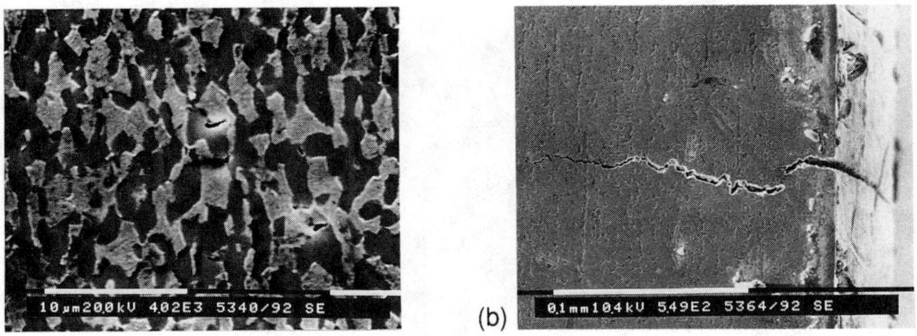

(b)

Fig. 1: (a) SEM micrograph of cracks in the microstructure from Fig. 2a at RT, (b) SEM micrograph of surface cracking in the microstructure from Fig. 2c at 650 ^{0}C.

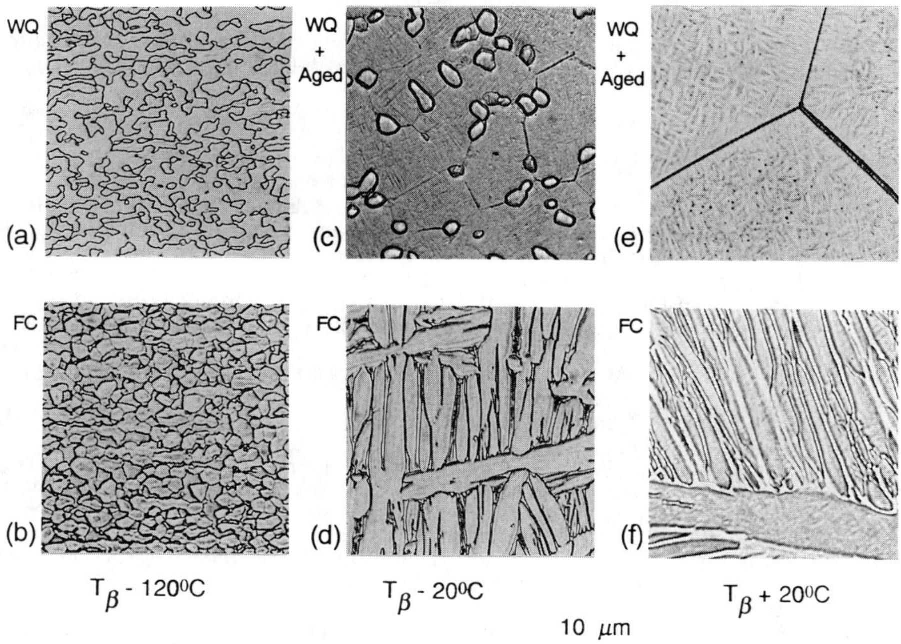

Fig. 2: Optical micrographs of Ti-25-10-3-1 subjected to different variations of SHT temperatures and cooling rates (WQ, FC).

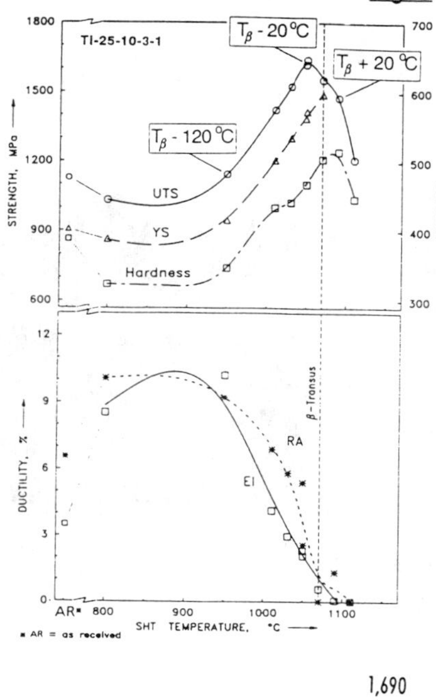

Fig. 3: Dependence of room temperature tensile properties and Vicker's Hardness of TMT Ti-25-10-3-1 on the SHT temperature at a constant cooling rate and aging treatment.

TMT: swaged 950°C/76% + SHT (15 min/WQ) + aged (700°C/4h).

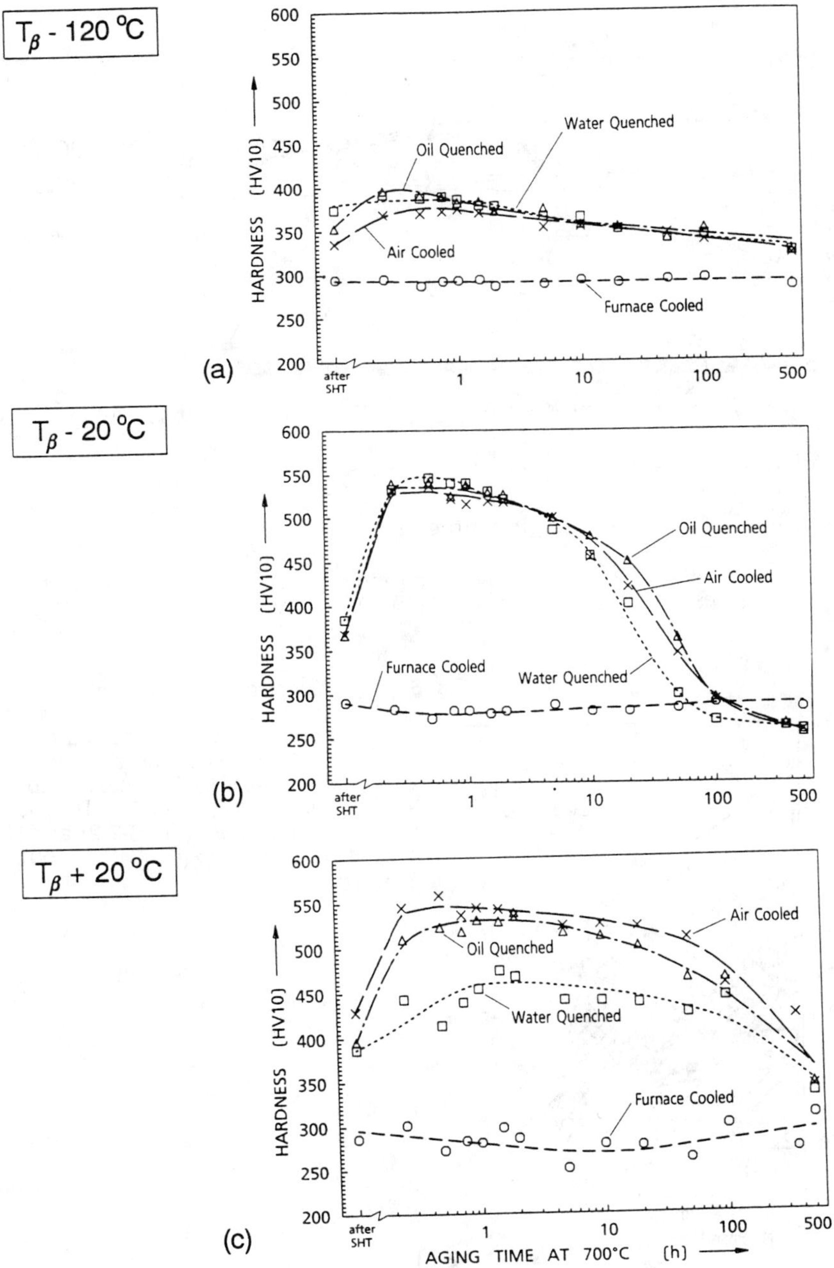

Fig. 4: The effect of cooling rate on the age hardening behavior at 700 °C in Ti-25-10-3-1 as a function of aging time for three SHT temperatures; (a) T_β-120 °C, (b) T_β-20 °C, (c) T+20 °C.

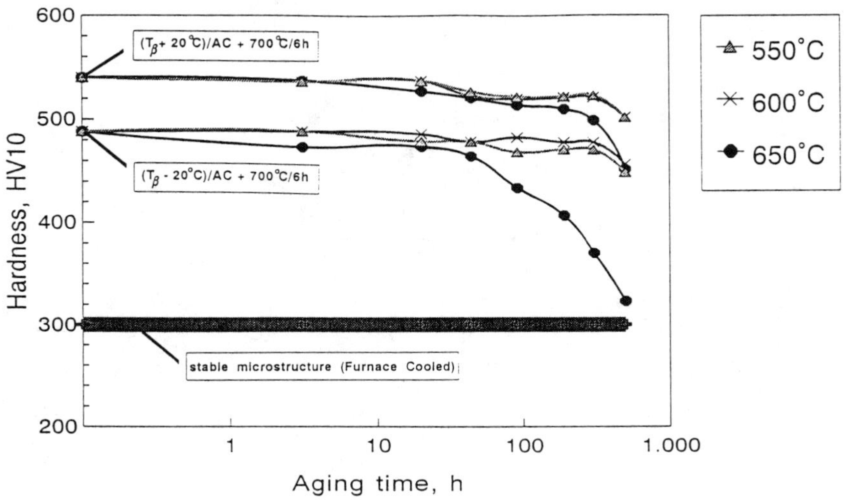

Fig. 5: The effect of long-term aging on room temperature Vicker's hardness for the β (T_β+20 ^{0}C) and $\alpha_2+\beta$ (T_β-20 ^{0}C) processed microstructures.

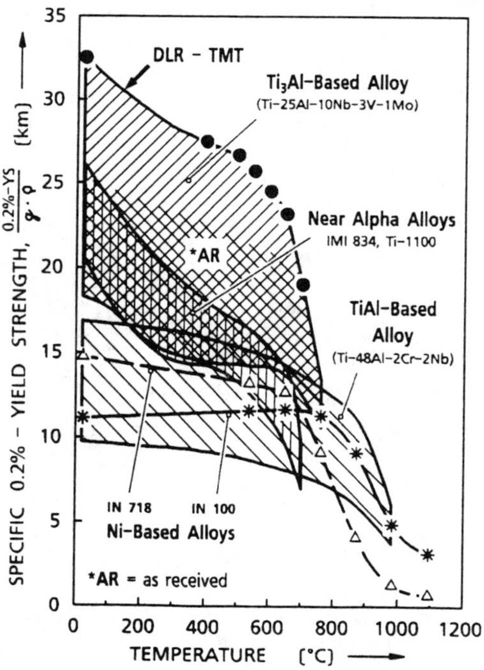

Fig. 6: Specific yield strength values of TMT Ti-25-10-3-1 as a function of temperature in comparison to the as received material (AR), near alpha alloys (IMI 834, Ti-1100), γ-TiAl alloys (Ti-48-2-2) and Ni based alloys (IN 718, IN 100).

THE ROLE OF HYDROGEN IN CYCLIC AND DWELL SENSITIVE FATIGUE OF A NEAR ALPHA TITANIUM ALLOY

W.J. Evans and M.R. Bache

I.R.C. in Materials for High Performance Applications,
University College of Swansea, U.K., SA2 8PP

Abstract

The paper demonstrates that hydrogen plays a significant role in the deformation and failure mechanisms in the titanium alloy IMI685. This role is complex and affects basketweave and aligned variants of the alloy to differing degrees. Strain-time records illustrate that hydrogen acts as a strengthening element upto concentrations around 60ppm. At higher levels, although strength is maintained premature failure occurs. Fractographic evidence is presented that suggests this reduction in fatigue life could be associated with the formation of brittle hydrides.

Introduction

Titanium alloys are used in engineering applications where a high specific strength is of paramount importance. A major factor in their popularity has been the development of the high strength near alpha compositions. These offer good creep properties due to solid solution strengthening and to a coarse grain, Widmanstatten or basketweave microstructure produced by the processing route. However, for thick sections where the cooling rate through the beta transus is not sufficiently high the alpha phase can develop an aligned morphology with a consequential adverse effect on mechanical properties.

A further problem that has taxed engineers in recent years is the fact that the near alpha alloys are sensitive to periods of dwell at peak stress during fatigue loading at near ambient temperatures. This phenomenon is addressed in a separate paper to be presented to this conference. Previous workers have suggested that hydrogen may play a critical role in the dwell fatigue behaviour but to date this has not been clearly established and the level of understanding of initiation and growth of cracks in these alloys is still a matter of some debate. This paper will focus on the role of hydrogen during a controlled study of dwell sensitive fatigue in IMI685 (Ti 6Al 5Zr 0.5Mo 0.2Si).

Titanium '92
Science and Technology
Edited by F.H. Froes and I. Caplan
The Minerals, Metals & Materials Society, 1993

Experimental Methods

The alloy was modified using a vacuum/hydrogenation technique to provide three classes of material based on their hydrogen content: "low hydrogen" material with a content of 20ppm hydrogen, "as received" at 60ppm and "high hydrogen" which encompassed concentrations between 100 and 275ppm. A test matrix was designed to explore the effects of cyclic and dwell loading on basketweave and aligned microstructures under uniaxial tension or torsion loading. The respective cyclic and dwell waveforms both consisted of a 2 second rise and fall time between minimum and maximum load, with either a 1 second or 120 second dwell at peak load. An R ratio of 0.1 was used for all testing. In addition to noting the life to failure each fatigue test was fully monitored to produce strain-time records. All specimens were examined by optical and scanning electron microscopy to provide information concerning the mechanisms of failure in the various materials.

Full details of specimen, pull rod and extensometer design plus other experimental techniques are awaiting publication [1].

Results

Stress dependence

The effects of hydrogen concentration on fatigue life are demonstrated by the S-N curves represented in Figures 1 to 4.

Figures 1 and 2 relate to tests on material in the basketweave condition. For comparison purposes, each graph contains a least squares best fit line from cyclic tension and torsion tests on basketweave material with a hydrogen content of 60ppm [2]. Figure 1 illustrates cyclic and dwell, tension and torsion lives for low hydrogen material. It is noted that the reduction in hydrogen from 60ppm to approximately 20ppm has had minimal effect on cyclic life but a significant effect on dwell lives under tensile loading. The effects of high hydrogen on basketweave material are summarised in Figure 2. A marked reduction in life for the dwell tension tests compared to the 60ppm data is clearly apparent.

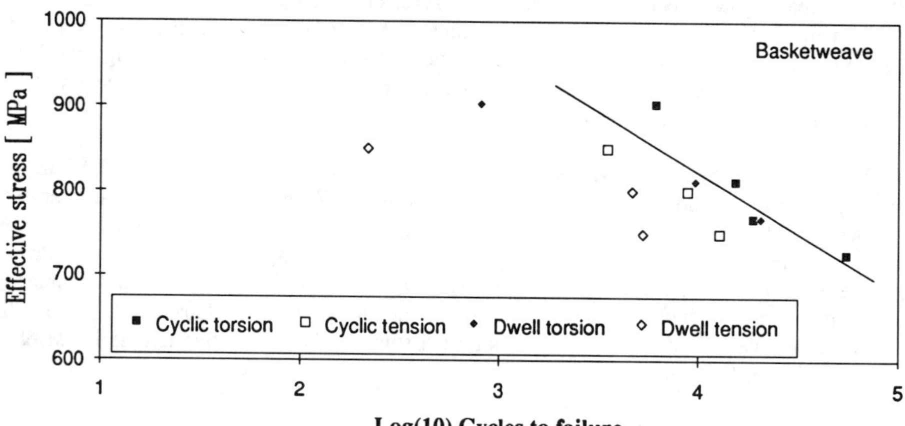

Figure 1 - Low hydrogen tests compared to the best fit line from 60ppm basketweave material.

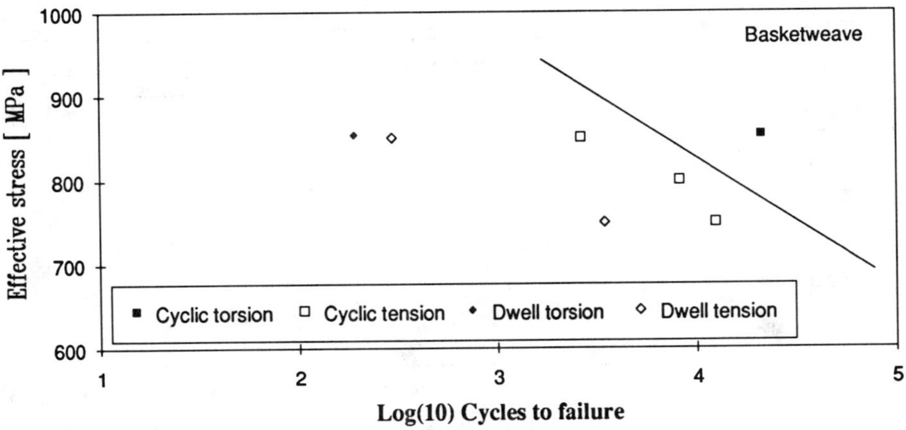

Figure 2 - High hydrogen tests compared to the best line fit for 60ppm basketweave material.

Figures 3 and 4 are the corresponding data for the aligned microstructure. The solid line is the best fit for cyclic torsion and cyclic tension tests in 60ppm aligned material [2]. (The cyclic tension data were obtained on solid rather than tubular specimens because, in the aligned material, they produced more reliable tensile life data due to the relatively large grainsize compared with the thin wall tubular testpieces). A reduction in life is apparent in Figure 3 for low hydrogen material under cyclic and dwell tension. The dwell torsion results also display a life reduction at the higher stress levels.

Figure 4 combines all high hydrogen data for the aligned condition. Under cyclic tension lives fall below the 60ppm trend by approximately one order of magnitude. However, even more dramatic de eases in life are observed for dwell tension. Typically the reduction is greater than two orders of magnitude. Repeat tests at a single stress level reinforce the validity of this result.

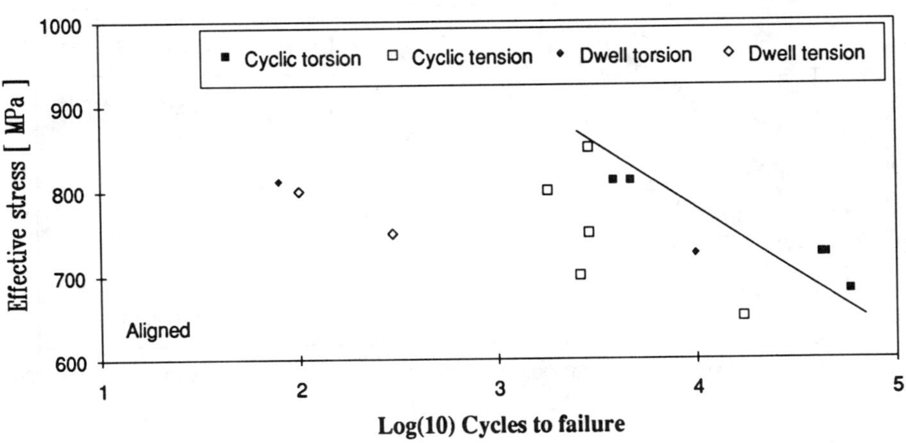

Figure 3 - Low hydrogen tests and best fit line from cyclic data on 60ppm aligned material.

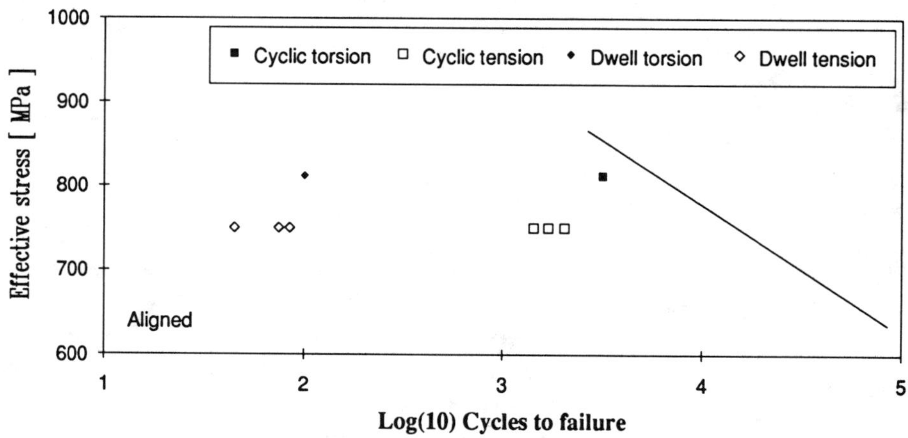

Figure 4 - High hydrogen tests and best fit line for cyclic data on 60ppm aligned material.

Strain-time behaviour

The role of hydrogen on bulk deformation is clearly demonstrated by strain-time curves from individual tests, Figure 5. For a given stress and microstructure, low hydrogen tests display relatively high strain accumulation compared to 60ppm material. The latter has a longer time to failure at a lower rate of strain accumulation. High hydrogen levels of 100ppm or greater virtually superimpose on the early stage of the 60ppm curve but then suffer premature failure. This trend was noted under tension and torsion for both cyclic and dwell loading. It suggests that hydrogen strengthens the alloy up to a certain concentration (around 60ppm) but further additions then lead to a brittle response.

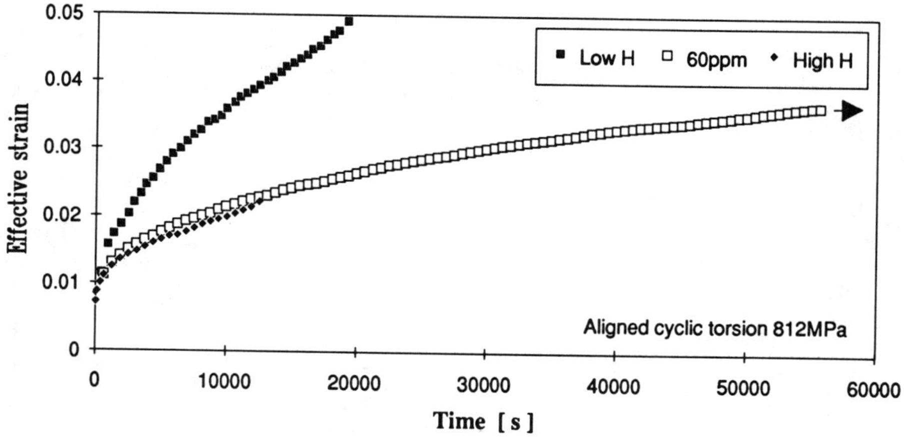

Figure 5 - Strain-time records for the aligned microstructure under cyclic tension at 750MPa.

Fractographic study

The strain-time response shows a transition from relatively ductile to virtually brittle behaviour with increasing hydrogen concentration. These changes were associated with characteristic features on the fracture surfaces. The differences were particularly well defined on failures in the aligned microstructure. At low hydrogen levels there was considerable evidence of slip bands on internal and external gauge surfaces in regions close to and well away from the main fracture. Under tension, the fracture surfaces themselves had a specific "rooftop" form strongly suggesting that shear on planes at $\pm 45^{o}$ to the tensile axis had been instrumental in this formation, Figure 6. At 60ppm, rooftop failures were again apparent at the highest stress levels. However, at lower stresses it was apparent that both cyclic and dwell fractures had initiated at large facets which were orientated approximately perpendicular to the tensile axis.

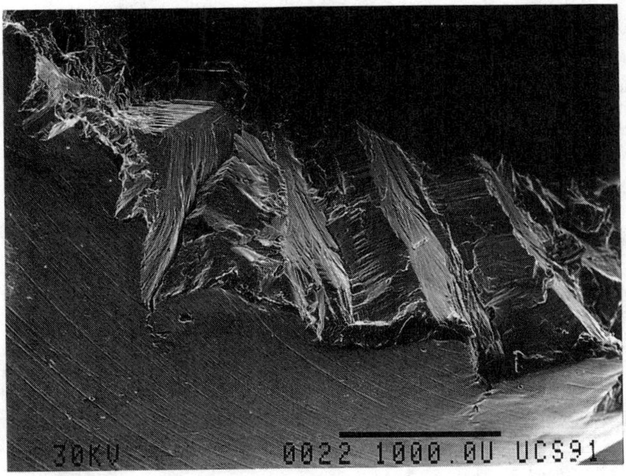

Figure 6 - "Rooftop" fracture in low hydrogen, aligned material under cyclic tension, 750MPa.

The high hydrogen failures had a particularly distinctive form which was totally different from the other levels studied. A typical example from a tensile test on aligned material is shown in Figure 7. The fracture has a marked step and plateau sequence. The steps appear to be comprised of a stack of planar features. Secondary cracking occurs at the foot of steps and along the interfaces within the stacked plates. Nelson [3] has proposed models to explain cracking in similar titanium alloys exposed to hydrogen gas atmosphere. He was able to show that there was a preference for cracking along alpha/beta phase boundaries. In the present example, this type of cracking can be related to the plateaux and interface undercuts. Other similarities with the fractures reported by Nelson were also noted. In particular, a terraced appearance was observed when viewing vertically downwards onto the plateaux regions, Figure 8. This example was found on a torsion fracture surface that had developed perpendicular to the maximum principal stress (mode I). It has been suggested [3] that these features may be the result of brittle hydrides formed by the migration of hydrogen along alpha/beta boundaries under the influence of a tensile stress.

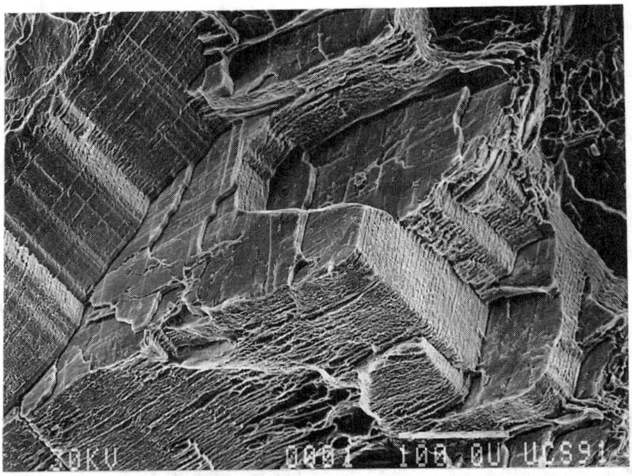

Figure 7 - Characteristic form of fracture for high hydrogen material.

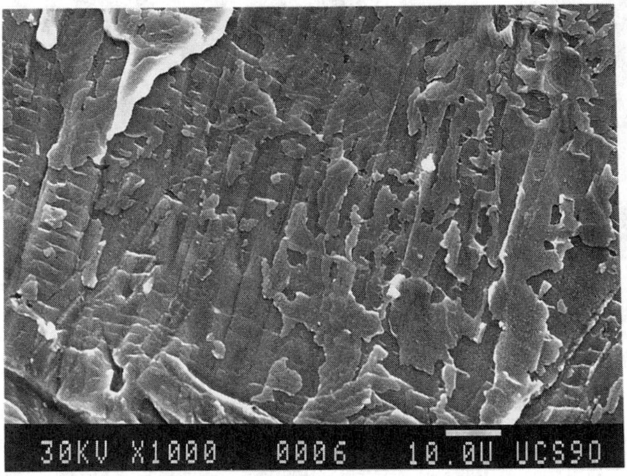

Figure 8 - Plan view of alpha / beta interface plateau.

Discussion

The results demonstrate that hydrogen plays a significant role in both deformation and failure mechanisms for this near alpha titanium alloy. However, this role is complex and affects aligned and basketweave microstructures to different degrees. Results from specimens that have initially been degassed but subsequently recharged to give hydrogen levels around 60ppm support the view that the optimum mechanical behaviour seen during this investigation is solely a function of hydrogen concentration.

In terms of bulk deformation, the hydrogen acts as a strengthening element. It reduces the rate of strain accumulation and improves fatigue lives for concentrations upto 60ppm. At higher hydrogen levels, however, the strength is maintained but there is a marked reduction in ductility and fatigue life. Fractographic evidence shows that these high hydrogen failures are dominated by tensile stress under tension or torsion loading with the cracks following alpha/beta interface planes. Features that might imply the formation of brittle hydrides are apparent on these planes. In fact, there are remarkable similarities between these failures and those for titanium alloys tested in gaseous hydrogen environments [3].

At lower hydrogen concentrations, two distinctive forms of fracture occured. At high stress, shear dominated failures are observed under both tension and torsion loading. The tendency towards shear related failures is greatest for the lowest hydrogen concentration. At lower stresses, crack initiation is associated with quasi-cleavage facet formation. These facets are orientated perpendicular to the dominant tensile stress. Previous work has shown that the facets have a basal plane orientation [4]. The formation of the facets has been attributed to stress redistribution associated with time dependent strain accumulation, such as that illustrated in Figure 5. It is likely that the reduced rate of strain accumulation at 60ppm contributes to the greater occurrence of this mode of failure at that hydrogen level.

Conclusions

1. Hydrogen in the titanium alloy IMI685 plays a vital role with regards to fatigue response.

2. Hydrogen is a strengthening element, improving cyclic and dwell lives up to a concentration of 60ppm. At higher levels, although strength is maintained, there is a reduction in ductility.

3. Features have been highlighted, that may result from the formation of brittle hydrides ahead of the tip of the advancing fatigue crack.

4. The combination of tensile dwell loading and high hydrogen content leads to significantly short fatigue lives particularly in material with an aligned microstructure.

References

1. Bache, M.R. and Evans, W.J., "Tension and torsion fatigue testing of a near alpha titanium alloy", to be published, Int. J. Fatigue.

2. Bache, M.R. and Evans, W.J., "Fatigue under tension/torsion loading in IMI685", this volume.

3. Nelson, H.G., "A film-rupture model of hydrogen induced, slow crack growth in acicular alpha-beta titanium", Met. Trans. A, **7A**, pp621, 1976.

4. Evans, W.J., "The influence of microstructure on dwell sensitive fatigue in a near alpha titanium alloy", Scripta Met., vol 21, No.4, pp469, 1987.

HOT STRENGTH OF Ti AND Ti ALLOYS DEFORMED IN AXIAL COMPRESSION

Ronaldo A.N.M. Barbosa[1], Henrique C. Braga[1]
and Jurgen Breme[2]

1- Dept. de Eng. Metalúrgica da Universidade Federal de Minas Gerais. R. Espírito Santo 35, s206- CENTRO. 30160- Belo Horizonte- MG- BRAZIL.

2- Dept. of Material Sciences, University of Saarbrücken, FB12.1-Werkstoffewissenschften, Gebaude 22, 5°-étage. 6600- Saarbrücken- GERMANY.

Abstract

Pure Ti and Ti6Al4V alloy have been often studied and there are many reports in the literature related to aspects of plastic deformation of these materials. Most published results are concerned with the relation between a given stress at constant strain and the hot deformation variables. Modeling of full stress-strain curves is rarely reported. However, knowledge of mean stresses during deformations is important in the design of hot forming processes.

This work analyses constitutive equations suitable to describe the relationship between stresses and strains at a given value of deformation. The expressions are then used in a published model to simulate stress-strain curves to equivalent strains up to approximately 0.8. The calculations are performed for hot deformed α and βTi as well as $\alpha+\beta$ Ti6Al4V. An analysis of the microstructure obtained from some hot deformed samples is also presented in this paper.

Introduction

Pure Ti and Ti6Al4V alloy have been frequently studied and there are many reports in the literature related to aspects of plastic deformation of these materials. Sargent and Ashby[1], for instance, have assembled a deformation map for pure α and β Ti using extensive data available in the literature. The map includes possible modes of deformation for a wide range of temperatures (from 73 to 1873K). This deformation diagram is valuable for predictions of mechanical behavior of Ti as a whole. However, it is somewhat theoretical in nature since all information was based in reports of the literature. Moreover, many interpolations had to be made in order to construct the map fully.

Several studies have also been concerned with the behavior of the stress at a particular strain during hot deformation. Most frequently the stress is related to ordinary hot processing variables. This is the case of Oikawa et al.[2] who have investigated the deformation characteristics of coarse grained βTi in the steady state. They found that the steady state stress is related to the strain through an equation of the Dorn type. Later, Oikawa et al.[3] extended their findings to cover αTi-Al alloys. However, their equations can not be used directly to simulate full stress-strain curves. A similar approach has also been taken by Sheppard and Norley[4], that is, they reported a

Titanium '92
Science and Technology
Edited by F.H. Froes and I. Caplan
The Minerals, Metals & Materials Society, 1993

dependence of the peak torque from torsion testing as a function of the temperature for Ti6Al4V. They also studied in detail the microstructures obtained from the deformed samples. No mention was made, however, to full simulation of the stress-strain curves.

The knowledge of mean stress during deformation is important in the design of hot forming processes. Nonetheless, few works have reported equations which are useful for load calculations, pass design or schedule optimization. This is not the case of Yamamoto and Ouchi(5). They employed empirical equations to predict mean stresses of Ti and Ti6Al4V. Nonetheless, their relationships are apparently not based on any fundamental concept but rather on regression analysis to obtain their coefficients. Another group of researchers(6) did simulate the stress-strain curves for Ti with several degrees of purity. The authors analyzed αTi deformed in the range of 873 to 1123K (600 to 850°C). The strain rates used in their experiments were in the range of 2 to $150s^{-1}$ and they complement the results obtained in the present work as will be seen later.

This work analyses constitutive equations suitable to describe the relationship between stresses and strains at a given value of deformation. The expressions are then used in Sellars´ model(7) to simulate stress-strain curves to equivalent strains up to approximately 0.8. The calculations are performed for hot deformed α and βTi as well as α+β Ti6Al4V. An analysis of the microstructure obtained from some hot deformed samples is also presented in this paper.

Experimental Technique

Material

The materials employed here were a commercially pure Ti grade 2 according to ASTM B265-79 and a Ti6Al4V. Details of the chemical compositions are given in Table I. Pure Ti presents a lamellar β (bcc) microstructure phase at temperatures higher than 1155K (882°C). Below this temperature, an α (hcp) phase exists. In the case of Grade 2 Ti however, there is a α + β dual phase region around 1155K (882°C). The Ti6Al4V, on the other hand, is an α + β mixture in the range of temperatures studied in this paper.

Table I Chemical compositions of the alloys used in this research. All numbers given in weight%.

Alloy	Ti	Al	V	C	Fe	O	H	N
Ti	b*	-	-	0.04	0.05	0.16	0.011	0.03
Ti6Al4V	b*	5.8	3.9	0.05	0.13	0.12	0.008	0.05

* b = balance

Isothermal Compression Testing

Isothermal compression testing of small cylinders was used to characterize the stress-strain behavior. Compression specimens were 5mm in diameter and 7.6mm in height. Tests were carried out at temperatures ranging from 1073 to 1223K (800 to 950°C) with intervals of 25K (25°C) and strain rates of 0.1, 1 and $3s^{-1}$. The strain rates were kept constant during compression. The specimens were deformed to equivalent strains in the range of 0.7 to 0.8 mostly. Testing was performed in a servo-hydraulic computer controlled machine. Samples were always heated up to compression temperature being homogenized for 300s before deformation was initiated.

Results and Discussion

Constitutive equations

The dependence of the peak stress on testing temperature is shown in Figures (1a-b) for Ti and Ti6Al4V respectively. The differences in stresses at approximately 1123K (900°C) are due to phase transformation in Ti. Ti6Al4V is $\alpha + \beta$ in all temperatures tested, therefore there is no sharp increase of stress as observed for pure Ti.

The analysis of the stress-strain curves was carried out using constitutive equations derived by Sellars and Tegart[8]. Similar treatment to stress-strain curves of a β Ti15V3Cr3Sn3Al has also been recently reported in the literature[9]. The constitutive equations can be written as follows

$$\dot{\varepsilon} = A\ F(\sigma)\ \exp\ (-Q/RT) \qquad (1)$$

where $F(\sigma)$ is a function of the applied stress with the following possible forms

$$\begin{aligned} F(\sigma) &= \sigma^n && \text{if } \alpha\sigma < 0.8 \\ &= \exp\ (\beta\ \sigma) && \text{if } \alpha\sigma > 1.2 \\ &= [\sinh\ (\alpha\ \sigma)]^n && \text{for all values of } \sigma \end{aligned} \qquad (2)$$

Here, $\beta = \alpha\ n$ and when $\alpha\ \sigma < 0.8$ equation (1) can be approximated by a power law and be re-written as

$$\sigma = B\ \dot{\varepsilon}^m \qquad (3)$$

The constant m, the strain rate sensitivity coefficient, is related to the constant n as m = 1/n. Therefore, if the dependence of the stress at a given strain on both strain rate and temperature is known, then the activation energy, Q, and the experimental constants α, β and n can be evaluated. Figure (2a) and (2b) show that Ti6Al4V clearly follows equation (2) for low (a) and high stresses (b). A similar pattern can be found for pure Ti. Note that the experimental data depart from a straight line at a value of $\alpha\sigma$ of about 1. Table II gives details of all experimental constants evaluated in the present work.

Table II Experimental constants evaluated at the peak stress for Ti and Ti6Al4V according to equations (1) and (2).

Alloy	Phases	n	α (Mpa^{-1})	β (Mpa^{-1})	Q (kJ/mol)
Ti	β	4.9	0.0480	0.233	179
	α	5.2	0.0104	0.054	346
Ti6Al4V	$\alpha + \beta$	3.7	0.0083	0.030	588

The values shown in Table II generally agree with the ones reported in the literature. Reca and Libanati[10] have reported an activation energy for βTi of 153kJ/mol. On the other hand, Q value as high as 253kJ/mol has been reported[5] for βTi alloyed with V. In the case of Ti6Al4V, Sheppard and Norley[4] have reported a value of 522kJ/mol.

The stress at a given strain can further be written as a function of the Zener-Hollomon parameter ($Z = \dot{\varepsilon} \exp(Q/R\,T)$) as

$$Z = A \left[\sinh \left(\alpha \sigma \right) \right]^{n} \tag{4}$$

where A is a constant which depends on the strain. Figure (3a) shows the dependence of ln (Z) on the ln (sinh ($\alpha\sigma$)) for pure Ti. All experimental data can be reasonably fit by a straight line in agreement with equation (4). Similar results are reported for Ti6Al4V alloys as shown in Figure (3b). The line for αTi in Figure (3a) can well explain the present data as well those of Senuma et al.[6]. Table III shows the values of A for strains of 0.05 ($\sigma_{0.05}$), peak stress (σ_p) and steady state stress (σ_{ss}).

Simulation of stress-strain curves

Equations described so far relate the stress at a given strain to the Zener-Hollomon parameter. Relationships connecting stress to strain are now required. In this work a model proposed by Sellars[7] has been used. Details of the model are shown below.

$$\sigma = \sigma_e \qquad \text{if} \quad \varepsilon \leq \varepsilon_c \tag{5}$$

$$\sigma = \sigma_e - \Delta\sigma \qquad \text{if} \quad \varepsilon > \varepsilon_c \tag{6}$$

$$\varepsilon_c = a\, \varepsilon_p \tag{7}$$

$$\sigma_e = \sigma_o + D \left[1 - \exp\left(-C\, \varepsilon \right) \right]^{p} \tag{8}$$

$$\Delta\sigma = D' \left\{ 1 - \left[\exp\left(-k \frac{\varepsilon - \varepsilon_c}{\varepsilon_p} \right) \right]^{q} \right\} \tag{9}$$

Here σ_o is the stress at $\varepsilon = 0.05$ and a, k, p and q are constants. D, D´ and C are a function of Z given by the expressions

$$D = \sigma_{ss(e)} - \sigma_o \tag{10}$$

$$D' = \sigma_{ss(e)} - \sigma_{ss} \tag{11}$$

$$C = X \ln\left[1 - \left[\frac{\sigma_x - \sigma_o}{D} \right]^{1/p} \right] \tag{12}$$

$$\sigma_{ss(e)} = b\, \sigma_p \tag{13}$$

In the case of Ti, X = 10 and σ_x is the stress at strain of 0.1. Moreover, $\sigma_{ss(e)} = \sigma_{ss} = \sigma_p$, which means that $\Delta\sigma$ is nil since there no dynamic recrystallization visible from the shape of the stress-strain curve. For Ti6Al4V X = 20 and σ_x is the stress at strain of 0.05. Additionally, the constant b is 1.05 for Ti6Al4V. Figures (4a) and (4b) show a comparison of measured and predicted stress-strain curves. The agreement between theoretical and experimental curves is reasonable regardless of testing temperature.

Microstructures Studies

Optical and transmission electron micrographs of various specimens are shown in Figures (5-7). With cp-Ti deformed at 800°C with 0.1 s^{-1} because of the

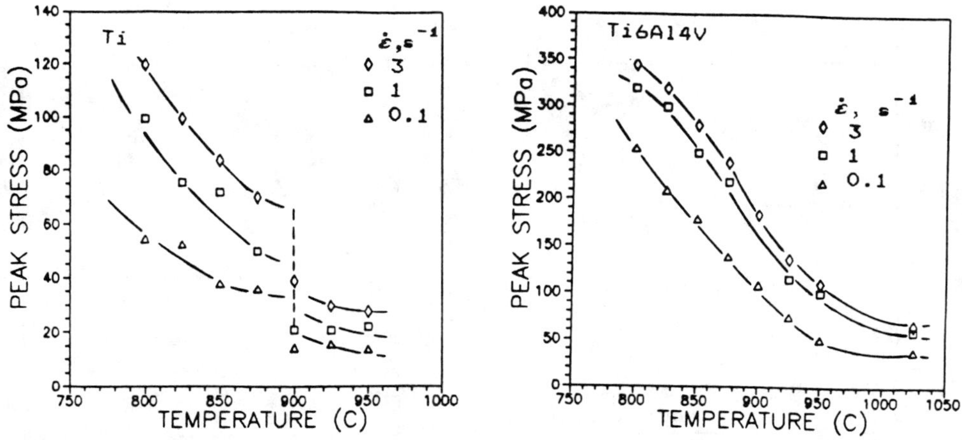

Figure 1: Dependence of σ_P, the peak stress, on testing temperature for Ti (a) and Ti6Al4V.

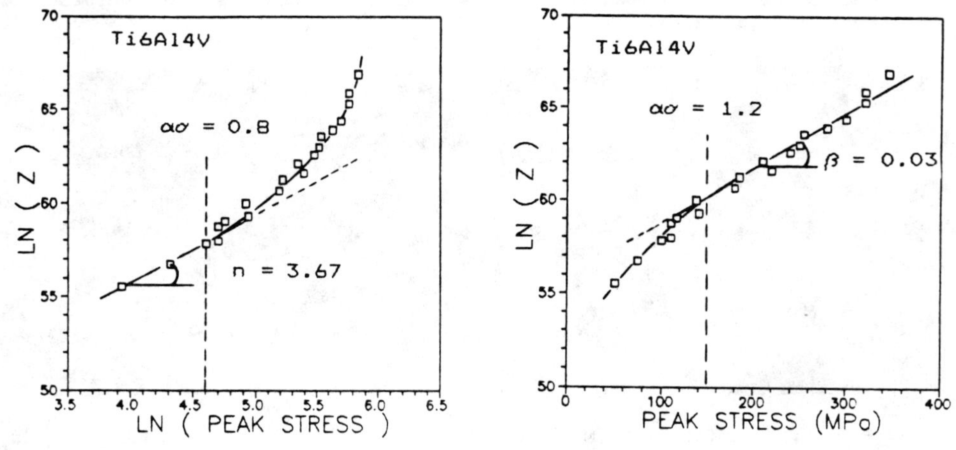

Figure 2: Dependence of Ln Z on Ln σ_P (a) and on σ_P (b) for Ti6Al4V alloy.

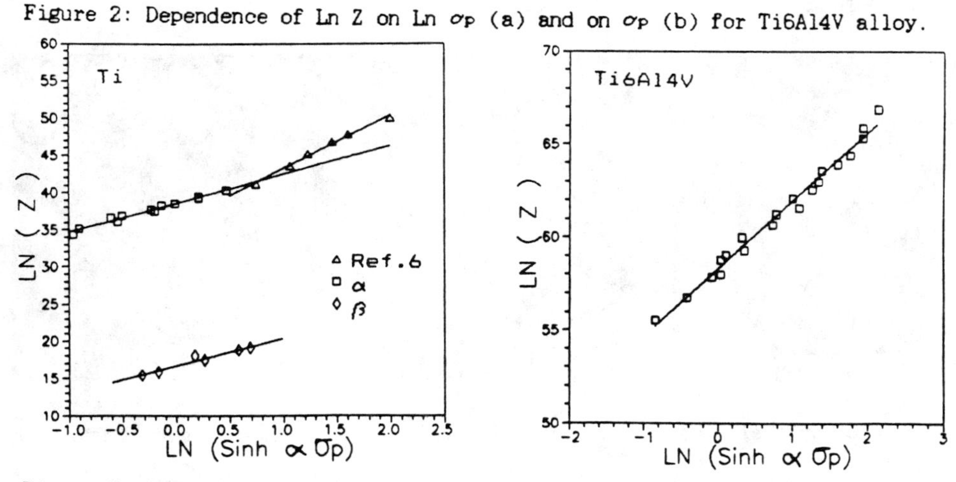

Figure 3: Correlation between Ln Z and Ln (sinh α σ_P) for Ti (a) and Ti6Al4V (b).

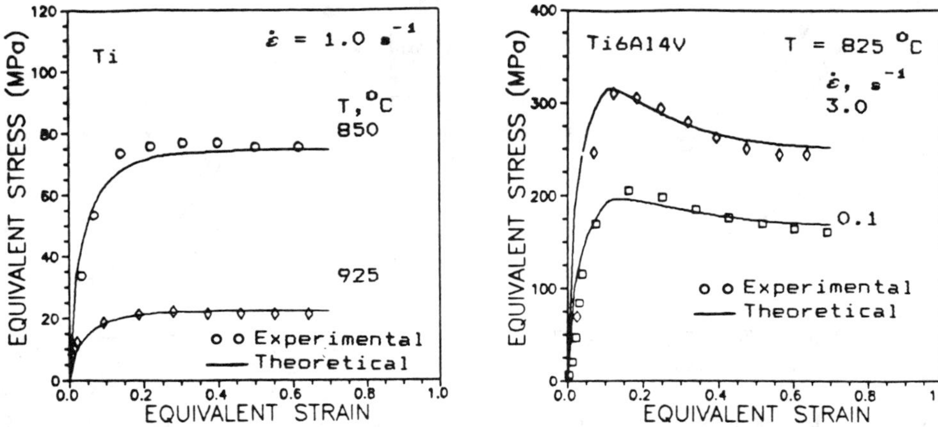

Figure 4: Comparison between predicted and measured stress-strain curves for Ti (a) and Ti6Al4V (b).

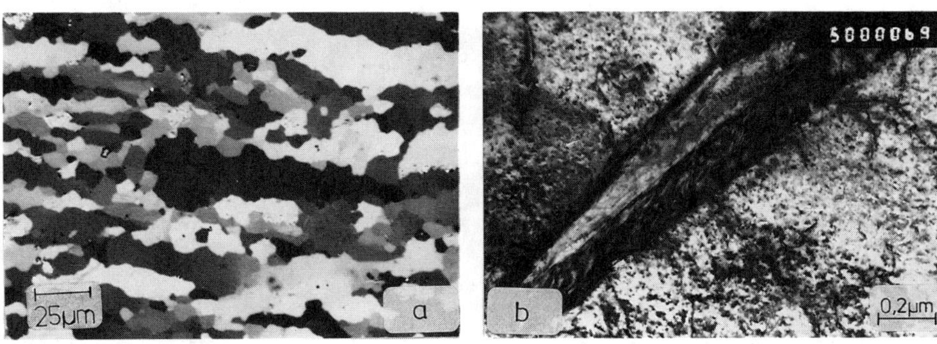

Figure 5: Optical (5a) and TEM (5b) micrographs of cp-Ti, deformed at 800°C; $\dot{\varepsilon}$ = 0.1s^{-1}.

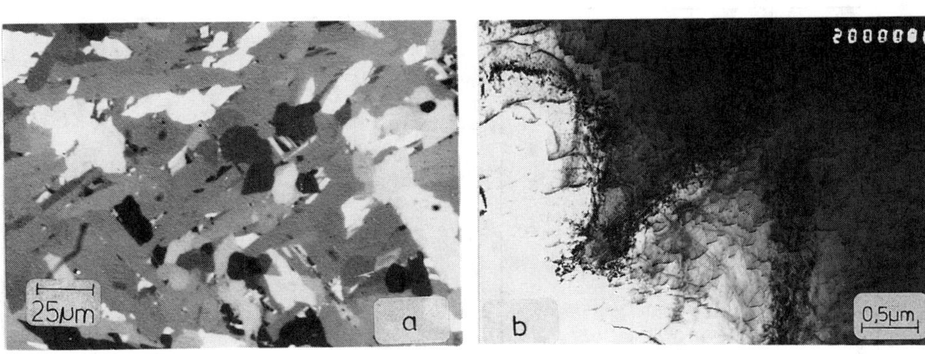

Figure 6: Optical (6a) and TEM (6b) micrographs of cp-Ti, deformed at 900°C; $\dot{\varepsilon}$ = 0.1s^{-1}.

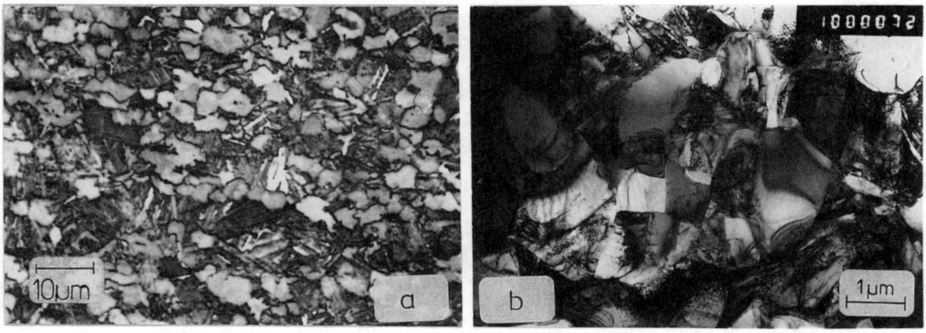

Figure 7: Optical (7a) and TEM (7b) micrographs of Ti6Al4V, deformed at 980°C; $\dot{\varepsilon}$ = 0.1s^{-1}.

high degree of deformation lengthened and elongated grains perpendicular to the deformation direction are observed (Figure 5a). According to their different color in polarized light the grains have different orientations. An extensive dislocation activity leading to a high density occurs (Figure 5b). In contrast the sample deformed in β-phase field at 900°C with 0.1 s^{-1} shows the typical microstructure of transformed β-phase without the elongation of the grains (fig. 6a) in a preferred direction. The dislocation density is less and subgrain formation because of dynamic recovery can be observed (Figure 6b). The microstructure of Ti6Al4V deformed at 950°C with 0.1 s^{-1} shows an amount of about 50% α- and 50% transformed β-phase according to the temperature of deformation. Deformation was not homogeneously distributed among the phases. This is the reason of the lower yield stress of the β-phase, which was highly deformed while α-phase was only slightly work hardened (Figure 7a). Nevertheless, both dynamic recrystallization and dynamic recovery occur (Figure 7b).

Table III Experimental constants evaluated for Ti and Ti6Al4V according to equation (4) at peak, ε_p, and steady state strains. ε_x is the strain of 0.10 for Ti and 0.05 for Ti6Al4V.

Alloy	Phases	Strains ε_x ln(A)	n	ε_p ln(A)	n	ε_{ss} ln(A)	n
Ti	β	17.6	4.82	16.6	3.68	16.6	3.68
	α	39.4	4.82	38.5	3.87	38.5	3.87
Ti6Al4V	$\alpha + \beta$	60.5	3.26	58.3	3.67	59.3	4.56

Conclusions

1- Stress at a given strain can be related to Z by equation (4).
2- Flow curves of α and β Ti as well as $\alpha + \beta$ Ti6Al4V can be simulated by equations (5) to (13).

Acknowledgements

RB and HB are grateful for the financial support received from CNPq (Conselho Nacional de Desenvolvimento Científico e Tecnológico) and GTZ (Deutsche Gesellschaft für Technische Zusammenarbeiten).

References

1- P.M. Sargent and M.F.Ashby, "Deformation Maps for Titanium and Zirconium", Scripta Metallurgica, 16 (1982), 1415-1422.

2- H. Oikawa et al., "High-Temperature Deformation of Polycrystalline Beta Titanium" , Scripta Metallurgica, 19 (1985), 825-828.

3- H. Oikawa et al., "Influence of Grain Size on High-Temperature Deformation of α-Ti-Al Alloys", Journal of Materials Science Letters, 7 (1988), 987-988.

4- T. Sheppard and J. Norley, "Deformation Characteristics of Ti-6Al-4V", Materials Science and Technology, 4 (1988), 903-908.

5- S. Yamamoto and C. Ouchi, "Deformation Resistence in Plate Rolling of Commercially Pure Titanium and Ti-6Al-4V Alloy", Tetsu-to-Hagane, 72 (1986), 595-602.

6- T. Senuma et al., "Deformation Resistance and Recrystallization of Commercially Pure Titanium in the High Strain Rate Hot Deformation", Tetsu-to-Hagane, 72 (1986), 321-328.

7- C.M. Sellars, "The Kinetics of Softening Processes during Hot Working of Austenite", Czecholosvak Journal of Physics, B35 (1985), 239-248.

8- C.M. Sellars and W.J. McG. Tegart, "La Relation entre la Resistance et la Structure dans la Deformation a Chaud", Memoires Scientifiques Revue de Metallurgie, 63 (1966), 731-746.

9- H.G. Suzuki and H. Fujii, "Deformation Characteristics of Ti-15V-3Cr-3Sn-3Al at Elevated Temperatue", ISIJ International, 31 (1991), 814-819.

10- N.E.W. de Reca and C.M. Libanati, "Autodifudion de Titanio Beta Y Hafnio Beta", Acta Metallurgica, 16 (1968), 1297-1305.

THE EFFECT OF BORON AND CERIUM ON THE CREEP BEHAVIOR OF A HIGH TEMPERATURE TITANIUM ALLOY

Zhang Jinxu, Xu Jialong, Chen Guoliang*, Chen Kexiu, Ge Honglin

Shanghai Iron and Steel Research Institute, Shanghai 200940, P. R. China
*Beijing University of Science and Technology, Beijing 100083, P. R. China

ABSTRACT

An investigation was conducted to examine the effect of boron and cerium addition on the mechanical properties of a new high temperature titanium alloy. Many kinds of advanced technology were used for study of the creep strengthening mechanism.

The results showed that micro-alloying with boron and cerium can extend greatly the creep life. The creep strengthening mechanisms are that distribution of Al, Mo and Nb elements in α phase and β phase is favorable to solid solution strengthening, that the segregation of boron and cerium at phase boundaries strengthens them and that the interface phase may be an obstacle to the movement of dislocation.

INTRODUCTION

Several countries have developed a near α type of titanium alloy system (Ti-Al-Sn-Zr-Mo-Nb), which is used at elevated temperature 500°C. Based on the composition of alloy Ti-6.5Al-2.2Sn-1.2Zr-2.0Mo-2.2Nb-0.2Si, a series of alloys was developed at the Shanghai Iron and Steel Research Institute over the past ten years. The alloys 7715A and 7715B have been used at 500°C for casting and forging respectively. Recently, forging 7715C alloy with boron and cerium has successfully passed a simulation test for a new type of plane at 500°C and 540°C for 200hr.

The beneficial effects of microaddition of elements on the mechanical behavior of titanium alloys have been reported in previous work[1-3] the addition of B in cast or welded forms of titanium alloys caused the refinement of grain and homogenization of structure but damaged the impact toughness[4-6]. The creep properties of castings can be improved by the addition of Ce, which resulted from stabilization of the dislocation substructure and strengthening

Titanium '92
Science and Technology
Edited by F.H. Froes and I. Caplan
The Minerals, Metals & Materials Society, 1993

of the grain boundary. However, there has been no concern about the effects of combined additions of B and Ce on forged titanium alloys.

In the present investigation, the effects of combined additions of B and Ce on the creep behaviour of a high temperature titanium alloy are investigated.

EXPERIMENTAL PROCEDURE

The compositions of the four alloys used in this investigation are given in Table 1. The alloys were vacuum melted by the self-consumable electrode process. The ingots were forged to Φ40mm bars and then rolled to Φ19mm rods. Under the same condition of heat treatment (980°C/1h A. C. + 560°C/2h A. C.) the four alloys have similar globular alpha structures.

The creep rupture time was measured at constant-load at 540°C.

The interfacial segregation and distribution of elements in alloys were analysed using both Auger Electron Spectroscopy (AES) and Electron Probe Microanalysis(EPM). Scanning and Transmission Electron Microscopy were used to observe the fracture surfaces and microstructures respectively.

Table. 1 Compositions of alloys investigated (wt%)

ALLOY	A	B	C	D
Al	6.05	6.54	6.61	6.64
Sn	2.17	2.23	2.17	2.29
Zr	1.27	1.22	1.20	1.25
Mo	2.07	2.07	2.00	2.10
Nb	2.32	2.32	2.30	2.24
Si	0.15	0.14	0.19	0.20
Ce	-	-	0.066	0.068
B	-	0.004	-	0.006

RESULTS

1. Creep properties

Table 2 shows the creep rupture times of four alloys under the five stress levels respectively. The changes in the creep rupture time on B or Ce addition are not noticeable, but the combined additions of B and Ce significantly increased the creep rupture time by approximately 10%, 136%, 132.1%, 177.7% and 137.5% for stress levels of 530MPa, 500MPa, 470MPa, 440MPa and 410MPa, respectively. This improvement resulted from both the reduction of steady state creep rate and the extension of the third stage of creep (shown in Fig. 1).

Table. 2 Creep rupture time of alloys (hr.)

ALLOY	530	500	470	440	410(MPa)
A	60	117	140	259	400
B	56	156	244	331	401
C	30	92	155	228	282
D	66	277	325	564	950

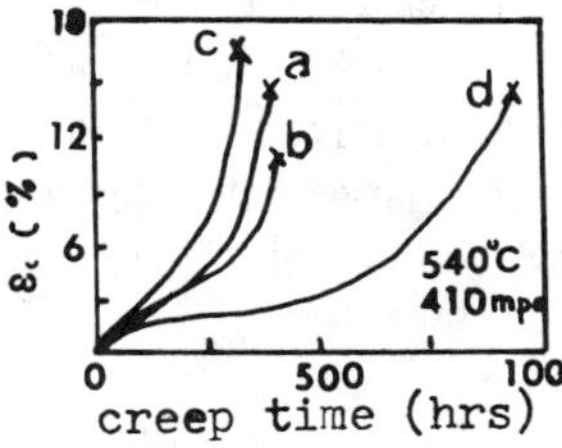

Figure 1
Creep curves of alloys

2. Interfacial segregation of B and Ce

Fig. 2 shows that the relative Ce level determined using EPMA (beam size is 1-2μm dia.) is significantly higher at the $\alpha+\beta/\alpha$ interface compared with the matrix. The Ce enrichment at the fracture surface, suggested to be $\alpha+\beta/\alpha$ interface, was determined also using AES analysis. The higher Ce peak in the spectrum is shown in Fig. 3. The relative content and thickness of Ce segregation in two alloys, one without B, and one with B, are shown in Fig. 4, which illustrates that the B addition prompted the Ce segregation.

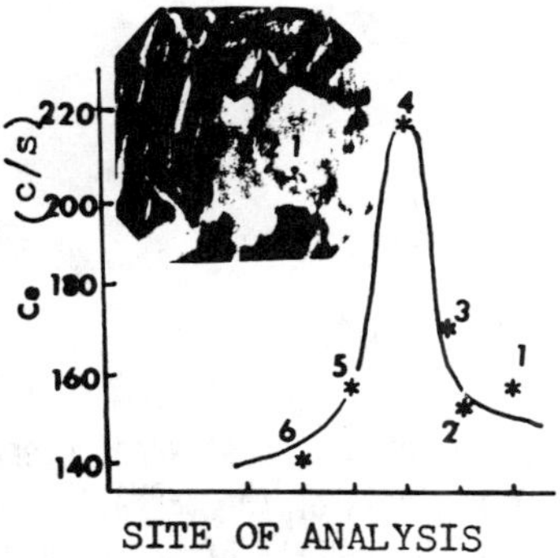

Figure 2
Relative Ce concentration determined by EPMA, showing + / interface Ce segregation.

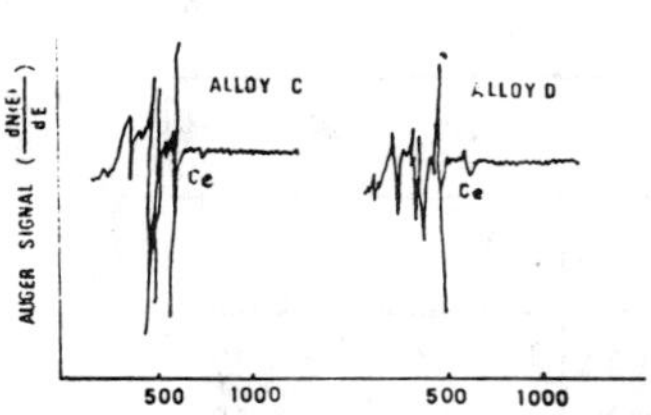

Figure 3
Auger spectra revealing Ce segregation

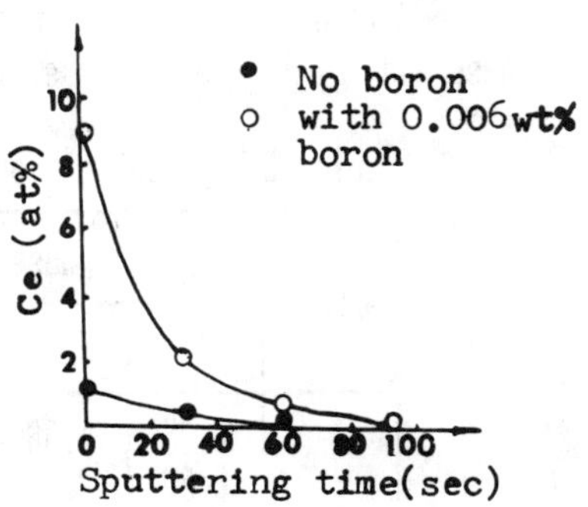

Figure 4
Relation Ce content vs. sputtering time determined by AES.

The examination of B segregation is difficult to determine due to the interference from the carbon peak in the AES analysis. We compared the AESt=0 spectrum (Fig. 5a) with AES t=30s spectrum (Fig. 5b) The B peak was not seen in the former but clearly appeared in the latter. It is evident that the B also has a propensity to segregate at the interface boundary.

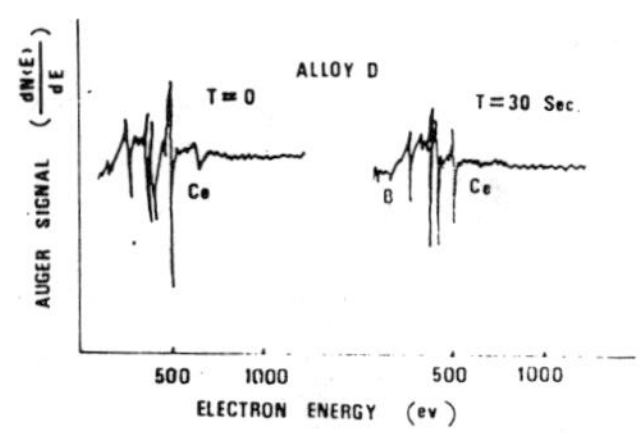

Figure 5

Auger spectra revealing B segreg[illegible]
T: sputtering time in sec.

3. The distribution of Al, Mo and Nb

The amounts of Al, Mo and Nb elements in the α or β phase were determined quantitatively using the wavelength dispersion analysis of the EPMA (Table 3). The addition of B shows a trend to increase the amounts of Mo and Nb in β phase and Al in the α phase. This trend was measured in the alloy D.

Table. 3 The Contents of elements in α and β phase (wt%)

ALLOY	α phase			β phase		
	Al	Nb	Mo	Al	Nb	Mo
A	6.95	1.66	0.83	7.56	2.13	1.97
B	6.89	2.04	1.58	5.89	2.74	2.83
C	6.54	1.87	1.71	6.50	1.84	1.91
D	7.05	1.54	0.33	5.46	2.99	3.43

4. Interface phase (IFP)

We examined several thin foils from the four alloys, which were prepared by electrothinning or ion-milling. The TEM study showed that IFP appeared in some α/β boundaries of alloy D(Fig. 6) while all of α/β boundaries of the other alloys were free from IFP.

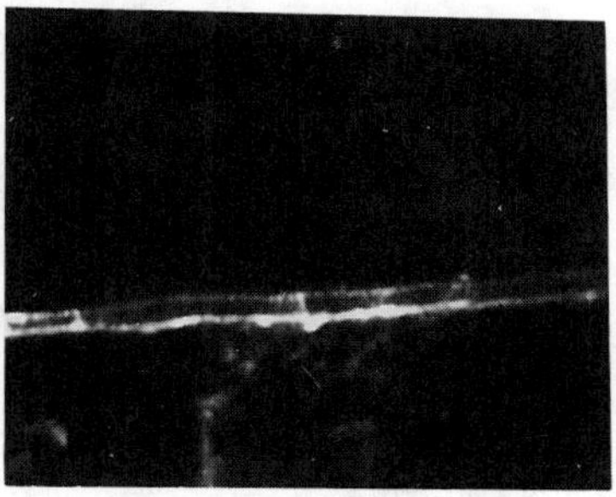

Figure 6

DF micrograph of alloy D. Note the thin layer of IFP at α/β boundaries.

DISCUSSION

The examination of the formation process of a creep crack is shown in Fig.7. When alloy A got to the end of the secondary stage of creep, voids formed at some phase boundaries and matrices (Fig. 7b). As creep continued, the voids combined to form the creep crack (Fig. 7c), then the growth of the crack resulted in rupture (Fig. 7d). Therefore, it is proposed that the increase of creep rupture time in alloy D can be attributed to the strengthening of the phase boundary due to the segregation of B and Ce, as can be seen when comparing the fractography of the creep specimen for alloy A and alloy D (Fig. 8). It is clearly shown that the patterns of creep fracture were changed from partially transgranular in alloy A to completely transgranular in alloy D.

At strain rates above $1.4\times10^{-9}s^{-1}$, creep in the α titanium alloy is probably controlled by the rate at which extended screw dislocation on basal planes can cross-slip onto prismatic or pyramidal planes[7]. Others reported that the α titanium alloy was strengthened by Al segregating to the stacking faults between partial dislocation and then the cross-slip of dislocation was impeded[8]. Because Mo and Nb are β stabilizing elements, the increase in the amount of Mo and Nb will make the β phase more stable and will add strengthening. It is suggested that the solution strengthening resulted from the increase in the amount of Al in the α phase. The amount of Mo and Nb in the β phase was due to the multialloying of B and Ce and may have contributed to the decrease of steady creep rate.

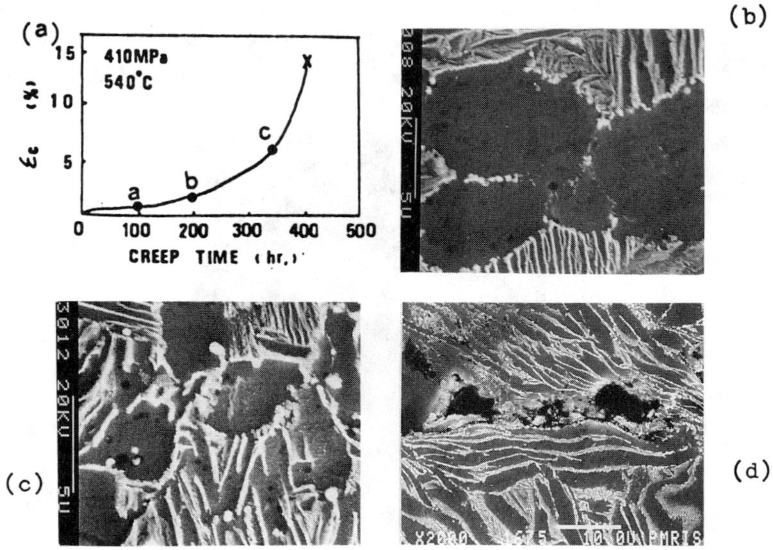

Figure 7

The formation process of creep crack

(a) creep curve of alloy A.
(b) The formation of creep voids ε_c=0.45%
(c) The formation of micro-crack ε_c=1.4%
(d) Showing crack propagation ε_c=6.6%

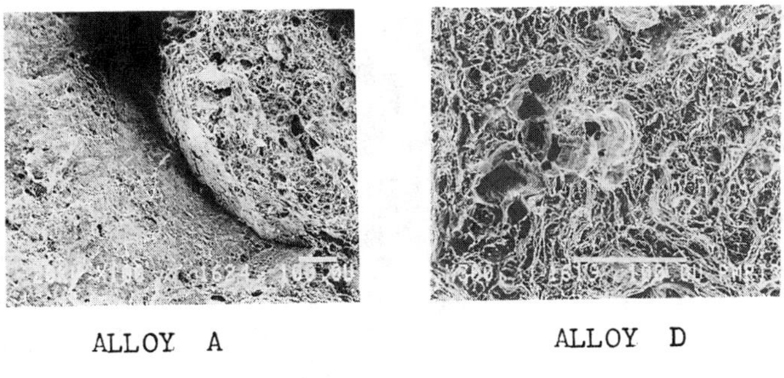

ALLOY A ALLOY D

Figure 8

The fractography of alloy A and alloy D.

In addition, comparing the boundaries of alloy C without IFP and alloy D with IFP (Fig. 9 and 10) after creep, we see that the interface phase may be an obstacle to the movement of dislocations.

Figure 9

TEM photograph of alloy C.

Figure 10

TEM photograph of alloy D.

CONCLUSIONS

The following conclusions can be drawn from the results of the present investigation.

1. Microalloying of boron and cerium is proved to be an efficient strengthening method for the titanium alloy Ti-6.5Al-2.2Sn-1.2Zr-2.0Mo-2.2Nb-0.2Si, especially for increasing the creep strength.

2. Boron and cerium have a high propensity to segregate to the $\alpha+\beta/\alpha$ interface, and also to change the distribution of Al, Mo and Nb elements in the α phase and β phase. When both boron and Cerium are added to one alloy, the trend will become stronger.

3. It is proposed that the increase of creep rupture life is mainly due to the $\alpha+\beta/\alpha$ interface strengthening, solution strengthening of the matrix and dislocation obstruction by the interface phase.

REFERENCES

1. M.R.Winstons, Reesd. Rawdings and D.R.F.West "Journal of the Less Common Metals" Vol.39 (1975) P205
2. M.R.Winston, Reesd. Rawdings and D.R.F.West "Journal of the Less Common Metals" v.31 (1973) No.1 P143
3. N.E.Daton, M.W.Mahorey,"Metall. Trans.," 7A(1976) P1685
4. H.Ikouye, S.A.David "Metall. Trans.," 15A(1984) P1505
5. H.Ikouye, S.A.David "Titanium Science and Technology" [Proc.] 1984 v.2 p791
6. R.E.Levis, ibid, p799
7. M.Kehoe and R.W.Broomfield "Titanium Science and Technology" [Proc.] 1972 p2167
8. Yu.N.Patron Phys. "Metall.," V.58 No.1 P67

SOME FACTORS AFFECTING THE CREEP PROPERTIES OF Ti 6242

K.E. Thiehsen and M.E. Kassner
Department of Mechanical Engineering and Center for Advanced Materials Research
Oregon State University
Corvallis, OR 97331-5001 USA

D.R. Hiatt and B.M. Bristow
Oregon Metallurgical Corporation
P.O. Box 580
Albany, OR 97321-0177 USA

High temperature tensile creep tests were performed on Ti 6242 0.1wt% Si at 510°C and an applied stress of 241 MPa. As the accumulated plastic strains, even after prolonged testing, were less than the elastic strains under these conditions careful alignment procedures were established. Creep specimens were solution annealed at different temperatures below the beta transus (T_β -5°C to T_β-50°C) which varied the volume fraction of primary alpha. The dependence of primary (Stage I) and steady-state (Stage II) creep rates on the volume fraction of primary alpha was determined. Relatively reproducible data was achieved in this study, and it was confirmed that increases in the volume fraction of primary alpha are associated with increased primary and steady state creep rates. The quantitative results are consistent with those of earlier investigations. The increased creep resistance with high solution annealing temperatures is primarily the result of a more refined microstructure.

Introduction

Ti 6Al-2Sn-4Zr-2Mo-0.1Si is a near alpha, alpha-beta alloy since it is composed primarily of the hcp alpha phase at ambient temperature, but some of the bcc beta phase still exists in the matrix. A typical microstructure for the alloy when annealed below the beta transus is globules of primary alpha in a matrix of basketweave, or Widmanstatten, lenticular alpha and beta platelets. The equiaxed, primary alpha phase tends to become more dominant in the microstructure as the solution anneal temperature is decreased below the beta transus.

Titanium 6242 must meet several stringent specifications before it is accepted for most applications. One of these specifications requires very low accumulated creep strains (e.g.<0.1% creep strain at 35 hrs, 241 Mpa and 510°C)[1], and this study attempts to identify some of the variables affecting creep rates. We particularly focused on low strain creep behavior.

An extensive search of the current literature identified possible sources of anomalously high creep in this alloy. First, the mechanical properties of the alloy change with the microstructure. Work done by Bania and Hall [2], Seagle et. al. [3], and Chen and Coyne [4] indicate that the steady state creep rate may be a function of the amount of primary alpha and hence solution annealing temperature. This investigation will attempt to confirm the work of these authors who suggested that by raising the solution annealing temperature toward the beta transus, thus reducing the amount of primary alpha, the steady state creep rate would decrease. Confirming the results was considered important since the data was limited in terms of both quantity and reproducibility. We also wished to extend the study to the primary creep range where these investigators report no (or inconsistent) data. This is considered important for low strain cases where primary creep may dominate. Furthermore, the composition of the Ti 6242 investigated in our study varied with respect to earlier work due to the presence of trace impurities often found in our samples. Thus, the results presented here may differ from earlier investigations.

Blenkinsop, Neal, and Goosey [5] also suggest that creep rate will be a function of cooling rate due to the changes in the microstructure associated with different cooling rates (for instance a slow cool will yield a coarser Widmanstatten structure). However, a further study of the effects of cooling rate on creep rates was not undertaken.

Work done by Semiatin and Lahoti [6] suggests that shear bands may form due to forging practice and the creep rate may, as a consequence, increase. These bands form as a result of localized deformation. Formation of shear bands is encouraged by a relatively long

Titanium '92
Science and Technology
Edited by F.H. Froes and I. Caplan
The Minerals, Metals & Materials Society, 1993

contact time between unheated dies and the workpiece causing chill zones near the dies which deform less easily than the material away from the die, a high strain rate which may not allow the heat generated during deformation to dissipate, a low forging temperature, and a low solution anneal temperature which does not restore banded substructures as readily as a high solution annealing temperature.

A third potential source of anomalously high creep could be bending moments in the creep sample caused by the testing apparatus. Bending stresses could create inhomogeneous axial stresses within the specimens. For cases, as in this study, where deflections, or strains, are measured only on one portion of the specimen, the observed strains and strain-rates may not represent the average or "true" mechanical behavior. Plasticity "homogenizes" the stress state, however, for cases in which the plastic strain ranges are less than the elastic strain on loading, such as in the present study, the inhomogeneous strains, and strain-rate may be present throughout the test, leading to anomalously high or low strain rates.

Experimental Procedure

All of the samples used in this testing program were forged from a billet of OREMET Ti-6242-0.1Si (heat # T 90847) which had a beta transus T_β=1004°C(determined by DTA). The composition is listed in Table 1.

TABLE I: COMPOSITION OF OREMET HEAT #T 90847 (wt%)

Al	Sn	Zr	Mo	Si	N	Cr	V	Ni	Fe	O
6.00	2.02	3.98	2.00	0.08	0.011	0.008	0.01	0.005	0.088	0.119

The billet was fabricated by forging a vacuum arc melted 914 mm diameter ingot to a 152 mm diameter billet in a two stage process. The initial breakdown was in the beta temperature range between 1065/1149°C (T_β+61/+145°C) where the ingot was forged down to a 254 mm octagon. The final forging step down to 152 mm was at T_β-36°C resulting in a microstructure of principally equiaxed primary alpha in a transformed beta matrix. This resulted in a microstructure of principally elongated primary alpha. Sample coupons were extracted from the billet and forged. The samples for the second set of tests had initial dimensions of 64 mm x 64 mm x 95 mm (the first, discarded, set being those tests done before the alignment of the creep testing system, the second set being those samples produced after the alignment corrections discussed in this paper). These coupons were heated for 2 hrs. at T_β-36°C +/- 14°C before forging. Hammer forging of the samples was performed at a strain rate of between 63 s^{-1} and 252 s^{-1}. Six forging passes were required to reduce the sample coupon to the desired cross sectional dimensions of 41mm x 16mm. The resultant true and engineering strains were 5.1 and 1.81 respectively. The samples were air cooled after forging and solution annealed at either T_β-6°C, T_β-15°C, T_β-28°C, or T_β-52°C for 1hr, air cooled again and then aged at 643°C for 8 hrs and air cooled to ambient temperature. Samples were cut from the coupons and machined in accordance with ASTM standard E 8.

Creep strain was measured using an Applied Test Systems model 4112 extensometer configured for 6.35mm dia. samples with a 25.4mm gage length, and a Measuretron model L1-12 linear variable capacitance transducer. The extensometer was determined to be accurate to +/-0.0013mm (a strain of +/- 50 microstrain).

The creep tests were performed at Oregon State University on an Arcweld (SATEC) model UC creep test machine. The test machine was used in its standard configuration for the first set of tests. However, due to anomalies in the data observed during this set of tests (i.e. significant variations in elastic strain on loading and poor reproducibility of the creep-rates), the presence of bending moments was suspected. For conventional (e.g. 10-20% plastic strain) creep tests the level of bending observed would be insignificant. However, at the very low plastic strains measured (<0.2%) bending in the system can significantly degrade the quality of the data because stress gradients can cause anomalously high or low creep rates as the sample deforms to "correct" the bending (the elastic strain on loading is about 0.3%). Therefore a systematic improvement of the testing system was initiated. The ASTM E606[8] and GE S400-

B[9] recommended practices for determining bending were consulted and a bending test was designed.

The following improvements were made for the second set of tests. The creep test machine pull rods were replaced with new rods that were precision ground straight to within 0.01 mm from stress relieved Inconel 600 rod. Since a substantial fraction of the bending originated at the sample/grip interface, unique high-temperature double clevis universal type grips were designed. Another universal was added to the lower part of the standard load train. The universal grips were lubricated with dry graphite.

We determined that Ti 6242-0.1Si did not measurably creep at ambient temperature at 241 MPa (the standard applied creep stress). Therefore, strain gages were mounted on all creep test samples for an ambient temperature pre-alignment at 241 MPa . Four strain gages were mounted at 90° intervals around the sample, and the bending in the sample was then measured for various rotations of the sample within the grip, and a configuration that minimized the bending at ambient temperature was located. Bending values were calculated according to the following formula:

$$\% \text{ bending} = \frac{\text{(average strain)-(max. or min. strain)}}{\text{(average strain)}} \times 100 \qquad (1)$$

The new system allowed the reduction of test sample bending at room temperature from an average of 8% before alignment (first set of tests) to less than 4% (and as low as 0.7%). The strain gages were then removed while the sample was still loaded to prevent misalignment during stripping. The adhesive was removed, and the sample was then given a final cleaning with methanol. Three thermocouples were attached, at the top, middle, and bottom sections of the gage length. This allowed for precise temperature monitoring and control. The extensometer was then attached and the sample was unloaded and brought to temperature. The creep tests were commenced once the temperature was allowed to stabilize at 510°C +/-2°C for 3.5-4.5 hours. A three zone furnace with Eurotherm power supplies and controllers was utilized. The samples were tested for 72 hours (or longer) to ensure that a mechanical steady state was achieved. The temperature typically did not vary by more than two degrees from 510°C at any point along the specimen throughout the test. After creep testing, the samples were sectioned and the microstructures were examined metallographically.

Results

The stress-strain data taken during the room temperature alignment procedure facilitated the calculation of ambient temperature elastic moduli, and Table 2 shows that there is a small but definite increase in elastic modulus with increasing solution anneal temperature.

Table II: Room Temperature Elastic Moduli of Ti 6242-0.1Si

Solution Annealing Temperature	Elastic Modulus (strain gage)
T_β-6°C	118 GPa
T_β-15°C	116 GPa
T_β-28°C	115 GPa
T_β-52°C	114 GPa

Figure 1 shows the typical microstructure of hammer forged samples solution annealed at T_β-6°C, T_β-15°C, T_β-28°C, and T_β-52°C. At lower annealing temperatures bands of elongated primary alpha are evident but diminish with higher solution annealing temperatures. A companion study by the authors suggests that the elongated primary alpha grains observed in the samples solution annealed at or below 1018°C may be remnants of the elongated alpha of the forged billets [10].

Figure 1. Typical Microstructures of Ti-6242-0.1Si Resulting From Various Solution Annealing Temperatures.

T_β-6°C T_β-15°C T_β-28°C T_β-52°C

It is apparent that the steady state creep rate varies inversely with the solution annealing temperature (and hence the microstructure) as shown in Figure 2. Since bending stresses decrease with plastic strain, the stage II or steady state creep value may be the most reliable indicator of creep resistance.

Figure 3 shows the primary creep strain as a function of solution annealing temperature relative to the beta transus. The onset of stage II creep typically occurred at a strain of 10^{-3} at about 10^5 seconds, however, it should be noted that the time of the onset of steady state creep varied from approximately $8x10^4$ sec to $1.6x10^6$ sec (possibly due to the presence of relatively small bending moments).

Figure 2. Steady State Creep Rate as a Function of Solution Annealing Temperature.

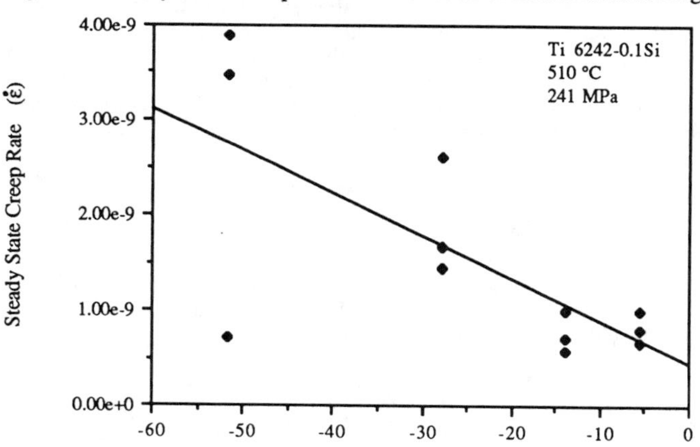

Figure 3. Primary Creep Strain as a Function of Solution Annealing Temperature.

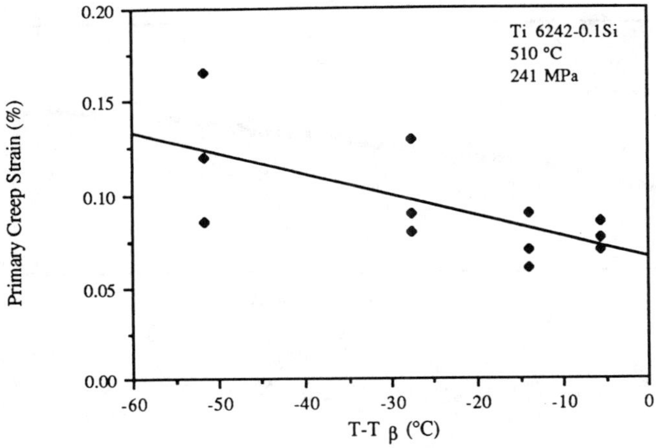

The dependence of the amount of primary alpha on solution annealing temperature (relative to the beta transus) is shown in Figure 4. As anticipated, the amount of primary alpha in the sample is a strong function of the solution annealing temperature. With the exception of the T_β-6°Csamples which contained slightly less primary alpha than predicted, the percentages of primary alpha all fall within the ranges reported by Seagle et. al.[3]. However, the percentages were generally higher than those predicted by Bania and Hall (who determined the beta transis micrographically rather than by DTA)[2]. Figure 5 shows a typical creep curve. It was found that fluctuations in ambient temperature caused fluctuations in the data which recovered when the ambient temperature returned to the baseline value. Such minor fluctuations were typical. Finally, Figure 6 shows that upon unloading the sample undergoes a significant anelastic recovery, often recovering up to one-third of the measured creep strain within 20hrs.

Figure 4. Amount of Primary Alpha as a Function of Solution Annealing Temperature.

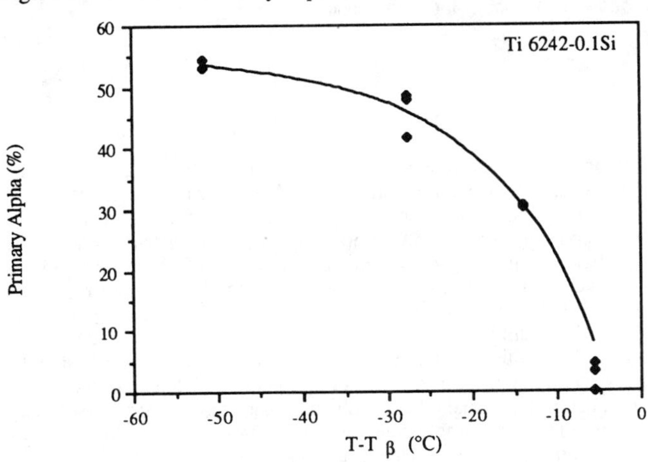

Figure 5. Typical Creep Curve For Ti 6242-0.1Si.

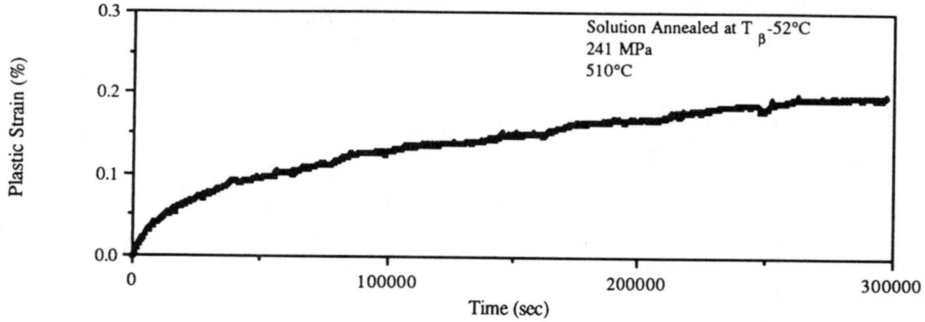

Figure 6. Typical Anelastic Recovery of Plastic Strain (Ti 6242-0.1 Si T_β-19°C Solution Anneal).

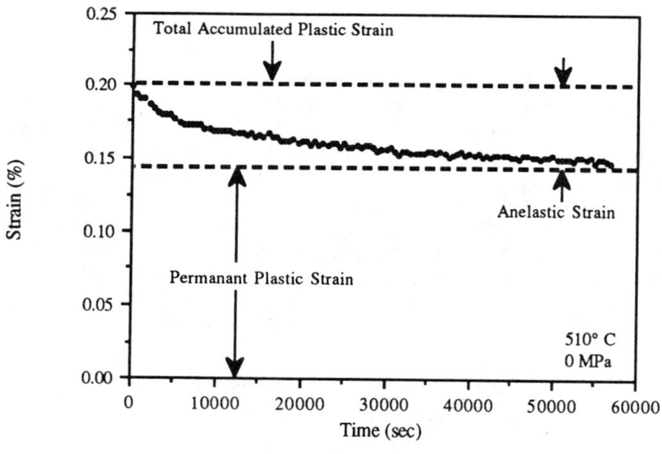

Discussion

Wide variations initially observed in elastic strain on loading and the bending observed in our early specimens at ambient temperature using strain gages have led to the conclusion that significant bending moments sometimes occur in the loaded specimens. The fluctuations in elastic strain on loading observed could partly be due to slippage of the extensometry system, especially since a cage type extensometer is used with rods that transmit the strain to the remote LVC. While some slippage will inevitably occur in a system like this, the observation of comparable bending measured directly at room temperature by strain gages leads to the conclusion that the extensometer is providing reliable data. The observed variation in elastic strain on loading even in the second set of tests (though substantially less than with the first tests) which were conducted after the new alignment procedures were implemented, leads to two other possible conclusions. First, with heating, thermal expansion occurs and this may create some misalignment. Second, the machining of the samples may introduce significant residual stresses in the sample which relax after the sample is heated to the test temperature, leading to some misalignment.

As is evident from Figure 2 there is a general trend toward lower steady state creep with increasing solution annealing temperature. With the increase in solution anneal temperature, the amount of primary alpha decreases. The observed increase in steady state creep rate with increasing amounts of primary alpha was expected based on the work of Bania and Hall [2], Chen and Coyne [4], and Seagle et. al. [3]. Chen and Coyne [4] report approximately a threefold increase in time to 0.2% total creep deformation for $\alpha+\beta$ forgings as the solution annealing temperature is raised from T_β-45°C to the beta transus (at 566°C and 211MPa). Seagle et. al. [3] reports an approximate 40% increase in total creep deformation as the solution annealing temperature is lowered from T_β-12°Cto T_β-42°C. Bania and Hall [2] found that a decrease in primary alpha from 35% to 0% resulted in a threefold decrease in steady state creep rate, while the effects on primary creep were unclear. In our work we found a linear decrease in creep rate of approximately fivefold upon reducing the amount of primary alpha from 53% to 3%.

Three possible explanations for the decrease in primary and steady state creep rate with increasing solution annealing temperature may be available. First, as reported in Table II at ambient temperature the elastic modulus decreases with increasing amounts of primary alpha. Sherby and Burke [11] proposed the classical empirical relation for strain rate (creep rate):

$$\varepsilon = A D_0 \exp(-Q/RT)(\sigma/E)^n(\gamma^{3.5}/L^2) \qquad (2)$$

D_0= diffusion constant, Q= activation energy of diffusion, σ= applied stress, E= elastic modulus, γ = stacking fault energy, A= constant, n= constant (4 to 5 above 0.6 T_m), R= gas constant, T= absolute temperature, L= grain size.

Assuming the relative difference between elastic modulus and and primary alpha persists at 510°C, then, using the preceding equation and the data from Table II, we would predict that the steady state creep rate for the T_β-6°C solution annealed samples would be 14% less than that of the T_β-52°C solution annealed samples. This difference, however would only explain part of the observed trend. The second possible explanation would be that the decrease in creep rate may be due to the diminishing of banding and texturing. Recovery and grain growth will increase with higher temperature annealing. Semiatin and Lahoti [5] suggest that both texturing and banding can as much as double the amount of creep observed. Although banding seems to disappear at about T_β-15°C and above, decreases in the creep rate are probably due to other effects. Finally, the basketweave structure may be intrinsically more creep resistant than one with equiaxed alpha. None of the presented explanations for the dependence of primary and steady state creep rate on the amount of primary alpha are mutually exclusive and all may be operating. However, the substructural explanation appears most attractive.

Summary

1) The bending that is inherent in the standard creep testing machines is probably causing substantial scatter in the small plastic strain creep data. However new methods described in this work reduce the bending moments and data scatter.

2) The new testing procedures show that raising the solution annealing temperature appears to increase the primary and steady state creep resistance of Ti 6242-0.1Si. This beneficial effect may be derived from a combination of several sources including increasing of the elastic modulus, the removal of such micrographic features such as banding or texturing, and especially the refining of the microstructure.

3) Current standard creep specifications require testing sensitivity that may extend beyond the current standard commercial state-of-the-art creep testing devices.

Acknowlegements

The authors would like to acknowledge the financial support provided by the U. S. Bureau of Mines, Oregon Metallurgical Corporation (OREMET), and the Oregon Economic Development Department through the Oregon Metals Initiative. We would also like to acknowledge the assistance of The U. S. Bureau of Mines for performing our press forging and of Greg Buchanan at OREMET who performed our sample forgings. Finally we would like to acknowledge the assistance of Utkarsh Kansal who provided the metallography appearing in this manuscript and Jeff Pollard who assisted us in the experimental portion of this investigation.

References

1) S.A.E. Aerospace Material Specification 4976D (1988).

2) P. J. Bania, J. A. Hall: Proceedings of the Fifth International Titanium Conference Munich. (1984) 2371.

3) S. R. Seagle, et. al.: Metals Engineering Quarterly. Feb. 1972, 48.

4) C. C. Chen, J. E. Coyne: Titanium 80 Science and Technology TMS-AIME, (1980) 1197.

5) P. A. Blenkinsop, et. al.: Titanium and Titanium Alloys, Plenum Press (1976) 2003.

6) S. L. Semiatin, G. D. Lahoti: Metallurgical Transactions A. 14A (1983) 743.

7) S. Ankem and S.R. Seagle: Proceedings of the Fifth International Titanium Conference Munich. (1984) 2411.

8) P. J. Bania, et. al.: Interim Technical Report, AFML Contract # F33615-75-C-5089. (1975).

9) ASTM Testing Standard E696. 3.1 (1980) 609.

10) M. E. Kassner, U. Kansal: To be published.

11) O. D. Sherby, P. M. Burke: Progress in Material Science. 13 (1968) 325.

Microstructure and Mechanical Properties for PREP / PM IMI829–1Gd and IMI829–1Er Alloys

Luo Guozhen, Deng Ju, Lan Tao, Zheng Yueqiu and Yang Guanjun

Northwest Institute for Nonferrous Metals Research

P.O.Box 71, Baoji, Shaanxi 721014 China

Abstract

The spherical powder for IMI829–1Gd and IMI829–1Er alloys produced by PREP was consolidated by HIP and HIP+hot swaging processing. The microstructure for the alloys and the distribution of RE oxides in the matrix were investigated by LM,SEM and TEM, and the tensile and creep properties of the alloys were determined at temperatures raging from 450 to 600℃. The room temperature tensile ductility of PREP HIPed materials can be improved by hot swaging after HIP, however, the creep properties are much lower than that of the I / M materials above 550℃. The creep resistances for both RS alloys are almost comparable to those for some commercial Ti alloys such as IMI679 and BT9.

Introduction

Recently, rapid solidification processing (RSP) has offered the opportunity to develop dispersoid strengthened Ti–RE alloys. In general, the plasma rotating electrode processing (PREP) is regarded as a simple RSP tecnique that is capable producing clean smooth spherical powders with excellent fluidity and tap density, which is well–suited to produce novel Ti alloys by powder metallurgy (PM). Furthermore, PREP is suitable for production of RS Ti–RE alloys on a large scale. Previous studies [1–2] on some Ti –RE alloys produced by IM had considered the role of RE oxide and intermetallic compound particles on dispersoid strenthening. It was demostrated that the rare earth elements Gd and Er appear to be promising dispersoid strengthening candidates. The key problems were that the RE oxide and intermetallic compound particles were too coarse and the density was too low. It is expected that PREP powders will produce fine and dense RE oxides. In recent years, many studies [3–6] have concentrated on evaluating the microstructure of binary or ternary RS Ti–RE alloys and paid less attention to high temperature properties for complicated RE–containing Ti alloys. A few works [7–8] examined Ti–RE alloys produced by PREP. The purpose of this study is to evaluate the microstructure and high temperature properties of RE containing Ti alloys IMI829–1Gd and IMI829–1Er produced by PREP.

Experimental proceduce

The prealloys for this study were melted in a vacuum consumble arc furnace by double melting into an ingot with weight of 15kg. The ingot was forged and machined into 30mm diameter by 500mm long rods as consumable electrodes for PREP. Then the rods were reduced to spherical alloy powders by PREP in Ar atmosphere with an electrode rotating speed of 11000 rpm. The cooling rate in this study was approximately 10^4k / s. The prealloy powders were packed into 16mm diameter CP Ti tubes with a 1mm wall thickness and sealed by eleceron beam welding in vacuum. Subsequently the canned Ti tubes containing the prealloy powders were HIPed by a QIH HIP machine at 900℃ / 196MPa / 2hrs, 950℃ / 147MPa / 2hrs and 1030℃ / 147MPa / 2hrs.

Titanium '92
Science and Technology
Edited by F.H. Froes and I. Caplan
The Minerals, Metals & Materials Society, 1993

Partial HIPed alloys were directly machined over the Ti can to examine the microstructure and mechanical properties after heat treatment. Another part of the HIPed alloys were hot swaged at 900℃ down to 6min diameter rods. The heat treatment were carried out in electronic resistance furnace at varied temperature.
The microsturcture of powder particles and consolidation products were examined by LM、 SEM and TEM. Tensile and creep tests of consolidated products were performed at room and elevated temperatures.

Results and Discussion

Prealloy Powder

The prealloy powders produced by PREP are typical smooth spherical particles. More than 80 percent of the powders have average diameter between 250–420μm according to sieve analysis. The powder alloy and the ingot compositions are listed in Table 1:

Table 1 The composition of ingot and prealloy powder

product	Alloy	Composition % wt											
		Al	Sn	Zr	Mo	Nb	Si	Gd	Er	O	C	H	N
Ingot	IMI829–1Gd	5.43 5.40	3.47 3.41	2.94 3.00	0.30 0.30	1.01 0.97	0.30 0.33	1.0 1.0	– –	0.08	0.03	0.002	0.01
Ingot	IMI829–1Er	5.46 5.30	3.54 3.50	2.93 2.97	0.30 0.30	1.05 1.04	0.36 0.36	–	0.9 0.9	0.09	0.03	0.002	0.011
Prealloy Powder	IMI829–1Gd	5.2	3.1	2.58	0.31	0.94	0.32	0.87	–	0.11	0.03	0.001	0.015
Prealloy Powder	IMI829–1Er	5.6	3.4	2.71	0.30	1.04	0.32	–	0.89	0.10	0.03	0.001	0.013

The results show that the compositions of prealloy powders are almost the same as those of the ingot compositions except for a slight decrease in Zr and Sn content. There was a little increase in oxygen and other interstitial elements in the prealloy powders.

Microstructure Examination

LM and SEM micrographs of the powder particles are shown in Fig 1. The original β grains in the individual powder particles are outlined by the dendrite arms. The grain sizes range from 11 to 30μm for IMI829–1Gd and 30 to 39μm for IMI829–1Er. These sizes decrease with decreasing the powder dimension. Compared to IMI829–1Er alloy powder, IMI829–1Gd alloy powder has a more refined grain size.
The microstructure for the HIP consolidated IMI829–1Gd alloy is shown in Fig.2 (a) and (b). In the as–HIPed condition, the interparticle bondings between the powder particles still exist although the shape of powder particles varied from spherical to ployhedral . The interparticle bonds seem to be outlined by many coarse quasi–grains boundaries and their size approaches the dimension of the powder particles. There are two typical quasi–grains with different contrast, bright and dark, under low magnification. The difference of composition between bright and dark quasi–grains was not detectable by EDAX . Howerer, the microstructure for both quasi–grains are quite different. The microstructure in the dark quasi–grain is coarser than that in bright one. This suggests that the powders formed during atomization have undergone a varied cooling rate . The boundaries between the dark and bright quasi–grains are visible and they appear to be not wettable. However, the boundary between the same species of the

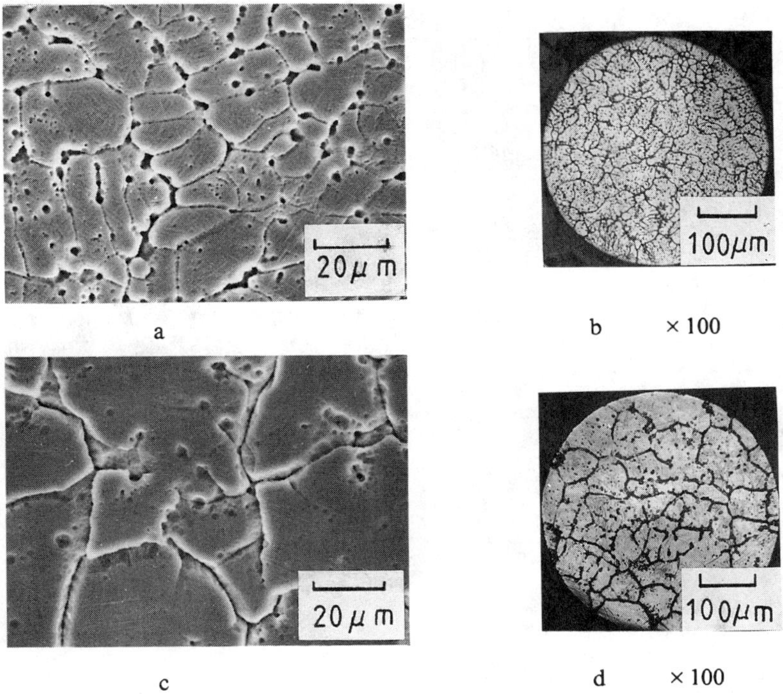

Fig.1 The microstructure of (a,b) IMI829−1Gd and (c,d) IMI829−1Er alloy powder

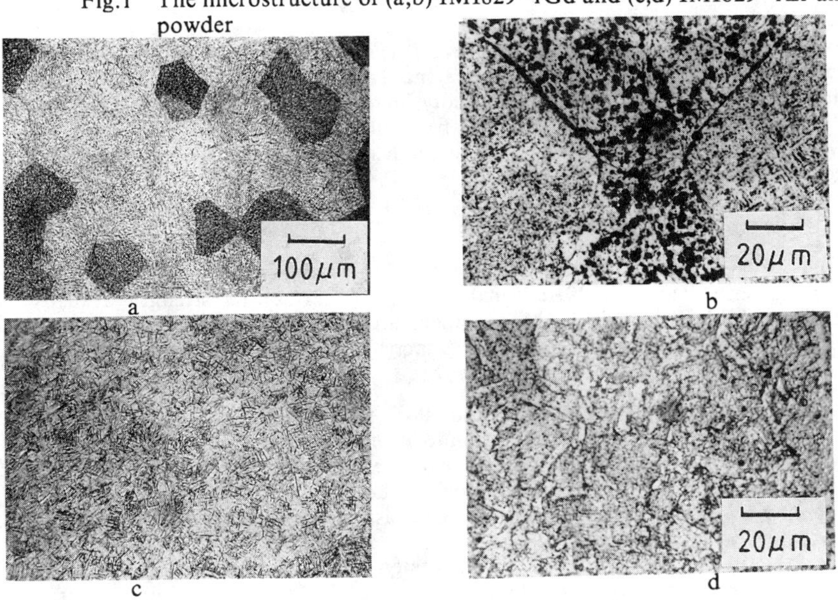

Fig.2 The microstructure for HIP consolidated IMI829−1Gd alloy (a,b) before and after (c,d) hot swagining

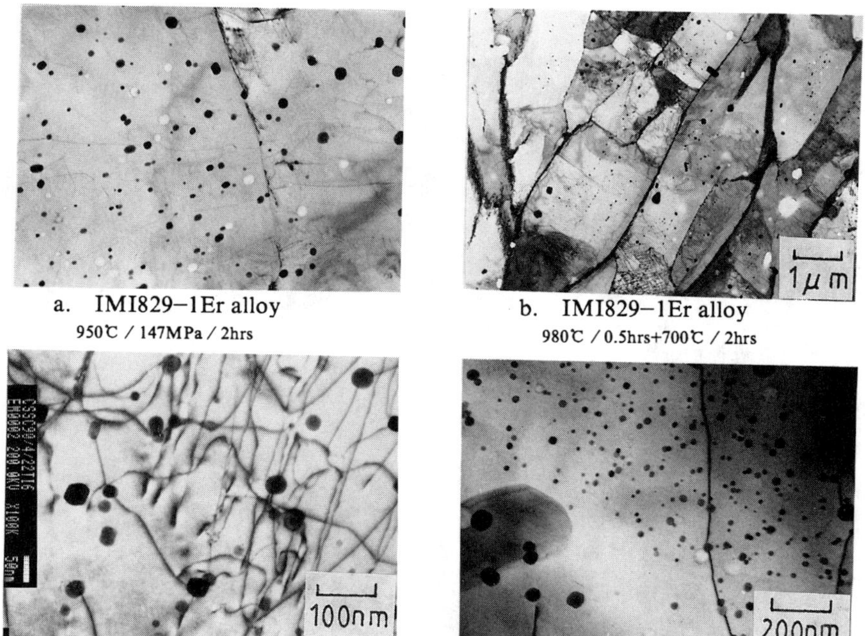

a. IMI829−1Er alloy
950℃ / 147MPa / 2hrs

b. IMI829−1Er alloy
980℃ / 0.5hrs+700℃ / 2hrs

c. IMI829−1Gd
950℃ / 147MPa+700℃ / 2hrs AC

b. IMI829−1Gd
900℃ / 196MPa / 2hrs+700℃ / 2hrs AC

Fig.3 TEM microstructure for both alloys
(a,b) IMI829−1Er and (c,d) IMI829−1Gd

quasi−grain boundaries have disappeared (Fig.2, c,d) This suggests that interparticle bonding has been strengthened by swaging. It should be noted that the tensile ductility for the $\alpha+\beta$ heat treated alloy markedly increases up to valued companable with IM IMI829−1Gd alloy. Compared to the basic alloy IMI829, both alloys have gained about 90−100MPa in strength from Orowan strengthening by the RE dispersoids, which is consistent with previous studies [3,4,8] .

B. High temperature tensile properties

The results of high temperature tensile test are also shown in Table 3.From this table , it can be seen that IMI829−1Er exhibited a higher tensile strength at 650℃ than IMI829−1Gd . However, the high temperature tensile strengths of both alloys are lower than the expected value probably due to increasing boundary slip above 550℃.

C. Creep properties

Fig.4 shows creep results at 550℃ and 600℃ for three alloys, and Table 3 also lists the resistance to creep fracture data at temperature between temperature 450−600℃ over loads of 200−384MPa for different heat treatment conditions. The creep test results show that RS processing reduced creep resistance to creep. From Fig.4 , it can be seen that the RS alloys exhibited much lower resistance to creep above 550℃ compared to IM alloys with similar compositions .

The data in table 3 show that β heat treatment can improve the resistance to creep fracture for both alloys compared to $\alpha+\beta$ heat treated alloys due to β grain coarsing.Compared to IMI829−1Gd , IMI829−1Er has a higher creep resistance in any condition quasi–grain are not visible and they are seem to be more wettable. The presence of these "boundaries" is the reason why the tensile ductility at room temperature is quite low. After hot swaging the quasi–grain boundaries seem to disappear (Fig 2 c, d) It is expected that the disappearance of these quasi–grain boundaries will result in an increase in tensile ductility of the alloys.

The TEM microstructures of as–HIP and after heat treated alloys at various temperatures are shown in Fig.3. TEM examination revealed many fine and homogeneous dispersed RE containing particles for all consolidation products. The RE particles are estimated to be average 18nm in diameter and 0.7 percent in volume fraction for IMI829–1Er alloy, and 20nm and 0.6 percent for IMI829–1Gd [10], which are almost the same as some recent reports [3,6]. Two populations of dispersed precipitates have been observed by TEM: one population of very densely distributed small precipitates with the size smaller than 0.1 μm which mostly distributed in the grain interior, and a second population of much less densely distributed coarse precipitates with a size larger than 0.1μm which is mostly distributed on the grain boundaries. It is suggested that the small precipitates have formed in the solid state during cooling after solidification or during hot consolidation and heat treatment. The coarse precipitates may have formed during solidification possibly due to interdendritic segregation on a powder particle scale. The present results of the TEM examination are simillar to those obtained by Naka [7], where most of both fine and coarse precipitates are sesquioxide RE_2O_3, e.g. Er_2O_3in IMI829–1Er alloy, and Gd_2O_3in IMI829–1Gd alloy. Other types of RE oxides have also been identified by TEM SAD, such as $RE_2Ti_2O_7$and $RE_2Sn_2O_7$ [10]. The coarse precipitates in IMI829–1Gd alloy include an intermetallic compound which suggests $GdSn_3$or Gd_3Sn according to the composition analysis by EDAX (Table 2). The stoichiometric composition Gd_3Sn was suggested by reference [9] and $GdSn_3$was suggested by a rough Gd:Sn ration of 3:1 (Table 2).

Table 2 The composition of intermetallic compound analyzed by EDXA

Alloy	chemical composition							
	Ti	Al	Si	Zr	Nb	Mo	Sn	Gd
RS IMI829–1Gd	60.598	3.109	0.626	1.351	0.269	0.052	8.756	25.234
1M IMI829–1Gd	4.963	–	–	–	–	–	23.121	71.915
	8.662	–	1.077	–	–	–	26.855	63.424

Mechanical Properties:

A. Room temperature (RT) tensile properties:

Table 3 lists the RT tensile properties for both PREP+HIPed alloys IMI829–1Gd and IMI829–1Er. The data include previous results obtained from IM alloys with similar comparisons as the alloy for this study for comparison . The present results can be summarized as following:

1) The tensile strength for PREP+HIPed IMI829–1Gd is almost equivalent to that of 1M alloy but has a lower tensile ductility. Compared to IMI829–1Gd , IMI829–1Er exhibited a slight higher tensile strenth and ductilitly . However the tensile ductilities for both alloys are much lower than expected value. The poor tensile ductilities for the RS alloys are probably associated with weak interparticle bonding.

2) After hot swaging , the tensile properties have improved for both IMI829–1Gd and IMI829–1Er, with fracture characterised by dimple failure . The LM picture reveals that

Table 3　The results of tensile and creep tests

Alloys	Processing and Heat treatment	RT Tensile Properties				650℃ Tensile Properties				Resistance to creep fracture			
		YS (MPa)	UTS (MPa)	e (%)	RA (%)	YS (MPa)	UTS (MPa)	e (%)	RA (%)	Temp (℃)	load (MPa)	Time (hrs)	Plastic strain
IMI829-1Gd	HIPed at 950℃ / 118MPa / 20min+700℃ / 2hrs AC	1005	1023	1.6	1.3	360	460	19	22	600	200	21.6	11.48
	HIPed at 900℃ / 196MPa / 2hrs+700℃ / 2hrs AC	–	1004	2.4	0.1	364	469	20	22				
	HIPed at 950℃ / 147MPa / 2hrs+700℃ / 2hrs AC	979	996	2.7	1.3	394	489	–	–				
	HIP* +HW** +650℃ / 8hr AC		1120	5.6	6.2					600	200	6.5	111
	HIP* +HW** +900℃ / 1hr AC+650℃ / 8hr AC		1006	17	21					600	200	6.3	25.52
	HIP* +HW** +980℃ / 1hr AC+650℃ / 8hr AC		1115	14	25					600	200	6.8	29.8
	HIP+HW+980℃ / 1hr AC+650℃ / 8hr AC									450	384	510***	2.47
	HIP+HW+980℃ / 1hr AC+650℃ / 8hr AC									500	350	510***	3.27
	HIP+HW+980℃ / 1hr AC+650℃ / 8hr AC									550	300	91	11.8
	HIP* +HW** +1050℃ / 1hr AC+650℃ / 8hr AC		1148	2.4	4.3					600	200	24.8	21.2
IMI829-1Er	HIPed at 950℃ / 118MPa / 20min+700℃ / 2hr AC	981	1011	3.3	2.0	361	467	33	31	600	200	66.4	34
	HIPed at 900℃ / 196MPa / 2hrs+700℃ / 2hr AC	981	1037	6.9	7.2	391	499	21	22				
	HIPed at 950℃ / 147MPa / 2hrs+700℃ / 2hr AC	979	996	2.7	1.3	420	496	5.3	–				
	HIPed at 1030℃ / 147MPa / 2hrs+700℃ / 2hr AC	870	943	3.3	6.9	421	461						
	HIP* +HW** +650℃ / 8hr AC		1095	11	14					600	200	23.3	70.5
	HIP* +HW** +900℃ / 1hrs+650℃ / 8hr AC		1062	11	11		490	53	84	600	200	10.9	99
	HIP* +HW** +980℃ / 1hrs+650℃ / 8hr AC		1061	6.7	7.9					600	200	38.6	55.5
	HIP* +HW** +1050℃ / 1hrs+650℃ / 8hr AC		1068	5.6	7.2		529	76	75	600	200	72.3	56.4
1M-IMI829	$\alpha+\beta$ rolled + $\alpha+\beta$ Heat treatment	1010	941	13.5	19.8								
		1000	917	10.8	12.0								
IMI829-1Gd	$\alpha+\beta$ rolled + $\alpha+\beta$ Heat treatment	950	890	13.6	23.8								
		1010	958	12.4	22.8								

* as HIPed at 900℃ / 196MPa / 2hrs　　** Hot swaged at 900℃　　*** not failure

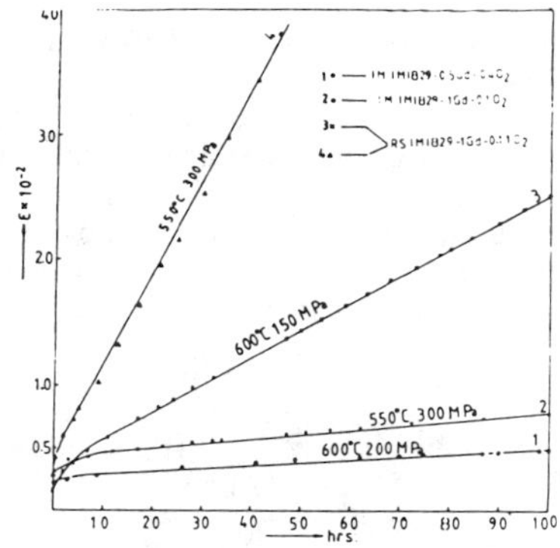

Fig.4 Creep curve for IMI829 containing Gd alloy

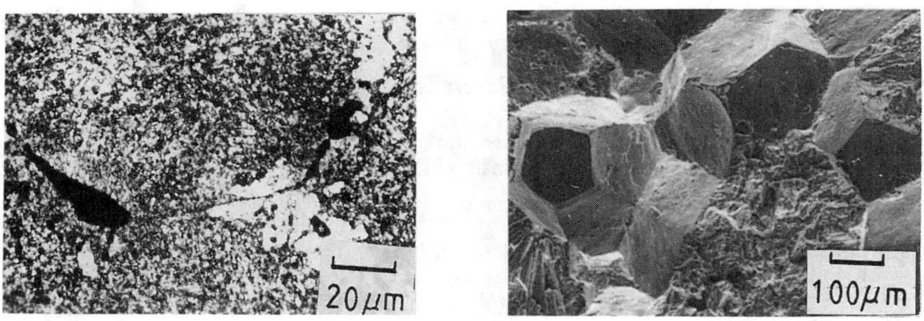

a b ×80

Fig.5 The microstructure(a) and fractography (b) for IMI829−1Gd alloy after creep testing.

due to the coarser grain size of IMI829−1Er . It should be noted that RS IMI829−1Gd significantly improved resistance to creep fracture in 450−500℃ temperature range. After thermal exposure at 450℃ / 384MPa and 500℃ / 350MPa for 510 hours, for example, the creep strain remained at 2.47% and 3.17% respectively. Unfortunately, no comparison between RS alloys IMI829−1Gd and IMI829−1Er has been made. However, it is expected that the creep resistance for both alloys approaches to that for some commercial high temperature titanium alloys such as IMI679 and BT9. By contrast, the creep resistance above 550℃ suddenly drops to a much lower level than that of IM alloy (as shown in Fig.4). This behavior can not be explained by RE dispersoids coarsing because most RE oxide dispersoids have much finer dimensions (18−20nm average diameter) even after HIPed or heat treated above 950℃ . Furthemore, after exposure at 700℃ for 100hrs, RE dispersoid particles coarsing in RS IMI829−1Gd alloy has not been detected by TEM [10] . These homogeneously distributed fine dispersoids would offer dispersion strengthening. However, the fine RE oxide dispersoids have not made any contribution to the creep resistance for either RS alloys. This contradiction phenomenon is probably due to the ultrafine grain size in RS alloys which offers much more opportunities for boundary sliding compared to coarse grain alloys obtained by IM. According to a report by Patore and Mahoney [11] , grain boundary sliding becames significant above 550℃ in some titanium alloys . It is suggested that the sudden decrease in creep resistance above 550℃ is probably due to grain boundary sliding and weak interparticle bonding. Fig.5 shows an example of this behavior. From fig.5, it can be seen that crack nucleation and growing initiated from the quasi–grain boundary in the creep specimen of RS IMI829–1Gd . Fractography (fig 5.b) revealed intergranular failure with a cleavage facet size almost equivalent to the size of powder particle. Furthermore, a diffusion band was observed around the quasi–grain boundary, EDAX analysis of the diffusion band indicated that alloying elements Mo and Nb are enriched and Gd is depleted in the band . This implies that the creep resistance above 550℃ for the RS alloys is controlled by a diffusion mechanism, which supports the view about boundary sliding above 550℃ in RS alloys.

Conclusion

Whether the RS alloys were consolidated by HIP or followed by heat treatment at various temperature, he RS alloys IMI829–1Gd and IMI829–1Er have dispersed RE oxide particles with an average diameter of 18nm–20nm for any consolidated products .
The creep resistance for RS IMI829–1Gd alloy in the temperature range of 450–500℃ is almost comparable to commercial high temperature alloys such as IMI679 and BT9.
With increasing temperature over 550℃ , the RS alloy IMI829–1Gd exhibited much lower creep resistance than IM alloy with the same composition probably due to weak interparticle bonding and grain boundary sliding.

Reference

1. Huang Hanlian, et al, Effect of rare earth element Gd and Oxygen content on microstructure and mechanical properties for IMI829 alloy . Rare Metal Materials and Engineering. No.1, 1991, China

2. Dan Ju, et al, A new high temperature titanium alloy containing Gd, Ti633G, Titanium Science and Engineering. Seventh Chinese Titanium Conference Proceeding. December, 1990, (China), P424–427

3. S.M.L.Sastry, P.J.Meschter and J.E.O′Neal, Structure and Properties of Rapidly Solidified Dispersion–Strengthening Titanium Alloys: Part Ⅰ , Characterization of Dispersoid Distribution, structure, and Chemistry, Met. Trans. A, Vol.15A, July 1984, 1451–1463

4. S.M.L.Sastry, T.C.Peng, and L.P.Beckerman, Structure and Properties of Rapidly Solidified Dispersion Strengthened Titanium Alloys: Part Ⅱ, Tensile and Creep Properties, Met. Trans A, Vol.15A, July 1984, 1465–1474

5. M.F.X. Gigliotti, C.E.Wasielewski, and R.G.Rowe Microstructure Development and High Temperature Mechanical characterization of A Dispersoid–Containing Titanium Alloy. Sixth World Conference on Titanium Proceeding ,France, 1988, 861–866

6. S.L.Kampe and D.A.Koss, Deformation of Rapidly Solidified Dispersion Strengthened Titanium Alloys. Met. Trans, Vol.20A, May 1989,875–887

7. S.Naka, M.Marty, H.Octor, Oxide–dispersed Titanium alloy Ti–Y prepared with the rotating electroed process, Journal of Materials Science 22 (1987) 887–895

8. C.Surganarayana, F.H.Froes, and R.G.Rowe, Rapid Solidification Processing of Titanium Alloys. International Material Reviews 1991, Vol.36, No3, 85–118

9. B.Φ. Tepexcba, E.M.Cafuykue. The Rare Earth Metallography, 1975,Chinese Translation, P140–141

10. Zheng Yuqiu, et al. The Microsturcture for RE dispersoid strengtheng Ti alloys unpublished Yeport. Dempt. 1991. Northwest Institute for Nonferrous Metals Research, Baoji, China.

High Temperature Deformation Behavior of α
$\alpha - \beta$ and β Titanium Alloys

M.N.Vijayshankar and S.Ankem
Department of Materials and Nuclear Engineering
University of Maryland, College Park
Maryland 20742 - 2115, USA

A systematic investigation was undertaken to determine the effect of temperature, strain rate and volume percents of phases on the high temperature deformation behavior of $\alpha, \alpha - \beta$ and β titanium alloys using Ti – Mn alloys as the model system. As expected, the flow stress of α – phase increased with strain rate and decreased with temperature. The apparent activation energies determined in the temperature range (923 K–1023 K) were found to be close to that of self diffusion of titanium. In regard to the β – phase, the strength increased with strain rate and decreased with temperature as expected. However, the β – phase exhibited sharp flow stress drops followed by steady state behavior. The extent of drop depended on strain rate, temperature and solute concentration. The apparent activation energies obtained for deformation of β – phase were found to be close to self and solute diffusion of β – phase. In regard to the $\alpha - \beta$ alloys, many anomalies were found. First of all, the strength of the two phase alloys did not vary linearly with volume percent of phases. The strength of two phase materials with nearly equal volume percents of phases were found to be much smaller than one would expect based on the law of mixture. Furthermore, the strength of two phase materials with nearly equal volume percents of phases did not decrease with an increase in temperature at the lower end of strain rates employed. These anomalies appeared to be related to the additional deformation mechanisms operating in two phase alloys. This suggestion is supported by the fact that the activation energies obtained for these $\alpha - \beta$ alloys, at lower strain rates, were found to be much smaller than the activation energies obtained for either α or β – phase alloys. At higher strain rates, the strength of the $\alpha - \beta$ alloys increased with volume percent of β and the activation energies were also found to be higher suggesting a change in the predominant deformation mechanism with increase in strain rate.

Introduction

A large number of technologically important titanium alloys consist of two phases (α and β) either at service temperatures and/or at processing temperatures. To design new titanium alloys with optimal properties or to improve the properties of existing titanium alloys, it is imperative to understand the deformation behavior of these alloys as a function of morphology (size, shape, distribution etc of the second phase) and testing variables. The properties of these materials cannot be predicted based on the simple rule of mixtures [1 - 4]. This is due to the fact that whenever a material consisting of two or more phases with different properties is subjected to stress, the deformation will be non-uniform resulting in interaction stresses. In addition, new deformation mechanisms such as interphase

Titanium '92
Science and Technology
Edited by F.H. Froes and I. Caplan
The Minerals, Metals & Materials Society, 1993

interface sliding can be operative which are otherwise absent in single phase materials.

In this investigation, an attempt has been made to systematically study the effect of the volume percent of phases, with equiaxed microstructures, on the high temperature deformation behavior of titanium alloys using Ti – Mn alloys.

Experimental Procedure

Six Ti – Mn alloys (designated as alloys 1 – 6) were used for this investigation. All the alloys were vacuum annealed for 200 hours at 923, 973 and 1023 K followed by water quenching. These heat treatments resulted in equiaxed microstructures [5]. Typical optical micrographs of alloys 1 through 6 in the as annealed condition are shown in Fig. 1. The particle and grain sizes were measured using standard techniques [6]. Typical microstructural parameters of the alloys which were annealed at 973 K for 200 hours followed by water quenching are shown in Fig. 2. It is to be noted that, at times, the particle size of a given phase is higher than the corresponding grain size because, often, the particles consist of two or more grains.

Tensile tests were conducted in an Instron machine. Tests were conducted in vacuum at temperatures in the range 923 – 1023 K, corresponding to the respective annealing temperatures in the strain rate range 0.00011/sec – 0.026/sec. Details of the test procedures are given elsewhere [4]. Typical true stress – true strain curves for alloys 1, 4 and 6 corresponding to the lowest strain rate (0.00011/sec) and highest strain rate (0.026 /sec) at 973 K are shown in Fig. 3. The curves corresponding to other testing conditions are not shown here [5].

Optical, SEM and TEM investigations were also carried out on the as annealed and deformed specimens by standard techniques [5, 6].

Results and Discussion

From the true stress – true strain curves, such as those shown in Fig. 3, flow stresses corresponding to 0.2, 0.5, 5, 10 and 20 % offset strains were determined. Variation of flow stress corresponding to 20 % offset strain with beta volume percent is shown in Fig. 4. Similar curves were obtained for other offsets at different temperatures [5].

To understand the deformation behavior, an attempt has been made to calculate approximate activation energies for the deformation mechanisms and the procedure is as follows.

The relationship between flow stress and strain rate has been analyzed using the Dorn equation [7]

$$\dot{\epsilon} = A\left(\frac{Gb}{kT}\right)\left(\frac{\sigma}{G}\right)^n \exp\left(\frac{-Q}{RT}\right) \qquad [1]$$

where G is the shear modulus, b the burgers vector, k the Boltzman constant, R the gas constant, T the absolute temperature, Q the activation energy, $\dot{\epsilon}$ the strain rate, σ the flow stress and A and n are dimensionless constants. The approximate shear modulii for different alloys were calculated from data available in literature [8].

For determination of activation energies the flow stresses corresponding to the 20 % offset strain were used because steady state behavior was observed for all of these alloys

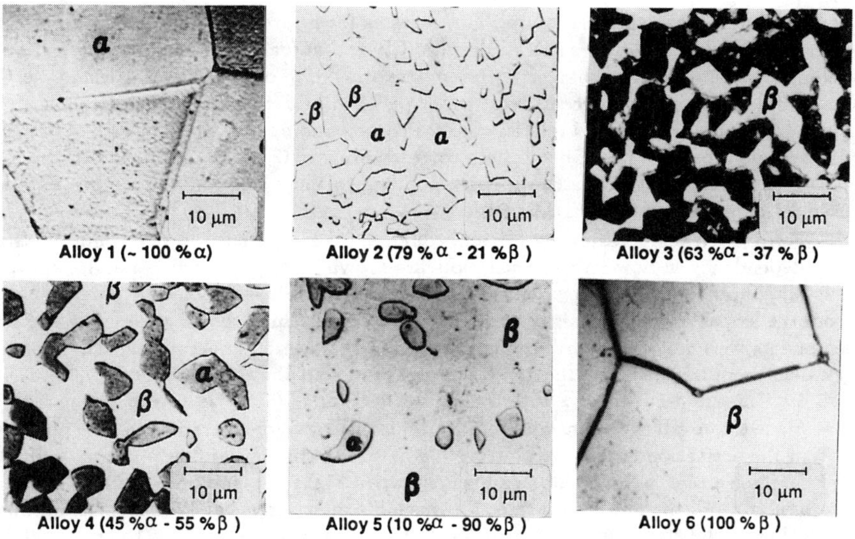

Fig. 1 : Typical microstructures of alpha, alpha - beta and beta titanium alloys which were annealed for 200 hours at 973 K followed by water quenching.

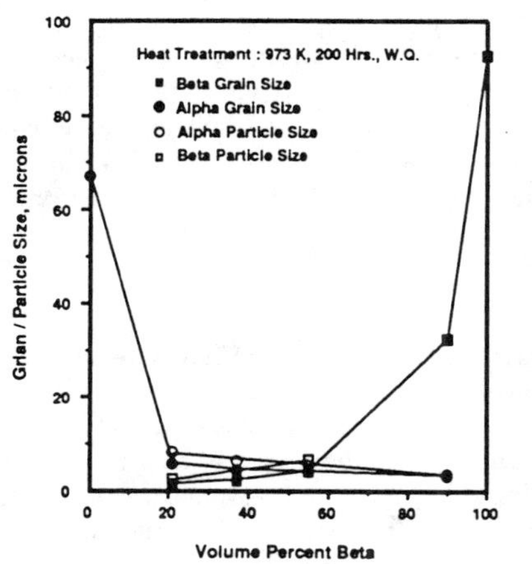

Fig. 2 : Effect of beta volume percent on grain and particle sizes in various Ti - Mn alloys.

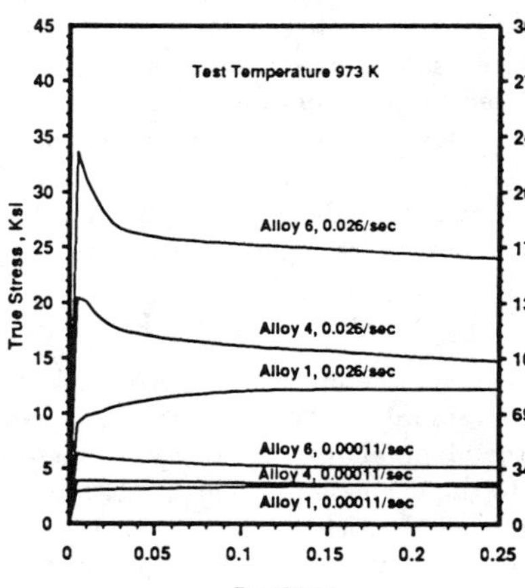

Fig. 3 : Effect of strain rate on the stress - strain behavior of alpha (alloy 1), alpha - beta (alloy 4) and beta (alloy 6) Ti - Mn alloys.

at this strain. As mentioned above, flow stresses for a given offset strain vary with beta volume percent, temperature and strain rate. Given that for a specific alloy composition the volume percent of phases changes with temperature [9], to obtain the relationship between flow stress and volume percent beta, the experimental data was fitted to a third degree polynomial. From this fitted relationship the corresponding flow stresses were obtained for different volume percents of beta phase corresponding to 0, 20, 40, 60, 80 and 100 % beta. It is to be noted that flow stresses corresponding to these selected volume percents were used for determining the activation energies.

For a given temperature and strain, equation [1] can be used to determine the strain rate dependency of flow stress. Such plots are shown in Fig. 5 for an all alpha alloy. For this alloy and the other alloys, not shown here, the stress exponents, n , were obtained from the slopes of these plots for the three test temperatures. For a given alloy the stress exponents were not significantly different for the three test temperatures. Therefore, the n values were averaged for the three temperatures and they were found to be 4.3, 4.75, 4.4, 4.1, 3.8 and 3.1 for alloys with 0, 20, 40, 60, 80 and 100 % beta, respectively.

Since the n values are known, from equation [1], for a given strain rate, it is possible to obtain the activation energy by plotting the natural logarithm of flow stress as a function of 1/T. Such a plot for an all alpha alloy is shown in Fig. 6. It is to be noted that the slopes are similar at all the four strain rates for this alloy indicating that similar mechanisms are operating in the strain rate range considered. However, the slopes for the other alpha – beta alloys were different for different strain rates indicating a change in the predominant deformation mechanism with strain rate.

It is to be noted that the stresses used in equation [1] are actually true stresses obtained by taking into consideration the instantaneous cross–sectional area of the specimen. However, similar corrections were not made for variation in strain rate as its effect is not expected to be significant [10] and hence would not significantly affect the activation energies obtained. Due to the number of assumptions made in estimating the activation energies, they are considered to be approximate apparent activation energies. Nevertheless, they give valuable information in the determination of the predominant deformation mechanisms operating in these alloys.

Deformation Behavior of α phase

Observation of Figs. 3 and 4 shows that the high temperature flow stresses increase with strain rate and decrease with temperature exhibiting mild strain hardening behavior for the alpha alloy. Part of the increase in flow stress with decrease in temperature is due to the decrease in grain size of the α – phase, not shown here. The stress exponent for this alloy which can be obtained from Fig. 5, is similar to the values reported earlier by other investigators for the alpha phase [7, 10, 12]. The activation energies determined from Fig. 6, are shown in Fig. 7 for various strain rates and they are in the range 210 – 250 kJ/mole for this alpha alloy. These values are close to the activation energy for self diffusion of alpha titanium as reported earlier by several other investigators [7, 10, 12]. This observation suggests that the deformation mechanism in alpha titanium, in the temperature range studied, is controlled by lattice diffusion. In addition, the range of stress exponents obtained further suggests that the deformation is predominantly climb controlled [10]. Further evidence for this suggestion comes from the TEM observations of the deformed specimens for this alloy where subgrain formation was observed [5].

Deformation Behavior of β phase

Unlike α – phase, the β – phase exhibits abrupt flow stress drops followed by steady state behavior, see alloy 6 in Fig. 3. As shown in this figure, the magnitude of the flow stress drop increases with strain rate. In addition, the extent of drop decreases with increase in temperature, not shown here. In the past, several investigators [11 - 16] have attributed both the flow stress drops and the subsequent steady state behavior to the process of dynamic recovery. Recently [4, 5, 16], the authors have considered the initial flow stress drops and the subsequent steady state behavior separately and have suggested that the former is due to the multiplication of mobile dislocations whereas the latter is due to dynamic recovery. The n values and the activation energies obtained in this investigation further support these earlier suggestions as explained below.

As mentioned above, the stress exponents for β – phase was found to be in the range 3 – 3.3. This is significantly lower than the values reported for pure β – titanium [15]. Given that the β – phase in the present investigation is a solid solution of manganese in titanium, the difference in n values appears to be related to the dislocation – solute atom interaction [7]. The low n values also suggest that the deformation in β – phase is predominantly glide controlled [7]. The activation energies shown in Fig. 7 for β – phase is in the range 210 – 280 kJ/mole. These values are close to the activation energies for self and solute lattice diffusion in β titanium [15]. This is also consistent with the suggestion that the deformation in β – titanium of Ti – Mn alloys is predominantly controlled by lattice diffusion. The microstructural observation of specimens after deformation also suggest that the flow stress drop is due to the multiplication of mobile dislocations and the subsequent steady state behavior is due to the process of dynamic recovery leading to the formation of subgrains [16].

Deformation Behavior of $\alpha - \beta$ alloys

As shown in Fig. 1, the $\alpha - \beta$ alloys consist of equiaxed type microstructures and the particle and grain sizes of the α and β phases are much smaller than the grain sizes of the terminal α and β phases. As shown in Fig. 3, for a given temperature and strain rate, the curves for a given $\alpha - \beta$ alloy are between those of the α and β curves. Furthermore, flow stress drops were also observed in $\alpha - \beta$ alloys and the extent of drop increased with beta volume percent and strain rate for a given temperature. In addition, the extent of drop decreased with an increase in temperature. For a given temperature, the flow stress did not vary linearly with beta volume percent, see Fig. 4. The actual variation depended on volume percent of phases and strain rate. Flow stresses for the lowest strain rate (0.00011/sec) were actually found to be lower for an alloy with about 50 % beta than for an alloy with 20 % beta alloy. Furthermore, the strength of this alloy is much lower than one would expect based on the law of mixtures. These observations suggest that a new deformation mechanism may be operating and such a mechanism is $\alpha - \beta$ interface sliding. Given that $\alpha - \beta$ interface sliding has been observed even at room temperature [17], it is

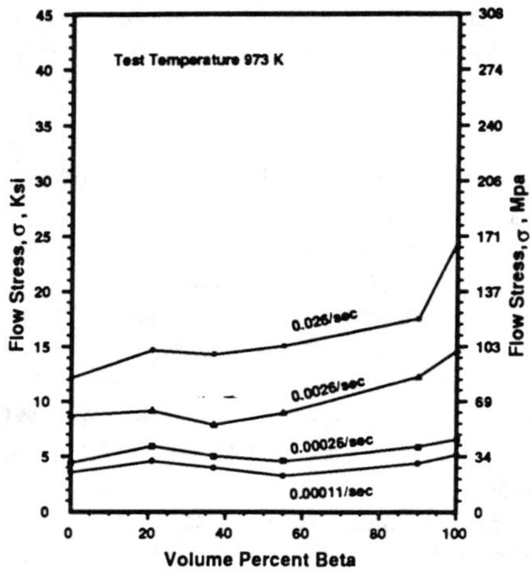

Fig 4: Offset flow stress vs. beta volume percent for different Ti - Mn alloys. The flow stress corresponds to 20 % plastic strain offset.

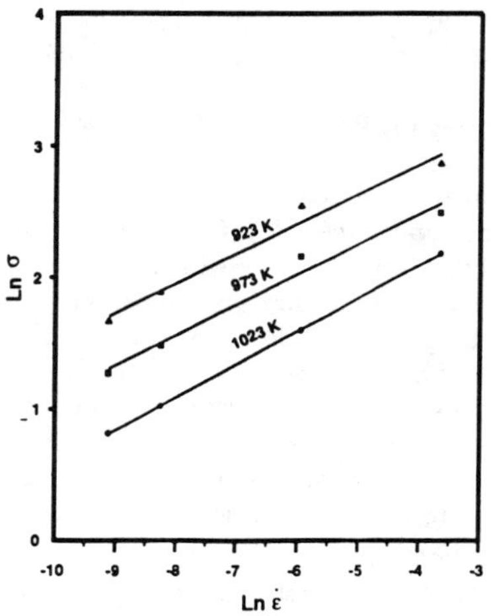

Fig 5: Effect of strain rate on flow stress for different temperatures for alpha titanium. The flow stress corresponds to 20 % plastic strain offset.

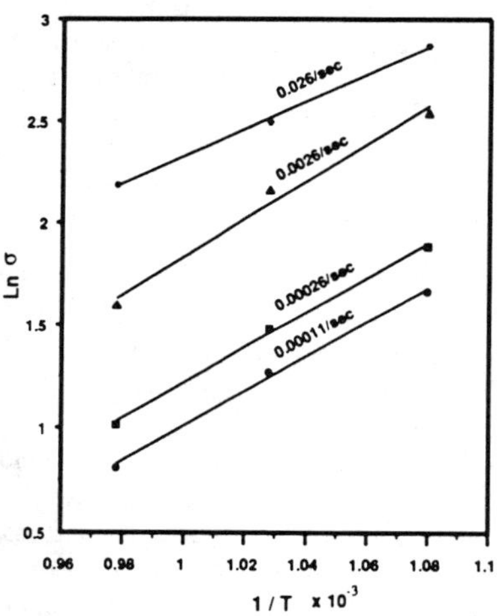

Fig 6: Effect of temperature on flow stress for different strain rates for alpha titanium. The flow stress corresponds to 20 % plastic strain offset.

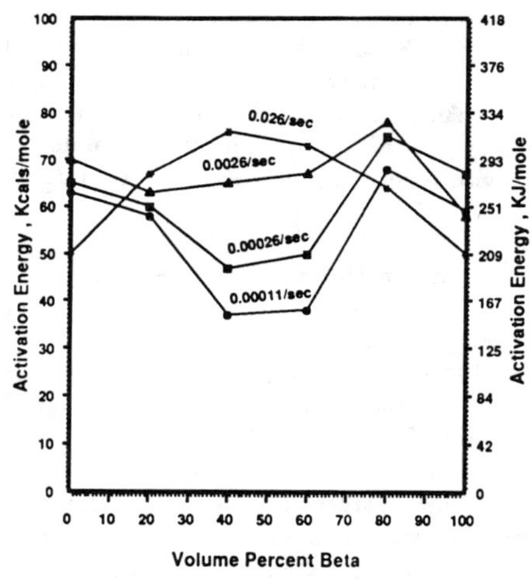

Fig 7: Effect of volume percent of beta phase on apparent activation energy for defor - mation for different strain rates.

reasonable to expect this mechanism to operate at high temperatures.

The activation energies shown in Fig. 7 are consistent with the suggestion that $\alpha - \beta$ interface sliding is a predominant deformation mechanism. It is to be noted that the activation energies obtained for an alloy with equal volume percents of α and β phases is about half of that obtained for the α or β phase. Observation of the microstructures of the undeformed and deformed specimens with fiducial lines indicated interface sliding [5]. Interface sliding was found to be much more significant at 923 K as compared to 973 K. This appears to be due to the much finer microstructures at 923 K. In fact, the flow stress of an alloy with almost equal volume percents of phases was found to be the same at 923 K and at 973 K. However, as the strain rate is increased from 0.00011/sec to 0.026/sec, the flow stress of this alloy increased and it was found to be about the same or slightly larger than that of an alloy with 20 volume percent beta. Also, the activation energies increased with strain rate and were close to that of the α and β phases suggesting that the deformation is controlled by lattice diffusion. The increase in activation energy with strain rate suggests that $\alpha - \beta$ interface sliding may not be the predominant mechanism at these high strain rates.

In regard to an alloy with 20 volume percent β phase or an alloy with 20 volume percent α - phase, the activation energies correspond to that for lattice diffusion which indicates that the deformation is predominantly lattice diffusion controlled. More work is needed to confirm these suggestions.

Conclusions

1. The flow stresses of the $\alpha - \beta$ Ti - Mn alloys, in the temperature range 923 - 1023 K and in strain rate range 0.00011/sec - 0.026/sec, did not vary linearly with beta volume percent. The deviation from linearity depended on strain rate, beta volume percent and temperature.
2. Stress exponents, n, for various $\alpha, \alpha - \beta$, and β alloys were determined and were found to be in the range 3.1 - 4.8. In addition, the corresponding activation energies for deformation were also determined. For α and β alloys the activation energies were found to be close to the activation energies for self and solute lattice diffusivities.
3. Based on the stress exponents, activation energies and microstructural studies of undeformed and deformed specimens, it is suggested that the deformation mechanism in alpha is controlled by lattice diffusion. For the beta phase also it is suggested that the deformation is controlled by lattice diffusion.
4. However, in $\alpha - \beta$ alloys, the deformation mechanism appears to depend on volume percent of beta phase and strain rate. For $\alpha - \beta$ alloys with 20 volume percent beta and 80 volume percent beta the deformation mechanism appears to be similar to the α and β phases i.e., lattice diffusion controlled. When the volume percent of α is about 50, the deformation mechanism appears to be predominantly $\alpha - \beta$ interface sliding at low strain rates and it gradually changes to lattice diffusion controlled at high strain rates. More work is needed to confirm this suggestion.

Acknowledgements

The authors would like to express their gratitude to Messrs. D.J. McNeish, D.E. Thomas, S.R. Seagle and their associates of the RMI company for making the alloys used in this investigation with great care at a nominal cost. The authors are indebted to Mr. I.L. Caplan of DTRC and Dr. A.H. Rosenstein of AFOSR for their keen interest and constant encouragement throughout the course of this investigation. This work was supported by the Air Force Office of Scientific Research on Grant AFOSR – 85 – 0367.

References

1. S.Ankem and H.Margolin, Metall. Trans., A, 13 (1982) 595–601.
2. S.Ankem and H.Margolin, Metall. Trans., A, 13 (1982) 603–609.
3. S.Ankem and H.Margolin, Metall. Trans., A, 17 (1986) 2209–2226.
4. S.Ankem, J.G.Shyue, M.N.Vijayshankar and R.J.Arsenault, Mat. Sci. and Engg., A111, April, (1989), 51–61.
5. M.N.Vijayshankar and S.Ankem, Unpublished research, University of Maryland, College Park, MD 20742, 1992.
6. Van Der Voort, Metallography Principles and Practice, McGraw–Hill, New York, 1984.
7. J.E.Bird, A.K.Mukherjee and J.E.Dorn, Quantitative Relation Between Properties and Microstructure, Eds. D.G.Brandon and A.Rosen, Israel University Press, Jerusalem, 1969.
8. Alloy Digest, Published by Engineering Alloys Digest, Inc., Upper Montclair, New Jersey, 1966.
9. R.P.Eliot, Constitution of Binary Alloys, First Supplement, McGraw–Hill, New York, 1965, 617.
10. H.Oikawa and M.X. Cui, Strength of Metals and Alloys, Proc. of ICSMA 7, Eds. H.J. McQueen, J.P. Bailon, J.I. Dickson, J.J. Jonas and M.G. Akben, Pergamon Press, (1985), 601–606.
11. P.Griffith and C.Hammond, Acta Metall., 20, (1972),935.
12. N.E.Paton and C.H.Hamilton, Titanium '84, Science and Technology, Eds. G.Lutjering, U. Zwicker and W.Bunk, AIME, (1984), 649.
13. S.M.L.Sastry, P.S.Pao and K.K. Sankaran, Titanium '84, Science and Technology, Eds. H. Kimura and O. Izumi, AIME, (1980), 874–886.
14. T. Sheppard and J. Norley, Mat. Sci. and Tech., Oct. (1988), Vol. 4, 903.
15. H. Oikawa, K. Nishimura and M.X. Cui, Scr. Metall., 19, (1985), 826.
16. M.N. Vijayshankar and S. Ankem, Mat. Sci. and Engg., A129, Nov., (1990), 229–237.
17. S.Ankem and H.Margolin, Metall. Trans., A, 14, (1983), 500–503.

The Microstructural Features Associated With the Growth of Short Fatigue Cracks in a Near-Alpha Ti Alloy.[1]

A L Dowson, C J Beevers and L Grabowski†

IRC in Materials for High Performance Applications,
The University of Birmingham, Birmingham, UK.

† Rolls Royce plc, P.O. Box 31, Elton Rd,
Derby, DE2 8BJ, UK.

Abstract

The formation and growth of fatigue cracks in a near alpha IMI 834 Ti alloy has been studied as a function of microstructure and stress ratio. In alpha plus transformed beta microstructures, cracks tended to form preferentially in the primary alpha phase along intense localised slip bands oriented between 30 and 70° to the principal stress direction. Subsequent crack propagation was found to be independent of the alpha phase volume fraction and relatively insensitive to variations in the prior beta grain size. Crack growth was observed to be highly irregular in nature and was accompanied by an intermittent mode of fracture with crack arrest, secondary cracking and crack tip deflection occurring within alpha colonies and at primary alpha and prior beta phase grain boundaries. The transition between short and long crack growth regimes was characterised by a progressive reduction in the frequency and duration of crack arrests and by an apparent build-up in fractographic features commonly associated with crack closure.

Introduction

The fatigue crack growth response of Ti and Ti alloys has been studied extensively and in most respects the microstructural features which influence the growth of long cracks have been well documented [1-3]. It is now widely accepted that acicular or colony based, Widmanstatten microstructures developed as a result of thermal processing or heat treatment within the beta phase field exhibit the greatest resistance to fatigue crack propagation. In general such improvements have been explained in terms of the crack path tortuosity and the alpha colony size. Crack propagation is invariably characterised by extensive crack tip bifurcation and by secondary cracking both parallel and perpendicular to alpha / beta platelet interfaces. The net result is an increased contribution from roughness induced crack closure and an associated reduction in the level of ΔK at the crack tip due to strain energy dispersal across the growing crack front. Whilst the mechanistic features associated with the growth of long cracks is well established, somewhat less attention has been directed at the growth of cracks in the short crack regime. In particular, experimental data relating to the effects of grain size and alpha phase distribution is lacking and where studies have been carried out the findings have often been inconclusive [4,5]. The principal experimental objective of the present study has been to examine the microstructural parameters (alpha phase volume fraction and prior beta phase grain size) which influence the formation and growth of short and long fatigue cracks in the near alpha IMI 834 Ti alloy with a view to developing an improved understanding of the mechanistic features which influence failure.

Experimental.

The principal objective of the experimental programme was to examine the effects of alpha phase volume fraction and prior beta phase grain size on the growth of long and short fatigue cracks in 'alpha plus beta' and 'fully beta' heat treated microstructures. Studies were conducted on a near alpha IMI 834 Ti alloy (nominal composition Ti-6Al-4Sn-3.5Zr-0.7Nb- 0.5Mo-0.35Si-0.05C)which had been heat treated at selected temperatures above and below the beta transus to yield alpha phase volume fractions of 5, 15 and 25% and prior beta grain sizes ranging from approximately 40 μm up to 1.0 mm. The various heat treatment conditions used in the fatigue crack growth programme are listed in Table I together with information relating to the variations in microstructure and

[1] The authors would like to acknowledge the support of Rolls Royce plc, Derby, DRA Aerospace Division, Pyestock and IMI Titanium, Birmingham. Thanks are also due to SERC and the director IRC for the provision of laboratory facilities and equipment.

Titanium '92
Science and Technology
Edited by F.H. Froes and I. Caplan
The Minerals, Metals & Materials Society, 1993

mechanical properties. Microstructures generated as a result of heat treatment below the beta transus were characterised by the presence of a dual phase structure consisting of a dispersion of primary alpha grains in a relatively fine grained, transformed beta matrix. In contrast material heat treated within the beta phase field exhibited a predominantly acicular microstructure consisting of non-aligned alpha colonies in a coarse grained matrix.

Table I. Heat treatment conditions, mechanical properties and microstructural characteristics of materials used in the fatigue crack growth programme.

Heat Treatment	Volume % Alpha	Primary Alpha GS (µm	Prior Beta GS (µm	0.2% PS (MPa)	UTS (MPa)	% EL	%RA
1/2 Hr 1100°C	0	-	420	1000	1100	4.7	-
2 Hrs 1044°C	5	16	120	984	1073	12.6	18.7
2 Hrs 1028°C	15	19	60	981	1069	13.2	23.8
2 Hrs 1015°C	25	22	46	976	1063	12.8	22.8

(All materials were subjected to a 2 hour age following solution treatment and oil-quenching.)

Short crack growth studies were conducted at R=0.1 and R=0.7 using axially loaded, hour-glass specimens of thickness 3 mm and Kt ≈ 1.03 and rectangular sectioned three and four point bend specimens of dimensions 5 mm x 15 mm. Each were loaded to a peak stress value equivalent to between 70 and 90% of the uniaxial yield stress. Long crack growth data was generated using axially aligned, corner notched specimens of dimensions 5 mm x 5 mm. In most cases short cracks were allowed to initiate naturally. However, because problems were encountered with the coarser grained beta heat treated microstructure it was found necessary to grow cracks from artificial defects introduced on the specimen surface by an EDM technique. All tests (long and short) were carried out in air at ambient temperature (≈ 25°C) using a Dartec, 50 kN capacity, servo-hydraulic testing machine operated under constant amplitude load control at a frequency of 25 Hz. The crack formation characteristics of the material and subsequent growth was monitored intermittently (≈ 1000 cycle intervals for short cracks) at mean load using the acetate replication technique. ΔK_I values for both short and long cracks were calculated on the basis of the projected crack length measured perpendicular to the stress axis using appropriate solutions derived by Pickard [6] and by Scott and Thorpe [7].

Results and Discussion.

Microstructural Features Associated With Crack Initiation.

The microstructural features associated with crack formation and subsequent propagation in a 15% alpha microstructure tested at R=0.1 using a peak stress equivalent to 80% of the uniaxial yield stress are illustrated in figure 1. Cracks tended to form preferentially in the primary alpha phase and were normally associated with the development of intense slip bands at angles of between 30 and 70° to the principal stress axis. Slip band cracking has been reported to occur predominantly in materials where cross slip is restricted and where there is an associated tendency towards strain localisation [6,7]. Since such effects tend to be pronounced in alpha Ti and in the near-alpha Ti alloys it follows that the crack initiation response of such materials will be significantly dependent on microstructural variations such as grain orientation, alpha phase volume fraction and the prior beta phase grain size since such parameters will alter the materials capacity for slip band formation and strain localisation. The possible effects of texture and grain orientation on crack initiation and subsequent stage I crack propagation is to some extent reflected in the crack path morphology. As indicated in figure 1, cracks on emerging from their grain of origin invariably maintained their directionality, propagating at an angle to the stress axis for a significant period before finally being deflected in towards the central axis, at a grain boundary or an alpha colony interface. Such observations would tend to suggest that slip compatibility between neighbouring alpha and transformed beta grains is an important consideration in the process of crack formation.

Although crack initiation in alpha plus transformed beta microstructures was observed to be specific to the alpha phase the overall resistance to crack formation and subsequent propagation was found to be relatively insensitive to variations in the alpha phase volume fraction within the range 5 to 25%. Whilst these observations are surprising they can be rationalised in terms of the relative changes in the prior beta grain size. In general, as the alpha phase volume fraction is increased there is a corresponding decrease in the prior beta phase grain size. Consequently, although the number of potential sites for crack initiation invariably increase as the alpha phase volume fraction is increased, the mean-free path for slip decreases by virtue of the decreasing beta phase grain size.

Whilst the resistance to crack initiation in alpha plus transformed beta microstructures was found to be insensitive to variations in the alpha phase volume fraction, the importance of the alpha phase in promoting crack formation is to some extent reflected in the difficulties which were experienced in initiating stable cracks in coarse grained, beta heat treated, fully transformed beta microstructures. Self initiated cracks were only observed in this material when the peak stress level was increased to in excess of 90% of the uniaxial yield stress. Slip band cracking was again the dominant mode of initiation with slip bands developing primarily within closely aligned alpha colonies at prior beta phase grain boundaries. Figure 2 shows a typical initiation site and demonstrates the early stages of development for a crack formed in material which had been heat treated for a half hour at 1100°C. In this particular instance crack initiation appeared to be associated with a shear displacement which resulted from slip within two closely spaced, parallel slip bands, the resultant crack extending rapidly across a prior beta phase grain boundary before being deflected at neighbouring beta phase grain boundary. Although slip band formation appeared to be a pre-requisite for crack formation such cracks did not necessarily propagate to failure. A feature specific to beta heat treated microstructures was the high incidence of non-propagating cracks. Such cracks invariably formed early in the specimen life and were normally associated with the development of planar or serrated slip bands either at prior beta phase grain boundaries or within closely aligned alpha colonies. In general such cracks tended to form in clusters within individual grains and in many cases were observed to extend across the full grain diameter, sometimes extending across grain boundaries.

Effect of Alpha Phase Volume Fraction on Crack Propagation.

Self-initiated short crack growth data obtained under tensile loading at R=0.1 for alpha phase volume fractions ranging from 5 up to 25% are presented in figure 3. The principal differences between long and short crack growth behaviour are demonstrated for a 15% alpha microstructure in figure 4. A classic short crack effect is observed with short cracks exhibiting faster growth rates and with crack growth persisting at ΔK_I values significantly below the long crack threshold at both R=0.1 and R=0.7. In general, crack propagation both within the long and the short crack growth regime was observed to be relatively insensitive to variations in the alpha phase volume fraction and hence, since the two parameters are inter-related, to small changes in the prior beta phase grain size. Crack propagation within the short crack regime was found to be highly irregular and was accompanied by an intermittent mode of growth with crack arrest ($da/dN \ll 10^{-8}$ mm/cycle) occurring both within alpha colonies and at primary alpha and prior beta phase grain boundaries. Considering the results presented in figure 4 it is apparent that, for alpha plus transformed beta microstructures, crack arrest tends to occur primarily within a narrow region of ΔK_I bounded by the long crack thresholds at R=0.1 and R=0.7. Within this region, as the level of ΔK_I is increased, there is a progressive reduction in both the frequency and duration of crack arrest points and a general convergence in the long and the short crack growth curves.

Figure 5 compares short crack growth data obtained under three point bend loading for a 15% alpha microstructure at R=0.1 and R=0.7. In contrast to observations made for long cracks (figure 4) the growth of short cracks was found to be independent of the stress ratio suggesting a possible reduction in crack closure effects. The potential absence of crack closure effects within this regime have been discussed by a number of workers [[10,11]. There is, however, little experimental evidence to-date to either confirm or deny such considerations. Recent in-situ SEM studies conducted by Halliday [12] of the crack opening displacemement at the centre of a 500 μm crack contained in a 15% alpha microstructure have demonstrated that crack closure only occurs at loads approaching zero. Such measurements, combined with the results presented in figure 5, would tend to suggest that crack closure plays only a minor role in short crack propagation; an observation which is consistent with finite element predictions made by Zhang [13]. Figure 6 illustrates some of the major fractographic variations observed as a function of crack depth for short, freely initiated cracks grown in a 15% alpha microstructure. Consistent with metallographic observations relating to the crack path morphology (i.e. figures 1 and 2), crack initiation and the early stages of crack propagation were characterised by a highly faceted mode of fracture with extensive secondary cracking and transgranular cleavage (figure 6a). Although such features were maintained throughout the short crack regime, as the crack length increased, an increasingly mixed mode of fracture was observed (figure 6b) with the fracture surface exhibiting increasing levels of plastic tearing. A gradual build-up in the level of crack face interactions (a feature normally associated with crack closure) was reflected by a general increase in the degree of asperity flattening and by the detection of rubbing damage on the fracture surface (figure 6c). Such features only really became apparent at crack depths greater than approximately 400 μm and were ultimately accompanied by a transition to a classic striated mode of fracture at crack depth in excess of ≈ 450 μm (figure 6d).

Crack Propagation in Beta Heat Treated Microstructures.

Figure 7 compares short and long crack growth data obtained at R=0.1 for a beta heat treated microstructure. In keeping with the results presented in figure 4 for alpha plus transformed beta microstructures a significant short crack effect is demonstrated with short cracks exhibiting significantly faster growth rates than their long crack counterparts for similar levels of ΔK_I. In figure 8 long and short crack growth data obtained at R=0.1 for beta heat treated material (1/2 hour 1100°C) are compared with similar results obtained for a 15% alpha microstructure. The absence of short crack data for beta heat treated microstructures at ΔK_I values below approximately 5.0 $MPa.m^{-1/2}$ and at surface crack lengths less than approximately 200 µm clearly precludes a detailed comparison of the short crack growth behaviour of the two microstructural conditions. Crack growth rates observed at higher values of ΔK_I are, however, comparable. Consequently for ΔK_I values within the range 5-10 $MPa.m^{-1/2}$ it may be concluded that the growth of short surface cracks is independent of changes in the prior beta phase grain size and associated variations in the microstructural morphology. The absence of a microstructural effect within the long crack regime is, perhaps, a little more surprising. The improved performance of colony-based, Widmanstatten-type microstructures is now widely accepted and has been attributed to increasing levels of secondary cracking, crack tip fragmentation and strain energy dispersal across a more complex crack front [1,2,3]. Although crack tip bifurcation was observed in the present microstructures (figure 9), with secondary cracking and crack branching occurring along the crack front, the contributions of such crack tip microstructural interactions in terms of the overall da/dN Vs ΔK_I curve cannot be quantified at this stage. Despite significant differences in the crack path morphology the rate of crack growth within the Paris and the near-threshold regions of the crack growth curve were essentially independent of the prior beta phase grain size and the alpha phase volume fraction.

Conclusions.

1. Crack initiation in alpha plus transformed beta and in beta heat treated microstructures was found to be associated with the development of intense slip bands either in primary alpha grains or in closely aligned alpha colonies at prior beta phase grain boundaries. Cracks tended to occur preferentially in the alpha phase and on this basis the intrinsic resistance of the material to crack formation increased in the fully beta heat treated microstructure.

2. Despite the general preference for cracks to form within the alpha phase the growth of short, self-initiated fatigue cracks was observed to be independent of the alpha phase volume fraction and the prior beta phase grain size. Crack growth rates observed for beta heat treated microstructures at ΔK_I values in excess of 5.0 $MPa.m^{-1/2}$ were found to be comparable with results obtained for alpha plus transformed beta microstructures.

3. The growth of short, self-initiated cracks in alpha plus transformed beta microstructures was found to be independent of the stress ratio. This is in direct contrast to observations made for long cracks and is consistent with a closure free mode of growth within the short crack regime.

4. The growth of long fatigue cracks was found to be independent of variations in the alpha phase volume fraction and changes in the prior beta phase grain size. Threshold ΔK_I values observed at R=0.1 for beta heat treated microstructures with prior beta phase grain sizes ranging up to approximately 650 µm were found to be comparable with those observed in alpha plus transformed beta microstructures.

5. The transition between short and long fatigue crack growth behaviour in alpha plus transformed beta microstructures was found to be characterised by a progressive reduction in the frequency and duration of crack arrest points and by an apparent build-up in fractographic features normally associated with crack closure.

References.

[1] G R Yoder, L A Cooley and T W Crooker, Metall Trans, Vol 9A, 1978, p1413.

[2] G T Gray and G Leutjering, "Titanium, Science and Technology.", Deutsche Gesellschaft fur Metallkunde, Munich, 1985, p2115.

[3] J C Chesnutt, "Titanium, Science and Technology.", Deutsche Gesellschaft fur Metallkunde, Munich, 1985, p227.

[4] L Wagner, J K Gregory, A Gysler and G Leutjering, "Small Fatigue Cracks.", ed. R O Ritchie and J Lankford, AIME, Warrendal, Pa., 1986, p171.

[5] C W Brown and D Taylor, "Fatigue Crack Growth Threshold Concepts.", ed. D Davidson and S Suresh, The Metallurgical Soc. AIME, 1984, p252.

[6] A C Pickard, The Application of 3-D Finite Element Methods to Fracture Mechanics and Fatigue Life Prediction., EMAS, 1986.

[7] P M Scott and T W Thorpe, Fatigue of Engng Mater Struct, Vol 4 (4), 1981, p291.

[8] C Laird and D J Ducquette, in Corrosion Fatigue, ed A J Mc Evily and R W Staehle, Nat Assoc Corr Eng, 1972, p88.

[9] J C Williams and G Leutjering, "Ti '80, Science and Technology.", ed. H Kimura and O Izumi, TMS-AIME, 1980.

[10] S Suresh and R O Ritchie, International Metals Review, Vol 29, 1984, p445.

[11] W N Sharpe and X Su, Engng Frac Mechanics, Vol 30, 1988, p275.

[12] M D Halliday, private communication, Univ Birmingham, Feb 1992.

[13] J Zhang, PhD Thesis, The University of Birmingham, 1992.

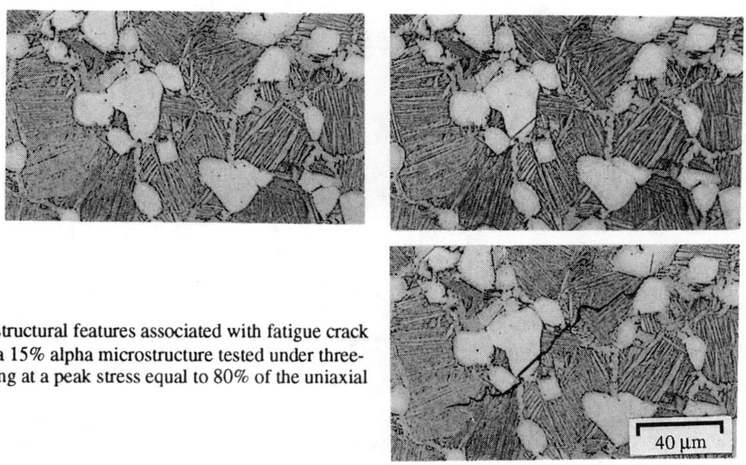

Figure 1. Microstructural features associated with fatigue crack development in a 15% alpha microstructure tested under three-point bend loading at a peak stress equal to 80% of the uniaxial yield stress.

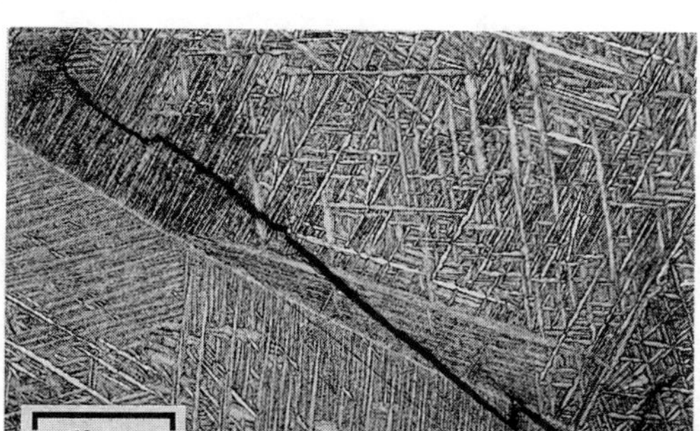

Figure 2. Slip band development and crack formation in a fully transformed beta microstructure tested under four-point bend loading (peak stress ≈ 90% of yield).

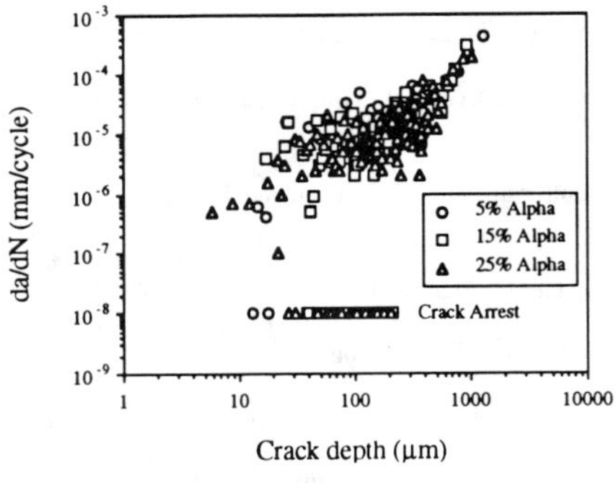

Figure 3 . Effect of alpha phase volume fraction on the growth of short, self-initiated cracks in the near alpha IMI 834 Ti alloy. All tests were conducted under tensile loading at R=0.1.

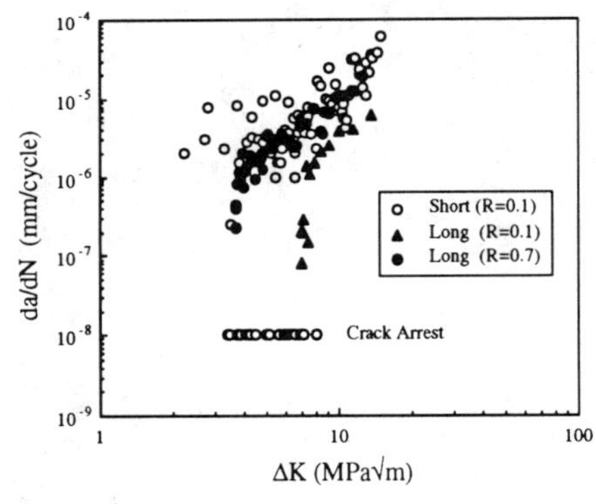

Figure 4. Comparison of long (R=0.1 and R=0.7) and short (R=0.1) fatigue crack growth data obtained for a 15% alpha containing microstructure tested under tensile load.

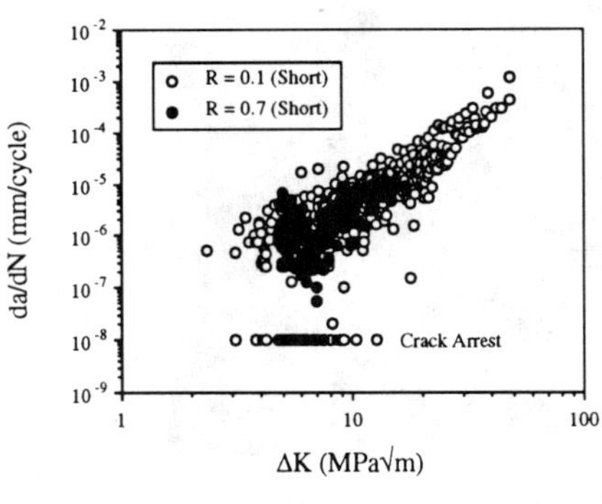

Figure 5. Effect of stress ratio on the growth of short fatigue cracks in a 15% alpha microstructure tested under three-point bend loading.

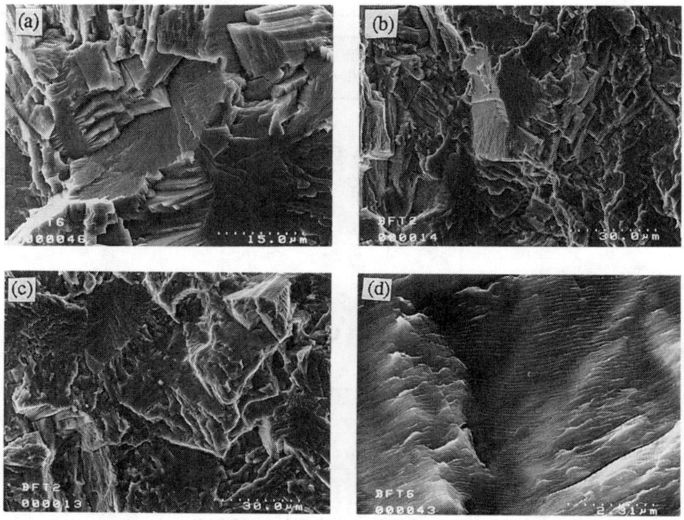

Figure 6 . Fractographic features associated with short fatigue cracks grown in a 15% alpha microstructure at R=0.1 under three-point bend loading (a) crack depth ≈ 30 μm, (b) crack depth ≈ 180 μm, (c) crack depth ≈ 430 μm and (d) ≥ 450 μm.

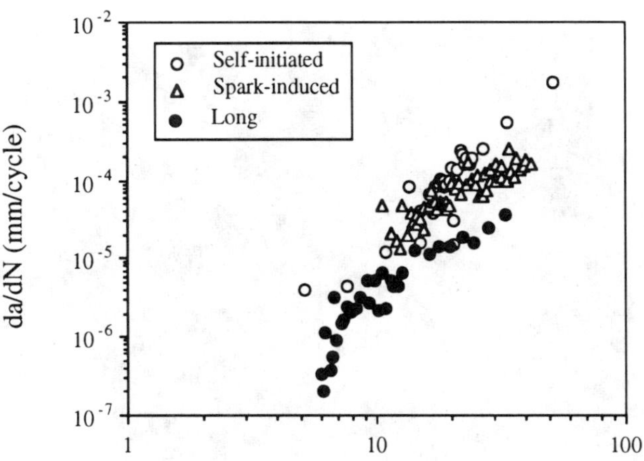

Figure 7 . Comparison of long and short fatigue crack growth data obtained at R=0.1 for material heat treated for a half hour at 1100°C and subsequently oil quenched and aged for two hours at 700°C. Short crack data was generated under four-point bend loading.

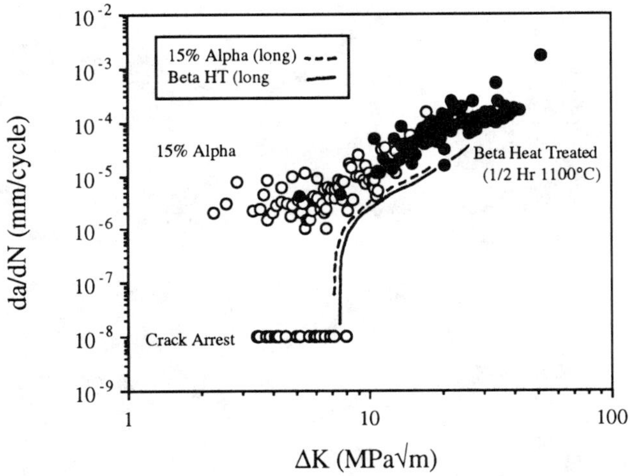

Figure 8. Comparison of long and short fatigue crack growth data obtained at R=0.1 for beta heat treated (half hour at 1100°C) and alpha (15%) plus transformed beta microstructures.

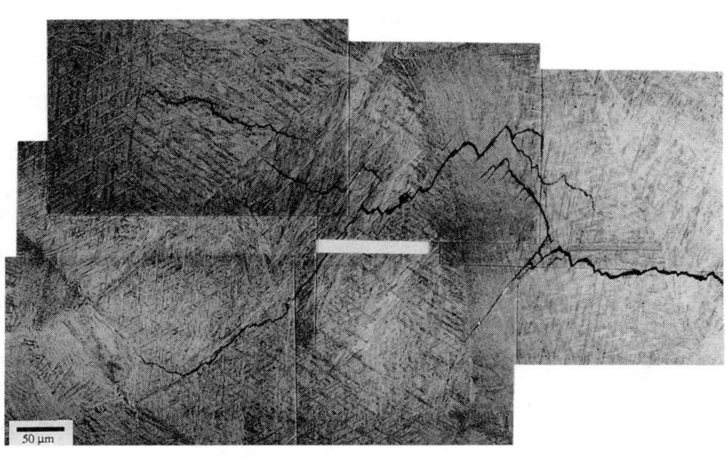

Figure 9 . Montage showing crack tip bifurcation and secondary cracking associated with long crack propagation in beta heat treated microstructures.

SMALL CRACK GROWTH AT ELEVATED TEMPERATURES

IN A NEAR ALPHA TITANIUM ALLOY.

S.H. Spence[+], W.J. Evans[*] and A. Goulding[+]

[*] I.R.C. in Materials for High Performance Applications,
University College of Swansea, U.K., SA2 8PP

[+] Department of Materials Engineering,
University College of Swansea, U.K., SA2 8PP

Abstract

Microstructurally small crack growth has been measured in the titanium alloy, IMI 829, both optically and *via* a d.c. potential drop system at temperatures of 20, 300 and 600°C. Part through small cracks were studied in a plain section specimen whilst long crack data were obtained from CT specimens. The programme explores the effect of temperature on small crack growth rates and compares and contrasts this with long crack behaviour.

Introduction

Many titanium based components in gas turbine compressors experience elevated temperatures during service. An ability, therefore, to understand and quantitatively assess fatigue response at temperature is essential for their safe service operation. Currently, design engineers are particularly interested in using fracture mechanics based criteria for lifing critical parts. A major point of debate, however, is the validity of such criteria for cracks that are comparable in size to alloy microstructural units and to design limiting structural features. Since cracks are likely to spend much of their total life in this regime it is obviously an issue that must be resolved if crack propagation based design methods are to be widely accepted. Any limitations associated with small crack growth will be magnified as temperatures increase due to interactions from other modes of deformation and failure such as creep and environmental effects.

The objective of the present programme was to characterise small crack growth behaviour over a range of temperatures and loading conditions for simple and complex stress fields. The paper sets out to describe the experimental facility and to present data obtained from the test programme on simple laboratory specimens. For the experimental evaluation, the near alpha titanium alloy, IMI 829, was specifically chosen. It is an important engineering alloy in the gas turbine industry and its comparatively coarse structure makes it an ideal vehicle for the study of microstructurally

Titanium '92
Science and Technology
Edited by F.H. Froes and I. Caplan
The Minerals, Metals & Materials Society, 1993

small cracks. Furthermore, previous work (1,2) provided a data base for room temperature crack growth characteristics for this alloy with which the behaviour at elevated temperatures can be compared.

Experimental Equipment and Procedure

The IMI 829 specimens were machined from bar stock which had been given a solution treatment at 1050°C for 30 minutes followed by air cooling and ageing at 625°C for two hours. This produced an essentially basketweave Widmanstätten alpha microstructure but with significant regions of aligned alpha platelets. The mean prior beta grain size was approximately 0.5 mm. This microstructure is typical for compressor discs currently in service.

Compact Tension, CT, specimens were used to obtain base data for long cracks. A Corner Crack, CC, specimen with a 10 mm square section was used to examine the behaviour of small part through cracks at plain surfaces. A starter slit was machined into one corner to a nominal depth and width of 0.25 mm and 0.15 mm respectively. This ensured initiation of a single crack at a known location and facilitated the use of crack monitoring equipment and subsequent growth rate analysis.

Constant load amplitude tests were conducted at two elevated temperatures in laboratory air using a Mayes 100 KN servo-hydraulic machine. Temperatures of 300 and 600°C were used. These are representative of bore and rim conditions in compressor discs. Loads were applied using a 1 Hz sine waveform. A range of stresses was applied with a load ratio (R value) of 0.1. A higher R value of 0.5 was also used in a limited number of tests.

Crack growth was monitored by a computer controlled system specifically developed for the project. It was comprised of a four port split furnace, optical microscopes and pulsed d.c. potential difference equipment. The working distance of the optical microscopes was sufficiently large as to permit surface crack monitoring from outside the furnace environment. The optical and p.d. techniques provide information that allows crack shape and morphology to be characterised. The observations were supplemented by fractographic studies using a scanning electron microscope.

Growth rates were calculated using a three point secant method and were assessed in terms of the stress intensity factor range, ΔK. Evaluations of K for CC and CT specimens were made using available polynomial approximations (3).

Results

Growth rate data from CC test pieces are illustrated in figures 1 and 2 for test temperatures of 300 and 600°C, respectively and R values of 0.1 and 0.5. In each case the best fit curve for room temperature data is included for comparison. A transition in the slope is apparent for data obtained from CC test pieces at 300°C. This transition occurs at a stress intensity value, ΔK_T, of approximately $11MPa\sqrt{m}$. Growth rates are similar for temperatures of 20 and $300°C$ for $\Delta K > \Delta K_T$, figure 1. However, rates are higher at $300°C$ for $\Delta K < \Delta K_T$. No significant difference in growth response was observed for R values of 0.1 and 0.5.

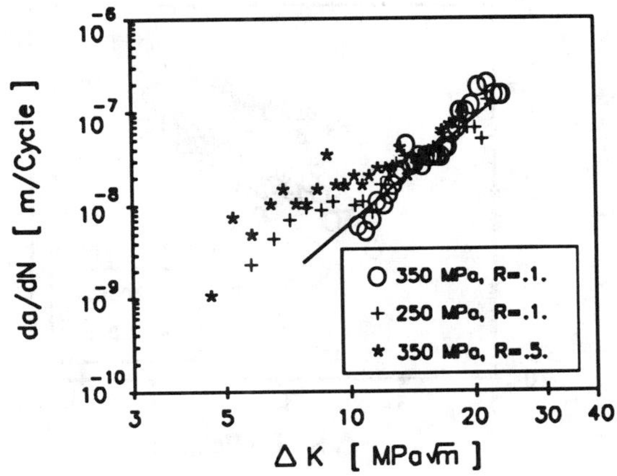

Figure 1. Growth rates in terms of ΔK for R values of 0.1 and 0.5 on CC test pieces at 300°C. The solid line included illustrates the best fit to data obtained at 20°C with R=0.1.

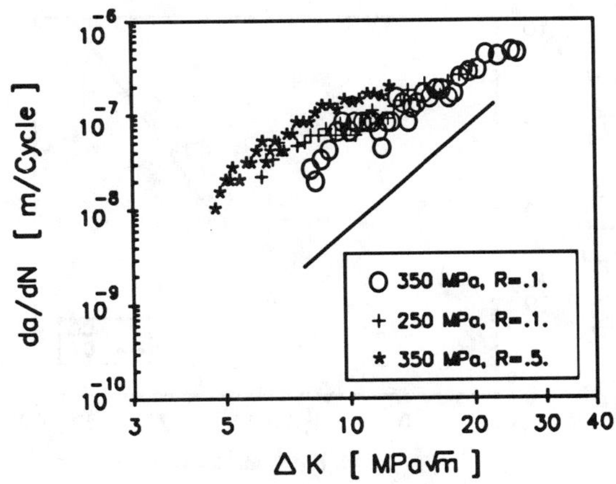

Figure 2. Growth rates in terms of ΔK for R values of 0.1 and 0.5 on CC test pieces at 600°C. The solid line illustrates the behaviour at room temperature.

Growth rates at $600°C$ are significantly higher than at ambient temperature throughout the range of ΔK values studied ($5-25MPa\sqrt{m}$), figure 2. The difference is most marked for lower values of ΔK being greater than an order of magnitude for $\Delta K < 10MPa\sqrt{m}$. No transitional value of ΔK was evident, the growth rate curve maintaining a constant slope within the range of ΔK

values studied. However the slope of the curve is similar to the $300°C$ data at values of ΔK less than ΔK_T, figure 3. At $600°C$ growth rate data for R values of 0.1 and 0.5 do not superimpose. Growth rates for R=0.5 are greater when compared on a ΔK basis, figure 2.

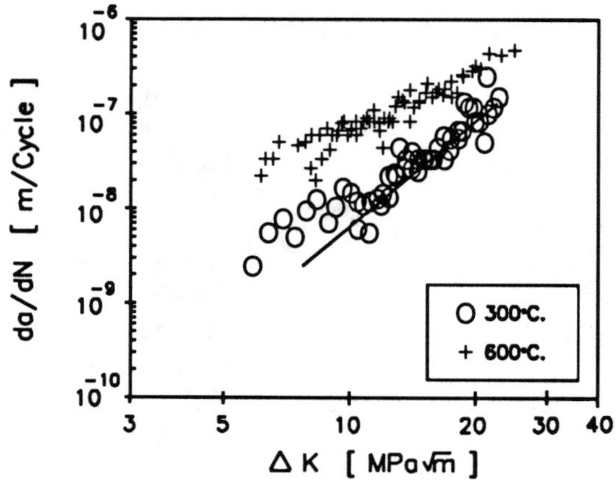

Figure 3. Growth rates for R=0.1 on CC specimens at 300 and 600°C, in terms of ΔK. The solid line illustrates room temperature behaviour.

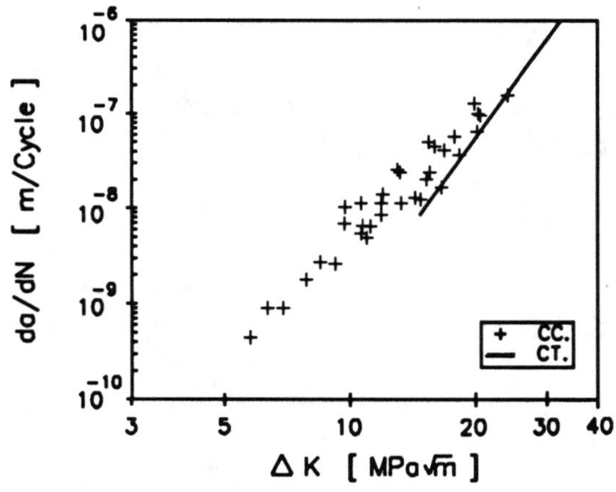

Figure 4. Growth rates in terms of ΔK from CC test pieces at 20°C with R=0.1. The solid line illustrates the best fit to data from CT specimens.

To assess small crack effects, growth rates from CC test pieces were compared with data from CT specimens for temperatures of 20, 300 and $600°C$ and an R value of 0.1. In figures 4, 5 and 6 the small crack growth data are plotted together with the best fit curves for CT specimens at the equivalent temperature. At ambient temperature and low values of ΔK, growth rates are significantly greater in CC test pieces, figure 4. However, the CT and CC data merge at $\Delta K \approx 23 MPa\sqrt{m}$.

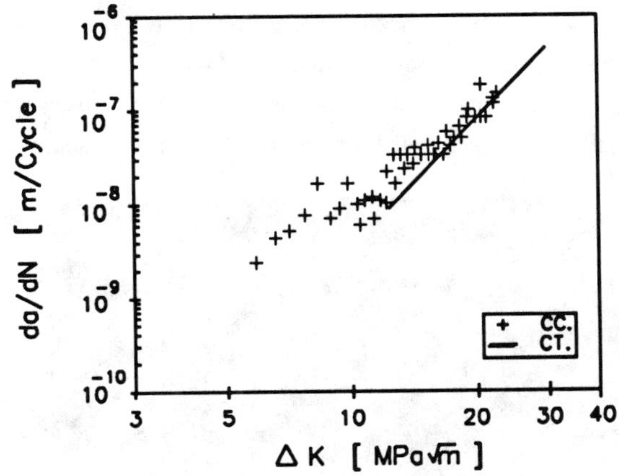

Figure 5. Crack growth behaviour at 300°C with R=0.1 on CC test pieces. Included is the best fit for CT specimens.

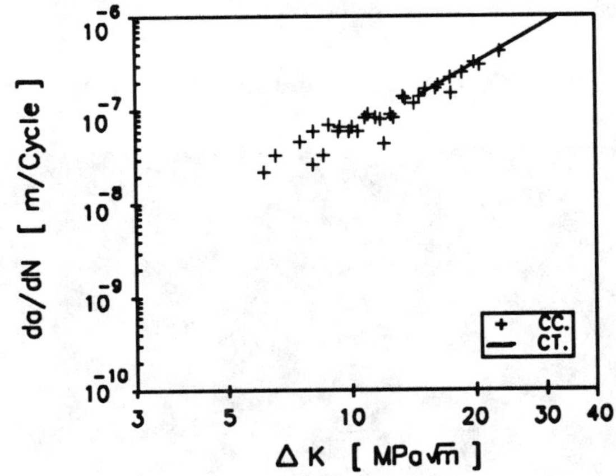

Figure 6. Crack growth rates at 600°C with R=0.1 on CC test pieces. Included is the best fit to CT data.

At 300°C small and long crack growth rates are similar for $\Delta K > \Delta K_T \approx 11 \quad MPa\sqrt{m}$, figure 5. For $\Delta K < \Delta K_T$ the small crack growth rates are significantly higher. For the highest temperature studied, 600°C, growth rates of small and long cracks are in good agreement over the range of ΔK studied. Both are significantly higher than the growth rates at 20 and 300°C.

A comprehensive fractographic study was carried out on the CC test pieces. The fracture surfaces from tests conducted at $300°C$ were similar to those found on room temperature specimens. Each was characterised by facets with interlinking regions of ductile tearing, figure 7. However, there was a marked change in fracture surface appearance for tests at $600°C$, figure 8. The facets are smaller and many have a much smoother appearance with less detail than those at lower temperatures. On changing R ratio from 0.1 to 0.5, at $600°C$, further differences in fracture surface appearance were apparent, figure 9. Macroscopically the surfaces were much flatter at R=0.5. Microscopically, facets were much smaller and less well defined.

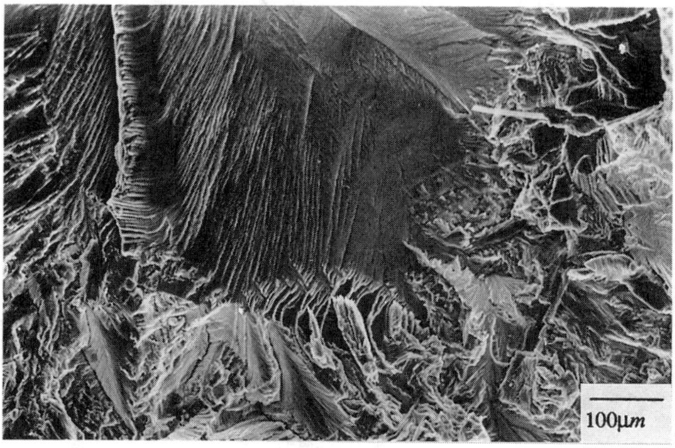

Figure 7. Fractograph of a CC specimen tested at 300°C, illustrating faceted surface.

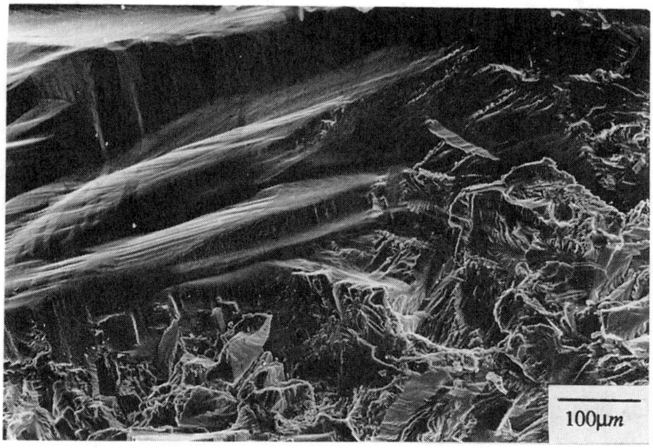

Figure 8. Fractograph of a CC specimen tested at 600°C with R=0.1.

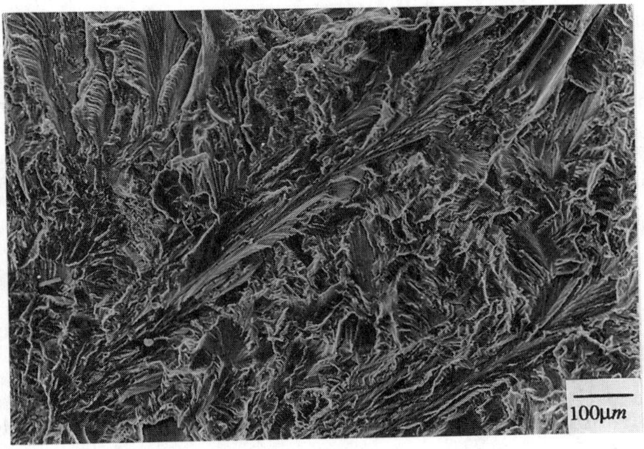

Figure 9. Typical fracture surface from a CC specimen tested at 600°C with R=0.5.

Discussion

This programme is still in its early stages, however, the preliminary results clearly reveal some important aspects of small crack growth at elevated temperatures. The transition in growth rate behaviour observed at 300°C and the significantly faster rates of growth at 600°C have important implications for fracture mechanics based lifing criteria. Similarly, the anomalously fast growth rates of small cracks, when compared with those for long, at 20 and 300°C further illustrate the problems associated with the use of parameters such as ΔK for component lifing criteria.

The transition in growth rate behaviour at 300°C could be associated with static and/or environmental mechanisms playing an important role. At 600°C a marked difference in growth rate behaviour was observed. The slope of the growth rate curve at 600°C (about 2) is approximately the same as for 300°C at $\Delta K < \Delta K_T$. This suggests that at 600°C static and/or environmental effects could play a significant role throughout the range of ΔK studied. Further, the significantly higher growth rates observed at 600°C could be due to these effects taking on a more dominant role. The effect of R value, observed only at 600°C, provides further evidence in support of this argument. The different fracture surface appearance at 600°C also suggests that different growth mechanisms are involved.

Conclusions

The research programme has highlighted major differences in the growth of small and long cracks at 20, 300 and 600°C:

1. A transition value of ΔK, ΔK_T, is associated with a change in slope of the growth rate curve for small cracks at 300°C.
2. Growth rates at 300°C are faster than those at 20°C for $\Delta K < \Delta K_T$.

3. Growth rates are significantly faster for $600°C$. At this temperature increased R values lead to increased growth rates.
4. Growth rates of small cracks at $20°C$ are faster than those for long cracks, for $\Delta K < 23MPa\sqrt{m}$, but merge at higher values.
5. Growth rates of small cracks at $300°C$ are faster than those for long cracks, for $\Delta K < \Delta K_T$.
6. Growth rates are similar for small and long cracks at $600°C$ and R=0.1.

Acknowledgements

This work has been supported by the Procurement Executive, Ministry of Defence.

References

1. S.H. Spence, "Small Fatigue Crack Propagation in a Near Alpha Titanium Alloy, IMI 829" (Ph.D. thesis, University of Wales, 1990).

2. W.J. Evans, S.H. Spence and A. Goulding, "Small Crack Growth at Stress Concentrations in a Near Alpha Titanium Alloy," Proc. This Conference.

3. A.J.A. Mom, "Revised Working Document for the Agard Cooperative Test Programme on Titanium Alloy Engine Disc Material" (Report NLR TR 86019 L, 1986).

LOW CYCLE FATIGUE BEHAVIOR OF Ti 6246.

A.S.BERANGER - X.FEAUGAS - D.LAI - M.CLAVEL.

Université de technologie de Compiègne. Département de Génie Mécanique.
Laboratoire de génie mécanique pour les matériaux et les structures.

Abstract :

The present study focused on the cyclic behavior of the Ti 6246 alloy, and more precisely on the cyclic softening mechanisms.
Measurements on the stress-strain hysteresis loops provided information on the origins of the phenomenon. Observations of the dislocation arrangements and investigations on the fine microstructure were undertaken using transmission electron microscopy (TEM). The effect of the strain incompatibilities between the two ductile phases α and β, on the macroscopic behavior, was studied, resorting to Finite Element Method calculations (FEM).

Introduction

In recent years, the aircraft industry has expanded the use of titanium alloys into the 500°C temperature range in order to produce lighter engines components. Within this framework, the applications of the Ti 6246 alloy, particularly as compressor discs, has generated a great interest in its low cycle fatigue behavior. Our purpose is to present experimental results of plastic strain control low cycle fatigue between 20°C and 500°C and to study in detail the mechanisms responsible for the macroscopic behavior. Mechanical analyses of the stress response of the material is related to microstructural investigations and FEM calculations to outline the different physical origins of the cyclic softening in different temperature ranges.

Experimental procedure

Material

The material to be investigated was Ti 6246, a two-phase α/β titanium alloy containing : 6 pct Al, 6 pct Mo, 4 pct Zr and 2 pct Sn by weight (Table 1). It was prepared as an ingot forged in the ß field at 965°C (the ß transus is approximately 950°C). Initial heat treatment was at 900°C, for 2 hours, following which the material was air quenched. It was then reheated to 595°C, held for 8 hours, and finally air quenched.

The process results in a Widmanstätten α, grain boundary α and transformed ß microstructure as shown on Figure 2. The phase percentages were measured by counting the respective areas on a superimposed millimeter net : the Widmanstätten α phase and the transformed ß phase both take up an equal volume fraction of 50%. TEM examinations showed that the transformed ß matrix is a mixture of retained ß and fine secondary α precipitates (Figure 3). Superlattice reflections were often observed in the SAD patterns of the α phase. The occurence of order in the form of Ti_3Al (α_2) fine precipitation within the α grains is responsible for these extra reflections. The Ti_3Al precipitates could be imaged in dark field (Figure 4). It was difficult to carry out exact measurements of these particles sizes. The diameter of the particles is approximately 15 Å, and the average inter-particle distance is about 30 Å.

Titanium '92
Science and Technology
Edited by F.H. Froes and I. Caplan
The Minerals, Metals & Materials Society, 1993

	Ti	Al	Zr	Sn	Mo	Si	Fe	O	H
Mini	Base	5.5	3.5	1.75	5.5				
Maxi		6.5	4.5	2.25	6.5	0.1	0.15	0.15	0.015

Table 1
Chemical composition of the 6246 alloy (weight %)

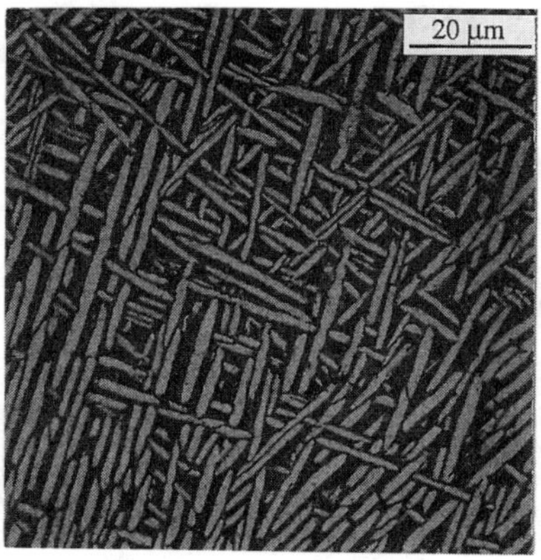

Figure 2
Optical micrograph : starting undeformed microstructure : Widmanstätten α phase and transformed ß matrix.

Figure 3
TEM micrograph (bright field) starting undeformed microstructure : secondary α precipitation within the transformed ß matrix.

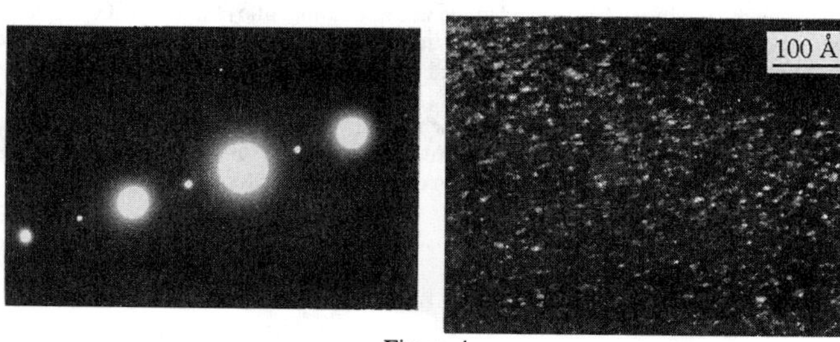

Figure 4
α SAD diffraction pattern : extra reflexions due to Ti_3Al
TEM dark-field micrograph : finely dispersed Ti_3Al in α.

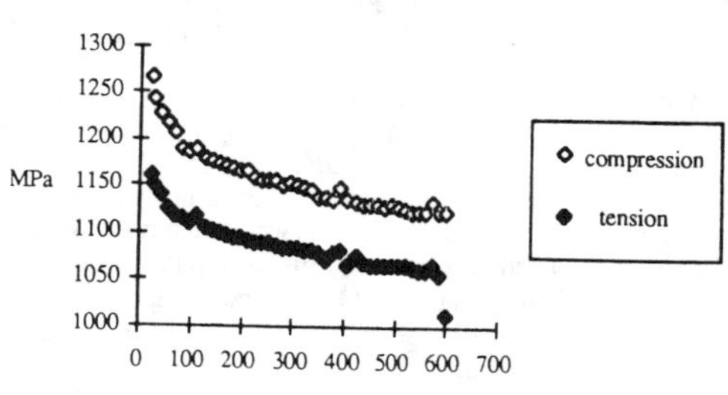

Figure 5
Compressive and tensile peak stresses as a function of p.

Tension-compression tests

Tension-compression specimens 6 mm diameter and 15 mm gage length were machined. Prior to testing, the specimens were prepared by consecutively mechanical polishing on a lathe through 4000 grit paper; then a operation was carried out on a cloth with a solution of 260 ml Al_2O_3 (4µm), 40 ml H_2O_2, 1 ml HNO_3 and 0,5 ml HF.

The fatigue tests were conducted on a Mayes 5t testing machine, under plastic strain control with R = -1. All the tests were carried out in laboratory air, between room temperature and 500°C. The axial strain level was controlled by extensometer arms tips located onto the gage length. The frequency was selected as 3 cycles/min and was identical for all the plastic strain range studied. The fatigue tests continued until the failure of the specimen occured. A triangular strain-time wave was used. A few additional tests were conducted, in slightly different conditions of frequency, or in some cases, fatigue tests were interrupted before the failure of the specimen.

TEM

Thin foils were prepared from 1 mm thick cross section slices taken from the gage length portion of the specimens. These slices were first mechanically thinned down to about 0,15 mm and then electropolished in a Strüers A3 electrolyte, at -15°C. All TEM work was done on a JEOL 100 CX electron microscope operating at 100kV.

Mechanical results

Cyclic deformation behavior

Within the range of plastic strain amplitudes $\frac{\Delta\varepsilon p}{2}$ (0,08% up to 1,2%), the 6246 alloy showed a pronounced difference between the compressive peak stress and the tensile peak stress. It was observed that the former is always much larger than the latter under cyclic loading, whatever the test temperature. In Figure 5, the compressive peak stress and the tensile peak stress were plotted as a function of cumulative plastic strain **p** ($p = 2N\,\Delta\varepsilon p - \frac{\Delta\varepsilon p}{2}$), for $\frac{\Delta\varepsilon p}{2} = 0{,}7\%$. A complementary test was conducted in the contrary way of cycling (compression first then tension). It showed that this phenomenon was not changed. It suggested that the hysteresis loops asymmetry is due to a different behavior in tension and in compression, rather than to an effect of e first loading on the subsequent ones.

The main feature of the cyclic behavior is that softening occured along the fatigue-life. Figure 6 shows cyclic softening curves for the range of plastic strain amplitudes used : the peak stress amplitudes (an average of the compresssive and tensile peak stresses) were plotted versus the cumulative plastic strains. It appears that softening is steep at the early stages of fatigue testing, then it slows down but does not cease until fracture. Such a continuous softening was observed at 20°C irrespective of the imposed plastic strain, in all the cases from 0,08% to 1,2%. It was also observed at higher temperatures, namely 300°C and 500°C, whatever the imposed plastic strain might be (Figure 7).

Internal hardening variables under plastic strain control

Two parameters are known to be of central significance for fatigue behavior of metals in relation to the dislocations behavior : the friction stress and the back stress. Quantitative values of the friction and the back stresses can be evaluated from the fatigue hysteresis loops by the Cottrell method **(1) (2)**. In order to investigate the cyclic softening of Ti 6246, analyses of the fatigue hysteresis loops were undertaken to extract the internal hardening variables **X** (corresponding to the back stress) and **R** (related to the friction stress). The connection between the yield stress **k**, the maximum stress and these internal variables in a fatigue loop is indicated in figure 8. The results are in fairly good agreement with those of a numerical procedure of identification of the internal variables (X(s) and R+k(s)) from experimental data **(3)**.

X is the kinematic hardening variable. It represents the translation of the yield surface in the stresses space. **R** is the isotropic hardening variable; it describes the expansion of the elastic domain **R+k**. For Ti 6246 at 20°C, **k** is about 1000 MPa.

In Figure 9, **X** and **R+k** were plotted together versus the number of cycles for a test under 0,7% plastic strain amplitude respectively, at 20°C. The measured values of **R+k** along the lifetime can be seen nearly constant, or slightly decreasing. Owing to the initial yield stress level, the isotropic hardening variable is

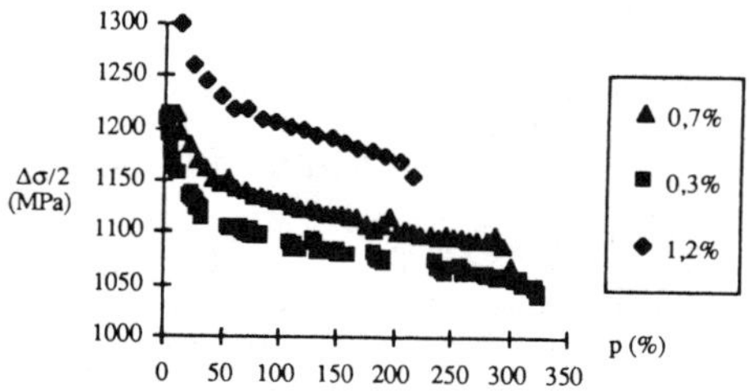

Figure 6
Cyclic softening curves at 20°C.

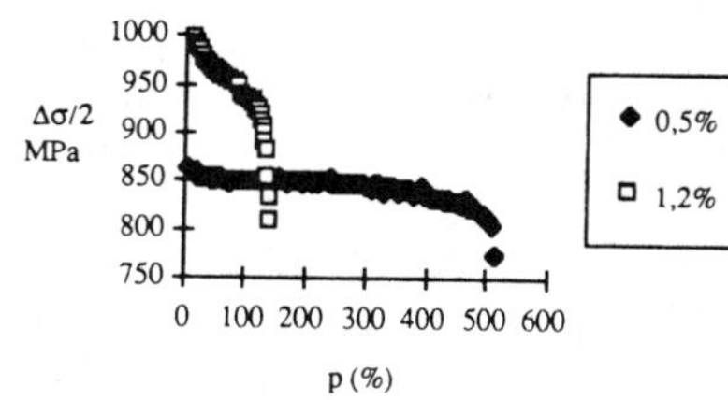

Figure 7
Cyclic softening curves at 500°C.

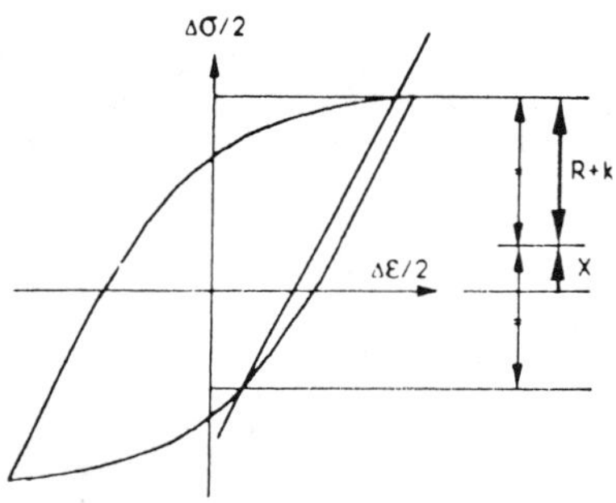

Figure 8
Connection between **k**, **X** and **R**.

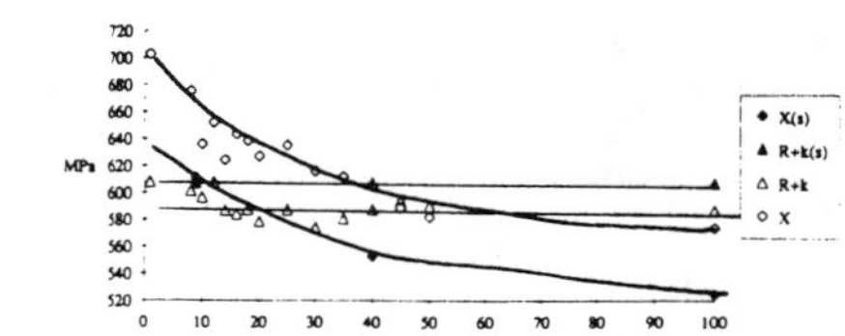

Figure 9
20°C : decrease of the kinematic component X.

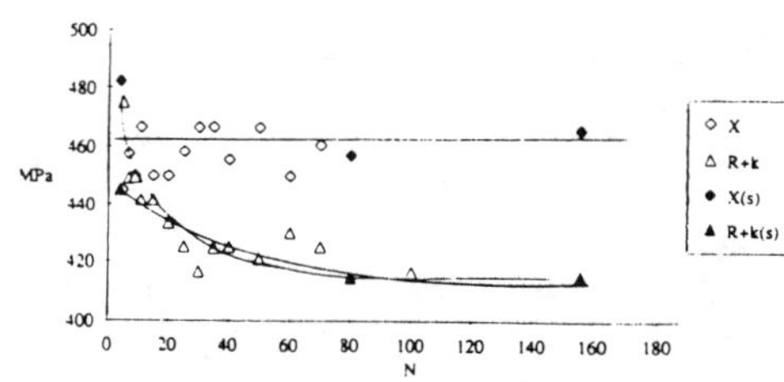

Figure 10
500°C : decrease of the isotropic component.

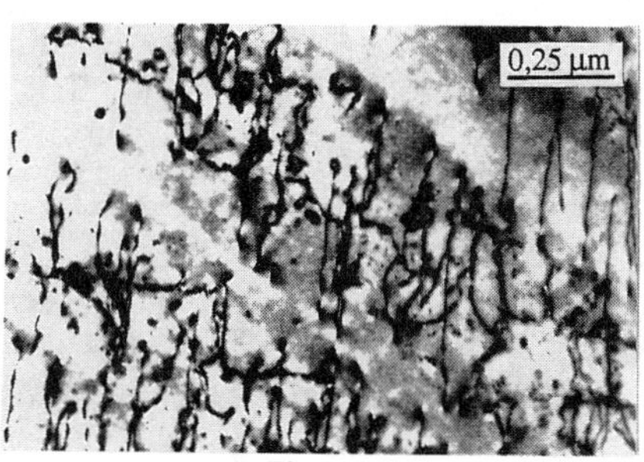

Figure 11
Distribution of the dislocations in the α phase after cyclic straining at 20°C (slip bands and loops).

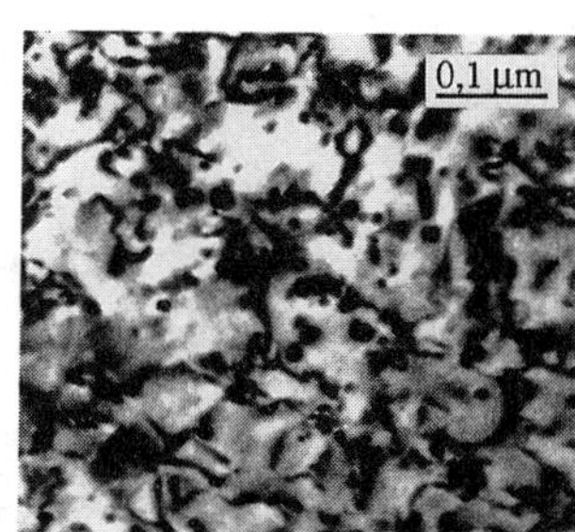

Figure 12
Random arrangement of dislocations in the α phase after cyclic straining at 500°C.

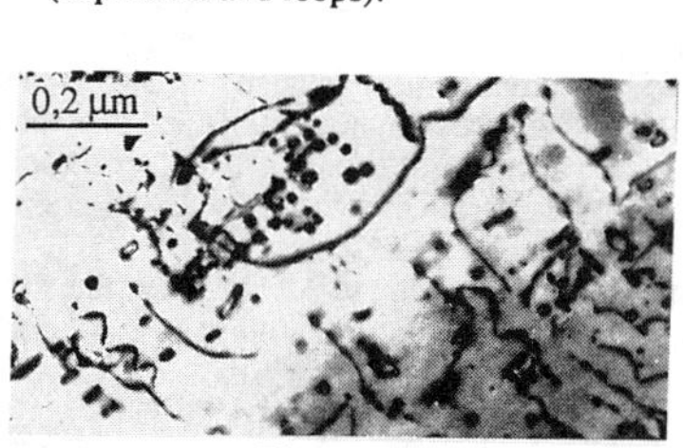

Figure 13
Elongated loops, helicoïds and jogs on screw dislocations.

negative. In other respects, the internal stress **X** is seen to decrease quickly at the early lifetime, and then more slowly. The evolution of the kinematic hardening variable **X** with the number of cycles is approximately similar to that of the average peak stress. Therefore, from these results, it seems fair to draw the conclusion that at 20°C, the major part of the cyclic softening can be described by the variations of **X**.

The analysis of hysteresis loops was also performed to investigate the high temperature cyclic softening. Figure 10 illustrates the values of **X** and **R+k** derived from a test at 500°C and under 0,7% plastic strain amplitude. The most striking observation is that, in these temperature conditions, the evolution of the isotropic variable coincides with that of the average peak stress, whereas the kinematic hardening variable keeps nearly constant. The initial yield stress at this temperature is about 700 MPa so that the **R** values are also negative in this context.

In terms of these results, the structural origins of the cyclic softening appear to be different depending on the test temperature.The cyclic softening is percieved to be a consequence of the decrease of **X** at 20°C and of **R** at 500°C.

Structural observations at the end of lifetime

The distributions of glide dislocations in the α phase, in specimens which have been cyclically deformed until failure, were studied by TEM on thin foils normal to the stress axis.

In the 20°C deformed material, inhomogeneous dislocation distribution was observed. The dislocations are arranged in localized planar slip bands. Only one slip system seems to be operating. Analyses of the Burgers vectors indicate that the arrays mainly consist of $\vec{a}$- type dislocations ($\frac{1}{3}$ <11$\bar{2}$0>), in accordance to the monotonic deformation modes in this alloy **(4)**. Correspondingly, prism planes {1$\bar{1}$00} appear to be the preferred glide planes. The dislocations are not so straight as in monotonically deformed samples, nevertheless they are still mainly in screw character. The major difference with monotonic deformation structures is the extensive amount of debris, especially dislocation loops. A thin foil micrograph of the typical configurations of dislocation is shown in Figure 11.

The deformation modes were also studied for Ti 6246 cylically deformed at high temperatures namely 300°C and 500°C. The deformation structures induced in these conditions differ from that in samples deformed at 20°C. Observations showed that slip becomes random as the temperature increases. At 500°C, there is no more planar slip, and the dislocations appear to be homogeneously distributed as shown in Figure 12. Numerous elongated loops can also be seen, as well as helicoidal dislocations. These particular configurations are the sign of cross slip mechanisms (Figure 13). Indeed, the motion of the dislocations out of their primary glide plane has been confirmed in **(5)**. As Naka mentioned **(6)**, the cross-slip mechanisms of $\vec{a}$- type dislocations seem to occur from prism planes into the corresponding pyramidal planes.

The deformation modes can be concluded to be strongly temperature dependent. As a matter of fact, localized planar slip in 20°C deformed samples demonstrates that cross-slip is inhibited at room temperature. On the contrary, at 500°C homogeneous distributions of dislocations, and typical configurations indicate that cross-slip is operating at elevated temperatures. This transition in the induced deformation structures from localized arrays to random arrangements with increasing temperature, is in good agreement with the different origins of the cyclic softening.

Discussion

Physical basis for X and R

The physical origins of the deformation behavior of alloys, interpreted in terms of friction and back stresses, are not well known yet, although several attempts were made to investigate this topic **(7)**. Generally, the friction stress (or isotropic strengthening) is attributed to interactions between dislocations and short-range obstacles by which their movement is resisted. Among these, solute atoms or short-range order precipitates can be a source of isotropic hardening. But forest dislocations cannot be eliminated from consideration either.

Moreover, in alloys, back stress hardening (or kinematic hardening) is ascribed to long-range obstacles such as large incoherent precipitates, interfaces, or typical dislocation substructures formed during deformation, including dislocation pile-ups or cells **(8)**.

Kinematic softening at room temperature

At 20°C, the mechanical cyclic behavior of the 6246 alloy mainly proceeds from the decrease of **X** with the accumulation of cycles. Corresponding microstructural observations clearly show dislocation pile-ups. When this type of arrays is not produced (at 500°C), then **X** remains constant. The dominant origin for the back stress decrease can be unambiguously concluded to be associated with the pile-ups arrangements. Furthermore, the slip multiplicity can also be suggested to contribute to the cyclic softening **(9)**. The multislip dislocation structure is generated at the very early stages of the cyclic loading, as confirmed by observations of the dislocation structures generated after half a cycle and after one cycle. After the first half-cycle, the α phase exhibits only one system of slip bands (on prism planes) (Figure 14). The subsequent compressive loading activates a secondary pyramidal glide (Figure 15), so that after the first tension-compression cycle, two slip systems are activated.

Saleh and Margolin **(10)** attributed the Bauschinger effect and high back stresses in two-ductile-phase alloys to interface effects. The nonuniformity of the strains make direct estimation of the local stresses or strains by laws of mixture inadequate. The FEM offers a viable tool for studying the deformation of such alloys. This procedure was used to evaluate the stress incompatibilities in the vicinity of the α/ß interfaces as a function of the volume fraction of deformed α phase (or slip bands) under strain control. The individual stress-strain curves for α, ß and the slip bands were approximated from Ankem and Margolin's study on α/ß Ti-Mn alloys **(11)** (Figure 16). The results of these calculations of a two dimensional (plane stress) elasto-plastic problem are shown on Figure 17. FEM modeling predicts a slight decrease of the external stress (σ_∞) as the volume fraction of deformed α phase (fv) increases. It also results from these calculations that the internal stress (the Von Mises stress gap ($\Delta\sigma$) at the interface between a slip band and the neighbour ß phase) decreases markedly while the volume fraction of deformed α phase increases, for a same amount of total deformation (4 %).

Another microstructural process can also be responsible for the decrease of the internal stress. Soon after they are generated in the well-oriented α grains, the dislocation pile up against the α/ß interfaces in such a manner as to cause an appreciable stress field ahead of the pile-up. This stress is added to the external applied stress and is expected to promote plastic flow in the neighbouring α phase, providing that ß remains quasi-elastic. Correspondingly, the back stress acting on the dislocations in the pile-up should necessarily be relieved, on the basis that it is partly dissipated for the spread of the plastic deformation.

Isotropic softening at 500°C

From the mechanical results, it is clear that **R** variations are responsible for the cyclic softening at 500°C. To understand the origins of these variations, one has to investigate first the short-range interactions in relation with the test temperature.

Whatever the test temperature may be, it was shown that the **R** values are negative, which suggest that a second-phase shearing mechanism is operating. The short-range obstacles which are expected to be sheared by the glide dislocations are the Ti_3Al precipitates scattered in α. These precipitates have an ordered hexagonal structure and it was shown that they are deformed by the glide of $\vec{a}$- type dislocations, particularly on prism planes **(12)**. Besides, numerous $\vec{a}$- type dislocation pairs were observed in α after cyclic deformation (figure 18), which confirm the shearing process. This is also in good agreement with the localization of slip at 20°C. As a result, the flow stress is lowered.

Microstructural investigations indicate that at least two slip systems are activated under cyclic loading. At 20°C, the two intersecting slip system necessarily strongly interact, owing to the planarity of slip; the interactions lead to the formation of small dislocation loops (Figure 11). Consequently, the associated flow stress component should increase. It seems that the negative contribution of the Ti_3Al shearing mechanism balancing the hardening caused by the forest dislocations interactions, the isotropic variable remains constant along the life-time, at room temperature.

On the other hand, at 500°C, the incidence of cross-slip was demonstrated. Therefore, the latent hardening contribution is no more linked to the effects of interactions but to the occurence of cross-slip, and is weaker. It was shown that the strain-hardening decreases when the moving dislocations can cross-slip **(13)**. Then, one can conclude that the role of the shearing process is decisive and actually determines the continuous decrease of the flow stress.

Conclusion

1. 6246 exhibited cyclic softening irrespective of the test temperature.

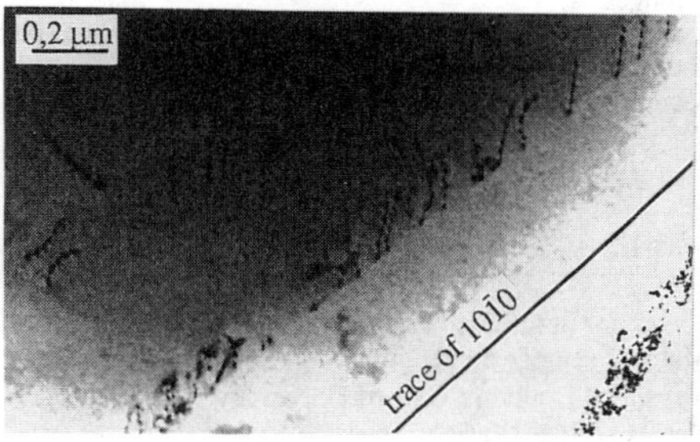

Figure 14
Dislocation arrays after half a cycle.

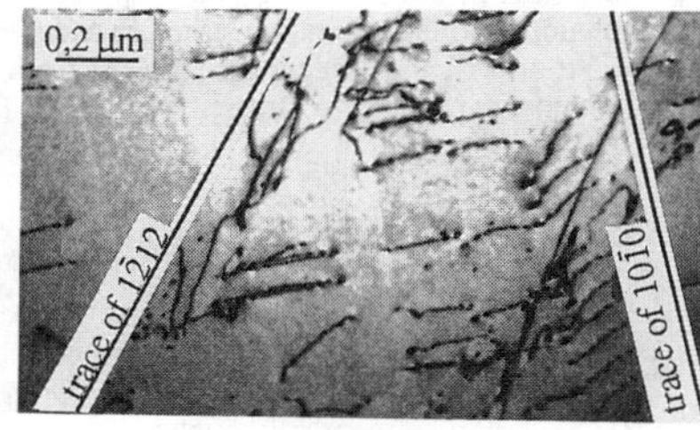

Figure 15
Dislocation arrays after one cycle.

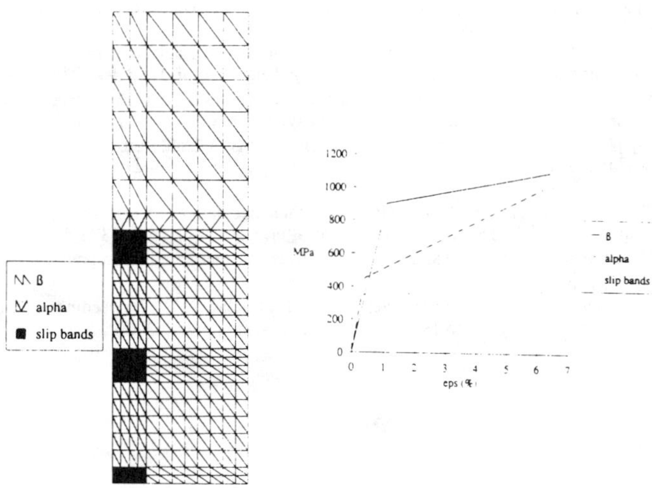

Figure 16
Mesh of triangular plate elements and stress-strain curves of α, ß and the slip bands used for the FEM calculations.

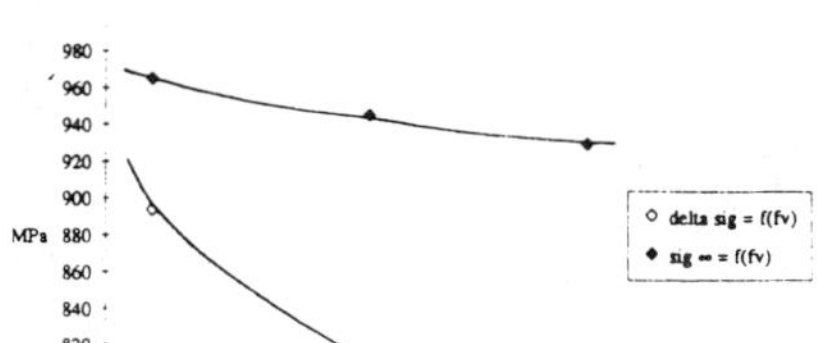

Figure 17
Calculated external stress (sig ∞) and difference of Von Mises stress (delta sig) at the interface between a slip band and the ß matrix as a function of the volume fraction of slip band (fv)

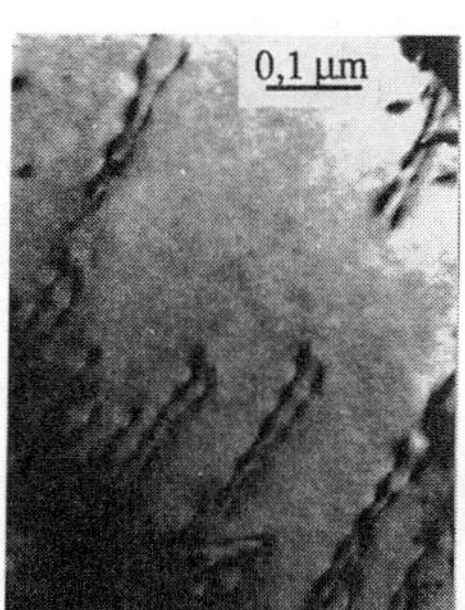

Figure 18
Dislocation pairs in the α phase .

2. Analyses of the stress-strain hysteresis loops illustrates that the softening is a consequence of different microstructural features in the different temperature ranges :
3. At 20°C, the softening mainly proceeds from the dominant role of long-range internal stress field, in relation with the two-phase structure and the slip localization in α.
4. At 500°C, the softening originates from the occurence of precipitate shearing mechanism (short-range component of the flow stress).

References

(1) D. Kuhlmann-Wilsdorf, C. Laird, "Dislocation behavior in fatigue : II. Friction stress and back stress as inferred from an analysis of hysteresis loops", Material science and engineering, 37 (1979), 111-120.
(2) A.H. Cottrell, Dislocations and plastic flow in crystals (London : Oxford University Press, 1953).
(3) J.Lemaitre, J.L. Chaboche, Mécanique des matériaux solides (Paris : Dunod, 1988).
(4) A.S. Béranger, M.Clavel, "Microstructure et ténacité de l'alliage de titane 6246" Mémoires et études scientifiques de la revue de métallurgie, march (1989), 145-152.
(5) A.S. Béranger, "Etude de l'alliage de titane 6246 : ténacité et comportement en fatigue oligocyclique en relation avec la microstructure" (thesis, University of technology of Compiègne, 1991).
(6) S. Naka, "Etude des mécanismes de déformation plastique à basse température de monocristaux de titane α" (thesis, Orsay, 1984).
(7) S.I. Hong, C.Laird, "Cyclic deformation behavior of Cu-16%at single crystals. Part III : Friction stress and back stress behavior" Material science and engineering, 128 (1990), 155-169.
(8) H. Mughrabi, "Internal stresses in wall and cell structures" Acta mettalurgica, 31 (1983), 1367-1379.
(9) S. Horibe, C. Laird, "Cyclic stress-strain behavior at a high strain amplitude of Al-Cu crystals containing partially coherent precipitates" Material science and engineering, 72 (1985), 149-153.
(10) Y. Saleh, H.Margolin, "Bauschinger effect during cyclic straining of two ductile phase alloys" Acta metallurgica, 27 (1979), 535-544.
(11) S. Ankem, H. Margolin, "Finite element method calculations of stress-strain behavior of alpha-beta Ti-Mn aloys : part I. Stress-strain relations" Metallurgical transactions, 13 (1982), 595-601.
(12) R.A. Varin, M.B. Winnicka, "Plasticity of structural intermetallic compounds" Material science and engineering, 137 (1991), 9-103.
(13) Z. Trojanova et al. "Strength of metals and alloys (ICSMA 7)", Proceedings of the 7th international conference, august the 12-16th 1985, (Montreal : Pergamon press, 1985).

FATIGUE UNDER TENSION/TORSION LOADING IN IMI685

W.J.Evans and M.R.Bache

I.R.C. in Materials for High Performance Applications
University College of Swansea, U.K., SA2 8PP

Abstract

Tension and torsion fatigue tests on the titanium alloy IMI685 are reported. Particular attention is paid to the effects of microstructure and the imposition of dwell periods at peak stress on fatigue life. Fractographic examinations reveal that two modes of failure occur in this alloy according to the magnitude of cyclic stress, namely a high strain/ductile mode at stresses above yield and quasi-cleavage facetting at lower stress levels. The implications of dwell sensitivity with respect to engineering design are discussed.

Introduction

The near alpha titanium alloy, IMI685 (Ti 6Al 5Zr 0.5Mo 0.2Si), is used in the gas turbine industry for compressor discs and blades. In these applications, it often experiences biaxial fatigue conditions. Over a number of years the alloy has been widely studied, not least because of an apparent sensitivity to dwell periods at peak stress. Little of this work, however, has explored the response of the alloy to biaxial loading. The present programme aimed to address this limitation. In particular, it set out to assess the response of the alloy to tension and torsion loading states and to establish the role of dwell periods under both conditions. It was anticipated that the information generated would not only provide biaxial failure criteria for the design engineer but also permit a clearer understanding of the mechanisms controlling cyclic and dwell fatigue in this important aerospace alloy.

Experimental Methods

The experimental evaluation was carried out on the standard commercial alloy in two heat treated conditions. A basketweave microstucture produced by rapid cooling from above the beta transus was representitive of the condition normally recommended for major rotating components. An aligned morphology, obtained by furnace cooling through the transus, was

Titanium '92
Science and Technology
Edited by F.H. Froes and I. Caplan
The Minerals, Metals & Materials Society, 1993

typical of the structure that can occur in thick sections. The latter tends to confer inferior fatigue properties and to display the greater dwell sensitivity. In both conditions the prior beta grain size was about 0.3 mm. Examples of these microstructures are included in Figure 1.

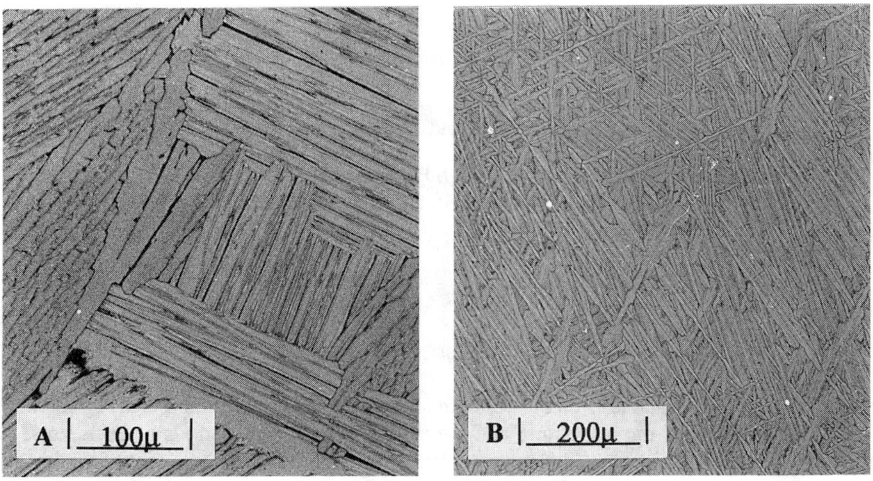

Figure 1 - Aligned (a) and basketweave (b) microstructures in IMI685.

A tension / torsion testing machine of ± 50KN axial load and ± 400Nm torque capacity was employed. The majority of the mechanical testing was carried out on tubular specimens with a wall thickness of 1.5 mm. The specimen design is illustrated in Figure 2. A small number of tests involved solid specimens in order to assess the extent to which the thin wall modified behaviour under tensile load. All testing was carried out at ambient temperature under load or torque control. The cyclic and dwell loading waveforms each employed a 2 second rise and fall time between minimum and maximum load. During cyclic tests a one second hold at peak stress facilitated data aquisition. For dwell tests the maximum load or torque was held constant for 2 minutes. In all tests a load ratio of R=0.1 was used. Schematic representations of the cyclic and dwell loading waveforms are included in Figure 3.

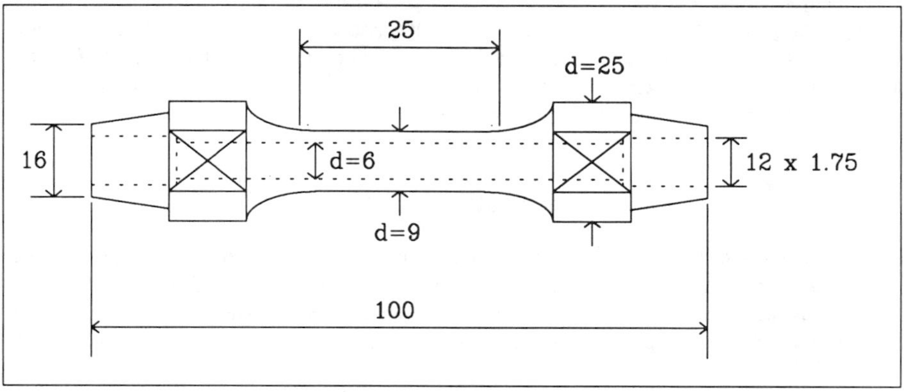

Figure 2 - Tension/torsion fatigue specimen design (dimensions in mm).

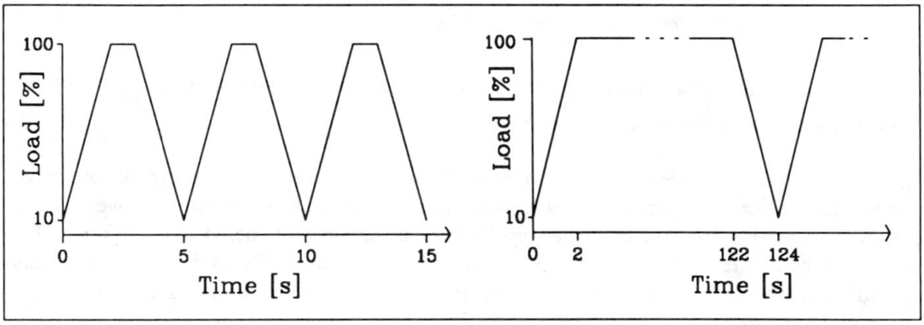

Figure 3 - Cyclic (left) and dwell loading waveforms.

In each test, the variation of extension or twist with cycles was monitored using a biaxial extensometer based on a design reported in [1]. For some specimens the development of surface features was followed by means of a replica technique. Detailed fractographic examinations were made for all specimens using optical and scanning electron microscopy.

Results

For tension fatigue tests, the definition of failure was taken as the complete separation of the specimen gauge section. However, such a definition under torsion was less specific as two distinct modes of fracture were noted. At high stresses (above yield) a near circumferential shear crack developed; at low stress helical fractures were observed, [2]. In the latter case, the growth of the fatigue crack coincided with a marked period of tertiary strain accumulation. On this basis, it has been argued [2] that a life to tertiary strain provides a consistent failure criterion for all torsion tests. This is the criterion used to construct the S-N curves below.

The effects of cyclic and dwell loading on fatigue life under tension and torsion are summarised by the S-N curves in Figures 4 to 8. The criteria for the calculation of effective stresses and strains in tension and torsion fatigue testing have been addressed previously [2]. In brief, for tension the effective stress and strain are simply calculated from the relative axial components:

$$\sigma_{eff} = \sigma_1 \qquad \text{and} \qquad \varepsilon_{eff} = \varepsilon_1$$

For torsion, it has been argued [2,3] that for tubular specimens with a relatively thin wall shear stress is constant across the section. This results in a modified Von Mises solution for effective stress of the form:

$$\sigma_{eff} = \sqrt{3}\,\frac{3T}{2\pi\,(r_o^3 - r_i^3)}$$

(T = torque, r_o = outer tube radius, r_i = inner radius)

Because the shear stress is no longer considered to reach a maximum on the outer surface, an average value of specimen radius is used in the calculation of effective strain:

$$\varepsilon_{eff} = \frac{1}{\sqrt{3}} \frac{(r_o + r_i)}{2} \frac{\theta}{l}$$

where θ is the angle of twist measured in radians. In the case of solid specimens r_i reduces to zero in the above equations.

Figure 4 illustrates the results from cyclic tension and torsion tests on the basketweave microstructure. The data compare favourably using the stress criterion outlined above. A least squares best fit line for this combined cyclic data is drawn and included in Figure 5 for comparison with data from dwell tests in the same microstructure. Dwell during torsion tests has had minimal effect on the measured lives. However, life reductions are significant in the case of dwell tension.

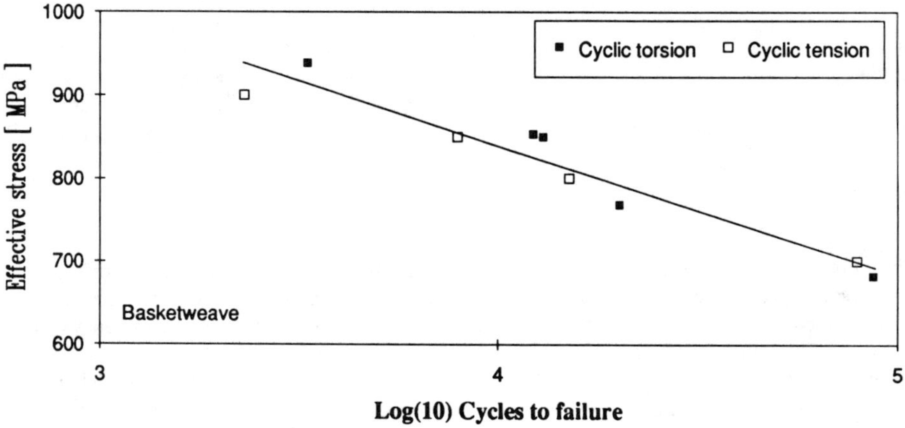

Figure 4 - Cyclic torsion and tension fatigue lives in the basketweave microstructure.

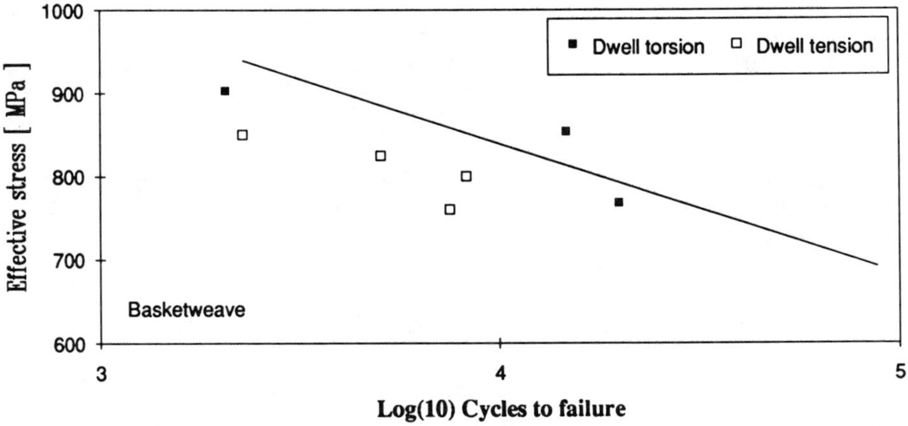

Figure 5 - Dwell torsion and tension for basketweave compared to the best fit line for cyclic basketweave results.

Figure 6 compares cyclic tension and torsion data on the aligned microstructure with the best fit line from the cyclic basketweave data. Two observations can be made. Firstly, under either mode of loading the aligned material has a weaker response. Secondly, the individual tension and torsion data sets do not compare as favourably as they did in the basketweave case. This can be attributed to the thin wall of the tubular specimen in relation to the coarse aligned microstructure. Plotting data obtained under cyclic tension on solid section aligned specimens together with the tubular torsion results provides a more favourable comparison, Figure 7. The best line fit from these data is then used in Figure 8 to illustrate the effect of dwell on the aligned material. Once again no effect is seen on torsion data but a marked reduction in life is noted under dwell tension. This reduction exceeds that noted for basketweave material.

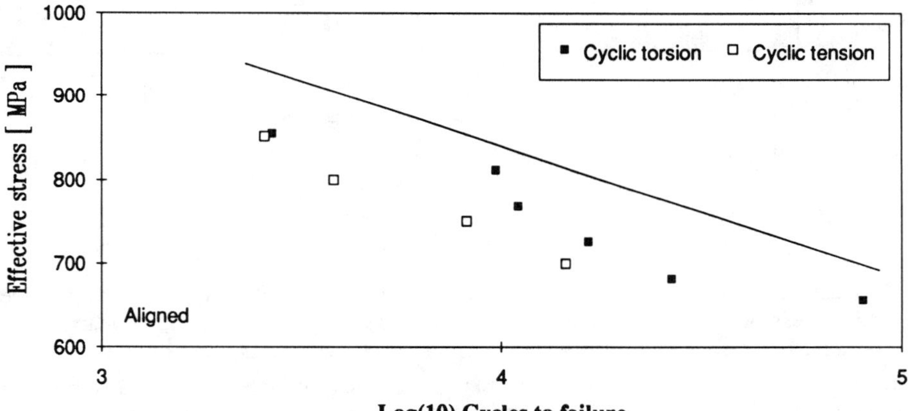

Figure 6 - Cyclic torsion and tension data for aligned tests compared to the best fit line for cyclic basketweave results.

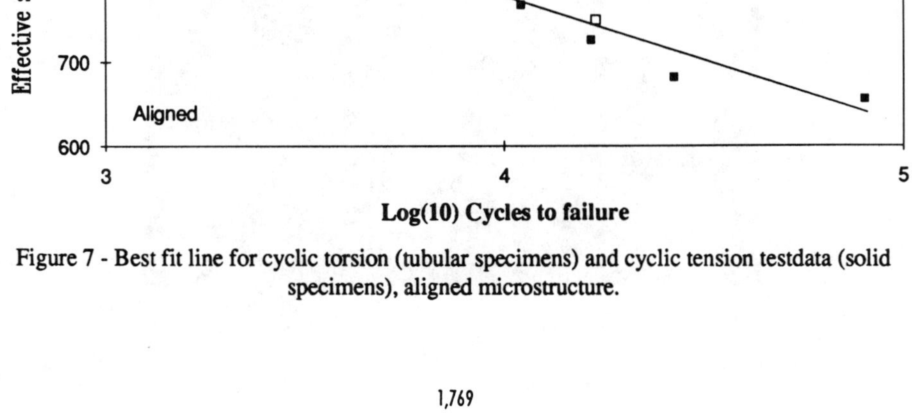

Figure 7 - Best fit line for cyclic torsion (tubular specimens) and cyclic tension testdata (solid specimens), aligned microstructure.

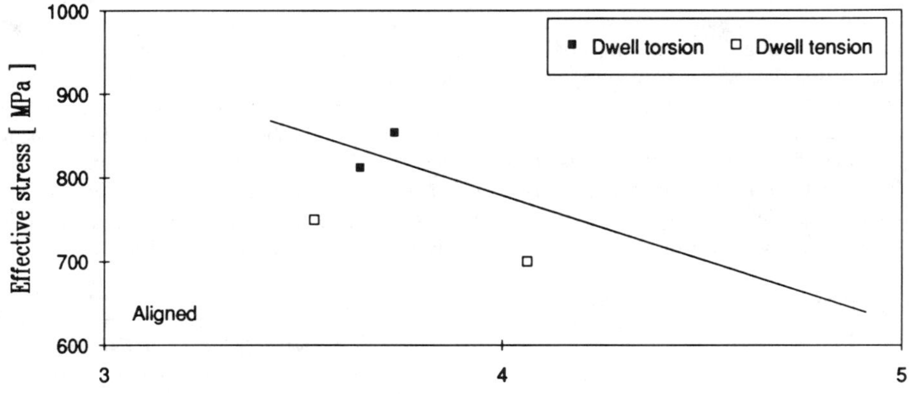

Figure 8 - Dwell torsion and tension results compared to the best fit line for cyclic aligned tests.

Fractography

Both tension and torsion testing modes demonstrate two regimes of fracture depending on the applied stress. At high stresses/low lives the failures are strain dominated. In tension this was reflected by extensive strain accumulation and a characteristic ductile form of fracture. Shear across planes at $\pm 45^{o}$ to the tensile axis produced "rooftop" features as illustrated in Figure 9 and reported elsewhere [2]. Specimen gauge surfaces often illustrated marked plastic deformation and secondary cracking. In torsion, shear on planes perpendicular and parallel to the longitudinal axis of the specimen was noted. The main fracture surface was usually obscurred due to inter-surface abrasion.

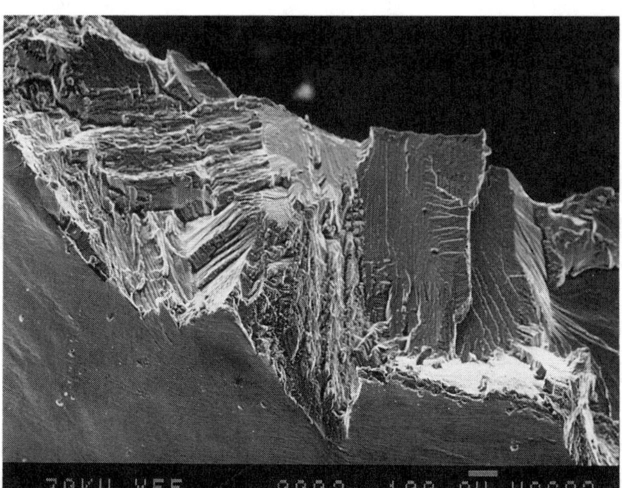

Figure 9 - "Rooftop" fracture morphology common in tension tests under high stress.

For lower stresses/longer lives, tensile tests were associated with the initiation of cracks from quasi-cleavage facets. These facets often traversed the complete wall thickness and possessed a fish eye appearance, Figure 10. Their orientation was perpendicular to the tensile stress axis. For torsion at relatively low stresses a helical mode of fracture was found. Facetting was still common but here they would lie at 45° to the longitudinal axis of the specimen and thus perpendicular to the maximum principal stress. There is evidence that the helical cracks initiated as Mode II cracks but then switched mode and propagated under the influence of the principal tensile stress.

Figure 10 - Quasi-cleavage facet formed in aligned material under cyclic tension, 750MPa.

Discussion

It has been established that tension and torsion cyclic fatigue data for the titanium alloy IMI685 correlate by using a modified form of the Von Mises stress criterion. This modified criterion is based on two assumptions. The first is that shear stress is constant across the thin walled section and the second is the use of an average radius for the calculation of effective strain. The success of the correlation supports the adoption of these assumptions. Their effectiveness must be due to the fact that both aligned and basketweave material experiences significant strain accumulation during all the tests reported and that there are a limited number of grains across the thin wall section.

The torsion tests are not sensitive to dwell fatigue. The tension tests show substantial reductions in life particularly in the aligned microstructure. The strain-time measurements and fracture surface studies suggest that cyclic and dwell fatigue in this alloy involves two competing processes. At high stress above yield, strain accumulation is extensive and shear failures dominate. For stresses below yield, quasi-cleavage facet formation is more evident. These facets are orientated perpendicular to the tensile stress. They have been shown to be more extensive under dwell conditions [4] and have been considered responsible for the dwell sensitivity.

The importance of each mode of failure must be related to the ratio of shear stress to tensile stress under each test condition. In torsion $\sigma_1=\tau$ but in tension $\sigma_1=2\tau$. On this basis, the tension test should be more favourable for facet formation. The fact that the maximum principal stress is less significant in torsion could account for the apparent lack of a dwell effect in those tests.

One other aspect of facet formation relates to the role of local constraint. It has been suggested that a biaxial or triaxial state of stress is a necessary condition. It is evident from the present work that extensive facetting occurs in thin sections. This observation appears to throw the constraint arguments into doubt. An alternative explanation has been offered [5] which attributes their formation to stress redistribution from weak to strong regions under the action of the time dependent deformation that is so evident in this material.

Conclusions

1. Room temperature tension and torsion cyclic fatigue data for the titanium alloy IMI685 compare favourably when using a modified solution of the Von Mises effective stress criterion.

2. Good correlation between tension and torsion data is achieved by adopting a failure criterion for the latter based on "life to tertiary strain".

3. Dwell sensitivity has been noted under tensile loading, particularly in the aligned microstructure. An equivilent effect was not seen in torsion tests.

4. Failures in this alloy are closely related to the magnitude of applied stress. At high stress/low lives shear deformation is dominant. At lower stress levels features characteristic of tensile stress are produced.

References

1. Brown, M.W. and Miller, K.J. "A biaxial fatigue machine for elevated temperature testing", Journal of Testing and Evaluation, **9** (4), pp202-208, July 1981.

2. Bache, M.R. and Evans, W.J., "Biaxial fatigue testing of a near alpha titanium alloy", to be published, Int. J. Fatigue.

3. Wang, C.H. and Miller, K.J., "The effect of mean shear stress on torsional fatigue behaviour", Fat. Fract. Eng. Mat. and Structures, vol 14, No. 2/3, 1991.

4. Evans, W.J. and Gostelow, C.R., "The effect of hold time on the fatigue properties of a β-processed titanium alloy", Met. Trans. A, vol 13A, pp1729, 1982.

5. Evans, W.J., "The influence of microstructure on dwell sensitive fatigue in a near alpha titanium alloy", Scripta Met., vol 21, No.4, pp469, 1987.

CREEP BEHAVIOR AND THERMAL STABILITY OF A NOVEL HIGH TEMPERATURE TITANIUM ALLOY

Han Chuanxi and Zhang Renqi

Northwest Institute for Non–ferrous Metal Research
P.O.Box 71, Baoji, Shaanxi 721014, P.R.China

ABSTRACT

The present paper introduces the creep behavior of a novel high temperature titanium alloy (Ti–Al–Sn–Zr–Nb–Mo–Si system) and the changes of its post–creep tensile properties. The influential factors on thermal stability were also discussed. From the room temperature properties of this alloy exposed at 550℃ –300MPa and the relations among plastic deformation, stress and temperature, which were measured from creep tests with different conditions (550℃ –300MPa–100~ 1000hrs), 550℃ –200~ 450MPa,400~ 650℃ –300MPa), it was indicated that the material should be used below 600℃ at 300MPa and the maximum working stress no more than 400MPa at 550℃. The primary factors which decrease thermal stability of the alloy have been determined by microstructure analysis and oxide surface measurement. All above provide the important foundation for parts designing of gas turbine.

INTRODUCTION

Ti–53311S is a near–α high temperature titanium alloy (Ti–5.4Al–3.4Sn–2.7Zr–0.9Mo–1.0Nb–0.25Si). The alloy can be used to manufacture vanes, compressor disc of air engine and astronautic structural details due to its good synthetic properties at 550℃.

In order to provide necessary design data for the material application departments, it is necessary to study creep behavior and thermal stability of the alloy. In this paper, the relations between creep strain of the alloy and time ,stress, temperature, as well as the tensile properties of the alloy exposed at 550℃ –300MPa for a long time were studied. Influences of oxygen–enriched brittle layer of surface and the formation of the Ti_3X phase on thermal stability are also discussed.

EXPERIMENTAL

Creep tests were carried out in a RD–2–3 creep test machine. Standard creep specimens of gauge section 5 mm in diameter were machined from hot rolled 14 mm diameter bars. The heat treatment was : 1050℃ / 0.5h / AC+580℃ / 4h / AC.Creep tests at 550℃ and variant stresses or at 300MPa and variant temperatures were proceeded with the single specimen method proposed byJ.E.Dorn [1]. INSTRON–1185 test machine was employed to measure tensile properties of the samples for certain stress exposure. Thickness of the oxygen–enriched brittle layer was examined by PSEM–500X scanning electron microscope. Microstructure and selected area diffraction analysis were carried out in JEM–200EX transmission electron microscope. Variances of β phase content were also determined by D–S X–ray diffractometer.

RESULTS AND DISCUSSION

Titanium '92
Science and Technology
Edited by F.H. Froes and I. Caplan
The Minerals, Metals & Materials Society, 1993

Creep properties of the alloy

Fig.1 shows creep cruves of the Ti−53311S alloy exposed at 550℃ −300MPa for 100, 200, 400, 600 and 1000hrs, respectively. All creep curves show two creep stages, which named decelerative creep and steady creep, respectively. Plastic deformations (ε%) increase with increase in creep time, which are 0.129%, 0.214%, 0.351%, 0.430% and 0.683% respectively. All Creep curves parallel proximately at steady creep stage, steady creep rate is $(8.1\sim 8.8)\times 10^{-6}$ / h. Average creep rates within the ranges of 50~100hrs, 50~200hrs, 50~400hrs, 50~600hrs and 50~1000hrs, however, are 25.8, 14.8, 10.2, 7.7, and 7.2×10^{-6} / h respectively. That is , the average creep rate decrease with prolonging in creep time.

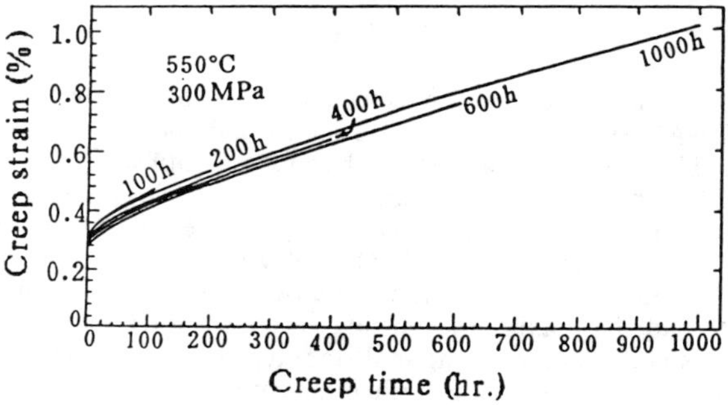

Fig.1 Creep curves of Ti−53311S alloy at 550℃−300MPa.

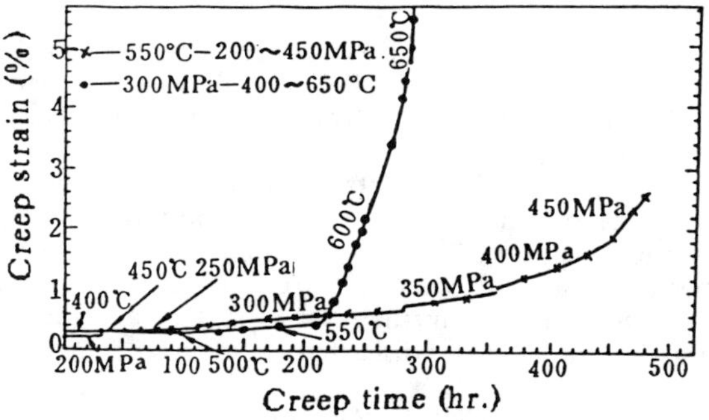

Fig.2 Plastic deformation vs. stress and temperature after creep exposure at constant temperature 550℃ or at constant stress 300MPa.

Relations among plastic deformation, stress and temperature at 550℃ and different stress or at 400~650℃ and constant stress are shown in Fig.2. Creep characteristics of the Ti−53311S alloy can be found from Fig.2 as follows : (1) Creep curve is nearly straight line at 550℃ −250MPa whereas accelerative creep stage occurs apparently at 400MPa. The creep curve accelerates rapidly when creep stress reaches 450MPa. (2) Creep curve at 600℃ and 300MPa belongs to accelerative creep stage, specimen occurs necking at 650℃ . (3) Under conditions of constant temperature with varying creep stress, the steady creep rate becomes double as creep stress increases per 50MPa, but the effects of stress on creep behavior are not more sensitive than temperature does. It is indicated that the Ti−53311S alloy should be used below 600℃ at 300MPa based on its creep properties.

The influences of temperature, stress and time on creep curves are different thoroughly, as shown in Fig. 1 and Fig.2, which means that the creep behavior of the

Ti–53311S alloy is controlled by distinct creep mechanisms under different creep conditions. The fact that steady creep rate rises with increases in stress at 550℃ indicates that dislocation movement obtains enough driving force with increment in stress, and thus it is easy to overcome obstacle and produce creep. Based on the results of the transformed microstructure, creep cruve at 550℃–300MPa for 1000 hrs is still at steady creep stage.

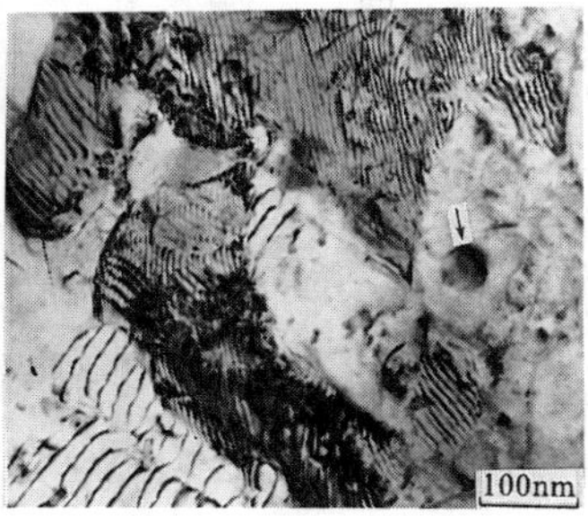

Fig.3 Ti_3Al particles resist dislocation motion and grain boundary sliding after creep exposure at 550℃–300MPa–1000hrs.

α / β interface has been strengthened by dot–shape α phase decomposed from β phase and " transitive phase " formed at α / β interface, therefore the interface sliding is blocked. Silicide precipitated from α / β interface further strengthens α / β interface after creep exposure for about 200hrs. Ti_3Al phase precipitated from grain boundary and α plates for 600 hrs exposure hinders slipping or sliding deformation of grain, grain boundary and α plates, leads to deceleration of creep rate (Fig.3).

Creep strain is sensitive to temperature at 300MPa. Creep activation energy of the Ti–53311S alloy at 500～550℃ is 174.3kJ / mol, which is higher than mutual diffusion activation energy (104.5kJ / mol) of α steady element aluminium in titanium, lower than self diffusion activation energy (250.8kJ / mol) of titanium. It can be regarded that both of slipping and diffusion act on creep behavior. Creep at above 550℃ is high temperatre creep. Slip systems of grain boundary, clony boundary and α / β interface increase with rising in temperature. Cross slip is easy to be carried out, cavities form, grow up and accumulate. All above cause the alloy to soften quickly. Density of samples decresaes during creep procedure, it descends violently at accelerative creep stage particularly [2], which relates to formation and growth of cavities. The order of the position in which cavities appear are α / β interface, colony boundary, grain boundary and interface between coarse silicide particle and matrix.

Thermal stability of the alloy

The room temperature tensile properties of the samples exposed at 550℃–300MPa are summarized in Table 1. From Table 1 it is known that tensile strength varies slightly with prolonging in exposure time, that is , the tensile strength increase with prolonging in time from 100 hrs to 600 hrs. Plasticities δ and ψ decrease considerably after exposure for 200 hrs. The creep life is longer than 100 hrs when reduction of area decreases 50% is under consideration.

Surface oxidation layer

An oxygen–enriched brittle layer will be formed when titanium alloy is used at elevated temperature. Depth of the oxygen–enriched layer is related to exposure time and temperature. Table 2 shows thickness of the oxygen–enriched brittle layer and α layer (polluted layer) of creep specimens. From the data it is known that the thickness of the oxygen–enriched brittle layer and α layer increase with increase in creep time, but total oxidation speed and total polluted speed are just on the contrary. The addition of thickness and net oxidation speed per 100hrs vary along saddle shape with increase in creep time. The lowest point of addition of thichness of oxygen–enriched brittle layer per 100 hrs appears at 400 hrs, while that of α layer at 600 hrs. Both of them indicate that the Ti–53311S alloy possesses sound oxidation resistance.

Table 1 Room Temperature Tensile Properties of Post–Creep Spceimens at 550℃–300MPa

Specimen State	UTS σ_b, MPa	YS $\sigma_{0.2}$, MPa	EL δ,%	RA ψ, %	Hardness Hv, MPa	Remarks
Original State	1065	974	10.3	19.3	3590.8	After 1050℃ × 0.5h • AC +580℃ × 4h • AC Heat Treatment
Creep for 100hrs	1118	1103	7.0	12.0	3680.4	
Creep for 200hrs	1134	1091	2.9	5.9	3551.1	
Creep for 400hrs	1137	1081	1.9	4.8	3570.9	
Creep for 600hrs	1169	1152	/	/	3630.6	
Creep for 1000hrs	1042	/	/	/	3610.7	
Creep at 550℃ with Variant Stresses	/	/	/	/	3471.5	To Observe Oxygen–enriched layer without tension
Creep at 300MPa with variant Temperature	/	/	/	/	3411.8	

Table 2 Thickness of Oxygen–enriched Brittle Layer and α Layer of Creep Samples

Samples	Thickness of Oxygen –enriched Brittle Layer μm		Oxidation Speed, $\mu m / h \times 10^{-3}$		Thickness of α Layer, μm		Pollution Speed, $\times 10^{-2} \mu m / h$	
	Total Thickness	Added Thickness Per 100hrs	Total	Net	Total Thickness	Added Thickness per 100 hrs	Total	Net
550℃–300MPa –100hrs	1.5	1.50	15.00	15.0	9.0	9.0	9.00	9.0
550℃–300MPa –200hrs	2.4	0.90	12.00	9.0	13.0	4.0	6.50	4.0
550℃–300MPa –400hrs	3.1	0.35	7.75	3.5	18.6	2.8	4.65	2.8
550℃–300MPa –600hrs	4.0	0.45	6.67	4.5	22.0	1.7	3.67	1.7
550℃–300MPa –1000hrs	6.0	0.50	6.00	5.0	31.3	2.3	3.13	2.3

Variations of metastable β phase

β phase contents of initial samples and samples creeped at 550℃ −300MPa for 100hrs, 200 hrs, 400 hrs, 600 hrs and 1000 hrs were determined by X-ray diffractometer, which are 7.2%, 5.7%, 5.6%, 5.5%, 6.3% and 9.2% respectively. It indicates that β phase content decreases with prolongation in creep time, the maximum reduction amount is about 23%, β phase contents have been improved after samples were exposured for 600 hrs, even over that of initial state, which imply precipitation of a new β phase. Phase composition analysis of creep specimens verifies that three kinds of β phase are in existence after 600 hrs exposure : molybdenum−enriched β phase (~ 3% Mo), molybdenum−poor β phase (1.0 % Mo) and tin−enriched β phase (12% Sn).

Precipitation of silicide and α_2 phase (Ti_3Al)

Silicide separates out prior in α / β interface of samples creeped at 550℃ −300MPa for 200 hrs, as shown in Fig.4.

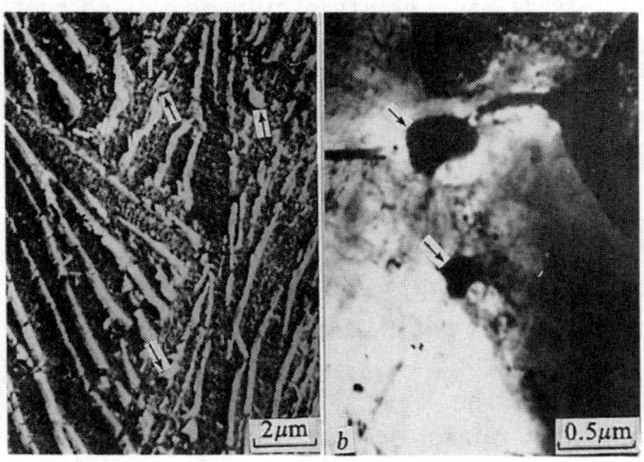

Fig.4 Precipitated silicide at α / β interface: (a) SEM−SEI; (b) TEM−BFI

Microanalysis showed that the composition of the silicide was $(Ti, Zr)_5Si_3$. Electron diffraction pattern of smaples treated at 550℃ −300MPa−600hrs shows that α_2 phase emerges from α phase. The direction correlation between α_2 phase and α phase is $[0001]\alpha_2 // [0001]\alpha$, as shown in Fig.5. Precipitation of silicide and $\alpha_2$phase improves room temperature strength of the alloy and descends plasticities, even causes decrease of thermal stability of the alloy.

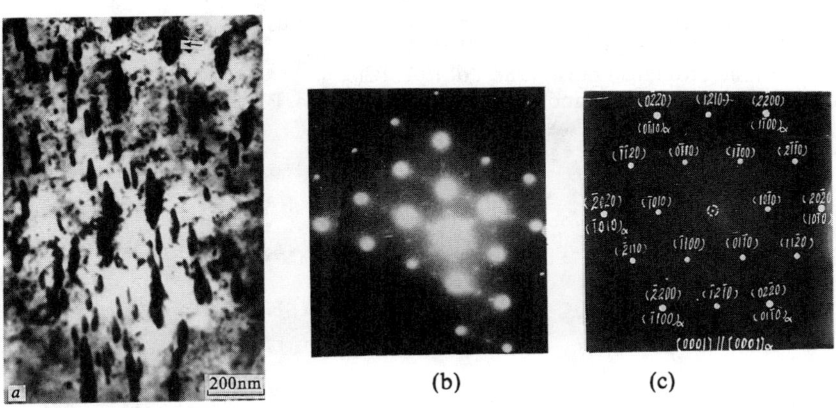

Fig.5. Structure of α_2 shown by TEM (a) BFI, morphology of α_2 phase; (b) Selected area diffraction pattern; (c) Indexation of diffraction pattern of α_2 phase.

Above–mentioned results demonstrate that decrease of room temperature tensile plasticities of the Ti–53311S alloy exposed for a long time is due to precipitation of α_2 phase, decomposition of metastable β phase, formation of surface oxygen–enriched layer and precipitation and growth of silicide, particularly due to Ti_3X phase and $(Ti,Zr)_5Si_3$ formed simultaneously and gathered at grain boundary or phase interface. It is observed by G. Scridher [3] that Ti_3Al phase distributated homogeneously in α matrix in IMI829 alloy after solution heat treatment and ageing, furthermore it had no correlation with dislocation, stacking fault or silicide. Tensile stress descends precipitation activation energy of Ti_3Al phase and accelerates decomposition of metastable β phase, precipitation and growth of silicide during heat exposure procedure. It also leads plasticities to go down obviously. D.Bonerjee [4] reported that S_2 type silicide might be formed in IMI829 alloy, its composition was situated between $(Ti,Zr)_5Si_3$ and $(Ti,Zr)_6Si_3$. Alloy plasticities could also worsen owing to accumulation of this kind of silicide at grain boundary and phase interface.

The oxygen–enriched layer formed on exposed sample surface makes it difficult for dislocations to move outside slip planes, thus many dislocations are within a few slip planes resulting in obstruction of dislocations. Stress concentration may be caused and fracture source can be formed once the obstruction reaches a certain degree. Consequently it leads plasticities to descent.

CONCLUSION

1. Ti–53311S alloy should be used below 600℃ at 300MPa or below 400MPa at 550℃, otherwise visible creep strain can be detected.

2. Ti–53311S alloy exposed at 550℃–300MPa for 600 hrs precipitates α_2 phase. α_2 phase improves creep properties meanwhile descends room temperature tensile plasticities.

3. Descent of room temperature plasticities of the Ti–53311S alloy exposed at 550℃ for a long time is due to precipitation of α_2 phase, decomposition of β phase, formation of surface oxygen–enriched layer and precipitation and growth of silicide. Stress accelerates the fluctuations above.

REFERENCES

1. J.E.Dorn, Cleveland Ohio, America Society for Metals 25
2. T.Nishimura, J.Jpn. Inst. Metals, 1975, 39(8), P825
3. G.Scridher, Metall. Trans., 1988 Vol 19A, P3025
4. D.Bonerjee, Ti Science and Technology, 1985, Vol 4, P151

DEFORMATION OF BETA TITANIUM SOLID SOLUTION

ALLOYS AT ELEVATED TEMPERATURES

Hiroshi OIKAWA, Yukio ISHIKAWA and Masahiro SEKI

Department of Materials Science
Faculty of Engineering
Tohoku University
Sendai 980, Japan

Abstract

Tensile test in an argon atmosphere was carried out on β-Ti-base binary solid solution alloys containing 1~7mol%Al or 5~30mol%V in the temperature range between 900 K and 1350 K and in the strain-rate range between $10^{-6}s^{-1}$ and $10^{-2}s^{-1}$. Deformation behavior can be classified into two types and the boundary condition expressed by the equation, $\sigma_u = 0.53\{(1+\nu)/(1-\nu)\pi\}\, e\, N(1-N)\, G$, where ν is Poisson's ratio, *e* the atom misfit factor, *N* the mole fraction of the solute, and *G* the shear modulus. In the low stress range, steady-state deformation behavior is of typical alloy-type and the relationship between stress σ and strain-rate $\dot{\varepsilon}$ can be represented by the equation, $\dot{\varepsilon} \approx \{D/N(1-N)e^2b^2\}(kT/Gb^3)(\sigma/G)^3$, where *D* is the diffusion coefficient of the solute element and *b* the Burgers vector.

Introduction

A large number of investigations of the mechanical behavior of titanium base alloys have been reported. However, the deformation characteristics of simple alloys at elevated temperatures have been investigated only to a limited extent, though these characteristics are important in the evaluation of materials for their hot-workability and/or heat resistivity.

A series of studies of high-temperature deformation behavior in Ti-base solid solutions has been made in the author's laboratory to establish a sound basis of knowledge of mechanical properties of Ti alloys. In this paper, some experimental results of the mechanical behavior obtained in two basic binary systems, β-Ti-Al and β-Ti-V, are reported and general deformation characteristics of bcc solid solutions of titanium are discussed, with a special reference to the steady-state deformation.

Experimental Procedure

Materials

Three series of materials were used and are listed in Table 1. In the first series the alloys contained 1~7mol%Al, which were double-arc-melted ingots. They were forged and hot-rolled to bars, from which rod-type specimens (the gage diameter = 4~7 mm) were machined.

Titanium '92
Science and Technology
Edited by F.H. Froes and I. Caplan
The Minerals, Metals & Materials Society, 1993

Table 1. Alloy composition and test conditions

Alloy designation	Main alloying content	Test conditions (Temp (K) - Strain rate ($10^{-4}s^{-1}$))
Ti- 1Al	1.0mol%Al	1250-(1.1~44), 1300-(0.22~56), 1350-(11)
Ti- 3Al	2.7mol%Al	1250-(0.2~44), 1300-(0.5~24), 1350-(2~11)
Ti- 5Al	5.6mol%Al	1250-(0.2~56), 1300-(2.2~11)
Ti- 7Al	6.9mol%Al	1300-(0.3~27), 1325-(1.1,11), 1350-(0.4~11)
Ti- 5V b	5.0mol%V	1164-(0.11~83)
Ti-10V b	10.1mol%V	1164-(0.17~83)
Ti-15V b	15.0mol%V	1164-(0.17~83)
Ti-20V b	20.1mol%V	1164-(0.17~83)
Ti-25V b	25.0mol%V	1164-(0.11~83)
Ti-30V b	29.6mol%V	1164-(0.06~8.3)
Ti-20V	19.6mol%V	900-(0.04~0.17), 950-(0.008~83), 1000-(0.04~83), 1025-(0.6~70), 1050-(0.017~83), 1100-(0.06~83), 1150-(0.04~83), 1200-(0.02~42), 1250-(0.02~17), 1300-(0.56~42), 1350-(0.56~43)
Ti-30V	31.4mol%V	1050-(0.008~33), 1215-(1.7), 1340-(1.7)

In the second series, which were multiple-arc-melted buttons, the alloys contained 5~30mol%V. They were hot-rolled to sheets, from which plate-type specimens (the cross section = 2×5 mm) were machined. In the third series, which were arc-plasma-melted ingots, the alloys contained 20 and 30mol%V. They were forged to bars, from which rod-type specimens were machined. The grain size was roughly 1 mm in all specimens tested.

Tensile Test

Mechanical behavior was investigated by conventional tensile testing in an argon atmosphere. The test conditions are listed also in Table 1.

The nominal stress - nominal strain (σ_0-ε_0) curves were converted into true stress - true strain (σ-ε) curves under the constant strain-rate condition assuming $\sigma = \sigma_0\ \varepsilon_0^{1+1/n}$, where n is the stress exponent of creep rate $\dot{\varepsilon}$ in the power-law creep equation.

Results

General

The true stress - true strain curve usually shows a broad maximum. The stress of this maximum level will be denoted as σ_s and used as a characteristic parameter throughout this paper. When the broad maximum could not be clearly recognized, σ at $\varepsilon = 0.20$ was used as σ_s.

In the following sections steady-state deformation characteristics will be shown using a traditional way of analysis,

$$\dot{\varepsilon} \propto (\sigma/G)^n \exp(-Q_c/RT), \qquad (1)$$

where $\dot{\varepsilon}$ is the (initial) strain rate, σ is the steady-state stress, or stress at 0.20 strain, G is the shear modulus, R is gas constant, and T is the test temperature.

Stress - Strain Rate Relationship

Typical examples of the σ_s- $\dot{\varepsilon}$ relationship are shown in Fig.1. In β-Ti-Al alloys the slope is about 4 and no bend can be seen under any test condition employed in this investigation.

In β-Ti-V alloys two regions with a unique slope in each region can be recognized. The low and high stress regions will be denoted as Regions M and H, respectively, referring to the case of Al-Mg solid solution alloys(3). The slopes, or the stress exponents, are about 3 and 4.5 in Regions M and H. The boundary condition (the upper transition stress σ_u) depends only slightly on temperature.

Temperature Dependence

Typical examples of the temperature dependence of $\dot{\varepsilon}$ under selected values of σ/G are shown in Fig.2 in each region. The apparent activation energies calculated from the slopes of data lines are 160 kJ/mol in Ti-Al alloys, and 180 and 160 kJ/mol in the low and high stress ranges of Ti-V alloys, respectively. These values are close to the activation energies for diffusion of Al in β-Ti-Al alloys(4), V and Ti in β-Ti-V alloys(5), respectively.

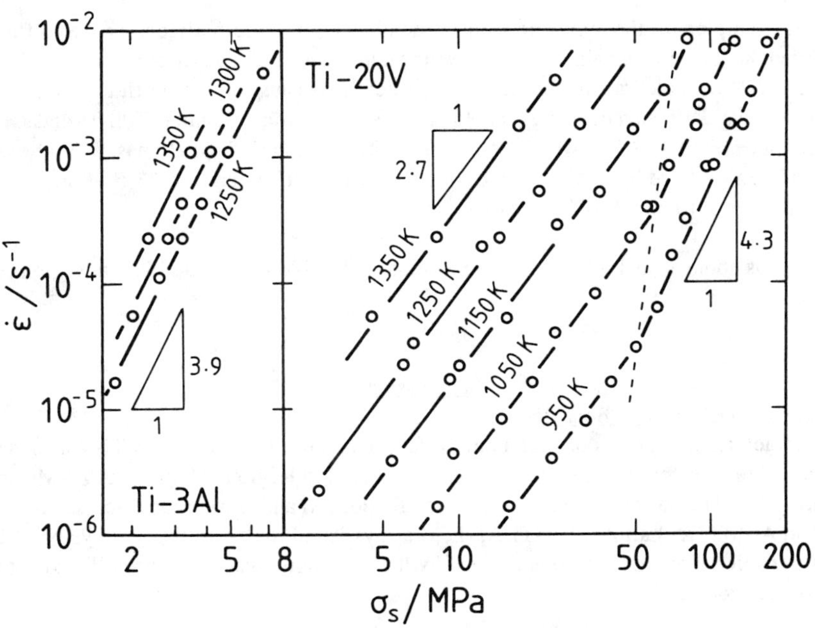

Fig.1 Typical examples of $\dot{\varepsilon}$-σ_s relationship obtained in β-Ti-3Al(1) and β-Ti-20V(2) alloys.

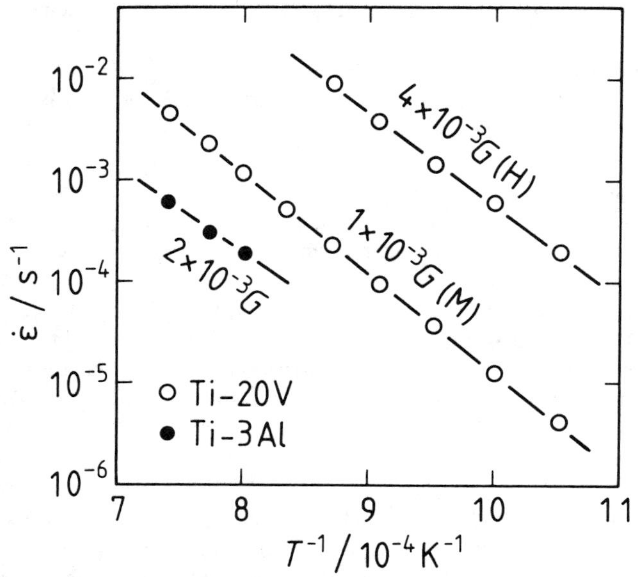

Fig.2
Typical examples of $\dot{\varepsilon}$-$1/T$ relationship under constant σ/G conditions in β-Ti-3Al[1] and β-Ti-20V[2] alloys.

Concentration Dependence

Examples of the influence of solute concentration on the $\dot{\varepsilon}$-σ relationship are shown in Fig.3. It is obvious in β-Ti-V alloys that the boundary condition between Regions M and H depends significantly on the composition.

Typical examples of the concentration dependence of $\dot{\varepsilon}$ normalized by the diffusion coefficient under selected values of σ/G are shown in Fig.4. The abscissa is taken as $N(1-N)$ instead of N (the mole fraction of the solute element), because of a better fit up to a concentrated range[7]. Diffusion coefficients used are those of interdiffusion in β-Ti/β-Ti-4mass%Al couples in cases of Ti-Al alloys, those of V self-diffusion in β-Ti-(0~30mass%)V alloys in cases of Region M of Ti-V alloys, and those of Ti self-diffusion in β-Ti-(0~30mass%)V alloys in cases of Region H of Ti-V alloys.

The slope is about 1 in Ti-Al alloys and Region M of Ti-V alloys, and 2.7 in Region H of Ti-V alloys.

Discussion

Transition in Deformation Behavior

The characteristic features of high temperature deformation observed in Ti-V alloys are considered to be essentially the same as those observed in Regions M and H of Al-Mg solid solutions[3]. The transition condition between Regions M and H has been discussed in some detail in Al-Mg (fcc) alloys[8], α-Ti-Al (hcp) alloys[9] and α-Fe-Mo (bcc) alloys[10]. The upper transition stress σ_u is well correlated by the condition given by Cottrell[11], except the numerical factor, i.e.,

$$\sigma_u \propto \{(1+\nu)/\pi(1-\nu)\}\, e\, N\, G. \qquad (2)$$

Here, ν is Poisson's ratio and e is the atom misfit parameter.

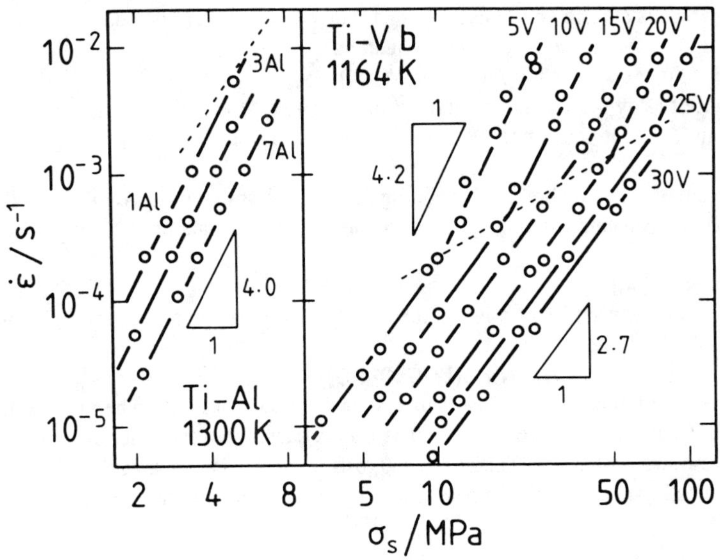

Fig.3 Typical examples of $\dot{\varepsilon}$-σ_s relationship in alloys containing various amounts of the solute in β-Ti-Al[1] and β-Ti-V[6] alloys.

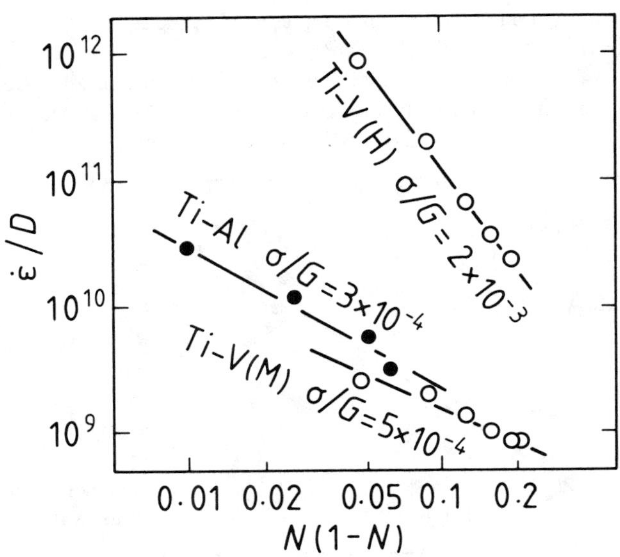

Fig.4 Concentration dependence of $\dot{\varepsilon}$[1,6] normalized by the diffusion coefficient, D, of the solute in Ti-Al and Ti-V(M) alloys and of the solvent in Ti-V(H) under given σ/G values.

The present results are shown in Fig.5 together with previously reported data in other systems. All results can be well represented by the equation,

$$\sigma_u/M = 0.2\{(1+\nu)/\pi(1-\nu)\}\ e\,N(1-N)\,G, \qquad (3)$$

independent of alloy system and crystal system. Here, M is the Taylor factor[12]: 3.06 for fcc lattice and 2.65 for bcc lattice.

This condition is drawn by broken lines in Fig.3. It is clear that in Ti-Al alloys all the experimental conditions locate within Region M and no bend on the data lines can be expected.

Creep Rate Equations

The relationship between $\dot{\varepsilon}$ and σ in Region M is analyzed using an equation[11] including the terms, $N(1\text{-}N)$ and e^2, i.e.,

$$\dot{\varepsilon} = A^M\{D/b^2N(1\text{-}N)e^2\}(kT/Gb^3)(\sigma/G)^n. \qquad (4)$$

This relationship holds very well in each system as shown in Fig.6, when diffusion coefficients of the solute are employed as D. The master curves, however, are obviously different between these two systems. This result suggests that there is still some missing factors for quantitative estimation of high-temperature deformation behavior in β-Ti-base solid solution alloys.

It was reported in Al-base ternary solid solutions[13] that a simple additive rule holds in high temperature strengthening, provided that the correction term, e^2/D, is applied for each solute element. In the Ti alloys, no such simple rule can be applied because of different master curves in the $\dot{\varepsilon}$-σ relation among alloys. One of the important problems remaining unsolved is the cause of differences in n and/or A^M among alloy systems.

The deformation behavior in Region H can be represented by the equation,

$$\dot{\varepsilon} = A^H\{N(1\text{-}N)\}^{-m}(Gb^3/kT)(D/b^2)(\sigma/G)^n, \qquad (5)$$

with $m\approx2.7$ and $n=4.2$, when the Ti tracer diffusion coefficients in each alloy are employed as D[5]. This relationship is essentially similar to that observed in (random) solid solution strengthening at low temperatures, i.e., $\sigma \propto N^{1/2\sim2/3}$ [14] at a given $\dot{\varepsilon}$.

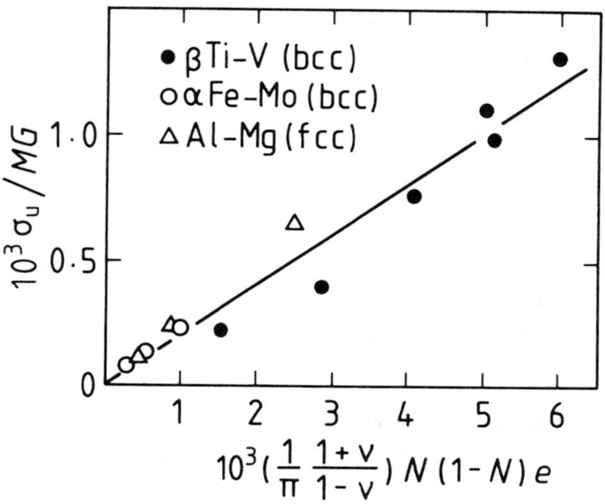

Fig.5
Relationship between the (upper) transition stress σ_u/G and the concentration $N(1\text{-}N)$ multiplized by factors according to the Cottrell model of the atmosphere breakaway[2,6].

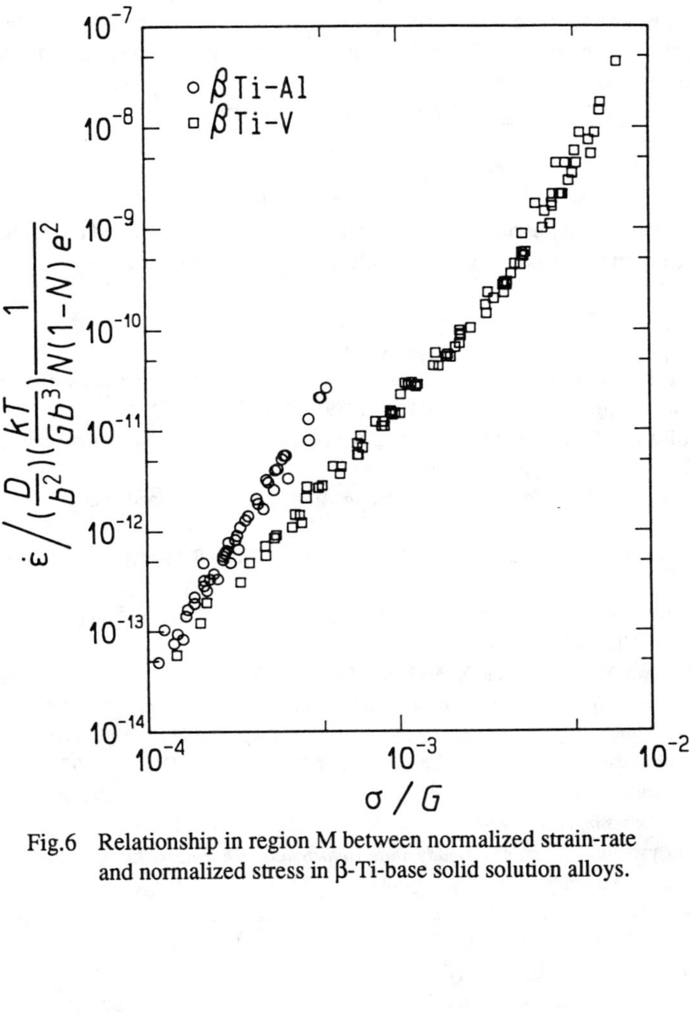

Fig.6 Relationship in region M between normalized strain-rate and normalized stress in β-Ti-base solid solution alloys.

Conclusions

Deformation behavior of β-Ti-base solid solution alloys is divided into two types and the boundary condition can be expressed by the equation,

$$\sigma_u = 0.53\{(1+\nu)/\pi(1-\nu)\}\ e\,N(1-N)\,G. \quad (6)$$

In the low stress range, the deformation of β-Ti-solid solution alloys is of the alloy-type, in which the glide motion of dislocations dragging solute atmosphere is the rate-controlling process. The relationship between stress and strain rate in the steady-state deformation can be represented by the equation,

$$\dot{\varepsilon}^M = A^M[D_{solute}/\{N(1-N)e^2\}](kT/Gb)(\sigma/G)^n. \quad (7)$$

The values of A^M and n depend slightly on the alloy system.

In the high stress range, the deformation of β-Ti solid solution alloys is of the metal-type, in which dislocations can glide with fairly high velocity and the climb motion of dislocations is the rate controlling process. The relationship between stress and strain-rate can be represented roughly by the equation,

$$\dot{\varepsilon}^{H} = A^{H}[D_{solvent}/\{N(1-N)\}^{2.5\sim3}](Gb/kT)(\sigma/G)^{\sim4.5}. \quad (8)$$

Acknowledgments

The work was supported partially by the Light Metal Educational Foundation, Incorp. The materials used were donated by Sumitomo Metal Industries, Ltd.

References

(1) M. Seki and H. Oikawa, *"High Temperature Deformation of Beta Titanium-Aluminum Solid Solution Alloys"*, Mater. Trans. JIM, 30(1989), 1027-32.

(2) H. Oikawa, M. Seki and M. Ohtsuka, *"High Temperature Deformation Behavior of Ti-20mol%V Alloy"* (Tohoku Univ., 1989).

(3) H. Oikawa, *"Hot Deformation of Aluminum Alloys"* eds. T.G. Langdon et al. (TMS,1991), 153-180.

(4) S.Y. Lee, Y. Iijima and K. Hirano, *"Interdiffusion in Dilute β-Ti-Al Alloys"* (Tohoku Univ., 1992).

(5) J.F. Murdock and C.J. McHarque, *"Self-Diffusion in Body-Centered Cubic Ti-V Alloys"*, Acta Metall., 16(1968), 493-500.

(6) H. Oikawa, Y. Ishikawa and K. Masada, *"Effects of the Solute Concentration on High Temperature Deformation Behavior of β-Ti-V Alloys"* (Tohoku Univ., 1988).

(7) M. Sakamoto, *"Diffusion Equation and Cottrell Atmosphere Dragging of Edge Dislocation in High Concentration Solid Solutions"*, Mater. Trans. JIM, 30(1989), 337-344.

(8) H. Sato and H. Oikawa, *"Further Experimental Study of Deformation Characteristics of Al-Mg Alloys in the Power-Law Creep Regime"*, Scr. Metall., 22(1988), 87-92.

(9) H. Oikawa and T. Oomori, *"Steady State Deformation Characteristics of α-Ti-Al Solid Solutions"*, Mater. Sci. Eng., A104(1988), 125-130.

(10) S. Nanba and H. Oikawa, *"Effects of the Solute Concentration on the Deformation Characteristics of Fe-Mo Solid Solutions in the Power-Law Creep Regime at 1100K"*, Mater. Sci. Eng., A101(1988), 87-92.

(11) A.H. Cottrell, *"Dislocations and Plastic Flow in Crystals"*, Oxford Univ. Press (Oxford, 1953), p.136.

(12) U.F. Kocks, *"The Relation between Polycrystal Deformation and Single-Crystal Deformation"*, Metall. Trans., 1(1970), 1121-1143.

(13) H. Nakashima, M. Watase and H. Yoshinaga, *"Experimental Examination of the Additivity of High Temperature Solution Hardening due to Different Solute Elements"*, Nippon Kinzoku Gakkai-shi, 53(1989), 28-33.

(14) P. Haasen, *"Dislocations in Solids"*, ed. F.R.N. Nabarro (North-Holland P., Amsterdam, 1979), vol.4, p.157-189.

STRENGTHENING MECHANISMS IN UNDERMATCHED

TITANIUM WELDS

M. E. Wells

Carderock Division, Naval Surface Warfare Center
Annapolis, Maryland 21402-5067

Abstract

This paper examines the strengthening mechanisms in lower yield strength (LYS) narrow groove weldments from both an experimental and analytical standpoint. All work was performed on nominal 25 mm thick Ti 6Al-4V weldments fabricated with commercially pure titanium filler wire.

Following radiographic examination of the completed weldments, microhardness measurements were taken across the baseplate, heat affected zone (HAZ) and weld metal. The weld metal hardness data was translated into tensile strengths using the strength-hardness relationships developed for titanium. Chemical and electron probe analyses of the weld metals were made to measure the extent of major and interstitial element pickup from the baseplate material during welding.

The tensile properties of the weldments were measured by full thickness transverse tensile tests. The results showed that as the weld width decreased, the transverse joint strength increased. At a weld width of 7.9 mm, the transverse joint strength matched the tensile strength of the baseplate material. Comparison of the transverse tensile test results with the estimated tensile strengths of the weld metal in the free state showed that contact strengthening was the dominant mechanism in producing higher strength in the weldments. Melting of the baseplate material into the weld deposit contributed to the joint strength at the narrowest weld width, but had a negligible effect at the wider weld widths. The results of compact tension tests showed that the increased joint strength was accompanied by a reduction in fracture toughness and resistance to crack propagation.

The numeric analysis of the LYS weldments was performed using the plane strain option of the PAPST finite element program. The results showed an expected build-up of axial, effective and hydrostatic stresses as the weld width decreased. The effects of contact strengthening were characterized by a decreasing ratio of effective stress to applied stress and an increasing ratio of hydrostatic stress to applied stress. The calculated values of effective stress at failure were found to closely match the tensile strengths of the weld metal in the free state.

Titanium '92
Science and Technology
Edited by F.H. Froes and I. Caplan
The Minerals, Metals & Materials Society, 1993

Introduction

As part of a reduced cost titanium alloy development program, the U.S. Navy is developing a lower yield strength (LYS) welding technology. This technology consists of two major factors. First, filler metals are employed that have yield and ultimate strength values lower than the baseplate. The use of LYS fillers in titanium fabrication offer the potential for improved toughness, increased tolerance for interstitial contamination and reduced cost due to lower alloy content and processing difficulty. Second, narrow groove joint designs are employed to provide required transverse weld strength properties. In addition to reducing the volume of weld metal required, the geometry of the narrow groove joint can be utilized with an LYS filler to increase the joint strength above that of the filler metal in the free state. When the weld width is less than the plate thickness, the easiest 45 degree slip planes are blocked by the harder baseplate material surrounding the weld. Geometric constraints on the plastic flow of the weld metal raise the yield and ultimate strength of the weld joint above that of the weld metal (1). As the joint strength exceeds the yield stress of the weld metal, a complex state of stress begins to develop at the center of the weld zone (2,3). The weld metal strain hardens, which in addition to strengthening the joint, can increase the susceptibility to brittle fracture (4,5,6). In the literature, this phenomenon is referred to as contact strengthening (7). While the majority of research into the strength of undermatched weldments has been accomplished through experimental programs, only a limited number of numerical studies appear in the open literature (8,9). The objective of this work was to examine the strengthening mechanisms in LYS titanium weldments from both an experimental and analytical standpoint.

Experimental Study

Procedure

The experimental work was focused on evaluating the effects of the joint geometry, as expressed by the aspect ratio of weld width to plate thickness, on the strength and toughness properties of narrow groove LYS weldments. A series of weldments were made in 25 mm thick Ti 6Al-4V plate with 1.6 mm diameter commercially pure (CP) wire. The chemical composition and tensile properties of the materials are provided in Tables I and II, respectively.

Table I. Chemical composition of baseplate and filler wire (weight %).

Material	Al	V	Fe	O ppm	N ppm	C ppm	H ppm
Baseplate	5.9	3.9	0.04	1220	100	210	39
Filler wire	-	-	-	590	100	80	46

Table II. Tensile properties of baseplate and weld metal.

Material	YS MPa	UTS MPa	El %	RA %
Baseplate	803	903	12	18
Weld Metal	422	516	29	50

Weldment Preparation. A 90 degree square groove was machined along the length of six plate sections, followed by tungsten carbide burring to a depth of 0.8 mm. The joint edges were then degreased with acetone and wiped dry with a clean disposable cloth. Three 25 x 610 x 914 mm weldments were fabricated by the GTAW cold-wire process in the flat position. A 6 mm thick Ti 6Al-4V backing bar was used to allow welding from one side. The weldments were identified as 1, 2 and 3 with joint openings of 19 mm, 11 mm and 6.5 mm, respectively. Arc current was maintained at 300-340 amperes, arc voltage at 14.5-18 volts, travel speed at 2.5-3.2 mm/sec, and wire feed speed of 23.7-30.0 mm/sec. The torch gas was a mixture of 75 percent helium and 25 percent argon with a flow rate of 28.3 l/min He and 9.4 l/min Ar. The trailing gas was 100 percent argon with a flow rate of 56.6 l/min. Between weld passes, the joint surface was cleaned with a rotary stainless steel wire brush and wiped with acetone.

Weldment Evaluation. After radiographic examination, microhardness measurements were taken on sections of each weldment using a diamond pyramid indentor with a 1 kg load. The measurements included baseplate, HAZ and weld metal and were performed to establish the width of the weld metal and corresponding aspect ratio.

Chemical analysis for interstitial pick-up was performed on weld metal from each weldment. Electron probe microanalysis was performed on the widest and narrowest weldments to measure aluminum and vanadium pick-up from the baseplate.

Full thickness transverse tensile specimens of 25 x 75 x 914 mm were removed from each weldment. Weld reinforcements were machined flush and small strain gages were attached to the baseplate and weld metal to measure surface strains during testing.

The fracture toughness of the weldments was measured by 1/2 T compact tension specimens using the unloading compliance J-integral test method. Specimens from the widest and narrowest weldments were notched in the weld metal with the notch normal to the plate surface.

Results and Discussion

Weld Hardness. Microhardness traverses across the fusion zones of each weldment provided the actual width of the weld metal and corresponding aspect ratio, Table III. The average microhardness values for the baseplate, HAZ and weld metal from each weldment are also presented in Table III. The weld metal hardness in weldments 1 and 2 were comparable and well below the baseplate hardness. Weld metal hardness in weldment 3 was significantly higher than the other weld deposits. Using the strength-hardness relationships developed for titanium in reference 10, the average weld metal hardness values were translated into ultimate tensile strength values. As expected, the estimated tensile strengths of weldments 1 and 2 were in close agreement with the Ti-CP all weld metal tensile strength value from Table II, and the weld metal tensile strength of weldment 3 was significantly higher than the other weld deposits.

Table III. Microhardness values for LYS weldments.

Weldment ID	Weld Metal Width	Aspect Ratio	Baseplate DPH	HAZ DPH	Weld Metal DPH	Estimated Weld Metal UTS MPa
1	20.6	0.70	329	337	161	490
2	11.9	0.39	331	330	176	517
3	7.9	0.30	338	340	233	655
Ti-CP Weld Metal UTS (Table II)						516

The strengthening effect observed in weldment 3 was attributed to pick-up from the baseplate during fabrication. Interstitial chemical analysis showed an oxygen content of 1180 ppm in this weldment, as compared with oxygen contents of 750 ppm and 730 ppm for the weld deposits of weldments 1 and 2, respectively. Electron probe analysis showed an increase in the weld metal aluminum content from 0.28 wt% in weldment 1 to 2.4 wt% in weldment 3. Using the strengthening rates of references 11 and 12, it is estimated that oxygen and aluminum pick-up increased the tensile strength of the weld deposit in weldment 3 by approximately 40 MPa and 117 MPa, respectively. The estimated tensile strength of the weld metal in weldment 3 from interstitial and aluminum pickup was in close agreement with the estimated tensile strength from microhardness measurements, 675 MPa versus 655 MPa, respectively.

Weldment Tensile Tests. The results of the full thickness tensile tests on each weldment are presented in Table IV. The data shows that as the weld width decreased, the transverse strength properties increased. At an weld width of 7.9 mm, the transverse ultimate tensile strength of the weldment matched the ultimate tensile strength of the Ti 6Al-4V baseplate. As noted in Table IV, all weldments fractured in the weld metal. The change in fracture appearance from 45 degree shear to a predominantly flat-rupture as the weld width decreases from 20.6 to 7.9 mm, is evidence of the increased constraint on plastic flow of the weld metal from the joint geometry.

Table IV. Tensile properties of LYS weldments.

Weldment ID	Weld Width mm	Aspect Ratio	Weld Metal YS MPa	UTS MPa	Fracture Location	Fracture Appearance
1	20.6	0.70	413	551	Weld	Shear
2	11.9	0.39	614	765	Weld	Shear/Flat
			655	765	Weld	Shear/Flat
3	7.9	0.30	689	930	Weld	Flat
			744	924	Weld	Flat

The transverse tensile strength values from Table IV are presented graphically in Fig. 3, along with the tensile strengths of the welds determined from the microhardness measurements of Table III. In this way, one can differentiate strengthening effects due to melting of the baseplate material into the weld deposit from contact strengthening due to joint geometry. At aspect ratios of 0.70 and 0.39, contact strengthening increased the joint strength by 13% and 48%, respectively. As the aspect ratio was further reduced to 0.30, major and interstitial element pickup from welding increased the strength of the weld metal by 27%. Contact strengthening accounted for an additional 42% increase in strength at this aspect ratio. From this graph, it is clear that contact strengthening is the dominant mechanism in increasing the joint strength above that of the weld metal in the free state. Melting of the higher strength baseplate material into the weld deposit increased the strength at the narrowest weld width, but had a negligible effect at the wider weld widths.

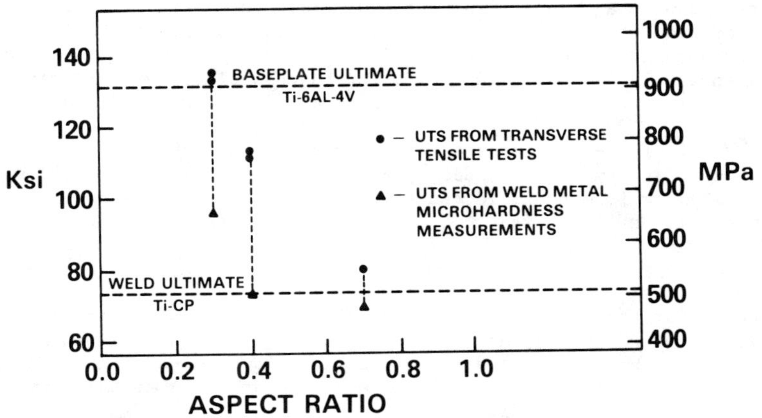

Fig. 3. Joint strength as a function of aspect ratio.

Fracture Toughness Tests. The results of compact tension tests on specimens from weldments 1 and 3 are provided in Table V. These data show that the fracture toughness J_{IC} decreased from 190 to 103 kJ/m^2 as the weld width decreased from 20.6 to 7.9 mm. These values were based upon a flow stress calculated from the Ti-CP all weld metal tensile test results of Table II. The fracture toughness was re-calculated using a flow stress based upon the transverse weld joint tensile properties of Table IV. For weldment 1, the J_{IC} and T values remained relatively unchanged. For weldment 3, the J_{IC} was reduced from 103 kJ/m^2 to 91 kJ/m^2, and T was reduced from 83 to 29. These results show that the increased joint strength is accompanied by a significant reduction in fracture toughness and tearing modulus.

Table V. Fracture toughness of LYS weldments.

Weldment ID	Weld Width	Aspect Ratio	Notch Orient.	Notch Location	E GPa	σ_f MPa	J_{IC} kJ/m^2	T
1	20.6	0.70	T-L	Weld	110.3	469	191	103
3	7.9	0.30	T-L	Weld	106.9	469	103	83
1						480	190	98
3						827	91	29

Note: E= Young's modulus
σ_f= flow stress

Analytical Study

Procedure

The experimental work provided narrow groove weldments with known failure loads for numerical analysis. The nonlinear finite element computer program PAPST was selected for the analysis. The analyses were performed on weldments 2 and 3, using the plane strain option of PAPST. This option was selected on the basis of previous in-house work showing that a 2-D plane strain model provided a good first-order approximation of the 3-D behavior in undermatched weldments. The material input data for the program included stress-strain data, Young's modulus and Poisson's ratio. The stress-strain curves of the baseplate and weld metal were divided into multiple straight line segments for input into the finite element analysis. Young's modulus was taken as 113.8 GPa and 106.2 GPa for the baseplate and weld metal, respectively. Poisson's ratio was taken as 0.3 for both materials.

One quadrant of the specimens was modeled using ten 12-nodal elements and 72 nodes. Each specimen was loaded by a uniform uniaxial tensile stress applied 12.7 cm from the weld centerline. Loads were increased incrementally until the fracture stress of each specimen was reached.

Results and Discussion

Examination of the output data showed that the extremes of stress-strain behavior occurred at the surface and center of the weld metal in each specimen. As expected, the highest stresses and strains occurred at the exact center of each weld. Pertinent stress quantities at these two locations are discussed below.

Calculated Stress Distributions. The calculated principal stresses in the weld metal of weldment 2 are shown in Fig. 3. The stresses at the surface remained in biaxial tension through the entire range of loading. The state of stress at the center of the weld metal began as one of plane stress but changed to a triaxial stress state as the load was increased above first yield of the weld metal. The effect on the von Mises effective and hydrostatic stresses is illustrated in Fig. 4. Above first yield, the development of high hydrostatic stresses accelerate, while the effective stress at the center increases more slowly. The ratio of the effective stress to applied axial stress dropped from a theoretical value of unity at first yield to a value on the order of 0.6 when the effective stress reached its free state tensile strength value of 516 MPa. This process is a manifestation of contact strengthening. At the same time, the ratio of hydrostatic stress to applied stress increased from a theoretical value of 1/3 at first yield to the order of 0.8 when the effective stress reached the tensile strength of the weld metal in the free state.

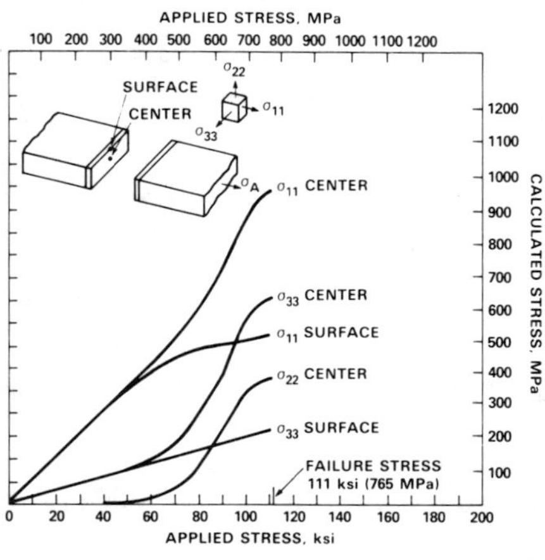

Fig. 3. Principal stresses in weld metal.

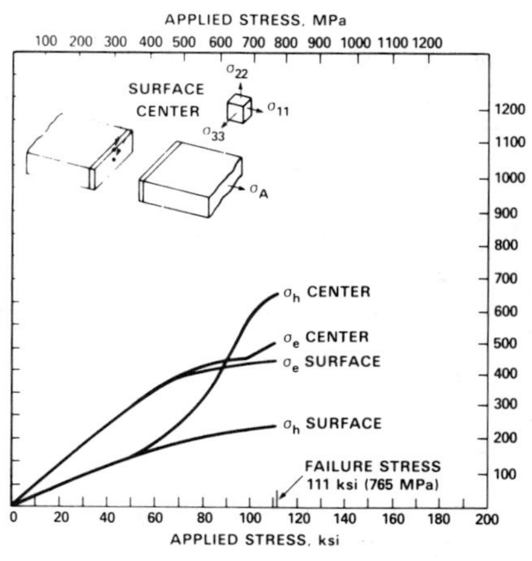

Fig. 4. Von Mises and hydrostatic stresses in weld metal.

The principal, effective and hydrostatic stresses at failure for both specimens are summarized in Table VI. While these stress values show that the narrowest weld had higher stresses within the weld at failure, there is nothing strongly evident in the calculated values to explain the difference in the joint strength of the two weldments. Since the program models only contact strengthening, the results indicate that other factors must contribute to the difference between load carrying capacity of the two weldments. Of these factors, the difference in weld metal ultimate tensile strength appears to be the most important. From Table III, the ultimate tensile strengths of the weld metal were estimated at 517 MPa for the 11.9 mm weld width and 655 MPa for the 7.9 mm width. These ultimate stress values match the calculated values of the center effective stresses at failure very closely. While based on limited experimental data, this result implies that the von Mises stress developed at the center of the weld can be used as a failure criterion for undermatched narrow gap welds, i.e. when the predicted von Mises effective stress at the center of the weld reaches the ultimate uniaxial tensile strength of the weld metal in the free state, failure occurs. In applying this criterion, allowance must be made for any increase in the weld metal ultimate tensile strength from welding.

Table VI. Computed stresses at failure for LYS weldments (MPa).

Weldment ID	2	3
Weld Width (mm)	11.9	7.9
σ_{11} @ surface	517	758
σ_{22} @ surface	0	0
σ_{33} @ surface	228	355
σ_{eff} (von Mises) @ surface	449	655
σ_h (hydrostatic) @ surface	248	372
σ_{11} @ center	955	1,200
σ_{22} @ center	379	441
σ_{33} @ center	627	776
σ_{eff} (von Mises) @ center	500	655
σ_h (hydrostatic) @ center	655	806
Applied stress at fracture	765	931

Conclusions

Based on the experimental and analytical results reported herein on narrow groove LYS weldments, the following conclusions are provided:

- As the weld width is reduced, the transverse yield and ultimate tensile strength of the weldments increase and the weld metal fracture toughness and resistance to crack propagation decrease.

- Contact strengthening is the dominant mechanism in producing higher strength in the LYS weldments. At the narrowest weld width tested in this work, additional strengthening was gained through melting of the baseplate material into the weld deposit.

- As the weld width decreases, the principal, von Mises effective and hydrostatic stresses increase. Contact strengthening is characterized by a decreasing ratio of effective stress to applied stress as the applied load increases.

- Failure of the welds is incipient when the von Mises effective stress approaches the ultimate tensile stress of the weld metal in the free state, adjusted upward for any strengthening from the welding process.

Work in Progress

Candidate LYS filler metals for the Ti-6211 baseplate system have been identified on the basis of cost, availability, and predicted metallurgical compatibility. Mechanical property and microstructural evaluation of the candidate filler metals are in progress. It is expected that the use of these filler metals, which more closely approximate the baseplate strength, will provide a less significant reduction in fracture toughness since the elevation in flow stress due to contact strengthening will be less severe.

Acknowledgements

The author gratefully acknowledges the assistance of Mr. Nash Gifford of the Carderock Division of the Naval Surface Warfare Center for his fine work on finite element analysis of the LYS weldments.

References

1. Hill, R., The Mathematical Theory of Plasticity, Oxford University Press, 1950, 226-236. (Note: The expressions developed for compression are reversible for the tensile case).

2. Erofeev, V.P. et al, "Calculation of the Parameters of the Stress State of Butt Welded Joints with Soft Areas," Welding Production, No. 7, 1979, 1-3.

3. Numes, A.C. et al, "Weld Width Indicates Weld Strength," Tech Brief, Vol. 6, No. 3, National Aeronautics and Space Administration, 1981.

4. Bakshi, O.A., "The Static Tensile Strength of Welded Joints with a Soft Interlayer," Welding Production, No. 5, 1962, 9-15.

5. Kharchenko, G. K., "The Strength of Joints with a Thin Soft Interlayer," Automatic Welding, No. 5, 1968, 31-33.

6. Karkhin, V.A. et al, "The Stress-Strain State of Welded Joints with Soft and Hard Interlayers," Welding Production, No. 3, 1984, 10-11.

7. Satoh, K. and M. Toyoda, "Joint Strength of Heavy Plates with Lower Strength Weld Metal," Welding Journal, No. 9, 1975, 311s-319s.

8. Agapakis, J., "Analytical and Numerical Evaluation of the Strength of Undermatched Butt Welded Joints in High Strength Steels," Special Research Project, Department of Ocean Engineering, MIT, 1980.

9. Jones, R.D., "The Effect of Undermatching in Shallow Cracked Specimens Containing Various Weld Geometries," University College of Swansea, Institute for Numerical Methods in Engineering, 1983.

10. Stark, L.E., "The Strength-Toughness Properties of Welds in Plates of Commercial Titanium Alloys," Welding Journal, No. 2, 1971, 67s.

11. Borisova, Ye.A. and I.I. Shashenkova, Titanium Alloys for Modern Technology, Nauka Press, Moscow, 1968, 202.

12. Vulf, B.K. and S.A. Yudina, Physical Metallurgy of Titanium, Moscow, 1964, 155.

Modelling of Texture Evolution during Thermomechanical Treatments for Different Titanium alloys.

M.J.PHILIPPE , F.WAGNER, M.SERGHAT, A.BOWEN+ and C.ESLING

LM2P - ISGMP -Université de METZ . F-57045 METZ CEDEX 01

+Materials. Structure Dept., Defence Research Agency; RAE Farnborough, Hampshire GU14 6TD, GB

Abstract

In hexagonal materials, plastic deformation is commonly accomodated by a combination of both dislocation glide and twinning . This is the case for titanium alloys, up to approximately a 50% reduction in area. For a higher degree of reduction, gliding is almost the only deformation mechanism, which may be accompanied by the formation of slip bands. In $\alpha+\beta$ alloys, the role of the β phase varies to a great extent with the rolling temperature and the percentage of the β phase.

Depending on the various stages of deformation, and in accordance with the evolution of the microstructure, different versions of the Taylor model were used for the modelling. The initial texture was the true hot rolling texture and the simulation of cold rolling required two or three different stages.

For two phase alloys (α and β phases as in TA6V), the modelling is restricted to the α phase. Indeed, at low temperatures, the percentage of the β phase is low and its deformation is negligible as it is much harder than the α phase. At a higher temperature, the volume fraction of the β phase is simultaneously increased and softened. In the latter case, it is necessary to resort to self consistent models which can be extended to multiphase versions.

Since the texture evolution during recrystallisation is well known, the modification of the deformation texture by different thermomechanical treatments offers promising prospects for the texture tayloring in view of the optimization of anisotropic properties.

Introduction

In order to calculate mechanical properties for hexagonal materials, such as yield stress or forming limit diagrams, it is necessary not only to identify the texture but also the deformation mechanisms with their critical resolved shear stresses. For titanium alloys, as for most hexagonal materials, it is difficult to find the appropriate literature since the results from one alloy to another can differ greatly (1).

With Transmission Electron Microscopy, we can attempt to determine the

Titanium '92
Science and Technology
Edited by F.H. Froes and I. Caplan
The Minerals, Metals & Materials Society, 1993

deformation mechanisms for the studied materials. For a better knowledge of the mechanisms and their critical shear stresses, it is interesting to observe the texture evolution in different conditions and to compare modelling and experiment.
If the texture evolution modelling is developed for cubic materials, it is not the case for hexagonal materials, and apart from our work, only one author has compared the results obtained with the Taylor to those obtained with the self-consistent models, for a defined set of CRSS' (2).
In a first approach we have worked with different versions of the Taylor model,taking twinning into account .
We have also chosen the version (full constrained, lath or pancake relaxed) according to the microstructure of the alloy, and the mechanisms observed into TEM have been introduced in the modelling.

Texture evolution by the cold rolling of α titanium or near α alloys

The textures produced by cold rolling of α titanium alloys have been listed (3). These textures can be classified in the range of two extreme boundaries :
- Type I or T-type (eg. T35, Fig. 1a) with C axes tilted at 40° to ND in TD and <10.0> directions in TD. Up to 50% deformation, this texture is not very sharp. However, after 50% deformation this texture becomes sharper with the increase of the deformation rate.
- Type II or C-type (eg. Ti + 2500ppmO_2; TA6V, Fig. 1b) with C axes tilted at 20° to ND in RD and with the <10.0> directions arranged randomly around C. These textures are never very pronounced whatever the deformation rate. This texture develops during the cold rolling. Generally, the hot rolling textures are either similar to type I or have their C-axes parallel to TD.

The type I textures are those of the α titanium alloys without a second phase and with an oxygen content of less than 1200 ppm. If the oxygen content increases, the C axes converge towards ND. Alloys with a large oxygen content (2500ppmO2) or which contain a second phase (TA6V for example), show again the rolling textures of type II (4).

Microstructure and deformation mechanisms

The titanium alloys with a low oxygen content (alloys with a type I texture), have an average grain size of between 10 and 30μm before cold rolling. During cold rolling, these alloys deform by gliding and twinning. For an alloy such as titanium T35, the twinned volume fraction increases with deformation and reaches 50% twinning at $\varepsilon = 0.5$ (Fig. 2).
The active deformation mechanisms, as determined by TEM during the first 20% of deformation (4), are as follows :
-prismatic glide {10.0}< 11.0 >
-deviated glide with the same direction <11.0> on a pyramidal or basal plane
-twinning on {10.2} and {11.2} planes.
During the first stage, no (or very little) gliding of type <c+a> is activated until 20% deformation.
After 20% deformation the nucleation of new twins decreases and becomes negligible at about 60% deformation (5). Thus, if the compression twinning {11.2} no longer appears, it is necessary to resort to pyramidal gliding <c+a> in order to adapt the deformation. Furthermore, the grain shape is very anisotropic, since the grains are very elongated.
For alloys such as TA6V or alloys with a large oxygen content, the grain sizes are in the range of 1 to 5μm. During cold rolling, the β phase (when it exists), does not participate directly in the deformation but only by limiting the grain size and consequently the twinning.
Furthermore, there is little twinning in these alloys (<5% twinned volume fraction up to a deformation of 50%). (Fig.3).

It can be considered therefore, that these alloys basically deform by gliding. The deformation mechanisms determined by TEM are:

- prismatic glide {10 .0} <11.0>
- cross slip with the same <11.0> direction
- pyramidal glide {10 .1} <11.3>

In this case only the α phase has been considered during the deformation.

Choice of model and conditions of use.

We used the Taylor Model because it is able to take twinning into account. Furthermore for cubic materials, this model gave comparably good results to those obtained from more sophisticated models (6). Moreover this model will only be used when the basic assumptions are fulfilled in practice. In addition, the best adapted version of the model (full constrained, lath or pancake-relaxed) for the microstructure and its evolution will be chosen.
Thus for the T type alloys like the T35 alloy, 3 different stages during the plastic deformation were distinguished: in the 1st stage, twinning was observed; in the second, twinning and pyramidal glide; and in the third, pyramidal glide.
Accordingly, the modelling was divided into three stages, in order to account of the changes in the deformation systems. In order to take into account the microstructure and its evolution, a fully constrained version of the model was used for the first stage, and a pancake-relaxed version for the following two stages.

In the case of TA6V, there is no change in the deformation mechanisms. We distinguished two stages in order to take into account the modification in the microstructure : a first stage up to 50% deformation with a full constrained version, and a second stage which ranges from 50 to 80% deformation, with a pancake-relaxed version (anisotropic grain shape).

Finally, the texture evolution was modelled by using as an initial texture the 'true' experimental texture, instead of an isotropic texture which is currently used.

Results of the modelling and comparison with the experimental textures.

In the case of the Titanium T 35 alloy, the procedure described above yielded the best results were obtained from a modelling in 3 stages, which allowed us to account for the evolution of the microstructure.
The ratios of critical resolved shear stresses have been assessed by means of TEM observations and been best adapted with respect to the results of the simulation.
For the first stage up to 20% deformation, the fully constrained version was used with the following relative critical shear stresses:

$$\frac{\tau\, pyr<11.0>}{\tau\, prism}=4{,}5 \qquad \frac{\tau\, twinning\,\{10.2\}}{\tau\, prism}=4{,}5 \qquad \frac{\tau\, twinning\,\{11.2\}}{\tau\, prism}=10$$

In a second stage ranging from 20 to 60% deformation, to account for the shape of the grains a pancake-relaxed version was used with the following relative critical shear stresses:

$$\frac{\tau\, pyr<11.0>}{\tau\, prism}=2 \qquad \frac{\tau\, twinning\,\{10.2\}}{\tau\, prism}=4{,}5 \qquad \frac{\tau\, twinning\,\{11.2\}}{\tau\, prism}=25 \qquad \frac{\tau\, pyr<1123>}{\tau\, prism}=1{,}5$$

In a third stage ranging from 60 to 80% deformation,we still used the pancake-relaxed version but the relative critical shear stresses were adapted to :

$$\frac{\tau\, pyr\langle 11.0\rangle}{\tau\, prism} = 1{,}5 \qquad \frac{\tau\, pyr\langle 11.3\rangle}{\tau\, prism} = 30$$

The modelling results are very similar to the experimental textures, as can be seen in figure 4 which corresponds to the result obtained at 80% deformation. These figures have to be compared with the experimental figures from figure 1a. The agreement of the modelled and experimental textures is quite good both when comparing the experimental pole figures and the theoretical pole figures calculated from simulations.
For the TA6V alloy there is only gliding. In the first stage from 0 to 50% deformation, a fully constrained version was used with the following critical resolved shear stress ratios :

$$\frac{\tau\, pyr\langle 11.0\rangle}{\tau\, prism} = 1{,}5 \qquad \frac{\tau\, pyr\langle 11.3\rangle}{\tau\, prism} = 10$$

In the second stage, with the same shear stresses ratios, a relaxed version was used to take account of the grain shape. In this case the modelling results, presented in fig. 5, correspond well to the experimental textures presented in figure 1b, where there is good agreement between the experimental pole figures and those calculated from the simulations.

Application to texture tayloring and conclusion.

The recrystallisation textures are well known, and have already been classified (2). The present work shows that the effect of the elementary deformation mechanisms (gliding, twinning) on the appearance of the various texture components by cold rolling are well understood. As the parameters which govern the activation of these elementary mechanisms such as alloying elements, second phase, grain size, deformation temperature etc. are understood too, it is possible to select the best adapted alloy and thermomechanical treatments to obtain a specific texture in view of the application of the sheets.
In particular, two applications can be mentioned :
1) if sheets are needed with a higher plane isotropy (similar properties in any direction in the sheet plane), it is necessary to start from a cold rolling texture with C axes in ND-RD plane (as for TA6V or Ti with 2500 ppmO_2). After recrystallisation the C axes are almost parallel to the normal direction - the 00.2 pole figure presents a central peak - whereas the <10.0> directions are almost randomly distributed in the sheet plane.
In order to obtain this kind of texture it is necessary to start with the kind of alloys which give this cold rolling texture and to submit it to the most suitable thermomechanical treatment. For T35 it is impossible to find this kind of texture in the classical sequence of cold roling and recrystallisation (without cross-rolling!).
2) if sheets are needed with the best formability in expansion and for cold drawing at room temperature, textures with a central pole (C parallel to ND) are not advised. C axes, tilted at about 40°(7) to the normal direction are preferable.

This is the case in T35 alloys and with very low addition of O_2 or Al. For the other alloys where the texture with C axes parallel to ND can appear it is necessary to control the deformation rate in cold rolling in order to keep the hot deformation texture.
A good knowledge of the texture evolution during cold rolling and recrystallization allows us to optimize the process to achieve texture control.

REFERENCES

(1) M.J. PHILIPPE, C. ESLING and B. HOCHEID
Textures and Microstructures 7 (1988) 265-303

(2) C. TOME, G. R. CANOVA, S. AHSI and A. MOLINARI
Proceedings of ICOTOM 8 ed. J. S. Kallend and G. Gottstein (1988) 395-401

(3) M.J. PHILIPPE, F. WAGNER and C. ESLING
Proceedings of ICOTOM 8 ed. J.S. Kallend and G. Gottstein (1988) 837-842

(4) M.J. PHILIPPE, F.E. MELLAB, F.W . WAGNER, C. ESLING and B. HOCHEID
Proceedings of 6 World Conference on Titanium ed. P. Lacombe, R. Tricot, C. Beranger
Les éditions de physique (1988) 135-140

(5) M. SERGHAT
Mid Term Report BRITE-EURAM 0117C (1991)

(6) F. WAGNER, G. CANOVA, P. VAN HOUTTE and A. MOLINARI
Proceedings of ICOTOM 9 Textures and Microstructures ed. H.J Bunge and C. Esling
(1990) 1135-1140

(7) M.J. PHILIPPE, C. ESLING and B. HOCHEID
Memoires scientifiques de le revue de Metallurgie 2 (1984) 7 1

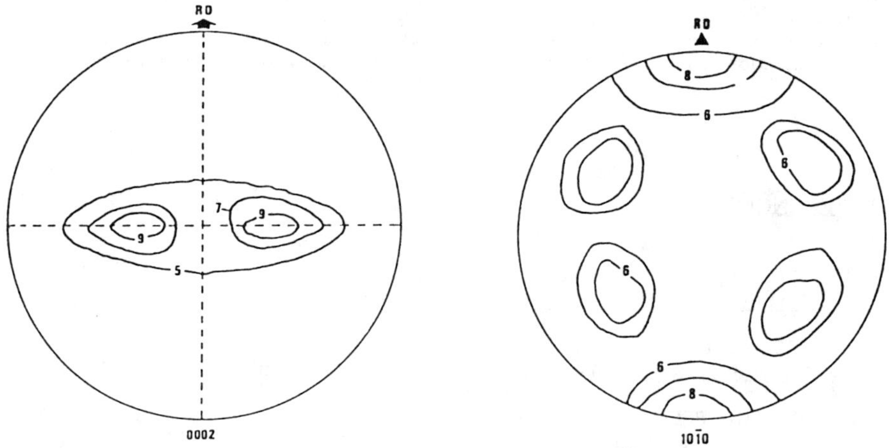

Fig. 1.a : Measured pole figures for T35 at 80% deformation (T texture)

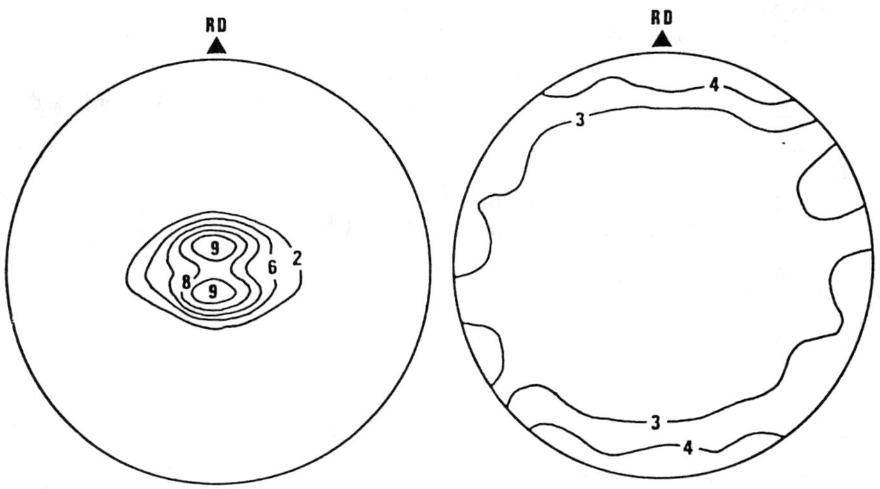

Fig. 1.b : Measured pole figures for TA6V at 80% deformation (L texture)

Fig. 2: Microstructure of T35 after 50 % deformation in cold rolling

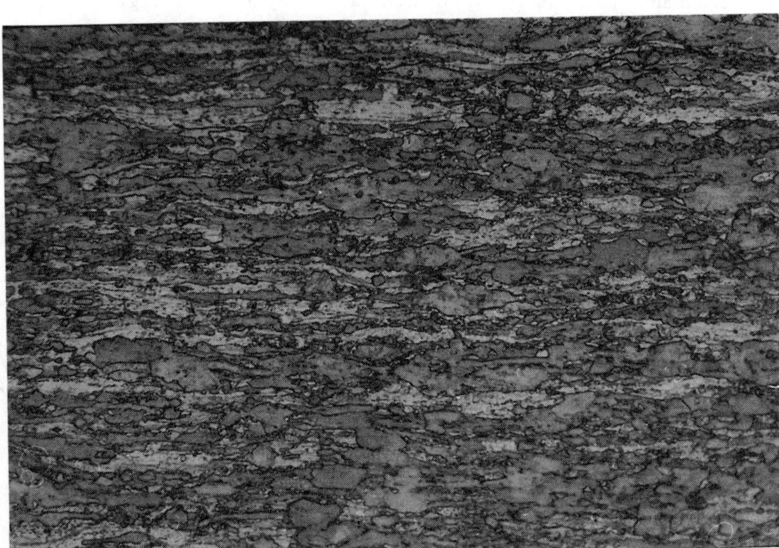

Fig. 3: Microstructure of TA6V after 50 % deformation in cold rolling

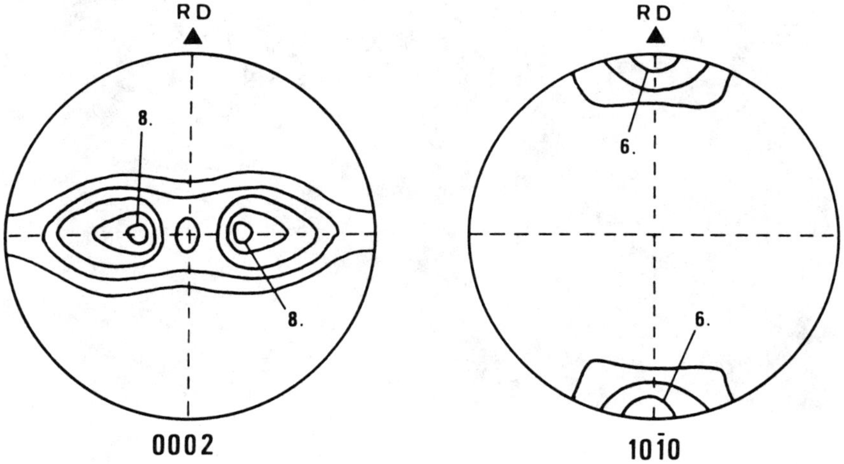

Fig. 4 : Theoretical pole figures for the T35 alloy at 80% deformation (calculated pole figures from ODF after Taylor modelling)

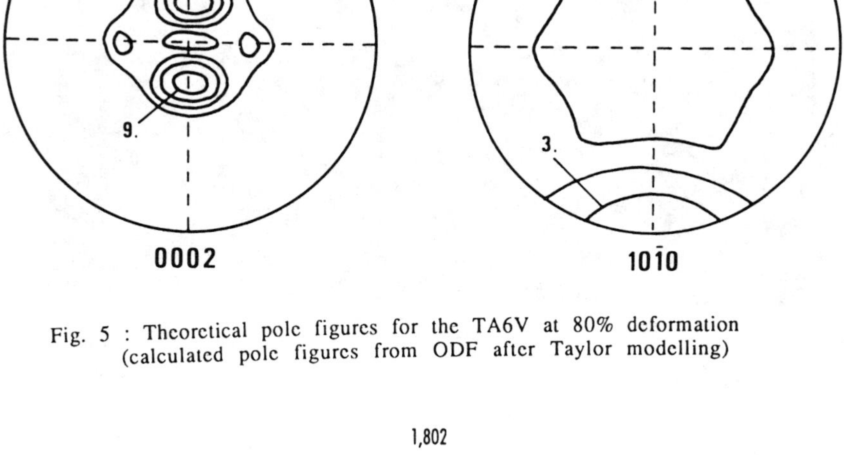

Fig. 5 : Theoretical pole figures for the TA6V at 80% deformation (calculated pole figures from ODF after Taylor modelling)

Fatigue Crack Growth Mechanisms of Two Titanium Alloys at Room and Elevated Temperatures

Y. Dai*, N. Marchand*, M. Hongoh** and J. I. Dickson*

*Dept. of Metallurgical Engng.
Ecole Polytechnique de Montreal
Canada H3C 3A7

**Pratt & Whitney Canada
LONGUEUIL (Quebec)
Canada J4G 1A1

ABSTRACT

Fatigue crack growth mechanisms at room and elevated temperatures of two titanium alloys, respectively Ti-6Al-4V and Ti-6Al-2Sn-4Zr-6Mo, were studied. Single edge notched specimens, which were fabricated from turbine compressor discs forgings, were used in this study. Crack initiation and crack growth rates (da/dN) for both small and long cracks emanating from the root of the notch were monitored using an advanced Alternating Current Potential Drop (ACPD) technique. The da/dN were plotted as a function of the stress intensity factors (ΔK_I) which were computed using respective short and long crack solutions. Correlation of da/dN data with ΔK_I suggest that fatigue crack growth is controlled by ΔK_I soon after nucleation ($a > 60\mu m$). The microscopic features of the fractured surfaces at different stages of fatigue crack growth are observed and analyzed. The results showed that the relationship between the crack front and the α/β interface controls the materials' resistance to fatigue crack growth.

1. INTRODUCTION

Titanium alloy forgings are extensively used in gas turbine engines for applications ranging from impellers, compressor discs and blades to sealings, cases and fan blades just to name a few. In recent years, the demand toward achieving higher thrust-to-weight ratios has led to increasing the temperatures of most stages (compressor, combustion and turbine) of the engines. As a result of operating at higher temperatures, compounded by the introduction of new regulations for safety, the accurate predictions of the low cycle fatigue life (i.e. the fatigue crack initiation life to 1/32" surface crack) as well as of the crack propagating lives (da/dN) has become paramount for Ti-based components. Furthermore, the accurate quantifications of the fatigue behaviour under realistic operating conditions including thermal-mechanical fatigue conditions must now taken into account in all the design stage of the components(1-2).

Fatigue cracks in a compressor disc usually start where some form of geometrical discontinuity exists such as fir tree root, slots and bolt holes. Recent advances in crack detection techniques such as ACPD techniques has rendered possible the detailed monitoring of crack initiation and growth mechanisms using laboratory specimen with notch features identical to those encountered in engine components(1,3). However, the study of cracks initiating from notches requires that 1) cracks with lengths (or depths) as small as 40μm be detected and their subsequent growth monitored with equal accuracy ($\Delta a \sim$ 1-2 μm), 2) the critical crack length for which further growth is mostly controlled by fracture mechanics must be determined, and 3) the proper stress intensity factors (ΔK)'s) that take account the stress field of the notch must be used when correlating the da/dN data.

Several mechanisms of fatigue crack initiation in $\alpha+\beta$ titanium alloys have been studied (3-5), including α/β interface cracking, primary α cracking at slip bands or slipless cracking, subsurface cracking by a cleavage-type, and grain egression. In addition to effects of microstructure, the operative mechanisms of crack initiation is also affected by a variety of mechanical and environmental conditions such as frequency of loading, R_ratio, temperature and corrosive environment.

2. MATERIALS AND EXPERIMENTAL PROCEDURES

Two titanium alloys were tested, both were fabricated from turbine compressor disc forgings. The microstructure of an α-ß forged Ti64 forging is shown in Figure 1. It can be seen that the equiaxed α particles, with average size about 32μm, are distributed in a transformed $\alpha+\beta$ matrix. The microstructure of an ß forged Ti6246 forging is also shown in Figure 1. The structure is featured by a coarse, acicular α-phase in the transformed ß phase. The average α lath size is about 2 X 60μm. Both alloys have volume fractions of α and ß phases of about 50%.

Composition analysis were performed using a JEOL 840 SEM with LINK spectrum analyzer. The chemical composition of beta stabilizers (V in Ti64 and Mo in Ti6246) varies from α to ß phases. These compositions were used as references to identify the fracture morphology associated with the different phases during the fractographic analysis.

The specimens used in this study were Single Edge Notched (SEN) specimens with rectangular cross section (W=10.67mm, B=5.33mm) and effective gauge length for extensometer equals to 25.4mm. A semi-circular notch (r/W = 0.1) was machined on the side of each specimen which matches the bolt hole geometries found in many engine compressor discs. The stress concentration factor (K_t) of the notch was found to be 2.12. The machining, cleaning and surface finishing procedures of the specimen were those that duplicate the actual manufacturing and conditions of the engine components. The tests were carried out under isothermal total strain controlled conditions using a high temperature extensometer attached to the back face of the specimen. Strain ranges investigated in this study varied between 0.2 ~ 0.4 % while the frequency was kept constant at 0.1667 Hz (10cpm). The specimens were heated using an induction heating system. The maximum temperatures for the tests were set at 400° C for Ti64 and 480°C for Ti6246.

The crack initiation and growth processes were monitored on-line using an advanced ACPD technique which has been described elsewhere(6). As compared with the Direct Current Potential Drop (DCPD) technique, the system sensitivity and linearity with respect to the measurement of crack length are enhanced by the skin effect which results from passing a high frequency AC current (30 KHz) through the specimen. Since the changes of potentials varied linearly with increasing crack length, the crack growth as a function of number of cycles was easily obtained from the ACPD curve which was continuously monitored during the test. With the present ACPD system, cracks as small as 20μm were detected with a probability of detection (POD) better than 95%. In turn, short and long crack growths (average) can be monitored with an accuracy better than 2μm. Furthermore, the ACPD system was shown to be promising for studying the closure mechanisms for short and long cracks.

3. K_I SOLUTIONS

The stress intensity factor (K_I) for small cracks emanating from the notch root were calculated by a model proposed by J. Schijve(4). This solution uses the notch peak stress and is written as

$$K_I = C\ (t/r)\ \sigma_{pk} \sqrt{\pi t} \tag{1}$$

Here σ_{pk} is the peak stress which is equal to $K_t \sigma_n$ with K_t the stress concentration factor and σ_n the far field nominal stress. C (t/r) is the notch geometry function which depends on t the crack depth measure from root surface and on r the radius of the notch. The function C(t/r) as derived by Schijve has the form

$$C(t/r) = 1.1215-3(t/r)+4(t/r)^{1.5}-1.7(t/r)^2 \tag{2}$$

When the crack grows longer as compared to r, i.e when a/w >0.14 (see Figure 2), the K_I's were calculated using the equation derived by Marchand et al (9) for a SEN specimen under fixed-ends displacement conditions. This solution can be briefly written as

$$K_I = (N/BW) \sqrt{\pi a}\ H(\ \zeta,\ \eta,\ \eta_b\ ) \tag{3}$$

Here N is the normal load *measured* during the test, B and W are the thickness and the width of the specimen, a is the crack length including the notch depth and $H(\zeta, \eta, \eta_b)$ is a geometrical correction factor which takes into account the actual loading conditions of the test (grips, no rotation ends, etc). For purpose of comparison, the K_I solution of Kujawaski(5) for a small notch crack in a semi-infinite domain was also used. When the cracks (a/W) are shorter than 0.14 (actual crack length t/W <0.04) the notch stress field controls the K_I's while for longer cracks the total crack length (notch plus actual crack length) and far field stress determine the K_I's. There exist a smooth transition between short and long crack solutions and thus a continuous solution for ΔK_I can be obtained to correlate the da/dN.

4. RESULTS

<u>Crack Growth Kinetics</u>

A typical fatige crack growth (FCG) rate vs. crack length curve for Ti64 tested at 400°C is shown in Figure 3. The following observations can be made from this figure. First, when the cracks are shorter (average depth from the root of the notch) than about 60μm, which is similar to the microstructural characteristic dimensions of the alloys, the da/dN varies erratically with increasing crack length, with da/dN spanning two orders of magnitude. This implies that these growth rates (for a<60μm) are (mostly) controlled by local microstructural events and thus are not and cannot be taken into account by a fracture mechanics parameter alone. When the crack grew longer than 60μm, their growth rates are well behaved (increased monotonically with increasing crack length) and thus can be correlated using a proper fracture mechanics parameter. Therefore it can be concluded that the da/dN's are controlled by fracture mechanics soon after initiation (a$\geqslant$60 μm) for Ti64. The same conclusion was also found to hold for the Ti6246 forgings tested.

The da/dN - ΔK curve pertaining to results shown in Figure 3 is presented in Figure 4. In this plot, the short crack K-solution (i.e Eq.(1)) was used for crack lengths ranging from 60 to 420μm while the long crack solution (i.e. Eq.(3)) was used for a > 420μm (a/W > 0.14, refer to Figure 2). For comparison, a set of long crack growth data is also plotted. Two conclusions can be drawn from this figure. Firstly, the da/dN for short cracks (60$\leq$a<130 μm) are faster than their long crack counterparts at the same level of ΔK. This is the result of the damage caused by inelastic deformation during cyclic straining. This was confirmed by carrying-out elasto-plastic finite element (FE) calculations which indicated the occurrence of small zones of inelastic deformation at the root of the notch. Secondly, as soon as the cracks grew outside the pre-damaged zone and into the elastic notch field (i.e 130 $\leq$a$\leq$ 420μm), the da/dN of the short crack became similar to that of the long cracks (at the same ΔK). This confirms that Schijive's K-solution takes properly into account the *elastic notch field*. Thus, short cracks emanating from notches can be considered as experiencing the following stages: (a) microstructural growth controlled, (b) inelastic damage growth controlled, (c) notch elastic field growth controlled, and (d) far-field stress growth controlled. As a result there exists three transitional (critical) crack lengths which are 60, 120 and 420μm for the Ti64 SEN tests. Similar results were also found for the Ti6246.

Fractographic observations

The Ti64 specimens tested at 400°C showed evidence of cleavage like facets morphology near the initiation sites at the root of the notch (see Figure 5a-5b). No cyclic cleavage striation was observed on these facets although fine and coarse fatigue striations were found near the cleaved facets. Compositional analyses showed these facets to be equiaxed α-phase particles, which is in agreement with other reported data (6-10). Analyses of sterofractographs of microregions at the initiation sites indicated that the crack nucleation planes coincide with the plane of maximum shear stresses obtained from the FEM results. Furthermore, see Figure 5c, analysis of several cleaved facets showed that fracture began near or at the α/β interface and proceeded toward the interior of the equiaxed α-phase particles. All this evidences suggest that the α-phase islands failed first and that the surrounding $\alpha+\beta$ matrix fractured under considerably higher cyclic strains as confirmed by the presence of coarse striations and faint cystallographic features in the $\alpha+\beta$ matrix. The above observations which pertains to the initiation zone were found to be more pronounced with increasing temperature. As expected, when the cracks grew outside the zone of influence of the notch, well defined fatigue striations are observed with secondary cracking occurring along the α/β interface and becoming more prevalent with increasing ΔK. For this region the measured microscopic da/dN are in agreement with those measured (bulk) by the ACPD technique. These results are also consistent with the observation of Rhodes et al.(7,9).

The fractured Ti6246 specimens also displayed numerous cleavage-like facets near the root of the notch, i.e. in the initiation zones (see Figure 6a). As expected, these facets were found to be the accicular α-phase grains. Subsurface cracking with clear cleavage-like facets were also observed in the 480°C tests. Due to the coarse microstructure of the material, the overall fractographic aspects of the surface appears relatively rougher as compared with Ti64. At moderate ΔK levels, but outside the zone of influence of the notch, delamination of acicular α-phase from the transformed β matrix is clearly visible (see Figure 6b). This indicates that crack growth proceeded through concurrent delamination of the α/β interfaces along the crack front. As a result, depending on the relative orientation of the α/β interfaces with respect to the loading axis, either striations along the delaminated areas (see Figure 6c) or secondary cracking along the α/β interfaces were produced. With increasing ΔK, interfacial secondary cracking was promoted because it is the most efficient mechanism to reduce the actual stress field at the head of the main cracks. This is supported by the fact that secondary cracks coincide with the striations at high increasing ΔK level.

5. DISCUSSION AND CONCLUSIONS

Fatigue cracks initiation (FCI) at the primary α-phase is also consistent with the other reported initiation sites in (Al)-rich reqions (6). The (Al) contents for both alloys were found higher in the α than in the transformend β matrix. The crack initiation and growth at the α+β interfaces (especially observed in Ti6246 tests at 480°C) indicate that elongated preferentially oriented α+β interfaces (with respected to the notch) can be detrimental to the FCI and FCG resistance. As suggested by D. Eylon (10), the surface connected interfacial cracking is an indication of environmental effects on the fatigue crack initiation processes. Diffusion of oxygen (at elevated temperatures) from the surface into the material is enhanced along the α/β interfaces either by lattice mismatch between the hexagonal α and cubic β phases or by the interface stresses developed during the load cycles. Interfacial diffusion is also promoted by the higher surface to volume ratio associated with the acicular α-phase and higher dislocation density. The resulting higher interstitial contents embrittle the interface zone (11) and reduce its ability to plastic deformation.

From this study, the following conclusions can be made:

- The crack growth rates in Ti64 and Ti6246 are controlled by ΔK soon after initiation at root of the notch when $a>60\mu m$.
- Three transitional (critical) crack lengths were indentified for the notched specimens using the ACPD crack growth measurement technique. These critical crack lengths separate microstructural growth controlled, inelastic notch field growth controlled, elastic notch field growth controlled, and far-field stress growth controlled regions.
- The crack initiation sites for both alloy were found to be at the α-phase particles or α/β interfaces, with cleavage-like facet fractured morphology.
- The α/β interfacial cracking, assisted by high temperature environmental embrittlement controls the FCI and FCG resistance of Ti6246.

ACKNOWLEDGEMENT

This research is supported by the Canadian National Science and Engineering Research Council (NSERC) and Pratt & Whitney Canada under R&D project 661-001/89.

REFERENCES

1. N. J. Marchand, Y. Dai and M. Hongoh, P&WC, WY-910069, 102 pages, 1991.
2. Engine Structural Integrity Program (ENSIP), MIL-STD-1783, Nov. 1984.
3. Y. Dai, et al, Proc. of 5th Int. on LCF, Berlin, Sept. 1992, in press.
4. J. Schijve, Fract. of Engng. Mater. & Struc., Vol.5, pp.77-90, 1982.
5. D. Kujiawski, Fract. of Engng. Mater. & Struc., Vol.14, pp.953-965, 1991.
6. J. L. Gilbert and H. R. Piehler, Met. Trans. 20A, pp. 1715-1725, 1989.
7. C. G. Rhodes, Proc. of 4th Int. Conf. on Titanium, pp. 1701-1707, 1980.
8. J. C. Chesnutt, et. al, ASTM STP 600, pp. 99-138, 1976.
9. C. G. Rhodes and J.C. Williams, Met. Trans. 6A, pp. 1670-1671, 1975.
10. D. Eylon et al, 4th Int. Conf. on Titanium, pp. 1845-1854, 1980.
11. G. T. Gray III, et. al, Met. Trans. 21A, pp.95-105, 1990.

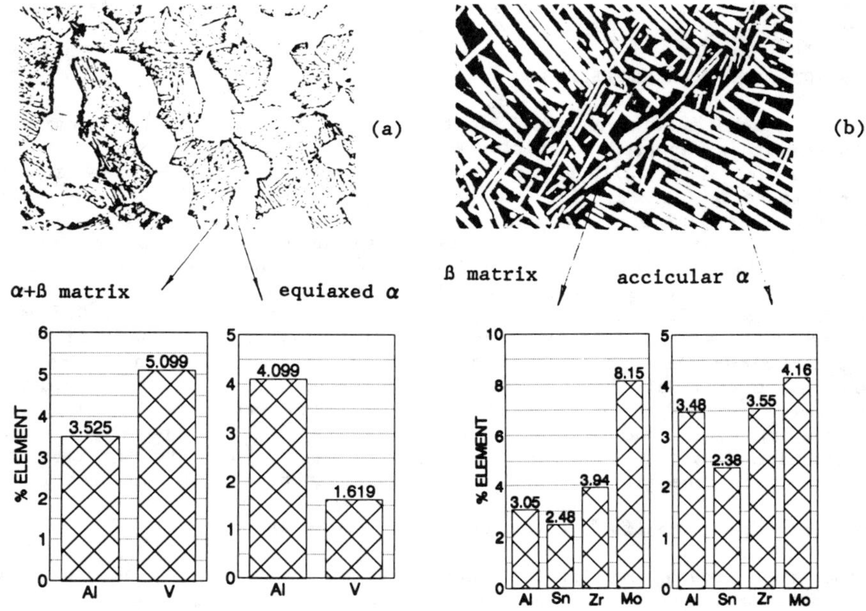

Fig.(1) Microstructure of the titanium forgings. (a) Ti64, (b) Ti6246. (500X)

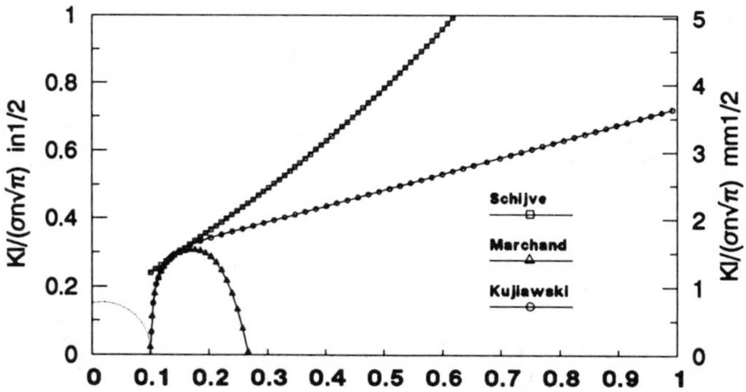

Fig.(2) K_I solutions for short and long cracks emanating at root of the notch.

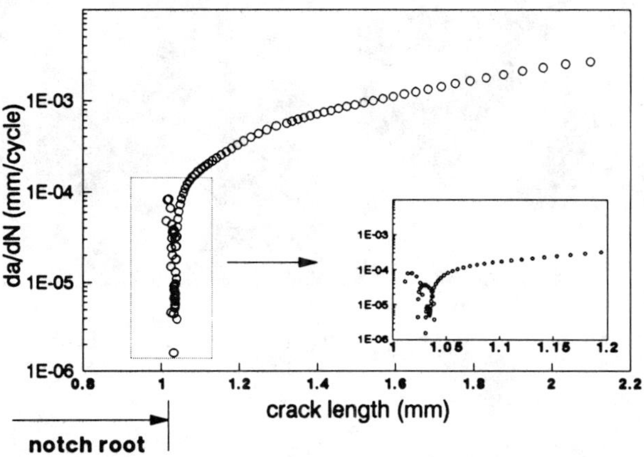

Fig.(3) A typical result for crack growth rate da/dN as a function of increasing crack length. Ti64 at 400°C.

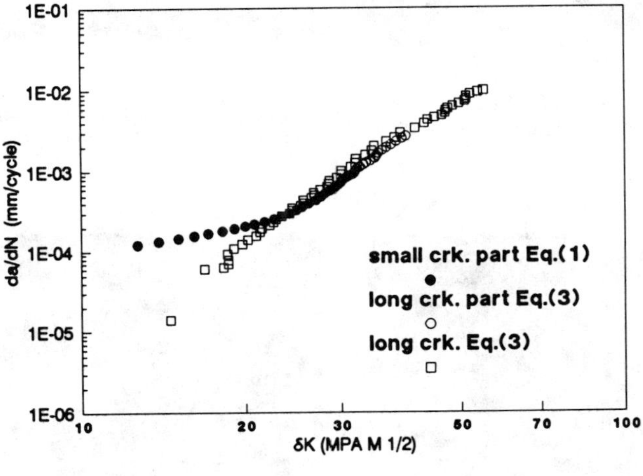

Fig.(4) Correlation of the da/dN data pertaining to Fig.(3) using ΔK computed for small and long cracks. Ti64 at 400°C.

Fig.(5.a) Crack initiation site at root of the notch. Ti64, 400°C.

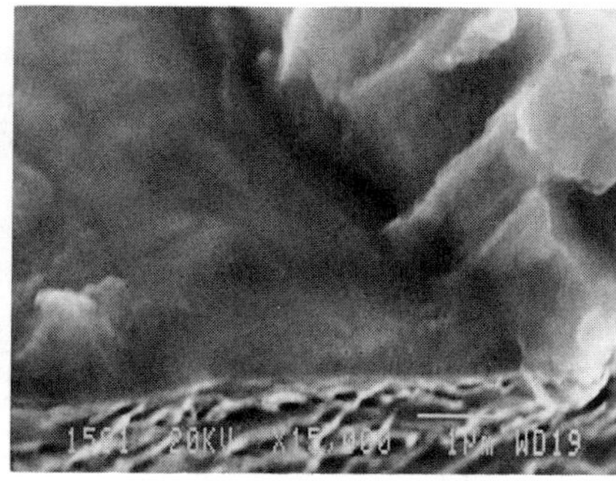

Fig.(5.b) Cleavage-like α facet at the Crack initiation site. Ti64, 400°C.

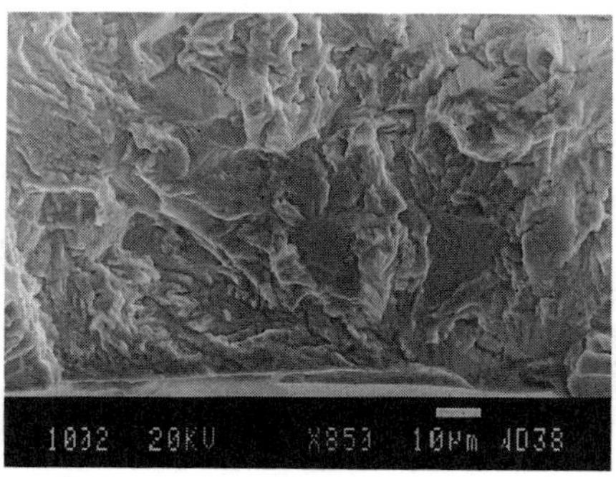

Fig.(5.c) Overview of the fracture surface near the notch. Ti64, 400°C.

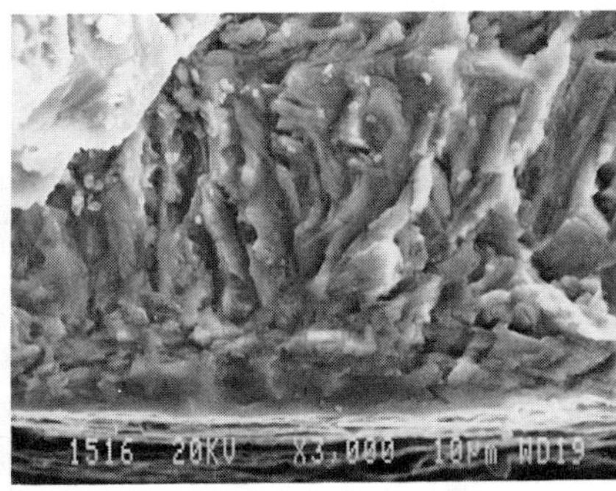

Fig.(6.a) A initiation site at the root of the notch. Ti6246, 480°C.

Fig.(6.b) Fatigue striations on the delaminated α-phases. Ti6246, 480°C.

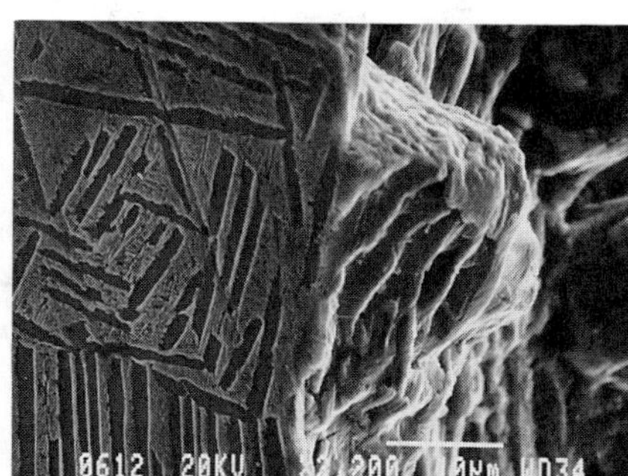

Fig.(6.c) Crack growth along the α/ß interfaces. Ti6246, 480°C.

THE MONITORING OF ANISOTROPY OF PHYSICO-MECHANICAL PROPERTIES OF FLAT SEMIPRODUCTS OF $\alpha+\beta$ TITANIUM ALLOYS

Igor Gorynin, Anatoliy Kudravtsev and R.A.Adamesku

CentralResearch Institute of Structural Materials
"Prometey", Ural Polytechnical Institute,
Russia

Abstract

The type and rate of anisotropy of physico-mechanical properties of the Ti-6Al-1Mo and Ti-4,7Al-1,8V alloys vs. crystallographic texture parameters has been studied. The effect of hot flat rolling schedules, texture condition of an original billet and heat treatment on the texture building and appearance of the property anisotropy in the indicated alloys are shown. The results indicate conditions to produce controlled anisotropy of the physico-mechanical properties of the flat semiproducts in $\alpha+\beta$ titanium alloys of the martensitic type.

Introduction

The anisotropy of the phisico-mechanical properties of the titanium alloy semiproducts is determined by a crystallographic texture /1-5/. Many works /4-12/ dealing with an investigation of the anisotropy of the titanium alloy properties, texture formation during a deformation and heat treatment have been published. However, the majority of the works were carried out on a commercially pure titanium or α-titanium alloys. In addition, sometimes the investigations were performed without precisely regulation of the composition of the alloys, the state of original parameter of deformation.This work aims to determine the conditions of the manufacture flat semiproducts in the $\alpha+\beta$ titanium alloys of the martensitic type of the desired anisotropy of the physico-mechanical properties based on the combined investigation of the type with desired anisotropy of the physico-mechanical properties.

Titanium '92
Science and Technology
Edited by F.H. Froes and I. Caplan
The Minerals, Metals & Materials Society, 1993

Material and Experimental Procedure

The Ti-Al-V and Ti-Al-V-Mo systems were investigated. The $\alpha+\beta\rightarrow\beta$ transformation temperatures (T_t) of the alloys are, ^{0}C: 1050 (5V), 980 (PT-3V) (Table1).

Table 1 The composition of the studied alloys

The alloy type	Content of the elements, wt.%									
	Aluminium	Vanadium	Molybdenum	Carbon	Iron	Silicon	Zirconium	Nitrogen	Oxygen	The total content of other impurities
PT-3V	4,50	1,80	-	0,060	0,080	0,017	0,018	0,011	0,094	0,066
5V	5,80	1,50	1,00	0,090	0,070	0,017	0,028	0,011	0,082	0,053

The crystallographic texture was studied by an analysis of direct (DPF) and opposite (OPF) pole figires obtained with an X-ray diffractometer. The servey was performed from the planes normal to ND (the plane parallel to a sheet plane), RD(longwise), TD(crosswise). The tensile properties were determined in accordance with the GOST 1497-84 on the 5-fold (l_0= 5d_0) cylindric specimens with a diameter of a gauge length of 3 and 5 mm. The coefficient of a strain hardening, $d\sigma/d\varepsilon$, uniform elongation, δ_u, and a coefficient of a normal plastic anisotropy, R, were evaluated at an uniform deformation stage/13/. $R = \ln(d_0/d_1) / \ln(d_0/d_2)$. The plane anisotropy rate was evaluated by a difference of the extremum (maximum-minimum) average values (by the test results of 3-5 specimens per point, not less) of the properties in the plane of the billet. The amount of the prismatic fraction in the texture in the dispersion field (0-45^{0}C) in accordance with the recommendations of the work /3/ was evaluated by the opposite pole figures: $f^{TD}_{<45^o} = \sum_{0-45^o} (I_{hkil}/I_R)\cdot A_{hkil} \cdot 100\%$.

Test Results

The investigation of the physico-mechanical properties as a function of a specimen cutting direction was carried out on a sheet billets of a texture that is most charactericitics of the $\alpha+\beta$ alloys (Figure 1):

- a single component texture $\{11\bar{2}0\}<10\bar{1}0>$, the c-axis of HCP is in a crosswise direction of the sheet, the $\{11\bar{2}0\}$ plane is parallel to a plane of a rolled billet,
- a biased basic or double basic (0001) $\pm\alpha$ ND-TD $[10\bar{1}0]$, the biased angle of the basic planes from the normal to the sheet plane α= 25-30^0;
- a two component $\{11\bar{2}0\}<10\bar{1}0>$ + (0001)$\pm\alpha$ ND-TD $[10\bar{1}0]$;
- a two component $\{11\bar{2}0\}<10\bar{1}0>$ + $\{11\bar{2}0\}<0001>$ texture with the c-axis in the rolling direction and in the crosswise direction.

The most appreciable anisotropy of the sheet plane (or plane anisotropy) was observed in the material of $\{11\bar{2}0\}$ $<10\bar{1}0>$ texture. By the type of variation of the properties in the sheet plane one can indicate three groups of variations:

- monotonic increase of the properties with the φ-angle variation in the 0-90^0 range (E, $\sigma_{0,2}$);
- the minimum variations at the 45^0 angle relative to the rolling direction σ_B, δ_u, $d\sigma/d\varepsilon$;
- the maximum variations at φ= 60-70^0 (δ_5, ψ).

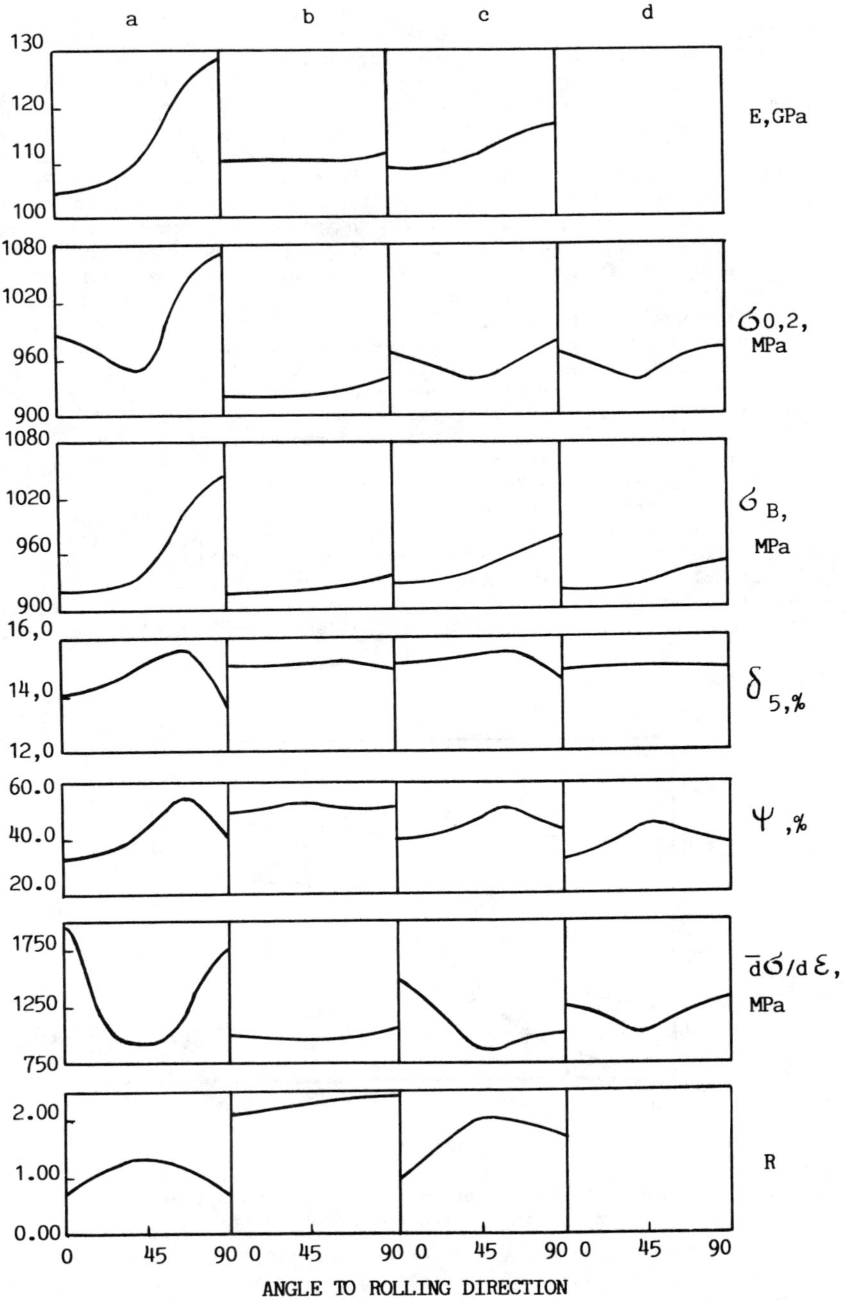

Figure 1 Anisotropy of mechanical properties of Ti-6Al-1Mo-1,5V alloy plates with different texture type:
a-{11$\bar{2}$0}<10$\bar{1}$0>; b-(0001)±αND-TD[10$\bar{1}$0];
c-{11$\bar{2}$0}<10$\bar{1}$0>±αND-TD[10$\bar{1}$0];
d-{11$\bar{2}$0}<10$\bar{1}$0> +{11$\bar{2}$0}<0001>

The plane anisotropic nature of the first group properties is determined by a deviation of a tensile axis relative to the c-axis of HCP, of the second group by the deformation mechanism under a plastic yield, of the third group by the type of an original texture variations with a deformation/13/. The sheet of (0001)$\pm\alpha$ NP-TD [10$\bar{1}$0] texture are characterized by a high **anisotro**py of the properties in the sheet plane and by an increase of a deformation in the thick direction of a sheet, which is indicated by a coefficient R>1 value and hardness values in the base sheet directions (HV_{ND}=357-370, HV_{RD}=324-326, HV_{TD}=322-340). The metal is characterized by low values of the deformation hardening coefficient and high values of a concentrated deformationThe plane anisotropy of the metal of the {11$\bar{2}$0}<10$\bar{1}$0>+ (0001)$\pm\alpha$ ND-TD [10$\bar{1}$0] texture is influenced by the both orientations. The distribution of the properties in the sheet plane is a function of the prismatic component of the texture. The basic orientation is resulted in a reduction of the plane anisotropy rate and in an increase of a deformation resistance in the thick direction of the sheet (R>1).The material of {11$\bar{2}$0}< 10$\bar{1}$0> +{11$\bar{2}$0}<0001> has a reduced anisotropy of the mechanical properties and the extremum values of the properties are observed at 45^0. The reversal rolling of the flat billets heated at about T_t and of a total reduction $\mathcal{E}$= ln H/h > 0,9 without a change of a drawing direction results in a formation of the {11$\bar{2}$0}<10$\bar{1}$0> texture in the studied alloys.

The generation of this texture type is induced by a generation of an intensive texture of β phase with the rolling in the β-range followed by a single alternative $\beta\to\alpha$(110) BCG(0001) HCP transformation /11,12/ occuring at the time of rolling (at the pauses between passes and at cooling, generally). This is to take place if the rolling is carried out at the temperature about the transformation temperature with high deformation rate (Figure 2).

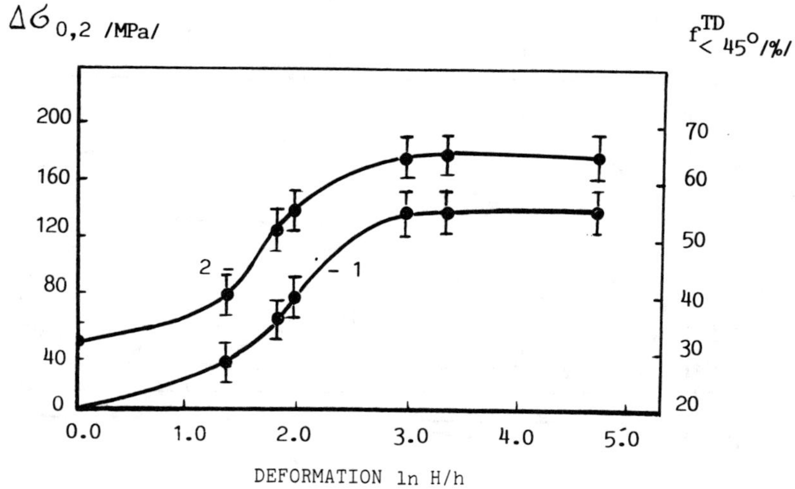

Figure 2 Effect of deformation on anisotropy of yield strength (1) and volume of fraction $f^{TD}_{<45^0}$ (2)

The range of the intensive formation of {11$\bar{2}$0}<10$\bar{1}$0> texture and plane anisotropy corresponds to 0,9-2,5 deformation (the reduction is 60-90 %). Effect of texture of an original material ,the cast including, is observed at the deformation rate = 1,6, not more. If a deformation rate is over 2,5,the rolling is carried out in a range of a preferencial existance of α- phase, which formation features are determined by the generated texture type. The existance of the original {11$\bar{2}$0}<10$\bar{1}$0> texture induces an unlimited plane deformation by the prismatic sliding which results in a stability of the original prismatic [11$\bar{2}$0] orientation in the normal direction and in decrease of [10$\bar{1}$0] orientation in the rolling direction. In the investigated parameter range the anisotropy of the yield strength and the modulus of the normal elasticity as a function of the prismatic fraction amount can be described with a reliable probability by a linear function. The limit values of the plane anisotropy of the flat bar properties of the each alloy with the described texture types is achieved by the deformation of H/h= 2.0-2,5.

The simplest realizable on a commercial base method to obtain the sheet materials of the basic texture (the biased basic texture for the investigated alloys) is the rolling of a billet of the original {11$\bar{2}$0}<10$\bar{1}$0> texture along c-axis in the temperature range of α-phase existance. The crosswise rolling of the 5V alloy billets of the indicated texture type at 850 and 950 ^{0}C results in a generation of the two component texture in a middle layer, where along with a weak {11$\bar{2}$0}<10$\bar{1}$0> orientation an inclined basic (0001) ± 20-30^0 ND-TD [10$\bar{1}$0] texture appears, which intensity increases with the deformation rate increase in the crosswise direction and with deformation temperature reduction (Table2). In a surface layer only the basic orientations appear and their intensity as a function of the temperature and deformation rate are similar.

Table 2 The variation of the texture and the anisotropy of the mechanical properties at the time of rolling

Parameter	The rolling schedule A	B	C	D	E	F	G	H	K
	The first rolling								
T, ^{0}C	950	950	950	900	900	900	850	850	850
ε, %	20	50	68	34	50	60	34	50	60
I_1	1,20	1,52	2,40	1.12	1,66	1,86	1,67	2,06	2,18
I_2	3,70	1,20	0,29	5,17	3,57	2,42	2,21	1,94	1,45
R	0,70	0,78	1,40	1,11	1,19	1,21	1,20	1,24	1,56
$\Delta\sigma_{0,2}$ MPa	- 88	-80	13	17	17	11	2	3	11
	The second rolling								
T, ^{0}C	-	-	-	900	900	900	850	850	850
ε, %	-	-	-	57	40	28	57	40	28
I_1	-	-	-	2,06	2,15	2,20	2.43	2,50	2,60
I_2	-	-	-	1,50	1,84	1,67	0,90	0,90	0,91
R	-	-	-	1,37	1,54	1,67	1,80	2,08	2,19
$\Delta\sigma_{0,2}$ MPa	-	-	-	51	5	4	37	15	2

I_1 is an average value of the intensity of the basic orientation;
I_2 is an average value of the intensity of the prismatic orientation.

There is no existance of the plane anisotropy at 34 % (ln H/h =0,42) deformation rate, actually, with temperature reduction and a crosswise deformation rate increase the plane anisotropy rate increases due to the strengthening of the basic orientations and weakening of the prismatic orientations. The deformation at 950 ^{0}C with deformation rate up to 50 % results in insignigicant weakening of the anisotropy of the properties due to large angle of bias of the basic planes to the direction of the rolling and holding of the prismatic orientations. The deformation rate increase up to 68 % results in the non-existance of the plane anisotropy induced by a generation of a sharp biased basic texture similar to the texture with the rolling at 850 and 900^{0}C.

The second rolling with the rechange of the deformation direction by 90^0 results in a further strengthening of the basic orientations and a weakening

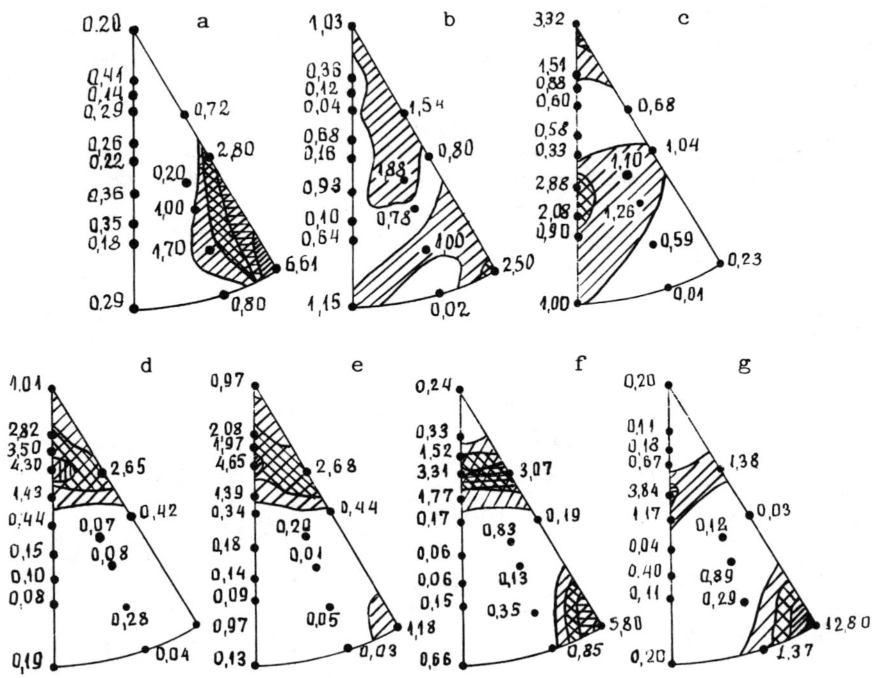

Figure 3 Experimental pole figures of Ti-6Al-1 Mo-1,5V alloy plates
Initial textures:
a,b,c -{11$\bar{2}$0}<10$\bar{1}$0> ; d,e,f,g {0001}±αND-TD [10$\bar{1}$0]
a,d -as hot rolled; solution treated at 850(e), 1000(f), 1040(g), 1060 °C (b,c)
Servey from ND(a,c-g), servey from TD (b)

of the prismatic. The rolling at 850 ^{0}C implies the disappearance of the prismatic component in a middle layer with the reduction of the inhomogeniety of the texture in the thick direction of the sheet. The inclination angle of the basic planes as a result of the second rolling does not change, in fact. The second rolling at 850 and 900 ^{0}C results in a subsequent increase of the plane anisotropy ,which is indicated by the increase of the R coefficient value.
The analysis of the evolution of a texture generation in a material with an original,multicomponent dispersed texture shows that during of the rolling at a temperature range where α-phase amounts to more 50% and the total reduction of up to 60 % the deformation is activated by a sliding in all deformation systems available in the titanium. It results in a small redistribution of the intensity of single component in the texture and non- generation of any definite texture type.
As a result of the billet heating the original $\{11\bar{2}0\}<10\bar{1}0>$ texture type and the anisotropy rate of the mechanical properties (the deformation hardening coefficient, excluding) is attained up to the temperature about $\alpha+\beta\rightarrow\beta$ transformation temperature (Figure 3) .The coefficient of the deformation hardening is very sensitive to the texture variations of the materials. Its value is reduced with anneal temperature increase ,which is due to the generation of an appreciable strong divergence of the basic orientations.
The effect of a postheating cooling rate on the texture and the anisotropy rate of all mechanical properties is observed at heating over $\alpha+\beta\rightarrow\beta$ transformation temperature only.
A β-anneal of the material with the sharp single component $\{11\bar{2}0\}<10\bar{1}0>$ or (0001) $\pm\alpha$ND-TD $[10\bar{1}0]$texture results in a generation of a sharp texture with a maximum pole density of the basic orientations in the rolling direction and in the crosswise direction. It induces the extremum appearance at the 45^0 to RD on the property level vs. direction of specimen cutting curve.
The accelerated cooling from the temperature over T_t weakens the anisotropy of the properties as compared to air cooling.
The (0001) $\pm\alpha$ND-TD$[10\bar{1}0]$texture type is held without an appreciable variations after the heating up to the temperature corresponding to 30-40 % αphase content. The heating up to a higher temperature (up to $\alpha+\beta\rightarrow\beta$ transformation temperature) results in a two component texture generation of the (0001) $\pm\alpha$ ND-TD $[10\bar{1}0]+\{11\bar{2}0\}<10\bar{1}0>$ type.
With the increasing of the heat treatment temperature (in the indicated range) the $\{11\bar{2}0\}<10\bar{1}0>$ component is strengthened, the amount of the basic fraction is reduced with the reduction of the range of dispersion of the basic orientations and with the increase of the biased angle of the basic poles fromthe rolling plane. The described texture variations result in the weakening of the normal and in the strengthening of the plane anisotropy of the properties. The average value of the coefficient of the normal plastic anisotropy is reduced to the values in 0-45 ^{0} range, mainly.

Conclusions

The results of the performed investigation of the effect of the parameters and schedules of the hot rolling, textire conditions of the billet and the heat treatment during of a texture generation process and the appearance of the anisotropy of the physico-mechanical properties in the Ti-6Al-1Mo-1,5V and Ti-4,5Al-1,8V alloys allow to determination of conditions to manufacture flat semiproducts in $\alpha+\beta$ titanium alloys of the martensitic type monitored by the anisotropy of the physico-mechanical properties.

References

1. G. Wassermann, J. Grewen, Textures of metallic materials (Moscow,Metallurgy,1969).

2. A.W. Bowen ,Scripta Metallurgica. 11,(1) (1977), 17-21.

3. I.D. Wishnjakov et al. Theory of texture formation in metals and alloys (Moscow, Science,1979),178-199.

4. Rbert Adamesku,Pavel Geld and Eugene Mitjushkov, Anisotropy of physical properties of metals (Moscow, Metallurgy, 1985), 118-130.

5. A.A. Babareko, " Textures of metals and alloys", The results of science and technics. Science of metals and heat treatment. VINITI,13 (1980), 79-148

6. A. Sommer et al. Titanium. Science of metals and technology. Proceedings of the 3th International Conference, vol. 3 (Moscow, VILS, 1978),87-96.

7. M. Peters and G. Luetjering, Titanium 80, Science and Technology. Proceedings 4th International Conference.(1980), vol. 2, 925

8. A.Tanabe et al. Titanium 80 ,Science and Technology. Proceedings 4th International Conference. vol. 2, 8(1980), 201.

9. A. Bowen, Titanium 80, Science and Technology. Proceedings 4th International Conference. vol. 2 (1980),215.

10. T.Nishimura et al. Titanium 80, Science and Technology. Proceedings 4th International Conference. vol. 2, (1980), 253,

11. N.B. Ageev, A.A.Babareko, Regularity of texture formation in technological processes in alloys with phase transformations, Physics of metals and science of metals.55,(1)(1983), 106-112.

12. A.A. Babareko,O.S. Belova, " Investigation of phase transformations in titanium alloys during deformation", Physics of metals and science of metals, 62,(5),(1986), 944-949.

13. Anisotropy of mechanical properties of $\alpha+\beta$ titanium alloy plates with transverse textures. Science of metals and technology of light alloys, ed.A.F.Belov (Moscow, VILS, 1990),122-127.

FATIGUE CRACK GROWTH BEHAVIOR OF Ti-6Al-4V

AT ELEVATED TEMPERATURE

J. Petit, W. Berata and B. Bouchet

Laboratoire de Mécanique et de Physique des Matériaux, URA CNRS 863
ENSMA , 86034 POITIERS Cedex, France

Abstract

A comprehensive study of fatigue crack propagation behaviour of Titanium alloys at elevated temperature has been undertaken. In a first stage, the near-threshold fatigue crack growth in a Ti-6Al-4V alloy has been studied at 300 °C. The specific influence of air environment has been investigated by comparing the crack propagation behaviour at room temperature and 300 °C in air and in high vacuum. The resistance of the material against crack propagation at elevated temperature is shown to be strongly affected by environment at growth rates lower than 10^{-8} m/cycle. Effective crack growth data (i.e. when crack closure is taken into account) bring out a detrimental environmental effect which can be associated to an accelerated oxygen assisted propagation, while nominal tests at low R ratio (0.1) show enhanced near-threshold oxide induced closure resulting in an increased threshold range. Results are discussed on the basis of microfractographic observations and Xray analysis of crack surfaces.

Introduction

Titanium alloys are particularly suited to help aeronautical industries which have to face two major problems, weight saving and increased operation temperatures in turbine engines. In such a context investigations of damage tolerance properties of Ti alloys experiencing high cycle fatigue is of a large interest. Recently some studies have been carried out on the near-threshold fatigue crack propagation behaviour at elevated temperature. The effect of R ratio has been investigated in ambient air (1-3) showing an increase in the near-threshold growth rates with increasing temperature in Ti-8Al-1Mo-1V alloy. When crack closure is taken into account (2), a temperature effect is still observed at 260 °C compared to Room Temperature (RT). So as to detect the specific influence of environment,from the intrinsic behaviour of the material in relation with its microstructure at high temperature, tests in vacuum were performed on a Ti-6Al-4V alloy to provide reference data in inert environmental condition (4). This paper deals with experiments performed in ambient air and a first analysis of the environmental influence at elevated temperature.

Experimental procedure

The material used was a forged Ti-6Al-4V alloy (wt % 6.27 Al, 3.86 V, 0.12 Fe, 0.18 O_2). After forging the alloy was heated at 965 °C for 1 h, water quenched, heat treated 2 h at 705 °C and cooled in air. The globular microstructure is shown in figure 1. It consists of colonies of aligned α platelets and of equiaxed fine α grains outlined by layers of enriched β phase. The Young modulus, yield stress, ultimate strength and elongation were respectively of 125.3 GPa, 975 MPa, 1035 MPa, 16 % at 20 °C and about 107 GPa, 650 MPa, 770 MPa , 20 % at 300 °C.

Fatigue crack propagation tests were carried out on a servohydraulic testing machine equipped with an environmental chamber providing high vacuum condition (< 5.10^{-4} Pa), on CT specimens 10 mm thick and 32 mm wide, at a test frequency of 35 Hz and a load ratio R = 0.1, or at constant K_{max} (variable R).

Titanium '92
Science and Technology
Edited by F.H. Froes and I. Caplan
The Minerals, Metals & Materials Society, 1993

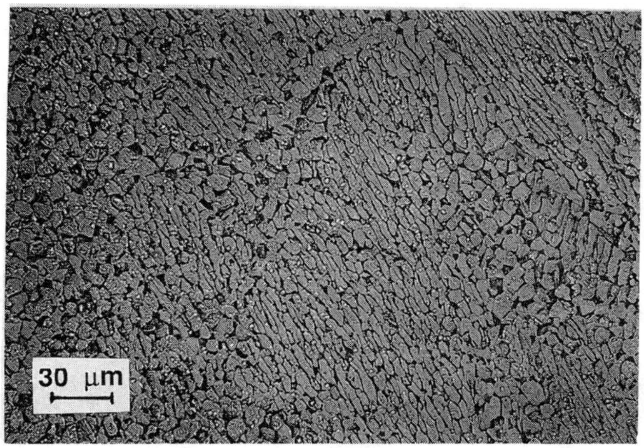

Figure 1 - Microstructure of Ti-6 Al-4V

Crack advance was monitored using a potential drop technique and crack closure was detected at room temperature by mean of the compliance technique with a back face strain gauge. The threshold was determined using a load shedding procedure in accordance with ASTM recommandation. At constant K_{max}, the R ratio was increased so as to eliminate crack closure and to provide the effective crack growth behaviour at room temperature and 300 °C. After threshold, the load amplitude was slightly increased, and subsequent crack growth behaviour was determined at increasing ΔK. Since such near-threshold experiments are highly time consuming, up to now, they have not been duplicated.

Experimental results

In figures 2 and 3 are presented the nominal crack growth data at R = 0.1 and the effective growth data at variable R (so as to eliminate crack closure) determined at 300 °C.

These results are compared to the mean curves of the nominal and effective propagation previously established in ambient air (typically 40 % R.H.) at room temperature (RT) and at a R ratio of 0.1.

Figure 2 shows results of tests performed at decreasing ΔK levels (threshold procedure). It can be observed that the nominal data at 300 °C fall onto the RT curve ; this would, at first sight, indicate a lack of temperature influence. But the effective near-threshold propagation at elevated temperature appears to be very much faster than at RT, which is consistent with a large detrimental influence of temperature. A very low effective threshold range is measured about 2 $MPa\sqrt{m}$. These results are in accordance with previous observations made elsewhere at 260 °C (2).
From comparing effective and nominal propagation curves, it can be assessed that, at low R ratio, an enhanced closure contribution to the nominal ΔK compensates actually the detrimental influence of temperature observed on effective curves.

Figure 3 shows the results of tests performed after threshold at increasing ΔK levels. The nominal crack growth at R = 0.1 and 300 °C is different from that obtained at decreasing ΔK. However the effective behaviour, determined at variable R, is not affected by the test procedure, and previous experiments have shown that the nominal and effective propagation at RT are also roughly independent on the procedure (fig. 2 and 3).Consequently the difference between the nominal curves at 300 °C and R = 0.1 for the two procedures, is stronghtly related to change in crack closure.

On the basis of the results presented in figure 2 and 3, three growth regimes can be considered :

- at mid-rates (> 10^{-8} m/cycle), the influence of temperature consists in a slight acceleration of the effective propagation and in an overlapping of the nominal curves ; for exemple the slightly slowered propagation at a ΔK range of about 12 $MPa\sqrt{m}$, suggests enhanced closure which could be related to enhanced crack surface oxidation at elevated temperature.

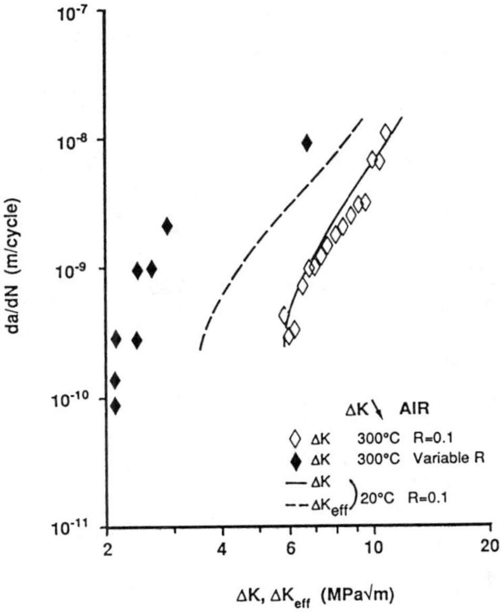

Figure 2 - Influence of temperature on fatigue crack propgation in air at R = 0.1, f = 35 Hz and decreasing ΔK.

- at about 10^{-8} m/cycle, a plateau range is observed on the effective curves suggesting a drastic change in the crack growth mechanism ; this behaviour is shaded off by crack closure an nominal curves.
- at low rates (< 10^{-8} m/cycle), a strong detrimental influence of temperature is brought out by the effective data, and the difference observed between the nominal curves at increasing and decreasing ΔK are typically related to substantial change in crack closure contribution which becomes smaller at increasing ΔK.

Figure 4 shows the evolution of the ratio K_{op}/K_{max}, (the stress intensity factor for crack opening against the maximum of the stress intensity factor), as evaluated from nominal and effective curves at 300 °C and compared to previous measurements at RT (5), and illustrates the variation in the closure contribution due to the test procedure. This result underlines the difficulty to have significant threshold measurements in an active environment because of the high sensibility of closure against test procedure due to oxidation.

Microfractographic observations of cracked surfaces (fig. 5) show enhanced oxidation and suggest a straight relationship between closure contribution and oxide thickening promoted by low R ratios. Such oxide thickening, as detected by mean of scanning microscope equipped with an Xray analyser, favorises a wedge effect as described for steels and Al alloys at room temperature in humid environments (6,7), and induces increased threshold range. Duplication of this kind of experiments is necessary to get a more precise picture of the influence of surface oxidation and on the probable related scatter on the nominal growth rates in the near threshold regime at elevated temperature.

<u>Crack propagation mechanism</u>

Previous results on the same alloy in high vacuum at RT and 300 °C and with the same other experimental conditions, have shown a large influence of the test procedure, which results in a wide hysterisis effect related to a fundamental change in the propagation mechanism when operating near threshold condition at decreasing ΔK (figure. 6) or increasing ΔK (figure. 7). A similar behaviour has been observed at RT on Al alloys tested in vacuum at low and ultra low growth rates (6).

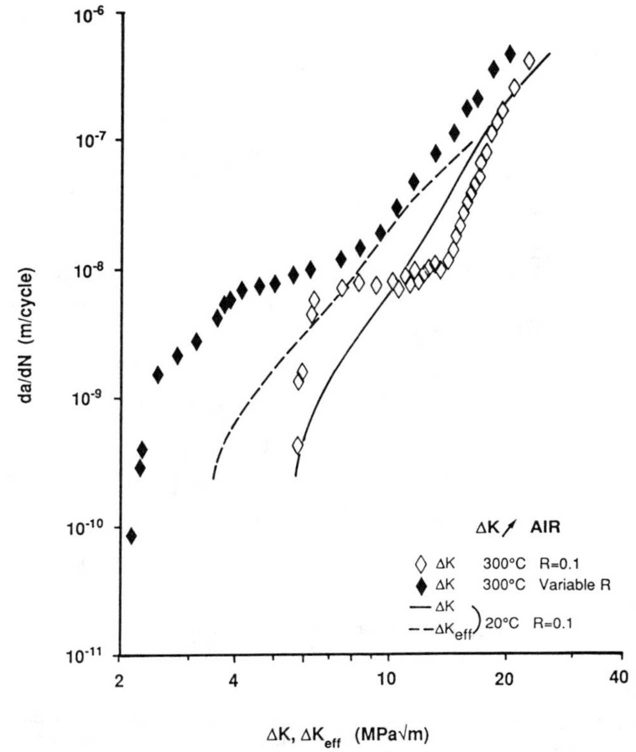

Figure 3 - Influence of temperature on fatigue crack propagation in air at R = 0.1, f = 35 Hz and increasing ΔK after threshold.

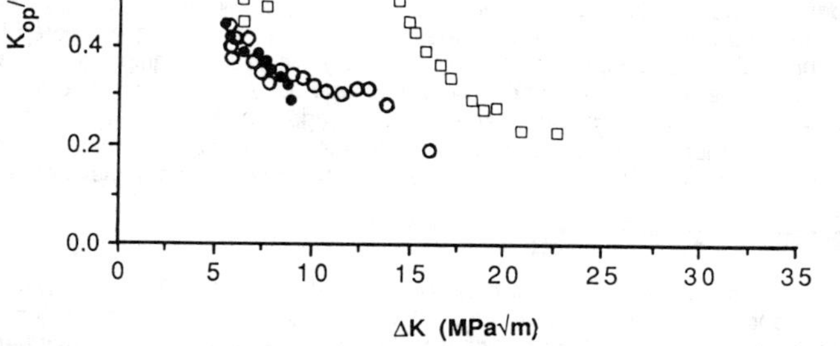

Figure 4 - Variation of the ratio K_{op}/K_{max} with respect to ΔK for tests performed in air at R = 0.1, RT and 300 °C.

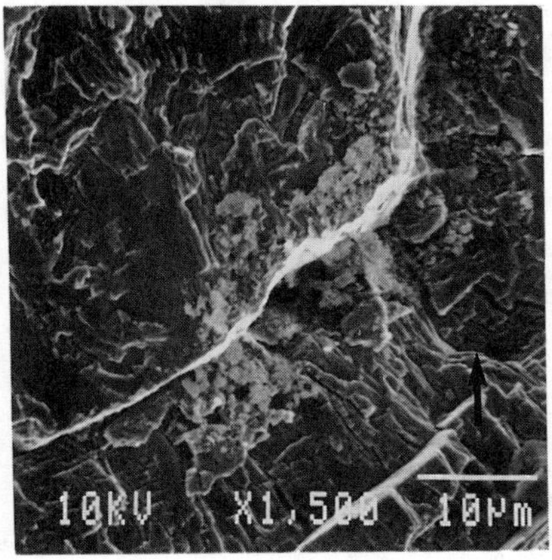

Fig. 5-a) Micrographic aspect of fracture surface at 300 °C in air (R = 0.1, da/dN = 7.10^{-10} m/cycle)

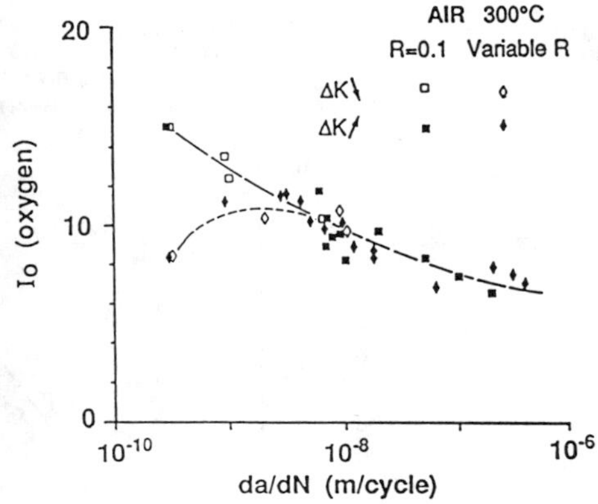

Fig.5-b) Oxygen intensity versus da/dN as detected using an Xray spectrometer (non calibrated intensity.

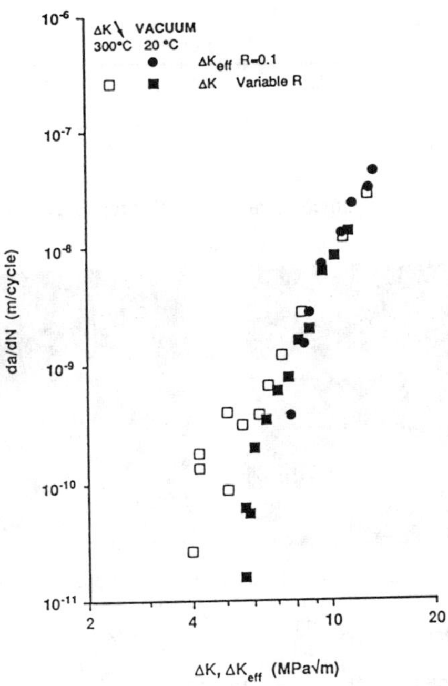

Figure 6 - Influence of temperature on the intrinsic propagation (i.e. in vacuum and after closure correction at decreasing ΔK)

At mid rates, a stage II regime has been identified from tests performed at variable R (no closure), with a flat growth path (figure 8a) resulting from an alternative slip mechanism which can develop along more or less symmetrical slip systems.

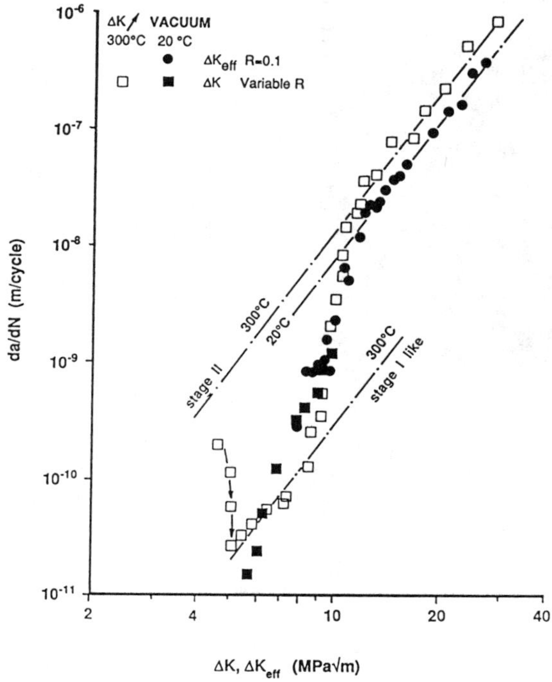

Figure 7 - Influence of temperature on intrinsic propagation at increasing ΔK.

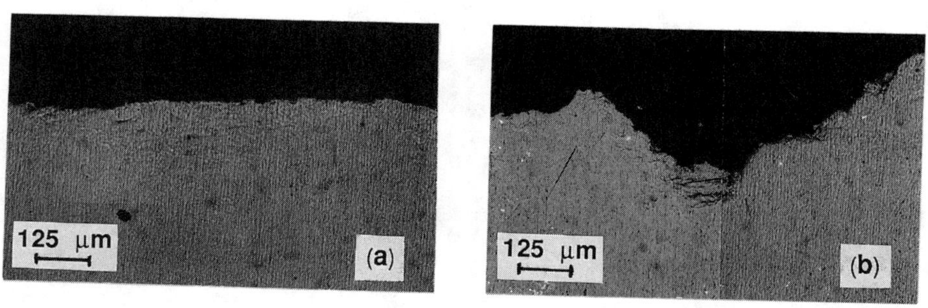

Figure 8 - Near-threshold crack profiles of specimens tested at 300 °C in vacuum at variable R and a frequency of 35 Hz :

a) decreasing ΔK : $\frac{da}{dN} = 2.10^{-10}$ m/cycle, $\Delta K = 4.2$ MPa$\sqrt{m}$;

b) increasing ΔK : $\frac{da}{dN} = 5.10^{-11}$ m/cycle, $\Delta K = 6.5$ MPa$\sqrt{m}$.

Near threshold, a localization of the deformation within a single slip system has been observed at the scale of each individual grain along the crack front. This process results in a retarded crystallographic propagation (figure 8b) due to barrier effects associated to crack deviation and branching at grain boundaries. it is of interest to notice that such very rough crystallographic crack path does not induce larger closure but instead smaller closer than the flat path of a stage II crack and, consequently, the slowed-down rates of a stage I like crack should not be attributed to enhanced closure contribution.

Figure 9 shows that the slight acceleration of the effective propagation observed in the mid-rate range (> 10^{-8} m/cycle) at 300 °C compared to RT can be explained by the diminution of the Young modulus E with temperature. The stage II intrinsic propagation in Ti-6AL-4V can be described using the relationship previously proposed for Al alloys and steels (9) :

$$\frac{da}{dN} = \frac{A}{D^*}\left(\frac{\Delta K_{eff}}{E}\right)^4 \qquad (1)$$

where A is dimensionless and D* a cumulated displacement. The acceleration in air compared to vacuum should result from a diminution of D* in an active environment. At grouwth rates lower that 10^{-8} m/cycle, the crack growth acceleration at 300 °C is not taken into account and the propagation cannot be described using relation (1). This is consistent with an environmentally assisted near threshold propagation.

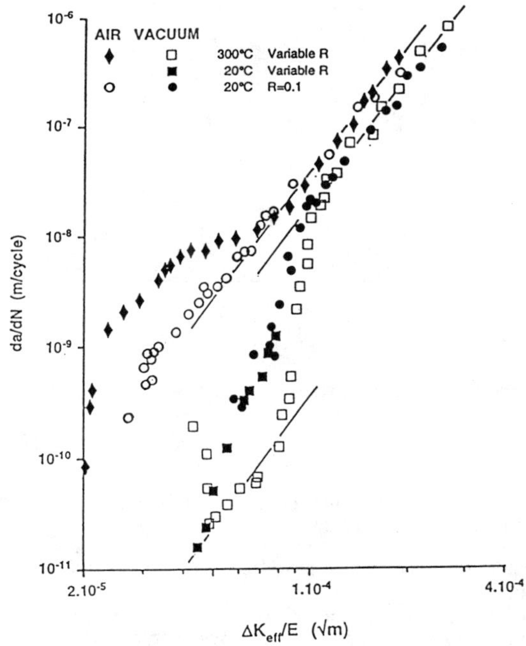

Figure 9 - da/dN vs $\Delta K_{eff}/E$ relationship for effective data at 20 °C and 300 °C in air and vacuum

Figure 10 compares the effective data in air to the intrinsic data in vacuum at elevated temperature. At mid rates, environment has little influence on the stage II propagation.

At low rates (< 10^{-8} m/cycle), the influence of air environment becomes very large due to the combination of two main factors :

- a strong acceleration of the crack growth rate which could be attributed to the development of an environmentally assisted crack growth mechanism involving an embrittling process probably due to oxygen which can penetrate into the plastic zone at the crack tip by dislocation dragging ;
- a change from stage II to stage I like regime which occurs near threshold condition in vacuum but not in air, and is more well marked at 300 °C than at RT.

In such conditions, the growth rate in air can be more than two orders higher than in vacuum as observed at ΔK ranges about 6 to 8 MPa$\sqrt{m}$.

Similar behaviour can be expected for surface microcracks. Ongoing experiments will document this crucial point.

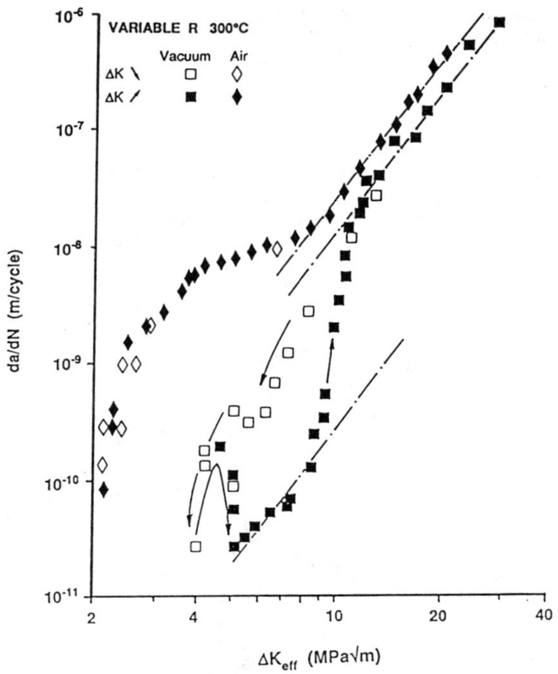

Figure 10 - Effective crack propagation in air and in vacuum at 300 °C (variable R, 35 Hz).

Conclusions

From fatigue crack propagation experiments conducted in air and in vacuum at 300 °C on a Ti-6Al-4V alloy, the following conclusions can be drawn :

- temperature and environment have little influence on stage II crack propagation at rates higher than 10^{-8} m/cycle ;
- compared to vacuum, a detrimental influence of oxygen is proposed to explain a very fast environmentally assisted crack growth observed at 300 °C, which leads to a very low effective threshold range in air ;
- when operating at a low R ratio, closure induced by oxide thickening can more or less compensate the detrimental effect of oxygen embrittlement.

References

1 - Salivar , C.G. and Heine, J.E., Engng Fract. Mech., 32, (5) (1989), 807-817.
2 - Salivar, G.C. and Hoak, F.K., Engng. Fract. Mech., 37, (3) (1990), 505-517.
3 - Vesier, L.S. and Antolovitch, S.D., Engng Fract. Mech., 37, (4) (1990), 753-775.
4 - Petit, J., Berata, W. and Bouchet, B., Scripta Metall. et Materialia, to be published in 1992.
5 - Petit, J., Mendez, J., Berata, W., Legendre, L. and Müller, C., "Short Fatigue Cracks", ESIS 13, (ed. : K.J. Miller, London, U.K. , Mech. Engng. Pub.) ,(1992), in press.
6 - Suresh, S., Zaminski, G.F., and Ritchie, R .O., Met. Trans. ,12 (1981), 1435-1443.
7 - Vasudevan, A.K. and Suresh, S., Met. Trans., 13 (1982), 2271-2279.
8 - Petit, J., Ranganathan, N. and Baïlon , J.P. "Fracture behaviour and desing of materials and structures", (ed. : D. Firao Warley, EMAS pub., U.K.), III, (1990) ,1263-1269.
9 - Petit, J. and Hénaff, G., Scripta Metall et Mater. , in press.

HIGH CYCLE FATIGUE PROPERTIES OF Ti-6Al-4V ALLOYS

AT CRYOGENIC TEMPERATURES

Kotobu NAGAI*, Tetsumi YURI*, Osamu UMEZAWA*, Toshio OGATA*, Keisuke ISHIKAWA*, Yoshimasa ITO**, and Takashi NISHIMURA***

* National Research Institute for Metals, Tsukuba Labs.,
1-2-1, Sengen, Tsukuba, Ibaraki 305, Japan
**Kobe Steel, Ltd., Technical Development Group
***Kobe Steel, Ltd., Iron and Steel Research Labs.,
1-3-18, Wakihamacho, Chuo-ku, Kobe, Hyogo 657, Japan

Abstract

Cryogenic systems require highly efficient structural materials. Titanium alloys are blessed with many advantages like low permeability and low thermal conductivity as well as high strength and low specific gravity. However, very few cryogenic fatigue data are available probably because the test is extremely difficult. The present paper describes the high cycle fatigue property of Ti-6Al-4V alloys at 293 K, 77 K and 4 K.

The Ti-6Al-4V alloys with 0.054, 0.104, and 0.135 mass% oxygen were tested. The ingots were forged(forged material) and part of them were further rolled (rolled material). Processing temperatures were in $\alpha+\beta$ region and finally mill-annealing was done. The fatigue strength increased as the test temperature decreased, which can be accounted for by increased strength at lower temperature. The rolled materials had higher fatigue strength than the forged materials over the temperatures range tested and at given strength level, independently of the oxygen content. This is believed to be due to the difference in the morphology of primary α grains, since the mean grain size itself was almost the same but the forged material had a "colony" structure in contrast with an equi-axed structure of the rolled one.

Introduction

Recent R&D projects based on superconductivity and cryogenics have a wide range of engineering applications like magnetic levitation car, electromagnetic thruster (ship), and superconducting generator etc. Cryogenic structural materials should have a high fracture toughness as well as a high yield strength [1]. In addition, a good fatigue strength is needed [2], since the machines experience stop-run load cycles and they often have "moving" components. And further some other properties are potentially demanded for better heat insulation and more sound operation under high magnetic field(static or alternate).

Titanium (Ti) alloy has many advantages for the cryogenic applications. Its low specific strength, strength-to-gravity ratio, and high yield strength are very favorable for high efficiency of the "moving" machines. And further, the alloy is more blessed with the low thermal conductivity, the extremely low magnetic permeability and the high electric resistivity [3], compared with austenitic stainless steels.

Some of the present authors previously reported the tensile properties, fracture toughness, and high cycle fatigue properties of a Ti-5Al-2.5Sn ELI (Extra- Low-Interstitials) alloy [4]. The reduction of oxygen content yielded the high fracture toughness at 4 K.

Accordingly the Ti-5Al-2.5Sn ELI alloy showed an excellent combination of yield strength and fracture toughness at 4 K and an increased fatigue strength at lower temperature.

Ti-6Al-4V alloy is one of the most popular Ti alloys. The principal interest of the present authors is the probability of cryogenic use of this alloy. The previous paper [5] pointed out that the low temperature fracture toughness of the Ti-6Al-4V alloy was also highly dependent on the oxygen content. Reduction of oxygen suppressed the drop of low temperature fracture toughness. In the lowest oxygen alloy, therefore, the fracture toughness was in the same level at all the temperatures. The present paper reports the high cycle fatigue properties at low temperatures of the Ti-6Al-4V alloys.

Experimental Procedure

Test materials

Three Ti-6Al-4V alloys with different impurity levels were melted; namely a normal-grade one (Normal), an extra-low-interstitial grade one (ELI), and an extremely-low-interstitial grade one (Special ELI, abbreviated as SpELI here). They had different oxygen contents and the nominal oxygen content was 0.15, 0.10 and 0.05 mass%. In the SpELI alloy, no iron was added expecting a better toughness [6], although 0.2% iron was conventionally added in other two alloys. The chemical compositions are listed in Table I.

Ingots were forged finally in $\alpha + \beta$ region, and then a part of them was further rolled also in $\alpha + \beta$ region. The ingot with or without rolling is called "forged material" or "rolled material", respectively. All the materials were finally heat-treated for 7.2 ks at 973 K and air-cooled. Some details of the processing history are shown in Table II.

Table I Chemical compositions of Ti-6Al-4V alloys tested in the present study in mass%

Alloy	Al	V	Fe	O	N	H	C
Normal	6.34	4.23	0.199	0.135	0.0071	0.0053	0.011
ELI	6.23	4.25	0.200	0.104	0.0035	0.0032	0.011
SpELI	5.97	4.12	0.028	0.054	0.0019	0.0055	0.024

Table II Process of forging, rolling, and heat treatment

Alloy	Forging	Rolling	Heat Treatment
Normal	$\alpha + \beta$ (75 x 85)→$\alpha + \beta$ (70 x 70)		
ELI	$\alpha + \beta$ (115 x 120)→$\alpha + \beta$ (70 x 70)	$\alpha + \beta$ (28ϕ)	973 K,7.2 ks Air Cooled
SpELI	β (170ϕ)→$\alpha + \beta$ (70 x 70)		

Forging or rolling temperatures are indicated as "$\alpha + \beta$ " which accounts for a temperature in the $\alpha + \beta$ region. The final section size in one heat series is shown in parenthesis (mm).

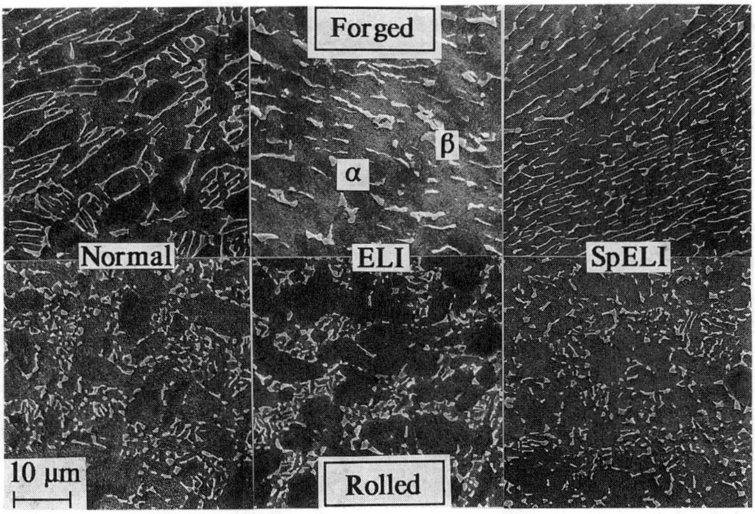

Figure 1 - SEM photographs of microstructure. The print planes are on the transverse section of the fatigue test pieces.

Microstructure

A finer and more homogeneous microstructure was expected for the rolled material because of the heavier reduction in forming. Figure 1 represents the SEM photographs of microstructure. The forged materials have lamellar microstructure principally composed of elongated or plate-like α and β (or transformed β) platelet. The formation of "colony", namely the region in which α plates are aligned, is seen especially in ELI and SpELI alloys. The mean width of α grain was 5.0, 4.0, and 1.9 μm in Normal, ELI, and SpELI, respectively. In the rolled materials, the α grain becomes globular. The mean diameter of α grain was 4.0, 4.0, and 2.8 μm in Normal, ELI, and SpELI, respectively.

Rolling did not always make α grains finer. The Normal and the ELI alloys had almost the same α grain size. The SpELI alloy had the finest microstructure. However, the reason is not made clear.

High cycle fatigue test

Hourglass type unnotched test pieces with a waist diameter of 4.5 mm were machined in the L-direction for both the forged and the rolled materials. S-N curves at 293, 77, and 4 K were determined using the cryogenic fatigue test machine [7]. In obtaining the S-N curves, the estimation of a million cycles fatigue strength (MFS) was intended. The test machine was servo-hydraulic and its dynamic capacity was ±50 kN. Load control test was done in a sinusoidal wave with a minimum-to-maximum load ratio, R=0.01 at 4 Hz at 4 K and at 10 - 20 Hz at 77 and 293 K.

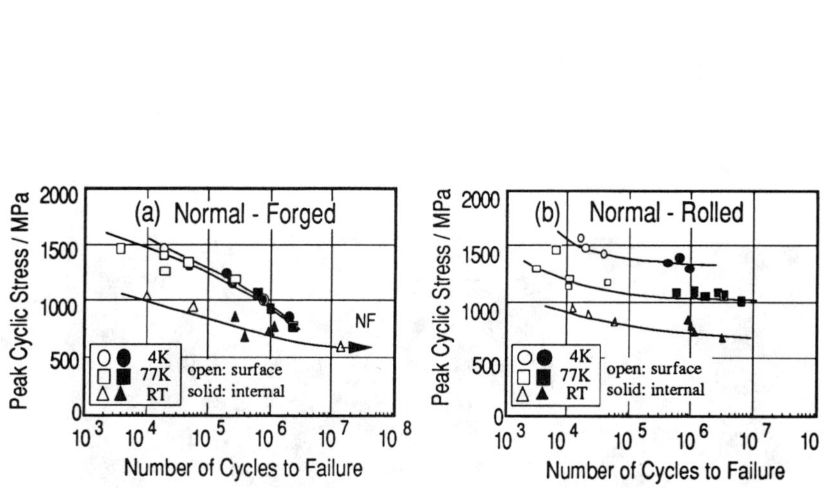

Figure 2 - S-N curves of Ti-6Al-4V Normal alloy: (a) forged, and (b) rolled.

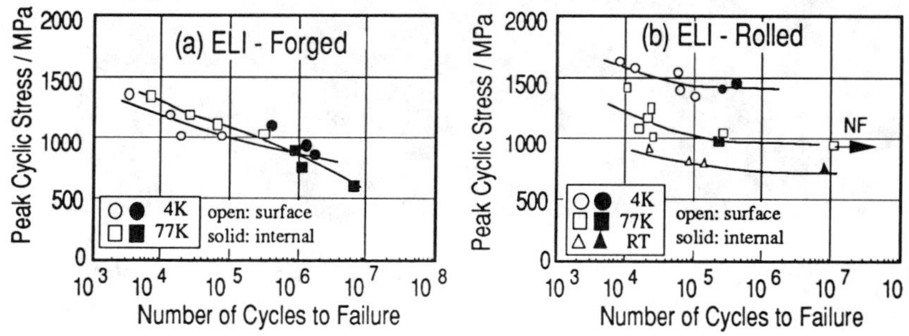

Figure 3 - S-N curves of Ti-6Al-4V ELI alloy: (a) forged and (b) rolled.

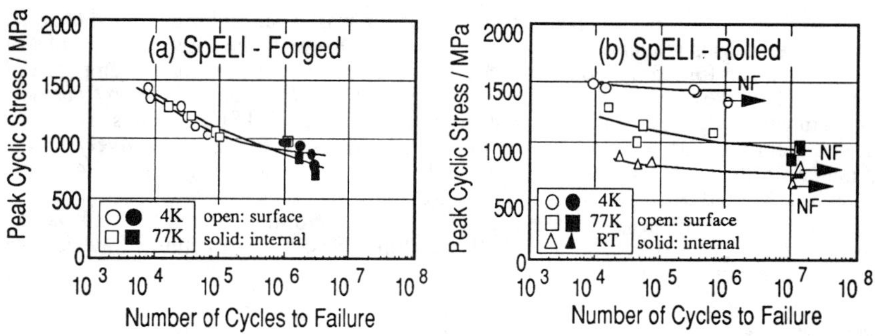

Figure 4 - S-N curves of T-6Al-4V SpELI alloy: (a) forged and (b) rolled.

Results and Discussion

High cycle fatigue property

All the S-N curves obtained in the present study are described in Figures 2 to 4. In these materials, two kinds of fatigue crack initiation site were observed. One was at the specimen surface, and the other was in the specimen interior. SEM photographs of the latter cases are presented in Figure 5. In Figs. 2 to 4, therefore, the crack initiation sites are classified into "Surface" or "Internal".

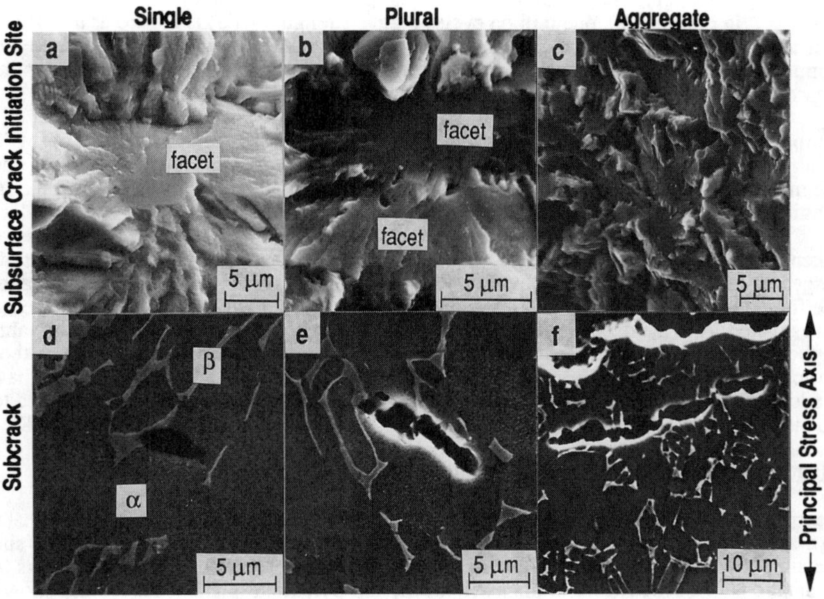

Figure 5 SEM micrographs of subsurface crack initiation sites and sub-cracks for Ti-6Al-4V forged materials : (a) Normal, 4K, 1,235MPa, (b) ELI, 4K, 944MPa, (c) ELI,77K, 601MPa, (d) Normal, 4K, 1,235MPa, (e) Normal, 4K, 1,058MPa, and (f) Normal, 4K, 1,147MPa.

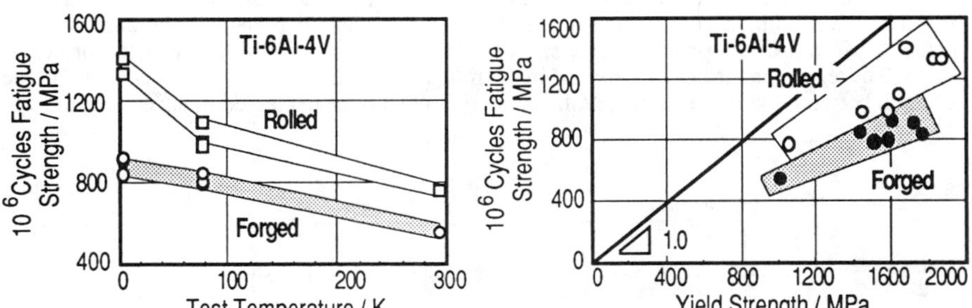

Figure 6 Comparison of a million-cycle-fatigue-strength between forged materials and rolled materials as a function of (a) test temperature and (b) yield strength.

Effect of test temperature and strength level

Temperature decrease produced an increase in strength, and it is generally said that the fatigue strength is proportional to the tensile strength. Hence a simple analogy leads to a speculation that the fatigue strength is increased at lower temperature. It is obviously true for the rolled materials irrespective of the oxygen level; the S-N curves shift to higher stress level at lower temperature and three temperature curves are nearly parallel. In the forged materials, on the other hand, there is almost no gap between the S-N curves at 77 K and 4 K. In the

Normal alloy, the gap becomes narrower as the number of cycles to failure increases and the three temperature curves are supposed to overlap at around 5 million cycles.

In Figure 6 (a), one million cycles fatigue strength (MFS) is plotted as a function of test temperature. As far as the MFS is concerned, the MFS increases with a decrease in temperature and the rolled material is superior to the forged material at all the temperatures. Especially at 4 K, the difference in the MFS between two materials is distinctly large.

Ratio of yield strength to ultimate tensile strength was higher than 95% at all the temperatures for three alloys. Therefore, the interrelation between strength and MFS is described in terms of yield strength vs. MFS as in Figure 6 (b), since yield strength is one of the most important measures in the selection of candidate materials. Anyway almost the same plotting was done when the tensile strength was taken as the abscissa.

As seen in Figure 6 (b), the plots form two separate groups when the processing is taken as a parameter. In other words, the MFS of the rolled material is higher than that of the forged material at a given yield strength. Roughly speaking, either in the rolled material or in the forged material, the MFS is proportional to the yield strength over the temperature range investigated. And the dependence of MFS on yield strength is less in the forged material than in the rolled material. This corresponds to the above-mentioned result in Figure 6. Although more detailed comparison leads to a different conclusion that at 4 K the SpELI alloy, having the lowest yield strength, showed a higher MFS than other two alloys, this is considered to be no major concern here except the fact that the SpELI alloy had the maximum MFS at 4 K.

Microstructural factors and fatigue strength

Texture. A.Sommer et al. said that in textured Ti-6Al-4V alloy, the alternating stress parallel to c-axis introduced longer fatigue lives than that perpendicular to c-axis[8]. Figure 7 shows the simplified X-rays analysis of texture, namely X-rays intensities of three major peaks from prismatic plane, $(10\bar{1}0)\alpha$ basal plane, $(0002)\alpha$, and pyramidal plane, $(10\bar{1}1)\alpha$ are shown in relative ratios to normalized total intensity,

$$I\alpha = 1/3\,(I_{(10\bar{1}0)\alpha} + I_{(0002)\alpha} + 1/4\, I_{(10\bar{1}1)\alpha})$$

This equation is based on an assumption that the ratio of the intensities is equal to 1 : 1 : 4 where the calculated ratio is 23.2 : 25.0 : 100 for randomly-oriented-polycrystalline pure-Ti (Cu-Kα). Therefore the maximum of relative intensity, $I/I\alpha$, is 3.0 and the larger $I/I\alpha$ shows the more preferred orientation. From Figure 7, it is concluded that all the materials have the similarly textured microstructure in which the prismatic plane is perpendicular and the basal plane is parallel to the principal stress, that is, the c-axis is perpendicular to the principal stress. The texture is further accentuated in the rolled materials. Hence the difference in fatigue strength between the forged and the rolled materials can not be explained in terms of texture.

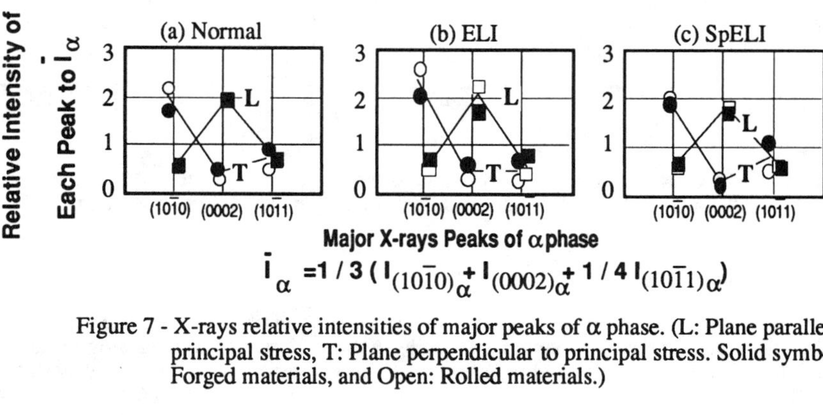

Figure 7 - X-rays relative intensities of major peaks of α phase. (L: Plane parallel to principal stress, T: Plane perpendicular to principal stress. Solid symbols: Forged materials, and Open: Rolled materials.)

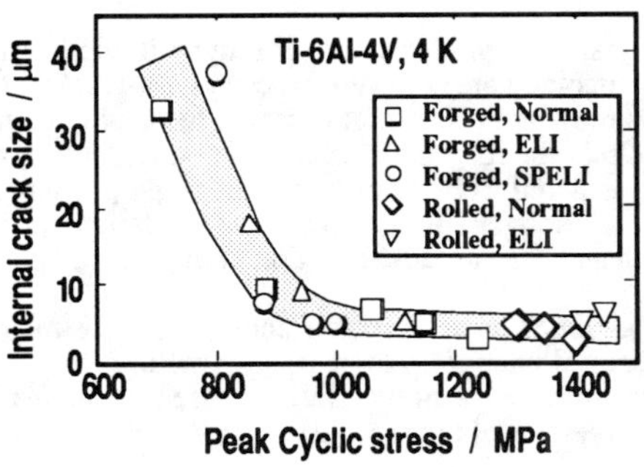

Figure 8 - Relationship between internal crack size and peak cyclic stress at 4K for Ti-6Al-4V alloys.

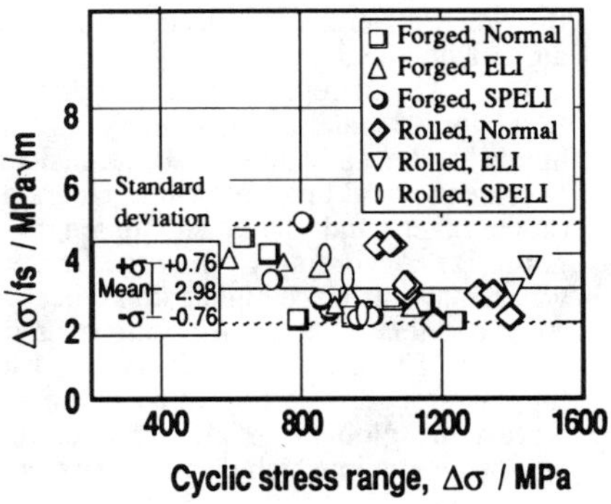

Figure 9 - Relationship between cyclic stress range 'Δσ' and Δσ√fs for Ti-6Al-4V alloys. 'fs' is the size of internal crack site.

Table III EDS analyses of α,β and internal initiation facet in the Normal alloy.

Region	Concentration (mass%)		
	Al	V	Fe
α phase	5.4 - 6.9	2.1 - 5.8	0.0 - 0.3
β phase	4.1 - 5.9	3.5 -11.8	0.1 - 1.2
initiation facet	4.8 - 7.3	1.6 - 3.0	0.0 - 0.1

Internal Initiation Site. Internal initiation prevailed more at lower cyclic stress as seen in Figs. 2 to 4. The site was composed of facet-like unit(s) and inclined by several ten degrees to the principal stress axis. The size became larger at lower stress as seen in Figure 5 (single, plural, and aggregate). Figure 8 shows the size varies between several μm and several ten μm. Here the size was defined as the minor axis of an orthographic projection of the initiation site on the main crack propagating plane, fs. The presence of subcracks near the main crack and its coalescence under lower stress, shown in Figure 5 (d)-(f), demonstrate that the first stage of fatigue is the subcracking of microstructure and the subcracks grow until a main crack forms originating at one of the subcracks. Figure 9 rearranges the data in Figure 8, when a critical condition like $\Delta K_I = k\,\Delta\sigma\sqrt{fs}$ is assumed where Δσ is cyclic stress range. The value of $\Delta\sigma\sqrt{fs}$ is in a narrow range around the mean value of 2.98 MPa√m. Strictly speaking, the coefficient k does not always take a constant value. The value depends on the shape and location of the initiation site. When the shape is taken into account, the smaller fs tends to have a higher k value due to its lower aspect ratio. However, the coefficient k can be roughly estimated to be around 2. Then ΔK_I becomes about 6 MPa√m. For the Ti-6Al-4V alloy, ΔK_{th} is slightly lower than 10MPa√m at 4K [9] and K_{IC} is greater than 30 MPa√m at 4K [10]. Hence it is considered that a main crack starts to grow when the initiation site grows up to a critical size determined for a given cyclic stress by a critical condition.

The α grain is a subcracking microstructure. As shown in Figure 5, subcracks were produced in α grains. Table III shows a comparison of point analysis data for α, β phases and facets in the initiation site. The chemical compositions of the facet are very similar to those of the α phase. In addition, the size and shape corresponded to those of the α grains. Namely, in the forged materials the sites were composed of elongated facet(s) and, on the other hand, in the rolled materials the facets were rather globular. Hence, the facet in the initiation site is attributed to an α grain.

Some reported that finer primary α grain size produced higher fatigue strength [11]. In the forged material, the primary α grains in a colony are believed to be crystallographically aligned and act as a single path for dislocation moving [12]. In that case, the mean slip length becomes several times of primary α grain width in the forged material. In the rolled material, on the other hand, the mean slip length is considered to be of an order of single primary α grain size. Accordingly, the forged material has a longer slip length than the rolled material, which may introduce higher stress localization and easier crack initiation at the same applied stress and this leads to lower fatigue strength.

From these consideration, it is concluded that the significant difference in the fatigue strength was caused by the morphological change in their microstructure. In other words, the globular α grain microstructure produces a higher fatigue strength than the microstructure in which plate-like or elongated α grains of the same size form "colonies".

Summary

The cryogenic high cycle fatigue properties were investigated for Ti-6Al-4V alloys with varied oxygen level; nominally 0.05, 0.10,and 0.15 mass %. The alloys were prepared both in the as-forged material and in the rolled material.

Mean primary α grain size of the lowest oxygen alloy was smallest. Although rolling did not always produce the finer α grain size, the process changed the morphology of α grains. Namely the globular grains were obtained in the rolled material in place of plate-like α grains forming "colony" in the forged one.

The fatigue strength increased as the test temperature decreased. This can be accounted for by increased strength at lower temperature. The fatigue strength of the rolled material was superior to that of the forged one especially at 4 K. At a given strength level, the former was higher than the latter over the temperature range investigated.This is believed to be ascribed to the difference in α grain morphology.

References

1. K.Yoshida, H.Nakajima, K.Koizumi, M.Shimada, Y.Sanada, Y.Takahashi, E.Tada, H.Tsuji, and S.Shimamoto, Austenitic Steels at Low Temperatures (New York, NY:Plenum Press, 1983), 29 - 39.
2. K.Hirano, New Aspects of Non-magnetic Steels (Tokyo, ISIJ, 1990), 85 - 94.
3. K.Nagai and K.Ishikawa, "Deformation and Fracture Characteristics of Titanium Alloys at Low Temperatures", Tetsu-to-Hagané, 75 (1989), 707 -715.
4. K.Nagai, T.Ogata, T.Yuri, K.Ishikawa, T.Nishimura, T.Mizoguchi, and Y.Ito, "Fatigue Fracture of Ti-5Al-2.5Sn ELI Alloy at Liquid Helium Temperature", Trans. ISIJ, 27(1987),376.
5. Y.Ito, T.Nishimura, K.Nagai and K.Ishikawa, "Cryogenic Properties of Extra-Low Oxygen Ti-6Al-4V Alloy", Proceedings of Sixth World Conference on Titanium, ed. P.Lacombe, R.Tricot and G.Beranger (Cannes, 6-9 June 1988), 87 -92
6. R.G.Broadwell and R.A.Wood, "Titanium Alloys for Cryogenic Service", Materials Research & Standards, 4(1964),549.
7. T.Ogata and K.Ishikawa, "A New Liquid Helium Temperature Fatigue Testing System", Trans. ISIJ, 26 (1986), 48 - 52.
8. A.Sommer, M.Creager, S.Fujishiro, and D.Eylon, Proc. of 3rd International Conference on Titanium, the Metall. Soc. of AIME, (1976), 1863.
9. R.L.Tobler and R.P.Reed, "Fatigue Crack Growth Resistance of Structural Alloys at Cryogenic Temperature", Advances in Cryogenic Engineering Materials, 24, (1978), 82-90.
10. K.Nagai, K.Hiraga, T.Ogata and K.Ishikawa, "Cryogenic Temperature Mechanical Properties of β-annealed Ti-6Al-4V Alloys", Trans. JIM, 26,(1985), 405-413.
11. J.J.Lucas and P.P.Konieczny , "Relationship between Alpha Grain Size and Crack Initiation Fatigue Strength in Ti-6Al-4V", Metall. Trans., 2(1971), 911 - 912.
12. R.Brown,"Fatigue of Ti-6Al-4V with a Widmanstatten Microstructure", Fracture and Fatigue, Elasto-Plasticity, Thin Sheet and Micromechanisms Problems, Proc. the 3rd Colloquium on Fracture, ed. J.C.Radon (London, Pergamon Press, 1980), 371-378.

FATIGUE CRACK PROPAGATION IN Ti-6Al-4V ALLOYS

CONTAINING RETAINED METASTABLE β PHASE

M. Niinomi *, T. Kobayashi *, and A. Shimokawa**

* Department of Production Systems Engineering, Toyohashi University of Technology, Toyohashi 441, Japan.
** Graduate School, Toyohashi University of Technology Toyohashi 441, Japan.

Abstract

Fatigue crack propagation characteristics in solutionized specimens and aged specimens of Ti-6Al-4V forged bars and rolled plates were investigated. Fatigue crack propagation rate of IIb region in solutionized specimens is greater than that in aged ones in, in particular, rolled Ti-6Al-4V plates. Crack closure of solutionized specimens is greater than that of aged ones in both forged bars and rolled plates of Ti-6Al-4V alloys. Deformation induced twins or martensite is consider to induce crack closure and retard the fatigue crack propagation rate in, in particular, rolled plates of Ti-6Al-4V alloys.

Introduction

Greater improvement in elongation, strength, toughness or fatigue life can be achieved in solutionized (ST) Ti-6Al-4V alloys under a certain condition compared with aged ones after solutionizing (STA) [1] - [2]. Such improvement in mechanical properties of Ti-6Al-4V alloys is attributed to the transformation of the retained metastable β phase to the martensite phase like orthorhombic martensite, α", due to the deformation.This phenomenon is so called " Deformation Induced Transformation ". This method for improving ductility or toughness is also well known in steels [3] , cast irons [4] and partially stabilized zirconia [5].

The deformation induced transformation is also expected to improve the fatigue crack propagation characteristics of Ti-6Al-4V alloys. However, the study on the fatigue crack propagation characteristics of Ti-6Al-4V alloys where the deformation induced transformation is highly expected is in lack while the study on the fatigue life on such alloy can be seen although the number of the reports is very ristricted [2].

Therefore, fatigue crack propagation characteristics of as solutionized forged bars and rolled plates of Ti-6Al-4V alloys were investigated related to the deformation induced transformation and microstructures in this study.

Experimental procedures

Materials

The materials used in this study were forged bars and rolled plates of Ti-6Al-4V alloys. Their chemical compositions are shown in Table I.

The forged bars and rolled plates were solutionized at both 1088K and 1173K for 3.6ks followed by water quenching (STQ). Some of the as solutionized bars at 1088K and plates at 1173K were aged at 813K for 14.4 ks followed by air cooling (STA).

Titanium '92
Science and Technology
Edited by F.H. Froes and I. Caplan
The Minerals, Metals & Materials Society, 1993

Table I Chemical Compositions of Ti-6Al-4V Alloys (mass%).

	Al	V	Fe	C	N	O	H
Forged Bar	6.03	4.12	0.19	0.01	0.01	0.203	0.0012
Rolled plate	6.37	4.01	0.15	0.08	0.01	0.101	0.0029

Table II Tensile properties of heat treated Ti-6Al-4V forged bars and rolled plates.

	Specimen	σ_y (MPa)	σ_B (MPa)	El (%)	ϕ (%)	E (GPa)
Forged bars	STQ1088F	800.3	1026.6	15.3	34.3	94.2
	STA1088F	996.4	1046.3	12.1	32.8	108.8
	STQ1173F	963.8	1092.3	11.3	32.2	96.0
Rolled plates	STQ1088R	862.3	1070.3	14.5	35.0	114.4
	STQ1173R	932.3	1084.1	9.2	9.4	118.8
	STA1173R	1049.5	1090.4	3.4	4.1	125.0

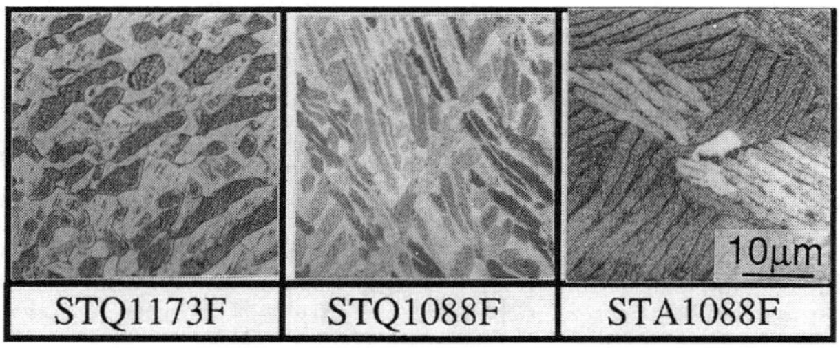

Figure 1 - Optical microstructures of forged Ti-6Al-4V bars conductedewith STQ or STA. The number indicates solutionized temperature in K. F indicate forged bar.

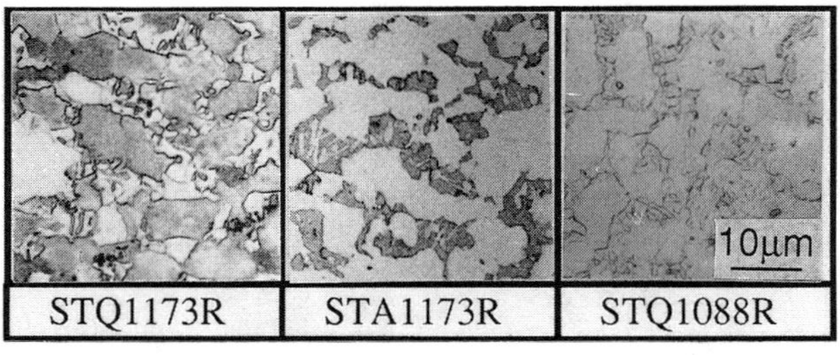

Figure 2 - Optical microstructures of rolleded Ti-6Al-4V plates conductedewith STQ or STA. The number indicates solutionized temperature in K. R indicates rolled plates.

The optical microstructures of forged bars and rolled plates conducted with heat treatments mentioned above are shown in Figure 1 and 2, respectively. The primary α morphology of

STQ1088F and STA1088F is elongated while that of STQ1173F is equiaxed in forged bars as shown in Figure 1. All the specimens exhibit equiaxed α structure in rolled plates as shown in Figure 2.

Tensile properties of heat treated Ti-6Al-4V forged bars and rolled plates are given in Table II where σ_y is yield strength, σ_B is ultimate tensile strength, El is elongation, Φ is reduction of area and E is Young's modulus.

Fatigue crack propagation test

Compact tension type (CT) specimens were machined from the heat treated rolled plates and forged bars so as for the fatigue crack propagation direction to be vertical to rolling and forging directions, respectively. Dimensions of CT specimens were W(width)=3.2×10^{-2}m, B(thickness)=8.0×10^{-3}m for forged bars and for rolled plate W=5.0×10^{-2}m , B=1.0×10^{-2}m for rolled plate, respectively.

Fatigue crack propagation tests were carried out on those CT specimens using electro - servo - hydraulic machine at a load ratio, R of 0.1, with a frequency of 30Hz (sine wave). A constant-load -amplitude method of ASTM E-647-81 [6] was adopted in the higher fatigue crack growth region above 10^{-8}m / cycle. ΔK decreasing method was adopted in the lower fatigue crack growth region than 10^{-8}m / cycle. Fatigue crack length measurements were done on the specimen surface using a light microscope. Crack closure measurements were carried out using back - face strain. Three specimens were used for obtaining one curve of fatigue crack growth rate, da / dN, and cyclic stress intensity factor range, ΔK, relation.

Fracture surfaces were characterized using a scanning electron microscope (SEM). The microstructure near the fracture surfaces were characterized using a transmission electron microscope (TEM).

Experimental results and discussion

Fatigue crack propagation characteristics in forged bars

The relationships between fatigue crack propagation rate, da/dN, and nominal cyclic stress intensity factor range, ΔK, divided by Young's modulus, E, in solutionized specimens (STQ1088F and STQ1173F) and aged specimens (STA1088F) of forged bars are shown in Figure 3 . The STQ1173F specimen exhibits the smallest da/dN, in particular, in Paris' region, that is, IIb region. A difference in da/dN of each specimen is not so big at greater $\Delta K/E$, that is, IIc region. da/dN of STQ1088F specimen is similar to that of STA1088F specimen although da/dN of STA1088F specimen is slightly greater than that of STQ1088F in IIb and IIc regions.

The relationships between da/dN and effective cyclic intensity factor range , ΔK_{eff} , divided by E, in solutionized specimens (STQ1088F and STQ1173F) and aged specimens (STA1088F) of forged bars are shown in Figure 4. The da/dN versus ΔK_{eff} /E curves of solutionized specimens (STQ1088F and STQ1173F) are shifted to much lower ΔK_{eff} /E region compared with that of aged specimens (STA1088F) in IIb region. The da/dN of STQ1088F specimens is much smaller than that of STA1088F specimens in IIb region although a difference in da/dN of STQ1088F and STA1088F specimens is not so big in da/dN and ΔK relation as described above. This result indicates that the crack closure of STQ1088F specimens is greater than that of STA1088F specimens.

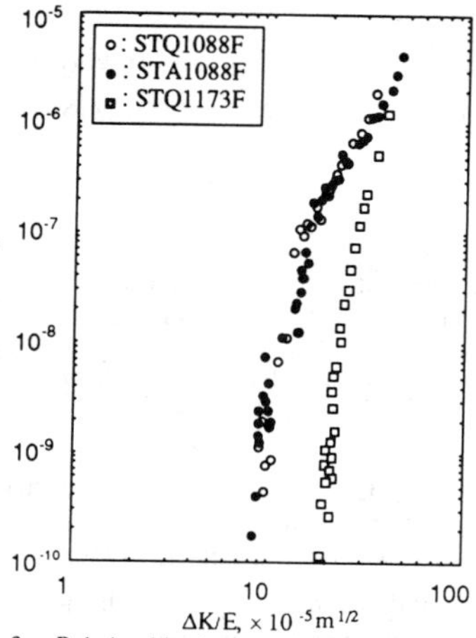

Figure 3 - Relationships between da/dN and ΔK/E in solution treated specimen and aged specimens of forged Ti-6Al-4V bars.

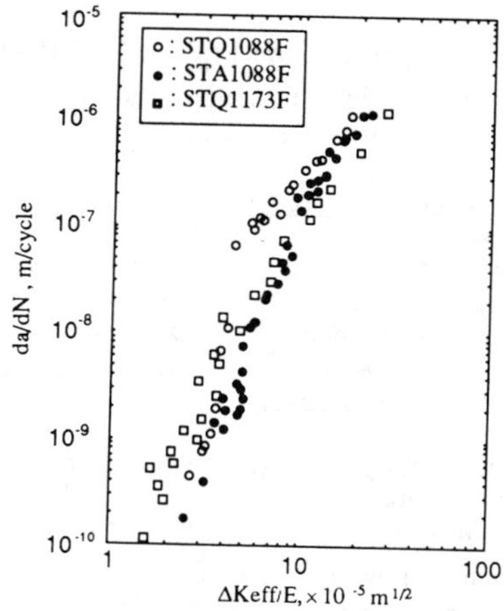

Figure 4 - Relationships between da/dN and ΔKeff/E in solution treated specimen and aged specimens of forged Ti-6Al-4V bars.

The relationships between Kop (stress intensity factor, K, at a crack opening load) / Kmax (K at a maximum load) and Kmax in solutionized specimens and aged specimens, which indicate the crack closure behavior, are shown in Figure 5 . STQ1173F specimens exhibits the greatest Kop/Kmax values compared with the other specimens. This crack closure suggests plastic deformation induced one because the fracture surfaces are fairly flat which will be shown in Figure 6.

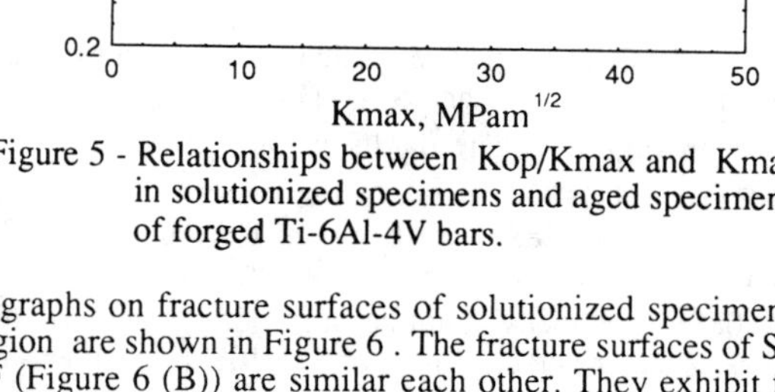

Figure 5 - Relationships between Kop/Kmax and Kmax in solutionized specimens and aged specimens of forged Ti-6Al-4V bars.

Typical SEM micrographs on fracture surfaces of solutionized specimens and aged ones of forged bars in IIb region are shown in Figure 6 . The fracture surfaces of STQ1088F (Figure 6 (A)) and STA1088F (Figure 6 (B)) are similar each other. They exhibit flat fracture surface which occupies a large portion of the fracture surface. Striation like patterns (Figure 6 (D))and intergranular cracking (Figure 6 (A)) can be seen on flat fracture surface. Calculating da/dN from striation spacing fitted with experimental da/dN. Intergranular cracking was found to occur between elongated primary α plates. On the other hand, facet like fracture surface occupies a large portion of the fracture surface of STQ1173F (Figure 6 (C)) .

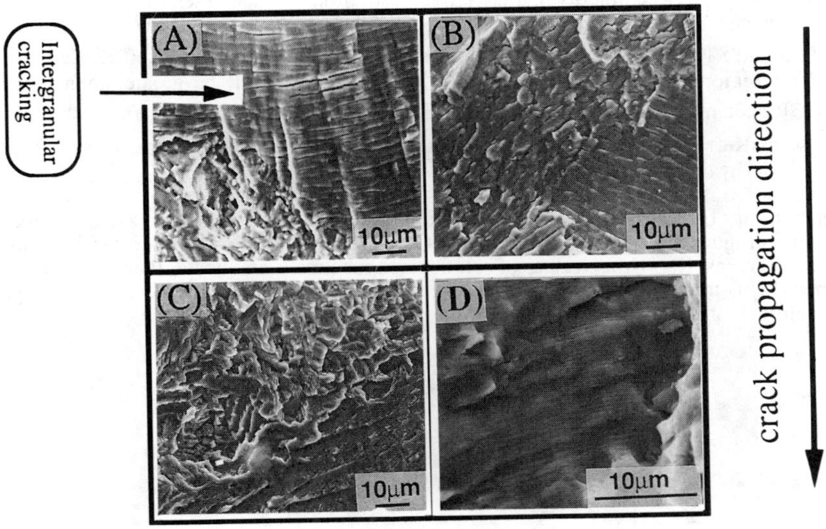

Figure 6 - Typical SEM of fracture surfaces solution treated specimens and aged specimens of forged Ti-6Al-4V bar.
(A) STQ1088F, $\Delta K/E=21 \times 10^{-5}$ $m^{1/2}$, (B) STA1088F, $\Delta K/E=20 \times 10^{-5}$ $m^{1/2}$ and (C) STQ1173F, $\Delta K/E=28 \times 10^{-5}$ $m^{1/2}$, (D) STQ1088F, $\Delta K/E=25 \times 10^{-5}$ $m^{1/2}$.

Fatigue crack propagation characteristics in rolled plates

The relationships between da/dN and $\Delta K/E$ in solutionized specimens (STQ1088R and STQ1173R) and aged specimens (STA1173R) of rolled plates are shown in Figure 7. STQ1088R specimens exhibit the smallest fatigue crack propagation rate in IIb and IIc regions among all the specimens in rolled plates. da/dN of STQ1173R specimens is greater than that of STA1173R specimens in IIb region. da/dN of as solutionized specimens were found to be considerably retarded at $\Delta K/E$ of approximately 20 and 30 $\times 10^{-5} m^{1/2}$ for STQ1173R and STQ1088R, respectively.

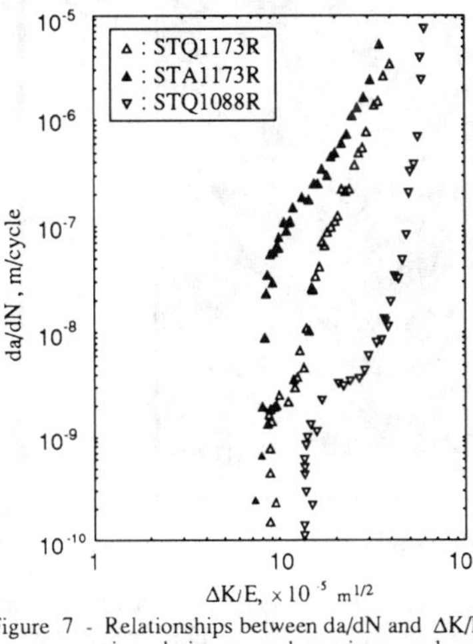

Figure 7 - Relationships between da/dN and $\Delta K/E$ in solution treated specimen and aged specimens of Rolled Ti-6Al-4V plates.

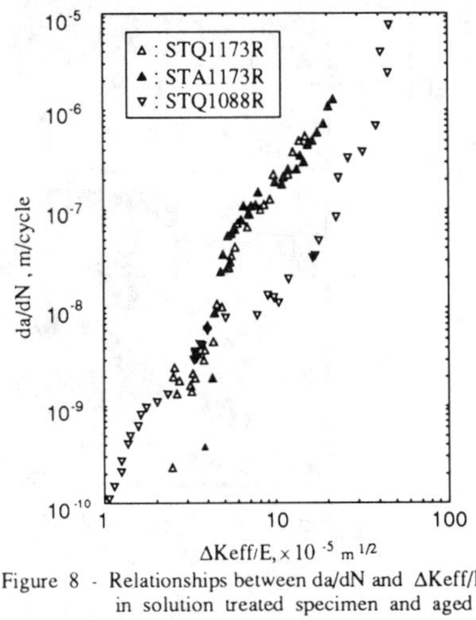

Figure 8 - Relationships between da/dN and $\Delta K_{eff}/E$ in solution treated specimen and aged specimens of rolled Ti-6Al-4V plates.

The relationships between da/dN and ΔKeff/E in solutionized specimens (STQ1088R and STQ1173R) and aged specimens (STA1173R) of rolled plates are shown in Figure 8. STQ1173R and STA1173R specimens exhibit a similar trend in da/dN and ΔKeff/E relations in IIb and IIc regions. Crack closure in STQ1173R specimens will be greater than that in STA1173R specimens. The da/dN and ΔKeff/E curve of STQ1088R specimens is shifted to much lower ΔKeff/E side compared with the da/dN and ΔK/E relation of those. This shift will be also caused by crack closure.

The relationships between Kop/Kmax and Kmax in solutionized specimens and aged specimens are shown in Figure 9. Kop/Kmax values of STQ1088R specimens is much greater than those of other specimens. Kop/Kmax values of solutionized specimens are greater than those of aged specimens. This trend is similar to that in forged bars. The crack closure here also suggests plastic deformation induced one.

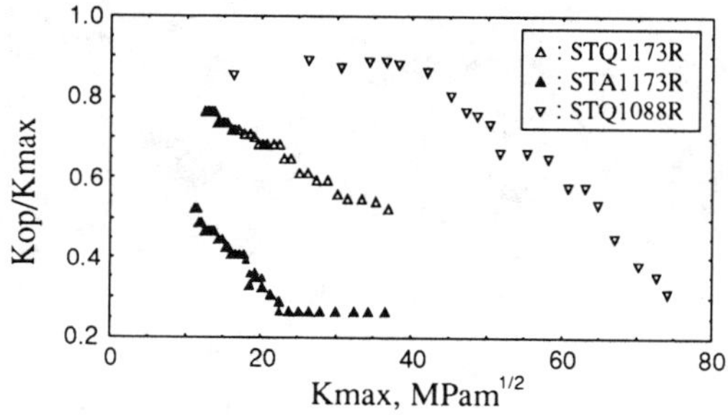

Figure 9 - Relationships between Kop/Kmax and Kmax in solutionized specimens and aged specimens of rolled Ti-6Al-4V plates.

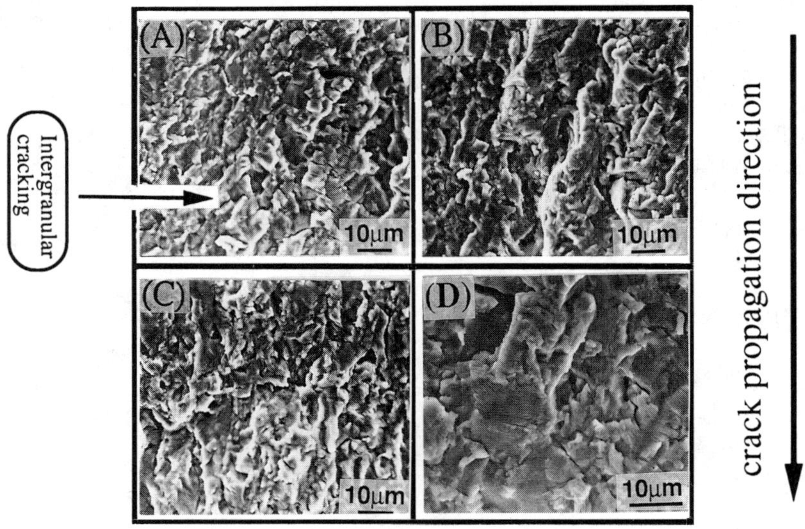

Figure 10 - Typical SEM of fracture surfaces solution treated specimens and aged specimens of rolled Ti-6Al-4V plates.
(A) STQ1173R, $\Delta K/E=17\times10^{-5}$ $m^{1/2}$, (B) STA1173R, $\Delta K/E=13\times10^{-5}$ $m^{1/2}$ and (C) STQ1088R, $\Delta K/E=47\times10^{-5}$ $m^{1/2}$, (D) STQ1173R, $\Delta K/E=22\times10^{-5}$ $m^{1/2}$.

Typical SEM micrographs on fracture surfaces of solutionized specimens and aged specimens of rolled plates in IIb region are shown in Figure 10 . The fracture surfaces of all the specimens are similar each other. Their fracture surfaces are, however, much different from those in forged specimens which have been shown in Figure 6. The fracture surface of each specimen is consists of facet like fracture surface. Striation like patterns (Figure 10 (D)) and intergranular crackings along primary α-β interface (Figure 10 (A)) are also observed on the fracture surface of each specimen.

TEM observation of microstructure near fracture surface

Figure 11 shows the microstructures of β regions in undeformed and deformed parts near the fracture surface formed in IIb region in the STQ1088R specimens. Deformation induced products are observed in the deformed β regions as shown in Figure 11 (b). These products were found to be deformation induced martensite by SADP analysis. This kind of deformation induced transformation is considered to induce crack closure and then retard the fatigue crack propagation rate of as solutionized specimens in, in particular, rolled plates.

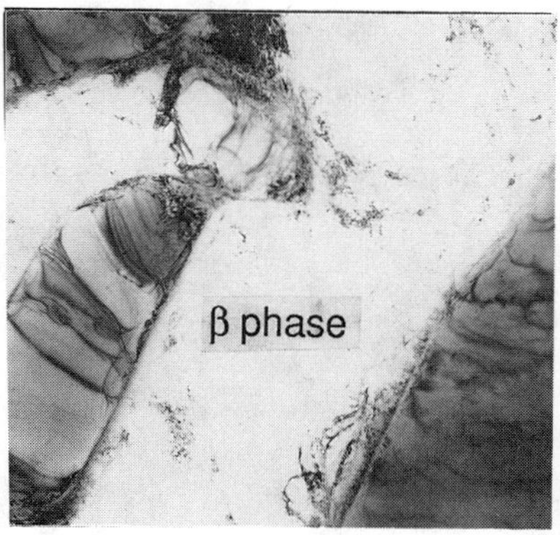

(A)

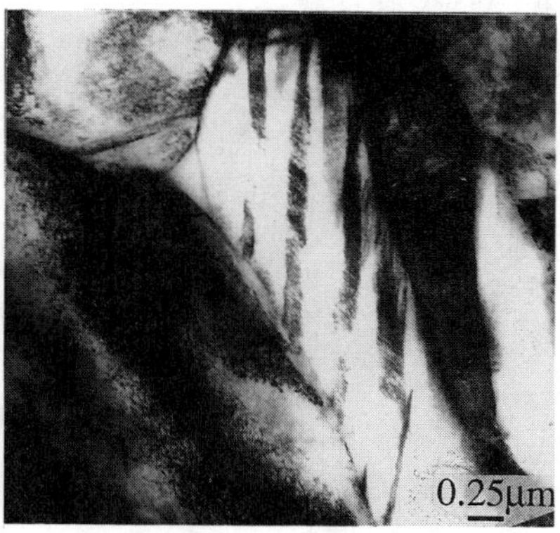

(B)

Figure 11 - Typical TEM micrographs of β regions in (A) undeformed and (B) deformed parts near the fracture surface formed in IIb region in the STQ1088R specimen.

Conclusions

Fatigue crack propagation characteristics in solutionized specimens and aged specimens of Ti-6Al-4V forged bars and rolled plates were investigated. The following results were obtained.

1. In the forged bars, fatigue crack propagation rate in IIb region of solutionized specimens (STQ1088F) is nearly equal to that of aged ones (STA1088F) when da/dN is related to ΔK/E.

2. In the rolled plates, fatigue crack propagation rate in IIb region of solutionized specimens (STQ1173R) is greater than that of aged ones (STA1173R) when da/dN is related to ΔK/E.

3. The fatigue crack propagation rate in IIb region of the specimens solutionized at 1088K is greater than that of the specimens solutionized at 1173K in rolled plates while a opposite trend is recognized in forged plates.

4. Crack closure in solutionized specimens is greater than that in aged specimens regardless of forged bars or rolled plates.

5. Deformation induced martensite is considered to induce crack closure and retard the fatigue crack propagation rate in, in particular, rolled plates.

References

[1] M. Niinomi , T. Kobayashi , I. Inagaki and A. W. Thompson ,"The Effect of Deformation - Induced Transformation on the Fracture Toughness of Commercial Titanium Alloys", Metall. Trans . A ,21A (6) (1990), 1733 - 1744 .

[2] M. A . Iman and C. M . Gilmore , "Fatigue and Microstructural Properties of Quenched Ti-6Al-4V", Metall . Trans . A , 14A (2) (1983) , 233 - 240.

[3] I. Tamura , " On the TRIP Steel (Strain - induced transformation and transformation - induced plasticity) ", J . Iron Steel Inst . ,56 (1970) , 429 - 445.

[4] T. Kobayashi , and H. Yamamoto , "Development of High Toughness in Austempered Type Ductile Cast Iron and Evaluation of Its Properties" , Metall . Trans . A ,19A (2) (1988) ,319 -327.

[5] T. Kobayashi , M. Niinomi , Y. Koide and I. Matsunuma , "Instrumented Impact Testing of Ceramics " , Trans . Jpn . Inst . Met . , 27 (10) (1986) ,775 - 783.

[6] ASTM E647-81 , 1981 Annual BooK of ASTM Standards, part 90, ASTM, Philadelphia, PA, (1981).

EFFECT OF HEAT TREATMENT ON THE CYCLIC SOFTENING

OF Ti-6Al-4V ALLOY.

F.J.Gil, J.M.Manero and J.A.Planell.

Dept. Ciencia de los Materiales e Ingeniería Metalúrgica.
ETSIIB. Universidad Politécnica de Cataluña.
Av. Diagonal 647. 08028-Barcelona. Spain.

Abstract

The Ti-6Al-4V alloy undergoes cyclic softening when cyclically deformed at room temperature at strain amplitudes over the yield strain. Such behaviour is observed for the "mill annealed", α-Widmanstatten and the martensitic microstructure. Softening takes place at an increasing rate as the cyclic strain amplitude is increased. The "mill annealed" microstructure sustains the longest cumulative plastic strain to fracture whilst the martensitic microstructure sustains the shortest. In the present work the dislocation substructures which give rise to such behaviour have been investigated.

Introduction

The high cycle fatigue behaviour of Ti-6Al-4V alloy has been extensively studied in the literature (1-5). However, there is not much information available related to its cyclic deformation response. It seems that the microstructure plays a strong role in the cyclic stress-strain behaviour (6,7). One of the main features, also observed in the Ti6242 near α alloy (8), is the cyclic softening which takes place at the different testing strain amplitudes. Such response seems also to be closely related to the microstructure of the alloy, and there is no clear interpretation accounting for this behaviour.

Although plasticity of high-purity titanium is heavily influenced by twinning at low plastic strains (9), it seems that this mechanism is less active for Ti-6Al-4V due to the effect of aluminium (10). It should be expected that the main plastic deformation mechanism in the α phase could be by glide of mobile dislocations on the different slip systems. It is well known that Ti has several slip systems which may accommodate plastic deformation by means mainly of a dislocations on the prismatic, basal and pyramidal planes, although $\vec{c}+\vec{a}$ dislocations can also become active (9,11,12,13).

In the present work the influence of the microstructure on the cyclic deformation behaviour of Ti-6Al-4V has been investigated. Three different microstructures have been prepared by conveniently controlling the heat treatment of the material. The mechanisms involved in the plastic deformation during the cyclic straining have been analyzed in terms of the unidirectional monotonic behaviour and by means of transmission electron microscopy (TEM) observations carried out in cyclically deformed specimens.

Titanium '92
Science and Technology
Edited by F.H. Froes and I. Caplan
The Minerals, Metals & Materials Society, 1993

Experimental methods and materials

The Ti-6Al-4V alloy used in the present work was kindly donated by Industrias Quirúrgicas de Levante S.A. The material came as cylindrical rods of 12 mm diameter, forged at 950°C, subsequently annealed at 700°C during 2 hours and then cooled in air.

The chemical composition of the alloy is shown in Table I. The microstructure, as observed using optical metallography, corresponds to that know as "mill-annealed" (MA), consisting of equiaxial α grains as well as grains constituted by small α Widmanstatten plates surrounded by residual ß phase. Both the composition and the microstructure satisfy the ASTM F136-84 standard for wrought Ti-6Al-4V ELI alloy for surgical applications.

Table I. Chemical composition of the Ti-6Al-4V alloy used in the present investigation (wt%).

Al	V	Fe	C	O_2	N_2	H_2
6.1	4.0	0.11	0.021	0.09	0.010	0.003

Tensile specimens with a ratio diameter to gauge length of 1/5 and fatigue specimens as shown in Fig. 1 were machined. A first batch of specimens was kept as the as received material (MA) whilst a second and third batch of specimens were heat treated. The specimens of the second batch were held 1 hour in a tubular furnace with argon atmosphere at 1050°C, which is a temperature just above the ß transus for the Ti-6Al-4V alloy, and they then cooled inside the furnace at a cooling rate of 17°C/min. The resulting microstructure (T) corresponds to the "basket weave" which consists in colonies of α Widmanstatten plates surrounded by residual ß phase. The specimens of the third batch were heat treated in the same way but rapidly quenched in water at room temperature. The resulting microstructure (M) corresponds to acicular α' martensite.

Two tensile specimens for each material were tested in a universal screwdriven testing machine of 100kN capacity at a cross-head speed of 1mm/min. The fatigue specimens were cyclically deformed in tension-compresion under strain control R_e =-1, in a servohydraulic testing machine of 100kN capacity. The strain rate was always kept constant at $6.5x10^{-3}$ s^{-1}.

The total strain amplitudes used were $\pm$ $5x10^{-3}$, $\pm$ $7x10^{-3}$ and $\pm$ $12x10^{-3}$. Tests were carried out at each total strain amplitude at different cumulative plastic strains. Two specimens for every microstructure were taken up to fracture at the different strain amplitudes tested.

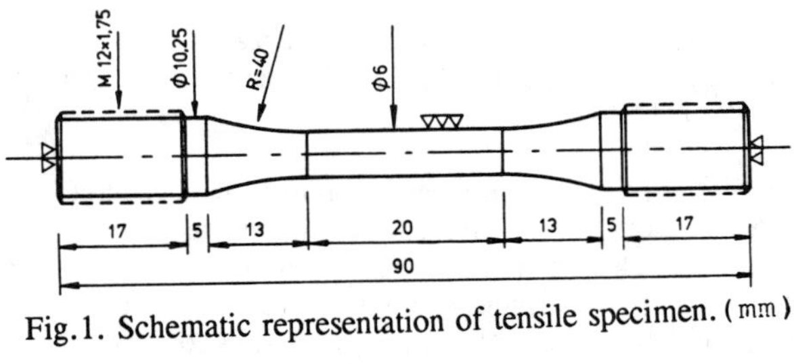

Fig.1. Schematic representation of tensile specimen. (mm).

The deformed specimens were transversely cut into discs of 0.5 mm thickness and 3 mm diameter. After grinding, TEM samples were prepared either using the ion beam thinning technique or by electropolishing the discs in a solution of butylcellusolve, perchloric acid and methanol at 10°C at a voltage of 40 V. Samples were loaded in a double tilt specimen holder and observed in a transmission electron microscope JEOL 1200 EXII.

Results and Discussion

The three materials, MA, T and M have been tested in unidirectional tension in order to obtain their monotonic stress-strain curves, which are shown in Fig.2. The relevant parameters are listed in Table II, where the values reported are the average of two different tests for each material.

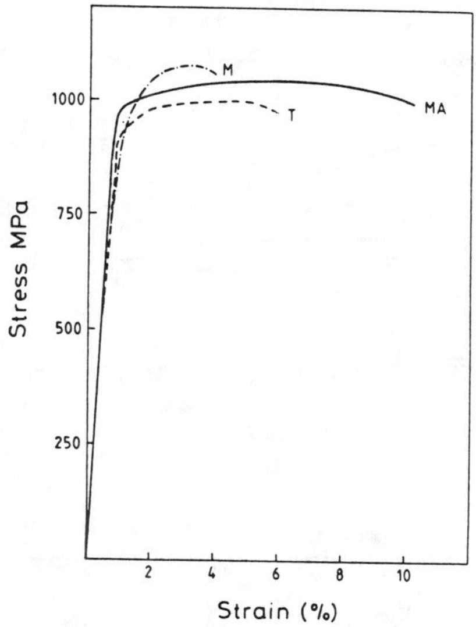

Fig.2. Monotonic stress-strain curves.

Table II. Tensile properties of the three microstructures studied of Ti-6Al-4V alloy.

Microstruct.	Proportional limit (MPa)	Yield Strength 0.2% (MPa)	Ultimate Strength (MPa)	% El	% RA
MA	837	986	1050	14	36
T	567	860	992	6	15
M	568	894	1080	4	7

It should be noticed that acicular microstructures reduce significantly the ductility of the alloy. It should be also pointed out that the proportional limit is much lower than the yield stress for the microstructures T and M. The proportional limit has been measured by means of an extensometer with a sensibility of 10^{-5} per mV. The microstructures T and M exhibit very similar proportional limits, 567 and 568 MPa respectively, and they yield macroscopically at 0.2% at stress of 860 and 894 MPa respectively. This means that irreversible plastic deformation is taking place between the proportional limit and the yield stress. In fact, plastic deformation starts to take place at much lower stresses than for the MA material.

When the three microstructures are cyclically deformed at the different total strain amplitudes, cyclic softening is observed in all cases. Fig.3 shows the cyclic softening curves for the three materials, MA, T and M, at the different total strain amplitudes, where the maximum stress has been plotted against the number of cycles. It should be mentioned that for the MA microstructure softening ocurred from the first cycle, whilst for the microstructure T and M hardening took place during the first two cycles which was followed by subsequent softening until fracture. It can be noticed that the cyclic softening increases as the total strain amplitude increases. In fact, not even at the lowest total strain amplitude, $\pm$ 5x10^{-3}, stress saturation could be reached. It should be pointed out that except for the highest strain amplitude studied, $\pm$ 12x10^{-3}, and only for a limited number of cycles, the maximum stress reached at every total strain amplitude is higher for the MA microstructure than for the other two. The highest rate of softening has been observed for the martensitic microstructure.

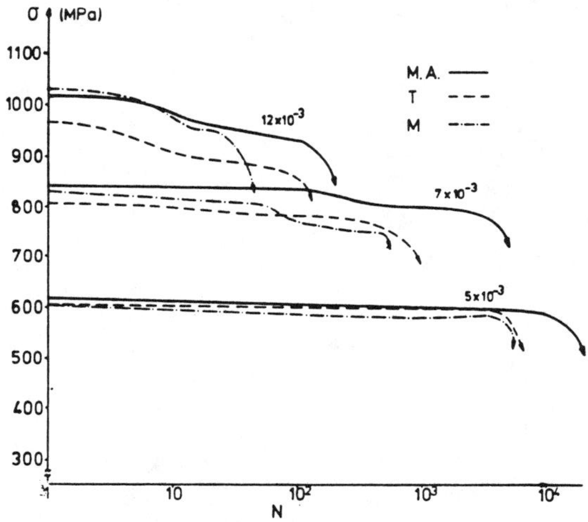

Fig.3. Cyclic softening curves.

Proper saturation is not achieved in any case for the different microstructures and for the different total strain amplitudes. In order to trace the cyclic stress-strain curves for each microstructure, saturation was taken as the stress level at which the rate of softening was the lowest. The trend of the cyclic stress-strain curves for the three microstructures is shown in Fig. 4, by plotting the saturation stress against the saturation plastic strain amplitude corresponding to each total strain amplitude. It can be noticed that the material M shows a higher cyclic hardening behaviour as plastic strain amplitude increases than materials MA and T. This fact agrees with the monotonic hardening behaviour exhibited by the tensile curves shown in Fig. 2. In order to define more exactly the cyclic stress-strain curves of the three microstructures, and more specifically their respective cyclic yield stresses, more tests in the range of total strain amplitudes between $\pm$ 3x10^{-3} and $\pm$ 7x10^{-3} should be conducted. Table III shows the plastic strain amplitudes and the maximum stresses reached at the different total strain amplitudes for the three microstructures at the first cycle.

It should be noticed that for the total strain amplitude of $\pm$ 5x10^{-3}, at the first cycle, a certain degree of plastic deformation has already taken place for the microstructures T and M, being their respective maximum stresses over their tensile proportional limits. At this strain amplitude, the microstructure MA is still being deformed inside its microstrain region. At the total strain amplitude of $\pm$ 7x10^{-3}, the MA microstructure is already being plastically deformed from the first cycle. Finally, at $\pm$ 12x10^{-3} the three microstructures are cyclically deformed well over their respective monotonic yield stresses.

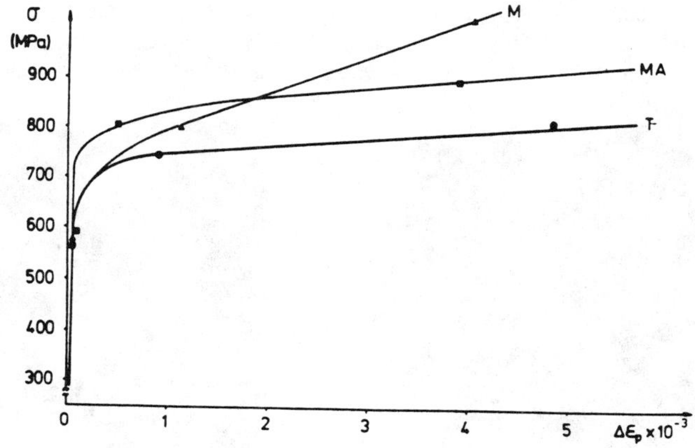

Fig.4. Cyclic stress-strain curves.

Table III. Plastic strain amplitudes and maximum stresses in the first cycle for the three microstructures deformed at the different total strain amplitudes.

Microst	MA		T		M	
$\Delta\epsilon_T$	$\Delta\epsilon_p$	σ (MPa)	$\Delta\epsilon_p$	σ (MPa)	$\Delta\epsilon_p$	σ (MPa)
$\pm 5\ 10^{-3}$	$\pm 3\ 10^{-5}$	628	$\pm 1\ 10^{-4}$	605	$\pm 8\ 10^{-4}$	595
$\pm 7\ 10^{-3}$	$\pm 1\ 10^{-4}$	845	$\pm 1.5\ 10^{-3}$	805	$\pm 1\ 10^{-3}$	843
$\pm 12\ 10^{-3}$	$\pm 3.7\ 10^{-3}$	1025	$\pm 4.1\ 10^{-3}$	965	$\pm 3.5\ 10^{-3}$	1050

The analysis of Tables II and III gives a good insight for the interpretation of the cyclic softening observed at each total strain amplitude and for every material. The MA microstructure corresponds to that of a wrought material at 950°C, inside the α+ß region. It can be easily understood that cyclic softening occurs for such a work hardened material. The microstructures T and M are obtained after fully annealing the material over its ß transus, and therefore cyclic softening can only be understood in terms of the initial cyclic hardening which takes place during the first few cycles. As the proportional limit for materials T and M is exceeded in the first cycle, plastic deformation mechanisms are activated up to the maximum stress reached. Such initial hardening should account for the subsequent softening observed.

In order to clarify such interpretation, a TEM stuy was carried out for the three microstructures both in their undeformed state and after cyclically deforming them up to fracture at a total strain amplitude of $\pm$ 7x10^{-3}.

The main features observed in the undeformed MA microstructure corresponded to high density dislocation substructures consisting mainly in planar bands along prismatic planes, as shown in Fig.5, both in α grains and in α Widmanstatten plates. The dislocation substructure corresponds to arrays of slip bands seen edge on in a $\{0001\}$ basal plane, running along a prismatic direction $<1\bar{1}00>$. In the cyclically deformed samples dislocation networks and pile-ups were observed, as shown in Fig.6. The dislocation density seemed to be lower than in the undeformed material.

Dislocations could scarcely be observed in samples of the undeformed T material. The residual ß phase surrounding the α Widmanstatten plates could be clearly identified in dark field, as shown in Fig.7. The width of the α Widmanstatten plates in the T microstructure ranges between 1 and 7 μm whilst in the MA microstructure it ranged between 0.1 up to 1 μm.

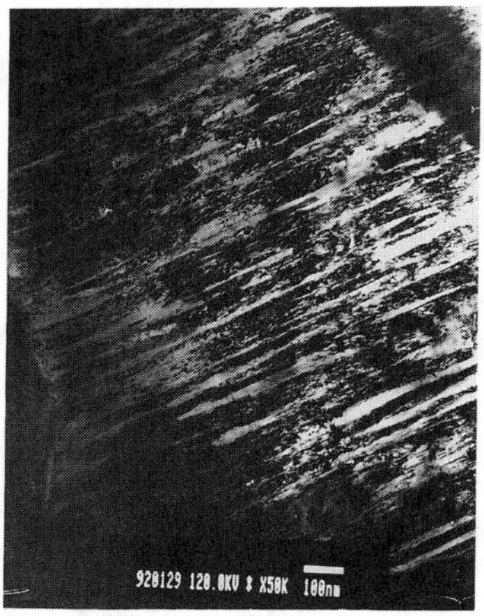

Fig.5. Planar bands along prismatic planes in α grains in the undeformed MA microstructure. Zone axis: $[0001]$

Fig.6. Network and pile-ups of dislocations in the cyclically deformed MA microstructure. Zone axis: $[01\bar{1}0]$

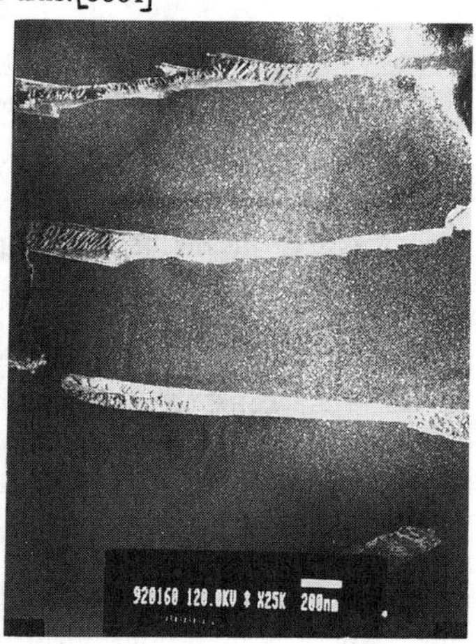

Fig.7. Undeformed T material microstructure. Zone axis: $[\bar{1}2\bar{1}3]$

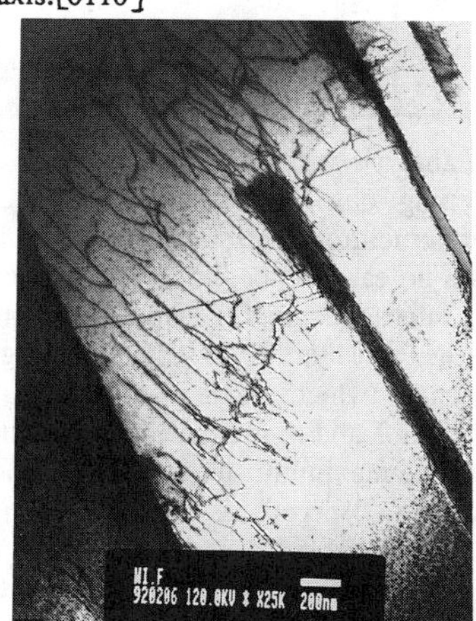

Fig.8. Parallel dislocation substructures extending along several α Widmanstatten plates in the cyclically deformed T material. Zone axis: $[\bar{1}2\bar{1}3]$

The crystallographic orientation of the α Widmanstatten plates within a colony of the T material is practically identical for all of them. This fact was clearly substantiated in the cyclically deformed samples where parallel dislocation substructures were seen to extend along several α Widmanstatten plates, as shown in Fig. 8. The Burgers vector of the dislocations is $\vec{c}+\vec{a}$. Long screw dislocations and debris can be observed.

In the undeformed M material an acicular microstructure can be observed as shown in Fig. 9. In some of the martensite plates $\vec{a}$ straight screw dislocations were observed as shown in Fig. 10. In the cyclically deformed M material it was noticed that both mechanisms, slip and twinning, accommodated the plastic deformation. Fig. 11 shows a martensitic plate with $\vec{a}$ edge dislocations forming loops and dipoles.

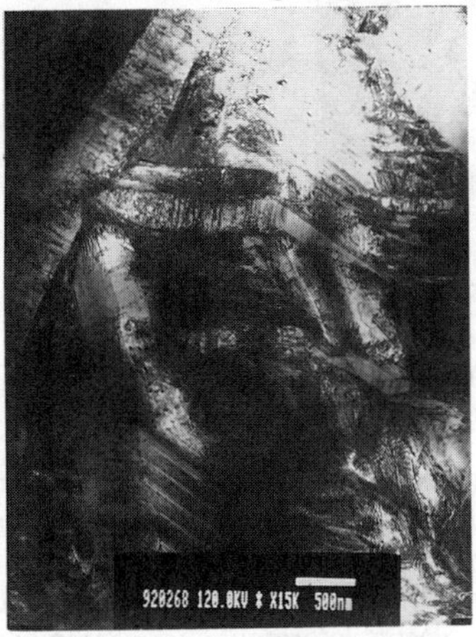

Fig.9. Acicular microstructure in the undeformed M material.

Fig.10. Straight screw dislocations in the undeformed M material. Zone axis:$[000\bar{1}]$

Parallel lenticular $\{11\bar{2}2\}$ twins were also observed as shown in dark field in Fig.12. Such twins are deformation twins since those appearing also in the undeformed specimens were $\{10\bar{1}1\}$, as shown in Fig.13. The deformation twins appeared in clusters in few areas of the samples. The fact that twinning is activated in the deformation of M material and not in the materials MA and T can be interpreted by the effect of aluminium which is more evenly distributed in the former material.

Fig.11. Loops and dipoles of a edge dislocations in a martensitic plate. Zone axis$[2\bar{1}\bar{1}0]$.

Fig.12. Parallel lenticular twins observed in dark field in the deformed M material. Zone axis:$[01\bar{1}0]$

Fig.13. $\{10\bar{1}1\}$ twins appearing both in the deformed and undeformed M material. Zone axis: $[2\bar{1}\bar{1}0]$

Although further work in TEM is being conducted in order to substantiate the mechanical results, the present evidence seems to show that the plastic deformation mechanisms activated during the first few cycles of deformation allow the subsequent softening. For the MA material, the initial high dislocation density can accommodate the low plastic strain amplitude considered. In the fully annealed T microstructure, the plastic deformation is accommodated by the to and fro movement of gliding dislocations generated during the first few hardening cycles. Such behaviour should lead to softening. Finally, for the M microstructure both dislocation movement and induced twinning activated during the first few cycles may explain the subsequent softening.

Acknowledgements

The authors are grateful to the CICYT for funding the present work through project MAT90-0755 and to Industrias Quirúrgicas de Levante S.A. for kindly donating the material.

References

1. P.E.Irving and C.J.Beevers. Mat. Sci. and Eng. 14, (1974), 229-238.
2. R.J.H.Wanhill, R.Galatolo and C.E.W.Looije. Int.J.Fatigue.11 (6), (1989), 407-416.
3. D.Eylon and C.M.Pierce. Metall. Trans. 7A, (1976), 111-121.
4. D.Eylon and P.J.Bania. Metall. Trans. 9A, (1978), 1273-1279.
5. K.S.Ravichandran. Acta Metall.Mater. 39(3), (1991) 401-410.
6. C.H.Wells and C.P.Sullivan. Trans. Am. Soc. Met., 62, (1969), 263-270
7. C.C. Wojcik and D.A.Koss. Mat. Sci. and Eng., 27, (1977), 97-99.
8. T.Bui-Quoc, R.Gomuc, A.Biron, H.L.Nguyen and R.N.Tadros. J. of Testing and Evaluation., 18(3), (1990), 160-171.
9. S.Nourbakhsh and T.D.O'Brien. Mat. Sci. and Eng. 100, (1988), 109-114.
10. D.K.Benson, J.C.Grosskreutz and J.J.Shaw. Metall. Trans. 3A, (1972), 1239-1248.
11. M.H.Yoo. Metall. Trans. 12A, (1981), 409-417.
12. S.Naka, L.P.Kubin and C.Perrier. Phil. Mag. A. 63(5), (1991), 1035-1043.
13. N.E.Paton and W.A.Backofen. Metall. Trans. 1, (1970), 2839-2847.

EFFECT OF HEAT TREATMENT ON THE MECHANICAL PROPERTIES OF Ti-10V-2Fe-3Al FOR DYNAMICALLY CRITICAL HELICOPTER COMPONENTS

D P Davies

Westland Helicopters Limited
Yeovil, Somerset, UK

Abstract

The requirement for enhanced handling characteristics, agility, low weight, high reliability and ease of maintenance were responsible for the adoption of the semi-rigid rotor head concept at Westland Helicopters Ltd (WHL) (1). Up to now the semi-rigid rotor head has been manufactured out of annealed Ti-6Al-4V, however, the desire to increase performance has lead to WHL undergoing an investigation into replacing this alloy with Ti-10V-2Fe-3Al. The paper will discuss the influence heat treatment has in determining the mechanical properties of Ti-10V-2Fe-3Al throughout a 120 mm cross section.

Introduction

In order to ensure the critical rotor head design requirements are consistently and confidently met throughout the forging, it is important that the desired ingot to billet thermomechanical processing route is stringently controlled. Of particular importance is establishing the fatigue properties in the billet material as this will greatly influence the properties attainable, in the forging, particularly in areas of low deformation (See Figure 1).

Consequently a billet was purchased of representative section size to that required to produce a rotor head forging, to enable the assessment and characterisation of Ti-10V-2Fe-3Al to be accomplished.

Titanium '92
Science and Technology
Edited by F.H. Froes and I. Caplan
The Minerals, Metals & Materials Society, 1993

Experimental Procedure

Material

The material chosen for this investigation was Ti-10V-2Fe-3Al supplied by TIMET and procured in accordance with AMS 4983, in which a summary of the chemical composition is shown in Table I.

Table I Chemical composition of the Ti-10V-2Fe-3Al billet

V	Fe	Al	C	H_2	N_2	O_2	Y	Ti
10.1	1.8	3.1	0.02	35 ppm	105 ppm	0.1	10 ppm	BAL

On-Receipt Testing

The billet stock was subjected to a detailed in release inspection which included (a) macroscopic examination to establish the general structure and (b) microscopic/segregation assessment at various locations throughout the billet cross section.

Trial Heat Treatments

A heat treatment programme was initiated in order to determine; (a) the transformation characteristics and (b) solution treatment and ageing response, as a precursor to establishing the optimum thermal processing parameters for the Ti-10V-2Fe-3Al billet.

Test Piece Removal and Design

Although, the billet size at 300 mm diameter was representative of the required starting stock, the maximum section thickness to be found on the Lynx rotor head is 120 mm. Consequently the central portion of the billet was trepanned out, leaving a ring with a more representative section thickness. After heat treatment the ring was sectioned to enable tensile, fatigue and fracture toughness property data to be collated and thereby enable sectional response to be determined. In the case of the tensile and fracture toughness specimens all data points were an average of at least two measurements.

Tensile, fatigue and fracture toughness specimens (i.e. B = 20 mm) were manufactured in the longitudinal direction to the relevant ASTM specification.

Experimental Results

The experimental programme was carried out in two stages. The first stage assessed the quality of the as received billet stock and devised the most suitable at treatment schedule for the Ti-10V-2Fe-3Al alloy, while the second stage determined the properties throughout the billet cross section.

On-Receipt Inspection

Macroscopic Examination. Macroslices taken from the end of the billet revealed a moderately coarse prior β grain structure, free of detrimental imperfections.

Microscopic Examination. A summary of the microstructural data obtained through the billet cross section, is given in Table II.

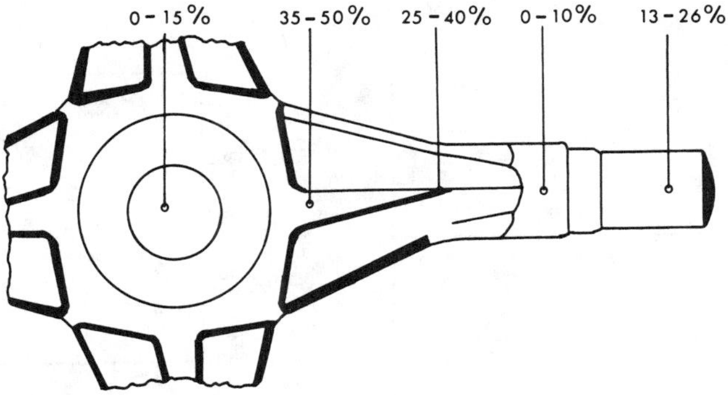

Figure 1 Variation in Deformation Rate in a Lynx Rotor Head Forging

Table II As received microstructural analysis through billet cross section

Distance from Surface mm	% Primary α	Primary α Grain Size μm	Prior β Grain Size mm
16	30.9 ± 1.3	4.6 ± 1.3	1.7 ± 1.0
96	30.3 ± 0.8	4.8 ± 1.4	1.8 ± 1.2
176	27.5 ± 2.2	4.7 ± 1.4	1.9 ± 0.9
256	27.6 ± 0.8	5.3 ± 1.6	2.0 ± 1.1

Elemental Analysis. Compositional differences through the billet cross section were considered negliable and well within the specified limits for Ti-10V-2Fe-3Al (See Figure 2). Furthermore the interstitial elements were consistently uniform throughout the billet cross section.

Heat Treatment Trials

Transformation Characteristics. The as received billet microstructure consisted of relatively small but highly elongated primary α grains in a β matrix. On raising the solution treatment temperatures above 700°C the amount of primary α decreased in relation to the solution treatment temperature. After exceeding 800°C (the β transus) a metastable β matrix completely devoid of primary α grains was produced.

Mechanical Properties. A summary of the properties obtained from the solution treatment/ageing response trials are shown in Figures 3 and 4. From this data the main trends can be summarised accordingly:

(a) On raising the solution treatment temperature, strength was found to gradually increase, reaching a plateau at ~ 800°C, with ductility following an inverse relationship to strength. The toughness data trend was more complicated, however, with toughness following an inverse relationship with strengths up to ~ 775°C, whereupon a further rise in strength produced a concomitant rise in toughness.

(b) By raising the ageing temperature from 475°C to 550°C, strength decreased, while ductility and toughness values increased.

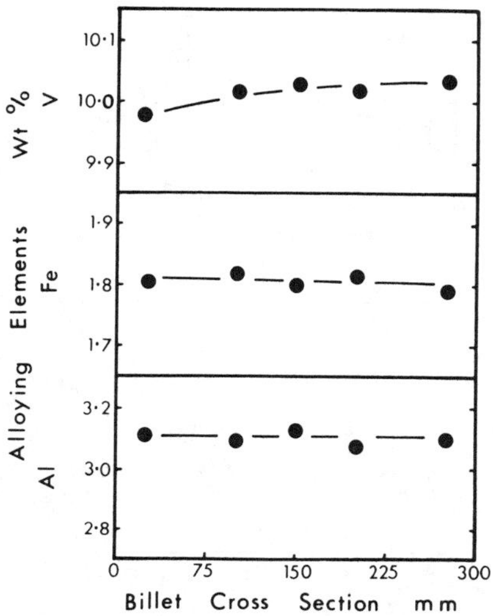

Figure 2 Variation in Chemical Composition Through Billet Cross Section

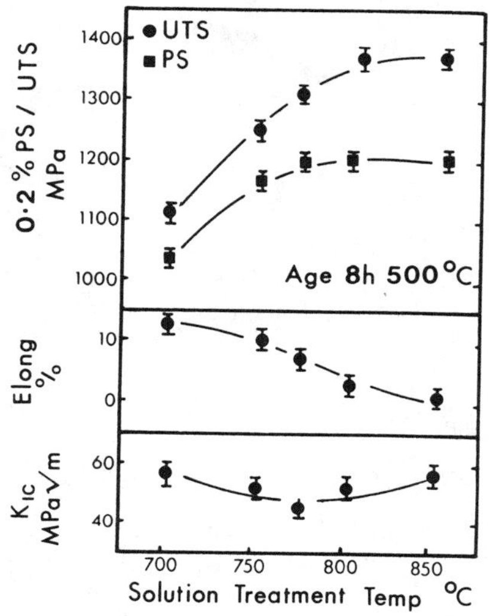

Figure 3 Influence of Solution Treatment Temperature on Static Mechanical Properties.

From the results of the heat treatment exercise and data extracted from the literature the billet was solution treated at 765°C, water quenched and aged at 500°C for 8 hours.

Billet Property Evaluation

Tensile Properties. The influence section thickness had on tensile properties is shown in Figure 5. From this diagram it can be seen that a reduction in strength on going from the inner edges of the ring towards its mid section of ~ 75 MPa was observed. Whereas the ductility values followed an inverse relationship to strength.

Fracture Toughness Properties. The influence that cooling rate and thereby section thickness had on fracture toughness properties can also be seen in Figure 5. Essentially no general trend in toughness values was observed, indicating Ti-10V-2Fe-3Al relative insensitivity to section size.

Fatigue Properties. An S/N curve generated for the Ti-10V-2Fe-3Al alloy at two mean stresses 0 MPa and 200 MPa and at two specified locations through the billet ring, is summarised in Figure 6. Furthermore, the endurance limits calculated for each of the loading conditions/billet ring position are plotted in the form of a Goodman Diagram (See Figure 7). Based on the information contained in Figures 6 and 7 the following observations can be made:

(a) By increasing the tensile mean stress a lower endurance limit is produced.

(b) The fatigue endurance limit is influenced by the position in the billet, i.e. superior fatigue endurance values were obtained from the outer portion more heavily worked area of the billet ring, although the effect was considered negligible.

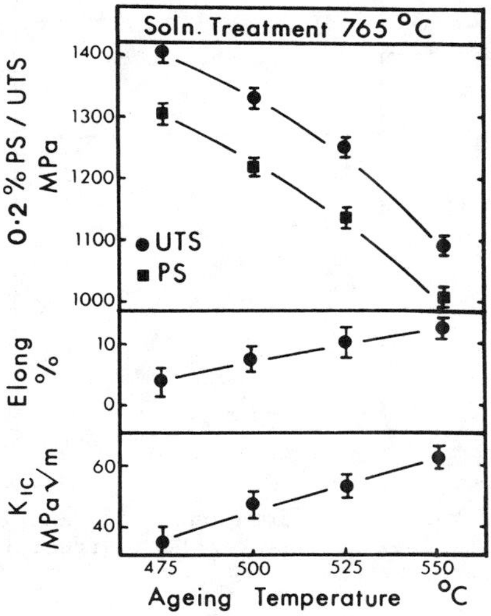

Figure 4 Influence of Ageing Temperature on Static Mechanical Properties

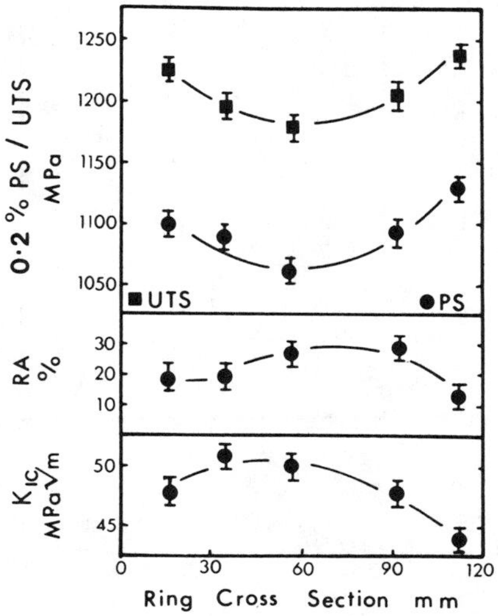

Figure 5 Influence of Section Thickness on Static Mechanical Properties

Microstructural Properties. The influence section size had on microstructure response can be seen schematically in Figure 8 and although in general the microstructural features were fairly insensitive to thickness slight differences were observed. On moving from the centre of the billet ring towards the outer edge the primary α content decreased, while the grain size increased. In all cases, however, the microstructure consisted of fine primary α grains in a β matrix.

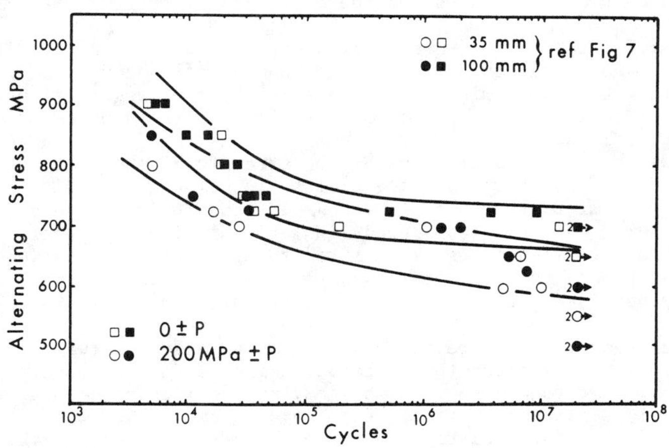

Figure 6 S/N Data Generated for Ti-10V-2Fe-3Al at 0 ± P and 200 MPa ± P

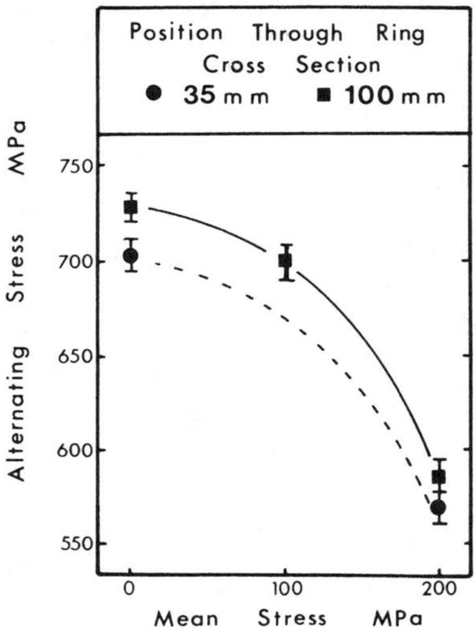

Figure 7 Goodman Diagram for Ti-10V-2Fe-3Al

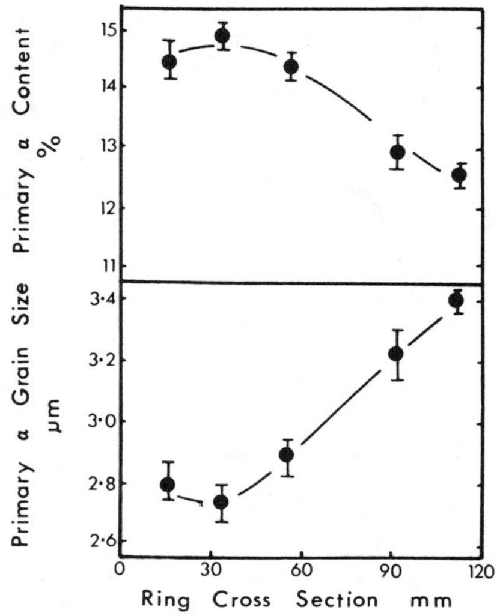

Figure 8 Influence of Section Thickness on Microstructural Response

Discussion

Influence of Microstructure on Static Mechanical Properties

By heat treating Ti-10V-2Fe-3Al below the β transus, the microstructure produced will consist of a bimodal distribution of α in a β matrix, with the relatively coarse primary α produced during solution treatment coexisting with the very fine acicular α precipitated during subsequent ageing. It has been reported by Duerig (2) that the primary α formed during solution treatment contributes little if anything to the strength of the alloy. Therefore, by lowering the solution treatment temperature lower strength levels will result after ageing, simply because the amount of fine acicular α that can be precipitated out of the matrix is decreased, (See Figure 3)

In general β titanium alloys exhibit a greater scatter in ductility than do α-β alloys and that this effect is strongly influenced by microstructure, yield strength and chemical inhomogeneity (3). As segregation was minimal the reported difference observed were considered to be due solely to microstructural and yield strength changes. There is, as with most alloys a general decrease in ductility as the strength of the alloy is increased, (See Figure 4). However, the rate of loss is not uniform, but also strongly dependent on microstructure. The relationship between ductility and microstructure can be considered in terms of slip homogeneity, i.e. with typical bimodal structures the strength differential between the soft primary α phase and the matrix (as related to primary α content) can encourage localised deformation and thereby lower ductility values.

In general an inverse relationship was observed between strength and toughness, i.e. as strength increases toughness falls. However, it should be realised that toughness is also sensitive to microstructural variations, i.e. previous investigations (4) have shown that the toughness of Ti-10V-2Fe-3Al is particularly sensitive to the morphology of the primary α. As the reported strength levels and microstructural variations were slight throughout the billet cross section, it would be expected that differences in toughness would be minimal, (See Figure 5).

As precipitated α acts as a barrier to dislocation movement its finesse is of paramount importance in determining strength levels, i.e. the higher the ageing temperature the coarser the secondary α precipitates and hence the lower the resultant strength levels. Consequently, great care must be taken to ensure the α phase precipitation is not too low, otherwise there is little or no plastic ductility, (See Figure 4).

<u>Influence of Microstructure on Fatigue Properties</u>

Although several investigators have shown that increases in tensile strength are likely to produce concomitant improvements in fatigue performance. Lucas (5) has shown that microstructural features is a more reliable indicator for assessing an alloy's fatigue performance. A number of these investigators (6) have found that a definite relationship exists between primary α grain size and fatigue strength, (i.e. by refining the primary α grain size improvements in fatigue strength is a consequence). Comparing the Ti-10V-2Fe-3Al results with previous α-β alloy billet data generated by Westland Helicopters (See Figure 9) that the relationship between primary α grain size and fatigue strength is clearly demonstrated. However, it is not a simple matter to determine primary α content in order to fully characterise fatigue strength, as early work by Stubbington et al (7) showed that the morphology of the primary α content is also of paramount importance. The most important microstructural feature with regard to fatigue resistance at endurance is absence of long α-β interfaces, either within the grains or at the grain boundaries. Consequently, in order to resist fatigue crack initiation it is necessary to minimise the amount of β transformation product which will reduce the number, length and continuity of the α-β interfaces. Furthermore, absence of grain boundary α is of prime importance as this can act as preferential nucleation sites. Examination of the microstructure in the billet ring revealed a homogenous bimodal distribution of equaxed primary α in a β matrix, with complete absence of grain boundary α. Consequently, it was considered the uniform microstructural features were mainly responsible for producing only a small reduction in fatigue properties on moving from the outer edge of the billet ring to the inner edge.

It can be seen in Figure 7 that by raising the mean stress the fatigue endurance is reduced, however the extent of this reduction can be related in part to the microstructural features. Lucas (5) has shown that by refining and increasing the amount of transformed β in the microstructure, the loss in fatigue endurance on moving to a more positive mean stress can be reduced. It would therefore be excepted from the microstructural features observed in this investigation, that the loss in fatigue on moving to a more positive mean stress would be minimal. Furthermore by comparing the loss in fatigue associated with Ti-6Al-4V in both the annealed and STOA condition, with that from the Ti-10V-2Fe-3Al data, see Table III a definite advantage for alloys solution treated high in the α-β region can be seen regardless of alloy composition.

Table III Relationship Between Primary α Content and Loss of Fatigue

Alloy	Condition	Primary α Content	Drop in Fatigue on moving from a 0 ± P to 200 MPa ± P
Ti-6Al-4V	Annealed	75%	~ 35%
Ti-6Al-4V	STOA	10 - 15%	~ 20%
Ti-10V-2Fe-3Al	STA	10 - 15%	~ 20%

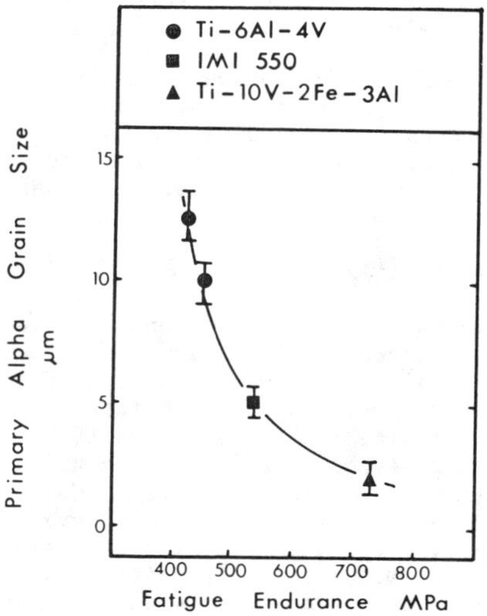

Figure 9 Relationship between Primary α Grain Size and Fatigue Resistance

Conclusions

(1) Ti-10V-2Fe-3Al exhibits a deep hardening capability, in which strengths in excess of 1150 MPa were achieved through a 120 mm billet cross section.

(2) Ti-10V-2Fe-3Al can be heat treated to a range of strength to toughness combinations, which appear to be superior to α-β alloys.

(3) The superior high cycle fatigue properties associated with Ti-10V-2Fe-3Al are considered to be due to the fine microstructure produced. Moreover the fatigue properties are capable of being sustained throughout a 120 mm cross section.

(4) There appears to be a relationship between the amount and finesse of the transformed β constituents and its resistance to fatigue under a positive mean stress.

References

(1) D. P. Davies, Metals and Materials, July 1988, p 417.

(2) T. W. Duerig, J Albrecht, D Richter, ACTA Met, 30, 1982, p 2161.

(3) C. C. Chen, J A Hall, 4th Int. Conf. on Titanium, 1980, p 457.

(4) T. W. Duering, J E Allison, J C Williams, Met Trans, 1983, p 1258.

(5) J. J. Lucas, 2nd Int. Conf. on Titanium, 1972, P 2081.

(6) A. W. Bowen, C A Stubbington, 2nd Int. Conf. on Titanium, 1972, p 1283.

(7) C. A. Stubbington, A W Bowen, Journal of Material Science, 1974, p 941.

Plastic Behavior of Ti−15−3 Alloy During Tensile and Deep Drawing Deformation

Luo Guozhen, Zhang Shuqi, Xie Liying, Wu Yiqing and Hong Quan

Northwest Institute for Nonferrous Metal Research, China

P.O.Box 71−15, Baoji, Shaanxi, 721014, P.R. China

Wang Shihong

Beijing University of Aeronautics and Astronautics

Abstract

The stress−strain response, texture features and certain plastic parameters (m,n, r and cup height) for Ti−15−3 sheet samples in various conditions were investigation. The heterogeneous thinning , earing and ellipticity of Ti−15−3 alloy hemispheres produced by hydropress drawing deforming were measured and the comparisons were made with stainless steel hemispheres. The relationship between the deep drawing formability and the plastic parameter are discussed. The effect of texture features on ears of the hemispheres of Ti−15−3 alloy is also discussed. The results show that the earing behavior of the alloy hemispheres is associated with the mechanical anisotropy and (001)[110] and (112)[111] texture feature of the alloy sheet.

Introduction

Ti−15V−3Cr−3Sn−3Al(Ti−15−3)is a superior new metastable β Ti alloy developed by TIMET,which has exhibited ideal strip−producibility, is age harderable and cold formable .

It is an economical and simple way to use cold−formed Ti−15−3 alloy instead of hot−formed Ti−6Al−4V for fabrication of certain aerospace components.The formability of Ti−15−3 alloy strongly depends on heat−treatment and processing condition and it is a function of texture , stress strain response and certain plastic parameters such as the plastic strain ratio r,strain hardening exponent n,strain rate sensitivity m and cup high H.Deep drawing is a traditional process,many studies [1−5] have concentrated on cold−formability for aluminium and steel.However,only a few studies have information about this process for titanium alloys,specialy for Ti−15−3 alloy,have been presented in the published literature [6−7]. The purpose of this study was to understand the relationship between the plastic parameters mentioned above and the plastic behavior during tensile and deep−drawing deformation.The effect of texture on mechanical anisotropy of Ti−15−3 alloy sheet is also discussed.

Procedure

The alloy used in this study was melted by double melting in a vacuum arc furnace into a ingot with weight of 24Kg and an average chemical composition of (wt%):Ti−15.08V−3.11Al−3.01Cr−3.19Sn−0.021N_2−0.04C−0.11O_2−0.10Fe−0.04Si−0.002H_2. The ingot was β forged into a slab with a size of $250 \times 200 \times 30$mm, then the slab was hot rolled in the β field into a plate with a thickness of 6mm and cold rolled into sheets with varied thickness from 0.6 to 2.5mm. In general, cold rolling reduction was

Titanium '92
Science and Technology
Edited by F.H. Froes and I. Caplan
The Minerals, Metals & Materials Society, 1993

controlled in a range 60–80%. The medial annealing and final solution treatment (ST) is at 800℃ / 20min followed by air cooling (AC). Tensile tests were preformed on an Instron Model 1185 by using extensometer and X–Y recorder to measures the stress–strain curvers and the plastic parameters r, n and m. The cross head speed varied from 0.1mm / min to 20mm / min, corresponding to strain rates of 1.7×10^{-4} / s––3.4 $\times 10^{-2}$ / s. The gauge length of the tensile specimens was $50mm$ or $5.65\sqrt{A}$.
Three types of hemispheres of Ti–15–3 alloy and stainless steel with diameter of 58, 76, 96mm and thickness of 1.0, 1.5, 2.5mm were successfully drawing–formed on a 63T hydropress in order to examine the deep–drawing formability of Ti–15–3 alloy sheets. The earing, heterogeneous thinning and ellipticity for these hemispheres was measured by using deepth gauger, optical micrometer and common micrometer tool.To understand the relationship between the earing of the hemispheres and texture , pole figures for Ti–15–3 sheets in various deformation and solution treatment conditions were automatically measured using a PW1700 Texture Diffactometer.

Results and Discussion

Typical plastic properties and stress strain response

It is well know that cold–deformed reduction, solution temperature and cooling rate all have a strong effect on the microstructure and plastic properties of metastable β Ti alloys in a given composition. The early work [8,9] reported that in order to gain more uniform refined microstructure and improved mechanical properties for the alloy sheets with 0.8–3mm thickness , the optimizing processing parameter were found to be as following: solution treated at 800℃ / 20min following by air cooling (AC) with a cooling rate of 1–3℃ / s and cold rolling reduction of 60–80%. As a result, typical plastic properties have been obtained in the optimized processing conditions as listed in Table 1.
The cup height reflects the deep drawing formability of the alloy, which is similar to the results as Rosenberg [10] reported. From Table 1, it can be seen that the cup height increases with increasing sheet thickness. However, this does not mean that the deep drawing formability increases with increasing thickness because of increase in resistance to fracture. Actually, r values decreased with the sheet thickness increasing according to an early report [8,9] . The lubricant is very important for making improvement in the cup height for the alloy sheets (as shown in Table 1).
The r value is very useful to evaluate the deep drawing formability of the alloy sheets. However, the measurement of r values is difficut to control precisely and it varied with the gauge distance of the specimen and specific strain to be used during tensile testing.
The plastic properties of the alloy solution treated at 800℃ / 20min, AC ,is not only associated with its microsturcture, which consisted of 100% β phase with fine grain size of 30–60μm, but also associated with its special inherent stress–strain response (as shown in Fig.1). It should be noted that only the engineering stress–strain curve for the alloy annealed at 800℃ / 20min. AC, is different from those for most common industry materials and also from those for the alloys in other conditions. Roseberg [10] reported that the spread between the yield stress and ultimate strength was quite small and it did not follow usual strain hardening laws. Using the following Equation : $n=\frac{d(log\sigma)}{d(log\varepsilon)}=\frac{d(ln\sigma)}{d(ln\varepsilon)}=\frac{\varepsilon}{\sigma}\cdot\frac{d\sigma}{d\varepsilon}$ [13–14] n values for eight specimens in ST condition were caculated by least square to be 0.0254–0.041. The n values are too small for the alloy .It has raised a question how the true uniform strain ε_u, r and n value are measured. There is no example to be followed so far. According to China's testing standard GB5027–85 and American ASTM E646–91, the n value must be determined in the region of uniform

Table 1 The typical plastic properties for Ti–15–3 sheets

Tensile Properties						Cup Hight	
Sheet Tickness	Testing Direction	YS MPa	UTS MPa	EL %	r Value	Sheet Tickness	Cup Hight mm
0.6mm*	LD	749	766	16	0.9516	0.6mm	7.4(1)
		783	802	18	1.161		
	TD	780	793	16	1.555	1.0mm	7.6(1)
		787	796	16	1.557		
	45°	810	816	10	0.694	1.5mm	8.8(2)
		791	804	11.5	0.749		
						2.0mm	8.5(1)
1.0mm	LD	796	802	27	0.685	2.5mm	10.2(2)
	TD	809	811	27	0.888		
	45°	799	813	22	0.532		9.2(1)
1.5mm	LD	777	792	28			10.5(1)
	TD	799	806	25			

* gauge distance is 50mm, others gauge distance is $5.65\sqrt{A}$.(1),(2) lubrication 1#, 2#.

plastic strain which is calculated from the yield point to the strain until maximum load. For most materials, necking and high localized strain begins at maximum load. However , it is not true for Ti–15–3 alloy in ST condition because the strain until maximum load is too small to identify the uniform strain. let us take the curve for the specimen annealed at 800℃ / 20min. AC, for an example to explain it in detail (Fig.1 B) . From Fig.1 B, the stress after yield point (A) quckly reaches the maximum stress(B=UTS) with strain increasing in a small strain range less than 8×10^{-3}, then it drops down through a short segment with larger slope within about 1.2×10^{-2} strain interval (BD). Subsequently, the stress very slowly decreases with strain increasing along a straight line DE until reaching point E at which point the curve deviates from a straight line. The results of tensile testing show that the necking begins at point E. It is suggested that the true uniform strain $\varepsilon_u$at E point is regarded as $\varepsilon_u = n$ [13–14] . Therefore, $\varepsilon_u$or n values listed in Table 2 were measured by this technique.

This special flow behavior of Ti–15–3 alloy is probably associated with its inherent complicated deformation mechanism. The sudden increasing and dropping in stress in the ABD segment in Fig.1 B is similar to the yielding phenomenon in a low carbon steel during tensile testing [13] probably due to interaction between dislocation and interstitial atmosphere.

Effect of strain rates.

Tensile tests for the alloy sheets in ST condition were performed at five different strain rates on an Instron 1185 model and the results are shown in Fig 2 and Table 2. The results show that the tensile properties of the alloy are sensitive to the strain rate. The tensile strength increases and ductility decreases with increasing strain rate . The strain rate sensitivity m was determined by measure a change in yield stress $\sigma_{0.2}$brought out by a change in strain rate at constant strain ε and room temperature by following equation :

$$m = \frac{D\log\sigma}{\delta\log\varepsilon} \approx \frac{\log(\sigma_2/\sigma_1)}{\log(\varepsilon_2/\varepsilon_1)} \approx 0.0231$$

The results show the m value is quite small and the n value or true uniform strain ε_u increases with decreasing strain rate . It means that slow strain rate is favorable to deep drawing deformation.

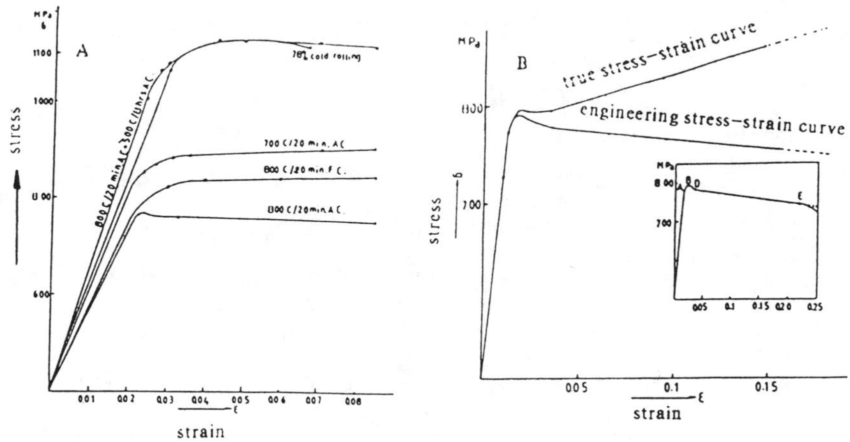

Fig.1 Stress–strain curve for Ti–15–3 alloy sheets in various condition

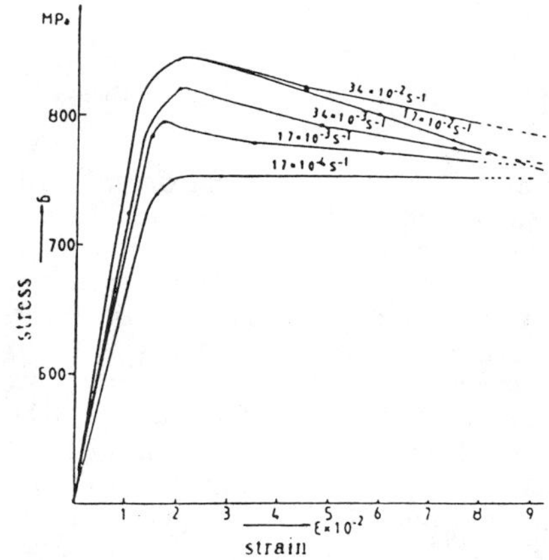

Fig.2 Effect of strain rate on tensile properties of Ti–15–3 alloy

Fig.3 Examples of Ti–15–3 alloy and stainless steel hemisphere

Deep drawing properties:

In considering optimization of the deep drawing process for Ti–15–3 alloy sheets, the cylinder cup drawing test using 1mm thickness sheets annealed at 800℃ / 20min AC,was successfully carried out at Beijing University of Aeronautics and Astronautics according to China's national standard HB6142–2–87.The results are shown in Table 3 .

Limited deep drawing ratio LDR which is defined as the ratio of the initial blank diameter D_0 to the inside diameter of the finished cup D was measured to be $LDR = D_0 / D_1 = 1.9098$ and the cup ears $\triangle H = 1.8–5.0$mm.

Afterwards, a series of Ti–15–3 and stainless steel hemispheres with 58, 76 and 96mm diameter and 1.0,1.5 and 2.5mm thickness were also successfully formed by 63T hydropress (Fig 3).The wall thinning variation and ears for the hemispheres of both materials are shown in Fig 4 and Fig.5 . It seems that both materials have similar deep–drawing behavior.Four high points symmetrically appeared at the edge of the hemisphere and the maximum wall thinning was observed at the same latitude 20mm from the pole point of the hemisphere for both materials (Fig 5).The earing position for Ti–15–3 is at 0 ° and 90 ° to rolling direction (Fig 4) but it is not clear for stainless steel. The average ear value and the maximum wall thinning of the hemispheres are 1.98mm and 9.7 percent for the Ti–15–3 alloy,0.98mm and 10.4 percent for stainless steel , respectively. However, Ti–15–3 hemispheres exhibited little higher ear values and a little more wall thinning heterogeneity. Furthermore the hemisphere of Ti–15–3 alloy also has a little larger ellipticity (0.22–0.93mm) than that of stainless steel hemispheres (0.09–0.23mm). In order to improve the deep drawing formability for Ti–15–3 alloy, must take care to consider optimization of the lubrication and the tool design.

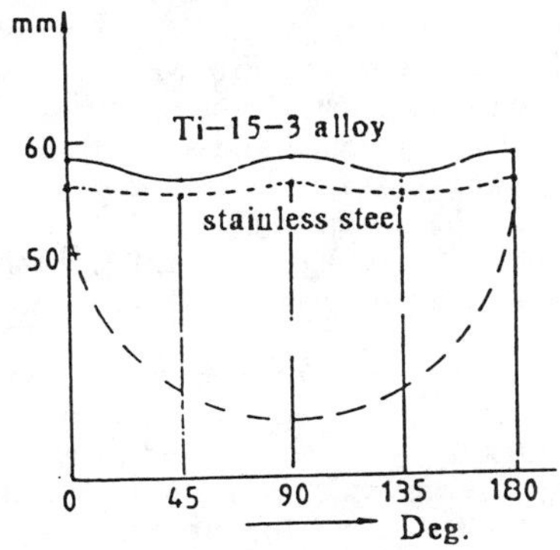

Fig.4 Earing dependence of the angle to rolling direction

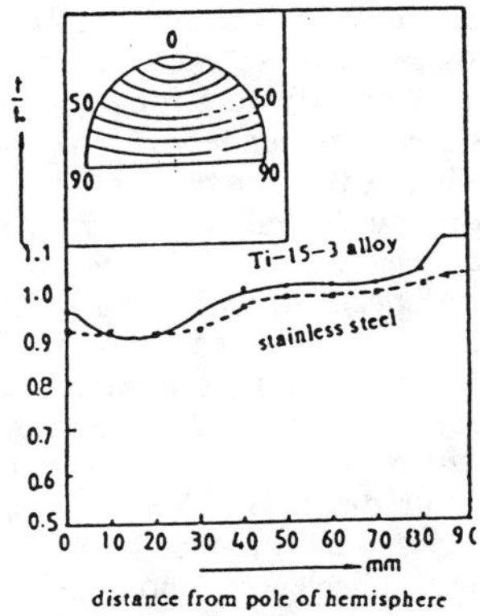

Fig.5 Wall thinning for Ti–15–3 and stainless steel hemisphere

Table 2 Effect of strain rate on tensile properties of the alloy sheet in ST condition

Sheet thickness	Cross head speed	Strain rate	tensile propeties			
			YS MPa	UTS MPa	EL %	$n=\varepsilon_u$
1.0mm (RD)	0.1mm	1.7×10^{-4}/s	740	750	31	0.188
	1.0mm	1.7×10^{-3}/s	778	788	30.5	0.167
	5mm	8.3×10^{-3}/s	805	820	21	0.156
	10mm	1.7×10^{-2}/s	835	845	19	0.143
	20mm	3.4×10^{-2}/s	830	845	19	0.122

Table 3

No.	D(mm)	P_{max}	H(mm)	Q	Status
1	97.80	1415.87	18.95	1022.60	BREAK
2	92.26	11880.97	46.56	1013.61	CUP
3	92.30	11919.98	46.49	1010.94	CUP
4	95.10	14112.07	49.60	1576.12	CUP
5	95.49	14080.86	35.14	1513.45	BREAK
6	95.49	13636.20	49.35	1500.66	CUP
7	96.50	12036.99	13.96	1494.27	BREAK

Mechanical anisotropy properties and texture

With 0 ° ,45 ° and 90 ° to rolling direction , the tensile specimens of Ti–15–3 alloy sheet annealed at 800℃ / 20min AC were tested by specified constant strain of 12 percent to measure the r value. In Fig.6, the r value and tensile properties were ploted as a function of the angle to the rolling direction. The results indicate that the tensile properties are not equivalent in 0 ° ,45 ° and 90 ° directions. It is clear that the spceimen in the 0 ° and 90 ° direction exhibited higher tensile ductility and r values as well as lower tensile strength, while the specimen in 45 ° direction has the highest yield stress and lowest r values and tensile ductility. This variation of r value and tensile properties within the plane of the alloy sheet is called mechanical anisotropy which is associated with crystallographic anisotropy resulting from the prefered orientation of the grains. Fig.7 show the texture feature for the alloy sheets in different condions. From Fig.7, the texture is strongly dependent on the cold rolling reduction and heat treatment. With small cold rolling reduction belows 32%, the sheet specimens in rolled and solution treated condition have a similar texture (001)[110]. With rolling reduction increasing over 65%, the specimens in rolled and solution treated condition have a strong $(112)[11\bar{1}]$ texture. The sheet specimen with 45% rolling reduction exhibited $(001)[110]+(112)[11\bar{1}]$ mixed texture. In spite of the variation of rolling reduction, the deformed and recrystallized texture of the alloy sheet involved (001)[110] or $(112)[11\bar{1}]$.

It is suggested that the mechanical anisotropy and the texture feature must be responsible for the ears of Ti–15–3 alloy hemisphere. Generally speaking, ears often occur at high r value direction. The formula for r value anisotropy is : $\triangle r=(r_0+r_{q0}-2r_{45})/2$. Here, $\triangle r$ is the difference between the $r_{45}$and the average value in $r_0$and r_{q0}direction. According to keeler [3] and kurtlange [11] state that when $\triangle r>0$ the ears are formed at angle 0 ° and 90 ° direcition, and if the r_{45}value is the largest ears are formed at 45 ° direction. The value of $\triangle r$ value in this study is $\triangle r=(r_0+r_{q0}-2r_{45})/2=0.585>0$, thus the ears of the alloy hemisphere occured at 0 ° and 90 ° direction. On the other hand, the highest yield stress and lowest tensile ductility for the specimen in 45 ° direction also means that the 45 ° direction exhibited higher resistance to deformation and lower deformability compared to 0 ° and 90 ° direction.

In general, the bcc crystal has (112), (110) and (123) slip planes and <111> slip direction with forty–eight possible slip systems [12] . Ti–15–3 alloy sheet annealed at 800℃ / 20min. AC, was approximately 100 percent β phase with a bcc crystal so that the alloy has the same slip systems. According to the formula: [13] $\tau_k=\sigma\cos\psi\cos\lambda$... where τ_k is critical shear stress, σ_s yield stress, ψ the angle between the principle stress and the normal of slip plane, λ the angle between the principle stress and the slip direction. The slip vector in 0 ° , 45 ° and 90 ° directions are not only dependent on the angle ψ and λ but are also dependent on the angle between the slip direction and each given direction. As Fig.7F shows the alloy sheets in any condition involved stronger (001) [110] and $(112)[11\bar{1}]$ textures. In this case, most (001) and (112) planes parallel to rolling planes, [110] and [111] direciton parallel to the rolling direction, as well as the principle stress parallel to the normal of (001) and (112) planes. According to the calculation of resolution shear stress for each of forty–eight possible slip systems as in a single bcc crystal, in the case of (001)[110] texture existing, $(1\bar{1}2)[\bar{1}11]$ and $(\bar{1}12)[1\bar{1}1]$ slip system have the highest resolution shear stress , e.g. $\tau=0.47\sigma_s$ and it can be expected to operate as the first slip system. The second one will be $(0\bar{1}1)[111]$, $(\bar{1}01)[111]$, $(011)[1\bar{1}1]$ and $(\bar{1}01)[1\bar{1}1]$ four slip system with $\tau=0.408\sigma_s$. In the case of $(112)[11\bar{1}]$ texture, the first one is $(213)[\bar{1}11]$ as $\tau=0.463\sigma_s$and second one will be $(121)[1\bar{1}1]$, $(211)[\bar{1}11]$, $(121)[\bar{1}11]$ as

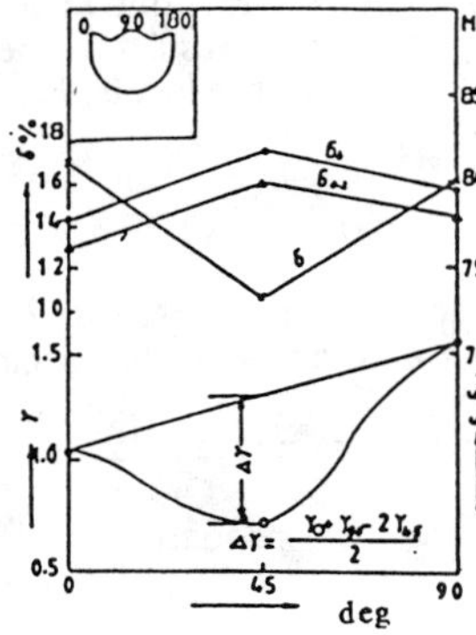

a. tensile properties dependence of the angle to rolling direction

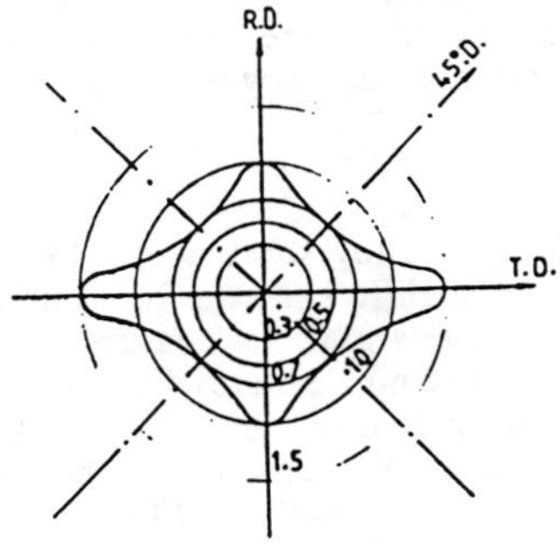

b. r value pole figure with angles to rolling direction

Fig.6 Mechanical anisotropy dependence of the angle to rolling direction

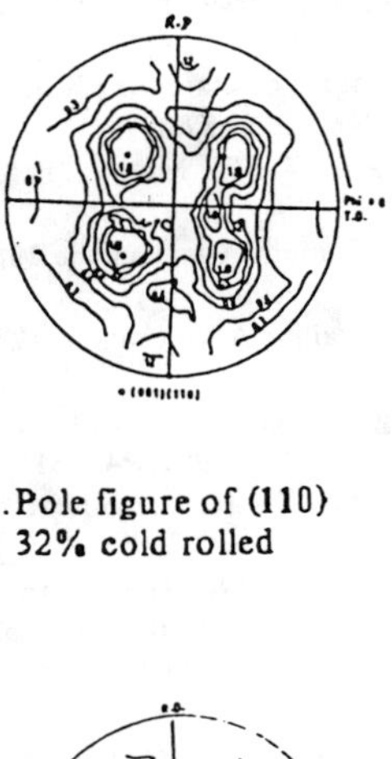

A. Pole figure of (110) 32% cold rolled

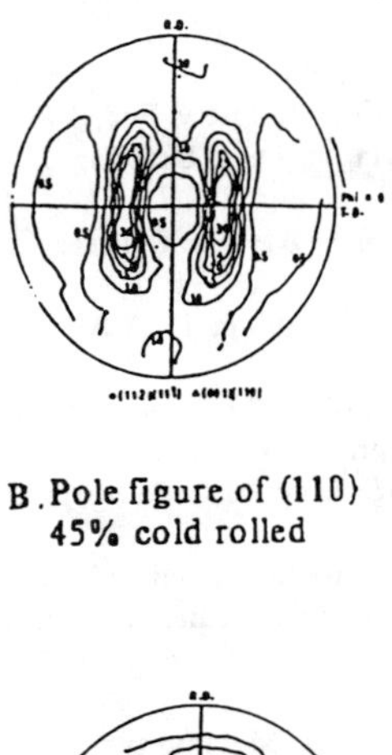

B. Pole figure of (110) 45% cold rolled

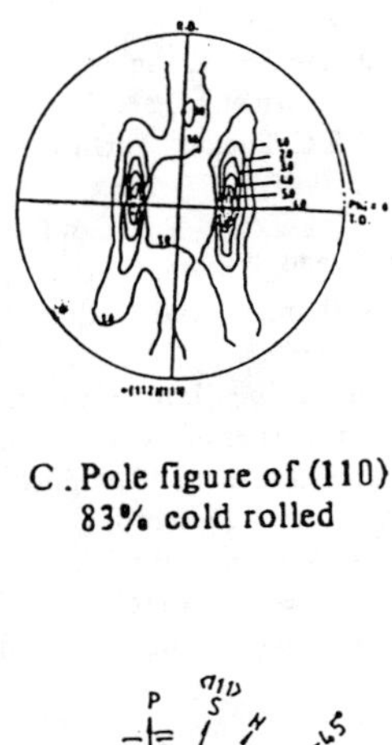

C. Pole figure of (110) 83% cold rolled

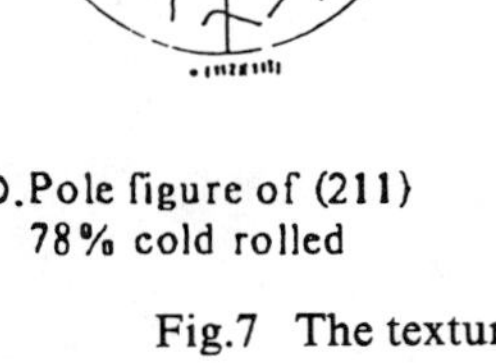

D. Pole figure of (211) 78% cold rolled

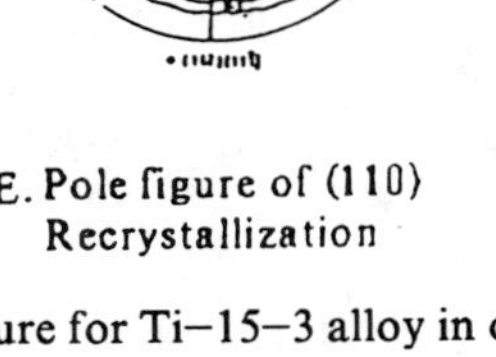

E. Pole figure of (110) Recrystallization

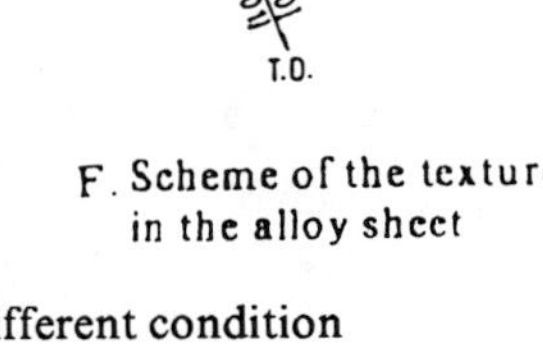

F. Scheme of the texture in the alloy sheet

Fig.7 The texture figure for Ti–15–3 alloy in different condition

$\tau = 0.392\sigma_s$. According to calculation of the orientation cosine of slip direction in 0 ° , 90 ° and 45 ° direction , the resolved shear stress in 0 ° , 90 ° direction are 1.41 times as large as that in 45 ° direction. It means that 0 ° , 90 ° directions have more slip vector than that of 45 ° direction. The application of the above results to the earing of polycrystalline sheet perhaps may explain 0 ° and 90 ° ears of the hemispheres and mechanical anisotropy for the alloy sheet.

Conclusion

Ti–15–3 alloy sheet in a solution–annealed condition has special stress–strain curve which is different from those of common engineering materials and the special method used in this study to measure r, n is available.
With strain rate increasing , n, $\varepsilon_u$value increase and slow strain rate is favorable to deep drawing deformation.
0 ° and 90 ° ears of the hemisphere for Ti–15–3 alloy are probably associated with (001) [110] and (112) [11T] texture.

Reference

1. J.C. Wright, ″ The phenomenon of Earing in Deep Drawing ″ Sheet Metal Industries, Nov. 1965
2. G.S.A.Shawki, et al ″ Assessign Deep–Drawing Qualities of Sheet ″, Sheet Metal Industries, May.1965, P363–368, 417–424
3. Stuart P.Keeler, ″ Properties Related to Forming ″, Sheet Metal Industries, July 1971, Vol 48, No7, P511–517
4. Dr.Ing Jam Galinowski, Using the planar amisotropy in sheet metal to facilitate the production of deep drawn cup. Part Ⅰ,Ⅱ, Sheet Metal Industry, Feb,1975, Vol52, No2, P79–87
5. G.E.G.Tucker, Texture and Earing in Deep Drawing of Aluminium, ACTA Metallurgica, Vol.9 April.1961, P275–286
6. A.E.Leach, ″ Formed Ti–15V–3Cr–3Al–3Sn Tankage ″ 《Beta Titaium Alloys in the 1980′s》 , P307–329
7. A.Shames, et al, ″ Cold Forming Titanium 15–3 Alloy 《Beta Titaium Alloys in the 1980′s》 , P307–329
8. Xie Li Ying, Luo GuoZhen, Effect of cold deformation and Aging on microstructure and Mechanical Properties of Ti–15–3 Alloy, Rare Metal Materials and Engineering. No1, 1991, P48–51
9. Luo Guozhen, 1990 Final Repore, Unpublished, Northwest Insititute for Nonferrous Metal Research, 1990, December.
10. H. W.Rosenberg, Ti–15–3 Properties Data, 《Beta Titaium Alloys in the 1980′s》 , P405–425
11. Kurt lauge, Handbook of Metal Forming, edited by Kurt lange. Mcgraw–Hibl Book company.
12. Xu Zu Yao, ed, Metallograph Principle (Science and Technique Publication Shanhai, China, 1965.4)
13. R.E.Smallman, ed, ″ Modern Physical Metallurgy ″ 3rd Edition. London, Bntteworths, Translation in Chinese 10,1978
14. George E. Dieter, Mechnical Metallurgy, Second Edition, 1976 by McGraw–Hill. Inc.

IMPROVEMENT OF HYDROGEN EMBRITTLEMENT OF BETA Ti ALLOY

Akira Nozue, Tadatsune Okubo, Yoji Saita and Hyoe Sakamaki

Department of Mechanical Engineering, Sophia University, Tokyo 102, JAPAN

Abstract

A special heat treatment process has been developed to reduce the susceptibility to hydrogen embrittlement in a beta Ti–15V–3Cr–3Sn–3Al alloy. Hydrogen embrittlement tests were carried out with CT specimens under sustained load conditions. Hydrogen was occluded for the specimens with a cathode charging method during the tests.

The beta Ti alloy of the good formability exhibited the extremely high susceptibility to hydrogen embrittlement compared with a widely used alpha – beta Ti–6Al–4V alloy. This susceptibility of the beta Ti alloy was found to be reduced by aging for a relatively short time. By this treatment, however, the strength of the present alloy was also reduced due to a decrease in the volume fraction of the alpha precipitates. Hence the special heat treatment process was carried out for the reduction of the susceptibility to hydrogen embrittlement and the retention of the high strength in the present alloy.

It is of importance both to form precipitate free zones almost around the beta grain boundary because of their extremely high fracture resistance and to retain an adequate fraction of the alpha precipitates within the grain. That is, this process is taken as follows; the specimens were solution–treated at relatively high temperature 1223K, aged at 783K for 14.4ks and then aged at 713K for 10.8ks. After the first aging, the precipitate free zones formed around grain boundaries and the specimens contained a small fraction of the alpha precipitates within the grains. After the secondary aging, the specimens contained a relatively high fraction of the alpha precipitates within the grains while still retaining the precipitate free zones. This process made it possible to obtain the desirable mechanical properties of the present alloy.

Introduction

A number of studies have been carried out on hydrogen embrittlement of widely used Ti–6Al–4V alloy [1–3]. We have reported that the resistance to hydrogen embrittlement of this alloy is enhanced by a decrease in the volume ratio of primary alpha phase, the discontinuity of beta phase and a decrease in the primary alpha grain size [4]. In addition, we have indicated that beta Ti alloys exhibit extremely high susceptibility to hydrogen embrittlement in comparison with the Ti–6Al–4V alloy under the same charging condition [5]. In the present study, a special heat treatment has been developed to reduce the susceptibility in the beta Ti–15V–3Cr–3Sn–3Al alloy.

Titanium '92
Science and Technology
Edited by F.H. Froes and I. Caplan
The Minerals, Metals & Materials Society, 1993

Experimental Procedure

Materials

The chemical composition of the Ti-15V-3Cr-3Sn-3Al alloy used in the present study is listed in **Table 1**. Compact tension (CT) specimens of 4mm thickness were machined after solution treatment, and then aged. A maximum stress intensity factor for fatigue pre-cracks was determined about 40% of an initial K_i in the present test. For the specimens solution-treated at 1123K for 3.6ks and aged at 783K for 50.4ks, the tensile strength is 1.27GPa and the elongation is 15%.

Table 1 Chemical composition of alloy used (mass %).

V	Cr	Sn	Al	Fe	C	N	O	H
14.90	2.89	2.90	3.14	0.21	0.011	0.0094	0.122	0.0013

Hydrogen embrittlement test

The CT specimens were cathodically charged in a 0.5% sulfuric acid solution under the condition of a current density of 200A/m^2 during sustained loading. The K_i value was determined about 50% of the fracture toughness. A crack growth rate was measured as a function of the stress intensity factor with a microscope and a compliance method.

Results and Discussion

Hydrogen embrittlement behavior of three types of Ti alloys

Hydrogen embrittlement tests of three types of Ti alloys were already carried out under the same condition of the present work [5]. These results give **Figure 1**. Three types of Ti alloys are a beta type of Ti-15V-3Cr-3Sn-3Al alloy, a near beta type of Ti-5Al-2Sn-2Zr-4Cr-4Mo alloy, and an alpha - beta type of Ti-6Al-4V alloy, which contain almost the same yield strength level of 1.2GPa. In Fig. 1, the beta and near beta types of Ti alloys exhibit extremely high crack growth rates compared with the alpha - beta type.

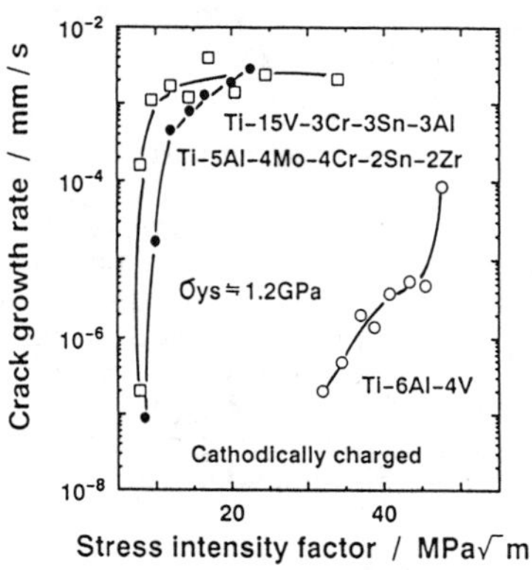

Fig.1 Crack growth rate vs.stress intensity factor under same cathode charging.

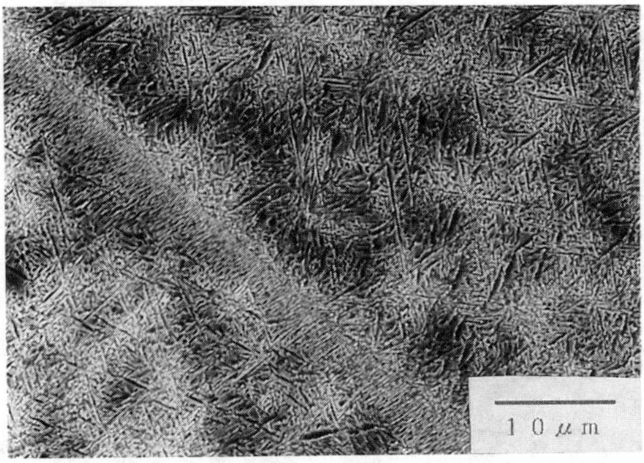

Fig. 2 Microstructure of specimen solution–treated at 1123K for 3.6ks and aged at 783K for 50.4ks.

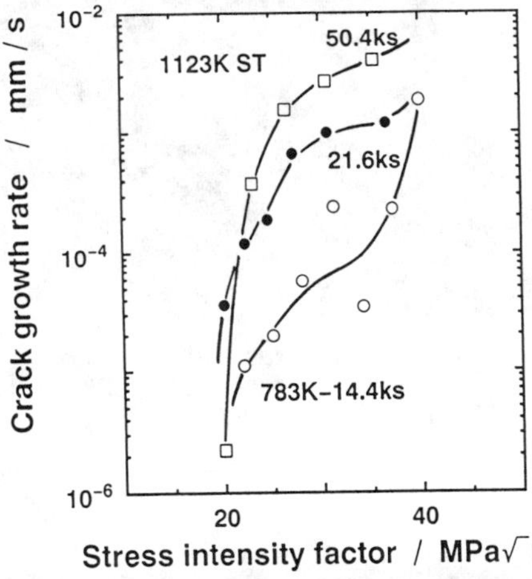

Fig. 3 Crack growth rate da/dt vs. stress intensity factor K for specimens aged at 783K for 14.4ks, 21.6ks and 50.4ks after solution–treated at 1123K for 3.6ks.

Effect of aging time on crack growth rate

Figure 2 provides a typical microstructure of the present alloy in a specimen solution–treated at 1123K for 3.6ks and aged at 783K for 50.4ks, which exhibits a close peak point at the relation between hardness and aging time. This specimen consists of very fine alpha precipitates within the beta matrix. The hydrogen embrittlement tests were made for specimens solution–treated at 1123K for 3.6ks and then aged at 783K for 14.4ks, 21.6ks and 50.4ks to clarify the effect of the aging time on the crack growth rate. As shown in **Figure 3**, the crack growth rate becomes low in specimens aged for 50.4ks, 21.6ks and 14.4ks in its turn. This result suggests that the alpha volume fraction and its distribution may control hydrogen embrittlement of the Ti–15V–3Cr–3Sn–3Al alloy.

Figure 4 shows microstructures of a specimen aged at 783K for 14.4ks after solution–treated at 1123K for 3.6ks. Figures 4(a) and 4(b) give low and high magnification photographs, respectively. A white area indicates an alpha precipitate. As shown in Fig. 4, a precipitate free zone, designated as PFZ in the present study, exists nearby the beta grain boundaries.

Figures 5(a) and 5(b) show fracture surfaces of specimens aged at 783K for 50.4ks and 14.4ks after solution-treated at 1123K for 3.6ks. In the specimen aged for 50.4ks, quasi-cleavage exists at almost all the surfaces, shown in Fig. 5(a). On the other hand, in the specimen aged for 14.4ks, which contains PFZ, the quasi-cleavage also exists only within the grain and a tear-ridge pattern appears almost around the grain boundary, shown in Fig. 5(b). This local ductile fracture pattern of the under-aged specimen may depend on the existence of PFZ which retains the high fracture resistance because of its low strength. As a result, the existence of PFZ reduces the crack growth rate, but its specimen also contains undesirable mechanical properties of the low tensile strength because of the low alpha volume ratio.

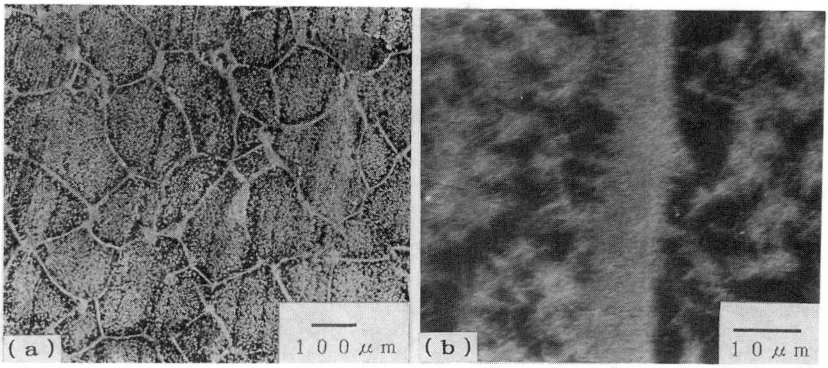

Fig. 4 Scanning electron micrographs of specimen aged at 783K for 14.4ks after solution-treated at 1123K for 3.6ks, shown with (a) low and (b) high magnification.

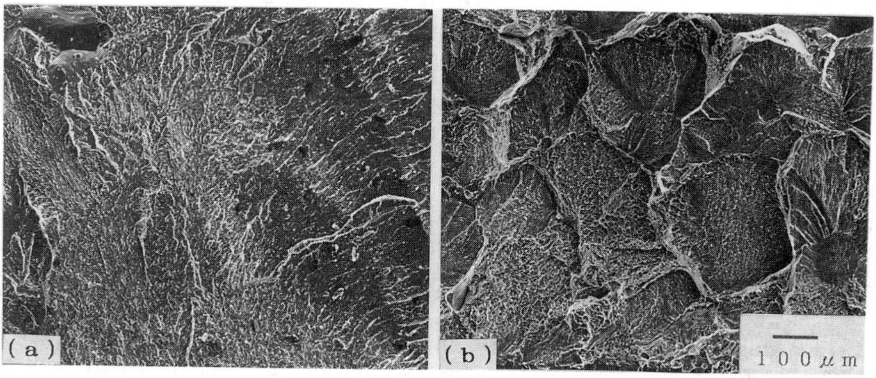

Fig. 5 Fracture surfaces of specimens aged at 783K for (a) 50.4ks and (b) 14.4ks after solution-treated at 1123K for 3.6ks.

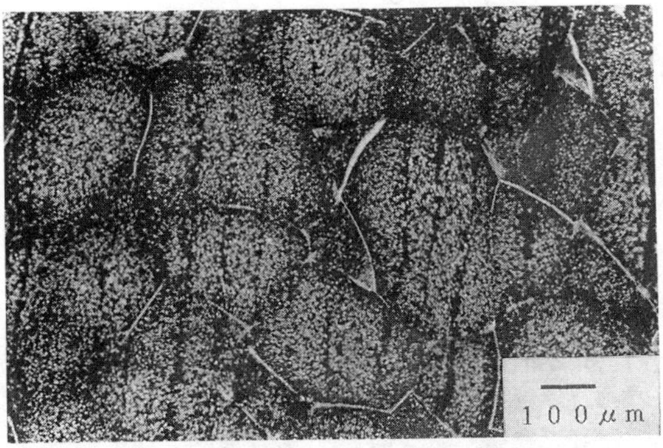

Fig. 6 Scanning electron micrograph of specimen solution-treated at 1223K for 3.6ks and aged at 783K for 14.4ks.

Effect of solution treatment temperature

The alpha precipitates on the grain boundary are widely known to reduce the elongation [6]. Therefore, to decrease the volume fraction of the alpha precipitates only on the grain boundary, specimens were solution-treated at relatively high temperature 1223K. **Figure 6** shows the microstructure of a specimen solution-treated at 1223K for 3.6ks and then aged at the same aging temperature 783K for the same time 14.4ks. The volume fraction of the alpha precipitate on the grain boundary becomes low in the specimen solution-treated at 1223K compared with the specimen shown in Fig. 4.

New heat treatment process

A special heat treatment process was carried out for the reduction of the susceptibility to hydrogen embrittlement and retention of the relatively high strength in the present alloy. It is of importance both to form PFZ almost around the beta grain boundaries because of the extremely high fracture resistance and to maintain an adequate fraction of the alpha precipitates within the grains. The heat treatment proposed in the present study is given in **Figure 7**. Specimens were solution-treated at relatively high temperature 1223K, aged at 783K for short time 14.4ks, and then aged at 713K for short time 10.8ks. After the first aging, the specimens exhibit PFZ around the grain boundaries and contain a small fraction of the alpha precipitates within the grains. After the secondary aging, the specimens exhibit a relatively high fraction of the alpha precipitates within the grains while still retaining PFZ. The microstructures of the specimens heat-treated with the new method maintain PFZ around the grain boundaries and the relatively high fraction of the alpha precipitates within the grains, shown in **Figures 8(a) and 8(b)**. **Figure 9** shows the crack growth rate vs. K for the specimens heat-treated with the new and conventional methods. In the specimens heat-treated with the new process, the crack growth rate becomes extremely low. As shown in **Figure 10**, its fracture surface is very similar to that of under-aged specimens, shown in Fig. 5(b). In addition, the specimen heat-treated with the new method retains a tensile strength of 1.23GPa and an elongation of 13 %. Therefore, the new process made it possible to obtain the desirable mechanical properties of the present alloy.

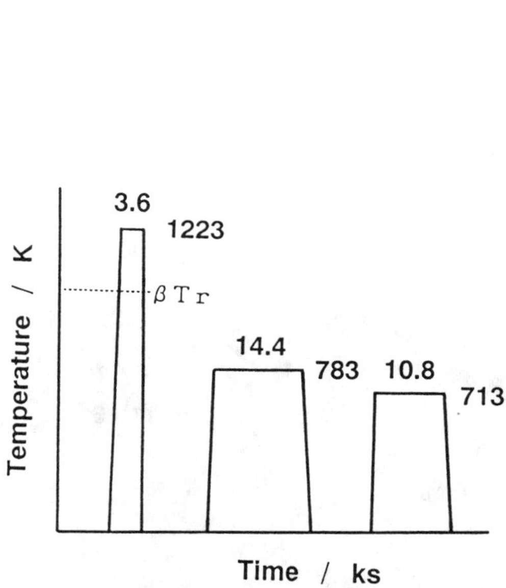

Fig. 7 New heat treatment proposed in the present study.

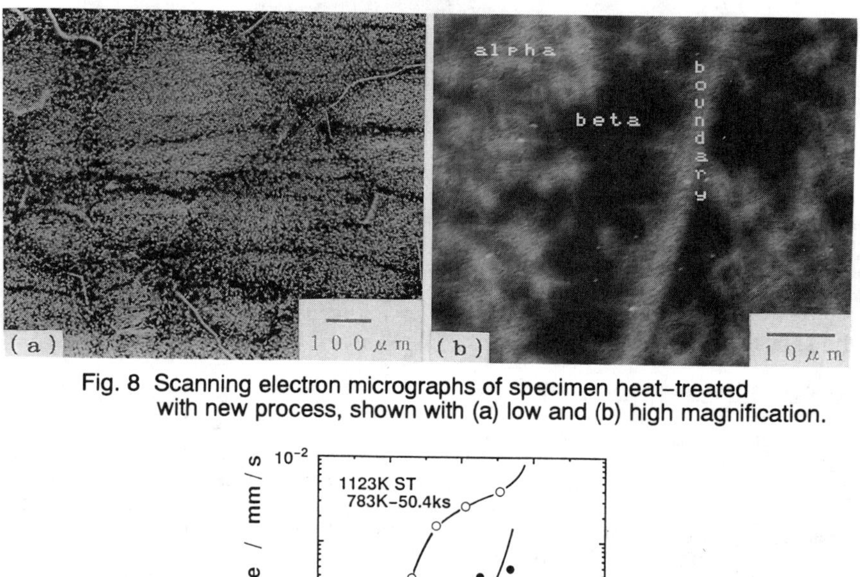

Fig. 8 Scanning electron micrographs of specimen heat-treated with new process, shown with (a) low and (b) high magnification.

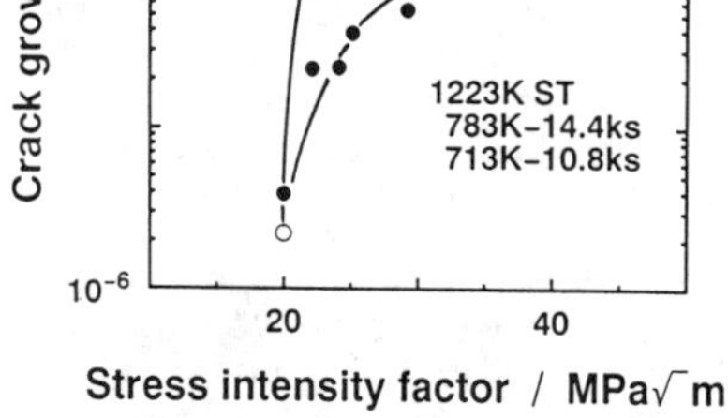

Fig. 9 Crack growth rate da/dt vs. stress intensity factor K for specimens with new and conventional heat treatments.

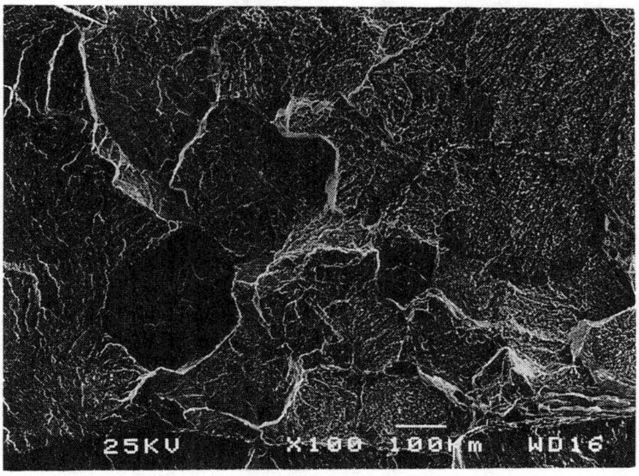

Fig. 10 Fracture surface of specimen heat-treated with new process.

Conclusion

A special heat treatment process has made it possible to enhance the hydrogen embrittlement resistance of a Ti-15V-3Cr-3Sn-3Al alloy.

Acknowledgment

The present work was partially supported by Grant-in-aid for Scientific Research from the Department of Education of Japan, and Iketani Science and Technology Foundation.

References

1. M. J. Blackburn, J. A. Feeney and T. R. Beck, "Stress-Corrosion Cracking of Titanium Alloys," Advances in Corrosion Science and Technology, Ed. by M. G. Fontana and R. W. Staehle, Plenum Press, **3**(1973), 67-292.
2. G. F. Pittinato and W. D. Hanna, "Hydrogen in Beta Transformed Ti-6Al-4V," Met. Trans., **3**(1972), 2905-2909.
3. Rodney R. Boyer and William F. Spurr, "Characteristics of Sustained-Load Cracking and Hydrogen Effects in Ti-6Al-4V," Met. Trans., **9A**(1978), 1443-1448.
4. Akira Nozue et al., "Hydrogen Embrittlement Behavior of Structure Controlled Ti-6Al-4V Alloy," 6th World Conference on Titanium, Ed. by P. Lacombe, R. Tricot and G. Beranger, part I (France, 1988), 223-228.
5. Akira Nozue et al., Proc. of Annual Meeting of The Japan Institute of Metals, (1988), 393-393.
6. J. C. Chesnutt, C. G. Rhodes and J. C. Williams, "Relationship Between Mechanical Properties, Microstructure, and Fracture Topography in alpha + beta Titanium Alloys," ASTM STP 600 (1976), 99-138.

FRACTURE CHARACTERISTICS OF SOLUTION-TREATED

Ti-15V-3Cr-3Sn-3Al ALLOYS

Kotobu NAGAI*, Keisuke ISHIKAWA*, Takao HORIYA**, and Hirowo SUZUKI**

* National Research Institute for Metals, Tsukuba Labs.,
1-2-1, Sengen, Tsukuba, Ibaraki 305, Japan
**Nippon Steel Corporation, Stainless Steel & Titanium Lab.,
20-1, Shintomi, Futtsu, Chiba 299-12, Japan

Abstract

The reason that β alloys show a poor toughness at low temperatures is unclear. The present paper elucidates the fracture characteristics of solution-treated Ti-15V-3Cr-3Sn-3Al alloy. The alloys with 0.17 weight% oxygen (High O alloy), 0.08 weight% oxygen (Low O alloy) and 0.04 weight% oxygen (Extra-low O alloy) were solution-treated for 900 s at 1073 K, and water-quenched. The mean grain diameter was about 50 μm for both. The material with larger grain diameter of 220 μm was made from the low O alloy. Charpy impact tests were done at temperatures between 77 K and 473 K. Tensile tests were also performed.

A marked energy transition occurred in every material. The upper-shelf fracture mode was featured by large equiaxed dimples of a size identical with the β grain size. On the other hand, the lower-shelf fracture was due to a flat fracture surface which was partly covered by micro-dimples on the order of 1 μm.

The Low O alloy had a lower transition temperature than the High O alloy. No difference in transition temperature existed between small grain and large grain size specimens; however, the large grain specimen had higher upper-shelf energy than the small grain one. These features are not always analogous to "cleavage fracture" in ferritic steel. The change of deformation mode is primarily responsible for this transition.

Introduction

Among titanium (Ti) and its alloys, only a few materials are specified for cryogenic temperature use by materials standards [1]. Low oxygen unalloyed Ti, Ti-5Al-2.5Sn, Ti-6Al-4V and others are listed in those standards, however no β alloys are listed, as they show a poor toughness at low temperature [2]. Figure 1 shows the Charpy energy change below room temperature for various types of Ti materials and Table I lists the fracture toughness for the alloys at 4 K [3]. Both the Charpy impact energy and the fracture toughness are lowest in the β alloys. Fracture in β alloys at low temperature is "brittle", compared with rather ductile mode in α and α-β alloys [3].

Beta alloys are conventionally classified as a bcc type alloy. Low temperature brittleness in bcc type ferrous alloys is a well-known phenomenon in which the fracture mode changes from ductile to brittle below a transition temperature and the brittle fracture surface is characterized by (quasi-) cleavage facets. The β alloys show a flat fracture surface resembling the " cleavage" of the ferrous alloys at low temperatures [3].

If the low toughness at low temperature of β alloys is analogous to that of ferrous alloys, the low toughness could be overcome by a reduction in the interstitial atom content, the refinement of microstructure and so on. Data exist for solution-treated-and-aged (STA'ed) alloys, which have α precipitates in β matrix. Hence as-ST'ed β alloys should be first

Titanium '92
Science and Technology
Edited by F.H. Froes and I. Caplan
The Minerals, Metals & Materials Society, 1993

investigated for a basic understanding of the fracture characteristic of the β alloys.

In the present study, the effect of oxygen content on the Charpy value was first investigated for a typical β alloy, Ti-15V-3Cr-3Sn-3Al (abbreviated as Ti-153 here). And further the β grain size effect was also examined. As a result, a localized deformation was considered to play an important role on the toughness deterioration. Hence the effect of temperature and strain rate on the tensile deformation behavior was examined.

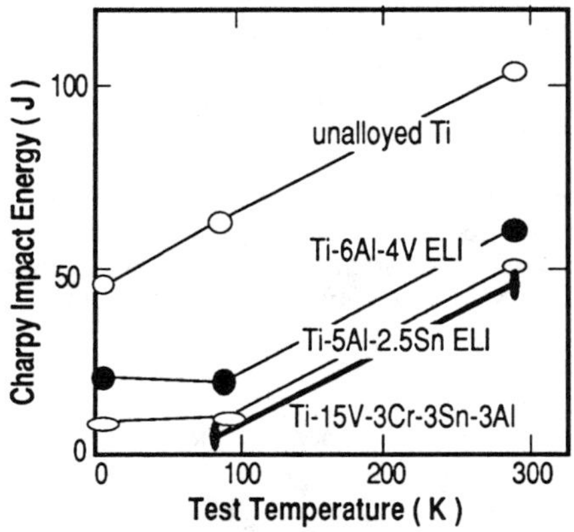

Figure 1 - Charpy impact energy change by test temperature for various titanium materials.

Table I Fracture toughness of various titanium alloys at 4 K

Alloy	Alloy type	Fracture toughness, K_{IC} (MPa $\sqrt{m}$)	Comment
Ti-5Al-2.5Sn ELI	α	94	mill-annealed
Ti-6Al-4V ELI	α−β	51	mill-anneal
Ti-15V-3Cr-3Sn-3Al	β	22	STA'ed
Ti-15Mo-5Zr-3Al	β	18	STA'ed

Experimental Procedure

Three Ti-153 alloys, namely 0.17 weight% oxygen (High O alloy), 0.08 weight% oxygen (Low O alloy), and 0.04 weight% oxygen (Extra-low O alloy) were melted. The chemical compositions are listed in Table I. All were ST'ed for 900 s at 1073 K. V-notched Charpy impact specimens (10 mm x 10 mm x 55 mm, 2 mm depth notch) were cut perpendicular to the hot-rolling direction and tested at temperatures between 77 K and 473 K to obtain energy transition curves. V-notched Charpy impact energy (CVN) was determined by duplicate tests at each temperature. In order to examine the grain size effect, Low and Extra-low O alloys were also ST'ed at 1223 K. Mean grain diameter was about 60 μm for the High O alloy ST'ed at 1073 K, 50 μm for the Low O alloy ST'ed at 1073 K, 220 μm for the Low O alloy ST'ed at 1223 K, 250 μm for the Extra-low O alloy ST'ed at 1073 K, and about 500 μm for the Extra-low O alloy ST'ed at 1223 K. Fracture surfaces were observed by SEM.

Cylindrical test pieces with 3.5 mm in gage diameter and 25 mm in gage length were also machined perpendicular to the hot-rolling direction only from the High O alloy.

Yield strength (principally 0.2% offset stress), tensile strength, elongation (fracture strain), and reduction of area were determined at crosshead speeds of 0.05, 0.5, 5, and 50 mm/min (3.3×10^{-5}, 3.3×10^{-4}, 3.3×10^{-3} , and 3.2×10^{-2} s^{-1}, respectively) using a screw-driven type tester. The tests were conducted at temperatures between 77 K and 355 K.

Table II Chemical compositions of materials studied in the present study in weight%

Alloy	V	Cr	Sn	Al	O	N	C	Fe	H
High O	15.7	3.4	3.3	2.9	0.17	0.01	0.006	0.06	0.007
Low O	15.3	3.3	3.4	3.3	0.08	0.01	0.005	0.04	0.008
Extra-low O	14.8	3.0	3.2	2.9	0.04	0.01	0.01	0.06	0.001

Table III Room temperature tensile properties of the specimens tested

Alloy	ST temperature (K)	Yield strength (MPa)	Elongation (%)	Reduction of Area (%)
High O	1073	818	27	63
Low O	1073	708	24	54
	1223	726	23	64
Extra-low O	1073	666	22	67
	1223	668	22	72

Results and Discussion

Effect of oxygen content

Charpy impact energy (CVN) at room temperature of as-ST'ed Ti-153 is as low as that of STA'ed alloy. However, the CVN increases at higher temperature and becomes higher than 150 J. Thus the CVN curve shows an obvious transition in energy just above room temperature as seen Figure 1. The energy transition temperature (ETT) is defined as the temperature at which the CVN is estimated to be an average of CVN at 77 K and at 473 K. For 1073 K ST, the transition occurred at lower temperature in the Low and the Extra-low O alloys than in the High O alloy. However no difference in the transition curves is seen between the Low O alloy and the Extra-low O alloy. For 1223 K ST, the Extra-low O alloy has also the same ETT as the Low O alloy.

Although the reduction of oxygen content had some effect in lowering the transition temperature, the effect on the CVN below room temperature was slight. The lowest oxygen content, 0.04 weight% is considered to be as low as possible in the current melting process. This means more improvement in the low temperature toughness is almost impossible through oxygen reduction in the Ti-153 system.

Effect of ST temperature

A higher ST temperature, 1223 K, produced a factor of four increase in grain diameter than a lower ST temperature, 1073 K. However, there was not a significant difference in yield strength for both ST temperatures. In Figure 3, the transition curves are compared between two ST temperatures. The upper shelf energy was obviously higher in larger grain specimens for both alloys, but the ETT was not changed. This differs from the low temperature brittleness in bcc ferrous alloys.

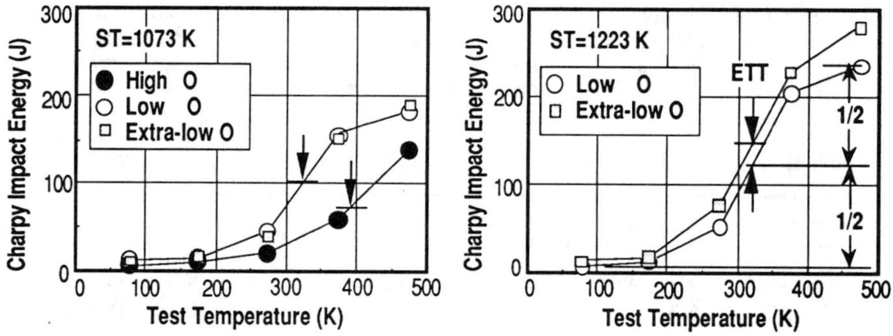

(a) (b)

Figure 2 - The effect of oxygen content on the Charpy impact energy transition :
(a) ST temperature = 1073 K and (b) ST temperature = 1223 K.

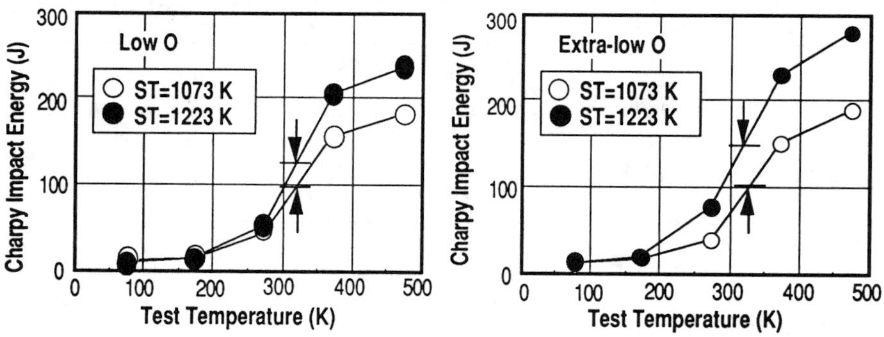

(a) (b)

Figure 3 - The effect of ST temperature on the Charpy impact energy transition :
(a) Low O alloy and (b) Extra-low O alloy.

Fracture Surface

In the specimens with a CVN higher than 100 J, equiaxed dimples of the size identical with β grain diameters cover the fracture surface. This fracture mode is a fully ductile mode. As the test temperature decreases, or the CVN decreases, arrays of straight chain of micro-dimples, on the order of 1 μm, is partly seen between large dimples, as shown in Figure 4 (the High O alloy, 373 K). In contrast, only large dimples are seen in the Low O alloy, 373 K, which showed the CVN of about 150 J. Below the room temperature, the micro-dimples become the main component of the fracture surface (Figure 4, 173 K), and at 77 K only the micro-dimples were seen. Hence a fracture mode of micro-dimples is believed to cause the poor CVN value at low temperatures. This also confirms the above-mentioned idea that the low temperature brittleness is not the mechanism for Ti-153 alloy.

The deformed zone near the notch root of fractured Low O alloy Charpy specimen was observed using optical microscope to detect the difference in deformation behavior between high and low CVN specimens as shown in Figure 5. In the high CVN specimen, the notch root area is highly deformed and some voids are seen. In the low CVN specimen, needle-like thin plates are produced in the deformed zone; none of such plates was seen in

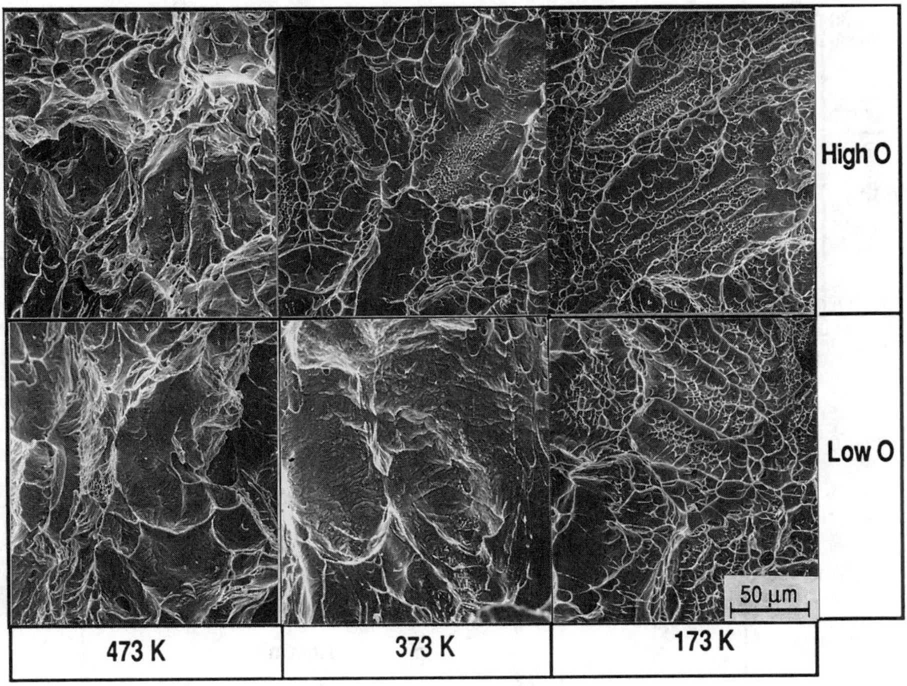

Figure 4 - SEM micrographs of fracture at various temperatures for the High O alloy and the Low O alloy solution-treated at 1073 K.

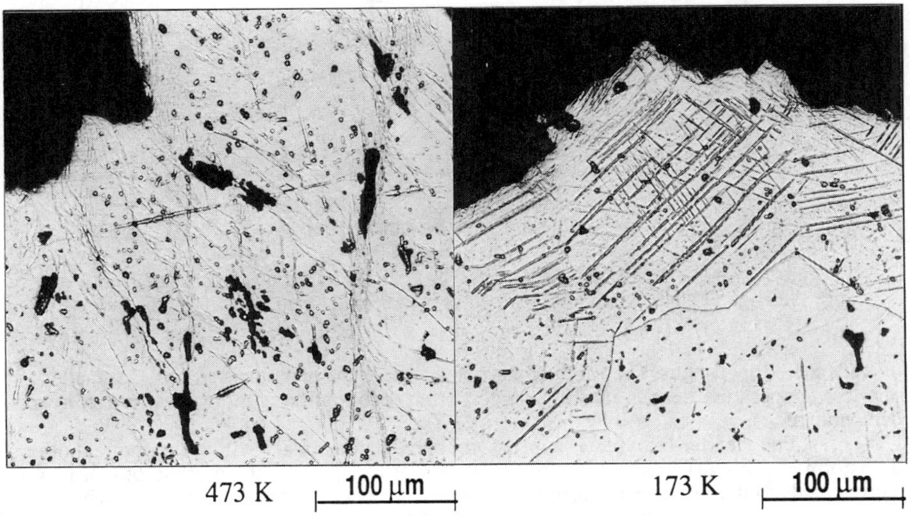

Figure 5 - Optical micrographs of the deformed area near the notch root of fractured Low O alloy Charpy specimen.

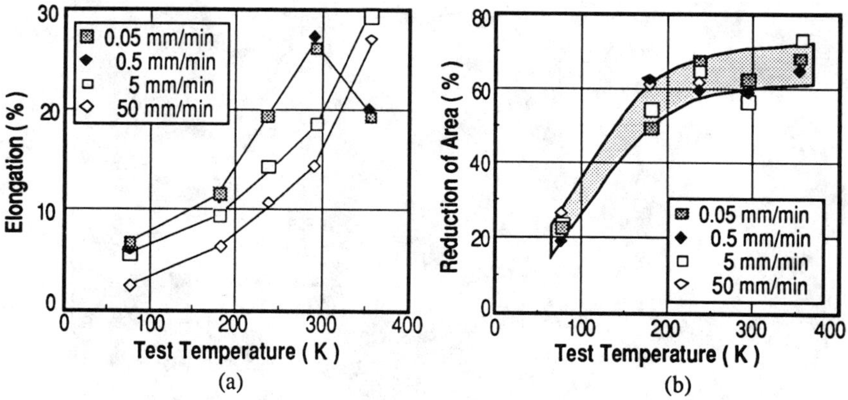

Figure 6 - Elongation (a) and reduction of area (b) as a function of test temperature for the High O alloy.

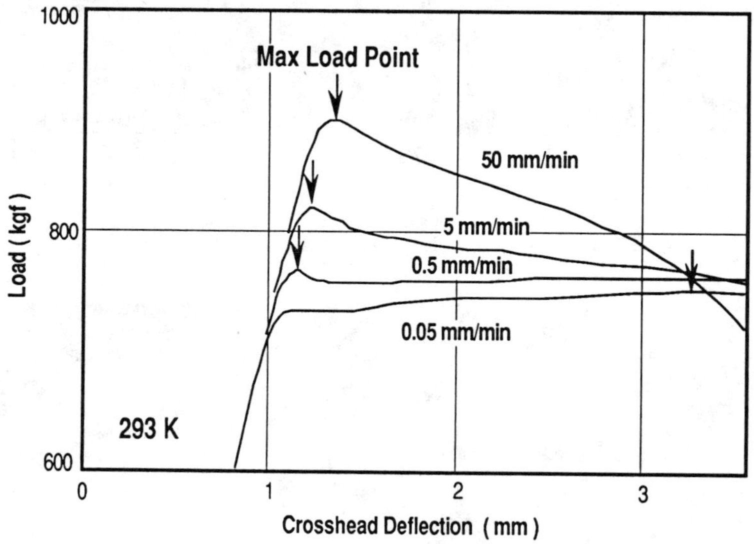

Figure 7 - Load-crosshead deflection curves at 293 K at various crosshead speeds for the High O alloy. Maximum load point is indicated by an arrow for each curve.

merely cooled specimens. The width of the zone in which the plates existed became smaller as the test temperature decreased. This suggests that deformation is more localized at lower temperature.

The needle-like plate appears to be a mechanical twin ratherthan another transformed phase. The plate width was of the order of 1 μm and the morphology of plates corresponds to the linear array of micro-dimples. The occurrence of mechanical twins may produce micro-voids which provide an easier path for fracture.

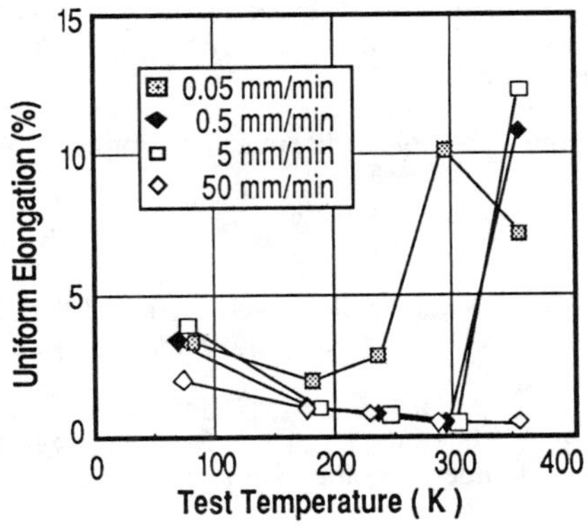

Figure 8 - Uniform elongation as a function of test temperature for the High O alloy.

Tensile behavior of High O alloy

Localized deformation is considered the key mechanism controlling the low temperature toughness of the High O alloy. Localized deformation can be largely influenced by strain rate in tensile deformation. Therefore, the effect of strain rate and test temperature on tensile behavior was studied for the High O alloy solution-treated at 1073 K.

Yield strength, as expected, had a linear relationship to the logarithm of strain rate between 181 K and 355 K. However, the yield strength showed no strain rate dependence at 77 K. Reduction of area (RA) was not significantly altered by strain rate and test temperature between 181 K and 355 K. The RA at 77 K was fairly low compared with others. In contrast, elongation was a function of strain rate and test temperature. The elongation was smaller with lower test temperature or with higher strain rate. As shown in Figure 6 (a), the relationship between elongation and test temperature reveals the curves similar to Charpy impact energy transition curve. No gap was seen between transition curves at lower two strain rates, but with higher strain rate, the curve shifted to higher temperature side.

It is strange that the elongation changed markedly while the RA did not vary between 181 K and 355 K. Most of load-displacement curves showed a quick necking as indicated in Figure 7. Only a small amount of uniform deformation was allowed before the necking started (This is another evidence that localized deformation prevailed in the present alloy). Most elongation resulted from a good resistance to failure during necking, producing a high value of RA. Hence uniform elongation, elongation before necking started or until maximum load point, was measured and plotted in Figure 8. In most cases, the uniform elongation was very small and less than 5 %. At 50 mm/min, the uniform elongation was smaller at higher test temperature. However, the uniform elongation jumped up to about 10% at a given test temperature at other strain rates. The jump occurred at lower temperature as the strain rate decreased. This result clearly shows that higher strain rate, or lower temperature, suppresses the gross uniform deformation by the occurrence of deformation localization.

Summary

The solution-treated Ti-15V-3Cr-3Sn-3Al alloy showed a marked energy transition at near room temperature in Charpy test. The upper-shelf fracture mode was characterized by large equiaxed dimples with a size similar to the β grain size. On the other hand, the lower-shelf fracture was due to the occurrence of micro-dimples of several μm. "Cleavage" was not the mechanism for the present alloy.

Reduction of oxygen from 0.17 weight% to 0.08 weight% gave a lower transition temperature, however a further reduction showed no effect. No difference in transition temperature existed between small grain and large grain specimens; however, the large grain specimen had a higher upper-shelf energy. These features are not always analogous to "cleavage fracture" of ferritic steel. Deformation localization is considered to be highly responsible for this transition.

This paper is supported by the Amada Foundation for Metal Work Technology.

References

1. ASM Metals Reference Book (American Society for Metals, 1983), 385.
2. R.G.Broadwell and R.A.Wood, "Titanium Alloys for Cryogenic Service", Materials Research & Standards, 4(1964),549.
3. K.Nagai, T.Yuri, O.Umezawa and K.Ishikawa, "The Cryogenic Mechanical Properties of Ti and Its Alloys", Cryogenic Materials '88, vol.2 Structural Materials, ed. R.P.Reed, Z.S.Xing and E.W.Collings (Boulder, CO:International Cryogenic Materials Conference, 1988), 727 -736.

Effect of Duplex-aging after Cold-working on the Mechanical Properties of Ti-15V-3Cr-3Sn-3Al

Naotake NIWA[1] and Hideo TAKATORI[2]

1) Faculty of Engineering, The University of Tokyo
7-3-1 Hongo Bunkyo-ku, Tokyo, 113 Japan

2) NIKKO ANCO-TECH INC.
2525 Beech Daly Road Dearborn Heights, Michigan 48125 U.S.A.

Abstract

The features of beta titanium alloys include their superior cold-formability in the solution-treated state and their increased strength after undergoing subsequent aging-treatment. Consequently, the method of aging after cold-forming is a valuable heat-treatment method which utilizes the features of beta titanium alloys. This research investigated the effects of aging at single temperature(CWA) and duplex-aging(CWDA) after cold-working on the mechanical properties of a Ti-15V-3Cr-3Sn-3Al alloy. In CWA, aging at higher temperature produces little gain in strength, on the other hand, aging at lower temperature reduced ductility while aiming to achieve high strength and the strength-ductility balance can not be improved. Duplex-aging after cold-working, i.e. combinations of short-time aging at 823K and 873K and re-aging at 673K have greatly improved the strength-ductility balance of this alloy, realizing high strength of about 1.8GPa with tensile elongation of 5%. This duplex-aging process further diminishes the effect of the amount of reduction by cold-swaging on the mechanical properties after aging.

Introduction

The features of beta titanium alloys include their superior cold-formability in the solution-treated state and their increased strength after undergoing subsequent aging. Consequently, the method of aging after cold-forming is a valuable heat-treatment method that utilizes the features of beta titanium alloys. This research investigated the effects of aging at single temperature and duplex-aging after cold-working on the mechanical properties of a Ti-15V-3Cr-3Sn-3Al alloy (hereafter abbreviated as Ti-15-3).

Experimental Procedures

The Ti-15-3 alloy bar was solution-treated at 1123K for 3.6ks in inert gas atmosphere and then air-cooled. After cold-working with 30%, 60%, 83% and 90% reductions of the bar using a rotary swaging machine, various types of aging-treatments were conducted on the specimens. Hereafter, the aging

Titanium '92
Science and Technology
Edited by F.H. Froes and I. Caplan
The Minerals, Metals & Materials Society, 1993

treatment of retention at a single temperature after cold-working is abbreviated as CWA, and the high-low temperature duplex-aging after cold-working is abbreviated as CWDA. Also, aging without cold-working after solution-treatment is abbreviated as STA. CWA was performed at aging temperatures; 973K, 923K, 873K, 823K, 773K, 673K and 573K. The second aging in the duplex-aging process for CWDA was performed at a single temperature of 673K. Of the aging-treatments after cold-working, short-time aging (60s and 180s) was performed in a salt bath, while the rest of the specimens were aged for longer times using a muffle furnace. After all the heat-treatments, specimens were air-cooled.

Table 1 shows the chemical composition of the Ti-15-3 alloy bar.

Hardness measurement, microstructure observations using an optical microscope and an electron microscope, as well as uniaxial tensile testing, were performed on the above specimens. Hardness value indicates the average value of five points on each specimen. Also, a X-ray diffraction method was used to detect the alpha phase precipitation.

Tensile testing was performed at room temperature at a cross-head speed of 1.7×10^{-2} mm/s. Tensile data was the average value of three specimens per each heat-treatment condition. The tensile specimens had a gauge size of 4.0mm in diameter and 15.0mm in length.

Table 1. Chemical composition (mass%).

V	Cr	Sn	Al	Fe	O	C	N	H	Ti
15.1	3.36	3.04	3.37	0.17	0.14	0.004	0.008	0.0061	bal.

Results

Aging Behavior in STA

Figure 1 shows the variation of hardness with aging time in STA. As for the specimens aged at 873K, even when the aging time was lengthened, little increase in hardness was observed. For the aging at 773K, the hardness attained almost the highest value after 30ks of aging, however, for the agings at 673K and 573K, the longer the aging time, the greater was the hardness. Within the aging time range of this experiment, the highest hardness of 490 was attained by the aging at 573K for 3Ms.

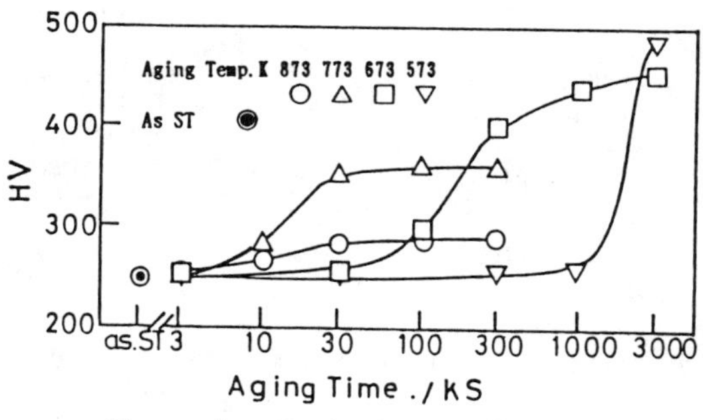

Figure 1 - Variation of hardness in STA.

Aging Behavior in CWA

Figure 2 shows the variation of hardness as a function of aging time and temperature for specimens aged after cold-swaging with 90% reduction from 28 mm to 9 mm in diameter of the bar. After 3.6ks aging, the hardness of all the specimens became higher than that before aging. The hardness of specimens aged at 873K and 573K were slightly higher than that of as cold-swaged, then, the hardness of aging at 873K decreased gradually and that of aging at 573K increased fast after 3.6ks. As for the specimens aged at 773K, the hardness became about 420 after 3.6ks and hardness decreased after 10.8ks. The hardness of the specimens aged at 673K became about 450 after 3.6ks and increased to about 510 until 360ks. For CWA, promotion of the aging response with cold-working and the recovery of the beta matrix can be considered to occur in parallel. Therefore, the variation of hardness with aging after cold-working should be controlled by rate of recovery and precipitation of alpha phase and size and distribution of alpha phase that strongly depend on aging temperature. The higher the retention temperature becomes, the greater the influence on recovery should be. The lack of hardening in the specimens aged at 873K would be due to the overgrowth of the alpha precipitates and to the rapid progress of recovery.

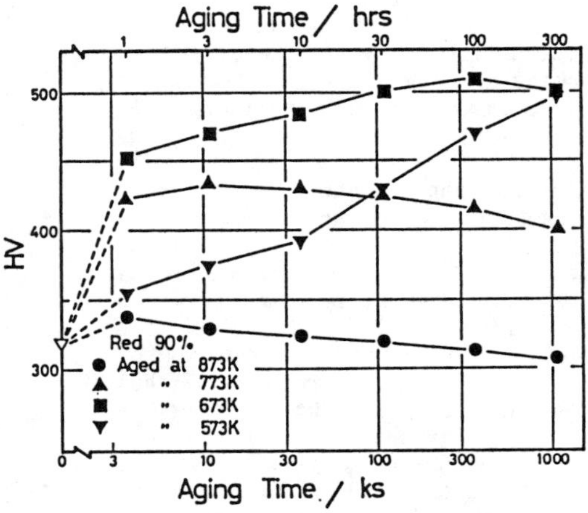

Figure 2 - Hardness variation with aging time and temperature in CWA.

Figure 3 shows the variation of hardness as a function of reduction in cold-swaging and aging time in the aging at 673K. This figure also shows the variation of hardness of STA specimens. It is indicated that the greater the reduction in cold-swaging, the greater is the hardness throughout aging and the promotion of the aging with cold-working is marked in comparison with the aging response in STA.

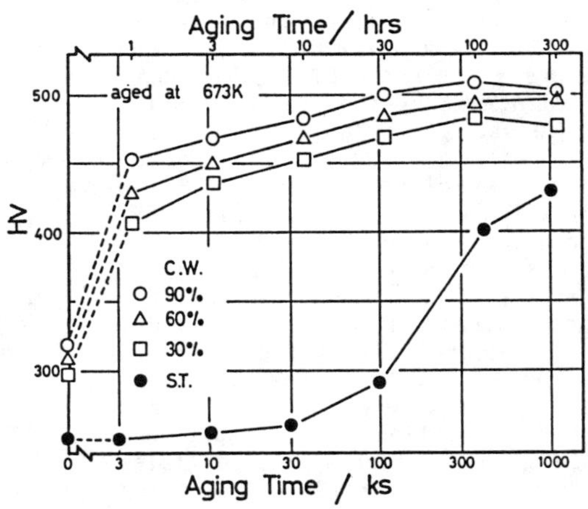

Figure 3 - Effect of reduction in cold-working on hardness variation in CWA.

Figure 4 shows the variation of tensile strength and reduction in area as a function of reduction in cold-swaging, aging temperature and aging time in CWA. The tendency in the variation of tensile strength corresponds to that in the variation of hardness as shown in Figure 3. Thus, these results indicate that difference of plastic strain in cold-working produces the difference in tensile strength after aging. Ductility in all aging condition of the CWA decreased greatly with aging.

<u>Aging Behavior in CWDA</u>

As mentioned above, cold-working before aging remarkably promoted the aging response and greatly reduced ductility while aiming to achieve high strength, therefore, the strength-ductility balance could not be improved by CWA.

The characteristic of CWDA[1] is adding short-time and high-temperature aging between cold-working and low temperature aging to get homogeneous recovered matrix based on the idea that highly strained area as grain-boundaries and shear-bands should recover more quickly than other areas.

Figure 5 shows the variation of hardness as a function of aging time and temperature for specimens aged after cold-swaging with 83% reduction. Heating temperatures of from 973K to 823K are higher than those adopted in CWA. After 60s aging, the hardness of all the aged specimens became lower than that before aging. The hardness of specimens aged at 973K is markedly low, being almost the same as that after solution-treatment. As for the specimens aged at 973K and 923K, even when the aging time was lengthened, little increase in hardness was observed. For the agings at 823K and 873K, the longer the aging time, the greater was the hardness. In particular, the increase in hardness of the specimens aged at 823K is significant. For the aging at 873K, hardness decreases after 1.8ks.

Microstructure observation revealed that for the specimens aged at 973K, the beta matrix completely recrystallized. The marked decrease in the hardness for these specimens is related to this recrystallization. For the specimens aged at 923K, recrystallized grains became more apparent as the aging time became longer. On the contrary, recrystallized grains were not observed under optical microscope even after 3.6ks aging for specimens aged at 823K and 873K. For specimens aged at 823K, 873K, as well as for those aged at 923K for 600s, precipitation of alpha phase was confirmed by X-ray diffraction. On the other hand, for the specimens aged at 973K, no precipitation was found even after 3.6ks aging. It was also found with TEM that the precipitated alpha phase grew larger and tended to become more spherical with increase in aging temperature.

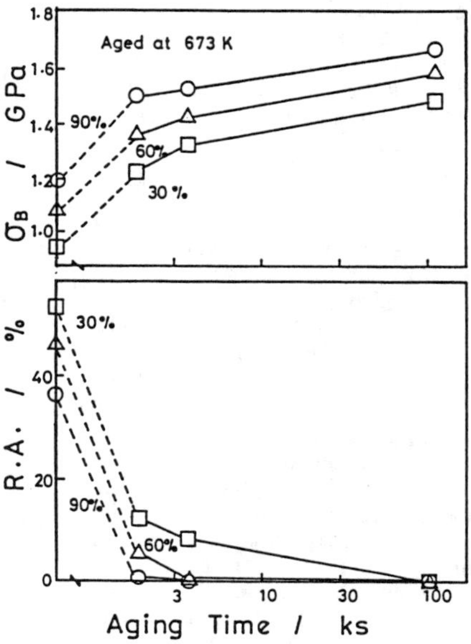

Figure 4 - Variation of tensile strength and reduction of area in CWA.

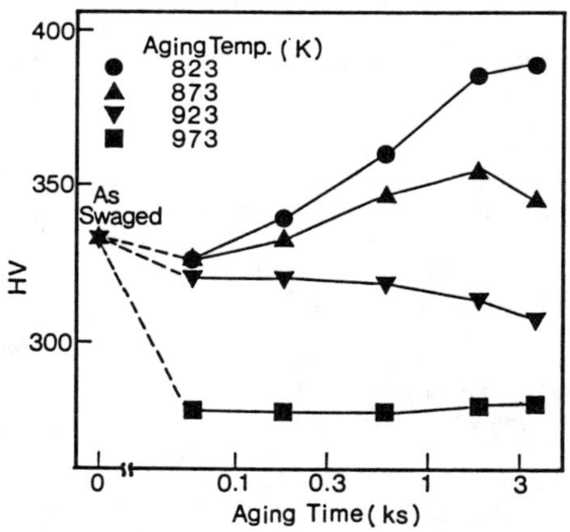

Figure 5 - Variation of hardness with aging time and temperature.

Under any aging condition, hardness increased with the progress of the second-stage aging. The shorter the time of the first aging, the more marked was the increase in hardness. Furthermore, for those specimens with the same first-stage aging times, the lower the aging temperature of the

first stage, the greater was the hardness during the initial period of the second aging. However, the hardness after 360ks of the second-stage aging was insensitive to the first-stage aging temperatures.
Variation of strength in CWDA corresponds to that of hardness. On the other hand, after 360ks aging, ductility decreased, in association with the increased strength. Changes in ductility with aging are dependent on the aging time of the first stage. However, despite the exhibition of increased strength, there were some cases in which total elongation also increased.

Discussion

Strength-ductility balance in CWA

Figure 6 shows the variation of the relation between ultimate tensile strength and reduction of area for specimens cold-worked with 90% reduction. STA. Compared with the strength-ductility balance in STA, CWA produced slightly worse balance. Rosenberg[2] also reported that aging after cold-working could not improve the strength-ductility balance of Ti-15-3. It would be originated from that, in CWA, preferential precipitation of alpha phase on highly strained areas would occur because aging enhancement strongly depends on accumulated strain, therefore, inhomogeneous aged microstructure with highly strained areas as shear bands and grain boundaries would be produced.

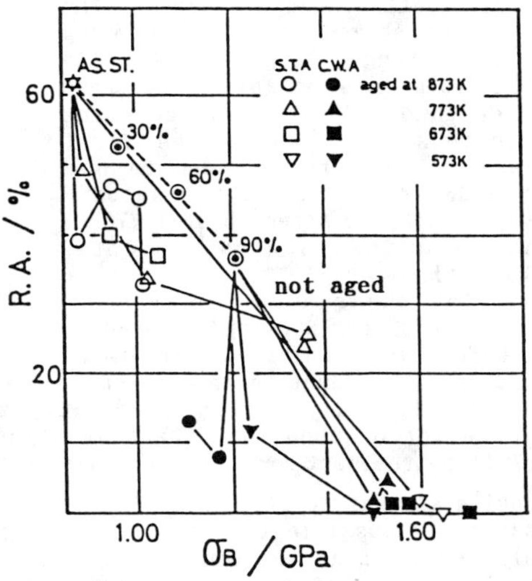

Figure 6 - Tensile strength and ductility in CWA and STA.

Figure 7 shows the variation of the relation between reduction of area and ultimate tensile strength in CWA for Ti-13V-11Cr-3Al alloy[3]. Specimens were cold-worked with from 90% to 30% reduction and aged at from 750K to 600K. In this figure, data obtained by STA is also shown. Compared with STA, CWA produced better balance of strength-ductility for this alloy. In STA, grain boundary cracking greatly reduced ductility with development of aging, which is peculiar to this alloy. In CWA, cold-working before aging increases precipitation sites and suppresses preferential precipitation of alpha phase on grain boundaries resulting in suppressing grain boundary cracking and improving strength-ductility balance.

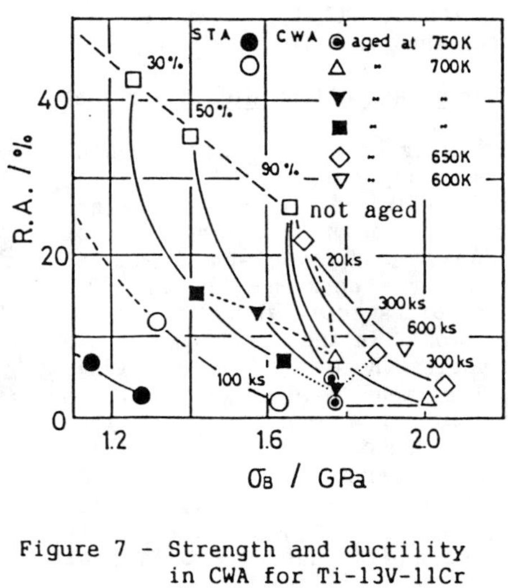

Figure 7 - Strength and ductility in CWA for Ti-13V-11Cr -3Al alloy.

Strength-ductility balance in CWDA

Figure 8 shows the relation between

reduction of area (RA) and ultimate tensile strength (σ_B) under the CWDA. The straight line indicates the relation between reduction of area and ultimate tensile strength obtained by STA on Ti-15-3. The shorter the first-stage aging time, the higher is the attainable strength. When strength reaches about 2GPa by CWDA, the ductility is significantly reduced. However, the specimens showed good ductility up to the strength level of 1.8GPa in CWDA. To give an example of the good strength-ductility balance, the specimens aged during the first stage at 823K for 600s and re-aged at 673K for 108ks in the second-stage exhibited strength of 1.77GPa, 31% reduction in area and total elongation of 5.7%. As another example, the specimens initially aged at 873K for 600s and re-aged at 673K for 108ks exhibited strength of 1.72GPa, 35% reduction in area and total elongation of 7.3%. Regarding the strength-ductility balance in Ti-15-3 with minimum requirement for total elongation of about 5%, the conventional heat-treatment of aging after solution-treatment or cold-working has been thought to give the maximum tensile strength of 1.5GPa[2,4-7]. However, the present results for CWDA show that it is possible to increase the tensile strength to up to about 1.8GPa while maintaining the same ductility. The specimens that were aged for a shorter period of time than 600s in the first stage and subjected to the longest aging in the second-stage exhibited increase in strength. On the contrary, the specimens that were initially aged for a longer period than 600s exhibited the decrease in strength. In both cases, the strength-ductility balance was poorer than that for 600s.

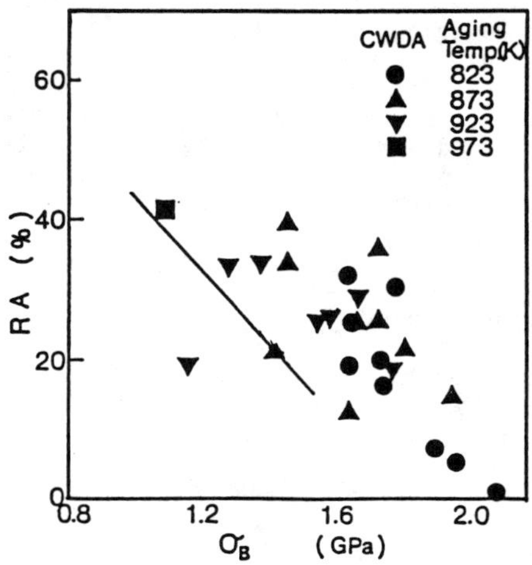

Figure 8 - Strength and ductility in CWDA.

Effect of Reduction in Cold-Working

The aging response after cold-working depends on the amount of strain caused by cold-working. Figure 9 shows the variation of hardness of specimens aged at temperatures between 973K and 823K after cold-swaging with 83% and 30% reduction. In the cold-worked state and during the early stage of aging, the specimens cold-swaged with 30% reduction show lower hardness than those with 83% reduction, except the specimens aged at 973K with 83% reduction. However, along with the progress of aging, differences in the hardness due to different reductions in cold-working for the respective aging temperatures

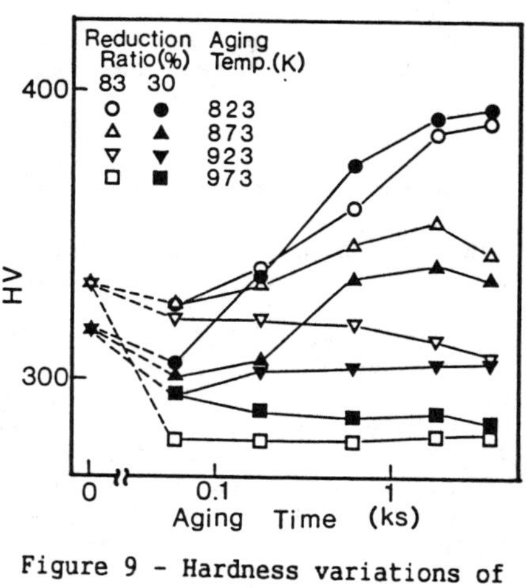

Figure 9 - Hardness variations of specimens aged after cold-swaging with 83% and 30% reduction.

decreased and the hardness became almost the same after 3.6ks. Strain accumulation caused by cold-working promotes the aging response and increases hardness. At the same time, it also promotes the recovery of the cold-worked matrix and reduces hardness. Therefore, the hardness is thought to be determined by the combination of the two effects. When the specimens cold-swaged with 83% reduction are compared with those cold-swaged with 30% reduction, the rate of precipitation is faster in the former because of their higher dislocation density. On the other hand, the rate of recovery in the former also becomes faster, besides the specimen heated at 973K. This must be the reason why hardness during the initial aging period is high in the former, but the difference between the hardness levels decreases with the progress of aging.

Figure 10 shows the relations between ultimate tensile strength (σ_B) and reduction of area (RA) of CWDA specimens cold-swaged by 83% and 30%. The specimens were first aged at either 823K or 873K for 600s. The first-stage aging conditions were chosen since they produce optimum ductility for specimens cold-worked with 83% reduction. The second aging was performed at 673K for times between 0 and 108ks. The results for specimens initially aged at 973K for 600s and were re-aged at 673K for between 0 and 108ks in the second aging are also shown. Under the same aging conditions, although the lower reduction specimens tended to exhibit lower strength and reduction of area, these tendencies became less pronounced as strength increased by the second-stage aging. For specimens aged at 973K in the first stage, strength after the second aging changed significantly depending on the amount of reduction in cold-working.

For all the CWDA specimens, except those initially aged at 973K, these results indicate that if the initial aging is performed over a certain length of time, it is possible to lessen the effect of reduction in cold-working on the hardening behavior during the second-stage of aging. In addition, the results also suggest that essentially the same mechanical properties can be achieved by the second-stage aging irrespective of the amount of cold-working; 83% or 30% reduction.

As mentioned above, the aging after cold-forming utilizes the features of this alloy. However, because of the inhomogeneous distribution of accumulated strain, attention should be paid to the local variation in the mechanical properties of a product that has been aged after cold-forming. As shown in Figure 10, since CWDA can decrease the effect of the amount of cold-working on mechanical properties. The advantages of this method are not only in the improved strength-ductility balance but also in the minimization of the influence of the amount of accumulated strain in cold-working on the mechanical properties after aging. Ti-15-3 was a beta titanium alloy developed on the premise of cold-forming[8]. It is thought that the present heat-treatment method has a superior potential for the manufacture of Ti-15-3 products that will undergo aging after cold-forming.

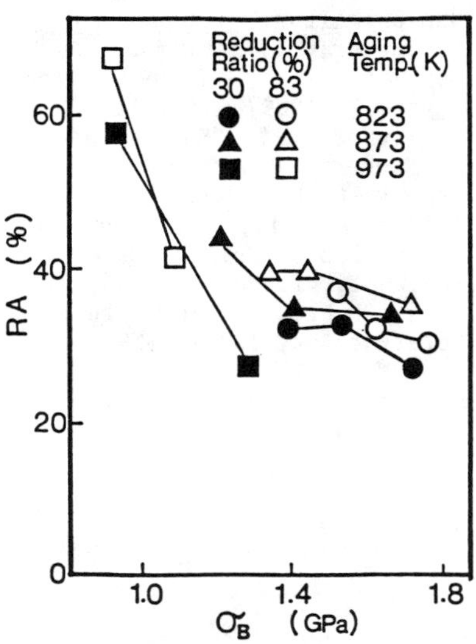

Figure 10 - Effect of reduction in cold-working on strength and ductility in CWDA.

Conclusion

This research investigated the mechanical properties of Ti-15-3 that underwent aging after cold-working. The main findings can be summarized as follows:

(1) By combining the first aging that was performed at a high temperature for a short period of time and the second aging that was performed at 673K, it was possible to improve the balance between the strength and ductility of this alloy. For example, while maintaining 5% total elongation, a tensile strength of approximately 1.8GPa was achieved.

(2) By this duplex aging method, the influence of accumulated strain differences caused by different reductions in cold-working on the mechanical properties after aging could be kept to a low level.

From these results, the duplex aging method using the features of Ti-15-3 is concluded to be an excellent method of heat-treatment that can overcome the problems of mechanical property unevenness and reduction of ductility caused by the conventional aging after cold-forming.

References

1. N. Niwa et al., "Mechanical Properties of Cold-worked and High-Low Temperature Duplex-aged Ti-15V-3Cr-3Sn-3Al Alloy," ISIJ International, 31(1991), 856-862
2. H.W. Rosenberg, "Ti-15-3: A New Cold-Formable Sheet Titanium Alloy," J. Met., 35(1983), 30-34
3. M. Gunji et al., "Effect of Degree of Cold Working and Aging Temperatures on Mechanical Properties of Aged Ti-13V-11Cr-3Al Alloy," Tetsu-to-Hagane, 72(1986), 610-616
4. M. Okada and T. Nishikawa, "The Effect of Cold Rolling and Heat Treatment Conditions on Microstructures and Mechanical Properties of Ti-15V-3Cr-3Sn-3Al Alloy," J.Japan Inst. Metals,50(1986),555-562
5. Y. Tsumori, T. Matsumoto and Y. Koyama, "Workability and Heat Treatment Characteristics of Beta Titanium Alloy, Ti-15V-3Cr-3Sn-3Al," Tetsu-to-Hagane,72(1986),603-609
6. N. Niwa et al., "Morphology of Microstructure and Mechanical Properties of Commercial Beta Titanium Alloys, Ti-15-3 and Ti-13-11-3," (Rep. Eng. Res. Inst. Fac. Eng. Univ. Tokyo, 47(1988),241-246)
7. N. Niwa et al., "Influence of Thermomechanical Treatment on Mechanical Properties of Commercial Beta Titanium Alloys" (Rep. Eng. Res. Inst. Fac. Eng. Univ. Tokyo, 47(1988),247-252)
8. J.A. Hall, The Metallurgy of Light Alloys(London:Institution of Metallurgist, 1983), 116

EFFECT OF Nb AND W ON MECHANICAL PROPERTIES AND OXIDATION RESISTANCE OF HIGH TEMPERATURE TITANIUM ALLOYS

Han Chuanxi and Liu Bingnan
Northwest Institute for Non–ferrous Metal Research
P.O. Box 71, Baoji, Shaanxi 721014, P.R. China

ABSTRACT

The effects of Nb and W contents on the tensile properties at room temperature and 550℃, the impact toughness, the creep and the oxidation resistance of the tested alloys has been investigated. The results indicate that alloying element Nb has no strong effect on the strength of Ti–alloy, but Nb can improve the surface stability and increase the oxdation resistance of Ti–alloy exposed at high temperature. Alloying element W can raise the strength and the resistance of creep in Ti–alloy, but has less contribution to the resistance of oxidation in Ti–alloy than Nb. The studies have provided necessary scientific data for choosing suitable β isomorphous adding elements in near α–Ti alloys.

INTRODUCTION

In titanium alloys, Nb and W are the important alloying elements. In recent years, several practical titanium alloys have been developed successfully by adding alloying elements Nb and W. Adding Nb abd W to intermetallic compounds of α_2–Ti_3Al seems to have retarded the ordering rate. The fine α_2 structure can improve strength and ductility at room temperature. TiAl–W alloy has a finer grain size than the stoichiometric gamma alloy. The fine structure of the alloy improves the creep resistance and the tensile strength without further loss in inductility [1].

This investigation was made to expound the correlation of mechanical properties and oxidation resistance with the amount of Nb and W, so as to provide necessary scientific basis for studying of selecting β isomorphous additive elements for near α–Ti alloys.

EXPERIMENTAL PROCEDURE

The chemical compositions of tested titanium alloys are shown in Table 1 and they will be referred to as alloys A (0.5% Nb), B (1% Nb), C (1.5% Nb), D (0% Nb), E (1%Nb, 0.5%W) and F (1%Nb, 1%W).

The specimens of these alloys were fabricated from ingots of 5kg melted in a consumable vacuum arc furnace. The ingots were β and $\alpha+\beta$ forged to Φ16mm × L mm bars which were followed by solution treatment and aging. The specimens were fabricated from Φ16mm × L mm heat–treated bars.

Titanium '92
Science and Technology
Edited by F.H. Froes and I. Caplan
The Minerals, Metals & Materials Society, 1993

Table 1. Chemical compositions (wt%) of six alloys under investigation

Alloy	Al	Sn	Zr	Mo	Nb	W	Si	Fe	C	N	O	H
A: Ti-5.5Al-3.5Sn-3Zr-0.25Mo-0.5Nb-0.3Si	5.51	3.41	2.72	0.26	0.50	–	0.26	0.047	0.02	0.009	0.092	<0.001
B: Ti-5.5Al-3.5Sn-3Zr-0.25Mo-1.0Nb-0.3Si	5.58	3.44	2.60	0.26	1.00	–	0.26	0.063	0.02	0.01	0.085	<0.001
C: Ti-5.5Al-3.5Sn-3Zr-0.25Mo-1.5Nb-0.3Si	5.52	3.41	2.66	0.26	1.48	–	0.26	0.047	0.02	0.01	0.09	<0.001
D: Ti-5.5Al-3.5Sn-3Zr-0.25Mo-0.3Si	5.57	3.51	2.89	0.26	–	–	0.26	0.045	0.02	0.01	0.08	<0.001
E: Ti-5.5Al-3.5Sn-3Zr-0.25Mo-0.5W-1Nb-0.3Si	5.50	3.51	2.65	0.26	1.01	0.50	0.26	0.053	0.025	0.01	0.075	<0.001
F: Ti-5.5Al-3.5Sn-3Zr-0.25Mo-1Nb-1W	5.56	3.55	2.67	0.25	1.00	0.99	0.26	0.046	0.025	0.01	0.08	<0.001

According to the $\alpha+\beta$ / β transus temperature of alloys, the heat treatment is carried out at 1035℃ for 1hr and at 620℃ for 2hrs. The solution treated temperature is 15–25℃ above β–transus.The mechanical properties studied include tensile, impact toughness and creep. The oxidation resistance of titanium alloys and thickness of the oxide layers of some of them were investigated by ion probe, scanning electron microscope and increasing weight in oxidation. First measuring the relation between oxygen ion current and the depth of bombarding pitch by using ion probe and then determining the depth of bombarding pitch by using scanning electron microscope of PSE 500X type as the thickness of the oxide layer. The samples used for increasing weight in oxidation are 10mm in diameter × 5mm in length. Before oxidation, the samples were mechanically polished with abrasive paper. Before and after oxidation, the weight of the sample was measured by using microbalance of 1.0μg maximum sensitivity. The oxidation was carried out at 550℃ for 100hrs in air and under the atmospheric pressure.

RESULTS AND DISCUSSION

1.The ways of adding Nb and W

There are great differences in density and melting point of metalic Nb,W,and Ti. It is a key technical problem for titanium melting to ensure the alloying elements Nb and W scattering uniformly in the basis metal without any unmelting bulks.In order to get a ingot with good quality the master alloys Al–Nb and Al–W are used as alloying additives, which are made by aluminothermics.The ingot obtained from double VAR melted electrode made of these two master alloys is approved to have homogenous contribution of Nb and W from the top to the bottom by chemical analysis, measuring Brinell hardness and dissecting a typical ingot.The variation of Nb and W contents is within the permissive range of commercial Ti–alloy. The analyses at ten points of a half ingot further indicate that the compositions of Nb and W in every part of the ingot are homogeneously distributed, i. e. without any metalic defects,such as inclusions and segregations, etc.

2. The influence of the contents of Nb and W on the mechnical properties of testing alloys

To make it easy for comparison, all the testing pieces were made by the same process, including hot forging and heat treatment ,etc. The microstructure obtained seems alike and all of them are acicular ($\alpha+\beta$).

(1)Effect of Nb on the alloys mechanical properties

Fig.1–3 show the effect of the amount of Nb in the alloy on the tensile properties,the tensile properties at 550℃ ,the impact toughness and the creep of the A,B,C,D alloys. From Fig.1 and 2,we can see that increasing the content of Nb among 0.5–1.5% doesn't make much difference on the tensile strength and the 0.2% proof stress,which is due to the little strengthening factor of Nb to Ti,since Nb is a soft strengthening element. As far as the reduction area and elongation are concerned,the reduction in area (RA) and elongation (EL) of the tensile at room and high temperature show good correlationship. The alloys with 1% Nb possess better plasticity comparing with that other Nb percentage. Fig.3. indicates that the general tendency of consumption of energy a_k, as fracturing, decreases with increasing the constents of Nb.The curve of the creep is made by measuring the plastic flow at constant temperature of 550℃ and constant stress of 304MPa for 100hrs. The curve show that the δ reaches its minimum value when the content of Nb is 0.5%, while for others, the increasing of Nb results in the increasing of δvalue. By comparing the alloys with 1% Nb with those without Nb, it seems that the δ value is almost the same.

The whole process of creep of the alloys A,B,C and D at 100hrs, may be devided into three stages i. e. , I , instantaneous deformation II , decelerative creep and III, stationary creep. At the III stage, the creep rate of alloys A,B,C and D are 6.4, 8.0, 11.0, 9.1×10^{-5} percent/ hr, respectively.

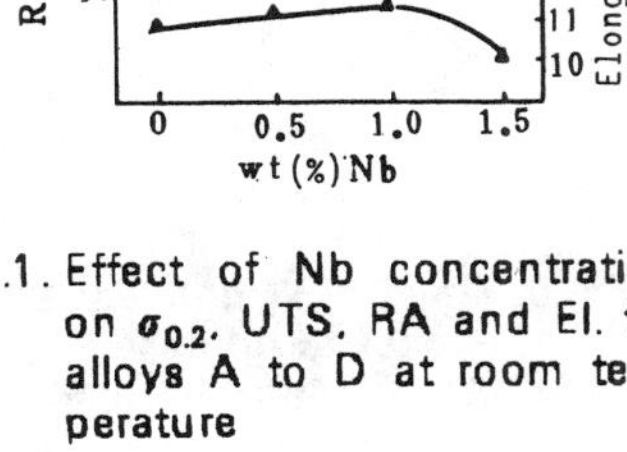

Fig.1. Effect of Nb concentration on $\sigma_{0.2}$, UTS, RA and El. for alloys A to D at room temperature

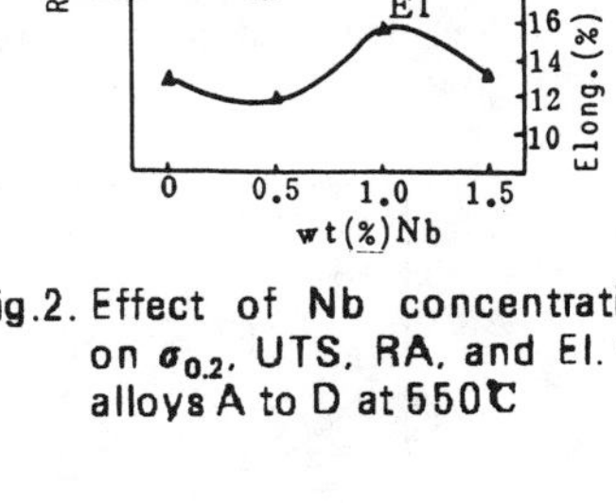

Fig.2. Effect of Nb concentration on $\sigma_{0.2}$, UTS, RA, and El. for alloys A to D at 550℃

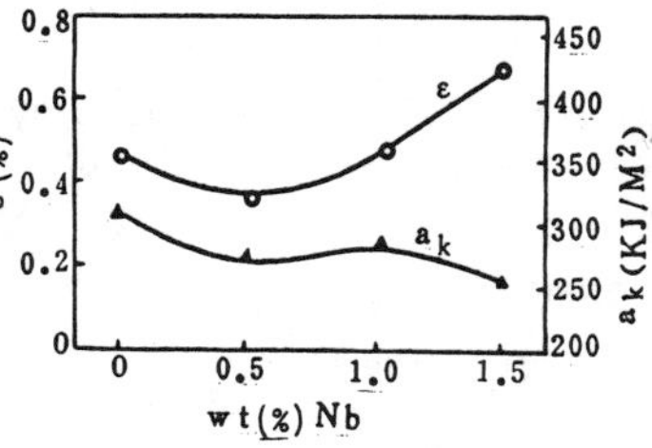

Fig.3. Effect of Nb concentration on impact toughness (a_k) and creep strain (ε) for alloys A to D.

(2) The effect of W on the mechanical properties of the alloys tested

Fig.4–6 show the effect of the amount of W added on the tensile strength, impact toughness and the creep of the B,E,F, alloys. From Fig.4 and Fig.5, we can see that the tensile strength and the 0.2% proof stress increase with increasing the content of W, while the reduction area and the elongation decrease at room temperature. At high temperature, the alloys with 1% W have higher RA and El comparing with those of 0.5% W.

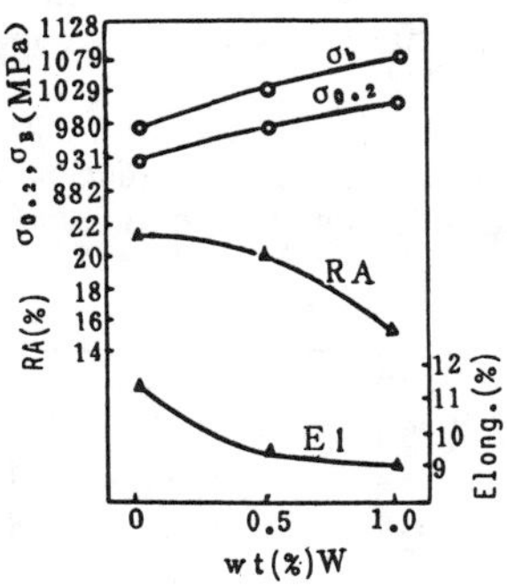

Fig.4. Effect of W concentration on $\sigma_{0.2}$, UTS, RA and El. for alloys B,E,F at room temperature

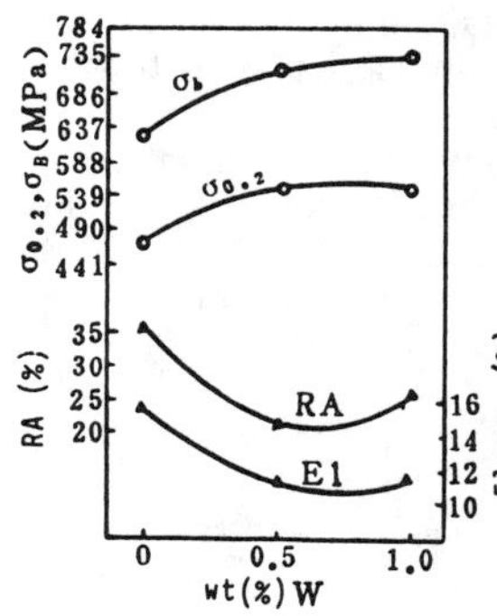

Fig.5. Effect of W concentration on $\sigma_{0.2}$, UTS, RA, and El. for alloys B,E,F at 550℃

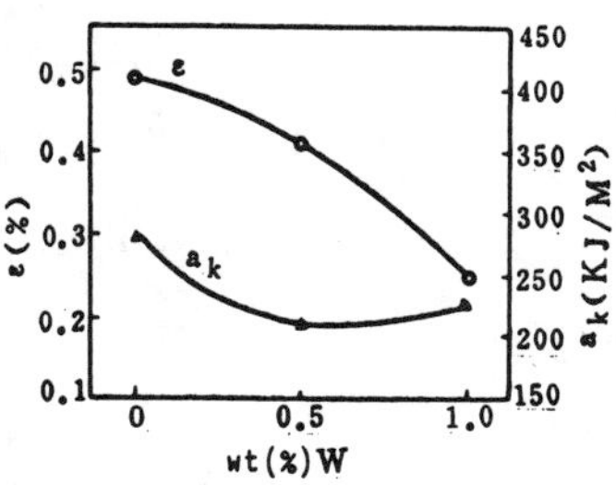

Fig.6. Effect of W concentration on impact toughness and creep strain for alloys B,E,F

Fig.6. shows that the special variasion of the impact toughness is like the changing feature of the RA and El during tensile process at high temperature .

For the creeps at 550℃ / 304MPa / 100hrs, as shown in Fig.6, the δ decrease obviously with increasing the content of W.

At the 1 stage, there is a big difference between the alloys with 0.5% W and those with 1% W. Alloy E turns into stationary creep stage rapidly after a little plastic flow at the 1 stage. The creep rate of the alloys E,F, at the Ⅲ stage are 6.4, 4.8 $\times 10^{-5}$ percent / hr, respectively.

From above results it can be easily seen that the major effect of W is to add strength and resistance to creep of the alloys. The mechanical properties of the alloys with Nb and W have much to do with the properties of the alloying elements. According to the strengthening mechanism of Ti–alloy, the factors which decide the solid solution strengthening effect include the size misfit δ_d and the elastic modulus misfit δ_G. The δ_d of Nb is −0.013 and δ_G is −0.081, the δ_d of W is −0.052 and δ_G is 2.74. Because of the difference is δ_d and δ_G of Nb and W, these two elements have different strengthening factors. The strengthening factors K_T of Nb is 1.75–3.5, while that of W is 5.5, therefore, the strenghthening effect of Nb is not so strong as that of W, which shows that Nb and W have different additional resistance of deformation τ_c. Besides, different elements have different alloying effects on the coefficient of elasticity, diffusivity, cohesion and energy of arranging defects which reflect on the mechanical properties of the alloys.

The effects of the mechanism of creep in the general creeping process depend upon the temperature and the stress conditions during the test. Creep temperature of 550℃ belongs to midtemperature, under this condition, the dynamic recoverying can be carried on obviously, but the diffusion is almost not obvious. The rate of creeping is usually described as the following experimental formula: $\delta = KD_S(\sigma / E)^N (\gamma)^{3.5} L^{-2}$ [2]. In order to make alloys with better resistance of creep, the stacking fault energy γ must be reduced while the self–diffusion activation energy Q_c which correlated with the diffusion factor D_s should be raised.

The D_s of W is much less than that of Nb at 900℃. Otherwise, Nb can reduce the diffusion of the alloying elements which may prevent the β phase from overgrowth. All these characteristics of Nb and W may exert effect on the γ and Ds in the formula, and furthermore, to determine the high temperature strength of the alloy. From the experimental results, it is known that the content of Nb may improve, in certain ranges, the resistance of creep of Ti–alloys. The more the content of W the better the alloys resistance of creep.

(3) The effects of alloying elements on the oxidation resistance

In Fig.7 the curve indicates the relationship between the oxygen ion current which is measured by ion probe and the depth of the bombarding pitch. The curve can be separated into three area, by ion probe and the depth of the bombarding pitch. The curve can be separated into three area, i.e., I exsisting in TiO_2 state, less 0^- is produced during ion bombarding, Ⅱ exsisting in solid solution state, the production of 0^- is decreases as the depth increases, Ⅲ exsisting in the base alloy, no changes occur in the production of 0^-.

The oxidation resistance and the thickness of oxide layer of the alloy are measured by the ion probe and increasing weight in oxidation. The results are shown in Table 2. From which we can see that the oxidation resistance of all the alloys is better than that of pure titanium. Nb has much stronger effect than W on oxidation resistance. Exposed at high temperature, the more Nb contain increases the better the surface stabilities. The external sub–layer of pure titanium which is much thicker is almost the sum of the external sub–layer and the internal sub–layer of another two alloys.

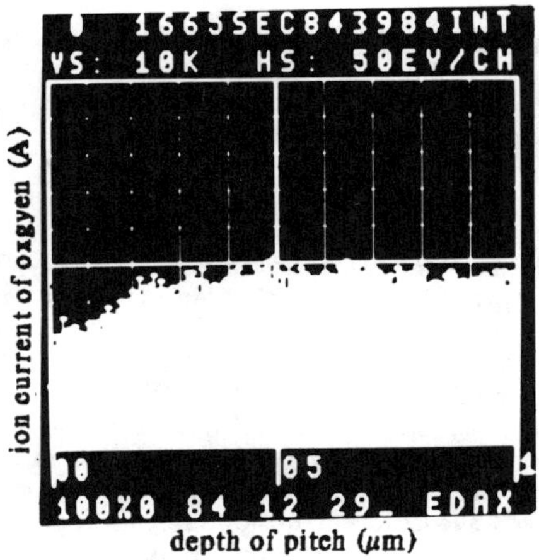

(a) pure titanium

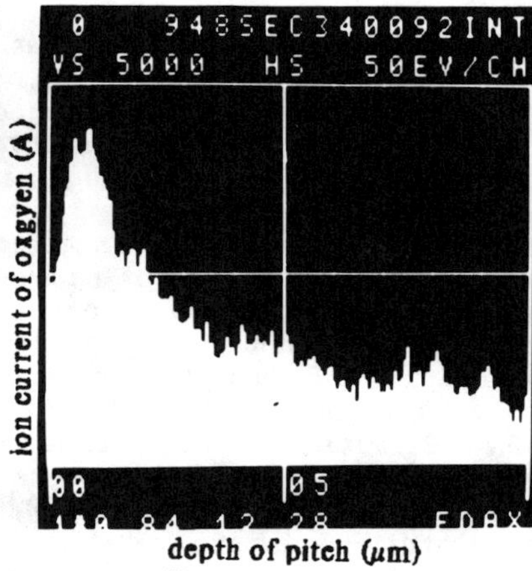

(b) alloy D+0.5%Nb

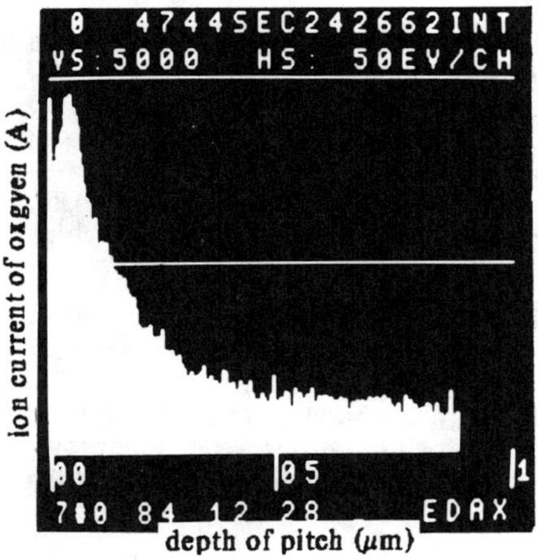

(c) alloy D+1%Nb+1%W

Fig.7. Relationships between the ion current of oxygen measured by ion probe and the depth of bombarding pitch (during the bombarding process the Ar ion current, 10^{-6} A, time, 800 second)

Table 2 Oxidation resistance and thickness of oxide layer of several Ti–alloys

Mathod	Material	Weight in–creasing in oxidation pre area mg / cm²	Method	Material	Depth of pitch (μm)	thickness of oxygen saturation layer (μm)	thickness of oxygen solid solution layer (μm)
increasing weight in oxida–tion	pure Ti	0.137	depth analysis using ion probe and SEM	prue–Ti	79.7	35.87	–
	D	0.060					
	A	0.034					
	B	0.015		A	82.6	8.26	24.78
	Ti–6242S	0.041					
	E	0.033					
	F	0.025		F	89.8	7.18	26.94

The oxidation resistance of Ti–alloy mainly depends upon the oxidation temperature, duration, the nature of atomosphere, properties of alloying elements and their amount [3]. The distribution of alloying elements in the oxide layer has much to do with their properties. in general, the solubility and the position of stack of the oxides formed in the titanium dioxied determines the porosity of the scale. It may reduce the oxidation rate to form stronger protecting oxide scale, increase the internal diffusion barrier of oxygen into the titanium dioxide. It seems that alloying elements Nb and W have different effects on that.

CONCLUSION

1. Titanium alloy ingot with homogeneously distributed chemical composition free from inclusion, segregation and other metalic defects can be made by adding master alloys Al–Nb, Al–W as the additives of β isomorphous alloying elements Nb and W.
2. Alloying element Nb has no strong effect on the strength of Ti–alloy. Considering the mechanical properties comprehensively, a proper content of Nb is 0.5–1% . Nb can improve the surface stability and increase the oxidation resistance of Ti–alloy exposed at high temperature.
3. Alloying element W can be raise the strength and the resistance of creep in Ti–alloy, but has less contribution to the resistance of oxidation in Ti–alloy than Nb.

REFERENCES

1. Yotaro Murakami, Titanium Science and Technology, Vol.3 DGM. 1984, P1403–1423
2. R.E.Slallman, Modern Physical Metallurgy, 3rd Edition, London, Butterworths.
3. A.M.Chaze, Titanium Science and Technology Vol.4. DGM.1984, P2665–2672.

The Tensile Properties of Post–Superplastically Formed

Ti–6Al–4V Fine Grain Sheet

Han Bing–Qiang and Lin Zhao–Rong
Nanjing Aeronautical Institute, Dept. of
Mechanical Engineering, Nanjing, JiangSu
210016,
P. R. China

Abstract

This paper reports on a study to determine the tensile properties of post–superplastically formed Ti–6Al–4V fine grain sheet. Superplastic tensile tests of specimens were carried out at various temperatures, initial strain rate, strains and cooling rates. The room temperature tensile properties for specimen of post–superplastically formed specimens were then measured. The effects of the different superplastic conditions on the room temperature tensile properties were explained through their effect on microstructure.

Key words: Titanium alloy, Superplastic tensile, tensile properties.

Introduction

Ti–6AL–4V alloy has a high strength to weight ratio, good corrosion resistance, heat resistance and excellent fatigue resistance as well as fracture toughness values. Although these properties are ideal for materials which can be used in the aerospace and chemical industries, it is difficult to manufacture parts from titanium alloys by traditional methods. However, superplastic forming recently has been developed as an advanced foming technigue.

Although research on the superplasticity of titanium alloys has been conducted for some time, there little data on the change of mechanical properties (such as tensile property, fracture toughness and fatigue strength) of post–superplastically fomed titanium alloys. In this paper, the tensile properties of post–superplastically formed titanium alloys have been studied under different conditions of superplasticity.

Experiment Procedure and Results

The material used in this study was 2mm thick Ti–6Al–4V fine grain sheet. Its as – received structure was equiaxed $\alpha+\beta$. The superplastic specimen gauge size was 15 × 6mm. Testing was carried out on 5 Ton material testing machine with an electrical furnace. The specimens were placed into the electric furnace, heated to temperature for 15 minutes, and then pulled to different elongation. When the superplastic tensile experiment was finished, the specimen was removed form furnace. A 15mm long gauge on the specimen surface was polished. The measurement of the room tensile properties

Titanium '92
Science and Technology
Edited by F.H. Froes and I. Caplan
The Minerals, Metals & Materials Society, 1993

was carried out on WD–1 type testing machine. A 10mm Instron displacement sensor was used to moniter room elongation. Analysis of the microstructure was performed using on Olympus metallographic microscopic and JEM–2000EX TEM.

The superplastic tensile experiments were conducted using different temperatures and different initial strain rates. After the specimens were pulled to certain elongations, the specimens were cooled in furnace at different rates and then removed. The tensile specimens were then pulled to fracture at room temperature. With this procedure we investigated the effects of temperature, initial strain rate, elongation and rate of furnace cooling on the tensile properties of post–superlastically formed material properties.

1. The effect of temperature on post–superplastically formed tensile properties.

The effect of temperature on tensile strength, yield strength and elongation at different temperature is shown in Fig.1. This figure shows that higher superplastic forming temperatures led to higher strength and lower elongation. Because β phase has higher stability, the structure has no change when it is cooled to the air from 820℃. The structure is softer, so the strength is lower and the elongation is higher. However, as tension temperature rising, some of β phase will change into sencondary platelet α phase when it is cooled. Then the microstructure at room temperature is primary α+transformed β phase in which the transformed β includes a sencond α phase and retained β. The volume fraction of primary α phase decreased and volume of transformed β increases at higher temperature. As a result, the strength will increase and the elongation will decrease. Fig.2 shows the microstructures at 850℃ and 920℃. We can see that the grain sizes at the two temperatures are similer. However, at 920℃, the grains are more equiaxed and the volume of transformed β is greater than that at 850℃.

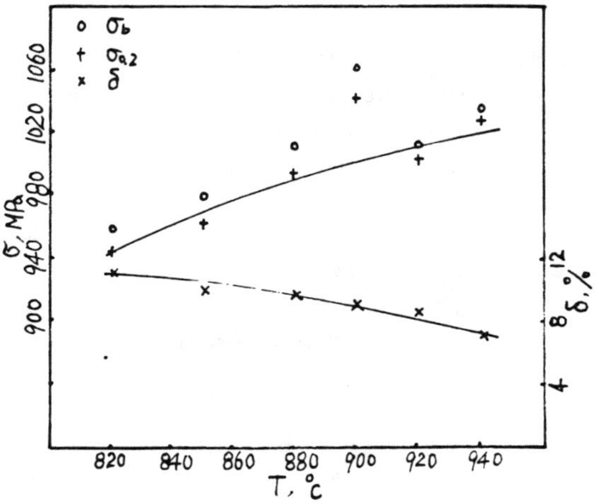

Fig.1 Under the superplastic conditions of 5.6×10^{-4} 1 / s initial strain rate, 100% elongation and air cooling, the effect of temperature on the tensile properties.

2. Effect of strain rate on post–superplastically fomed tensile properties

The effect of the initial strain rate on the strength and elongation of post–superplastically formed Ti–6Al–4V is illustrated in Fig.3. The specimen were pulled at high temperature and show microstructure changes along with the property differences. At lower strain rates, the specimens spend more time at high temperature. The grain size increases which leads to a decrease in yield

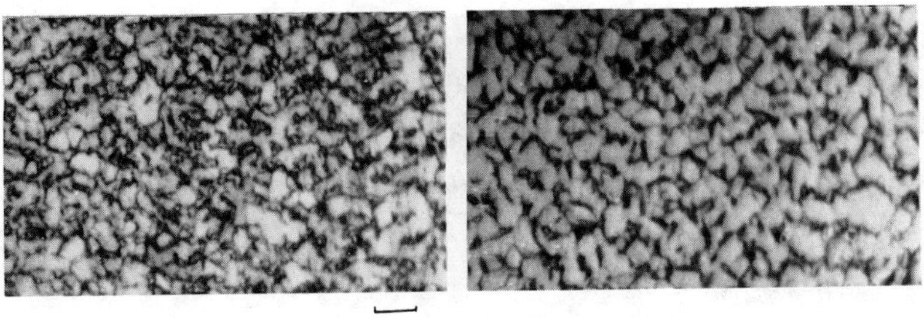

(a) 850℃ 10μm (b) 920℃

Fig.2 Under the superplastic conditions of 5.6×10^{-4} 1/s initial strain rate, 100% elongation and air cooling, the effect of temperature on microstructure, 650X

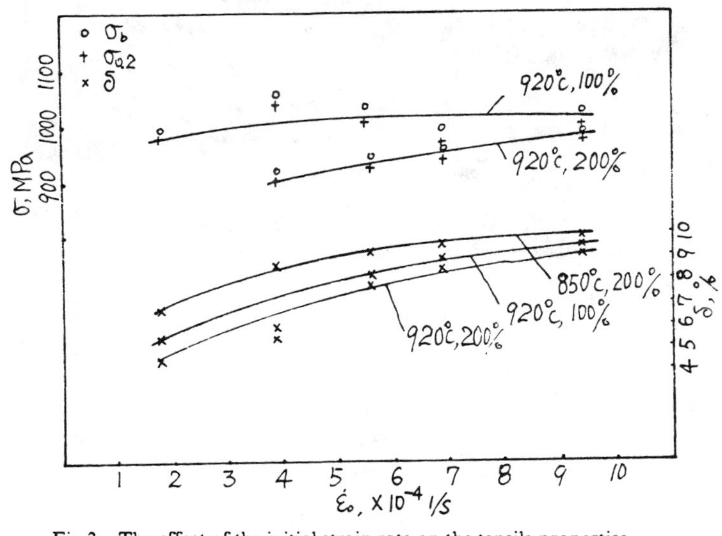

Fig.3 The effect of the initial strain rate on the tensile properties of post–superplastic forming

strength and elongation. Higher initial strain rates result in shorter times at high temperatures. The grain size remains small and the yield strength and elongation increase.

Under the superplastic forming condition, properly increasing strain rate will be beneficial to post–superplastically formed tensile properties. If the strain rate exceeds the optimumrange of superplastic strain rates, more Cavitations form. The intergranular strength will be weaked. So the yield strength and elongation of post–superplastic forming will be decreased.

At a superplastic elongation of 200%, the effect of the initial strain rate on the microstructure is shown in Fig.4. The grain sizes vary inversely with the strain rates as shown in Fig.5.

3. The effect of elongation on post–super plastically formed tensile properties

At the 5.6×10^{-4} 1/s initial strain rate, the effect of superplastic elongation on the tensile properties is showed in Fig.6. The greater the superplastic elongation, the lower the yield strength and the smaller the elongation. The effect of the superplastic elongation on microstructure is shown in Fig.7. The grain sizes change with elongation as seen in Fig.8. As elongation increases, the specimens spend

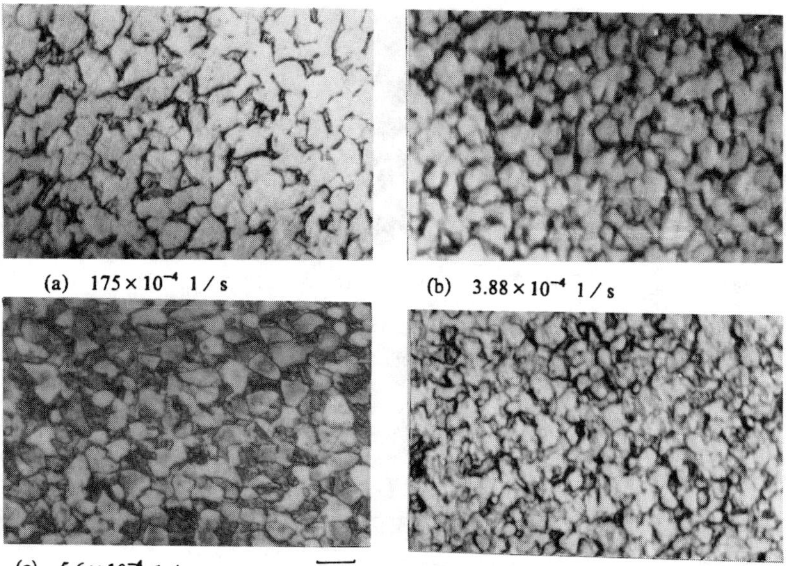

Fig.4 Under the condition of 920℃, 200% and air cooling, the effect of the initial strains rates on the microstructure, 650X.

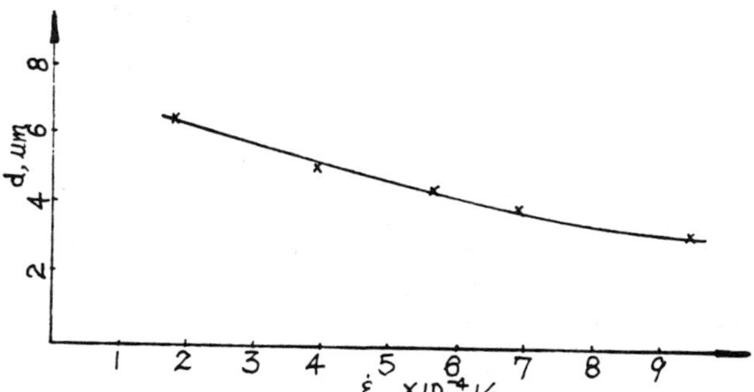

Fig.5 Under the condition of 920℃, 200% and cooling in air, the effect of the initial strain rates on the sizes of grains

more time at high temperation and the grains grow larger. Therefore the tensile strengths and elongations are reduced.

The microstructure of deformed zone (gauge) and undefomed zone (shoulder) were examined for the specimen deformed at 920℃, an initial strain rate of 5.6×10^{-4} 1/s, 320% elongation and air cooled. The microstructure (see Fig.9) show that the grain size of superplastically deformed zone is clearly larger than that of undeformed zone. Although the size of the undeformed zone has recrystallized into an exquiaxed fine grain microstructure, the grain size has not increased. Fig.7 and Fig.8 show that superplastic forming stimulates grain growth.

4. Effect of the time of furnace cooling on post–superplastically formed tensile properties.

When superplastic forming process was finished, the specimens were removed and either air

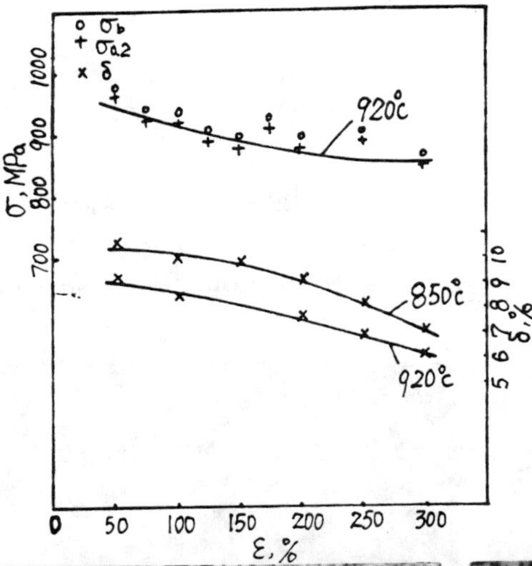

Fig.6 Under the condition of the 5.6 $\times 10^{-4}$ 1 / s initial strain rate, the effect of superplastic tensile elongations on post–superplastically formed tensile properties.

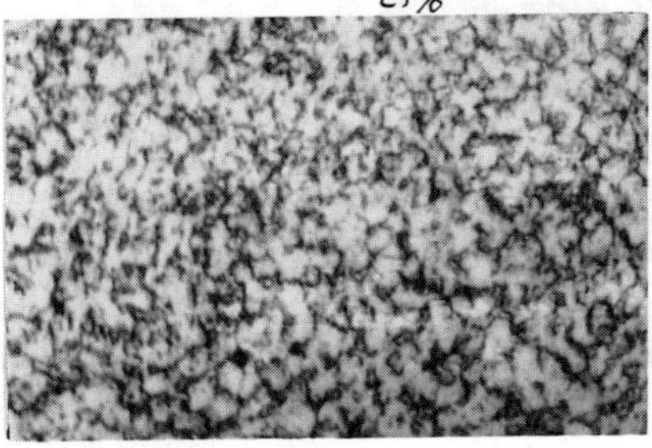

(a) 50%

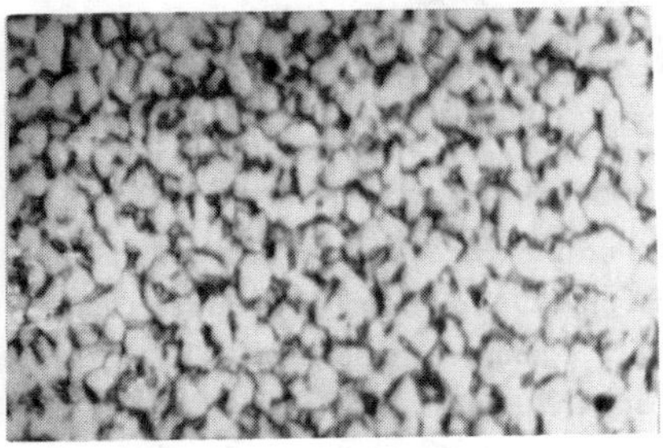

(b) 100%

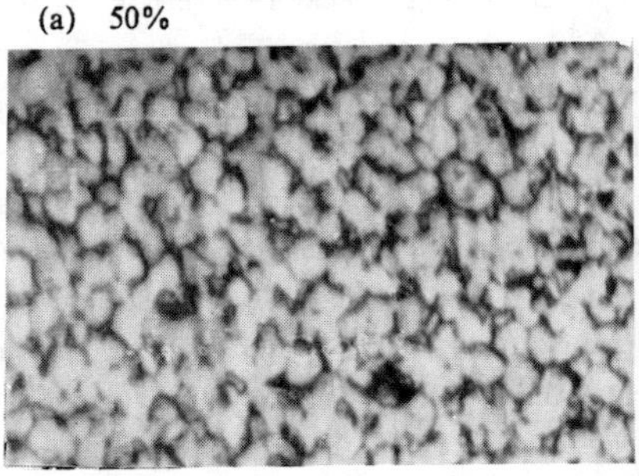

(c) 250%

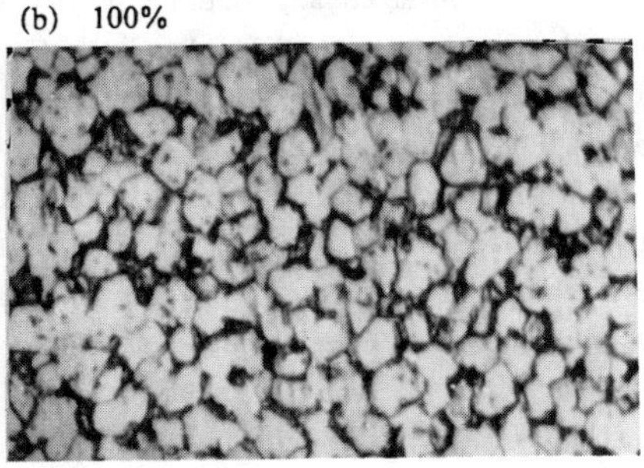

(d) 300%

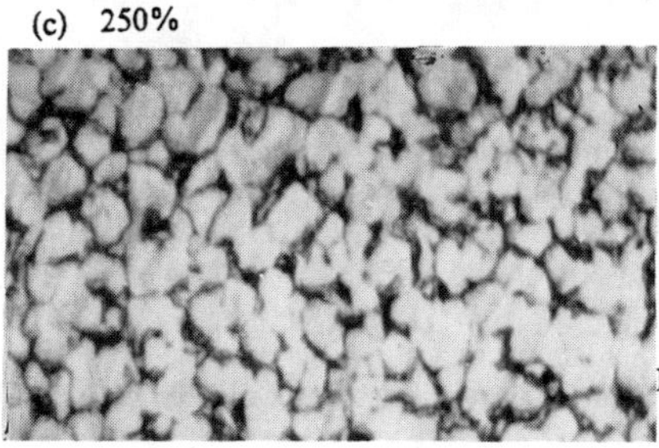

(e) 350%

10μm

Fig.7 Under the condition of 920℃, 5.6×10^{-4} 1 / s and air cooling, the effect of superplastic elongations on microstructure, 650X.

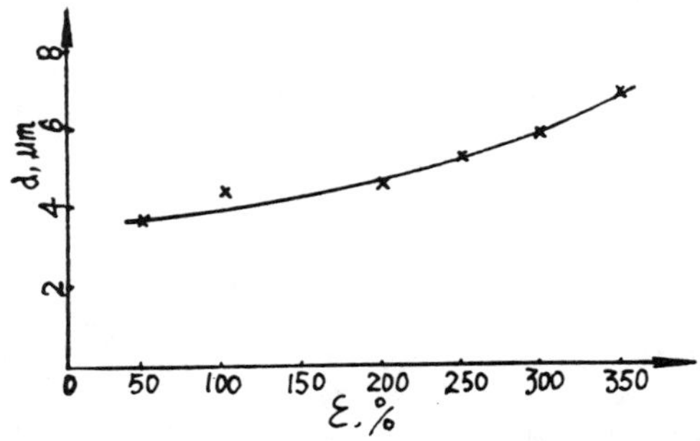

Fig.8 Under the condition of 920℃, 5.6 × 10⁻⁴1 / s and air cooling, the effect of superplastic elongation on the grain sizes.

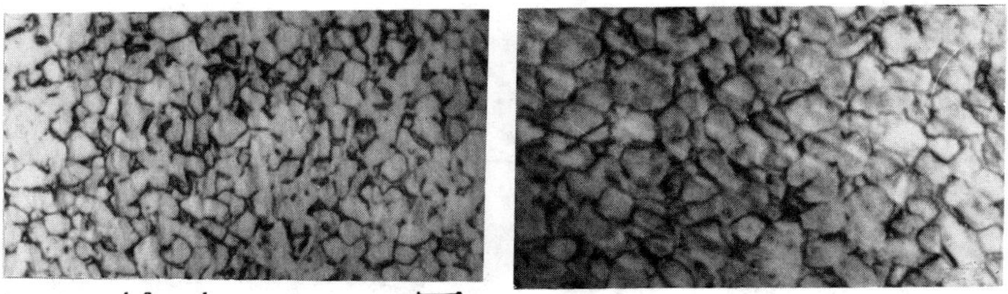

(a) undefomed zone 10μm (b) deformed zone

Fig.9 Under the condition of 920℃, 5.6 × 10⁻⁴1 / s, 320% and aircooling, 650X.

cooled or furnace cooled for a period of time and then air cooled. In our test, furnace cooling for 0.5 hour from 920℃ corresponds to furnaces cooling to 640℃ and then air cooling. 1 hour corresponds to 460℃. 2 hours corresponds to 200℃. 4 hours corresponds to 50℃. 6 hours corresponds to room temperature.

The effect of the cooling condition on yield strength and elongation can been seen in Fig.10. It shows that the cooling condition has little effect on the post-superplasticcilly formed. The yield strength and elongation are higher after air cooling while furnace cooling produces only small changes. Fig.11 show the effects of the cooling condition on the microstructures. The grain sizes shows little change under the furnace cooling condition. While β phase between α phases increases with the furnace cooling. A comparision with the corresponding air cooled microstructure in Fig.4(c) shows that the grain size is smaller after air cooling. The air cooled microstructure of post-superplastically formed consists of primary α+transformed β (platelet sencondary α+retained β). Fig.12(a) and (b) are TEM structures for different forming conditions following air cooling, in which the large grains are primary α phase, the platelets are sencondary α phase and remainer is β

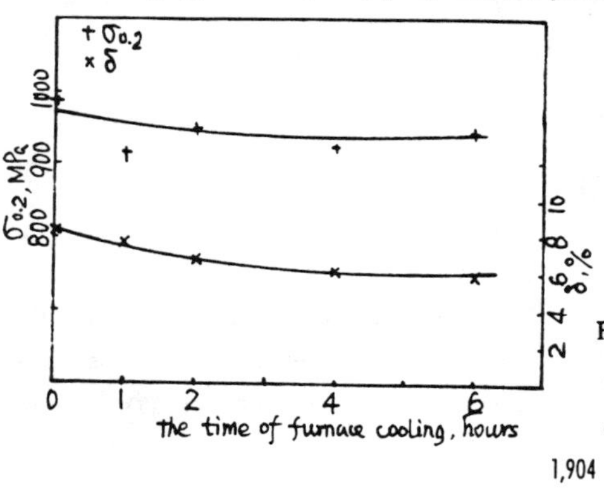

Fig.10 After the superplastic tension of 920℃, 5.6 × 10⁻⁴1 / s and 100%, the effect of time of furnace cooling on the tensile properties.

phase. There are many dislocations in the air cooled samples. Fig.12(c) and (d) are TEM structures following furnace cooling, in which the sencondary α phase in transformed β is globuler like the primary α phase. The retained β phase lies in the α phase. The dislocations in α phase decrease gradually while the boundaries of α and β phase become straight with increased time for furnace cooling. After 4 hours furnace cooling the boundaries become very straight with few dislocation in the grains.

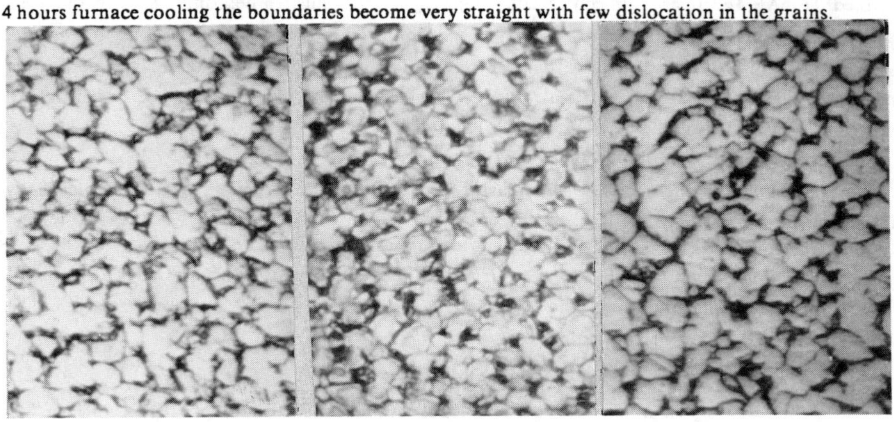

(a) 0.5 hour (b) 1 hour (c) 4 hours 10μm

Fig.11. After the superplastic stension of 920℃, 5.6×10^{-4}1 / s and 200%, the effect of time of furnace cooling on microstructure, 650X.

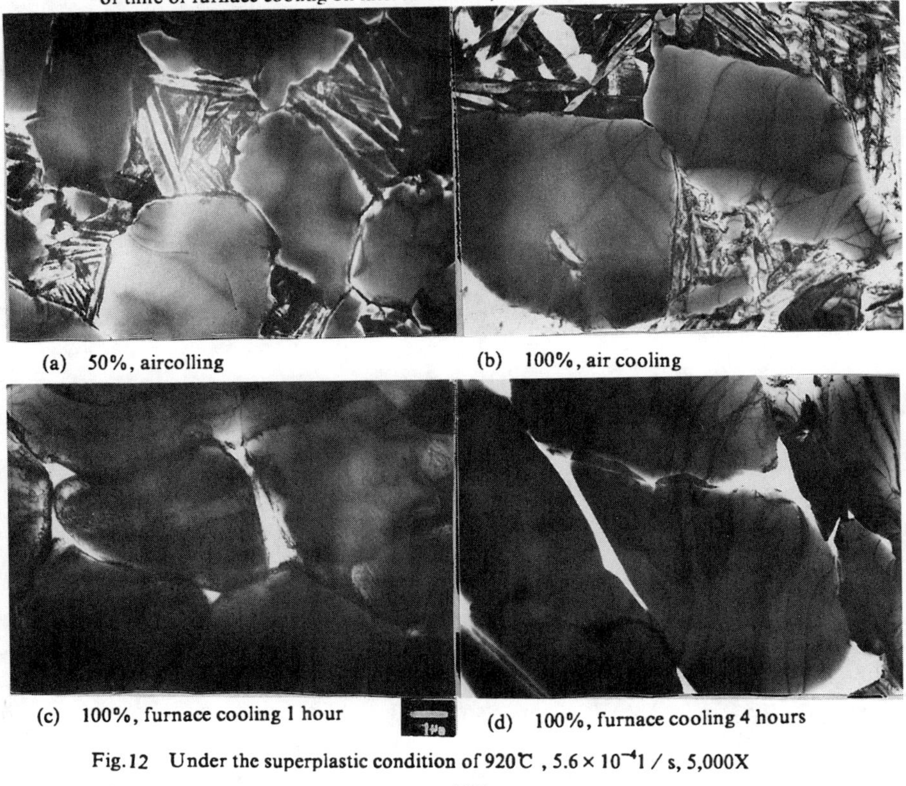

(a) 50%, aircolling (b) 100%, air cooling

(c) 100%, furnace cooling 1 hour 1μm (d) 100%, furnace cooling 4 hours

Fig.12 Under the superplastic condition of 920℃ , 5.6×10^{-4}1 / s, 5,000X

Conclusions

We summerise the experimental results as follow:

1. The superplastic forming temperatures affect the tensile properties of post–superplastically formed Ti–6Al–4V. In the air cooled condition, higher forming temperatures increase the yield strength and decrease elongation.

2. The initial superplastic forming strain rates also affect the tensile properties of post–superplastically formed material. Properly increasing strain rates will be beneficial to post–superplastically formed tensile properties.

3. The larger the superplastic elongation, the lower the tensile perperties.

4. The cooling rate following superplastic forming has little effect on post–superplastically formed tensile properties.

Acknowledgement

This work was supported by a grant from Chinese Ministry of Aerospace. The authors are grateful to Sun Jing–ping, Zao Hui, Chen Ming–He, Xu Lian–Zen, Zhou Jian–Hua and Yu Ya–Ping for their help in the experiment.

References

1. Lin Zhao–Rong, et al: Principle and Application of Metal Superplastic forming, 1990, Chinese Aerospace Industry Publishing House.

2. M.T. Coper, N.Ridley: Materials Science and Technology, Feb.1986, Vol.2, P.140.

3. M.T. Coper, D.R.Evetts, N.Ridley: Materials Science and Technology, June 1987, Vol.3, P.455.

4. B.P. Kashyap, et al: Journal of Materials Science, 20(1985), 2661–2686.

5.C.C. Bampton, J.W.Edington: Journal of Engineering Materials and Technology, Jan.1983, Vol.105 / 55.

A State Variable Material Model

for Superplastic Titanium-6Al-4V

Daniel A. Mosher and Paul R. Dawson

Sibley School of Mechanical and Aerospace Engineering
Cornell University
Ithaca, New York 14853

Abstract

Analysis of superplastic forming requires that the material behavior be accurately quantified. A model has been developed which captures the changing strain rate sensitivity occurring between superplastic regions II and III and which accounts for strain path dependence of the flow stress. Isothermal compression tests have been conducted for strain rates of 10^{-5} to 10^{-3} s^{-1} and temperatures of 875 to 950 °C for nine different equiaxed microstructural states. Tension and compression tests were found to give the same stress response. Static and deformation enhanced grain growth have been measured and their relation to strain hardening discussed.

Titanium '92
Science and Technology
Edited by F.H. Froes and I. Caplan
The Minerals, Metals & Materials Society, 1993

Introduction

The application of superplastic material behavior to the production of complex parts requires that pressure cycles, temperatures, and other processing conditions be designed to minimize tensile localization. Computer simulation allows one to efficiently examine the effect of changing these conditions and to determine process variable cycles which produce optimum workpieces. For the results of simulations to be meaningful, it is essential that the material behavior be described by an accurate model. According to state variable methodology, the current stress response of a material should depend only on the current conditions such as imposed strain rate, temperature and its current microstructural state. Some models have used variables related to the material's history such as plastic strain or time successfully because over limited paths, such variables closely correlate with the state. However, this is not the case for superplastic forming of titanium-6Al-4V since the material typically undergoes a complex strain path and the predominant state variable, grain size, does not depend on strain or time in a path independent manner.

Characterization of the material behavior can be divided into two parts: the kinetics of flow at constant microstructural state and the kinetics of the evolution of state. The kinetics at fixed state gives the instantaneous effective stress, σ, for various strain rates, $\dot{\epsilon}$, temperatures, T, and sets of relevant state variables, $\{S\}$.

$$\sigma = \sigma(\dot{\epsilon}, T, \{S\}) \tag{1}$$

The evolution of state describes how the microstructure will change over large strains, and it is cast in rate form so that it will apply for any strain rate or temperature history.

$$\dot{S}_i = \dot{S}_i(\dot{\epsilon}, T, \{S\}) \tag{2}$$

Modeling of the Microstructural State

The choice of proper state variables to be included in $\{S\}$ is an important step in the development of an accurate material model. A number of authors [1,2,3,4] have presented a comprehensive list of state variables for superplastic titanium alloys, the most important of which are α/β phase volume fractions, average grain sizes, grain size distributions, grain shape and phase contiguity.

The temperature dependent volume fraction of the phases naturally effects the flow stress and could be the cause of the abnormally high thermal activation energy [3,4]. However, because the equilibrium volume fraction depends exclusively on the current temperature and not on the thermomechanical history, its effect may be incorporated implicitly through the temperature dependence of the kinetics of flow. This approach is an approximation since for large, rapid changes in temperature the volume fraction may differ from equilibrium.

The importance of grain size distribution for the kinetics of flow can be estimated from the dependence on average grain size, $\bar{L}_3$. From studies in the literature [2,5] as well as this work, the stress in region II has a grain size exponent of approximately 1 while deformation rate has an exponent of approximately -2.

$$\sigma \approx \sigma(\dot{\epsilon}, T)(\bar{L}_3)^1 \; ; \qquad \dot{\epsilon} \approx \dot{\epsilon}(\sigma, T)(\bar{L}_3)^{-2} \tag{3}$$

If one applies these relations for individual grains of different sizes and assumes an iso-strain rate condition on a microscopic scale, then the exponent of approximately 1 results in the average stress depending only on average grain size. If one assumes an iso-stress condition then the exponent of -2 on grain size causes the relationship between average deformation rate and stress to depend not only on the grain size average but also on its distribution. Raj and Ghosh [6] have demonstrated the importance of grain size distribution, but this modeling was conducted for diffusional creep which has a stress grain size exponent of 3. Thus, if the iso-strain rate condition is dominant for a material with a stress grain size exponent of approximately 1, grain size distribution should have little effect on the average kinetics of flow.

In this work as in many others, the important state variable was hypothesized to be the average grain size of the α phase. While the size of the β grains is also important, it is assumed that β grain size is proportional to α grain size. Since the function of the α grains is to retard grain growth of the β phase, the size of the β grains should be related to the α mean free path which is proportional to the α average grain size through the phase volume fraction.

Average grain size of the α phase was evaluated by measuring the mean linear intercept, $\bar{L}_3$, in three orthogonal directions, the rolling direction, transverse direction and normal direction (RD, TD and ND) of the cross rolled titanium plate. It was estimated that these directions corresponded to the principal directions of the orthotropic material properties based on examination of the microstructure. The use of directional grain sizes allows average grain shape to be computed which can be represented as a second rank tensor [7] for use in an anisotropic model. When the material response is approximated as isotropic, an effective grain size is used, defined as

$$\bar{L}_3^{eff} = \frac{\bar{L}_3^{TD} + \bar{L}_3^{RD} + \bar{L}_3^{ND}}{3} \tag{4}$$

Experimental Method and Results

The experimental apparatus consisted of an MTS servo-hydraulic load frame, a three zone split tube furnace and MAR-M246 alloy push/pull rods. Samples with six different equiaxed microstructures were obtained from the RMI Titanium Company and the General Electric Company. Isothermal compression and tension tests were conducted in an air atmosphere; to decrease the diffusion rate of impurities, the titanium specimens were coated with a glass barrier.[1] The diameter of the compression specimens varied between 4 and 10 mm depending on the thickness of the rolled plate, while the height was 1.5 times the diameter. The axisymmetric tensile specimens had a gauge diameter of 4.5 mm with a gauge length of 38 mm. Grooves were machined on the ends of the cylindrical compression specimens to retain and distribute a second glass coating[2] used as a lubricant. Silicon nitride platens proved to remain smooth during the testing, resulting in only slight barreling of the samples.

The furnace and load train were preheated to the test temperature. The compression specimen/platen or tensile specimen/collar assembly was placed into the furnace and allowed to sit for fifteen minutes to retrieve temperature uniformity. After testing the samples were removed from the furnace and air cooled. The specimens were typically small enough so that cooling rates were adequate to produce a fine transformed α phase which was easily distinguishable from the primary α during metallographic examination.

To obtain data for the kinetics of flow at fixed state, isothermal strain rate jump tests were performed. For a typical test the stress could be measured for seven strain rates between $10^{-3.0}$ and $10^{-5.0}$ s^{-1} in as little as 7% strain, resulting in little change of the microstructure. Tests were repeated for temperatures of 875, 900, 925 and 950 °C. The strain rate and temperature dependencies were obtained for the six sources of titanium in their *as received* state as well as for three of them deformed to a strain of 0.4 at a strain rate of $10^{-4.0}$ s^{-1} and temperature of 900 °C. The *deformed* specimens were remachined before performing strain rate jump tests.

The directional mean linear intercepts were measured on planes with RD and TD normals. Five micrographs were taken for each plane with mean linear intercepts being measured in two perpendicular directions. In this manner approximately 500 intercepts were counted for each directional grain size resulting in a standard deviation of 2–4% of the average for the effective grain size. The measurement of the directional $\bar{L}_3$ and the computation of $\bar{L}_3^{eff}$ are given in Table I.

[1]Deltaglaze 13, Acheson Colloids Company
[2]TK 13073, Acheson Colloids Company

State	$\bar{L}_3^{TD}$	$\bar{L}_3^{ND}$	$\bar{L}_3^{RD}$	$\bar{L}_3^{eff}$
1. As Received #1A	17.5	13.3	20.6	17.1
(As Received #1B)	17.6	13.3	18.5	16.5
2. As Received #2	11.9	10.3	13.7	12.0
3. As Received #3	6.3	4.6	6.1	5.7
4. As Received #4	6.4	4.7	6.8	6.0
5. As Received #5	7.9	6.7	9.2	7.9
6. As Received #6	13.5	11.7	16.1	13.8
7. Deformed #1B	18.3	14.5	17.2	16.7
8. Deformed #2	13.4	11.8	14.0	13.1
9. Deformed #3	7.4	6.1	7.3	6.9

Table I: Mean linear intercepts of the α grains in the transverse, normal and rolling directions at 925 °C and the computed effective grain sizes; States #1A and #1B are different heats of the same plate product; units are micrometers.

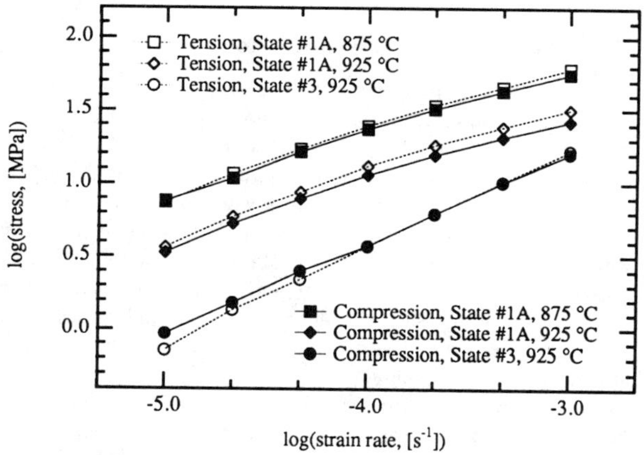

Figure 1: Comparison of Tension and Compression

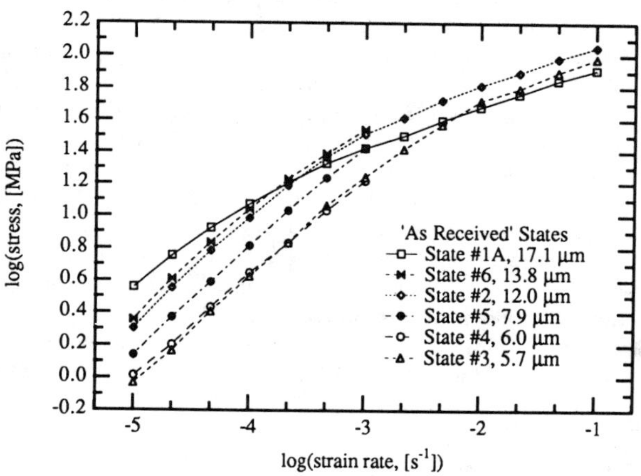

Figure 2: Microstructural State Dependence of Flow Kinetics

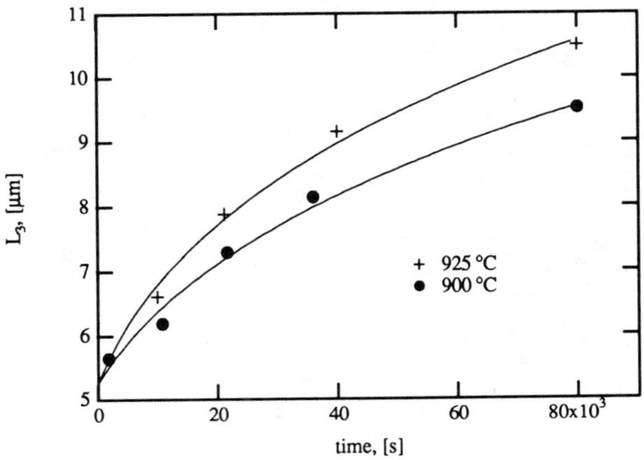

Figure 3: Static Grain Growth Data and Model

Comparison of tension and compression showed no distinguishable effect of the mean stress[3] on the kinetics of flow as shown in Figure 1. Thus, the use of compression tests to obtain data for tensile superplastic models is justified. All subsequent data are from compression tests.

The strain rate dependence for various states at constant temperature is shown in Figure 2. The presence of two regions is apparent. At lower strain rates and smaller grain sizes, grain boundary sliding of region II produces a higher slope or strain rate sensitivity, m. At higher strain rates and grain sizes, dislocation creep of region III becomes activated and m decreases. The strain rate at which the region II–III transition takes place is strongly dependent on the grain size with smaller grains having a higher transition strain rate.

At high temperatures and small grain sizes, the temperature dependence was noted to be abnormal. For the *as received* States #3 and #4, the stresses at 950 °C were equal to or greater than those at 925 °C. The *deformed* State #3 had stresses at 950 °C which were only slightly lower than at 925 °C, while all larger grain size states had normal temperature dependence.

Grain size evolution experiments were conducted in compression for 900 and 925 °C primarily for State # 3. Grain sizes, measured for static conditions as well as for three strain rates, are plotted in Figures 3 and 4. The rate of growth was practically the same for all three directional grain sizes, $\bar{L}_3^{TD}$, $\bar{L}_3^{ND}$ and $\bar{L}_3^{RD}$. The different slopes for different strain rates in Figure 4 demonstrate that the use of strain to predict the grain size dependent flow stress is inadequate.

Material Model Functional Forms

Kinetics at Fixed State

The hyperbolic sine has been used to match creep data over a wide range of strain rates [8]. The following modified functional form was used to describe the kinetics of flow at fixed state with $\bar{L}_3^0 = 1\ \mu m$ being present purely to simplify units.

$$\sigma = \sigma_0(\bar{L}_3^{eff}) \left[\sinh^{-1}\left(\frac{\dot{\epsilon}}{\dot{\epsilon}_0(\bar{L}_3^{eff})} \right) \right]^{1/n} \exp\left(\frac{\tilde{Q}}{RT} \right) \tag{5}$$

$$\dot{\epsilon}_0(\bar{L}_3^{eff}) = A_{\dot{\epsilon}} \left(\frac{\bar{L}_3^{eff}}{\bar{L}_3^0} \right)^{p_{\dot{\epsilon}}} \quad ; \quad \sigma_0(\bar{L}_3^{eff}) = A_{\sigma} \left(\frac{\bar{L}_3^{eff}}{\bar{L}_3^0} \right)^{p_{\sigma}} \tag{6}$$

The inverse hyperbolic sine factor captures the strain rate dependence of the data in Figure 2; the high, constant m of region II goes through a transition to lower m values at approximately

[3] $\sigma_{mean} = (\sigma_{11} + \sigma_{22} + \sigma_{33})/3$

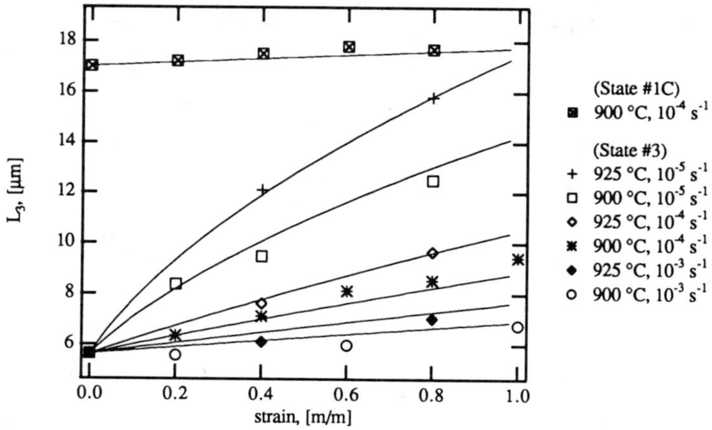

Figure 4: Grain Growth Data and Model at Strain Rates of 10^{-3}, 10^{-4} and 10^{-5} s^{-1}

n	$A_{\dot{\epsilon}}$	$p_{\dot{\epsilon}}$	$\tilde{Q}/R$	A_{σ}	p_{σ}	$\bar{L}_3^0$
1.55	$0.251\ s^{-1}$	-2.73	$19,250\ K$	$9.41 * 10^{-6}$ MPa	-0.643	$1\ \mu m$

Table II: Parameters for Kinetics of Flow

a strain rate of $\dot{\epsilon}_0(\bar{L}_3^{eff})$. Combres and Levaillant [5] have discussed a transition grain size as a function of strain rate and its derivation from mechanisms. Their results estimate a grain size exponent between -2.5 and -4.0 which agrees with $p_{\dot{\epsilon}} = -2.73$ from the fits below. Note that $\tilde{Q}$ is not a true activation energy since its exponential factor normalizes stress not strain rate. However, in region II for $\dot{\epsilon}/\dot{\epsilon}_0 < 1$, the activation energy will be approximately $n\tilde{Q}$.

Another possible approach to capture the transition between regions II and III is to add the strain rate contributions of grain boundary sliding and dislocation creep [9,10]. For the velocity based finite element method, this requires an iterative solution for the stress given the imposed or guessed strain rate, whereas the above inverse hyperbolic sine model gives the stress directly.

The parameters were computed by nonlinear regression one at a time with the computed values being fixed and used in subsequent parameter fits. They were fit in the order, $n \rightarrow \dot{\epsilon}_0 \rightarrow \tilde{Q} \rightarrow \sigma_0$ with the results being given in Table II. The parameters, n, $\dot{\epsilon}_0$ and $\tilde{Q}$ showed no distinguishable trend when comparing *as received* versus *deformed*. The parameter σ_0, however, was smaller for the *deformed* materials. Other experiments in this work have also revealed a relative softening of the material when superplastically deformed from the mill annealed state. Possible causes are recrystallization and refinement of the β grains relative to the α mean free path and α contiguity which decreases with deformation as clusters of α are separated by β phase [11]. This potentially important effect has not been accounted for in the present model.

A comparison of the model for flow kinetics with the experimental data is shown in Figure 5 for four states at 925 °C. It should be noted that the parameter fits were conducted for significantly more data than is shown in Figure 5 including five additional states and three other temperatures. In general, the basic trends of decreasing slope at the region II–III transition and the grain size dependence of the stress are captured. The precise shape at the transition is not captured well since the data for State #1 actually crosses over States #2 and #6. This may be caused by the influence of state variables other than average α grain size which are not constant for the different materials tested here.

Evolution of State

To model the grain growth phenomenon, the static and deformation enhanced grain growth

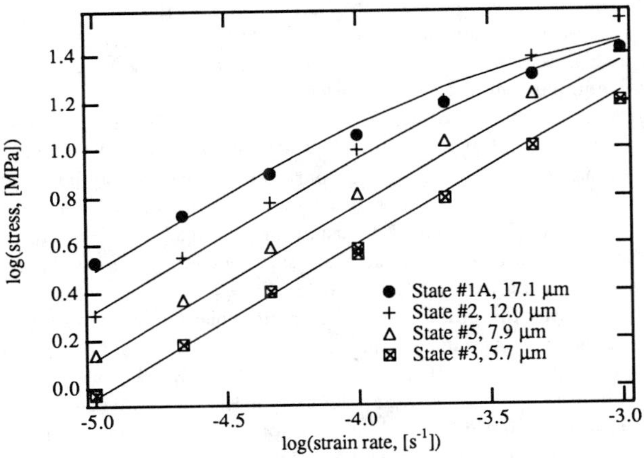

Figure 5: Comparison of the Kinetics of Flow Model with Experiment, 925 °C

Static				
B	$\bar{L}_3^0$	q	Q_s/R	
$3.67 * 10^7\ s^{-1}$	$1\ \mu m$	3.78	$23,800$ K	
Deformation Enhanced				
$\dot{\bar{L}}_3^{ref}$	$\dot{\epsilon}_d$	p	Q_d/R	$\bar{L}_3^s$
$4.71 * 10^9\ \mu m/s$	$1\ s^{-1}$	0.7	$28,000$ K	$22\ \mu m$

Table III: Parameters for Grain Size Evolution

portions were assumed to take place independently. An additive decomposition of the growth rate of the effective grain size was used.

$$\dot{\bar{L}}_3^{total} = \dot{\bar{L}}_3^{static}(T, \bar{L}_3) + \dot{\bar{L}}_3^{def}(\dot{\epsilon}, T, \bar{L}_3) \tag{7}$$

The functional form for static grain growth has been derived and discussed in the literature [12].

$$\left(\frac{\bar{L}_3}{\bar{L}_3^0}\right)^q - \left(\frac{\bar{L}_3^{init}}{\bar{L}_3^0}\right)^q = B\exp\left(-\frac{Q_s}{RT}\right)t\ ; \qquad \dot{\bar{L}}_3^{static} = \frac{\bar{L}_3^0}{q}B\exp\left(-\frac{Q_s}{RT}\right)\left(\frac{\bar{L}_3}{\bar{L}_3^0}\right)^{1-q} \tag{8}$$

Once the static grain growth component is estimated, it can be subtracted from the total grain growth to obtain the deformation enhanced portion. Grain growth was noted to be considerably slower for larger grain sizes and so a saturation factor was used in the functional form for deformation enhanced grain growth.

$$\dot{\bar{L}}_3^{def} = \dot{\bar{L}}_3^{ref}\left(\frac{\dot{\epsilon}}{\dot{\epsilon}_d}\right)^p \exp\left(-\frac{Q_d}{RT}\right)\left(1 - \frac{\bar{L}_3}{\bar{L}_3^s}\right) \tag{9}$$

Least squares regression gave the parameters in Table III and integration of $\dot{\bar{L}}_3^{total}$ gave the curves in Figures 3 and 4.

Discussion and Summary

To achieve the large tensile strains often necessary during superplastic forming the ability to inhibit necking is provided through high strain rate sensitivity and strain hardening. Examination of the trends indicated by the present model and data helps illuminate the interaction of these two effects.

The strain rate sensitivity, m, is relatively constant in the grain boundary sliding regime but begins to decrease as the strain rate exceeds $\dot{\epsilon}_0(\bar{L}_3^{eff})$. Thus, $\dot{\epsilon}_0(\bar{L}_3^{eff})$ can be thought of approximately as the highest strain rate for which m is still acceptable for superplasticity. Grain growth has a negative impact on m as it will lower $\dot{\epsilon}_0(\bar{L}_3^{eff})$.

The microstructural foundation of strain hardening is predominantly grain growth. The static grain growth portion is not expected to help stabilize flow localization since the hardening of the material occurs uniformly throughout the part [13]. However, the deformation enhanced portion of grain growth occurs at a faster rate for higher strain rates. Thus, grains will grow faster in the necking region. The impact of deformation enhanced grain growth on stabilization will be greater for lower strain rates since the grain size dependence of the flow stress is higher for grain boundary sliding than for dislocation creep.

Thus, at low strain rates the grain growth has a significant static component, which contributes no strain hardening. At high strain rates, deformation enhanced grain growth may occur, but the grain size dependence of the flow stress will be low. At intermediate strain rates both deformation enhanced grain growth and high grain size dependence of the stress occur, resulting in the greatest contribution to stabilization for a given amount of grain growth.

The interaction of strain rate sensitivity and strain hardening results in complex behavior with the net effect on localization resistance depending on the particular quantitative dependencies. Because of this complexity and the strain path dependence, the need for an accurate state variable material model is apparent.

Acknowledgements

This work was supported by the General Electric Company. The authors would like to thank H. A. Nied and R. Sundell for their discussions on the modeling and experimental method as well as RMI Titanium for supplying material and helpful information.

References

[1] A. K. Ghosh and C. H. Hamilton, *Metallurgical Transactions A,* **10A** (1979), 699–706.

[2] N. E. Paton and C. H. Hamilton, *Metallurgical Transactions A*, **10A** (1979), 241–250.

[3] A. K. Ghosh and C. H. Hamilton, *Metallurgical Transactions A*, **13A** (1982), 733–743.

[4] C. H. Hamilton, *Superplastic Forming*, (Proceedings of a Symposium by the Los Angeles Chapter of ASM) (1984), 13–22.

[5] Y. Combres and C. Levaillant, *International Journal of Plasticity*, **6** (1990), 505–519.

[6] R. Raj and A. K. Ghosh, *Acta Metallurgica*, **29** (1981), 283–292.

[7] C. H. Turner and S. C. Cowin, *Journal of Materials Science*, **22** (1987), 3178–3184.

[8] F. Garofalo, *Trans. Metall. Soc. AIME*, **227** (1963), 351–356.

[9] M. F. Ashby and R. A. Verrall, *Acta Metallurgica*, **21** (1973), 149–163.

[10] C. H. Hamilton, H. M. Zbib, C. H. Johnson and S. K. Richter, *International Conference on Superplasticity in Advanced Materials*, 1991, Osaka, Japan (in press).

[11] P. G. Partridge, D. S. McDarmaid, and A. W. Bowen, *Acta Metallurgica*, **33** (4) (1985), 571–577.

[12] A. J. Ardell, *Acta Metallurgica*, **20** (1972), 601–609.

[13] C. H. Hamilton, *Metallurgical Transactions A*, **20A** (1989), 2783–2792.

SMALL CRACK GROWTH AT STRESS CONCENTRATIONS

IN A NEAR ALPHA TITANIUM ALLOY.

W.J. Evans*, S.H. Spence+ and A. Goulding+

* I.R.C. in Materials for High Performance Applications,
University College of Swansea, U.K., SA2 8PP

+ Department of Materials Engineering,
University College of Swansea, U.K., SA2 8PP

Abstract

Small crack growth data, measured by optical and potential difference techniques, are reported for the titanium alloy IMI 829. Part-through crack development has been studied in the uniform stress field of a Corner Crack (CC) specimen and in the complex stress field of a thick section (10 mm) double edge notch (DEN) test piece. Through cracks have been assessed in a thin section (2 mm) version of the DEN geometry. The programme explores the role of section thickness, microstructure, applied stress and R value in the development of small cracks.

Introduction

Stress concentration features such as holes and machined radii are unavoidable design limitations in most engineering applications. Knowledge of fatigue crack growth behaviour at notches, therefore, is of paramount importance to designers. In many applications most of the service life of a component can be spent in the small crack regime where growth may not be readily quantified by linear elastic fracture mechanics. In such cases small fatigue crack propagation at notches must be characterised for accurate component life predictions.

The present programme of work set out to compare and contrast the development of small fatigue cracks at a stress concentration feature with the growth of an equivalent crack at a plain surface. It concentrated particularly on the role of applied stress, covering a range that encompassed both elastic and elasto-plastic regimes of behaviour, and of mean stress through a range of R values. For the experimental evaluation, the near alpha titanium alloy IMI 829 (Ti,5.5Al,3.5Sn,3Zr,1Nb,0.25Mo,0.35Si) was specifically chosen. It is an important engineering alloy in the gas turbine industry being used in a number of primary engine applications as compressor discs and blades. It has, however, a comparatively coarse structure thereby making it an ideal vehicle for evaluating interactions between crack growth, microstructure and notch geometry.

Experimental Equipment and Procedure

The IMI 829 specimens were machined from bar stock which had been given a solution treatment at 1050°C for 30 minutes followed by air cooling and ageing at 625°C for two hours. This produced an essentially basketweave Widmanstätten alpha microstructure with a mean prior beta grain size of 0.5 mm and an alpha colony size of 0.2 mm.

Titanium '92
Science and Technology
Edited by F.H. Froes and I. Caplan
The Minerals, Metals & Materials Society, 1993

A Corner Crack (CC) specimen with a 10 mm square section was used to obtain data on part-through cracks at plain surfaces. A starter slit was machined into one corner to a nominal depth and width of 0.25 mm and 0.15 mm respectively. Small crack behaviour at stress concentrations was studied using double edge notch (DEN) test pieces. A thick section DEN was used to evaluate part-through cracks. It had a cross section at the root of the notch identical to that of the CC specimen. The notch radius was 3 mm resulting in a gross section of 10 x 16 mm and a k_t of approximately 1.9 with respect to the net section nominal stress. A starter slit, identical to that for the CC specimen, was machined at the root of one notch to facilitate crack growth monitoring.

A thin section DEN specimen was used to evaluate through crack behaviour. This measured only 2 mm along the notch root. A through geometry starter slit was machined at the root of one notch to a nominal depth of 0.25 mm. In all other respects this geometry was the same as the thicker section test piece. The fine starter slits ensured initiation of a single crack at a known location.

Constant load amplitude tests were conducted at room temperature in laboratory air at a frequency of 1 Hz and a range of stresses covering several R values. Computer automated d.c. potential drop and photomicroscopic systems were used to monitor crack growth. The p.d. system provided average crack length data whilst surface lengths were recorded by the optical system. The p.d. system was also used to record the voltage across the crack throughout selected load cycles in order to detect the occurrence of crack closure. Crack shape development was assessed in terms of pairs of surface crack lengths. Crack shapes were obtained at the end of propagation tests by examining the fracture surfaces optically and via a scanning electron microscope. Growth rates were evaluated in terms of the stress intensity factor range, ΔK. Evaluations of K for CC and DEN test pieces were made using an in-house finite element programme (1).

Results

For all test piece geometries, crack paths were tortuous and growth rate curves in the form of da/dN against surface crack length showed extensive accelerations and decelerations in rate. This is similar to the work reported previously (2).

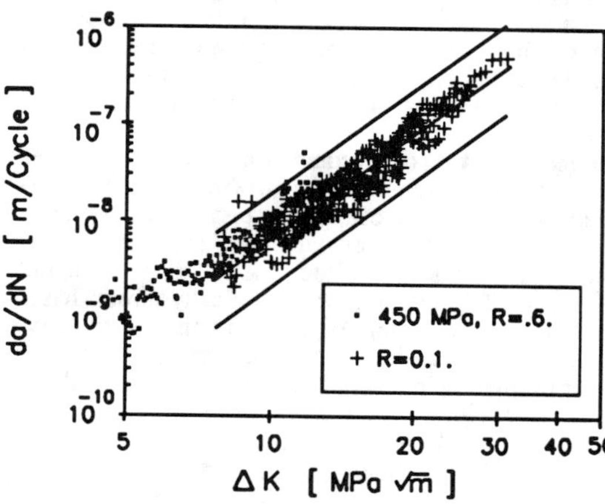

Figure 1. Growth rates in terms of ΔK for CC specimens, R=0.1 and 0.6. Solid lines illustrate the best fit and tolerance bands to the data for R=0.1

Crack growth rates derived from p.d. data on CC specimens are plotted against ΔK in figure 1. These were used for comparison with growth rates at stress concentrations. The figure includes best fit curves and 95% tolerance bands for the R=0.1 tests plus data at 450 MPa and R=0.6. There is a linear relationship between growth rates and ΔK for the range of stresses considered, $250 \rightarrow 600MPa$, suggesting that a Paris relationship is appropriate. The scatter is relatively low as illustrated by the spacing of the tolerance bands. It is noticeable that the R=0.6 data are well within these bands. The optical data showed a similar dependence on ΔK except that the scatter was considerably larger.

The growth rate data from thick DEN specimens are more scattered than the CC results. These data are shown in figure 2 together with the tolerance bands for the CC tests. It is evident that the DEN growth rates lie within the CC tolerance bands. Furthermore, values of R between 0.1 and 0.6 appear to have little effect on the growth response, figure 3.

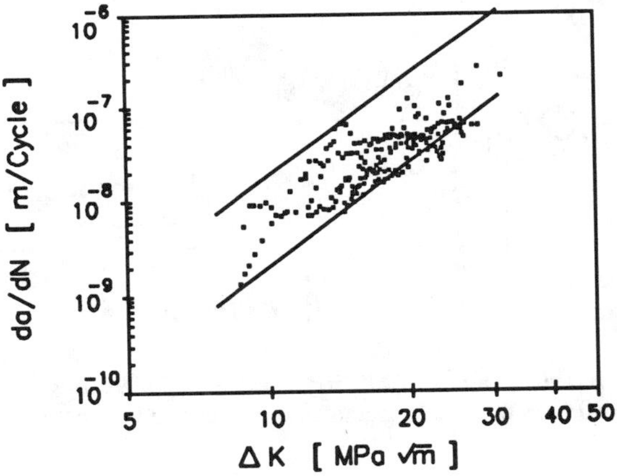

Figure 2. Growth rates in terms of ΔK for thick DEN specimens, R=0.1. Included are the tolerance bands for data from CC test pieces.

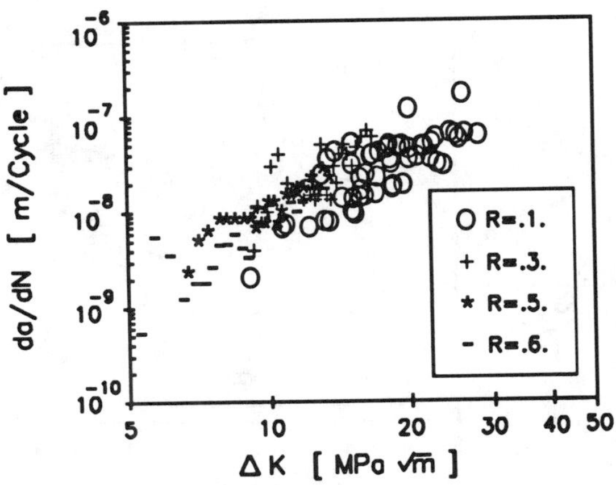

Figure 3. Growth rate data, in terms of ΔK, for a maximum applied stress of 350 MPa and R values of 0.1, 0.3, 0.5 and 0.6 on thick DEN test pieces.

Figure 4 shows that growth rate data for through cracks in thin DEN specimens were significantly different from the thick DEN and CC test pieces. It is evident that stress intensity factors alone are not sufficient to describe the behaviour. There is a marked dependence on nominal applied stress. A stepped response is apparent as illustrated in figure 4 for stress levels of 150 and 350 MPa and R=0.1. For a given ΔK, growth rates are higher at larger applied stresses. The dependence of growth rates on applied stress range, when R is constant, complicates the assessment of mean stress effects. However, a dependence on R value can also be found within the data.

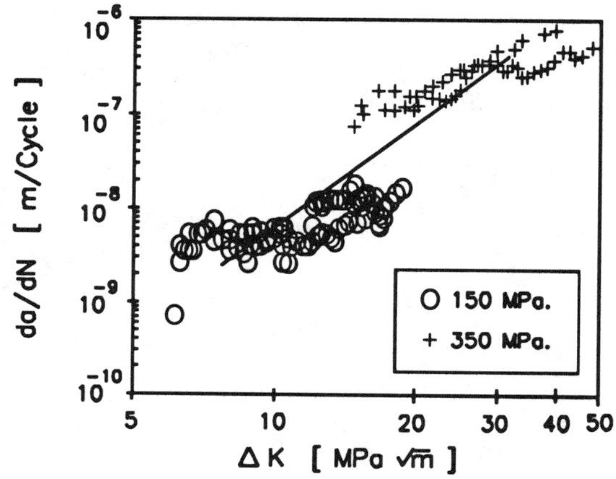

Figure 4. Growth rates, in terms of ΔK, for maximum applied stresses of 150 and 350 MPa and R=0.1 on thin DEN specimens. Included is the best fit line to data from CC specimens.

In figure 5 growth rates are plotted for a maximum applied stress of 83 MPa and R=0.1 together with data for a stress of 250 MPa and R=0.7. The two sets of data have the same stress range ($\Delta\sigma = 75 \quad MPa$) but they do not behave in the same way suggesting that the growth rates of these through cracks are also mean stress dependent.

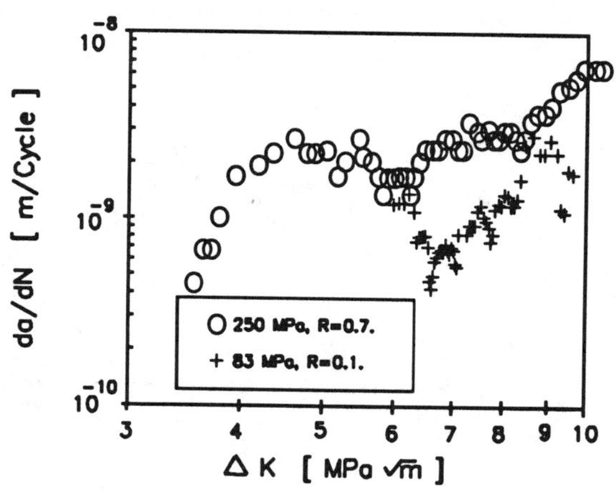

Figure 5. Growth rates, in terms of ΔK, for 250 MPa, R=0.7 and 83 MPa, R=0.1 on thin DEN test pieces. Δσ=75 MPa for both sets of data.

Crack shape development was studied in some detail on the thick DEN specimen used to construct figures 2 and 3. The shapes were found to be dependent on both the level of applied stress and R values. The cracks can be represented as ideal quarter ellipses. Each quarter ellipse has a major axis, b, and minor axis, a, where a and b are the lengths of the crack along the plain surface and notch root respectively. For low levels of applied stress ($\leq 350 MPa$) and R=0.1, cracks are largely quarter elliptical, figure 6. However, for higher levels of stress ($\geq 450 MPa$) the aspect ratio is lower. At 550 MPa the cracks begin to approach a quarter circular shape, figure 7, but with a bowed crack front. However, when R is increased to 0.6 at 550 MPa, the crack returns to a quarter elliptical shape with a similar aspect ratio to the low stress tests at R=0.1.

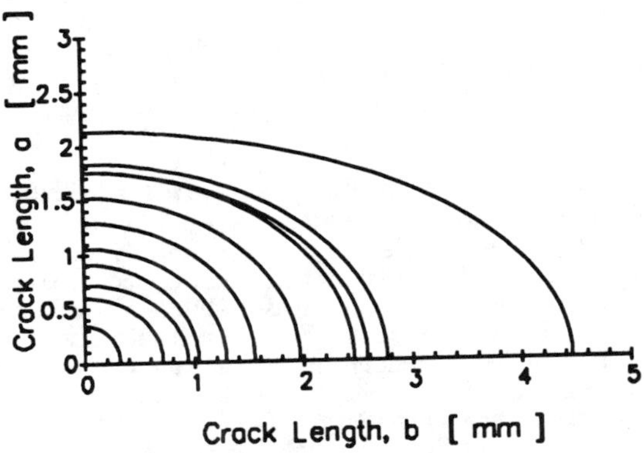

Figure 6. Crack shape development for 250 MPa on a thick DEN test piece.

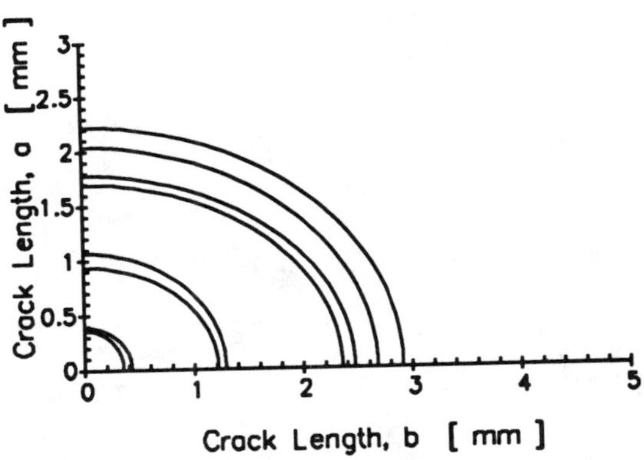

Figure 7. Crack shape development for 550 MPa on a thick DEN specimen.

Discussion.

Growth rate curves from p.d. data on CC specimens show less variation than the optical data due to the averaging effect inherent in the technique. The p.d. detects crack growth along the whole crack front rather than just at the two surfaces. Once the crack front spans a few grains, localised growth within one grain will have less effect upon the overall crack length measurement.

The growth rate data for CC specimens can be described by a Paris relationship:

$$\frac{da}{dN} = C\Delta K^{m}, \tag{1}$$

with $C=1.4x10^{-12}$ and m=3.6.

The crack growth rates for the thick DEN specimens lie within the tolerance limits for the CC specimens. However, there are differences in the growth characteristics which suggest that the two sets of results might not be entirely comparable. In the first place, the scatter for the DEN results was greater. Secondly, it is evident from the crack shape studies that applied stress and R can significantly influence the way in which the cracks in the DEN develop. Similar changes in shape were not evident in the CC specimens where all the cracks were quarter circular. The difference can be attributed to the occurrence of crack closure at the notch root.

Plasticity induced crack closure has been invoked to account for many aspects of crack growth behaviour by a large number of authors (3-5). At stress levels ≤ 350 *MPa* and R=0.1, the notch cracks are highly elliptical in shape consistent with a predominantly elastic stress field. Closure due to crack tip plasticity is likely to be similar for DEN and CC specimens at these stress levels. For stresses >400 MPa the degree of ellipticity decreases while at 550 MPa the crack is virtually quarter circular. The different crack shapes strongly suggest that the crack fronts are being held up at the notch surface. This might be due simply to increased plasticity and stress redistribution at the surface layers, since the K_t=1.9 factor ensured that notch root plasticity occurs for applied stresses >400 MPa. However, when a high R value (=0.6) test is carried out at 550 MPa, the crack shape returns to elliptical. This change can only be attributed to the occurrence of closure in the equivalent R=0.1 test. The p.d. technique was used to assess the likelihood of closure. If surfaces are in contact, resistance is lowered and large voltage changes should be observed. On this basis, p.d. measurements were made throughout selected load cycles and for equivalent crack sizes in R=0.1 and 0.6 tests at 550 MPa. The resultant outputs are shown in figure 8. It is evident that substantial changes in voltage occur at R=0.1 but that there is no change at R=0.6. The measurements are consistent with closure occurring during the test at R=0.1.

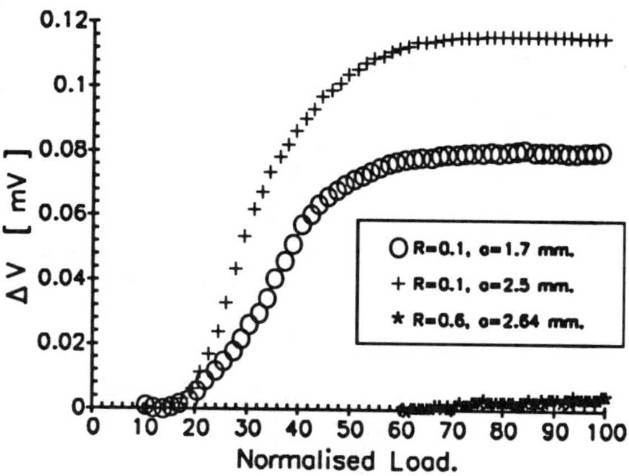

Figure 8. Variation of normalised crack voltage, ΔV, with load on a thick DEN specimen. σ_{max} = 550 MPa.

The closure is highly localised being concentrated in surface regions (5) but predominantly at the notch root for stress levels that induce significant levels of plasticity. This localisation is responsible for the fact that R value has little effect on the crack growth response in DEN, figure 3. It also explains the fact that the data for DEN tests correlate well with the CC behaviour, figure 2. The high level of scatter in DEN growth rates in this graph, however, suggests that the localised closure at the notch has a second order influence on crack development.

The crack growth behaviour of the thin DEN is significantly different from the CC or thick DEN response. In this case a through thickness crack is growing away from the notch root. A closure effect due to notch root plasticity is unlikely because of the crack geometry and the fact that the applied stress levels ≤ 350 MPa are insufficient for the macroscopic yield strength to be exceeded. The different response appears to be associated with other factors. In particular, the coarse microstructure of the alloy means that the 2 mm long crack front is addressing only 3-5 prior beta grains or approximately 10 alpha colonies. Chan and Lankford (6) have suggested that such conditions promote a small crack growth response in which local, and possibly single crystal, deformation characteristics become important. Recently, a detailed assessment of crack growth response in this particular crack/specimen geometry has been carried out (2). It was observed that K_{max} as well as the applied stress range, $\Delta\sigma$, affects growth rate. To asses the relative importance of K_{max} and $\Delta\sigma$ an empirical evaluation of the growth rate dependence was carried out. It was found that a relationship of the form:

$$da/dN = AK_{max}^{m}\Delta\sigma^{n}, \qquad (2)$$

described the full range of data with A=1.4x10^{-17}, m=1 and n=3.6.

The dependence of growth rates on both $\Delta\sigma$ and K_{max} strongly suggests an interaction between cyclic and static modes of failure. The static processes are associated with the accumulation of plastic strain at the crack tip and may also involve other factors such as environment. These processes will obviously be encouraged by the lack of constraint in the thin section. The cyclic mode of growth is clearly associated with the development of reversed plastic zones and the occurrence of factors such as crack tip closure. The extent to which each dominates will depend on loading conditions and test piece geometry. A clear task for the future is the definition of the bounds for each of these modes of failure and for the region of their interaction.

The present work has clear implications for component life prediction calculations. These techniques often rely on the use of data generated on simple test pieces to predict the behaviour of more complex stress concentration features on components. It is apparent that a simple relationship between the two situations might not exist. Major differences due to a range of factors including degree of plasticity, closure and mode of fracture can arise. It is evident, therefore, that despite our joint efforts in recent years there is still much to be done to optimise design and life prediction methods for major engineering applications.

Conclusions

Crack growth behaviour of corner crack, thick double edge notch and thin DEN test pieces has been compared.

1. The da/dN against ΔK results for the thick DEN tests lie within the 95% tolerance limits of the CC data but display a higher degree of scatter.

2. The DEN crack shapes are influenced by stress level and R value. At low stress and R, the cracks are elliptical. For high stress and low R they become more circular but then at high stress and high R return to the elliptical form.

3. The crack shape changes have been attributed to enhanced closure at the notch root. P.D. measurements confirm this belief. The increased scatter in DEN data has been ascribed to the closure influence.

4. Crack growth in the thin DEN specimens displays the small crack, microstructural effects associated with the crack front addressing a small number of prior beta or alpha colony boundaries.

Acknowledgements

This work has been supported by the Procurement Executive, Ministry of Defence.

References

1. S.H. Spence, W.J. Evans and P.J. Nicholas, "The Growth of Microstructurally Small Cracks in a Near Alpha Titanium Alloy," Proc. Int. Conf., Fracture of Engineering Materials and Structures, Singapore, 1991, Joint FEFG/ICF.

2. W.J. Evans and S.H. Spence, "Small Crack Growth at Notches in IMI 829," Short Fatigue Cracks, ESIS 13 (Edited by K.J. Miller) (1992), MEP, London.

3. W. Elber, "Fatigue Crack Closure Under Cyclic Tension," Engng. Fract. Mechs., 2 (1970), 37-45.

4. S. Chand and S.B.L. Garg, "Crack Closure Studies Under Constant Amplitude Loading," Engng. Fract. Mechs., 18 (2) (1983), 333-347.

5. T.C. Lindley and C.E. Richards,"The Relevance of Crack Closure to Fatigue Crack Propagation," Mets. Sci. and Engng., 14 (1974), 281-293.

6. K.S. Chan and J. Lankford, "The Role of Microstructure Dissimilitude in Fatigue and Fracture of Small Cracks," Acta Metall., 36 (1) (1988), 193-206.

INFLUENCE OF PROCESSING VARIABLES ON THE MECHANICAL PROPERTIES OF Ti-15V-3Cr-3Al-3Sn SHEET FOR HELICOPTER EROSION SHIELD APPLICATIONS

D P Davies

Westland Helicopters Limited
Yeovil, Somerset, UK

Abstract

Although Ti-15V-3Cr-3Al-3Sn was originally designed as a cold formable alloy, the paper will discuss the influence elevated forming temperature has on the resultant mechanical properties. Furthermore the effect of temperature on the forming/processing characteristics will also be discussed.

Introduction

Although an erosion shield provides damage protection to the composite blade it also requires two additional factors, namely (i) it must be electrically and thermally conducting and (ii) it must have specific engineering properties (i.e. adequate strength and fatigue resistance) to resist the induced strains from the composite blade.

In order to bond the erosion shields to the complex aerofoil profile, See Figure 1, it is necessary to either hot form (~ 700°C) or superplastically form (~ 900°C) the current Ti-6Al-4V material. However, the desire to enhance mechanical performance, especially fatigue, and to reduce material and processing costs were responsible for Westland Helicopters evaluating Ti-15V-3Cr-3Al-3Sn as a potential helicopter erosion shield material (1).

Titanium '92
Science and Technology
Edited by F.H. Froes and I. Caplan
The Minerals, Metals & Materials Society, 1993

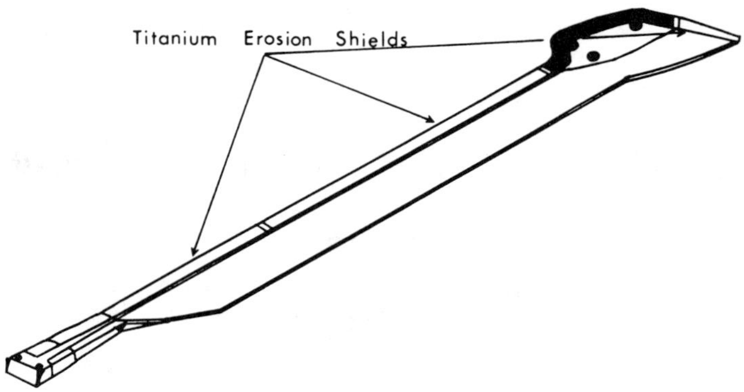

Figure 1 Schematic Representation of Helicopter Erosion Shield

Experimental Procedure

Material

The material chosen for this investigation was Ti-15V-3Cr-3Al-3Sn supplied by TIMET, in the solution annealed condition, in accordance with AMS 4914.

Forming Trials

Forming trials were performed on material subjected to the processing parameters defined in Table I.

Table I Summary of the Forming Temperature/Method

Forming Temperature	Shield Type	Forming Method
450 ± 10°C	Outboard	Stretch Forming
575 ± 10°C	Outboard	Stretch Forming
675 ± 10°C	Outboard	Stretch Forming
675 ± 15°C	Inboard	Blow Forming
725 ± 15°C	Inboard	Blow Forming
755 ± 15°C	Inboard	Blow Forming

Inspection and Testing

Prior to undergoing a detailed investigation each shield was assessed for degree of alpha case contamination and extent of hydrogen absorption after forming and chemical etching.

In order to facilitate shield evaluation and determine the effect of strain rate on mechanical performance, test pieces were removed from both the upper and lower surfaces at three separate chordwise locations. Tensile, fatigue and crack propagation specimens, were manufactured in accordance with the relevant ASTM specifications.

Fatigue testing was carried out using an axially loaded unnotched specimen at a stress ratio of R = 0.05. Fatigue endurance limits were calculated using Weibull's emperically derived formula, in which the material constants were derived from a computer program which fits the four point parameter S/N curve equation through the fatigue test data points.

In the case of crack propagation, testing was carried out using a M(T) specimen at a stress ratio of R = 0.1. The da/dN Vs ΔK data appropriate to each forming scenario was calculated using a computer program devised from the Paris equation.

Experimental Results

The experimental programme was carried out in two stages. The first stage established the forming/processing characteristics (i.e. formability, susceptibility to contamination etc.), while the second stage determined the mechanical properties within each shield.

Process Evaluation

Alpha Case Contamination. The degree of alpha case contamination exhibited with each of the forming trials can be seen in Figure 2, on which representative Ti-6Al-4V data has been superimposed. From the results it can be seen that by raising the forming temperature the depth of alpha case contamination increases, regardless of alloy composition, although Ti-15V-3Cr-3Al-3Sn is also more susceptible to oxidation that Ti-6Al-4V.

Absorbed Hydrogen. Analysis of the hydrogen content, both before and after pickling, is shown in Table II. From these results at can be seen that increases in hydrogen levels are a direct result of the forming temperature and hence the amount of time in the pickling bath.

Table II Effect of Forming Temperature/Pickling on Hydrogen Content

Forming Temperature °C	Hydrogen Content ppm	
	After Forming	After Pickling
575	20	110
675	25	155
725	35	175

Property Levels. Tensile data was obtained from all the forming trials, regardless of whether forming was successful, in producing an acceptable shield profile. These results are shown in Figure 3. It is interesting to note that by raising the forming temperature reductions in tensile strength were observed. In addition, by forming at temperatures $\leqslant$ 675°C the strengths so produced were higher than is the as received sheet, inferring some form of strain or precipitation hardening is occurring. Moreover, the difference in strength levels between the inboard and outboard erosion shields formed at ~ 675°C (i.e. the inboard erosion shield was nominally 4% lower) was considered a direct result of the lower strain rate attributable to blow forming operation.

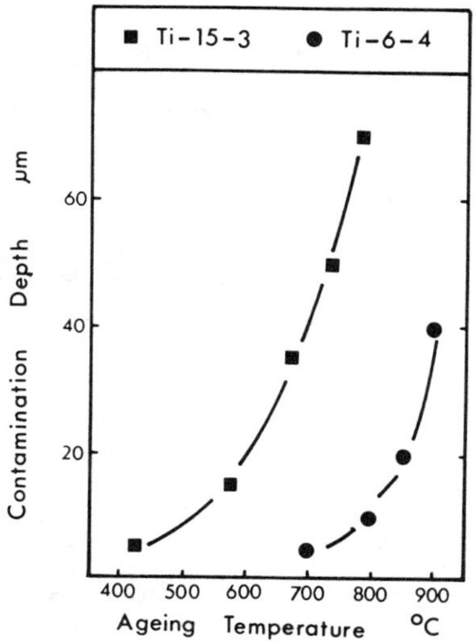

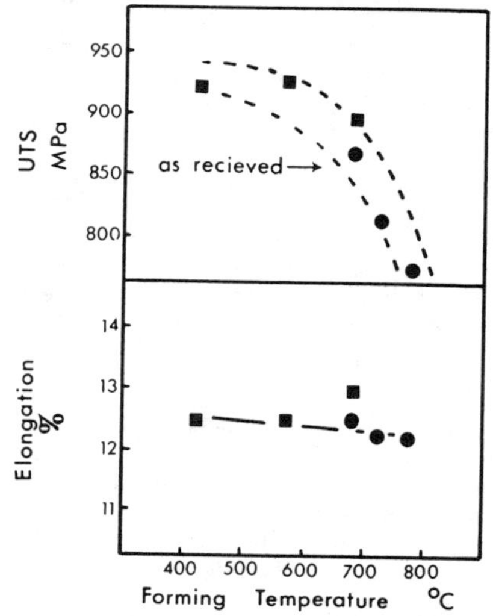

Figure 2 Influence of Forming Temperature on Alpha Case Contamination

Figure 3 Influence of Forming Temperature on Tensile Properties.

Mechanical Property Response

Tensile Properties. A comparison of the mean tensile properties obtained from each of the satisfactory forming trials is shown in Figure 4. From this data it can be seen that by raising the forming temperature a reduction in the resultant 0.2% PS and UTS values were observed while ductility levels were unaffected.

In order to study the effect of variations in forming strain on mechanical performance, tensile samples were removed from three separate chordwise locations, the results of which are shown in Figure 5. From the data it can be seen that strengths decreased on moving away from the trailing edge of the shield (i.e. region of highest strain) regardless of forming temperature. However, the chordwise differences declined with increasing forming temperature.

Fatigue Properties. A comparison of the S/N data generated from each of the successful forming trials, is shown in Figure 6. Furthermore the endurance limits determined from each S/N curve (i.e. representing life at 2 x 10^7 cycles) is shown schematically in Figure 7.

From Figures 6 and 7 it can be seen that by raising the forming temperature a reduction in the high cycle fatigue performance was apparent, although the general shape of the curve was unaffected.

Crack Growth Properties. A comparison of the crack growth rate data for the various forming regimes is plotted in Figure 8, in which data trends have been approximated by a straight line for clarity. In general there does not appear to be a direct correlation between fatigue crack growth and forming temperature (i.e. by raising the forming temperature), crack propagation rates were found to be essentially similar.

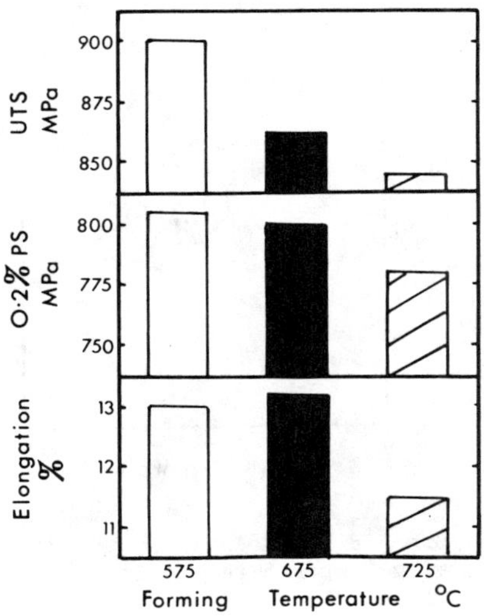

Figure 4 Comparison of Mean Tensile Data with Forming Temperature

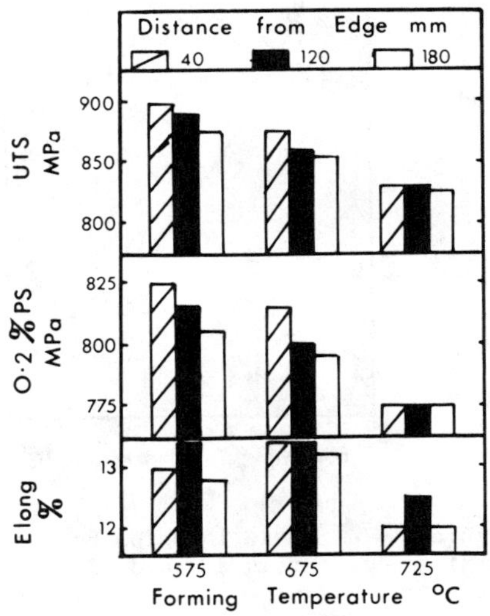

Figure 5 Inter-Relationship between Forming Temperature, Test Piece Position and Tensile Properties

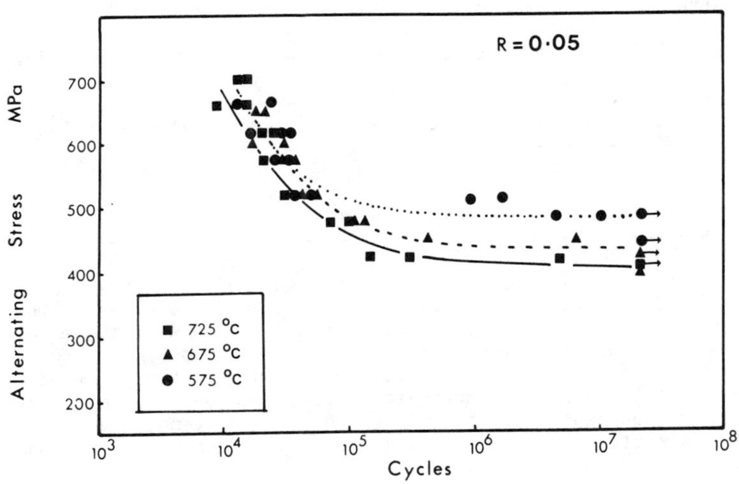

Figure 6 S/N Data Generated for Ti-15V-3Cr-3Al-3Sn from Various Forming Trials

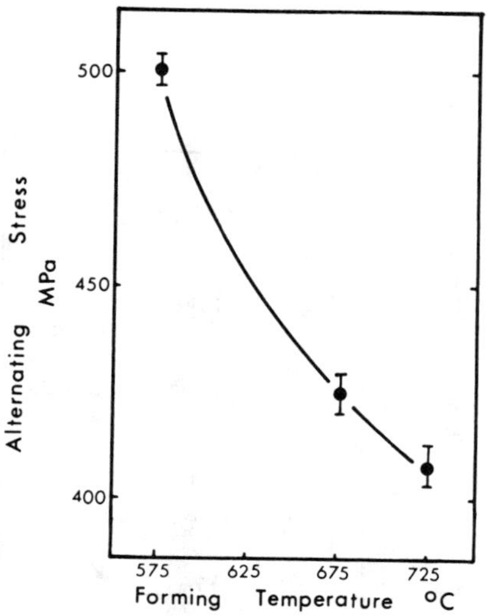

Figure 7 Relationship Between Forming Temperature and Fatigue Performance

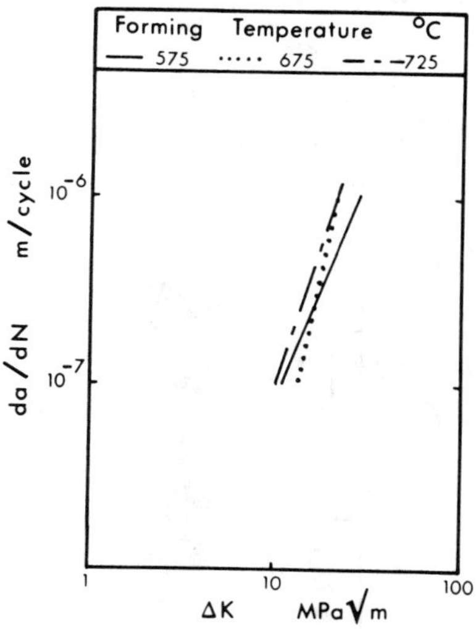

Figure 8 Relationship between Forming Temperature and Crack Growth Resistance

Microstructural Features. Photomicrographs of the general grain structure of each of forming scenarios is shown in Figures 9(a) to 9(d). On comparing the microstructural features, differences were apparent as summarised accordingly:

(i) On raising the forming temperature, increases in the mean beta grain size was observed, as summarised in Table III.

Table III Influence of Forming Temperature on Beta Grain Size

Forming Temperature °C	Mean Beta Grain Size μm
725	30
675	25
575	23

(ii) By reducing the forming temperature below 725°C strain-induced martensite and thermally precipitated alpha was observed. The strain induced martensite ostensively occurred around the nose of the shield, (See Figure 9(c)) while the thermally precipitated alpha was more evenly distributed around the grain boundaries and throughout the prior beta grains (See Figures 9(b) and 9(d)). Moreover, differences between the precipitated alpha was also apparent, i.e. the higher the forming temperature the coarser was the alpha particles, and the thicker was the grain boundary alpha.

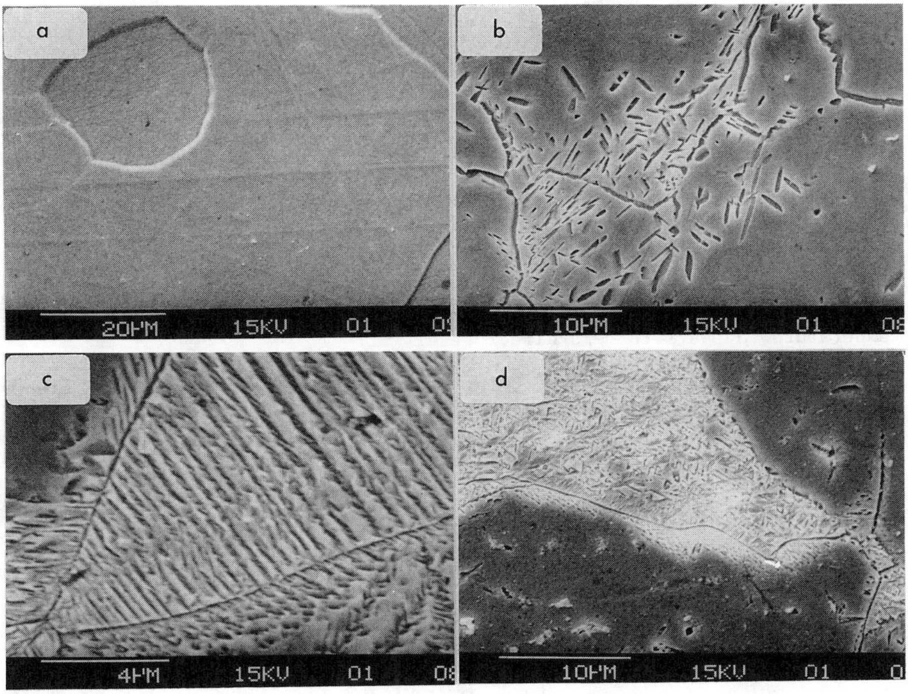

Figure 9 Influence of Forming Temperature on Microstructural Response:

(a) 725°C (General) (b) 675°C (General) (c) 575°C (Nose) (d) 575°C (General)

Discussion

Processing

Contamination During Heat Treatment. As with all titanium alloys when exposed to air at elevated temperature, Ti-15V-3Cr-3Al-3Sn will absorb oxygen at the surface. The degree of oxide formation and hence alpha case contamination, will depend on the alloy content, i.e. Ti-15V-3Cr- 3Al-3Sn is more susceptible to oxidation due to the presence of Chromium in the chemical composition (2). If Ti-15V-3Cr-3Al-3Sn is to fully develop its high cycle fatigue properties, it is essential that the oxide scale and the contaminated metal beneath is completely removed.

In addition to oxidation, hydrogen absorption and its avoidance is also of great importance. Current specification limit for hydrogen content is between 125 - 200 ppm, depending on the alloy and mill form. Generally beta alloys have a high tolerance for hydrogen (i.e. up to 2000 ppm) without detrimentally affecting mechanical properties (4). However, the aim was to limit the absorbed hydrogen using adequate buffer solutions in the chemical etching bath, to below 250 ppm. The trials carried out have been successful in limiting the degree of absorbed hydrogen to below 200 ppm.

As oxidation (alpha case contamination) resistance and its proclivity of rapid hydrogen absorption is more severe with Ti-15V-3Cr-3Al-3Sn than Ti-6Al-4V and since the alloy has a lower beta transformation temperature, Ti-15V-3Cr-3Al-3Sn should not be formed above 750°C.

Mechanical Properties

Influence of Test Piece Position. There appears to be a direct relationship between test piece position and resultant tensile properties, as shown in Figure 5. Specifically, the tensile strengths increase on moving towards the trailing edge and this is considered to be primarily due to the amount of strain induced into the shields during forming. Furthermore, as the reported differences appear to decrease with forming temperature, this would suggest that the induced thermal strains are either (i) being removed by the higher forming temperature or (ii) a lower level of induced strain is occurring due to the blow forming operation.

Influence of Forming Temperature. The reductions in tensile and fatigue properties observed (See Figures 4 and 7) on raising the forming temperature, appears to be a direct result of:

(i) less strain hardening being induced into the shield at the higher working temperatures,

(ii) coarsening and eventual elimination of the alpha phase attributable to higher thermal temperatures (3), and

(iii) beta grain growth (4).

These hypothese are strengthened by comparing the tensile properties of the as received sheet with the resultant as formed tensile properties, See Figure 3. From this it can be seen that only shields that were formed above 700°C exhibited inferior tensile strengths, and this is considered a direct result of the lack of alpha phase precipitation and resultant grain growth.

As crack growth rates are relatively insensitive to small changes in microstructure, it is considered that the reported differences would be unlikely to produce substantial changes in the propagation rate, which was in fact observed (See Figure 8).

Conclusions

Although Ti-15V-3Cr-3Al-3Sn offers a far greater level of formability than is capable with Ti-6Al-4V it must be remembered that its inferior oxidation resistance and lower beta transformation temperature, should limit its maximum processing temperature to below 750°C. Furthermore, if fatigue resistance is of primary importance, then forming should not be carried out above 650°C.

References

(1) D P Davies, B C Gittos, WHL Contract MMB/EA621T40W/F01, 1990, pp69

(2) Properties and Processing of Ti-15-3, Timet Data Sheet

(3) M Okade, D Banerjee, J C Williams, 5th Int. Conf. on Titanium, 1984, p 1835

(4) Titanium and Titanium Alloys, Source Book ASTM 1982.

INFLUENCE OF THERMOMECHANICAL PROCESSING ON THE MECHANICAL PROPERTIES OF Ti-6Al-4V FOR HELICOPTER DYNAMIC COMPONENTS

D P Davies

Westland Helicopters Limited
Yeovil, Somerset, UK

Abstract

The mechanical properties of an annealed Ti-6Al-4V forging are greatly influenced by the type of thermomechanical working the billet receives. Hence, the paper will discuss the most relevant of these billet processing steps which directly affect the properties in the forging. In addition, the paper will also make reference to the influence specific working/post forging heat treatments have on the resultant forging properties.

Introduction

The purpose of the work was to (a) determine the optimum billet processing route for large annealed titanium Ti-6Al-4V forgings against present and future helicopter design requirements and (b) assess the effect thermomechanical processing steps have on the resultant mechanical properties for dynamically critical rotating components.

Titanium '92
Science and Technology
Edited by F.H. Froes and I. Caplan
The Minerals, Metals & Materials Society, 1993

Experimental Procedure

Material Procurement

The material chosen for this investigation was Ti-6Al-4V supplied by TIMET, IMI and CEZUS and procured in accordance with WHMS 462.

On-Receipt Inspection and Testing

Billet (Processing Parameters). Each of the materials selected for the ingot to billet conversion processing routes were heat treated (i.e. annealed at 700°C per 1 hour of section thickness, followed by air cooling) prior to undergoing a full metallurgical/mechanical investigation. The billets were examined by a combination of macroscopic/microscopic examination and by static/dynamic mechanical testing. Macroscopic evaluation consisted of establishing the general grain flow per billet route, whereas relative microstructural differences (i.e. primary α grain size, % primary α, etc.) were determined at various locations throughout the billet cross section.

Mechanical property determinations were carried out by trepanning out of each billet two separate rings approximately 15 mm and 85 mm from the billet surface.

Fatigue testing was carried out using a simple unnotched direct stress fatigue specimen, the dimensions of which is given in reference (3). The specimens were tested over a range of loading conditions, particular to helicopter requirements, (i.e. from a zero mean stress 0 ± P to a maximum applied tensile stress of 300 MPa ± P). For each loading condition a S/N curve was generated. Fatigue endurance limits were calculated using Weibull's empirically derived formula, in which the material constants were determined from a computer programme which fits the four point parameter S/N curve equation through the fatigue test data points.

Forging (Processing Parameters). A Lynx sleeve forging, shown schematically in Figure 1, was used to determine the effect various heat treatment processing routes, had on the resultant mechanical/microstructure parameters. Each forging was subjected to individual post forging thermal treatments as described in Table I.

Table I Summary of the Thermomechanical Processing Routes

Forging Identity	Thermomechanical Processing Information
A	Forged at 930°C, Air cooled (AC) Overaged at 730°C, AC
B	Forged at 1040°C, Water Quenched (WC), Overaged at 730°C, AC
C	Forged at 930°C, AC, Solution treated at 970°C, Overaged at 730°C, AC

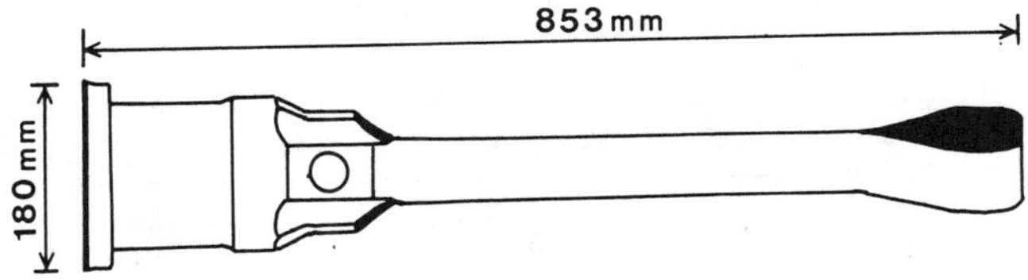

Figure 1 Schematic Diagram of a Lynx Sleeve Forging

A number of macrosections/microsections were taken from each forging in order to determine (a) the relative grain flow throughout the forged shape and (b) thermomechanical response on microstructural features. Mechanical test specimens were obtained at similar depths to those obtained from the billet investigation, in order to relate the sectional response of the forging to the degree of working it had received. Tensile and Fatigue specimens were manufactured in accordance with the relevant ASTM specifications. Moreover, fatigue testing and analysis was carried out in a similar manner to that used in the billet investigation.

Results/Discussion

Billet Material

Although a number of ingot/billet conversions were evaluated, in order to determine the preferred route, the major processing steps can be summarised accordingly:

(a) Ingot (800 mm diameter)
(b) β working
(c) α-β working
(d) β recrystallisation/β Working
(e) Cooling Rate
(f) α-β Working (265 mm diameter)

It is clear from the results obtained, that similarly processed billet stock which had been processed through a number of ingot/billet manufacturing routes exhibited significantly different fatigue and tensile properties. In particular, two processing parameters whose effects can be determined in the microstructure of the billet material and which appear to significantly affect the fatigue and tensile properties are:

(i) The rate of cooling through the β phase field prior to the final α-β working.

(ii) The amount of working, particularly the final α-β working that the billet received.

Cooling Rate. The effect of introducing a more rapid rate of cooling from the β phase prior to final α-β working, can be seen in Figure 2, in which higher fatigue endurance values are produced by water quenching. It is interesting to note that enhanced fatigue properties are not only limited to the surface areas of the billet, but are also apparent in the more centralised regions. Moreover, water quenching also produces improvements in tensile properties and this is not at the expense of ductility (See Figure 3). The improvement in mechanical properties appear to be due to microstructural effects, in that water quenching produces a finer acicular structure, thereby arresting β grain growth and ensuring the absence of grain boundaries which are rich in the α phase. This results in a finer primary α grain size during the final α-β working sequence (Table II).

Table II Influence of Cooling Rate on the Primary α Grain Size of Ti-6Al-4V Billet

Cooling Medium	Mean Primary α Grain Size μm		
	Transverse	Radial	Longitudinal
Air Water	10.7 9.0	13.7 10.0	10.8 10.5

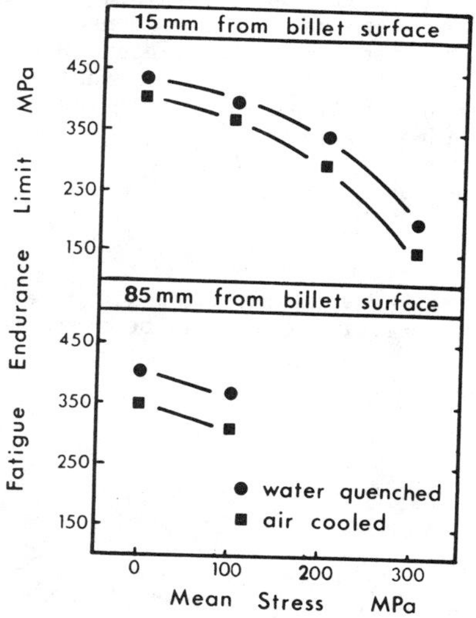

Figure 2 Influence of Cooling Rate on the Fatigue Properties of a Ti-6Al-4V Billet.

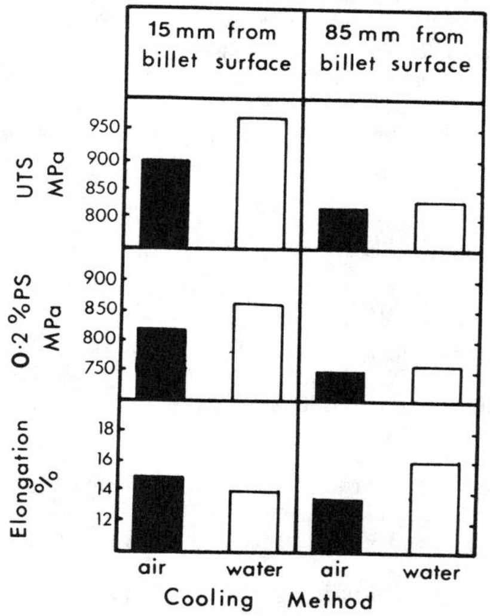

Figure 3 Influence of Cooling Rate on the Tensile Properties of a Ti-6Al-4V Billet.

The results obtained by Westland Helicopters are consistent with the work of previous investigators (1) who notice that refinement of the primary α grain size accompanied water quenching.

Degree of α-β Working. There appears to be a direct relationship between the degree of α-β working and mechanical properties, in that, by increasing the amount of α-β working a billet receives, improvements in fatigue and tensile properties are a consequence (See Figures 4 and 5). This is in agreement with the work of Bowen et al (2) on smaller diameter sections, who concluded that working in the α-β phase field contributed to the resultant fatigue and tensile strength of that material.

The degree of working that a billet receives during its processing route contributes not only to the general refinement of the equiaxed primary α grain size but also to the breakdown and eventual elimination of the prior β grains and the preferentially aligned colonies of elongated α. An inverse relationship exists between the primary α grain size and the degree of working (See Figure 6), i.e. the greater the degree of working, the smaller the primary α grain size and the better the resulting fatigue and tensile properties of the material. The refinement and deformation of the microstructure was, however, more apparent near the surface of the billet. Whereas as one approached the more central regions, the primary α grain size, the number of undeformed preferentially aligned colonies of primary α and the number of grain boundary α films all tended to increase, resulting in lower fatigue and tensile properties.

However, excessively high degrees of α-β working (i.e. > 85%) are to be avoided, as banding and texturing can be introduced into the billet, thereby detrimentally affecting fatigue performance.

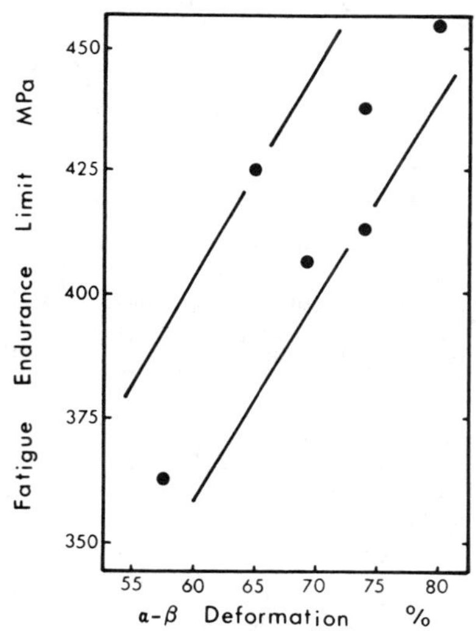

Figure 4 Influence of α-β Working on the Fatigue Properties of Numerous Ti-6Al-4V Processed Billet

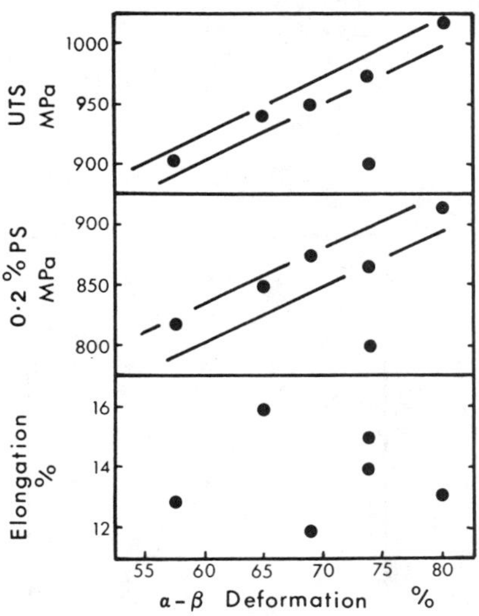

Figure 5 Influence of α-β Working on the Tensile Properties of Numerous Ti-6Al-4V Processed Billets

Forging Material

Effect of α-β Phase Working. Deformation in the α-β field has been shown to produce increases in fatigue strength (3), and this improvement can be related to the microstructural characteristics produced by the forging sequence. However, the fatigue strength of the forging, particularly in areas where a low degree of forging reduction has occurred, is determined to some extent by the original billet stock fatigue properties. This can be seen in Figure 7 in which the fatigue strength in areas of the forging that had received the lowest degree of working, was only slightly higher than the original billet properties. In contrast, substantial improvements in fatigue were observed in areas on the forging that had received the heaviest degree of working. The general shape of the original billets Goodman Diagram has only be slightly altered in the final forged item. This suggests that at zero mean stress some effect of the billet processing route on the microstructure of the original billet material has been maintained throughout the subsequent forging operation. Whereas, at higher mean stresses the forging has enhanced fatigue performance as a result of the α-β interface refinement. It is therefore apparent from these results that, although improvements in fatigue strength can be achieved by subsequent working during the forging operation, it is the fatigue strength exhibited by the original billet material in low worked areas which can profoundly influence the fatigue properties of the forging. It is interesting to note that even in low worked areas the forging can produce significantly improved tensile properties when compared to the original billet properties.

Effect of β Phase Working. Although α-β titanium alloys are most often used in the annealed condition, substantial differences in both the microstructure and mechanical properties can be achieved depending upon whether or not forging was carried out above or below the β Transus (4).

Table III compares the properties of two Lynx sleeve forgings deformed in the α-β phase field (i.e. 930°C) and the β phase field (i.e. 1040°C) respectively.

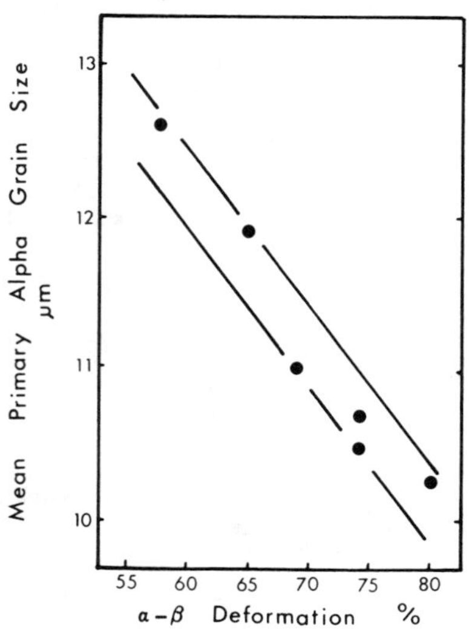

Figure 6 Influence of α-β Working on the Primary α Grain Size of Numerous Ti-6Al-4V Processed Billet

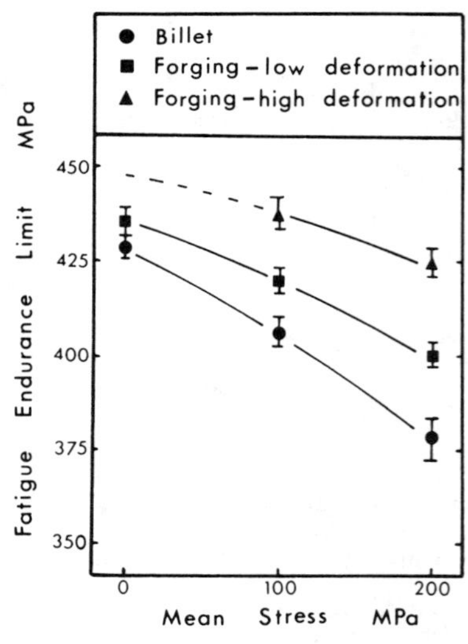

Figure 7 Fatigue Property Enhancement in an Annealed Forging as a Result of α-β Deformation

Table III Comparison of the Mechanical Property Levels due to α-β and β Processing

Property	α-β Forging		β Forging	
	15 mm below surface	85 mm below surface	15 mm below surface	85 mm below surface
Tensile Strength (MPa)	901	885	1056	1023
0.2% Proof Stress (MPa)	818	807	961	927
Elongation (%)	13.5	12.5	6.5	7.5
RA (%)	40	38	10	9
Fracture Toughness (MPa√m)	62	73	81	85
Fatigue Limit (MPa) mean stress 200 MPa ± P	± 256	± 206	± 441	± 370

From this data it can be seen that the tensile, fracture toughness and fatigue strength are all notably higher in the β forged/overage condition (containing a Widmanstatten structure) than in the α-β forged, annealed material, which is more ductile. The reason for the higher fatigue resistance with the β annealed structure is due to the fact that cracks propagate at a slower rate though a Widmanstatten microstructure due to crack branching within the α laths (4). However, while β forgings appears to offer considerable improvements in properties over conventional α-β forgings, the Widmanstatten microstructure is less resistant to low cycle fatigue and is considerably less ductile.

Solution Treatment and Over Aged (STOA) Processing. Heat treatment of α-β processed material at temperatures high in the α-β phase field produces mixed microstructures consisting of islands of primary α in a matrix of transformed β (5). The proportion of primary α depends on the proximity of the treatment temperature to the β transus temperature, while the morphology of the transformed β is a function of the rate of cooling.

The properties achieved with a forging subjected to a STOA thermal treatment, designed to produce a duplex structure containing ~ 10 - 25% primary α in a matrix of transformed β, can be seen in Table IV.

Table IV Influence of an STOA treatment on the Mechanical Properties of an α-β Forging

Property	α-β Solution Treatment	
	15 mm below surface	85 mm below surface
Tensile Strength (MPa)	1004	951
0.2% Proof Stress (MPa)	917	865
Elongation (%)	13.5	14.5
RA (%)	41	39
Fracture Toughness (MPa√m)	75	78
Fatigue Limit (MPa) mean stress 200 MPa ± P	± 464	± 421

By comparing the mechanical property data for the α–β processed and β worked forging shown in Table III, it can be seen that the α–β STOA forging offers comparable strength, toughness, and fatigue resistance to those produced by β forging while still retaining a level of ductility only experienced with the α–β annealed forging.

As with all water quenching operations, there will be a limiting sectional size response, above which reductions in tensile strength and fatigue are a likely consequence. This effect appears to be mainly microstructural, in that during water quenching the rapid rate of cooling experienced on the surface restricts the α grain size and inhibits grain growth (particularly in the fine acicular α of the transformed β matrix), thereby increasing tensile strength and fatigue properties. Nearer the centre of the forging, however, the cooling rate is much slower, allowing time for diffusion controlled reactions within the microstructure to occur, resulting in a general coarsening of the transformed β and hence lower tensile and fatigue properties, (See Tables III and IV). It is interesting to note that relative amounts of equiaxed α in the microstructure has an important influence on the ductility of α–β alloys, i.e. ductility increases with increasing amounts of equiaxed α up to ~ 15%, whereupon it levels off.

Conclusions

Billet

To ensure optimum properties in the billet, the ingot to billet conversion should include (a) a water quenching operation prior to the final α–β working stage, and (b) a high degree of α–β working (i.e. in excess of 60%) as the final deformation step.

Forging

In order to obtain optimum properties in the final forging, the processing route should include (a) a water quench operation after the final deformation step, and (b) a solution treatment process high in the α–β region.

References

(1) J. J. Lucas, 2nd Int. Conf. on Titanium, 1972, p 2081.

(2) A. W. Bowen, CA Stubbington, 2nd Int. Conf. on Titanium, 1972, p 1769.

(3) D. P. Davies, New Developments in Aerospace Materials II, AEROTECH 89, 1989.

(4) I. J. Polmear, Light Alloys, Edward Arnold, 1980.

(5) D. P. Davies, Metals and Materials, July 1988, p 417.

ADIABATIC SHEAR FAILURE OF THE TITANIUM ALLOY TI6AL4V

UNDER BIAXIAL DYNAMIC COMPRESSION/SHEAR LOADING

L. W. Meyer and E. Staskewitsch

IFAM-Fraunhofer-Institut für Angewandte
Materialforschung, Bremen, Germany

Abstract

In order to study the propensity of metallic materials to adiabatic shear failure under multiaxial stresses a new testing method was developed to investigate the dynamic failure behaviour under combined compression/shear loading. The combined simultaneous loading is applied by a drop weight hammer to a cylindrical specimen whose axis is inclined to the hammer axis.

The testing results on the titanium alloy Ti6Al4V show that the onset of an adiabatic shear failure is promoted strongly even under small amounts of a shear component to the main compression stress.

Introduction

Adiabatic shear is one of the most important failure mechanisms in impact loaded structures because the failure acts like a brittle fracture even when the fracture is microscopically a ductile one.

In conventional testing and evaluation the material behaviour under monoaxial stress states in tensile or compressive loading is often used to calculate the permitted load of a structure. For a multiaxial stress condition Mises' or Tresca's laws are applicable to reduce a multiaxial loading to an equivalent monoaxial state of stress.

Mises' and Tresca's laws can also be applied for dynamic biaxial loading by using some corrections / 1/ , if no adiabatic shear failure occurs. This failure behaviour indeed cannot be predicted, it has to be investigated under the real conditions including biaxial loading. This is especially important for titanium alloys which tend to fail under adiabatic shear conditions in the case of impact loading /2/.

Investigations of the susceptibility of metallic materials to adiabatic shear failure can be done in a pure shear condition by dynamic torsion testing / 3,4 / or by using special shear specimens / 5,6/ loaded in a split Hopkinson bar. Both techniques have the benefit to study the shear strain and strain-rate conditions which initiate adiabatic shear failure / 7/ . In order to investigate the propensity of metallic materials to adiabatic

Titanium '92
Science and Technology
Edited by F.H. Froes and I. Caplan
The Minerals, Metals & Materials Society, 1993

shear failure under multiaxial loading a new test procedure was developed. The testing method for combined compression/shear loading is described in this paper and first results of an investigation program for titanium alloys are presented.

Test procedure

The testing system (Fig. 1) consists essentially of a heavy drop weight hammer, which is mounted in a special ball-bearing system in order to guarantee a very precise fall motion, an anvil and a machine frame. The frame and all other parts of the test setup are of high stiffness to prevent noticeable elastic deformation. Under the hammer a small punch is fixed to give a precise loading condition and to measure the load time history by calibrated strain gage records. To reach a combined compression/shear loading the cylindrical specimen were manufactured with an inclination up to an angle of 10 °. The punch surface as well as the anvil surface is shaped by laser treatment for keeping a lubricant to enable an unrestrained lateral deformation of the specimen.

Adjustable stiff stopping devices make it possible to arrest the test at given points of deformation. This is important for the following metallurgical analysis of the shear band forming and the failure development.

Besides the force measurement at the tup strain measurements directly at the specimen were undertaken by using strain gages and additionally by using an optoelectrical camera. A light gate system measures exactly the velocity of the drop hammer respectively the loading rate. All measured data were registered and stored in a transient recorder.

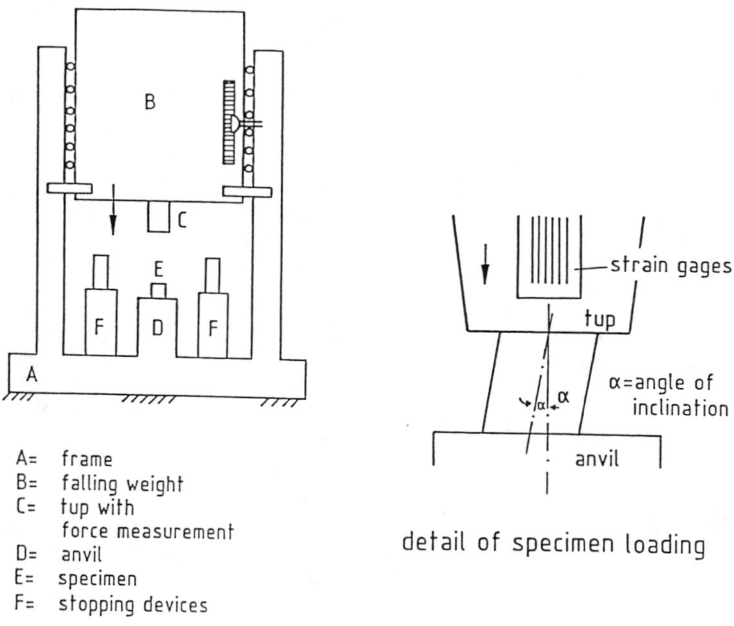

Figure 1 - Schematic representation of the test system

Results

To evaluate the influence of the biaxial shear/compression loading the commercial titanium alloy Ti6Al4V was used, which was already investigated under pure uniaxial compression loading at various temperatures and various strains /2/. With the relative small inclination of the specimen of three, six and in the maximum of ten degrees nearly the same compressive stress distribution is reached as in the uninclined specimen, but this inclination implies shear stresses of up to 15 % of the compressive stress.

Under quasistatic loading (strain rate $\dot{\varepsilon} \approx 5 \cdot 10^{-4}\ s^{-1}$) and room temperature the material failed at about 50 % compressive deformation with a nearly homogeneous distributed deformation without shearbands. With increasing strain rates first slightly higher deformability is noticed, but under high rates of strain in the range of 200 to 250 s^{-1} the deformability is limited by the occurance of adiabatic shear failure. In this case the compressive deformation is reduced to half of the deformation under quasistatic loading, Fig. 2, where each test is represented by a data print.

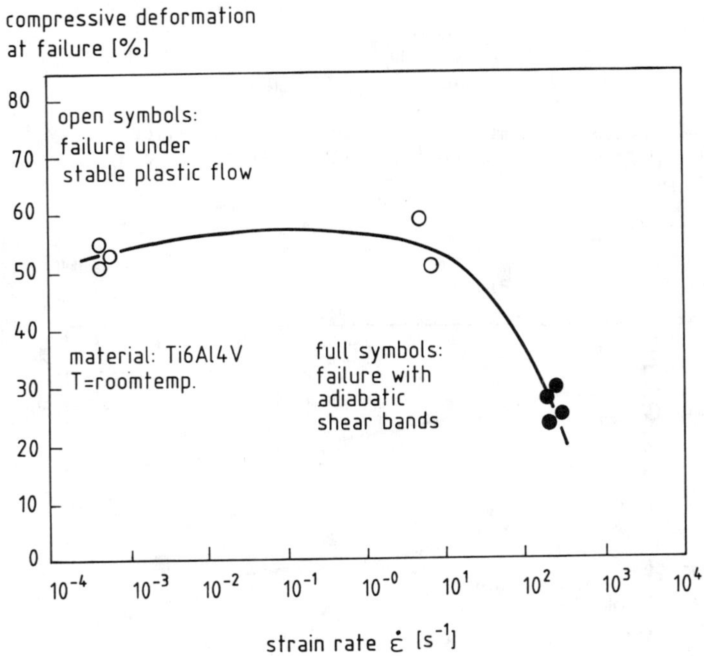

Figure 2 - Deformability behaviour under compressive loading as a function of strain rate

With increasing inclination of the specimen the adiabatic shear failure occured at lower strains and lower load as the averaged curves of several tests show, Fig. 3. That means that an increasing ratio of shear/compression stress leads to a strong reduction of the deformability and, more important, to a distinct reduction of the ultimate compressive stress. In Fig. 4 all results of the dynamical tests are put together. The marked

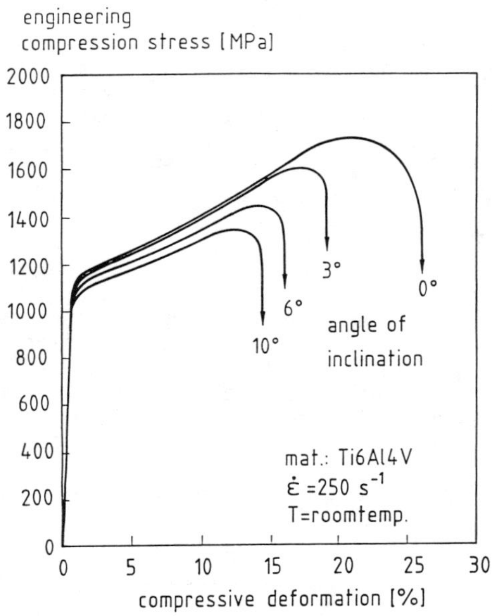

Figure 3 - Engineering compressive stress/strain behaviour of cylindrical specimen with varied inclination between 0° and 10° under dynamical loading

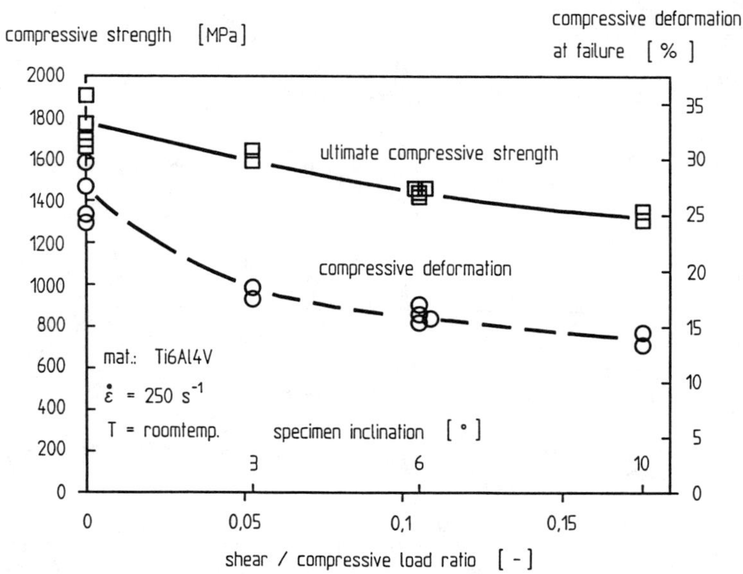

Figure 4 - The effect of shear/compression stress ratio on ultimate compressive strength and deformability

data points show the scattering of the test results. A reduction of the ultimate compressive strength of about 25 % - from 1750 MPa sloping to 1330 MPa - was found. Before the onset of adiabatic shearing the compressive yield stress undergoes only a small reduction caused by the increasing shear stress in the specimens with increasing inclination, according to the Mises' law.

In one dynamic test series specimens with an inclination of 6° were used to stop the test at several states of deformation. Metallographic investigations were performed to find out the initiation and the development of adiabatic shearing. These investigations showed that the beginning of the adiabatic shear formation leads to a decrease of the slope of the load strain curve, specimen No. 5 in Fig. 5. In this case the micrograph (Fig.6) shows small shear bands starting in the specimen edges.

In the specimen No. 7 which was loaded respectively deformed until the load began to diminish, the shear-band is full established and initiates the failure, as shown in the micrograph, Fig. 7. In the last state of failure - specimen No. 1 in Fig. 5 - some secondary effects lead to small shear zones at the specimen edges, Fig. 8.

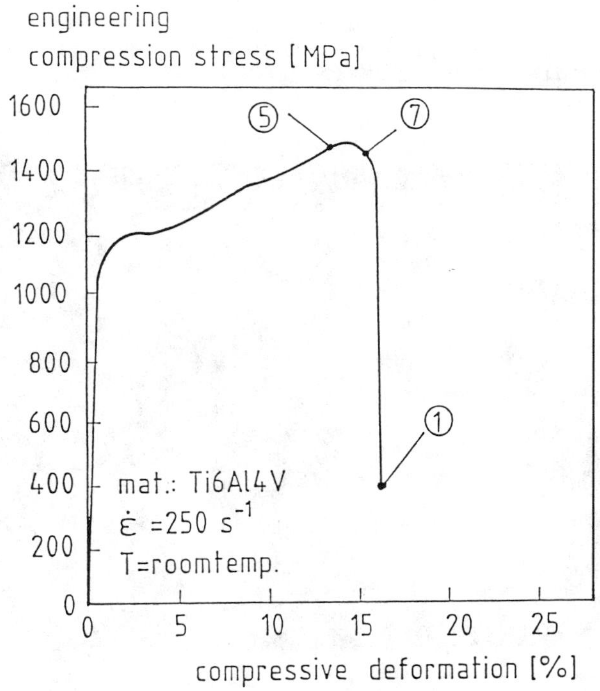

Figure 5 - Initiation and development of adiabatic shear failure in specimens with 6° inclination under dynamic compression loading

Figure 6 - Micrograph of specimen No. 5 showing the beginning of the adiabatic shear formation

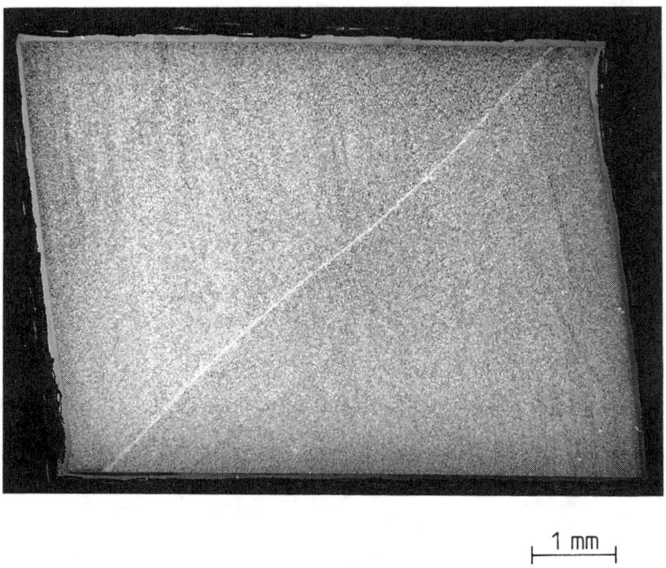

Figure 7 - Micrograph of specimen No. 7 showing the full established shear-band

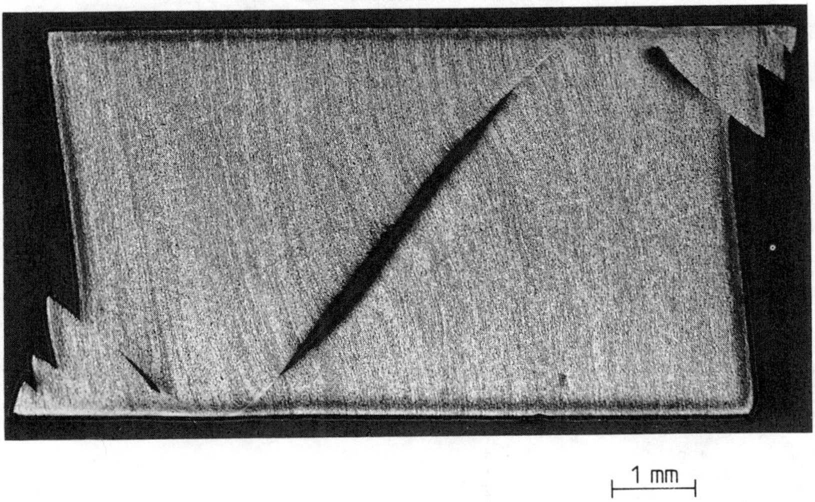

Figure 8 - Micrograph of specimen No. 1 showing the failure and secondary effects at the specimen edges

Conclusion

The results of this study show the influence of adiabatic shear failure on the ultimate strength of the titanium alloy Ti6Al4V under dynamical loading. This effect exists generally for all metallic materials under high loading rates and will be intensified by increasing shear stress under multiaxial loading condition.

The propensity to this failure mode seems to be high for titanium alloys. More investigations are needed to improve the knowledge in this area.

References

1. Ewald Staskewitsch and Klemens Stiebler, "Mechanical Behaviour of a High-Strength Austenitic Steel under Dynamical Biaxial Loading", Shock-Wave on High Strain-Rate Phenomena in Materials, ed. M.A. Meyers, L.Murr and K.P. Staudhammer (New York, NY: Marcel Dekker, Inc. 1992) 107-115.

2. Lothar W. Meyer and Chi Y. Chiem, "Conditions of Adiabatic Shear in Dynamic Torsional and Compressive Loading of TiAl6V4 Titanium Alloy", Titanium, Science and Technology, ed. G. Lütjering, U. Zwicker and W. Bunk (Oberursel, Germany: DGM Informationsgesellschaft-Verlag, 1985) 1907-1914.

3. L.S. Costin et al., "On the Localisation of Plastic Flow in Mild Steel Tubes under Dynamic Torsional Loading", Mechanical Properties at High Rates of Strain, ed. John Harding (London, Great Britain: The Institute of Physics, Conf. Series No. 47, 1979) 90 - 100.

4. J.H. Giovanola, "Observation of Adiabatic Shear Banding in Simple Torsion", Impact Loading and Dynamic Behaviour of Materials", ed. C.Y. Chiem, H.-D. Kunze and L.W. Meyer (Oberursel, Germany: DGM Informationsgesellschaft-Verlag, 1988) 705-710.

5. Karl-Heinz-Hartmann, Hans-Dieter Kunze and Lothar W. Meyer, "Metallurgical Effects on Impact Loaded Materials", Shock Waves and High-Strain-Rate Phenomena in Metals, ed. Marc A. Meyers and Lawrence E. Murr (New York, NY: Plenum Press 1981) 325-337.

6. Lothar W. Meyer and Sara Manwaring, "Critical Adiabatic Shear Strength of Low Alloyed Steel under Compressive Loading", Metallurgical Applications of Shock-Wave and High-Strain-Rate Phenomena, ed. Lawrence E. Murr, Karl P. Staudhammer and Marc A. Meyers (New York, NY: Marcel Dekker, Inc. 1986) 657-673.

7. Jacques Duffy, "Experimental Studies of Shear Band Formation Through Temperature Measurements and High Speed Photography", Proceed. 3rd Intern. Conf. on Mechanical and Physical Behaviour of Materials under Dynamic Loading; (Paris, France: DYMAT Association 1991) 645-652.

EFFECT Of Zr, Sn AND Al ADDITION ON MECHANICAL PROPERTIES OF METASTABLE β TITANIUM ALLOYS

S.Isiyama*, S.Hanada**

* Nippon Stainless Steel Co., Ltd.
Minatocho, Jyoetsu, Niigata-ken, 942 Japan
** Institute for Materials Research, Tohoku University
Katahira, Aoba-ku, Sendai, Miyagi-ken, 980 Japan

Abstract

Mechanical properties of metastable β titaniumu alloys were investigated using Ti-16V and Ti-7Cr base alloys doped with Sn,Zr and Al as ternary or quaternary additions. Zr addition had little influence on the mechanical properties, while Sn addition decreased flow stress and increased elongation markedly. In Al added alloys, rapid work hardening followed by secondary yielding was observed after extremely low yielding. This tendency became clear with increasing Al content, which resulted in remarkable deterioration in elongation. These results were explained by considering the effect of Zr,Sn and Al addition on deformation modes and β phase stability.

Introduction

It has been reported in metastable β titanium alloys that different types of deformation modes,slip,(332)twinning and deformation induced martensitic transformation operate depending on alloy compositions. The present authors[1] showed that the dominant deformation mode of metastable β titanium alloys was controlled by β phase stability referred to martensitic transformation and ω phase transformation which varied with concentration and type of alloying element. On the other hand, on the mechanical properties of metastable β titanium alloys, Hanada et al.[2-4] found in mainly binary alloys that (332)twinning which operated predominantly in the alloys with large instability of β phase led to low yield strength and large elongation, although crystallographic slip in the alloys with small instability of β phase resulted in high yield strength and small elongation. Moreover, Duerig et al.[5,6] and Flower[7] investigated the mechanical properties of the alloys such as Ti-10V-2Fe-3Al in which deformation induced α " martensitic transformation occurred as dominant deformation mode. They also observed the tensile properties characterized by the similar large elongation and low yield strength to the alloys deform predominantly by (332)twinning. From these results, it seems apparent that both (332)twinning and deformation induced martensitic transformation in metastable β titanium alloys cause lower yield strength and higher elongation than slip does. However, the difference between the effect of

Titanium '92
Science and Technology
Edited by F.H. Froes and I. Caplan
The Minerals, Metals & Materials Society, 1993

(332)twinning on the mechanical properties and that of deformation induced martensitic transformation is left unknown. Further, the effect of ternary or quaternary alloying elements on the mechanical properties of metastable β titanium alloys is also unclear.
The object of this study is to make clear the correlation of mechanical properties, deformation modes, and ternary or quaternary alloying elements in metastable β titanium alloys. In this study, therefore, tensile tests were conducted for a series of Ti-16V-(Zr,Sn,Al) and a series of Ti-7Cr-(Zr,Sn,Al) alloys, and the effect of Zr, Sn and Al on the mechanical properties was discussed in connection with the the deformation modes presented in our previous paper[1] in detail. V and Cr were selected as representatives of isomorphous and eutectoid β stabilizer, respectively. Both 16wt%V and 7wt%Cr correspond to a minimum content preventing martensitic transformation during quenching[8]. Zr, Sn and Al were chose as ternary or quaternary additions, since they are extensively used as important alloying elements in most commercial titanium alloys.

Procedure

Ti-16V-(Zr,Sn,Al) and Ti-7Cr-(Zr,Sn,Al) alloys containing 0/6wt% Zr, 0/6wt% Sn and 0/3wt% Al were prepared by arc melting in an argon atmosphere. The arc melted buttons were hot rolled and cold rolled to 1mm thick sheets. Tensile test specimens with the gage size of 1mm × 3mm × 16mm were obtained from the sheets by spark-cutting, and were sealed in a quartz tube under a vacuum. They were solution-treated at 1023-1123K for 1.8ks, quenched into ice water and then descaled by both mechanical and chemical polishing. Tensile test were performed on Instron type testing machine at an initial strain rate of 1.04×10^{-4} s^{-1}. All the alloys had an oxygen content in the range of 0.04 to 0.06wt%.

Results

Tensile Properties of Zr Added Alloys and Sn Added Ones

The effect of Zr addition on the stress-strain curves of Ti-16V and Ti-7Cr alloy systems is shown in Fig.1 and the effect of Sn addition in Fig.2. Although Ti-16V and Ti-7Cr systems show the similar dependence of the stress-strain curves upon Zr and Sn addition, the effect of Zr addition on tensile properties and that of Sn addition are quite different each other. Namely, Zr addition up to 6wt% has no significant influence on the tensile properties and thus Zr added alloys in Fig.1 all show the similar stress-strain curves characterized by 600/650MPa of yield strength, linear work hardening and 25/30% elongation in engineering strain. In contrast with Zr addition, Sn addition changes tensile properties drastically, as shown in Fig.2. Although Sn added alloys also show the similar linear work hardening to Zr added alloys, increase in Sn content up to 4/5wt% decreases yield stress and flow stress and increases elongation markedly. The alloys containing 4/5wt% Sn show the low yield strength of about 250MPa and the excellent elongations above 50% in engineering strain. Further Sn addition, however, tends to increase yield strength and decrease elongation. The effect of Sn addition on tensile properties is more distinct in Ti-16V systems than in Ti-7Cr ones.

Tensile Properties of Al Added Alloys

β phase could not be retained at room temperature in a series of Ti-16V-Al and a series of Ti-7Cr-Al alloys because of martensitic transformation during quenching. The tensile properties of the alloys doped both with Zr and Al were controlled only by Al content, which may be understood from the results on the effect of Zr addition on the tensile properties as mentioned above. Thus the effect of Al addition was investigated using a series of Ti-16V-6Zr-Al and a series of Ti-7Cr-4Zr-Al alloys in which β phase was retained under solution treated conditions except for Ti-16V-6Zr-1.5Al.

Fig.3 shows the effect of Al addition on the stress-strain curves. The similar dependence of the tensile properties on Al content is also found in Ti-16V and Ti-7Cr systems. The yield stress is lowered extremely by Al addition as in the case of Sn addition. However, different from Sn addition, rapid work hardening followed by second yielding can be seen after first yielding in stress-strain curves of Al added alloys shown in Fig.3. This tendency becomes clear with increasing Al content, which results in remarkable deterioration in elongation.

Tensile Properties of The Alloys Doped Both with Sn and Al

Under a small amount of Sn content, tensile properties of the alloys containing both Sn and Al were analogous to those of the alloys doped both with Zr and Al stated earlier. However, increase in Al content increased yield strength and decreased elongation in the alloys with a large amount of Sn content. This situation is indicated in the stress-strain curves of a series of Ti-16V-6Sn-Al and a series of Ti-7Cr-6Sn-Al alloys in Fig.4. In these alloys, rapid work hardening after first yielding and second yielding observed in Al added alloys in Fig.3 disappear.

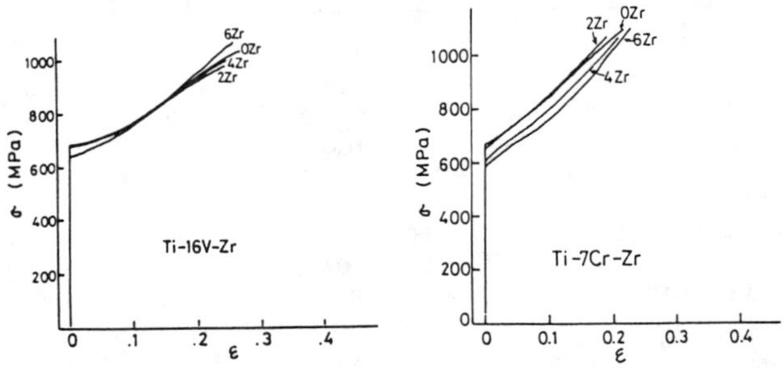

Fig. 1 Stress-strain curves of the alloys doped with Zr.

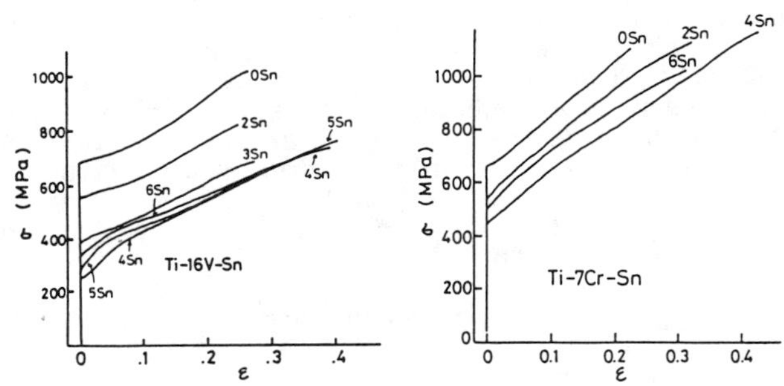

Fig. 2 Stress-strain curves of the alloys doped with Sn.

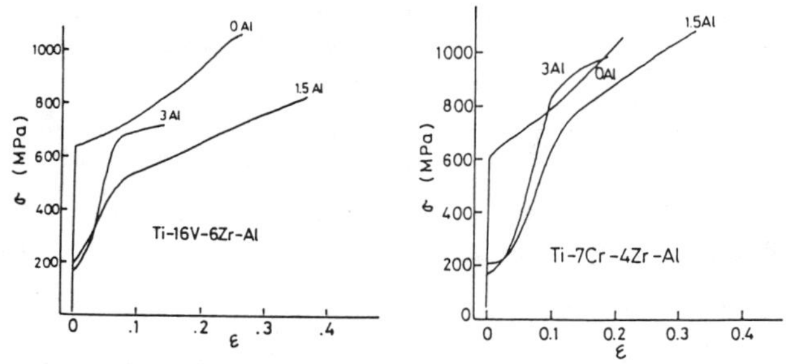

Fig. 3 Stress-strain curves of the alloys doped both with Al and Zr.

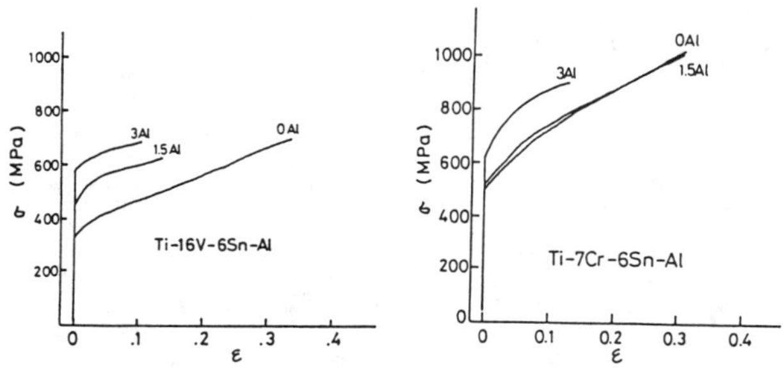

Fig. 4 Stress-strain curves of the alloys doped both with Al and Sn.

Discussion

Effect of Alloying Elements on Deformation Modes and β Phase Stability

It is pertinent to summarize the results on the effect of Zr, Sn and Al addition on the deformation mode and β phase stability determined by X-ray diffraction and selected area electron diffraction described in our previous paper[1] before discussing the present results on the mechanical properties.

Zr addition has little influence on deformation modes and β phase stability and thus the dominant deformation mode in Ti-16V-Zr and Ti-7Cr-Zr systems is (332)twinning in which β phase is transformed to ω phase, as in the case of many binary alloy systems[2,9,10]. On the other hand, Sn and Al addition change dominant deformation mode from (332)twinning to deformation induced α" or α' martensitic transformation depending on alloy systems, which is ascribed to the suppression of athermal ω phase transformation by these elements addition. Sn addition above 4/5wt% and Al addition above 1.5wt%, however, have a tendency to restrain the deformation induced martensitic transformation. This is considered to be attributed to increase in shear stress for martensitic transformation resulting from solution hardening with higher Sn and Al content. As a result, addition of a large amount of Sn and Al changes dominant deformation mode from deformation induced martensitic transformation to slip.

Correlation of Alloying Elements, Deformation Modes and Mechanical Properties

As shown in Fig 1, Zr addition has no significant influence on tensile properties of metastable β titanium alloys. This may be understood by the fact that Zr addition does not change β phase stability and thus deformation mode. On the other hand, since Sn addition varies dominant deformation mode from (332)twinning to deformation induced martensitic transformation, tensile properties change drastically with Sn content. It is reported[5-7] that yield stress decreases remarkably when deformation occurs by deformation induced martensitic transformation, which is good agreement with the present results on the tensile properties of Sn added alloys. Both (332)twinning and deformation induced martensitic transformation have higher work hardenability than slip[2-7]. Therefore, these deformation modes may increase elongation by means of inhibiting early necking. In this study, while high work hardening rate is also found in both Zr added alloys deforming predominantly by (332)twinning and Sn added alloys deforming preferentially by deformation induced α'' martensitic transformation, the elongations of the former are obviously smaller than those of the latter. The lower ductility of the former may be attributed to higher sensitivity to cracking resulting from a large amount of ω phase within β phase induced both by quenching and deformation. The decrease in yield stress and flow stress with Sn addition in Ti-7Cr alloy systems is smaller than in Ti-16V ones. This is probably due to the smaller amount of athermal ω phase in binary Ti-7Cr alloy than in binary Ti-16V one[1], which may result in smaller decrease in yield strength and flow stress with depression of athermal ω phase by Sn addition. While both Sn added alloys and Al added alloys deform by deformation induced martensitic transformation, the tensile properties are quite different each other. Al added alloys show the "double yielding" and have smaller elongation than Sn added alloys, which becomes clear with increase in Al content. A similar phenomenon was observed in a Ti-10V-2Fe-3Al alloy by Duerig et al.[5,6] who explained the stress-strain curves as follows: first yielding is due to occurrence of α'' transformation, flat portion after first yielding is ascribed to α'' transformation continuing under fairly constant stress, rapid work hardening is nearly elastic deformation because of difficulty in producing additional α'' phase and second yielding is caused by operation of slip resulting from increase in applied stress. The difference of the effect of Al addition on tensile properties from that of Sn addition is apparently shown in Fig.5 and Fig.6. The stress-strain curve of Ti-16V-6Zr-3Al alloy obtained by alternating loading and unloading and optical microstructures under unloading after applying a given value of strain are shown in Fig.5. It should be noted that pseudoelasticity is apparently observed in Fig.5. Namely, when the applied stress is below the second yielding, the strain under loading is restored by unloading, which results in no permanent strain in the stress-strain curve and no deformation structure in optical micrographs(A,B). Permanent strain and deformation structure probably involving slip line in addition to α'' plates (micrograph:C,D) appear after loading exceeding the second yielding. The stress-strain curve during loading in Fig.5, therefore, can be interpreted according to the explanations presented by Duerig et al. mentioned above, although they do not describe any pseudoelastic behavior in Ti-10V-2Fe-3Al. The pseudoelasticity in Fig.5 is considered to occur in a similar manner to transformation pseudoelasticity observed in Cu-Al-Ni alloy[11]. Namely, in Ti-16V-6Zr-3Al alloy, the deformation induced α'' phase is stable only under loading unless slip is introduced, and thus the strain under loading below second yielding is restored by unloading according to the reversion of α'' phase to β phase. In any event, it is presumed that the "double yielding" in Al added alloys occurs because of the difference in shear stress for martensitic transformation and for slip and becomes clear due to increase in shear stress for slip with increasing Al content. Fig.6 shows the stress-strain curve and optical microstructures of Ti-16V-4Sn alloy obtained in the same manner as Ti-16V-6Zr-3Al alloy in Fig.5. Ti-16V-4Sn alloy shows conventionally shaped stress-strain curve without the pseudoelastic behavior found in Ti-16V-6Zr-3Al alloy, and thus deformation structures in response to applied strain values are observed in micrographs. This is considered to mean that slip as well as α " martensitic transformation also operates simultaneously after yielding. Unfortunately, obvious evidence of slip deformation can not be obtained in optical observation because of difficulty in distinguishing slip lines from a lot of deformation induced α " plates. However, since Sn added alloys have extremely large elongations which can not be explained only by deformation induced martensitic transformation[7], it is likely that slip occurs with the martensitic transformation in these alloys. If a combination of slip and α " martensitic transformation operates in Sn added alloys,

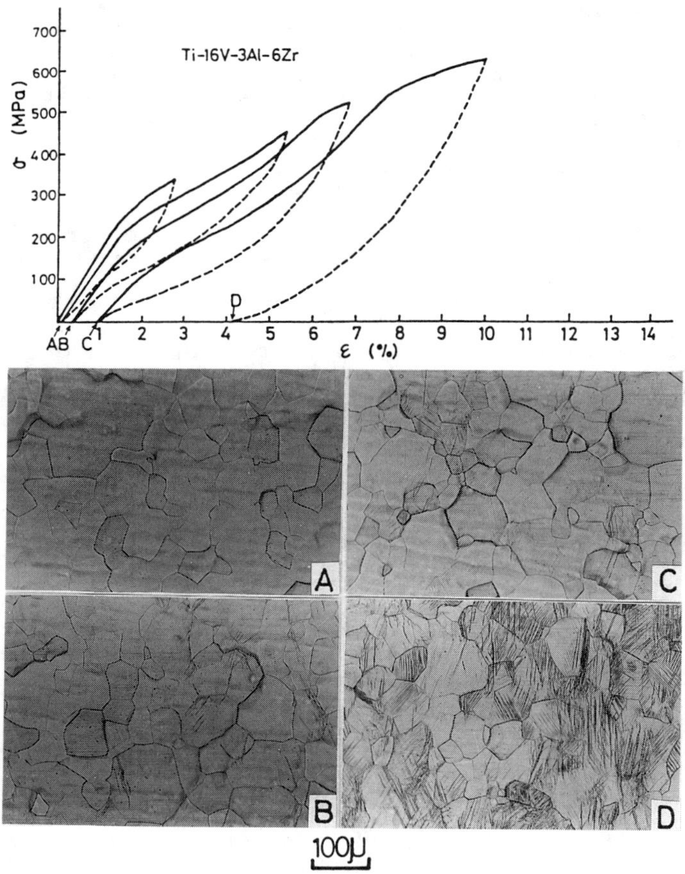

Fig. 5 Stress-strain curve of Ti-16V-6Zr-3Al alloy obtained by alternating loading and unloading and microstructures under unloading.

high work hardening rate attributed to work hardening of martensite itself and martensite boundaries acting as barriers to dislocations movement may continue until failure to result in large elongation. Relatively large elongations observed in 1.5wt% Al added alloys in Fig.3 are considered to be obtained in a manner similar to Sn added alloys. However, since higher Al addition restrains deformation induced martensitic transformation, the alloys containing 3wt% Al deform only by slip with poor work hardenability after second yielding, and thus the elongations of these alloys decrease markedly as shown in Fig.3.

Since both Sn and Al addition above a given value suppress deformation induced martensitic transformation and change dominant deformation mode to slip, the alloys doped both with Sn and Al tend to increase yield strength and decrease elongation with increasing Sn and Al content, as shown in Fig.4. Therefore, Ti-16V-6Sn-3Al and Ti-7Cr-6Sn-3Al alloy containing the largest amount of Sn and Al in present work deform almost only by slip, and the stress-strain curves of these alloys are in close agreement with

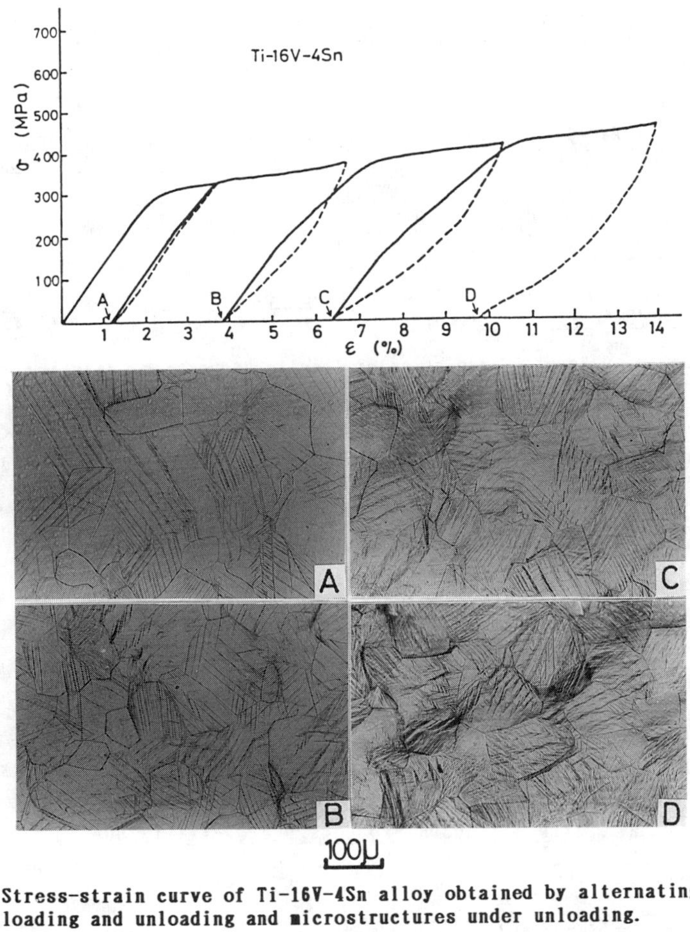

Fig. 6 Stress-strain curve of Ti-16V-4Sn alloy obtained by alternating loading and unloading and microstructures under unloading.

those of β titanium alloys deform only by slip reported by Hanada et al.[4].

Summary

The effect of Zr, Sn and Al addition as ternary or quaternary alloying element on the mechanical properties of Ti-16V and Ti-7Cr metastable β titanium alloy systems was investigated.
Zr addition up to 6wt% did not change significantly the tensile properties. Yield strength and flow stress were decreased and elongation was increased markedly with increasing Sn addition up to 4/5wt%, but further addition showed opposite tendency. Al addition up to 1.5wt% indicated the similar effect to Sn addition up to 4/5wt% on tensile properties. However, further Al addition brought about "double yielding", pseudoelasticity, and remarkable deterioration of elongation. A large amount of both Sn and Al addition caused increase in yield strength and decrease in elongation. The similar dependence of tensile properties on Zr, Sn and Al addition was observed in Ti-16V alloy systems and Ti-7Cr ones. These results were explained by considering the effect of Zr, Sn and Al addition on deformation modes and β phase stability.

Reference

1. S. Isiyama, S. Hanada and O. Izumi, "Effect of Zr, Sn and Al Addition on Deformation Mode and Beta Phase Stability of Metastable Beta Ti Alloys," ISIJ Int., 31(1991), 807-813

2. S. Hanada and O. Izumi, "Deformation Behaviour of Retained β Phase in β-eutectoid Ti-Cr Alloys," J. Mater. Sci., 21(1986), 4131-4139

3. S. Hanada, Y. Yoshio and O. Izumi, "Effect of Deformation Modes on Tensile Properties of Beta Titanium Alloys," Trans. Jpn. Inst. Met., 27(1986), 496-503

4. S. Hanada and O. Izumi, "Correlation of Tensile Properties, Deformation Modes, and Phase Stability in Commercial β-Phase Titanium Alloys," Metall. Trans., 18A(1987), 265-271

5. T. W. Duerig, G. T. Terlinde and J. C. Williams, "Phase Transformations and Tensile Properties of Ti-10V-2Fe-3Al," Metall. Trans., 11A(1980), 1987-1998

6. T. W. Duerig et al., "Stress Assisted Transformation in Ti-10V-2Fe-3Al," Proc. 4th Inter. Conf. on Titanium, (1980), 1503-1512

7. H. M. Flower, "The Plastic Deformation of Metastable β Titanium Alloys," Proc. 5th Inter. Conf. on Titanium, (1984), 1651-1658

8. S. Ankem and S. R. Seagle: Beta Titanium Alloys in the 1980's (ed. by R. R. Boyer and H. W. Rosenberg, AIME, New York, 1984), 107-126

9. S. Hanada and O. Izumi, "Transmission Electron Microscopic Observation of Mechanical Twinning in Metastable Beta Titanium Alloys," Metall. Trans., 17A(1986), 1409-1420

10. M. Hida et al., "Stress Induced Products and Ductility due to Lattice Instability of β Phase Single Crystal of Ti-Mo Alloys," Acta Metall., 30(1982), 1471-1479

11. K. Otsuka, H. Sakamoto and K. Shimizu, "Successive Stress-Induced Martensitic Transformations and Associated Transformation Pseudoelasticity in Cu-Al-Ni Alloys," Acta Metall., 27(1979), 585-601

CREEP-FATIGUE INTERACTION IN TITANIUM ALLOY

Ti-6Al-1.6Zr-3.3Mo-0.3Si AT ROOM TEMPERATURE

M. Tavafoghi, C. Ramachandra, A.K. Singh, and V. Singh

Centre of Advanced Study
Department of Metallurgical Engineering
Institute of Technology
Banaras Hindu University
Varanasi-221 005, India

Abstract

Low cycle fatigue tests are conducted under strain control at $\Delta\varepsilon_t/2$ from ± 0.75 to ± 1.65% and cyclic frequency of 0.2 Hz with and without tensile hold of 2 minutes after each cycle, at RT, for β solution treated (ST) as well as solution treated & 973K-12h aged (ST-A) condition of the alloy Ti-6Al-1.6Zr-3.3Mo-0.3Si (VT9). Dual slopes are exhibited in Coffin-Manson plots for both the conditions. Fatigue life in ST-A condition was far less than in ST. The tensile hold of 2 minutes had detrimental effect on fatigue life of both ST and ST-A materials.

Introduction

Titanium and its alloys are known to exhibit appreciable amounts of creep at ambient temperature (T< 0.2 Tm) even at stress levels of ≃ 80% of tensile yield strength [1-4]. Thus, creep-fatigue interaction which usually occurs in other alloys at elevated temperature and seriously affects fatigue life, may occur even at room temperature in titanium alloys under cyclic loading. However, only limited investigations have recently been made on this aspect of titanium alloys [5-7].

The alloy Ti-6Al-1.6Zr-3.3Mo-0.3Si (VT9) is a commercial $\alpha+\beta$ alloy designed for high temperature application as disc material in jet engines. Presence of small amount of Si(< 0.5 wt.%) in this alloy and other high temperature titanium alloys is essential for their creep resistance at elevated temperatures. There have been differing views about the mechanism of creep strengthening due to silicon; silicon in solid solution as well as in the form of silicide precipitates has been considered effective in creep strengthening of titanium alloys, depending upon the test conditions [8-11]. Silicide precipitates, however, are reported to promote crack initiation in the structural components of the near α alloy 685 [12].

The present investigation is concerned with low cycle fatigue behaviour of the titanium alloy VT9 in β solution treated & oil quenched as well as solution treated and aged condition, without and with hold period of 2 minutes at peak tensile strain.

Titanium '92
Science and Technology
Edited by F.H. Froes and I. Caplan
The Minerals, Metals & Materials Society, 1993

Material and Experimental Procedure

The alloy VT9 was obtained from Mishra Dhatu Nigam, Hyderabad, India in the form of hot rolled and mill annealed rods of 15mm diameter. The alloy contained by wt.pct. 6.5Al, 3.3Mo, 1.75Zr, 0.3Si, 21ppm H_2, 12ppm N_2, 574ppm O_2 and balance Titanium. Blanks of 120mm length and 13mm diameter were solution treated in the β phase field at 1293K for 1/2 hour, under vacuum ($\simeq 10^{-5}$ torr) and quenched in oil using vacuum quenching furnace. This heat treatment is designated as ST. Half of the total stock of the oil quenched blanks were aged at 973K for 12 hours under the same level of vacuum and designated as ST-A. Small pieces cut from the ST and ST-A blanks were mechanically polished and etched by 10HF + $5HNO_3$ + $85H_2O$ (vol.%) for optical metallography. Thin foils for transmission electron microscopy (TEM) were prepared by electrothinning of 3mm discs at 12.5V, using twin jet electropolisher and electrolyte 59% methanol, 35% n-butanol and 6% perchloric acid cooled to 233K. TEM was carried out using JEOL 200CX transmission electron microscope.

Low cycle fatigue specimens, with gauge length of 14mm, gauge diameter of 6mm and shoulder radii of 26mm were machined from the heat treated blanks and mechanically polished. Tension-compression LCF tests were conducted in total strain controlled mode at different total strain amplitudes ($\Delta\varepsilon_t/2$) from ± 0.75 to 1.65%, at constant cyclic frequency of 0.2Hz, with triangular Waveform, using MTS servo hydraulic machine. Hysteresis loops were recorded at frequent intervals and tests were continued till failure. Tensile tests were conducted with cylindrical specimens of 15.4mm gauge length, 4.5mm gauge diameter, at a strain rate of $4 \times 10^{-4} s^{-1}$ using Instron machine.

Results and Discussion

Microstructure

The optical micrograph in Fig. 1 shows microstructure of the alloy in ST condition. The criss-cross morphology of the transformed β structure may be seen in the micrograph. No observable difference could be seen in the microstructure on aging the ST specimen i.e. ST-A condition.

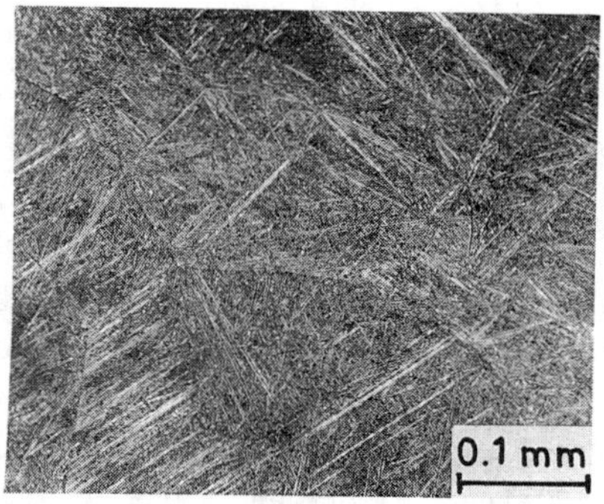

Figure 1 – Optical micrograph showing microstructure in ST condition.

Bright field transmission electron micrograph (BFTEM) in Fig. 2a for the ST condition reveals thin films of β sandwiched between interface phases which in turn are sandwiched between α' platelets. The α' platelets show dislocations and stacking faults. These microstructural features are typical of the alloys like VT9 and near α alloys resulting from β solution treatment and less severe cooling [13-15]. The interface phase is a subject of much controversy in the literature [13-20]. While Banerjee et al. [16,17] described it an artifact resulting from electrothinning, very recently Servant et al. [18] have reported it a metastable phase existing under certain thermal and thermomechanical treatments. Fig. 2b shows the BFTEM in the ST-A condition. It may be observed that silicides are precipitated predominantly at the interplatelet boundaries of α' and β [21]. A few silicides are seen inside β and also in α' . In an earlier investigation [21] these silicides were confirmed to be hexagonal S_2 silicides with lattice parameters a = 0.70 nm and c = 0.36 nm.

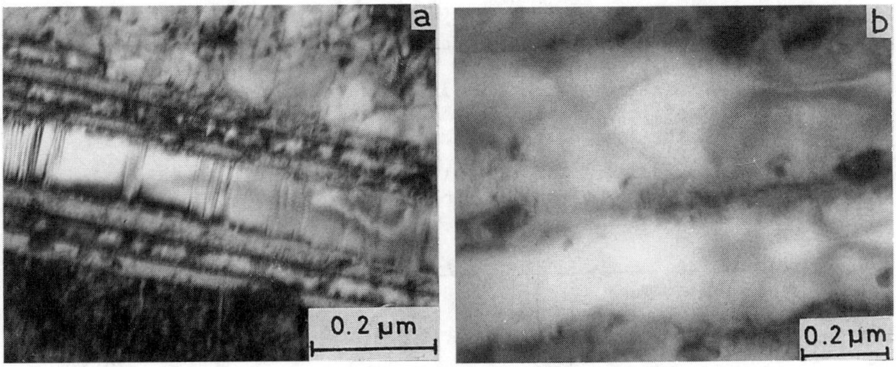

Figure 2 - Bright field transmission electron micrographs for (a) ST (b) ST-A conditions.

Low cycle fatigue behaviour

The dependence of fatigue life, expressed as number of reversals to failure, $2N_f$, on plastic strain amplitude, $\Delta\varepsilon_p/2$, is shown by Coffin-Manson plot in Fig. 3. Since the tests were conducted under constant total strain

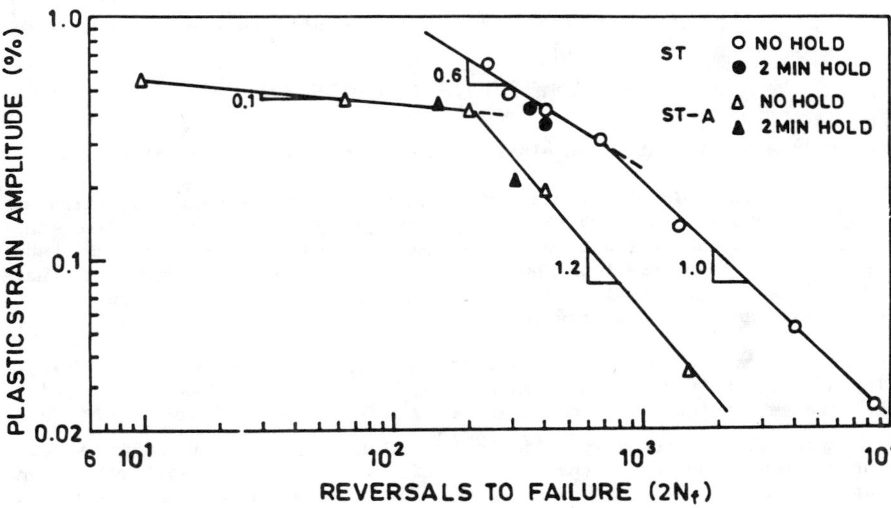

Figure 3 - Coffin-Manson plot.

amplitude ($\Delta\varepsilon_t/2$) and the plastic strain amplitude continually increased with number of cycles due to cyclic softening, both in ST and ST-A conditions (Fig. 4), the values of plastic strain amplitude for Coffin-Manson plot, were taken from the hysteresis loops corresponding to half fatigue life ($N_f/2$). It is obvious that fatigue life in ST-A condition is much lower than that in ST condition over the entire range of strain amplitude investigated. The detrimental effect of silicides on LCF life of the alloy VT9 is in line with that observed in the near α alloy IMI 685 [22]. The drastic reduction in fatigue life resulting from precipitation of silicide particles may be attributed to enhanced planarity of slip [23] and the loss of tensile ductility ($\varepsilon_p \simeq 0.6\%$). While the planarity of slip enhances the initiation of surface cracks [24], the loss in ductility increases the rate of propagation of the small surface cracks [25] during cyclic loading.

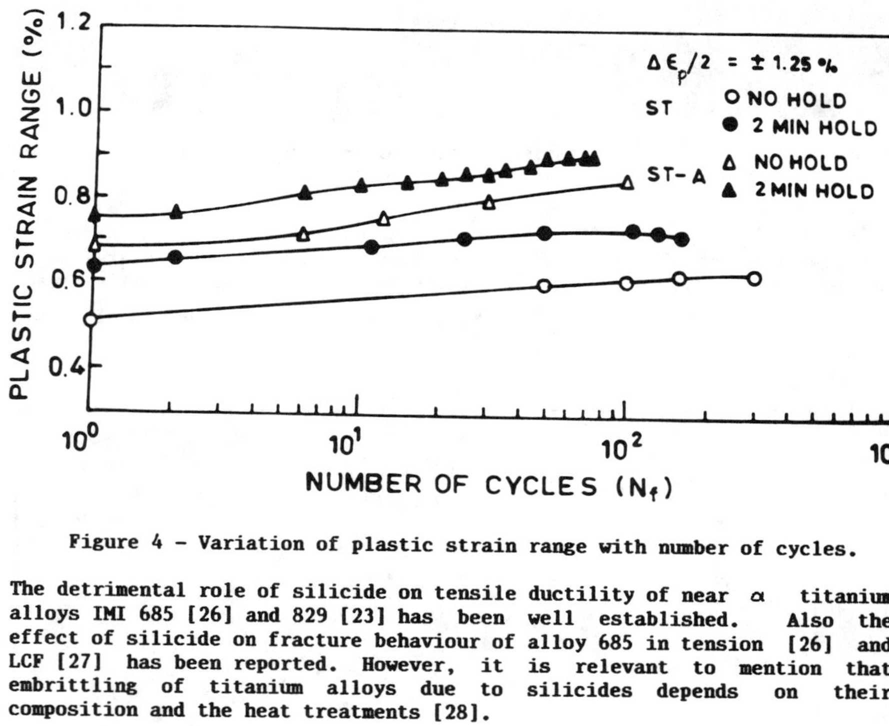

Figure 4 - Variation of plastic strain range with number of cycles.

The detrimental role of silicide on tensile ductility of near α titanium alloys IMI 685 [26] and 829 [23] has been well established. Also the effect of silicide on fracture behaviour of alloy 685 in tension [26] and LCF [27] has been reported. However, it is relevant to mention that embrittling of titanium alloys due to silicides depends on their composition and the heat treatments [28].

Further, it may be seen in Fig. 3 that there are two segments of different slopes in the Coffin-Manson plots of both ST as well as ST-A materials. The change in the slope occurs at $\Delta\varepsilon_p/2 \simeq 0.28\%$ and 0.4% for the ST and ST-A materials respectively. Similar two slope behaviour in Coffin-Manson plot has been reported in the $\alpha+\beta$ titanium alloy Ti-6Al-4V with change in slope at $\Delta\varepsilon_p/2 \simeq 0.45\%$ and 0.75% for the acicular and equiaxed microstructures respectively [29]. However, no explanation has been offered for the dual slope in Coffin-Manson plot of the alloy Ti-6Al-4V [29]. Mediratta et al. [30] have observed two slope behaviour in the dual phase steel and attributed it to difference in the cyclic work hardening behaviour in the lower and upper regions of the strain amplitude. Recently, Singh et al. [31] have observed change in the slope of the Coffin-Manson plot in the Nimonic Alloy PE16 in solution treated and aged condition. They have established through detailed TEM examination that change in the slope of Coffin-Manson plot occurs essentially due to the change in the mode of deformation from inhomogeneous planar slip to homogeneous deformation from the lower to higher region of strain.

The planarity of deformation at lower strain amplitude leads to drastic reduction in the number of cycles to crack initiation [24], the major component of the fatigue life [31]. The slopes of the two segments of Coffin-Manson plots for ST and ST-A conditions are given in Fig. 3. The slope of the lower segments are considerably higher than those of upper segments. The slope of the upper segment for ST condition (≃ 0.60) is comparable to the common value of 0.5 for many metals and alloys [32]. The slope of the upper segment for ST-A condition, however, is much lower, and this may be attributed to much wider range of fatigue life, in the region of high plastic strain amplitude as compared to that in the ST condition. It is relevant to mention that estimation of fatigue life in the region of low $\Delta\varepsilon_p/2$ from extrapolation of the upper segment of the Coffin-Manson plot would give much higher life than the actual life and hence it is essential to establish the Coffin-Manson plot over a wide range of strain amplitude for such materials.

Effect of tensile hold on fatigue life

The effect of creep-fatigue interaction, resulting from the tensile hold of 2 minutes, after each cycle of loading, on fatigue life of the alloy VT9 in ST and ST-A conditions may be seen in Table I. It is obvious that fatigue life is reduced due to tensile hold in both the conditions. The data points of hold time tests are also shown on the Coffin-Manson plots for the respective conditions (Fig. 3). Although they are seen to lie on the respective plots, it should be pointed out that they are at relatively higher levels of plastic strain amplitudes as compared to those of the companion specimens tested without hold, at identical total strain amplitudes (Table I). The increase in the plastic strain amplitude of the hold time specimens is due to creep resulting from stress relaxation during the tensile hold.

Table I. Effect of hold period of two minutes at peak tensile strain on fatigue life

Condition	Total strain amplitude $\pm \Delta\varepsilon_t/2$ (%)	Plastic strain amplitude at $N_f/2$ $\pm \Delta\varepsilon_p/2$ (%) No hold	Plastic strain amplitude at $N_f/2$ $\pm \Delta\varepsilon_p/2$ (%) hold	Fatigue life (cycles) No hold	Fatigue life (cycles) hold	Reduction in fatigue life (%)
ST	1.25	0.313	0.368	344	206	40.11
	1.35	0.415	0.428	206	182	11.65
ST-A	1.00	0.187	0.215	201	154	23.38
	1.25	0.403	0.437	102	74	27.45

The effect of tensile hold on fatigue resistance of titanium alloys has been found to depend on a number of factors like the mode of cyclic loading [7], number of cycles between the hold periods [7], Al and Mo equivalents [33] and texture [34] of the alloys.

Earlier investigators have emphasized the detrimental effect of hydrogen during tensile hold at hydrogen content of ≃ 40ppm [6], however, recent investigations in a number of titanium alloys have revealed that there is no detrimental effect of hydrogen even at a much higher concentration (≃ 190ppm) [33]. Thus decrease in fatigue life of the alloy VT9 (H_2 ≃ 21ppm) resulting from tensile hold may not be attributed essentially to hydrogen effect.

The observed decrease in fatigue life of the alloy VT9 due to tensile hold in the present investigation, is in agreement with the compositional criterion >6 wt.% Al and < 4wt.% Mo equivalents for the detrimental effect of hold time on fatigue life of titanium alloys at room temperature [33]. The decrease in fatigue life of both ST and ST-A materials due to tensile hold is essentially due to increase in the plastic strain component from creep. It may be seen that pct. reduction in fatigue life of ST and ST-A materials (Table I) due to hold effect, tested at identical total strain amplitude of ±1.25% is nearly comparable and thus there is not any significant influence of the form of silicon on degradation of fatigue resistance of the alloy VT9 resulting from creep-fatigue interaction at room temperature under the test conditions investigated.

Conclusions

1. Low cycle fatigue life of the alloy VT9 is drastically reduced due to precipitation of silicides.

2. A hold period of 2 minutes at peak tensile strain, after each cycle, has detrimental effect on LCF life in both ST and ST-A conditions.

3. There is not much influence of the form of silicon, present in the alloy VT9, on degradation of its fatigue resistance, due to creep-fatigue interaction at RT.

References

1. W.H. Reiman, "Room temperature creep in Ti-6Al-4V," J. Mater., 6 (1971), 926-940.

2. A.W. Thompson and B.C. Odegard, "The Influence of Microstructure on Low Temperature Creep of Ti-5Al-2.5Sn," Metall. Trans., 4A (1973), 899-908.

3. M.A. Imam and C.M. Gilmore, "Room Temperature Creep of Ti-6Al-4V," Metall. Trans., 10A (1979), 419-425.

4. R.T. Chen, W.H. Miller, Jr. and E.A. Starke, Jr. "Effect of Microstructure on Creep Behaviour of Ti-6211," Titanium Science and Technology, ed. G. Lütjering, U. Zwicker, W. Bunk (Munich, PA:DGM, Sept. 1984), 2363-2370.

5. D. Eylon and J.A. Hall, "Fatigue Behaviour of Beta Processed Titanium Alloy IMI 685," Metall. Trans., 8A (1977), 981-990.

6. J. White, M.H. Loretto and R.E. Smallmen, "The Effect of Dwell on Low Cycle Fatigue in IMI 829 (Ti 5331S)," Titanium Science and Technology, ed. G. Lütjering, U. Zwicker, W. Bunk (Munich, PA:DGM, Sept. 1984), 2297-2304.

7. H.P. Chu, B.A. MacDonold, O.P. Arora, "Creep-Fatigue Interaction of Titanium Alloy Ti-6Al-2Cb-1Ta-0.8Mo At Room Temperature," ibid, 2395-2402.

8. Y. Imbert, "Creep-Resistance and Embrittlement of a Ti-6%Al-5%Zr-1%W-0.4%Si Alloy," J. of Less Common Metals, 37 (1974), 71-89.

9. N.E. Paton and M.W. Mahoney, "Creep of Titanium-Silicon Alloys," Metall. Trans., 7A (1976), 1685-1694.

10. A.T.K. Assadi, H.M. Flower and D.R.F. West, "Microstructure and Strength of Alloys of the Ti-Al-Zr-Mo-Si Systems," Metals Tech., (1979), 8-15.

11. M.W. Mahoney and N.E. Paton, "Fatigue and Fracture Characteristics of Silicon-Bearing Titanium Alloy," Metall. Trans., 9A (1978) 1497-1501.

12. D. Hoeppner, "A Fractographic Analysis of Flaw Growth in a High Strength Titanium Alloys," Metallography, 11 (1978), 129-154.

13. D. Banerjee et al., "Microstructure and Tensile Ductility of β Heat Treated Titanium Alloy," Metall. Trans., 14A (1983), 413-420.

14. C. Ramachandra and V. Singh, "Silicide Precipitation in Alloy Ti-6Al-5Zr-0.5Mo-0.25Si," Metall. Trans., 13A (1982), 771-775.

15. G. Sridhar, V.V. Kutumba Rao and D.S. Sarma, "The Influence of Heat Treatment on the Structure and Properties of Near-α Titanium Alloy," Metall. Trans., 18A (1987), 877-891.

16. D. Banerjee and J.C. Williams, "The Effect of Foil Preparation Technique on Interface Phase Formation in Ti Alloys," Scripta Metall., 17 (1983), 1125-1128.

17. D. Banerjee et al., "A Resolution of the Interface Phase Problem in Titanium Alloys," Acta Metall., 36(1) (1988), 125-141.

18. C. Servant et al., "Contribution to the Analysis of the α/β Interface in Some Titanium Alloys," J. Mater. Res., 6(5) (1991), 987-998.

19. N.E. Paton and H.L. Fraser, "Phase Transformations in Titanium Alloys," Sixth World Conference on Titanium (Proc.Conf.), Cannes, France, Les Editions de Physique, (1988), 1469-1479.

20. T.T. Ye and H. Ling, "A New Mechanism of the Hexagonal Interface Phase Formation," Scripta Metall., 23 (1989), 1755-1759.

21. A.K. Singh et al., "Microstructure of β-Solution-Treated Quenched and Aged $\alpha + \beta$ Titanium Alloy Ti-6Al-1.6Zr-3.3Mo-0.3Si," J. of Alloys and Componds, (1991) (in press).

22. C. Ramachandra and Vakil Singh, "Effect of Silicide Precipitation on the Low Cycle Fatigue Behaviour of Alloy Ti-6Al-5Zr-0.5Mo-0.25Si," Scripta Metall., 21 (1987), 633-636.

23. A.P. Woodfield et al., "The Effect of Long-Term High Temperature Exposure on the Strength and Properties of the Titanium Alloy Ti 5331S," Acta Metall., 36(3) (1988), 507-515.

24. E. Hornbogen and M. Graf, "The Effect of Inhomogeneity of Cyclic Strain on Initiation of Cracks," Scripta Metall., 12(2) (1978), 147-150.

25. G. Lütjering, A. Gysler and L. Wagner, "Crack Propagation in Ti Alloys," Sixth World Conference on Titanium (Proc.Conf.), Cannes, France, Les Editions de Physique, (1988), 71-80.

26. C. Ramachandra and Vakil Singh, "Effect of Silicide Precipitation on Tensile Properties and Fracture of Alloy Ti-6Al-5Zr-0.5Mo-0.25Si," Metall. Trans., 16A (1985), 227-231.

27. C. Ramachandra and Vakil Singh, "Fracture of Titanium Alloy IMI 685 During Low Cycle Fatigue," Bull. of Electron Microscope Society of India, 7 (1983), 95-100.

28. M. Tavafoghi et al., "Effect of Heat Treatment on Tensile Properties and Fracture Behaviour of Titanium Alloys VT9 and IMI 685," Fracture of Engineering Materials & Structure, ed. S.H. Teoh and K.H. Lee (London, PA: Elsvier Applied Science, 1991), 720-725.

29. B. Vittemant and G. Thauvin, "Low-Cycle Fatigue in a Ti-6Al-4V Alloys with Equiaxed or Acicular Structure,"Sixth World Conference on Titanium (Proc.Conf.), Cannes, France, Les Editions de Physique, (1988), 319-324.

30. S.R. Mediratta, V. Ramaswamy and P. Rama Rao, "Two Stage Cyclic Work Hardening and two Slope Coffin-Manson Relationship in Dual Phase Steels," Scripta Metall., 20 (1986), 555-558.

31. V. Singh et al., "Low Cycle Fatigue Behaviour of Nimonic PE16 at Room Temperature," Metall. Trans., 22A (1991), 499-506.

32. J.E. Tavernells and L.F. Coffin, "A Compilation and Interpretation of Cyclic Strain Fatigue Tests on Metals," ASM Trans., 51 (1959), 438-450.

33. D.F. Neal,"Creep-Fatigue Interactions in Titanium Alloys,"Sixth World Conference on Titanium (Proc.Conf.), Cannes, France, Les Editions de Physique, (1988), 175-180.

34. M.R. Winstone, "Effect of Texture on the Dwell Fatigue Properties of a Near Alpha Titanium Alloys," ibid, 169-173.

The Behaviors of Fretting and Fretting Fatigue

of Ti–10V–2Fe–3Al Alloy

Wang shihong   Liang youming   Shen guiqin

(Beijing University of Aeronautics and Astronautics)
Beijing   100083   China

Abstract

To evaluate the influence of fretting damage on fatigue properties of Ti–1023 alloy, the fretting and fretting fatigue behaviors of the alloy have been systematically studied.The experimental results showed that Ti–1023 alloy was very sensitive to fretting damage, the fretting fatigue strength at room temperature was reduced by 50%. But, the fretting damage was decreased with increased testing temperature and fretting wear volume was reached minium at 250℃. The SEM observations showed that the basic fretting damage machanism was fatigue delamilation caused by alternate shear stress at the sub–surface of the material. Adhension and metal transfer phenomenon were very slight.

Introduction

The phenomenon of fretting can be detected in many practical applications, such as, at the bearing surface of roller bearing races and in shaft–hub combination, and also at loosely jointed component combination, for example, at the blade foot of tubine blade. The characteristic of this is the appearance of abrasive and friction weld marks, which are caused by oscillation sliding movements, of very small amplitude–usually only a few μm. This surface damage leads to the formation of microcracks, which spread and finally lead to the fatigue failure of the component.

The Ti–10V–2Fe–3AL alloy is a relatively new near beta titanium alloy. It offers excellent strength / ductility / toughness combination, deep hardenability and good forgeability. Over the last few years, extensive work has been carried out on the metallurgical characteristics and mechanical properties of the alloy, but nonthing is known about its fretting fatigue behaviors. Titanium and its alloys are particularly susceptible to fretting fatigue and therefore it is essential that fretting fatigue data be acquired for this high strength beta titanium alloy.

The aim of the investigation presented here is to describe the fretting fatigue behaviors of the material combination Ti–1023 / Ti–1023, effect factors and expore improvement methods.

Titanium '92
Science and Technology
Edited by F.H. Froes and I. Caplan
The Minerals, Metals & Materials Society, 1993

Experimental

Table 1 shows the chemical analysis of the test alloy. The alloy was solution treated by water cooling from 740℃ (i.e. from below the β transus) followed by aging 8 h at 510℃. This produced a uniform microstructure which was consisted of primary α and aging β matrix. Typical michanical properties of the alloy in this condition are as following:

$\sigma_b=1200$Mpa, $\sigma_{0.2}=1120$Mpa, $\delta=10\%$, $\Psi=40\%$.

Table 1 The chemical compositions of Ti–1023 alloy

Elements	Al	V	Fe	Si	N	C	H	O
wt%	3.06	10.65	2.13	10.1	0.011	0.023	0.009	0.073

The fretting fatigue tests were performed under axial fluctuating tension. The fatigue specimen had parallel flats machined on its surface and fretting was induced by clamping a pair of bridges on to the parallel flats. The bridges were of the same material as the fatigue specimen and had been given the same heat–treatment. For high temperature test, the specimens and bridges were surrounded by a small eletrical furnace and tests conducted at 100, 250 and 350℃ The main parameters of tests are listed in table 2. The fretting wear was carried out with semiball and flat specimen which were both made of same material. The contact pressure P=49N, frequence of test f=98 Hz and amplitude A=60μm.

Table 2 Testing parameters of fretting fatigue

clamping load P(N)	500 1000 1500
testing temperature t℃	18 100 250 350
frequence f(Hz)	98
atmosphere	laboratory air

Results and Discussion

S–N curves of fretting fatigue tests and the usual fatigue tests without fretting at room temperature are shown in Fig.1. It can be seen that the fretting fatigue strength of Ti–1023 at 10^7cycles as compared with usual fatigue is rapidly decreased, from 695 Mpa to 345 Mpa, the percentage of reduction is about 50%. The visual examination showed the fretting fatigue specimens are always fractured in contact region between fretting pad and surface of fatigue specimens. To observing the features of fretting damage all the specimens were ultrasonically cleaned to remove the loose debris and examined in the scanning electron microscope. Fig 2 shows the appearance of original region of fretting fatigue fracture. The fretting pitting and oxide debres filled up in it can be observed. Fig.3 shows the morphology of scar of contact surface, from this, the outline of

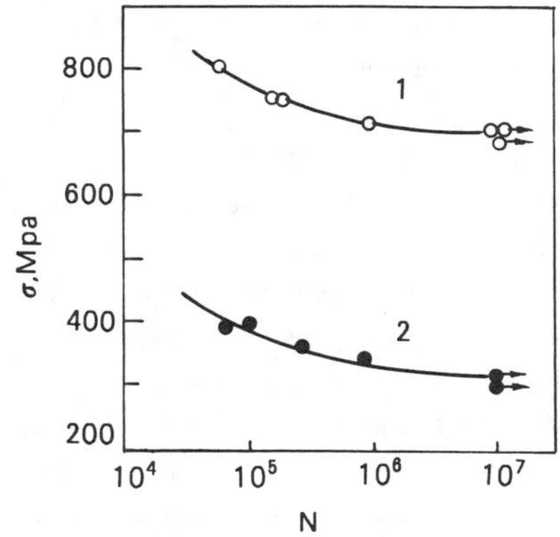

Fig.1 S–N curves at room temperature for Ti–1023

1. usual fatigue 2. Fretting fatigue

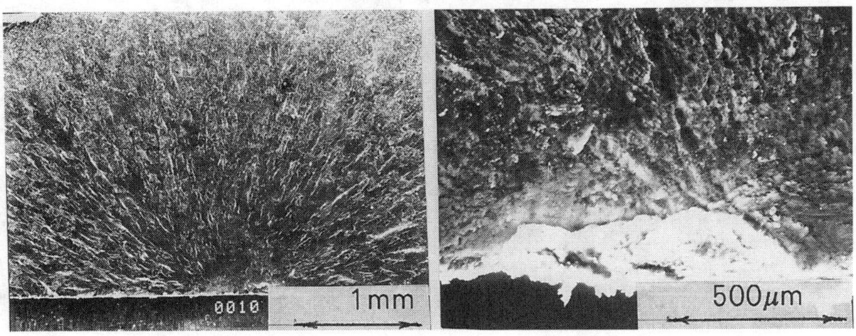

a b

Fig.2 Original region of fretting fatigue fracture

a.Origin region of fatigue fracture b.Oxide debres in origin region

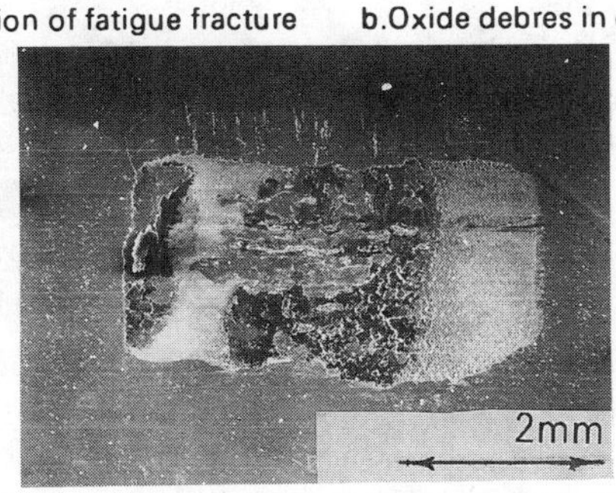

Fig.3 Scar of contact surface of fretting fatigue specimen

slip regions and non–slip regions is clearly discernible. The slip regions are characterised by the production of fretting debris and pitting, whereas non–slip regions by slight abrasion. It was founded that the fretting fatigue cracks were initiated in the boundary between the slip and non–slip regions. This phenomenon is in keeping with general pattern for fretting fatigue fracture. According to fundamental investigations of fretting fatigue[1,2], the maximum repeated shearing stress, which is the combination of repeated stress and the tangential fretting stress, is known to reach a maximum value at the boundary of slip region and therefore the fatigue crack is easily initiated in this areas. It was also founded that the basic fretting damage mechanism was fatigue delamilation caused by alternative shear stress at the sub–surface of the material (Fig.4). Adhension and metal transfer phenomenon were very slight. In later period of fretting fatigue, oxide film is ruptured and removed from surface of specimen by delamilation (Fig.4b), a large number of pitting is left on contact surface which results in stress concentration and induced initiation of fatigue cracks (Fig.5). Fig.6 shows the direction of fatigue crack propagation in a cross section of specimen. At first the crack propagetes in the zone influenced by the alternating stress arising from the fretting action. The crack at this stage is oblique to the surface. When it passes out of the this zone the crack changes direction and propagates perpendicular to the surface, i.e. right angles to the applied alternating stress in the specimen.

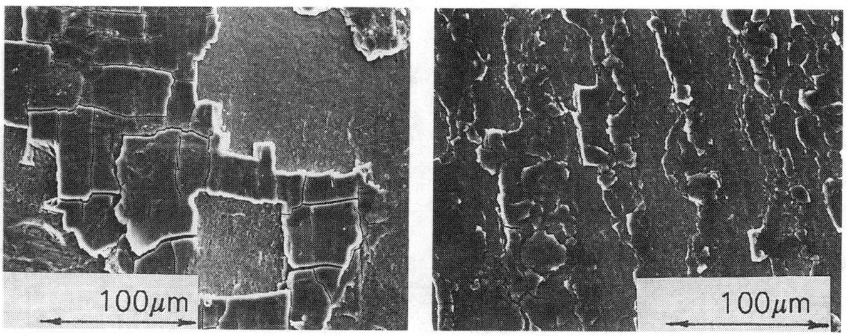

a b

Fig.4 Appearance of fretting damage

a,fretting damage at early period b,fretting damage at later period

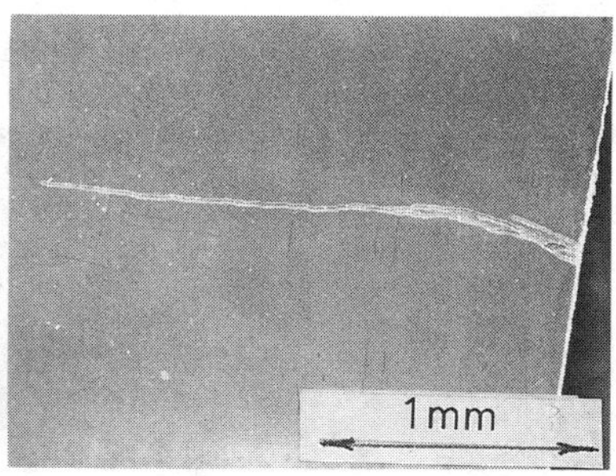

Fig.5 Morphology of fretting fatigue fracture Fig.6 Direction of fretting crack propagation

In general, if the amplitude of slip is maintained constant the volume of fretting damage increases linerly with the applied normal load [3]. But, in the present study, the fretting fatigue strength of

Ti–1023 alloy is increased with normal load (Fig.7). That means the actual amplitued is reduced as normal load is increased.

Fig.8 shows the effect of temperature on fretting fatgiue, the fretting fatigue strength decreased with temperature. But, this reduction is more likely caused by softening of material matrix with temperature increasing, rather than by fretting. This is because the fretting wear of Ti–1023 alloy is gradually reduced when testing temperature raise from room temperature to 350℃ (Fig.9). This result indicated suitable oxide film on contact surface can be acted as lubricant and reduce metal

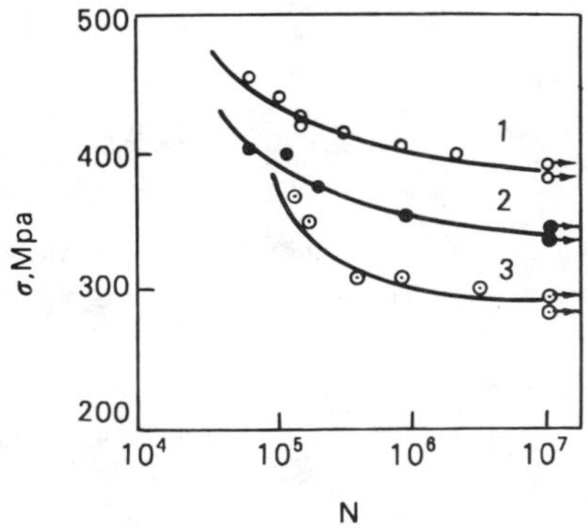

Fig.7 Influence of the normal load on fretting fatigue strength of Ti–1023 alloy
1.1500N 2.1000N 3.500N.

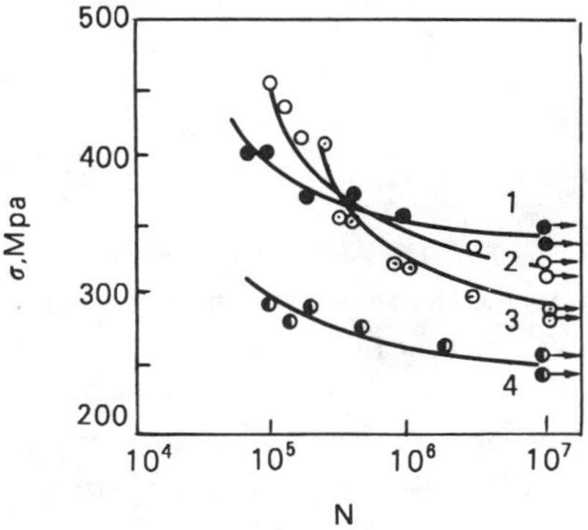

Fig.8 Effect of testing temperature on fretting fatigue strength of Ti–1023 alloy
1.18℃ 2.100℃ 3.250℃ 4.350℃

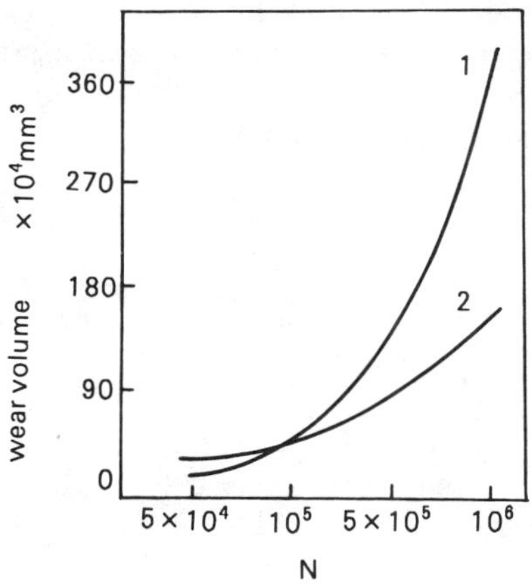

Fig.9 Effect of testing temperature on fretting wear of Ti–1023
1.shot peening, 2.non–shot peening

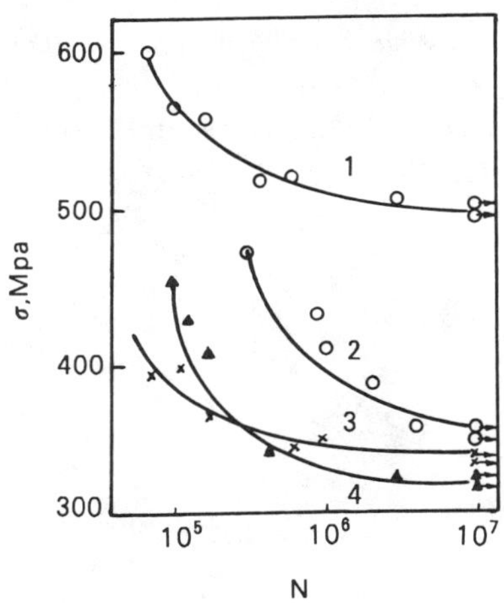

Fig.10 Effect of shot–peening on fretting fatgiue strength
1.18℃ shot peening, 2.250℃ shot peening
3.18℃ non–shot peening, 4.250℃ non–shot peening

to metal contact.

The development of surface compressive stresses by cold working the surface of a component has a beneficial effect on the fretting fatigue strength. Shot peening is the simplest of the methods to apply. Fig.10 shows the experiment result for Ti–1023 alloy and indicates that the fretting fatgiue strength of shot–peened specimen at room temperature was increased by 45%. This beneficial effect was even remained at 250℃ . The residual surface stress in the shot–peened specimens measured by X–ray diffraction was 950 Mpa, i.e. approching yield strength value of Ti–1023 alloy.

Conclusion

1. The Ti–1023 alloy is very sensitive to fretting damage, the fretting fatigue strengthh at room temperature was reduced by 50%, But, at elevated temperature the effect of fretting was weakened.
2. The basic fretting damage in Ti–1023 alloy was fatigue delamilation.
3. Shot peening process significantly improved the fatigue strength of the alloy, as a result of inducing residual compressive stress in specimens.

References

1. R.B.Waterhouse, Fretting Fatigue, (London, Applied science, 1981) P.87–149.
2. Kunio Nishioka and Kenji Hinakawa, "Fundamental Investigations of Fretting Fatigue", Bulletin of JSME 12(1969), 397–407.
3. R.B.Waterhouse, Fretting Corrosion, (Oxford, Pergamon, 1972) P.50–94.